W9-AFA-769

Astronomy

Senior Contributing Authors

Andrew Fraknoi, Foothill College
David Morrison, National Aeronautics and Space Administration
Sidney C. Wolff, National Optical Astronomy Observatories (Emeritus)

Download for free at http://cnx.org/content/col11992/latest/

© 2017 Rice University

Rice University
6100 Main Street MS-375
Houston, Texas 77005

Textbook content produced by OpenStax is licensed under a
Creative Commons Attribution 4.0 International License.
https://creativecommons.org/licenses/by/4.0/
https://creativecommons.org/licenses/by/4.0/legalcode

Changes to original document:
Non-numbered pages 1, 2, 3 and 4 were removed. The previous title
page, this page, and the last two blank pages have been added.

Download the original document for free at:
http://cnx.org/content/col11992/latest/

☰ TABLE OF CONTENTS

Download for free at http://cnx.org/content/col11992/latest/

This OpenStax book is available for free at http://cnx.org/content/col11992/1.8

Download for free at http://cnx.org/content/col11992/latest/

This OpenStax book is available for free at http://cnx.org/content/col11992/1.8

Download for free at http://cnx.org/content/col11992/latest/

This OpenStax book is available for free at http://cnx.org/content/col11992/1.8

Download for free at http://cnx.org/content/col11992/latest/

PREFACE

Welcome to *Astronomy*, an OpenStax resource. This textbook was written to increase student access to high-quality learning materials, maintaining highest standards of academic rigor at little to no cost.

About OpenStax

OpenStax is a nonprofit based at Rice University, and it's our mission to improve student access to education. Our first openly licensed college textbook was published in 2012 and our library has since scaled to over 20 books for college and AP courses used by hundreds of thousands of students. Our adaptive learning technology, designed to improve learning outcomes through personalized educational paths, is being piloted in college courses throughout the country. Through our partnerships with philanthropic foundations and our alliance with other educational resource organizations, OpenStax is breaking down the most common barriers to learning and empowering students and instructors to succeed.

About OpenStax Resources

Customization

Astronomy is licensed under a Creative Commons Attribution 4.0 International (CC BY) license, which means that you can distribute, remix, and build upon the content, as long as you provide attribution to OpenStax and its content contributors.

Because our books are openly licensed, you are free to use the entire book or pick and choose the sections that are most relevant to the needs of your course. Feel free to remix the content by assigning your students certain chapters and sections in your syllabus, in the order that you prefer. You can even provide a direct link in your syllabus to the sections in the web view of your book.

Faculty also have the option of creating a customized version of their OpenStax book through the aerSelect platform. The custom version can be made available to students in low-cost print or digital form through your campus bookstore. Visit your book page on openstax.org for a link to your book on aerSelect.

Errata

All OpenStax textbooks undergo a rigorous review process. However, like any professional-grade textbook, errors sometimes occur. Since our books are web based, we can make updates periodically when deemed pedagogically necessary. If you have a correction to suggest, submit it through the link on your book page on openstax.org. Subject-matter experts review all errata suggestions. OpenStax is committed to remaining transparent about all updates, so you will also find a list of past errata changes on your book page on openstax.org.

Format

You can access this textbook for free in web view or PDF through openstax.org, and for a low cost in print.

About *Astronomy*

Astronomy is written in clear non-technical language, with the occasional touch of humor and a wide range of clarifying illustrations. It has many analogies drawn from everyday life to help non-science majors appreciate,

Download for free at http://cnx.org/content/col11992/latest/

on their own terms, what our modern exploration of the universe is revealing. The book can be used for either a one-semester or two-semester introductory course (bear in mind, you can customize your version and include only those chapters or sections you will be teaching.) It is made available free of charge in electronic form (and low cost in printed form) to students around the world. If you have ever thrown up your hands in despair over the spiraling cost of astronomy textbooks, you owe your students a good look at this one.

Coverage and Scope

Astronomy was written, updated, and reviewed by a broad range of astronomers and astronomy educators in a strong community effort. It is designed to meet scope and sequence requirements of introductory astronomy courses nationwide.

Chapter 1: Science and the Universe: A Brief Tour

Chapter 2: Observing the Sky: The Birth of Astronomy

Chapter 3: Orbits and Gravity

Chapter 4: Earth, Moon, and Sky

Chapter 5: Radiation and Spectra

Chapter 6: Astronomical Instruments

Chapter 7: Other Worlds: An Introduction to the Solar System

Chapter 8: Earth as a Planet

Chapter 9: Cratered Worlds

Chapter 10: Earthlike Planets: Venus and Mars

Chapter 11: The Giant Planets

Chapter 12: Rings, Moons, and Pluto

Chapter 13: Comets and Asteroids: Debris of the Solar System

Chapter 14: Cosmic Samples and the Origin of the Solar System

Chapter 15: The Sun: A Garden-Variety Star

Chapter 16: The Sun: A Nuclear Powerhouse

Chapter 17: Analyzing Starlight

Chapter 18: The Stars: A Celestial Census

Chapter 19: Celestial Distances

Chapter 20: Between the Stars: Gas and Dust in Space

Chapter 21: The Birth of Stars and the Discovery of Planets outside the Solar System

Chapter 22: Stars from Adolescence to Old Age

Chapter 23: The Death of Stars

Chapter 24: Black Holes and Curved Spacetime

Chapter 25: The Milky Way Galaxy

Chapter 26: Galaxies

Chapter 27: Active Galaxies, Quasars, and Supermassive Black Holes

Download for free at http://cnx.org/content/col11992/latest/

Currency and Accuracy

Astronomy has information and images from the New Horizons exploration of Pluto, the discovery of gravitational waves, the Rosetta Mission to Comet C-G, and many other recent projects in astronomy. The discussion of exoplanets has been updated with recent information—indicating not just individual examples, but trends in what sorts of planets seem to be most common. Black holes receive their own chapter, and the role of supermassive black holes in active galaxies and galaxy evolution is clearly explained. Chapters have been reviewed by subject-matter experts for accuracy and currency.

Flexibility

Because there are many different ways to teach introductory astronomy, we have made the text as flexible as we could. Math examples are shown in separate sections throughout, so that you can leave out the math or require it as you deem best. Each section of a chapter treats a different aspect of the topic being covered; a number of sections could be omitted in shorter overview courses and can be included where you need more depth. And, as we have already discussed, you can customize the book in a variety of ways that have never been possible in traditional textbooks.

Student-Centered Focus

This book is written to help students understand the big picture rather than get lost in random factoids to memorize. The language is accessible and inviting. Helpful diagrams and summary tables review and encapsulate the ideas being covered. Each chapter contains interactive group activities you can assign to help students work in teams and pool their knowledge.

Download for free at http://cnx.org/content/col11992/latest/

Interactive Online Resources

Interesting "Links to Learning" are scattered throughout the chapters, which direct students to online animations, short videos, or enrichment readings to enhance their learning. Also, the resources listed at the end of each chapter include links to websites and other useful educational videos.

Feature Boxes That Help Students Think outside the Box

A variety of feature boxes within the chapters connect astronomy to the students' other subjects and humanize the face of astronomy by highlighting the lives of the men and women who have been key to its progress. Besides the math examples that we've already mentioned, the boxes include:

Making Connections. This feature connects the chapter topic to students' experiences with other fields, from poetry to engineering, popular culture, and natural disasters.

Voyagers in Astronomy. This feature presents brief and engaging biographies of the people behind historically significant discoveries, as well as emerging research.

Astronomy Basics. This feature explains basic science concepts that we often (incorrectly) assume students know from earlier classes.

Seeing for Yourself. This feature provides practical ways that students can make astronomical observations on their own.

End-of-Chapter Materials to Extend Students' Learning

Chapter Summaries. Summaries give the gist of each section for easy review.

For Further Exploration. This section offers a list of suggested articles, websites, and videos so students can delve into topics of interest, whether for their own learning, for homework, extra credit, or papers.

Review Questions. Review questions allow students to show you (or themselves) how well they understood the chapter.

Thought Questions. Thought questions help students assess their learning by asking for critical reflection on principles or ideas in the chapter.

Figuring For Yourself. Mathematical questions, using only basic algebra and arithmetic, allow students to apply the math principles given in the example boxes throughout the chapter.

Collaborative Group Activities. This section suggests ideas for group discussion, research, or reports.

Beautiful Art Program

Our comprehensive art program is designed to enhance students' understanding of concepts through clear and effective illustrations, diagrams, and photographs. Here are a few examples.

This OpenStax book is available for free at http://cnx.org/content/col11992/1.8

Download for free at http://cnx.org/content/col11992/latest/

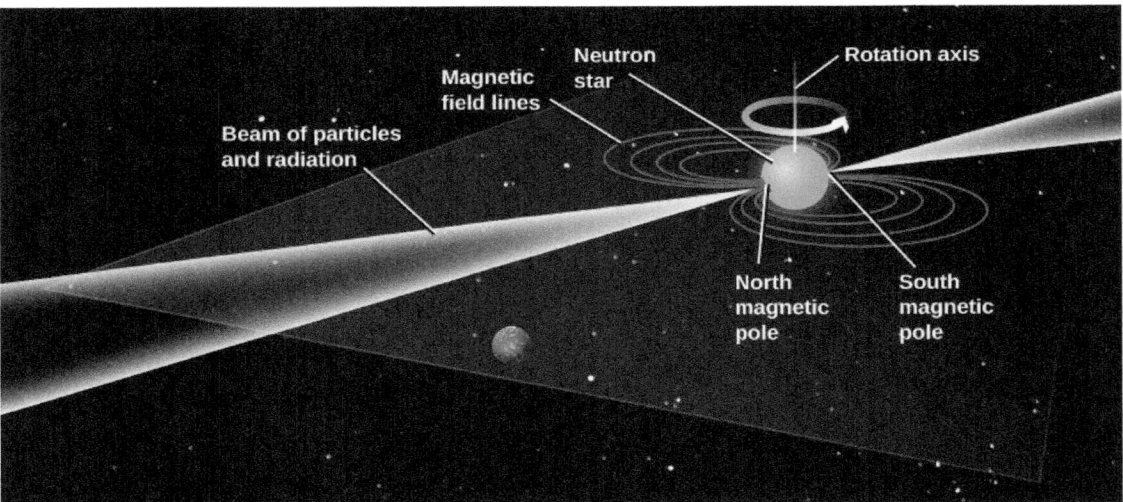

Figure 1. How a Pulsar Beam Sweeps over Earth.

Figure 2. Structure of the Milky Way Galaxy.

Download for free at http://cnx.org/content/col11992/latest/

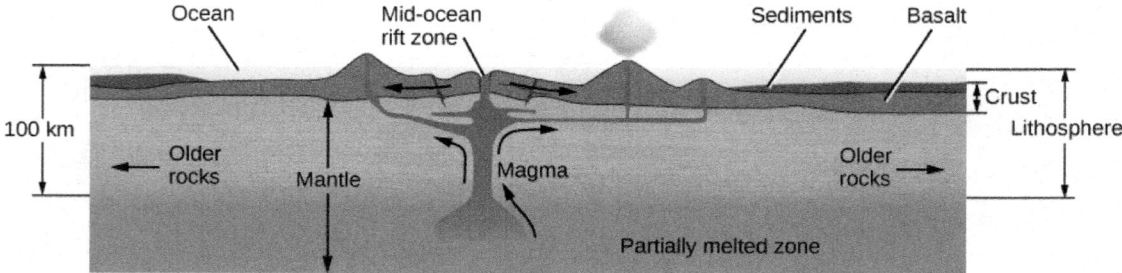

Rift zone

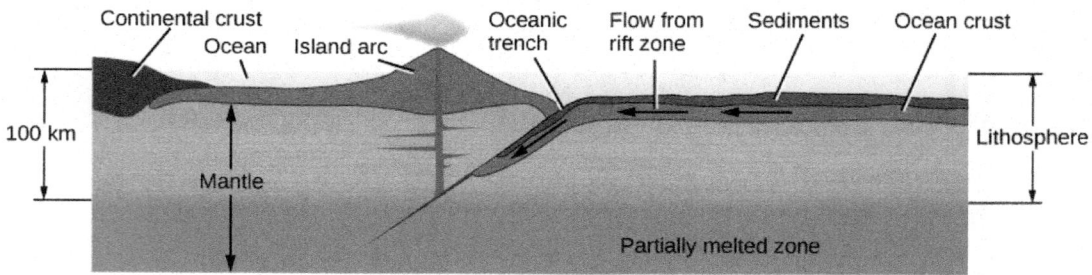

Subduction zone

Figure 3. Two Aspects of Plate Tectonics.

This OpenStax book is available for free at http://cnx.org/content/col11992/1.8

Download for free at http://cnx.org/content/col11992/latest/

Figure 4. Pluto Close Up.

Additional Resources

Student and Instructor Resources

We've compiled additional resources for both students and instructors, including Getting Started Guides, PowerPoint slides, and an instructor answer guide. Instructor resources require a verified instructor account, which can be requested on your openstax.org log-in. Take advantage of these resources to supplement your OpenStax book.

Partner Resources

OpenStax Partners are our allies in the mission to make high-quality learning materials affordable and accessible to students and instructors everywhere. Their tools integrate seamlessly with our OpenStax titles at a low cost. To access the partner resources for your text, visit your book page on openstax.org.

About the Authors

Senior Contributing Authors

Andrew Fraknoi, Foothill College

Download for free at http://cnx.org/content/col11992/latest/

Andrew Fraknoi is Chair of the Astronomy Department at Foothill College and served as the Executive Director of the Astronomical Society of the Pacific from 1978–1992. His work with the society included editing *Mercury* Magazine, *Universe in the Classroom*, and *Astronomy Beat*. He's taught at San Francisco State University, Canada College, and the University of California Extension. He is editor/co-author of *The Universe at Your Fingertips 2.0*, a collection of teaching activities, and co-author of *Solar Science,* a book for middle-school teachers. He was co-author of a syndicated newspaper column on astronomy, and appears regularly on local and national radio. With Sidney Wolff, he was founder of *Astronomy Education Review*. He serves on the Board of Trustees of the SETI Institute and on the Lick Observatory Council. In addition, he has organized six national symposia on teaching introductory astronomy. He received the Klumpke-Roberts Prize of the ASP, the Gemant Award of the American Institute of Physics, and the Faraday Award of the NSTA.

David Morrison, National Aeronautics and Space Administration

David Morrison is a Senior Scientist at NASA Ames Research Center. He received his PhD in astronomy from Harvard, where he was one of Carl Sagan's graduate students. He is a founder of the field of astrobiology and is known for research on small bodies in the solar system. He spent 17 years at University of Hawaii's Institute for Astronomy and the Department of Physics and Astronomy. He was Director of the IRTF at Mauna Kea Observatory. Morrison has held senior NASA positions including Chief of the Ames Space Science Division and founding Director of the Lunar Science Institute. He's been on science teams for the Voyager, Galileo, and Kepler missions. Morrison received NASA Outstanding Leadership Medals and the NASA Exceptional Achievement Medal. He was awarded the AAS Carl Sagan medal and the ASP Klumpke-Roberts prize. Committed to the struggle against pseudoscience, he serves as Contributing Editor of *Skeptical Inquirer* and on the Advisory Council of the National Center for Science Education.

Sidney C. Wolff, National Optical Astronomy Observatories (Emeritus)

After receiving her PhD from the UC Berkeley, Dr. Wolff was involved with the astronomical development of Mauna Kea. In 1984, she became the Director of Kitt Peak National Observatory, and was director of National Optical Astronomy Observatory. Most recently, she led the design and development of the 8.4-meter Large Synoptic Survey Telescope. Dr. Wolff has published over ninety refereed papers on star formation and stellar atmospheres. She has served as President of the AAS and the ASP. Her recently published book, *The Boundless Universe: Astronomy in the New Age of Discovery*, won the 2016 IPPY (Independent Publisher Book Awards) Silver Medal in Science.

All three senior contributing authors have received the Education Prize of the American Astronomical Society and have had an asteroid named after them by the International Astronomical Union. They have worked together on a series of astronomy textbooks over the past two decades.

Contributing Authors

John Beck, Stanford University
Susan D. Benecchi, Planetary Science Institute
John Bochanski, Rider University
Howard Bond, Pennsylvania State University, Emeritus, Space Telescope Science Institute
Jennifer Carson, Occidental College
Bryan Dunne, University of Illinois at Urbana-Champaign
Martin Elvis, Harvard-Smithsonian Center for Astrophysics
Debra Fischer, Yale University
Heidi Hammel, Association of Universities for Research in Astronomy
Tori Hoehler, NASA Ames Research Center

This OpenStax book is available for free at http://cnx.org/content/col11992/1.8

Douglas Ingram, Texas Christian University

Steven Kawaler, Iowa State University

Lloyd Knox, University of California, Davis

Mark Krumholz, Australian National University

James Lowenthal, Smith College

Siobahn Morgan, University of Northern Iowa

Daniel Perley, California Institute of Technology

Claire Raftery, National Solar Observatory

Deborah Scherrer, retired, Stanford University

Phillip Scherrer, Stanford University

Sanjoy Som, Blue Marble Space Institute of Science, NASA Ames Research Center

Wes Tobin, Indiana University East

William H. Waller, retired, Tufts University, Rockport (MA) Public Schools

Todd Young, Wayne State College

Reviewers

Elisabeth R. Adams, Planetary Science Institute

Alfred N. Alaniz, San Antonio College

Charles Allison, Texas A&M University–Kingsville

Douglas Arion, Carthage College

Timothy Barker, Wheaton College

Marshall Bartlett, The Hockaday School

Charles Benesh, Wesleyan College

Gerald B. Cleaver, Baylor University

Kristi Concannon, King's College

Anthony Crider, Elon University

Scott Engle, Villanova University

Matthew Fillingim, University of California, Berkeley

Robert Fisher, University of Massachusetts, Dartmouth

Carrie Fitzgerald, Montgomery College

Christopher Fuse, Rollins College

Shila Garg, Emeritus, The College of Wooster

Richard Gelderman, Western Kentucky University

Lee Hartman, University of Michigan

Beth Hufnagel, Anne Arundel Community College

Francine Jackson, Brown University

Joseph Jensen, Utah Valley University

John Kielkopf, University of Louisville

James C. Lombardi, Jr., Allegheny College

Amy Lovell, Agnes Scott College

Charles Niederriter, Gustavus Adolphus College

Richard Olenick, University of Dallas

Matthew Olmstead, King's College

Zoran Pazameta, Eastern Connecticut State University

David Quesada, Saint Thomas University

Valerie A. Rapson, Dudley Observatory

Download for free at http://cnx.org/content/col11992/latest/

Joseph Ribaudo, Utica College
Dean Richardson, Xavier University of Louisiana
Andrew Rivers, Northwestern University
Marc Sher, College of William & Mary
Christopher Sirola, University of Southern Mississippi
Ran Sivron, Baker University
J. Allyn Smith, Austin Peay State University
Jason Smolinski, Calvin College
Michele Thornley, Bucknell University
Richard Webb, Union College
Terry Willis, Chesapeake College
David Wood, San Antonio College
Jeremy Wood, Hazard Community and Technical College
Jared Workman, Colorado Mesa University
Kaisa E. Young, Nicholls State University

This OpenStax book is available for free at http://cnx.org/content/col11992/1.8

Download for free at http://cnx.org/content/col11992/latest/

SCIENCE AND THE UNIVERSE: A BRIEF TOUR

Figure 1.1. Distant Galaxies. These two interacting islands of stars (galaxies) are so far away that their light takes hundreds of millions of years to reach us on Earth (photographed with the Hubble Space Telescope). (credit: modification of work by NASA, ESA, the Hubble Heritage (STScI/AURA)-ESA/Hubble Collaboration, and K. Noll (STScI))

Chapter Outline

Introduction

We invite you to come along on a series of voyages to explore the universe as astronomers understand it today. Beyond Earth are vast and magnificent realms full of objects that have no counterpart on our home planet. Nevertheless, we hope to show you that the evolution of the universe has been directly responsible for your presence on Earth today.

Along your journey, you will encounter:

- a canyon system so large that, on Earth, it would stretch from Los Angeles to Washington, DC (Figure 1.2).

Download for free at http://cnx.org/content/col11992/latest/

Figure 1.2. Mars Mosaic. This image of Mars is centered on the Valles Marineris (Mariner Valley) complex of canyons, which is as long as the United States is wide. (credit: modification of work by NASA)

- a crater and other evidence on Earth that tell us that the dinosaurs (and many other creatures) died because of a cosmic collision.
- a tiny moon whose gravity is so weak that one good throw from its surface could put a baseball into orbit.
- a collapsed star so dense that to duplicate its interior we would have to squeeze every human being on Earth into a single raindrop.
- exploding stars whose violent end could wipe clean all of the life-forms on a planet orbiting a neighboring star (Figure 1.3).
- a "cannibal galaxy" that has already consumed a number of its smaller galaxy neighbors and is not yet finished finding new victims.
- a radio echo that is the faint but unmistakable signal of the creation event for our universe.

Figure 1.3. Stellar Corpse. We observe the remains of a star that was seen to explode in our skies in 1054 (and was, briefly, bright enough to be visible during the daytime). Today, the remnant is called the Crab Nebula and its central region is seen here. Such exploding stars are crucial to the development of life in the universe. (credit: NASA, ESA, J. Hester (Arizona State University))

This OpenStax book is available for free at http://cnx.org/content/col11992/1.8

Such discoveries are what make astronomy such an exciting field for scientists and many others—but you will explore much more than just the objects in our universe and the latest discoveries about them. We will pay equal attention to the *process* by which we have come to understand the realms beyond Earth and the tools we use to increase that understanding.

We gather information about the cosmos from the messages the universe sends our way. Because the stars are the fundamental building blocks of the universe, decoding the message of starlight has been a central challenge and triumph of modern astronomy. By the time you have finished reading this text, you will know a bit about how to read that message and how to understand what it is telling us.

1.1 | THE NATURE OF ASTRONOMY

Astronomy is defined as the study of the objects that lie beyond our planet Earth and the processes by which these objects interact with one another. We will see, though, that it is much more. It is also humanity's attempt to organize what we learn into a clear history of the universe, from the instant of its birth in the Big Bang to the present moment. Throughout this book, we emphasize that science is a *progress report*—one that changes constantly as new techniques and instruments allow us to probe the universe more deeply.

In considering the history of the universe, we will see again and again that the cosmos *evolves*; it changes in profound ways over long periods of time. For example, the universe made the carbon, the calcium, and the oxygen necessary to construct something as interesting and complicated as you. Today, many billions of years later, the universe has evolved into a more hospitable place for life. Tracing the evolutionary processes that continue to shape the universe is one of the most important (and satisfying) parts of modern astronomy.

1.2 | THE NATURE OF SCIENCE

The ultimate judge in science is always what nature itself reveals based on observations, experiments, models, and testing. Science is not merely a body of knowledge, but a *method* by which we attempt to understand nature and how it behaves. This method begins with many observations over a period of time. From the trends found through observations, scientists can *model* the particular phenomena we want to understand. Such models are always approximations of nature, subject to further testing.

As a concrete astronomical example, ancient astronomers constructed a model (partly from observations and partly from philosophical beliefs) that Earth was the center of the universe and everything moved around it in circular orbits. At first, our available observations of the Sun, Moon, and planets did fit this model; however, after further observations, the model had to be updated by adding circle after circle to represent the movements of the planets around Earth at the center. As the centuries passed and improved instruments were developed for keeping track of objects in the sky, the old model (even with a huge number of circles) could no longer explain all the observed facts. As we will see in the chapter on Observing the Sky: The Birth of Astronomy, a new model, with the Sun at the center, fit the experimental evidence better. After a period of philosophical struggle, it became accepted as our view of the universe.

When they are first proposed, new models or ideas are sometimes called *hypotheses*. You may think there can be no new hypotheses in a science such as astronomy—that everything important has already been learned. Nothing could be further from the truth. Throughout this textbook you will find discussions of recent, and occasionally still controversial, hypotheses in astronomy. For example, the significance that the huge chunks of rock and ice that hit Earth have for life on Earth itself is still debated. And while the evidence is strong that vast quantities of invisible "dark energy" make up the bulk of the universe, scientists have no convincing explanation for what the dark energy actually is. Resolving these issues will require difficult observations done

Download for free at http://cnx.org/content/col11992/latest/

at the forefront of our technology, and all such hypotheses need further testing before we incorporate them fully into our standard astronomical models.

This last point is crucial: a hypothesis must be a proposed explanation that can be *tested*. The most straightforward approach to such testing in science is to perform an experiment. If the experiment is conducted properly, its results either will agree with the predictions of the hypothesis or they will contradict it. If the experimental result is truly inconsistent with the hypothesis, a scientist must discard the hypothesis and try to develop an alternative. If the experimental result agrees with predictions, this does not necessarily prove that the hypothesis is absolutely correct; perhaps later experiments will contradict crucial parts of the hypothesis. But, the more experiments that agree with the hypothesis, the more likely we are to accept the hypothesis as a useful description of nature.

One way to think about this is to consider a scientist who was born and lives on an island where only black sheep live. Day after day the scientist encounters black sheep only, so he or she hypothesizes that all sheep are black. Although every observed sheep adds confidence to the theory, the scientist only has to visit the mainland and observe one white sheep to prove the hypothesis wrong.

When you read about experiments, you probably have a mental picture of a scientist in a laboratory conducting tests or taking careful measurements. This is certainly the case for a biologist or a chemist, but what can astronomers do when our laboratory is the universe? It's impossible to put a group of stars into a test tube or to order another comet from a scientific supply company.

As a result, astronomy is sometimes called an *observational* science; we often make our tests by observing many samples of the kind of object we want to study and noting carefully how different samples vary. New instruments and technology can let us look at astronomical objects from new perspectives and in greater detail. Our hypotheses are then judged in the light of this new information, and they pass or fail in the same way we would evaluate the result of a laboratory experiment.

Much of astronomy is also a *historical* science—meaning that what we observe has already happened in the universe and we can do nothing to change it. In the same way, a geologist cannot alter what has happened to our planet, and a paleontologist cannot bring an ancient animal back to life. While this can make astronomy challenging, it also gives us fascinating opportunities to discover the secrets of our cosmic past.

You might compare an astronomer to a detective trying to solve a crime that occurred before the detective arrived at the scene. There is lots of evidence, but both the detective and the scientist must sift through and organize the evidence to test various hypotheses about what actually happened. And there is another way in which the scientist is like a detective: they both must prove their case. The detective must convince the district attorney, the judge, and perhaps ultimately the jury that his hypothesis is correct. Similarly, the scientist must convince colleagues, editors of journals, and ultimately a broad cross-section of other scientists that her hypothesis is provisionally correct. In both cases, one can only ask for evidence "beyond a reasonable doubt." And sometimes new evidence will force both the detective and the scientist to revise their last hypothesis.

This self-correcting aspect of science sets it off from most human activities. Scientists spend a great deal of time questioning and challenging one another, which is why applications for project funding—as well as reports for publication in academic journals—go through an extensive process of *peer review*, which is a careful examination by other scientists in the same field. In science (after formal education and training), everyone is encouraged to improve upon experiments and to challenge any and all hypotheses. New scientists know that one of the best ways to advance their careers is to find a weakness in our current understanding of something and to correct it with a new or modified hypothesis.

This is one of the reasons science has made such dramatic progress. An undergraduate science major today knows more about science and did math than Sir Isaac Newton, one of the most renowned scientists who ever

This OpenStax book is available for free at http://cnx.org/content/col11992/1.8

lived. Even in this introductory astronomy course, you will learn about objects and processes that no one a few generations ago even dreamed existed.

1.3 THE LAWS OF NATURE

Over centuries scientists have extracted various *scientific laws* from countless observations, hypotheses, and experiments. These scientific laws are, in a sense, the "rules" of the game that nature plays. One remarkable discovery about nature—one that underlies everything you will read about in this text—is that the same laws apply everywhere in the universe. The rules that determine the motion of stars so far away that your eye cannot see them are the same laws that determine the arc of a baseball after a batter has hit it out of the park.

Note that without the existence of such universal laws, we could not make much headway in astronomy. If each pocket of the universe had different rules, we would have little chance of interpreting what happened in other "neighborhoods." But, the consistency of the laws of nature gives us enormous power to understand distant objects without traveling to them and learning the local laws. In the same way, if every region of a country had completely different laws, it would be very difficult to carry out commerce or even to understand the behavior of people in those different regions. A consistent set of laws, though, allows us to apply what we learn or practice in one state to any other state.

This is not to say that our current scientific models and laws cannot change. New experiments and observations can lead to new, more sophisticated models—models that can include new phenomena and laws about their behavior. The general theory of relativity proposed by Albert Einstein is a perfect example of such a transformation that took place about a century ago; it led us to predict, and eventually to observe, a strange new class of objects that astronomers call *black holes*. Only the patient process of observing nature ever more carefully and precisely can demonstrate the validity of such new scientific models.

One important problem in describing scientific models has to do with the limitations of language. When we try to describe complex phenomena in everyday terms, the words themselves may not be adequate to do the job. For example, you may have heard the structure of the atom likened to a miniature solar system. While some aspects of our modern model of the atom do remind us of planetary orbits, many other of its aspects are fundamentally different.

This problem is the reason scientists often prefer to describe their models using equations rather than words. In this book, which is designed to introduce the field of astronomy, we use mainly words to discuss what scientists have learned. We avoid complex math, but if this course piques your interest and you go on in science, more and more of your studies will involve the precise language of mathematics.

1.4 NUMBERS IN ASTRONOMY

In astronomy we deal with distances on a scale you may never have thought about before, with numbers larger than any you may have encountered. We adopt two approaches that make dealing with astronomical numbers a little bit easier. First, we use a system for writing large and small numbers called *scientific notation* (or sometimes *powers-of-ten notation*). This system is very appealing because it eliminates the many zeros that can seem overwhelming to the reader. In scientific notation, if you want to write a number such as 500,000,000, you express it as 5×10^8. The small raised number after the 10, called an *exponent*, keeps track of the number of places we had to move the decimal point to the left to convert 500,000,000 to 5. If you are encountering this system for the first time or would like a refresher, we suggest you look at Appendix C and Example 1.1 for more information. The second way we try to keep numbers simple is to use a consistent set of units—the metric International System of Units, or SI (from the French *Système International d'Unités*). The metric system is summarized in Appendix D (see Example 1.2).

Download for free at http://cnx.org/content/col11992/latest/

LINK TO LEARNING

Watch this brief PBS animation (https://openstax.org/l/30scinotation) that explains how scientific notation works and why it's useful.

A common unit astronomers use to describe distances in the universe is a light-year, which is the distance light travels during one year. Because light always travels at the same speed, and because its speed turns out to be the fastest possible speed in the universe, it makes a good standard for keeping track of distances. You might be confused because a "light-year" seems to imply that we are measuring time, but this mix-up of time and distance is common in everyday life as well. For example, when your friend asks where the movie theater is located, you might say "about 20 minutes from downtown."

So, how many kilometers are there in a light-year? Light travels at the amazing pace of 3×10^5 kilometers per second (km/s), which makes a light-year 9.46×10^{12} kilometers. You might think that such a large unit would reach the nearest star easily, but the stars are far more remote than our imaginations might lead us to believe. Even the nearest star is 4.3 light-years away—more than 40 trillion kilometers. Other stars visible to the unaided eye are hundreds to thousands of light-years away (Figure 1.4).

Figure 1.4. Orion Nebula. This beautiful cloud of cosmic raw material (gas and dust from which new stars and planets are being made) called the Orion Nebula is about 1400 light-years away. That's a distance of roughly 1.34×10^{16} kilometers—a pretty big number. The gas and dust in this region are illuminated by the intense light from a few extremely energetic adolescent stars. (credit: NASA, ESA, M. Robberto (Space Telescope Science Institute/ESA) and the Hubble Space Telescope Orion Treasury Project Team)

EXAMPLE 1.1

Scientific Notation

This OpenStax book is available for free at http://cnx.org/content/col11992/1.8

In 2015, the richest human being on our planet had a net worth of $79.2 billion. Some might say this is an astronomical sum of money. Express this amount in scientific notation.

Solution

$79.2 billion can be written $79,200,000,000. Expressed in scientific notation it becomes 7.92×10^{10}.

EXAMPLE 1.2

Getting Familiar with a Light-Year

How many kilometers are there in a light-year?

Solution

Light travels 3×10^5 km in 1 s. So, let's calculate how far it goes in a year:

- There are 60 (6×10^1) s in 1 min, and 6×10^1 min in 1 h.
- Multiply these together and you find that there are 3.6×10^3 s/h.
- Thus, light covers 3×10^5 km/s $\times$ 3.6×10^3 s/h = 1.08×10^9 km/h.
- There are 24 or 2.4×10^1 h in a day, and 365.24 (3.65×10^2) days in 1 y.
- The product of these two numbers is 8.77×10^3 h/y.
- Multiplying this by 1.08×10^9 km/h gives 9.46×10^{12} km/light-year.

That's almost 10,000,000,000,000 km that light covers in a year. To help you imagine how long this distance is, we'll mention that a string 1 light-year long could fit around the circumference of Earth 236 million times.

1.5 CONSEQUENCES OF LIGHT TRAVEL TIME

There is another reason the speed of light is such a natural unit of distance for astronomers. Information about the universe comes to us almost exclusively through various forms of light, and all such light travels at the speed of light—that is, 1 light-year every year. This sets a limit on how quickly we can learn about events in the universe. If a star is 100 light-years away, the light we see from it tonight left that star 100 years ago and is just now arriving in our neighborhood. The soonest we can learn about any changes in that star is 100 years after the fact. For a star 500 light-years away, the light we detect tonight left 500 years ago and is carrying 500-year-old news.

Because many of us are accustomed to instant news from the Internet, some might find this frustrating.

"You mean, when I see that star up there," you ask, "I won't know what's actually happening there for another 500 years?"

But this isn't the most helpful way to think about the situation. For astronomers, *now* is when the light reaches us here on Earth. There is no way for us to know anything about that star (or other object) until its light reaches us.

Download for free at http://cnx.org/content/col11992/latest/

But what at first may seem a great frustration is actually a tremendous benefit in disguise. If astronomers really want to piece together what has happened in the universe since its beginning, they must find evidence about each epoch (or period of time) of the past. Where can we find evidence today about cosmic events that occurred billions of years ago?

The delay in the arrival of light provides an answer to this question. The farther out in space we look, the longer the light has taken to get here, and the longer ago it left its place of origin. By looking billions of light-years out into space, astronomers are actually seeing billions of years into the past. In this way, we can reconstruct the history of the cosmos and get a sense of how it has evolved over time.

This is one reason why astronomers strive to build telescopes that can collect more and more of the faint light in the universe. The more light we collect, the fainter the objects we can observe. On average, fainter objects are farther away and can, therefore, tell us about periods of time even deeper in the past. Instruments such as the Hubble Space Telescope (Figure 1.5) and the Very Large Telescope in Chile (which you will learn about in the chapter on Astronomical Instruments), are giving astronomers views of deep space and deep time better than any we have had before.

Figure 1.5. Telescope in Orbit. The Hubble Space Telescope, shown here in orbit around Earth, is one of many astronomical instruments in space. (credit: modification of work by European Space Agency)

1.6 A TOUR OF THE UNIVERSE

We can now take a brief introductory tour of the universe as astronomers understand it today to get acquainted with the types of objects and distances you will encounter throughout the text. We begin at home with Earth, a nearly spherical planet about 13,000 kilometers in diameter (Figure 1.6). A space traveler entering our planetary system would easily distinguish Earth from the other planets in our solar system by the large amount of liquid water that covers some two thirds of its crust. If the traveler had equipment to receive radio or television signals, or came close enough to see the lights of our cities at night, she would soon find signs that this watery planet has sentient life.

This OpenStax book is available for free at http://cnx.org/content/col11992/1.8

Download for free at http://cnx.org/content/col11992/latest/

Figure 1.6. Humanity's Home Base. This image shows the Western hemisphere as viewed from space 35,400 kilometers (about 22,000 miles) above Earth. Data about the land surface from one satellite was combined with another satellite's data about the clouds to create the image. (credit: modification of work by R. Stockli, A. Nelson, F. Hasler, NASA/ GSFC/ NOAA/ USGS)

Our nearest astronomical neighbor is Earth's satellite, commonly called the *Moon*. Figure 1.7 shows Earth and the Moon drawn to scale on the same diagram. Notice how small we have to make these bodies to fit them on the page with the right scale. The Moon's distance from Earth is about 30 times Earth's diameter, or approximately 384,000 kilometers, and it takes about a month for the Moon to revolve around Earth. The Moon's diameter is 3476 kilometers, about one fourth the size of Earth.

Figure 1.7. Earth and Moon, Drawn to Scale. This image shows Earth and the Moon shown to scale for both size and distance. (credit: modification of work by NASA)

Light (or radio waves) takes 1.3 seconds to travel between Earth and the Moon. If you've seen videos of the Apollo flights to the Moon, you may recall that there was a delay of about 3 seconds between the time Mission Control asked a question and the time the astronauts responded. This was not because the astronomers were thinking slowly, but rather because it took the radio waves almost 3 seconds to make the round trip.

Earth revolves around our star, the Sun, which is about 150 million kilometers away—approximately 400 times as far away from us as the Moon. We call the average Earth–Sun distance an *astronomical unit* (AU) because, in the early days of astronomy, it was the most important measuring standard. Light takes slightly more than 8 minutes to travel 1 astronomical unit, which means the latest news we receive from the Sun is always 8 minutes old. The diameter of the Sun is about 1.5 million kilometers; Earth could fit comfortably inside one of the minor eruptions that occurs on the surface of our star. If the Sun were reduced to the size of a basketball, Earth would be a small apple seed about 30 meters from the ball.

It takes Earth 1 year (3×10^7 seconds) to go around the Sun at our distance; to make it around, we must travel at approximately 110,000 kilometers per hour. (If you, like many students, still prefer miles to kilometers, you might find the following trick helpful. To convert kilometers to miles, just multiply kilometers by 0.6. Thus,

Download for free at http://cnx.org/content/col11992/latest/

110,000 kilometers per hour becomes 66,000 miles per hour.) Because gravity holds us firmly to Earth and there is no resistance to Earth's motion in the vacuum of space, we participate in this extremely fast-moving trip without being aware of it day to day.

Earth is only one of eight planets that revolve around the Sun. These planets, along with their moons and swarms of smaller bodies such as dwarf planets, make up the solar system (Figure 1.8). A planet is defined as a body of significant size that orbits a star and does not produce its own light. (If a large body consistently produces its own light, it is then called a *star*.) Later in the book this definition will be modified a bit, but it is perfectly fine for now as you begin your voyage.

Figure 1.8. Our Solar Family. The Sun, the planets, and some dwarf planets are shown with their sizes drawn to scale. The orbits of the planets are much more widely separated than shown in this drawing. Notice the size of Earth compared to the giant planets. (credit: modification of work by NASA)

We are able to see the nearby planets in our skies only because they reflect the light of our local star, the Sun. If the planets were much farther away, the tiny amount of light they reflect would usually not be visible to us. The planets we have so far discovered orbiting other stars were found from the pull their gravity exerts on their parent stars, or from the light they block from their stars when they pass in front of them. We can't see most of these planets directly, although a few are now being imaged directly.

The Sun is our local star, and all the other stars are also enormous balls of glowing gas that generate vast amounts of energy by nuclear reactions deep within. We will discuss the processes that cause stars to shine in more detail later in the book. The other stars look faint only because they are so very far away. If we continue our basketball analogy, Proxima Centauri, the nearest star beyond the Sun, which is 4.3 light-years away, would be almost 7000 kilometers from the basketball.

When you look up at a star-filled sky on a clear night, all the stars visible to the unaided eye are part of a single collection of stars we call the *Milky Way Galaxy*, or simply the *Galaxy*. (When referring to the Milky Way, we capitalize *Galaxy*; when talking about other galaxies of stars, we use lowercase *galaxy*.) The Sun is one of hundreds of billions of stars that make up the Galaxy; its extent, as we will see, staggers the human imagination. Within a sphere 10 light-years in radius centered on the Sun, we find roughly ten stars. Within a sphere 100 light-years in radius, there are roughly 10,000 (10^4) stars—far too many to count or name—but we have still

This OpenStax book is available for free at http://cnx.org/content/col11992/1.8

traversed only a tiny part of the Milky Way Galaxy. Within a 1000-light-year sphere, we find some ten million (10^7) stars; within a sphere of 100,000 light-years, we finally encompass the entire Milky Way Galaxy.

Our Galaxy looks like a giant disk with a small ball in the middle. If we could move outside our Galaxy and look down on the disk of the Milky Way from above, it would probably resemble the galaxy in Figure 1.9, with its spiral structure outlined by the blue light of hot adolescent stars.

Figure 1.9. Spiral Galaxy. This galaxy of billions of stars, called by its catalog number NGC 1073, is thought to be similar to our own Milky Way Galaxy. Here we see the giant wheel-shaped system with a bar of stars across its middle. (credit: NASA, ESA)

The Sun is somewhat less than 30,000 light-years from the center of the Galaxy, in a location with nothing much to distinguish it. From our position inside the Milky Way Galaxy, we cannot see through to its far rim (at least not with ordinary light) because the space between the stars is not completely empty. It contains a sparse distribution of gas (mostly the simplest element, hydrogen) intermixed with tiny solid particles that we call *interstellar dust*. This gas and dust collect into enormous clouds in many places in the Galaxy, becoming the raw material for future generations of stars. Figure 1.10 shows an image of the disk of the Galaxy as seen from our vantage point.

Download for free at http://cnx.org/content/col11992/latest/

Figure 1.10. Milky Way Galaxy. Because we are inside the Milky Way Galaxy, we see its disk in cross-section flung across the sky like a great milky white avenue of stars with dark "rifts" of dust. In this dramatic image, part of it is seen above Trona Pinnacles in the California desert. (credit: Ian Norman)

Typically, the interstellar material is so extremely sparse that the space between stars is a much better vacuum than anything we can produce in terrestrial laboratories. Yet, the dust in space, building up over thousands of light-years, can block the light of more distant stars. Like the distant buildings that disappear from our view on a smoggy day in Los Angeles, the more distant regions of the Milky Way cannot be seen behind the layers of interstellar smog. Luckily, astronomers have found that stars and raw material shine with various forms of light, some of which do penetrate the smog, and so we have been able to develop a pretty good map of the Galaxy.

Recent observations, however, have also revealed a rather surprising and disturbing fact. There appears to be more—much more—to the Galaxy than meets the eye (or the telescope). From various investigations, we have evidence that much of our Galaxy is made of material we cannot currently observe directly with our instruments. We therefore call this component of the Galaxy *dark matter*. We know the dark matter is there by the pull its gravity exerts on the stars and raw material we can observe, but what this dark matter is made of and how much of it exists remain a mystery. Furthermore, this dark matter is not confined to our Galaxy; it appears to be an important part of other star groupings as well.

By the way, not all stars live by themselves, as the Sun does. Many are born in double or triple systems with two, three, or more stars revolving about each other. Because the stars influence each other in such close systems, multiple stars allow us to measure characteristics that we cannot discern from observing single stars. In a number of places, enough stars have formed together that we recognized them as star clusters (Figure

This OpenStax book is available for free at http://cnx.org/content/col11992/1.8

1.11). Some of the largest of the star clusters that astronomers have cataloged contain hundreds of thousands of stars and take up volumes of space hundreds of light-years across.

Figure 1.11. Star Cluster. This large star cluster is known by its catalog number, M9. It contains some 250,000 stars and is seen more clearly from space using the Hubble Space Telescope. It is located roughly 25,000 light-years away. (credit: NASA, ESA)

You may hear stars referred to as "eternal," but in fact no star can last forever. Since the "business" of stars is making energy, and energy production requires some sort of fuel to be used up, eventually all stars run out of fuel. This news should not cause you to panic, though, because our Sun still has at least 5 or 6 billion years to go. Ultimately, the Sun and all stars will die, and it is in their death throes that some of the most intriguing and important processes of the universe are revealed. For example, we now know that many of the atoms in our bodies were once inside stars. These stars exploded at the ends of their lives, recycling their material back into the reservoir of the Galaxy. In this sense, all of us are literally made of recycled "star dust."

1.7 THE UNIVERSE ON THE LARGE SCALE

In a very rough sense, you could think of the solar system as your house or apartment and the Galaxy as your town, made up of many houses and buildings. In the twentieth century, astronomers were able to show that, just as our world is made up of many, many towns, so the universe is made up of enormous numbers of galaxies. (We define the universe to be everything that exists that is accessible to our observations.) Galaxies stretch as far into space as our telescopes can see, many billions of them within the reach of modern instruments. When they were first discovered, some astronomers called galaxies *island universes*, and the term is aptly descriptive; galaxies do look like islands of stars in the vast, dark seas of intergalactic space.

The nearest galaxy, discovered in 1993, is a small one that lies 75,000 light-years from the Sun in the direction of the constellation Sagittarius, where the smog in our own Galaxy makes it especially difficult to discern. (A constellation, we should note, is one of the 88 sections into which astronomers divide the sky, each named after a prominent star pattern within it.) Beyond this Sagittarius dwarf galaxy lie two other small galaxies, about 160,000 light-years away. First recorded by Magellan's crew as he sailed around the world, these are called the *Magellanic Clouds* (Figure 1.12). All three of these small galaxies are satellites of the Milky Way Galaxy, interacting with it through the force of gravity. Ultimately, all three may even be swallowed by our much larger Galaxy, as other small galaxies have been over the course of cosmic time.

Download for free at http://cnx.org/content/col11992/latest/

Figure 1.12. Neighbor Galaxies. This image shows both the Large Magellanic Cloud and the Small Magellanic Cloud above the telescopes of the Atacama Large Millimeter/Submillimeter Array (ALMA) in the Atacama Desert of northern Chile. (credit: ESO, C. Malin)

The nearest large galaxy is a spiral quite similar to our own, located in the constellation of Andromeda, and is thus called the Andromeda galaxy; it is also known by one of its catalog numbers, M31 (Figure 1.13). M31 is a little more than 2 million light-years away and, along with the Milky Way, is part of a small cluster of more than 50 galaxies referred to as the *Local Group*.

This OpenStax book is available for free at http://cnx.org/content/col11992/1.8

Download for free at http://cnx.org/content/col11992/latest/

Figure 1.13. Closest Spiral Galaxy. The Andromeda galaxy (M31) is a spiral-shaped collection of stars similar to our own Milky Way. (credit: Adam Evans)

At distances of 10 to 15 million light-years, we find other small galaxy groups, and then at about 50 million light-years there are more impressive systems with thousands of member galaxies. We have discovered that galaxies occur mostly in clusters, both large and small (Figure 1.14).

Download for free at http://cnx.org/content/col11992/latest/

Figure 1.14. Fornax Cluster of Galaxies. In this image, you can see part of a cluster of galaxies located about 60 million light-years away in the constellation of Fornax. All the objects that are not pinpoints of light in the picture are galaxies of billions of stars. (credit: ESO, J. Emerson, VISTA. Acknowledgment: Cambridge Astronomical Survey Unit)

Some of the clusters themselves form into larger groups called *superclusters*. The Local Group is part of a supercluster of galaxies, called the Virgo Supercluster, which stretches over a diameter of 110 million light-years. We are just beginning to explore the structure of the universe at these enormous scales and are already encountering some unexpected findings.

At even greater distances, where many ordinary galaxies are too dim to see, we find *quasars*. These are brilliant centers of galaxies, glowing with the light of an extraordinarily energetic process. The enormous energy of the quasars is produced by gas that is heated to a temperature of millions of degrees as it falls toward a massive black hole and swirls around it. The brilliance of quasars makes them the most distant beacons we can see in the dark oceans of space. They allow us to probe the universe 10 billion light-years away or more, and thus 10 billion years or more in the past.

With quasars we can see way back close to the Big Bang explosion that marks the beginning of time. Beyond the quasars and the most distant visible galaxies, we have detected the feeble glow of the explosion itself, filling the universe and thus coming to us from all directions in space. The discovery of this "afterglow of creation" is considered to be one of the most significant events in twentieth-century science, and we are still exploring the many things it has to tell us about the earliest times of the universe.

Measurements of the properties of galaxies and quasars in remote locations require large telescopes, sophisticated light-amplifying devices, and painstaking labor. Every clear night, at observatories around the world, astronomers and students are at work on such mysteries as the birth of new stars and the large-scale structure of the universe, fitting their results into the tapestry of our understanding.

This OpenStax book is available for free at http://cnx.org/content/col11992/1.8

Download for free at http://cnx.org/content/col11992/latest/

1.8 THE UNIVERSE OF THE VERY SMALL

The foregoing discussion has likely impressed on you that the universe is extraordinarily large and extraordinarily empty. On average, it is 10,000 times more empty than our Galaxy. Yet, as we have seen, even the Galaxy is mostly empty space. The air we breathe has about 10^{19} atoms in each cubic centimeter—and we usually think of air as empty space. In the interstellar gas of the Galaxy, there is about one atom in every cubic centimeter. Intergalactic space is filled so sparsely that to find one atom, on average, we must search through a cubic meter of space. Most of the universe is fantastically empty; places that are dense, such as the human body, are tremendously rare.

Even our most familiar solids are mostly space. If we could take apart such a solid, piece by piece, we would eventually reach the tiny molecules from which it is formed. Molecules are the smallest particles into which any matter can be divided while still retaining its chemical properties. A molecule of water (H_2O), for example, consists of two hydrogen atoms and one oxygen atom bonded together.

Molecules, in turn, are built of atoms, which are the smallest particles of an element that can still be identified as that element. For example, an atom of gold is the smallest possible piece of gold. Nearly 100 different kinds of atoms (elements) exist in nature. Most of them are rare, and only a handful account for more than 99% of everything with which we come in contact. The most abundant elements in the cosmos today are listed in Table 1.1; think of this table as the "greatest hits" of the universe when it comes to elements.

The Cosmically Abundant Elements

Element[1]	Symbol	Number of Atoms per Million Hydrogen Atoms
Hydrogen	H	1,000,000
Helium	He	80,000
Carbon	C	450
Nitrogen	N	92
Oxygen	O	740
Neon	Ne	130
Magnesium	Mg	40
Silicon	Si	37
Sulfur	S	19
Iron	Fe	32

Table 1.1

---

1   This list of elements is arranged in order of the atomic number, which is the number of protons in each nucleus.

Download for free at http://cnx.org/content/col11992/latest/

All atoms consist of a central, positively charged nucleus surrounded by negatively charged electrons. The bulk of the matter in each atom is found in the nucleus, which consists of positive protons and electrically neutral neutrons all bound tightly together in a very small space. Each element is defined by the number of protons in its atoms. Thus, any atom with 6 protons in its nucleus is called *carbon*, any with 50 protons is called *tin*, and any with 70 protons is called *ytterbium*. (For a list of the elements, see Appendix K.)

The distance from an atomic nucleus to its electrons is typically 100,000 times the size of the nucleus itself. This is why we say that even solid matter is mostly space. The typical atom is far emptier than the solar system out to Neptune. (The distance from Earth to the Sun, for example, is only 100 times the size of the Sun.) This is one reason atoms are not like miniature solar systems.

Remarkably, physicists have discovered that everything that happens in the universe, from the smallest atomic nucleus to the largest superclusters of galaxies, can be explained through the action of only four forces: gravity, electromagnetism (which combines the actions of electricity and magnetism), and two forces that act at the nuclear level. The fact that there are four forces (and not a million, or just one) has puzzled physicists and astronomers for many years and has led to a quest for a unified picture of nature.

LINK TO LEARNING

To construct an atom, particle by particle, check out this guided animation (https://openstax.org/l/30buildanatom) for building an atom.

1.9 A CONCLUSION AND A BEGINNING

If you are new to astronomy, you have probably reached the end of our brief tour in this chapter with mixed emotions. On the one hand, you may be fascinated by some of the new ideas you've read about and you may be eager to learn more. On the other hand, you may be feeling a bit overwhelmed by the number of topics we have covered, and the number of new words and ideas we have introduced. Learning astronomy is a little like learning a new language: at first it seems there are so many new expressions that you'll never master them all, but with practice, you soon develop facility with them.

At this point you may also feel a bit small and insignificant, dwarfed by the cosmic scales of distance and time. But, there is another way to look at what you have learned from our first glimpses of the cosmos. Let us consider the history of the universe from the Big Bang to today and compress it, for easy reference, into a single year. (We have borrowed this idea from Carl Sagan's 1997 Pulitzer Prize-winning book, *The Dragons of Eden*.)

On this scale, the Big Bang happened at the first moment of January 1, and this moment, when you are reading this chapter would be the end of the very last second of December 31. When did other events in the development of the universe happen in this "cosmic year?" Our solar system formed around September 10, and the oldest rocks we can date on Earth go back to the third week in September (Figure 1.15).

Download for free at http://cnx.org/content/col11992/latest/

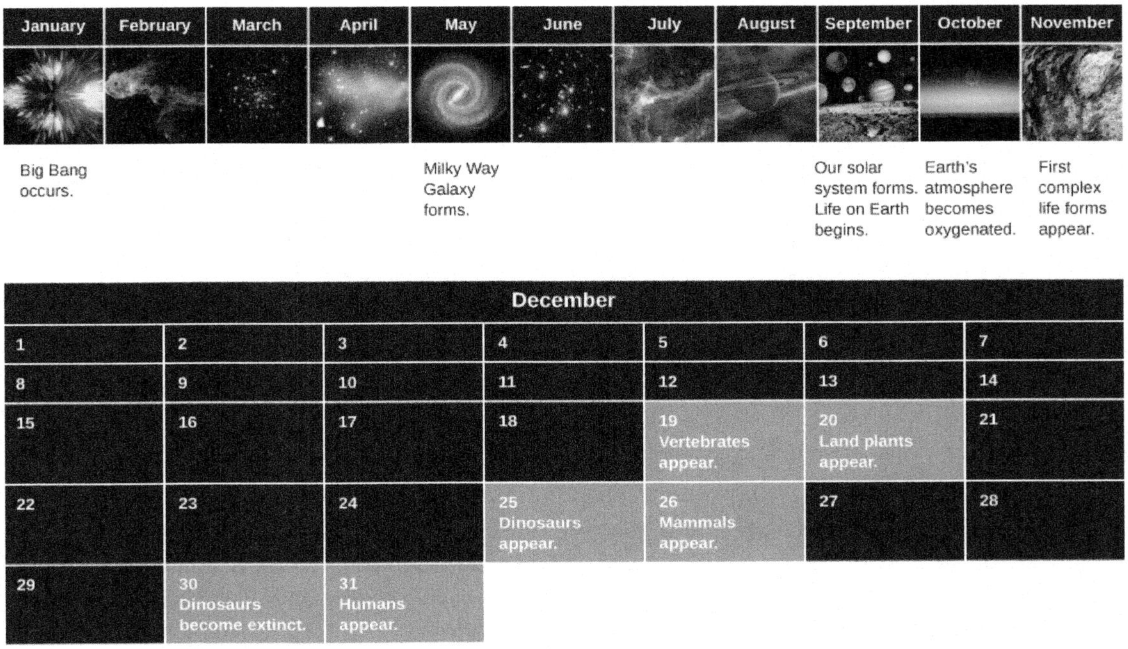

January	February	March	April	May	June	July	August	September	October	November

Big Bang occurs.

Milky Way Galaxy forms.

Our solar system forms. Life on Earth begins.

Earth's atmosphere becomes oxygenated.

First complex life forms appear.

December						
1	2	3	4	5	6	7
8	9	10	11	12	13	14
15	16	17	18	19 Vertebrates appear.	20 Land plants appear.	21
22	23	24	25 Dinosaurs appear.	26 Mammals appear.	27	28
29	30 Dinosaurs become extinct.	31 Humans appear.				

Figure 1.15. Charting Cosmic Time. On a cosmic calendar, where the time since the Big Bang is compressed into 1 year, creatures we would call human do not emerge on the scene until the evening of December 31. (credit: February: modification of work by NASA, JPL-Caltech, W. Reach (SSC/Caltech); March: modification of work by ESA, Hubble and NASA, Acknowledgement: Giles Chapdelaine; April: modification of work by NASA, ESA, CFHT, CXO, M.J. Jee (University of California, Davis), A. Mahdavi (San Francisco State University); May: modification of work by NASA, JPL-Caltech; June: modification of work by NASA/ESA; July: modification of work by NASA, JPL-Caltech, Harvard-Smithsonian; August: modification of work by NASA, JPL-Caltech, R. Hurt (SSC-Caltech); September: modification of work by NASA; October: modification of work by NASA; November: modification of work by Dénes Emőke)

Where does the origin of human beings fall during the course of this cosmic year? The answer turns out to be the evening of December 31. The invention of the alphabet doesn't occur until the fiftieth second of 11:59 p.m. on December 31. And the beginnings of modern astronomy are a mere fraction of a second before the New Year. Seen in a cosmic context, the amount of time we have had to study the stars is minute, and our success in piecing together as much of the story as we have is remarkable.

Certainly our attempts to understand the universe are not complete. As new technologies and new ideas allow us to gather more and better data about the cosmos, our present picture of astronomy will very likely undergo many changes. Still, as you read our current progress report on the exploration of the universe, take a few minutes every once in a while just to savor how much you have already learned.

 FOR FURTHER EXPLORATION

Books

Miller, Ron, and William Hartmann. *The Grand Tour: A Traveler's Guide to the Solar System*. 3rd ed. Workman, 2005. This volume for beginners is a colorfully illustrated voyage among the planets.

Sagan, Carl. *Cosmos.* Ballantine, 2013 [1980]. This tome presents a classic overview of astronomy by an astronomer who had a true gift for explaining things clearly. (You can also check out Sagan's television series *Cosmos: A Personal Voyage* and Neil DeGrasse Tyson's current series *Cosmos: A Spacetime Odyssey.*)

Download for free at http://cnx.org/content/col11992/latest/

Tyson, Neil DeGrasse, and Don Goldsmith. *Origins: Fourteen Billion Years of Cosmic Evolution.* Norton, 2004. This book provides a guided tour through the beginnings of the universe, galaxies, stars, planets, and life.

Websites

If you enjoyed the beautiful images in this chapter (and there are many more fabulous photos to come in other chapters), you may want to know where you can obtain and download such pictures for your own enjoyment. (Many astronomy images are from government-supported instruments or projects, paid for by tax dollars, and therefore are free of copyright laws.) Here are three resources we especially like:

- Astronomy Picture of the Day: apod.nasa.gov/apod/astropix.html. Two space scientists scour the Internet and select one beautiful astronomy image to feature each day. Their archives range widely, from images of planets and nebulae to rockets and space instruments; they also have many photos of the night sky. The search function (see the menu on the bottom of the page) works quite well for finding something specific among the many years' worth of daily images.

- Hubble Space Telescope Images: www.hubblesite.org/newscenter/archive/browse/images. Starting at this page, you can select from among hundreds of Hubble pictures by subject or by date. Note that many of the images have supporting pictures with them, such as diagrams, animations, or comparisons. Excellent captions and background information are provided. Other ways to approach these images are through the more public-oriented Hubble Gallery (www.hubblesite.org/gallery) and the European homepage (www.spacetelescope.org/images).

- National Aeronautics and Space Administration's (NASA's) Planetary Photojournal: photojournal.jpl.nasa.gov. This site features thousands of images from planetary exploration, with captions of varied length. You can select images by world, feature name, date, or catalog number, and download images in a number of popular formats. However, only NASA mission images are included. Note the Photojournal Search option on the menu at the top of the homepage to access ways to search their archives.

Videos

Cosmic Voyage: www.youtube.com/watch?v=qxXf7AJZ73A. This video presents a portion of Cosmic Voyage, narrated by Morgan Freeman (8:34).

Powers of Ten: www.youtube.com/watch?v=0fKBhvDjuy0. This classic short video is a much earlier version of Powers of Ten, narrated by Philip Morrison (9:00).

The Known Universe: www.youtube.com/watch?v=17jymDn0W6U. This video tour from the American Museum of Natural History has realistic animation, music, and captions (6:30).

Wanderers: apod.nasa.gov/apod/ap141208.html. This video provides a tour of the solar system, with narrative by Carl Sagan, imagining other worlds with dramatically realistic paintings (3:50).

This OpenStax book is available for free at http://cnx.org/content/col11992/1.8

2 OBSERVING THE SKY: THE BIRTH OF ASTRONOMY

Figure 2.1. Night Sky. In this panoramic photograph of the night sky from the Atacama Desert in Chile, we can see the central portion of the Milky Way Galaxy arcing upward in the center of the frame. On the left, the Large Magellanic Cloud and the Small Magellanic Cloud (smaller galaxies that orbit the Milky Way Galaxy) are easily visible from the Southern Hemisphere. (credit: modification of work by ESO/Y. Beletsky)

Chapter Outline

✎ Thinking Ahead

Much to your surprise, a member of the Flat Earth Society moves in next door. He believes that Earth is flat and all the NASA images of a spherical Earth are either faked or simply show the round (but flat) disk of Earth from above. How could you prove to your new neighbor that Earth really is a sphere? (When you've thought about this on your own, you can check later in the chapter for some suggested answers.)

Today, few people really spend much time looking at the night sky. In ancient days, before electric lights robbed so many people of the beauty of the sky, the stars and planets were an important aspect of everyone's daily life. All the records that we have—on paper and in stone—show that ancient civilizations around the world noticed, worshipped, and tried to understand the lights in the sky and fit them into their own view of the world. These ancient observers found both majestic regularity and never-ending surprise in the motions of the heavens. Through their careful study of the planets, the Greeks and later the Romans laid the foundation of the science of astronomy.

Download for free at http://cnx.org/content/col11992/latest/

2.1 THE SKY ABOVE

Learning Objectives

By the end of this section, you will be able to:

> Define the main features of the celestial sphere
> Explain the system astronomers use to describe the sky
> Describe how motions of the stars appear to us on Earth
> Describe how motions of the Sun, Moon, and planets appear to us on Earth
> Understand the modern meaning of the term *constellation*

Our senses suggest to us that Earth is the center of the universe—the hub around which the heavens turn. This **geocentric** (Earth-centered) view was what almost everyone believed until the European Renaissance. After all, it is simple, logical, and seemingly self-evident. Furthermore, the geocentric perspective reinforced those philosophical and religious systems that taught the unique role of human beings as the central focus of the cosmos. However, the geocentric view happens to be wrong. One of the great themes of our intellectual history is the overthrow of the geocentric perspective. Let us, therefore, take a look at the steps by which we reevaluated the place of our world in the cosmic order.

The Celestial Sphere

If you go on a camping trip or live far from city lights, your view of the sky on a clear night is pretty much identical to that seen by people all over the world before the invention of the telescope. Gazing up, you get the impression that the sky is a great hollow dome with you at the center (Figure 2.2), and all the stars are an equal distance from you on the surface of the dome. The top of that dome, the point directly above your head, is called the **zenith**, and where the dome meets Earth is called the **horizon**. From the sea or a flat prairie, it is easy to see the horizon as a circle around you, but from most places where people live today, the horizon is at least partially hidden by mountains, trees, buildings, or smog.

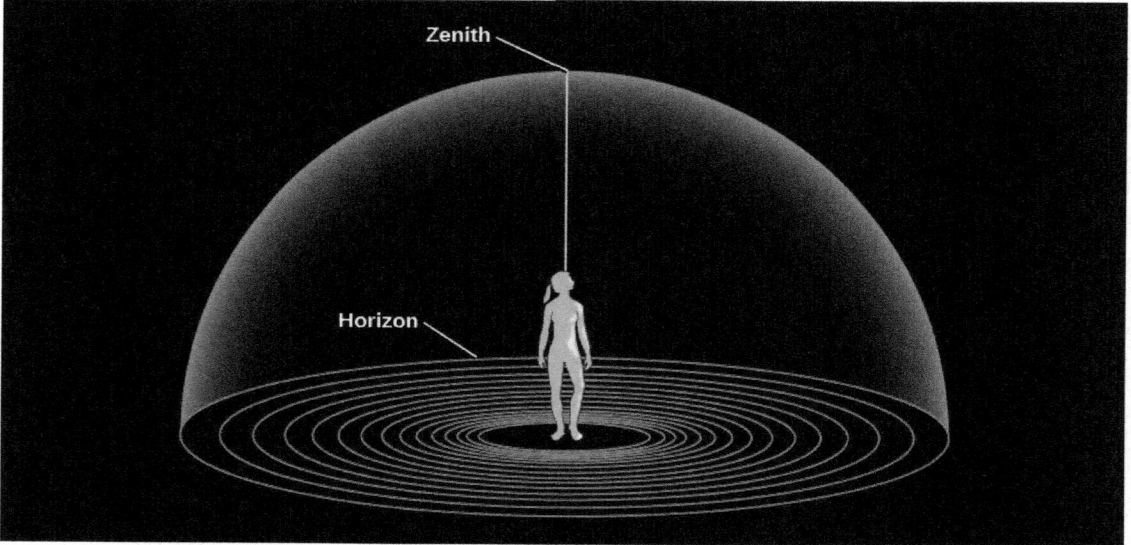

Figure 2.2. The Sky around Us. The horizon is where the sky meets the ground; an observer's zenith is the point directly overhead.

This OpenStax book is available for free at http://cnx.org/content/col11992/1.8

Download for free at http://cnx.org/content/col11992/latest/

If you lie back in an open field and observe the night sky for hours, as ancient shepherds and travelers regularly did, you will see stars rising on the eastern horizon (just as the Sun and Moon do), moving across the dome of the sky in the course of the night, and setting on the western horizon. Watching the sky turn like this night after night, you might eventually get the idea that the dome of the sky is really part of a great sphere that is turning around you, bringing different stars into view as it turns. The early Greeks regarded the sky as just such a **celestial sphere** (Figure 2.3). Some thought of it as an actual sphere of transparent crystalline material, with the stars embedded in it like tiny jewels.

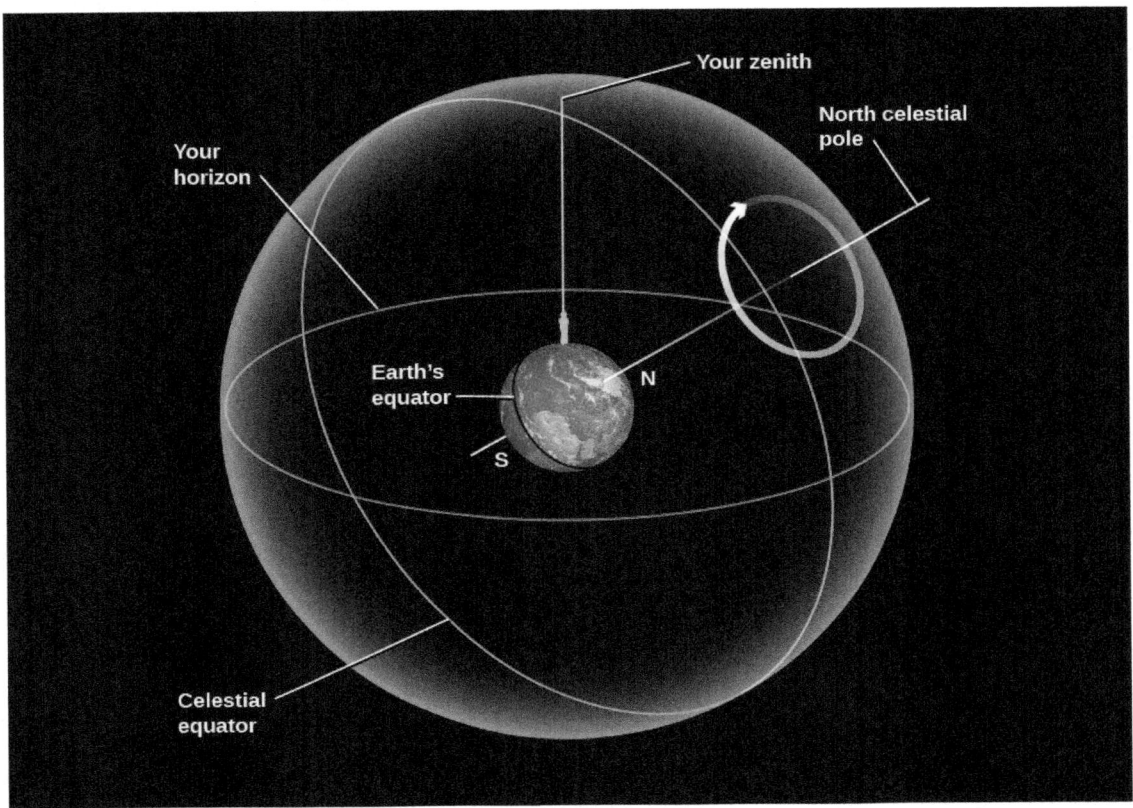

Figure 2.3. Circles on the Celestial Sphere. Here we show the (imaginary) celestial sphere around Earth, on which objects are fixed, and which rotates around Earth on an axis. In reality, it is Earth that turns around this axis, creating the illusion that the sky revolves around us. Note that Earth in this picture has been tilted so that your location is at the top and the North Pole is where the N is. The apparent motion of celestial objects in the sky around the pole is shown by the circular arrow.

Today, we know that it is not the celestial sphere that turns as night and day proceed, but rather the planet on which we live. We can put an imaginary stick through Earth's North and South Poles, representing our planet's axis. It is because Earth turns on this axis every 24 hours that we see the Sun, Moon, and stars rise and set with clockwork regularity. Today, we know that these celestial objects are not really on a dome, but at greatly varying distances from us in space. Nevertheless, it is sometimes still convenient to talk about the celestial dome or sphere to help us keep track of objects in the sky. There is even a special theater, called a *planetarium*, in which we project a simulation of the stars and planets onto a white dome.

As the celestial sphere rotates, the objects on it maintain their positions with respect to one another. A grouping of stars such as the Big Dipper has the same shape during the course of the night, although it turns with the

Download for free at http://cnx.org/content/col11992/latest/

sky. During a single night, even objects we know to have significant motions of their own, such as the nearby planets, seem fixed relative to the stars. Only meteors—brief "shooting stars" that flash into view for just a few seconds—move appreciably with respect to other objects on the celestial sphere. (This is because they are not stars at all. Rather, they are small pieces of cosmic dust, burning up as they hit Earth's atmosphere.) We can use the fact that the entire celestial sphere seems to turn together to help us set up systems for keeping track of what things are visible in the sky and where they happen to be at a given time.

Celestial Poles and Celestial Equator

To help orient us in the turning sky, astronomers use a system that extends Earth's axis points into the sky. Imagine a line going through Earth, connecting the North and South Poles. This is Earth's axis, and Earth rotates about this line. If we extend this imaginary line outward from Earth, the points where this line intersects the celestial sphere are called the *north celestial pole* and the *south celestial pole*. As Earth rotates about its axis, the sky appears to turn in the opposite direction around those **celestial poles** (Figure 2.4). We also (in our imagination) throw Earth's equator onto the sky and call this the **celestial equator**. It lies halfway between the celestial poles, just as Earth's equator lies halfway between our planet's poles.

Figure 2.4. Circling the South Celestial Pole. This long-exposure photo shows trails left by stars as a result of the apparent rotation of the celestial sphere around the south celestial pole. (In reality, it is Earth that rotates.) (Credit: ESO/Iztok Bončina)

Now let's imagine how riding on different parts of our spinning Earth affects our view of the sky. The apparent motion of the celestial sphere depends on your latitude (position north or south of the equator). First of all, notice that Earth's axis is pointing at the celestial poles, so these two points in the sky do not appear to turn.

If you stood at the North Pole of Earth, for example, you would see the north celestial pole overhead, at your zenith. The celestial equator, 90° from the celestial poles, would lie along your horizon. As you watched the stars during the course of the night, they would all circle around the celestial pole, with none rising or setting. Only that half of the sky north of the celestial equator is ever visible to an observer at the North Pole. Similarly, an observer at the South Pole would see only the southern half of the sky.

If you were at Earth's equator, on the other hand, you see the celestial equator (which, after all, is just an "extension" of Earth's equator) pass overhead through your zenith. The celestial poles, being 90° from the celestial equator, must then be at the north and south points on your horizon. As the sky turns, all stars rise

This OpenStax book is available for free at http://cnx.org/content/col11992/1.8

and set; they move straight up from the east side of the horizon and set straight down on the west side. During a 24-hour period, all stars are above the horizon exactly half the time. (Of course, during some of those hours, the Sun is too bright for us to see them.)

What would an observer in the latitudes of the United States or Europe see? Remember, we are neither at Earth's pole nor at the equator, but in between them. For those in the continental United States and Europe, the north celestial pole is neither overhead nor on the horizon, but in between. It appears above the northern horizon at an angular height, or altitude, equal to the observer's latitude. In San Francisco, for example, where the latitude is 38° N, the north celestial pole is 38° above the northern horizon.

For an observer at 38° N latitude, the south celestial pole is 38° below the southern horizon and, thus, never visible. As Earth turns, the whole sky seems to pivot about the north celestial pole. For this observer, stars within 38° of the North Pole can never set. They are always above the horizon, day and night. This part of the sky is called the north **circumpolar zone**. For observers in the continental United States, the Big Dipper, Little Dipper, and Cassiopeia are examples of star groups in the north circumpolar zone. On the other hand, stars within 38° of the south celestial pole never rise. That part of the sky is the south circumpolar zone. To most U.S. observers, the Southern Cross is in that zone. (Don't worry if you are not familiar with the star groups just mentioned; we will introduce them more formally later on.)

LINK TO LEARNING

The Rotating Sky Lab (https://openstaxcollege.org/l/30rotatingsky) created by the University of Nebraska–Lincoln provides an interactive demonstration that introduces the horizon coordinate system, the apparent rotation of the sky, and allows for exploration of the relationship between the horizon and celestial equatorial coordinate systems.

At this particular time in Earth's history, there happens to be a star very close to the north celestial pole. It is called Polaris, the pole star, and has the distinction of being the star that moves the least amount as the northern sky turns each day. Because it moved so little while the other stars moved much more, it played a special role in the mythology of several Native American tribes, for example (some called it the "fastener of the sky").

ASTRONOMY BASICS

What's Your Angle?

Astronomers measure how far apart objects appear in the sky by using angles. By definition, there are 360° in a circle, so a circle stretching completely around the celestial sphere contains 360°. The half-sphere or dome of the sky then contains 180° from horizon to opposite horizon. Thus, if two stars are 18° apart, their separation spans about 1/10 of the dome of the sky. To give you a sense of how big a degree is, the full Moon is about half a degree across. This is about the width of your smallest finger (pinkie) seen at arm's length.

Download for free at http://cnx.org/content/col11992/latest/

Rising and Setting of the Sun

We described the movement of stars in the night sky, but what about during the daytime? The stars continue to circle during the day, but the brilliance of the Sun makes them difficult to see. (The Moon can often be seen in the daylight, however.) On any given day, we can think of the Sun as being located at some position on the hypothetical celestial sphere. When the Sun rises—that is, when the rotation of Earth carries the Sun above the horizon—sunlight is scattered by the molecules of our atmosphere, filling our sky with light and hiding the stars above the horizon.

For thousands of years, astronomers have been aware that the Sun does more than just rise and set. It changes position gradually on the celestial sphere, moving each day about 1° to the east relative to the stars. Very reasonably, the ancients thought this meant the Sun was slowly moving around Earth, taking a period of time we call 1 **year** to make a full circle. Today, of course, we know it is Earth that is going around the Sun, but the effect is the same: the Sun's position in our sky changes day to day. We have a similar experience when we walk around a campfire at night; we see the flames appear in front of each person seated about the fire in turn.

The path the Sun appears to take around the celestial sphere each year is called the **ecliptic** (Figure 2.5). Because of its motion on the ecliptic, the Sun rises about 4 minutes later each day with respect to the stars. Earth must make just a bit more than one complete rotation (with respect to the stars) to bring the Sun up again.

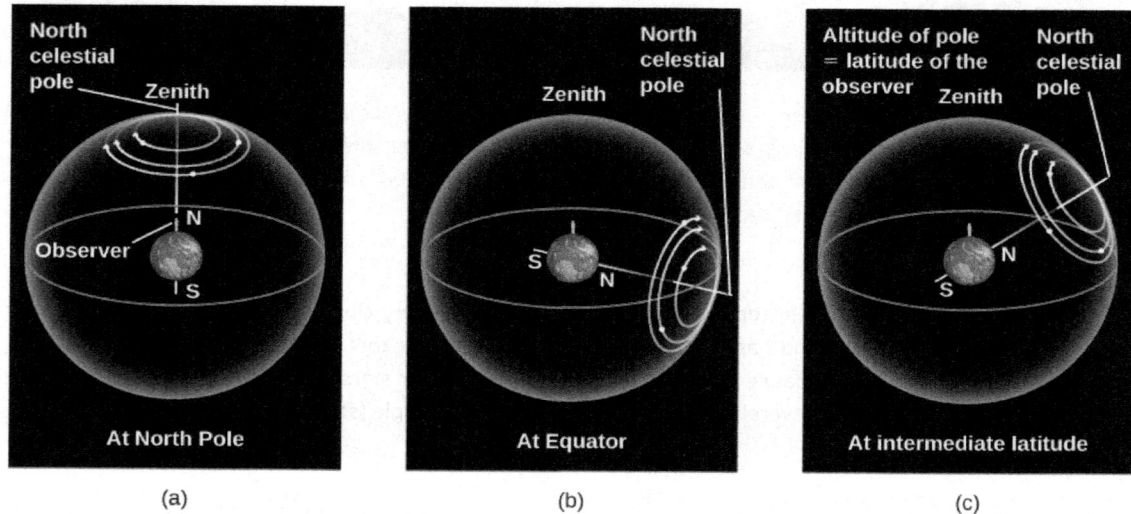

Figure 2.5. Star Circles at Different Latitudes. The turning of the sky looks different depending on your latitude on Earth. (a) At the North Pole, the stars circle the zenith and do not rise and set. (b) At the equator, the celestial poles are on the horizon, and the stars rise straight up and set straight down. (c) At intermediate latitudes, the north celestial pole is at some position between overhead and the horizon. Its angle above the horizon turns out to be equal to the observer's latitude. Stars rise and set at an angle to the horizon.

As the months go by and we look at the Sun from different places in our orbit, we see it projected against different places in our orbit, and thus against different stars in the background (Figure 2.6 and Table 2.1)—or we would, at least, if we could see the stars in the daytime. In practice, we must deduce which stars lie behind and beyond the Sun by observing the stars visible in the opposite direction at night. After a year, when Earth has completed one trip around the Sun, the Sun will appear to have completed one circuit of the sky along the ecliptic.

This OpenStax book is available for free at http://cnx.org/content/col11992/1.8

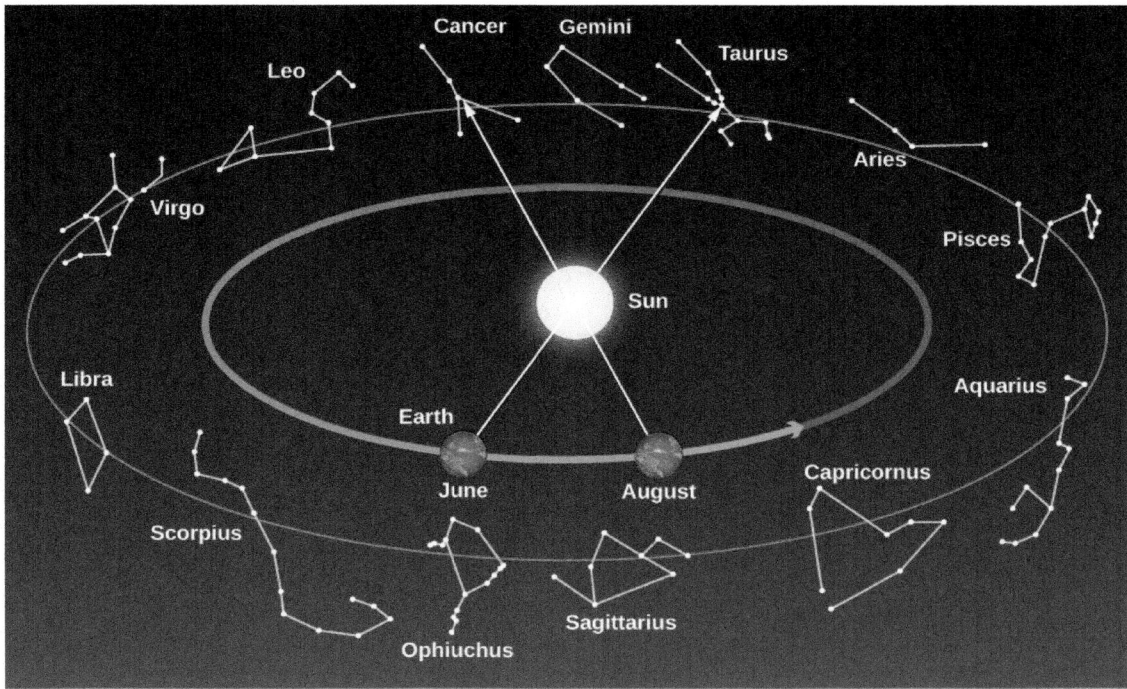

Figure 2.6. Constellations on the Ecliptic. As Earth revolves around the Sun, we sit on "platform Earth" and see the Sun moving around the sky. The circle in the sky that the Sun appears to make around us in the course of a year is called the *ecliptic*. This circle (like all circles in the sky) goes through a set of constellations. The ancients thought these constellations, which the Sun (and the Moon and planets) visited, must be special and incorporated them into their system of astrology. Note that at any given time of the year, some of the constellations crossed by the ecliptic are visible in the night sky; others are in the day sky and are thus hidden by the brilliance of the Sun.

Constellations on the Ecliptic

Constellation on the Ecliptic	Dates When the Sun Crosses It
Capricornus	January 21–February 16
Aquarius	February 16–March 11
Pisces	March 11–April 18
Aries	April 18–May 13
Taurus	May 13–June 22
Gemini	June 22–July 21
Cancer	July 21–August 10
Leo	August 10–September 16
Virgo	September 16–October 31

Table 2.1

Download for free at http://cnx.org/content/col11992/latest/

Constellations on the Ecliptic

Constellation on the Ecliptic	Dates When the Sun Crosses It
Libra	October 31–November 23
Scorpius	November 23–November 29
Ophiuchus	November 29–December 18
Sagittarius	December 18–January 21

Table 2.1

The ecliptic does not lie along the celestial equator but is inclined to it at an angle of about 23.5°. In other words, the Sun's annual path in the sky is not linked with Earth's equator. This is because our planet's axis of rotation is tilted by about 23.5° from a vertical line sticking out of the plane of the ecliptic (Figure 2.7). Being tilted from "straight up" is not at all unusual among celestial bodies; Uranus and Pluto are actually tilted so much that they orbit the Sun "on their side."

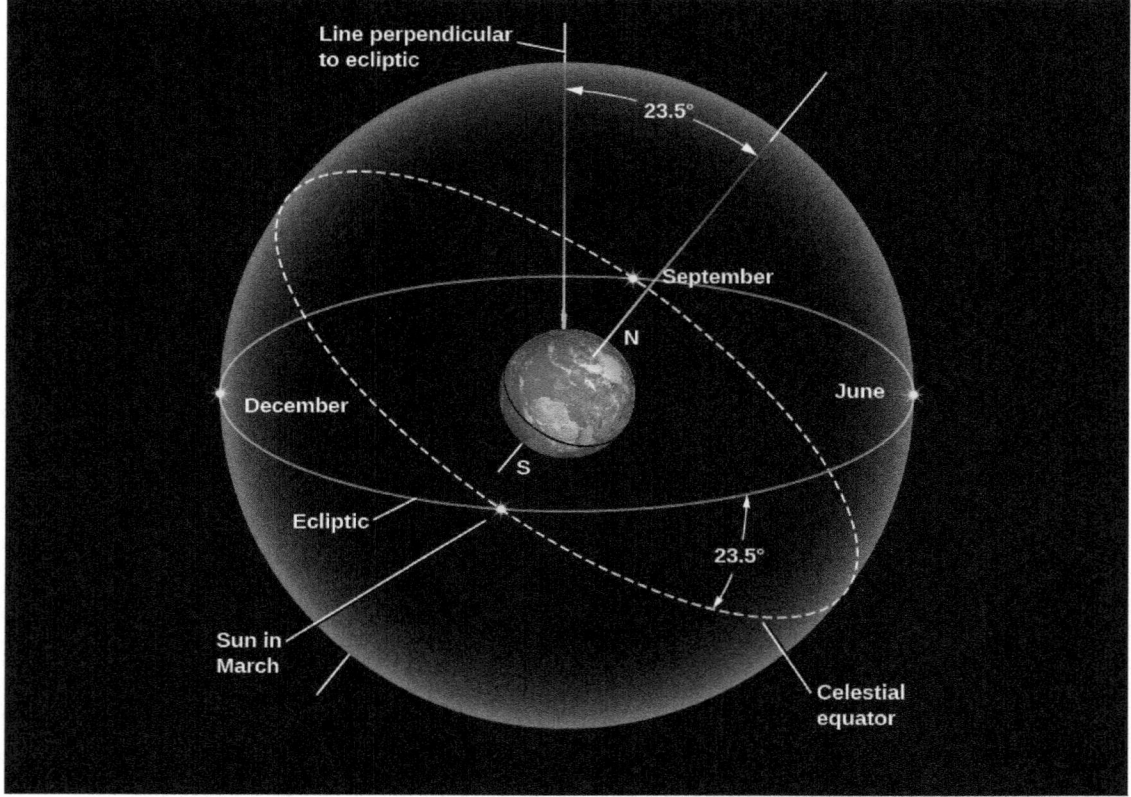

Figure 2.7. The Celestial Tilt. The celestial equator is tilted by 23.5° to the ecliptic. As a result, North Americans and Europeans see the Sun north of the celestial equator and high in our sky in June, and south of the celestial equator and low in the sky in December.

This OpenStax book is available for free at http://cnx.org/content/col11992/1.8

The inclination of the ecliptic is the reason the Sun moves north and south in the sky as the seasons change. In Earth, Moon, and Sky, we discuss the progression of the seasons in more detail.

Fixed and Wandering Stars

The Sun is not the only object that moves among the fixed stars. The Moon and each of the planets that are visible to the unaided eye—Mercury, Venus, Mars, Jupiter, Saturn, and Uranus (although just barely)—also change their positions slowly from day to day. During a single day, the Moon and planets all rise and set as Earth turns, just as the Sun and stars do. But like the Sun, they have independent motions among the stars, superimposed on the daily rotation of the celestial sphere. Noticing these motions, the Greeks of 2000 years ago distinguished between what they called the *fixed stars*—those that maintain fixed patterns among themselves through many generations—and the *wandering stars*, or **planets**. The word "planet," in fact, means "wanderer" in ancient Greek.

Today, we do not regard the Sun and Moon as planets, but the ancients applied the term to all seven of the moving objects in the sky. Much of ancient astronomy was devoted to observing and predicting the motions of these celestial wanderers. They even dedicated a unit of time, the week, to the seven objects that move on their own; that's why there are 7 days in a week. The Moon, being Earth's nearest celestial neighbor, has the fastest apparent motion; it completes a trip around the sky in about 1 month (or *moonth*). To do this, the Moon moves about 12°, or 24 times its own apparent width on the sky, each day.

EXAMPLE 2.1

Angles in the Sky

A circle consists of 360 degrees (°). When we measure the angle in the sky that something moves, we can use this formula:

$$\text{speed} = \frac{\text{distance}}{\text{time}}$$

This is true whether the motion is measured in kilometers per hour or degrees per hour; we just need to use consistent units.

As an example, let's say you notice the bright star Sirius due south from your observing location in the Northern Hemisphere. You note the time, and then later, you note the time that Sirius sets below the horizon. You find that Sirius has traveled an angular distance of about 75° in 5 h. About how many hours will it take for Sirius to return to its original location?

Solution

The speed of Sirius is $\frac{75°}{5\,\text{h}} = \frac{15°}{1\,\text{h}}$. If we want to know the time required for Sirius to return to its original location, we need to wait until it goes around a full circle, or 360°. Rearranging the formula for speed we were originally given, we find:

$$\text{time} = \frac{\text{distance}}{\text{speed}} = \frac{360°}{15°/\text{h}} = 24\,\text{h}$$

The actual time is a few minutes shorter than this, and we will explore why in a later chapter.

Download for free at http://cnx.org/content/col11992/latest/

Check Your Learning

The Moon moves in the sky relative to the background stars (in addition to moving with the stars as a result of Earth's rotation.) Go outside at night and note the position of the Moon relative to nearby stars. Repeat the observation a few hours later. How far has the Moon moved? (For reference, the diameter of the Moon is about 0.5°.) Based on your estimate of its motion, how long will it take for the Moon to return to the position relative to the stars in which you first observed it?

Answer:

The speed of the moon is 0.5°/1 h. To move a full 360°, the moon needs 720 h: $\frac{0.5°}{1 \text{ h}} = \frac{360°}{720 \text{ h}}$. Dividing 720 h by the conversion factor of 24 h/day reveals the lunar cycle is about 30 days.

The individual paths of the Moon and planets in the sky all lie close to the ecliptic, although not exactly on it. This is because the paths of the planets about the Sun, and of the Moon about Earth, are all in nearly the same plane, as if they were circles on a huge sheet of paper. The planets, the Sun, and the Moon are thus always found in the sky within a narrow 18-degree-wide belt, centered on the ecliptic, called the **zodiac** (Figure 2.6). (The root of the term "zodiac" is the same as that of the word "zoo" and means a collection of animals; many of the patterns of stars within the zodiac belt reminded the ancients of animals, such as a fish or a goat.)

How the planets appear to move in the sky as the months pass is a combination of their actual motions plus the motion of Earth about the Sun; consequently, their paths are somewhat complex. As we will see, this complexity has fascinated and challenged astronomers for centuries.

Constellations

The backdrop for the motions of the "wanderers" in the sky is the canopy of stars. If there were no clouds in the sky and we were on a flat plain with nothing to obstruct our view, we could see about 3000 stars with the unaided eye. To find their way around such a multitude, the ancients found groupings of stars that made some familiar geometric pattern or (more rarely) resembled something they knew. Each civilization found its own patterns in the stars, much like a modern Rorschach test in which you are asked to discern patterns or pictures in a set of inkblots. The ancient Chinese, Egyptians, and Greeks, among others, found their own groupings—or constellations—of stars. These were helpful in navigating among the stars and in passing their star lore on to their children.

You may be familiar with some of the old star patterns we still use today, such as the Big Dipper, Little Dipper, and Orion the hunter, with his distinctive belt of three stars (Figure 2.8). However, many of the stars we see are not part of a distinctive star pattern at all, and a telescope reveals millions of stars too faint for the eye to see. Therefore, during the early decades of the 20th century, astronomers from many countries decided to establish a more formal system for organizing the sky.

This OpenStax book is available for free at http://cnx.org/content/col11992/1.8

Download for free at http://cnx.org/content/col11992/latest/

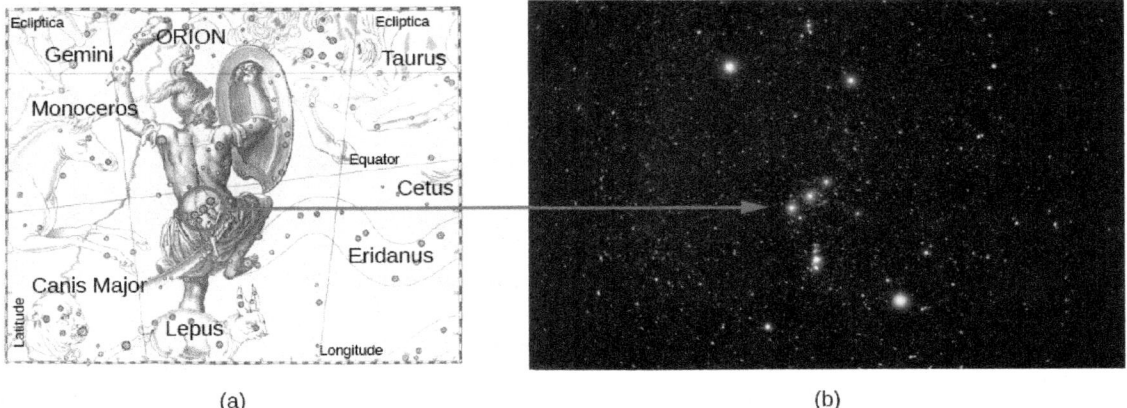

(a)   (b)

Figure 2.8. Orion. (a) The winter constellation of Orion, the hunter, is surrounded by neighboring constellations, as illustrated in the seventeenth-century atlas by Hevelius. (b) A photograph shows the Orion region in the sky. Note the three blue stars that make up the belt of the hunter. The bright red star above the belt denotes his armpit and is called Betelgeuse (pronounced "Beetel-juice"). The bright blue star below the belt is his foot and is called Rigel. (credit a: modification of work by Johannes Hevelius; b: modification of work by Matthew Spinelli)

Today, we use the term *constellation* to mean one of 88 sectors into which we divide the sky, much as the United States is divided into 50 states. The modern boundaries between the constellations are imaginary lines in the sky running north–south and east–west, so that each point in the sky falls in a specific constellation, although, like the states, not all constellations are the same size. All the constellations are listed in Appendix L. Whenever possible, we have named each modern constellation after the Latin translations of one of the ancient Greek star patterns that lies within it. Thus, the modern constellation of Orion is a kind of box on the sky, which includes, among many other objects, the stars that made up the ancient picture of the hunter. Some people use the term *asterism* to denote an especially noticeable star pattern within a constellation (or sometimes spanning parts of several constellations). For example, the Big Dipper is an asterism within the constellation of Ursa Major, the Big Bear.

Students are sometimes puzzled because the constellations seldom resemble the people or animals for which they were named. In all likelihood, the Greeks themselves did not name groupings of stars because they looked like actual people or subjects (any more than the outline of Washington state resembles George Washington). Rather, they named sections of the sky in honor of the characters in their mythology and then fit the star configurations to the animals and people as best they could.

LINK TO LEARNING

This website about objects in the sky (https://openstaxcollege.org/l/30heavensabove) allows users to construct a detailed sky map showing the location and information about the Sun, Moon, planets, stars, constellations, and even satellites orbiting Earth. Begin by setting your observing location using the option in the menu in the upper right corner of the screen.

Download for free at http://cnx.org/content/col11992/latest/

2.2 | ANCIENT ASTRONOMY

Learning Objectives

By the end of this section, you will be able to:

> Describe early examples of astronomy around the world
> Explain how Greek astronomers were able to deduce that Earth is spherical
> Explain how Greek astronomers were able to calculate Earth's size
> Describe the motion of Earth called precession
> Describe Ptolemy's geocentric system of planetary motion

Let us now look briefly back into history. Much of modern Western civilization is derived in one way or another from the ideas of the ancient Greeks and Romans, and this is true in astronomy as well. However, many other ancient cultures also developed sophisticated systems for observing and interpreting the sky.

Astronomy around the World

Ancient Babylonian, Assyrian, and Egyptian astronomers knew the approximate length of the year. The Egyptians of 3000 years ago, for example, adopted a calendar based on a 365-day year. They kept careful track of the rising time of the bright star Sirius in the predawn sky, which has a yearly cycle that corresponded with the flooding of the Nile River. The Chinese also had a working calendar; they determined the length of the year at about the same time as the Egyptians. The Chinese also recorded comets, bright meteors, and dark spots on the Sun. (Many types of astronomical objects were introduced in Science and the Universe: A Brief Tour. If you are not familiar with terms like *comets* and *meteors*, you may want to review that chapter.) Later, Chinese astronomers kept careful records of "guest stars"—those that are normally too faint to see but suddenly flare up to become visible to the unaided eye for a few weeks or months. We still use some of these records in studying stars that exploded a long time ago.

The Mayan culture in Mexico and Central America developed a sophisticated calendar based on the planet Venus, and they made astronomical observations from sites dedicated to this purpose a thousand years ago. The Polynesians learned to navigate by the stars over hundreds of kilometers of open ocean—a skill that enabled them to colonize new islands far away from where they began.

In Britain, before the widespread use of writing, ancient people used stones to keep track of the motions of the Sun and Moon. We still find some of the great stone circles they built for this purpose, dating from as far back as 2800 BCE. The best known of these is Stonehenge, which is discussed in Earth, Moon, and Sky.

Early Greek and Roman Cosmology

Our concept of the cosmos—its basic structure and origin—is called **cosmology**, a word with Greek roots. Before the invention of telescopes, humans had to depend on the simple evidence of their senses for a picture of the universe. The ancients developed cosmologies that combined their direct view of the heavens with a rich variety of philosophical and religious symbolism.

At least 2000 years before Columbus, educated people in the eastern Mediterranean region knew Earth was round. Belief in a spherical Earth may have stemmed from the time of Pythagoras, a philosopher and mathematician who lived 2500 years ago. He believed circles and spheres to be "perfect forms" and suggested that Earth should therefore be a sphere. As evidence that the gods liked spheres, the Greeks cited the fact that the Moon is a sphere, using evidence we describe later.

This OpenStax book is available for free at http://cnx.org/content/col11992/1.8

The writings of Aristotle (384–322 BCE), the tutor of Alexander the Great, summarize many of the ideas of his day. They describe how the progression of the Moon's phases—its apparent changing shape—results from our seeing different portions of the Moon's sunlit hemisphere as the month goes by (see Earth, Moon, and Sky). Aristotle also knew that the Sun has to be farther away from Earth than is the Moon because occasionally the Moon passed exactly between Earth and the Sun and hid the Sun temporarily from view. We call this a *solar eclipse*.

Aristotle cited convincing arguments that Earth must be round. First is the fact that as the Moon enters or emerges from Earth's shadow during an eclipse of the Moon, the shape of the shadow seen on the Moon is always round (Figure 2.9). Only a spherical object always produces a round shadow. If Earth were a disk, for example, there would be some occasions when the sunlight would strike it edge-on and its shadow on the Moon would be a line.

Figure 2.9. Earth's Round Shadow. A lunar eclipse occurs when the Moon moves into and out of Earth's shadow. Note the curved shape of the shadow—evidence for a spherical Earth that has been recognized since antiquity. (credit: modification of work by Brian Paczkowski)

As a second argument, Aristotle explained that travelers who go south a significant distance are able to observe stars that are not visible farther north. And the height of the North Star—the star nearest the north celestial pole—decreases as a traveler moves south. On a flat Earth, everyone would see the same stars overhead. The only possible explanation is that the traveler must have moved over a curved surface on Earth, showing stars from a different angle. (See the How Do We Know Earth Is Round? feature for more ideas on proving Earth is round.)

One Greek thinker, Aristarchus of Samos (310–230 BCE), even suggested that Earth was moving around the Sun, but Aristotle and most of the ancient Greek scholars rejected this idea. One of the reasons for their conclusion was the thought that if Earth moved about the Sun, they would be observing the stars from different places along Earth's orbit. As Earth moved along, nearby stars should shift their positions in the sky relative to more distant stars. In a similar way, we see foreground objects appear to move against a more distant background whenever we are in motion. When we ride on a train, the trees in the foreground appear to shift their position relative to distant hills as the train rolls by. Unconsciously, we use this phenomenon all of the time to estimate distances around us.

The apparent shift in the direction of an object as a result of the motion of the observer is called **parallax**. We call the shift in the apparent direction of a star due to Earth's orbital motion *stellar parallax*. The Greeks made dedicated efforts to observe stellar parallax, even enlisting the aid of Greek soldiers with the clearest vision, but to no avail. The brighter (and presumably nearer) stars just did not seem to shift as the Greeks observed them in the spring and then again in the fall (when Earth is on the opposite side of the Sun).

This meant either that Earth was not moving or that the stars had to be so tremendously far away that the parallax shift was immeasurably small. A cosmos of such enormous extent required a leap of imagination that most ancient philosophers were not prepared to make, so they retreated to the safety of the Earth-centered view, which would dominate Western thinking for nearly two millennia.

Download for free at http://cnx.org/content/col11992/latest/

ASTRONOMY BASICS

How Do We Know Earth Is Round?

In addition to the two ways (from Aristotle's writings) discussed in this chapter, you might also reason as follows:

1. Let's watch a ship leave its port and sail into the distance on a clear day. On a flat Earth, we would just see the ship get smaller and smaller as it sails away. But this isn't what we actually observe. Instead, ships sink below the horizon, with the hull disappearing first and the mast remaining visible for a while longer. Eventually, only the top of the mast can be seen as the ship sails around the curvature of Earth. Finally, the ship disappears under the horizon.

2. The International Space Station circles Earth once every 90 minutes or so. Photographs taken from the shuttle and other satellites show that Earth is round from every perspective.

3. Suppose you made a friend in each time zone of Earth. You call all of them at the same hour and ask, "Where is the Sun?" On a flat Earth, each caller would give you roughly the same answer. But on a round Earth you would find that, for some friends, the Sun would be high in the sky whereas for others it would be rising, setting, or completely out of sight (and this last group of friends would be upset with you for waking them up).

Measurement of Earth by Eratosthenes

The Greeks not only knew Earth was round, but also they were able to measure its size. The first fairly accurate determination of Earth's diameter was made in about 200 BCE by Eratosthenes (276–194 BCE), a Greek living in Alexandria, Egypt. His method was a geometric one, based on observations of the Sun.

The Sun is so distant from us that all the light rays that strike our planet approach us along essentially parallel lines. To see why, look at Figure 2.10. Take a source of light near Earth—say, at position A. Its rays strike different parts of Earth along diverging paths. From a light source at B, or at C (which is still farther away), the angle between rays that strike opposite parts of Earth is smaller. The more distant the source, the smaller the angle between the rays. For a source infinitely distant, the rays travel along parallel lines.

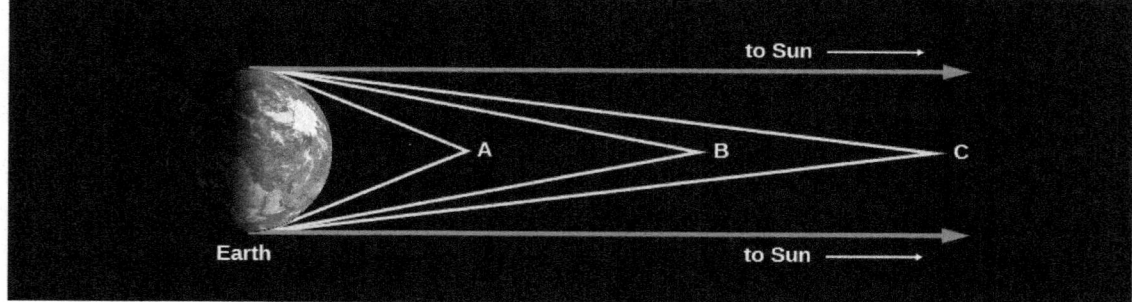

Figure 2.10. Light Rays from Space. The more distant an object, the more nearly parallel the rays of light coming from it.

Of course, the Sun is not infinitely far away, but given its distance of 150 million kilometers, light rays striking Earth from a point on the Sun diverge from one another by an angle far too small to be observed with the unaided eye. As a consequence, if people all over Earth who could see the Sun were to point at it, their fingers

This OpenStax book is available for free at http://cnx.org/content/col11992/1.8

Download for free at http://cnx.org/content/col11992/latest/

would, essentially, all be parallel to one another. (The same is also true for the planets and stars—an idea we will use in our discussion of how telescopes work.)

Eratosthenes was told that on the first day of summer at Syene, Egypt (near modern Aswan), sunlight struck the bottom of a vertical well at noon. This indicated that the Sun was directly over the well—meaning that Syene was on a direct line from the center of Earth to the Sun. At the corresponding time and date in Alexandria, Eratosthenes observed the shadow a column made and saw that the Sun was not directly overhead, but was slightly south of the zenith, so that its rays made an angle with the vertical equal to about 1/50 of a circle (7°). Because the Sun's rays striking the two cities are parallel to one another, why would the two rays not make the same angle with Earth's surface? Eratosthenes reasoned that the curvature of the round Earth meant that "straight up" was not the same in the two cities. And the measurement of the angle in Alexandria, he realized, allowed him to figure out the size of Earth. Alexandria, he saw, must be 1/50 of Earth's circumference north of Syene (Figure 2.11). Alexandria had been measured to be 5000 stadia north of Syene. (The *stadium* was a Greek unit of length, derived from the length of the racetrack in a stadium.) Eratosthenes thus found that Earth's circumference must be 50 × 5000, or 250,000 stadia.

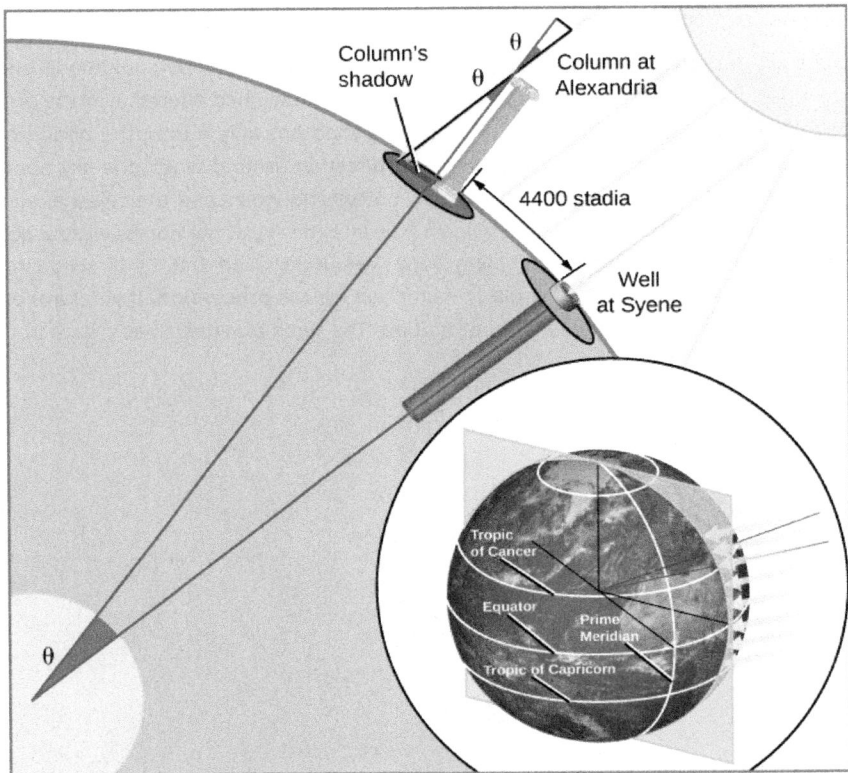

Figure 2.11. How Eratosthenes Measured the Size of Earth. Eratosthenes measured the size of Earth by observing the angle at which the Sun's rays hit our planet's surface. The Sun's rays come in parallel, but because Earth's surface curves, a ray at Syene comes straight down whereas a ray at Alexandria makes an angle of 7° with the vertical. That means, in effect, that at Alexandria, Earth's surface has curved away from Syene by 7° of 360°, or 1/50 of a full circle. Thus, the distance between the two cities must be 1/50 the circumference of Earth. (credit: modification of work by NOAA Ocean Service Education)

It is not possible to evaluate precisely the accuracy of Eratosthenes solution because there is doubt about which of the various kinds of Greek stadia he used as his unit of distance. If it was the common Olympic stadium, his result is about 20% too large. According to another interpretation, he used a stadium equal to

Download for free at http://cnx.org/content/col11992/latest/

about 1/6 kilometer, in which case his figure was within 1% of the correct value of 40,000 kilometers. Even if his measurement was not exact, his success at measuring the size of our planet by using only shadows, sunlight, and the power of human thought was one of the greatest intellectual achievements in history.

Hipparchus and Precession

Perhaps the greatest astronomer of antiquity was Hipparchus, born in Nicaea in what is present-day Turkey. He erected an observatory on the island of Rhodes around 150 BCE, when the Roman Republic was expanding its influence throughout the Mediterranean region. There he measured, as accurately as possible, the positions of objects in the sky, compiling a pioneering star catalog with about 850 entries. He designated celestial coordinates for each star, specifying its position in the sky, just as we specify the position of a point on Earth by giving its latitude and longitude.

He also divided the stars into **apparent magnitudes** according to their apparent brightness. He called the brightest ones "stars of the first magnitude"; the next brightest group, "stars of the second magnitude"; and so forth. This rather arbitrary system, in modified form, still remains in use today (although it is less and less useful for professional astronomers).

By observing the stars and comparing his data with older observations, Hipparchus made one of his most remarkable discoveries: the position in the sky of the north celestial pole had altered over the previous century and a half. Hipparchus deduced correctly that this had happened not only during the period covered by his observations, but was in fact happening all the time: the direction around which the sky appears to rotate changes slowly but continuously. Recall from the section on celestial poles and the celestial equator that the north celestial pole is just the projection of Earth's North Pole into the sky. If the north celestial pole is wobbling around, then Earth itself must be doing the wobbling. Today, we understand that the direction in which Earth's axis points does indeed change slowly but regularly—a motion we call **precession**. If you have ever watched a spinning top wobble, you observed a similar kind of motion. The top's axis describes a path in the shape of a cone, as Earth's gravity tries to topple it (Figure 2.12).

This OpenStax book is available for free at http://cnx.org/content/col11992/1.8

Download for free at http://cnx.org/content/col11992/latest/

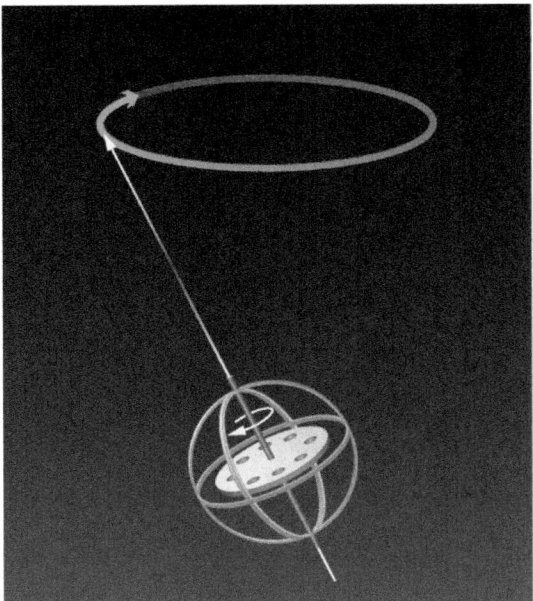 

Figure 2.12. Precession. Just as the axis of a rapidly spinning top wobbles slowly in a circle, so the axis of Earth wobbles in a 26,000-year cycle. Today the north celestial pole is near the star Polaris, but about 5000 years ago it was close to a star called Thuban, and in 14,000 years it will be closest to the star Vega.

Because our planet is not an exact sphere, but bulges a bit at the equator, the pulls of the Sun and Moon cause it to wobble like a top. It takes about 26,000 years for Earth's axis to complete one circle of precession. As a result of this motion, the point where our axis points in the sky changes as time goes on. While Polaris is the star closest to the north celestial pole today (it will reach its closest point around the year 2100), the star Vega in the constellation of Lyra will be the North Star in 14,000 years.

Ptolemy's Model of the Solar System

The last great astronomer of the Roman era was Claudius Ptolemy (or Ptolemaeus), who flourished in Alexandria in about the year 140. He wrote a mammoth compilation of astronomical knowledge, which today is called by its Arabic name, *Almagest* (meaning "The Greatest"). *Almagest* does not deal exclusively with Ptolemy's own work; it includes a discussion of the astronomical achievements of the past, principally those of Hipparchus. Today, it is our main source of information about the work of Hipparchus and other Greek astronomers.

Ptolemy's most important contribution was a geometric representation of the solar system that predicted the positions of the planets for any desired date and time. Hipparchus, not having enough data on hand to solve the problem himself, had instead amassed observational material for posterity to use. Ptolemy supplemented this material with new observations of his own and produced a cosmological model that endured more than a thousand years, until the time of Copernicus.

The complicating factor in explaining the motions of the planets is that their apparent wandering in the sky results from the combination of their own motions with Earth's orbital revolution. As we watch the planets from our vantage point on the moving Earth, it is a little like watching a car race while you are competing in it. Sometimes opponents' cars pass you, but at other times you pass them, making them appear to move backward for a while with respect to you.

Download for free at http://cnx.org/content/col11992/latest/

Figure 2.13 shows the motion of Earth and a planet farther from the Sun—in this case, Mars. Earth travels around the Sun in the same direction as the other planet and in nearly the same plane, but its orbital speed is faster. As a result, it overtakes the planet periodically, like a faster race car on the inside track. The figure shows where we see the planet in the sky at different times. The path of the planet among the stars is illustrated in the star field on the right side of the figure.

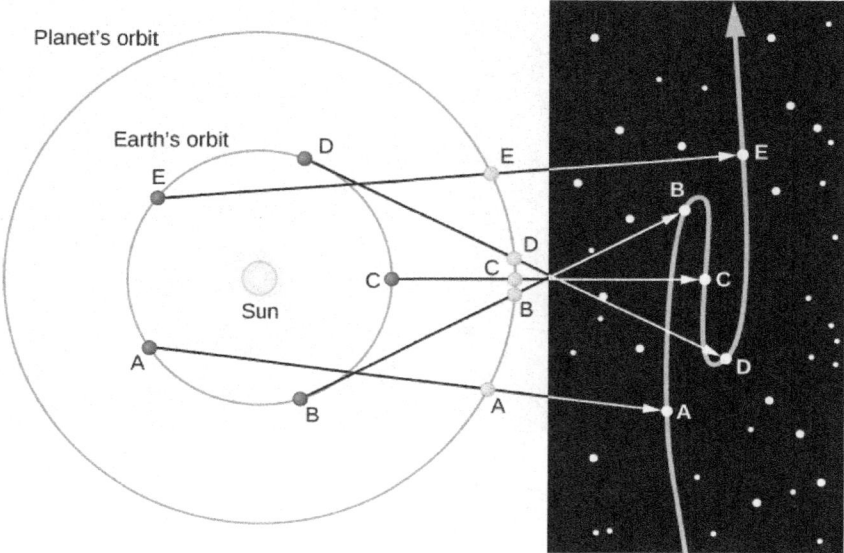

Figure 2.13. Retrograde Motion of a Planet beyond Earth's Orbit. The letters on the diagram show where Earth and Mars are at different times. By following the lines from each Earth position through each corresponding Mars position, you can see how the retrograde path of Mars looks against the background stars.

LINK TO LEARNING

This retrograde simulation of Mars (https://openstaxcollege.org/l/30marsretrograd) illustrates the motion of Mars as seen from Earth as well as Earth's retrograde motion as seen from Mars. There is also an animation of the movement of the two planets relative to each other that creates the appearance of this motion.

Normally, planets move eastward in the sky over the weeks and months as they orbit the Sun, but from positions B to D in Figure 2.13, as Earth passes the planets in our example, it appears to drift backward, moving west in the sky. Even though it is actually moving to the east, the faster-moving Earth has overtaken it and seems, from our perspective, to be leaving it behind. As Earth rounds its orbit toward position E, the planet again takes up its apparent eastward motion in the sky. The temporary apparent westward motion of a planet as Earth swings between it and the Sun is called **retrograde motion**. Such backward motion is much easier for us to understand today, now that we know Earth is one of the moving planets and not the unmoving center of all creation. But Ptolemy was faced with the far more complex problem of explaining such motion while assuming a stationary Earth.

Furthermore, because the Greeks believed that celestial motions had to be circles, Ptolemy had to construct his model using circles alone. To do it, he needed dozens of circles, some moving around other circles, in a complex

This OpenStax book is available for free at http://cnx.org/content/col11992/1.8

structure that makes a modern viewer dizzy. But we must not let our modern judgment cloud our admiration for Ptolemy's achievement. In his day, a complex universe centered on Earth was perfectly reasonable and, in its own way, quite beautiful. However, as Alfonso X, the King of Castile, was reported to have said after having the Ptolemaic system of planet motions explained to him, "If the Lord Almighty had consulted me before embarking upon Creation, I should have recommended something simpler."

Ptolemy solved the problem of explaining the observed motions of planets by having each planet revolve in a small orbit called an **epicycle**. The center of the epicycle then revolved about Earth on a circle called a *deferent* (Figure 2.14). When the planet is at position *x* in Figure 2.14 on the epicycle orbit, it is moving in the same direction as the center of the epicycle; from Earth, the planet appears to be moving eastward. When the planet is at *y*, however, its motion is in the direction opposite to the motion of the epicycle's center around Earth. By choosing the right combination of speeds and distances, Ptolemy succeeded in having the planet moving westward at the correct speed and for the correct interval of time, thus replicating retrograde motion with his model.

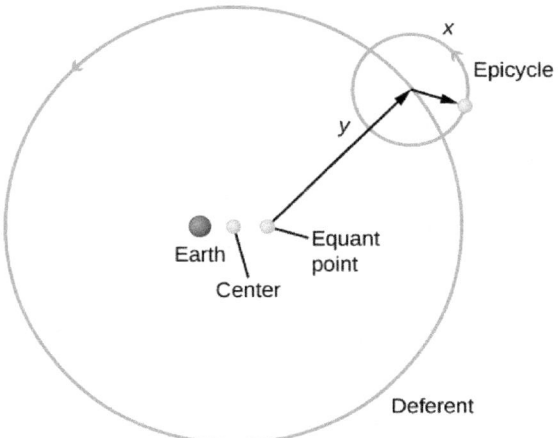

Figure 2.14. Ptolemy's Complicated Cosmological System. Each planet orbits around a small circle called an *epicycle*. Each epicycle orbits on a larger circle called the *deferent*. This system is not centered exactly on Earth but on an offset point called the *equant*. The Greeks needed all this complexity to explain the actual motions in the sky because they believed that Earth was stationary and that all sky motions had to be circular.

However, we shall see in Orbits and Gravity that the planets, like Earth, travel about the Sun in orbits that are ellipses, not circles. Their actual behavior cannot be represented accurately by a scheme of uniform circular motions. In order to match the observed motions of the planets, Ptolemy had to center the deferent circles, not on Earth, but at points some distance from Earth. In addition, he introduced uniform circular motion around yet another axis, called the *equant point*. All of these considerably complicated his scheme.

It is a tribute to the genius of Ptolemy as a mathematician that he was able to develop such a complex system to account successfully for the observations of planets. It may be that Ptolemy did not intend for his cosmological model to describe reality, but merely to serve as a mathematical representation that allowed him to predict the positions of the planets at any time. Whatever his thinking, his model, with some modifications, was eventually accepted as authoritative in the Muslim world and (later) in Christian Europe.

2.3 ASTROLOGY AND ASTRONOMY

Learning Objectives

By the end of this section, you will be able to:

Download for free at http://cnx.org/content/col11992/latest/

> Explain the origins of astrology
> Explain what a horoscope is
> Summarize the arguments that invalidate astrology as a scientific practice

Many ancient cultures regarded the planets and stars as representatives or symbols of the gods or other supernatural forces that controlled their lives. For them, the study of the heavens was not an abstract subject; it was connected directly to the life-and-death necessity of understanding the actions of the gods and currying favor with them. Before the time of our scientific perspectives, everything that happened in nature—from the weather, to diseases and accidents, to celestial surprises such as eclipses or new comets—was thought to be an expression of the whims or displeasure of the gods. Any signs that helped people understand what these gods had in mind were considered extremely important.

The movements of the seven objects that had the power to "wander" through the realm of the sky—the Sun, the Moon, and five planets visible to the unaided eye—clearly must have special significance in such a system of thinking.

Most ancient cultures associated these seven objects with various supernatural rulers in their pantheon and kept track of them for religious reasons. Even in the comparatively sophisticated Greece of antiquity, the planets had the names of gods and were credited with having the same powers and influences as the gods whose names they bore. From such ideas was born the ancient system called **astrology**, still practiced by some people today, in which the positions of these bodies among the stars of the zodiac are thought to hold the key to understanding what we can expect from life.

The Beginnings of Astrology

Astrology began in Babylonia about two and half millennia ago. The Babylonians, believing the planets and their motions influenced the fortunes of kings and nations, used their knowledge of astronomy to guide their rulers. When the Babylonian culture was absorbed by the Greeks, astrology gradually came to influence the entire Western world and eventually spread to Asia as well.

By the 2nd century BCE the Greeks democratized astrology by developing the idea that the planets influence every individual. In particular, they believed that the configuration of the Sun, Moon, and planets at the moment of birth affected a person's personality and fortune—a doctrine called *natal astrology*. Natal astrology reached its peak with Ptolemy 400 years later. As famous for his astrology as for his astronomy, Ptolemy compiled the *Tetrabiblos*, a treatise on astrology that remains the "bible" of the subject. It is essentially this ancient religion, older than Christianity or Islam, that is still practiced by today's astrologers.

The Horoscope

The key to natal astrology is the **horoscope**, a chart showing the positions of the planets in the sky at the moment of an individual's birth. The word "horoscope" comes from the Greek words *hora* (meaning "time") and *skopos* (meaning a "watcher" or "marker"), so "horoscope" can literally be translated as "marker of the hour." When a horoscope is charted, the planets (including the Sun and Moon, classed as *wanderers* by the ancients) must first be located in the zodiac. At the time astrology was set up, the zodiac was divided into 12 sectors called *signs* (Figure 2.15), each 30° long. Each sign was named after a constellation in the sky through which the Sun, Moon, and planets were seen to pass—the sign of Virgo after the constellation of Virgo, for example.

This OpenStax book is available for free at http://cnx.org/content/col11992/1.8

Download for free at http://cnx.org/content/col11992/latest/

Figure 2.15. Zodiac Signs. The signs of the zodiac are shown in a medieval woodcut.

When someone today casually asks you your "sign," they are asking for your "sun sign"—which zodiac sign the Sun was in at the moment you were born. However, more than 2000 years have passed since the signs received their names from the constellations. Because of precession, the constellations of the zodiac slide westward along the ecliptic, going once around the sky in about 26,000 years. Thus, today the real stars have slipped around by about 1/12 of the zodiac—about the width of one sign.

In most forms of astrology, however, the signs have remained assigned to the dates of the year they had when astrology was first set up. This means that the astrological signs and the real constellations are out of step; the sign of Aries, for example, now occupies the constellation of Pisces. When you look up your sun sign in a newspaper astrology column, the name of the sign associated with your birthday is no longer the name of the constellation in which the Sun was actually located when you were born. To know that constellation, you must look for the sign before the one that includes your birthday.

A complete horoscope shows the location of not only the Sun, but also the Moon and each planet in the sky by indicating its position in the appropriate sign of the zodiac. However, as the celestial sphere turns (owing to the rotation of Earth), the entire zodiac moves across the sky to the west, completing a circuit of the heavens each day. Thus, the position in the sky (or "house" in astrology) must also be calculated. There are more or less standardized rules for the interpretation of the horoscope, most of which (at least in Western schools of astrology) are derived from the *Tetrabiblos* of Ptolemy. Each sign, each house, and each planet—the last acting as a center of force—is supposed to be associated with particular matters in a person's life.

The detailed interpretation of a horoscope is a very complicated business, and there are many schools of astrological thought on how it should be done. Although some of the rules may be standardized, how each rule is to be weighed and applied is a matter of judgment—and "art." It also means that it is very difficult to tie down astrology to specific predictions or to get the same predictions from different astrologers.

Download for free at http://cnx.org/content/col11992/latest/

Astrology Today

Astrologers today use the same basic principles laid down by Ptolemy nearly 2000 years ago. They cast horoscopes (a process much simplified by the development of appropriate computer programs) and suggest interpretations. Sun sign astrology (which you read in the newspapers and many magazines) is a recent, simplified variant of natal astrology. Although even professional astrologers do not place much trust in such a limited scheme, which tries to fit everyone into just 12 groups, sun sign astrology is taken seriously by many people (perhaps because it is discussed so commonly in the media).

Today, we know much more about the nature of the planets as physical bodies, as well as about human genetics, than the ancients could. It is hard to imagine how the positions of the Sun, Moon, or planets in the sky at the moment of our birth could have anything to do with our personality or future. There are no known forces, not gravity or anything else, that could cause such effects. (For example, a straightforward calculation shows that the gravitational pull of the obstetrician delivering a newborn baby is greater than that of Mars.) Astrologers thus have to argue there must be unknown forces exerted by the planets that depend on their configurations with respect to one another and that do not vary according to the distance of the planet—forces for which there is no shred of evidence.

Another curious aspect of astrology is its emphasis on planet configurations at birth. What about the forces that might influence us at conception? Isn't our genetic makeup more important for determining our personality than the circumstances of our birth? Would we really be a different person if we had been born a few hours earlier or later, as astrology claims? (Back when astrology was first conceived, birth was thought of as a moment of magic significance, but today we understand a lot more about the long process that precedes it.)

Actually, very few well-educated people today buy the claim that our entire lives are predetermined by astrological influences at birth, but many people apparently believe that astrology has validity as an indicator of affinities and personality. A surprising number of Americans make judgments about people—whom they will hire, associate with, and even marry—on the basis of astrological information. To be sure, these are difficult decisions, and you might argue that we should use any relevant information that might help us to make the right choices. But does astrology actually provide any useful information on human personality? This is the kind of question that can be tested using the scientific method (see Testing Astrology).

The results of hundreds of tests are all the same: there is no evidence that natal astrology has any predictive power, even in a statistical sense. Why, then, do people often seem to have anecdotes about how well their own astrologer advised them? Effective astrologers today use the language of the zodiac and the horoscope only as the outward trappings of their craft. Mostly they work as amateur therapists, offering simple truths that clients like or need to hear. (Recent studies have shown that just about any sort of short-term therapy makes people feel a little better because the very act of talking about our problems with someone who listens attentively is, in itself, beneficial.)

The scheme of astrology has no basis in scientific fact, however; at best, it can be described as a pseudoscience. It is an interesting historical system, left over from prescientific days and best remembered for the impetus it gave people to learn the cycles and patterns of the sky. From it grew the science of astronomy, which is our main subject for discussion.

Download for free at http://cnx.org/content/col11992/latest/

MAKING CONNECTIONS

Testing Astrology

In response to modern public interest in astrology, scientists have carried out a wide range of statistical tests to assess its predictive power. The simplest of these examine sun sign astrology to determine whether—as astrologers assert—some signs are more likely than others to be associated with some objective measure of success, such as winning Olympic medals, earning high corporate salaries, or achieving elective office or high military rank. (You can devise such a test yourself by looking up the birth dates of all members of Congress, for example, or all members of the U.S. Olympic team.) Are our political leaders somehow selected at birth by their horoscopes and thus more likely to be Leos, say, than Scorpios?

You do not even need to be specific about your prediction in such tests. After all, many schools of astrology disagree about which signs go with which personality characteristics. To demonstrate the validity of the astrological hypothesis, it would be sufficient if the birthdays of all our leaders clustered in any one or two signs in some statistically significant way. Dozens of such tests have been performed, and all have come up completely negative: the birth dates of leaders in all fields tested have been found to be distributed randomly among *all* the signs. Sun sign astrology does not predict anything about a person's future occupation or strong personality traits.

In a fine example of such a test, two statisticians examined the reenlistment records of the United States Marine Corps. We suspect you will agree that it takes a certain kind of personality not only to enlist, but also to reenlist in the Marines. If sun signs can predict strong personality traits—as astrologers claim—then those who reenlisted (with similar personalities) should have been distributed preferentially in those one or few signs that matched the personality of someone who loves being a Marine. However, the reenlisted were distributed randomly among all the signs.

More sophisticated studies have also been done, involving full horoscopes calculated for thousands of individuals. The results of all these studies are also negative: none of the systems of astrology has been shown to be at all effective in connecting astrological aspects to personality, success, or finding the right person to love.

Other tests show that it hardly seems to matter what a horoscope interpretation says, as long as it is vague enough, and as long as each subject feels it was prepared personally just for him or her. The French statistician Michel Gauquelin, for example, sent the horoscope interpretation for one of the worst mass murderers in history to 150 people, but told each recipient that it was a "reading" prepared exclusively for him or her. Ninety-four percent of the readers said they recognized themselves in the interpretation of the mass murderer's horoscope.

Geoffrey Dean, an Australian researcher, reversed the astrological readings of 22 subjects, substituting phrases that were the opposite of what the horoscope actually said. Yet, his subjects said that the resulting readings applied to them just as often (95%) as the people to whom the original phrases were given.

Download for free at http://cnx.org/content/col11992/latest/

LINK TO LEARNING

For more on astrology and science from an astronomer's point of view, read this article (https://openstaxcollege.org/l/30astrosociety) that shines light on the topic through an accessible Q&A.

2.4 THE BIRTH OF MODERN ASTRONOMY

Learning Objectives

By the end of this section, you will be able to:

> Explain how Copernicus developed the heliocentric model of the solar system
> Explain the Copernican model of planetary motion and describe evidence or arguments in favor of it
> Describe Galileo's discoveries concerning the study of motion and forces
> Explain how Galileo's discoveries tilted the balance of evidence in favor of the Copernican model

Astronomy made no major advances in strife-torn medieval Europe. The birth and expansion of Islam after the seventh century led to a flowering of Arabic and Jewish cultures that preserved, translated, and added to many of the astronomical ideas of the Greeks. Many of the names of the brightest stars, for example, are today taken from the Arabic, as are such astronomical terms as "zenith."

As European culture began to emerge from its long, dark age, trading with Arab countries led to a rediscovery of ancient texts such as *Almagest* and to a reawakening of interest in astronomical questions. This time of rebirth (in French, "*renaissance*") in astronomy was embodied in the work of Copernicus (Figure 2.16).

Figure 2.16. Nicolaus Copernicus (1473–1543). Copernicus was a cleric and scientist who played a leading role in the emergence of modern science. Although he could not prove that Earth revolves about the Sun, he presented such compelling arguments for this idea that he turned the tide of cosmological thought and laid the foundations upon which Galileo and Kepler so effectively built in the following century.

Copernicus

One of the most important events of the Renaissance was the displacement of Earth from the center of the universe, an intellectual revolution initiated by a Polish cleric in the sixteenth century. Nicolaus Copernicus was

This OpenStax book is available for free at http://cnx.org/content/col11992/1.8

born in Torun, a mercantile town along the Vistula River. His training was in law and medicine, but his main interests were astronomy and mathematics. His great contribution to science was a critical reappraisal of the existing theories of planetary motion and the development of a new Sun-centered, or **heliocentric**, model of the solar system. Copernicus concluded that Earth is a planet and that all the planets circle the Sun. Only the Moon orbits Earth (Figure 2.17).

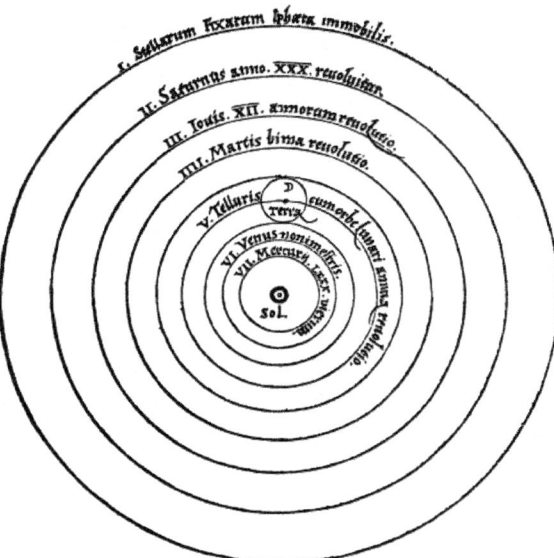

Figure 2.17. Copernicus' System. Copernicus developed a heliocentric plan of the solar system. This system was published in the first edition of *De Revolutionibus Orbium Coelestium*. Notice the word *Sol* for "Sun" in the middle. (credit: Nicolai Copernici)

Copernicus described his ideas in detail in his book *De Revolutionibus Orbium Coelestium* (*On the Revolution of Celestial Orbs*), published in 1543, the year of his death. By this time, the old Ptolemaic system needed significant adjustments to predict the positions of the planets correctly. Copernicus wanted to develop an improved theory from which to calculate planetary positions, but in doing so, he was himself not free of all traditional prejudices.

He began with several assumptions that were common in his time, such as the idea that the motions of the heavenly bodies must be made up of combinations of uniform circular motions. But he did not assume (as most people did) that Earth had to be in the center of the universe, and he presented a defense of the heliocentric system that was elegant and persuasive. His ideas, although not widely accepted until more than a century after his death, were much discussed among scholars and, ultimately, had a profound influence on the course of world history.

One of the objections raised to the heliocentric theory was that if Earth were moving, we would all sense or feel this motion. Solid objects would be ripped from the surface, a ball dropped from a great height would not strike the ground directly below it, and so forth. But a moving person is not necessarily aware of that motion. We have all experienced seeing an adjacent train, bus, or ship appear to move, only to discover that it is we who are moving.

Copernicus argued that the apparent motion of the Sun about Earth during the course of a year could be represented equally well by a motion of Earth about the Sun. He also reasoned that the apparent rotation of the celestial sphere could be explained by assuming that Earth rotates while the celestial sphere is stationary. To the objection that if Earth rotated about an axis it would fly into pieces, Copernicus answered that if such motion

Download for free at http://cnx.org/content/col11992/latest/

would tear Earth apart, the still faster motion of the much larger celestial sphere required by the geocentric hypothesis would be even more devastating.

The Heliocentric Model

The most important idea in Copernicus' *De Revolutionibus* is that Earth is one of six (then-known) planets that revolve about the Sun. Using this concept, he was able to work out the correct general picture of the solar system. He placed the planets, starting nearest the Sun, in the correct order: Mercury, Venus, Earth, Mars, Jupiter, and Saturn. Further, he deduced that the nearer a planet is to the Sun, the greater its orbital speed. With his theory, he was able to explain the complex retrograde motions of the planets without epicycles and to work out a roughly correct scale for the solar system.

Copernicus could not prove that Earth revolves about the Sun. In fact, with some adjustments, the old Ptolemaic system could have accounted, as well, for the motions of the planets in the sky. But Copernicus pointed out that the Ptolemaic cosmology was clumsy and lacking the beauty and symmetry of its successor.

In Copernicus' time, in fact, few people thought there were ways to prove whether the heliocentric or the older geocentric system was correct. A long philosophical tradition, going back to the Greeks and defended by the Catholic Church, held that pure human thought combined with divine revelation represented the path to truth. Nature, as revealed by our senses, was suspect. For example, Aristotle had reasoned that heavier objects (having more of the quality that made them heavy) must fall to Earth faster than lighter ones. This is absolutely incorrect, as any simple experiment dropping two balls of different weights shows. However, in Copernicus' day, experiments did not carry much weight (if you will pardon the expression); Aristotle's reasoning was more convincing.

In this environment, there was little motivation to carry out observations or experiments to distinguish between competing cosmological theories (or anything else). It should not surprise us, therefore, that the heliocentric idea was debated for more than half a century without any tests being applied to determine its validity. (In fact, in the North American colonies, the older geocentric system was still taught at Harvard University in the first years after it was founded in 1636.)

Contrast this with the situation today, when scientists rush to test each new hypothesis and do not accept any ideas until the results are in. For example, when two researchers at the University of Utah announced in 1989 that they had discovered a way to achieve nuclear fusion (the process that powers the stars) at room temperature, other scientists at more than 25 laboratories around the United States attempted to duplicate "cold fusion" within a few weeks—without success, as it turned out. The cold fusion theory soon went down in flames.

How would we look at Copernicus' model today? When a new hypothesis or theory is proposed in science, it must first be checked for consistency with what is already known. Copernicus' heliocentric idea passes this test, for it allows planetary positions to be calculated at least as well as does the geocentric theory. The next step is to determine which predictions the new hypothesis makes that differ from those of competing ideas. In the case of Copernicus, one example is the prediction that, if Venus circles the Sun, the planet should go through the full range of phases just as the Moon does, whereas if it circles Earth, it should not (Figure 2.18). Also, we should not be able to see the full phase of Venus from Earth because the Sun would then be between Venus and Earth. But in those days, before the telescope, no one imagined testing these predictions.

This OpenStax book is available for free at http://cnx.org/content/col11992/1.8

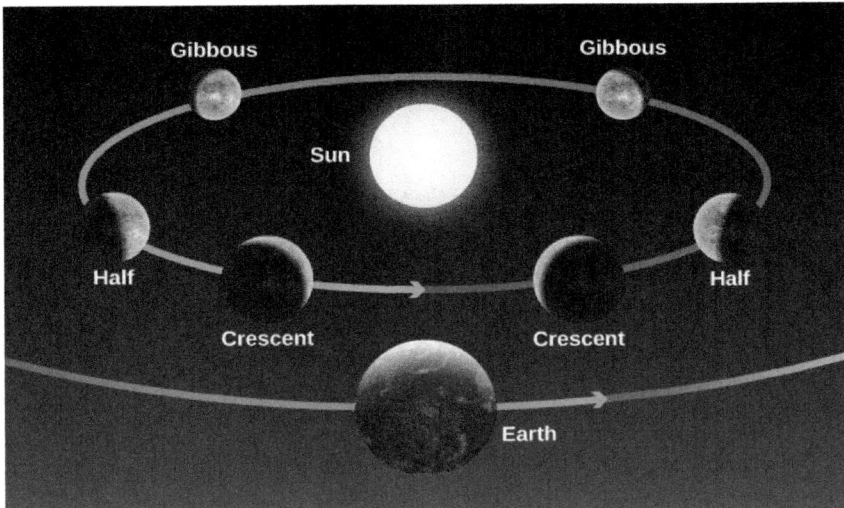

Figure 2.18. Phases of Venus. As Venus moves around the Sun, we see changing illumination of its surface, just as we see the face of the Moon illuminated differently in the course of a month.

LINK TO LEARNING

This animation (https://openstaxcollege.org/l/30venusphases) shows the phases of Venus. You can also see its distance from Earth as it orbits the Sun.

Galileo and the Beginning of Modern Science

Many of the modern scientific concepts of observation, experimentation, and the testing of hypotheses through careful quantitative measurements were pioneered by a man who lived nearly a century after Copernicus. Galileo Galilei (Figure 2.19), a contemporary of Shakespeare, was born in Pisa. Like Copernicus, he began training for a medical career, but he had little interest in the subject and later switched to mathematics. He held faculty positions at the University of Pisa and the University of Padua, and eventually became mathematician to the Grand Duke of Tuscany in Florence.

Download for free at http://cnx.org/content/col11992/latest/

Figure 2.19. Galileo Galilei (1564–1642). Galileo advocated that we perform experiments or make observations to ask nature its ways. When Galileo turned the telescope to the sky, he found things were not the way philosophers had supposed.

Galileo's greatest contributions were in the field of mechanics, the study of motion and the actions of forces on bodies. It was familiar to all persons then, as it is to us now, that if something is at rest, it tends to remain at rest and requires some outside influence to start it in motion. Rest was thus generally regarded as the natural state of matter. Galileo showed, however, that rest is no more natural than motion.

If an object is slid along a rough horizontal floor, it soon comes to rest because friction between it and the floor acts as a retarding force. However, if the floor and the object are both highly polished, the object, given the same initial speed, will slide farther before stopping. On a smooth layer of ice, it will slide farther still. Galileo reasoned that if all resisting effects could be removed, the object would continue in a steady state of motion indefinitely. He argued that a force is required not only to start an object moving from rest but also to slow down, stop, speed up, or change the direction of a moving object. You will appreciate this if you have ever tried to stop a rolling car by leaning against it, or a moving boat by tugging on a line.

Galileo also studied the way objects **accelerate**—change their speed or direction of motion. Galileo watched objects as they fell freely or rolled down a ramp. He found that such objects accelerate uniformly; that is, in equal intervals of time they gain equal increments in speed. Galileo formulated these newly found laws in precise mathematical terms that enabled future experimenters to predict how far and how fast objects would move in various lengths of time.

LINK TO LEARNING

In theory, if Galileo is right, a feather and a hammer, dropped at the same time from a height, should land at the same moment. On Earth, this experiment is not possible because air resistance and air movements make the feather flutter, instead of falling straight down, accelerated only by the force of gravity. For generations, physics teachers had said that the place to try this experiment is somewhere where there is no air, such as the Moon. In 1971, *Apollo 15* astronaut David Scott took a hammer and feather to the Moon and tried it, to the delight of physics nerds everywhere. NASA provides the video of

This OpenStax book is available for free at http://cnx.org/content/col11992/1.8

the hammer and feather (https://openstaxcollege.org/l/30HamVsFeath) as well as a brief explanation.

Sometime in the 1590s, Galileo adopted the Copernican hypothesis of a heliocentric solar system. In Roman Catholic Italy, this was not a popular philosophy, for Church authorities still upheld the ideas of Aristotle and Ptolemy, and they had powerful political and economic reasons for insisting that Earth was the center of creation. Galileo not only challenged this thinking but also had the audacity to write in Italian rather than scholarly Latin, and to lecture publicly on those topics. For him, there was no contradiction between the authority of the Church in matters of religion and morality, and the authority of nature (revealed by experiments) in matters of science. It was primarily because of Galileo and his "dangerous" opinions that, in 1616, the Church issued a prohibition decree stating that the Copernican doctrine was "false and absurd" and not to be held or defended.

Galileo's Astronomical Observations

It is not certain who first conceived of the idea of combining two or more pieces of glass to produce an instrument that enlarged images of distant objects, making them appear nearer. The first such "spyglasses" (now called *telescopes*) that attracted much notice were made in 1608 by the Dutch spectacle maker Hans Lippershey (1570–1619). Galileo heard of the discovery and, without ever having seen an assembled telescope, constructed one of his own with a three-power magnification (3×), which made distant objects appear three times nearer and larger (Figure 2.20).

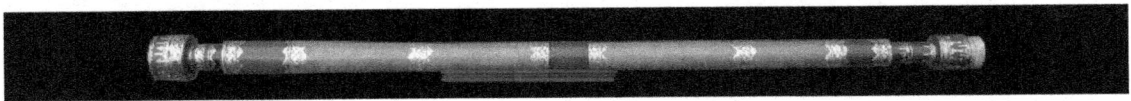

Figure 2.20. Telescope Used by Galileo. The telescope has a wooden tube covered with paper and a lens 26 millimeters across.

On August 25, 1609, Galileo demonstrated a telescope with a magnification of 9× to government officials of the city-state of Venice. By a magnification of 9×, we mean the linear dimensions of the objects being viewed appeared nine times larger or, alternatively, the objects appeared nine times closer than they really were. There were obvious military advantages associated with a device for seeing distant objects. For his invention, Galileo's salary was nearly doubled, and he was granted lifetime tenure as a professor. (His university colleagues were outraged, particularly because the invention was not even original.)

Others had used the telescope before Galileo to observe things on Earth. But in a flash of insight that changed the history of astronomy, Galileo realized that he could turn the power of the telescope toward the heavens. Before using his telescope for astronomical observations, Galileo had to devise a stable mount and improve the optics. He increased the magnification to 30×. Galileo also needed to acquire confidence in the telescope.

At that time, human eyes were believed to be the final arbiter of truth about size, shape, and color. Lenses, mirrors, and prisms were known to distort distant images by enlarging, reducing, or inverting them, or spreading the light into a spectrum (rainbow of colors). Galileo undertook repeated experiments to convince himself that what he saw through the telescope was identical to what he saw up close. Only then could he begin to believe that the miraculous phenomena the telescope revealed in the heavens were real.

Beginning his astronomical work late in 1609, Galileo found that many stars too faint to be seen with the unaided eye became visible with his telescope. In particular, he found that some nebulous blurs resolved into

Download for free at http://cnx.org/content/col11992/latest/

many stars, and that the Milky Way—the strip of whiteness across the night sky—was also made up of a multitude of individual stars.

Examining the planets, Galileo found four moons revolving about Jupiter in times ranging from just under 2 days to about 17 days. This discovery was particularly important because it showed that not everything has to revolve around Earth. Furthermore, it demonstrated that there could be centers of motion that are themselves in motion. Defenders of the geocentric view had argued that if Earth was in motion, then the Moon would be left behind because it could hardly keep up with a rapidly moving planet. Yet, here were Jupiter's moons doing exactly that. (To recognize this discovery and honor his work, NASA named a spacecraft that explored the Jupiter system Galileo.)

With his telescope, Galileo was able to carry out the test of the Copernican theory mentioned earlier, based on the phases of Venus. Within a few months, he had found that Venus goes through phases like the Moon, showing that it must revolve about the Sun, so that we see different parts of its daylight side at different times (see Figure 2.18.) These observations could not be reconciled with Ptolemy's model, in which Venus circled about Earth. In Ptolemy's model, Venus could also show phases, but they were the wrong phases in the wrong order from what Galileo observed.

Galileo also observed the Moon and saw craters, mountain ranges, valleys, and flat, dark areas that he thought might be water. These discoveries showed that the Moon might be not so dissimilar to Earth—suggesting that Earth, too, could belong to the realm of celestial bodies.

LINK TO LEARNING

For more information about the life and work of Galileo, see the Galileo Project (https://openstaxcollege.org/l/30GalProj) at Rice University.

After Galileo's work, it became increasingly difficult to deny the Copernican view, and Earth was slowly dethroned from its central position in the universe and given its rightful place as one of the planets attending the Sun. Initially, however, Galileo met with a great deal of opposition. The Roman Catholic Church, still reeling from the Protestant Reformation, was looking to assert its authority and chose to make an example of Galileo. He had to appear before the Inquisition to answer charges that his work was heretical, and he was ultimately condemned to house arrest. His books were on the Church's forbidden list until 1836, although in countries where the Roman Catholic Church held less sway, they were widely read and discussed. Not until 1992 did the Catholic Church admit publicly that it had erred in the matter of censoring Galileo's ideas.

The new ideas of Copernicus and Galileo began a revolution in our conception of the cosmos. It eventually became evident that the universe is a vast place and that Earth's role in it is relatively unimportant. The idea that Earth moves around the Sun like the other planets raised the possibility that they might be worlds themselves, perhaps even supporting life. As Earth was demoted from its position at the center of the universe, so, too, was humanity. The universe, despite what we may wish, does not revolve around us.

Most of us take these things for granted today, but four centuries ago such concepts were frightening and heretical for some, immensely stimulating for others. The pioneers of the Renaissance started the European world along the path toward science and technology that we still tread today. For them, nature was rational and ultimately knowable, and experiments and observations provided the means to reveal its secrets.

This OpenStax book is available for free at http://cnx.org/content/col11992/1.8

Download for free at http://cnx.org/content/col11992/latest/

Observing the Planets

At most any time of the night, and at any season, you can spot one or more bright planets in the sky. All five of the planets known to the ancients—Mercury, Venus, Mars, Jupiter, and Saturn—are more prominent than any but the brightest stars, and they can be seen even from urban locations if you know where and when to look. One way to tell planets from bright stars is that planets twinkle less.

Venus, which stays close to the Sun from our perspective, appears either as an "evening star" in the west after sunset or as a "morning star" in the east before sunrise. It is the brightest object in the sky after the Sun and Moon. It far outshines any real star, and under the most favorable circumstances, it can even cast a visible shadow. Some young military recruits have tried to shoot Venus down as an approaching enemy craft or UFO.

Mars, with its distinctive red color, can be nearly as bright as Venus is when close to Earth, but normally it remains much less conspicuous. Jupiter is most often the second-brightest planet, approximately equaling in brilliance the brightest stars. Saturn is dimmer, and it varies considerably in brightness, depending on whether its large rings are seen nearly edge-on (faint) or more widely opened (bright).

Mercury is quite bright, but few people ever notice it because it never moves very far from the Sun (it's never more than 28° away in the sky) and is always seen against bright twilight skies.

True to their name, the planets "wander" against the background of the "fixed" stars. Although their apparent motions are complex, they reflect an underlying order upon which the heliocentric model of the solar system, as described in this chapter, was based. The positions of the planets are often listed in newspapers (sometimes on the weather page), and clear maps and guides to their locations can be found each month in such magazines as *Sky & Telescope* and *Astronomy* (available at most libraries and online). There are also a number of computer programs and phone and tablet apps that allow you to display where the planets are on any night.

Download for free at http://cnx.org/content/col11992/latest/

CHAPTER 2 REVIEW

 KEY TERMS

accelerate to change velocity; to speed up, slow down, or change direction.

apparent magnitude a measure of how bright a star looks in the sky; the larger the number, the dimmer the star appears to us

astrology the pseudoscience that deals with the supposed influences on human destiny of the configurations and locations in the sky of the Sun, Moon, and planets

celestial equator a great circle on the celestial sphere 90° from the celestial poles; where the celestial sphere intersects the plane of Earth's equator

celestial poles points about which the celestial sphere appears to rotate; intersections of the celestial sphere with Earth's polar axis

celestial sphere the apparent sphere of the sky; a sphere of large radius centered on the observer; directions of objects in the sky can be denoted by their position on the celestial sphere

circumpolar zone those portions of the celestial sphere near the celestial poles that are either always above or always below the horizon

cosmology the study of the organization and evolution of the universe

ecliptic the apparent annual path of the Sun on the celestial sphere

epicycle the circular orbit of a body in the Ptolemaic system, the center of which revolves about another circle (the deferent)

geocentric centered on Earth

heliocentric centered on the Sun

horizon (astronomical) a great circle on the celestial sphere 90° from the zenith; more popularly, the circle around us where the dome of the sky meets Earth

horoscope a chart used by astrologers that shows the positions along the zodiac and in the sky of the Sun, Moon, and planets at some given instant and as seen from a particular place on Earth—usually corresponding to the time and place of a person's birth

parallax the apparent displacement of a nearby star that results from the motion of Earth around the Sun

planet today, any of the larger objects revolving about the Sun or any similar objects that orbit other stars; in ancient times, any object that moved regularly among the fixed stars

precession (of Earth) the slow, conical motion of Earth's axis of rotation caused principally by the gravitational pull of the Moon and Sun on Earth's equatorial bulge

retrograde motion the apparent westward motion of a planet on the celestial sphere or with respect to the stars

year the period of revolution of Earth around the Sun

This OpenStax book is available for free at http://cnx.org/content/col11992/1.8

zenith the point on the celestial sphere opposite the direction of gravity; point directly above the observer

zodiac a belt around the sky about 18° wide centered on the ecliptic

 ## SUMMARY

2.1 The Sky Above

The direct evidence of our senses supports a geocentric perspective, with the celestial sphere pivoting on the celestial poles and rotating about a stationary Earth. We see only half of this sphere at one time, limited by the horizon; the point directly overhead is our zenith. The Sun's annual path on the celestial sphere is the ecliptic—a line that runs through the center of the zodiac, which is the 18-degree-wide strip of the sky within which we always find the Moon and planets. The celestial sphere is organized into 88 constellations, or sectors.

2.2 Ancient Astronomy

Ancient Greeks such as Aristotle recognized that Earth and the Moon are spheres, and understood the phases of the Moon, but because of their inability to detect stellar parallax, they rejected the idea that Earth moves. Eratosthenes measured the size of Earth with surprising precision. Hipparchus carried out many astronomical observations, making a star catalog, defining the system of stellar magnitudes, and discovering precession from the apparent shift in the position of the north celestial pole. Ptolemy of Alexandria summarized classic astronomy in his *Almagest*; he explained planetary motions, including retrograde motion, with remarkably good accuracy using a model centered on Earth. This geocentric model, based on combinations of uniform circular motion using epicycles, was accepted as authority for more than a thousand years.

2.3 Astrology and Astronomy

The ancient religion of astrology, with its main contribution to civilization a heightened interest in the heavens, began in Babylonia. It reached its peak in the Greco-Roman world, especially as recorded in the *Tetrabiblos* of Ptolemy. Natal astrology is based on the assumption that the positions of the planets at the time of our birth, as described by a horoscope, determine our future. However, modern tests clearly show that there is no evidence for this, even in a broad statistical sense, and there is no verifiable theory to explain what might cause such an astrological influence.

2.4 The Birth of Modern Astronomy

Nicolaus Copernicus introduced the heliocentric cosmology to Renaissance Europe in his book *De Revolutionibus*. Although he retained the Aristotelian idea of uniform circular motion, Copernicus suggested that Earth is a planet and that the planets all circle about the Sun, dethroning Earth from its position at the center of the universe. Galileo was the father of both modern experimental physics and telescopic astronomy. He studied the acceleration of moving objects and, in 1610, began telescopic observations, discovering the nature of the Milky Way, the large-scale features of the Moon, the phases of Venus, and four moons of Jupiter. Although he was accused of heresy for his support of heliocentric cosmology, Galileo is credited with observations and brilliant writings that convinced most of his scientific contemporaries of the reality of the Copernican theory.

 ## FOR FURTHER EXPLORATION

Articles

Ancient Astronomy

Gingerich, O. "From Aristarchus to Copernicus." *Sky & Telescope* (November 1983): 410.

Download for free at http://cnx.org/content/col11992/latest/

Gingerich, O. "Islamic Astronomy." *Scientific American* (April 1986): 74.

Astronomy and Astrology

Fraknoi, A. "Your Astrology Defense Kit." *Sky & Telescope* (August 1989): 146.

Copernicus and Galileo

Gingerich, O. "Galileo and the Phases of Venus." *Sky & Telescope* (December 1984): 520.

Gingerich, O. "How Galileo Changed the Rules of Science." *Sky & Telescope* (March 1993): 32.

Maran, S., and Marschall, L. "The Moon, the Telescope, and the Birth of the Modern World." *Sky & Telescope* (February 2009): 28.

Sobel, D. "The Heretic's Daughter: A Startling Correspondence Reveals a New Portrait of Galileo." *The New Yorker* (September 13, 1999): 52.

Websites

Ancient Astronomy

Aristarchos of Samos: http://adsabs.harvard.edu//full/seri/JRASC/0075//0000029.000.html. By Dr. Alan Batten.

Claudius Ptolemy: http://www-history.mcs.st-and.ac.uk/Biographies/Ptolemy.html. An interesting biography.

Hipparchus of Rhodes: http://www-history.mcs.st-andrews.ac.uk/Biographies/Hipparchus.html. An interesting biography.

Astronomy and Astrology

Astrology and Science: http://www.astrology-and-science.com/hpage.htm. The best site for a serious examination of the issues with astrology and the research on whether it works.

Real Romance in the Stars: http://www.independent.co.uk/voices/the-real-romance-in-the-stars-1527970.html. 1995 newspaper commentary attacking astrology.

Copernicus and Galileo

Galileo Galilei: http://www-history.mcs.st-andrews.ac.uk/Biographies/Galileo.html. A good biography with additional links.

Galileo Project: http://galileo.rice.edu/. Rice University's repository of information on Galileo.

Nicolaus Copernicus: http://www-groups.dcs.st-and.ac.uk/~history/Biographies/Copernicus.html. A biography including links to photos about his life.

Videos

Astronomy and Astrology

Astrology Debunked: https://www.youtube.com/watch?v=y84HX2pMo5U. A compilation of scientists and magicians commenting skeptically on astrology (9:09).

Copernicus and Galileo

Galileo: http://www.biography.com/people/galileo-9305220. A brief biography (2:51).

Galileo's Battle for the Heavens: https://www.youtube.com/watch?v=VnEH9rbrIkk. A NOVA episode on PBS (1:48:55)

Nicolaus Copernicus: http://www.biography.com/people/nicolaus-copernicus-9256984. An overview of his life and work (2:41).

This OpenStax book is available for free at http://cnx.org/content/col11992/1.8

COLLABORATIVE GROUP ACTIVITIES

A. With your group, consider the question with which we began this chapter. How many ways can you think of to prove to a member of the "Flat Earth Society" that our planet is, indeed, round?

B. Make a list of ways in which a belief in astrology (the notion that your life path or personality is controlled by the position of the Sun, Moon, and planets at the time of your birth) might be harmful to an individual or to society at large.

C. Have members of the group compare their experiences with the night sky. Did you see the Milky Way? Can you identify any constellations? Make a list of reasons why you think so many fewer people know the night sky today than at the time of the ancient Greeks. Discuss reasons for why a person, today, may want to be acquainted with the night sky.

D. Constellations commemorate great heroes, dangers, or events in the legends of the people who name them. Suppose we had to start from scratch today, naming the patterns of stars in the sky. Whom or what would you choose to commemorate by naming a constellation after it, him, or her and why (begin with people from history; then if you have time, include living people as well)? Can the members of your group agree on any choices?

E. Although astronomical mythology no longer holds a powerful sway over the modern imagination, we still find proof of the power of astronomical images in the number of products in the marketplace that have astronomical names. How many can your group come up with? (Think of things like Milky Way candy bars, Eclipse and Orbit gum, or Comet cleanser.)

 EXERCISES

Review Questions

1. From where on Earth could you observe all of the stars during the course of a year? What fraction of the sky can be seen from the North Pole?

2. Give four ways to demonstrate that Earth is spherical.

3. Explain, according to both geocentric and heliocentric cosmologies, why we see retrograde motion of the planets.

4. In what ways did the work of Copernicus and Galileo differ from the views of the ancient Greeks and of their contemporaries?

5. What were four of Galileo's discoveries that were important to astronomy?

6. Explain the origin of the magnitude designation for determining the brightness of stars. Why does it seem to go backward, with smaller numbers indicating brighter stars?

7. Ursa Minor contains the pole star, Polaris, and the asterism known as the Little Dipper. From most locations in the Northern Hemisphere, all of the stars in Ursa Minor are circumpolar. Does that mean these stars are also above the horizon during the day? Explain.

Download for free at http://cnx.org/content/col11992/latest/

8. How many degrees does the Sun move per day relative to the fixed stars? How many days does it take for the Sun to return to its original location relative to the fixed stars?

9. How many degrees does the Moon move per day relative to the fixed stars? How many days does it take for the Moon to return to its original location relative to the fixed stars?

10. Explain how the zodiacal constellations are different from the other constellations.

11. The Sun was once thought to be a planet. Explain why.

12. Is the ecliptic the same thing as the celestial equator? Explain.

13. What is an asterism? Can you name an example?

14. Why did Pythagoras believe that Earth should be spherical?

15. How did Aristotle deduce that the Sun is farther away from Earth than the Moon?

16. What are two ways in which Aristotle deduced that Earth is spherical?

17. How did Hipparchus discover the wobble of Earth's axis, known as *precession*?

18. Why did Ptolemy have to introduce multiple circles of motion for the planets instead of a single, simple circle to represent the planet's motion around the Sun?

19. Why did Copernicus want to develop a completely new system for predicting planetary positions? Provide two reasons.

20. What two factors made it difficult, at first, for astronomers to choose between the Copernican heliocentric model and the Ptolemaic geocentric model?

21. What phases would Venus show if the geocentric model were correct?

Thought Questions

22. Describe a practical way to determine in which constellation the Sun is found at any time of the year.

23. What is a constellation as astronomers define it today? What does it mean when an astronomer says, "I saw a comet in Orion last night"?

24. Draw a picture that explains why Venus goes through phases the way the Moon does, according to the heliocentric cosmology. Does Jupiter also go through phases as seen from Earth? Why?

25. Show with a simple diagram how the lower parts of a ship disappear first as it sails away from you on a spherical Earth. Use the same diagram to show why lookouts on old sailing ships could see farther from the masthead than from the deck. Would there be any advantage to posting lookouts on the mast if Earth were flat? (Note that these nautical arguments for a spherical Earth were quite familiar to Columbus and other mariners of his time.)

26. Parallaxes of stars were not observed by ancient astronomers. How can this fact be reconciled with the heliocentric hypothesis?

27. Why do you think so many people still believe in astrology and spend money on it? What psychological needs does such a belief system satisfy?

28. Consider three cosmological perspectives—the geocentric perspective, the heliocentric perspective, and the modern perspective—in which the Sun is a minor star on the outskirts of one galaxy among billions. Discuss some of the cultural and philosophical implications of each point of view.

This OpenStax book is available for free at http://cnx.org/content/col11992/1.8

29. The north celestial pole appears at an altitude above the horizon that is equal to the observer's latitude. Identify Polaris, the North Star, which lies very close to the north celestial pole. Measure its altitude. (This can be done with a protractor. Alternatively, your fist, extended at arm's length, spans a distance approximately equal to 10°.) Compare this estimate with your latitude. (Note that this experiment cannot be performed easily in the Southern Hemisphere because Polaris itself is not visible in the south and no bright star is located near the south celestial pole.)

30. What were two arguments or lines of evidence in support of the geocentric model?

31. Although the Copernican system was largely correct to place the Sun at the center of all planetary motion, the model still gave inaccurate predictions for planetary positions. Explain the flaw in the Copernican model that hindered its accuracy.

32. During a retrograde loop of Mars, would you expect Mars to be brighter than usual in the sky, about average in brightness, or fainter than usual in the sky? Explain.

33. The Great Pyramid of Giza was constructed nearly 5000 years ago. Within the pyramid, archaeologists discovered a shaft leading from the central chamber out of the pyramid, oriented for favorable viewing of the bright star Thuban at that time. Thinking about Earth's precession, explain why Thuban might have been an important star to the ancient Egyptians.

34. Explain why more stars are circumpolar for observers at higher latitudes.

35. What is the altitude of the north celestial pole in the sky from your latitude? If you do not know your latitude, look it up. If you are in the Southern Hemisphere, answer this question for the south celestial pole, since the north celestial pole is not visible from your location.

36. If you were to drive to some city south of your current location, how would the altitude of the celestial pole in the sky change?

37. Hipparchus could have warned us that the dates associated with each of the natal astrology sun signs would eventually be wrong. Explain why.

38. Explain three lines of evidence that argue against the validity of astrology.

39. What did Galileo discover about the planet Jupiter that cast doubt on exclusive geocentrism?

40. What did Galileo discover about Venus that cast doubt on geocentrism?

Figuring For Yourself

41. Suppose Eratosthenes had found that, in Alexandria, at noon on the first day of summer, the line to the Sun makes an angle 30° with the vertical. What, then, would he have found for Earth's circumference?

42. Suppose Eratosthenes' results for Earth's circumference were quite accurate. If the diameter of Earth is 12,740 km, what is the length of his stadium in kilometers?

43. Suppose you are on a strange planet and observe, at night, that the stars do not rise and set, but circle parallel to the horizon. Next, you walk in a constant direction for 8000 miles, and at your new location on the planet, you find that all stars rise straight up in the east and set straight down in the west, perpendicular to the horizon. How could you determine the circumference of the planet without any further observations? What is the circumference, in miles, of the planet?

Download for free at http://cnx.org/content/col11992/latest/

This OpenStax book is available for free at http://cnx.org/content/col11992/1.8

3

ORBITS AND GRAVITY

Figure 3.1. International Space Station. This space habitat and laboratory orbits Earth once every 90 minutes. (credit: modification of work by NASA)

Chapter Outline

✎ Thinking Ahead

How would you find a new planet at the outskirts of our solar system that is too dim to be seen with the unaided eye and is so far away that it moves very slowly among the stars? This was the problem confronting astronomers during the nineteenth century as they tried to pin down a full inventory of our solar system.

If we could look down on the solar system from somewhere out in space, interpreting planetary motions would be much simpler. But the fact is, we must observe the positions of all the other planets from our own moving planet. Scientists of the Renaissance did not know the details of Earth's motions any better than the motions of the other planets. Their problem, as we saw in Observing the Sky: The Birth of Astronomy, was that they had to deduce the nature of all planetary motion using only their earthbound observations of the other planets' positions in the sky. To solve this complex problem more fully, better observations and better models of the planetary system were needed.

Download for free at http://cnx.org/content/col11992/latest/

3.1 THE LAWS OF PLANETARY MOTION

Learning Objectives

By the end of this section, you will be able to:

> Describe how Tycho Brahe and Johannes Kepler contributed to our understanding of how planets move around the Sun
> Explain Kepler's three laws of planetary motion

At about the time that Galileo was beginning his experiments with falling bodies, the efforts of two other scientists dramatically advanced our understanding of the motions of the planets. These two astronomers were the observer Tycho Brahe and the mathematician Johannes Kepler. Together, they placed the speculations of Copernicus on a sound mathematical basis and paved the way for the work of Isaac Newton in the next century.

Tycho Brahe's Observatory

Three years after the publication of Copernicus' *De Revolutionibus*, Tycho Brahe was born to a family of Danish nobility. He developed an early interest in astronomy and, as a young man, made significant astronomical observations. Among these was a careful study of what we now know was an exploding star that flared up to great brilliance in the night sky. His growing reputation gained him the patronage of the Danish King Frederick II, and at the age of 30, Brahe was able to establish a fine astronomical observatory on the North Sea island of Hven (Figure 3.2). Brahe was the last and greatest of the pre-telescopic observers in Europe.

(a)   (b)

Figure 3.2. Tycho Brahe (1546–1601) and Johannes Kepler (1571–1630).   (a) A stylized engraving shows Tycho Brahe using his instruments to measure the altitude of celestial objects above the horizon. The large curved instrument in the foreground allowed him to measure precise angles in the sky. Note that the scene includes hints of the grandeur of Brahe's observatory at Hven. (b) Kepler was a German mathematician and astronomer. His discovery of the basic laws that describe planetary motion placed the heliocentric cosmology of Copernicus on a firm mathematical basis.

This OpenStax book is available for free at http://cnx.org/content/col11992/1.8

At Hven, Brahe made a continuous record of the positions of the Sun, Moon, and planets for almost 20 years. His extensive and precise observations enabled him to note that the positions of the planets varied from those given in published tables, which were based on the work of Ptolemy. These data were extremely valuable, but Brahe didn't have the ability to analyze them and develop a better model than what Ptolemy had published. He was further inhibited because he was an extravagant and cantankerous fellow, and he accumulated enemies among government officials. When his patron, Frederick II, died in 1597, Brahe lost his political base and decided to leave Denmark. He took up residence in Prague, where he became court astronomer to Emperor Rudolf of Bohemia. There, in the year before his death, Brahe found a most able young mathematician, Johannes Kepler, to assist him in analyzing his extensive planetary data.

Johannes Kepler

Johannes Kepler was born into a poor family in the German province of Württemberg and lived much of his life amid the turmoil of the Thirty Years' War (see Figure 3.2). He attended university at Tubingen and studied for a theological career. There, he learned the principles of the Copernican system and became converted to the heliocentric hypothesis. Eventually, Kepler went to Prague to serve as an assistant to Brahe, who set him to work trying to find a satisfactory theory of planetary motion—one that was compatible with the long series of observations made at Hven. Brahe was reluctant to provide Kepler with much material at any one time for fear that Kepler would discover the secrets of the universal motion by himself, thereby robbing Brahe of some of the glory. Only after Brahe's death in 1601 did Kepler get full possession of the priceless records. Their study occupied most of Kepler's time for more than 20 years.

Through his analysis of the motions of the planets, Kepler developed a series of principles, now known as *Kepler's three laws,* which described the behavior of planets based on their paths through space. The first two laws of planetary motion were published in 1609 in *The New Astronomy*. Their discovery was a profound step in the development of modern science.

The First Two Laws of Planetary Motion

The path of an object through space is called its **orbit**. Kepler initially assumed that the orbits of planets were circles, but doing so did not allow him to find orbits that were consistent with Brahe's observations. Working with the data for Mars, he eventually discovered that the orbit of that planet had the shape of a somewhat flattened circle, or **ellipse**. Next to the circle, the ellipse is the simplest kind of closed curve, belonging to a family of curves known as *conic sections* (Figure 3.3).

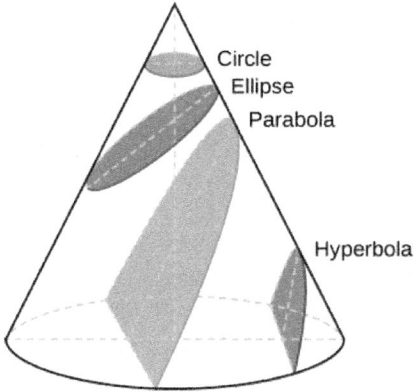

Figure 3.3. Conic Sections. The circle, ellipse, parabola, and hyperbola are all formed by the intersection of a plane with a cone. This is why such curves are called conic sections.

Download for free at http://cnx.org/content/col11992/latest/

You might recall from math classes that in a circle, the center is a special point. The distance from the center to anywhere on the circle is exactly the same. In an ellipse, the sum of the distance from two special points inside the ellipse to any point on the ellipse is always the same. These two points inside the ellipse are called its foci (singular: **focus**), a word invented for this purpose by Kepler.

This property suggests a simple way to draw an ellipse (Figure 3.4). We wrap the ends of a loop of string around two tacks pushed through a sheet of paper into a drawing board, so that the string is slack. If we push a pencil against the string, making the string taut, and then slide the pencil against the string all around the tacks, the curve that results is an ellipse. At any point where the pencil may be, the sum of the distances from the pencil to the two tacks is a constant length—the length of the string. The tacks are at the two foci of the ellipse.

The widest diameter of the ellipse is called its **major axis**. Half this distance—that is, the distance from the center of the ellipse to one end—is the **semimajor axis**, which is usually used to specify the size of the ellipse. For example, the semimajor axis of the orbit of Mars, which is also the planet's average distance from the Sun, is 228 million kilometers.

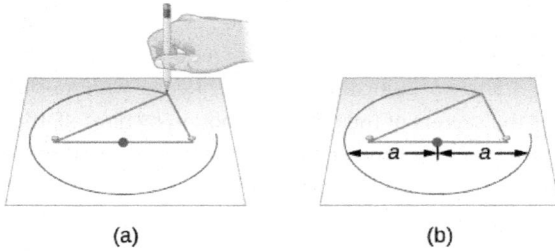

Figure 3.4. Drawing an Ellipse. (a) We can construct an ellipse by pushing two tacks (the white objects) into a piece of paper on a drawing board, and then looping a string around the tacks. Each tack represents a focus of the ellipse, with one of the tacks being the Sun. Stretch the string tight using a pencil, and then move the pencil around the tacks. The length of the string remains the same, so that the sum of the distances from any point on the ellipse to the foci is always constant. (b) In this illustration, each semimajor axis is denoted by *a*. The distance 2*a* is called the major axis of the ellipse.

The shape (roundness) of an ellipse depends on how close together the two foci are, compared with the major axis. The ratio of the distance between the foci to the length of the major axis is called the **eccentricity** of the ellipse.

If the foci (or tacks) are moved to the same location, then the distance between the foci would be zero. This means that the eccentricity is zero and the ellipse is just a circle; thus, a circle can be called an ellipse of zero eccentricity. In a circle, the semimajor axis would be the radius.

Next, we can make ellipses of various elongations (or extended lengths) by varying the spacing of the tacks (as long as they are not farther apart than the length of the string). The greater the eccentricity, the more elongated is the ellipse, up to a maximum eccentricity of 1.0, when the ellipse becomes "flat," the other extreme from a circle.

The size and shape of an ellipse are completely specified by its semimajor axis and its eccentricity. Using Brahe's data, Kepler found that Mars has an elliptical orbit, with the Sun at one focus (the other focus is empty). The eccentricity of the orbit of Mars is only about 0.1; its orbit, drawn to scale, would be practically indistinguishable from a circle, but the difference turned out to be critical for understanding planetary motions.

Kepler generalized this result in his first law and said that *the orbits of all the planets are ellipses*. Here was a decisive moment in the history of human thought: it was not necessary to have only circles in order to have an acceptable cosmos. The universe could be a bit more complex than the Greek philosophers had wanted it to be.

This OpenStax book is available for free at http://cnx.org/content/col11992/1.8

Download for free at http://cnx.org/content/col11992/latest/

Kepler's second law deals with the speed with which each planet moves along its ellipse, also known as its **orbital speed**. Working with Brahe's observations of Mars, Kepler discovered that the planet speeds up as it comes closer to the Sun and slows down as it pulls away from the Sun. He expressed the precise form of this relationship by imagining that the Sun and Mars are connected by a straight, elastic line. When Mars is closer to the Sun (positions 1 and 2 in Figure 3.5), the elastic line is not stretched as much, and the planet moves rapidly. Farther from the Sun, as in positions 3 and 4, the line is stretched a lot, and the planet does not move so fast. As Mars travels in its elliptical orbit around the Sun, the elastic line sweeps out areas of the ellipse as it moves (the colored regions in our figure). Kepler found that in equal intervals of time (t), the areas swept out in space by this imaginary line are always equal; that is, the area of the region B from 1 to 2 is the same as that of region A from 3 to 4.

If a planet moves in a circular orbit, the elastic line is always stretched the same amount and the planet moves at a constant speed around its orbit. But, as Kepler discovered, in most orbits that speed of a planet orbiting its star (or moon orbiting its planet) tends to vary because the orbit is elliptical.

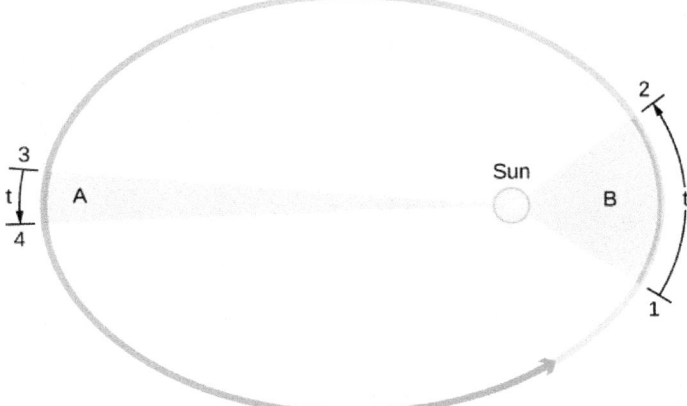

Figure 3.5. Kepler's Second Law: The Law of Equal Areas. The orbital speed of a planet traveling around the Sun (the circular object inside the ellipse) varies in such a way that in equal intervals of time (t), a line between the Sun and a planet sweeps out equal areas (A and B). Note that the eccentricities of the planets' orbits in our solar system are substantially less than shown here.

Kepler's Third Law

Kepler's first two laws of planetary motion describe the shape of a planet's orbit and allow us to calculate the speed of its motion at any point in the orbit. Kepler was pleased to have discovered such fundamental rules, but they did not satisfy his quest to fully understand planetary motions. He wanted to know why the orbits of the planets were spaced as they are and to find a mathematical pattern in their movements—a "harmony of the spheres" as he called it. For many years he worked to discover mathematical relationships governing planetary spacing and the time each planet took to go around the Sun.

In 1619, Kepler discovered a basic relationship to relate the planets' orbits to their relative distances from the Sun. We define a planet's **orbital period**, (**P**), as the time it takes a planet to travel once around the Sun. Also, recall that a planet's semimajor axis, a, is equal to its average distance from the Sun. The relationship, now known as *Kepler's third law*, says that a planet's orbital period squared is proportional to the semimajor axis of its orbit cubed, or

$$P^2 \propto a^3$$

Download for free at http://cnx.org/content/col11992/latest/

When *P* (the orbital period) is measured in years, and *a* is expressed in a quantity known as an **astronomical unit (AU)**, the two sides of the formula are not only proportional but equal. One AU is the average distance between Earth and the Sun and is approximately equal to 1.5×10^8 kilometers. In these units,

$$P^2 = a^3$$

Kepler's third law applies to all objects orbiting the Sun, including Earth, and provides a means for calculating their relative distances from the Sun from the time they take to orbit. Let's look at a specific example to illustrate how useful Kepler's third law is.

For instance, suppose you time how long Mars takes to go around the Sun (in Earth years). Kepler's third law can then be used to calculate Mars' average distance from the Sun. Mars' orbital period (1.88 Earth years) squared, or P^2, is $1.88^2 = 3.53$, and according to the equation for Kepler's third law, this equals the cube of its semimajor axis, or a^3. So what number must be cubed to give 3.53? The answer is 1.52 (since $1.52 \times 1.52 \times 1.52 = 3.53$). Thus, Mars' semimajor axis in astronomical units must be 1.52 AU. In other words, to go around the Sun in a little less than two years, Mars must be about 50% (half again) as far from the Sun as Earth is.

EXAMPLE 3.1

Calculating Periods

Imagine an object is traveling around the Sun. What would be the orbital period of the object if its orbit has a semimajor axis of 50 AU?

Solution

From Kepler's third law, we know that (when we use units of years and AU)

$$P^2 = a^3$$

If the object's orbit has a semimajor axis of 4 AU (*a* = 50), we can cube 50 and then take the square root of the result to get P:

$$P = \sqrt{a^3}$$
$$P = \sqrt{50 \times 50 \times 50} = \sqrt{125,000} = 353.6 \text{ years}$$

Therefore, the orbital period of the object is about 350 years. This would place our hypothetical object beyond the orbit of Pluto.

Check Your Learning

What would be the orbital period of an asteroid (a rocky chunk between Mars and Jupiter) with a semimajor axis of 3 AU?

Answer:

$$P = \sqrt{3 \times 3 \times 3} = \sqrt{27} = 5.2 \text{ years}$$

Kepler's three laws of planetary motion can be summarized as follows:

- **Kepler's first law**: Each planet moves around the Sun in an orbit that is an ellipse, with the Sun at one focus of the ellipse.

This OpenStax book is available for free at http://cnx.org/content/col11992/1.8

- **Kepler's second law**: The straight line joining a planet and the Sun sweeps out equal areas in space in equal intervals of time.
- **Kepler's third law**: The square of a planet's orbital period is directly proportional to the cube of the semimajor axis of its orbit.

Kepler's three laws provide a precise geometric description of planetary motion within the framework of the Copernican system. With these tools, it was possible to calculate planetary positions with greatly improved precision. Still, Kepler's laws are purely descriptive: they do not help us understand what forces of nature constrain the planets to follow this particular set of rules. That step was left to Isaac Newton.

EXAMPLE 3.2

Applying Kepler's Third Law

Using the orbital periods and semimajor axes for Venus and Earth that are provided here, calculate P^2 and a^3, and verify that they obey Kepler's third law. Venus' orbital period is 0.62 year, and its semimajor axis is 0.72 AU. Earth's orbital period is 1.00 year, and its semimajor axis is 1.00 AU.

Solution

We can use the equation for Kepler's third law, $P^2 \propto a^3$. For Venus, $P^2 = 0.62 \times 0.62 = 0.38$ year and $a^3 = 0.72 \times 0.72 \times 0.72 = 0.37$ AU (rounding numbers sometimes causes minor discrepancies like this). The orbital period (0.38 year) approximates the semimajor axis (0.37 AU). Therefore, Venus obeys Kepler's third law. For Earth, $P^2 = 1.00 \times 1.00 = 1.00$ year and $a^3 = 1.00 \times 1.00 \times 1.00 = 1.00$ AU. The orbital period (1.00 year) approximates (in this case, equals) the semimajor axis (1.00 AU). Therefore, Earth obeys Kepler's third law.

Check Your Learning

Using the orbital periods and semimajor axes for Saturn and Jupiter that are provided here, calculate P^2 and a^3, and verify that they obey Kepler's third law. Saturn's orbital period is 29.46 years, and its semimajor axis is 9.54 AU. Jupiter's orbital period is 11.86 years, and its semimajor axis is 5.20 AU.

Answer:

For Saturn, $P^2 = 29.46 \times 29.46 = 867.9$ years and $a^3 = 9.54 \times 9.54 \times 9.54 = 868.3$ AU. The orbital period (867.9 years) approximates the semimajor axis (868.3 AU). Therefore, Saturn obeys Kepler's third law.

LINK TO LEARNING

In honor of the scientist who first devised the laws that govern the motions of planets, the team that built the first spacecraft to search for planets orbiting other stars decided to name the probe "Kepler." To learn more about Johannes Kepler's life and his laws of planetary motion, as well as lots of information on the Kepler Mission, visit NASA's Kepler website (https://openstaxcollege.org/l/30nasakepmiss) and follow the links that interest you.

Download for free at http://cnx.org/content/col11992/latest/

3.2 NEWTON'S GREAT SYNTHESIS

Learning Objectives

By the end of this section, you will be able to:

> Describe Newton's three laws of motion
> Explain how Newton's three laws of motion relate to momentum
> Define mass, volume, and density and how they differ
> Define angular momentum

It was the genius of Isaac Newton that found a conceptual framework that completely explained the observations and rules assembled by Galileo, Brahe, Kepler, and others. Newton was born in Lincolnshire, England, in the year after Galileo's death (Figure 3.6). Against the advice of his mother, who wanted him to stay home and help with the family farm, he entered Trinity College at Cambridge in 1661 and eight years later was appointed professor of mathematics. Among Newton's contemporaries in England were architect Christopher Wren, authors Aphra Behn and Daniel Defoe, and composer G. F. Handel.

Figure 3.6. Isaac Newton (1643–1727), 1689 Portrait by Sir Godfrey Kneller. Isaac Newton's work on the laws of motion, gravity, optics, and mathematics laid the foundations for much of physical science.

Newton's Laws of Motion

As a young man in college, Newton became interested in natural philosophy, as science was then called. He worked out some of his first ideas on machines and optics during the plague years of 1665 and 1666, when students were sent home from college. Newton, a moody and often difficult man, continued to work on his ideas in private, even inventing new mathematical tools to help him deal with the complexities involved. Eventually, his friend Edmund Halley (profiled in Comets and Asteroids: Debris of the Solar System) prevailed on him to collect and publish the results of his remarkable investigations on motion and gravity. The result was a volume that set out the underlying system of the physical world, *Philosophiae Naturalis Principia Mathematica*. The *Principia*, as the book is generally known, was published at Halley's expense in 1687.

At the very beginning of the *Principia*, Newton proposes three laws that would govern the motions of all objects:

This OpenStax book is available for free at http://cnx.org/content/col11992/1.8

- **Newton's first law**: Every object will continue to be in a state of rest or move at a constant speed in a straight line unless it is compelled to change by an outside force.

- **Newton's second law**: The change of motion of a body is proportional to and in the direction of the force acting on it.

- **Newton's third law**: For every action there is an equal and opposite reaction (*or:* the mutual actions of two bodies upon each other are always equal and act in opposite directions).

In the original Latin, the three laws contain only 59 words, but those few words set the stage for modern science. Let us examine them more carefully.

Interpretation of Newton's Laws

Newton's first law is a restatement of one of Galileo's discoveries, called the *conservation of momentum*. The law states that in the absence of any outside influence, there is a measure of a body's motion, called its **momentum**, that remains unchanged. You may have heard the term momentum used in everyday expressions, such as "This bill in Congress has a lot of momentum; it's going to be hard to stop."

Newton's first law is sometimes called the *law of inertia*, where inertia is the tendency of objects (and legislatures) to keep doing what they are already doing. In other words, a stationary object stays put, and a moving object keeps moving unless some force intervenes.

Let's define the precise meaning of momentum—it depends on three factors: (1) speed—how fast a body moves (zero if it is stationary), (2) the direction of its motion, and (3) its mass—a measure of the amount of matter in a body, which we will discuss later. Scientists use the term **velocity** to describe the speed and direction of motion. For example, 20 kilometers per hour due south is velocity, whereas 20 kilometers per hour just by itself is speed. Momentum then can be defined as an object's mass times its velocity.

It's not so easy to see this rule in action in the everyday world because of the many forces acting on a body at any one time. One important force is friction, which generally slows things down. If you roll a ball along the sidewalk, it eventually comes to a stop because the sidewalk exerts a rubbing force on the ball. But in the space between the stars, where there is so little matter that friction is insignificant, objects can in fact continue to move (to coast) indefinitely.

The momentum of a body can change only under the action of an outside influence. Newton's second law expresses *force* in terms of its ability to change momentum with time. A force (a push or a pull) has both size and direction. When a force is applied to a body, the momentum changes in the direction of the applied force. This means that a force is required to change either the speed or the direction of a body, or both—that is, to start it moving, to speed it up, to slow it down, to stop it, or to change its direction.

As you learned in Observing the Sky: The Birth of Astronomy, the rate of change in an object's velocity is called *acceleration*. Newton showed that the acceleration of a body was proportional to the force being applied to it. Suppose that after a long period of reading, you push an astronomy book away from you on a long, smooth table. (We use a smooth table so we can ignore friction.) If you push the book steadily, it will continue to speed up as long as you are pushing it. The harder you push the book, the larger its acceleration will be. How much a force will accelerate an object is also determined by the object's mass. If you kept pushing a pen with the same force with which you pushed the textbook, the pen—having less mass—would be accelerated to a greater speed.

Newton's third law is perhaps the most profound of the rules he discovered. Basically, it is a generalization of the first law, but it also gives us a way to define mass. If we consider a system of two or more objects isolated from outside influences, Newton's first law says that the total momentum of the objects should remain

Download for free at http://cnx.org/content/col11992/latest/

constant. Therefore, any change of momentum within the system must be balanced by another change that is equal and opposite so that the momentum of the entire system is not changed.

This means that forces in nature do not occur alone: we find that in each situation there is always a *pair* of forces that are equal to and opposite each other. If a force is exerted on an object, it must be exerted by something else, and the object will exert an equal and opposite force back on that something. We can look at a simple example to demonstrate this.

Suppose that a daredevil astronomy student—and avid skateboarder—wants to jump from his second-story dorm window onto his board below (we don't recommend trying this!). The force pulling him down after jumping (as we will see in the next section) is the force of gravity between him and Earth. Both he and Earth must experience the same total change of momentum because of the influence of these mutual forces. So, both the student and Earth are accelerated by each other's pull. However, the student does much more of the moving. Because Earth has enormously greater mass, it can experience the same change of momentum by accelerating only a very small amount. Things fall toward Earth all the time, but the acceleration of our planet as a result is far too small to be measured.

A more obvious example of the mutual nature of forces between objects is familiar to all who have batted a baseball. The recoil you feel as you swing your bat shows that the ball exerts a force on it during the impact, just as the bat does on the ball. Similarly, when a rifle you are bracing on your shoulder is discharged, the force pushing the bullet out of the muzzle is equal to the force pushing backward upon the gun and your shoulder.

This is the principle behind jet engines and rockets: the force that discharges the exhaust gases from the rear of the rocket is accompanied by the force that pushes the rocket forward. The exhaust gases need not push against air or Earth; a rocket actually operates best in a vacuum (Figure 3.7).

Figure 3.7. Demonstrating Newton's Third Law. The U.S. Space Shuttle (here launching *Discovery*), powered by three fuel engines burning liquid oxygen and liquid hydrogen, with two solid fuel boosters, demonstrates Newton's third law. (credit: modification of work by NASA)

LINK TO LEARNING

For more about Isaac Newton's life and work, check out this timeline page (https://openstaxcollege.org/l/30IsaacNewTime) with snapshots from his career, produced by the British Broadcasting Corporation (BBC).

This OpenStax book is available for free at http://cnx.org/content/col11992/1.8

Mass, Volume, and Density

Before we go on to discuss Newton's other work, we want to take a brief look at some terms that will be important to sort out clearly. We begin with *mass,* which is a measure of the amount of material within an object.

The *volume* of an object is the measure of the physical space it occupies. Volume is measured in cubic units, such as cubic centimeters or liters. The volume is the "size" of an object. A penny and an inflated balloon may both have the same mass, but they have very different volumes. The reason is that they also have very different *densities*, which is a measure of how much mass there is per unit volume. Specifically, **density** is the mass divided by the volume. Note that in everyday language we often use "heavy" and "light" as indications of density (rather than weight) as, for instance, when we say that iron is heavy or that whipped cream is light.

The units of density that will be used in this book are grams per cubic centimeter (g/cm^3).[1] If a block of some material has a mass of 300 grams and a volume of 100 cm^3, its density is 3 g/cm^3. Familiar materials span a considerable range in density, from artificial materials such as plastic insulating foam (less than 0.1 g/cm^3) to gold (19.3 g/cm^3). Table 3.1 gives the densities of some familiar materials. In the astronomical universe, much more remarkable densities can be found, all the way from a comet's tail (10^{-16} g/cm^3) to a collapsed "star corpse" called a neutron star (10^{15} g/cm^3).

Densities of Common Materials

Material	Density (g/cm³)
Gold	19.3
Lead	11.3
Iron	7.9
Earth (bulk)	5.5
Rock (typical)	2.5
Water	1
Wood (typical)	0.8
Insulating foam	0.1
Silica gel	0.02

Table 3.1

To sum up, mass is *how much*, volume is *how big*, and density is *how tightly packed*.

---

1   Generally we use standard metric (or SI) units in this book. The proper metric unit of density in that system is kg/m^3. But to most people, g/cm^3 provides a more meaningful unit because the density of water is exactly 1 g/cm^3, and this is useful information for comparison. Density expressed in g/cm^3 is sometimes called specific density or specific weight.

Download for free at http://cnx.org/content/col11992/latest/

LINK TO LEARNING

You can play with a simple animation (https://openstaxcollege.org/l/30phetsimdenmas)
demonstrating the relationship between the concepts of density, mass, and volume, and find out why
objects like wood float in water.

Angular Momentum

A concept that is a bit more complex, but important for understanding many astronomical objects, is **angular
momentum**, which is a measure of the rotation of a body as it revolves around some fixed point (an example is
a planet orbiting the Sun). The angular momentum of an object is defined as the product of its mass, its velocity,
and its distance from the fixed point around which it revolves.

If these three quantities remain constant—that is, if the motion of a particular object takes place at a constant
velocity at a fixed distance from the spin center—then the angular momentum is also a constant. Kepler's
second law is a consequence of the *conservation of angular momentum*. As a planet approaches the Sun on its
elliptical orbit and the distance to the spin center decreases, the planet speeds up to conserve the angular
momentum. Similarly, when the planet is farther from the Sun, it moves more slowly.

The conservation of angular momentum is illustrated by figure skaters, who bring their arms and legs in to spin
more rapidly, and extend their arms and legs to slow down (Figure 3.8). You can duplicate this yourself on a
well-oiled swivel stool by starting yourself spinning slowly with your arms extended and then pulling your arms
in. Another example of the conservation of angular momentum is a shrinking cloud of dust or a star collapsing
on itself (both are situations that you will learn about as you read on). As material moves to a lesser distance
from the spin center, the speed of the material increases to conserve angular momentum.

Figure 3.8. Conservation of Angular Momentum. When a spinning figure skater brings in her arms, their distance from her spin center is
smaller, so her speed increases. When her arms are out, their distance from the spin center is greater, so she slows down.

This OpenStax book is available for free at http://cnx.org/content/col11992/1.8

3.3 NEWTON'S UNIVERSAL LAW OF GRAVITATION

Learning Objectives

By the end of this section, you will be able to:

> Explain what determines the strength of gravity
> Describe how Newton's universal law of gravitation extends our understanding of Kepler's laws

Newton's laws of motion show that objects at rest will stay at rest and those in motion will continue moving uniformly in a straight line unless acted upon by a force. Thus, it is the *straight line* that defines the most natural state of motion. But the planets move in ellipses, not straight lines; therefore, some force must be bending their paths. That force, Newton proposed, was **gravity**.

In Newton's time, gravity was something associated with Earth alone. Everyday experience shows us that Earth exerts a gravitational force upon objects at its surface. If you drop something, it accelerates toward Earth as it falls. Newton's insight was that Earth's gravity might extend as far as the Moon and produce the force required to curve the Moon's path from a straight line and keep it in its orbit. He further hypothesized that gravity is not limited to Earth, but that there is a general force of attraction between all material bodies. If so, the attractive force between the Sun and each of the planets could keep them in their orbits. (This may seem part of our everyday thinking today, but it was a remarkable insight in Newton's time.)

Once Newton boldly hypothesized that there was a universal attraction among all bodies everywhere in space, he had to determine the exact nature of the attraction. The precise mathematical description of that gravitational force had to dictate that the planets move exactly as Kepler had described them to (as expressed in Kepler's three laws). Also, that gravitational force had to predict the correct behavior of falling bodies on Earth, as observed by Galileo. How must the force of gravity depend on distance in order for these conditions to be met?

The answer to this question required mathematical tools that had not yet been developed, but this did not deter Isaac Newton, who invented what we today call calculus to deal with this problem. Eventually he was able to conclude that the magnitude of the force of gravity must decrease with increasing distance between the Sun and a planet (or between any two objects) in proportion to the inverse square of their separation. In other words, if a planet were twice as far from the Sun, the force would be $(1/2)^2$, or 1/4 as large. Put the planet three times farther away, and the force is $(1/3)^2$, or 1/9 as large.

Newton also concluded that the gravitational attraction between two bodies must be proportional to their masses. The more mass an object has, the stronger the pull of its gravitational force. The gravitational attraction between any two objects is therefore given by one of the most famous equations in all of science:

$$F_{\text{gravity}} = G\frac{M_1 M_2}{R^2}$$

where F_{gravity} is the gravitational force between two objects, M_1 and M_2 are the masses of the two objects, and R is their separation. G is a constant number known as the *universal gravitational constant*, and the equation itself symbolically summarizes Newton's *universal law of gravitation*. With such a force and the laws of motion, Newton was able to show mathematically that the only orbits permitted were exactly those described by Kepler's laws.

Newton's universal law of gravitation works for the planets, but is it really universal? The gravitational theory should also predict the observed acceleration of the Moon toward Earth as it orbits Earth, as well as of any

Download for free at http://cnx.org/content/col11992/latest/

object (say, an apple) dropped near Earth's surface. The falling of an apple is something we can measure quite easily, but can we use it to predict the motions of the Moon?

Recall that according to Newton's second law, forces cause acceleration. Newton's universal law of gravitation says that the force acting upon (and therefore the acceleration of) an object toward Earth should be inversely proportional to the square of its distance from the center of Earth. Objects like apples at the surface of Earth, at a distance of one Earth-radius from the center of Earth, are observed to accelerate downward at 9.8 meters per second per second (9.8 m/s^2).

It is this force of gravity on the surface of Earth that gives us our sense of *weight*. Unlike your mass, which would remain the same on any planet or moon, your weight depends on the local force of gravity. So you would weigh less on Mars and the Moon than on Earth, even though there is no change in your mass. (Which means you would still have to go easy on the desserts in the college cafeteria when you got back!)

The Moon is 60 Earth radii away from the center of Earth. If gravity (and the acceleration it causes) gets weaker with distance squared, the acceleration the Moon experiences should be a lot less than for the apple. The acceleration should be $(1/60)^2 = 1/3600$ (or 3600 times less—about 0.00272 m/s^2. This is precisely the observed acceleration of the Moon in its orbit. (As we shall see, the Moon does not fall *to* Earth with this acceleration, but falls *around* Earth.) Imagine the thrill Newton must have felt to realize he had discovered, and verified, a law that holds for Earth, apples, the Moon, and, as far as he knew, everything in the universe.

EXAMPLE 3.3

Calculating Weight

By what factor would a person's weight at the surface of Earth change if Earth had its present mass but eight times its present volume?

Solution

With eight times the volume, Earth's radius would double. This means the gravitational force at the surface would reduce by a factor of $(1/2)^2 = 1/4$, so a person would weigh only one-fourth as much.

Check Your Learning

By what factor would a person's weight at the surface of Earth change if Earth had its present size but only one-third its present mass?

Answer:

With one-third its present mass, the gravitational force at the surface would reduce by a factor of 1/3, so a person would weight only one-third as much.

Gravity is a "built-in" property of mass. Whenever there are masses in the universe, they will interact via the force of gravitational attraction. The more mass there is, the greater the force of attraction. Here on Earth, the largest concentration of mass is, of course, the planet we stand on, and its pull dominates the gravitational interactions we experience. But everything with mass attracts everything else with mass anywhere in the universe.

This OpenStax book is available for free at http://cnx.org/content/col11992/1.8

Download for free at http://cnx.org/content/col11992/latest/

Newton's law also implies that gravity never becomes zero. It quickly gets weaker with distance, but it continues to act to some degree no matter how far away you get. The pull of the Sun is stronger at Mercury than at Pluto, but it can be felt far beyond Pluto, where astronomers have good evidence that it continuously makes enormous numbers of smaller icy bodies move around huge orbits. And the Sun's gravitational pull joins with the pull of billions of others stars to create the gravitational pull of our Milky Way Galaxy. That force, in turn, can make other smaller galaxies orbit around the Milky Way, and so on.

Why is it then, you may ask, that the astronauts aboard the Space Shuttle appear to have no gravitational forces acting on them when we see images on television of the astronauts and objects floating in the spacecraft? After all, the astronauts in the shuttle are only a few hundred kilometers above the surface of Earth, which is not a significant distance compared to the size of Earth, so gravity is certainly not a great deal weaker that much farther away. The astronauts feel "weightless" (meaning that they don't feel the gravitational force acting on them) for the same reason that passengers in an elevator whose cable has broken or in an airplane whose engines no longer work feel weightless: they are falling (Figure 3.9).[2]

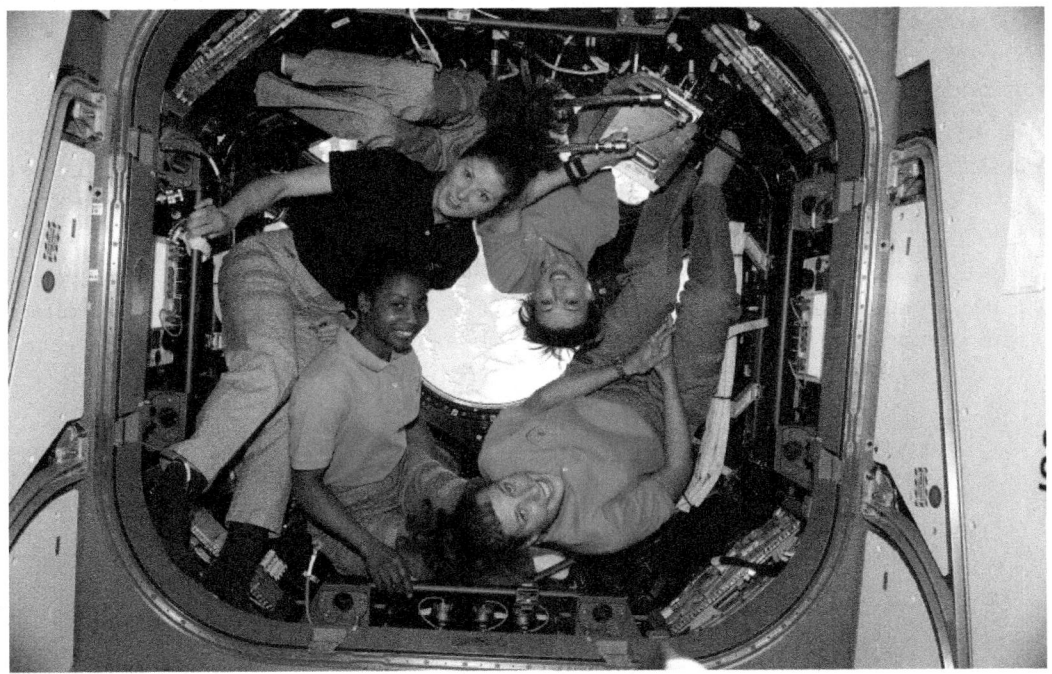

Figure 3.9. Astronauts in Free Fall. While in space, astronauts are falling freely, so they experience "weightlessness." Clockwise from top left: Tracy Caldwell Dyson (NASA), Naoko Yamzaki (JAXA), Dorothy Metcalf-Lindenburger (NASA), and Stephanie Wilson (NASA). (credit: NASA)

When *falling*, they are in free fall and accelerate at the same rate as everything around them, including their spacecraft or a camera with which they are taking photographs of Earth. When doing so, astronauts experience no additional forces and therefore feel "weightless." Unlike the falling elevator passengers, however, the astronauts are falling *around* Earth, not *to* Earth; as a result they will continue to fall and are said to be "in orbit" around Earth (see the next section for more about orbits).

---

2   In the film *Apollo 13*, the scenes in which the astronauts were "weightless" were actually filmed in a falling airplane. As you might imagine, the plane fell for only short periods before the engines engaged again.

Download for free at http://cnx.org/content/col11992/latest/

Orbital Motion and Mass

Kepler's laws describe the orbits of the objects whose motions are described by Newton's laws of motion and the law of gravity. Knowing that gravity is the force that attracts planets toward the Sun, however, allowed Newton to rethink Kepler's third law. Recall that Kepler had found a relationship between the orbital period of a planet's revolution and its distance from the Sun. But Newton's formulation introduces the additional factor of the masses of the Sun (M_1) and the planet (M_2), both expressed in units of the Sun's mass. Newton's universal law of gravitation can be used to show mathematically that this relationship is actually

$$a^3 = \left(M_1 + M_2\right) \times P^2$$

where a is the semimajor axis and P is the orbital period.

How did Kepler miss this factor? In units of the Sun's mass, the mass of the Sun is 1, and in units of the Sun's mass, the mass of a typical planet is a negligibly small factor. This means that the sum of the Sun's mass and a planet's mass, ($M_1 + M_2$), is very, very close to 1. This makes Newton's formula appear almost the same as Kepler's; the tiny mass of the planets compared to the Sun is the reason that Kepler did not realize that both masses had to be included in the calculation. There are many situations in astronomy, however, in which we *do* need to include the two mass terms—for example, when two stars or two galaxies orbit each other.

Including the mass term allows us to use this formula in a new way. If we can measure the motions (distances and orbital periods) of objects acting under their mutual gravity, then the formula will permit us to deduce their masses. For example, we can calculate the mass of the Sun by using the distances and orbital periods of the planets, or the mass of Jupiter by noting the motions of its moons.

Indeed, Newton's reformulation of Kepler's third law is one of the most powerful concepts in astronomy. Our ability to deduce the masses of objects from their motions is key to understanding the nature and evolution of many astronomical bodies. We will use this law repeatedly throughout this text in calculations that range from the orbits of comets to the interactions of galaxies.

EXAMPLE 3.4

Calculating the Effects of Gravity

A planet like Earth is found orbiting its star at a distance of 1 AU in 0.71 Earth-year. Can you use Newton's version of Kepler's third law to find the mass of the star? (Remember that compared to the mass of a star, the mass of an earthlike planet can be considered negligible.)

Solution

In the formula $a^3 = (M_1 + M_2) \times P^2$, the factor $M_1 + M_2$ would now be approximately equal to M_1 (the mass of the star), since the planet's mass is so small by comparison. Then the formula becomes $a^3 = M_1 \times P^2$, and we can solve for M_1:

$$M_1 = \frac{a^3}{P^2}$$

Since $a = 1$, $a^3 = 1$, so

$$M_1 = \frac{1}{P^2} = \frac{1}{0.71^2} = \frac{1}{0.5} = 2$$

This OpenStax book is available for free at http://cnx.org/content/col11992/1.8

Download for free at http://cnx.org/content/col11992/latest/

So the mass of the star is twice the mass of our Sun. (Remember that this way of expressing the law has units in terms of Earth and the Sun, so masses are expressed in units of the mass of our Sun.)

Check Your Learning

Suppose a star with twice the mass of our Sun had an earthlike planet that took 4 years to orbit the star. At what distance (semimajor axis) would this planet orbit its star?

Answer:

Again, we can neglect the mass of the planet. So $M_1 = 2$ and $P = 4$ years. The formula is $a^3 = M_1 \times P^2$, so $a^3 = 2 \times 4^2 = 2 \times 16 = 32$. So a is the cube root of 32. To find this, you can just ask Google, "What is the cube root of 32?" and get the answer 3.2 AU.

LINK TO LEARNING

You might like to try a simulation (https://openstaxcollege.org/l/30phetsimsunear) that lets you move the Sun, Earth, Moon, and space station to see the effects of changing their distances on their gravitational forces and orbital paths. You can even turn off gravity and see what happens.

3.4 ORBITS IN THE SOLAR SYSTEM

Learning Objectives

By the end of this section, you will be able to:

> Compare the orbital characteristics of the planets in the solar system
> Compare the orbital characteristics of asteroids and comets in the solar system

Recall that the path of an object under the influence of gravity through space is called its orbit, whether that object is a spacecraft, planet, star, or galaxy. An orbit, once determined, allows the future positions of the object to be calculated.

Two points in any orbit in our solar system have been given special names. The place where the planet is closest to the Sun (*helios* in Greek) and moves the fastest is called the **perihelion** of its orbit, and the place where it is farthest away and moves the most slowly is the **aphelion**. For the Moon or a satellite orbiting Earth (*gee* in Greek), the corresponding terms are **perigee** and **apogee**. (In this book, we use the word *moon* for a natural object that goes around a planet and the word **satellite** to mean a human-made object that revolves around a planet.)

Orbits of the Planets

Today, Newton's work enables us to calculate and predict the orbits of the planets with marvelous precision. We know eight planets, beginning with Mercury closest to the Sun and extending outward to Neptune. The average orbital data for the planets are summarized in Table 3.2. (Ceres is the largest of the *asteroids,* now considered a dwarf planet.)

Download for free at http://cnx.org/content/col11992/latest/

According to Kepler's laws, Mercury must have the shortest orbital period (88 Earth-days); thus, it has the highest orbital speed, averaging 48 kilometers per second. At the opposite extreme, Neptune has a period of 165 years and an average orbital speed of just 5 kilometers per second.

All the planets have orbits of rather low eccentricity. The most eccentric orbit is that of Mercury (0.21); the rest have eccentricities smaller than 0.1. It is fortunate that among the rest, Mars has an eccentricity greater than that of many of the other planets. Otherwise the pre-telescopic observations of Brahe would not have been sufficient for Kepler to deduce that its orbit had the shape of an ellipse rather than a circle.

The planetary orbits are also confined close to a common plane, which is near the plane of Earth's orbit (called the ecliptic). The strange orbit of the dwarf planet Pluto is inclined about 17° to the ecliptic, and that of the dwarf planet Eris (orbiting even farther away from the Sun than Pluto) by 44°, but all the major planets lie within 10° of the common plane of the solar system.

LINK TO LEARNING

You can use an orbital simulator (https://openstaxcollege.org/l/30phetorbsim) to design your own mini solar system with up to four bodies. Adjust masses, velocities, and positions of the planets, and see what happens to their orbits as a result.

Orbits of Asteroids and Comets

In addition to the eight planets, there are many smaller objects in the solar system. Some of these are moons (natural satellites) that orbit all the planets except Mercury and Venus. In addition, there are two classes of smaller objects in heliocentric orbits: *asteroids* and *comets*. Both asteroids and comets are believed to be small chunks of material left over from the formation process of the solar system.

In general, asteroids have orbits with smaller semimajor axes than do comets (Figure 3.10). The majority of them lie between 2.2 and 3.3 AU, in the region known as the **asteroid belt** (see Comets and Asteroids: Debris of the Solar System). As you can see in Table 3.2, the asteroid belt (represented by its largest member, Ceres) is in the middle of a gap between the orbits of Mars and Jupiter. It is because these two planets are so far apart that stable orbits of small bodies can exist in the region between them.

This OpenStax book is available for free at http://cnx.org/content/col11992/1.8

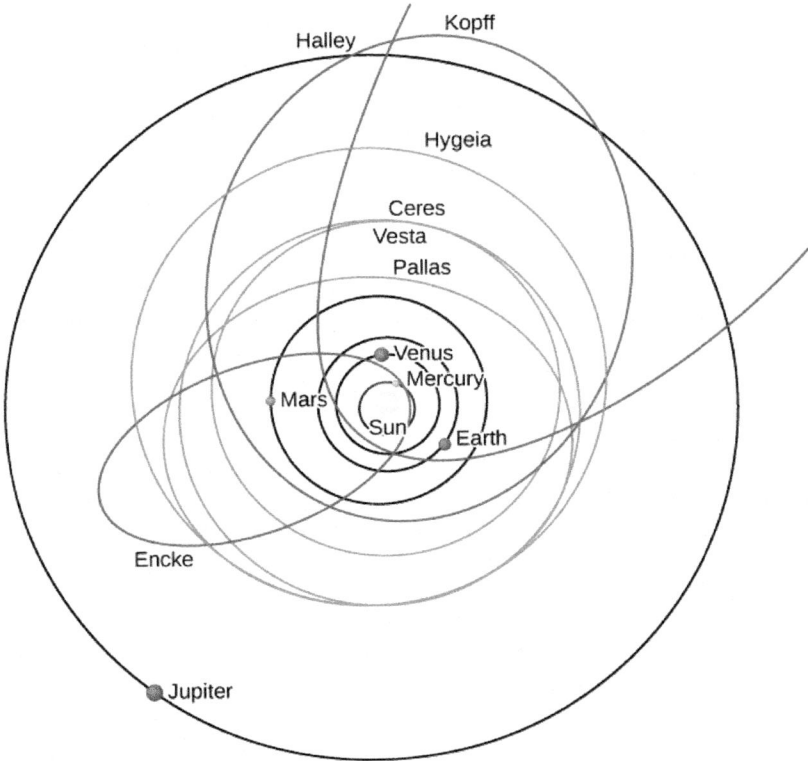

Figure 3.10. Solar System Orbits. We see the orbits of typical comets and asteroids compared with those of the planets Mercury, Venus, Earth, Mars, and Jupiter (black circles). Shown in red are three comets: Halley, Kopff, and Encke. In blue are the four largest asteroids: Ceres, Pallas, Vesta, and Hygeia.

Orbital Data for the Planets

Planet	Semimajor Axis (AU)	Period (y)	Eccentricity
Mercury	0.39	0.24	0.21
Venus	0.72	0.6	0.01
Earth	1	1.00	0.02
Mars	1.52	1.88	0.09
(Ceres)	2.77	4.6	0.08
Jupiter	5.20	11.86	0.05
Saturn	9.54	29.46	0.06
Uranus	19.19	84.01	0.05

Table 3.2

Download for free at http://cnx.org/content/col11992/latest/

Orbital Data for the Planets

Planet	Semimajor Axis (AU)	Period (y)	Eccentricity
Neptune	30.06	164.82	0.01

Table 3.2

Comets generally have orbits of larger size and greater eccentricity than those of the asteroids. Typically, the eccentricity of their orbits is 0.8 or higher. According to Kepler's second law, therefore, they spend most of their time far from the Sun, moving very slowly. As they approach perihelion, the comets speed up and whip through the inner parts of their orbits more rapidly.

3.5 MOTIONS OF SATELLITES AND SPACECRAFT

Learning Objectives

By the end of this section, you will be able to:

> Explain how an object (such as a satellite) can be put into orbit around Earth
> Explain how an object (such as a planetary probe) can escape from orbit

Newton's universal law of gravitation and Kepler's laws describe the motions of Earth satellites and interplanetary spacecraft as well as the planets. Sputnik, the first artificial Earth satellite, was launched by what was then called the Soviet Union on October 4, 1957. Since that time, thousands of satellites have been placed into orbit around Earth, and spacecraft have also orbited the Moon, Venus, Mars, Jupiter, Saturn, and a number of asteroids and comets.

Once an artificial satellite is in orbit, its behavior is no different from that of a natural satellite, such as our Moon. If the satellite is high enough to be free of atmospheric friction, it will remain in orbit forever. However, although there is no difficulty in maintaining a satellite once it is in orbit, a great deal of energy is required to lift the spacecraft off Earth and accelerate it to orbital speed.

To illustrate how a satellite is launched, imagine a gun firing a bullet horizontally from the top of a high mountain, as in Figure 3.11, which has been adapted from a similar diagram by Newton. Imagine, further, that the friction of the air could be removed and that nothing gets in the bullet's way. Then the only force that acts on the bullet after it leaves the muzzle is the gravitational force between the bullet and Earth.

This OpenStax book is available for free at http://cnx.org/content/col11992/1.8

Download for free at http://cnx.org/content/col11992/latest/

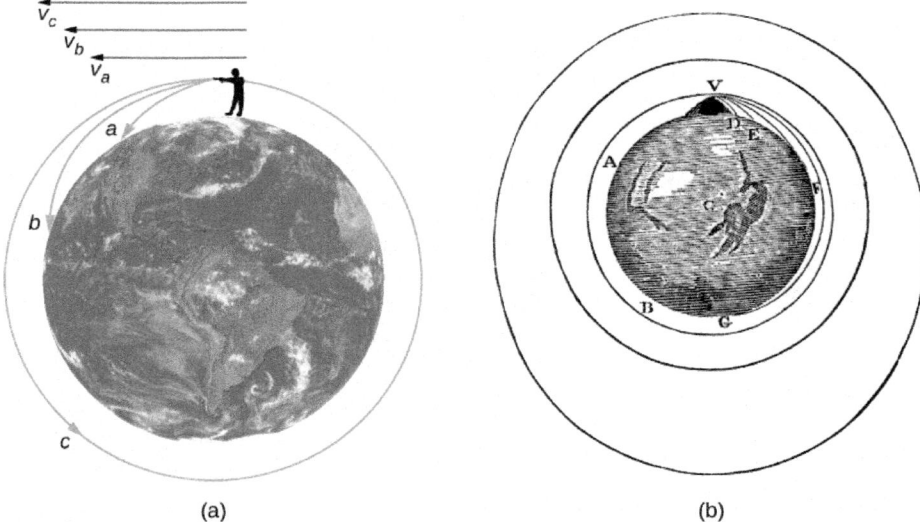

Figure 3.11. Firing a Bullet into Orbit. (a) For paths *a* and *b*, the velocity is not enough to prevent gravity from pulling the bullet back to Earth; in case *c*, the velocity allows the bullet to fall completely around Earth. (b) This diagram by Newton in his *De Mundi Systemate*, 1731 edition, illustrates the same concept shown in (a).

If the bullet is fired with a velocity we can call v_a, the gravitational force acting upon it pulls it downward toward Earth, where it strikes the ground at point *a*. However, if it is given a higher muzzle velocity, v_b, its higher speed carries it farther before it hits the ground at point *b*.

If our bullet is given a high enough muzzle velocity, v_c, the curved surface of Earth causes the ground to remain the same distance from the bullet so that the bullet falls *around* Earth in a complete circle. The speed needed to do this—called the circular satellite velocity—is about 8 kilometers per second, or about 17,500 miles per hour in more familiar units.

Each year, more than 50 new satellites are launched into orbit by such nations as Russia, the United States, China, Japan, India, and Israel, as well as by the European Space Agency (ESA), a consortium of European nations (Figure 3.12). Today, these satellites are used for weather tracking, ecology, global positioning systems, communications, and military purposes, to name a few uses. Most satellites are launched into low Earth orbit, since this requires the minimum launch energy. At the orbital speed of 8 kilometers per second, they circle the planet in about 90 minutes. Some of the very low Earth orbits are not indefinitely stable because, as Earth's atmosphere swells from time to time, a frictional drag is generated by the atmosphere on these satellites, eventually leading to a loss of energy and "decay" of the orbit.

Download for free at http://cnx.org/content/col11992/latest/

Figure 3.12. Satellites in Earth Orbit. This figure shows the larger pieces of orbital debris that are being tracked by NASA in Earth's orbit. (credit: NASA/JSC)

Interplanetary Spacecraft

The exploration of the solar system has been carried out largely by robot spacecraft sent to the other planets. To escape Earth, these craft must achieve **escape speed**, the speed needed to move away from Earth forever, which is about 11 kilometers per second (about 25,000 miles per hour). After escaping Earth, these craft coast to their targets, subject only to minor trajectory adjustments provided by small thruster rockets on board. In interplanetary flight, these spacecraft follow orbits around the Sun that are modified only when they pass near one of the planets.

As it comes close to its target, a spacecraft is deflected by the planet's gravitational force into a modified orbit, either gaining or losing energy in the process. Spacecraft controllers have actually been able to use a planet's gravity to redirect a flyby spacecraft to a second target. For example, Voyager 2 used a series of gravity-assisted encounters to yield successive flybys of Jupiter (1979), Saturn (1980), Uranus (1986), and Neptune (1989). The Galileo spacecraft, launched in 1989, flew past Venus once and Earth twice to gain the energy required to reach its ultimate goal of orbiting Jupiter.

If we wish to orbit a planet, we must slow the spacecraft with a rocket when the spacecraft is near its destination, allowing it to be captured into an elliptical orbit. Additional rocket thrust is required to bring a vehicle down from orbit for a landing on the surface. Finally, if a return trip to Earth is planned, the landed payload must include enough propulsive power to repeat the entire process in reverse.

This OpenStax book is available for free at http://cnx.org/content/col11992/1.8

GRAVITY WITH MORE THAN TWO BODIES

Learning Objectives

By the end of this section, you will be able to:

> Explain how the gravitational interactions of many bodies can causes perturbations in their motions
> Explain how the planet Neptune was discovered

Until now, we have considered the Sun and a planet (or a planet and one of its moons) as nothing more than a pair of bodies revolving around each other. In fact, all the planets exert gravitational forces upon one another as well. These interplanetary attractions cause slight variations from the orbits than would be expected if the gravitational forces between planets were neglected. The motion of a body that is under the gravitational influence of two or more other bodies is very complicated and can be calculated properly only with large computers. Fortunately, astronomers have such computers at their disposal in universities and government research institutes.

The Interactions of Many Bodies

As an example, suppose you have a cluster of a thousand stars all orbiting a common center (such clusters are quite common, as we shall see in Star Clusters). If we know the exact position of each star at any given instant, we can calculate the combined gravitational force of the entire group on any one member of the cluster. Knowing the force on the star in question, we can therefore find how it will accelerate. If we know how it was moving to begin with, we can then calculate how it will move in the next instant of time, thus tracking its motion.

However, the problem is complicated by the fact that the other stars are also moving and thus changing the effect they will have on our star. Therefore, we must simultaneously calculate the acceleration of each star produced by the combination of the gravitational attractions of all the others in order to track the motions of all of them, and hence of any one. Such complex calculations have been carried out with modern computers to track the evolution of hypothetical clusters of stars with up to a million members (Figure 3.13).

Figure 3.13. Modern Computing Power. These supercomputers at NASA's Ames Research Center are capable of tracking the motions of more than a million objects under their mutual gravitation. (credit: NASA Ames Research Center/Tom Trower)

Within the solar system, the problem of computing the orbits of planets and spacecraft is somewhat simpler. We have seen that Kepler's laws, which do not take into account the gravitational effects of the other planets on an orbit, really work quite well. This is because these additional influences are very small in comparison with

Download for free at http://cnx.org/content/col11992/latest/

the dominant gravitational attraction of the Sun. Under such circumstances, it is possible to treat the effects of other bodies as small **perturbations** (or disturbances). During the eighteenth and nineteenth centuries, mathematicians developed many elegant techniques for calculating perturbations, permitting them to predict very precisely the positions of the planets. Such calculations eventually led to the prediction and discovery of a new planet in 1846.

The Discovery of Neptune

The discovery of the eighth planet, Neptune, was one of the high points in the development of gravitational theory. In 1781, William Herschel, a musician and amateur astronomer, accidentally discovered the seventh planet, Uranus. It happens that Uranus had been observed a century before, but in none of those earlier sightings was it recognized as a planet; rather, it was simply recorded as a star. Herschel's discovery showed that there could be planets in the solar system too dim to be visible to the unaided eye, but ready to be discovered with a telescope if we just knew where to look.

By 1790, an orbit had been calculated for Uranus using observations of its motion in the decade following its discovery. Even after allowance was made for the perturbing effects of Jupiter and Saturn, however, it was found that Uranus did not move on an orbit that exactly fit the earlier observations of it made since 1690. By 1840, the discrepancy between the positions observed for Uranus and those predicted from its computed orbit amounted to about $0.03°$—an angle barely discernable to the unaided eye but still larger than the probable errors in the orbital calculations. In other words, Uranus just did not seem to move on the orbit predicted from Newtonian theory.

In 1843, John Couch Adams, a young Englishman who had just completed his studies at Cambridge, began a detailed mathematical analysis of the irregularities in the motion of Uranus to see whether they might be produced by the pull of an unknown planet. He hypothesized a planet more distant from the Sun than Uranus, and then determined the mass and orbit it had to have to account for the departures in Uranus' orbit. In October 1845, Adams delivered his results to George Airy, the British Astronomer Royal, informing him where in the sky to find the new planet. We now know that Adams' predicted position for the new body was correct to within $2°$, but for a variety of reasons, Airy did not follow up right away.

Meanwhile, French mathematician Urbain Jean Joseph Le Verrier, unaware of Adams or his work, attacked the same problem and published its solution in June 1846. Airy, noting that Le Verrier's predicted position for the unknown planet agreed to within $1°$ with that of Adams, suggested to James Challis, Director of the Cambridge Observatory, that he begin a search for the new object. The Cambridge astronomer, having no up-to-date star charts of the Aquarius region of the sky where the planet was predicted to be, proceeded by recording the positions of all the faint stars he could observe with his telescope in that location. It was Challis' plan to repeat such plots at intervals of several days, in the hope that the planet would distinguish itself from a star by its motion. Unfortunately, he was negligent in examining his observations; although he had actually seen the planet, he did not recognize it.

About a month later, Le Verrier suggested to Johann Galle, an astronomer at the Berlin Observatory, that he look for the planet. Galle received Le Verrier's letter on September 23, 1846, and, possessing new charts of the Aquarius region, found and identified the planet that very night. It was less than a degree from the position Le Verrier predicted. The discovery of the eighth planet, now known as Neptune (the Latin name for the god of the sea), was a major triumph for gravitational theory for it dramatically confirmed the generality of Newton's laws. The honor for the discovery is properly shared by the two mathematicians, Adams and Le Verrier (Figure 3.14).

This OpenStax book is available for free at http://cnx.org/content/col11992/1.8

Download for free at http://cnx.org/content/col11992/latest/

(a) (b)

Figure 3.14. Mathematicians Who Discovered a Planet. (a) John Couch Adams (1819–1892) and (b) Urbain J. J. Le Verrier (1811–1877) share the credit for discovering the planet Neptune.

We should note that the discovery of Neptune was not a complete surprise to astronomers, who had long suspected the existence of the planet based on the "disobedient" motion of Uranus. On September 10, 1846, two weeks before Neptune was actually found, John Herschel, son of the discoverer of Uranus, remarked in a speech before the British Association, "We see [the new planet] as Columbus saw America from the shores of Spain. Its movements have been felt trembling along the far-reaching line of our analysis with a certainty hardly inferior to ocular demonstration."

This discovery was a major step forward in combining Newtonian theory with painstaking observations. Such work continues in our own times with the discovery of planets around other stars.

LINK TO LEARNING

For the fuller story of how Neptune was predicted and found (and the effect of the discovery on the search for Pluto), you can read this page (https://openstaxcollege.org/l/30nepplumatdis) on the mathematical discovery of planets.

MAKING CONNECTIONS

Astronomy and the Poets

When Copernicus, Kepler, Galileo, and Newton formulated the fundamental rules that underlie everything in the physical world, they changed much more than the face of science. For some, they gave humanity the courage to let go of old superstitions and see the world as rational and manageable; for

Download for free at http://cnx.org/content/col11992/latest/

others, they upset comforting, ordered ways that had served humanity for centuries, leaving only a dry, "mechanical clockwork" universe in their wake.

Poets of the time reacted to such changes in their work and debated whether the new world picture was an appealing or frightening one. John Donne (1573–1631), in a poem called "Anatomy of the World," laments the passing of the old certainties:

> The new Philosophy [science] calls all in doubt,
> The element of fire is quite put out;
> The Sun is lost, and th' earth, and no man's wit
> Can well direct him where to look for it.

(Here the "element of fire" refers also to the sphere of fire, which medieval thought placed between Earth and the Moon.)

By the next century, however, poets like Alexander Pope were celebrating Newton and the Newtonian world view. Pope's famous couplet, written upon Newton's death, goes

> Nature, and nature's laws lay hid in night.
> God said, Let Newton be! And all was light.

In his 1733 poem, *An Essay on Man*, Pope delights in the complexity of the new views of the world, incomplete though they are:

> Of man, what see we, but his station here,
> From which to reason, to which refer? . . .
> He, who thro' vast immensity can pierce,
> See worlds on worlds compose one universe,
> Observe how system into system runs,
> What other planets circle other suns,
> What vary'd being peoples every star,
> May tell why Heav'n has made us as we are . . .
> All nature is but art, unknown to thee;
> All chance, direction, which thou canst not see;
> All discord, harmony not understood;
> All partial evil, universal good:
> And, in spite of pride, in erring reason's spite,
> One truth is clear, whatever is, is right.

Poets and philosophers continued to debate whether humanity was exalted or debased by the new views of science. The nineteenth-century poet Arthur Hugh Clough (1819–1861) cries out in his poem "The New Sinai":

> And as old from Sinai's top God said that God is one,
> By science strict so speaks He now to tell us, there is None!
> Earth goes by chemic forces; Heaven's a Mécanique Celeste!
> And heart and mind of humankind a watchwork as the rest!

This OpenStax book is available for free at http://cnx.org/content/col11992/1.8

Download for free at http://cnx.org/content/col11992/latest/

(A "mécanique celeste" is a clockwork model to demonstrate celestial motions.)

The twentieth-century poet Robinson Jeffers (whose brother was an astronomer) saw it differently in a poem called "Star Swirls":

> There is nothing like astronomy to pull the stuff out of man.
> His stupid dreams and red-rooster importance:
> Let him count the star-swirls.

Download for free at http://cnx.org/content/col11992/latest/

CHAPTER 3 REVIEW

🔑 KEY TERMS

angular momentum the measure of the motion of a rotating object in terms of its speed and how widely the object's mass is distributed around its axis

aphelion the point in its orbit where a planet (or other orbiting object) is farthest from the Sun

apogee the point in its orbit where an Earth satellite is farthest from Earth

asteroid belt the region of the solar system between the orbits of Mars and Jupiter in which most asteroids are located; the main belt, where the orbits are generally the most stable, extends from 2.2 to 3.3 AU from the Sun

astronomical unit (AU) the unit of length defined as the average distance between Earth and the Sun; this distance is about 1.5×10^8 kilometers

density the ratio of the mass of an object to its volume

eccentricity in an ellipse, the ratio of the distance between the foci to the major axis

ellipse a closed curve for which the sum of the distances from any point on the ellipse to two points inside (called the foci) is always the same

escape speed the speed a body must achieve to break away from the gravity of another body

focus (plural: foci) one of two fixed points inside an ellipse from which the sum of the distances to any point on the ellipse is constant

gravity the mutual attraction of material bodies or particles

Kepler's first law each planet moves around the Sun in an orbit that is an ellipse, with the Sun at one focus of the ellipse

Kepler's second law the straight line joining a planet and the Sun sweeps out equal areas in space in equal intervals of time

Kepler's third law the square of a planet's orbital period is directly proportional to the cube of the semimajor axis of its orbit

major axis the maximum diameter of an ellipse

momentum the measure of the amount of motion of a body; the momentum of a body is the product of its mass and velocity; in the absence of an unbalanced force, momentum is conserved

Newton's first law every object will continue to be in a state of rest or move at a constant speed in a straight line unless it is compelled to change by an outside force

Newton's second law the change of motion of a body is proportional to and in the direction of the force acting on it

Newton's third law for every action there is an equal and opposite reaction (*or:* the mutual actions of two bodies upon each other are always equal and act in opposite directions)

This OpenStax book is available for free at http://cnx.org/content/col11992/1.8

orbit the path of an object that is in revolution about another object or point

orbital period (P) the time it takes an object to travel once around the Sun

orbital speed the speed at which an object (usually a planet) orbits around the mass of another object; in the case of a planet, the speed at which each planet moves along its ellipse

perigee the point in its orbit where an Earth satellite is closest to Earth

perihelion the point in its orbit where a planet (or other orbiting object) is nearest to the Sun

perturbation a small disturbing effect on the motion or orbit of a body produced by a third body

satellite an object that revolves around a planet

semimajor axis half of the major axis of a conic section, such as an ellipse

velocity the speed and direction a body is moving—for example, 44 kilometers per second toward the north galactic pole

 # SUMMARY

3.1 The Laws of Planetary Motion

Tycho Brahe's accurate observations of planetary positions provided the data used by Johannes Kepler to derive his three fundamental laws of planetary motion. Kepler's laws describe the behavior of planets in their orbits as follows: (1) planetary orbits are ellipses with the Sun at one focus; (2) in equal intervals, a planet's orbit sweeps out equal areas; and (3) the relationship between the orbital period (P) and the semimajor axis (a) of an orbit is given by $P^2 = a^3$ (when a is in units of AU and P is in units of Earth years).

3.2 Newton's Great Synthesis

In his *Principia*, Isaac Newton established the three laws that govern the motion of objects: (1) objects continue to be at rest or move with a constant velocity unless acted upon by an outside force; (2) an outside force causes an acceleration (and changes the momentum) for an object; and (3) for every action there is an equal and opposite reaction. Momentum is a measure of the motion of an object and depends on both its mass and its velocity. Angular momentum is a measure of the motion of a spinning or revolving object and depends on its mass, velocity, and distance from the point around which it revolves. The density of an object is its mass divided by its volume.

3.3 Newton's Universal Law of Gravitation

Gravity, the attractive force between all masses, is what keeps the planets in orbit. Newton's universal law of gravitation relates the gravitational force to mass and distance:

$$F_{\text{gravity}} = G\frac{M_1 M_2}{R^2}$$

The force of gravity is what gives us our sense of weight. Unlike mass, which is constant, weight can vary depending on the force of gravity (or acceleration) you feel. When Kepler's laws are reexamined in the light of Newton's gravitational law, it becomes clear that the masses of both objects are important for the third law, which becomes $a^3 = (M_1 + M_2) \times P^2$. Mutual gravitational effects permit us to calculate the masses of astronomical objects, from comets to galaxies.

Download for free at http://cnx.org/content/col11992/latest/

3.4 Orbits in the Solar System

The closest point in a satellite orbit around Earth is its perigee, and the farthest point is its apogee (corresponding to perihelion and aphelion for an orbit around the Sun). The planets follow orbits around the Sun that are nearly circular and in the same plane. Most asteroids are found between Mars and Jupiter in the asteroid belt, whereas comets generally follow orbits of high eccentricity.

3.5 Motions of Satellites and Spacecraft

The orbit of an artificial satellite depends on the circumstances of its launch. The circular satellite velocity needed to orbit Earth's surface is 8 kilometers per second, and the escape speed from our planet is 11 kilometers per second. There are many possible interplanetary trajectories, including those that use gravity-assisted flybys of one object to redirect the spacecraft toward its next target.

3.6 Gravity with More Than Two Bodies

Calculating the gravitational interaction of more than two objects is complicated and requires large computers. If one object (like the Sun in our solar system) dominates gravitationally, it is possible to calculate the effects of a second object in terms of small perturbations. This approach was used by John Couch Adams and Urbain Le Verrier to predict the position of Neptune from its perturbations of the orbit of Uranus and thus discover a new planet mathematically.

FOR FURTHER EXPLORATION

Articles

Brahe and Kepler

Christianson, G. "The Celestial Palace of Tycho Brahe." *Scientific American* (February 1961): 118.

Gingerich, O. "Johannes Kepler and the Rudolphine Tables." *Sky & Telescope* (December 1971): 328. Brief article on Kepler's work.

Wilson, C. "How Did Kepler Discover His First Two Laws?" *Scientific American* (March 1972): 92.

Newton

Christianson, G. "Newton's *Principia*: A Retrospective." *Sky & Telescope* (July 1987): 18.

Cohen, I. "Newton's Discovery of Gravity." *Scientific American* (March 1981): 166.

Gingerich, O. "Newton, Halley, and the Comet." *Sky & Telescope* (March 1986): 230.

Sullivant, R. "When the Apple Falls." *Astronomy* (April 1998): 55. Brief overview.

The Discovery of Neptune

Sheehan, W., et al. "The Case of the Pilfered Planet: Did the British Steal Neptune?" *Scientific American* (December 2004): 92.

Websites

Brahe and Kepler

Johannes Kepler: His Life, His Laws, and Time: http://kepler.nasa.gov/Mission/JohannesKepler/. From NASA's Kepler mission.

Johannes Kepler: http://www.britannica.com/biography/Johannes-Kepler. Encyclopedia Britannica article.

This OpenStax book is available for free at http://cnx.org/content/col11992/1.8

Johannes Kepler: http://www-history.mcs.st-andrews.ac.uk/Biographies/Kepler.html. MacTutor article with additional links.

Noble Dane: Images of Tycho Brahe: http://www.mhs.ox.ac.uk/tycho/index.htm. A virtual museum exhibit from Oxford.

Newton

Sir Isaac Newton: http://www-groups.dcs.st-and.ac.uk/~history//Biographies/Newton.html. MacTutor article with additional links.

Sir Isaac Newton: http://www.luminarium.org/sevenlit/newton/newtonbio.htm. Newton Biography at the Luminarium.

The Discovery of Neptune

Adams, Airy, and the Discovery of Neptune: http://www.mikeoates.org/lassell/adams-airy.htm. A defense of Airy's role by historian Alan Chapman.

Mathematical Discovery of Planets: http://www-groups.dcs.st-and.ac.uk/~history/HistTopics/Neptune_and_Pluto.html. MacTutor article.

Videos

Brahe and Kepler

"Harmony of the Worlds." This third episode of Carl Sagan's TV series Cosmos focuses on Kepler and his life and work.

Tycho Brahe, Johannes Kepler, and Planetary Motion: https://www.youtube.com/watch?v=x3ALuycrCwI. German-produced video, in English (14:27).

Newton

Beyond the Big Bang: Sir Isaac Newton's Law of Gravity: http://www.history.com/topics/enlightenment/videos/beyond-the-big-bang-sir-isaac-newtons-law-of-gravity. From the History Channel (4:35).

Sir Isaac Newton versus Bill Nye: Epic Rap Battles of History: https://www.youtube.com/watch?v=8yis7GzlXNM. (2:47).

The Discovery of Neptune

Richard Feynman: On the Discovery of Neptune: https://www.youtube.com/watch?v=FgXQffVgZRs. A brief black-and-white Caltech lecture (4:33).

COLLABORATIVE GROUP ACTIVITIES

A. An eccentric, but very rich, alumnus of your college makes a bet with the dean that if you drop a baseball and a bowling ball from the tallest building on campus, the bowling ball would hit the ground first. Have your group discuss whether you would make a side bet that the alumnus is right. How would you decide who is right?

B. Suppose someone in your astronomy class was unhappy about his or her weight. Where could a person go to weigh one-fourth as much as he or she does now? Would changing the unhappy person's weight have any effect on his or her mass?

Download for free at http://cnx.org/content/col11992/latest/

C. When the Apollo astronauts landed on the Moon, some commentators commented that it ruined the mystery and "poetry" of the Moon forever (and that lovers could never gaze at the full moon in the same way again). Others felt that knowing more about the Moon could only enhance its interest to us as we see it from Earth. How do the various members of your group feel? Why?

D. Figure 3.12 shows a swarm of satellites in orbit around Earth. What do you think all these satellites do? How many categories of functions for Earth satellites can your group come up with?

E. The Making Connections feature box Astronomy and the Poets discusses how poets included the most recent astronomical knowledge in their poetry. Is this still happening today? Can your group members come up with any poems or songs that you know that deal with astronomy or outer space? If not, perhaps you could find some online, or by asking friends or roommates who are into poetry or music.

🖱 EXERCISES

Review Questions

1. State Kepler's three laws in your own words.

2. Why did Kepler need Tycho Brahe's data to formulate his laws?

3. Which has more mass: an armful of feathers or an armful of lead? Which has more volume: a kilogram of feathers or a kilogram of lead? Which has higher density: a kilogram of feathers or a kilogram of lead?

4. Explain how Kepler was able to find a relationship (his third law) between the orbital periods and distances of the planets that did not depend on the masses of the planets or the Sun.

5. Write out Newton's three laws of motion in terms of what happens with the momentum of objects.

6. Which major planet has the largest . . .
   A. semimajor axis?

   B. average orbital speed around the Sun?

   C. orbital period around the Sun?

   D. eccentricity?

7. Why do we say that Neptune was the first planet to be discovered through the use of mathematics?

8. Why was Brahe reluctant to provide Kepler with all his data at one time?

9. According to Kepler's second law, where in a planet's orbit would it be moving fastest? Where would it be moving slowest?

10. The gas pedal, the brakes, and the steering wheel all have the ability to accelerate a car—how?

11. Explain how a rocket can propel itself using Newton's third law.

12. A certain material has a mass of 565 g while occupying 50 cm^3 of space. What is this material? (Hint: Use Table 3.1.)

13. To calculate the momentum of an object, which properties of an object do you need to know?

This OpenStax book is available for free at http://cnx.org/content/col11992/1.8

14. To calculate the angular momentum of an object, which properties of an object do you need to know?

15. What was the great insight Newton had regarding Earth's gravity that allowed him to develop the universal law of gravitation?

16. Which of these properties of an object best quantifies its inertia: velocity, acceleration, volume, mass, or temperature?

17. Pluto's orbit is more eccentric than any of the major planets. What does that mean?

18. Why is Tycho Brahe often called "the greatest naked-eye astronomer" of all time?

Thought Questions

19. Is it possible to escape the force of gravity by going into orbit around Earth? How does the force of gravity in the International Space Station (orbiting an average of 400 km above Earth's surface) compare with that on the ground?

20. What is the momentum of an object whose velocity is zero? How does Newton's first law of motion include the case of an object at rest?

21. Evil space aliens drop you and your fellow astronomy student 1 km apart out in space, very far from any star or planet. Discuss the effects of gravity on each of you.

22. A body moves in a perfectly circular path at constant speed. Are there forces acting in such a system? How do you know?

23. As friction with our atmosphere causes a satellite to spiral inward, closer to Earth, its orbital speed increases. Why?

24. Use a history book, an encyclopedia, or the internet to find out what else was happening in England during Newton's lifetime and discuss what trends of the time might have contributed to his accomplishments and the rapid acceptance of his work.

25. Two asteroids begin to gravitationally attract one another. If one asteroid has twice the mass of the other, which one experiences the greater force? Which one experiences the greater acceleration?

26. How does the mass of an astronaut change when she travels from Earth to the Moon? How does her weight change?

27. If there is gravity where the International Space Station (ISS) is located above Earth, why doesn't the space station get pulled back down to Earth?

28. Compare the density, weight, mass, and volume of a pound of gold to a pound of iron on the surface of Earth.

29. If identical spacecraft were orbiting Mars and Earth at identical radii (distances), which spacecraft would be moving faster? Why?

Figuring For Yourself

30. By what factor would a person's weight be increased if Earth had 10 times its present mass, but the same volume?

31. Suppose astronomers find an earthlike planet that is twice the size of Earth (that is, its radius is twice that of Earth's). What must be the mass of this planet such that the gravitational force ($F_{gravity}$) at the surface would be identical to Earth's?

Download for free at http://cnx.org/content/col11992/latest/

32. What is the semimajor axis of a circle of diameter 24 cm? What is its eccentricity?

33. If 24 g of material fills a cube 2 cm on a side, what is the density of the material?

34. If 128 g of material is in the shape of a brick 2 cm wide, 4 cm high, and 8 cm long, what is the density of the material?

35. If the major axis of an ellipse is 16 cm, what is the semimajor axis? If the eccentricity is 0.8, would this ellipse be best described as mostly circular or very elongated?

36. What is the average distance from the Sun (in astronomical units) of an asteroid with an orbital period of 8 years?

37. What is the average distance from the Sun (in astronomical units) of a planet with an orbital period of 45.66 years?

38. In 1996, astronomers discovered an icy object beyond Pluto that was given the designation 1996 TL 66. It has a semimajor axis of 84 AU. What is its orbital period according to Kepler's third law?

This OpenStax book is available for free at http://cnx.org/content/col11992/1.8

4

EARTH, MOON, AND SKY

Figure 4.1. Southern Summer. As captured with a fish-eye lens aboard the Atlantis Space Shuttle on December 9, 1993, Earth hangs above the Hubble Space Telescope as it is repaired. The reddish continent is Australia, its size and shape distorted by the special lens. Because the seasons in the Southern Hemisphere are opposite those in the Northern Hemisphere, it is summer in Australia on this December day. (credit: modification of work by NASA)

Chapter Outline

📝 Thinking Ahead

If Earth's orbit is nearly a perfect circle (as we saw in earlier chapters), why is it hotter in summer and colder in winter in many places around the globe? And why are the seasons in Australia or Peru the opposite of those in the United States or Europe?

The story is told that Galileo, as he left the Hall of the Inquisition following his retraction of the doctrine that Earth rotates and revolves about the Sun, said under his breath, "But nevertheless it moves." Historians are not sure whether the story is true, but certainly Galileo knew that Earth was in motion, whatever church authorities said.

It is the motions of Earth that produce the seasons and give us our measures of time and date. The Moon's motions around us provide the concept of the month and the cycle of lunar phases. In this chapter we examine some of the basic phenomena of our everyday world in their astronomical context.

Download for free at http://cnx.org/content/col11992/latest/

4.1 EARTH AND SKY

Learning Objectives

By the end of this section, you will be able to:

> Describe how latitude and longitude are used to map Earth
> Explain how right ascension and declination are used to map the sky

In order to create an accurate map, a mapmaker needs a way to uniquely and simply identify the location of all the major features on the map, such as cities or natural landmarks. Similarly, astronomical mapmakers need a way to uniquely and simply identify the location of stars, galaxies, and other celestial objects. On Earth maps, we divide the surface of Earth into a grid, and each location on that grid can easily be found using its *latitude* and *longitude* coordinate. Astronomers have a similar system for objects on the sky. Learning about these can help us understand the apparent motion of objects in the sky from various places on Earth.

Locating Places on Earth

Let's begin by fixing our position on the surface of planet Earth. As we discussed in Observing the Sky: The Birth of Astronomy, Earth's axis of rotation defines the locations of its North and South Poles and of its equator, halfway between. Two other directions are also defined by Earth's motions: east is the direction toward which Earth rotates, and west is its opposite. At almost any point on Earth, the four directions—north, south, east, and west—are well defined, despite the fact that our planet is round rather that flat. The only exceptions are exactly at the North and South Poles, where the directions east and west are ambiguous (because points exactly at the poles do not turn).

We can use these ideas to define a system of coordinates attached to our planet. Such a system, like the layout of streets and avenues in Manhattan or Salt Lake City, helps us find where we are or want to go. Coordinates on a sphere, however, are a little more complicated than those on a flat surface. We must define circles on the sphere that play the same role as the rectangular grid that you see on city maps.

A **great circle** is any circle on the surface of a sphere whose center is at the center of the sphere. For example, Earth's equator is a great circle on Earth's surface, halfway between the North and South Poles. We can also imagine a series of great circles that pass through both the North and South Poles. Each of this circles is called a **meridian**; they are each perpendicular to the equator, crossing it at right angles.

Any point on the surface of Earth will have a meridian passing through it (Figure 4.2). The meridian specifies the east-west location, or longitude, of the place. By international agreement (and it took many meetings for the world's countries to agree), longitude is defined as the number of degrees of arc along the equator between your meridian and the one passing through Greenwich, England, which has been designated as the Prime Meridian. The longitude of the Prime Meridian is defined as 0°.

This OpenStax book is available for free at http://cnx.org/content/col11992/1.8

Download for free at http://cnx.org/content/col11992/latest/

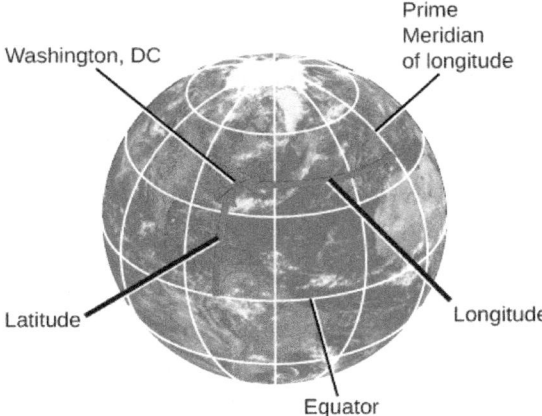

Figure 4.2. Latitude and Longitude of Washington, DC. We use latitude and longitude to find cities like Washington, DC, on a globe. Latitude is the number of degrees north or south of the equator, and longitude is the number of degrees east or west of the Prime Meridian. Washington, DC's coordinates are 38° N and 77° W.

Why Greenwich, you might ask? Every country wanted 0° longitude to pass through its own capital. Greenwich, the site of the old Royal Observatory (Figure 4.3), was selected because it was between continental Europe and the United States, and because it was the site for much of the development of the method to measure longitude at sea. Longitudes are measured either to the east or to the west of the Greenwich meridian from 0° to 180°. As an example, the longitude of the clock-house benchmark of the U.S. Naval Observatory in Washington, DC, is 77.066° W.

Figure 4.3. Royal Observatory in Greenwich, England. At the internationally agreed-upon zero point of longitude at the Royal Observatory Greenwich, tourists can stand and straddle the exact line where longitude "begins."(credit left: modification of work by "pdbreen"/Flickr; credit right: modification of work by Ben Sutherland)

Your latitude (or north-south location) is the number of degrees of arc you are away from the equator along your meridian. Latitudes are measured either north or south of the equator from 0° to 90°. (The latitude of the equator is 0°.) As an example, the latitude of the previously mentioned Naval Observatory benchmark is 38.921° N. The latitude of the South Pole is 90° S, and the latitude of the North Pole is 90° N.

Download for free at http://cnx.org/content/col11992/latest/

Locating Places in the Sky

Positions in the sky are measured in a way that is very similar to the way we measure positions on the surface of Earth. Instead of latitude and longitude, however, astronomers use coordinates called **declination** and **right ascension**. To denote positions of objects in the sky, it is often convenient to make use of the fictitious celestial sphere. We saw in Observing the Sky: The Birth of Astronomy that the sky appears to rotate about points above the North and South Poles of Earth—points in the sky called the north celestial pole and the south celestial pole. Halfway between the celestial poles, and thus 90° from each pole, is the *celestial equator,* a great circle on the celestial sphere that is in the same plane as Earth's equator. We can use these markers in the sky to set up a system of celestial coordinates.

Declination on the celestial sphere is measured the same way that latitude is measured on the sphere of Earth: from the celestial equator toward the north (positive) or south (negative). So Polaris, the star near the north celestial pole, has a declination of almost +90°.

Right ascension (RA) is like longitude, except that instead of Greenwich, the arbitrarily chosen point where we start counting is the *vernal equinox*, a point in the sky where the *ecliptic* (the Sun's path) crosses the celestial equator. RA can be expressed either in units of angle (degrees) or in units of time. This is because the celestial sphere appears to turn around Earth once a day as our planet turns on its axis. Thus the 360° of RA that it takes to go once around the celestial sphere can just as well be set equal to 24 hours. Then each 15° of arc is equal to 1 hour of time. For example, the approximate celestial coordinates of the bright star Capella are RA 5h = 75° and declination +50°.

One way to visualize these circles in the sky is to imagine Earth as a transparent sphere with the terrestrial coordinates (latitude and longitude) painted on it with dark paint. Imagine the celestial sphere around us as a giant ball, painted white on the inside. Then imagine yourself at the center of Earth, with a bright light bulb in the middle, looking out through its transparent surface to the sky. The terrestrial poles, equator, and meridians will be projected as dark shadows on the celestial sphere, giving us the system of coordinates in the sky.

LINK TO LEARNING

You can explore a variety of basic animations about coordinates and motions in the sky at this interactive site (https://openstaxcollege.org/l/30anicoormot) from ClassAction. Click on the "Animations" tab for a list of options. If you choose the second option in the menu, you can play with the celestial sphere and see RA and declination defined visually.

The Turning Earth

Why do many stars rise and set each night? Why, in other words, does the night sky seem to turn? We have seen that the apparent rotation of the celestial sphere could be accounted for either by a daily rotation of the sky around a stationary Earth or by the rotation of Earth itself. Since the seventeenth century, it has been generally accepted that it is Earth that turns, but not until the nineteenth century did the French physicist Jean Foucault provide an unambiguous demonstration of this rotation. In 1851, he suspended a 60-meter pendulum weighing about 25 kilograms from the dome of the Pantheon in Paris and started the pendulum swinging evenly. If Earth had not been turning, there would have been no alteration of the pendulum's plane of oscillation, and so it would have continued tracing the same path. Yet after a few minutes Foucault could see that the pendulum's plane of motion was turning. Foucault explained that it was not the pendulum that was shifting, but rather

This OpenStax book is available for free at http://cnx.org/content/col11992/1.8

Earth that was turning beneath it (Figure 4.4). You can now find such pendulums in many science centers and planetariums around the world.

Figure 4.4. Foucault's Pendulum. As Earth turns, the plane of oscillation of the Foucault pendulum shifts gradually so that over the course of 12 hours, all the targets in the circle at the edge of the wooden platform are knocked over in sequence. (credit: Manuel M. Vicente)

Can you think of other pieces of evidence that indicate that it is Earth and not the sky that is turning? (See Collaborative Group Activity A at the end of this chapter.)

4.2 THE SEASONS

Learning Objectives

By the end of this section, you will be able to:

> Describe how the tilt of Earth's axis causes the seasons
> Explain how seasonal differences on Earth vary with latitude

One of the fundamental facts of life at Earth's midlatitudes, where most of this book's readers live, is that there are significant variations in the heat we receive from the Sun during the course of the year. We thus divide the year into *seasons*, each with its different amount of sunlight. The difference between seasons gets more pronounced the farther north or south from the equator we travel, and the seasons in the Southern Hemisphere are the opposite of what we find on the northern half of Earth. With these observed facts in mind, let us ask what causes the seasons.

Many people have believed that the seasons were the result of the changing distance between Earth and the Sun. This sounds reasonable at first: it should be colder when Earth is farther from the Sun. But the facts don't bear out this hypothesis. Although Earth's orbit around the Sun is an ellipse, its distance from the Sun varies by only about 3%. That's not enough to cause significant variations in the Sun's heating. To make matters worse for people in North America who hold this hypothesis, Earth is actually closest to the Sun in January, when the Northern Hemisphere is in the middle of winter. And if distance were the governing factor, why would the

Download for free at http://cnx.org/content/col11992/latest/

two hemispheres have opposite seasons? As we shall show, the seasons are actually caused by the 23.5° tilt of Earth's axis.

The Seasons and Sunshine

Figure 4.5 shows Earth's annual path around the Sun, with Earth's axis tilted by 23.5°. Note that our axis continues to point the same direction in the sky throughout the year. As Earth travels around the Sun, in June the Northern Hemisphere "leans into" the Sun and is more directly illuminated. In December, the situation is reversed: the Southern Hemisphere leans into the Sun, and the Northern Hemisphere leans away. In September and March, Earth leans "sideways"—neither into the Sun nor away from it—so the two hemispheres are equally favored with sunshine.

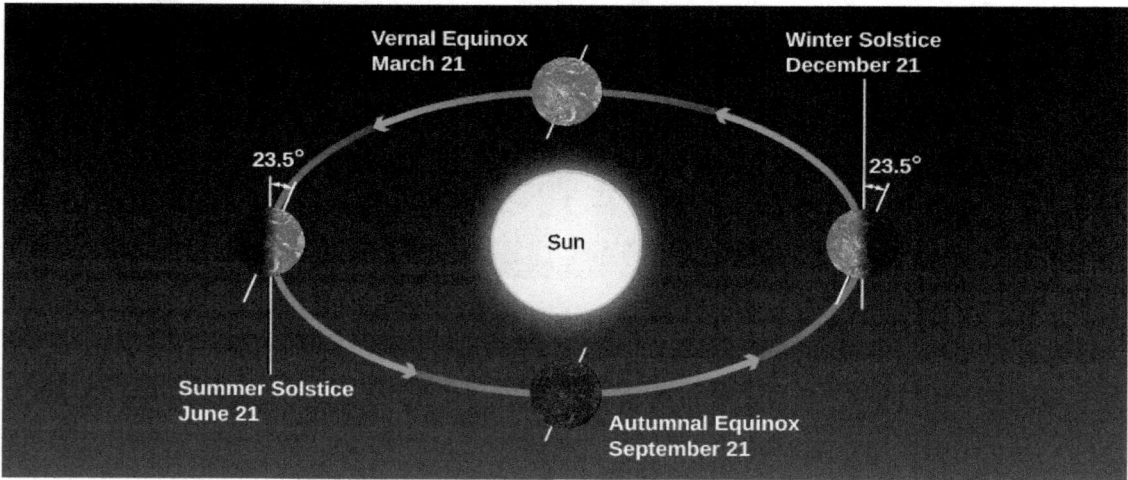

Figure 4.5. Seasons. We see Earth at different seasons as it circles the Sun. In June, the Northern Hemisphere "leans into" the Sun, and those in the North experience summer and have longer days. In December, during winter in the Northern Hemisphere, the Southern Hemisphere "leans into" the Sun and is illuminated more directly. In spring and autumn, the two hemispheres receive more equal shares of sunlight.[1]

How does the Sun's favoring one hemisphere translate into making it warmer for us down on the surface of Earth? There are two effects we need to consider. When we lean into the Sun, sunlight hits us at a more direct angle and is more effective at heating Earth's surface (Figure 4.6). You can get a similar effect by shining a flashlight onto a wall. If you shine the flashlight straight on, you get an intense spot of light on the wall. But if you hold the flashlight at an angle (if the wall "leans out" of the beam), then the spot of light is more spread out. Like the straight-on light, the sunlight in June is more direct and intense in the Northern Hemisphere, and hence more effective at heating.

---

1   Note that the dates indicated for the solstices and equinoxes are approximate; depending on the year, they may occur a day or two earlier or later.

This OpenStax book is available for free at http://cnx.org/content/col11992/1.8

Download for free at http://cnx.org/content/col11992/latest/

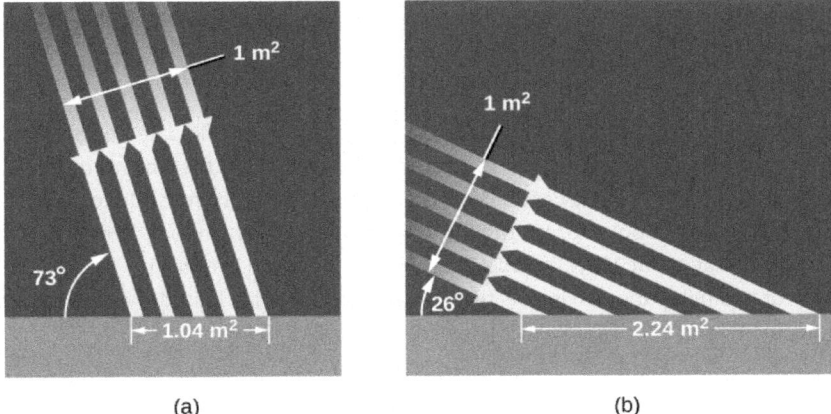

Figure 4.6. The Sun's Rays in Summer and Winter. (a) In summer, the Sun appears high in the sky and its rays hit Earth more directly, spreading out less. (b) In winter, the Sun is low in the sky and its rays spread out over a much wider area, becoming less effective at heating the ground.

The second effect has to do with the length of time the Sun spends above the horizon (Figure 4.7). Even if you've never thought about astronomy before, we're sure you have observed that the hours of daylight increase in summer and decrease in winter. Let's see why this happens.

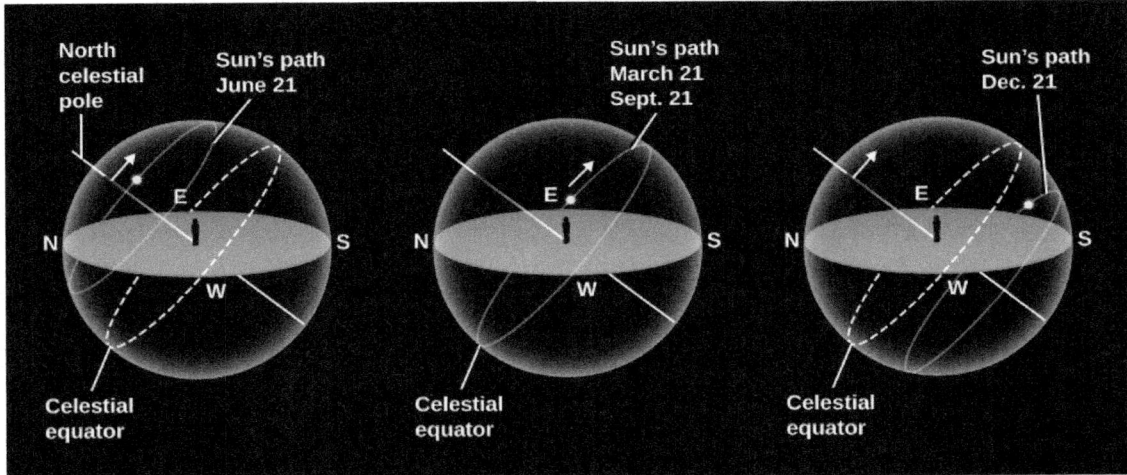

Figure 4.7. The Sun's Path in the Sky for Different Seasons. On June 21, the Sun rises north of east and sets north of west. For observers in the Northern Hemisphere of Earth, the Sun spends about 15 hours above the horizon in the United States, meaning more hours of daylight. On December 21, the Sun rises south of east and sets south of west. It spends 9 hours above the horizon in the United States, which means fewer hours of daylight and more hours of night in northern lands (and a strong need for people to hold celebrations to cheer themselves up). On March 21 and September 21, the Sun spends equal amounts of time above and below the horizon in both hemispheres.

As we saw in Observing the Sky: The Birth of Astronomy, an equivalent way to look at our path around the Sun each year is to pretend that the Sun moves around Earth (on a circle called the ecliptic). Because Earth's axis is tilted, the ecliptic is tilted by about 23.5° relative to the celestial equator (review Figure 2.7). As a result, where we see the Sun in the sky changes as the year wears on.

In June, the Sun is north of the celestial equator and spends more time with those who live in the Northern Hemisphere. It rises high in the sky and is above the horizon in the United States for as long as 15 hours. Thus, the Sun not only heats us with more direct rays, but it also has more time to do it each day. (Notice in Figure 4.7

Download for free at http://cnx.org/content/col11992/latest/

that the Northern Hemisphere's gain is the Southern Hemisphere's loss. There the June Sun is low in the sky, meaning fewer daylight hours. In Chile, for example, June is a colder, darker time of year.) In December, when the Sun is south of the celestial equator, the situation is reversed.

Let's look at what the Sun's illumination on Earth looks like at some specific dates of the year, when these effects are at their maximum. On or about June 21 (the date we who live in the Northern Hemisphere call the *summer solstice* or sometimes the first day of summer), the Sun shines down most directly upon the Northern Hemisphere of Earth. It appears about 23° north of the equator, and thus, on that date, it passes through the zenith of places on Earth that are at 23° N latitude. The situation is shown in detail in Figure 4.8. To a person at 23° N (near Hawaii, for example), the Sun is directly overhead at noon. This latitude, where the Sun can appear at the zenith at noon on the first day of summer, is called the *Tropic of Cancer*.

We also see in Figure 4.8 that the Sun's rays shine down all around the North Pole at the solstice. As Earth turns on its axis, the North Pole is continuously illuminated by the Sun; all places within 23° of the pole have sunshine for 24 hours. The Sun is as far north on this date as it can get; thus, 90° – 23° (or 67° N) is the southernmost latitude where the Sun can be seen for a full 24-hour period (sometimes called the "land of the midnight Sun"). That circle of latitude is called the *Arctic Circle*.

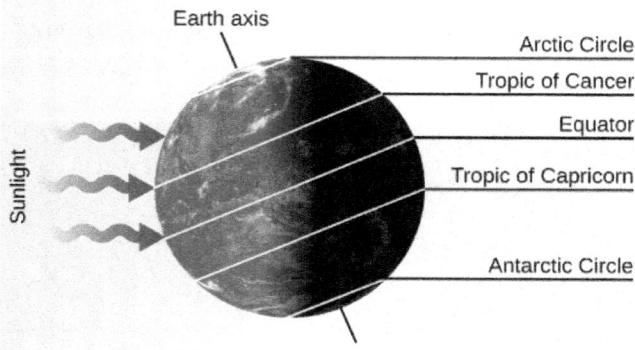

Figure 4.8. Earth on June 21. This is the date of the summer solstice in the Northern Hemisphere. Note that as Earth turns on its axis (the line connecting the North and South Poles), the North Pole is in constant sunlight while the South Pole is veiled in 24 hours of darkness. The Sun is at the zenith for observers on the Tropic of Cancer.

Many early cultures scheduled special events around the summer solstice to celebrate the longest days and thank their gods for making the weather warm. This required people to keep track of the lengths of the days and the northward trek of the Sun in order to know the right day for the "party." (You can do the same thing by watching for several weeks, from the same observation point, where the Sun rises or sets relative to a fixed landmark. In spring, the Sun will rise farther and farther north of east, and set farther and farther north of west, reaching the maximum around the summer solstice.)

Now look at the South Pole in Figure 4.8. On June 21, all places within 23° of the South Pole—that is, south of what we call the *Antarctic Circle*—do not see the Sun at all for 24 hours.

The situation is reversed 6 months later, about December 21 (the date of the *winter solstice*, or the first day of winter in the Northern Hemisphere), as shown in Figure 4.9. Now it is the Arctic Circle that has the 24-hour night and the Antarctic Circle that has the midnight Sun. At latitude 23° S, called the *Tropic of Capricorn*, the Sun passes through the zenith at noon. Days are longer in the Southern Hemisphere and shorter in the north. In the United States and Southern Europe, there may be only 9 or 10 hours of sunshine during the day. It is winter in the Northern Hemisphere and summer in the Southern Hemisphere.

This OpenStax book is available for free at http://cnx.org/content/col11992/1.8

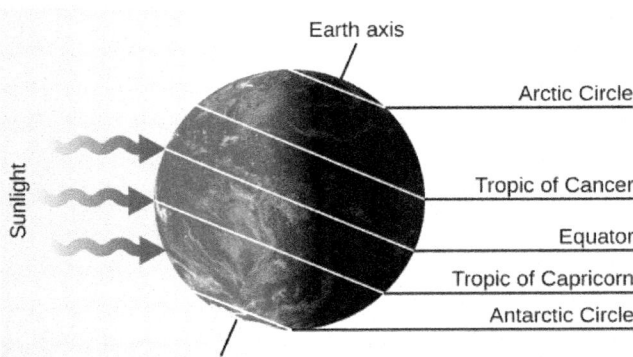

Figure 4.9. Earth on December 21. This is the date of the winter solstice in the Northern Hemisphere. Now the North Pole is in darkness for 24 hours and the South Pole is illuminated. The Sun is at the zenith for observers on the Tropic of Capricorn and thus is low in the sky for the residents of the Northern Hemisphere.

EXAMPLE 4.1

Seasonal Variations

As you can see in Figure 4.8, the Tropic of Cancer is the latitude for which the Sun is directly overhead on the summer solstice. At this time, the Sun is at a declination of 23° N of the celestial equator, and the corresponding latitude on Earth is 23° N of the equator. If Earth were tilted a bit less, then the Tropic of Cancer would be at a lower latitude, closer to the equator.

The Arctic Circle marks the southernmost latitude for which the day length is 24 hours on the day of the summer solstice. This is located at 90° – 23° = 67° N of Earth's equator. If Earth were tilted a bit less, then the Arctic Circle would move farther North. In the limit at which Earth is not tilted at all (its axis is perpendicular to the ecliptic), the Tropic of Cancer would be right on Earth's equator, and the Arctic Circle would simply be the North Pole. Suppose the tilt of Earth's axis were tilted only 5°. What would be the effect on the seasons and the locations of the Tropic of Cancer and Arctic Circle?

Solution

If Earth were tilted less, the seasons would be less extreme. The variation in day length and direct sunlight would be very small over the course of a year, and the Sun's daily path in the sky would not vary much. If Earth were tilted by 5°, the Sun's position on the day of the summer solstice would be 5° N of the celestial equator, so the Tropic of Cancer would be at the corresponding latitude on Earth of 5° N of the Equator. The Arctic Circle would be located at 90° – 5° = 85° N of the equator.

Check Your Learning

Suppose the tilt of Earth's axis were 16°. What, then, would be the difference in latitude between the Arctic Circle and the Tropic of Cancer? What would be the effect on the seasons compared with that produced by the actual tilt of 23°?

Download for free at http://cnx.org/content/col11992/latest/

Answer:

The Tropic of Cancer is at a latitude equal to Earth's tilt, so in this case, it would be at 16° N latitude. The Arctic Circle is at a latitude equal to 90° minus Earth's tilt, or 90° – 16° = 74°. The difference between these two latitudes is 74° – 16° = 58°. Since the tilt of Earth is less, there would be less variation in the tilt of Earth and less variation in the Sun's paths throughout the year, so there would be milder seasonal changes.

LINK TO LEARNING

You can see an animation (https://openstaxcollege.org/l/30anisunpath) of the Sun's path during the seasons alongside a time-lapse view of light and shadow from a camera set up on the University of Nebraska campus.

Many cultures that developed some distance north of the equator have a celebration around December 21 to help people deal with the depressing lack of sunlight and the often dangerously cold temperatures. Originally, this was often a time for huddling with family and friends, for sharing the reserves of food and drink, and for rituals asking the gods to return the light and heat and turn the cycle of the seasons around. Many cultures constructed elaborate devices for anticipating when the shortest day of the year was coming. Stonehenge in England, built long before the invention of writing, is probably one such device. In our own time, we continue the winter solstice tradition with various holiday celebrations around that December date.

Halfway between the solstices, on about March 21 and September 21, the Sun is on the celestial equator. From Earth, it appears above our planet's equator and favors neither hemisphere. Every place on Earth then receives roughly 12 hours of sunshine and 12 hours of night. The points where the Sun crosses the celestial equator are called the *vernal* (spring) and *autumnal* (fall) *equinoxes*.

The Seasons at Different Latitudes

The seasonal effects are different at different latitudes on Earth. Near the equator, for instance, all seasons are much the same. Every day of the year, the Sun is up half the time, so there are approximately 12 hours of sunshine and 12 hours of night. Local residents define the seasons by the amount of rain (wet season and dry season) rather than by the amount of sunlight. As we travel north or south, the seasons become more pronounced, until we reach extreme cases in the Arctic and Antarctic.

At the North Pole, all celestial objects that are north of the celestial equator are always above the horizon and, as Earth turns, circle around parallel to it. The Sun is north of the celestial equator from about March 21 to September 21, so at the North Pole, the Sun rises when it reaches the vernal equinox and sets when it reaches the autumnal equinox. Each year there are 6 months of sunshine at each pole, followed by 6 months of darkness.

This OpenStax book is available for free at http://cnx.org/content/col11992/1.8

EXAMPLE 4.2

The Position of the Sun in the Sky

The Sun's coordinates on the celestial sphere range from a declination of 23° N of the celestial equator (or +23°) to a declination 23° S of the celestial equator (or –23°). So, the Sun's altitude at noon, when it crosses the meridian, varies by a total of 46°. What is the altitude of the Sun at noon on March 21, as seen from a place on Earth's equator? What is its altitude on June 21, as seen from a place on Earth's equator?

Solution

On Earth's equator, the celestial equator passes through the zenith. On March 21, the Sun is crossing the celestial equator, so it should be found at the zenith (90°) at noon. On June 21, the Sun is 23° N of the celestial equator, so it will be 23° away from the zenith at noon. The altitude above the horizon will be 23° less than the altitude of the zenith (90°), so it is 90° – 23° = 67° above the horizon.

Check Your Learning

What is the altitude of the Sun at noon on December 21, as seen from a place on the Tropic of Cancer?

Answer:

On the day of the winter solstice, the Sun is located about 23° S of the celestial equator. From the Tropic of Cancer, a latitude of 23° N, the zenith would be a declination of 23° N. The difference in declination between zenith and the position of the Sun is 46°, so the Sun would be 46° away from the zenith. That means it would be at an altitude of 90° – 46° = 44°.

Clarifications about the Real World

In our discussions so far, we have been describing the rising and setting of the Sun and stars as they would appear if Earth had little or no atmosphere. In reality, however, the atmosphere has the curious effect of allowing us to see a little way "over the horizon." This effect is a result of *refraction*, the bending of light passing through air or water, something we will discuss in Astronomical Instruments. Because of this atmospheric refraction (and the fact that the Sun is not a point of light but a disk), the Sun appears to rise earlier and to set later than it would if no atmosphere were present.

In addition, the atmosphere scatters light and provides some twilight illumination even when the Sun is below the horizon. Astronomers define morning twilight as beginning when the Sun is 18° below the horizon, and evening twilight extends until the Sun sinks more than 18° below the horizon.

These atmospheric effects require small corrections in many of our statements about the seasons. At the equinoxes, for example, the Sun appears to be above the horizon for a few minutes longer than 12 hours, and below the horizon for fewer than 12 hours. These effects are most dramatic at Earth's poles, where the Sun actually can be seen more than a week before it reaches the celestial equator.

You probably know that the summer solstice (June 21) is not the warmest day of the year, even if it is the longest. The hottest months in the Northern Hemisphere are July and August. This is because our weather involves the air and water covering Earth's surface, and these large reservoirs do not heat up instantaneously. You have probably observed this effect for yourself; for example, a pond does not get warm the moment the Sun rises but is warmest late in the afternoon, after it has had time to absorb the Sun's heat. In the same way, Earth

Download for free at http://cnx.org/content/col11992/latest/

gets warmer after it has had a chance to absorb the extra sunlight that is the Sun's summer gift to us. And the coldest times of winter are a month or more after the winter solstice.

4.3 KEEPING TIME

Learning Objectives

By the end of this section, you will be able to:

> Explain the difference between the solar day and the sidereal day
> Explain mean solar time and the reason for time zones

The measurement of time is based on the rotation of Earth. Throughout most of human history, time has been reckoned by positions of the Sun and stars in the sky. Only recently have mechanical and electronic clocks taken over this function in regulating our lives.

The Length of the Day

The most fundamental astronomical unit of time is the day, measured in terms of the rotation of Earth. There is, however, more than one way to define the day. Usually, we think of it as the rotation period of Earth with respect to the Sun, called the **solar day**. After all, for most people sunrise is more important than the rising time of Arcturus or some other star, so we set our clocks to some version of Sun-time. However, astronomers also use a **sidereal day**, which is defined in terms of the rotation period of Earth with respect to the stars.

A solar day is slightly longer than a sidereal day because (as you can see from Figure 4.10) Earth not only turns but also moves along its path around the Sun in a day. Suppose we start when Earth's orbital position is at day 1, with both the Sun and some distant star (located in the direction indicated by the long white arrow pointing left), directly in line with the zenith for the observer on Earth. When Earth has completed one rotation with respect to the distant star and is at day 2, the long arrow again points to the same distant star. However, notice that because of the movement of Earth along its orbit from day 1 to 2, the Sun has not yet reached a position above the observer. To complete a solar day, Earth must rotate an additional amount, equal to 1/365 of a full turn. The time required for this extra rotation is 1/365 of a day, or about 4 minutes. So the solar day is about 4 minutes longer than the sidereal day.

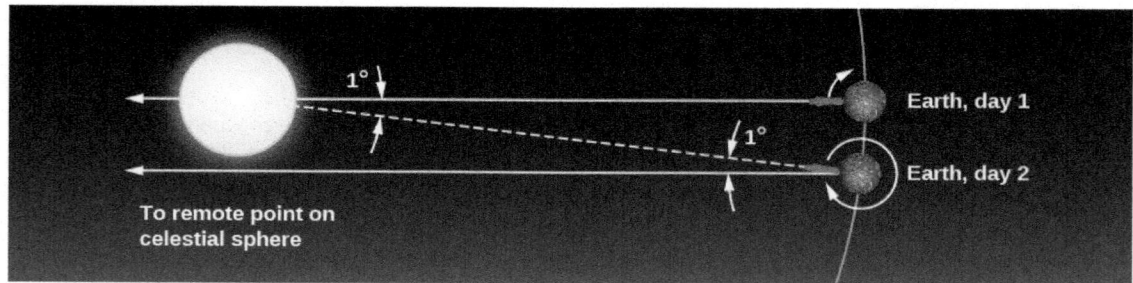

Figure 4.10. Difference Between a Sidereal Day and a Solar Day. This is a top view, looking down as Earth orbits the Sun. Because Earth moves around the Sun (roughly 1° per day), after one complete rotation of Earth relative to the stars, we do not see the Sun in the same position.

Because our ordinary clocks are set to solar time, stars rise 4 minutes earlier each day. Astronomers prefer sidereal time for planning their observations because in that system, a star rises at the same time every day.

This OpenStax book is available for free at http://cnx.org/content/col11992/1.8

Download for free at http://cnx.org/content/col11992/latest/

EXAMPLE 4.3

Sidereal Time and Solar Time

The Sun makes a complete circle in the sky approximately every 24 hours, while the stars make a complete circle in the sky in 4 minutes less time, or 23 hours and 56 minutes. This causes the positions of the stars at a given time of day or night to change slightly each day. Since stars rise 4 minutes earlier each day, that works out to about 2 hours per month (4 minutes × 30 = 120 minutes or 2 hours). So, if a particular constellation rises at sunset during the winter, you can be sure that by the summer, it will rise about 12 hours earlier, with the sunrise, and it will not be so easily visible in the night sky. Let's say that tonight the bright star Sirius rises at 7:00 p.m. from a given location so that by midnight, it is very high in the sky. At what time will Sirius rise in three months?

Solution

In three months' time, Sirius will be rising earlier by:

$$90 \text{ days} \times \frac{4 \text{ minutes}}{\text{day}} = 360 \text{ minutes or 6 hours}$$

It will rise at about 1:00 p.m. and be high in the sky at around sunset instead of midnight. Sirius is the brightest star in the constellation of Canis Major (the big dog). So, some other constellation will be prominently visible high in the sky at this later date.

Check Your Learning

If a star rises at 8:30 p.m. tonight, approximately what time will it rise two months from now?

Answer:

In two months, the star will rise:

$$60 \text{ days} \times \frac{4 \text{ minutes}}{\text{day}} = 240 \text{ minutes or 4 hours earlier.}$$

This means it will rise at 4:30 p.m.

Apparent Solar Time

We can define **apparent solar time** as time reckoned by the actual position of the Sun in the sky (or, during the night, its position below the horizon). This is the kind of time indicated by sundials, and it probably represents the earliest measure of time used by ancient civilizations. Today, we adopt the middle of the night as the starting point of the day and measure time in hours elapsed since midnight.

During the first half of the day, the Sun has not yet reached the meridian (the great circle in the sky that passes through our zenith). We designate those hours as before midday (*ante meridiem*, or a.m.), before the Sun reaches the local meridian. We customarily start numbering the hours after noon over again and designate them by p.m. (*post meridiem*), after the Sun reaches the local meridian.

Although apparent solar time seems simple, it is not really very convenient to use. The exact length of an apparent solar day varies slightly during the year. The eastward progress of the Sun in its annual journey around the sky is not uniform because the speed of Earth varies slightly in its elliptical orbit. Another complication is that Earth's axis of rotation is not perpendicular to the plane of its revolution. Thus, apparent

Download for free at http://cnx.org/content/col11992/latest/

solar time does not advance at a uniform rate. After the invention of mechanical clocks that run at a uniform rate, it became necessary to abandon the apparent solar day as the fundamental unit of time.

Mean Solar Time and Standard Time

Instead, we can consider the **mean solar time**, which is based on the average value of the solar day over the course of the year. A mean solar day contains exactly 24 hours and is what we use in our everyday timekeeping. Although mean solar time has the advantage of progressing at a uniform rate, it is still inconvenient for practical use because it is determined by the position of the Sun. For example, noon occurs when the Sun is overhead. But because we live on a round Earth, the exact time of noon is different as you change your longitude by moving east or west.

If mean solar time were strictly observed, people traveling east or west would have to reset their watches continually as the longitude changed, just to read the local mean time correctly. For instance, a commuter traveling from Oyster Bay on Long Island to New York City would have to adjust the time on the trip through the East River tunnel because Oyster Bay time is actually about 1.6 minutes more advanced than that of Manhattan. (Imagine an airplane trip in which an obnoxious flight attendant gets on the intercom every minute, saying, "Please reset your watch for local mean time.")

Until near the end of the nineteenth century, every city and town in the United States kept its own local mean time. With the development of railroads and the telegraph, however, the need for some kind of standardization became evident. In 1883, the United States was divided into four standard time zones (now six, including Hawaii and Alaska), each with one system of time within that zone.

By 1900, most of the world was using the system of 24 standardized global time zones. Within each zone, all places keep the same *standard time*, with the local mean solar time of a standard line of longitude running more or less through the middle of each zone. Now travelers reset their watches only when the time change has amounted to a full hour. Pacific standard time is 3 hours earlier than eastern standard time, a fact that becomes painfully obvious in California when someone on the East Coast forgets and calls you at 5:00 a.m.

Globally, almost all countries have adopted one or more standard time zones, although one of the largest nations, India, has settled on a half-zone, being 5.5 hours from Greenwich standard. Also, several large countries (Russia, China) officially use only one time zone, so all the clocks in that country keep the same time. In Tibet, for example, the Sun rises while the clocks (which keep Beijing time) say it is midmorning already.

Daylight saving time is simply the local standard time of the place plus 1 hour. It has been adopted for spring and summer use in most states in the United States, as well as in many countries, to prolong the sunlight into evening hours, on the apparent theory that it is easier to change the time by government action than it would be for individuals or businesses to adjust their own schedules to produce the same effect. It does not, of course, "save" any daylight at all—because the amount of sunlight is not determined by what we do with our clocks—and its observance is a point of legislative debate in some states.

The International Date Line

The fact that time is always advancing as you move toward the east presents a problem. Suppose you travel eastward around the world. You pass into a new time zone, on the average, about every 15° of longitude you travel, and each time you dutifully set your watch ahead an hour. By the time you have completed your trip, you have set your watch ahead a full 24 hours and thus gained a day over those who stayed at home.

The solution to this dilemma is the **International Date Line**, set by international agreement to run approximately along the 180° meridian of longitude. The date line runs down the middle of the Pacific Ocean, although it jogs a bit in a few places to avoid cutting through groups of islands and through Alaska (Figure

This OpenStax book is available for free at http://cnx.org/content/col11992/1.8

4.11). By convention, at the date line, the date of the calendar is changed by one day. Crossing the date line from west to east, thus advancing your time, you compensate by decreasing the date; crossing from east to west, you increase the date by one day. To maintain our planet on a rational system of timekeeping, we simply must accept that the date will differ in different cities at the same time. A good example is the date when the Imperial Japanese Navy bombed Pearl Harbor in Hawaii, known in the United States as Sunday, December 7, 1941, but taught to Japanese students as Monday, December 8.

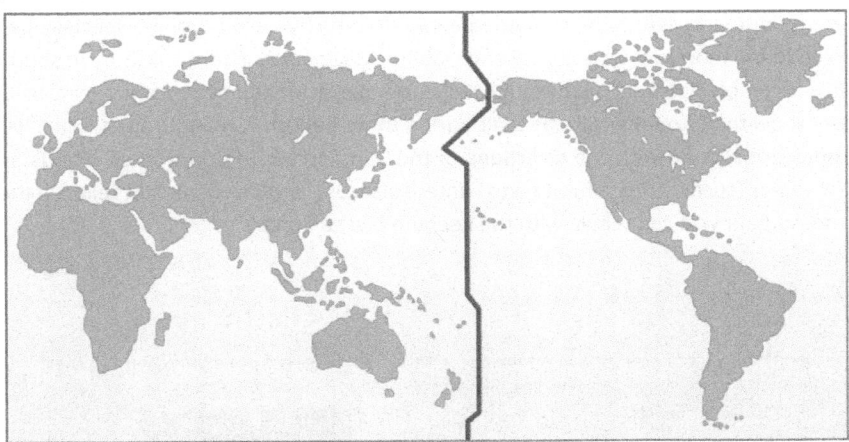

Figure 4.11. Where the Date Changes. The International Date Line is an arbitrarily drawn line on Earth where the date changes. So that neighbors do not have different days, the line is located where Earth's surface is mostly water.

4.4 | THE CALENDAR

Learning Objectives

By the end of this section, you will be able to:

> › Understand how calendars varied among different cultures
> › Explain the origins of our modern calendar

"What's today's date?" is one of the most common questions you can ask (usually when signing a document or worrying about whether you should have started studying for your next astronomy exam). Long before the era of digital watches, smartphones, and fitness bands that tell the date, people used calendars to help measure the passage of time.

The Challenge of the Calendar

There are two traditional functions of any calendar. First, it must keep track of time over the course of long spans, allowing people to anticipate the cycle of the seasons and to honor special religious or personal anniversaries. Second, to be useful to a large number of people, a calendar must use natural time intervals that everyone can agree on—those defined by the motions of Earth, the Moon, and sometimes even the planets. The natural units of our calendar are the *day*, based on the period of rotation of Earth; the *month*, based on the cycle of the Moon's phases (see later in this chapter) about Earth; and the year, based on the period of revolution of Earth about the Sun. Difficulties have resulted from the fact that these three periods are not commensurable; that's a fancy way of saying that one does not divide evenly into any of the others.

The rotation period of Earth is, by definition, 1.0000 day (and here the solar day is used, since that is the basis of human experience). The period required by the Moon to complete its cycle of phases, called the *lunar month*,

Download for free at http://cnx.org/content/col11992/latest/

is 29.5306 days. The basic period of revolution of Earth, called the *tropical year*, is 365.2422 days. The ratios of these numbers are not convenient for calculations. This is the historic challenge of the calendar, dealt with in various ways by different cultures.

Early Calendars

Even the earliest cultures were concerned with the keeping of time and the calendar. Some interesting examples include monuments left by Bronze Age people in northwestern Europe, especially the British Isles. The best preserved of the monuments is Stonehenge, about 13 kilometers from Salisbury in southwest England (Figure 4.12). It is a complex array of stones, ditches, and holes arranged in concentric circles. Carbon dating and other studies show that Stonehenge was built during three periods ranging from about 2800 to 1500 BCE. Some of the stones are aligned with the directions of the Sun and Moon during their risings and settings at critical times of the year (such as the summer and winter solstices), and it is generally believed that at least one function of the monument was connected with the keeping of a calendar.

Figure 4.12. Stonehenge. The ancient monument known as Stonehenge was used to keep track of the motions of the Sun and Moon. (credit: modification of work by Adriano Aurelio Araujo)

The Maya in Central America, who thrived more than a thousand years ago, were also concerned with the keeping of time. Their calendar was as sophisticated as, and perhaps more complex than, contemporary calendars in Europe. The Maya did not attempt to correlate their calendar accurately with the length of the year or lunar month. Rather, their calendar was a system for keeping track of the passage of days and for counting time far into the past or future. Among other purposes, it was useful for predicting astronomical events, such as the position of Venus in the sky (Figure 4.13).

This OpenStax book is available for free at http://cnx.org/content/col11992/1.8

Figure 4.13. El Caracol. This Mayan observatory at Chichen Itza in the Yucatan, Mexico, dates from around the year 1000. (credit: "wiredtourist.com"/Flickr)

The ancient Chinese developed an especially complex calendar, largely limited to a few privileged hereditary court astronomer-astrologers. In addition to the motions of Earth and the Moon, they were able to fit in the approximately 12-year cycle of Jupiter, which was central to their system of astrology. The Chinese still preserve some aspects of this system in their cycle of 12 "years"—the Year of the Dragon, the Year of the Pig, and so on—that are defined by the position of Jupiter in the zodiac.

Our Western calendar derives from a long history of timekeeping beginning with the Sumerians, dating back to at least the second millennium BCE, and continuing with the Egyptians and the Greeks around the eighth century BCE. These calendars led, eventually, to the *Julian calendar*, introduced by Julius Caesar, which approximated the year at 365.25 days, fairly close to the actual value of 365.2422. The Romans achieved this approximation by declaring years to have 365 days each, with the exception of every fourth year. The *leap year* was to have one extra day, bringing its length to 366 days, and thus making the average length of the year in the Julian calendar 365.25 days.

In this calendar, the Romans had dropped the almost impossible task of trying to base their calendar on the Moon as well as the Sun, although a vestige of older lunar systems can be seen in the fact that our months have an average length of about 30 days. However, lunar calendars remained in use in other cultures, and Islamic calendars, for example, are still primarily lunar rather than solar.

The Gregorian Calendar

Although the Julian calendar (which was adopted by the early Christian Church) represented a great advance, its average year still differed from the true year by about 11 minutes, an amount that accumulates over the centuries to an appreciable error. By 1582, that 11 minutes per year had added up to the point where the first day of spring was occurring on March 11, instead of March 21. If the trend were allowed to continue, eventually the Christian celebration of Easter would be occurring in early winter. Pope Gregory XIII, a contemporary of Galileo, felt it necessary to institute further calendar reform.

The Gregorian calendar reform consisted of two steps. First, 10 days had to be dropped out of the calendar to bring the vernal equinox back to March 21; by proclamation, the day following October 4, 1582, became October 15. The second feature of the new Gregorian calendar was a change in the rule for leap year, making the average length of the year more closely approximate the tropical year. Gregory decreed that three of every four century years—all leap years under the Julian calendar—would be common years henceforth. The rule was that only century years divisible by 400 would be leap years. Thus, 1700, 1800, and 1900—all divisible by 4 but not by 400—were not leap years in the Gregorian calendar. On the other hand, the years 1600 and 2000, both

Download for free at http://cnx.org/content/col11992/latest/

divisible by 400, were leap years. The average length of this Gregorian year, 365.2425 mean solar days, is correct to about 1 day in 3300 years.

The Catholic countries immediately put the Gregorian reform into effect, but countries of the Eastern Church and most Protestant countries did not adopt it until much later. It was 1752 when England and the American colonies finally made the change. By parliamentary decree, September 2, 1752, was followed by September 14. Although special laws were passed to prevent such abuses as landlords collecting a full month's rent for September, there were still riots, and people demanded their 12 days back. Russia did not abandon the Julian calendar until the time of the Bolshevik revolution. The Russians then had to omit 13 days to come into step with the rest of the world. The anniversary of the October Revolution (old calendar) of 1917, bringing the communists to power, thus ended up being celebrated in November (new calendar), a difference that is perhaps not so important since the fall of communism.

4.5 PHASES AND MOTIONS OF THE MOON

Learning Objectives

By the end of this section, you will be able to:

> Explain the cause of the lunar phases
> Understand how the Moon rotates and revolves around Earth

After the Sun, the Moon is the brightest and most obvious object in the sky. Unlike the Sun, it does not shine under its own power, but merely glows with reflected sunlight. If you were to follow its progress in the sky for a month, you would observe a cycle of **phases** (different appearances), with the Moon starting dark and getting more and more illuminated by sunlight over the course of about two weeks. After the Moon's disk becomes fully bright, it begins to fade, returning to dark about two weeks later.

These changes fascinated and mystified many early cultures, which came up with marvelous stories and legends to explain the cycle of the Moon. Even in the modern world, many people don't understand what causes the phases, thinking that they are somehow related to the shadow of Earth. Let us see how the phases can be explained by the motion of the Moon relative to the bright light source in the solar system, the Sun.

Lunar Phases

Although we know that the Sun moves 1/12 of its path around the sky each month, for purposes of explaining the phases, we can assume that the Sun's light comes from roughly the same direction during the course of a four-week lunar cycle. The Moon, on the other hand, moves completely around Earth in that time. As we watch the Moon from our vantage point on Earth, how much of its face we see illuminated by sunlight depends on the angle the Sun makes with the Moon.

Here is a simple experiment to show you what we mean: stand about 6 feet in front of a bright electric light in a completely dark room (or outdoors at night) and hold in your hand a small round object such as a tennis ball or an orange. Your head can then represent Earth, the light represents the Sun, and the ball the Moon. Move the ball around your head (making sure you don't cause an eclipse by blocking the light with your head). You will see phases just like those of the Moon on the ball. (Another good way to get acquainted with the phases and motions of the Moon is to follow our satellite in the sky for a month or two, recording its shape, its direction from the Sun, and when it rises and sets.)

Let's examine the Moon's cycle of phases using Figure 4.14, which depicts the Moon's behavior for the entire month. The trick to this figure is that you must imagine yourself standing on Earth, facing the Moon in each of

This OpenStax book is available for free at http://cnx.org/content/col11992/1.8

its phases. So, for the position labeled "New," you are on the right side of Earth and it's the middle of the day; for the position "Full," you are on the left side of Earth in the middle of the night. Note that in every position on Figure 4.14, the Moon is half illuminated and half dark (as a ball in sunlight should be). The difference at each position has to do with what part of the Moon faces Earth.

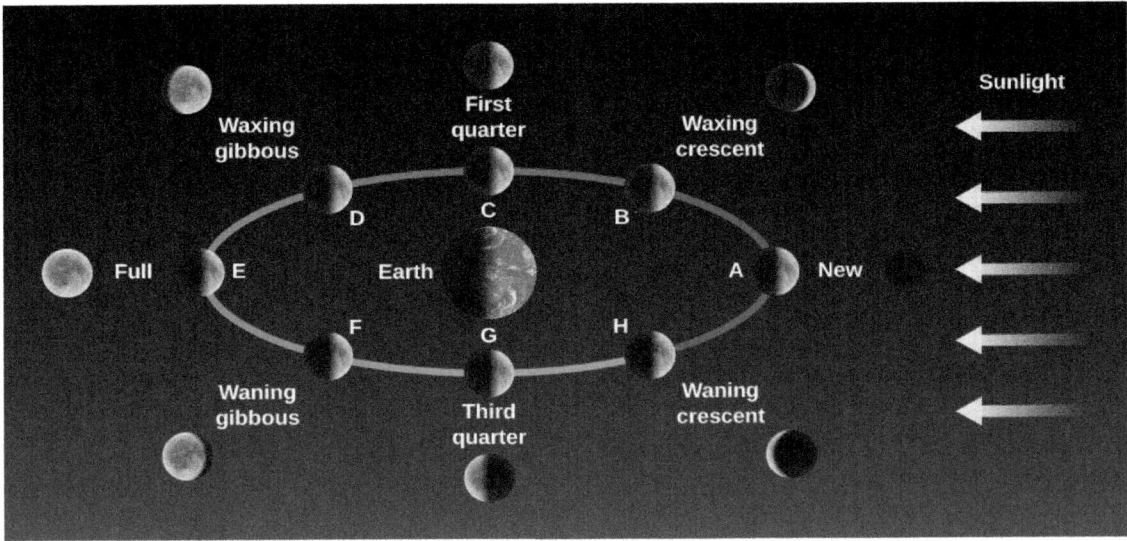

Figure 4.14. Phases of the Moon. The appearance of the Moon changes over the course of a complete monthly cycle. The pictures of the Moon on the white circle show the perspective from space, with the Sun off to the right in a fixed position. The outer images show how the Moon appears to you in the sky from each point in the orbit. Imagine yourself standing on Earth, facing the Moon at each stage. In the position "New," for example, you are facing the Moon from the right side of Earth in the middle of the day. (Note that the distance of the Moon from Earth is not to scale in this diagram: the Moon is roughly 30 Earth-diameters away from us.) (credit: modification of work by NASA)

The Moon is said to be *new* when it is in the same general direction in the sky as the Sun (position A). Here, its illuminated (bright) side is turned away from us and its dark side is turned toward us. You might say that the Sun is shining on the "wrong " side of the Moon from our perspective. In this phase the Moon is invisible to us; its dark, rocky surface does not give off any light of its own. Because the new moon is in the same part of the sky as the Sun, it rises at sunrise and sets at sunset.

But the Moon does not remain in this phase long because it moves eastward each day in its monthly path around us. Since it takes about 30 days to orbit Earth and there are 360° in a circle, the Moon will move about 12° in the sky each day (or about 24 times its own diameter). A day or two after the new phase, the thin *crescent* first appears, as we begin to see a small part of the Moon's illuminated hemisphere. It has moved into a position where it now reflects a little sunlight toward us along one side. The bright crescent increases in size on successive days as the Moon moves farther and farther around the sky away from the direction of the Sun (position B). Because the Moon is moving eastward away from the Sun, it rises later and later each day (like a student during summer vacation).

After about one week, the Moon is one-quarter of the way around its orbit (position C) and so we say it is at the *first quarter* phase. Half of the Moon's illuminated side is visible to Earth observers. Because of its eastward motion, the Moon now lags about one-quarter of the day behind the Sun, rising around noon and setting around midnight.

During the week after the first quarter phase, we see more and more of the Moon's illuminated hemisphere (position D), a phase that is called *waxing* (or growing) gibbous (from the Latin *gibbus*, meaning hump).

Download for free at http://cnx.org/content/col11992/latest/

Eventually, the Moon arrives at position E in our figure, where it and the Sun are opposite each other in the sky. The side of the Moon turned toward the Sun is also turned toward Earth, and we have the *full* phase.

When the Moon is full, it is opposite the Sun in the sky. The Moon does the opposite of what the Sun does, rising at sunset and setting at sunrise. Note what that means in practice: the completely illuminated (and thus very noticeable) Moon rises just as it gets dark, remains in the sky all night long, and sets as the Sun's first rays are seen at dawn. Its illumination throughout the night helps lovers on a romantic stroll and students finding their way back to their dorms after a long night in the library or an off-campus party.

And when is the full moon highest in the sky and most noticeable? At midnight, a time made famous in generations of horror novels and films. (Note how the behavior of a vampire like Dracula parallels the behavior of the full Moon: Dracula rises at sunset, does his worst mischief at midnight, and must be back down in his coffin by sunrise. The old legends were a way of personifying the behavior of the Moon, which was a much more dramatic part of people's lives in the days before electric lights and television.)

Folklore has it that more crazy behavior is seen during the time of the full moon (the Moon even gives a name to crazy behavior—"lunacy"). But, in fact, statistical tests of this "hypothesis" involving thousands of records from hospital emergency rooms and police files do not reveal any correlation of human behavior with the phases of the Moon. For example, homicides occur at the same rate during the new moon or the crescent moon as during the full moon. Most investigators believe that the real story is not that more crazy behavior happens on nights with a full moon, but rather that we are more likely to notice or remember such behavior with the aid of a bright celestial light that is up all night long.

During the two weeks following the full moon, the Moon goes through the same phases again in reverse order (points F, G, and H in Figure 4.14), returning to new phase after about 29.5 days. About a week after the full moon, for example, the Moon is at *third quarter*, meaning that it is three-quarters of the way around (not that it is three-quarters illuminated—in fact, half of the visible side of the Moon is again dark). At this phase, the Moon is now rising around midnight and setting around noon.

Note that there is one thing quite misleading about Figure 4.14. If you look at the Moon in position E, although it is full in theory, it appears as if its illumination would in fact be blocked by a big fat Earth, and hence we would not see anything on the Moon except Earth's shadow. In reality, the Moon is nowhere near as close to Earth (nor is its path so identical with the Sun's in the sky) as this diagram (and the diagrams in most textbooks) might lead you to believe.

The Moon is actually 30 *Earth-diameters* away from us; Science and the Universe: A Brief Tour contains a diagram that shows the two objects to scale. And, since the Moon's orbit is tilted relative to the path of the Sun in the sky, Earth's shadow misses the Moon most months. That's why we regularly get treated to a full moon. The times when Earth's shadow does fall on the Moon are called lunar eclipses and are discussed in Eclipses of the Sun and Moon.

MAKING CONNECTIONS

Astronomy and the Days of the Week

The week seems independent of celestial motions, although its length may have been based on the time between quarter phases of the Moon. In Western culture, the seven days of the week are named after the seven "wanderers" that the ancients saw in the sky: the Sun, the Moon, and the five planets visible to the unaided eye (Mercury, Venus, Mars, Jupiter, and Saturn).

This OpenStax book is available for free at http://cnx.org/content/col11992/1.8

In English, we can easily recognize the names Sun-day (Sunday), Moon-day (Monday), and Saturn-day (Saturday), but the other days are named after the Norse equivalents of the Roman gods that gave their names to the planets. In languages more directly related to Latin, the correspondences are clearer. Wednesday, Mercury's day, for example, is *mercoledi* in Italian, *mercredi* in French, and *miércoles* in Spanish. Mars gives its name to Tuesday (*martes* in Spanish), Jupiter or Jove to Thursday (*giovedi* in Italian), and Venus to Friday (*vendredi* in French).

There is no reason that the week has to have seven days rather than five or eight. It is interesting to speculate that if we had lived in a planetary system where more planets were visible without a telescope, the Beatles could have been right and we might well have had "Eight Days a Week."

LINK TO LEARNING

View this animation (https://openstaxcollege.org/l/30phamoonearth) to see the phases of the Moon as it orbits Earth and as Earth orbits the Sun.

The Moon's Revolution and Rotation

The Moon's sidereal period—that is, the period of its revolution about Earth measured with respect to the stars—is a little over 27 days: the **sidereal month** is 27.3217 days to be exact. The time interval in which the phases repeat—say, from full to full—is the **solar month**, 29.5306 days. The difference results from Earth's motion around the Sun. The Moon must make more than a complete turn around the moving Earth to get back to the same phase with respect to the Sun. As we saw, the Moon changes its position on the celestial sphere rather rapidly: even during a single evening, the Moon creeps visibly eastward among the stars, traveling its own width in a little less than 1 hour. The delay in moonrise from one day to the next caused by this eastward motion averages about 50 minutes.

The Moon *rotates* on its axis in exactly the same time that it takes to *revolve* about Earth. As a consequence, the Moon always keeps the same face turned toward Earth (Figure 4.15). You can simulate this yourself by "orbiting" your roommate or another volunteer. Start by facing your roommate. If you make one rotation (spin) with your shoulders in the exact same time that you revolve around him or her, you will continue to face your roommate during the whole "orbit." As we will see in coming chapters, our Moon is not the only world that exhibits this behavior, which scientists call **synchronous rotation**.

Download for free at http://cnx.org/content/col11992/latest/

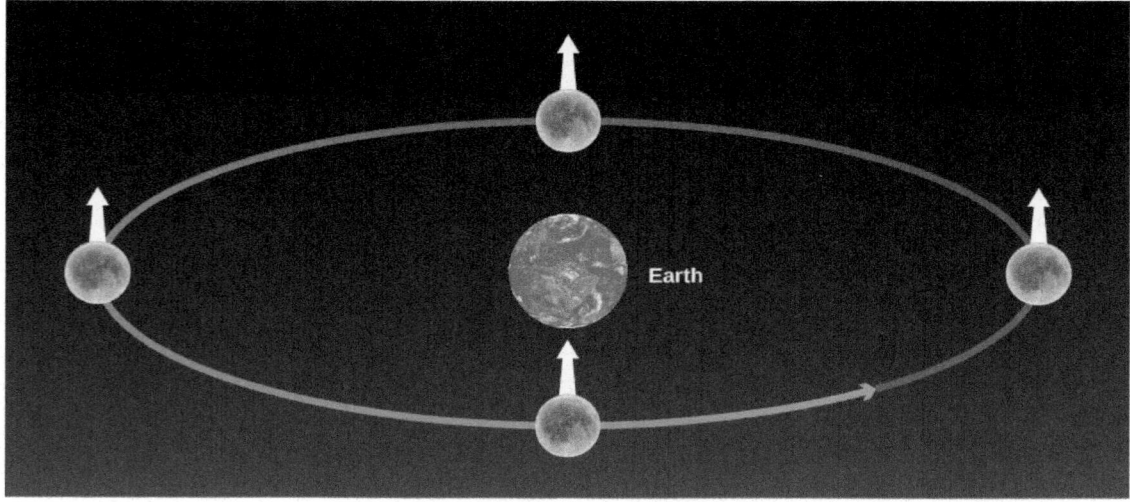

(a)

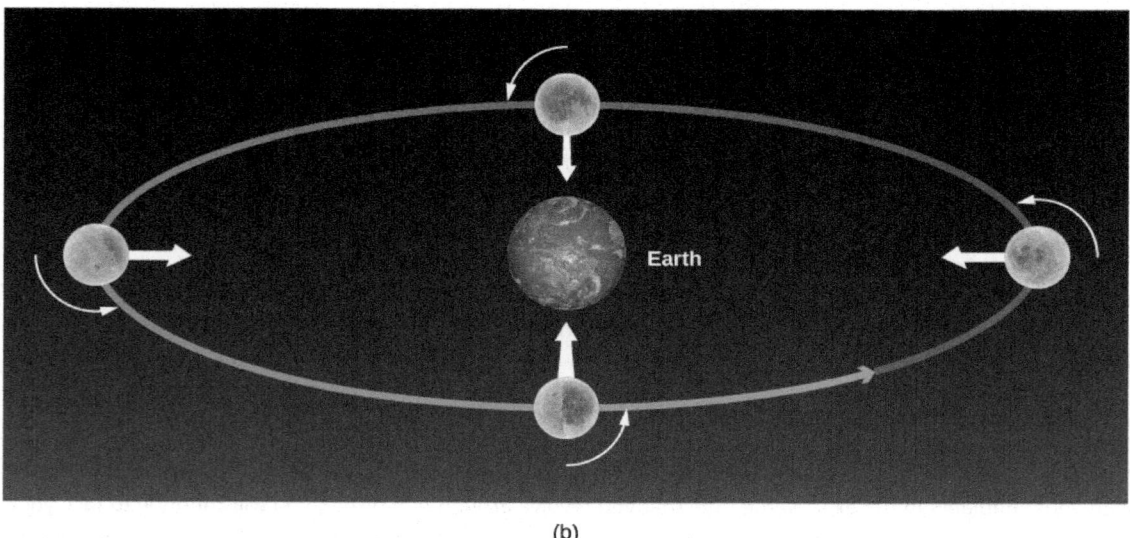

(b)

Figure 4.15. The Moon without and with Rotation. In this figure, we stuck a white arrow into a fixed point on the Moon to keep track of its sides. (a) If the Moon did not rotate as it orbited Earth, it would present all of its sides to our view; hence the white arrow would point directly toward Earth only in the bottom position on the diagram. (b) Actually, the Moon rotates in the same period that it revolves, so we always see the same side (the white arrow keeps pointing to Earth).

The differences in the Moon's appearance from one night to the next are due to changing illumination by the Sun, not to its own rotation. You sometimes hear the back side of the Moon (the side we never see) called the "dark side." This is a misunderstanding of the real situation: which side is light and which is dark changes as the Moon moves around Earth. The back side is dark no more frequently than the front side. Since the Moon rotates, the Sun rises and sets on all sides of the Moon. With apologies to Pink Floyd, there is simply no regular "Dark Side of the Moon."

This OpenStax book is available for free at http://cnx.org/content/col11992/1.8

4.6 OCEAN TIDES AND THE MOON

Learning Objectives

By the end of this section, you will be able to:

> Describe what causes tides on Earth
> Explain why the amplitude of tides changes during the course of a month

Anyone living near the sea is familiar with the twice-daily rising and falling of the **tides**. Early in history, it was clear that tides must be related to the Moon because the daily delay in high tide is the same as the daily delay in the Moon's rising. A satisfactory explanation of the tides, however, awaited the theory of gravity, supplied by Newton.

The Pull of the Moon on Earth

The gravitational forces exerted by the Moon at several points on Earth are illustrated in Figure 4.16. These forces differ slightly from one another because Earth is not a point, but has a certain size: all parts are not equally distant from the Moon, nor are they all in exactly the same direction from the Moon. Moreover, Earth is not perfectly rigid. As a result, the differences among the forces of the Moon's attraction on different parts of Earth (called *differential forces*) cause Earth to distort slightly. The side of Earth nearest the Moon is attracted toward the Moon more strongly than is the center of Earth, which in turn is attracted more strongly than is the side opposite the Moon. Thus, the differential forces tend to stretch Earth slightly into a *prolate spheroid* (a football shape), with its long diameter pointed toward the Moon.

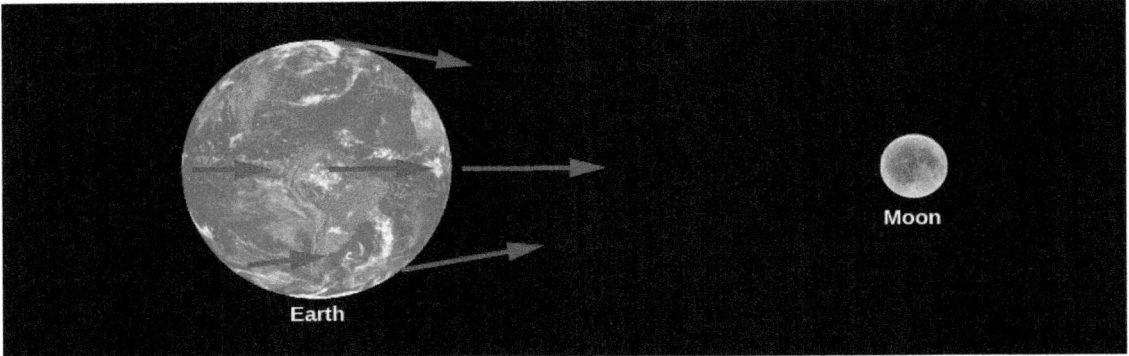

Figure 4.16. Pull of the Moon. The Moon's differential attraction is shown on different parts of Earth. (Note that the differences have been exaggerated for educational purposes.)

If Earth were made of water, it would distort until the Moon's differential forces over different parts of its surface came into balance with Earth's own gravitational forces pulling it together. Calculations show that in this case, Earth would distort from a sphere by amounts ranging up to nearly 1 meter. Measurements of the actual deformation of Earth show that the solid Earth does distort, but only about one-third as much as water would, because of the greater rigidity of Earth's interior.

Because the tidal distortion of the solid Earth amounts—at its greatest—to only about 20 centimeters, Earth does not distort enough to balance the Moon's differential forces with its own gravity. Hence, objects at Earth's surface experience tiny horizontal tugs, tending to make them slide about. These *tide-raising forces* are too

Download for free at http://cnx.org/content/col11992/latest/

insignificant to affect solid objects like astronomy students or rocks in Earth's crust, but they do affect the waters in the oceans.

The Formation of Tides

The tide-raising forces, acting over a number of hours, produce motions of the water that result in measurable tidal bulges in the oceans. Water on the side of Earth facing the Moon flows toward it, with the greatest depths roughly at the point below the Moon. On the side of Earth opposite the Moon, water also flows to produce a tidal bulge (Figure 4.17).

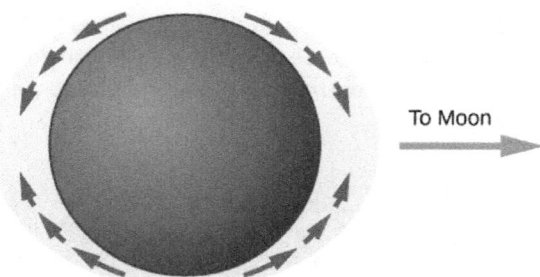

Figure 4.17. Tidal Bulges in an "Ideal" Ocean. Differences in gravity cause tidal forces that push water in the direction of tidal bulges on Earth.

LINK TO LEARNING

You can run this animation (https://openstaxcollege.org/l/30visdemotidal) for a visual demonstration of the tidal bulge.

Note that the tidal bulges in the oceans do not result from the Moon's compressing or expanding the water, nor from the Moon's lifting the water "away from Earth." Rather, they result from an actual flow of water over Earth's surface toward the two regions below and opposite the Moon, causing the water to pile up to greater depths at those places (Figure 4.18).

Figure 4.18. High and Low Tides. This is a side-by-side comparison of the Bay of Fundy in Canada at high and low tides. (credit a, b: modification of work by Dylan Kereluk)

This OpenStax book is available for free at http://cnx.org/content/col11992/1.8

In the idealized (and, as we shall see, oversimplified) model just described, the height of the tides would be only a few feet. The rotation of Earth would carry an observer at any given place alternately into regions of deeper and shallower water. An observer being carried toward the regions under or opposite the Moon, where the water was deepest, would say, "The tide is coming in"; when carried away from those regions, the observer would say, "The tide is going out." During a day, the observer would be carried through two tidal bulges (one on each side of Earth) and so would experience two high tides and two low tides.

The Sun also produces tides on Earth, although it is less than half as effective as the Moon at tide raising. The actual tides we experience are a combination of the larger effect of the Moon and the smaller effect of the Sun. When the Sun and Moon are lined up (at new moon or full moon), the tides produced reinforce each other and so are greater than normal (Figure 4.19). These are called spring tides (the name is connected not to the season but to the idea that higher tides "spring up"). Spring tides are approximately the same, whether the Sun and Moon are on the same or opposite sides of Earth, because tidal bulges occur on both sides. When the Moon is at first quarter or last quarter (at right angles to the Sun's direction), the tides produced by the Sun partially cancel the tides of the Moon, making them lower than usual. These are called neap tides.

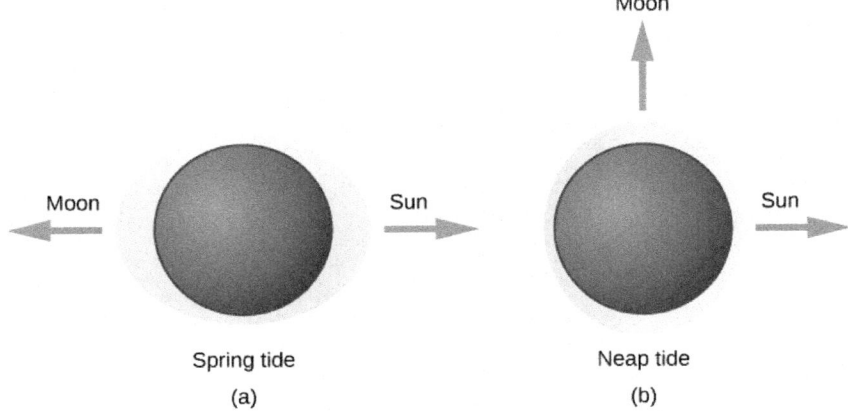

Figure 4.19. Tides Caused by Different Alignments of the Sun and Moon. (a) In spring tides, the Sun's and Moon's pulls reinforce each other. (b) In neap tides, the Sun and the Moon pull at right angles to each other and the resulting tides are lower than usual.

The "simple" theory of tides, described in the preceding paragraphs, would be sufficient if Earth rotated very slowly and were completely surrounded by very deep oceans. However, the presence of land masses stopping the flow of water, the friction in the oceans and between oceans and the ocean floors, the rotation of Earth, the wind, the variable depth of the ocean, and other factors all complicate the picture. This is why, in the real world, some places have very small tides while in other places huge tides become tourist attractions. If you have been in such places, you may know that "tide tables" need to be computed and published for each location; one set of tide predictions doesn't work for the whole planet. In this introductory chapter, we won't delve further into these complexities.

Download for free at http://cnx.org/content/col11992/latest/

VOYAGERS IN ASTRONOMY

George Darwin and the Slowing of Earth

The rubbing of water over the face of Earth involves an enormous amount of energy. Over long periods of time, the friction of the tides is slowing down the rotation of Earth. Our day gets longer by about 0.002 second each century. That seems very small, but such tiny changes can add up over millions and billions of years.

Although Earth's spin is slowing down, the angular momentum (see Orbits and Gravity) in a system such as the Earth-Moon system cannot change. Thus, some other spin motion must speed up to take the extra angular momentum. The details of what happens were worked out over a century ago by George Darwin, the son of naturalist Charles Darwin. George Darwin (see Figure 4.20) had a strong interest in science but studied law for six years and was admitted to the bar. However, he never practiced law, returning to science instead and eventually becoming a professor at Cambridge University. He was a protégé of Lord Kelvin, one of the great physicists of the nineteenth century, and he became interested in the long-term evolution of the solar system. He specialized in making detailed (and difficult) mathematical calculations of how orbits and motions change over geologic time.

Figure 4.20. George Darwin (1845–1912). George Darwin is best known for studying Earth's spin in relation to angular momentum.

What Darwin calculated for the Earth-Moon system was that the Moon will slowly spiral outward, away from Earth. As it moves farther away, it will orbit less quickly (just as planets farther from the Sun move more slowly in their orbits). Thus, the month will get longer. Also, because the Moon will be more distant, total eclipses of the Sun will no longer be visible from Earth.

Both the day and the month will continue to get longer, although bear in mind that the effects are very gradual. Darwin's calculations were confirmed by mirrors placed on the Moon by Apollo 11 astronauts. These show that the Moon is moving away by 3.8 centimeters per year, and that ultimately—billions of years in the future—the day and the month will be the same length (about 47 of our present days). At this point the Moon will be stationary in the sky over the same spot on Earth, meaning some parts of Earth will see the Moon and its phases and other parts will never see them. This kind of alignment is

This OpenStax book is available for free at http://cnx.org/content/col11992/1.8

already true for Pluto's moon Charon (among others). Its rotation and orbital period are the same length as a day on Pluto.

4.7 ECLIPSES OF THE SUN AND MOON

Learning Objectives

By the end of this section, you will be able to:

> Describe what causes lunar and solar eclipses
> Differentiate between a total and partial solar eclipse
> Explain why lunar eclipses are much more common than solar eclipses

One of the coincidences of living on Earth at the present time is that the two most prominent astronomical objects, the Sun and the Moon, have nearly the same apparent size in the sky. Although the Sun is about 400 times larger in diameter than the Moon, it is also about 400 times farther away, so both the Sun and the Moon have the same angular size—about 1/2°. As a result, the Moon, as seen from Earth, can appear to cover the Sun, producing one of the most impressive events in nature.

Any solid object in the solar system casts a shadow by blocking the light of the Sun from a region behind it. This shadow in space becomes apparent whenever another object moves into it. In general, an *eclipse* occurs whenever any part of either Earth or the Moon enters the shadow of the other. When the Moon's shadow strikes Earth, people within that shadow see the Sun at least partially covered by the Moon; that is, they witness a **solar eclipse**. When the Moon passes into the shadow of Earth, people on the night side of Earth see the Moon darken in what is called a **lunar eclipse**. Let's look at how these happen in more detail.

The shadows of Earth and the Moon consist of two parts: a cone where the shadow is darkest, called the *umbra*, and a lighter, more diffuse region of darkness called the *penumbra*. As you can imagine, the most spectacular eclipses occur when an object enters the umbra. Figure 4.21 illustrates the appearance of the Moon's shadow and what the Sun and Moon would look like from different points within the shadow.

Download for free at http://cnx.org/content/col11992/latest/

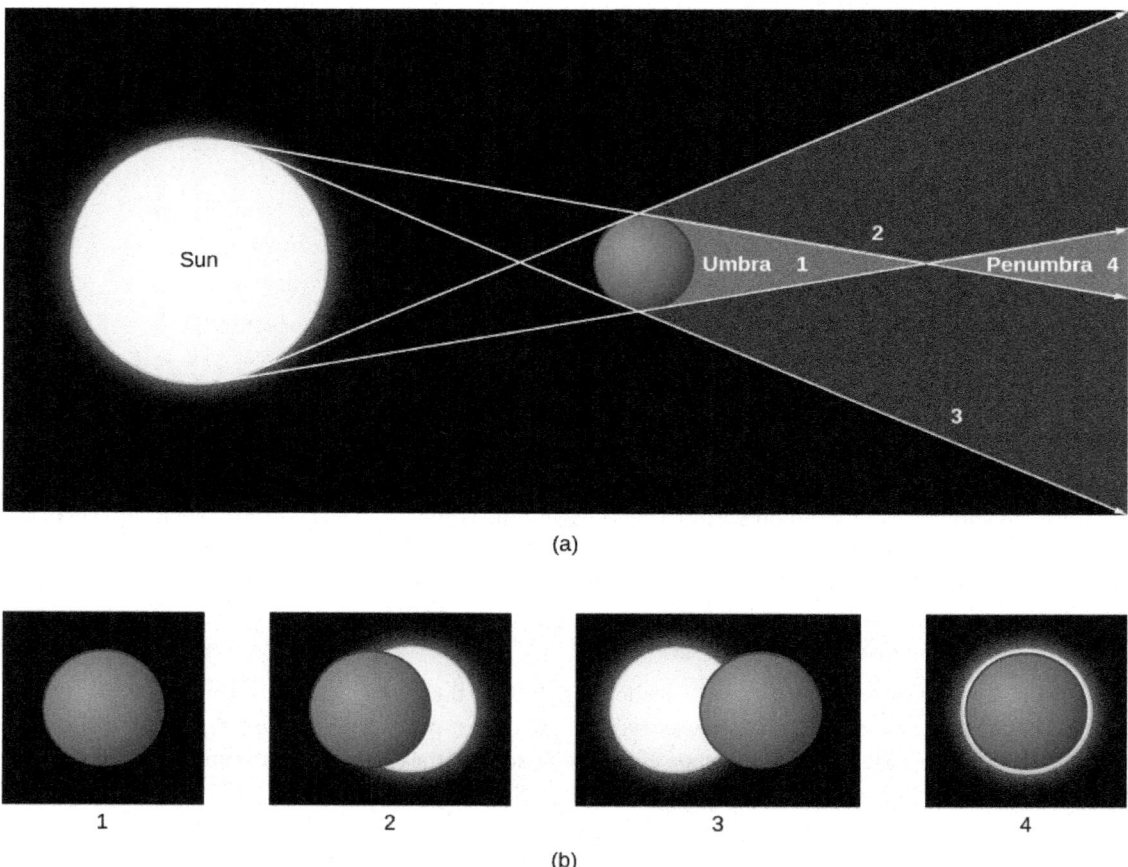

Figure 4.21. Solar Eclipse. (a) The shadow cast by a spherical body (the Moon, for example) is shown. Notice the dark umbra and the lighter penumbra. Four points in the shadow are labeled with numbers. In (b) you see what the Sun and Moon would look like in the sky at the four labeled points. At position 1, you see a total eclipse. At positions 2 and 3, the eclipse is partial. At position 4, the Moon is farther away and thus cannot cover the Sun completely; a ring of light thus shows around the Sun, creating what is called an "annular" eclipse.

If the path of the Moon in the sky were identical to the path of the Sun (the ecliptic), we might expect to see an eclipse of the Sun and the Moon each month—whenever the Moon got in front of the Sun or into the shadow of Earth. However, as we mentioned, the Moon's orbit is tilted relative to the plane of Earth's orbit about the Sun by about 5° (imagine two hula hoops with a common center, but tilted a bit). As a result, during most months, the Moon is sufficiently above or below the ecliptic plane to avoid an eclipse. But when the two paths cross (twice a year), it is then "eclipse season" and eclipses are possible.

Eclipses of the Sun

The apparent or angular sizes of both the Sun and Moon vary slightly from time to time as their distances from Earth vary. (Figure 4.21 shows the distance of the observer varying at points A–D, but the idea is the same.) Much of the time, the Moon looks slightly smaller than the Sun and cannot cover it completely, even if the two are perfectly aligned. In this type of "annular eclipse," there is a ring of light around the dark sphere of the Moon.

However, if an eclipse of the Sun occurs when the Moon is somewhat nearer than its average distance, the Moon can completely hide the Sun, producing a *total* solar eclipse. Another way to say it is that a total eclipse of the Sun occurs at those times when the umbra of the Moon's shadow reaches the surface of Earth.

This OpenStax book is available for free at http://cnx.org/content/col11992/1.8

The geometry of a total solar eclipse is illustrated in Figure 4.22. If the Sun and Moon are properly aligned, then the Moon's darkest shadow intersects the ground at a small point on Earth's surface. Anyone on Earth within the small area covered by the tip of the Moon's shadow will, for a few minutes, be unable to see the Sun and will witness a total eclipse. At the same time, observers on a larger area of Earth's surface who are in the penumbra will see only a part of the Sun eclipsed by the Moon: we call this a *partial* solar eclipse.

Between Earth's rotation and the motion of the Moon in its orbit, the tip of the Moon's shadow sweeps eastward at about 1500 kilometers per hour along a thin band across the surface of Earth. The thin zone across Earth within which a total solar eclipse is visible (weather permitting) is called the eclipse path. Within a region about 3000 kilometers on either side of the eclipse path, a partial solar eclipse is visible. It does not take long for the Moon's shadow to sweep past a given point on Earth. The duration of totality may be only a brief instant; it can never exceed about 7 minutes.

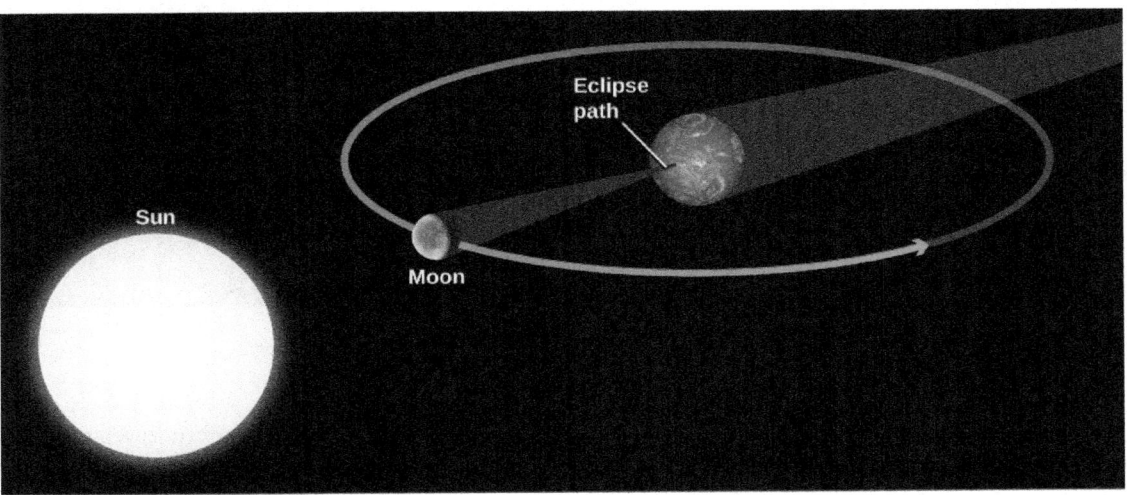

Figure 4.22. Geometry of a Total Solar Eclipse. Note that our diagram is not to scale. The Moon blocks the Sun during new moon phase as seen from some parts of Earth and casts a shadow on our planet.

Because a total eclipse of the Sun is so spectacular, it is well worth trying to see one if you can. There are some people whose hobby is "eclipse chasing" and who brag about how many they have seen in their lifetimes. Because much of Earth's surface is water, eclipse chasing can involve lengthy boat trips (and often requires air travel as well). As a result, eclipse chasing is rarely within the budget of a typical college student. Nevertheless, a list of future eclipses is given for your reference in Appendix H, just in case you strike it rich early. (And, as you can see in the Appendix, there will be total eclipses visible in the United States in 2017 and 2024, to which even college students may be able to afford travel.)

Appearance of a Total Eclipse

What can you see if you are lucky enough to catch a total eclipse? A solar eclipse starts when the Moon just begins to silhouette itself against the edge of the Sun's disk. A partial phase follows, during which more and more of the Sun is covered by the Moon. About an hour after the eclipse begins, the Sun becomes completely hidden behind the Moon. In the few minutes immediately before this period of totality begins, the sky noticeably darkens, some flowers close up, and chickens may go to roost. As an eerie twilight suddenly descends during the day, other animals (and people) may get disoriented. During totality, the sky is dark enough that planets become visible in the sky, and usually the brighter stars do as well.

Download for free at http://cnx.org/content/col11992/latest/

As the bright disk of the Sun becomes entirely hidden behind the Moon, the Sun's remarkable corona flashes into view (Figure 4.23). The *corona* is the Sun's outer atmosphere, consisting of sparse gases that extend for millions of miles in all directions from the apparent surface of the Sun. It is ordinarily not visible because the light of the corona is feeble compared with the light from the underlying layers of the Sun. Only when the brilliant glare from the Sun's visible disk is blotted out by the Moon during a total eclipse is the pearly white corona visible. (We'll talk more about the corona in the chapter on The Sun: A Garden-Variety Star.)

Figure 4.23. The Sun's Corona. The corona (thin outer atmosphere) of the Sun is visible during a total solar eclipse. (It looks more extensive in photographs than it would to the unaided eye.) (credit: modification of work by Lutfar Rahman Nirjhar)

The total phase of the eclipse ends, as abruptly as it began, when the Moon begins to uncover the Sun. Gradually, the partial phases of the eclipse repeat themselves, in reverse order, until the Moon has completely uncovered the Sun. We should make one important safety point here: while the few minutes of the *total* eclipse are safe to look at, if any part of the Sun is uncovered, you must protect your eyes with safe eclipse glasses[2] or by projecting an image of the Sun (instead of looking at it directly). For more, read the How to Observe Solar Eclipses box in this chapter.

Eclipses of the Moon

A lunar eclipse occurs when the Moon enters the shadow of Earth. The geometry of a lunar eclipse is shown in Figure 4.24. Earth's dark shadow is about 1.4 million kilometers long, so at the Moon's distance (an average of 384,000 kilometers), it could cover about four full moons. Unlike a solar eclipse, which is visible only in certain local areas on Earth, a lunar eclipse is visible to everyone who can see the Moon. Because a lunar eclipse can be seen (weather permitting) from the entire night side of Earth, lunar eclipses are observed far more frequently from a given place on Earth than are solar eclipses.

---

2   Eclipse glasses are available in many planetarium and observatory gift stores, and also from the two main U.S. manufacturers: American Paper Optics and Rainbow Symphony.

This OpenStax book is available for free at http://cnx.org/content/col11992/1.8

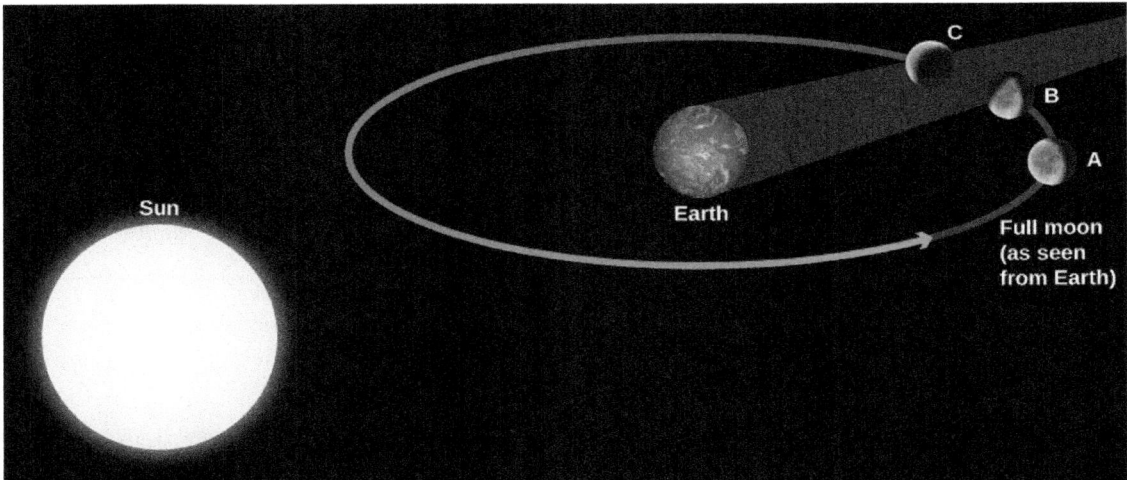

Figure 4.24. Geometry of a Lunar Eclipse. The Moon is shown moving through the different parts of Earth's shadow during a total lunar eclipse. Note that the distance the Moon moves in its orbit during the eclipse has been exaggerated here for clarity.

An eclipse of the Moon is total only if the Moon's path carries it though Earth's umbra. If the Moon does not enter the umbra completely, we have a partial eclipse of the Moon. But because Earth is larger than the Moon, its umbra is larger, so that lunar eclipses last longer than solar eclipses, as we will discuss below.

A lunar eclipse can take place only when the Sun, Earth, and Moon are in a line. The Moon is opposite the Sun, which means the Moon will be in full phase before the eclipse, making the darkening even more dramatic. About 20 minutes before the Moon reaches the dark shadow, it dims somewhat as Earth partly blocks the sunlight. As the Moon begins to dip into the shadow, the curved shape of Earth's shadow upon it soon becomes apparent.

Even when totally eclipsed, the Moon is still faintly visible, usually appearing a dull coppery red. The illumination on the eclipsed Moon is sunlight that has been bent into Earth's shadow by passing through Earth's atmosphere.

After totality, the Moon moves out of the shadow and the sequence of events is reversed. The total duration of the eclipse depends on how closely the Moon's path approaches the axis of the shadow. For an eclipse where the Moon goes through the center of Earth's shadow, each partial phase consumes at least 1 hour, and totality can last as long as 1 hour and 40 minutes. Eclipses of the Moon are much more "democratic" than solar eclipses. Since the full moon is visible on the entire night side of Earth, the lunar eclipse is visible for all those who live in that hemisphere. (Recall that a total eclipse of the Sun is visible only in a narrow path where the shadow of the umbra falls.) Total eclipses of the Moon occur, on average, about once every two or three years. A list of future total eclipses of the Moon is in Appendix H. In addition, since the lunar eclipse happens to a full moon, and a full moon is not dangerous to look at, everyone can look at the Moon during all the parts of the eclipse without worrying about safety.

Thanks to our understanding of gravity and motion (see Orbits and Gravity), eclipses can now be predicted centuries in advance. We've come a long way since humanity stood frightened by the darkening of the Sun or the Moon, fearing the displeasure of the gods. Today, we enjoy the sky show with a healthy appreciation of the majestic forces that keep our solar system running.

Download for free at http://cnx.org/content/col11992/latest/

SEEING FOR YOURSELF

How to Observe Solar Eclipses

A total eclipse of the Sun is a spectacular sight and should not be missed. However, it is extremely dangerous to look directly at the Sun: even a brief exposure can damage your eyes. Normally, few rational people are tempted to do this because it is painful (and something your mother told you never to do!). But during the partial phases of a solar eclipse, the temptation to take a look is strong. Think before you give in. The fact that the Moon is covering part of the Sun doesn't make the uncovered part any less dangerous to look at. Still, there are perfectly safe ways to follow the course of a solar eclipse, if you are lucky enough to be in the path of the shadow.

The easiest technique is to make a pinhole projector. Take a piece of cardboard with a small (1 millimeter) hole punched in it, and hold it several feet above a light surface, such as a concrete sidewalk or a white sheet of paper, so that the hole is "aimed" at the Sun. The hole produces a fuzzy but adequate image of the eclipsed Sun. Alternatively, if it's the right time of year, you can let the tiny spaces between a tree's leaves form multiple pinhole images against a wall or sidewalk. Watching hundreds of little crescent Suns dancing in the breeze can be captivating. A kitchen colander also makes an excellent pinhole projector.

Although there are safe filters for looking at the Sun directly, people have suffered eye damage by looking through improper filters, or no filter at all. For example, neutral density photographic filters are not safe because they transmit infrared radiation that can cause severe damage to the retina. Also unsafe are smoked glass, completely exposed color film, sunglasses, and many other homemade filters. Safe filters include welders' goggles and specially designed eclipse glasses.

You should certainly look at the Sun directly when it is totally eclipsed, even through binoculars or telescopes. Unfortunately, the total phase, as we discussed, is all too brief. But if you know when it is coming and going, be sure you look, for it's an unforgettably beautiful sight. And, despite the ancient folklore that presents eclipses as dangerous times to be outdoors, the partial phases of eclipses—as long as you are not looking directly at the Sun—are not any more dangerous than being out in sunlight.

During past eclipses, unnecessary panic has been created by uninformed public officials acting with the best intentions. There were two marvelous total eclipses in Australia in the twentieth century during which townspeople held newspapers over their heads for protection and schoolchildren cowered indoors with their heads under their desks. What a pity that all those people missed what would have been one of the most memorable experiences of their lifetimes.

On August 21, 2017, there will be a total solar eclipse visible across a large swath of the continental United States. The path the Moon's shadow will cast is shown in Figure 4.25.

This OpenStax book is available for free at http://cnx.org/content/col11992/1.8

Download for free at http://cnx.org/content/col11992/latest/

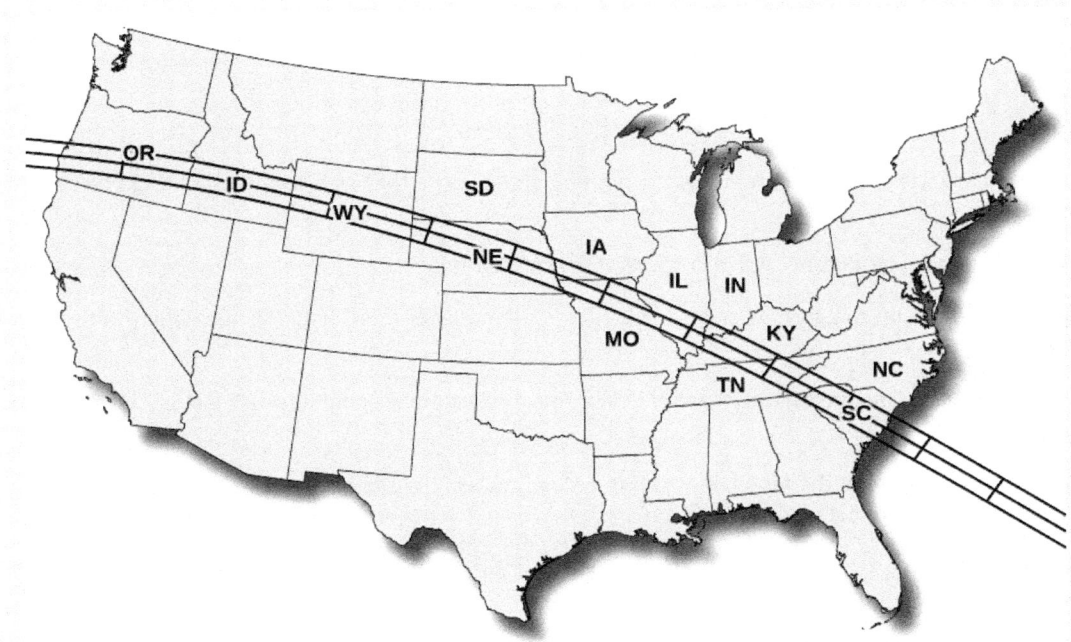

Figure 4.25. 2017 Total Solar Eclipse. This map of the United States shows the path of the total solar eclipse of 2017. On August 21, 2017, the shadow will first cross onto the West Coast near Portland, Oregon, traversing the United States and exiting the East Coast in South Carolina approximately 90 minutes later, covering about 3000 miles in the process. (credit: modification of work by NASA)

Since the eclipse path is not more than a one-day drive for most people in the United States, this would be a prime opportunity to witness this extraordinary spectacle.

LINK TO LEARNING

Check out this useful booklet (https://openstaxcollege.org/l/302017ecliboo) about the 2017 eclipse (with specific times in different locations).

Download for free at http://cnx.org/content/col11992/latest/

CHAPTER 4 REVIEW

 KEY TERMS

apparent solar time time as measured by the position of the Sun in the sky (the time that would be indicated by a sundial)

declination the angular distance north or south of the celestial equator

great circle a circle on the surface of a sphere that is the curve of intersection of the sphere with a plane passing through its center

International Date Line an arbitrary line on the surface of Earth near longitude 180° across which the date changes by one day

lunar eclipse an eclipse of the Moon, in which the Moon moves into the shadow of Earth; lunar eclipses can occur only at the time of full moon

mean solar time time based on the rotation of Earth; mean solar time passes at a constant rate, unlike apparent solar time

meridian a great circle on the terrestrial or celestial sphere that passes through the poles

phases of the Moon the different appearance of light and dark on the Moon as seen from Earth during its monthly cycle, from new moon to full moon and back to new moon

right ascension the coordinate for measuring the east-west positions of celestial bodies; the angle measured eastward along the celestial equator from the vernal equinox to the hour circle passing through a body

sidereal day Earth's rotation period as defined by the positions of the stars in the sky; the time between successive passages of the same star through the meridian

sidereal month the period of the Moon's revolution about Earth measured with respect to the stars

solar day Earth's rotation period as defined by the position of the Sun in the sky; the time between successive passages of the Sun through the meridian

solar eclipse an eclipse of the Sun by the Moon, caused by the passage of the Moon in front of the Sun; solar eclipses can occur only at the time of the new moon

solar month the time interval in which the phases repeat—say, from full to full phase

synchronous rotation when a body (for example, the Moon) rotates at the same rate that it revolves around another body

tides alternate rising and falling of sea level caused by the difference in the strength of the Moon's gravitational pull on different parts of Earth

This OpenStax book is available for free at http://cnx.org/content/col11992/1.8

SUMMARY

4.1 Earth and Sky

The terrestrial system of latitude and longitude makes use of the great circles called meridians. Longitude is arbitrarily set to 0° at the Royal Observatory at Greenwich, England. An analogous celestial coordinate system is called right ascension (RA) and declination, with 0° of declination starting at the vernal equinox. These coordinate systems help us locate any object on the celestial sphere. The Foucault pendulum is a way to demonstrate that Earth is turning.

4.2 The Seasons

The familiar cycle of the seasons results from the 23.5° tilt of Earth's axis of rotation. At the summer solstice, the Sun is higher in the sky and its rays strike Earth more directly. The Sun is in the sky for more than half of the day and can heat Earth longer. At the winter solstice, the Sun is low in the sky and its rays come in at more of an angle; in addition, it is up for fewer than 12 hours, so those rays have less time to heat. At the vernal and autumnal equinoxes, the Sun is on the celestial equator and we get about 12 hours of day and night. The seasons are different at different latitudes.

4.3 Keeping Time

The basic unit of astronomical time is the day—either the solar day (reckoned by the Sun) or the sidereal day (reckoned by the stars). Apparent solar time is based on the position of the Sun in the sky, and mean solar time is based on the average value of a solar day during the year. By international agreement, we define 24 time zones around the world, each with its own standard time. The convention of the International Date Line is necessary to reconcile times on different parts of Earth.

4.4 The Calendar

The fundamental problem of the calendar is to reconcile the incommensurable lengths of the day, month, and year. Most modern calendars, beginning with the Roman (Julian) calendar of the first century BCE, neglect the problem of the month and concentrate on achieving the correct number of days in a year by using such conventions as the leap year. Today, most of the world has adopted the Gregorian calendar established in 1582 while finding ways to coexist with the older lunar calendars' system of months.

4.5 Phases and Motions of the Moon

The Moon's monthly cycle of phases results from the changing angle of its illumination by the Sun. The full moon is visible in the sky only during the night; other phases are visible during the day as well. Because its period of revolution is the same as its period of rotation, the Moon always keeps the same face toward Earth.

4.6 Ocean Tides and the Moon

The twice-daily ocean tides are primarily the result of the Moon's differential force on the material of Earth's crust and ocean. These tidal forces cause ocean water to flow into two tidal bulges on opposite sides of Earth; each day, Earth rotates through these bulges. Actual ocean tides are complicated by the additional effects of the Sun and by the shape of the coasts and ocean basins.

4.7 Eclipses of the Sun and Moon

The Sun and Moon have nearly the same angular size (about 1/2°). A solar eclipse occurs when the Moon moves between the Sun and Earth, casting its shadow on a part of Earth's surface. If the eclipse is total, the light from the bright disk of the Sun is completely blocked, and the solar atmosphere (the corona) comes into view. Solar eclipses take place rarely in any one location, but they are among the most spectacular sights in nature. A lunar

Download for free at http://cnx.org/content/col11992/latest/

eclipse takes place when the Moon moves into Earth's shadow; it is visible (weather permitting) from the entire night hemisphere of Earth.

 FOR FURTHER EXPLORATION

Articles

Bakich, M. "Your Twenty-Year Solar Eclipse Planner." *Astronomy* (October 2008): 74. Describes the circumstances of upcoming total eclipses of the Sun.

Coco, M. "Not Just Another Pretty Phase." *Astronomy* (July 1994): 76. Moon phases explained.

Espenak, F., & Anderson, J. "Get Ready for America's Coast to Coast Experience." *Sky & Telescope* (February 2016): 22.

Gingerich, O. "Notes on the Gregorian Calendar Reform." *Sky & Telescope* (December 1982): 530.

Kluepfel, C. "How Accurate Is the Gregorian Calendar?" *Sky & Telescope* (November 1982): 417.

Krupp, E. "Calendar Worlds." *Sky & Telescope* (January 2001): 103. On how the days of the week got their names.

Krupp, E. "Behind the Curve." *Sky & Telescope* (September 2002): 68. On the reform of the calendar by Pope Gregory XIII.

MacRobert, A., & Sinnott, R. "Young Moon Hunting." *Sky & Telescope* (February 2005): 75. Hints for finding the Moon as soon after its new phase as possible.

Pasachoff, J. "Solar Eclipse Science: Still Going Strong." *Sky & Telescope* (February 2001): 40. On what we have learned and are still learning from eclipses.

Regas, D. "The Quest for Totality." *Sky & Telescope* (July 2012): 36. On eclipse chasing as a hobby.

Schaefer, B. "Lunar Eclipses That Changed the World." *Sky & Telescope* (December 1992): 639.

Schaefer, B. "Solar Eclipses That Changed the World." *Sky & Telescope* (May 1994): 36.

Websites

Ancient Observatories, Timeless Knowledge (Stanford Solar Center): http://solar-center.stanford.edu/AO/. An introduction to ancient sites where the movements of celestial objects were tracked over the years (with a special focus on tracking the Sun).

Astronomical Data Services: http://aa.usno.navy.mil/data/index.php. This rich site from the U.S. Naval Observatory has information about Earth, the Moon, and the sky, with tables and online calculators.

Calendars through the Ages: http://www.webexhibits.org/calendars/index.html. Like a good museum exhibit on the Web.

Calendar Zone: http://www.calendarzone.com/. Everything you wanted to ask or know about calendars and timekeeping, with links from around the world.

Eclipse 2017 Information and Safe Viewing Instructions: http://www.nsta.org/publications/press/extras/files/solarscience/SolarScienceInsert.pdf.

Eclipse Maps: http://www.eclipse-maps.com/Eclipse-Maps/Welcome.html. Michael Zeiler specializes in presenting helpful and interactive maps of where solar eclipses will be visible

This OpenStax book is available for free at http://cnx.org/content/col11992/1.8

Eclipse Predictions: http://astro.unl.edu/classaction/animations/lunarcycles/eclipsetable.html. This visual calendar provides dates for upcoming solar and lunar eclipses through 2029.EclipseWise: http://www.eclipsewise.com/intro.html. An introductory site on future eclipses and eclipse observing by NASA's Fred Espenak.

History of the International Date Line: http://www.staff.science.uu.nl/~gent0113/idl/idl.htm. From R. H. van Gent at Utrecht University in the Netherlands.

Lunacy and the Full Moon: http://www.scientificamerican.com/article/lunacy-and-the-full-moon/. This *Scientific American* article explores whether the Moon's phase is related to strange behavior.

Moon Phase Calculator: https://stardate.org/nightsky/moon. Keep track of the phases of the Moon with this calendar.

NASA Eclipse Website: http://eclipse.gsfc.nasa.gov/eclipse.html. This site, by NASA's eclipse expert Fred Espenak, contains a wealth of information on lunar and solar eclipses, past and future, as well as observing and photography links.

Phases of the Moon Gallery and Information: http://astropixels.com/moon/phases/phasesgallery.html. Photographs and descriptions presented by NASA's Fred Espenak.

Time and Date Website: http://www.timeanddate.com/. Comprehensive resource about how we keep time on Earth; has time zone converters and many other historical and mathematical tools.

Walk through Time: The Evolution of Time Measurement through the Ages (National Institute of Standards and Technology): http://www.nist.gov/pml/general/time/.

Videos

Bill Nye, the Science Guy, Explains the Seasons: https://www.youtube.com/watch?v=KUU7IyfR34o. For kids, but college students can enjoy the bad jokes, too (4:45).

Geography Lesson Idea: Time Zones: https://www.youtube.com/watch?v=-j-SWKtWEcU. (3:11).

How to View a Solar Eclipse: http://www.exploratorium.edu/eclipse/how-to-view-eclipse. (1:35).

Shadow of the Moon: https://www.youtube.com/watch?v=XNcfKUJwnjM. This NASA video explains eclipses of the Sun, with discussion and animation, focusing on a 2015 eclipse, and shows what an eclipse looks like from space (1:54).

Strangest Time Zones in the World: https://www.youtube.com/watch?v=uW6QqcmCfm8. (8:38).

Understanding Lunar Eclipses: https://www.youtube.com/watch?v=lNi5UFpales. This NASA video explains why there isn't an eclipse every month, with good animation (1:58).

🏛 COLLABORATIVE GROUP ACTIVITIES

A. Have your group brainstorm about other ways (besides the Foucault pendulum) you could prove that it is our Earth that is turning once a day, and not the sky turning around us. (Hint: How does the spinning of Earth affect the oceans and the atmosphere?)

B. What would the seasons on Earth be like if Earth's axis were not tilted? Discuss with your group how many things about life on Earth you think would be different.

Download for free at http://cnx.org/content/col11992/latest/

C. After college and graduate training, members of your U.S. student group are asked to set up a school in New Zealand. Describe some ways your yearly school schedule in the Southern Hemisphere would differ from what students are used to in the Northern Hemisphere.

D. During the traditional U.S. Christmas vacation weeks, you are sent to the vicinity of the South Pole on a research expedition (depending on how well you did on your astronomy midterm, either as a research assistant or as a short-order cook!). Have your group discuss how the days and nights will be different there and how these differences might affect you during your stay.

E. Discuss with your group all the stories you have heard about the full moon and crazy behavior. Why do members of your group think people associate crazy behavior with the full moon? What other legends besides vampire stories are connected with the phases of the Moon? (Hint: Think Professor Lupin in the Harry Potter stories, for example.)

F. Your college town becomes the founding site for a strange new cult that worships the Moon. These true believers gather regularly around sunset and do a dance in which they must extend their arms in the direction of the Moon. Have your group discuss which way their arms will be pointing at sunset when the Moon is new, first quarter, full, and third quarter.

G. Changes of the seasons play a large part in our yearly plans and concerns. The seasons have inspired music, stories, poetry, art, and much groaning from students during snowstorms. Search online to come up with some examples of the seasons being celebrated or overcome in fields other than science.

H. Use the information in Appendix H and online to figure out when the next eclipse of the Sun or eclipse of the Moon will be visible from where your group is going to college or from where your group members live. What time of day will the eclipse be visible? Will it be a total or partial eclipse? What preparations can you make to have an enjoyable and safe eclipse experience? How do these preparations differ between a solar and lunar eclipse?

I. On Mars, a day (often called a sol) is 24 hours and 40 minutes. Since Mars takes longer to go around the Sun, a year is 668.6 sols. Mars has two tiny moons, Phobos and Deimos. Phobos, the inner moon, rises in the west and sets in the east, taking 11 hours from moonrise to the next moonrise. Using your calculators and imaginations, have your group members come up with a calendar for Mars. (After you do your own, and only after, you can search online for the many suggestions that have been made for a martian calendar over the years.)

⎙ EXERCISES

Review Questions

1. Discuss how latitude and longitude on Earth are similar to declination and right ascension in the sky.

2. What is the latitude of the North Pole? The South Pole? Why does longitude have no meaning at the North and South Poles?

3. Make a list of each main phase of the Moon, describing roughly when the Moon rises and sets for each phase. During which phase can you see the Moon in the middle of the morning? In the middle of the afternoon?

This OpenStax book is available for free at http://cnx.org/content/col11992/1.8

4. What are advantages and disadvantages of apparent solar time? How is the situation improved by introducing mean solar time and standard time?

5. What are the two ways that the tilt of Earth's axis causes the summers in the United States to be warmer than the winters?

6. Why is it difficult to construct a practical calendar based on the Moon's cycle of phases?

7. Explain why there are two high tides and two low tides each day. Strictly speaking, should the period during which there are two high tides be 24 hours? If not, what should the interval be?

8. What is the phase of the Moon during a total solar eclipse? During a total lunar eclipse?

9. On a globe or world map, find the nearest marked latitude line to your location. Is this an example of a great circle? Explain.

10. Explain three lines of evidence that indicate that the seasons in North America are not caused by the changing Earth-Sun distance as a result of Earth's elliptical orbit around the Sun.

11. What is the origin of the terms "a.m." and "p.m." in our timekeeping?

12. Explain the origin of the leap year. Why is it necessary?

13. Explain why the year 1800 was not a leap year, even though years divisible by four are normally considered to be leap years.

14. What fraction of the Moon's visible face is illuminated during first quarter phase? Why is this phase called first quarter?

15. Why don't lunar eclipses happen during every full moon?

16. Why does the Moon create tidal bulges on both sides of Earth instead of only on the side of Earth closest to the Moon?

17. Why do the heights of the tides change over the course of a month?

18. Explain how tidal forces are causing Earth to slow down.

19. Explain how tidal forces are causing the Moon to slowly recede from Earth.

20. Explain why the Gregorian calendar modified the nature of the leap year from its original definition in the Julian calendar.

21. The term *equinox* translates as "equal night." Explain why this translation makes sense from an astronomical point of view.

22. The term *solstice* translates as "Sun stop." Explain why this translation makes sense from an astronomical point of view.

23. Why is the warmest day of the year in the United States (or in the Northern Hemisphere temperate zone) usually in August rather than on the day of the summer solstice, in late June?

Thought Questions

24. When Earth's Northern Hemisphere is tilted toward the Sun during June, some would argue that the cause of our seasons is that the Northern Hemisphere is physically closer to the Sun than the Southern Hemisphere, and this is the primary reason the Northern Hemisphere is warmer. What argument or line of evidence could contradict this idea?

Download for free at http://cnx.org/content/col11992/latest/

25. Where are you on Earth if you experience each of the following? (Refer to the discussion in Observing the Sky: The Birth of Astronomy as well as this chapter.)
    A. The stars rise and set perpendicular to the horizon.

    B. The stars circle the sky parallel to the horizon.

    C. The celestial equator passes through the zenith.

    D. In the course of a year, all stars are visible.

    E. The Sun rises on March 21 and does not set until September 21 (ideally).

26. In countries at far northern latitudes, the winter months tend to be so cloudy that astronomical observations are nearly impossible. Why can't good observations of the stars be made at those places during the summer months?

27. What is the phase of the Moon if it . . .
    A. rises at 3:00 p.m.?

    B. is highest in the sky at sunrise?

    C. sets at 10:00 a.m.?

28. A car accident occurs around midnight on the night of a full moon. The driver at fault claims he was blinded momentarily by the Moon rising on the eastern horizon. Should the police believe him?

29. The secret recipe to the ever-popular veggie burgers in the college cafeteria is hidden in a drawer in the director's office. Two students decide to break in to get their hands on it, but they want to do it a few hours before dawn on a night when there is no Moon, so they are less likely to be caught. What phases of the Moon would suit their plans?

30. Your great-great-grandfather, who often exaggerated events in his own life, once told your relatives about a terrific adventure he had on February 29, 1900. Why would this story make you suspicious?

31. One year in the future, when money is no object, you enjoy your birthday so much that you want to have another one right away. You get into your supersonic jet. Where should you and the people celebrating with you travel? From what direction should you approach? Explain.

32. Suppose you lived in the crater Copernicus on the side of the Moon facing Earth.
    A. How often would the Sun rise?

    B. How often would Earth set?

    C. During what fraction of the time would you be able to see the stars?

33. In a lunar eclipse, does the Moon enter the shadow of Earth from the east or west side? Explain.

34. Describe what an observer at the crater Copernicus would see while the Moon is eclipsed on Earth. What would the same observer see during what would be a total solar eclipse as viewed from Earth?

35. The day on Mars is 1.026 Earth-days long. The martian year lasts 686.98 Earth-days. The two moons of Mars take 0.32 Earth-day (for Phobos) and 1.26 Earth-days (for Deimos) to circle the planet. You are given the task of coming up with a martian calendar for a new Mars colony. Would a solar or lunar calendar be better for tracking the seasons?

36. What is the right ascension and declination of the vernal equinox?

This OpenStax book is available for free at http://cnx.org/content/col11992/1.8

37. What is the right ascension and declination of the autumnal equinox?

38. What is the right ascension and declination of the Sun at noon on the summer solstice in the Northern Hemisphere?

39. During summer in the Northern Hemisphere, the North Pole is illuminated by the Sun 24 hours per day. During this time, the temperature often does not rise above the freezing point of water. Explain why.

40. On the day of the vernal equinox, the day length for all places on Earth is actually slightly longer than 12 hours. Explain why.

41. Regions north of the Arctic Circle are known as the "land of the midnight Sun." Explain what this means from an astronomical perspective.

42. In a part of Earth's orbit where Earth is moving faster than usual around the Sun, would the length of the sidereal day change? If so, how? Explain.

43. In a part of Earth's orbit where Earth is moving faster than usual around the Sun, would the length of the solar day change? If so, how? Explain.

44. If Sirius rises at 8:00 p.m. tonight, at what time will it rise tomorrow night, to the nearest minute? Explain.

45. What are three lines of evidence you could use to indicate that the phases of the Moon are not caused by the shadow of Earth falling on the Moon?

46. If the Moon rises at a given location at 6:00 p.m. today, about what time will it rise tomorrow night?

47. Explain why some solar eclipses are total and some are annular.

48. Why do lunar eclipses typically last much longer than solar eclipses?

Figuring For Yourself

49. Suppose Earth took exactly 300.0 days to go around the Sun, and everything else (the day, the month) was the same. What kind of calendar would we have? How would this affect the seasons?

50. Consider a calendar based entirely on the day and the month (the Moon's period from full phase to full phase). How many days are there in a month? Can you figure out a scheme analogous to leap year to make this calendar work?

51. If a star rises at 8:30 p.m. tonight, approximately what time will it rise two months from now?

52. What is the altitude of the Sun at noon on December 22, as seen from a place on the Tropic of Cancer?

53. Show that the Gregorian calendar will be in error by 1 day in about 3300 years.

Download for free at http://cnx.org/content/col11992/latest/

This OpenStax book is available for free at http://cnx.org/content/col11992/1.8

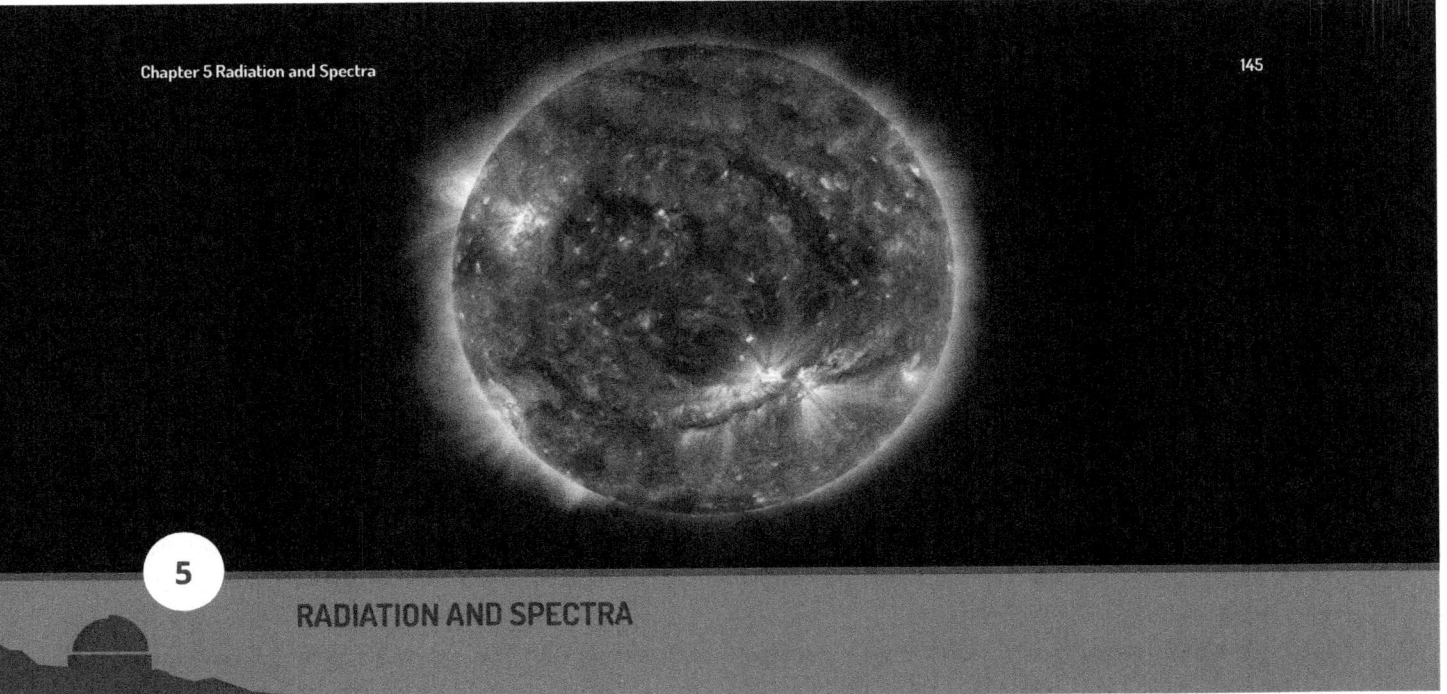

Figure 5.1. Our Sun in Ultraviolet Light. This photograph of the Sun was taken at several different wavelengths of ultraviolet, which our eyes cannot see, and then color coded so it reveals activity in our Sun's atmosphere that cannot be observed in visible light. This is why it is important to observe the Sun and other astronomical objects in wavelengths other than the visible band of the spectrum. This image was taken by a satellite from above Earth's atmosphere, which is necessary since Earth's atmosphere absorbs much of the ultraviolet light coming from space. (credit: modification of work by NASA)

Chapter Outline

5.1 The Behavior of Light

5.2 The Electromagnetic Spectrum

5.3 Spectroscopy in Astronomy

5.4 The Structure of the Atom

5.5 Formation of Spectral Lines

5.6 The Doppler Effect

Thinking Ahead

The nearest star is so far away that the fastest spacecraft humans have built would take almost 100,000 years to get there. Yet we very much want to know what material this neighbor star is composed of and how it differs from our own Sun. How can we learn about the chemical makeup of stars that we cannot hope to visit or sample?

In astronomy, most of the objects that we study are completely beyond our reach. The temperature of the Sun is so high that a spacecraft would be fried long before it reached it, and the stars are much too far away to visit in our lifetimes with the technology now available. Even light, which travels at a speed of 300,000 kilometers per second (km/s), takes more than 4 years to reach us from the nearest star. If we want to learn about the Sun and stars, we must rely on techniques that allow us to analyze them from a distance.

Download for free at http://cnx.org/content/col11992/latest/

5.1   THE BEHAVIOR OF LIGHT

Learning Objectives

By the end of this section, you will be able to:

> Explain the evidence for Maxwell's electromagnetic model of light
> Describe the relationship between wavelength, frequency, and speed of light
> Discuss the particle model of light and the definition of photon
> Explain how and why the amount of light we see from an object depends upon its distance

Coded into the light and other kinds of radiation that reach us from objects in the universe is a wide range of information about what those objects are like and how they work. If we can decipher this code and read the messages it contains, we can learn an enormous amount about the cosmos without ever having to leave Earth or its immediate environment.

The visible light and other radiation we receive from the stars and planets is generated by processes at the atomic level—by changes in the way the parts of an atom interact and move. Thus, to appreciate how light is generated, we must explore how atoms work. There is a bit of irony in the fact that in order to understand some of the largest structures in the universe, we must become acquainted with some of the smallest.

Notice that we have twice used the phrase "light and other radiation." One of the key ideas explored in this chapter is that visible light is not unique; it is merely the most familiar example of a much larger family of radiation that can carry information to us.

The word "radiation" will be used frequently in this book, so it is important to understand what it means. In everyday language, "radiation" is often used to describe certain kinds of energetic subatomic particles released by radioactive materials in our environment. (An example is the kind of radiation used to treat some cancers.) But this is not what we mean when we use the word "radiation" in an astronomy text. *Radiation*, as used in this book, is a general term for waves (including light waves) that *radiate* outward from a source.

As we saw in Orbits and Gravity, Newton's theory of gravity accounts for the motions of planets as well as objects on Earth. Application of this theory to a variety of problems dominated the work of scientists for nearly two centuries. In the nineteenth century, many physicists turned to the study of electricity and magnetism, which are intimately connected with the production of light.

The scientist who played a role in this field comparable to Newton's role in the study of gravity was physicist James Clerk Maxwell, born and educated in Scotland (Figure 5.2). Inspired by a number of ingenious experiments that showed an intimate relationship between electricity and magnetism, Maxwell developed a theory that describes both electricity and magnetism with only a small number of elegant equations. It is this theory that gives us important insights into the nature and behavior of light.

This OpenStax book is available for free at http://cnx.org/content/col11992/1.8

Figure 5.2. James Clerk Maxwell (1831–1879). Maxwell unified the rules governing electricity and magnetism into a coherent theory.

Maxwell's Theory of Electromagnetism

We will look at the structure of the atom in more detail later, but we begin by noting that the typical atom consists of several types of particles, a number of which have not only mass but an additional property called electric charge. In the nucleus (central part) of every atom are *protons*, which are positively charged; outside the nucleus are electrons, which have a negative charge.

Maxwell's theory deals with these electric charges and their effects, especially when they are moving. In the vicinity of an electron charge, another charge feels a force of attraction or repulsion: opposite charges attract; like charges repel. When charges are not in motion, we observe only this electric attraction or repulsion. If charges are in motion, however (as they are inside every atom and in a wire carrying a current), then we measure another force called *magnetism*.

Magnetism was well known for much of recorded human history, but its cause was not understood until the nineteenth century. Experiments with electric charges demonstrated that magnetism was the result of moving charged particles. Sometimes, the motion is clear, as in the coils of heavy wire that make an industrial electromagnet. Other times, it is more subtle, as in the kind of magnet you buy in a hardware store, in which many of the electrons inside the atoms are spinning in roughly the same direction; it is the alignment of their motion that causes the material to become magnetic.

Physicists use the word *field* to describe the action of forces that one object exerts on other distant objects. For example, we say the Sun produces a *gravitational field* that controls Earth's orbit, even though the Sun and Earth do not come directly into contact. Using this terminology, we can say that stationary electric charges produce *electric fields*, and moving electric charges also produce *magnetic fields*.

Actually, the relationship between electric and magnetic phenomena is even more profound. Experiments showed that changing magnetic fields could produce electric currents (and thus changing electric fields), and changing electric currents could in turn produce changing magnetic fields. So once begun, electric and magnetic field changes could continue to trigger each other.

Maxwell analyzed what would happen if electric charges were oscillating (moving constantly back and forth) and found that the resulting pattern of electric and magnetic fields would spread out and travel rapidly through space. Something similar happens when a raindrop strikes the surface of water or a frog jumps into a pond. The disturbance moves outward and creates a pattern we call a *wave* in the water (Figure 5.3). You might, at first, think that there must be very few situations in nature where electric charges oscillate, but this is not at all the case. As we shall see, atoms and molecules (which consist of charged particles) oscillate back and forth all the time. The resulting electromagnetic disturbances are among the most common phenomena in the universe.

Download for free at http://cnx.org/content/col11992/latest/

Figure 5.3. Making Waves. An oscillation in a pool of water creates an expanding disturbance called a wave. (credit: modification of work by "vastateparksstaff"/Flickr)

Maxwell was able to calculate the speed at which an electromagnetic disturbance moves through space; he found that it is equal to the speed of light, which had been measured experimentally. On that basis, he speculated that light was one form of a family of possible electromagnetic disturbances called **electromagnetic radiation**, a conclusion that was again confirmed in laboratory experiments. When light (reflected from the pages of an astronomy textbook, for example) enters a human eye, its changing electric and magnetic fields stimulate nerve endings, which then transmit the information contained in these changing fields to the brain. The science of astronomy is primarily about analyzing radiation from distant objects to understand what they are and how they work.

The Wave-Like Characteristics of Light

The changing electric and magnetic fields in light are similar to the waves that can be set up in a quiet pool of water. In both cases, the disturbance travels rapidly outward from the point of origin and can use its energy to disturb other things farther away. (For example, in water, the expanding ripples moving away from our frog could disturb the peace of a dragonfly resting on a leaf in the same pool.) In the case of electromagnetic waves, the radiation generated by a transmitting antenna full of charged particles and moving electrons at your local radio station can, sometime later, disturb a group of electrons in your car radio antenna and bring you the news and weather while you are driving to class or work in the morning.

The waves generated by charged particles differ from water waves in some profound ways, however. Water waves require water to travel in. The sound waves we hear, to give another example, are pressure disturbances that require air to travel though. But electromagnetic waves do not require water or air: the fields generate each other and so can move through a vacuum (such as outer space). This was such a disturbing idea to nineteenth-century scientists that they actually made up a substance to fill all of space—one for which there was not a single shred of evidence—just so light waves could have something to travel through: they called it the *aether*. Today, we know that there is no aether and that electromagnetic waves have no trouble at all moving through empty space (as all the starlight visible on a clear night must surely be doing).

This OpenStax book is available for free at http://cnx.org/content/col11992/1.8

Download for free at http://cnx.org/content/col11992/latest/

The other difference is that *all* electromagnetic waves move at the same speed in empty space (the speed of light—approximately 300,000 kilometers per second, or 300,000,000 meters per second, which can also be written as 3×10^8 m/s), which turns out to be the fastest possible speed in the universe. No matter where electromagnetic waves are generated from and no matter what other properties they have, when they are moving (and not interacting with matter), they move at the speed of light. Yet you know from everyday experience that there are different kinds of light. For example, we perceive that light waves differ from one another in a property we call color. Let's see how we can denote the differences among the whole broad family of electromagnetic waves.

The nice thing about a wave is that it is a repeating phenomenon. Whether it is the up-and-down motion of a water wave or the changing electric and magnetic fields in a wave of light, the pattern of disturbance repeats in a cyclical way. Thus, any wave motion can be characterized by a series of crests and troughs (Figure 5.4). Moving from one crest through a trough to the next crest completes one cycle. The horizontal length covered by one cycle is called the **wavelength**. Ocean waves provide an analogy: the wavelength is the distance that separates successive wave crests.

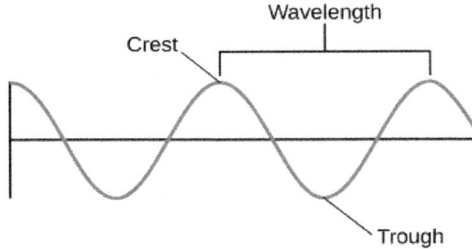

Figure 5.4. Characterizing Waves. Electromagnetic radiation has wave-like characteristics. The wavelength (λ) is the distance between crests, the frequency (f) is the number of cycles per second, and the speed (c) is the distance the wave covers during a specified period of time (e.g., kilometers per second).

For visible light, our eyes perceive different wavelengths as different colors: red, for example, is the longest visible wavelength, and violet is the shortest. The main colors of visible light from longest to shortest wavelength can be remembered using the mnemonic ROY G BIV—for Red, Orange, Yellow, Green, Blue, Indigo, and Violet. Other invisible forms of electromagnetic radiation have different wavelengths, as we will see in the next section.

We can also characterize different waves by their **frequency**, the number of wave cycles that pass by per second. If you count 10 crests moving by each second, for example, then the frequency is 10 cycles per second (cps). In honor of Heinrich Hertz, the physicist who—inspired by Maxwell's work—discovered radio waves, a cps is also called a *hertz* (Hz). Take a look at your radio, for example, and you will see the channel assigned to each radio station is characterized by its frequency, usually in units of KHz (kilohertz, or thousands of hertz) or MHz (megahertz, or millions of hertz).

Wavelength (λ) and frequency (f) are related because all electromagnetic waves travel at the same speed. To see how this works, imagine a parade in which everyone is forced by prevailing traffic conditions to move at exactly the same speed. You stand on a corner and watch the waves of marchers come by. First you see row after row of miniature ponies. Because they are not very large and, therefore, have a shorter wavelength, a good number of the ponies can move past you each minute; we can say they have a high frequency. Next, however, come several rows of circus elephants. The elephants are large and marching at the same speed as the ponies, so far fewer of them can march past you per minute: Because they have a wider spacing (longer wavelength), they represent a lower frequency.

Download for free at http://cnx.org/content/col11992/latest/

The formula for this relationship can be expressed as follows: for any wave motion, the speed at which a wave moves equals the frequency times the wavelength. Waves with longer wavelengths have lower frequencies. Mathematically, we can express this as

$$c = \lambda f$$

where the Greek letter for "l"—lambda, λ—is used to denote wavelength and c is the scientific symbol for the speed of light. Solving for the wavelength, this is expressed as:

$$\lambda = \frac{c}{f}.$$

EXAMPLE 5.1

Deriving and Using the Wave Equation

The equation for the relationship between the speed and other characteristics of a wave can be derived from our basic understanding of motion. The average speed of anything that is moving is:

$$\text{average speed} = \frac{\text{distance}}{\text{time}}$$

(So, for example, a car on the highway traveling at a speed of 100 km/h covers 100 km during the time of 1 h.) For an electromagnetic wave to travel the distance of one of its wavelengths, λ, at the speed of light, c, we have $c = \lambda/t$. The frequency of a wave is the number of cycles per second. If a wave has a frequency of a million cycles per second, then the time for each cycle to go by is a millionth of a second. So, in general, $t = 1/f$. Substituting into our wave equation, we get $c = \lambda \times f$. Now let's use this to calculate an example. What is the wavelength of visible light that has a frequency of 5.66×10^{14} Hz?

Solution

Solving the wave equation for wavelength, we find:

$$\lambda = \frac{c}{f}$$

Substituting our values gives:

$$\lambda = \frac{3.00 \times 10^8 \text{ m/s}}{5.66 \times 10^{14} \text{ Hz}} = 5.30 \times 10^{-7} \text{ m}$$

This answer can also be written as 530 nm, which is in the yellow-green part of the visible spectrum (nm stands for nanometers, where the term "nano" means "billionths").

Check Your Learning

"Tidal waves," or tsunamis, are waves caused by earthquakes that travel rapidly through the ocean. If a tsunami travels at the speed of 600 km/h and approaches a shore at a rate of one wave crest every 15 min (4 waves/h), what would be the distance between those wave crests at sea?

Answer:

$$\lambda = \frac{600 \text{ km/h}}{4 \text{ waves/h}} = 150 \text{ km}$$

This OpenStax book is available for free at http://cnx.org/content/col11992/1.8

Light as a Photon

The electromagnetic wave model of light (as formulated by Maxwell) was one of the great triumphs of nineteenth-century science. In 1887, when Heinrich Hertz actually made invisible electromagnetic waves (what today are called radio waves) on one side of a room and detected them on the other side, it ushered in a new era that led to the modern age of telecommunications. His experiment ultimately led to the technologies of television, cell phones, and today's wireless networks around the globe.

However, by the beginning of the twentieth century, more sophisticated experiments had revealed that light behaves in certain ways that cannot be explained by the wave model. Reluctantly, physicists had to accept that sometimes light behaves more like a "particle"—or at least a self-contained packet of energy—than a wave. We call such a packet of electromagnetic energy a **photon**.

The fact that light behaves like a wave in certain experiments and like a particle in others was a very surprising and unlikely idea. After all, our common sense says that waves and particles are opposite concepts. On one hand, a wave is a repeating disturbance that, by its very nature, is not in only one place, but spreads out. A particle, on the other hand, is something that can be in only one place at any given time. Strange as it sounds, though, countless experiments now confirm that electromagnetic radiation can sometimes behave like a wave and at other times like a particle.

Then, again, perhaps we shouldn't be surprised that something that always travels at the "speed limit" of the universe and doesn't need a medium to travel through might not obey our everyday common sense ideas. The confusion that this wave-particle duality of light caused in physics was eventually resolved by the introduction of a more complicated theory of waves and particles, now called quantum mechanics. (This is one of the most interesting fields of modern science, but it is mostly beyond the scope of our book. If you are interested in it, see some of the suggested resources at the end of this chapter.)

In any case, you should now be prepared when scientists (or the authors of this book) sometimes discuss electromagnetic radiation as if it consisted of waves and at other times refer to it as a stream of photons. A photon (being a packet of energy) carries a specific amount of energy. We can use the idea of energy to connect the photon and wave models. How much energy a photon has depends on its frequency when you think about it as a wave. A low-energy radio wave has a low frequency as a wave, while a high-energy X-ray at your dentist's office is a high-frequency wave. Among the colors of visible light, violet-light photons have the highest energy and red-light photons have the lowest.

Test whether the connection between photons and waves is clear to you. In the above example, which photon would have the longer wavelength as a wave: the radio wave or the X-ray? If you answered the radio wave, you are correct. Radio waves have a lower frequency, so the wave cycles are longer (they are elephants, not miniature ponies).

Propagation of Light

Let's think for a moment about how light from a lightbulb moves through space. As waves expand, they travel away from the bulb, not just toward your eyes but in all directions. They must therefore cover an ever-widening space. Yet the total amount of light available can't change once the light has left the bulb. This means that, as the same expanding shell of light covers a larger and larger area, there must be less and less of it in any given place. Light (and all other electromagnetic radiation) gets weaker and weaker as it gets farther from its source.

The increase in the area that the light must cover is proportional to the square of the distance that the light has traveled (Figure 5.5). If we stand twice as far from the source, our eyes will intercept two-squared (2 × 2), or four times less light. If we stand 10 times farther from the source, we get 10-squared, or 100 times less light. You can see how this weakening means trouble for sources of light at astronomical distances. One of the nearest stars,

Download for free at http://cnx.org/content/col11992/latest/

Alpha Centauri A, emits about the same total energy as the Sun. But it is about 270,000 times farther away, and so it appears about 73 billion times fainter. No wonder the stars, which close-up would look more or less like the Sun, look like faint pinpoints of light from far away.

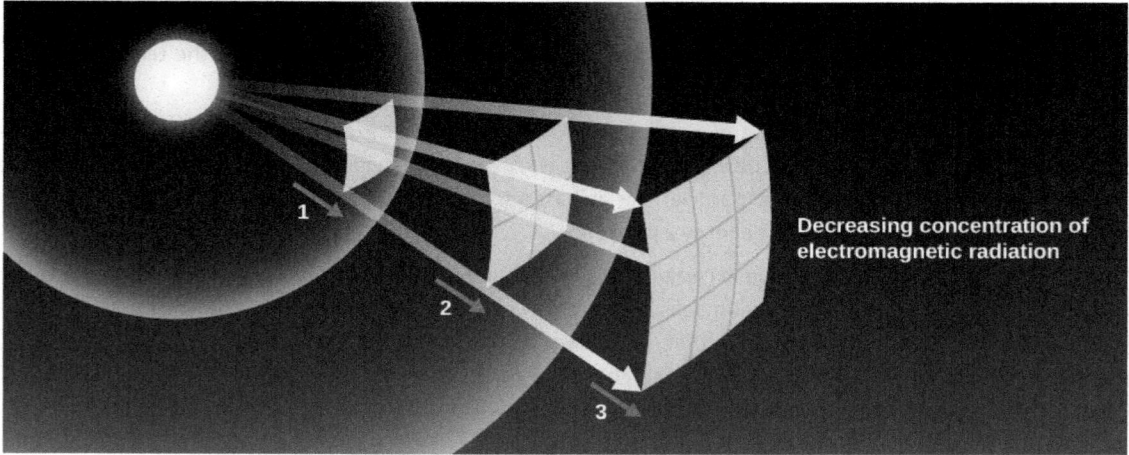

Figure 5.5. Inverse Square Law for Light. As light radiates away from its source, it spreads out in such a way that the energy per unit area (the amount of energy passing through one of the small squares) decreases as the square of the distance from its source.

This idea—that the apparent brightness of a source (how bright it looks to us) gets weaker with distance in the way we have described—is known as the **inverse square law** for light propagation. In this respect, the propagation of light is similar to the effects of gravity. Remember that the force of gravity between two attracting masses is also inversely proportional to the square of their separation.

EXAMPLE 5.2

The Inverse Square Law for Light

The intensity of a 120-W lightbulb observed from a distance 2 m away is 2.4 W/m^2. What would be the intensity if this distance was doubled?

Solution

If we move twice as far away, then the answer will change according to the inverse square of the distance, so the new intensity will be $(1/2)^2$ = 1/4 of the original intensity, or 0.6 W/m^2.

Check Your Learning

How many times brighter or fainter would a star appear if it were moved to:

a. twice its present distance?

b. ten times its present distance?

c. half its present distance?

This OpenStax book is available for free at http://cnx.org/content/col11992/1.8

Answer:

a. $\left(\frac{1}{2}\right)^2 = \frac{1}{4}$; b. $\left(\frac{1}{10}\right)^2 = \frac{1}{100}$; c. $\left(\frac{1}{1/2}\right)^2 = 4$

5.2 THE ELECTROMAGNETIC SPECTRUM

Learning Objectives

By the end of this section, you will be able to:

> Understand the bands of the electromagnetic spectrum and how they differ from one another
> Understand how each part of the spectrum interacts with Earth's atmosphere
> Explain how and why the light emitted by an object depends on its temperature

Objects in the universe send out an enormous range of electromagnetic radiation. Scientists call this range the **electromagnetic spectrum**, which they have divided into a number of categories. The spectrum is shown in Figure 5.6, with some information about the waves in each part or band.

Download for free at http://cnx.org/content/col11992/latest/

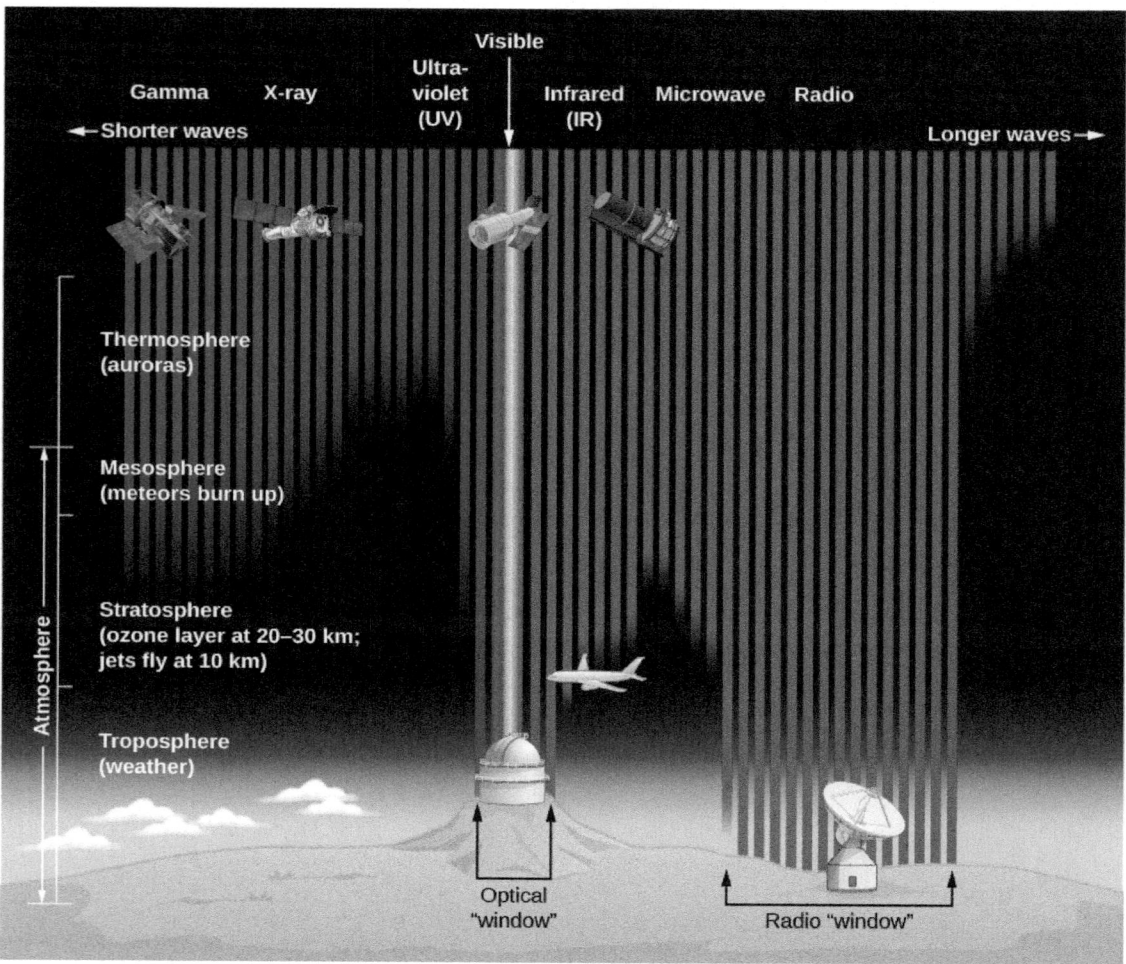

Figure 5.6. Radiation and Earth's Atmosphere. This figure shows the bands of the electromagnetic spectrum and how well Earth's atmosphere transmits them. Note that high-frequency waves from space do not make it to the surface and must therefore be observed from space. Some infrared and microwaves are absorbed by water and thus are best observed from high altitudes. Low-frequency radio waves are blocked by Earth's ionosphere. (credit: modification of work by STScI/JHU/NASA)

Types of Electromagnetic Radiation

Electromagnetic radiation with the shortest wavelengths, no longer than 0.01 nanometer, is categorized as **gamma rays** (1 nanometer = 10^{-9} meters; see Appendix D). The name *gamma* comes from the third letter of the Greek alphabet: gamma rays were the third kind of radiation discovered coming from radioactive atoms when physicists first investigated their behavior. Because gamma rays carry a lot of energy, they can be dangerous for living tissues. Gamma radiation is generated deep in the interior of stars, as well as by some of the most violent phenomena in the universe, such as the deaths of stars and the merging of stellar corpses. Gamma rays coming to Earth are absorbed by our atmosphere before they reach the ground (which is a good thing for our health); thus, they can only be studied using instruments in space.

Electromagnetic radiation with wavelengths between 0.01 nanometer and 20 nanometers is referred to as **X-rays**. Being more energetic than visible light, X-rays are able to penetrate soft tissues but not bones, and so allow us to make images of the shadows of the bones inside us. While X-rays can penetrate a short length of human flesh, they are stopped by the large numbers of atoms in Earth's atmosphere with which they interact.

This OpenStax book is available for free at http://cnx.org/content/col11992/1.8

Thus, X-ray astronomy (like gamma-ray astronomy) could not develop until we invented ways of sending instruments above our atmosphere (Figure 5.7).

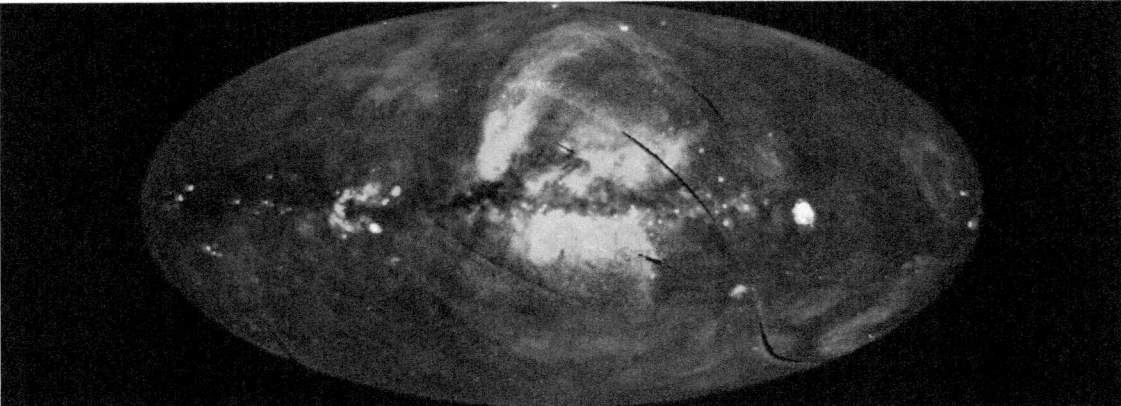

Figure 5.7. X-Ray Sky. This is a map of the sky tuned to certain types of X-rays (seen from above Earth's atmosphere). The map tilts the sky so that the disk of our Milky Way Galaxy runs across its center. It was constructed and artificially colored from data gathered by the European ROSAT satellite. Each color (red, yellow, and blue) shows X-rays of different frequencies or energies. For example, red outlines the glow from a hot local bubble of gas all around us, blown by one or more exploding stars in our cosmic vicinity. Yellow and blue show more distant sources of X-rays, such as remnants of other exploded stars or the active center of our Galaxy (in the middle of the picture). (credit: modification of work by NASA)

Radiation intermediate between X-rays and visible light is **ultraviolet** (meaning higher energy than violet). Outside the world of science, ultraviolet light is sometimes called "black light" because our eyes cannot see it. Ultraviolet radiation is mostly blocked by the ozone layer of Earth's atmosphere, but a small fraction of ultraviolet rays from our Sun do penetrate to cause sunburn or, in extreme cases of overexposure, skin cancer in human beings. Ultraviolet astronomy is also best done from space.

Electromagnetic radiation with wavelengths between roughly 400 and 700 nm is called **visible light** because these are the waves that human vision can perceive. This is also the band of the electromagnetic spectrum that most readily reaches Earth's surface. These two observations are not coincidental: human eyes evolved to see the kinds of waves that arrive from the Sun most effectively. Visible light penetrates Earth's atmosphere effectively, except when it is temporarily blocked by clouds.

Between visible light and radio waves are the wavelengths of **infrared** or heat radiation. Astronomer William Herschel first discovered infrared in 1800 while trying to measure the temperatures of different colors of sunlight spread out into a spectrum. He noticed that when he accidently positioned his thermometer beyond the reddest color, it still registered heating due to some invisible energy coming from the Sun. This was the first hint about the existence of the other (invisible) bands of the electromagnetic spectrum, although it would take many decades for our full understanding to develop.

A heat lamp radiates mostly infrared radiation, and the nerve endings in our skin are sensitive to this band of the electromagnetic spectrum. Infrared waves are absorbed by water and carbon dioxide molecules, which are more concentrated low in Earth's atmosphere. For this reason, infrared astronomy is best done from high mountaintops, high-flying airplanes, and spacecraft.

After infrared comes the familiar **microwave**, used in short-wave communication and microwave ovens. (Wavelengths vary from 1 millimeter to 1 meter and are absorbed by water vapor, which makes them effective in heating foods.) The "micro-" prefix refers to the fact that microwaves are small in comparison to radio waves, the next on the spectrum. You may remember that tea—which is full of water—heats up quickly in

Download for free at http://cnx.org/content/col11992/latest/

your microwave oven, while a ceramic cup—from which water has been removed by baking—stays cool in comparison.

All electromagnetic waves longer than microwaves are called **radio waves**, but this is so broad a category that we generally divide it into several subsections. Among the most familiar of these are radar waves, which are used in radar guns by traffic officers to determine vehicle speeds, and AM radio waves, which were the first to be developed for broadcasting. The wavelengths of these different categories range from over a meter to hundreds of meters, and other radio radiation can have wavelengths as long as several kilometers.

With such a wide range of wavelengths, not all radio waves interact with Earth's atmosphere in the same way. FM and TV waves are not absorbed and can travel easily through our atmosphere. AM radio waves are absorbed or reflected by a layer in Earth's atmosphere called the ionosphere (the ionosphere is a layer of charged particles at the top of our atmosphere, produced by interactions with sunlight and charged particles that are ejected from the Sun).

We hope this brief survey has left you with one strong impression: although visible light is what most people associate with astronomy, the light that our eyes can see is only a tiny fraction of the broad range of waves generated in the universe. Today, we understand that judging some astronomical phenomenon by using only the light we can see is like hiding under the table at a big dinner party and judging all the guests by nothing but their shoes. There's a lot more to each person than meets our eye under the table. It is very important for those who study astronomy today to avoid being "visible light chauvinists"—to respect only the information seen by their eyes while ignoring the information gathered by instruments sensitive to other bands of the electromagnetic spectrum.

Table 5.1 summarizes the bands of the electromagnetic spectrum and indicates the temperatures and typical astronomical objects that emit each kind of electromagnetic radiation. While at first, some of the types of radiation listed in the table may seem unfamiliar, you will get to know them better as your astronomy course continues. You can return to this table as you learn more about the types of objects astronomers study.

Types of Electromagnetic Radiation

Type of Radiation	Wavelength Range (nm)	Radiated by Objects at This Temperature	Typical Sources
Gamma rays	Less than 0.01	More than 10^8 K	Produced in nuclear reactions; require very high-energy processes
X-rays	0.01–20	10^6–10^8 K	Gas in clusters of galaxies, supernova remnants, solar corona
Ultraviolet	20–400	10^4–10^6 K	Supernova remnants, very hot stars
Visible	400–700	10^3–10^4 K	Stars
Infrared	10^3–10^6	10–10^3 K	Cool clouds of dust and gas, planets, moons
Microwave	10^6–10^9	Less than 10 K	Active galaxies, pulsars, cosmic background radiation

Table 5.1

This OpenStax book is available for free at http://cnx.org/content/col11992/1.8

Types of Electromagnetic Radiation

Type of Radiation	Wavelength Range (nm)	Radiated by Objects at This Temperature	Typical Sources
Radio	More than 10^9	Less than 10 K	Supernova remnants, pulsars, cold gas

Table 5.1

Radiation and Temperature

Some astronomical objects emit mostly infrared radiation, others mostly visible light, and still others mostly ultraviolet radiation. What determines the type of electromagnetic radiation emitted by the Sun, stars, and other dense astronomical objects? The answer often turns out to be their *temperature*.

At the microscopic level, everything in nature is in motion. A solid is composed of molecules and atoms in continuous vibration: they move back and forth in place, but their motion is much too small for our eyes to make out. A gas consists of atoms and/or molecules that are flying about freely at high speed, continually bumping into one another and bombarding the surrounding matter. The hotter the solid or gas, the more rapid the motion of its molecules or atoms. The temperature of something is thus a measure of the average motion energy of the particles that make it up.

This motion at the microscopic level is responsible for much of the electromagnetic radiation on Earth and in the universe. As atoms and molecules move about and collide, or vibrate in place, their electrons give off electromagnetic radiation. The characteristics of this radiation are determined by the temperature of those atoms and molecules. In a hot material, for example, the individual particles vibrate in place or move rapidly from collisions, so the emitted waves are, on average, more energetic. And recall that higher energy waves have a higher frequency. In very cool material, the particles have low-energy atomic and molecular motions and thus generate lower-energy waves.

LINK TO LEARNING

Check out the NASA briefing (https://openstax.org/l/30elmagsp1) or NASA's 5-minute introductory video (https://openstax.org/l/30elmagsp2) to learn more about the electromagnetic spectrum.

Radiation Laws

To understand, in more quantitative detail, the relationship between temperature and electromagnetic radiation, we imagine an idealized object called a **blackbody**. Such an object (unlike your sweater or your astronomy instructor's head) does not reflect or scatter any radiation, but absorbs all the electromagnetic energy that falls onto it. The energy that is absorbed causes the atoms and molecules in it to vibrate or move around at increasing speeds. As it gets hotter, this object will radiate electromagnetic waves until absorption and radiation are in balance. We want to discuss such an idealized object because, as you will see, stars behave in very nearly the same way.

The radiation from a blackbody has several characteristics, as illustrated in Figure 5.8. The graph shows the power emitted at each wavelength by objects of different temperatures. In science, the word *power* means the

Download for free at http://cnx.org/content/col11992/latest/

energy coming off per second (and it is typically measured in *watts*, which you are probably familiar with from buying lightbulbs).

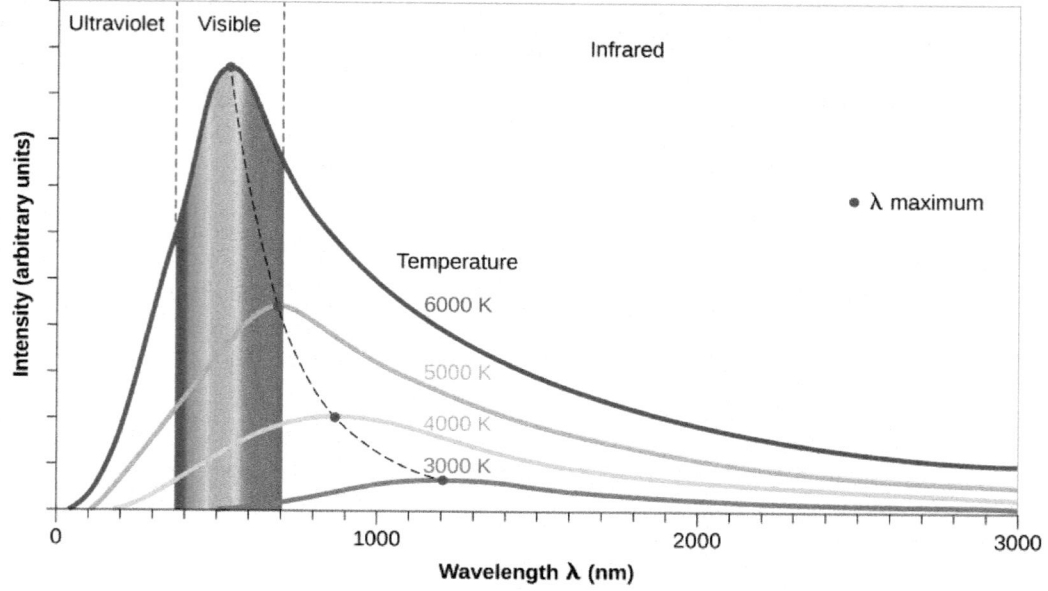

Figure 5.8. Radiation Laws Illustrated. This graph shows in arbitrary units how many photons are given off at each wavelength for objects at four different temperatures. The wavelengths corresponding to visible light are shown by the colored bands. Note that at hotter temperatures, more energy (in the form of photons) is emitted at all wavelengths. The higher the temperature, the shorter the wavelength at which the peak amount of energy is radiated (this is known as Wien's law).

First of all, notice that the curves show that, at each temperature, our blackbody object emits radiation (photons) at all wavelengths (all colors). This is because in any solid or denser gas, some molecules or atoms vibrate or move between collisions slower than average and some move faster than average. So when we look at the electromagnetic waves emitted, we find a broad range, or spectrum, of energies and wavelengths. More energy is emitted at the average vibration or motion rate (the highest part of each curve), but if we have a large number of atoms or molecules, some energy will be detected at each wavelength.

Second, note that an object at a higher temperature emits more power at all wavelengths than does a cooler one. In a hot gas (the taller curves in Figure 5.8), for example, the atoms have more collisions and give off more energy. In the real world of stars, this means that hotter stars give off more energy at every wavelength than do cooler stars.

Third, the graph shows us that the higher the temperature, the shorter the wavelength at which the maximum power is emitted. Remember that a shorter wavelength means a higher frequency and energy. It makes sense, then, that hot objects give off a larger fraction of their energy at shorter wavelengths (higher energies) than do cool objects. You may have observed examples of this rule in everyday life. When a burner on an electric stove is turned on low, it emits only heat, which is infrared radiation, but does not glow with visible light. If the burner is set to a higher temperature, it starts to glow a dull red. At a still-higher setting, it glows a brighter orange-red (shorter wavelength). At even higher temperatures, which cannot be reached with ordinary stoves, metal can appear brilliant yellow or even blue-white.

We can use these ideas to come up with a rough sort of "thermometer" for measuring the temperatures of stars. Because many stars give off most of their energy in visible light, the color of light that dominates a star's appearance is a rough indicator of its temperature. If one star looks red and another looks blue, which one

This OpenStax book is available for free at http://cnx.org/content/col11992/1.8

has the higher temperature? Because blue is the shorter-wavelength color, it is the sign of a hotter star. (Note that the temperatures we associate with different colors in science are not the same as the ones artists use. In art, red is often called a "hot" color and blue a "cool" color. Likewise, we commonly see red on faucet or air conditioning controls to indicate hot temperatures and blue to indicate cold temperatures. Although these are common uses to us in daily life, in nature, it's the other way around.)

We can develop a more precise star thermometer by measuring how much energy a star gives off at each wavelength and by constructing diagrams like Figure 5.8. The location of the peak (or maximum) in the power curve of each star can tell us its temperature. The average temperature at the surface of the Sun, which is where the radiation that we see is emitted, turns out to be 5800 K. (Throughout this text, we use the kelvin or absolute temperature scale. On this scale, water freezes at 273 K and boils at 373 K. All molecular motion ceases at 0 K. The various temperature scales are described in Appendix D.) There are stars cooler than the Sun and stars hotter than the Sun.

The wavelength at which maximum power is emitted can be calculated according to the equation

$$\lambda_{\text{max}} = \frac{3 \times 10^6}{T}$$

where the wavelength is in nanometers (one billionth of a meter) and the temperature is in K. This relationship is called **Wien's law**. For the Sun, the wavelength at which the maximum energy is emitted is 520 nanometers, which is near the middle of that portion of the electromagnetic spectrum called visible light. Characteristic temperatures of other astronomical objects, and the wavelengths at which they emit most of their power, are listed in Table 5.1.

EXAMPLE 5.3

Calculating the Temperature of a Blackbody

We can use Wien's law to calculate the temperature of a star provided we know the wavelength of peak intensity for its spectrum. If the emitted radiation from a red dwarf star has a wavelength of maximum power at 1200 nm, what is the temperature of this star, assuming it is a blackbody?

Solution

Solving Wien's law for temperature gives:

$$T = \frac{3 \times 10^6 \text{ nm K}}{\lambda_{\text{max}}} = \frac{3 \times 10^6 \text{ nm K}}{1200 \text{ nm}} = 2500 \text{ K}$$

Check Your Learning

What is the temperature of a star whose maximum light is emitted at a much shorter wavelength of 290 nm?

Answer:

$$T = \frac{3 \times 10^6 \text{ nm K}}{\lambda_{\text{max}}} = \frac{3 \times 10^6 \text{ nm K}}{290 \text{ nm}} = 10,300 \text{ K}$$

Download for free at http://cnx.org/content/col11992/latest/

Since this star has a peak wavelength that is at a shorter wavelength (in the ultraviolet part of the spectrum) than that of our Sun (in the visible part of the spectrum), it should come as no surprise that its surface temperature is much hotter than our Sun's.

We can also describe our observation that hotter objects radiate more power at all wavelengths in a mathematical form. If we sum up the contributions from all parts of the electromagnetic spectrum, we obtain the total energy emitted by a blackbody. What we usually measure from a large object like a star is the **energy flux**, the power emitted per square meter. The word *flux* means "flow" here: we are interested in the flow of power into an area (like the area of a telescope mirror). It turns out that the energy flux from a blackbody at temperature T is proportional to the fourth power of its absolute temperature. This relationship is known as the **Stefan-Boltzmann law** and can be written in the form of an equation as

$$F = \sigma T^4$$

where F stands for the energy flux and σ (Greek letter sigma) is a constant number (5.67×10^8).

Notice how impressive this result is. Increasing the temperature of a star would have a tremendous effect on the power it radiates. If the Sun, for example, were twice as hot—that is, if it had a temperature of 11,600 K—it would radiate 2^4, or 16 times more power than it does now. Tripling the temperature would raise the power output 81 times. Hot stars really shine away a tremendous amount of energy.

EXAMPLE 5.4

Calculating the Power of a Star

While energy flux tells us how much power a star emits per square meter, we would often like to know how much total power is emitted by the star. We can determine that by multiplying the energy flux by the number of square meters on the surface of the star. Stars are mostly spherical, so we can use the formula $4\pi R^2$ for the surface area, where R is the radius of the star. The total power emitted by the star (which we call the star's "absolute luminosity") can be found by multiplying the formula for energy flux and the formula for the surface area:

$$L = 4\pi R^2 \sigma T^4$$

Two stars have the same size and are the same distance from us. Star A has a surface temperature of 6000 K, and star B has a surface temperature twice as high, 12,000 K. How much more luminous is star B compared to star A?

Solution

$$L_A = 4\pi R_A{}^2 \sigma T_A{}^4 \text{ and } L_B = 4\pi R_B{}^2 \sigma T_B{}^4$$

Take the ratio of the luminosity of Star A to Star B:

$$\frac{L_B}{L_A} = \frac{4\pi R_B{}^2 \sigma T_B{}^4}{4\pi R_A{}^2 \sigma T_A{}^4} = \frac{R_B{}^2 T_B{}^4}{R_A{}^2 T_A{}^4}$$

Because the two stars are the same size, $R_A = R_B$, leaving

$$\frac{T_B{}^4}{T_A{}^4} = \frac{(12{,}000 \text{ K})^4}{(8{,}000 \text{ K})^4} = 2^4 = 16$$

This OpenStax book is available for free at http://cnx.org/content/col11992/1.8

Check Your Learning

Two stars with identical diameters are the same distance away. One has a temperature of 8700 K and the other has a temperature of 2900 K. Which is brighter? How much brighter is it?

Answer:

The 5800 K star has triple the temperature, so it is 3^4 = 81 times brighter.

5.3 SPECTROSCOPY IN ASTRONOMY

Learning Objectives

By the end of this section, you will be able to:

> Describe the properties of light
> Explain how astronomers learn the composition of a gas by examining its spectral lines
> Discuss the various types of spectra

Electromagnetic radiation carries a lot of information about the nature of stars and other astronomical objects. To extract this information, however, astronomers must be able to study the amounts of energy we receive at different wavelengths of light in fine detail. Let's examine how we can do this and what we can learn.

Properties of Light

Light exhibits certain behaviors that are important to the design of telescopes and other instruments. For example, light can be *reflected* from a surface. If the surface is smooth and shiny, as with a mirror, the direction of the reflected light beam can be calculated accurately from knowledge of the shape of the reflecting surface. Light is also bent, or *refracted*, when it passes from one kind of transparent material into another—say, from the air into a glass lens.

Reflection and refraction of light are the basic properties that make possible all *optical* instruments (devices that help us to see things better)—from eyeglasses to giant astronomical telescopes. Such instruments are generally combinations of glass lenses, which bend light according to the principles of refraction, and curved mirrors, which depend on the properties of reflection. Small optical devices, such as eyeglasses or binoculars, generally use lenses, whereas large telescopes depend almost entirely on mirrors for their main optical elements. We will discuss astronomical instruments and their uses more fully in Astronomical Instruments. For now, we turn to another behavior of light, one that is essential for the decoding of light.

In 1672, in the first paper that he submitted to the Royal Society, Sir Isaac Newton described an experiment in which he permitted sunlight to pass through a small hole and then through a prism. Newton found that sunlight, which looks white to us, is actually made up of a mixture of all the colors of the rainbow (Figure 5.9).

Download for free at http://cnx.org/content/col11992/latest/

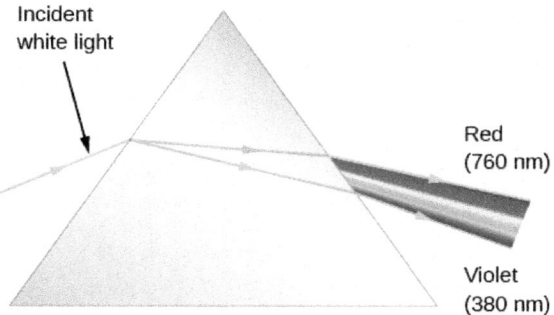

Figure 5.9. Action of a Prism. When we pass a beam of white sunlight through a prism, we see a rainbow-colored band of light that we call a continuous spectrum.

Figure 5.9 shows how light is separated into different colors with a prism—a piece of glass in the shape of a triangle with refracting surfaces. Upon entering one face of the prism, the path of the light is refracted (bent), but not all of the colors are bent by the same amount. The bending of the beam depends on the wavelength of the light as well as the properties of the material, and as a result, different wavelengths (or colors of light) are bent by different amounts and therefore follow slightly different paths through the prism. The violet light is bent more than the red. This phenomenon is called **dispersion** and explains Newton's rainbow experiment.

Upon leaving the opposite face of the prism, the light is bent again and further dispersed. If the light leaving the prism is focused on a screen, the different wavelengths or colors that make up white light are lined up side by side just like a rainbow (Figure 5.10). (In fact, a rainbow is formed by the dispersion of light though raindrops; see The Rainbow feature box.) Because this array of colors is a spectrum of light, the instrument used to disperse the light and form the spectrum is called a **spectrometer**.

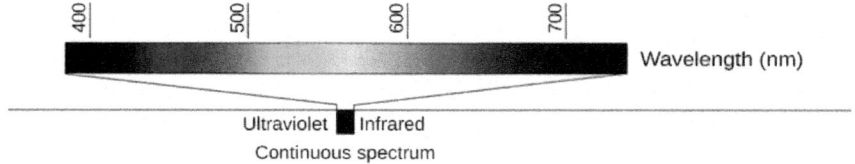

Figure 5.10. Continuous Spectrum. When white light passes through a prism, it is dispersed and forms a continuous spectrum of all the colors. Although it is hard to see in this printed version, in a well-dispersed spectrum, many subtle gradations in color are visible as your eye scans from one end (violet) to the other (red).

The Value of Stellar Spectra

When Newton described the laws of refraction and dispersion in optics, and observed the solar spectrum, all he could see was a continuous band of colors. If the spectrum of the white light from the Sun and stars were simply a continuous rainbow of colors, astronomers would have little interest in the detailed study of a star's spectrum once they had learned its average surface temperature. In 1802, however, William Wollaston built an improved spectrometer that included a lens to focus the Sun's spectrum on a screen. With this device, Wollaston saw that the colors were not spread out uniformly, but instead, some ranges of color were missing, appearing as dark bands in the solar spectrum. He mistakenly attributed these lines to natural boundaries between the colors. In 1815, German physicist Joseph Fraunhofer, upon a more careful examination of the solar spectrum, found about 600 such dark lines (missing colors), which led scientists to rule out the boundary hypothesis (Figure 5.11).

This OpenStax book is available for free at http://cnx.org/content/col11992/1.8

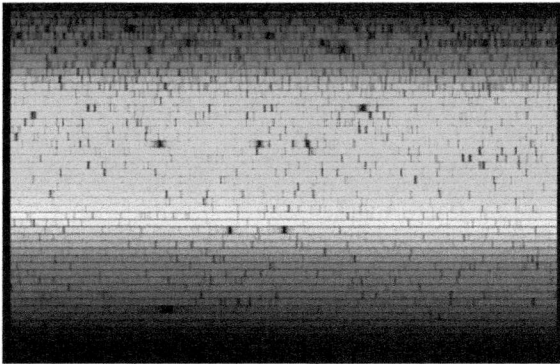

Figure 5.11. Visible Spectrum of the Sun. Our star's spectrum is crossed by dark lines produced by atoms in the solar atmosphere that absorb light at certain wavelengths. (credit: modification of work by Nigel Sharp, NOAO/National Solar Observatory at Kitt Peak/AURA, and the National Science Foundation)

Later, researchers found that similar dark lines could be produced in the spectra ("spectra" is the plural of "spectrum") of artificial light sources. They did this by passing their light through various apparently transparent substances—usually containers with just a bit of thin gas in them.

These gases turned out not to be transparent at *all* colors: they were quite opaque at a few sharply defined wavelengths. Something in each gas had to be absorbing just a few colors of light and no others. All gases did this, but each different element absorbed a different set of colors and thus showed different dark lines. If the gas in a container consisted of two elements, then light passing through it was missing the colors (showing dark lines) for both of the elements. So it became clear that certain lines in the spectrum "go with" certain elements. This discovery was one of the most important steps forward in the history of astronomy.

What would happen if there were no continuous spectrum for our gases to remove light from? What if, instead, we heated the same thin gases until they were hot enough to glow with their own light? When the gases were heated, a spectrometer revealed no continuous spectrum, but several separate bright lines. That is, these hot gases emitted light only at certain specific wavelengths or colors.

When the gas was pure hydrogen, it would emit one pattern of colors; when it was pure sodium, it would emit a different pattern. A mixture of hydrogen and sodium emitted both sets of spectral lines. The colors the gases emitted when they were heated were the very same colors as those they had absorbed when a continuous source of light was behind them. From such experiments, scientists began to see that different substances showed distinctive *spectral signatures* by which their presence could be detected (Figure 5.12). Just as your signature allows the bank to identify you, the unique pattern of colors for each type of atom (its spectrum) can help us identify which element or elements are in a gas.

Download for free at http://cnx.org/content/col11992/latest/

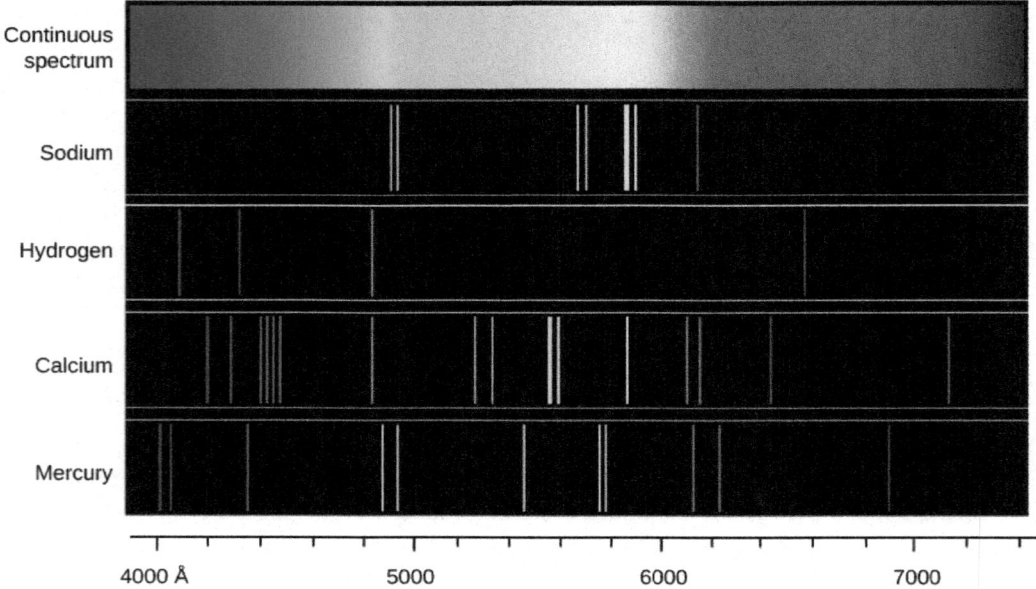

Figure 5.12. Continuous Spectrum and Line Spectra from Different Elements. Each type of glowing gas (each element) produces its own unique pattern of lines, so the composition of a gas can be identified by its spectrum. The spectra of sodium, hydrogen, calcium, and mercury gases are shown here.

Types of Spectra

In these experiments, then, there were three different types of spectra. A **continuous spectrum** (formed when a solid or very dense gas gives off radiation) is an array of all wavelengths or colors of the rainbow. A continuous spectrum can serve as a backdrop from which the atoms of much less dense gas can absorb light. A dark line, or **absorption spectrum**, consists of a series or pattern of dark lines—missing colors—superimposed upon the continuous spectrum of a source. A bright line, or **emission spectrum**, appears as a pattern or series of bright lines; it consists of light in which only certain discrete wavelengths are present. (Figure 5.11 shows an absorption spectrum, whereas Figure 5.12 shows the emission spectrum of a number of common elements along with an example of a continuous spectrum.)

When we have a hot, thin gas, each particular chemical element or compound produces its own characteristic pattern of spectral lines—its spectral signature. No two types of atoms or molecules give the same patterns. In other words, each particular gas can absorb or emit only certain wavelengths of the light peculiar to that gas. In contrast, absorption spectra occur when passing white light through a cool, thin gas. The temperature and other conditions determine whether the lines are bright or dark (whether light is absorbed or emitted), but the wavelengths of the lines for any element are the same in either case. It is the precise pattern of wavelengths that makes the signature of each element unique. Liquids and solids can also generate spectral lines or bands, but they are broader and less well defined—and hence, more difficult to interpret. Spectral analysis, however, can be quite useful. It can, for example, be applied to light reflected off the surface of a nearby asteroid as well as to light from a distant galaxy.

The dark lines in the solar spectrum thus give evidence of certain chemical elements between us and the Sun absorbing those wavelengths of sunlight. Because the space between us and the Sun is pretty empty, astronomers realized that the atoms doing the absorbing must be in a thin atmosphere of cooler gas around the Sun. This outer atmosphere is not all that different from the rest of the Sun, just thinner and cooler. Thus, we can use what we learn about its composition as an indicator of what the whole Sun is made of. Similarly, we

This OpenStax book is available for free at http://cnx.org/content/col11992/1.8

can use the presence of absorption and emission lines to analyze the composition of other stars and clouds of gas in space.

Such analysis of spectra is the key to modern astronomy. Only in this way can we "sample" the stars, which are too far away for us to visit. Encoded in the electromagnetic radiation from celestial objects is clear information about the chemical makeup of these objects. Only by understanding what the stars were made of could astronomers begin to form theories about what made them shine and how they evolved.

In 1860, German physicist Gustav Kirchhoff became the first person to use spectroscopy to identify an element in the Sun when he found the spectral signature of sodium gas. In the years that followed, astronomers found many other chemical elements in the Sun and stars. In fact, the element helium was found first in the Sun from its spectrum and only later identified on Earth. (The word "helium" comes from *helios*, the Greek name for the Sun.)

Why are there specific lines for each element? The answer to that question was not found until the twentieth century; it required the development of a model for the atom. We therefore turn next to a closer examination of the atoms that make up all matter.

MAKING CONNECTIONS

The Rainbow

Rainbows are an excellent illustration of the dispersion of sunlight. You have a good chance of seeing a rainbow any time you are between the Sun and a rain shower, as illustrated in Figure 5.13. The raindrops act like little prisms and break white light into the spectrum of colors. Suppose a ray of sunlight encounters a raindrop and passes into it. The light changes direction—is refracted—when it passes from air to water; the blue and violet light are refracted more than the red. Some of the light is then reflected at the backside of the drop and reemerges from the front, where it is again refracted. As a result, the white light is spread out into a rainbow of colors.

(a)   (b)   (c)

Figure 5.13. Rainbow Refraction. (a) This diagram shows how light from the Sun, which is located behind the observer, can be refracted by raindrops to produce (b) a rainbow. (c) Refraction separates white light into its component colors.

Note that violet light lies above the red light after it emerges from the raindrop. When you look at a rainbow, however, the red light is higher in the sky. Why? Look again at Figure 5.13. If the observer looks at a raindrop that is high in the sky, the violet light passes over her head and the red light enters her eye. Similarly, if the observer looks at a raindrop that is low in the sky, the violet light reaches her eye and the drop appears violet, whereas the red light from that same drop strikes the ground and is not seen. Colors of intermediate wavelengths are refracted to the eye by drops that are intermediate in altitude between

Download for free at http://cnx.org/content/col11992/latest/

the drops that appear violet and the ones that appear red. Thus, a single rainbow always has red on the outside and violet on the inside.

5.4 THE STRUCTURE OF THE ATOM

Learning Objectives

By the end of this section, you will be able to:

> Describe the structure of atoms and the components of nuclei
> Explain the behavior of electrons within atoms and how electrons interact with light to move among energy levels

The idea that matter is composed of tiny particles called atoms is at least 25 centuries old. It took until the twentieth century, however, for scientists to invent instruments that permitted them to probe inside an atom and find that it is not, as had been thought, hard and indivisible. Instead, the atom is a complex structure composed of still smaller particles.

Probing the Atom

The first of these smaller particles was discovered by British physicist James (J. J.) Thomson in 1897. Named the *electron*, this particle is negatively charged. (It is the flow of these particles that produces currents of electricity, whether in lightning bolts or in the wires leading to your lamp.) Because an atom in its normal state is electrically neutral, each electron in an atom must be balanced by the same amount of positive charge.

The next step was to determine where in the atom the positive and negative charges are located. In 1911, British physicist Ernest Rutherford devised an experiment that provided part of the answer to this question. He bombarded an extremely thin piece of gold foil, only about 400 atoms thick, with a beam of alpha particles (Figure 5.14). *Alpha particles* (α particles) are helium atoms that have lost their electrons and thus are positively charged. Most of these particles passed though the gold foil just as if it and the atoms in it were nearly empty space. About 1 in 8000 of the alpha particles, however, completely reversed direction and bounced backward from the foil. Rutherford wrote, "It was quite the most incredible event that has ever happened to me in my life. It was almost as incredible as if you fired a 15-inch shell at a piece of tissue paper and it came back and hit you."

This OpenStax book is available for free at http://cnx.org/content/col11992/1.8

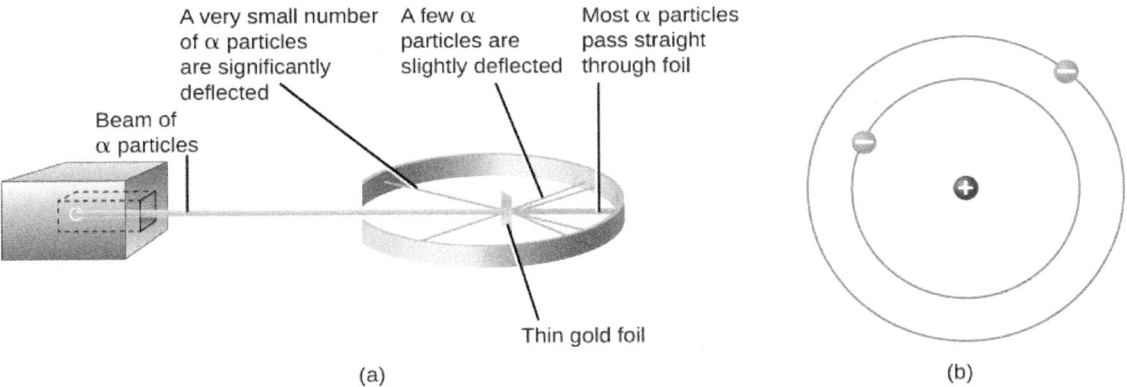

Figure 5.14. Rutherford's Experiment. (a) When Rutherford allowed α particles from a radioactive source to strike a target of gold foil, he found that, although most of them went straight through, some rebounded back in the direction from which they came. (b) From this experiment, he concluded that the atom must be constructed like a miniature solar system, with the positive charge concentrated in the nucleus and the negative charge orbiting in the large volume around the nucleus. Note that this drawing is not to scale; the electron orbits are much larger relative to the size of the nucleus.

The only way to account for the particles that reversed direction when they hit the gold foil was to assume that nearly all of the mass, as well as all of the positive charge in each individual gold atom, is concentrated in a tiny center or **nucleus**. When a positively charged alpha particle strikes a nucleus, it reverses direction, much as a cue ball reverses direction when it strikes another billiard ball. Rutherford's model placed the other type of charge—the negative electrons—in orbit around this nucleus.

Rutherford's model required that the electrons be in motion. Positive and negative charges attract each other, so stationary electrons would fall into the positive nucleus. Also, because both the electrons and the nucleus are extremely small, most of the atom is empty, which is why nearly all of Rutherford's particles were able to pass right through the gold foil without colliding with anything. Rutherford's model was a very successful explanation of the experiments he conducted, although eventually scientists would discover that even the nucleus itself has structure.

The Atomic Nucleus

The simplest possible atom (and the most common one in the Sun and stars) is hydrogen. The nucleus of ordinary hydrogen contains a single proton. Moving around this proton is a single electron. The mass of an electron is nearly 2000 times smaller than the mass of a proton; the electron carries an amount of charge exactly equal to that of the proton but opposite in sign (Figure 5.15). Opposite charges attract each other, so it is an electromagnetic force that holds the proton and electron together, just as gravity is the force that keeps planets in orbit around the Sun.

Download for free at http://cnx.org/content/col11992/latest/

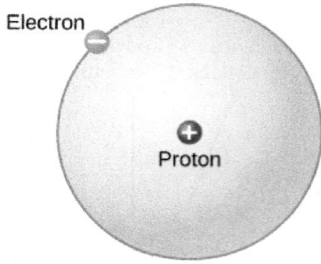

Figure 5.15. Hydrogen Atom. This is a schematic diagram of a hydrogen atom in its lowest energy state, also called the ground state. The proton and electron have equal but opposite charges, which exert an electromagnetic force that binds the hydrogen atom together. In the illustration, the size of the particles is exaggerated so that you can see them; they are not to scale. They are also shown much closer than they would actually be as it would take more than an entire page to show their actual distance to scale.

There are many other types of atoms in nature. Helium, for example, is the second-most abundant element in the Sun. Helium has two protons in its nucleus instead of the single proton that characterizes hydrogen. In addition, the helium nucleus contains two neutrons, particles with a mass comparable to that of the proton but with no electric charge. Moving around this nucleus are two electrons, so the total net charge of the helium atom is also zero (Figure 5.16).

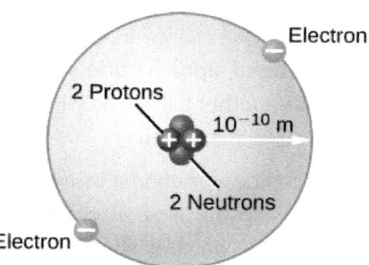

Figure 5.16. Helium Atom. Here we see a schematic diagram of a helium atom in its lowest energy state. Two protons are present in the nucleus of all helium atoms. In the most common variety of helium, the nucleus also contains two neutrons, which have nearly the same mass as the proton but carry no charge. Two electrons orbit the nucleus.

From this description of hydrogen and helium, perhaps you have guessed the pattern for building up all the elements (different types of atoms) that we find in the universe. The type of element is determined by the number of protons in the nucleus of the atom. For example, any atom with six protons is the element carbon, with eight protons is oxygen, with 26 is iron, and with 92 is uranium. On Earth, a typical atom has the same number of electrons as protons, and these electrons follow complex orbital patterns around the nucleus. Deep inside stars, however, it is so hot that the electrons get loose from the nucleus and (as we shall see) lead separate yet productive lives.

The ratio of neutrons to protons increases as the number of protons increases, but each element is unique. The number of neutrons is not necessarily the same for all atoms of a given element. For example, most hydrogen atoms contain no neutrons at all. There are, however, hydrogen atoms that contain one proton and one neutron, and others that contain one proton and two neutrons. The various types of hydrogen nuclei with different numbers of neutrons are called **isotopes** of hydrogen (Figure 5.17), and all other elements have isotopes as well. You can think of isotopes as siblings in the same element "family"—closely related but with different characteristics and behaviors.

This OpenStax book is available for free at http://cnx.org/content/col11992/1.8

Download for free at http://cnx.org/content/col11992/latest/

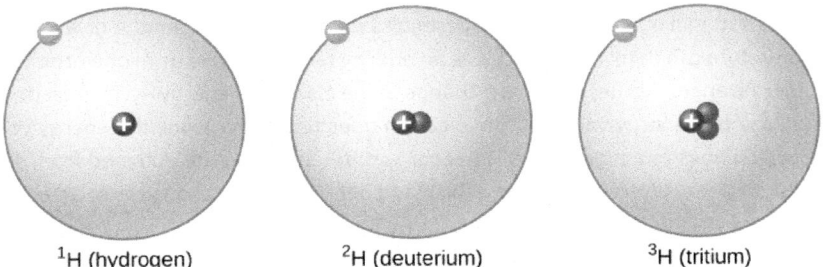

^{1}H (hydrogen) ^{2}H (deuterium) ^{3}H (tritium)

Figure 5.17. Isotopes of Hydrogen. A single proton in the nucleus defines the atom to be hydrogen, but there may be zero, one, or two neutrons. The most common isotope of hydrogen is the one with only a single proton and no neutrons.

LINK TO LEARNING

To explore the structure of atoms, go to the PhET Build and Atom website (https://openstax.org/l/ 30atombld) where you can add protons, neutrons, or electrons to a model and the name of the element you have created will appear. You can also see the net charge, the mass number, whether it is stable or unstable, and whether it is an ion or a neutral atom.

The Bohr Atom

Rutherford's model for atoms has one serious problem. Maxwell's theory of electromagnetic radiation says that when electrons change either speed or the direction of motion, they must emit energy. Orbiting electrons constantly change their direction of motion, so they should emit a constant stream of energy. Applying Maxwell's theory to Rutherford's model, all electrons should spiral into the nucleus of the atom as they lose energy, and this collapse should happen very quickly—in about 10^{-16} seconds.

It was Danish physicist Niels Bohr (1885–1962) who solved the mystery of how electrons remain in orbit. He was trying to develop a model of the atom that would also explain certain regularities observed in the spectrum of hydrogen. He suggested that the spectrum of hydrogen can be understood if we assume that orbits of only certain sizes are possible for the electron. Bohr further assumed that as long as the electron moves in only one of these allowed orbits, it radiates no energy: its energy would change only if it moved from one orbit to another.

This suggestion, in the words of science historian Abraham Pais, was "one of the most audacious hypotheses ever introduced in physics." If something equivalent were at work in the everyday world, you might find that, as you went for a walk after astronomy class, nature permitted you to walk two steps per minute, five steps per minute, and 12 steps per minute, but no speeds in between. No matter how you tried to move your legs, only certain walking speeds would be permitted. To make things more bizarre, it would take no effort to walk at any one of the allowed speeds, but it would be difficult to change from one speed to another. Luckily, no such rules apply at the level of human behavior. But at the microscopic level of the atom, experiment after experiment has confirmed the validity of Bohr's strange idea. Bohr's suggestions became one of the foundations of the new (and much more sophisticated) model of the subatomic world called quantum mechanics.

In Bohr's model, if the electron moves from one orbit to another closer to the atomic nucleus, it must give up some energy in the form of electromagnetic radiation. If the electron goes from an inner orbit to one farther from the nucleus, however, it requires some additional energy. One way to obtain the necessary energy is to absorb electromagnetic radiation that may be streaming past the atom from an outside source.

Download for free at http://cnx.org/content/col11992/latest/

A key feature of Bohr's model is that each of the permitted electron orbits around a given atom has a certain energy value; we therefore can think of each orbit as an **energy level**. To move from one orbit to another (which will have its own specific energy value) requires a change in the electron's energy—a change determined by the difference between the two energy values. If the electron goes to a lower level, the energy difference will be given off; if the electron goes to a higher level, the energy difference must be obtained from somewhere else. Each jump (or transition) to a different level has a fixed and definite energy change associated with it.

A crude analogy for this situation might be life in a tower of luxury apartments where the rent is determined by the quality of the view. Such a building has certain, definite numbered levels or floors on which apartments are located. No one can live on floor 5.37 or 22.5. In addition, the rent gets higher as you go up to higher floors. If you want to exchange an apartment on the twentieth floor for one on the second floor, you will not owe as much rent. However, if you want to move from the third floor to the twenty-fifth floor, your rent will increase. In an atom, too, the "cheapest" place for an electron to live is the lowest possible level, and energy is required to move to a higher level.

Here we have one of the situations where it is easier to think of electromagnetic radiation as particles (photons) rather than as waves. As electrons move from one level to another, they give off or absorb little packets of energy. When an electron moves to a higher level, it absorbs a photon of just the right energy (provided one is available). When it moves to a lower level, it emits a photon with the exact amount of energy it no longer needs in its "lower-cost living situation."

The photon and wave perspectives must be equivalent: light is light, no matter how we look at it. Thus, each photon carries a certain amount of energy that is proportional to the frequency (f) of the wave it represents. The value of its energy (E) is given by the formula

$$E = hf$$

where the constant of proportionality, h, is called Planck's constant.

The constant is named for Max Planck, the German physicist who was one of the originators of the quantum theory (Figure 5.18). If metric units are used (that is, if energy is measured in joules and frequency in hertz), then Planck's constant has the value $h = 6.626 \times 10^{-34}$ joule-seconds (J-s). Higher-energy photons correspond to higher-frequency waves (which have a shorter wavelength); lower-energy photons are waves of lower frequency.

This OpenStax book is available for free at http://cnx.org/content/col11992/1.8

Download for free at http://cnx.org/content/col11992/latest/

(a)   (b)

Figure 5.18. Niels Bohr (1885–1962) and Max Planck (1858–1947). (a) Bohr, shown at his desk in this 1935 photograph, and (b) Planck helped us understand the energy behavior of photons.

To take a specific example, consider a calcium atom inside the Sun's atmosphere in which an electron jumps from a lower level to a higher level. To do this, it needs about 5×10^{-19} joules of energy, which it can conveniently obtain by absorbing a passing photon of that energy coming from deeper inside the Sun. This photon is equivalent to a wave of light whose frequency is about 7.5×10^{14} hertz and whose wavelength is about 3.9×10^{-7} meters (393 nanometers), in the deep violet part of the visible light spectrum. Although it may seem strange at first to switch from picturing light as a photon (or energy packet) to picturing it as a wave, such switching has become second nature to astronomers and can be a handy tool for doing calculations about spectra.

EXAMPLE 5.5

The Energy of a Photon

Now that we know how to calculate the wavelength and frequency of a photon, we can use this information, along with Planck's constant, to determine how much energy each photon carries. How much energy does a red photon of wavelength 630 nm have?

Solution

First, as we learned earlier, we can find the frequency of the photon:

$$f = \frac{c}{\lambda} = \frac{3 \times 10^8 \text{ m/s}}{630 \times 10^{-9} \text{ m}} = 4.8 \times 10^{14} \text{ Hz}$$

Next, we can use Planck's constant to determine the energy (remember that a Hz is the same as 1/s):

$$E = hf = \left(6.626 \times 10^{-34} \text{ J-s}\right)\left(4.8 \times 10^{14} \text{ (1/s)}\right) = 3.2 \times 10^{-19} \text{ J}$$

Check Your Learning

What is the energy of a yellow photon with a frequency of 5.5×10^{14} Hz?

Download for free at http://cnx.org/content/col11992/latest/

Answer:

$$E = hf = \left(6.626 \times 10^{-34}\right)\left(5.5 \times 10^{14}\right) = 3.6 \times 10^{-19} \text{ J}$$

5.5 FORMATION OF SPECTRAL LINES

Learning Objectives

By the end of this section, you will be able to:

> Explain how emission line spectra and absorption line spectra are formed
> Describe what ions are and how they are formed
> Explain how spectral lines and ionization levels in a gas can help us determine its temperature

We can use Bohr's model of the atom to understand how spectral lines are formed. The concept of energy levels for the electron orbits in an atom leads naturally to an explanation of why atoms absorb or emit only specific energies or wavelengths of light.

The Hydrogen Spectrum

Let's look at the hydrogen atom from the perspective of the Bohr model. Suppose a beam of white light (which consists of photons of all visible wavelengths) shines through a gas of atomic hydrogen. A photon of wavelength 656 nanometers has just the right energy to raise an electron in a hydrogen atom from the second to the third orbit. Thus, as all the photons of different energies (or wavelengths or colors) stream by the hydrogen atoms, photons with *this* particular wavelength can be absorbed by those atoms whose electrons are orbiting on the second level. When they are absorbed, the electrons on the second level will move to the third level, and a number of the photons of this wavelength and energy will be missing from the general stream of white light.

Other photons will have the right energies to raise electrons from the second to the fourth orbit, or from the first to the fifth orbit, and so on. Only photons with these exact energies can be absorbed. All of the other photons will stream past the atoms untouched. Thus, hydrogen atoms absorb light at only certain wavelengths and produce dark lines at those wavelengths in the spectrum we see.

Suppose we have a container of hydrogen gas through which a whole series of photons is passing, allowing many electrons to move up to higher levels. When we turn off the light source, these electrons "fall" back down from larger to smaller orbits and emit photons of light—but, again, only light of those energies or wavelengths that correspond to the energy difference between permissible orbits. The orbital changes of hydrogen electrons that give rise to some spectral lines are shown in Figure 5.19.

This OpenStax book is available for free at http://cnx.org/content/col11992/1.8

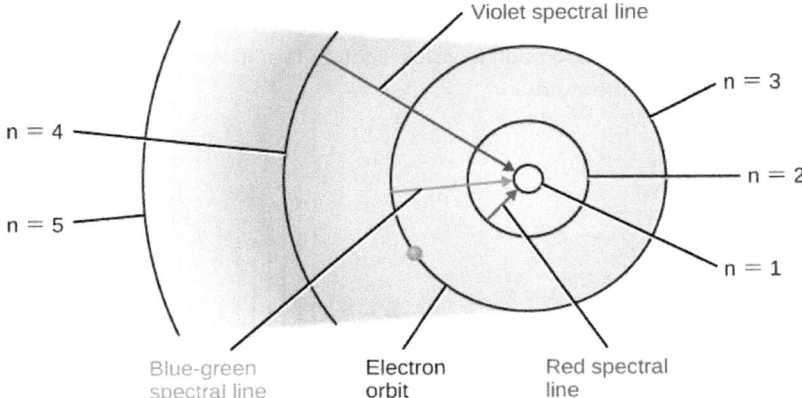

Figure 5.19. Bohr Model for Hydrogen. In this simplified model of a hydrogen atom, the concentric circles shown represent permitted orbits or energy levels. An electron in a hydrogen atom can only exist in one of these energy levels (or states). The closer the electron is to the nucleus, the more tightly bound the electron is to the nucleus. By absorbing energy, the electron can move to energy levels farther from the nucleus (and even escape if enough energy is absorbed).

Similar pictures can be drawn for atoms other than hydrogen. However, because these other atoms ordinarily have more than one electron each, the orbits of their electrons are much more complicated, and the spectra are more complex as well. For our purposes, the key conclusion is this: *each type of atom has its own unique pattern of electron orbits, and no two sets of orbits are exactly alike.* This means that each type of atom shows its own unique set of spectral lines, produced by electrons moving between its unique set of orbits.

Astronomers and physicists have worked hard to learn the lines that go with each element by studying the way atoms absorb and emit light in laboratories here on Earth. Then they can use this knowledge to identify the elements in celestial bodies. In this way, we now know the chemical makeup of not just any star, but even galaxies of stars so distant that their light started on its way to us long before Earth had even formed.

Energy Levels and Excitation

Bohr's model of the hydrogen atom was a great step forward in our understanding of the atom. However, we know today that atoms cannot be represented by quite so simple a picture. For example, the concept of sharply defined electron orbits is not really correct; however, at the level of this introductory course, the notion that only certain discrete energies are allowable for an atom is very useful. The energy levels we have been discussing can be thought of as representing certain average distances of the electron's possible orbits from the atomic nucleus.

Ordinarily, an atom is in the state of lowest possible energy, its **ground state**. In the Bohr model of the hydrogen atom, the ground state corresponds to the electron being in the innermost orbit. An atom can absorb energy, which raises it to a higher energy level (corresponding, in the simple Bohr picture, to an electron's movement to a larger orbit)—this is referred to as **excitation**. The atom is then said to be in an *excited state*. Generally, an atom remains excited for only a very brief time. After a short interval, typically a hundred-millionth of a second or so, it drops back spontaneously to its ground state, with the simultaneous emission of light. The atom may return to its lowest state in one jump, or it may make the transition in steps of two or more jumps, stopping at intermediate levels on the way down. With each jump, it emits a photon of the wavelength that corresponds to the energy difference between the levels at the beginning and end of that jump.

An energy-level diagram for a hydrogen atom and several possible atomic transitions are shown in Figure 5.20. When we measure the energies involved as the atom jumps between levels, we find that the transitions to or from the ground state, called the *Lyman series* of lines, result in the emission or absorption of ultraviolet

Download for free at http://cnx.org/content/col11992/latest/

photons. But the transitions to or from the first excited state (labeled n = 2 in part (a) of Figure 5.20), called the Balmer series, produce emission or absorption in visible light. In fact, it was to explain this Balmer series that Bohr first suggested his model of the atom.

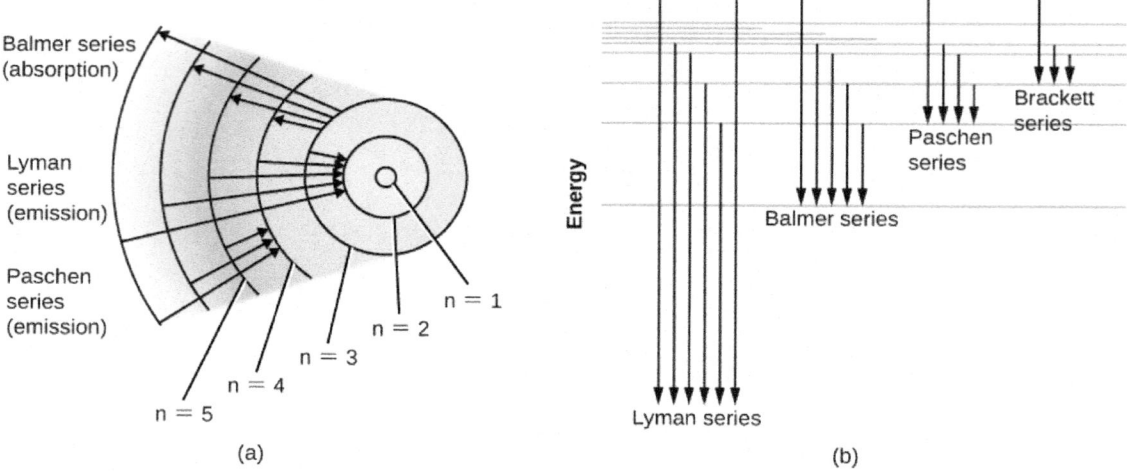

Figure 5.20. Energy-Level Diagrams for Hydrogen. (a) Here we follow the emission or absorption of photons by a hydrogen atom according to the Bohr model. Several different series of spectral lines are shown, corresponding to transitions of electrons from or to certain allowed orbits. Each series of lines that terminates on a specific inner orbit is named for the physicist who studied it. At the top, for example, you see the Balmer series, and arrows show electrons jumping from the second orbit (n = 2) to the third, fourth, fifth, and sixth orbits. Each time a "poor" electron from a lower level wants to rise to a higher position in life, it must absorb energy to do so. It can absorb the energy it needs from passing waves (or photons) of light. The next set of arrows (Lyman series) show electrons falling down to the first orbit from different (higher) levels. Each time a "rich" electron goes downward toward the nucleus, it can afford to give off (emit) some energy it no longer needs. (b) At higher and higher energy levels, the levels become more and more crowded together, approaching a limit. The region above the top line represents energies at which the atom is ionized (the electron is no longer attached to the atom). Each series of arrows represents electrons falling from higher levels to lower ones, releasing photons or waves of energy in the process.

Atoms that have absorbed specific photons from a passing beam of white light and have thus become excited generally de-excite themselves and emit that light again in a very short time. You might wonder, then, why *dark* spectral lines are ever produced. In other words, why doesn't this reemitted light quickly "fill in" the darker absorption lines?

Imagine a beam of white light coming toward you through some cooler gas. Some of the reemitted light *is* actually returned to the beam of white light you see, but this fills in the absorption lines only to a slight extent. The reason is that the atoms in the gas reemit light *in all directions*, and only a small fraction of the reemitted light is in the direction of the original beam (toward you). In a star, much of the reemitted light actually goes in directions leading back into the star, which does observers outside the star no good whatsoever.

Figure 5.21 summarizes the different kinds of spectra we have discussed. An incandescent lightbulb produces a continuous spectrum. When that continuous spectrum is viewed through a thinner cloud of gas, an absorption line spectrum can be seen superimposed on the continuous spectrum. If we look only at a cloud of excited gas atoms (with no continuous source seen behind it), we see that the excited atoms give off an emission line spectrum.

This OpenStax book is available for free at http://cnx.org/content/col11992/1.8

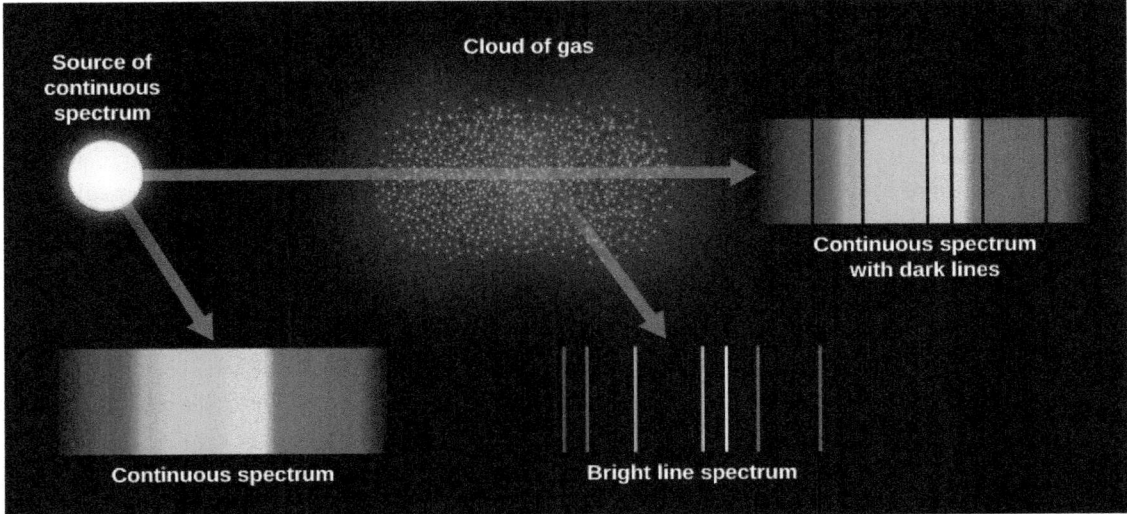

Figure 5.21. Three Kinds of Spectra. When we see a lightbulb or other source of continuous radiation, all the colors are present. When the continuous spectrum is seen through a thinner gas cloud, the cloud's atoms produce absorption lines in the continuous spectrum. When the excited cloud is seen without the continuous source behind it, its atoms produce emission lines. We can learn which types of atoms are in the gas cloud from the pattern of absorption or emission lines.

Atoms in a hot gas are moving at high speeds and continually colliding with one another and with any loose electrons. They can be excited (electrons moving to a higher level) and de-excited (electrons moving to a lower level) by these collisions as well as by absorbing and emitting light. The speed of atoms in a gas depends on the temperature. When the temperature is higher, so are the speed and energy of the collisions. The hotter the gas, therefore, the more likely that electrons will occupy the outermost orbits, which correspond to the highest energy levels. This means that the level where electrons *start* their upward jumps in a gas can serve as an indicator of how hot that gas is. In this way, the absorption lines in a spectrum give astronomers information about the temperature of the regions where the lines originate.

LINK TO LEARNING

Use this simulation (https://openstax.org/l/30Hatom) to play with a hydrogen atom and see what happens when electrons move to higher levels and then give off photons as they go to a lower level.

Ionization

We have described how certain discrete amounts of energy can be absorbed by an atom, raising it to an excited state and moving one of its electrons farther from its nucleus. If enough energy is absorbed, the electron can be completely removed from the atom—this is called **ionization**. The atom is then said to be ionized. The minimum amount of energy required to remove one electron from an atom in its ground state is called its ionization energy.

Still-greater amounts of energy must be absorbed by the now-ionized atom (called an **ion**) to remove an additional electron deeper in the structure of the atom. Successively greater energies are needed to remove the third, fourth, fifth—and so on—electrons from the atom. If enough energy is available, an atom can become completely ionized, losing all of its electrons. A hydrogen atom, having only one electron to lose, can be ionized

Download for free at http://cnx.org/content/col11992/latest/

only once; a helium atom can be ionized twice; and an oxygen atom up to eight times. When we examine regions of the cosmos where there is a great deal of energetic radiation, such as the neighborhoods where hot young stars have recently formed, we see a lot of ionization going on.

An atom that has become positively ionized has lost a negative charge—the missing electron—and thus is left with a net positive charge. It therefore exerts a strong attraction on any free electron. Eventually, one or more electrons will be captured and the atom will become neutral (or ionized to one less degree) again. During the electron-capture process, the atom emits one or more photons. Which photons are emitted depends on whether the electron is captured at once to the lowest energy level of the atom or stops at one or more intermediate levels on its way to the lowest available level.

Just as the excitation of an atom can result from a collision with another atom, ion, or electron (collisions with electrons are usually most important), so can ionization. The rate at which such collisional ionizations occur depends on the speeds of the atoms and hence on the temperature of the gas—the hotter the gas, the more of its atoms will be ionized.

The rate at which ions and electrons recombine also depends on their relative speeds—that is, on the temperature. In addition, it depends on the density of the gas: the higher the density, the greater the chance for recapture, because the different kinds of particles are crowded more closely together. From a knowledge of the temperature and density of a gas, it is possible to calculate the fraction of atoms that have been ionized once, twice, and so on. In the Sun, for example, we find that most of the hydrogen and helium atoms in its atmosphere are neutral, whereas most of the calcium atoms, as well as many other heavier atoms, are ionized once.

The energy levels of an ionized atom are entirely different from those of the same atom when it is neutral. Each time an electron is removed from the atom, the energy levels of the ion, and thus the wavelengths of the spectral lines it can produce, change. This helps astronomers differentiate the ions of a given element. Ionized hydrogen, having no electron, can produce no absorption lines.

5.6 THE DOPPLER EFFECT

Learning Objectives

By the end of this section, you will be able to:

> Explain why the spectral lines of photons we observe from an object will change as a result of the object's motion toward or away from us
> Describe how we can use the Doppler effect to deduce how astronomical objects are moving through space

The last two sections introduced you to many new concepts, and we hope that through those, you have seen one major idea emerge. Astronomers can learn about the elements in stars and galaxies by decoding the information in their spectral lines. There is a complicating factor in learning how to decode the message of starlight, however. If a star is moving toward or away from us, its lines will be in a slightly different place in the spectrum from where they would be in a star at rest. And most objects in the universe do have some motion relative to the Sun.

Motion Affects Waves

In 1842, Christian Doppler first measured the effect of motion on waves by hiring a group of musicians to play on an open railroad car as it was moving along the track. He then applied what he learned to all waves, including

This OpenStax book is available for free at http://cnx.org/content/col11992/1.8

light, and pointed out that if a light source is approaching or receding from the observer, the light waves will be, respectively, crowded more closely together or spread out. The general principle, now known as the **Doppler effect**, is illustrated in Figure 5.22.

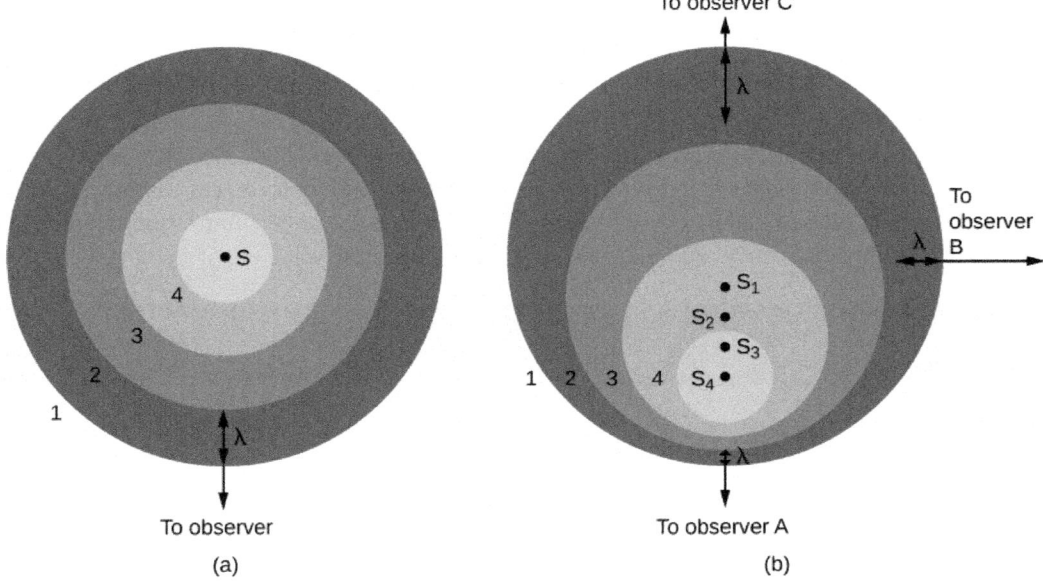

Figure 5.22. Doppler Effect. (a) A source, S, makes waves whose numbered crests (1, 2, 3, and 4) wash over a stationary observer. (b) The source S now moves toward observer A and away from observer C. Wave crest 1 was emitted when the source was at position S4, crest 2 at position S2, and so forth. Observer A sees waves compressed by this motion and sees a blueshift (if the waves are light). Observer C sees the waves stretched out by the motion and sees a redshift. Observer B, whose line of sight is perpendicular to the source's motion, sees no change in the waves (and feels left out).

In part (a) of the figure, the light source (S) is at rest with respect to the observer. The source gives off a series of waves, whose crests we have labeled 1, 2, 3, and 4. The light waves spread out evenly in all directions, like the ripples from a splash in a pond. The crests are separated by a distance, λ, where λ is the wavelength. The observer, who happens to be located in the direction of the bottom of the image, sees the light waves coming nice and evenly, one wavelength apart. Observers located anywhere else would see the same thing.

On the other hand, if the source of light is moving with respect to the observer, as seen in part (b), the situation is more complicated. Between the time one crest is emitted and the next one is ready to come out, the source has moved a bit, toward the bottom of the page. From the point of view of observer A, this motion of the source has decreased the distance between crests—it's squeezing the crests together, this observer might say.

In part (b), we show the situation from the perspective of three observers. The source is seen in four positions, S_1, S_2, S_3, and S_4, each corresponding to the emission of one wave crest. To observer A, the waves seem to follow one another more closely, at a decreased wavelength and thus increased frequency. (Remember, all light waves travel at the speed of light through empty space, no matter what. This means that motion cannot affect the speed, but only the wavelength and the frequency. As the wavelength decreases, the frequency must increase. If the waves are shorter, more will be able to move by during each second.)

The situation is not the same for other observers. Let's look at the situation from the point of view of observer C, located opposite observer A in the figure. For her, the source is moving away from her location. As a result, the waves are not squeezed together but instead are spread out by the motion of the source. The crests arrive with an increased wavelength and decreased frequency. To observer B, in a direction at right angles to the motion

Download for free at http://cnx.org/content/col11992/latest/

of the source, no effect is observed. The wavelength and frequency remain the same as they were in part (a) of the figure.

We can see from this illustration that the Doppler effect is produced only by a motion toward or away from the observer, a motion called **radial velocity**. Sideways motion does not produce such an effect. Observers between *A* and *B* would observe some shortening of the light waves for that part of the motion of the source that is along their line of sight. Observers between *B* and *C* would observe lengthening of the light waves that are along their line of sight.

You may have heard the Doppler effect with sound waves. When a train whistle or police siren approaches you and then moves away, you will notice a decrease in the pitch (which is how human senses interpret sound wave frequency) of the sound waves. Compared to the waves at rest, they have changed from slightly more frequent when coming toward you, to slightly less frequent when moving away from you.

LINK TO LEARNING

A nice example of this change in the sound of a train whistle can be heard at the end of the classic Beach Boys song "Caroline, No" on their album *Pet Sounds*. To hear this sound, go to this YouTube (https://openstax.org/l/30BBtrain) version of the song. The sound of the train begins at approximately 2:20.

Color Shifts

When the source of waves moves toward you, the wavelength decreases a bit. If the waves involved are visible light, then the colors of the light change slightly. As wavelength decreases, they shift toward the blue end of the spectrum: astronomers call this a *blueshift* (since the end of the spectrum is really violet, the term should probably be *violetshift*, but blue is a more common color). When the source moves away from you and the wavelength gets longer, we call the change in colors a *redshift*. Because the Doppler effect was first used with visible light in astronomy, the terms " blueshift" and " redshift" became well established. Today, astronomers use these words to describe changes in the wavelengths of radio waves or X-rays as comfortably as they use them to describe changes in visible light.

The greater the motion toward or away from us, the greater the Doppler shift. If the relative motion is entirely along the line of sight, the formula for the Doppler shift of light is

$$\frac{\Delta\lambda}{\lambda} = \frac{v}{c}$$

where λ is the wavelength emitted by the source, $\Delta\lambda$ is the difference between λ and the wavelength measured by the observer, c is the speed of light, and v is the relative speed of the observer and the source in the line of sight. The variable v is counted as positive if the velocity is one of recession, and negative if it is one of approach. Solving this equation for the velocity, we find $v = c \times \Delta\lambda/\lambda$.

If a star approaches or recedes from us, the wavelengths of light in its continuous spectrum appear shortened or lengthened, respectively, as do those of the dark lines. However, unless its speed is tens of thousands of kilometers per second, the star does not appear noticeably bluer or redder than normal. The Doppler shift is thus not easily detected in a continuous spectrum and cannot be measured accurately in such a spectrum. The wavelengths of the absorption lines can be measured accurately, however, and their Doppler shift is relatively simple to detect.

This OpenStax book is available for free at http://cnx.org/content/col11992/1.8

EXAMPLE 5.6

The Doppler Effect

We can use the Doppler effect equation to calculate the radial velocity of an object if we know three things: the speed of light, the original (unshifted) wavelength of the light emitted, and the difference between the wavelength of the emitted light and the wavelength we observe. For particular absorption or emission lines, we usually know exactly what wavelength the line has in our laboratories on Earth, where the source of light is not moving. We can measure the new wavelength with our instruments at the telescope, and so we know the difference in wavelength due to Doppler shifting. Since the speed of light is a universal constant, we can then calculate the radial velocity of the star.

A particular emission line of hydrogen is originally emitted with a wavelength of 656.3 nm from a gas cloud. At our telescope, we observe the wavelength of the emission line to be 656.6 nm. How fast is this gas cloud moving toward or away from Earth?

Solution

Because the light is shifted to a longer wavelength (redshifted), we know this gas cloud is moving away from us. The speed can be calculated using the Doppler shift formula:

$$v = c \times \frac{\Delta\lambda}{\lambda} = \left(3.0 \times 10^8 \text{ m/s}\right)\left(\frac{0.3 \text{ nm}}{656.3 \text{ nm}}\right) = \left(3.0 \times 10^8 \text{ m/s}\right)\left(\frac{0.3 \times 10^{-9} \text{ m}}{656.3 \times 10^{-9} \text{ m}}\right)$$
$$= 140{,}000 \text{ m/s} = 140 \text{ km/s}$$

Check Your Learning

Suppose a spectral line of hydrogen, normally at 500 nm, is observed in the spectrum of a star to be at 500.1 nm. How fast is the star moving toward or away from Earth?

Answer:

Because the light is shifted to a longer wavelength, the star is moving away from us:

$$v = c \times \frac{\Delta\lambda}{\lambda} = \left(3.0 \times 10^8 \text{ m/s}\right)\left(\frac{0.1 \text{ nm}}{500 \text{ nm}}\right) = \left(3.0 \times 10^8 \text{ m/s}\right)\left(\frac{0.1 \times 10^{-9} \text{ m}}{500 \times 10^{-9} \text{ m}}\right) = 60{,}000 \text{ m/s}. \text{ Its speed is}$$

60,000 m/s.

You may now be asking: if all the stars are moving and motion changes the wavelength of each spectral line, won't this be a disaster for astronomers trying to figure out what elements are present in the stars? After all, it is the precise wavelength (or color) that tells astronomers which lines belong to which element. And we first measure these wavelengths in containers of gas in our laboratories, which are not moving. If every line in a star's spectrum is now shifted by its motion to a different wavelength (color), how can we be sure which lines and which elements we are looking at in a star whose speed we do not know?

Take heart. This situation sounds worse than it really is. Astronomers rarely judge the presence of an element in an astronomical object by a single line. It is the *pattern* of lines unique to hydrogen or calcium that enables us to determine that those elements are part of the star or galaxy we are observing. The Doppler effect does not change the pattern of lines from a given element—it only shifts the whole pattern slightly toward redder

Download for free at http://cnx.org/content/col11992/latest/

or bluer wavelengths. The shifted pattern is still quite easy to recognize. Best of all, when we do recognize a familiar element's pattern, we get a bonus: the amount the pattern is shifted can enable us to determine the speed of the objects in our line of sight.

The training of astronomers includes much work on learning to decode light (and other electromagnetic radiation). A skillful "decoder" can learn the temperature of a star, what elements are in it, and even its speed in a direction toward us or away from us. That's really an impressive amount of information for stars that are light-years away.

This OpenStax book is available for free at http://cnx.org/content/col11992/1.8

CHAPTER 5 REVIEW

KEY TERMS

absorption spectrum a series or pattern of dark lines superimposed on a continuous spectrum

blackbody an idealized object that absorbs all electromagnetic energy that falls onto it

continuous spectrum a spectrum of light composed of radiation of a continuous range of wavelengths or colors, rather than only certain discrete wavelengths

dispersion separation of different wavelengths of white light through refraction of different amounts

Doppler effect the apparent change in wavelength or frequency of the radiation from a source due to its relative motion away from or toward the observer

electromagnetic radiation radiation consisting of waves propagated through regularly varying electric and magnetic fields and traveling at the speed of light

electromagnetic spectrum the whole array or family of electromagnetic waves, from radio to gamma rays

emission spectrum a series or pattern of bright lines superimposed on a continuous spectrum

energy flux the amount of energy passing through a unit area (for example, 1 square meter) per second; the units of flux are watts per square meter

energy level a particular level, or amount, of energy possessed by an atom or ion above the energy it possesses in its least energetic state; also used to refer to the states of energy an electron can have in an atom

excitation the process of giving an atom or an ion an amount of energy greater than it has in its lowest energy (ground) state

frequency the number of waves that cross a given point per unit time (in radiation)

gamma rays photons (of electromagnetic radiation) of energy with wavelengths no longer than 0.01 nanometer; the most energetic form of electromagnetic radiation

ground state the lowest energy state of an atom

infrared electromagnetic radiation of wavelength 10^3–10^6 nanometers; longer than the longest (red) wavelengths that can be perceived by the eye, but shorter than radio wavelengths

inverse square law (for light) the amount of energy (light) flowing through a given area in a given time decreases in proportion to the square of the distance from the source of energy or light

ion an atom that has become electrically charged by the addition or loss of one or more electrons

ionization the process by which an atom gains or loses electrons

isotope any of two or more forms of the same element whose atoms have the same number of protons but different numbers of neutrons

microwave electromagnetic radiation of wavelengths from 1 millimeter to 1 meter; longer than infrared but shorter than radio waves

Download for free at http://cnx.org/content/col11992/latest/

nucleus (of an atom) the massive part of an atom, composed mostly of protons and neutrons, and about which the electrons revolve

photon a discrete unit (or "packet") of electromagnetic energy

radial velocity motion toward or away from the observer; the component of relative velocity that lies in the line of sight

radio waves all electromagnetic waves longer than microwaves, including radar waves and AM radio waves

spectrometer an instrument for obtaining a spectrum; in astronomy, usually attached to a telescope to record the spectrum of a star, galaxy, or other astronomical object

Stefan-Boltzmann law a formula from which the rate at which a blackbody radiates energy can be computed; the total rate of energy emission from a unit area of a blackbody is proportional to the fourth power of its absolute temperature: $F = \sigma T^4$

ultraviolet electromagnetic radiation of wavelengths 10 to 400 nanometers; shorter than the shortest visible wavelengths

visible light electromagnetic radiation with wavelengths of roughly 400–700 nanometers; visible to the human eye

wavelength the distance from crest to crest or trough to trough in a wave

Wien's law formula that relates the temperature of a blackbody to the wavelength at which it emits the greatest intensity of radiation

X-rays electromagnetic radiation with wavelengths between 0.01 nanometer and 20 nanometers; intermediate between those of ultraviolet radiation and gamma rays

 # SUMMARY

5.1 The Behavior of Light

James Clerk Maxwell showed that whenever charged particles change their motion, as they do in every atom and molecule, they give off waves of energy. Light is one form of this electromagnetic radiation. The wavelength of light determines the color of visible radiation. Wavelength (λ) is related to frequency (f) and the speed of light (c) by the equation $c = \lambda f$. Electromagnetic radiation sometimes behaves like waves, but at other times, it behaves as if it were a particle—a little packet of energy, called a photon. The apparent brightness of a source of electromagnetic energy decreases with increasing distance from that source in proportion to the square of the distance—a relationship known as the inverse square law.

5.2 The Electromagnetic Spectrum

The electromagnetic spectrum consists of gamma rays, X-rays, ultraviolet radiation, visible light, infrared, and radio radiation. Many of these wavelengths cannot penetrate the layers of Earth's atmosphere and must be observed from space, whereas others—such as visible light, FM radio and TV—can penetrate to Earth's surface. The emission of electromagnetic radiation is intimately connected to the temperature of the source. The higher the temperature of an idealized emitter of electromagnetic radiation, the shorter is the wavelength at which the maximum amount of radiation is emitted. The mathematical equation describing this relationship is known as Wien's law: $\lambda_{max} = (3 \times 10^6)/T$. The total power emitted per square meter increases with increasing temperature. The relationship between emitted energy flux and temperature is known as the Stefan-Boltzmann law: $F = \sigma T^4$.

This OpenStax book is available for free at http://cnx.org/content/col11992/1.8

5.3 Spectroscopy in Astronomy

A spectrometer is a device that forms a spectrum, often utilizing the phenomenon of dispersion. The light from an astronomical source can consist of a continuous spectrum, an emission (bright line) spectrum, or an absorption (dark line) spectrum. Because each element leaves its spectral signature in the pattern of lines we observe, spectral analyses reveal the composition of the Sun and stars.

5.4 The Structure of the Atom

Atoms consist of a nucleus containing one or more positively charged protons. All atoms except hydrogen can also contain one or more neutrons in the nucleus. Negatively charged electrons orbit the nucleus. The number of protons defines an element (hydrogen has one proton, helium has two, and so on) of the atom. Nuclei with the same number of protons but different numbers of neutrons are different isotopes of the same element. In the Bohr model of the atom, electrons on permitted orbits (or energy levels) don't give off any electromagnetic radiation. But when electrons go from lower levels to higher ones, they must absorb a photon of just the right energy, and when they go from higher levels to lower ones, they give off a photon of just the right energy. The energy of a photon is connected to the frequency of the electromagnetic wave it represents by Planck's formula, $E = hf$.

5.5 Formation of Spectral Lines

When electrons move from a higher energy level to a lower one, photons are emitted, and an emission line can be seen in the spectrum. Absorption lines are seen when electrons absorb photons and move to higher energy levels. Since each atom has its own characteristic set of energy levels, each is associated with a unique pattern of spectral lines. This allows astronomers to determine what elements are present in the stars and in the clouds of gas and dust among the stars. An atom in its lowest energy level is in the ground state. If an electron is in an orbit other than the least energetic one possible, the atom is said to be excited. If an atom has lost one or more electrons, it is called an ion and is said to be ionized. The spectra of different ions look different and can tell astronomers about the temperatures of the sources they are observing.

5.6 The Doppler Effect

If an atom is moving toward us when an electron changes orbits and produces a spectral line, we see that line shifted slightly toward the blue of its normal wavelength in a spectrum. If the atom is moving away, we see the line shifted toward the red. This shift is known as the Doppler effect and can be used to measure the radial velocities of distant objects.

 FOR FURTHER EXPLORATION

Articles

Augensen, H. & Woodbury, J. "The Electromagnetic Spectrum." *Astronomy* (June 1982): 6.

Darling, D. "Spectral Visions: The Long Wavelengths." *Astronomy* (August 1984): 16; "The Short Wavelengths." *Astronomy* (September 1984): 14.

Gingerich, O. "Unlocking the Chemical Secrets of the Cosmos." *Sky & Telescope* (July 1981): 13.

Stencil, R. et al. "Astronomical Spectroscopy." *Astronomy* (June 1978): 6.

Download for free at http://cnx.org/content/col11992/latest/

Websites

Doppler Effect: http://www.physicsclassroom.com/class/waves/Lesson-3/The-Doppler-Effect. A shaking bug and the Doppler Effect explained.

Electromagnetic Spectrum: http://imagine.gsfc.nasa.gov/science/toolbox/emspectrum1.html. An introduction to the electromagnetic spectrum from NASA's *Imagine the Universe*; note that you can click the "Advanced" button near the top and get a more detailed discussion.

Rainbows: How They Form and How to See Them: http://www.livescience.com/30235-rainbows-formation-explainer.html. By meteorologist and amateur astronomer Joe Rao.

Videos

Doppler Effect: http://www.esa.int/spaceinvideos/Videos/2014/07/Doppler_effect_-_classroom_demonstration_video_VP05. ESA video with Doppler ball demonstration and Doppler effect and satellites (4:48).

How a Prism Works to Make Rainbow Colors: https://www.youtube.com/watch?v=JGqsi_LDUn0. Short video on how a prism bends light to make a rainbow of colors (2:44).

Tour of the Electromagnetic Spectrum: https://www.youtube.com/watch?v=HPcAWNlVl-8. *NASA Mission Science* video tour of the bands of the electromagnetic spectrum (eight short videos).

Introductions To Quantum Mechanics

Ford, Kenneth. *The Quantum World*. 2004. A well-written recent introduction by a physicist/educator.

Gribbin, John. *In Search of Schroedinger's Cat*. 1984. Clear, very basic introduction to the fundamental ideas of quantum mechanics, by a British physicist and science writer.

Rae, Alastair. *Quantum Physics: A Beginner's Guide*. 2005. Widely praised introduction by a British physicist.

ᗊ COLLABORATIVE GROUP ACTIVITIES

A. Have your group make a list of all the electromagnetic wave technology you use during a typical day.

B. How many applications of the Doppler effect can your group think of in everyday life? For example, why would the highway patrol find it useful?

C. Have members of your group go home and "read" the face of your radio set and then compare notes. If you do not have a radio, research "broadcast radio frequencies" to find answers to the following questions. What do all the words and symbols mean? What frequencies can your radio tune to? What is the frequency of your favorite radio station? What is its wavelength?

D. If your instructor were to give you a spectrometer, what kind of spectra does your group think you would see from each of the following: (1) a household lightbulb, (2) the Sun, (3) the "neon lights of Broadway," (4) an ordinary household flashlight, and (5) a streetlight on a busy shopping street?

E. Suppose astronomers want to send a message to an alien civilization that is living on a planet with an atmosphere very similar to that of Earth's. This message must travel through space, make it through the other planet's atmosphere, and be noticeable to the residents of that planet. Have your group discuss

This OpenStax book is available for free at http://cnx.org/content/col11992/1.8

Download for free at http://cnx.org/content/col11992/latest/

what band of the electromagnetic spectrum might be best for this message and why. (Some people, including noted physicist Stephen Hawking, have warned scientists not to send such messages and reveal the presence of our civilization to a possible hostile cosmos. Do you agree with this concern?)

⌐ EXERCISES

Review Questions

1. What distinguishes one type of electromagnetic radiation from another? What are the main categories (or bands) of the electromagnetic spectrum?

2. What is a wave? Use the terms *wavelength* and *frequency* in your definition.

3. Is your textbook the kind of idealized object (described in section on radiation laws) that absorbs all the radiation falling on it? Explain. How about the black sweater worn by one of your classmates?

4. Where in an atom would you expect to find electrons? Protons? Neutrons?

5. Explain how emission lines and absorption lines are formed. In what sorts of cosmic objects would you expect to see each?

6. Explain how the Doppler effect works for sound waves and give some familiar examples.

7. What kind of motion for a star does not produce a Doppler effect? Explain.

8. Describe how Bohr's model used the work of Maxwell.

9. Explain why light is referred to as electromagnetic radiation.

10. Explain the difference between radiation as it is used in most everyday language and radiation as it is used in an astronomical context.

11. What are the differences between light waves and sound waves?

12. Which type of wave has a longer wavelength: AM radio waves (with frequencies in the kilohertz range) or FM radio waves (with frequencies in the megahertz range)? Explain.

13. Explain why astronomers long ago believed that space must be filled with some kind of substance (the "aether") instead of the vacuum we know it is today.

14. Explain what the ionosphere is and how it interacts with some radio waves.

15. Which is more dangerous to living things, gamma rays or X-rays? Explain.

16. Explain why we have to observe stars and other astronomical objects from above Earth's atmosphere in order to fully learn about their properties.

17. Explain why hotter objects tend to radiate more energetic photons compared to cooler objects.

18. Explain how we can deduce the temperature of a star by determining its color.

19. Explain what dispersion is and how astronomers use this phenomenon to study a star's light.

20. Explain why glass prisms disperse light.

21. Explain what Joseph Fraunhofer discovered about stellar spectra.

Download for free at http://cnx.org/content/col11992/latest/

22. Explain how we use spectral absorption and emission lines to determine the composition of a gas.

23. Explain the results of Rutherford's gold foil experiment and how they changed our model of the atom.

24. Is it possible for two different atoms of carbon to have different numbers of neutrons in their nuclei? Explain.

25. What are the three isotopes of hydrogen, and how do they differ?

26. Explain how electrons use light energy to move among energy levels within an atom.

27. Explain why astronomers use the term "blueshifted" for objects moving toward us and "redshifted" for objects moving away from us.

28. If spectral line wavelengths are changing for objects based on the radial velocities of those objects, how can we deduce which type of atom is responsible for a particular absorption or emission line?

Thought Questions

29. Make a list of some of the many practical consequences of Maxwell's theory of electromagnetic waves (television is one example).

30. With what type of electromagnetic radiation would you observe:
   A. A star with a temperature of 5800 K?

   B. A gas heated to a temperature of one million K?

   C. A person on a dark night?

31. Why is it dangerous to be exposed to X-rays but not (or at least much less) dangerous to be exposed to radio waves?

32. Go outside on a clear night, wait 15 minutes for your eyes to adjust to the dark, and look carefully at the brightest stars. Some should look slightly red and others slightly blue. The primary factor that determines the color of a star is its temperature. Which is hotter: a blue star or a red one? Explain

33. Water faucets are often labeled with a red dot for hot water and a blue dot for cold. Given Wien's law, does this labeling make sense?

34. Suppose you are standing at the exact center of a park surrounded by a circular road. An ambulance drives completely around this road, with siren blaring. How does the pitch of the siren change as it circles around you?

35. How could you measure Earth's orbital speed by photographing the spectrum of a star at various times throughout the year? (Hint: Suppose the star lies in the plane of Earth's orbit.)

36. Astronomers want to make maps of the sky showing sources of X-rays or gamma rays. Explain why those X-rays and gamma rays must be observed from above Earth's atmosphere.

37. The greenhouse effect can be explained easily if you understand the laws of blackbody radiation. A greenhouse gas blocks the transmission of infrared light. Given that the incoming light to Earth is sunlight with a characteristic temperature of 5800 K (which peaks in the visible part of the spectrum) and the outgoing light from Earth has a characteristic temperature of about 300 K (which peaks in the infrared part of the spectrum), explain how greenhouse gases cause Earth to warm up. As part of your answer, discuss that greenhouse gases block both incoming and outgoing infrared light. Explain why these two effects don't simply cancel each other, leading to no net temperature change.

This OpenStax book is available for free at http://cnx.org/content/col11992/1.8

38. An idealized radiating object does not reflect or scatter any radiation but instead absorbs all of the electromagnetic energy that falls on it. Can you explain why astronomers call such an object a blackbody? Keep in mind that even stars, which shine brightly in a variety of colors, are considered blackbodies. Explain why.

39. Why are ionized gases typically only found in very high-temperature environments?

40. Explain why each element has a unique spectrum of absorption or emission lines.

Figuring For Yourself

41. What is the wavelength of the carrier wave of a campus radio station, broadcasting at a frequency of 97.2 MHz (million cycles per second or million hertz)?

42. What is the frequency of a red laser beam, with a wavelength of 670 nm, which your astronomy instructor might use to point to slides during a lecture on galaxies?

43. You go to a dance club to forget how hard your astronomy midterm was. What is the frequency of a wave of ultraviolet light coming from a blacklight in the club, if its wavelength is 150 nm?

44. What is the energy of the photon with the frequency you calculated in Exercise 5.43?

45. If the emitted infrared radiation from Pluto, has a wavelength of maximum intensity at 75,000 nm, what is the temperature of Pluto assuming it follows Wien's law?

46. What is the temperature of a star whose maximum light is emitted at a wavelength of 290 nm?

Download for free at http://cnx.org/content/col11992/latest/

This OpenStax book is available for free at http://cnx.org/content/col11992/1.8

Download for free at http://cnx.org/content/col11992/latest/

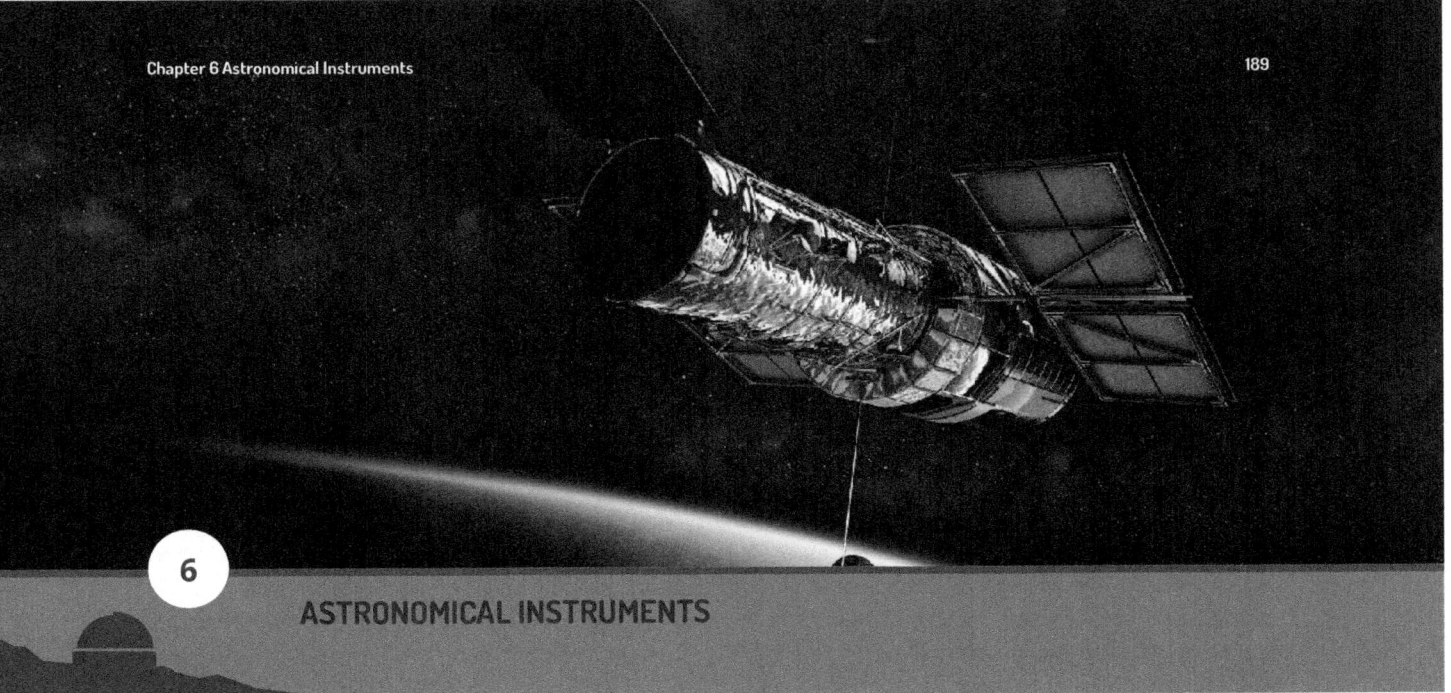

Figure 6.1. Hubble Space Telescope (HST). This artist's impression shows the Hubble above Earth, with the rectangular solar panels that provide it with power seen to the left and right.

Chapter Outline

Thinking Ahead

If you look at the sky when you are far away from city lights, there seem to be an overwhelming number of stars up there. In reality, only about 9000 stars are visible to the unaided eye (from both hemispheres of our planet). The light from most stars is so weak that by the time it reaches Earth, it cannot be detected by the human eye. How can we learn about the vast majority of objects in the universe that our unaided eyes simply cannot see?

In this chapter, we describe the tools astronomers use to extend their vision into space. We have learned almost everything we know about the universe from studying electromagnetic radiation, as discussed in the chapter on Radiation and Spectra. In the twentieth century, our exploration of space made it possible to detect electromagnetic radiation at all wavelengths, from gamma rays to radio waves. The different wavelengths carry different kinds of information, and the appearance of any given object often depends on the wavelength at which the observations are made.

Download for free at http://cnx.org/content/col11992/latest/

6.1 TELESCOPES

Learning Objectives

By the end of this section, you will be able to:

> Describe the three basic components of a modern system for measuring astronomical sources
> Describe the main functions of a telescope
> Describe the two basic types of visible-light telescopes and how they form images

Systems for Measuring Radiation

There are three basic components of a modern system for measuring radiation from astronomical sources. First, there is a **telescope**, which serves as a "bucket" for collecting visible light (or radiation at other wavelengths, as shown in (Figure 6.2). Just as you can catch more rain with a garbage can than with a coffee cup, large telescopes gather much more light than your eye can. Second, there is an instrument attached to the telescope that sorts the incoming radiation by wavelength. Sometimes the sorting is fairly crude. For example, we might simply want to separate blue light from red light so that we can determine the temperature of a star. But at other times, we want to see individual spectral lines to determine what an object is made of, or to measure its speed (as explained in the Radiation and Spectra chapter). Third, we need some type of **detector**, a device that senses the radiation in the wavelength regions we have chosen and permanently records the observations.

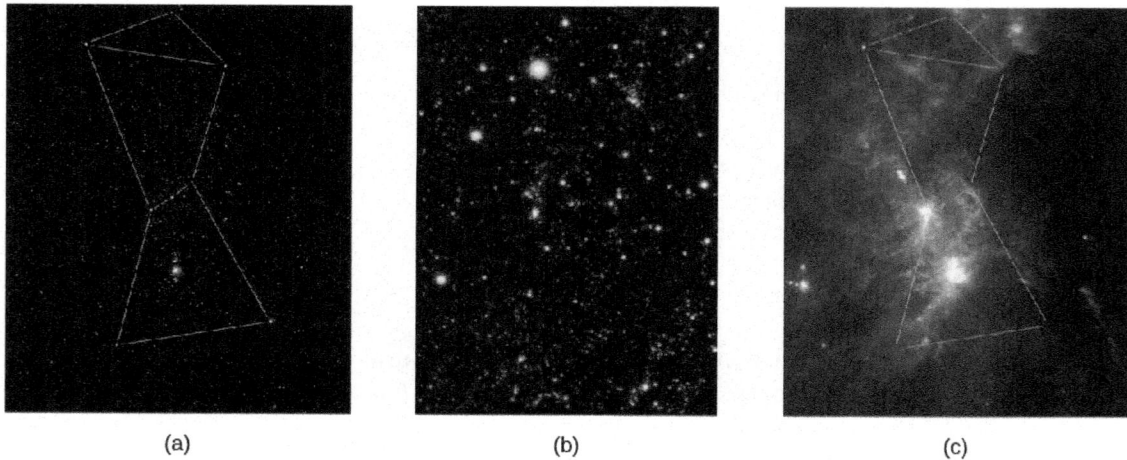

(a)   (b)   (c)

Figure 6.2. Orion Region at Different Wavelengths. The same part of the sky looks different when observed with instruments that are sensitive to different bands of the spectrum. (a) Visible light: this shows part of the Orion region as the human eye sees it, with dotted lines added to show the figure of the mythical hunter, Orion. (b) X-rays: here, the view emphasizes the point-like X-ray sources nearby. The colors are artificial, changing from yellow to white to blue with increasing energy of the X-rays. The bright, hot stars in Orion are still seen in this image, but so are many other objects located at very different distances, including other stars, star corpses, and galaxies at the edge of the observable universe. (c) Infrared radiation: here, we mainly see the glowing dust in this region. (credit a: modification of work by Howard McCallon/NASA/ IRAS; credit b: modification of work by Howard McCallon/NASA/IRAS; credit c: modification of work by Michael F. Corcoran)

The history of the development of astronomical telescopes is about how new technologies have been applied to improve the efficiency of these three basic components: the telescopes, the wavelength-sorting device, and the detectors. Let's first look at the development of the telescope.

Many ancient cultures built special sites for observing the sky (Figure 6.3). At these ancient *observatories*, they could measure the positions of celestial objects, mostly to keep track of time and date. Many of these ancient

This OpenStax book is available for free at http://cnx.org/content/col11992/1.8

observatories had religious and ritual functions as well. The eye was the only device available to gather light, all of the colors in the light were observed at once, and the only permanent record of the observations was made by human beings writing down or sketching what they saw.

(a)

(b)

Figure 6.3. Two Pre-Telescopic Observatories. (a) Machu Picchu is a fifteenth century Incan site located in Peru. (b) Stonehenge, a prehistoric site (3000–2000 BCE), is located in England. (credit a: modification of work by Allard Schmidt)

While Hans Lippershey, Zaccharias Janssen, and Jacob Metius are all credited with the invention of the telescope around 1608—applying for patents within weeks of each other—it was Galileo who, in 1610, used this simple tube with lenses (which he called a spyglass) to observe the sky and gather more light than his eyes alone could. Even his small telescope—used over many nights—revolutionized ideas about the nature of the planets and the position of Earth.

How Telescopes Work

Telescopes have come a long way since Galileo's time. Now they tend to be huge devices; the most expensive cost hundreds of millions to billions of dollars. (To provide some reference point, however, keep in mind that just renovating college football stadiums typically costs hundreds of millions of dollars—with the most expensive recent renovation, at Texas A&M University's Kyle Field, costing $450 million.) The reason astronomers keep building bigger and bigger telescopes is that celestial objects—such as planets, stars, and galaxies—send much more light to Earth than any human eye (with its tiny opening) can catch, and bigger telescopes can detect fainter objects. If you have ever watched the stars with a group of friends, you know that there's plenty of starlight to go around; each of you can see each of the stars. If a thousand more people were watching, each of them would also catch a bit of each star's light. Yet, as far as you are concerned, the light not shining into your eye is wasted. It would be great if some of this "wasted" light could also be captured and brought to your eye. This is precisely what a telescope does.

The most important functions of a telescope are (1) to *collect* the faint light from an astronomical source and (2) to *focus* all the light into a point or an image. Most objects of interest to astronomers are extremely faint: the more light we can collect, the better we can study such objects. (And remember, even though we are focusing on visible light first, there are many telescopes that collect other kinds of electromagnetic radiation.)

Telescopes that collect visible radiation use a lens or mirror to gather the light. Other types of telescopes may use collecting devices that look very different from the lenses and mirrors with which we are familiar, but they serve the same function. In all types of telescopes, the light-gathering ability is determined by the area of the device acting as the light-gathering "bucket." Since most telescopes have mirrors or lenses, we can compare

Download for free at http://cnx.org/content/col11992/latest/

their light-gathering power by comparing the **apertures**, or diameters, of the opening through which light travels or reflects.

The amount of light a telescope can collect increases with the size of the aperture. A telescope with a mirror that is 4 meters in diameter can collect 16 times as much light as a telescope that is 1 meter in diameter. (The diameter is squared because the area of a circle equals $\pi d^2/4$, where d is the diameter of the circle.)

EXAMPLE 6.1

Calculating the Light-Collecting Area

What is the area of a 1-m diameter telescope? A 4-m diameter one?

Solution

Using the equation for the area of a circle,

$$A = \frac{\pi d^2}{4}$$

the area of a 1-m telescope is

$$\frac{\pi d^2}{4} = \frac{\pi (1 \text{ m})^2}{4} = 0.79 \text{ m}^2$$

and the area of a 4-m telescope is

$$\frac{\pi d^2}{4} = \frac{\pi (4 \text{ m})^2}{4} = 12.6 \text{ m}^2$$

Check Your Learning

Show that the ratio of the two areas is 16:1.

Answer:

$\dfrac{12.6 \text{ m}^2}{0.79 \text{ m}^2} = 16.$ Therefore, with 16 times the area, a 4-m telescope collects 16 times the light of a 1-m

telescope.

After the telescope forms an image, we need some way to detect and record it so that we can measure, reproduce, and analyze the image in various ways. Before the nineteenth century, astronomers simply viewed images with their eyes and wrote descriptions of what they saw. This was very inefficient and did not lead to a very reliable long-term record; you know from crime shows on television that eyewitness accounts are often inaccurate.

In the nineteenth century, the use of photography became widespread. In those days, photographs were a chemical record of an image on a specially treated glass plate. Today, the image is generally detected with sensors similar to those in digital cameras, recorded electronically, and stored in computers. This permanent record can then be used for detailed and quantitative studies. Professional astronomers rarely look through the large telescopes that they use for their research.

This OpenStax book is available for free at http://cnx.org/content/col11992/1.8

Formation of an Image by a Lens or a Mirror

Whether or not you wear glasses, you see the world through lenses; they are key elements of your eyes. A lens is a transparent piece of material that bends the rays of light passing through it. If the light rays are parallel as they enter, the lens brings them together in one place to form an image (Figure 6.4). If the curvatures of the lens surfaces are just right, all parallel rays of light (say, from a star) are bent, or *refracted*, in such a way that they converge toward a point, called the **focus** of the lens. At the focus, an image of the light source appears. In the case of parallel light rays, the distance from the lens to the location where the light rays focus, or image, behind the lens is called the *focal length* of the lens.

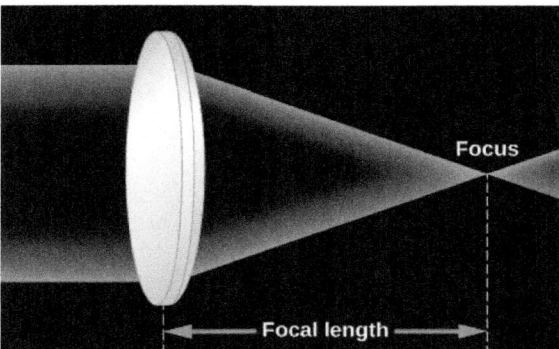

Figure 6.4. Formation of an Image by a Simple Lens. Parallel rays from a distant source are bent by the convex lens so that they all come together in a single place (the focus) to form an image.

As you look at Figure 6.4, you may ask why two rays of light from the same star would be parallel to each other. After all, if you draw a picture of star shining in all directions, the rays of light coming from the star don't look parallel at all. But remember that the stars (and other astronomical objects) are all extremely far away. By the time the few rays of light pointed toward us actually arrive at Earth, they are, for all practical purposes, parallel to each other. Put another way, any rays that were *not* parallel to the ones pointed at Earth are now heading in some very different direction in the universe.

To view the image formed by the lens in a telescope, we use an additional lens called an **eyepiece**. The eyepiece focuses the image at a distance that is either directly viewable by a human or at a convenient place for a detector. Using different eyepieces, we can change the *magnification* (or size) of the image and also redirect the light to a more accessible location. Stars look like points of light, and magnifying them makes little difference, but the image of a planet or a galaxy, which has structure, can often benefit from being magnified.

Many people, when thinking of a telescope, picture a long tube with a large glass lens at one end. This design, which uses a lens as its main optical element to form an image, as we have been discussing, is known as a *refractor* (Figure 6.5), and a telescope based on this design is called a **refracting telescope**. Galileo's telescopes were refractors, as are today's binoculars and field glasses. However, there is a limit to the size of a refracting telescope. The largest one ever built was a 49-inch refractor built for the Paris 1900 Exposition, and it was dismantled after the Exposition. Currently, the largest refracting telescope is the 40-inch refractor at Yerkes Observatory in Wisconsin.

Download for free at http://cnx.org/content/col11992/latest/

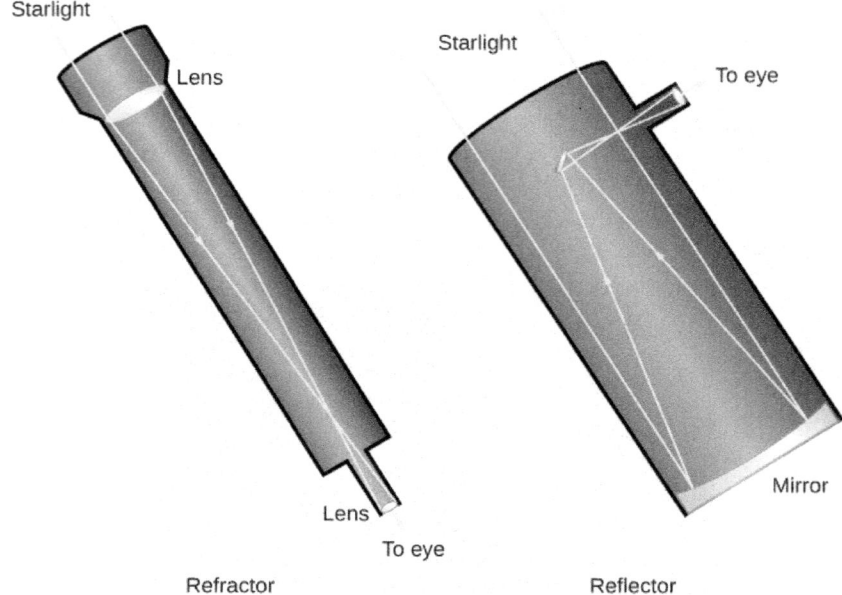

Figure 6.5. Refracting and Reflecting Telescopes. Light enters a refracting telescope through a lens at the upper end, which focuses the light near the bottom of the telescope. An eyepiece then magnifies the image so that it can be viewed by the eye, or a detector like a photographic plate can be placed at the focus. The upper end of a reflecting telescope is open, and the light passes through to the mirror located at the bottom of the telescope. The mirror then focuses the light at the top end, where it can be detected. Alternatively, as in this sketch, a second mirror may reflect the light to a position outside the telescope structure, where an observer can have easier access to it. Professional astronomers' telescopes are more complicated than this, but they follow the same principles of reflection and refraction.

One problem with a refracting telescope is that the light must pass *through* the lens of a refractor. That means the glass must be perfect all the way through, and it has proven very difficult to make large pieces of glass without flaws and bubbles in them. Also, optical properties of transparent materials change a little bit with the wavelengths (or colors) of light, so there is some additional distortion, known as **chromatic aberration**. Each wavelength focuses at a slightly different spot, causing the image to appear blurry.

In addition, since the light must pass through the lens, the lens can only be supported around its edges (just like the frames of our eyeglasses). The force of gravity will cause a large lens to sag and distort the path of the light rays as they pass through it. Finally, because the light passes through it, both sides of the lens must be manufactured to precisely the right shape in order to produce a sharp image.

A different type of telescope uses a concave *primary mirror* as its main optical element. The mirror is curved like the inner surface of a sphere, and it reflects light in order to form an image (Figure 6.5). Telescope mirrors are coated with a shiny metal, usually silver, aluminum, or, occasionally, gold, to make them highly reflective. If the mirror has the correct shape, all parallel rays are reflected back to the same point, the focus of the mirror. Thus, images are produced by a mirror exactly as they are by a lens.

Telescopes designed with mirrors avoid the problems of refracting telescopes. Because the light is reflected from the front surface only, flaws and bubbles within the glass do not affect the path of the light. In a telescope designed with mirrors, only the front surface has to be manufactured to a precise shape, and the mirror can be supported from the back. For these reasons, most astronomical telescopes today (both amateur and professional) use a mirror rather than a lens to form an image; this type of telescope is called a **reflecting telescope**. The first successful reflecting telescope was built by Isaac Newton in 1668.

In a reflecting telescope, the concave mirror is placed at the bottom of a tube or open framework. The mirror reflects the light back up the tube to form an image near the front end at a location called the **prime focus**.

This OpenStax book is available for free at http://cnx.org/content/col11992/1.8

The image can be observed at the prime focus, or additional mirrors can intercept the light and redirect it to a position where the observer can view it more easily (Figure 6.6). Since an astronomer at the prime focus can block much of the light coming to the main mirror, the use of a small *secondary mirror* allows more light to get through the system.

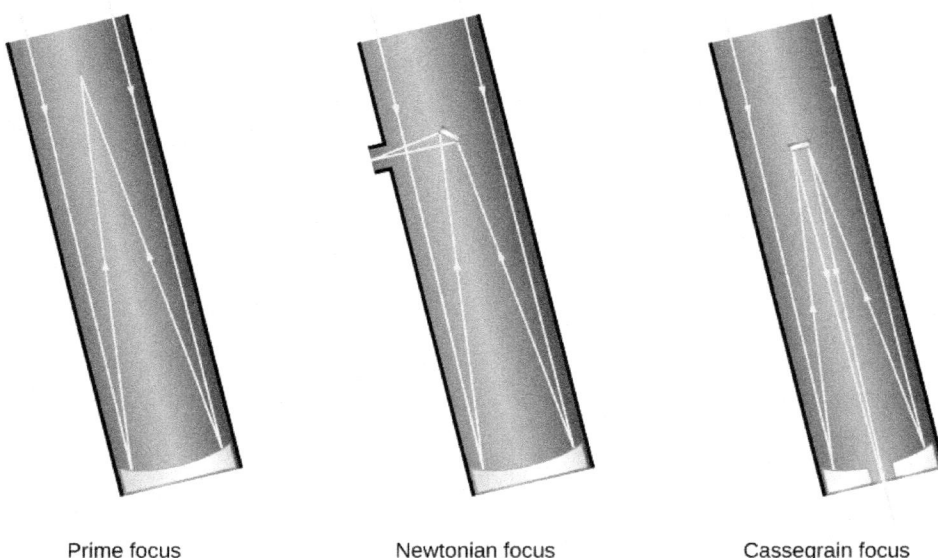

Prime focus   Newtonian focus   Cassegrain focus

Figure 6.6. Focus Arrangements for Reflecting Telescopes. Reflecting telescopes have different options for where the light is brought to a focus. With prime focus, light is detected where it comes to a focus after reflecting from the primary mirror. With Newtonian focus, light is reflected by a small secondary mirror off to one side, where it can be detected (see also Figure 6.5). Most large professional telescopes have a Cassegrain focus in which light is reflected by the secondary mirror down through a hole in the primary mirror to an observing station below the telescope.

MAKING CONNECTIONS

Choosing Your Own Telescope

If the astronomy course you are taking whets your appetite for exploring the sky further, you may be thinking about buying your own telescope. Many excellent amateur telescopes are available, and some research is required to find the best model for your needs. Some good sources of information about personal telescopes are the two popular US magazines aimed at amateur astronomers: *Sky & Telescope* and *Astronomy*. Both carry regular articles with advice, reviews, and advertisements from reputable telescope dealers.

Some of the factors that determine which telescope is right for you depend upon your preferences:

- Will you be setting up the telescope in one place and leaving it there, or do you want an instrument that is portable and can come with you on outdoor excursions? How portable should it be, in terms of size and weight?

- Do you want to observe the sky with your eyes only, or do you want to take photographs? (Long-exposure photography, for example, requires a good clock drive to turn your telescope to compensate for Earth's rotation.)

Download for free at http://cnx.org/content/col11992/latest/

- What types of objects will you be observing? Are you interested primarily in comets, planets, star clusters, or galaxies, or do you want to observe all kinds of celestial sights?

You may not know the answers to some of these questions yet. For this reason, you may want to "test-drive" some telescopes first. Most communities have amateur astronomy clubs that sponsor star parties open to the public. The members of those clubs often know a lot about telescopes and can share their ideas with you. Your instructor may know where the nearest amateur astronomy club meets; or, to find a club near you, use the websites suggested in Appendix B.

Furthermore, you may already have an instrument like a telescope at home (or have access to one through a relative or friend). Many amateur astronomers recommend starting your survey of the sky with a good pair of binoculars. These are easily carried around and can show you many objects not visible (or clear) to the unaided eye.

When you are ready to purchase a telescope, you might find the following ideas useful:

- The key characteristic of a telescope is the aperture of the main mirror or lens; when someone says they have a 6-inch or 8-inch telescope, they mean the diameter of the collecting surface. The larger the aperture, the more light you can gather, and the fainter the objects you can see or photograph.

- Telescopes of a given aperture that use lenses (refractors) are typically more expensive than those using mirrors (reflectors) because both sides of a lens must be polished to great accuracy. And, because the light passes through it, the lens must be made of high-quality glass throughout. In contrast, only the front surface of a mirror must be accurately polished.

- Magnification is not one of the criteria on which to base your choice of a telescope. As we discussed, the magnification of the image is done by a smaller eyepiece, so the magnification can be adjusted by changing eyepieces. However, a telescope will magnify not only the astronomical object you are viewing but also the turbulence of Earth's atmosphere. If the magnification is too high, your image will shimmer and shake and be difficult to view. A good telescope will come with a variety of eyepieces that stay within the range of useful magnification.

- The mount of a telescope (the structure on which it rests) is one of its most critical elements. Because a telescope shows a tiny field of view, which is magnified significantly, even the smallest vibration or jarring of the telescope can move the object you are viewing around or out of your field of view. A sturdy and stable mount is essential for serious viewing or photography (although it clearly affects how portable your telescope can be).

- A telescope requires some practice to set up and use effectively. Don't expect everything to go perfectly on your first try. Take some time to read the instructions. If a local amateur astronomy club is nearby, use it as a resource.

6.2 TELESCOPES TODAY

Learning Objectives

By the end of this section, you will be able to:

> Recognize the largest visible-light and infrared telescopes in operation today

This OpenStax book is available for free at http://cnx.org/content/col11992/1.8

Download for free at http://cnx.org/content/col11992/latest/

> Discuss the factors relevant to choosing an appropriate telescope site
> Define the technique of adaptive optics and describe the effects of the atmosphere on astronomical observations

Since Newton's time, when the sizes of the mirrors in telescopes were measured in inches, reflecting telescopes have grown ever larger. In 1948, US astronomers built a telescope with a 5-meter (200-inch) diameter mirror on Palomar Mountain in Southern California. It remained the largest visible-light telescope in the world for several decades. The giants of today, however, have primary mirrors (the largest mirrors in the telescope) that are 8- to 10-meters in diameter, and larger ones are being built (Figure 6.7).

Figure 6.7. Large Telescope Mirror. This image shows one of the primary mirrors of the European Southern Observatory's Very Large Telescope, named Yepun, just after it was recoated with aluminum. The mirror is a little over 8 meters in diameter. (credit: ESO/G. Huedepohl)

Modern Visible-Light and Infrared Telescopes

The decades starting in 1990 saw telescope building around the globe grow at an unprecedented rate. (See Table 6.1, which also includes websites for each telescope in case you want to visit or learn more about them.) Technological advancements had finally made it possible to build telescopes significantly larger than the 5-meter telescope at Palomar at a reasonable cost. New technologies have also been designed to work well in the infrared, and not just visible, wavelengths.

Large Single-Dish Visible-Light and Infrared Telescopes

Aperture (m)	Telescope Name	Location	Status	Website
39	European Extremely Large Telescope (E-ELT)	Cerro Armazonas, Chile	First light 2025 (estimated)	www.eso.org/sci/facilities/eelt
30	Thirty-Meter Telescope (TMT)	Mauna Kea, HI	First light 2025 (estimated)	www.tmt.org
24.5	Giant Magellan Telescope (GMT)	Las Campanas Observatory, Chile	First light 2025 (estimated)	www.gmto.org

Table 6.1

Download for free at http://cnx.org/content/col11992/latest/

Large Single-Dish Visible-Light and Infrared Telescopes

Aperture (m)	Telescope Name	Location	Status	Website
11.1 × 9.9	Southern African Large Telescope (SALT)	Sutherland, South Africa	2005	www.salt.ac.za
10.4	Gran Telescopio Canarias (GTC)	La Palma, Canary Islands	First light 2007	http://www.gtc.iac.es
10.0	Keck I and II (two telescopes)	Mauna Kea, HI	Completed 1993–96	www.keckobservatory.org
9.1	Hobby–Eberly Telescope (HET)	Mount Locke, TX	Completed 1997	www.as.utexas.edu/ mcdonald/het
8.4	Large Binocular Telescope (LBT) (two telescopes)	Mount Graham, AZ	First light 2004	www.lbto.org
8.4	Large Synoptic Survey Telescope (LSST)	The Cerro Pachón, Chile	First light 2021	www.lsst.org
8.3	Subaru Telescope	Mauna Kea, HI	First light 1998	www.naoj.org
8.2	Very Large Telescope (VLT)	Cerro Paranal, Chile	All four telescopes completed 2000	www.eso.org/public/ teles-instr/paranal
8.1	Gemini North and Gemini South	Mauna Kea, HI (North) and Cerro Pachón, Chile (South)	First light 1999 (North), First light 2000 (South)	www.gemini.edu
6.5	Magellan Telescopes (two telescopes: Baade and Landon Clay)	Las Campanas, Chile	First light 2000 and 2002	obs.carnegiescience.edu/ Magellan
6.5	Multi-Mirror Telescope (MMT)	Mount Hopkins, AZ	Completed 1979	www.mmto.org
6.0	Big Telescope Altazimuth (BTA-6)	Mount Pastukhov, Russia	Completed 1976	w0.sao.ru/Doc-en/ Telescopes/bta/ descrip.html

Table 6.1

This OpenStax book is available for free at http://cnx.org/content/col11992/1.8

Download for free at http://cnx.org/content/col11992/latest/

Large Single-Dish Visible-Light and Infrared Telescopes

Aperture (m)	Telescope Name	Location	Status	Website
5.1	Hale Telescope	Mount Palomar, CA	Completed 1948	www.astro.caltech.edu/palomar/about/telescopes/hale.html

Table 6.1

The differences between the Palomar telescope and the modern Gemini North telescope (to take an example) are easily seen in Figure 6.8. The Palomar telescope is a massive steel structure designed to hold the 14.5-ton primary mirror with a 5-meter diameter. Glass tends to sag under its own weight; hence, a huge steel structure is needed to hold the mirror. A mirror 8 meters in diameter, the size of the Gemini North telescope, if it were built using the same technology as the Palomar telescope, would have to weigh at least eight times as much and would require an enormous steel structure to support it.

(a)   (b)

Figure 6.8. Modern Reflecting Telescopes. (a) The Palomar 5-meter reflector: The Hale telescope on Palomar Mountain has a complex mounting structure that enables the telescope (in the open "tube" pointing upward in this photo) to swing easily into any position. (b) The Gemini North 8-meter telescope: The Gemini North mirror has a larger area than the Palomar mirror, but note how much less massive the whole instrument seems. (credit a: modification of work by Caltech/Palomar Observatory; credit b: modification of work by Gemini Observatory/AURA)

The 8-meter Gemini North telescope looks like a featherweight by contrast, and indeed it is. The mirror is only about 8 inches thick and weighs 24.5 tons, less than twice as much as the Palomar mirror. The Gemini North telescope was completed about 50 years after the Palomar telescope. Engineers took advantage of new technologies to build a telescope that is much lighter in weight relative to the size of the primary mirror. The

Download for free at http://cnx.org/content/col11992/latest/

Gemini mirror does sag, but with modern computers, it is possible to measure that sag many times each second and apply forces at 120 different locations to the back of the mirror to correct the sag, a process called *active control*. Seventeen telescopes with mirrors 6.5 meters in diameter and larger have been constructed since 1990.

The twin 10-meter Keck telescopes on Mauna Kea, which were the first of these new-technology instruments, use precision control in an entirely novel way. Instead of a single primary mirror 10 meters in diameter, each Keck telescope achieves its larger aperture by combining the light from 36 separate hexagonal mirrors, each 1.8 meters wide (Figure 6.9). Computer-controlled actuators (motors) constantly adjust these 36 mirrors so that the overall reflecting surface acts like a single mirror with just the right shape to collect and focus the light into a sharp image.

Figure 6.9. Thirty-Six Eyes Are Better Than One. The mirror of the 10-meter Keck telescope is composed of 36 hexagonal sections. (credit: NASA)

LINK TO LEARNING

Learn more about the Keck Observatory on Mauna Kea (https://openstaxcollege.org/l/30KeckObserv) through this History Channel clip on the telescopes and the work that they do.

In addition to holding the mirror, the steel structure of a telescope is designed so that the entire telescope can be pointed quickly toward any object in the sky. Since Earth is rotating, the telescope must have a motorized drive system that moves it very smoothly from east to west at exactly the same rate that Earth is rotating from west to east, so it can continue to point at the object being observed. All this machinery must be housed in a dome to protect the telescope from the elements. The dome has an opening in it that can be positioned in front of the telescope and moved along with it, so that the light from the objects being observed is not blocked.

This OpenStax book is available for free at http://cnx.org/content/col11992/1.8

VOYAGERS IN ASTRONOMY

George Ellery Hale: Master Telescope Builder

George Ellery Hale (Figure 6.10) was a giant among early telescope builders. Not once, but four times, he initiated projects that led to the construction of what was the world's largest telescope at the time. And he was a master at winning over wealthy benefactors to underwrite the construction of these new instruments.

Figure 6.10. George Ellery Hale (1868–1938). Hale's work led to the construction of several major telescopes, including the 40-inch refracting telescope at Yerkes Observatory, and three reflecting telescopes: the 60-inch Hale and 100-inch Hooker telescopes at Mount Wilson Observatory, and the 200-inch Hale Telescope at Palomar Observatory.

Hale's training and early research were in solar physics. In 1892, at age 24, he was named associate professor of astral physics and director of the astronomical observatory at the University of Chicago. At the time, the largest telescope in the world was the 36-inch refractor at the Lick Observatory near San Jose, California. Taking advantage of an existing glass blank for a 40-inch telescope, Hale set out to raise money for a larger telescope than the one at Lick. One prospective donor was Charles T. Yerkes, who, among other things, ran the trolley system in Chicago.

Hale wrote to Yerkes, encouraging him to support the construction of the giant telescope by saying that "the donor could have no more enduring monument. It is certain that Mr. Lick's name would not have been nearly so widely known today were it not for the famous observatory established as a result of his munificence." Yerkes agreed, and the new telescope was completed in May 1897; it remains the largest refractor in the world (Figure 6.11).

Download for free at http://cnx.org/content/col11992/latest/

Figure 6.11. World's Largest Refractor. The Yerkes 40-inch (1-meter) telescope.

Even before the completion of the Yerkes refractor, Hale was not only dreaming of building a still larger telescope but was also taking concrete steps to achieve that goal. In the 1890s, there was a major controversy about the relative quality of refracting and reflecting telescopes. Hale realized that 40 inches was close to the maximum feasible aperture for refracting telescopes. If telescopes with significantly larger apertures were to be built, they would have to be reflecting telescopes.

Using funds borrowed from his own family, Hale set out to construct a 60-inch reflector. For a site, he left the Midwest for the much better conditions on Mount Wilson—at the time, a wilderness peak above the small city of Los Angeles. In 1904, at the age of 36, Hale received funds from the Carnegie Foundation to establish the Mount Wilson Observatory. The 60-inch mirror was placed in its mount in December 1908.

Two years earlier, in 1906, Hale had already approached John D. Hooker, who had made his fortune in hardware and steel pipe, with a proposal to build a 100-inch telescope. The technological risks were substantial. The 60-inch telescope was not yet complete, and the usefulness of large reflectors for astronomy had yet to be demonstrated. George Ellery Hale's brother called him "the greatest gambler in the world." Once again, Hale successfully obtained funds, and the 100-inch telescope was completed in November 1917. (It was with this telescope that Edwin Hubble was able to establish that the spiral nebulae were separate islands of stars—or galaxies—quite removed from our own Milky Way.)

Hale was not through dreaming. In 1926, he wrote an article in *Harper's Magazine* about the scientific value of a still larger telescope. This article came to the attention of the Rockefeller Foundation, which granted $6 million for the construction of a 200-inch telescope. Hale died in 1938, but the 200-inch (5-meter) telescope on Palomar Mountain was dedicated 10 years later and is now named in Hale's honor.

Picking the Best Observing Sites

A telescope like the Gemini or Keck telescope costs about $100 million to build. That kind of investment demands that the telescope be placed in the best possible site. Since the end of the nineteenth century, astronomers have realized that the best observatory sites are on mountains, far from the lights and pollution of cities. Although a number of urban observatories remain, especially in the large cities of Europe, they have

This OpenStax book is available for free at http://cnx.org/content/col11992/1.8

Download for free at http://cnx.org/content/col11992/latest/

become administrative centers or museums. The real action takes place far away, often on desert mountains or isolated peaks in the Atlantic and Pacific Oceans, where we find the staff's living quarters, computers, electronic and machine shops, and of course the telescopes themselves. A large observatory today requires a supporting staff of 20 to 100 people in addition to the astronomers.

The performance of a telescope is determined not only by the size of its mirror but also by its location. Earth's atmosphere, so vital to life, presents challenges for the observational astronomer. In at least four ways, our air imposes limitations on the usefulness of telescopes:

1. The most obvious limitation is weather conditions such as clouds, wind, and rain. At the best sites, the weather is clear as much as 75% of the time.

2. Even on a clear night, the atmosphere filters out a certain amount of starlight, especially in the infrared, where the absorption is due primarily to water vapor. Astronomers therefore prefer dry sites, generally found at high altitudes.

3. The sky above the telescope should be dark. Near cities, the air scatters the glare from lights, producing an illumination that hides the faintest stars and limits the distances that can be probed by telescopes. (Astronomers call this effect *light pollution*.) Observatories are best located at least 100 miles from the nearest large city.

4. Finally, the air is often unsteady; light passing through this turbulent air is disturbed, resulting in blurred star images. Astronomers call these effects "bad **seeing**." When seeing is bad, images of celestial objects are distorted by the constant twisting and bending of light rays by turbulent air.

The best observatory sites are therefore high, dark, and dry. The world's largest telescopes are found in such remote mountain locations as the Andes Mountains of Chile (Figure 6.12), the desert peaks of Arizona, the Canary Islands in the Atlantic Ocean, and Mauna Kea in Hawaii, a dormant volcano with an altitude of 13,700 feet (4200 meters).

LINK TO LEARNING

Light pollution is a problem not just for professional astronomers but for everyone who wants to enjoy the beauty of the night sky. In addition research is now showing that it can disrupt the life cycle of animals with whom we share the urban and suburban landscape. And the light wasted shining into the sky leads to unnecessary municipal expenses and use of fossil fuels. Concerned people have formed an organization, the International Dark-Sky Association, whose website (https://openstaxcollege.org/l/30IntDSA) is full of good information. A citizen science project called Globe at Night (https://openstaxcollege.org/l/30GlbatNght) allows you to measure the light levels in your community by counting stars and to compare it to others around the world. And, if you get interested in this topic and want to do a paper for your astronomy course or another course while you are in college, the Dark Night Skies guide (https://openstaxcollege.org/l/30DNSGuide) can point you to a variety of resources on the topic.

Download for free at http://cnx.org/content/col11992/latest/

Figure 6.12. High and Dry Site. Cerro Paranal, a mountain summit 2.7 kilometers above sea level in Chile's Atacama Desert, is the site of the European Southern Observatory's Very Large Telescope. This photograph shows the four 8-meter telescope buildings on the site and vividly illustrates that astronomers prefer high, dry sites for their instruments. The 4.1-meter Visible and Infrared Survey Telescope for Astronomy (VISTA) can be seen in the distance on the next mountain peak. (credit: ESO)

The Resolution of a Telescope

In addition to gathering as much light as they can, astronomers also want to have the sharpest images possible. **Resolution** refers to the precision of detail present in an image: that is, the smallest features that can be distinguished. Astronomers are always eager to make out more detail in the images they study, whether they are following the weather on Jupiter or trying to peer into the violent heart of a "cannibal galaxy" that recently ate its neighbor for lunch.

One factor that determines how good the resolution will be is the size of the telescope. Larger apertures produce sharper images. Until very recently, however, visible-light and infrared telescopes on Earth's surface could not produce images as sharp as the theory of light said they should.

The problem—as we saw earlier in this chapter—is our planet's atmosphere, which is turbulent. It contains many small-scale blobs or cells of gas that range in size from inches to several feet. Each cell has a slightly different temperature from its neighbor, and each cell acts like a lens, bending (refracting) the path of the light by a small amount. This bending slightly changes the position where each light ray finally reaches the detector in a telescope. The cells of air are in motion, constantly being blown through the light path of the telescope by winds, often in different directions at different altitudes. As a result, the path followed by the light is constantly changing.

For an analogy, think about watching a parade from a window high up in a skyscraper. You decide to throw some confetti down toward the marchers. Even if you drop a handful all at the same time and in the same direction, air currents will toss the pieces around, and they will reach the ground at different places. As we described earlier, we can think of the light from the stars as a series of parallel beams, each making its way through the atmosphere. Each path will be slightly different, and each will reach the detector of the telescope at a slightly different place. The result is a blurred image, and because the cells are being blown by the wind, the nature of the blur will change many times each second. You have probably noticed this effect as the "twinkling" of stars seen from Earth. The light beams are bent enough that part of the time they reach your eye, and part of the time some of them miss, thereby making the star seem to vary in brightness. In space, however, the light of the stars is steady.

Astronomers search the world for locations where the amount of atmospheric blurring, or turbulence, is as small as possible. It turns out that the best sites are in coastal mountain ranges and on isolated volcanic peaks in the middle of an ocean. Air that has flowed long distances over water before it encounters land is especially stable.

This OpenStax book is available for free at http://cnx.org/content/col11992/1.8

Download for free at http://cnx.org/content/col11992/latest/

The resolution of an image is measured in units of angle on the sky, typically in units of arcseconds. One arcsecond is 1/3600 degree, and there are 360 degrees in a full circle. So we are talking about tiny angles on the sky. To give you a sense of just how tiny, we might note that 1 arcsecond is how big a quarter would look when seen from a distance of 5 kilometers. The best images obtained from the ground with traditional techniques reveal details as small as several tenths of an arcsecond across. This image size is remarkably good. One of the main reasons for launching the Hubble Space Telescope was to escape Earth's atmosphere and obtain even sharper images.

But since we can't put every telescope into space, astronomers have devised a technique called **adaptive optics** that can beat Earth's atmosphere at its own game of blurring. This technique (which is most effective in the infrared region of the spectrum with our current technology) makes use of a small flexible mirror placed in the beam of a telescope. A sensor measures how much the atmosphere has distorted the image, and as often as 500 times per second, it sends instructions to the flexible mirror on how to change shape in order to compensate for distortions produced by the atmosphere. The light is thus brought back to an almost perfectly sharp focus at the detector. Figure 6.13 shows just how effective this technique is. With adaptive optics, ground-based telescopes can achieve resolutions of 0.1 arcsecond or a little better in the infrared region of the spectrum. This impressive figure is the equivalent of the resolution that the Hubble Space Telescope achieves in the visible-light region of the spectrum.

Figure 6.13. Power of Adaptive Optics. One of the clearest pictures of Jupiter ever taken from the ground, this image was produced with adaptive optics using an 8-meter-diameter telescope at the Very Large Telescope in Chile. Adaptive optics uses infrared wavelengths to remove atmospheric blurring, resulting in a much clearer image. (credit: modification of work by ESO, F.Marchis, M.Wong (UC Berkeley); E.Marchetti, P.Amico, S.Tordo (ESO))

ASTRONOMY BASICS

How Astronomers Really Use Telescopes

In the popular view (and some bad movies), an astronomer spends most nights in a cold observatory peering through a telescope, but this is not very accurate today. Most astronomers do not live at

Download for free at http://cnx.org/content/col11992/latest/

observatories, but near the universities or laboratories where they work. An astronomer might spend only a week or so each year observing at the telescope and the rest of the time measuring or analyzing the data acquired from large project collaborations and dedicated surveys. Many astronomers use radio telescopes for space experiments, which work just as well during the daylight hours. Still others work at purely theoretical problems using supercomputers and never observe at a telescope of any kind.

Even when astronomers are observing with large telescopes, they seldom peer through them. Electronic detectors permanently record the data for detailed analysis later. At some observatories, observations may be made remotely, with the astronomer sitting at a computer thousands of miles away from the telescope.

Time on major telescopes is at a premium, and an observatory director will typically receive many more requests for telescope time than can be accommodated during the year. Astronomers must therefore write a convincing proposal explaining how they would like to use the telescope and why their observations will be important to the progress of astronomy. A committee of astronomers is then asked to judge and rank the proposals, and time is assigned only to those with the greatest merit. Even if your proposal is among the high-rated ones, you may have to wait many months for your turn. If the skies are cloudy on the nights you have been assigned, it may be more than a year before you get another chance.

Some older astronomers still remember long, cold nights spent alone in an observatory dome, with only music from a tape recorder or an all-night radio station for company. The sight of the stars shining brilliantly hour after hour through the open slit in the observatory dome was unforgettable. So, too, was the relief as the first pale light of dawn announced the end of a 12-hour observation session. Astronomy is much easier today, with teams of observers working together, often at their computers, in a warm room. Those who are more nostalgic, however, might argue that some of the romance has gone from the field, too.

6.3 VISIBLE-LIGHT DETECTORS AND INSTRUMENTS

Learning Objectives

By the end of this section, you will be able to:

> Describe the difference between photographic plates and charge-coupled devices
> Describe the unique difficulties associated with infrared observations and their solutions
> Describe how a spectrometer works

After a telescope collects radiation from an astronomical source, the radiation must be *detected* and measured. The first detector used for astronomical observations was the human eye, but it suffers from being connected to an imperfect recording and retrieving device—the human brain. Photography and modern electronic detectors have eliminated the quirks of human memory by making a permanent record of the information from the cosmos.

The eye also suffers from having a very short *integration time*; it takes only a fraction of a second to add light energy together before sending the image to the brain. One important advantage of modern detectors is that the light from astronomical objects can be collected by the detector over longer periods of time; this technique is called "taking a long exposure." Exposures of several hours are required to detect very faint objects in the cosmos.

This OpenStax book is available for free at http://cnx.org/content/col11992/1.8

Before the light reaches the detector, astronomers today normally use some type of instrument to sort the light according to wavelength. The instrument may be as simple as colored filters, which transmit light within a specified range of wavelengths. A red transparent plastic is an everyday example of a filter that transmits only the red light and blocks the other colors. After the light passes through a filter, it forms an image that astronomers can then use to measure the apparent brightness and color of objects. We will show you many examples of such images in the later chapters of this book, and we will describe what we can learn from them.

Alternatively, the instrument between telescope and detector may be one of several devices that spread the light out into its full rainbow of colors so that astronomers can measure individual lines in the spectrum. Such an instrument (which you learned about in the chapter on Radiation and Spectra) is called a *spectrometer* because it allows astronomers to measure (to meter) the spectrum of a source of radiation. Whether a filter or a spectrometer, both types of wavelength-sorting instruments still have to use detectors to record and measure the properties of light.

Photographic and Electronic Detectors

Throughout most of the twentieth century, photographic film or *glass plates* served as the prime astronomical detectors, whether for photographing spectra or direct images of celestial objects. In a photographic plate, a light-sensitive chemical coating is applied to a piece of glass that, when developed, provides a lasting record of the image. At observatories around the world, vast collections of photographs preserve what the sky has looked like during the past 100 years. Photography represents a huge improvement over the human eye, but it still has limitations. Photographic films are inefficient: only about 1% of the light that actually falls on the film contributes to the chemical change that makes the image; the rest is wasted.

Astronomers today have much more efficient electronic detectors to record astronomical images. Most often, these are **charge-coupled devices (CCDs)**, which are similar to the detectors used in video camcorders or in digital cameras (like the one more and more students have on their cell phones) (see Figure 6.14). In a CCD, photons of radiation hitting any part of the detector generate a stream of charged particles (electrons) that are stored and counted at the end of the exposure. Each place where the radiation is counted is called a pixel (picture element), and modern detectors can count the photons in millions of pixels (megapixels, or MPs).

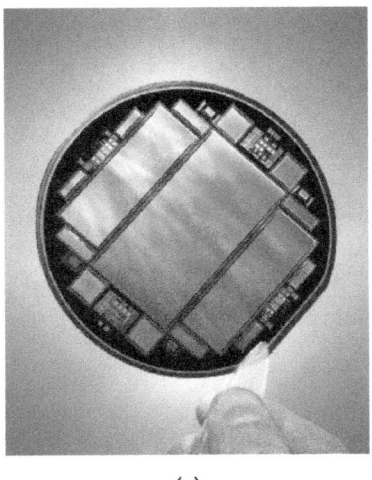
(a)

(b)

Figure 6.14. Charge-Coupled Devices (CCDs). (a) This CCD is a mere 300-micrometers thick (thinner than a human hair) yet holds more than 21 million pixels. (b) This matrix of 42 CCDs serves the Kepler telescope. (credit a: modification of work by US Department of Energy; credit b: modification of work by NASA and Ball Aerospace)

Download for free at http://cnx.org/content/col11992/latest/

Because CCDs typically record as much as 60–70% of all the photons that strike them, and the best silicon and infrared CCDs exceed 90% sensitivity, we can detect much fainter objects. Among these are many small moons around the outer planets, icy dwarf planets beyond Pluto, and dwarf galaxies of stars. CCDs also provide more accurate measurements of the brightness of astronomical objects than photography, and their output is digital—in the form of numbers that can go directly into a computer for analysis.

Infrared Observations

Observing the universe in the infrared band of the spectrum presents some additional challenges. The infrared region extends from wavelengths near 1 micrometer (μm), which is about the long wavelength sensitivity limit of both CCDs and photography, to 100 micrometers or longer. Recall from the discussion on radiation and spectra that infrared is "heat radiation" (given off at temperatures that we humans are comfortable with). The main challenge to astronomers using infrared is to distinguish between the tiny amount of heat radiation that reaches Earth from stars and galaxies, and the much greater heat radiated by the telescope itself and our planet's atmosphere.

Typical temperatures on Earth's surface are near 300 K, and the atmosphere through which observations are made is only a little cooler. According to Wien's law (from the chapter on Radiation and Spectra), the telescope, the observatory, and even the sky are radiating infrared energy with a peak wavelength of about 10 micrometers. To infrared eyes, everything on Earth is brightly aglow—including the telescope and camera (Figure 6.15). The challenge is to detect faint cosmic sources against this sea of infrared light. Another way to look at this is that an astronomer using infrared must always contend with the situation that a visible-light observer would face if working in broad daylight with a telescope and optics lined with bright fluorescent lights.

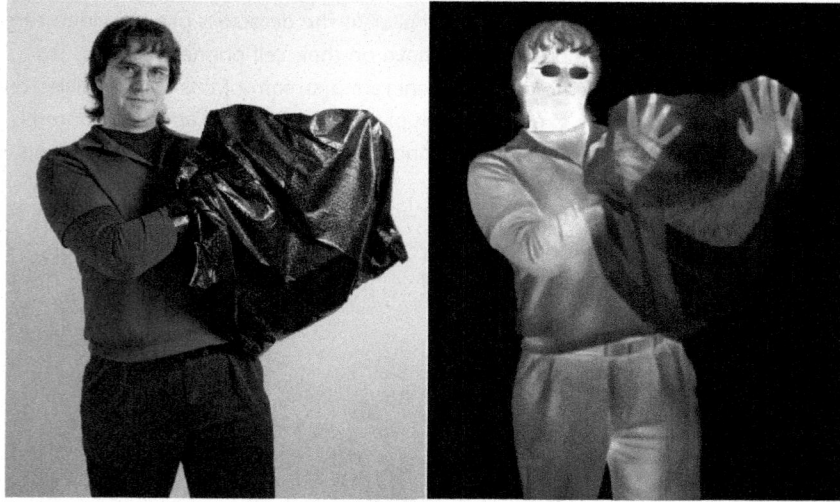

Figure 6.15. Infrared Eyes. Infrared waves can penetrate places in the universe from which light is blocked, as shown in this infrared image where the plastic bag blocks visible light but not infrared. (credit: NASA/JPL-Caltech/R. Hurt (SSC))

To solve this problem, astronomers must protect the infrared detector from nearby radiation, just as you would shield photographic film from bright daylight. Since anything warm radiates infrared energy, the detector must be isolated in very cold surroundings; often, it is held near absolute zero (1 to 3 K) by immersing it in liquid helium. The second step is to reduce the radiation emitted by the telescope structure and optics, and to block this heat from reaching the infrared detector.

This OpenStax book is available for free at http://cnx.org/content/col11992/1.8

Download for free at http://cnx.org/content/col11992/latest/

LINK TO LEARNING

Check out The Infrared Zoo (https://openstaxcollege.org/l/30IFZoo) to get a sense of what familiar objects look like with infrared radiation. Slide the slider to change the wavelength of radiation for the picture, and click the arrow to see other animals.

Spectroscopy

Spectroscopy is one of the astronomer's most powerful tools, providing information about the composition, temperature, motion, and other characteristics of celestial objects. More than half of the time spent on most large telescopes is used for spectroscopy.

The many different wavelengths present in light can be separated by passing them through a spectrometer to form a spectrum. The design of a simple spectrometer is illustrated in Figure 6.16. Light from the source (actually, the image of a source produced by the telescope) enters the instrument through a small hole or narrow slit, and is collimated (made into a beam of parallel rays) by a lens. The light then passes through a prism, producing a spectrum: different wavelengths leave the prism in different directions because each wavelength is bent by a different amount when it enters and leaves the prism. A second lens placed behind the prism focuses the many different images of the slit or entrance hole onto a CCD or other detecting device. This collection of images (spread out by color) is the spectrum that astronomers can then analyze at a later point. As spectroscopy spreads the light out into more and more collecting bins, fewer photons go into each bin, so either a larger telescope is needed or the integration time must be greatly increased—usually both.

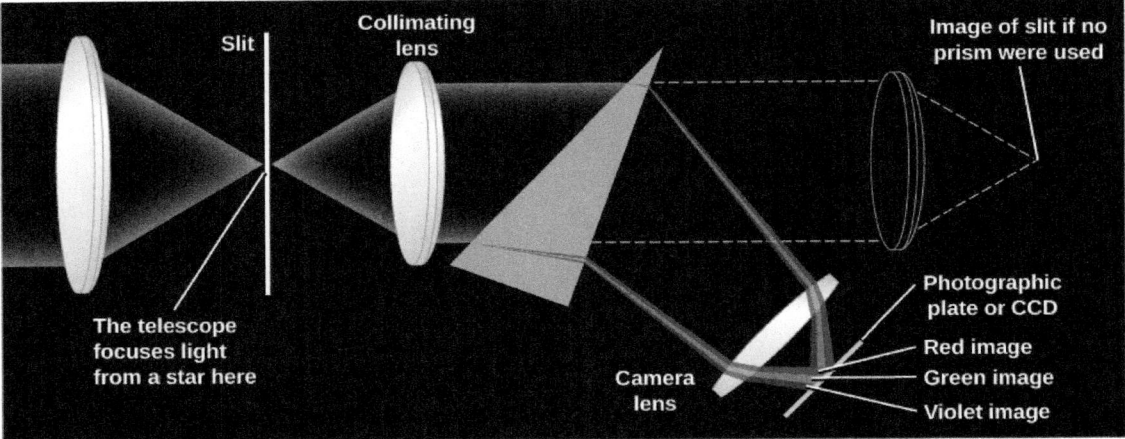

Figure 6.16. Prism Spectrometer. The light from the telescope is focused on a slit. A prism (or grating) disperses the light into a spectrum, which is then photographed or recorded electronically.

In practice, astronomers today are more likely to use a different device, called a *grating*, to disperse the spectrum. A grating is a piece of material with thousands of grooves on its surface. While it functions completely differently, a grating, like a prism, also spreads light out into a spectrum.

Download for free at http://cnx.org/content/col11992/latest/

6.4 | RADIO TELESCOPES

Learning Objectives

By the end of this section, you will be able to:

> Describe how radio waves from space are detected
> Identify the world's largest radio telescopes
> Define the technique of interferometry and discuss the benefits of interferometers over single-dish telescopes

In addition to visible and infrared radiation, radio waves from astronomical objects can also be detected from the surface of Earth. In the early 1930s, Karl G. Jansky, an engineer at Bell Telephone Laboratories, was experimenting with antennas for long-range radio communication when he encountered some mysterious static—radio radiation coming from an unknown source (Figure 6.17). He discovered that this radiation came in strongest about four minutes earlier on each successive day and correctly concluded that since Earth's sidereal rotation period (how long it takes us to rotate relative to the stars) is four minutes shorter than a solar day, the radiation must be originating from some region fixed on the celestial sphere. Subsequent investigation showed that the source of this radiation was part of the Milky Way Galaxy; Jansky had discovered the first source of cosmic radio waves.

Figure 6.17. First Radio Telescope. This rotating radio antenna was used by Jansky in his serendipitous discovery of radio radiation from the Milky Way.

In 1936, Grote Reber, who was an amateur astronomer interested in radio communications, used galvanized iron and wood to build the first antenna specifically designed to receive cosmic radio waves. Over the years, Reber built several such antennas and used them to carry out pioneering surveys of the sky for celestial radio sources; he remained active in radio astronomy for more than 30 years. During the first decade, he worked practically alone because professional astronomers had not yet recognized the vast potential of radio astronomy.

Detection of Radio Energy from Space

It is important to understand that radio waves cannot be "heard": they are not the sound waves you hear coming out of the radio receiver in your home or car. Like light, radio waves are a form of electromagnetic radiation, but unlike light, we cannot detect them with our senses—we must rely on electronic equipment to

This OpenStax book is available for free at http://cnx.org/content/col11992/1.8

pick them up. In commercial radio broadcasting, we encode sound information (music or a newscaster's voice) into radio waves. These must be decoded at the other end and then turned back into sound by speakers or headphones.

The radio waves we receive from space do not, of course, have music or other program information encoded in them. If cosmic radio signals were translated into sound, they would sound like the static you hear when scanning between stations. Nevertheless, there is information in the radio waves we receive—information that can tell us about the chemistry and physical conditions of the sources of the waves.

Just as vibrating charged particles can produce electromagnetic waves (see the Radiation and Spectra chapter), electromagnetic waves can make charged particles move back and forth. Radio waves can produce a current in conductors of electricity such as metals. An antenna is such a conductor: it intercepts radio waves, which create a feeble current in it. The current is then amplified in a radio receiver until it is strong enough to measure or record. Like your television or radio, receivers can be tuned to select a single frequency (channel). In astronomy, however, it is more common to use sophisticated data-processing techniques that allow thousands of separate frequency bands to be detected simultaneously. Thus, the astronomical radio receiver operates much like a spectrometer on a visible-light or infrared telescope, providing information about how much radiation we receive at each wavelength or frequency. After computer processing, the radio signals are recorded on magnetic disks for further analysis.

Radio waves are reflected by conducting surfaces, just as light is reflected from a shiny metallic surface, and according to the same laws of optics. A radio-reflecting telescope consists of a concave metal reflector (called a *dish*), analogous to a telescope mirror. The radio waves collected by the dish are reflected to a focus, where they can then be directed to a receiver and analyzed. Because humans are such visual creatures, radio astronomers often construct a pictorial representation of the radio sources they observe. Figure 6.18 shows such a radio image of a distant galaxy, where radio telescopes reveal vast jets and complicated regions of radio emissions that are completely invisible in photographs taken with light.

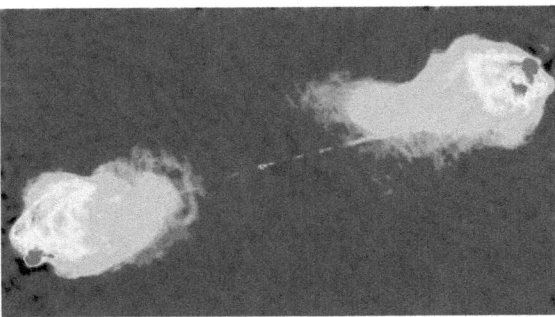

Figure 6.18. Radio Image. This image has been constructed of radio observations at the Very Large Array of a galaxy called Cygnus A. Colors have been added to help the eye sort out regions of different radio intensities. Red regions are the most intense, blue the least. The visible galaxy would be a small dot in the center of the image. The radio image reveals jets of expelled material (more than 160,000 light-years long) on either side of the galaxy. (credit: NRAO/AUI)

Radio astronomy is a young field compared with visible-light astronomy, but it has experienced tremendous growth in recent decades. The world's largest radio reflectors that can be pointed to any direction in the sky have apertures of 100 meters. One of these has been built at the US National Radio Astronomy Observatory in West Virginia (Figure 6.19). Table 6.2 lists some of the major radio telescopes of the world.

Download for free at http://cnx.org/content/col11992/latest/

Figure 6.19. Robert C. Byrd Green Bank Telescope. This fully steerable radio telescope in West Virginia went into operation in August 2000. Its dish is about 100 meters across. (credit: modification of work by "b3nscott"/Flickr)

Major Radio Observatories of the World

Observatory	Location	Description	Website
Individual Radio Dishes			
Arecibo Observatory	Arecibo, Puerto Rico	305-m fixed dish	www.naic.edu
Green Bank Telescope (GBT)	Green Bank, WV	110 × 100-m steerable dish	www.science.nrao.edu/facilities/gbt
Effelsberg 100-m Telescope	Bonn, Germany	100-m steerable dish	www.mpifr-bonn.mpg.de/en/effelsberg
Lovell Telescope	Manchester, England	76-m steerable dish	www.jb.man.ac.uk/aboutus/lovell
Canberra Deep Space Communication Complex (CDSCC)	Tidbinbilla, Australia	70-m steerable dish	www.cdscc.nasa.gov
Goldstone Deep Space Communications Complex (GDSCC)	Barstow, CA	70-m steerable dish	www.gdscc.nasa.gov
Parkes Observatory	Parkes, Australia	64-m steerable dish	www.parkes.atnf.csiro.au
Arrays of Radio Dishes			

Table 6.2

This OpenStax book is available for free at http://cnx.org/content/col11992/1.8

Major Radio Observatories of the World

Observatory	Location	Description	Website
Square Kilometre Array (SKA)	South Africa and Western Australia	Thousands of dishes, km^2 collecting area, partial array in 2020	www.skatelescope.org
Atacama Large Millimeter/ submillimeter Array (ALMA)	Atacama desert, Northern Chile	66 7-m and 12-m dishes	www.almaobservatory.org
Very Large Array (VLA)	Socorro, New Mexico	27-element array of 25-m dishes (36-km baseline)	www.science.nrao.edu/facilities/ vla
Westerbork Synthesis Radio Telescope (WSRT)	Westerbork, the Netherlands	12-element array of 25-m dishes (1.6-km baseline)	www.astron.nl/radio-observatory/public/public-0
Very Long Baseline Array (VLBA)	Ten US sites, HI to the Virgin Islands	10-element array of 25-m dishes (9000 km baseline)	www.science.nrao.edu/facilities/ vlba
Australia Telescope Compact Array (ATCA)	Several sites in Australia	8-element array (seven 22-m dishes plus Parkes 64 m)	www.narrabri.atnf.csiro.au
Multi-Element Radio Linked Interferometer Network (MERLIN)	Cambridge, England, and other British sites	Network of seven dishes (the largest is 32 m)	www.e-merlin.ac.uk
Millimeter-wave Telescopes			
IRAM	Granada, Spain	30-m steerable mm-wave dish	www.iram-institute.org
James Clerk Maxwell Telescope (JCMT)	Mauna Kea, HI	15-m steerable mm-wave dish	www.eaobservatory.org/jcmt
Nobeyama Radio Observatory (NRO)	Minamimaki, Japan	6-element array of 10-m wave dishes	www.nro.nao.ac.jp/en
Hat Creek Radio Observatory (HCRO)	Cassel, CA	6-element array of 5-m wave dishes	www.sri.com/research-development/specialized-facilities/hat-creek-radio-observatory

Table 6.2

Download for free at http://cnx.org/content/col11992/latest/

Radio Interferometry

As we discussed earlier, a telescope's ability to show us fine detail (its resolution) depends upon its aperture, but it also depends upon the wavelength of the radiation that the telescope is gathering. The longer the waves, the harder it is to resolve fine detail in the images or maps we make. Because radio waves have such long wavelengths, they present tremendous challenges for astronomers who need good resolution. In fact, even the largest radio dishes on Earth, operating alone, cannot make out as much detail as the typical small visible-light telescope used in a college astronomy lab. To overcome this difficulty, radio astronomers have learned to sharpen their images by linking two or more radio telescopes together electronically. Two or more telescopes linked together in this way are called an **interferometer**.

"Interferometer" may seem like a strange term because the telescopes in an interferometer work cooperatively; they don't "interfere" with each other. **Interference**, however, is a technical term for the way that multiple waves interact with each other when they arrive in our instruments, and this interaction allows us to coax more detail out of our observations. The resolution of an interferometer depends upon the separation of the telescopes, not upon their individual apertures. Two telescopes separated by 1 kilometer provide the same resolution as would a single dish 1 kilometer across (although they are not, of course, able to collect as much radiation as a radio-wave bucket that is 1 kilometer across).

To get even better resolution, astronomers combine a large number of radio dishes into an **interferometer array**. In effect, such an array works like a large number of two-dish interferometers, all observing the same part of the sky together. Computer processing of the results permits the reconstruction of a high-resolution radio image. The most extensive such instrument in the United States is the National Radio Astronomy Observatory's Very Large Array (VLA) near Socorro, New Mexico. It consists of 27 movable radio telescopes (on railroad tracks), each having an aperture of 25 meters, spread over a total span of about 36 kilometers. By electronically combining the signals from all of its individual telescopes, this array permits the radio astronomer to make pictures of the sky at radio wavelengths comparable to those obtained with a visible-light telescope, with a resolution of about 1 arcsecond.

The Atacama Large Millimeter/submillimeter array (ALMA) in the Atacama Desert of Northern Chile (Figure 6.20), at an altitude of 16,400 feet, consists of 12 7-meter and 54 12-meter telescopes, and can achieve baselines up to 16 kilometers. Since it became operational in 2013, it has made observations at resolutions down to 6 milliarcseconds (0.006 arcseconds), a remarkable achievement for radio astronomy.

Figure 6.20. Atacama Large Millimeter/Submillimeter Array (ALMA). Located in the Atacama Desert of Northern Chile, ALMA currently provides the highest resolution for radio observations. (credit: ESO/S. Guisard)

This OpenStax book is available for free at http://cnx.org/content/col11992/1.8

LINK TO LEARNING

Watch this documentary (https://openstax.org/l/30ALMAdoc) that explains the work that went into designing and building ALMA, discusses some of its first images, and explores its future.

Initially, the size of interferometer arrays was limited by the requirement that all of the dishes be physically wired together. The maximum dimensions of the array were thus only a few tens of kilometers. However, larger interferometer separations can be achieved if the telescopes do not require a physical connection. Astronomers, with the use of current technology and computing power, have learned to time the arrival of electromagnetic waves coming from space very precisely at each telescope and combine the data later. If the telescopes are as far apart as California and Australia, or as West Virginia and Crimea in Ukraine, the resulting resolution far surpasses that of visible-light telescopes.

The United States operates the Very Long Baseline Array (VLBA), made up of 10 individual telescopes stretching from the Virgin Islands to Hawaii (Figure 6.21). The VLBA, completed in 1993, can form astronomical images with a resolution of 0.0001 arcseconds, permitting features as small as 10 astronomical units (AU) to be distinguished at the center of our Galaxy.

Figure 6.21. Very Long Baseline Array. This map shows the distribution of 10 antennas that constitute an array of radio telescopes stretching across the United States and its territories.

Recent advances in technology have also made it possible to do interferometry at visible-light and infrared wavelengths. At the beginning of the twenty-first century, three observatories with multiple telescopes each began using their dishes as interferometers, combining their light to obtain a much greater resolution. In addition, a dedicated interferometric array was built on Mt. Wilson in California. Just as in radio arrays, these observations allow astronomers to make out more detail than a single telescope could provide.

Visible-Light Interferometers

Longest Baseline (m)	Telescope Name	Location	Mirrors	Status
400	CHARA Array (Center for High Angular Resolution Astronomy)	Mount Wilson, CA	Six 1-m telescopes	Operational since 2004

Table 6.3

Download for free at http://cnx.org/content/col11992/latest/

Visible-Light Interferometers

Longest Baseline (m)	Telescope Name	Location	Mirrors	Status
200	Very Large Telescope	Cerro Paranal, Chile	Four 8.2-m telescopes	Completed 2000
85	Keck I and II telescopes	Mauna Kea, HI	Two 10-m telescopes	Operated from 2001 to 2012
22.8	Large Binocular Telescope	Mount Graham, AZ	Two 8.4-m telescopes	First light 2004

Table 6.3

Radar Astronomy

Radar is the technique of transmitting radio waves to an object in our solar system and then detecting the radio radiation that the object reflects back. The time required for the round trip can be measured electronically with great precision. Because we know the speed at which radio waves travel (the speed of light), we can determine the distance to the object or a particular feature on its surface (such as a mountain).

Radar observations have been used to determine the distances to planets and how fast things are moving in the solar system (using the Doppler effect, discussed in the Radiation and Spectra chapter). Radar waves have played important roles in navigating spacecraft throughout the solar system. In addition, as will be discussed in later chapters, radar observations have determined the rotation periods of Venus and Mercury, probed tiny Earth-approaching asteroids, and allowed us to investigate the mountains and valleys on the surfaces of Mercury, Venus, Mars, and the large moons of Jupiter.

Any radio dish can be used as a radar telescope if it is equipped with a powerful transmitter as well as a receiver. The most spectacular facility in the world for radar astronomy is the 1000-foot (305-meter) telescope at Arecibo in Puerto Rico (Figure 6.22). The Arecibo telescope is too large to be pointed directly at different parts of the sky. Instead, it is constructed in a huge natural "bowl" (more than a mere dish) formed by several hills, and it is lined with reflecting metal panels. A limited ability to track astronomical sources is achieved by moving the receiver system, which is suspended on cables 100 meters above the surface of the bowl. An even larger (500-meter) radar telescope is currently under construction. It is the Five-hundred-meter Aperture Spherical Telescope (FAST) in China and is expected to be completed in 2016.

This OpenStax book is available for free at http://cnx.org/content/col11992/1.8

Figure 6.22. Largest Radio and Radar Dish. The Arecibo Observatory, with its 1000-foot radio dish-filling valley in Puerto Rico, is part of the National Astronomy and Ionosphere Center, operated by SRI International, USRA, and UMET under a cooperative agreement with the National Science Foundation. (credit: National Astronomy and Ionosphere Center, Cornell U., NSF)

6.5 OBSERVATIONS OUTSIDE EARTH'S ATMOSPHERE

Learning Objectives

By the end of this section, you will be able to:

> List the advantages of making astronomical observations from space
> Explain the importance of the Hubble Space Telescope
> Describe some of the major space-based observatories astronomers use

Earth's atmosphere blocks most radiation at wavelengths shorter than visible light, so we can only make direct ultraviolet, X-ray, and gamma ray observations from space (though indirect gamma ray observations can be made from Earth). Getting above the distorting effects of the atmosphere is also an advantage at visible and infrared wavelengths. The stars don't "twinkle" in space, so the amount of detail you can observe is limited only by the size of your instrument. On the other hand, it is expensive to place telescopes into space, and repairs can present a major challenge. This is why astronomers continue to build telescopes for use on the ground as well as for launching into space.

Airborne and Space Infrared Telescopes

Water vapor, the main source of atmospheric interference for making infrared observations, is concentrated in the lower part of Earth's atmosphere. For this reason, a gain of even a few hundred meters in elevation can make an important difference in the quality of an infared observatory site. Given the limitations of high mountains, most of which attract clouds and violent storms, and the fact that the ability of humans to perform complex tasks degrades at high altitudes, it was natural for astronomers to investigate the possibility of observing infrared waves from airplanes and ultimately from space.

Infrared observations from airplanes have been made since the 1960s, starting with a 15-centimeter telescope on board a Learjet. From 1974 through 1995, NASA operated a 0.9-meter airborne telescope flying regularly out of the Ames Research Center south of San Francisco. Observing from an altitude of 12 kilometers, the telescope was above 99% of the atmospheric water vapor. More recently, NASA (in partnership with the German Aerospace Center) has constructed a much larger 2.5-meter telescope, called the Stratospheric Observatory for Infrared Astronomy (SOFIA), which flies in a modified Boeing 747SP (Figure 6.23).

Download for free at http://cnx.org/content/col11992/latest/

Figure 6.23. Stratospheric Observatory for Infrared Astronomy (SOFIA). SOFIA allows observations to be made above most of Earth's atmospheric water vapor. (credit: NASA)

LINK TO LEARNING

To find out more about SOFIA, watch this video (https://openstaxcollege.org/l/30SOFIAvid) provided by NASA's Armstrong Flight Research Center.

Getting even higher and making observations from space itself have important advantages for infrared astronomy. First is the elimination of all interference from the atmosphere. Equally important is the opportunity to cool the entire optical system of the instrument in order to nearly eliminate infrared radiation from the telescope itself. If we tried to cool a telescope within the atmosphere, it would quickly become coated with condensing water vapor and other gases, making it useless. Only in the vacuum of space can optical elements be cooled to hundreds of degrees below freezing and still remain operational.

The first orbiting infrared observatory, launched in 1983, was the Infrared Astronomical Satellite (IRAS), built as a joint project by the United States, the Netherlands, and Britain. IRAS was equipped with a 0.6-meter telescope cooled to a temperature of less than 10 K. For the first time, the infrared sky could be seen as if it were night, rather than through a bright foreground of atmospheric and telescope emissions. IRAS carried out a rapid but comprehensive survey of the entire infrared sky over a 10-month period, cataloging about 350,000 sources of infrared radiation. Since then, several other infrared telescopes have operated in space with much better sensitivity and resolution due to improvements in infrared detectors. The most powerful of these infrared telescopes is the 0.85-meter Spitzer Space Telescope, which launched in 2003. A few of its observations are shown in Figure 6.24. With infrared observations, astronomers can detect cooler parts of cosmic objects, such as the dust clouds around star nurseries and the remnants of dying stars, that visible-light images don't reveal.

This OpenStax book is available for free at http://cnx.org/content/col11992/1.8

Flame nebula

Cassiopeia A

Helix nebula

Figure 6.24. Observations from the Spitzer Space Telescope (SST). These infrared images—a region of star formation, the remnant of an exploded star, and a region where an old star is losing its outer shell—show just a few of the observations made and transmitted back to Earth from the SST. Since our eyes are not sensitive to infrared rays, we don't perceive colors from them. The colors in these images have been selected by astronomers to highlight details like the composition or temperature in these regions. (credit "Flame nebula": modification of work by NASA (X-ray: NASA/CXC/PSU/K.Getman, E.Feigelson, M.Kuhn & the MYStIX team; Infrared:NASA/JPL-Caltech); credit "Cassiopeia A": modification of work by NASA/JPL-Caltech; credit "Helix nebula": modification of work by NASA/JPL-Caltech)

Hubble Space Telescope

In April 1990, a great leap forward in astronomy was made with the launch of the Hubble Space Telescope (HST). With an aperture of 2.4 meters, this is the largest telescope put into space so far. (Its aperture was limited by the size of the payload bay in the Space Shuttle that served as its launch vehicle.) It was named for Edwin Hubble, the astronomer who discovered the expansion of the universe in the 1920s (whose work we will discuss in the chapters on Galaxies).

HST is operated jointly by NASA's Goddard Space Flight Center and the Space Telescope Science Institute in Baltimore. It was the first orbiting observatory designed to be serviced by Shuttle astronauts and, over the years since it was launched, they made several visits to improve or replace its initial instruments and to repair some of the systems that operate the spacecraft (Figure 6.1)—though this repair program has now been discontinued, and no more visits or improvements will be made.

With the Hubble, astronomers have obtained some of the most detailed images of astronomical objects from the solar system outward to the most distant galaxies. Among its many great achievements is the Hubble Ultra-Deep Field, an image of a small region of the sky observed for almost 100 hours. It contains views of about 10,000 galaxies, some of which formed when the universe was just a few percent of its current age (Figure 6.25).

Download for free at http://cnx.org/content/col11992/latest/

Figure 6.25. Hubble Ultra-Deep Field (HUDF). The Hubble Space Telescope has provided an image of a specific region of space built from data collected between September 24, 2003, and January 16, 2004. These data allow us to search for galaxies that existed approximately 13 billion years ago. (credit: modification of work by NASA)

The HST's mirror was ground and polished to a remarkable degree of accuracy. If we were to scale up its 2.4-meter mirror to the size of the entire continental United States, there would be no hill or valley larger than about 6 centimeters in its smooth surface. Unfortunately, after it was launched, scientists discovered that the primary mirror had a slight error in its *shape,* equal to roughly 1/50 the width of a human hair. Small as that sounds, it was enough to ensure that much of the light entering the telescope did not come to a clear focus and that all the images were blurry. (In a misplaced effort to save money, a complete test of the optical system had not been carried out before launch, so the error was not discovered until HST was in orbit.)

The solution was to do something very similar to what we do for astronomy students with blurry vision: put corrective optics in front of their eyes. In December 1993, in one of the most exciting and difficult space missions ever flown, astronauts captured the orbiting telescope and brought it back into the shuttle payload bay. There they installed a package containing compensating optics as well as a new, improved camera before releasing HST back into orbit. The telescope now works as it was intended to, and further missions to it were able to install even more advanced instruments to take advantage of its capabilities.

High-Energy Observatories

Ultraviolet, X-ray, and direct gamma-ray (high-energy electromagnetic wave) observations can be made only from space. Such observations first became possible in 1946, with V2 rockets captured from Germany after World War II. The US Naval Research Laboratory put instruments on these rockets for a series of pioneering flights, used initially to detect ultraviolet radiation from the Sun. Since then, many other rockets have been launched to make X-ray and ultraviolet observations of the Sun, and later of other celestial objects.

Beginning in the 1960s, a steady stream of high-energy observatories has been launched into orbit to reveal and explore the universe at short wavelengths. Among recent X-ray telescopes is the Chandra X-ray Observatory, which was launched in 1999 (Figure 6.26). It is producing X-ray images with unprecedented resolution and sensitivity. Designing instruments that can collect and focus energetic radiation like X-rays and gamma rays

This OpenStax book is available for free at http://cnx.org/content/col11992/1.8

is an enormous technological challenge. The 2002 Nobel Prize in physics was awarded to Riccardo Giacconi, a pioneer in the field of building and launching sophisticated X-ray instruments. In 2008, NASA launched the Fermi Gamma-ray Space Telescope, designed to measure cosmic gamma rays at energies greater than any previous telescope, and thus able to collect radiation from some of the most energetic events in the universe.

Figure 6.26. Chandra X-Ray Satellite. Chandra, the world's most powerful X-ray telescope, was developed by NASA and launched in July 1999. (credit: modification of work by NASA)

One major challenge is to design "mirrors" to reflect such penetrating radiation as X-rays and gamma rays, which normally pass straight through matter. However, although the technical details of design are more complicated, the three basic components of an observing system, as we explained earlier in this chapter, are the same at all wavelengths: a telescope to gather up the radiation, filters or instruments to sort the radiation according to wavelength, and some method of detecting and making a permanent record of the observations. Table 6.4 lists some of the most important active space observatories that humanity has launched.

Gamma-ray detections can also be made from Earth's surface by using the atmosphere as the primary detector. When a gamma ray hits our atmosphere, it accelerates charged particles (mostly electrons) in the atmosphere. Those energetic particles hit other particles in the atmosphere and give off their own radiation. The effect is a cascade of light and energy that can be detected on the ground. The VERITAS array in Arizona and the H.E.S.S. array in Namibia are two such ground-based gamma-ray observatories.

Recent Observatories in Space

Observatory	Date Operation Began	Bands of the Spectrum	Notes	Website
Hubble Space Telescope (HST)	1990	visible, UV, IR	2.4-m mirror; images and spectra	www.hubblesite.org

Table 6.4

Download for free at http://cnx.org/content/col11992/latest/

Recent Observatories in Space

Observatory	Date Operation Began	Bands of the Spectrum	Notes	Website
Chandra X-Ray Observatory	1999	X-rays	X-ray images and spectra	www.chandra.si.edu
XMM-Newton	1999	X-rays	X-ray spectroscopy	http://www.cosmos.esa.int/web/xmm-newton
International Gamma-Ray Astrophysics Laboratory (INTEGRAL)	2002	X- and gamma-rays	higher resolution gamma-ray images	http://sci.esa.int/integral/
Spitzer Space Telescope	2003	IR	0.85-m telescope	www.spitzer.caltech.edu
Fermi Gamma-ray Space Telescope	2008	gamma-rays	first high-energy gamma-ray observations	fermi.gsfc.nasa.gov
Kepler	2009	visible-light	planet finder	http://kepler.nasa.gov
Wide-field Infrared Survey Explorer (WISE)	2009	IR	whole-sky map, asteroid searches	www.nasa.gov/mission_pages/WISE/main
Gaia	2013	visible-light	Precise map of the Milky Way	http://sci.esa.int/gaia/

Table 6.4

6.6 | THE FUTURE OF LARGE TELESCOPES

Learning Objectives

By the end of this section, you will be able to:

> Describe the next generation of ground- and space-based observatories
> Explain some of the challenges involved in building these observatories

If you've ever gone on a hike, you have probably been eager to see what lies just around the next bend in the path. Researchers are no different, and astronomers and engineers are working on the technologies that will allow us to explore even more distant parts of the universe and to see them more clearly.

The premier space facility planned for the next decade is the James Webb Space Telescope (Figure 6.27), which (in a departure from tradition) is named after one of the early administrators of NASA instead of a scientist. This telescope will have a mirror 6 meters in diameter, made up, like the Keck telescopes, of 36 small hexagons. These will have to unfold into place once the telescope reaches its stable orbit point, some 1.5 million kilometers

This OpenStax book is available for free at http://cnx.org/content/col11992/1.8

from Earth (where no astronauts can currently travel if it needs repair.) The telescope is scheduled for launch in 2018 and should have the sensitivity needed to detect the very first generation of stars, formed when the universe was only a few hundred million years old. With the ability to measure both visible and infrared wavelengths, it will serve as the successor to both HST and the Spitzer Space Telescope.

Figure 6.27. James Webb Space Telescope (JWST). This image shows some of the mirrors of the JWST as they underwent cryogenic testing. The mirrors were exposed to extreme temperatures in order to gather accurate measurements on changes in their shape as they heated and cooled. (credit: NASA/MSFC/David Higginbotham/Emmett Given)

LINK TO LEARNING

Watch this video (https://openstaxcollege.org/l/30JWSTvid) to learn more about the James Webb Space Telescope and how it will build upon the work that Hubble has allowed us to begin in exploring the universe.

On the ground, astronomers have started building the Large Synoptic Survey Telescope (LSST), an 8.4-meter telescope with a significantly larger field of view than any existing telescopes. It will rapidly scan the sky to find *transients*, phenomena that change quickly, such as exploding stars and chunks of rock that orbit near Earth. The LSST is expected to see first light in 2021.

The international gamma-ray community is planning the Cherenkov Telescope Array (CTA), two arrays of telescopes, one in each hemisphere, which will indirectly measure gamma rays from the ground. The CTA will measure gamma-ray energies a thousand times as great as the Fermi telescope can detect.

Several groups of astronomers around the globe interested in studying visible light and infrared are exploring the feasibility of building ground-based telescopes with mirrors larger than 30 meters across. Stop and think what this means: 30 meters is one-third the length of a football field. It is technically impossible to build and transport a single astronomical mirror that is 30 meters or larger in diameter. The primary mirror of these giant telescopes will consist of smaller mirrors, all aligned so that they act as a very large mirror in combination. These include the Thirty-Meter Telescope for which construction has begun at the top of Mauna Kea in Hawaii.

The most ambitious of these projects is the European Extremely Large Telescope (E-ELT) (Figure 6.28). (Astronomers try to outdo each other not only with the size of these telescopes, but also their names!) The design of the E-ELT calls for a 39.3-meter primary mirror, which will follow the Keck design and be made up of 798 hexagonal mirrors, each 1.4 meters in diameter and all held precisely in position so that they form a continuous surface.

Download for free at http://cnx.org/content/col11992/latest/

Construction on the site in the Atacama Desert in Northern Chile started in 2014. The E-ELT, along with the Thirty Meter Telescope and the Giant Magellan Telescope, which are being built by international consortia led by US astronomers, will combine light-gathering power with high-resolution imaging. These powerful new instruments will enable astronomers to tackle many important astronomical problems. For example, they should be able to tell us when, where, and how often planets form around other stars. They should even be able to provide us images and spectra of such planets and thus, perhaps, give us the first real evidence (from the chemistry of these planets' atmospheres) that life exists elsewhere.

Figure 6.28. Artist's Conception of the European Extremely Large Telescope. The primary mirror in this telescope is 39.3 meters across. The telescope is under construction in the Atacama Desert in Northern Chile. (credit: ESO/L. Calçada)

LINK TO LEARNING

Check out this fun diagram (https://openstaxcollege.org/l/30JWSTdiag) comparing the sizes of the largest planned and existing telescopes to a regulation basketball and tennis court.

This OpenStax book is available for free at http://cnx.org/content/col11992/1.8

CHAPTER 6 REVIEW

 KEY TERMS

adaptive optics systems used with telescopes that can compensate for distortions in an image introduced by the atmosphere, thus resulting in sharper images

aperture diameter of the primary lens or mirror of a telescope

charge-coupled device (CCD) array of high-sensitivity electronic detectors of electromagnetic radiation, used at the focus of a telescope (or camera lens) to record an image or spectrum

chromatic aberration distortion that causes an image to appear fuzzy when each wavelength coming into a transparent material focuses at a different spot

detector device sensitive to electromagnetic radiation that makes a record of astronomical observations

eyepiece magnifying lens used to view the image produced by the objective lens or primary mirror of a telescope

focus (of telescope) point where the rays of light converged by a mirror or lens meet

interference process in which waves mix together such that their crests and troughs can alternately reinforce and cancel one another

interferometer instrument that combines electromagnetic radiation from one or more telescopes to obtain a resolution equivalent to what would be obtained with a single telescope with a diameter equal to the baseline separating the individual separate telescopes

interferometer array combination of multiple radio dishes to, in effect, work like a large number of two-dish interferometers

prime focus point in a telescope where the objective lens or primary mirror focuses the light

radar technique of transmitting radio waves to an object and then detecting the radiation that the object reflects back to the transmitter; used to measure the distance to, and motion of, a target object or to form images of it

reflecting telescope telescope in which the principal light collector is a concave mirror

refracting telescope telescope in which the principal light collector is a lens or system of lenses

resolution detail in an image; specifically, the smallest angular (or linear) features that can be distinguished

seeing unsteadiness of Earth's atmosphere, which blurs telescopic images; good seeing means the atmosphere is steady

telescope instrument for collecting visible-light or other electromagnetic radiation

Download for free at http://cnx.org/content/col11992/latest/

📑 SUMMARY

6.1 Telescopes

A telescope collects the faint light from astronomical sources and brings it to a focus, where an instrument can sort the light according to wavelength. Light is then directed to a detector, where a permanent record is made. The light-gathering power of a telescope is determined by the diameter of its aperture, or opening—that is, by the area of its largest or primary lens or mirror. The primary optical element in a telescope is either a convex lens (in a refracting telescope) or a concave mirror (in a reflector) that brings the light to a focus. Most large telescopes are reflectors; it is easier to manufacture and support large mirrors because the light does not have to pass through glass.

6.2 Telescopes Today

New technologies for creating and supporting lightweight mirrors have led to the construction of a number of large telescopes since 1990. The site for an astronomical observatory must be carefully chosen for clear weather, dark skies, low water vapor, and excellent atmospheric seeing (low atmospheric turbulence). The resolution of a visible-light or infrared telescope is degraded by turbulence in Earth's atmosphere. The technique of adaptive optics, however, can make corrections for this turbulence in real time and produce exquisitely detailed images.

6.3 Visible-Light Detectors and Instruments

Visible-light detectors include the human eye, photographic film, and charge-coupled devices (CCDs). Detectors that are sensitive to infrared radiation must be cooled to very low temperatures since everything in and near the telescope gives off infrared waves. A spectrometer disperses the light into a spectrum to be recorded for detailed analysis.

6.4 Radio Telescopes

In the 1930s, radio astronomy was pioneered by Karl G. Jansky and Grote Reber. A radio telescope is basically a radio antenna (often a large, curved dish) connected to a receiver. Significantly enhanced resolution can be obtained with interferometers, including interferometer arrays like the 27-element VLA and the 66-element ALMA. Expanding to very long baseline interferometers, radio astronomers can achieve resolutions as precise as 0.0001 arcsecond. Radar astronomy involves transmitting as well as receiving. The largest radar telescope currently in operation is a 305-meter bowl at Arecibo.

6.5 Observations outside Earth's Atmosphere

Infrared observations are made with telescopes aboard aircraft and in space, as well as from ground-based facilities on dry mountain peaks. Ultraviolet, X-ray, and gamma-ray observations must be made from above the atmosphere. Many orbiting observatories have been flown to observe in these bands of the spectrum in the last few decades. The largest-aperture telescope in space is the Hubble Space telescope (HST), the most significant infrared telescope is Spitzer, and Chandra and Fermi are the premier X-ray and gamma-ray observatories, respectively.

6.6 The Future of Large Telescopes

New and even larger telescopes are on the drawing boards. The James Webb Space Telescope, a 6-meter successor to Hubble, is currently scheduled for launch in 2018. Gamma-ray astronomers are planning to build the CTA to measure very energetic gamma rays. Astronomers are building the LSST to observe with an unprecedented field of view and a new generation of visible-light/infrared telescopes with apertures of 24.5 to 39 meters in diameter.

This OpenStax book is available for free at http://cnx.org/content/col11992/1.8

FOR FURTHER EXPLORATION

Articles

Blades, J. C. "Fixing the Hubble One Last Time." *Sky & Telescope* (October 2008): 26. On the last Shuttle service mission and what the Hubble was then capable of doing.

Brown, A. "How Gaia will Map a Billion Stars." *Astronomy* (December 2014): 32. Nice review of the mission to do photometry and spectroscopy of all stars above a certain brightness.

Irion, R. "Prime Time." *Astronomy* (February 2001): 46. On how time is allotted on the major research telescopes.

Jedicke, Peter & Robert. "The Coming Giant Sky Patrols." *Sky & Telescope* (September 2008): 30. About giant telescopes to survey the sky continuously.

Lazio, Joseph, et al. "Tuning in to the Universe: 21st Century Radio Astronomy." *Sky & Telescope* (July 2008): 21. About ALMA and the Square Kilometer Array.

Lowe, Jonathan. "Mirror, Mirror." *Sky & Telescope* (December 2007): 22. On the Large Binocular Telescope in Arizona.

Lowe, Jonathan. "Next Light: Tomorrow's Monster Telescopes." *Sky & Telescope* (April 2008): 20. About plans for extremely large telescopes on the ground.

Mason, Todd & Robin. "Palomar's Big Eye." *Sky & Telescope* (December 2008): 36. On the Hale 200-inch telescope.

Subinsky, Raymond. "Who Really Invented the Telescope." *Astronomy* (August 2008): 84. Brief historical introduction, focusing on Hans Lippershey.

Websites

Websites for major telescopes are given in Table 6.1, Table 6.2, Table 6.3, and Table 6.4.

Videos

Astronomy from the Stratosphere: SOFIA: https://www.youtube.com/watch?v=NV98BcBBA9c. A talk by Dr. Dana Backman (1:15:32)

Galaxies Viewed in Full Spectrum of Light: https://www.youtube.com/watch?v=368K0iQv8nE. Scientists with the Spitzer Observatory show how a galaxy looks different at different wavelengths (6:22)

Lifting the Cosmic Veil: Highlights from a Decade of the Spitzer Space Telescope: https://www.youtube.com/watch?v=nkrNQcwkY78. A talk by Dr. Michael Bicay (1:42:44)

 COLLABORATIVE GROUP ACTIVITIES

A. Most large telescopes get many more proposals for observing projects than there is night observing time available in a year. Suppose your group is the telescope time allocation committee reporting to an observatory director. What criteria would you use in deciding how to give out time on the telescope? What steps could you take to make sure all your colleagues thought the process was fair and people would still talk to you at future astronomy meetings?

Download for free at http://cnx.org/content/col11992/latest/

B. Your group is a committee of nervous astronomers about to make a proposal to the government ministers of your small European country to chip in with other countries to build the world's largest telescope in the high, dry desert of the Chilean Andes Mountains. You expect the government ministers to be very skeptical about supporting this project. What arguments would you make to convince them to participate?

C. The same government ministers we met in the previous activity ask you to draw up a list of the pros and cons of having the world's largest telescope in the mountains of Chile (instead of a mountain in Europe). What would your group list in each column?

D. Your group should discuss and make a list of all the ways in which an observing session at a large visible-light telescope and a large radio telescope might differ. (Hint: Bear in mind that because the Sun is not especially bright at many radio wavelengths, observations with radio telescopes can often be done during the day.)

E. Another "environmental threat" to astronomy (besides light pollution) comes from the spilling of terrestrial communications into the "channels"—wavelengths and frequencies—previously reserved for radio astronomy. For example, the demand for cellular phones means that more and more radio channels will be used for this purpose. The faint signals from cosmic radio sources could be drowned in a sea of earthly conversation (translated and sent as radio waves). Assume your group is a congressional committee being lobbied by both radio astronomers, who want to save some clear channels for doing astronomy, and the companies that stand to make a lot of money from expanding cellular phone use. What arguments would sway you to each side?

F. When the site for the new Thirty-Meter Telescope on Hawaii's Mauna Kea was dedicated, a group of native Hawaiians announced opposition to the project because astronomers were building too many telescopes on a mountain that native Hawaiians consider a sacred site. You can read more about this controversy at http://www.nytimes.com/2015/12/04/science/space/hawaii-court-rescinds-permit-to-build-thirty-meter-telescope.html?_r=0 and at http://www.nature.com/news/the-mountain-top-battle-over-the-thirty-meter-telescope-1.18446. Once your group has the facts, discuss the claims of each side in the controversy. How do you think it should be resolved?

G. If you could propose to use a large modern telescope, what would you want to find out? What telescope would you use and why?

H. Light pollution (spilled light in the night sky making it difficult to see the planets and stars) used to be an issue that concerned mostly astronomers. Now spilled light at night is also of concern to environmentalists and those worrying about global warming. Can your group come up with some non-astronomical reasons to be opposed to light pollution?

🗏 EXERCISES

Review Questions

1. What are the three basic components of a modern astronomical instrument? Describe each in one to two sentences.

2. Name the two spectral windows through which electromagnetic radiation easily reaches the surface of Earth and describe the largest-aperture telescope currently in use for each window.

This OpenStax book is available for free at http://cnx.org/content/col11992/1.8

3. List the largest-aperture single telescope currently in use in each of the following bands of the electromagnetic spectrum: radio, X-ray, gamma ray.

4. When astronomers discuss the apertures of their telescopes, they say bigger is better. Explain why.

5. The Hooker telescope at Palomar Observatory has a diameter of 5 m, and the Keck I telescope has a diameter of 10 m. How much more light can the Keck telescope collect than the Hooker telescope in the same amount of time?

6. What is meant by "reflecting" and "refracting" telescopes?

7. Why are the largest visible-light telescopes in the world made with mirrors rather than lenses?

8. Compare the eye, photographic film, and CCDs as detectors for light. What are the advantages and disadvantages of each?

9. What is a charge-coupled device (CCD), and how is it used in astronomy?

10. Why is it difficult to observe at infrared wavelengths? What do astronomers do to address this difficulty?

11. Radio and radar observations are often made with the same antenna, but otherwise they are very different techniques. Compare and contrast radio and radar astronomy in terms of the equipment needed, the methods used, and the kind of results obtained.

12. Look back at Figure 6.18 of Cygnus A and read its caption again. The material in the giant lobes at the edges of the image had to have been ejected from the center *at least* how many years ago?

13. Why do astronomers place telescopes in Earth's orbit? What are the advantages for the different regions of the spectrum?

14. What was the problem with the Hubble Space Telescope and how was it solved?

15. Describe the techniques radio astronomers use to obtain a resolution comparable to what astronomers working with visible light can achieve.

16. What kind of visible-light and infrared telescopes on the ground are astronomers planning for the future? Why are they building them on the ground and not in space?

17. Describe one visible-light or infrared telescope that astronomers are planning to launch into space in the future.

Thought Questions

18. What happens to the image produced by a lens if the lens is "stopped down" (the aperture reduced, thereby reducing the amount of light passing through the lens) with an iris diaphragm—a device that covers its periphery?

19. What would be the properties of an ideal astronomical detector? How closely do the actual properties of a CCD approach this ideal?

20. Many decades ago, the astronomers on the staff of Mount Wilson and Palomar Observatories each received about 60 nights per year for their observing programs. Today, an astronomer feels fortunate to get 10 nights per year on a large telescope. Can you suggest some reasons for this change?

21. The largest observatory complex in the world is on Mauna Kea, the tallest mountain on Earth. What are some factors astronomers consider when selecting an observatory site? Don't forget practical ones. Should astronomers, for example, consider building an observatory on Denali (Mount McKinley) or Mount Everest?

Download for free at http://cnx.org/content/col11992/latest/

22. Suppose you are looking for sites for a visible-light observatory, an infrared observatory, and a radio observatory. What are the main criteria of excellence for each? What sites are actually considered the best today?

23. Radio astronomy involves wavelengths much longer than those of visible light, and many orbiting observatories have probed the universe for radiation of very short wavelengths. What sorts of objects and physical conditions would you expect to be associated with emission of radiation at very long and very short wavelengths?

24. The dean of a university located near the ocean (who was not a science major in college) proposes building an infrared telescope right on campus and operating it in a nice heated dome so that astronomers will be comfortable on cold winter nights. Criticize this proposal, giving your reasoning.

Figuring For Yourself

25. What is the area, in square meters, of a 10-m telescope?

26. Approximately 9000 stars are visible to the naked eye in the whole sky (imagine that you could see around the entire globe and both the northern and southern hemispheres), and there are about 41,200 square degrees on the sky. How many stars are visible per square degree? Per square arcsecond?

27. Theoretically (that is, if seeing were not an issue), the resolution of a telescope is inversely proportional to its diameter. How much better is the resolution of the ALMA when operating at its longest baseline than the resolution of the Arecibo telescope?

28. In broad daylight, the size of your pupil is typically 3 mm. In dark situations, it expands to about 7 mm. How much more light can it gather?

29. How much more light can be gathered by a telescope that is 8 m in diameter than by your fully dark-adapted eye at 7 mm?

30. How much more light can the Keck telescope (with its 10-m diameter mirror) gather than an amateur telescope whose mirror is 25 cm (0.25 m) across?

31. People are often bothered when they discover that reflecting telescopes have a second mirror in the middle to bring the light out to an accessible focus where big instruments can be mounted. "Don't you lose light?" people ask. Well, yes, you do, but there is no better alternative. You can estimate how much light is lost by such an arrangement. The primary mirror (the one at the bottom in Figure 6.6) of the Gemini North telescope is 8 m in diameter. The secondary mirror at the top is about 1 m in diameter. Use the formula for the area of a circle to estimate what fraction of the light is blocked by the secondary mirror.

32. Telescopes can now be operated remotely from a warm room, but until about 25 years ago, astronomers worked at the telescope to guide it so that it remained pointed in exactly the right place. In a large telescope, like the Palomar 200-inch telescope, astronomers sat in a cage at the top of the telescope, where the secondary mirror is located, as shown in Figure 6.6. Assume for the purpose of your calculation that the diameter of this cage was 40 inches. What fraction of the light is blocked?

33. The HST cost about $1.7 billion for construction and $300 million for its shuttle launch, and it costs $250 million per year to operate. If the telescope lasts for 20 years, what is the total cost per year? Per day? If the telescope can be used just 30% of the time for actual observations, what is the cost per hour and per minute for the astronomer's observing time on this instrument? What is the cost per person in the United States? Was your investment in the Hubble Space telescope worth it?

34. How much more light can the James Webb Space Telescope (with its 6-m diameter mirror) gather than the Hubble Space Telescope (with a diameter of 2.4 m)?

This OpenStax book is available for free at http://cnx.org/content/col11992/1.8

35. The Palomar telescope's 5-m mirror weighs 14.5 tons. If a 10-m mirror were constructed of the same thickness as Palomar's (only bigger), how much would it weigh?

Download for free at http://cnx.org/content/col11992/latest/

This OpenStax book is available for free at http://cnx.org/content/col11992/1.8

Download for free at http://cnx.org/content/col11992/latest/

Figure 7.1. "Self-Portrait" of Mars. This picture was taken by the *Curiosity* Rover on Mars in 2012. The image is reconstructed digitally from 55 different images taken by a camera on the rover's extended mast, so that the many positions of the mast (which acted like a selfie stick) are edited out. (credit: modification of work by NASA/JPL-Caltech/MSSS)

Chapter Outline

7.1 Overview of Our Planetary System

7.2 Composition and Structure of Planets

7.3 Dating Planetary Surfaces

7.4 Origin of the Solar System

Thinking Ahead

Surrounding the Sun is a complex system of worlds with a wide range of conditions: eight major planets, many dwarf planets, hundreds of moons, and countless smaller objects. Thanks largely to visits by spacecraft, we can now envision the members of the solar system as other worlds like our own, each with its own chemical and geological history, and unique sights that interplanetary tourists may someday visit. Some have called these past few decades the "golden age of planetary exploration," comparable to the golden age of exploration in the fifteenth century, when great sailing ships plied Earth's oceans and humanity became familiar with our own planet's surface.

In this chapter, we discuss our planetary system and introduce the idea of comparative planetology—studying how the planets work by comparing them with one another. We want to get to know the planets not only for what we can learn about them, but also to see what they can tell us about the origin and evolution of the entire solar system. In the upcoming chapters, we describe the better-known members of the solar system and begin to compare them to the thousands of planets that have been discovered recently, orbiting other stars.

Download for free at http://cnx.org/content/col11992/latest/

7.1 | OVERVIEW OF OUR PLANETARY SYSTEM

Learning Objectives

By the end of this section, you will be able to:

> Describe how the objects in our solar system are identified, explored, and characterized
> Describe the types of small bodies in our solar system, their locations, and how they formed
> Model the solar system with distances from everyday life to better comprehend distances in space

The solar system[1] consists of the Sun and many smaller objects: the planets, their moons and rings, and such "debris" as asteroids, comets, and dust. Decades of observation and spacecraft exploration have revealed that most of these objects formed together with the Sun about 4.5 billion years ago. They represent clumps of material that condensed from an enormous cloud of gas and dust. The central part of this cloud became the Sun, and a small fraction of the material in the outer parts eventually formed the other objects.

During the past 50 years, we have learned more about the solar system than anyone imagined before the space age. In addition to gathering information with powerful new telescopes, we have sent spacecraft directly to many members of the planetary system. (Planetary astronomy is the only branch of our science in which we can, at least vicariously, travel to the objects we want to study.) With evocative names such as Voyager, Pioneer, *Curiosity*, and Pathfinder, our robot explorers have flown past, orbited, or landed on every planet, returning images and data that have dazzled both astronomers and the public. In the process, we have also investigated two dwarf planets, hundreds of fascinating moons, four ring systems, a dozen asteroids, and several comets (smaller members of our solar system that we will discuss later).

Our probes have penetrated the atmosphere of Jupiter and landed on the surfaces of Venus, Mars, our Moon, Saturn's moon Titan, the asteroids Eros and Itokawa, and the Comet Churyumov-Gerasimenko (usually referred to as 67P). Humans have set foot on the Moon and returned samples of its surface soil for laboratory analysis (Figure 7.2). We have even discovered other places in our solar system that might be able to support some kind of life.

---

1   The generic term for a group of planets and other bodies circling a star is *planetary system*. Ours is called the *solar system* because our Sun is sometimes called *Sol*. Strictly speaking, then, there is only one solar system; planets orbiting other stars are in planetary systems.

This OpenStax book is available for free at http://cnx.org/content/col11992/1.8

Download for free at http://cnx.org/content/col11992/latest/

Figure 7.2. Astronauts on the Moon. The lunar lander and surface rover from the Apollo 15 mission are seen in this view of the one place beyond Earth that has been explored directly by humans. (credit: modification of work by David R. Scott, NASA)

LINK TO LEARNING

View this gallery of NASA images (https://openstaxcollege.org/l/30projapolloarc) that trace the history of the Apollo mission.

An Inventory

The Sun, a star that is brighter than about 80% of the stars in the Galaxy, is by far the most massive member of the solar system, as shown in Table 7.1. It is an enormous ball about 1.4 million kilometers in diameter, with surface layers of incandescent gas and an interior temperature of millions of degrees. The Sun will be discussed in later chapters as our first, and best-studied, example of a star.

Mass of Members of the Solar System

Object	Percentage of Total Mass of Solar System
Sun	99.80
Jupiter	0.10
Comets	0.0005–0.03 (estimate)
All other planets and dwarf planets	0.04
Moons and rings	0.00005

Table 7.1

Download for free at http://cnx.org/content/col11992/latest/

Mass of Members of the Solar System

Object	Percentage of Total Mass of Solar System
Asteroids	0.000002 (estimate)
Cosmic dust	0.0000001 (estimate)

Table 7.1

Table 7.1 also shows that most of the material of the planets is actually concentrated in the largest one, Jupiter, which is more massive than all the rest of the planets combined. Astronomers were able to determine the masses of the planets centuries ago using Kepler's laws of planetary motion and Newton's law of gravity to measure the planets' gravitational effects on one another or on moons that orbit them (see Orbits and Gravity). Today, we make even more precise measurements of their masses by tracking their gravitational effects on the motion of spacecraft that pass near them.

Beside Earth, five other planets were known to the ancients—Mercury, Venus, Mars, Jupiter, and Saturn—and two were discovered after the invention of the telescope: Uranus and Neptune. The eight planets all revolve in the same direction around the Sun. They orbit in approximately the same plane, like cars traveling on concentric tracks on a giant, flat racecourse. Each planet stays in its own "traffic lane," following a nearly circular orbit about the Sun and obeying the "traffic" laws discovered by Galileo, Kepler, and Newton. Besides these planets, we have also been discovering smaller worlds beyond Neptune that are called trans-Neptunian objects or TNOs (see Figure 7.3). The first to be found, in 1930, was Pluto, but others have been discovered during the twenty-first century. One of them, Eris, is about the same size as Pluto and has at least one moon (Pluto has five known moons.) The largest TNOs are also classed as *dwarf planets,* as is the largest asteroid, Ceres. (Dwarf planets will be discussed further in the chapter on Rings, Moons, and Pluto). To date, more than 1750 of these TNOs have been discovered.

This OpenStax book is available for free at http://cnx.org/content/col11992/1.8

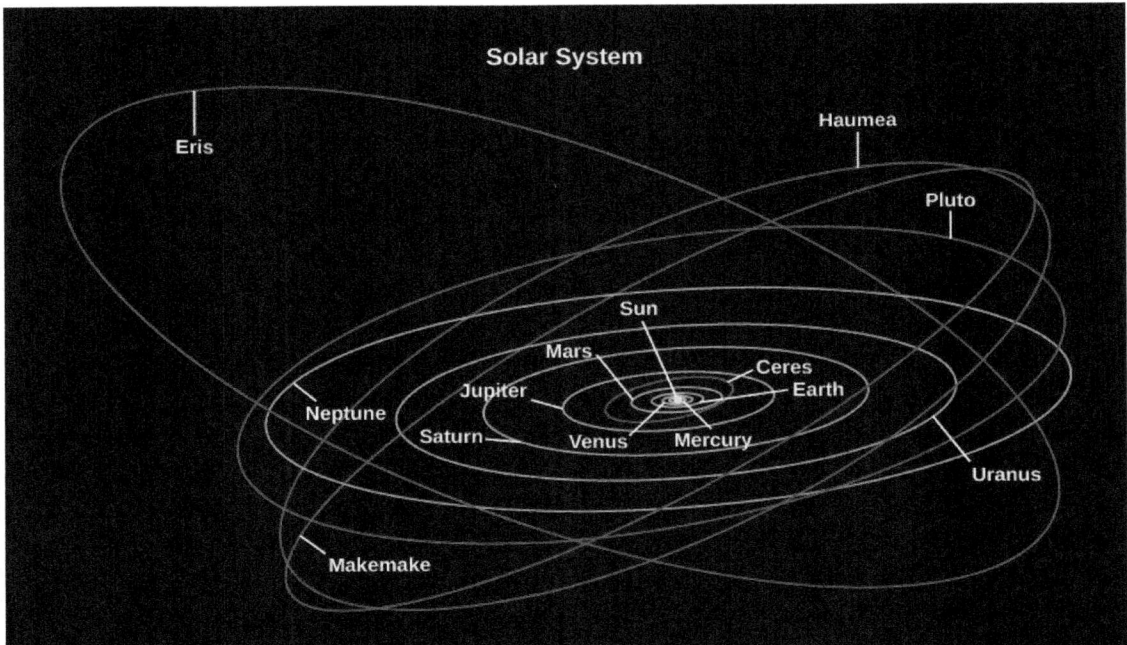

Figure 7.3. Orbits of the Planets. All eight major planets orbit the Sun in roughly the same plane. The five currently known dwarf planets are also shown: Eris, Haumea, Pluto, Ceres, and Makemake. Note that Pluto's orbit is not in the plane of the planets.

Each of the planets and dwarf planets also rotates (spins) about an axis running through it, and in most cases the direction of rotation is the same as the direction of revolution about the Sun. The exceptions are Venus, which rotates backward very slowly (that is, in a retrograde direction), and Uranus and Pluto, which also have strange rotations, each spinning about an axis tipped nearly on its side. We do not yet know the spin orientations of Eris, Haumea, and Makemake.

The four planets closest to the Sun (Mercury through Mars) are called the inner or **terrestrial planets**. Often, the Moon is also discussed as a part of this group, bringing the total of terrestrial objects to five. (We generally call Earth's satellite "the Moon," with a capital M, and the other satellites "moons," with lowercase m's.) The terrestrial planets are relatively small worlds, composed primarily of rock and metal. All of them have solid surfaces that bear the records of their geological history in the forms of craters, mountains, and volcanoes (Figure 7.4).

Download for free at http://cnx.org/content/col11992/latest/

Figure 7.4. Surface of Mercury. The pockmarked face of the terrestrial world of Mercury is more typical of the inner planets than the watery surface of Earth. This black-and-white image, taken with the Mariner 10 spacecraft, shows a region more than 400 kilometers wide. (credit: modification of work by NASA/John Hopkins University Applied Physics Laboratory/Carnegie Institution of Washington)

The next four planets (Jupiter through Neptune) are much larger and are composed primarily of lighter ices, liquids, and gases. We call these four the jovian planets (after "Jove," another name for Jupiter in mythology) or **giant planets**—a name they richly deserve (Figure 7.5). More than 1400 Earths could fit inside Jupiter, for example. These planets do not have solid surfaces on which future explorers might land. They are more like vast, spherical oceans with much smaller, dense cores.

Figure 7.5. The Four Giant Planets. This montage shows the four giant planets: Jupiter, Saturn, Uranus, and Neptune. Below them, Earth is shown to scale. (credit: modification of work by NASA, Solar System Exploration)

Near the outer edge of the system lies Pluto, which was the first of the distant icy worlds to be discovered beyond Neptune (Pluto was visited by a spacecraft, the NASA New Horizons mission, in 2015 [see Figure 7.6]). Table 7.2 summarizes some of the main facts about the planets.

This OpenStax book is available for free at http://cnx.org/content/col11992/1.8

Figure 7.6. Pluto Close-up. This intriguing image from the New Horizons spacecraft, taken when it flew by the dwarf planet in July 2015, shows some of its complex surface features. The rounded white area is temporarily being called the Sputnik Plain, after humanity's first spacecraft. (credit: modification of work by NASA/Johns Hopkins University Applied Physics Laboratory/Southwest Research Institute)

The Planets

Name	Distance from Sun (AU)[2]	Revolution Period (y)	Diameter (km)	Mass (10^{23} kg)	Density (g/cm^3)[3]
Mercury	0.39	0.24	4,878	3.3	5.4
Venus	0.72	0.62	12,120	48.7	5.2
Earth	1.00	1.00	12,756	59.8	5.5
Mars	1.52	1.88	6,787	6.4	3.9
Jupiter	5.20	11.86	142,984	18,991	1.3
Saturn	9.54	29.46	120,536	5686	0.7
Uranus	19.18	84.07	51,118	866	1.3
Neptune	30.06	164.82	49,660	1030	1.6

Table 7.2

EXAMPLE 7.1

Comparing Densities

Let's compare the densities of several members of the solar system. The density of an object equals its mass divided by its volume. The volume (*V*) of a sphere (like a planet) is calculated using the equation

---

2   An AU (or astronomical unit) is the distance from Earth to the Sun.
3   We give densities in units where the density of water is 1 g/cm^3. To get densities in units of kg/m^3, multiply the given value by 1000.

Download for free at http://cnx.org/content/col11992/latest/

$$V = \frac{4}{3}\pi R^3$$

where π (the Greek letter pi) has a value of approximately 3.14. Although planets are not perfect spheres, this equation works well enough. The masses and diameters of the planets are given in Table 7.2. For data on selected moons, see Appendix G. Let's use Saturn's moon Mimas as our example, with a mass of 4×10^{19} kg and a diameter of approximately 400 km (radius, 200 km = 2×10^5 m).

Solution

The volume of Mimas is

$$\frac{4}{3} \times 3.14 \times \left(2 \times 10^5 \text{ m}\right)^3 = 3.3 \times 10^{16} \text{ m}^3.$$

Density is mass divided by volume:

$$\frac{4 \times 10^{19} \text{ kg}}{3.3 \times 10^{16} \text{ m}^3} = 1.2 \times 10^3 \text{ kg/m}^3.$$

Note that the density of water in these units is 1000 kg/m^3, so Mimas must be made mainly of ice, not rock. (Note that the density of Mimas given in Appendix G is 1.2, but the units used there are different. In that table, we give density in units of
g/cm^3, for which the density of water equals 1. Can you show, by converting units, that 1 g/cm^3 is the same as 1000 kg/m^3?)

Check Your Learning

Calculate the average density of our own planet, Earth. Show your work. How does it compare to the density of an ice moon like Mimas? See Table 7.2 for data.

Answer:

For a sphere,
$$\text{density} = \frac{\text{mass}}{\left(\frac{4}{3}\pi R^3\right)} \text{ kg/m}^3.$$

For Earth, then,
$$\text{density} = \frac{6 \times 10^{24} \text{ kg}}{4.2 \times 2.6 \times 10^{20} \text{ m}^3} = 5.5 \times 10^3 \text{ kg/m}^3.$$

This density is four to five times greater than Mimas'. In fact, Earth is the densest of the planets.

LINK TO LEARNING

Learn more about NASA's mission to Pluto (https://openstaxcollege.org/l/30NASAmisspluto) and see high-resolution images of Pluto's moon Charon.

Smaller Members of the Solar System

Most of the planets are accompanied by one or more moons; only Mercury and Venus move through space alone. There are more than 180 known moons orbiting planets and dwarf planets (see Appendix G for a listing

Download for free at http://cnx.org/content/col11992/latest/

of the larger ones), and undoubtedly many other small ones remain undiscovered. The largest of the moons are as big as small planets and just as interesting. In addition to our Moon, they include the four largest moons of Jupiter (called the Galilean moons, after their discoverer) and the largest moons of Saturn and Neptune (confusingly named Titan and Triton).

Each of the giant planets also has rings made up of countless small bodies ranging in size from mountains to mere grains of dust, all in orbit about the equator of the planet. The bright rings of Saturn are, by far, the easiest to see. They are among the most beautiful sights in the solar system (Figure 7.7). But, all four ring systems are interesting to scientists because of their complicated forms, influenced by the pull of the moons that also orbit these giant planets.

Figure 7.7. Saturn and Its Rings. This 2007 Cassini image shows Saturn and its complex system of rings, taken from a distance of about 1.2 million kilometers. This natural-color image is a composite of 36 images taken over the course of 2.5 hours. (credit: modification of work by NASA/JPL/Space Science Institute)

The solar system has many other less-conspicuous members. Another group is the **asteroids**, rocky bodies that orbit the Sun like miniature planets, mostly in the space between Mars and Jupiter (although some do cross the orbits of planets like Earth—see Figure 7.8). Most asteroids are remnants of the initial population of the solar system that existed before the planets themselves formed. Some of the smallest moons of the planets, such as the moons of Mars, are very likely captured asteroids.

Figure 7.8. Asteroid Eros. This small Earth-crossing asteroid image was taken by the NEAR-Shoemaker spacecraft from an altitude of about 100 kilometers. This view of the heavily cratered surface is about 10 kilometers wide. The spacecraft orbited Eros for a year before landing gently on its surface. (credit: modification of work by NASA/JHUAPL)

Another class of small bodies is composed mostly of ice, made of frozen gases such as water, carbon dioxide, and carbon monoxide; these objects are called **comets** (see Figure 7.9). Comets also are remnants from the formation of the solar system, but they were formed and continue (with rare exceptions) to orbit the Sun in

Download for free at http://cnx.org/content/col11992/latest/

distant, cooler regions—stored in a sort of cosmic deep freeze. This is also the realm of the larger icy worlds, called dwarf planets.

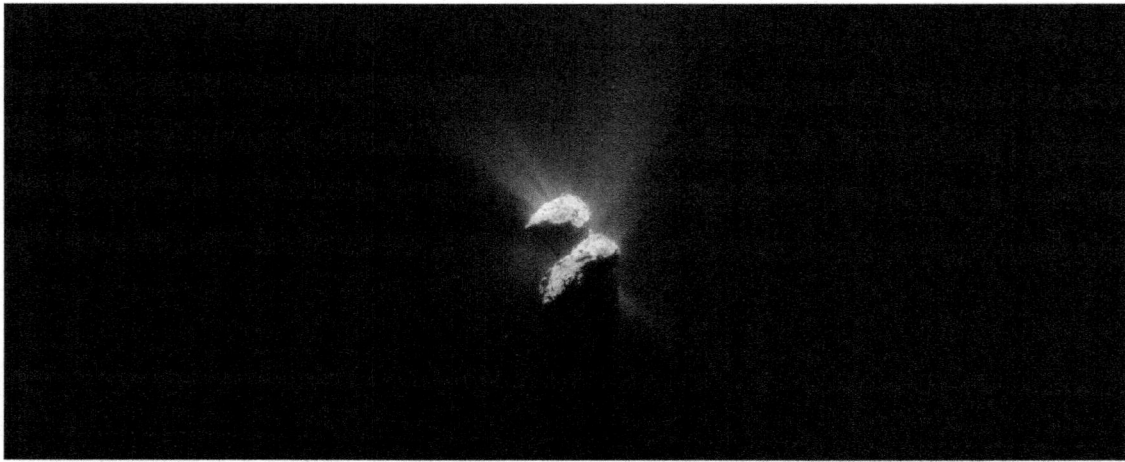

Figure 7.9. Comet Churyumov-Gerasimenko (67P). This image shows Comet Churyumov-Gerasimenko, also known as 67P, near its closest approach to the Sun in 2015, as seen from the *Rosetta* spacecraft. Note the jets of gas escaping from the solid surface. (credit: modification of work by ESA/Rosetta/NAVACAM, CC BY-SA IGO 3.0 (http://creativecommons.org/licenses/by-sa/3.0/igo/))

Finally, there are countless grains of broken rock, which we call cosmic dust, scattered throughout the solar system. When these particles enter Earth's atmosphere (as millions do each day) they burn up, producing a brief flash of light in the night sky known as a **meteor** (meteors are often referred to as shooting stars). Occasionally, some larger chunk of rocky or metallic material survives its passage through the atmosphere and lands on Earth. Any piece that strikes the ground is known as a **meteorite**. (You can see meteorites on display in many natural history museums and can sometimes even purchase pieces of them from gem and mineral dealers.)

VOYAGERS IN ASTRONOMY

Carl Sagan: Solar System Advocate

The best-known astronomer in the world during the 1970s and 1980s, Carl Sagan devoted most of his professional career to studying the planets and considerable energy to raising public awareness of what we can learn from exploring the solar system (see Figure 7.10). Born in Brooklyn, New York, in 1934, Sagan became interested in astronomy as a youngster; he also credits science fiction stories for sustaining his fascination with what's out in the universe.

This OpenStax book is available for free at http://cnx.org/content/col11992/1.8

Download for free at http://cnx.org/content/col11992/latest/

Figure 7.10. Carl Sagan (1934–1996) and Neil deGrasse Tyson. Sagan was Tyson's inspiration to become a scientist. (credit "Sagan": modification of work by NASA, JPL; credit "Tyson": modification of work by Bruce F. Press)

In the early 1960s, when many scientists still thought Venus might turn out to be a hospitable place, Sagan calculated that the thick atmosphere of Venus could act like a giant greenhouse, keeping the heat in and raising the temperature enormously. He showed that the seasonal changes astronomers had seen on Mars were caused, not by vegetation, but by wind-blown dust. He was a member of the scientific teams for many of the robotic missions that explored the solar system and was instrumental in getting NASA to put a message-bearing plaque aboard the Pioneer spacecraft, as well as audio-video records on the Voyager spacecraft—all of them destined to leave our solar system entirely and send these little bits of Earth technology out among the stars.

To encourage public interest and public support of planetary exploration, Sagan helped found The Planetary Society, now the largest space-interest organization in the world. He was a tireless and eloquent advocate of the need to study the solar system close-up and the value of learning about other worlds in order to take better care of our own.

Sagan simulated conditions on early Earth to demonstrate how some of life's fundamental building blocks might have formed from the "primordial soup" of natural compounds on our planet. In addition, he and his colleagues developed computer models showing the consequences of nuclear war for Earth would be even more devastating than anyone had thought (this is now called the nuclear winter hypothesis) and demonstrating some of the serious consequences of continued pollution of our atmosphere.

Sagan was perhaps best known, however, as a brilliant popularizer of astronomy and the author of many books on science, including the best-selling *Cosmos*, and several evocative tributes to solar system exploration such as *The Cosmic Connection* and *Pale Blue Dot*. His book *The Demon Haunted World*, completed just before his death in 1996, is perhaps the best antidote to fuzzy thinking about pseudo-science and irrationality in print today. An intriguing science fiction novel he wrote, titled *Contact*, which became a successful film as well, is still recommended by many science instructors as a scenario for making contact with life elsewhere that is much more reasonable than most science fiction.

Sagan was a master, too, of the television medium. His 13-part public television series, *Cosmos*, was seen by an estimated 500 million people in 60 countries and has become one of the most-watched series in the

Download for free at http://cnx.org/content/col11992/latest/

history of public broadcasting. A few astronomers scoffed at a scientist who spent so much time in the public eye, but it is probably fair to say that Sagan's enthusiasm and skill as an explainer won more friends for the science of astronomy than anyone or anything else in the second half of the twentieth century.

In the two decades since Sagan's death, no other scientist has achieved the same level of public recognition. Perhaps closest is the director of the Hayden Planetarium, Neil deGrasse Tyson, who followed in Sagan's footsteps by making an updated version of the *Cosmos* program in 2014. Tyson is quick to point out that Sagan was his inspiration to become a scientist, telling how Sagan invited him to visit for a day at Cornell when he was a high school student looking for a career. However, the media environment has fragmented a great deal since Sagan's time. It is interesting to speculate whether Sagan could have adapted his communication style to the world of cable television, Twitter, Facebook, and podcasts.

LINK TO LEARNING

Two imaginative videos provide a tour of the solar system objects we have been discussing. Shane Gellert's I Need Some Space (https://openstaxcollege.org/l/30needsomespace) uses NASA photography and models to show the various worlds with which we share our system. In the more science fiction-oriented Wanderers (https://openstaxcollege.org/l/30wanderers) video, we see some of the planets and moons as tourist destinations for future explorers, with commentary taken from recordings by Carl Sagan.

A Scale Model of the Solar System

Astronomy often deals with dimensions and distances that far exceed our ordinary experience. What does 1.4 billion kilometers—the distance from the Sun to Saturn—really mean to anyone? It can be helpful to visualize such large systems in terms of a scale model.

In our imaginations, let us build a scale model of the solar system, adopting a scale factor of 1 billion (10^9)—that is, reducing the actual solar system by dividing every dimension by a factor of 10^9. Earth, then, has a diameter of 1.3 centimeters, about the size of a grape. The Moon is a pea orbiting this at a distance of 40 centimeters, or a little more than a foot away. The Earth-Moon system fits into a standard backpack.

In this model, the Sun is nearly 1.5 meters in diameter, about the average height of an adult, and our Earth is at a distance of 150 meters—about one city block—from the Sun. Jupiter is five blocks away from the Sun, and its diameter is 15 centimeters, about the size of a very large grapefruit. Saturn is 10 blocks from the Sun; Uranus, 20 blocks; and Neptune, 30 blocks. Pluto, with a distance that varies quite a bit during its 249-year orbit, is currently just beyond 30 blocks and getting farther with time. Most of the moons of the outer solar system are the sizes of various kinds of seeds orbiting the grapefruit, oranges, and lemons that represent the outer planets.

In our scale model, a human is reduced to the dimensions of a single atom, and cars and spacecraft to the size of molecules. Sending the Voyager spacecraft to Neptune involves navigating a single molecule from the Earth–grape toward a lemon 5 kilometers away with an accuracy equivalent to the width of a thread in a spider's web.

This OpenStax book is available for free at http://cnx.org/content/col11992/1.8

If that model represents the solar system, where would the nearest stars be? If we keep the same scale, the closest stars would be tens of thousands of kilometers away. If you built this scale model in the city where you live, you would have to place the representations of these stars on the other side of Earth or beyond.

By the way, model solar systems like the one we just presented have been built in cities throughout the world. In Sweden, for example, Stockholm's huge Globe Arena has become a model for the Sun, and Pluto is represented by a 12-centimeter sculpture in the small town of Delsbo, 300 kilometers away. Another model solar system is in Washington on the Mall between the White House and Congress (perhaps proving they are worlds apart?).

MAKING CONNECTIONS

Names in the Solar System

We humans just don't feel comfortable until something has a name. Types of butterflies, new elements, and the mountains of Venus all need names for us to feel we are acquainted with them. How do we give names to objects and features in the solar system?

Planets and moons are named after gods and heroes in Greek and Roman mythology (with a few exceptions among the moons of Uranus, which have names drawn from English literature). When William Herschel, a German immigrant to England, first discovered the planet we now call Uranus, he wanted to name it Georgium Sidus (George's star) after King George III of his adopted country. This caused such an outcry among astronomers in other nations, however, that the classic tradition was upheld—and has been maintained ever since. Luckily, there were a lot of minor gods in the ancient pantheon, so plenty of names are left for the many small moons we are discovering around the giant planets. (Appendix G lists the larger moons).

Comets are often named after their discoverers (offering an extra incentive to comet hunters). Asteroids are named by their discoverers after just about anyone or anything they want. Recently, asteroid names have been used to recognize people who have made significant contributions to astronomy, including the three original authors of this book.

That was pretty much all the naming that was needed while our study of the solar system was confined to Earth. But now, our spacecraft have surveyed and photographed many worlds in great detail, and each world has a host of features that also need names. To make sure that naming things in space remains multinational, rational, and somewhat dignified, astronomers have given the responsibility of approving names to a special committee of the International Astronomical Union (IAU), the body that includes scientists from every country that does astronomy.

This IAU committee has developed a set of rules for naming features on other worlds. For example, craters on Venus are named for women who have made significant contributions to human knowledge and welfare. Volcanic features on Jupiter's moon Io, which is in a constant state of volcanic activity, are named after gods of fire and thunder from the mythologies of many cultures. Craters on Mercury commemorate famous novelists, playwrights, artists, and composers. On Saturn's moon Tethys, all the features are named after characters and places in Homer's great epic poem, *The Odyssey*. As we explore further, it may well turn out that more places in the solar system need names than Earth history can provide. Perhaps by then, explorers and settlers on these worlds will be ready to develop their own names for the places they may (if but for a while) call home.

Download for free at http://cnx.org/content/col11992/latest/

You may be surprised to know that the meaning of the word *planet* has recently become controversial because we have discovered many other planetary systems that don't look very much like our own. Even within our solar system, the planets differ greatly in size and chemical properties. The biggest dispute concerns Pluto, which is much smaller than the other eight major planets. The category of dwarf planet was invented to include Pluto and similar icy objects beyond Neptune. But is a dwarf planet also a planet? Logically, it should be, but even this simple issue of grammar has been the subject of heated debate among both astronomers and the general public.

7.2 COMPOSITION AND STRUCTURE OF PLANETS

Learning Objectives

By the end of this section, you will be able to:

> Describe the characteristics of the giant planets, terrestrial planets, and small bodies in the solar system
> Explain what influences the temperature of a planet's surface
> Explain why there is geological activity on some planets and not on others

The fact that there are two distinct kinds of planets—the rocky terrestrial planets and the gas-rich jovian planets—leads us to believe that they formed under different conditions. Certainly their compositions are dominated by different elements. Let us look at each type in more detail.

The Giant Planets

The two largest planets, Jupiter and Saturn, have nearly the same chemical makeup as the Sun; they are composed primarily of the two elements hydrogen and helium, with 75% of their mass being hydrogen and 25% helium. On Earth, both hydrogen and helium are gases, so Jupiter and Saturn are sometimes called gas planets. But, this name is misleading. Jupiter and Saturn are so large that the gas is compressed in their interior until the hydrogen becomes a liquid. Because the bulk of both planets consists of compressed, liquefied hydrogen, we should really call them liquid planets.

Under the force of gravity, the heavier elements sink toward the inner parts of a liquid or gaseous planet. Both Jupiter and Saturn, therefore, have cores composed of heavier rock, metal, and ice, but we cannot see these regions directly. In fact, when we look down from above, all we see is the atmosphere with its swirling clouds (Figure 7.11). We must infer the existence of the denser core inside these planets from studies of each planet's gravity.

This OpenStax book is available for free at http://cnx.org/content/col11992/1.8

Figure 7.11. Jupiter. This true-color image of Jupiter was taken from the Cassini spacecraft in 2000. (credit: modification of work by NASA/JPL/ University of Arizona)

Uranus and Neptune are much smaller than Jupiter and Saturn, but each also has a core of rock, metal, and ice. Uranus and Neptune were less efficient at attracting hydrogen and helium gas, so they have much smaller atmospheres in proportion to their cores.

Chemically, each giant planet is dominated by hydrogen and its many compounds. Nearly all the oxygen present is combined chemically with hydrogen to form water (H_2O). Chemists call such a hydrogen-dominated composition *reduced*. Throughout the outer solar system, we find abundant water (mostly in the form of ice) and reducing chemistry.

The Terrestrial Planets

The terrestrial planets are quite different from the giants. In addition to being much smaller, they are composed primarily of rocks and metals. These, in turn, are made of elements that are less common in the universe as a whole. The most abundant rocks, called silicates, are made of silicon and oxygen, and the most common metal is iron. We can tell from their densities (see Table 7.2) that Mercury has the greatest proportion of metals (which are denser) and the Moon has the lowest. Earth, Venus, and Mars all have roughly similar bulk compositions: about one third of their mass consists of iron-nickel or iron-sulfur combinations; two thirds is made of silicates. Because these planets are largely composed of oxygen compounds (such as the silicate minerals of their crusts), their chemistry is said to be *oxidized*.

When we look at the internal structure of each of the terrestrial planets, we find that the densest metals are in a central core, with the lighter silicates near the surface. If these planets were liquid, like the giant planets, we could understand this effect as the result the sinking of heavier elements due to the pull of gravity. This leads us to conclude that, although the terrestrial planets are solid today, at one time they must have been hot enough to melt.

Differentiation is the process by which gravity helps separate a planet's interior into layers of different compositions and densities. The heavier metals sink to form a core, while the lightest minerals float to the surface to form a crust. Later, when the planet cools, this layered structure is preserved. In order for a rocky planet to differentiate, it must be heated to the melting point of rocks, which is typically more than 1300 K.

Download for free at http://cnx.org/content/col11992/latest/

Moons, Asteroids, and Comets

Chemically and structurally, Earth's Moon is like the terrestrial planets, but most moons are in the outer solar system, and they have compositions similar to the cores of the giant planets around which they orbit. The three largest moons—Ganymede and Callisto in the jovian system, and Titan in the saturnian system—are composed half of frozen water, and half of rocks and metals. Most of these moons differentiated during formation, and today they have cores of rock and metal, with upper layers and crusts of very cold and—thus very hard—ice (Figure 7.12).

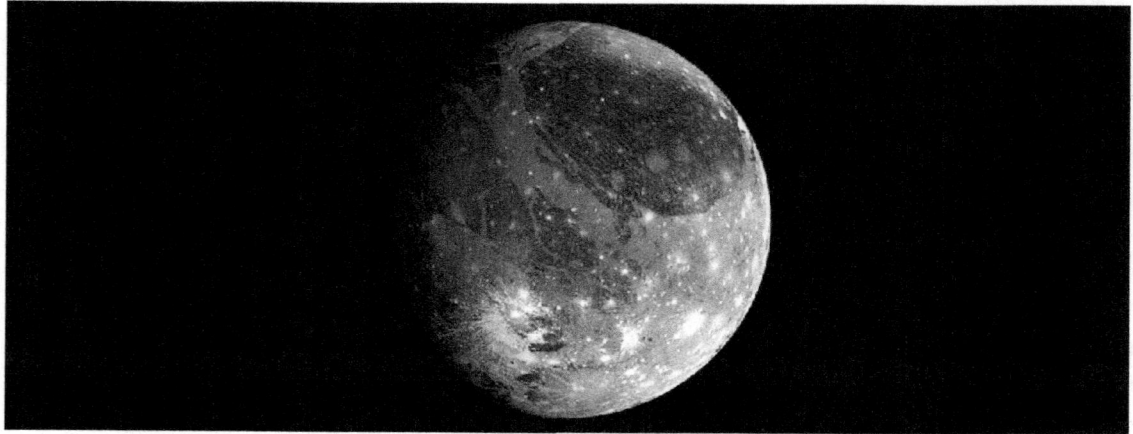

Figure 7.12. Ganymede. This view of Jupiter's moon Ganymede was taken in June 1996 by the Galileo spacecraft. The brownish gray color of the surface indicates a dusty mixture of rocky material and ice. The bright spots are places where recent impacts have uncovered fresh ice from underneath. (credit: modification of work by NASA/JPL)

Most of the asteroids and comets, as well as the smallest moons, were probably never heated to the melting point. However, some of the largest asteroids, such as Vesta, appear to be differentiated; others are fragments from differentiated bodies. Because most asteroids and comets retain their original composition, they represent relatively unmodified material dating back to the time of the formation of the solar system. In a sense, they act as chemical fossils, helping us to learn about a time long ago whose traces have been erased on larger worlds.

Temperatures: Going to Extremes

Generally speaking, the farther a planet or moon is from the Sun, the cooler its surface. The planets are heated by the radiant energy of the Sun, which gets weaker with the square of the distance. You know how rapidly the heating effect of a fireplace or an outdoor radiant heater diminishes as you walk away from it; the same effect applies to the Sun. Mercury, the closest planet to the Sun, has a blistering surface temperature that ranges from 280–430 °C on its sunlit side, whereas the surface temperature on Pluto is only about –220 °C, colder than liquid air.

Mathematically, the temperatures decrease approximately in proportion to the square root of the distance from the Sun. Pluto is about 30 AU at its closest to the Sun (or 100 times the distance of Mercury) and about 49 AU at its farthest from the Sun. Thus, Pluto's temperature is less than that of Mercury by the square root of 100, or a factor of 10: from 500 K to 50 K.

In addition to its distance from the Sun, the surface temperature of a planet can be influenced strongly by its atmosphere. Without our atmospheric insulation (the greenhouse effect, which keeps the heat in), the oceans of Earth would be permanently frozen. Conversely, if Mars once had a larger atmosphere in the past,

This OpenStax book is available for free at http://cnx.org/content/col11992/1.8

it could have supported a more temperate climate than it has today. Venus is an even more extreme example, where its thick atmosphere of carbon dioxide acts as insulation, reducing the escape of heat built up at the surface, resulting in temperatures greater than those on Mercury. Today, Earth is the only planet where surface temperatures generally lie between the freezing and boiling points of water. As far as we know, Earth is the only planet to support life.

ASTRONOMY BASICS

There's No Place Like Home

In the classic film *The Wizard of Oz*, Dorothy, the heroine, concludes after her many adventures in "alien" environments that "there's no place like home." The same can be said of the other worlds in our solar system. There are many fascinating places, large and small, that we might like to visit, but humans could not survive on any without a great deal of artificial assistance.

A thick carbon dioxide atmosphere keeps the surface temperature on our neighbor Venus at a sizzling 700 K (near 900 °F). Mars, on the other hand, has temperatures generally below freezing, with air (also mostly carbon dioxide) so thin that it resembles that found at an altitude of 30 kilometers (100,000 feet) in Earth's atmosphere. And the red planet is so dry that it has not had any rain for billions of years.

The outer layers of the jovian planets are neither warm enough nor solid enough for human habitation. Any bases we build in the systems of the giant planets may well have to be in space or one of their moons—none of which is particularly hospitable to a luxury hotel with a swimming pool and palm trees. Perhaps we will find warmer havens deep inside the clouds of Jupiter or in the ocean under the frozen ice of its moon Europa.

All of this suggests that we had better take good care of Earth because it is the only site where life as we know it could survive. Recent human activity may be reducing the habitability of our planet by adding pollutants to the atmosphere, especially the potent greenhouse gas carbon dioxide. Human civilization is changing our planet dramatically, and these changes are not necessarily for the better. In a solar system that seems unready to receive us, making Earth less hospitable to life may be a grave mistake.

Geological Activity

The crusts of all of the terrestrial planets, as well as of the larger moons, have been modified over their histories by both internal and external forces. Externally, each has been battered by a slow rain of projectiles from space, leaving their surfaces pockmarked by impact craters of all sizes (see Figure 7.4). We have good evidence that this bombardment was far greater in the early history of the solar system, but it certainly continues to this day, even if at a lower rate. The collision of more than 20 large pieces of Comet Shoemaker–Levy 9 with Jupiter in the summer of 1994 (see Figure 7.13) is one dramatic example of this process.

Download for free at http://cnx.org/content/col11992/latest/

Figure 7.13. Comet Shoemaker–Levy 9. In this image of Comet Shoemaker–Levy 9 taken on May 17, 1994, by NASA's Hubble Space Telescope, you can see about 20 icy fragments into which the comet broke. The comet was approximately 660 million kilometers from Earth, heading on a collision course with Jupiter. (credit: modification of work by NASA, ESA, H. Weaver (STScI), E. Smith (STScI))

Figure 7.14 shows the aftermath of these collisions, when debris clouds larger than Earth could be seen in Jupiter's atmosphere.

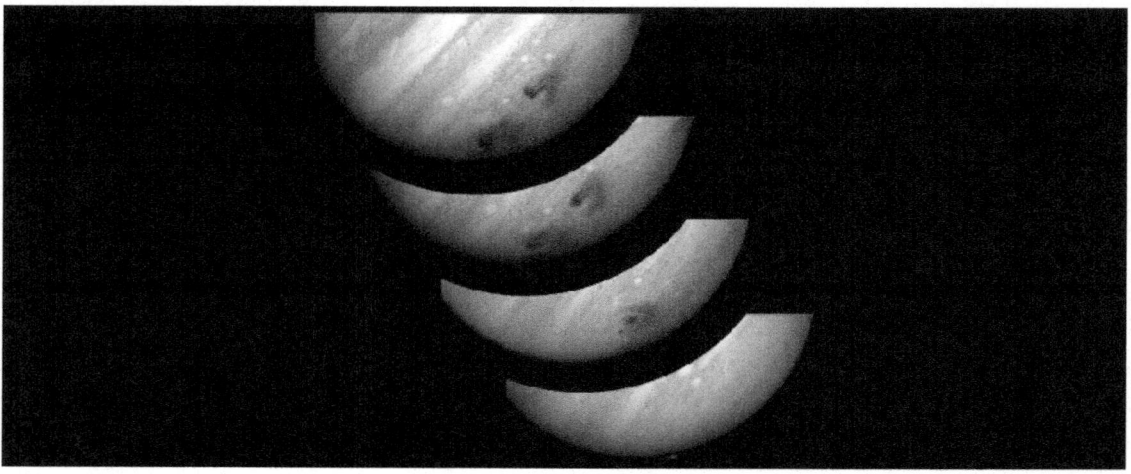

Figure 7.14. Jupiter with Huge Dust Clouds. The Hubble Space Telescope took this sequence of images of Jupiter in summer 1994, when fragments of Comet Shoemaker–Levy 9 collided with the giant planet. Here we see the site hit by fragment G, from five minutes to five days after impact. Several of the dust clouds generated by the collisions became larger than Earth. (credit: modification of work by H. Hammel, NASA)

During the time all the planets have been subject to such impacts, internal forces on the terrestrial planets have buckled and twisted their crusts, built up mountain ranges, erupted as volcanoes, and generally reshaped the surfaces in what we call geological activity. (The prefix *geo* means "Earth," so this is a bit of an "Earth-chauvinist" term, but it is so widely used that we bow to tradition.) Among the terrestrial planets, Earth and Venus have experienced the most geological activity over their histories, although some of the moons in the outer solar system are also surprisingly active. In contrast, our own Moon is a dead world where geological activity ceased billions of years ago.

Geological activity on a planet is the result of a hot interior. The forces of volcanism and mountain building are driven by heat escaping from the interiors of planets. As we will see, each of the planets was heated at the time of its birth, and this primordial heat initially powered extensive volcanic activity, even on our Moon. But, small objects such as the Moon soon cooled off. The larger the planet or moon, the longer it retains its internal heat, and therefore the more we expect to see surface evidence of continuing geological activity. The effect is similar to our own experience with a hot baked potato: the larger the potato, the more slowly it cools. If we want a potato to cool quickly, we cut it into small pieces.

This OpenStax book is available for free at http://cnx.org/content/col11992/1.8

Download for free at http://cnx.org/content/col11992/latest/

For the most part, the history of volcanic activity on the terrestrial planets conforms to the predictions of this simple theory. The Moon, the smallest of these objects, is a geologically dead world. Although we know less about Mercury, it seems likely that this planet, too, ceased most volcanic activity about the same time the Moon did. Mars represents an intermediate case. It has been much more active than the Moon, but less so than Earth. Earth and Venus, the largest terrestrial planets, still have molten interiors even today, some 4.5 billion years after their birth.

7.3 | DATING PLANETARY SURFACES

Learning Objectives

By the end of this section, you will be able to:

> › Explain how astronomers can tell whether a planetary surface is geologically young or old
> › Describe different methods for dating planets

How do we know the age of the surfaces we see on planets and moons? If a world has a surface (as opposed to being mostly gas and liquid), astronomers have developed some techniques for estimating how long ago that surface solidified. Note that the age of these surfaces is not necessarily the age of the planet as a whole. On geologically active objects (including Earth), vast outpourings of molten rock or the erosive effects of water and ice, which we call planet weathering, have erased evidence of earlier epochs and present us with only a relatively young surface for investigation.

Counting the Craters

One way to estimate the age of a surface is by counting the number of impact craters. This technique works because the rate at which impacts have occurred in the solar system has been roughly constant for several billion years. Thus, in the absence of forces to eliminate craters, the number of craters is simply proportional to the length of time the surface has been exposed. This technique has been applied successfully to many solid planets and moons (Figure 7.15).

Figure 7.15. Our Cratered Moon. This composite image of the Moon's surface was made from many smaller images taken between November 2009 and February 2011 by the Lunar Reconnaissance Orbiter (LRO) and shows craters of many different sizes. (credit: modification of work by NASA/GSFC/Arizona State University)

Download for free at http://cnx.org/content/col11992/latest/

Bear in mind that crater counts can tell us only the time since the surface experienced a major change that could modify or erase preexisting craters. Estimating ages from crater counts is a little like walking along a sidewalk in a snowstorm after the snow has been falling steadily for a day or more. You may notice that in front of one house the snow is deep, while next door the sidewalk may be almost clear. Do you conclude that less snow has fallen in front of Ms. Jones' house than Mr. Smith's? More likely, you conclude that Jones has recently swept the walk clean and Smith has not. Similarly, the numbers of craters indicate how long it has been since a planetary surface was last "swept clean" by ongoing lava flows or by molten materials ejected when a large impact happened nearby.

Still, astronomers can use the numbers of craters on different parts of the same world to provide important clues about how regions on that world evolved. On a given planet or moon, the more heavily cratered terrain will generally be older (that is, more time will have elapsed there since something swept the region clean).

Radioactive Rocks

Another way to trace the history of a solid world is to measure the age of individual rocks. After samples were brought back from the Moon by Apollo astronauts, the techniques that had been developed to date rocks on Earth were applied to rock samples from the Moon to establish a geological chronology for the Moon. Furthermore, a few samples of material from the Moon, Mars, and the large asteroid Vesta have fallen to Earth as meteorites and can be examined directly (see the chapter on Cosmic Samples and the Origin of the Solar System).

Scientists measure the age of rocks using the properties of natural **radioactivity**. Around the beginning of the twentieth century, physicists began to understand that some atomic nuclei are not stable but can split apart (decay) spontaneously into smaller nuclei. The process of radioactive decay involves the emission of particles such as electrons, or of radiation in the form of gamma rays (see the chapter on Radiation and Spectra).

For any one radioactive nucleus, it is not possible to predict when the decay process will happen. Such decay is random in nature, like the throw of dice: as gamblers have found all too often, it is impossible to say just when the dice will come up 7 or 11. But, for a very large number of dice tosses, we can calculate the odds that 7 or 11 will come up. Similarly, if we have a very large number of radioactive atoms of one type (say, uranium), there is a specific time period, called its **half-life**, during which the chances are fifty-fifty that decay will occur for any of the nuclei.

A particular nucleus may last a shorter or longer time than its half-life, but in a large sample, almost exactly half of the nuclei will have decayed after a time equal to one half-life. Half of the remaining nuclei will have decayed after two half-lives pass, leaving only one half of a half—or one quarter—of the original sample (Figure 7.16).

This OpenStax book is available for free at http://cnx.org/content/col11992/1.8

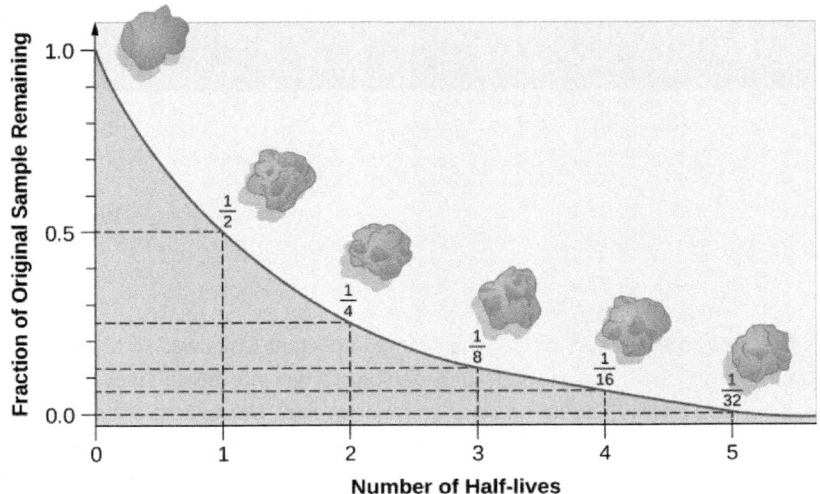

Figure 7.16. Radioactive Decay. This graph shows (in pink) the amount of a radioactive sample that remains after several half-lives have passed. After one half-life, half the sample is left; after two half-lives, one half of the remainder (or one quarter) is left; and after three half-lives, one half of that (or one eighth) is left. Note that, in reality, the decay of radioactive elements in a rock sample would not cause any visible change in the appearance of the rock; the splashes of color are shown here for conceptual purposes only.

If you had 1 gram of pure radioactive nuclei with a half-life of 100 years, then after 100 years you would have 1/2 gram; after 200 years, 1/4 gram; after 300 years, only 1/8 gram; and so forth. However, the material does not disappear. Instead, the radioactive atoms are replaced with their decay products. Sometimes the radioactive atoms are called *parents* and the decay products are called *daughter* elements.

In this way, radioactive elements with half-lives we have determined can provide accurate nuclear clocks. By comparing how much of a radioactive parent element is left in a rock to how much of its daughter products have accumulated, we can learn how long the decay process has been going on and hence how long ago the rock formed. Table 7.3 summarizes the decay reactions used most often to date lunar and terrestrial rocks.

Radioactive Decay Reaction Used to Date Rocks[4]

Parent	Daughter	Half-Life (billions of years)
Samarium-147	Neodymium-143	106
Rubidium-87	Strontium-87	48.8
Thorium-232	Lead-208	14.0
Uranium-238	Lead-206	4.47
Potassium-40	Argon-40	1.31

Table 7.3

4   The number after each element is its atomic weight, equal to the number of protons plus neutrons in its nucleus. This specifies the *isotope* of the element: different isotopes of the same element differ in the number of neutrons.

Download for free at http://cnx.org/content/col11992/latest/

LINK TO LEARNING

PBS provides an evolution series excerpt (https://openstaxcollege.org/l/30pbsradiomat) that explains how we use radioactive elements to date Earth.

This Science Channel video (https://openstaxcollege.org/l/30billnyevideo) features Bill Nye the Science Guy showing how scientists have used radioactive dating to determine the age of Earth.

When astronauts first flew to the Moon, one of their most important tasks was to bring back lunar rocks for radioactive age-dating. Until then, astronomers and geologists had no reliable way to measure the age of the lunar surface. Counting craters had let us calculate relative ages (for example, the heavily cratered lunar highlands were older than the dark lava plains), but scientists could not measure the actual age in years. Some thought that the ages were as young as those of Earth's surface, which has been resurfaced by many geological events. For the Moon's surface to be so young would imply active geology on our satellite. Only in 1969, when the first Apollo samples were dated, did we learn that the Moon is an ancient, geologically dead world. Using such dating techniques, we have been able to determine the ages of both Earth and the Moon: each was formed about 4.5 billion years ago (although, as we shall see, Earth probably formed earlier).

We should also note that the decay of radioactive nuclei generally releases energy in the form of heat. Although the energy from a single nucleus is not very large (in human terms), the enormous numbers of radioactive nuclei in a planet or moon (especially early in its existence) can be a significant source of internal energy for that world. Geologists estimate that about half of Earth's current internal heat budget comes from the decay of radioactive isotopes in its interior.

7.4 ORIGIN OF THE SOLAR SYSTEM

Learning Objectives

By the end of this section, you will be able to:

> Describe the characteristics of planets that are used to create formation models of the solar system
> Describe how the characteristics of extrasolar systems help us to model our own solar system
> Explain the importance of collisions in the formation of the solar system

Much of astronomy is motivated by a desire to understand the origin of things: to find at least partial answers to age-old questions of where the universe, the Sun, Earth, and we ourselves came from. Each planet and moon is a fascinating place that may stimulate our imagination as we try to picture what it would be like to visit. Taken together, the members of the solar system preserve patterns that can tell us about the formation of the entire system. As we begin our exploration of the planets, we want to introduce our modern picture of how the solar system formed.

The recent discovery of hundreds of planets in orbit around other stars has shown astronomers that many exoplanetary systems can be quite different from our own solar system. For example, it is common for these systems to include planets intermediate in size between our terrestrial and giant planets. These are often called *superearths*. Some exoplanet systems even have giant planets close to the star, reversing the order we see in our system. In The Birth of Stars and the Discovery of Planets outside the Solar System, we will look at

Download for free at http://cnx.org/content/col11992/latest/

these exoplanet systems. But for now, let us focus on theories of how our own particular system has formed and evolved.

Looking for Patterns

One way to approach our question of origin is to look for regularities among the planets. We found, for example, that all the planets lie in nearly the same plane and revolve in the same direction around the Sun. The Sun also spins in the same direction about its own axis. Astronomers interpret this pattern as evidence that the Sun and planets formed together from a spinning cloud of gas and dust that we call the **solar nebula** (Figure 7.17).

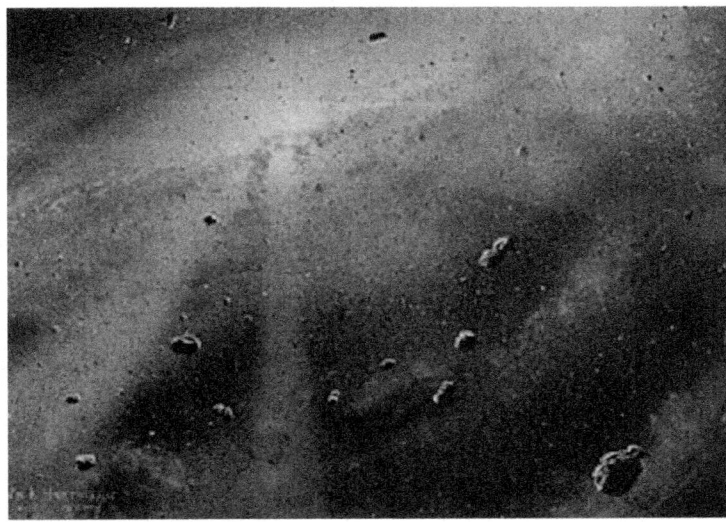

Figure 7.17. Solar Nebula. This artist's conception of the solar nebula shows the flattened cloud of gas and dust from which our planetary system formed. Icy and rocky planetesimals (precursors of the planets) can be seen in the foreground. The bright center is where the Sun is forming. (credit: William K. Hartmann, Planetary Science Institute)

The composition of the planets gives another clue about origins. Spectroscopic analysis allows us to determine which elements are present in the Sun and the planets. The Sun has the same hydrogen-dominated composition as Jupiter and Saturn, and therefore appears to have been formed from the same reservoir of material. In comparison, the terrestrial planets and our Moon are relatively deficient in the light gases and the various ices that form from the common elements oxygen, carbon, and nitrogen. Instead, on Earth and its neighbors, we see mostly the rarer heavy elements such as iron and silicon. This pattern suggests that the processes that led to planet formation in the inner solar system must somehow have excluded much of the lighter materials that are common elsewhere. These lighter materials must have escaped, leaving a residue of heavy stuff.

The reason for this is not hard to guess, bearing in mind the heat of the Sun. The inner planets and most of the asteroids are made of rock and metal, which can survive heat, but they contain very little ice or gas, which evaporate when temperatures are high. (To see what we mean, just compare how long a rock and an ice cube survive when they are placed in the sunlight.) In the outer solar system, where it has always been cooler, the planets and their moons, as well as icy dwarf planets and comets, are composed mostly of ice and gas.

Download for free at http://cnx.org/content/col11992/latest/

The Evidence from Far Away

A second approach to understanding the origins of the solar system is to look outward for evidence that other systems of planets are forming elsewhere. We cannot look back in time to the formation of our own system, but many stars in space are much younger than the Sun. In these systems, the processes of planet formation might still be accessible to direct observation. We observe that there are many other "solar nebulas" or *circumstellar disks*—flattened, spinning clouds of gas and dust surrounding young stars. These disks resemble our own solar system's initial stages of formation billions of years ago (Figure 7.18).

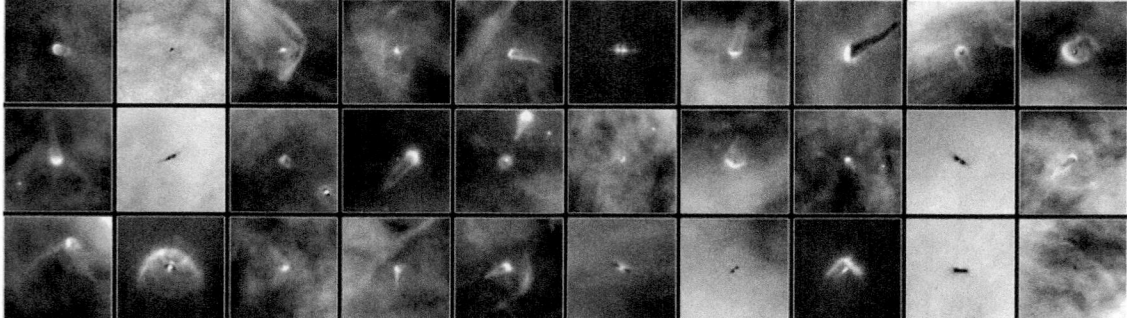

Figure 7.18. Atlas of Planetary Nurseries. These Hubble Space Telescope photos show sections of the Orion Nebula, a relatively close-by region where stars are currently forming. Each image shows an embedded circumstellar disk orbiting a very young star. Seen from different angles, some are energized to glow by the light of a nearby star while others are dark and seen in silhouette against the bright glowing gas of the Orion Nebula. Each is a contemporary analog of our own solar nebula—a location where planets are probably being formed today. (credit: modification of work by NASA/ESA, L. Ricci (ESO))

Building Planets

Circumstellar disks are a common occurrence around very young stars, suggesting that disks and stars form together. Astronomers can use theoretical calculations to see how solid bodies might form from the gas and dust in these disks as they cool. These models show that material begins to coalesce first by forming smaller objects, precursors of the planets, which we call **planetesimals**.

Today's fast computers can simulate the way millions of planetesimals, probably no larger than 100 kilometers in diameter, might gather together under their mutual gravity to form the planets we see today. We are beginning to understand that this process was a violent one, with planetesimals crashing into each other and sometimes even disrupting the growing planets themselves. As a consequence of those violent impacts (and the heat from radioactive elements in them), all the planets were heated until they were liquid and gas, and therefore differentiated, which helps explain their present internal structures.

The process of impacts and collisions in the early solar system was complex and, apparently, often random. The solar nebula model can explain many of the regularities we find in the solar system, but the random collisions of massive collections of planetesimals could be the reason for some exceptions to the "rules" of solar system behavior. For example, why do the planets Uranus and Pluto spin on their sides? Why does Venus spin slowly and in the opposite direction from the other planets? Why does the composition of the Moon resemble Earth in many ways and yet exhibit substantial differences? The answers to such questions probably lie in enormous collisions that took place in the solar system long before life on Earth began.

Today, some 4.5 billion years after its origin, the solar system is—thank goodness—a much less violent place. As we will see, however, some planetesimals have continued to interact and collide, and their fragments move about the solar system as roving "transients" that can make trouble for the established members of the Sun's family, such as our own Earth. (We discuss this "troublemaking" in Comets and Asteroids: Debris of the Solar System.)

This OpenStax book is available for free at http://cnx.org/content/col11992/1.8

LINK TO LEARNING

A great variety of infographics (https://openstaxcollege.org/l/30worldsinsolar) at space.com let you explore what it would be like to live on various worlds in the solar system.

Download for free at http://cnx.org/content/col11992/latest/

CHAPTER 7 REVIEW

 KEY TERMS

asteroid a stony or metallic object orbiting the Sun that is smaller than a major planet but that shows no evidence of an atmosphere or of other types of activity associated with comets

comet a small body of icy and dusty matter that revolves about the Sun; when a comet comes near the Sun, some of its material vaporizes, forming a large head of tenuous gas and often a tail

differentiation gravitational separation of materials of different density into layers in the interior of a planet or moon

giant planet any of the planets Jupiter, Saturn, Uranus, and Neptune in our solar system, or planets of roughly that mass and composition in other planetary systems

half-life time required for half of the radioactive atoms in a sample to disintegrate

meteor a small piece of solid matter that enters Earth's atmosphere and burns up, popularly called a *shooting star* because it is seen as a small flash of light

meteorite a portion of a meteor that survives passage through an atmosphere and strikes the ground

planetesimals objects, from tens to hundreds of kilometers in diameter, that formed in the solar nebula as an intermediate step between tiny grains and the larger planetary objects we see today; the comets and some asteroids may be leftover planetesimals

radioactivity process by which certain kinds of atomic nuclei decay naturally, with the spontaneous emission of subatomic particles and gamma rays

solar nebula the cloud of gas and dust from which the solar system formed

terrestrial planet any of the planets Mercury, Venus, Earth, or Mars; sometimes the Moon is included in the list

 SUMMARY

7.1 Overview of Our Planetary System

Our solar system currently consists of the Sun, eight planets, five dwarf planets, nearly 200 known moons, and a host of smaller objects. The planets can be divided into two groups: the inner terrestrial planets and the outer giant planets. Pluto, Eris, Haumea, and Makemake do not fit into either category; as icy dwarf planets, they exist in an ice realm on the fringes of the main planetary system. The giant planets are composed mostly of liquids and gases. Smaller members of the solar system include asteroids (including the dwarf planet Ceres), which are rocky and metallic objects found mostly between Mars and Jupiter; comets, which are made mostly of frozen gases and generally orbit far from the Sun; and countless smaller grains of cosmic dust. When a meteor survives its passage through our atmosphere and falls to Earth, we call it a meteorite.

7.2 Composition and Structure of Planets

The giant planets have dense cores roughly 10 times the mass of Earth, surrounded by layers of hydrogen and helium. The terrestrial planets consist mostly of rocks and metals. They were once molten, which allowed their structures to differentiate (that is, their denser materials sank to the center). The Moon resembles the terrestrial

This OpenStax book is available for free at http://cnx.org/content/col11992/1.8

planets in composition, but most of the other moons—which orbit the giant planets—have larger quantities of frozen ice within them. In general, worlds closer to the Sun have higher surface temperatures. The surfaces of terrestrial planets have been modified by impacts from space and by varying degrees of geological activity.

7.3 Dating Planetary Surfaces

The ages of the surfaces of objects in the solar system can be estimated by counting craters: on a given world, a more heavily cratered region will generally be older than one that is less cratered. We can also use samples of rocks with radioactive elements in them to obtain the time since the layer in which the rock formed last solidified. The half-life of a radioactive element is the time it takes for half the sample to decay; we determine how many half-lives have passed by how much of a sample remains the radioactive element and how much has become the decay product. In this way, we have estimated the age of the Moon and Earth to be roughly 4.5 billion years.

7.4 Origin of the Solar System

Regularities among the planets have led astronomers to hypothesize that the Sun and the planets formed together in a giant, spinning cloud of gas and dust called the solar nebula. Astronomical observations show tantalizingly similar circumstellar disks around other stars. Within the solar nebula, material first coalesced into planetesimals; many of these gathered together to make the planets and moons. The remainder can still be seen as comets and asteroids. Probably all planetary systems have formed in similar ways, but many exoplanet systems have evolved along quite different paths, as we will see in Cosmic Samples and the Origin of the Solar System.

 FOR FURTHER EXPLORATION

Articles

Davidson, K. "Carl Sagan's Coming of Age." *Astronomy*. (November 1999): 40. About the noted popularizer of science and how he developed his interest in astronomy.

Garget, J. "Mysterious Microworlds." *Astronomy*. (July 2005): 32. A quick tour of a number of the moons in the solar system.

Hartmann, W. "The Great Solar System Revision." *Astronomy*. (August 1998): 40. How our views have changed over the past 25 years.

Kross, J. "What's in a Name?" *Sky & Telescope*. (May 1995): 28. How worlds are named.

Rubin, A. "Secrets of Primitive Meteorites." *Scientific American*. (February 2013): 36. What meteorites can teach us about the environment in which the solar system formed.

Soter, S. "What Is a Planet?" *Scientific American*. (January 2007): 34. The IAU's new definition of a planet in our solar system, and what happened to Pluto as a result.

Talcott, R. "How the Solar System Came to Be." *Astronomy*. (November 2012): 24. On the formation period of the Sun and the planets.

Wood, J. "Forging the Planets: The Origin of our Solar System." *Sky & Telescope*. (January 1999): 36. Good overview.

Download for free at http://cnx.org/content/col11992/latest/

Websites

Gazetteer of Planetary Nomenclature: http://planetarynames.wr.usgs.gov/. Outlines the rules for naming bodies and features in the solar system.

Planetary Photojournal: http://photojournal.jpl.nasa.gov/index.html. This NASA site features thousands of the best images from planetary exploration, with detailed captions and excellent indexing. You can find images by world, feature name, or mission, and download them in a number of formats. And the images are copyright-free because your tax dollars paid for them.

The following sites present introductory information and pictures about each of the worlds of our solar system:

- NASA/JPL Solar System Exploration pages: http://solarsystem.nasa.gov/index.cfm.
- National Space Science Data Center Lunar and Planetary Science pages: http://nssdc.gsfc.nasa.gov/planetary/.
- Nine [now 8] Planets Solar System Tour: http://www.nineplanets.org/.
- Planetary Society solar system pages: http://www.planetary.org/explore/space-topics/compare/.
- Views of the Solar System by Calvin J. Hamilton: http://www.solarviews.com/eng/homepage.htm.

Videos

Brown Dwarfs and Free Floating Planets: When You Are Just Too Small to Be a Star: https://www.youtube.com/watch?v=zXCDsb4n4KU. A nontechnical talk by Gibor Basri of the University of California at Berkeley, discussing some of the controversies about the meaning of the word "planet" (1:32:52).

In the Land of Enchantment: The Epic Story of the Cassini Mission to Saturn: https://www.youtube.com/watch?v=Vx135n8VFxY. A public lecture by Dr. Carolyn Porco that focuses mainly on the exploration of Saturn and its moons, but also presents an eloquent explanation of why we explore the solar system (1:37:52).

Origins of the Solar System: http://www.pbs.org/wgbh/nova/space/origins-solar-system.html. A video from PBS that focuses on the evidence from meteorites, narrated by Neil deGrasse Tyson (13:02).

To Scale: The Solar System: https://www.youtube.com/watch?t=84&v=zR3Igc3Rhfg. Constructing a scale model of the solar system in the Nevada desert (7:06).

 COLLABORATIVE GROUP ACTIVITIES

A. Discuss and make a list of the reasons why we humans might want to explore the other worlds in the solar system. Does your group think such missions of exploration are worth the investment? Why?

B. Your instructor will assign each group a world. Your task is to think about what it would be like to be there. (Feel free to look ahead in the book to the relevant chapters.) Discuss where on or around your world we would establish a foothold and what we would need to survive there.

C. In the There's No Place Like Home feature, we discuss briefly how human activity is transforming our planet's overall environment. Can you think of other ways that this is happening?

D. Some scientists criticized Carl Sagan for "wasting his research time" popularizing astronomy. To what extent do you think scientists should spend their time interpreting their field of research for the public?

This OpenStax book is available for free at http://cnx.org/content/col11992/1.8

Why or why not? Are there ways that scientists who are not as eloquent or charismatic as Carl Sagan or Neil deGrasse Tyson can still contribute to the public understanding of science?

E. Your group has been named to a special committee by the International Astronomical Union to suggest names of features (such as craters, trenches, and so on) on a newly explored asteroid. Given the restriction that any people after whom features are named must no longer be alive, what names or types of names would you suggest? (Keep in mind that you are not restricted to names of people, by the way.)

F. A member of your group has been kidnapped by a little-known religious cult that worships the planets. They will release him only if your group can tell which of the planets are currently visible in the sky during the evening and morning. You are forbidden from getting your instructor involved. How and where else could you find out the information you need? (Be as specific as you can. If your instructor says it's okay, feel free to answer this question using online or library resources.)

G. In the Carl Sagan: Solar System Advocate feature, you learned that science fiction helped spark and sustain his interest in astronomy. Did any of the members of your group get interested in astronomy as a result of a science fiction story, movie, or TV show? Did any of the stories or films you or your group members saw take place on the planets of our solar system? Can you remember any specific ones that inspired you? If no one in the group is into science fiction, perhaps you can interview some friends or classmates who are and report back to the group.

H. A list of NASA solar system spacecraft missions can be found at http://www.nasa.gov/content/solar-missions-list. Your instructor will assign each group a mission. Look up when the mission was launched and executed, and describe the mission goals, the basic characteristics of the spacecraft (type of instruments, propellant, size, and so on), and what was learned from the mission. If time allows, each group should present its findings to the rest of the class.

I. What would be some of the costs or risks of developing a human colony or base on another planetary body? What technologies would need to be developed? What would people need to give up to live on a different world in our solar system?

▣ EXERCISES

Review Questions

1. Venus rotates backward and Uranus and Pluto spin about an axis tipped nearly on its side. Based on what you learned about the motion of small bodies in the solar system and the surfaces of the planets, what might be the cause of these strange rotations?

2. What is the difference between a differentiated body and an undifferentiated body, and how might that influence a body's ability to retain heat for the age of the solar system?

3. What does a planet need in order to retain an atmosphere? How does an atmosphere affect the surface of a planet and the ability of life to exist?

4. Which type of planets have the most moons? Where did these moons likely originate?

5. What is the difference between a meteor and a meteorite?

6. Explain our ideas about why the terrestrial planets are rocky and have less gas than the giant planets.

Download for free at http://cnx.org/content/col11992/latest/

7. Do all planetary systems look the same as our own?

8. What is comparative planetology and why is it useful to astronomers?

9. What changed in our understanding of the Moon and Moon-Earth system as a result of humans landing on the Moon's surface?

10. If Earth was to be hit by an extraterrestrial object, where in the solar system could it come from and how would we know its source region?

11. List some reasons that the study of the planets has progressed more in the past few decades than any other branch of astronomy.

12. Imagine you are a travel agent in the next century. An eccentric billionaire asks you to arrange a "Guinness Book of Solar System Records" kind of tour. Where would you direct him to find the following (use this chapter and Appendix F and Appendix G):

    A. the least-dense planet

    B. the densest planet

    C. the largest moon in the solar system

    D. excluding the jovian planets, the planet where you would weigh the most on its surface (Hint: Weight is directly proportional to surface gravity.)

    E. the smallest planet

    F. the planet that takes the longest time to rotate

    G. the planet that takes the shortest time to rotate

    H. the planet with a diameter closest to Earth's

    I. the moon with the thickest atmosphere

    J. the densest moon

    K. the most massive moon

13. What characteristics do the worlds in our solar system have in common that lead astronomers to believe that they all formed from the same "mother cloud" (solar nebula)?

14. How do terrestrial and giant planets differ? List as many ways as you can think of.

15. Why are there so many craters on the Moon and so few on Earth?

16. How do asteroids and comets differ?

17. How and why is Earth's Moon different from the larger moons of the giant planets?

18. Where would you look for some "original" planetesimals left over from the formation of our solar system?

19. Describe how we use radioactive elements and their decay products to find the age of a rock sample. Is this necessarily the age of the entire world from which the sample comes? Explain.

20. What was the solar nebula like? Why did the Sun form at its center?

This OpenStax book is available for free at http://cnx.org/content/col11992/1.8

Thought Questions

21. What can we learn about the formation of our solar system by studying other stars? Explain.

22. Earlier in this chapter, we modeled the solar system with Earth at a distance of about one city block from the Sun. If you were to make a model of the distances in the solar system to match your height, with the Sun at the top of your head and Pluto at your feet, which planet would be near your waist? How far down would the zone of the terrestrial planets reach?

23. Seasons are a result of the inclination of a planet's axial tilt being inclined from the normal of the planet's orbital plane. For example, Earth has an axis tilt of 23.4° (Appendix F). Using information about just the inclination alone, which planets might you expect to have seasonal cycles similar to Earth, although different in duration because orbital periods around the Sun are different?

24. Again using Appendix F, which planet(s) might you expect not to have significant seasonal activity? Why?

25. Again using Appendix F, which planets might you expect to have extreme seasons? Why?

26. Using some of the astronomical resources in your college library or the Internet, find five names of features on each of three other worlds that are named after real people. In a sentence or two, describe each of these people and what contributions they made to the progress of science or human thought.

27. Explain why the planet Venus is differentiated, but asteroid Fraknoi, a very boring and small member of the asteroid belt, is not.

28. Would you expect as many impact craters per unit area on the surface of Venus as on the surface of Mars? Why or why not?

29. Interview a sample of 20 people who are not taking an astronomy class and ask them if they can name a living astronomer. What percentage of those interviewed were able to name one? Typically, the two living astronomers the public knows these days are Stephen Hawking and Neil deGrasse Tyson. Why are they better known than most astronomers? How would your result have differed if you had asked the same people to name a movie star or a professional basketball player?

30. Using Appendix G, complete the following table that describes the characteristics of the Galilean moons of Jupiter, starting from Jupiter and moving outward in distance.

Moon	Semimajor Axis (km³)	Diameter	Density (g/cm³)
Io			
Europa			
Ganymede			
Callisto			

Table A

This system has often been described as a mini solar system. Why might this be so? If Jupiter were to represent the Sun and the Galilean moons represented planets, which moons could be considered more terrestrial in nature and which ones more like gas/ice giants? Why? (Hint: Use the values in your table to help explain your categorization.)

Figuring For Yourself

31. Calculate the density of Jupiter. Show your work. Is it more or less dense than Earth? Why?

Download for free at http://cnx.org/content/col11992/latest/

32. Calculate the density of Saturn. Show your work. How does it compare with the density of water? Explain how this can be.

33. What is the density of Jupiter's moon Europa (see Appendix G for data on moons)? Show your work.

34. Look at Appendix F and Appendix G and indicate the moon with a diameter that is the largest fraction of the diameter of the planet or dwarf planet it orbits.

35. Barnard's Star, the second closest star to us, is about 56 trillion (5.6×10^{12}) km away. Calculate how far it would be using the scale model of the solar system given in Overview of Our Planetary System.

36. A radioactive nucleus has a half-life of 5×10^8 years. Assuming that a sample of rock (say, in an asteroid) solidified right after the solar system formed, approximately what fraction of the radioactive element should be left in the rock today?

This OpenStax book is available for free at http://cnx.org/content/col11992/1.8

Figure 8.1. Active Geology. This image, taken from the International Space Station in 2006, shows a plume of ash coming from the Cleveland Volcano in the Aleutian Islands. Although the plume was only visible for around two hours, such events are a testament to the dynamic nature of Earth's crust. (credit: modification of work by NASA)

Chapter Outline

Thinking Ahead

Airless worlds in our solar system seem peppered with craters large and small. Earth, on the other hand, has few craters, but a thick atmosphere and much surface activity. Although impacts occurred on Earth at the same rate, craters have since been erased by forces in the planet's crust and atmosphere. What can the comparison between the obvious persistent cratering on so many other worlds, and the different appearance of Earth, tell us about the history of our planet?

As our first step in exploring the solar system in more detail, we turn to the most familiar planet, our own Earth. The first humans to see Earth as a blue sphere floating in the blackness of space were the astronauts who made the first voyage around the Moon in 1968. For many people, the historic images showing our world as a small, distant globe represent a pivotal moment in human history, when it became difficult for educated human beings to view our world without a global perspective. In this chapter, we examine the composition and structure of our planet with its envelope of ocean and atmosphere. We ask how our terrestrial environment came to be the way it is today, and how it compares with other planets.

Download for free at http://cnx.org/content/col11992/latest/

8.1 THE GLOBAL PERSPECTIVE

Learning Objectives

By the end of this section, you will be able to:

> Describe the components of Earth's interior and explain how scientists determined its structure
> Specify the origin, size, and extent of Earth's magnetic field

Earth is a medium-size planet with a diameter of approximately 12,760 kilometers (Figure 8.2). As one of the inner or terrestrial planets, it is composed primarily of heavy elements such as iron, silicon, and oxygen—very different from the composition of the Sun and stars, which are dominated by the light elements hydrogen and helium. Earth's orbit is nearly circular, and Earth is warm enough to support liquid water on its surface. It is the only planet in our solar system that is neither too hot nor too cold, but "just right" for the development of life as we know it. Some of the basic properties of Earth are summarized in Table 8.1.

Figure 8.2. Blue Marble. This image of Earth from space, taken by the Apollo 17 astronauts, is known as the "Blue Marble." This is one of the rare images of a full Earth taken during the Apollo program; most images show only part of Earth's disk in sunlight. (credit: modification of work by NASA)

Some Properties of Earth

Property	Measurement
Semimajor axis	1.00 AU
Period	1.00 year
Mass	5.98×10^{24} kg
Diameter	12,756 km
Radius	6378 km

Table 8.1

This OpenStax book is available for free at http://cnx.org/content/col11992/1.8

Some Properties of Earth

Property	Measurement
Escape velocity	11.2 km/s
Rotational period	23 h 56 m 4 s
Surface area	5.1×10^8 km^2
Density	5.514 g/cm^3
Atmospheric pressure	1.00 bar

Table 8.1

Earth's Interior

The interior of a planet—even our own Earth—is difficult to study, and its composition and structure must be determined indirectly. Our only direct experience is with the outermost skin of Earth's crust, a layer no more than a few kilometers deep. It is important to remember that, in many ways, we know less about our own planet 5 kilometers beneath our feet than we do about the surfaces of Venus and Mars.

Earth is composed largely of metal and silicate rock (see the Composition and Structure of Planets section). Most of this material is in a solid state, but some of it is hot enough to be molten. The structure of material in Earth's interior has been probed in considerable detail by measuring the transmission of **seismic waves** through Earth. These are waves that spread through the interior of Earth from earthquakes or explosion sites.

Seismic waves travel through a planet rather like sound waves through a struck bell. Just as the sound frequencies vary depending on the material the bell is made of and how it is constructed, so a planet's response depends on its composition and structure. By monitoring the seismic waves in different locations, scientists can learn about the layers through which the waves have traveled. Some of these vibrations travel along the surface; others pass directly through the interior. Seismic studies have shown that Earth's interior consists of several distinct layers with different compositions, illustrated in Figure 8.3. As waves travel through different materials in Earth's interior, the waves—just like light waves in telescope lenses—bend (or refract) so that some seismic stations on Earth receive the waves and others are in "shadows." Detecting the waves in a network of seismographs helps scientists construct a model of Earth's interior, showing liquid and solid layers. This type of seismic imaging is not unlike that used in ultrasound, a type of imaging used to see inside the body.

Download for free at http://cnx.org/content/col11992/latest/

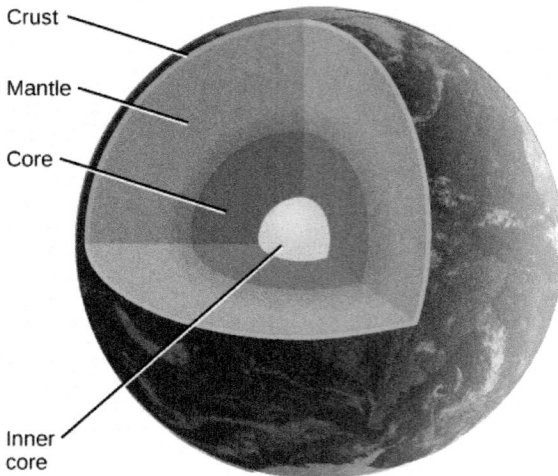

Figure 8.3. Interior Structure of Earth. The crust, mantle, and inner and outer cores (liquid and solid, respectively) as shown as revealed by seismic studies.

The top layer is the **crust**, the part of Earth we know best (Figure 8.4). Oceanic crust covers 55% of Earth's surface and lies mostly submerged under the oceans. It is typically about 6 kilometers thick and is composed of volcanic rocks called **basalt**. Produced by the cooling of volcanic lava, basalts are made primarily of the elements silicon, oxygen, iron, aluminum, and magnesium. The continental crust covers 45% of the surface, some of which is also beneath the oceans. The continental crust is 20 to 70 kilometers thick and is composed predominantly of a different volcanic class of silicates (rocks made of silicon and oxygen) called **granite**. These crustal rocks, both oceanic and continental, typically have densities of about 3 g/cm^3. (For comparison, the density of water is 1 g/cm^3.) The crust is the easiest layer for geologists to study, but it makes up only about 0.3% of the total mass of Earth.

Figure 8.4. Earth's Crust. This computer-generated image shows the surface of Earth's crust as determined from satellite images and ocean floor radar mapping. Oceans and lakes are shown in blue, with darker areas representing depth. Dry land is shown in shades of green and brown, and the Greenland and Antarctic ice sheets are depicted in shades of white. (credit: modification of work by C. Amante, B. W. Eakins, National Geophysical Data Center, NOAA)

This OpenStax book is available for free at http://cnx.org/content/col11992/1.8

Download for free at http://cnx.org/content/col11992/latest/

The largest part of the solid Earth, called the **mantle**, stretches from the base of the crust downward to a depth of 2900 kilometers. The mantle is more or less solid, but at the temperatures and pressures found there, mantle rock can deform and flow slowly. The density in the mantle increases downward from about 3.5 g/cm^3 to more than 5 g/cm^3 as a result of the compression produced by the weight of overlying material. Samples of upper mantle material are occasionally ejected from volcanoes, permitting a detailed analysis of its chemistry.

Beginning at a depth of 2900 kilometers, we encounter the dense metallic **core** of Earth. With a diameter of 7000 kilometers, our core is substantially larger than the entire planet Mercury. The outer core is liquid, but the innermost part of the core (about 2400 kilometers in diameter) is probably solid. In addition to iron, the core probably also contains substantial quantities of nickel and sulfur, all compressed to a very high density.

The separation of Earth into layers of different densities is an example of *differentiation,* the process of sorting the major components of a planet by density. The fact that Earth is differentiated suggests that it was once warm enough for its interior to melt, permitting the heavier metals to sink to the center and form the dense core. Evidence for differentiation comes from comparing the planet's bulk density (5.5 g/cm^3) with the surface materials (3 g/cm^3) to suggest that denser material must be buried in the core.

Magnetic Field and Magnetosphere

We can find additional clues about Earth's interior from its magnetic field. Our planet behaves in some ways as if a giant bar magnet were inside it, aligned approximately with the rotational poles of Earth. This magnetic field is generated by moving material in Earth's liquid metallic core. As the liquid metal inside Earth circulates, it sets up a circulating electric current. When many charged particles are moving together like that—in the laboratory or on the scale of an entire planet—they produce a magnetic field.

Earth's magnetic field extends into surrounding space. When a charged particle encounters a magnetic field in space, it becomes trapped in the magnetic zone. Above Earth's atmosphere, our field is able to trap small quantities of electrons and other atomic particles. This region, called the **magnetosphere**, is defined as the zone within which Earth's magnetic field dominates over the weak interplanetary magnetic field that extends outward from the Sun (Figure 8.5).

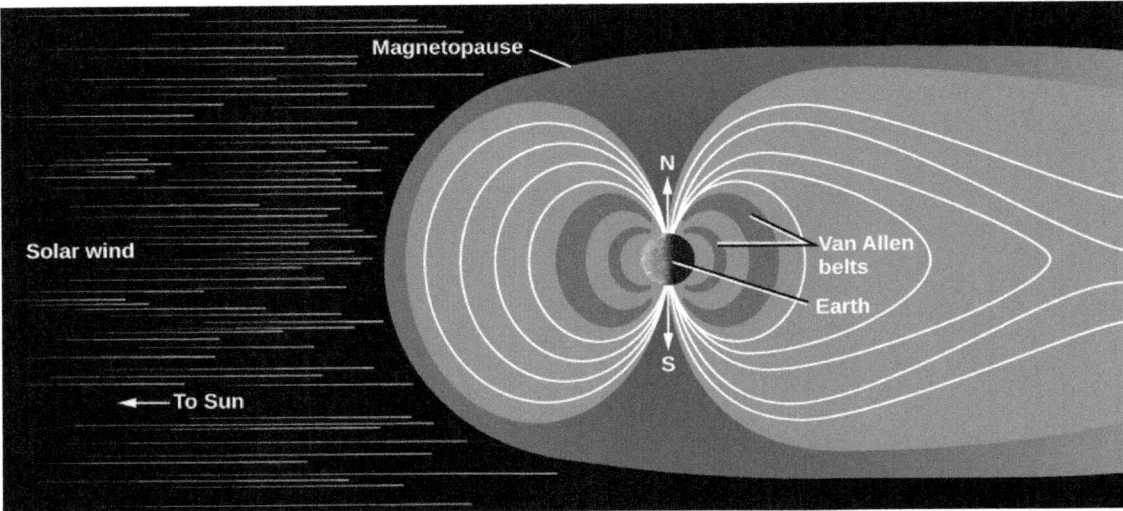

Figure 8.5. Earth's Magnetosphere. A cross-sectional view of our magnetosphere (or zone of magnetic influence), as revealed by numerous spacecraft missions. Note how the wind of charged particles from the Sun "blows" the magnetic field outward like a wind sock.

Download for free at http://cnx.org/content/col11992/latest/

Where do the charged particles trapped in our magnetosphere come from? They flow outward from the hot surface of the Sun; this is called the *solar wind*. It not only provides particles for Earth's magnetic field to trap, it also stretches our field in the direction pointing away from the Sun. Typically, Earth's magnetosphere extends about 60,000 kilometers, or 10 Earth radii, in the direction of the Sun. But, in the direction away from the Sun, the magnetic field can reach as far as the orbit of the Moon, and sometimes farther.

The magnetosphere was discovered in 1958 by instruments on the first US Earth satellite, *Explorer 1*, which recorded the ions (charged particles) trapped in its inner part. The regions of high-energy ions in the magnetosphere are often called the *Van Allen belts* in recognition of the University of Iowa professor who built the scientific instrumentation for *Explorer 1*. Since 1958, hundreds of spacecraft have explored various regions of the magnetosphere. You can read more about its interaction with the Sun in a later chapter.

8.2 EARTH'S CRUST

Learning Objectives

By the end of this section, you will be able to:

> Denote the primary types of rock that constitute Earth's crust
> Explain the theory of plate tectonics
> Describe the difference between rift and subduction zones
> Describe the relationship between fault zones and mountain building
> Explain the various types of volcanic activity occurring on Earth

Let us now examine our planet's outer layers in more detail. Earth's crust is a dynamic place. Volcanic eruptions, erosion, and large-scale movements of the continents rework the surface of our planet constantly. Geologically, ours is the most active planet. Many of the geological processes described in this section have taken place on other planets as well, but usually in their distant pasts. Some of the moons of the giant planets also have impressive activity levels. For example, Jupiter's moon Io has a remarkable number of active volcanoes.

Composition of the Crust

Earth's crust is largely made up of oceanic basalt and continental granite. These are both **igneous rock**, the term used for any rock that has cooled from a molten state. All volcanically produced rock is igneous (Figure 8.6).

Figure 8.6. Formation of Igneous Rock as Liquid Lava Cools and Freezes. This is a lava flow from a basaltic eruption. Basaltic lava flows quickly and can move easily over distances of more than 20 kilometers. (credit: USGS)

This OpenStax book is available for free at http://cnx.org/content/col11992/1.8

Download for free at http://cnx.org/content/col11992/latest/

Two other kinds of rock are familiar to us on Earth, although it turns out that neither is common on other planets. **Sedimentary rocks** are made of fragments of igneous rock or the shells of living organisms deposited by wind or water and cemented together without melting. On Earth, these rocks include the common sandstones, shales, and limestones. **Metamorphic rocks** are produced when high temperature or pressure alters igneous or sedimentary rock physically or chemically (the word *metamorphic* means "changed in form"). Metamorphic rocks are produced on Earth because geological activity carries surface rocks down to considerable depths and then brings them back up to the surface. Without such activity, these changed rocks would not exist at the surface.

There is a fourth very important category of rock that can tell us much about the early history of the planetary system: **primitive rock**, which has largely escaped chemical modification by heating. Primitive rock represents the original material out of which the planetary system was made. No primitive material is left on Earth because the entire planet was heated early in its history. To find primitive rock, we must look to smaller objects such as comets, asteroids, and small planetary moons. We can sometimes see primitive rock in samples that fall to Earth from these smaller objects.

A block of quartzite on Earth is composed of materials that have gone through all four of these states. Beginning as primitive material before Earth was born, it was heated in the early Earth to form igneous rock, transformed chemically and redeposited (perhaps many times) to form sedimentary rock, and finally changed several kilometers below Earth's surface into the hard, white metamorphic stone we see today.

Plate Tectonics

Geology is the study of Earth's crust and the processes that have shaped its surface throughout history. (Although *geo-* means "related to Earth," astronomers and planetary scientists also talk about the geology of other planets.) Heat escaping from the interior provides energy for the formation of our planet's mountains, valleys, volcanoes, and even the continents and ocean basins themselves. But not until the middle of the twentieth century did geologists succeed in understanding just how these landforms are created.

Plate tectonics is a theory that explains how slow motions within the mantle of Earth move large segments of the crust, resulting in a gradual "drifting" of the continents as well as the formation of mountains and other large-scale geological features. Plate tectonics is a concept as basic to geology as evolution by natural selection is to biology or gravity is to understanding the orbits of planets. Looking at it from a different perspective, plate tectonics is a mechanism for Earth to transport heat efficiently from the interior, where it has accumulated, out to space. It is a cooling system for the planet. All planets develop a heat transfer process as they evolve; mechanisms may differ from that on Earth as a result of chemical makeup and other constraints.

Earth's crust and upper mantle (to a depth of about 60 kilometers) are divided into about a dozen tectonic plates that fit together like the pieces of a jigsaw puzzle (Figure 8.7). In some places, such as the Atlantic Ocean, the plates are moving apart; in others, such as off the western coast of South America, they are being forced together. The power to move the plates is provided by slow **convection** of the mantle, a process by which heat escapes from the interior through the upward flow of warmer material and the slow sinking of cooler material. (Convection, in which energy is transported from a warm region, such as the interior of Earth, to a cooler region, such as the upper mantle, is a process we encounter often in astronomy—in stars as well as planets. It is also important in boiling water for coffee while studying for astronomy exams.)

Download for free at http://cnx.org/content/col11992/latest/

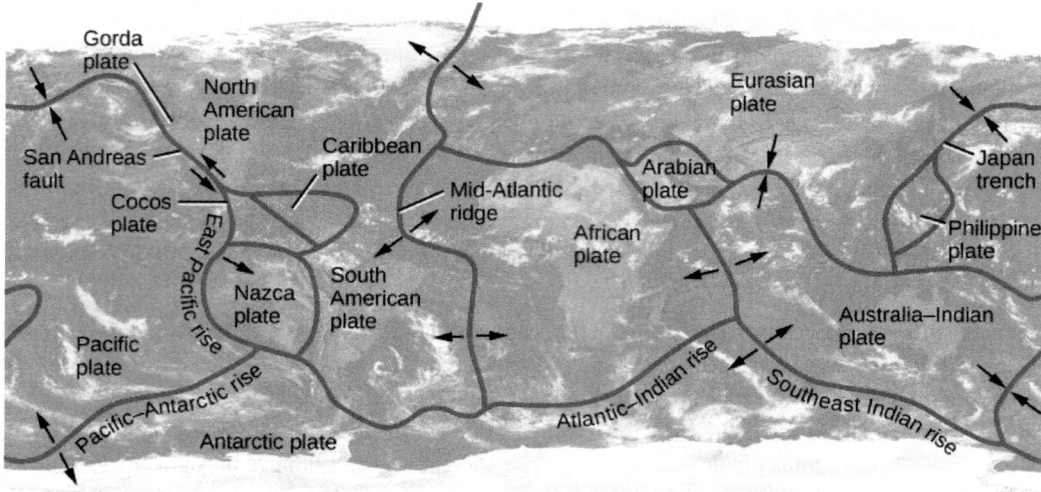

Figure 8.7. Earth's Continental Plates. This map shows the major plates into which the crust of Earth is divided. Arrows indicate the motion of the plates at average speeds of 4 to 5 centimeters per year, similar to the rate at which your hair grows.

LINK TO LEARNING

The US Geological Survey provides a map of recent earthquakes (https://openstax.org/l/ 30geosurmapeart) and shows the boundaries of the tectonic plates and where earthquakes occur in relation to these boundaries. You can look close-up at the United States or zoom out for a global view.

As the plates slowly move, they bump into each other and cause dramatic changes in Earth's crust over time. Four basic kinds of interactions between crustal plates are possible at their boundaries: (1) they can pull apart, (2) one plate can burrow under another, (3) they can slide alongside each other, or (4) they can jam together. Each of these activities is important in determining the geology of Earth.

VOYAGERS IN ASTRONOMY

Alfred Wegener: Catching the Drift of Plate Tectonics

When studying maps or globes of Earth, many students notice that the coast of North and South America, with only minor adjustments, could fit pretty well against the coast of Europe and Africa. It seems as if these great landmasses could once have been together and then were somehow torn apart. The same idea had occurred to others (including Francis Bacon as early as 1620), but not until the twentieth century could such a proposal be more than speculation. The scientist who made the case for continental drift in 1920 was a German meteorologist and astronomer named Alfred Wegener (Figure 8.8).

This OpenStax book is available for free at http://cnx.org/content/col11992/1.8

Download for free at http://cnx.org/content/col11992/latest/

Figure 8.8. Alfred Wegener (1880–1930). Wegener proposed a scientific theory for the slow shifting of the continents.

Born in Berlin in 1880, Wegener was, from an early age, fascinated by Greenland, the world's largest island, which he dreamed of exploring. He studied at the universities in Heidelberg, Innsbruck, and Berlin, receiving a doctorate in astronomy by reexamining thirteenth-century astronomical tables. But, his interests turned more and more toward Earth, particularly its weather. He carried out experiments using kites and balloons, becoming so accomplished that he and his brother set a world record in 1906 by flying for 52 hours in a balloon.

Wegener first conceived of continental drift in 1910 while examining a world map in an atlas, but it took 2 years for him to assemble sufficient data to propose the idea in public. He published the results in book form in 1915. Wegener's evidence went far beyond the congruence in the shapes of the continents. He proposed that the similarities between fossils found only in South America and Africa indicated that these two continents were joined at one time. He also showed that resemblances among living animal species on different continents could best be explained by assuming that the continents were once connected in a supercontinent he called *Pangaea* (from Greek elements meaning "all land").

Wegener's suggestion was met with a hostile reaction from most scientists. Although he had marshaled an impressive list of arguments for his hypothesis, he was missing a *mechanism*. No one could explain *how* solid continents could drift over thousands of miles. A few scientists were sufficiently impressed by Wegener's work to continue searching for additional evidence, but many found the notion of moving continents too revolutionary to take seriously. Developing an understanding of the mechanism (plate tectonics) would take decades of further progress in geology, oceanography, and geophysics.

Wegener was disappointed in the reception of his suggestion, but he continued his research and, in 1924, he was appointed to a special meteorology and geophysics professorship created especially for him at the University of Graz (where he was, however, ostracized by most of the geology faculty). Four years later, on his fourth expedition to his beloved Greenland, he celebrated his fiftieth birthday with colleagues and then set off on foot toward a different camp on the island. He never made it; he was found a few days later, dead of an apparent heart attack.

Download for free at http://cnx.org/content/col11992/latest/

Critics of science often point to the resistance to the continental drift hypothesis as an example of the flawed way that scientists regard new ideas. (Many people who have advanced crackpot theories have claimed that they are being ridiculed unjustly, just as Wegener was.) But we think there is a more positive light in which to view the story of Wegener's suggestion. Scientists in his day maintained a skeptical attitude because they needed more evidence and a clear mechanism that would fit what they understood about nature. Once the evidence and the mechanism were clear, Wegener's hypothesis quickly became the centerpiece of our view of a dynamic Earth.

LINK TO LEARNING

See how the drift of the continents (https://openstax.org/l/30contintdrift) has changed the appearance of our planet's crust.

Rift and Subduction Zones

Plates pull apart from each other along **rift zones**, such as the Mid-Atlantic ridge, driven by upwelling currents in the mantle (Figure 8.9). A few rift zones are found on land. The best known is the central African rift—an area where the African continent is slowly breaking apart. Most rift zones, however, are in the oceans. Molten rock rises from below to fill the space between the receding plates; this rock is basaltic lava, the kind of igneous rock that forms most of the ocean basins.

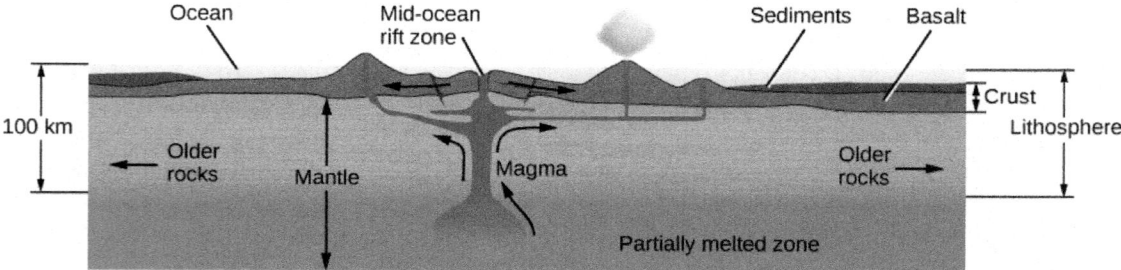

Rift zone

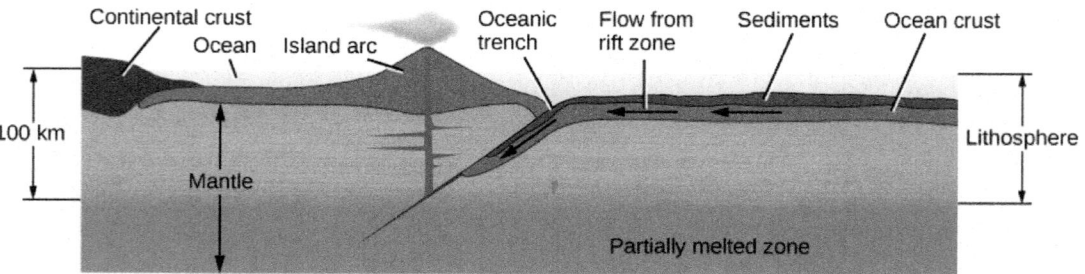

Subduction zone

Figure 8.9. Rift Zone and Subduction Zone. Rift and subduction zones are the regions (mostly beneath the oceans) where new crust is formed and old crust is destroyed as part of the cycle of plate tectonics.

This OpenStax book is available for free at http://cnx.org/content/col11992/1.8

From a knowledge of how the seafloor is spreading, we can calculate the average age of the oceanic crust. About 60,000 kilometers of active rifts have been identified, with average separation rates of about 4 centimeters per year. The new area added to Earth each year is about 2 square kilometers, enough to renew the entire oceanic crust in a little more than 100 million years. This is a very short interval in geological time—less than 3% of the age of Earth. The present ocean basins thus turn out to be among the youngest features on our planet.

As new crust is added to Earth, the old crust must go somewhere. When two plates come together, one plate is often forced beneath another in what is called a **subduction** zone (Figure 8.9). In general, the thick continental masses cannot be subducted, but the thinner oceanic plates can be rather readily thrust down into the upper mantle. Often a subduction zone is marked by an ocean trench; a fine example of this type of feature is the deep Japan trench along the coast of Asia. The subducted plate is forced down into regions of high pressure and temperature, eventually melting several hundred kilometers below the surface. Its material is recycled into a downward-flowing convection current, ultimately balancing the flow of material that rises along rift zones. The amount of crust destroyed at subduction zones is approximately equal to the amount formed at rift zones.

All along the subduction zone, earthquakes and volcanoes mark the death throes of the plate. Some of the most destructive earthquakes in history have taken place along subduction zones, including the 1923 Yokohama earthquake and fire that killed 100,000 people, the 2004 Sumatra earthquake and tsunami that killed more than 200,000 people, and the 2011 Tohoku earthquake that resulted in the meltdown of three nuclear power reactors in Japan.

Fault Zones and Mountain Building

Along much of their length, the crustal plates slide parallel to each other. These plate boundaries are marked by cracks or **faults**. Along active fault zones, the motion of one plate with respect to the other is several centimeters per year, about the same as the spreading rates along rifts.

One of the most famous faults is the San Andreas Fault in California, which lies at the boundary between the Pacific plate and the North American plate (Figure 8.10). This fault runs from the Gulf of California to the Pacific Ocean northwest of San Francisco. The Pacific plate, to the west, is moving northward, carrying Los Angeles, San Diego, and parts of the southern California coast with it. In several million years, Los Angeles may be an island off the coast of San Francisco.

Download for free at http://cnx.org/content/col11992/latest/

Figure 8.10. San Andreas Fault. We see part of a very active region in California where one crustal plate is sliding sideways with respect to the other. The fault is marked by the valley running up the right side of the photo. Major slippages along this fault can produce extremely destructive earthquakes. (credit: John Wiley)

Unfortunately for us, the motion along fault zones does not take place smoothly. The creeping motion of the plates against each other builds up stresses in the crust that are released in sudden, violent slippages that generate earthquakes. Because the average motion of the plates is constant, the longer the interval between earthquakes, the greater the stress and the more energy released when the surface finally moves.

For example, the part of the San Andreas Fault near the central California town of Parkfield has slipped every 25 years or so during the past century, moving an average of about 1 meter each time. In contrast, the average interval between major earthquakes in the Los Angeles region is about 150 years, and the average motion is about 7 meters. The last time the San Andreas fault slipped in this area was in 1857; tension has been building ever since, and sometime soon it is bound to be released. Sensitive instruments placed within the Los Angeles basin show that the basin is distorting and contracting in size as these tremendous pressures build up beneath the surface.

EXAMPLE 8.1

Fault Zones and Plate Motion

After scientists mapped the boundaries between tectonic plates in Earth's crust and measured the annual rate at which the plates move (which is about 5 cm/year), we could estimate quite a lot about the rate at which the geology of Earth is changing. As an example, let's suppose that the next slippage along the San Andreas Fault in southern California takes place in the year 2017 and that it completely relieves the accumulated strain in this region. How much slippage is required for this to occur?

Solution

The speed of motion of the Pacific plate relative to the North American plate is 5 cm/y. That's 500 cm (or 5 m) per century. The last southern California earthquake was in 1857. The time from 1857 to 2017 is

This OpenStax book is available for free at http://cnx.org/content/col11992/1.8

160 y, or 1.6 centuries, so the slippage to relieve the strain completely would be
5 m/century × 1.6 centuries = 8.0 m.

Check Your Learning

If the next major southern California earthquake occurs in 2047 and only relieves one-half of the accumulated strain, how much slippage will occur?

Answer:

The difference in time from 1857 to 2047 is 190 y, or 1.9 centuries. Because only half the strain is released, this is equivalent to half the annual rate of motion. The total slippage comes to
0.5 × 5 m/century × 1.9 centuries = 4.75 m.

When two continental masses are moving on a collision course, they push against each other under great pressure. Earth buckles and folds, dragging some rock deep below the surface and raising other folds to heights of many kilometers. This is the way many, but not all, of the mountain ranges on Earth were formed. The Alps, for example, are a result of the African plate bumping into the Eurasian plate. As we will see, however, quite different processes produced the mountains on other planets.

Once a mountain range is formed by upthrusting of the crust, its rocks are subject to erosion by water and ice. The sharp peaks and serrated edges have little to do with the forces that make the mountains initially. Instead, they result from the processes that tear down mountains. Ice is especially effective sculptor of rock (Figure 8.11). In a world without moving ice or running water (such as the Moon or Mercury), mountains remain smooth and dull.

Figure 8.11. Mountains on Earth. The Torres del Paine are a young region of Earth's crust where sharp mountain peaks are being sculpted by glaciers. We owe the beauty of our young, steep mountains to the erosion by ice and water. (credit: David Morrison)

Volcanoes

Volcanoes mark locations where lava rises to the surface. One example is mid ocean ridges, which are long undersea mountain ranges formed by lava rising from Earth's mantle at plate boundaries. A second major kind of volcanic activity is associated with subduction zones, and volcanoes sometimes also appear in regions where continental plates are colliding. In each case, the volcanic activity gives us a way to sample some of the material from deeper within our planet.

Download for free at http://cnx.org/content/col11992/latest/

Other volcanic activity occurs above mantle "hot spots"—areas far from plate boundaries where heat is nevertheless rising from the interior of Earth. One of the best-known hot spot is under the island of Hawaii, where it currently supplies the heat to maintain three active volcanoes, two on land and one under the ocean. The Hawaii hot spot has been active for at least 100 million years. As Earth's plates have moved during that time, the hot spot has generated a 3500-kilometer-long chain of volcanic islands. The tallest Hawaiian volcanoes are among the largest individual mountains on Earth, more than 100 kilometers in diameter and rising 9 kilometers above the ocean floor. One of the Hawaiian volcanic mountains, the now-dormant Mauna Kea, has become one of the world's great sites for doing astronomy.

LINK TO LEARNING

The US Geological Service provides an interactive map (https://openstax.org/l/30mapringoffire) of the famous "ring of fire," which is the chain of volcanoes surrounding the Pacific Ocean, and shows the Hawaiian "hot spot" enclosed within.

Not all volcanic eruptions produce mountains. If lava flows rapidly from long cracks, it can spread out to form lava plains. The largest known terrestrial eruptions, such as those that produced the Snake River basalts in the northwestern United States or the Deccan plains in India, are of this type. Similar lava plains are found on the Moon and the other terrestrial planets.

8.3 EARTH'S ATMOSPHERE

Learning Objectives

By the end of this section, you will be able to:

> Differentiate between Earth's various atmospheric layers
> Describe the chemical composition and possible origins of our atmosphere
> Explain the difference between weather and climate

We live at the bottom of the ocean of air that envelops our planet. The atmosphere, weighing down upon Earth's surface under the force of gravity, exerts a pressure at sea level that scientists define as 1 **bar** (a term that comes from the same root as *barometer*, an instrument used to measure atmospheric pressure). A bar of pressure means that each square centimeter of Earth's surface has a weight equivalent to 1.03 kilograms pressing down on it. Humans have evolved to live at this pressure; make the pressure a lot lower or higher and we do not function well.

The total mass of Earth's atmosphere is about 5×10^{18} kilograms. This sounds like a large number, but it is only about a millionth of the total mass of Earth. The atmosphere represents a smaller fraction of Earth than the fraction of your mass represented by the hair on your head.

Structure of the Atmosphere

The structure of the atmosphere is illustrated in Figure 8.12. Most of the atmosphere is concentrated near the surface of Earth, within about the bottom 10 kilometers where clouds form and airplanes fly. Within this region—called the **troposphere**—warm air, heated by the surface, rises and is replaced by descending currents of cooler air; this is an example of convection. This circulation generates clouds and wind. Within

This OpenStax book is available for free at http://cnx.org/content/col11992/1.8

the troposphere, temperature decreases rapidly with increasing elevation to values near 50 °C below freezing at its upper boundary, where the **stratosphere** begins. Most of the stratosphere, which extends to about 50 kilometers above the surface, is cold and free of clouds.

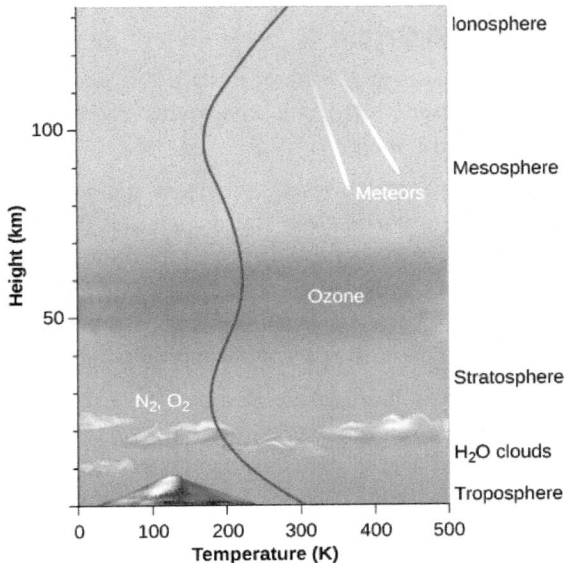

Figure 8.12. Structure of Earth's Atmosphere. Height increases up the left side of the diagram, and the names of the different atmospheric layers are shown at the right. In the upper ionosphere, ultraviolet radiation from the Sun can strip electrons from their atoms, leaving the atmosphere ionized. The curving red line shows the temperature (see the scale on the *x*-axis).

Near the top of the stratosphere is a layer of **ozone** (O_3), a heavy form of oxygen with three atoms per molecule instead of the usual two. Because ozone is a good absorber of ultraviolet light, it protects the surface from some of the Sun's dangerous ultraviolet radiation, making it possible for life to exist on Earth. The breakup of ozone adds heat to the stratosphere, reversing the decreasing temperature trend in the troposphere. Because ozone is essential to our survival, we reacted with justifiable concern to evidence that became clear in the 1980s that atmospheric ozone was being destroyed by human activities. By international agreement, the production of industrial chemicals that cause ozone depletion, called chlorofluorocarbons, or CFCs, has been phased out. As a result, ozone loss has stopped and the "ozone hole" over the Antarctic is shrinking gradually. This is an example of how concerted international action can help maintain the habitability of Earth.

LINK TO LEARNING

Visit NASA's scientific visualization studio for a short video (https://openstax.org/l/302065regcfc) of what would have happened to Earth's ozone layer by 2065 if CFCs had not been regulated.

At heights above 100 kilometers, the atmosphere is so thin that orbiting satellites can pass through it with very little friction. Many of the atoms are ionized by the loss of an electron, and this region is often called the ionosphere. At these elevations, individual atoms can occasionally escape completely from the gravitational field of Earth. There is a continuous, slow leaking of atmosphere—especially of lightweight atoms, which move faster than heavy ones. Earth's atmosphere cannot, for example, hold on for long to hydrogen or helium, which

Download for free at http://cnx.org/content/col11992/latest/

escape into space. Earth is not the only planet to experience atmosphere leakage. Atmospheric leakage also created Mars' thin atmosphere. Venus' dry atmosphere evolved because its proximity to the Sun vaporized and dissociated any water, with the component gases lost to space.

Atmospheric Composition and Origin

At Earth's surface, the atmosphere consists of 78% nitrogen (N_2), 21% oxygen (O_2), and 1% argon (Ar), with traces of water vapor (H_2O), carbon dioxide (CO_2), and other gases. Variable amounts of dust particles and water droplets are also found suspended in the air.

A complete census of Earth's volatile materials, however, should look at more than the gas that is now present. *Volatile* materials are those that evaporate at a relatively low temperature. If Earth were just a little bit warmer, some materials that are now liquid or solid might become part of the atmosphere. Suppose, for example, that our planet were heated to above the boiling point of water (100 °C, or 373 K); that's a large change for humans, but a small change compared to the range of possible temperatures in the universe. At 100 °C, the oceans would boil and the resulting water vapor would become a part of the atmosphere.

To estimate how much water vapor would be released, note that there is enough water to cover the entire Earth to a depth of about 300 meters. Because the pressure exerted by 10 meters of water is equal to about 1 bar, the average pressure at the ocean floor is about 300 bars. Water weighs the same whether in liquid or vapor form, so if the oceans boiled away, the atmospheric pressure of the water would still be 300 bars. Water would therefore greatly dominate Earth's atmosphere, with nitrogen and oxygen reduced to the status of trace constituents.

On a warmer Earth, another source of additional atmosphere would be found in the sedimentary carbonate rocks of the crust. These minerals contain abundant carbon dioxide. If all these rocks were heated, they would release about 70 bars of CO_2, far more than the current CO_2 pressure of only 0.0005 bar. Thus, the atmosphere of a warm Earth would be dominated by water vapor and carbon dioxide, with a surface pressure nearing 400 bars.

Several lines of evidence show that the composition of Earth's atmosphere has changed over our planet's history. Scientists can infer the amount of atmospheric oxygen, for example, by studying the chemistry of minerals that formed at various times. We examine this issue in more detail later in this chapter.

Today we see that CO_2, H_2O, sulfur dioxide (SO_2), and other gases are released from deeper within Earth through the action of volcanoes. (For CO_2, the primary source today is the burning of fossil fuels, which releases far more CO_2 than that from volcanic eruptions.) Much of this apparently new gas, however, is recycled material that has been subducted through plate tectonics. But where did our planet's original atmosphere come from?

Three possibilities exist for the original source of Earth's atmosphere and oceans: (1) the atmosphere could have been formed with the rest of Earth as it accumulated from debris left over from the formation of the Sun; (2) it could have been released from the interior through volcanic activity, subsequent to the formation of Earth; or (3) it may have been derived from impacts by comets and asteroids from the outer parts of the solar system. Current evidence favors a combination of the interior and impact sources.

Weather and Climate

All planets with atmospheres have *weather*, which is the name we give to the circulation of the atmosphere. The energy that powers the weather is derived primarily from the sunlight that heats the surface. Both the rotation of the planet and slower seasonal changes cause variations in the amount of sunlight striking different parts of Earth. The atmosphere and oceans redistribute the heat from warmer to cooler areas. Weather on any planet

This OpenStax book is available for free at http://cnx.org/content/col11992/1.8

represents the response of its atmosphere to changing inputs of energy from the Sun (see Figure 8.13 for a dramatic example).

Figure 8.13. Storm from Space. This satellite image shows Hurricane Irene in 2011, shortly before the storm hit land in New York City. The combination of Earth's tilted axis of rotation, moderately rapid rotation, and oceans of liquid water can lead to violent weather on our planet. (credit: NASA/NOAA GOES Project)

Climate is a term used to refer to the effects of the atmosphere that last through decades and centuries. Changes in climate (as opposed to the random variations in weather from one year to the next) are often difficult to detect over short time periods, but as they accumulate, their effect can be devastating. One saying is that "Climate is what you expect, and weather is what you get." Modern farming is especially sensitive to temperature and rainfall; for example, calculations indicate that a drop of only 2 °C throughout the growing season would cut the wheat production by half in Canada and the United States. At the other extreme, an increase of 2 °C in the average temperature of Earth would be enough to melt many glaciers, including much of the ice cover of Greenland, raising sea level by as much as 10 meters, flooding many coastal cities and ports, and putting small islands completely under water.

The best documented changes in Earth's climate are the great ice ages, which have lowered the temperature of the Northern Hemisphere periodically over the past half million years or so (Figure 8.14). The last ice age, which ended about 14,000 years ago, lasted some 20,000 years. At its height, the ice was almost 2 kilometers thick over Boston and stretched as far south as New York City.

Download for free at http://cnx.org/content/col11992/latest/

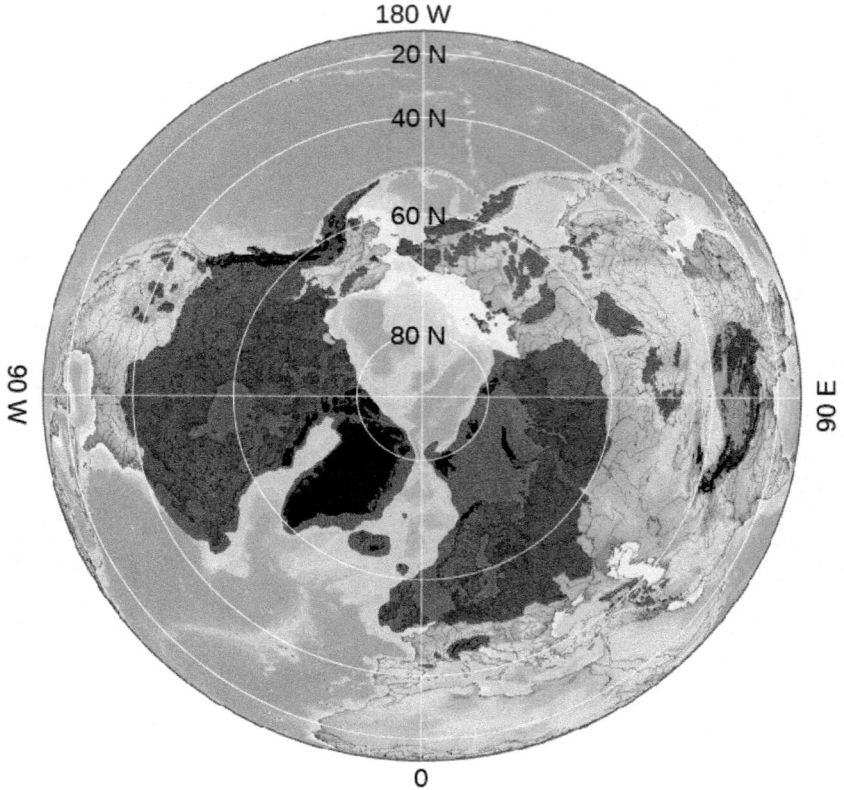

Figure 8.14. Ice Age. This computer-generated image shows the frozen areas of the Northern Hemisphere during past ice ages from the vantage point of looking down on the North Pole. The area in black indicates the most recent glaciation (coverage by glaciers), and the area in gray shows the maximum level of glaciation ever reached. (credit: modification of work by Hannes Grobe/AWI)

These ice ages were primarily the result of changes in the tilt of Earth's rotational axis, produced by the gravitational effects of the other planets. We are less certain about evidence that at least once (and perhaps twice) about a billion years ago, the entire ocean froze over, a situation called *snowball Earth*.

The development and evolution of life on Earth has also produced changes in the composition and temperature of our planet's atmosphere, as we shall see in the next section.

LINK TO LEARNING

Watch this short excerpt (https://openstax.org/l/30natgeoearth) from the National Geographic documentary *Earth: The Biography*. In this segment, Dr. Iain Stewart explains the fluid nature of our atmosphere.

This OpenStax book is available for free at http://cnx.org/content/col11992/1.8

8.4 | LIFE, CHEMICAL EVOLUTION, AND CLIMATE CHANGE

Learning Objectives

By the end of this section, you will be able to:

> Outline the origins and subsequent diversity of life on Earth
> Explain the ways that life and geological activity have influenced the evolution of the atmosphere
> Describe the causes and effects of the atmospheric greenhouse effect and global warming
> Describe the impact of human activity on our planet's atmosphere and ecology

As far as we know, Earth seems to be the only planet in the solar system with life. The origin and development of life are an important part of our planet's story. Life arose early in Earth's history, and living organisms have been interacting with their environment for billions of years. We recognize that life-forms have evolved to adapt to the environment on Earth, and we are now beginning to realize that Earth itself has been changed in important ways by the presence of living matter. The study of the coevolution of life and our planet is one of the subjects of the modern science of *astrobiology*.

The Origin of Life

The record of the birth of life on Earth has been lost in the restless motions of the crust. According to chemical evidence, by the time the oldest surviving rocks were formed about 3.9 billion years ago, life already existed. At 3.5 billion years ago, life had achieved the sophistication to build large colonies called *stromatolites*, a form so successful that stromatolites still grow on Earth today (Figure 8.15). But, few rocks survive from these ancient times, and abundant fossils have been preserved only during the past 600 million years—less than 15% of our planet's history.

Figure 8.15. Cross-Sections of Fossil Stromatolites. This polished cross-section of a fossilized colony of stromatolites dates to the Precambrian Era. The layered, domelike structures are mats of sediment trapped in shallow waters by large numbers of blue-green bacteria that can photosynthesize. Such colonies of microorganisms date back more than 3 billion years. (credit: James St. John)

There is little direct evidence about the actual origin of life. We know that the atmosphere of early Earth, unlike today's, contained abundant carbon dioxide and some methane, but no oxygen gas. In the absence of oxygen, many complex chemical reactions are possible that lead to the production of amino acids, proteins, and other chemical building blocks of life. Therefore, it seems likely that these chemical building blocks were available very early in Earth's history and they would have combined to make living organisms.

Download for free at http://cnx.org/content/col11992/latest/

For tens of millions of years after Earth's formation, life (perhaps little more than large molecules, like the viruses of today) probably existed in warm, nutrient-rich seas, living off accumulated organic chemicals. When this easily accessible food became depleted, life began the long evolutionary road that led to the vast numbers of different organisms on Earth today. As it did so, life began to influence the chemical composition of the atmosphere.

In addition to the study of life's history as revealed by chemical and fossil evidence in ancient rocks, scientists use tools from the rapidly advancing fields of genetics and *genomics*—the study of the genetic code that is shared by all life on Earth. While each individual has a unique set of genes (which is why genetic "fingerprinting" is so useful for the study of crime), we also have many genetic traits in common. Your *genome*, the complete map of the DNA in your body, is identical at the 99.9% level to that of Julius Caesar or Marie Curie. At the 99% level, human and chimpanzee genomes are the same. By looking at the gene sequences of many organisms, we can determine that all life on Earth is descended from a common ancestor, and we can use the genetic variations among species as a measure of how closely different species are related.

These genetic analysis tools have allowed scientists to construct what is called the " tree of life" (Figure 8.16). This diagram illustrates the way organisms are related by examining one sequence of the nucleic acid RNA that all species have in common. This figure shows that life on Earth is dominated by microscopic creatures that you have probably never heard of. Note that the plant and animal kingdoms are just two little branches at the far right. Most of the diversity of life, and most of our evolution, has taken place at the microbial level. Indeed, it may surprise you to know that there are more microbes in a bucket of soil than there are stars in the Galaxy. You may want to keep this in mind when, later in this book, we turn to the search for life on other worlds. The "aliens" that are most likely to be out there are microbes.

This OpenStax book is available for free at http://cnx.org/content/col11992/1.8

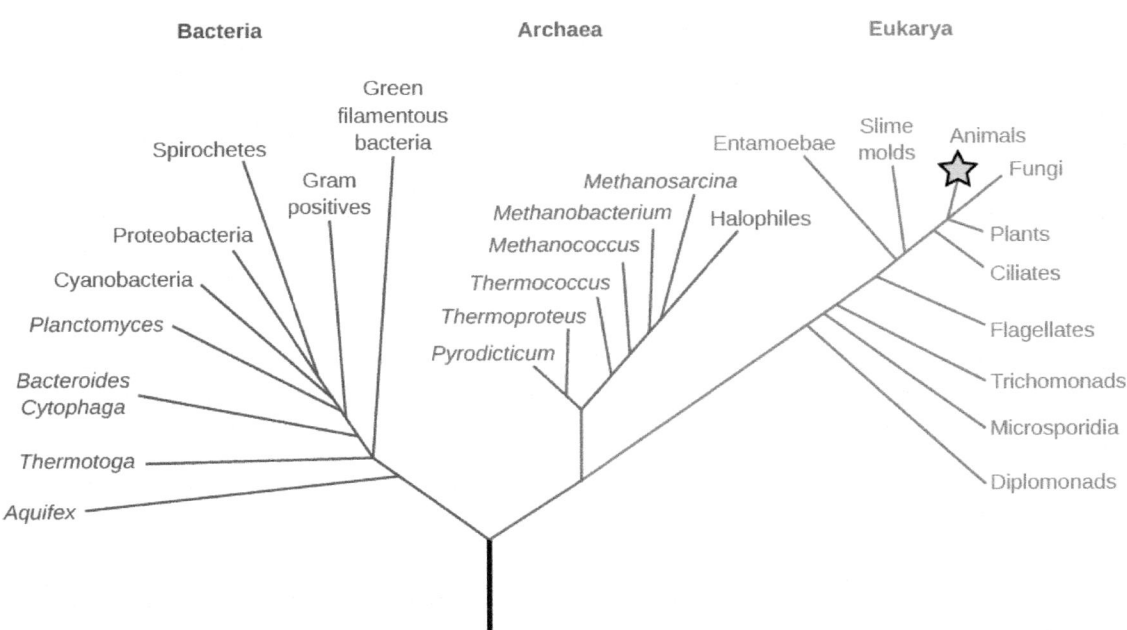

Figure 8.16. Tree of Life. This chart shows the main subdivisions of life on Earth and how they are related. Note that the animal and plant kingdoms are just short branches on the far right, along with the fungi. The most fundamental division of Earth's living things is onto three large domains called bacteria, archaea, and eukarya. Most of the species listed are microscopic. (credit: modification of work by Eric Gaba)

Such genetic studies lead to other interesting conclusions as well. For example, it appears that the earliest surviving terrestrial life-forms were all adapted to live at high temperatures. Some biologists think that life might actually have begun in locations on our planet that were extremely hot. Yet another intriguing possibility is that life began on Mars (which cooled sooner) rather than Earth and was "seeded" onto our planet by meteorites traveling from Mars to Earth. Mars rocks are still making their way to Earth, but so far none has shown evidence of serving as a "spaceship" to carry microorganisms from Mars to Earth.

The Evolution of the Atmosphere

One of the key steps in the evolution of life on Earth was the development of blue-green algae, a very successful life-form that takes in carbon dioxide from the environment and releases oxygen as a waste product. These successful microorganisms proliferated, giving rise to all the lifeforms we call plants. Since the energy for making new plant material from chemical building blocks comes from sunlight, we call the process **photosynthesis**.

Studies of the chemistry of ancient rocks show that Earth's atmosphere lacked abundant free oxygen until about 2 billion years ago, despite the presence of plants releasing oxygen by photosynthesis. Apparently, chemical reactions with Earth's crust removed the oxygen gas as quickly as it formed. Slowly, however, the increasing evolutionary sophistication of life led to a growth in the plant population and thus increased oxygen production. At the same time, it appears that increased geological activity led to heavy erosion on our planet's surface. Tthis buried much of the plant carbon before it could recombine with oxygen to form CO_2.

Download for free at http://cnx.org/content/col11992/latest/

Free oxygen began accumulating in the atmosphere about 2 billion years ago, and the increased amount of this gas led to the formation of Earth's ozone layer (recall that ozone is a triple molecule of oxygen, O_3), which protects the surface from deadly solar ultraviolet light. Before that, it was unthinkable for life to venture outside the protective oceans, so the landmasses of Earth were barren.

The presence of oxygen, and hence ozone, thus allowed colonization of the land. It also made possible a tremendous proliferation of animals, which lived by taking in and using the organic materials produced by plants as their own energy source.

As animals evolved in an environment increasingly rich in oxygen, they were able to develop techniques for breathing oxygen directly from the atmosphere. We humans take it for granted that plenty of free oxygen is available in Earth's atmosphere, and we use it to release energy from the food we take in. Although it may seem funny to think of it this way, we are lifeforms that have evolved to breathe in the waste product of plants. It is plants and related microbes that are the primary producers, using sunlight to create energy-rich "food" for the rest of us.

On a planetary scale, one of the consequences of life has been a decrease in atmospheric carbon dioxide. In the absence of life, Earth would probably have an atmosphere dominated by CO_2, like Mars or Venus. But living things, in combination with high levels of geological activity, have effectively stripped our atmosphere of most of this gas.

The Greenhouse Effect and Global Warming

We have a special interest in the carbon dioxide content of the atmosphere because of the key role this gas plays in retaining heat from the Sun through a process called the **greenhouse effect**. To understand how the greenhouse effect works, consider the fate of sunlight that strikes the surface of Earth. The light penetrates our atmosphere, is absorbed by the ground, and heats the surface layers. At the temperature of Earth's surface, that energy is then reemitted as infrared or heat radiation (Figure 8.17). However, the molecules of our atmosphere, which allow visible light through, are good at absorbing infrared energy. As a result, CO_2 (along with methane and water vapor) acts like a blanket, trapping heat in the atmosphere and impeding its flow back to space. To maintain an energy balance, the temperature of the surface and lower atmosphere must increase until the total energy radiated by Earth to space equals the energy received from the Sun. The more CO_2 there is in our atmosphere, the higher the temperature at which Earth's surface reaches a new balance.

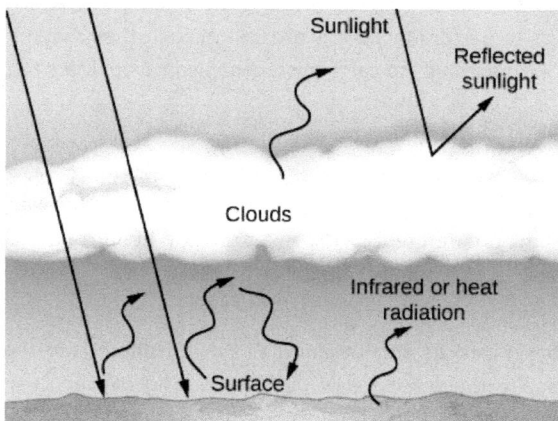

Figure 8.17. How the Greenhouse Effect Works. Sunlight that penetrates to Earth's lower atmosphere and surface is reradiated as infrared or heat radiation, which is trapped by greenhouse gases such as water vapor, methane, and CO_2 in the atmosphere. The result is a higher surface temperature for our planet.

This OpenStax book is available for free at http://cnx.org/content/col11992/1.8

Download for free at http://cnx.org/content/col11992/latest/

The greenhouse effect in a planetary atmosphere is similar to the heating of a gardener's greenhouse or the inside of a car left out in the Sun with the windows rolled up. In these examples, the window glass plays the role of **greenhouse gases**, letting sunlight in but reducing the outward flow of heat radiation. As a result, a greenhouse or car interior winds up much hotter than would be expected from the heating of sunlight alone. On Earth, the current greenhouse effect elevates the surface temperature by about 23 °C. Without this greenhouse effect, the average surface temperature would be well below freezing and Earth would be locked in a global ice age.

That's the good news; the bad news is that the heating due to the greenhouse effect is increasing. Modern industrial society depends on energy extracted from burning fossil fuels. In effect, we are exploiting the energy-rich material created by photosynthesis tens of millions of years ago. As these ancient coal and oil deposits are oxidized (burned using oxygen), large quantities of carbon dioxide are released into the atmosphere. The problem is exacerbated by the widespread destruction of tropical forests, which we depend on to extract CO_2 from the atmosphere and replenish our supply of oxygen. In the past century of increased industrial and agricultural development, the amount of CO_2 in the atmosphere increased by about 30% and continues to rise at more than 0.5% per year.

Before the end of the present century, Earth's CO_2 level is predicted to reach twice the value it had before the industrial revolution (Figure 8.18). The consequences of such an increase for Earth's surface and atmosphere (and the creatures who live there) are likely to be complex changes in climate, and may be catastrophic for many species. Many groups of scientists are now studying the effects of such global warming with elaborate computer models, and climate change has emerged as the greatest known threat (barring nuclear war) to both industrial civilization and the ecology of our planet.

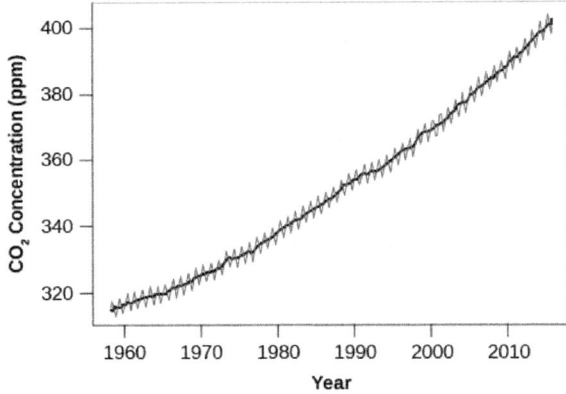

Figure 8.18. Increase of Atmospheric Carbon Dioxide over Time. Scientists expect that the amount of CO_2 will double its preindustrial level before the end of the twenty-first century. Measurements of the isotopic signatures of this added CO_2 demonstrate that it is mostly coming from burning fossil fuels. (credit: modification of work by NOAA)

LINK TO LEARNING

This short PBS video (https://openstax.org/l/30pbsgreengas) explains the physics of the greenhouse effect.

Download for free at http://cnx.org/content/col11992/latest/

Already climate change is widely apparent. Around the world, temperature records are constantly set and broken; all but one of the hottest recorded years have taken place since 2000. Glaciers are retreating, and the Arctic Sea ice is now much thinner than when it was first explored with nuclear submarines in the 1950s. Rising sea levels (from both melting glaciers and expansion of the water as its temperature rises) pose one of the most immediate threats, and many coastal cities have plans to build dikes or seawalls to hold back the expected flooding. The rate of temperature increase is without historical precedent, and we are rapidly entering "unknown territory" where human activities are leading to the highest temperatures on Earth in more than 50 million years.

Human Impacts on Our Planet

Earth is so large and has been here for so long that some people have trouble accepting that humans are really changing the planet, its atmosphere, and its climate. They are surprised to learn, for example, that the carbon dioxide released from burning fossil fuels is 100 times greater than that emitted by volcanoes. But, the data clearly tell the story that our climate is changing rapidly, and that almost all of the change is a result of human activity.

This is not the first time that humans have altered our environment dramatically. Some of the greatest changes were caused by our ancestors, before the development of modern industrial society. If aliens had visited Earth 50,000 years ago, they would have seen much of the planet supporting large animals of the sort that now survive only in Africa. The plains of Australia were occupied by giant marsupials such as diprododon and zygomaturus (the size of our elephants today), and a species of kangaroo that stood 10 feet high. North America and North Asia hosted mammoths, saber tooth cats, mastodons, giant sloths, and even camels. The Islands of the Pacific teemed with large birds, and vast forests covered what are now the farms of Europe and China. Early human hunters killed many large mammals and marsupials, early farmers cut down most of the forests, and the Polynesian expansion across the Pacific doomed the population of large birds.

An even greater mass extinction is underway as a result of rapid climate change. In recognition of our impact on the environment, scientists have proposed giving a new name to the current epoch, the *anthropocine*, when human activity started to have a significant global impact. Although not an officially approved name, the concept of "anthropocine" is useful for recognizing that we humans now represent the dominant influence on our planet's atmosphere and ecology, for better or for worse.

8.5 COSMIC INFLUENCES ON THE EVOLUTION OF EARTH

Learning Objectives

By the end of this section, you will be able to:

> Explain the scarcity of impact craters on Earth compared with other planets and moons
> Describe the evidence for recent impacts on Earth
> Detail how a massive impact changed the conditions for life on Earth, leading to the extinction of the dinosaurs
> Describe how impacts have influenced the evolution of life on Earth
> Discuss the search for objects that could potentially collide with our planet

In discussing Earth's geology earlier in this chapter, we dealt only with the effects of internal forces, expressed through the processes of plate tectonics and volcanism. On the Moon, in contrast, we see primarily craters, produced by the impacts of interplanetary debris such as asteroids and comets. Why don't we see more evidence on Earth of the kinds of impact craters that are so prominent on the Moon and other worlds?

This OpenStax book is available for free at http://cnx.org/content/col11992/1.8

Where Are the Craters on Earth?

It is not possible that Earth escaped being struck by the interplanetary debris that has pockmarked the Moon. From a cosmic perspective, the Moon is almost next door. Our atmosphere does make small pieces of cosmic debris burn up (which we see as *meteors*—commonly called shooting stars). But, the layers of our air provide no shield against the large impacts that form craters several kilometers in diameter and are common on the Moon.

In the course of its history, Earth must therefore have been impacted as heavily as the Moon. The difference is that, on Earth, these craters are destroyed by our active geology before they can accumulate. As plate tectonics constantly renews our crust, evidence of past cratering events is slowly erased. Only in the past few decades have geologists succeeded in identifying the eroded remnants of many impact craters (Figure 8.19). Even more recent is our realization that, over the history of Earth, these impacts have had an important influence on the evolution of life.

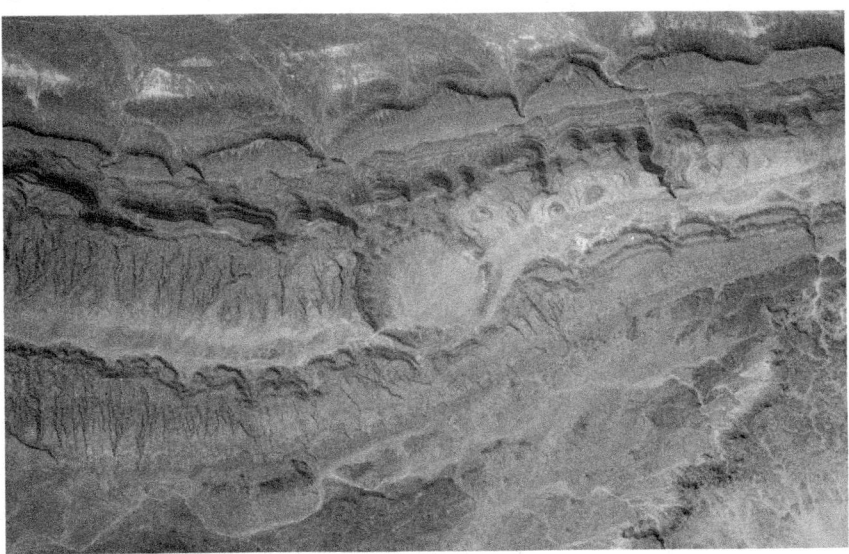

Figure 8.19. Ouarkziz Impact Crater. Located in Algeria, this crater (the round feature in the center) is the result of a meteor impact during the Cretaceous period. Although the crater has experienced heavy erosion, this image from the International Space Station shows the circular pattern resulting from impact. (credit: modification of work by NASA)

Recent Impacts

The collision of interplanetary debris with Earth is not a hypothetical idea. Evidence of relatively recent impacts can be found on our planet's surface. One well-studied historic collision took place on June 30, 1908, near the Tunguska River in Siberia. In this desolate region, there was a remarkable explosion in the atmosphere about 8 kilometers above the surface. The shock wave flattened more than a thousand square kilometers of forest (Figure 8.20). Herds of reindeer and other animals were killed, and a man at a trading post 80 kilometers from the blast was thrown from his chair and knocked unconscious. The blast wave spread around the world, as recorded by instruments designed to measure changes in atmospheric pressure.

Download for free at http://cnx.org/content/col11992/latest/

Figure 8.20. Aftermath of the Tunguska Explosion. This photograph, taken 21 years after the blast, shows a part of the forest that was destroyed by the 5-megaton explosion, resulting when a stony projectile about the size of a small office building (40 meters in diameter) collided with our planet. (credit: modification of work by Leonid Kulik)

Despite this violence, no craters were formed by the Tunguska explosion. Shattered by atmospheric pressure, the stony projectile with a mass of approximately 10,000 tons disintegrated above our planet's surface to create a blast equivalent to a 5-megaton nuclear bomb. Had it been smaller or more fragile, the impacting body would have dissipated its energy at high altitude and probably attracted no attention. Today, such high-altitude atmospheric explosions are monitored regularly by military surveillance systems.

If it had been larger or made of stronger material (such as metal), the Tunguska projectile would have penetrated all the way to the surface of Earth and exploded to form a crater. Instead, only the heat and shock of the atmospheric explosion reached the surface, but the devastation it left behind in Siberia bore witness to the power of such impacts. Imagine if the same rocky impactor had exploded over New York City in 1908; history books might today record it as one of the most deadly events in human history.

Tens of thousands of people witnessed directly the explosion of a smaller (20-meter) projectile over the Russian city of Chelyabinsk on an early winter morning in 2013. It exploded at a height of 21 kilometers in a burst of light brighter than the Sun, and the shockwave of the 0.5-megaton explosion broke tens of thousands of windows and sent hundreds of people to the hospital. Rock fragments (meteorites) were easily collected by people in the area after the blast because they landed on fresh snow.

LINK TO LEARNING

Dr. David Morrison, one of the original authors of this textbook, provides a nontechnical talk (https://openstax.org/l/30chelyabinskex) about the Chelyabinsk explosion, and impacts in general.

The best-known recent crater on Earth was formed about 50,000 years ago in Arizona. The projectile in this case was a lump of iron about 40 meters in diameter. Now called *Meteor Crater* and a major tourist attraction on the way to the Grand Canyon, the crater is about a mile across and has all the features associated with similar-

This OpenStax book is available for free at http://cnx.org/content/col11992/1.8

size lunar impact craters (Figure 8.21). Meteor Crater is one of the few impact features on Earth that remains relatively intact; some older craters are so eroded that only a trained eye can distinguish them. Nevertheless, more than 150 have been identified. (See the list of suggested online sites at the end of this chapter if you want to find out more about these other impact scars.)

Figure 8.21. Meteor Crater in Arizona. Here we see a 50,000-year-old impact crater made by the collision of a 40-meter lump of iron with our planet. Although impact craters are common on less active bodies such as the Moon, this is one of the very few well-preserved craters on Earth. (modification of work by D. Roddy/USGS)

Mass Extinction

The impact that produced Meteor Crater would have been dramatic indeed to any humans who witnessed it (from a safe distance) since the energy release was equivalent to a 10-megaton nuclear bomb. But such explosions are devastating only in their local areas; they have no *global* consequences. Much larger (and rarer) impacts, however, can disturb the ecological balance of the entire planet and thus influence the course of evolution.

The best-documented large impact took place 65 million years ago, at the end of what is now called the Cretaceous period of geological history. This time in the history of life on Earth was marked by a **mass extinction**, in which more than half of the species on our planet died out. There are a dozen or more mass extinctions in the geological record, but this particular event (nicknamed the "great dying") has always intrigued paleontologists because it marks the end of the dinosaur age. For tens of millions of years these great creatures had flourished and dominated. Then, they suddenly disappeared (along with many other species), and thereafter mammals began the development and diversification that ultimately led to all of us.

The object that collided with Earth at the end of the Cretaceous period struck a shallow sea in what is now the Yucatán peninsula of Mexico. Its mass must have been more than a trillion tons, determined from study of a worldwide layer of sediment deposited from the dust cloud that enveloped the planet after its impact. First identified in 1979, this sediment layer is rich in the rare metal iridium and other elements that are relatively abundant in asteroids and comets, but exceedingly rare in Earth's crust. Even though it was diluted by the material that the explosion excavated from the surface of Earth, this cosmic component can still be identified. In addition, this layer of sediment contains many minerals characteristic of the temperatures and pressures of a gigantic explosion.

Download for free at http://cnx.org/content/col11992/latest/

The impact that led to the extinction of dinosaurs released energy equivalent to 5 billion Hiroshima-size nuclear bombs and excavated a crater 200 kilometers across and deep enough to penetrate through Earth's crust. This large crater, named Chicxulub for a small town near its center, has subsequently been buried in sediment, but its outlines can still be identified (Figure 8.22). The explosion that created the Chicxulub crater lifted about 100 trillion tons of dust into the atmosphere. We can determine this amount by measuring the thickness of the sediment layer that formed when this dust settled to the surface.

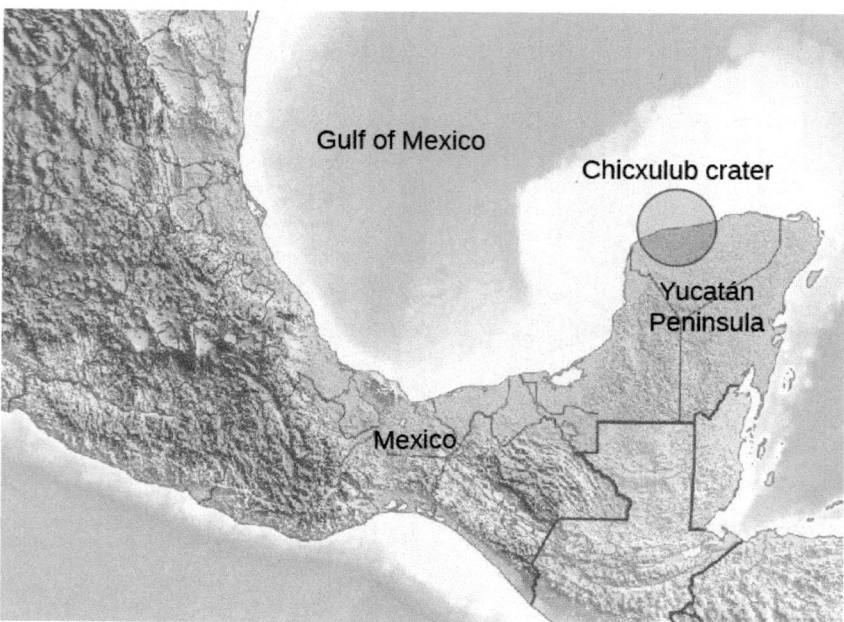

Figure 8.22. Site of the Chicxulub Crater. This map shows the location of the impact crater created 65 million years ago on Mexico's Yucatán peninsula. The crater is now buried under more than 500 meters of sediment. (credit: modification of work by "Carport"/Wikimedia)

Such a quantity of airborne material would have blocked sunlight completely, plunging Earth into a period of cold and darkness that lasted several months. Many plants dependent on sunlight would have died, leaving plant-eating animals without a food supply. Other worldwide effects included large-scale fires (started by the hot, flying debris from the explosion) that destroyed much of the planet's forests and grasslands, and a long period in which rainwater around the globe was acidic. It was these environmental effects, rather than the explosion itself, that were responsible for the mass extinction, including the demise of the dinosaurs.

Impacts and the Evolution of Life

It is becoming clear that many—perhaps most—mass extinctions in Earth's long history resulted from a variety of other causes, but in the case of the dinosaur killer, the cosmic impact certainly played a critical role and may have been the "final straw" in a series of climactic disturbances that resulted in the "great dying."

A catastrophe for one group of living things, however, may create opportunities for another group. Following each mass extinction, there is a sudden evolutionary burst as new species develop to fill the ecological niches opened by the event. Sixty-five million years ago, our ancestors, the mammals, began to thrive when so many other species died out. We are the lucky beneficiaries of this process.

Impacts by comets and asteroids represent the only mechanisms we know of that could cause truly global catastrophes and seriously influence the evolution of life all over the planet. As paleontologist Stephen Jay Gould of Harvard noted, such a perspective changes fundamentally our view of biological evolution. The central

This OpenStax book is available for free at http://cnx.org/content/col11992/1.8

issues for the survival of a species must now include more than just its success in competing with other species and adapting to slowly changing environments, as envisioned by Darwin's idea of natural selection. Also required is an ability to survive random global catastrophes due to impacts.

Still earlier in its history, Earth was subject to even larger impacts from the leftover debris of planet formation. We know that the Moon was struck repeatedly by objects larger than 100 kilometers in diameter—1000 times more massive than the object that wiped out most terrestrial life 65 million years ago. Earth must have experienced similar large impacts during its first 700 million years of existence. Some of them were probably violent enough to strip the planet of most its atmosphere and to boil away its oceans. Such events would sterilize the planet, destroying any life that had begun. Life may have formed and been wiped out several times before our own microbial ancestors took hold sometime about 4 billion years ago.

The fact that the oldest surviving microbes on Earth are thermophiles (adapted to very high temperatures) can also be explained by such large impacts. An impact that was just a bit too small to sterilize the planet would still have destroyed anything that lived in what we consider "normal" environments, and only the creatures adapted to high temperatures would survive. Thus, the oldest surviving terrestrial lifeforms are probably the remnants of a sort of evolutionary bottleneck caused by repeated large impacts early in the planet's history.

Impacts in Our Future?

The impacts by asteroids and comets that have had such a major influence on life are not necessarily a thing of the past. In the full scope of planetary history, 65 million years ago was just yesterday. Earth actually orbits the Sun within a sort of cosmic shooting gallery, and although major impacts are rare, they are by no means over. Humanity could suffer the same fate as the dinosaurs, or lose a city to the much more frequent impacts like the one over Tunguska, unless we figure out a way to predict the next big impact and to protect our planet. The fact that our solar system is home to some very large planets in outer orbits may be beneficial to us; the gravitational fields of those planets can be very effective at pulling in cosmic debris and shielding us from larger, more frequent impacts.

Beginning in the 1990s, a few astronomers began to analyze the cosmic impact hazard and to persuade the government to support a search for potentially hazardous asteroids. Several small but sophisticated wide-field telescopes are now used for this search, which is called the NASA Spaceguard Survey. Already we know that there are currently no asteroids on a collision course with Earth that are as big (10–15 kilometers) as the one that killed the dinosaurs. The Spaceguard Survey now concentrates on finding smaller potential impactors. By 2015, the search had netted more than 15,000 near-Earth-asteroids, including most of those larger than 1 kilometer. None of those discovered so far poses any danger to us. Of course, we cannot make a similar statement about the asteroids that have not yet been discovered, but these will be found and evaluated one by one for their potential hazard. These asteroid surveys are one of the few really life-and-death projects carried out by astronomers, with a potential to help to save our planet from future major impacts.

LINK TO LEARNING

The Torino Impact Hazard Scale (https://openstax.org/l/30torhazscale) is a method for categorizing the impact hazard associated with near-Earth objects such as asteroids and comets. It is a communication tool for astronomers and the public to assess the seriousness of collision predictions by combining probability statistics and known kinetic damage potentials into a single threat value.

Download for free at http://cnx.org/content/col11992/latest/

Purdue University's "Impact: Earth" calculator (https://openstax.org/l/30purimpearcal) lets you input the characteristics of an approaching asteroid to determine the effect of its impact on our planet.

This OpenStax book is available for free at http://cnx.org/content/col11992/1.8

CHAPTER 8 REVIEW

 KEY TERMS

bar a force of 100,000 Newtons acting on a surface area of 1 square meter; the average pressure of Earth's atmosphere at sea level is 1.013 bars

basalt igneous rock produced by the cooling of lava; makes up most of Earth's oceanic crust and is found on other planets that have experienced extensive volcanic activity

convection movement caused within a gas or liquid by the tendency of hotter, and therefore less dense material, to rise and colder, denser material to sink under the influence of gravity, which consequently results in transfer of heat

core the central part of the planet; consists of higher density material

crust the outer layer of a terrestrial planet

fault in geology, a crack or break in the crust of a planet along which slippage or movement can take place, accompanied by seismic activity

granite a type of igneous silicate rock that makes up most of Earth's continental crust

greenhouse effect the blanketing (absorption) of infrared radiation near the surface of a planet—for example, by CO_2 in its atmosphere

greenhouse gas a gas in an atmosphere that absorbs and emits radiation within the thermal infrared range; on Earth, these atmospheric gases primarily include carbon dioxide, methane, and water vapor

igneous rock rock produced by cooling from a molten state

magnetosphere the region around a planet in which its intrinsic magnetic field dominates the interplanetary field carried by the solar wind; hence, the region within which charged particles can be trapped by the planetary magnetic field

mantle the largest part of Earth's interior; lies between the crust and the core

mass extinction the sudden disappearance in the fossil record of a large number of species of life, to be replaced by fossils of new species in subsequent layers; mass extinctions are indicators of catastrophic changes in the environment, such as might be produced by a large impact on Earth

metamorphic rock rock produced by physical and chemical alteration (without melting) under high temperature and pressure

ozone (O_3) a heavy molecule of oxygen that contains three atoms rather than the more normal two

photosynthesis a complex sequence of chemical reactions through which some living things can use sunlight to manufacture products that store energy (such as carbohydrates), releasing oxygen as one by-product

plate tectonics the motion of segments or plates of the outer layer of a planet over the underlying mantle

primitive rock rock that has not experienced great heat or pressure and therefore remains representative of the original condensed materials from the solar nebula

Download for free at http://cnx.org/content/col11992/latest/

rift zone in geology, a place where the crust is being torn apart by internal forces generally associated with the injection of new material from the mantle and with the slow separation of tectonic plates

sedimentary rock rock formed by the deposition and cementing of fine grains of material, such as pieces of igneous rock or the shells of living things

seismic wave a vibration that travels through the interior of Earth or any other object; on Earth, these are generally caused by earthquakes

stratosphere the layer of Earth's atmosphere above the troposphere and below the ionosphere

subduction the sideways and downward movement of the edge of a plate of Earth's crust into the mantle beneath another plate

troposphere the lowest level of Earth's atmosphere, where most weather takes place

volcano a place where material from a planet's mantle erupts on its surface

SUMMARY

8.1 The Global Perspective

Earth is the prototype terrestrial planet. Its interior composition and structure are probed using seismic waves. Such studies reveal that Earth has a metal core and a silicate mantle. The outer layer, or crust, consists primarily of oceanic basalt and continental granite. A global magnetic field, generated in the core, produces Earth's magnetosphere, which can trap charged atomic particles.

8.2 Earth's Crust

Terrestrial rocks can be classified as igneous, sedimentary, or metamorphic. A fourth type, primitive rock, is not found on Earth. Our planet's geology is dominated by plate tectonics, in which crustal plates move slowly in response to mantle convection. The surface expression of plate tectonics includes continental drift, recycling of the ocean floor, mountain building, rift zones, subduction zones, faults, earthquakes, and volcanic eruptions of lava from the interior.

8.3 Earth's Atmosphere

The atmosphere has a surface pressure of 1 bar and is composed primarily of N_2 and O_2, plus such important trace gases as H_2O, CO_2, and O_3. Its structure consists of the troposphere, stratosphere, mesosphere, and ionosphere. Changing the composition of the atmosphere also influences the temperature. Atmospheric circulation (weather) is driven by seasonally changing deposition of sunlight. Many longer term climate variations, such as the ice ages, are related to changes in the planet's orbit and axial tilt.

8.4 Life, Chemical Evolution, and Climate Change

Life originated on Earth at a time when the atmosphere lacked O_2 and consisted mostly of CO_2. Later, photosynthesis gave rise to free oxygen and ozone. Modern genomic analysis lets us see how the wide diversity of species on the planet are related to each other. CO_2 and methane in the atmosphere heat the surface through the greenhouse effect; today, increasing amounts of atmospheric CO_2 are leading to the global warming of our planet.

8.5 Cosmic Influences on the Evolution of Earth

Earth, like the Moon and other planets, has been influenced by the impacts of cosmic debris, including such recent examples as Meteor Crater and the Tunguska explosion. Larger past impacts are implicated in some mass extinctions, including the large impact 65 million years ago at the end of the Cretaceous period that

This OpenStax book is available for free at http://cnx.org/content/col11992/1.8

Download for free at http://cnx.org/content/col11992/latest/

wiped out the dinosaurs and many other species. Today, astronomers are working to predict the next impact in advance, while other scientists are coming to grips with the effect of impacts on the evolution and diversity of life on Earth.

 FOR FURTHER EXPLORATION

Articles

Earth

Collins, W., et al. "The Physical Science behind Climate Change." *Scientific American* (August 2007): 64. Why scientists are now confident that human activities are changing our planet's climate.

Glatzmaier, G., & Olson, P. "Probing the Geodynamo." *Scientific American* (April 2005): 50. Experiments and modeling that tell us about the source and reversals of Earth's magnetic field.

Gurnis, M. "Sculpting the Earth from Inside Out." *Scientific American* (March 2001): 40. On motions that lift and lower the continents.

Hartmann, W. "Piecing Together Earth's Early History." *Astronomy* (June 1989): 24.

Jewitt, D., & Young, E. "Oceans from the Skies." *Scientific American* (March 2015): 36. How did Earth get its water after its initial hot period?

Impacts

Boslaugh, M. "In Search of Death-Plunge Asteroids." *Astronomy* (July 2015): 28. On existing and proposed programs to search for earth-crossing asteroids.

Brusatte, S. "What Killed the Dinosaurs?" *Scientific American* (December 2015): 54. The asteroid hit Earth at an already vulnerable time.

Chyba, C. "Death from the Sky: Tunguska." *Astronomy* (December 1993): 38. Excellent review article.

Durda, D. "The Chelyabinsk Super-Meteor." *Sky & Telescope* (June 2013): 24. A nice summary with photos and eyewitness reporting.

Gasperini, L., et al. "The Tunguska Mystery." *Scientific American* (June 2008): 80. A more detailed exploration of the site of the 1908 impact over Siberia.

Kring, D. "Blast from the Past." *Astronomy* (August 2006): 46. Six-page introduction to Arizona's meteor crater.

Websites

Earth

Astronaut Photography of Earth from Space: http://earth.jsc.nasa.gov/. A site with many images and good information.

Exploration of the Earth's Magnetosphere: http://phy6.org/Education/Intro.html. An educational website by Dr. Daniel Stern.

NASA Goddard: Earth from Space: Fifteen Amazing Things in 15 Years: https://www.nasa.gov/content/goddard/earth-from-space-15-amazing-things-in-15-years. Images and videos that reveal things about our planet and its atmosphere.

U.S. Geological Survey: Earthquake Information Center: http://earthquake.usgs.gov/learn/

Views of the Solar System: http://www.solarviews.com/eng/earth.htm. Overview of Earth.

Download for free at http://cnx.org/content/col11992/latest/

Impacts

B612 Foundation : https://b612foundation.org/. Set up by several astronauts for research and education about the asteroid threat to Earth and to build a telescope in space to search for dangerous asteroids.

Lunar and Planetary Institute: Introduction to Terrestrial Impact Craters: http://www.lpi.usra.edu/publications/slidesets/craters/. Includes images.

Meteor Crater Tourist Site: http://meteorcrater.com/.

NASA/Jet Propulsion Lab Near Earth Object Program: http://neo.jpl.nasa.gov/neo/.

What Are Near-Earth-Objects: http://spaceguardcentre.com/what-are-neos/. From the British Spaceguard Centre.

Videos

Earth

All Alone in the Night: http://apod.nasa.gov/apod/ap120305.html. Flying over Earth at night (2:30).

Earth Globes Movies (including Earth at night): http://astro.uchicago.edu/cosmus/projects/earth/.

Earth: The Operator's Manual: http://earththeoperatorsmanual.com/feature-video/earth-the-operators-manual. A National Science Foundation–sponsored miniseries on climate change and energy, with geologist Richard Alley (53:43).

PBS NOVA Videos about Earth: http://www.pbs.org/wgbh/nova/earth/. Programs and information about planet Earth. Click full episodes on the menu at left to be taken to a nice array of videos.

U. S. National Weather Service: http://earth.nullschool.net. Real Time Globe of Earth showing wind patterns which can be zoomed and moved to your preferred view.

Impacts

Chelyabinsk Meteor: Can We Survive a Bigger Impact?: https://www.youtube.com/watch?v=Y-e6xyUZLLs . Talk by Dr. David Morrison (1:34:48).

Large Asteroid Impact Simulation: https://www.youtube.com/watch?v=bU1QPtOZQZU. Large asteroid impact simulation from the Discovery Channel (4:45).

Meteor Hits Russia February 15, 2013: https://www.youtube.com/watch?v=dpmXyJrs7iU. Archive of eyewitness footage (10:11).

Sentinel Mission: Finding an Asteroid Headed for Earth: https://www.youtube.com/watch?v=efz8c3ijD_A. Public lecture by astronaut Ed Lu (1:08:57).

🛉 COLLABORATIVE GROUP ACTIVITIES

A. If we can predict that lots of ground movement takes place along subduction zones and faults, then why do so many people live there? Should we try to do anything to discourage people from living in these areas? What inducement would your group offer people to move? Who would pay for the relocation? (Note that two of the original authors of this book live quite close to the San Andreas and Hayward faults. If they wrote this chapter and haven't moved, what are the chances others living in these kinds of areas will move?)

This OpenStax book is available for free at http://cnx.org/content/col11992/1.8

B. After your group reads the feature box on Alfred Wegener: Catching the Drift of Plate Tectonics, discuss some reasons his idea did not catch on right away among scientists. From your studies in this course and in other science courses (in college and before), can you cite other scientific ideas that we now accept but that had controversial beginnings? Can you think of any scientific theories that are still controversial today? If your group comes up with some, discuss ways scientists could decide whether each theory on your list is right.

C. Suppose we knew that a large chunk of rock or ice (about the same size as the one that hit 65 million years ago) will impact Earth in about 5 years. What could or should we do about it? (The film *Deep Impact* dealt with this theme.) Does your group think that the world as a whole should spend more money to find and predict the orbits of cosmic debris near Earth?

D. Carl Sagan pointed out that any defensive weapon that we might come up with to deflect an asteroid *away* from Earth could be used as an offensive weapon by an unstable dictator in the future to cause an asteroid not heading our way to come toward Earth. The history of human behavior, he noted, has shown that most weapons that are built (even with the best of motives) seem to wind up being used. Bearing this in mind, does your group think we should be building weapons to protect Earth from asteroid or comet impact? Can we afford not to build them? How can we safeguard against these collisions?

E. Is there evidence of climate change in your area over the past century? How would you distinguish a true climate change from the random variations in weather that take place from one year to the next?

⎙ EXERCISES

Review Questions

1. What is the thickest interior layer of Earth? The thinnest?

2. What are Earth's core and mantle made of? Explain how we know.

3. Describe the differences among primitive, igneous, sedimentary, and metamorphic rock, and relate these differences to their origins.

4. Explain briefly how the following phenomena happen on Earth, relating your answers to the theory of plate tectonics
   - **A.** earthquakes
   - **B.** continental drift
   - **C.** mountain building
   - **D.** volcanic eruptions
   - **E.** creation of the Hawaiian island chain

5. What is the source of Earth's magnetic field?

6. Why is the shape of the magnetosphere not spherical like the shape of Earth?

7. Although he did not present a mechanism, what were the key points of Alfred Wegener's proposal for the concept of continental drift?

Download for free at http://cnx.org/content/col11992/latest/

8. List the possible interactions between Earth's crustal plates that can occur at their boundaries.

9. List, in order of decreasing altitude, the principle layers of Earth's atmosphere.

10. In which atmospheric layer are almost all water-based clouds formed?

11. What is, by far, the most abundant component of Earth's atmosphere?

12. In which domain of living things do you find humankind?

13. Describe three ways in which the presence of life has affected the composition of Earth's atmosphere.

14. Briefly describe the greenhouse effect.

15. How do impacts by comets and asteroids influence Earth's geology, its atmosphere, and the evolution of life?

16. Why are there so many impact craters on our neighbor world, the Moon, and so few on Earth?

17. Detail some of the anthropogenic changes to Earth's climate and their potential impact on life.

Thought Questions

18. If you wanted to live where the chances of a destructive earthquake were small, would you pick a location near a fault zone, near a mid ocean ridge, near a subduction zone, or on a volcanic island such as Hawaii? What are the relative risks of earthquakes at each of these locations?

19. Which type of object would likely cause more damage if it struck near an urban area: a small metallic object or a large stony/icy one?

20. If all life were destroyed on Earth by a large impact, would new life eventually form to take its place? Explain how conditions would have to change for life to start again on our planet.

21. Why is a decrease in Earth's ozone harmful to life?

22. Why are we concerned about the increases in CO_2 and other gases that cause the greenhouse effect in Earth's atmosphere? What steps can we take in the future to reduce the levels of CO_2 in our atmosphere? What factors stand in the way of taking the steps you suggest? (You may include technological, economic, and political factors in your answer.)

23. Do you think scientists should make plans to defend Earth from future asteroid impacts? Is it right to intervene in the same evolutionary process that made the development of mammals (including us) possible after the big impact 65 million years ago?

Figuring For Yourself

24. Europe and North America are moving apart by about 5 m per century. As the continents separate, new ocean floor is created along the mid-Atlantic Rift. If the rift is 5000 km long, what is the total area of new ocean floor created in the Atlantic each century? (Remember that 1 km = 1000 m.)

25. Over the entire Earth, there are 60,000 km of active rift zones, with average separation rates of 5 m/century. How much area of new ocean crust is created each year over the entire planet? (This area is approximately equal to the amount of ocean crust that is subducted since the total area of the oceans remains about the same.)

This OpenStax book is available for free at http://cnx.org/content/col11992/1.8

26. With the information from Exercise 8.25, you can calculate the average age of the ocean floor. First, find the total area of the ocean floor (equal to about 60% of the surface area of Earth). Then compare this with the area created (or destroyed) each year. The average lifetime is the ratio of these numbers: the total area of ocean crust compared to the amount created (or destroyed) each year.

27. What is the volume of new oceanic basalt added to Earth's crust each year? Assume that the thickness of the new crust is 5 km, that there are 60,000 km of rifts, and that the average speed of plate motion is 4 cm/y. What fraction of Earth's entire volume does this annual addition of new material represent?

28. Suppose a major impact that produces a mass extinction takes place on Earth once every 5 million years. Suppose further that if such an event occurred today, you and most other humans would be killed (this would be true even if the human species as a whole survived). Such impact events are random, and one could take place at any time. Calculate the probability that such an impact will occur within the next 50 years (within your lifetime).

29. How do the risks of dying from the impact of an asteroid or comet compare with other risks we are concerned about, such as dying in a car accident or from heart disease or some other natural cause? (Hint: To find the annual risk, go to the library or internet and look up the annual number of deaths from a particular cause in a particular country, and then divide by the population of that country.)

30. What fraction of Earth's volume is taken up by the core?

31. Approximately what percentage of Earth's radius is represented by the crust?

32. What is the drift rate of the Pacific plate over the Hawaiian hot spot?

33. What is the percent increase of atmospheric CO_2 in the past 20 years?

34. Estimate the mass of the object that formed Meteor Crater in Arizona.

Download for free at http://cnx.org/content/col11992/latest/

This OpenStax book is available for free at http://cnx.org/content/col11992/1.8

Download for free at http://cnx.org/content/col11992/latest/

9

CRATERED WORLDS

Figure 9.1. Apollo 11 Astronaut Edwin "Buzz" Aldrin on the Surface of the Moon. Because there is no atmosphere, ocean, or geological activity on the Moon today, the footprints you see in the image will likely be preserved in the lunar soil for millions of years (credit: modification of work by NASA/ Neil A. Armstrong).

Chapter Outline

✎ Thinking Ahead

The Moon is the only other world human beings have ever visited. What is it like to stand on the surface of our natural satellite? And what can we learn from going there and bringing home pieces of a different world?

We begin our discussion of the planets as cratered worlds with two relatively simple objects: the Moon and Mercury. Unlike Earth, the Moon is geologically dead, a place that has exhausted its internal energy sources. Because its airless surface preserves events that happened long ago, the Moon provides a window on earlier epochs of solar system history. The planet Mercury is in many ways similar to the Moon, which is why the two are discussed together: both are relatively small, lacking in atmospheres, deficient in geological activity, and dominated by the effects of impact cratering. Still, the processes that have molded their surfaces are not unique to these two worlds. We shall see that they have acted on many other members of the planetary system as well.

9.1 GENERAL PROPERTIES OF THE MOON

Learning Objectives

By the end of this section, you will be able to:

Download for free at http://cnx.org/content/col11992/latest/

> Discuss what has been learned from both manned and robotic lunar exploration
> Describe the composition and structure of the Moon

The Moon has only one-eightieth the mass of Earth and about one-sixth Earth's surface gravity—too low to retain an atmosphere (Figure 9.2). Moving molecules of a gas can escape from a planet just the way a rocket does, and the lower the gravity, the easier it is for the gas to leak away into space. While the Moon can acquire a temporary atmosphere from impacting comets, this atmosphere is quickly lost by freezing onto the surface or by escape to surrounding space. The Moon today is dramatically deficient in a wide range of *volatiles*, those elements and compounds that evaporate at relatively low temperatures. Some of the Moon's properties are summarized in Table 9.1, along with comparative values for Mercury.

Figure 9.2. Two Sides of the Moon. The left image shows part of the hemisphere that faces Earth; several dark maria are visible. The right image shows part of the hemisphere that faces away from Earth; it is dominated by highlands. The resolution of this image is several kilometers, similar to that of high-powered binoculars or a small telescope. (credit: modification of work by NASA/GSFC/Arizona State University)

Properties of the Moon and Mercury

Property	Moon	Mercury
Mass (Earth = 1)	0.0123	0.055
Diameter (km)	3476	4878
Density (g/cm^3)	3.3	5.4
Surface gravity (Earth = 1)	0.17	0.38
Escape velocity (km/s)	2.4	4.3
Rotation period (days)	27.3	58.65

Table 9.1

This OpenStax book is available for free at http://cnx.org/content/col11992/1.8

Download for free at http://cnx.org/content/col11992/latest/

Properties of the Moon and Mercury

Property	Moon	Mercury
Surface area (Earth = 1)	0.27	0.38

Table 9.1

Exploration of the Moon

Most of what we know about the Moon today derives from the US Apollo program, which sent nine piloted spacecraft to our satellite between 1968 and 1972, landing 12 astronauts on its surface (Figure 9.1). Before the era of spacecraft studies, astronomers had mapped the side of the Moon that faces Earth with telescopic resolution of about 1 kilometer, but lunar geology hardly existed as a scientific subject. All that changed beginning in the early 1960s. Initially, Russia took the lead in lunar exploration with Luna 3, which returned the first photos of the lunar far side in 1959, and then with Luna 9, which landed on the surface in 1966 and transmitted pictures and other data to Earth. However, these efforts were overshadowed on July 20, 1969, when the first American astronaut set foot on the Moon.

Table 9.2 summarizes the nine Apollo flights: six that landed and three others that circled the Moon but did not land. The initial landings were on flat plains selected for safety reasons. But with increasing experience and confidence, NASA targeted the last three missions to more geologically interesting locales. The level of scientific exploration also increased with each mission, as the astronauts spent longer times on the Moon and carried more elaborate equipment. Finally, on the last Apollo landing, NASA included one scientist, geologist Jack Schmitt, among the astronauts (Figure 9.3).

Apollo Flights to the Moon

Flight	Date	Landing Site	Main Accomplishment
Apollo 8	Dec. 1968	—	First humans to fly around the Moon
Apollo 10	May 1969	—	First spacecraft rendezvous in lunar orbit
Apollo 11	July 1969	Mare Tranquillitatis	First human landing on the Moon; 22 kilograms of samples returned
Apollo 12	Nov. 1969	Oceanus Procellarum	First Apollo Lunar Surface Experiment Package (ALSEP); visit to Surveyor 3 lander
Apollo 13	Apr. 1970	—	Landing aborted due to explosion in command module
Apollo 14	Jan. 1971	Mare Nubium	First "rickshaw" on the Moon
Apollo 15	July 1971	Mare Imbrium/ Hadley	First "rover;" visit to Hadley Rille; astronauts traveled 24 kilometers

Table 9.2

Download for free at http://cnx.org/content/col11992/latest/

Apollo Flights to the Moon

Flight	Date	Landing Site	Main Accomplishment
Apollo 16	Apr. 1972	Descartes	First landing in highlands; 95 kilograms of samples returned
Apollo 17	Dec. 1972	Taurus-Littrow highlands	Geologist among the crew; 111 kilograms of samples returned

Table 9.2

Figure 9.3. Scientist on the Moon. Geologist (and later US senator) Harrison "Jack" Schmitt in front of a large boulder in the Littrow Valley at the edge of the lunar highlands. Note how black the sky is on the airless Moon. No stars are visible because the surface is brightly lit by the Sun, and the exposure therefore is not long enough to reveal stars.

In addition to landing on the lunar surface and studying it at close range, the Apollo missions accomplished three objectives of major importance for lunar science. First, the astronauts collected nearly 400 kilograms of samples for detailed laboratory analysis on Earth (Figure 9.4). These samples have revealed as much about the Moon and its history as all other lunar studies combined. Second, each Apollo landing after the first one deployed an Apollo Lunar Surface Experiment Package (ALSEP), which continued to operate for years after the astronauts departed. Third, the orbiting Apollo command modules carried a wide range of instruments to photograph and analyze the lunar surface from above.

This OpenStax book is available for free at http://cnx.org/content/col11992/1.8

Figure 9.4. Handling Moon Rocks. Lunar samples collected in the Apollo Project are analyzed and stored in NASA facilities at the Johnson Space Center in Houston, Texas. Here, a technician examines a rock sample using gloves in a sealed environment to avoid contaminating the sample. (credit: NASA JSC)

The last human left the Moon in December 1972, just a little more than three years after Neil Armstrong took his "giant leap for mankind." The program of lunar exploration was cut off midstride due to political and economic pressures. It had cost just about $100 per American, spread over 10 years—the equivalent of one large pizza per person per year. Yet for many people, the Moon landings were one of the central events in twentieth-century history.

The giant Apollo rockets built to travel to the Moon were left to rust on the lawns of NASA centers in Florida, Texas, and Alabama, although recently, some have at least been moved indoors to museums (Figure 9.5). Today, neither NASA nor Russia have plans to send astronauts to the Moon, and China appears to be the nation most likely to attempt this feat. (In a bizarre piece of irony, a few people even question whether we went to the Moon at all, proposing instead that the Apollo program was a fake, filmed on a Hollywood sound stage. See the Link to Learning box below for some scientists' replies to such claims.) However, scientific interest in the Moon is stronger than ever, and more than half a dozen scientific spacecraft—sent from NASA, ESA, Japan, India, and China—have orbited or landed on our nearest neighbor during the past decade.

LINK TO LEARNING

Read The Great Moon Hoax (https://openstax.org/l/30greatmoonhoax) about the claim that NASA never succeeded in putting people on the Moon.

Download for free at http://cnx.org/content/col11992/latest/

Figure 9.5. Moon Rocket on Display. One of the unused Saturn 5 rockets built to go to the Moon is now a tourist attraction at NASA's Johnson Space Center in Houston, although it has been moved indoors since this photo was taken. (credit: modification of work by David Morrison)

Lunar exploration has become an international enterprise with many robotic spacecraft focusing on lunar science. The USSR sent a number in the 1960s, including robot sample returns. Table 9.3 lists some of the most recent lunar missions.

Some International Missions to the Moon

Launch Year	Spacecraft	Type of Mission	Agency
1994	Clementine	Orbiter	US (USAF/NASA)
1998	Lunar Prospector	Orbiter	US (NASA)
2003	SMART-1	Orbiter	Europe (ESA)
2007	SELENE 1	Orbiter	Japan (JAXA)
2007	Chang'e 1	Orbiter	China (CNSA)
2008	Chandrayaan-1	Orbiter	India (ISRO)
2009	LRO	Orbiter	US (NASA)
2009	LCROSS	Impactor	US (NASA)

Table 9.3

This OpenStax book is available for free at http://cnx.org/content/col11992/1.8

Download for free at http://cnx.org/content/col11992/latest/

Some International Missions to the Moon

Launch Year	Spacecraft	Type of Mission	Agency
2010	Chang'e 2	Orbiter	China (CNSA)
2011	GRAIL	Twin orbiters	US (NASA)
2013	LADEE	Orbiter	US (NASA)
2013	Chang'e 3	Lander/Rover	China (CNSA)

Table 9.3

Composition and Structure of the Moon

The composition of the Moon is not the same as that of Earth. With an average density of only 3.3 g/cm^3, the Moon must be made almost entirely of silicate rock. Compared to Earth, it is depleted in iron and other metals. It is as if the Moon were composed of the same silicates as Earth's mantle and crust, with the metals and the volatiles selectively removed. These differences in composition between Earth and Moon provide important clues about the origin of the Moon, a topic we will cover in detail later in this chapter.

Studies of the Moon's interior carried out with seismometers taken to the Moon as part of the Apollo program confirm the absence of a large metal core. The twin GRAIL spacecraft launched into lunar orbit in 2011 provided even more precise tracking of the interior structure. We also know from the study of lunar samples that water and other volatiles have been depleted from the lunar crust. The tiny amounts of water detected in these samples were originally attributed to small leaks in the container seal that admitted water vapor from Earth's atmosphere. However, scientists have now concluded that some chemically bound water is present in the lunar rocks.

Most dramatically, water ice has been detected in permanently shadowed craters near the lunar poles. In 2009, NASA crashed a small spacecraft called the Lunar Crater Observation and Sensing Satellite (LCROSS) into the crater Cabeus near the Moon's south pole. The impact at 9,000 kilometers per hour released energy equivalent to 2 tons of dynamite, blasting a plume of water vapor and other chemicals high above the surface. This plume was visible to telescopes in orbit around the Moon, and the LCROSS spacecraft itself made measurements as it flew through the plume. A NASA spacecraft called the Lunar Reconnaissance Orbiter (LRO) also measured the very low temperatures inside several lunar craters, and its sensitive cameras were even able to image crater interiors by starlight.

The total quantity of water ice in the Moon's polar craters is estimated to be hundreds of billions of tons. As liquid, this would only be enough water to fill a lake 100 miles across, but compared with the rest of the dry lunar crust, so much water is remarkable. Presumably, this polar water was carried to the Moon by comets and asteroids that hit its surface. Some small fraction of the water froze in a few extremely cold regions (cold traps) where the Sun never shines, such as the bottom of deep craters at the Moon's poles. One reason this discovery could be important is that it raises the possibility of future human habitation near the lunar poles, or even of a lunar base as a way-station on routes to Mars and the rest of the solar system. If the ice could be mined, it would yield both water and oxygen for human support, and it could be broken down into hydrogen and oxygen, a potent rocket fuel.

Download for free at http://cnx.org/content/col11992/latest/

9.2 THE LUNAR SURFACE

Learning Objectives

By the end of this section, you will be able to:

> Differentiate between the major surface features of the Moon
> Describe the history of the lunar surface
> Describe the properties of the lunar "soil"

General Appearance

If you look at the Moon through a telescope, you can see that it is covered by impact craters of all sizes. The most conspicuous of the Moon's surface features—those that can be seen with the unaided eye and that make up the feature often called "the man in the Moon"—are vast splotches of darker lava flows.

Centuries ago, early lunar observers thought that the Moon had continents and oceans and that it was a possible abode of life. They called the dark areas "seas" (*maria* in Latin, or *mare* in the singular, pronounced "mah ray"). Their names, Mare Nubium (Sea of Clouds), Mare Tranquillitatis (Sea of Tranquility), and so on, are still in use today. In contrast, the "land" areas between the seas are not named. Thousands of individual craters have been named, however, mostly for great scientists and philosophers (Figure 9.6). Among the most prominent craters are those named for Plato, Copernicus, Tycho, and Kepler. Galileo only has a small crater, however, reflecting his low standing among the Vatican scientists who made some of the first lunar maps.

We know today that the resemblance of lunar features to terrestrial ones is superficial. Even when they look somewhat similar, the origins of lunar features such as craters and mountains are very different from their terrestrial counterparts. The Moon's relative lack of internal activity, together with the absence of air and water, make most of its geological history unlike anything we know on Earth.

Figure 9.6. Sunrise on the Central Mountain Peaks of Tycho Crater, as Imaged by the NASA Lunar Reconnaissance Orbiter. Tycho, about 82 kilometers in diameter, is one of the youngest of the very large lunar craters. The central mountain rises 12 kilometers above the crater floor. (credit: modification of work by NASA/Goddard/Arizona State University)

Lunar History

To trace the detailed history of the Moon or of any planet, we must be able to estimate the ages of individual rocks. Once lunar samples were brought back by the Apollo astronauts, the radioactive dating techniques that had been developed for Earth were applied to them. The solidification ages of the samples ranged from about

This OpenStax book is available for free at http://cnx.org/content/col11992/1.8

3.3 to 4.4 billion years old, substantially older than most of the rocks on Earth. For comparison, as we saw in the chapter on Earth, Moon, and Sky, both Earth and the Moon were formed between 4.5 and 4.6 billion years ago.

Most of the crust of the Moon (83%) consists of silicate rocks called *anorthosites*; these regions are known as the lunar **highlands**. They are made of relatively low-density rock that solidified on the cooling Moon like slag floating on the top of a smelter. Because they formed so early in lunar history (between 4.1 and 4.4 billion years ago), the highlands are also extremely heavily cratered, bearing the scars of all those billions of years of impacts by interplanetary debris (Figure 9.7).

Figure 9.7. Lunar Highlands. The old, heavily cratered lunar highlands make up 83% of the Moon's surface. (credit: Apollo 11 Crew, NASA)

Unlike the mountains on Earth, the Moon's highlands do not have any sharp folds in their ranges. The highlands have low, rounded profiles that resemble the oldest, most eroded mountains on Earth (Figure 9.8). Because there is no atmosphere or water on the Moon, there has been no wind, water, or ice to carve them into cliffs and sharp peaks, the way we have seen them shaped on Earth. Their smooth features are attributed to gradual erosion, mostly due to impact cratering from meteorites.

Download for free at http://cnx.org/content/col11992/latest/

Figure 9.8. Lunar Mountain. This photo of Mt. Hadley on the edge of Mare Imbrium was taken by Dave Scott, one of the Apollo 15 astronauts. Note the smooth contours of the lunar mountains, which have not been sculpted by water or ice. (credit: NASA/Apollo Lunar Surface Journal)

The maria are much less cratered than the highlands, and cover just 17% of the lunar surface, mostly on the side of the Moon that faces Earth (Figure 9.9).

Figure 9.9. Lunar Maria. About 17% of the Moon's surface consists of the maria—flat plains of basaltic lava. This view of Mare Imbrium also shows numerous secondary craters and evidence of material ejected from the large crater Copernicus on the upper horizon. Copernicus is an impact crater almost 100 kilometers in diameter that was formed long after the lava in Imbrium had already been deposited. (credit: NASA, Apollo 17)

Today, we know that the maria consist mostly of dark-colored basalt (volcanic lava) laid down in volcanic eruptions billions of years ago. Eventually, these lava flows partly filled the huge depressions called *impact basins*, which had been produced by collisions of large chunks of material with the Moon relatively early in its history. The basalt on the Moon (Figure 9.10) is very similar in composition to the crust under the oceans of

This OpenStax book is available for free at http://cnx.org/content/col11992/1.8

Earth or to the lavas erupted by many terrestrial volcanoes. The youngest of the lunar impact basins is Mare Orientale, shown in Figure 9.11.

Figure 9.10. Rock from a Lunar Mare. In this sample of basalt from the mare surface, you can see the holes left by gas bubbles, which are characteristic of rock formed from lava. All lunar rocks are chemically distinct from terrestrial rocks, a fact that has allowed scientists to identify a few lunar samples among the thousands of meteorites that reach Earth. (credit: modification of work by NASA)

Figure 9.11. Mare Orientale. The youngest of the large lunar impact basins is Orientale, formed 3.8 billion years ago. Its outer ring is about 1000 kilometers in diameter, roughly the distance between New York City and Detroit, Michigan. Unlike most of the other basins, Orientale has not been completely filled in with lava flows, so it retains its striking "bull's-eye" appearance. It is located on the edge of the Moon as seen from Earth. (credit: NASA)

Volcanic activity may have begun very early in the Moon's history, although most evidence of the first half billion years is lost. What we do know is that the major mare volcanism, which involved the release of lava from hundreds of kilometers below the surface, ended about 3.3 billion years ago. After that, the Moon's interior cooled, and volcanic activity was limited to a very few small areas. The primary forces altering the surface come from the outside, not the interior.

Download for free at http://cnx.org/content/col11992/latest/

On the Lunar Surface

"The surface is fine and powdery. I can pick it up loosely with my toe. But I can see the footprints of my boots and the treads in the fine sandy particles." —Neil Armstrong, Apollo 11 astronaut, immediately after stepping onto the Moon for the first time.

The surface of the Moon is buried under a fine-grained soil of tiny, shattered rock fragments. The dark basaltic dust of the lunar maria was kicked up by every astronaut footstep, and thus eventually worked its way into all of the astronauts' equipment. The upper layers of the surface are porous, consisting of loosely packed dust into which their boots sank several centimeters (Figure 9.12). This lunar dust, like so much else on the Moon, is the product of impacts. Each cratering event, large or small, breaks up the rock of the lunar surface and scatters the fragments. Ultimately, billions of years of impacts have reduced much of the surface layer to particles about the size of dust or sand.

Figure 9.12. Footprint on Moon Dust. Apollo photo of an astronaut's boot print in the lunar soil. (credit: NASA)

In the absence of any air, the lunar surface experiences much greater temperature extremes than the surface of Earth, even though Earth is virtually the same distance from the Sun. Near local noon, when the Sun is highest in the sky, the temperature of the dark lunar soil rises above the boiling point of water. During the long lunar night (which, like the lunar day, lasts two Earth weeks[1]), the temperature drops to about 100 K (–173 °C). The extreme cooling is a result not only of the absence of air but also of the porous nature of the Moon's dusty soil, which cools more rapidly than solid rock would.

LINK TO LEARNING

Learn how the moon's craters and maria were formed by watching a video produced by NASA's Lunar Reconnaissance Orbiter (LRO) team (https://openstax.org/l/30mooncratersfo) about the evolution of the Moon, tracing it from its origin about 4.5 billion years ago to the Moon we see today. See a simulation of how the Moon's craters and maria were formed through periods of impact, volcanic activity, and heavy bombardment.

---

1   You can see the cycle of day and night on the side of the Moon facing us in the form of the Moon's phases. It takes about 14 days for the side of the Moon facing us to go from full moon (all lit up) to new moon (all dark). There is more on this in **Chapter 4: Earth, Moon, and Sky**.

This OpenStax book is available for free at http://cnx.org/content/col11992/1.8

Download for free at http://cnx.org/content/col11992/latest/

9.3 IMPACT CRATERS

Learning Objectives

By the end of this section, you will be able to:

> Compare and contrast ideas about how lunar craters form
> Explain the process of impact crater formation
> Discuss the use of crater counts to determine relative ages of lunar landforms

The Moon provides an important benchmark for understanding the history of our planetary system. Most solid worlds show the effects of impacts, often extending back to the era when a great deal of debris from our system's formation process was still present. On Earth, this long history has been erased by our active geology. On the Moon, in contrast, most of the impact history is preserved. If we can understand what has happened on the Moon, we may be able to apply this knowledge to other worlds. The Moon is especially interesting because it is not just any moon, but *our* Moon—a nearby world that has shared the history of Earth for more than 4 billion years and preserved a record that, for Earth, has been destroyed by our active geology.

Volcanic Versus Impact Origin of Craters

Until the middle of the twentieth century, scientists did not generally recognize that lunar craters were the result of impacts. Since impact craters are extremely rare on Earth, geologists did not expect them to be the major feature of lunar geology. They reasoned (perhaps unconsciously) that since the craters we have on Earth are volcanic, the lunar craters must have a similar origin.

One of the first geologists to propose that lunar craters were the result of impacts was Grove K. Gilbert, a scientist with the US Geological Survey in the 1890s. He pointed out that the large lunar craters—mountain-rimmed, circular features with floors generally below the level of the surrounding plains—are larger and have different shapes from known volcanic craters on Earth. Terrestrial volcanic craters are smaller and deeper and almost always occur at the tops of volcanic mountains (Figure 9.13). The only alternative to explain the Moon's craters was an impact origin. His careful reasoning, although not accepted at the time, laid the foundations for the modern science of lunar geology.

Terrestrial volcano   Lunar impact crater

Figure 9.13. Volcanic and Impact Craters. Profiles of a typical terrestrial volcanic crater and a typical lunar impact crater are quite different.

Gilbert concluded that the lunar craters were produced by impacts, but he didn't understand why all of them were circular and not oval. The reason lies in the escape velocity, the minimum speed that a body must reach to permanently break away from the gravity of another body; it is also the minimum speed that a projectile approaching Earth or the Moon will hit with. Attracted by the gravity of the larger body, the incoming chunk strikes with at least escape velocity, which is 11 kilometers per second for Earth and 2.4 kilometers per second (5400 miles per hour) for the Moon. To this escape velocity is added whatever speed the projectile already had with respect to Earth or Moon, typically 10 kilometers per second or more.

At these speeds, the energy of impact produces a violent *explosion* that excavates a large volume of material in a symmetrical way. Photographs of bomb and shell craters on Earth confirm that explosion craters are always

Download for free at http://cnx.org/content/col11992/latest/

essentially circular. Only following World War I did scientists recognize the similarity between impact craters and explosion craters, but, sadly, Gilbert did not live to see his impact hypothesis widely accepted.

The Cratering Process

Let's consider how an impact at these high speeds produces a crater. When such a fast projectile strikes a planet, it penetrates two or three times its own diameter before stopping. During these few seconds, its energy of motion is transferred into a shock wave (which spreads through the target body) and into heat (which vaporizes most of the projectile and some of the surrounding target). The shock wave fractures the rock of the target, while the expanding silicate vapor generates an explosion similar to that of a nuclear bomb detonated at ground level (Figure 9.14). The size of the excavated crater depends primarily on the speed of impact, but generally it is 10 to 15 times the diameter of the projectile.

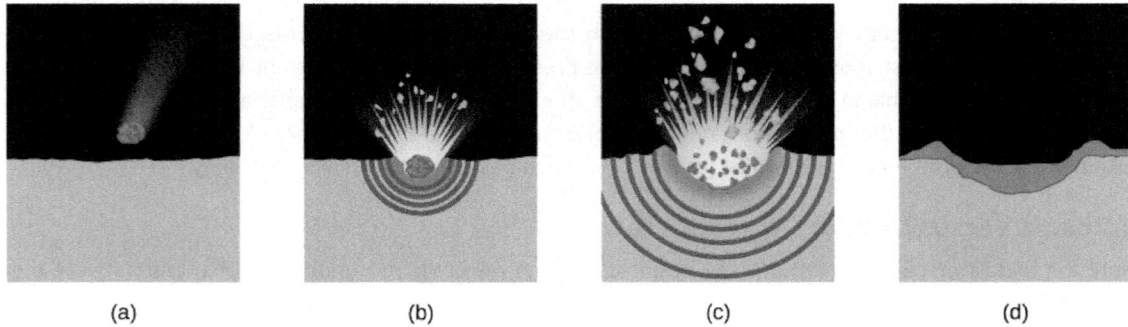

    (a)   (b)   (c)   (d)

Figure 9.14. Stages in the Formation of an Impact Crater. (a) The impact occurs. (b) The projectile vaporizes and a shock wave spreads through the lunar rock. (c) Ejecta are thrown out of the crater. (d) Most of the ejected material falls back to fill the crater, forming an ejecta blanket.

An impact explosion of the sort described above leads to a characteristic kind of crater, as shown in Figure 9.15. The central cavity is initially bowl-shaped (the word "crater" comes from the Greek word for "bowl"), but the rebound of the crust partially fills it in, producing a flat floor and sometimes creating a central peak. Around the rim, landslides create a series of terraces.

Figure 9.15. Typical Impact Crater. King Crater on the far side of the Moon, a fairly recent lunar crater 75 kilometers in diameter, shows most of the features associated with large impact structures. (credit: NASA/JSC/Arizona State University)

The rim of the crater is turned up by the force of the explosion, so it rises above both the floor and the adjacent terrain. Surrounding the rim is an *ejecta blanket* consisting of material thrown out by the explosion. This debris falls back to create a rough, hilly region, typically about as wide as the crater diameter. Additional, higher-

This OpenStax book is available for free at http://cnx.org/content/col11992/1.8

speed ejecta fall at greater distances from the crater, often digging small *secondary craters* where they strike the surface (Figure 9.9).

Some of these streams of ejecta can extend for hundreds or even thousands of kilometers from the crater, creating the bright *crater rays* that are prominent in lunar photos taken near full phase. The brightest lunar crater rays are associated with large young craters such as Kepler and Tycho.

SEEING FOR YOURSELF

Observing the Moon

The Moon is one of the most beautiful sights in the sky, and it is the only object close enough to reveal its *topography* (surface features such as mountains and valleys) without a visit from a spacecraft. A fairly small amateur telescope easily shows craters and mountains on the Moon as small as a few kilometers across.

Even as seen through a good pair of binoculars, we can observe that the appearance of the Moon's surface changes dramatically with its phase. At full phase, it shows almost no topographic detail, and you must look closely to see more than a few craters. This is because sunlight illuminates the surface straight on, and in this flat lighting, no shadows are cast. Much more revealing is the view near first or third quarter, when sunlight streams in from the side, causing topographic features to cast sharp shadows. It is almost always more rewarding to study a planetary surface under such oblique lighting, when the maximum information about surface relief can be obtained.

The flat lighting at full phase does, however, accentuate brightness contrasts on the Moon, such as those between the maria and highlands. Notice in Figure 9.16 that several of the large mare craters seem to be surrounded by white material and that the light streaks or rays that can stretch for hundreds of kilometers across the surface are clearly visible. These lighter features are ejecta, splashed out from the crater-forming impact.

Download for free at http://cnx.org/content/col11992/latest/

(a)   (b)

Figure 9.16. Appearance of the Moon at Different Phases. (a) Illumination from the side brings craters and other topographic features into sharp relief, as seen on the far left side. (b) At full phase, there are no shadows, and it is more difficult to see such features. However, the flat lighting at full phase brings out some surface features, such as the bright rays of ejecta that stretch out from a few large young craters. (credit: modification of work by Luc Viatour)

By the way, there is no danger in looking at the Moon with binoculars or telescopes. The reflected sunlight is never bright enough to harm your eyes. In fact, the sunlit surface of the Moon has about the same brightness as a sunlit landscape of dark rock on Earth. Although the Moon looks bright in the night sky, its surface is, on average, much less reflective than Earth's, with its atmosphere and white clouds. This difference is nicely illustrated by the photo of the Moon passing in front of Earth taken from the Deep Space Climate Observatory spacecraft (Figure 9.17). Since the spacecraft took the image from a position inside the orbit of Earth, we see both objects fully illuminated (full Moon and full Earth). By the way, you cannot see much detail on the Moon because the exposure has been set to give a bright image of Earth, not the Moon.

Figure 9.17. The Moon Crossing the Face of Earth. In this 2015 image from the Deep Space Climate Observatory spacecraft, both objects are fully illuminated, but the Moon looks darker because it has a much lower average reflectivity than Earth. (credit: modification of work by NASA, DSCOVR EPIC team)

This OpenStax book is available for free at http://cnx.org/content/col11992/1.8

Download for free at http://cnx.org/content/col11992/latest/

One interesting thing about the Moon that you can see without binoculars or telescopes is popularly called "the new Moon in the old Moon's arms." Look at the Moon when it is a thin crescent, and you can often make out the faint circle of the entire lunar disk, even though the sunlight shines on only the crescent. The rest of the disk is illuminated not by sunlight but by earthlight—sunlight reflected from Earth. The light of the full Earth on the Moon is about 50 times brighter than that of the full Moon shining on Earth.

Using Crater Counts

If a world has had little erosion or internal activity, like the Moon during the past 3 billion years, it is possible to use the number of impact craters on its surface to estimate the age of that surface. By "age" here we mean the time since a major disturbance occurred on that surface (such as the volcanic eruptions that produced the lunar maria).

We cannot directly measure the rate at which craters are being formed on Earth and the Moon, since the average interval between large crater-forming impacts is longer than the entire span of human history. Our best-known example of such a large crater, Meteor Crater in Arizona (Figure 9.18), is about 50,000 years old. However, the cratering rate can be estimated from the number of craters on the lunar maria or calculated from the number of potential "projectiles" (asteroids and comets) present in the solar system today. Both lines of reasoning lead to about the same estimations.

Figure 9.18. Meteor Crater. This aerial photo of Meteor Crater in Arizona shows the simple form of a meteorite impact crater. The crater's rim diameter is about 1.2 kilometers. (credit: Shane Torgerson)

For the Moon, these calculations indicate that a crater 1 kilometer in diameter should be produced about every 200,000 years, a 10-kilometer crater every few million years, and one or two 100-kilometer craters every billion years. If the cratering rate has stayed the same, we can figure out how long it must have taken to make all the craters we see in the lunar maria. Our calculations show that it would have taken several billion years. This result is similar to the age determined for the maria from radioactive dating of returned samples—3.3 to 3.8 billion years old.

The fact that these two calculations agree suggests that astronomers' original assumption was right: comets and asteroids in approximately their current numbers have been impacting planetary surfaces for billions of years. Calculations carried out for other planets (and their moons) indicate that they also have been subject to about the same number of interplanetary impacts during this time.

Download for free at http://cnx.org/content/col11992/latest/

We have good reason to believe, however, that earlier than 3.8 billion years ago, the impact rates must have been a great deal higher. This becomes immediately evident when comparing the numbers of craters on the lunar highlands with those on the maria. Typically, there are 10 times more craters on the highlands than on a similar area of maria. Yet the radioactive dating of highland samples showed that they are only a little older than the maria, typically 4.2 billion years rather than 3.8 billion years. If the rate of impacts had been constant throughout the Moon's history, the highlands would have had to be at least 10 times older. They would thus have had to form 38 billion years ago—long before the universe itself began.

In science, when an assumption leads to an implausible conclusion, we must go back and re-examine that assumption—in this case, the constant impact rate. The contradiction is resolved if the impact rate varied over time, with a much heavier bombardment earlier than 3.8 billion years ago (Figure 9.19). This "heavy bombardment" produced most of the craters we see today in the highlands.

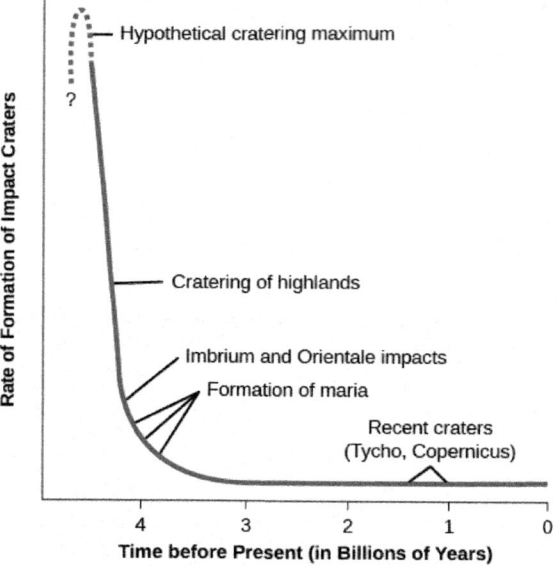

Figure 9.19. Cratering Rates over Time. The number of craters being made on the Moon's surface has varied with time over the past 4.3 billion years.

This idea we have been exploring—that large impacts (especially during the early history of the solar system) played a major role in shaping the worlds we see—is not unique to our study of the Moon. As you read through the other chapters about the planets, you will see further indications that a number of the present-day characteristics of our system may be due to its violent past.

9.4 THE ORIGIN OF THE MOON

Learning Objectives

By the end of this section, you will be able to:

> Describe the top three early hypotheses of the formation of the Moon
> Summarize the current "giant impact" concept of how the Moon formed

It is characteristic of modern science to ask how things originated. Understanding the origin of the Moon has proven to be challenging for planetary scientists, however. Part of the difficulty is simply that we know so much

This OpenStax book is available for free at http://cnx.org/content/col11992/1.8

about the Moon (quite the opposite of our usual problem in astronomy). As we will see, one key problem is that the Moon is both tantalizingly similar to Earth and frustratingly different.

Ideas for the Origin of the Moon

Most of the earlier hypotheses for the Moon's origin followed one of three general ideas:

1. The fission theory—the Moon was once part of Earth, but somehow separated from it early in their history.

2. The sister theory—the Moon formed together with (but independent of) Earth, as we believe many moons of the outer planets formed.

3. The capture theory—the Moon formed elsewhere in the solar system and was captured by Earth.

Unfortunately, there seem to be fundamental problems with each of these ideas. Perhaps the easiest hypothesis to reject is the capture theory. Its primary drawback is that no one knows of any way that early Earth could have captured such a large moon from elsewhere. One body approaching another cannot go into orbit around it without a substantial loss of energy; this is the reason that spacecraft destined to orbit other planets are equipped with retro-rockets. Furthermore, if such a capture did take place, the captured object would go into a very eccentric orbit rather than the nearly circular orbit our Moon occupies today. Finally, there are too many compositional similarities between Earth and the Moon, particularly an identical fraction of the major isotopes[2] of oxygen, to justify seeking a completely independent origin.

The fission hypothesis, which states that the Moon separated from Earth, was suggested in the late nineteenth century. Modern calculations have shown that this sort of spontaneous fission or splitting is impossible. Furthermore, it is difficult to understand how a Moon made out of terrestrial material in this way could have developed the many distinctive chemical differences now known to characterize our neighbor.

Scientists were therefore left with the sister hypothesis—that the Moon formed alongside Earth—or with some modification of the fission hypothesis that can find a more acceptable way for the lunar material to have separated from Earth. But the more we learned about our Moon, the less these old ideas seem to fit the bill.

The Giant Impact Hypothesis

In an effort to resolve these apparent contradictions, scientists developed a fourth hypothesis for the origin of the Moon, one that involves a giant impact early in Earth's history. There is increasing evidence that large chunks of material—objects of essentially planetary mass—were orbiting in the inner solar system at the time that the terrestrial planets formed. The giant impact hypothesis envisions Earth being struck obliquely by an object approximately one-tenth Earth's mass—a "bullet" about the size of Mars. This is very nearly the largest impact Earth could experience without being shattered.

Such an impact would disrupt much of Earth and eject a vast amount of material into space, releasing almost enough energy to break the planet apart. Computer simulations indicate that material totaling several percent of Earth's mass could be ejected in such an impact. Most of this material would be from the stony mantles of Earth and the impacting body, not from their metal cores. This ejected rock vapor then cooled and formed a ring of material orbiting Earth. It was this ring that ultimately condensed into the Moon.

While we do not have any current way of showing that the giant impact hypothesis is the correct model of the Moon's origin, it does offer potential solutions to most of the major problems raised by the chemistry of the Moon. First, since the Moon's raw material is derived from the mantles of Earth and the projectile, the absence of metals is easily understood. Second, most of the volatile elements would have been lost during the high-

---

2   Remember from the **Radiation and Spectra** chapter that the term isotope means a different "version" of an element. Specifically, different isotopes of the same element have equal numbers of protons but different numbers of neutrons (as in carbon-12 versus carbon-14.)

Download for free at http://cnx.org/content/col11992/latest/

temperature phase following the impact, explaining the lack of these materials on the Moon. Yet, by making the Moon primarily of terrestrial mantle material, it is also possible to understand similarities such as identical abundances of various oxygen isotopes.

9.5 MERCURY

Learning Objectives

By the end of this section, you will be able to:

> Characterize the orbit of Mercury around the Sun
> Describe Mercury's structure and composition
> Explain the relationship between Mercury's orbit and rotation
> Describe the topography and features of Mercury's surface
> Summarize our ideas about the origin and evolution of Mercury

The planet Mercury is similar to the Moon in many ways. Like the Moon, it has no atmosphere, and its surface is heavily cratered. As described later in this chapter, it also shares with the Moon the likelihood of a violent birth.

Mercury's Orbit

Mercury is the nearest planet to the Sun, and, in accordance with Kepler's third law, it has the shortest period of revolution about the Sun (88 of our days) and the highest average orbital speed (48 kilometers per second). It is appropriately named for the fleet-footed messenger god of the Romans. Because Mercury remains close to the Sun, it can be difficult to pick out in the sky. As you might expect, it's best seen when its eccentric orbit takes it as far from the Sun as possible.

The semimajor axis of Mercury's orbit—that is, the planet's average distance from the Sun—is 58 million kilometers, or 0.39 AU. However, because its orbit has the high eccentricity of 0.206, Mercury's actual distance from the Sun varies from 46 million kilometers at perihelion to 70 million kilometers at aphelion (the ideas and terms that describe orbits were introduced in Orbits and Gravity).

Composition and Structure

Mercury's mass is one-eighth that of Earth, making it the smallest terrestrial planet. Mercury is the smallest planet (except for the dwarf planets), having a diameter of 4878 kilometers, less than half that of Earth. Mercury's density is 5.4 g/cm³, much greater than the density of the Moon, indicating that the composition of those two objects differs substantially.

Mercury's composition is one of the most interesting things about it and makes it unique among the planets. Mercury's high density tells us that it must be composed largely of heavier materials such as metals. The most likely models for Mercury's interior suggest a metallic iron-nickel core amounting to 60% of the total mass, with the rest of the planet made up primarily of silicates. The core has a diameter of 3500 kilometers and extends out to within 700 kilometers of the surface. We could think of Mercury as a metal ball the size of the Moon surrounded by a rocky crust 700 kilometers thick (Figure 9.20). Unlike the Moon, Mercury does have a weak magnetic field. The existence of this field is consistent with the presence of a large metal core, and it suggests that at least part of the core must be liquid in order to generate the observed magnetic field.[3]

---

3   Recall from the **Radiation and Spectra** chapter that magnetism is an effect of moving electric charges. In atoms of metals, the outer electrons are easier to dislodge and they can form a current when the metal is in liquid form and can flow.

This OpenStax book is available for free at http://cnx.org/content/col11992/1.8

Figure 9.20. Mercury's Internal Structure. The interior of Mercury is dominated by a metallic core about the same size as our Moon.

EXAMPLE 9.1

Densities of Worlds

The average density of a body equals its mass divided by its volume. For a sphere, density is:

$$\text{density} = \frac{\text{mass}}{\frac{4}{3}\pi R^3}$$

Astronomers can measure both mass and radius accurately when a spacecraft flies by a body.

Using the information in this chapter, we can calculate the approximate average density of the Moon.

Solution

For a sphere,

$$\text{density} = \frac{\text{mass}}{\frac{4}{3}\pi R^3} = \frac{7.35 \times 10^{22} \text{ kg}}{4.2 \times 5.2 \times 10^{18} \text{ m}^3} = 3.4 \times 10^3 \text{ kg/m}^3$$

Table 9.1 gives a value of 3.3 g/cm^3, which is 3.3 × 10^3 kg/m^3.

Check Your Learning

Using the information in this chapter, calculate the average density of Mercury. Show your work. Does your calculation agree with the figure we give in this chapter?

Answer:

$$\text{density} = \frac{\text{mass}}{\frac{4}{3}\pi R^3} = \frac{3.3 \times 10^{23} \text{ kg}}{4.2 \times 1.45 \times 10^{19} \text{ m}^3} = 5.4 \times 10^3 \text{ kg/m}^3$$

That matches the value given in Table 9.1 when g/cm^3 is converted into kg/m^3.

Download for free at http://cnx.org/content/col11992/latest/

Mercury's Strange Rotation

Visual studies of Mercury's indistinct surface markings were once thought to indicate that the planet kept one face to the Sun (as the Moon does to Earth). Thus, for many years, it was widely believed that Mercury's rotation period was equal to its revolution period of 88 days, making one side perpetually hot while the other was always cold.

Radar observations of Mercury in the mid-1960s, however, showed conclusively that Mercury does not keep one side fixed toward the Sun. If a planet is turning, one side seems to be approaching Earth while the other is moving away from it. The resulting Doppler shift spreads or broadens the precise transmitted radar-wave frequency into a range of frequencies in the reflected signal (Figure 9.21). The degree of broadening provides an exact measurement of the rotation rate of the planet.

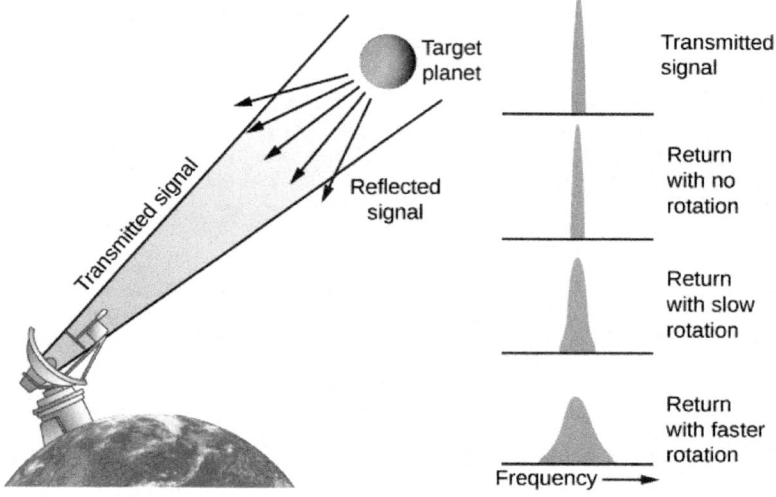

Figure 9.21. Doppler Radar Measures Rotation. When a radar beam is reflected from a rotating planet, the motion of one side of the planet's disk toward us and the other side away from us causes Doppler shifts in the reflected signal. The effect is to cause both a redshift and a blueshift, widening the spread of frequencies in the radio beam.

Mercury's period of rotation (how long it takes to turn with respect to the distant stars) is 59 days, which is just two-thirds of the planet's period of revolution. Subsequently, astronomers found that a situation where the spin and the orbit of a planet (its year) are in a 2:3 ratio turns out to be stable. (See What a Difference a Day Makes for more on the effects of having such a long day on Mercury.)

Mercury, being close to the Sun, is very hot on its daylight side; but because it has no appreciable atmosphere, it gets surprisingly cold during the long nights. The temperature on the surface climbs to 700 K (430 °C) at noontime. After sunset, however, the temperature drops, reaching 100 K (–170 °C) just before dawn. (It is even colder in craters near the poles that receive no sunlight at all.) The range in temperature on Mercury is thus 600 K (or 600 °C), a greater difference than on any other planet.

This OpenStax book is available for free at http://cnx.org/content/col11992/1.8

MAKING CONNECTIONS

What a Difference a Day Makes

Mercury rotates three times for each two orbits around the Sun. It is the only planet that exhibits this relationship between its spin and its orbit, and there are some interesting consequences for any observers who might someday be stationed on the surface of Mercury.

Here on Earth, we take for granted that days are much shorter than years. Therefore, the two astronomical ways of defining the local "day"—how long the planet takes to rotate and how long the Sun takes to return to the same position in the sky—are the same on Earth for most practical purposes. But this is not the case on Mercury. While Mercury rotates (spins once) in 59 Earth days, the time for the Sun to return to the same place in Mercury's sky turns out to be two Mercury years, or 176 Earth days. (Note that this result is not intuitively obvious, so don't be upset if you didn't come up with it.) Thus, if one day at noon a Mercury explorer suggests to her companion that they should meet at noon the next day, this could mean a very long time apart!

To make things even more interesting, recall that Mercury has an eccentric orbit, meaning that its distance from the Sun varies significantly during each mercurian year. By Kepler's law, the planet moves fastest in its orbit when closest to the Sun. Let's examine how this affects the way we would see the Sun in the sky during one 176-Earth-day cycle. We'll look at the situation as if we were standing on the surface of Mercury in the center of a giant basin that astronomers call Caloris (Figure 9.23).

At the location of Caloris, Mercury is most distant from the Sun at sunrise; this means the rising Sun looks smaller in the sky (although still more than twice the size it appears from Earth). As the Sun rises higher and higher, it looks bigger and bigger; Mercury is now getting closer to the Sun in its eccentric orbit. At the same time, the apparent motion of the Sun slows down as Mercury's faster motion in orbit begins to catch up with its rotation.

At noon, the Sun is now three times larger than it looks from Earth and hangs almost motionless in the sky. As the afternoon wears on, the Sun appears smaller and smaller, and moves faster and faster in the sky. At sunset, a full Mercury year (or 88 Earth days after sunrise), the Sun is back to its smallest apparent size as it dips out of sight. Then it takes another Mercury year before the Sun rises again. (By the way, sunrises and sunsets are much more sudden on Mercury, since there is no atmosphere to bend or scatter the rays of sunlight.)

Astronomers call locations like the Caloris Basin the "hot longitudes" on Mercury because the Sun is closest to the planet at noon, just when it is lingering overhead for many Earth days. This makes these areas the hottest places on Mercury.

We bring all this up not because the exact details of this scenario are so important but to illustrate how many of the things we take for granted on Earth are not the same on other worlds. As we've mentioned before, one of the best things about taking an astronomy class should be ridding you forever of any "Earth chauvinism" you might have. The way things are on our planet is just one of the many ways nature can arrange reality.

Download for free at http://cnx.org/content/col11992/latest/

The Surface of Mercury

The first close-up look at Mercury came in 1974, when the US spacecraft Mariner 10 passed 9500 kilometers from the surface of the planet and transmitted more than 2000 photographs to Earth, revealing details with a resolution down to 150 meters. Subsequently, the planet was mapped in great detail by the MESSENGER spacecraft, which was launched in 2004 and made multiple flybys of Earth, Venus, and Mercury before settling into orbit around Mercury in 2011. It ended its life in 2015, when it was commanded to crash into the surface of the planet.

Mercury's surface strongly resembles the Moon in appearance (Figure 9.22 and Figure 9.23). It is covered with thousands of craters and larger basins up to 1300 kilometers in diameter. Some of the brighter craters are rayed, like Tycho and Copernicus on the Moon, and many have central peaks. There are also *scarps* (cliffs) more than a kilometer high and hundreds of kilometers long, as well as ridges and plains.

MESSENGER instruments measured the surface composition and mapped past volcanic activity. One of its most important discoveries was the verification of water ice (first detected by radar) in craters near the poles, similar to the situation on the Moon, and the unexpected discovery of organic (carbon-rich) compounds mixed with the water ice.

LINK TO LEARNING

Scientists working with data from the MESSENGER mission (https://openstax.org/l/30MESSmercuryrt) put together a rotating globe of Mercury, in false color, showing some of the variations in the composition of the planet's surface. You can watch it spin.

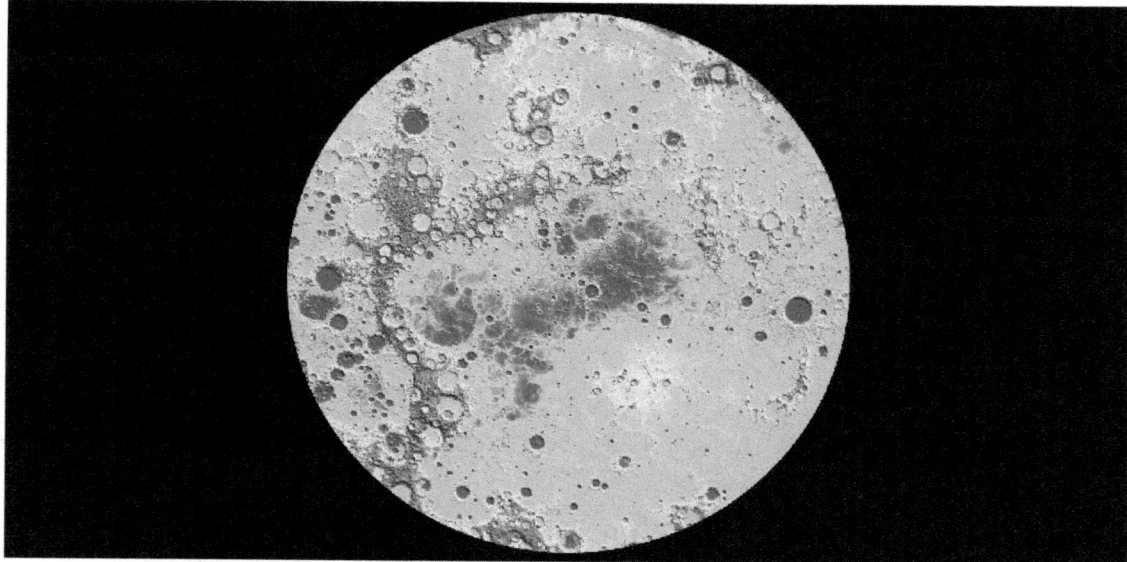

Figure 9.22. Mercury's Topography. The topography of Mercury's northern hemisphere is mapped in great detail from MESSENGER data. The lowest regions are shown in purple and blue, and the highest regions are shown in red. The difference in elevation between the lowest and highest regions shown here is roughly 10 kilometers. The permanently shadowed low-lying craters near the north pole contain radar-bright water ice. (credit: modification of work by NASA/Johns Hopkins University Applied Physics Laboratory/Carnegie Institution of Washington)

This OpenStax book is available for free at http://cnx.org/content/col11992/1.8

Download for free at http://cnx.org/content/col11992/latest/

Figure 9.23. Caloris Basin. This partially flooded impact basin is the largest known structural feature on Mercury. The smooth plains in the interior of the basin have an area of almost two million square kilometers. Compare this photo with Figure 9.11, the Orientale Basin on the Moon. (credit: NASA/Johns Hopkins University Applied Physics Laboratory/Carnegie Institution of Washington)

Most of the mercurian features have been named in honor of artists, writers, composers, and other contributors to the arts and humanities, in contrast with the scientists commemorated on the Moon. Among the named craters are Bach, Shakespeare, Tolstoy, Van Gogh, and Scott Joplin.

There is no evidence of plate tectonics on Mercury. However, the planet's distinctive long scarps can sometimes be seen cutting across craters; this means the scarps must have formed later than the craters (Figure 9.24). These long, curved cliffs appear to have their origin in the slight compression of Mercury's crust. Apparently, at some point in its history, the planet shrank, wrinkling the crust, and it must have done so after most of the craters on its surface had already formed.

If the standard cratering chronology applies to Mercury, this shrinkage must have taken place during the last 4 billion years and not during the solar system's early period of heavy bombardment.

Figure 9.24. Discovery Scarp on Mercury. This long cliff, nearly 1 kilometer high and more than 100 kilometers long, cuts across several craters. Astronomers conclude that the compression that made "wrinkles" like this in the plank's surface must have taken place after the craters were formed. (credit: modification of work by NASA/JPL/Northwestern University)

Download for free at http://cnx.org/content/col11992/latest/

The Origin of Mercury

The problem with understanding how Mercury formed is the reverse of the problem posed by the composition of the Moon. We have seen that, unlike the Moon, Mercury is composed mostly of metal. However, astronomers think that Mercury should have formed with roughly the same ratio of metal to silicate as that found on Earth or Venus. How did it lose so much of its rocky material?

The most probable explanation for Mercury's silicate loss may be similar to the explanation for the Moon's lack of a metal core. Mercury is likely to have experienced several giant impacts very early in its youth, and one or more of these may have torn away a fraction of its mantle and crust, leaving a body dominated by its iron core.

LINK TO LEARNING

You can follow some of NASA's latest research on Mercury (https://openstax.org/l/30NASAresmercu) and see some helpful animations on the MESSENGER web page.

Today, astronomers recognize that the early solar system was a chaotic place, with the final stages of planet formation characterized by impacts of great violence. Some objects of planetary mass have been destroyed, whereas others could have fragmented and then re-formed, perhaps more than once. Both the Moon and Mercury, with their strange compositions, bear testimony to the catastrophes that must have characterized the solar system during its youth.

This OpenStax book is available for free at http://cnx.org/content/col11992/1.8

CHAPTER 9 REVIEW

 ## KEY TERMS

highlands the lighter, heavily cratered regions of the Moon, which are generally several kilometers higher than the maria

mare (plural: maria) Latin for "sea;" the name applied to the dark, relatively smooth features that cover 17% of the Moon's surface

 ## SUMMARY

9.1 General Properties of the Moon

Most of what we know about the Moon derives from the Apollo program, including 400 kilograms of lunar samples still being intensively studied. The Moon has one-eightieth the mass of Earth and is severely depleted in both metals and volatile materials. It is made almost entirely of silicates like those in Earth's mantle and crust. However, more recent spacecraft have found evidence of a small amount of water near the lunar poles, most likely deposited by comet and asteroid impacts.

9.2 The Lunar Surface

The Moon, like Earth, was formed about 4.5 billion year ago. The Moon's heavily cratered highlands are made of rocks more than 4 billion years old. The darker volcanic plains of the maria were erupted primarily between 3.3 and 3.8 billion years ago. Generally, the surface is dominated by impacts, including continuing small impacts that produce its fine-grained soil.

9.3 Impact Craters

A century ago, Grove Gilbert suggested that the lunar craters were caused by impacts, but the cratering process was not well understood until more recently. High-speed impacts produce explosions and excavate craters 10 to 15 times the size of the impactor with raised rims, ejecta blankets, and often central peaks. Cratering rates have been roughly constant for the past 3 billion years but earlier were much greater. Crater counts can be used to derive approximate ages for geological features on the Moon and other worlds with solid surfaces.

9.4 The Origin of the Moon

The three standard hypotheses for the origin of the Moon were the fission hypothesis, the sister hypothesis, and the capture hypothesis. All have problems, and they have been supplanted by the giant impact hypothesis, which ascribes the origin of the Moon to the impact of a Mars-sized projectile with Earth 4.5 billion years ago. The debris from the impact made a ring around Earth which condensed and formed the Moon.

9.5 Mercury

Mercury is the nearest planet to the Sun and the fastest moving. Mercury is similar to the Moon in having a heavily cratered surface and no atmosphere, but it differs in having a very large metal core. Early in its evolution, it apparently lost part of its silicate mantle, probably due to one or more giant impacts. Long scarps on its surface testify to a global compression of Mercury's crust during the past 4 billion years.

Download for free at http://cnx.org/content/col11992/latest/

FOR FURTHER EXPLORATION

Articles

The Moon

Bakich, Michael. "Asia's New Assault on the Moon." *Astronomy* (August 2009): 50. The Japanese Selene and Chinese Chang'e 1 missions.

Beatty, J. "NASA Slams the Moon." *Sky & Telescope* (February 2010): 28. The impact of the LCROSS mission on the Moon and what we learned from it.

Bell, T. "Warning: Dust Ahead." *Astronomy* (March 2006): 46. What we know about lunar dust and the problems it can cause.

Dorminey, B. "Secrets beneath the Moon's Surface." *Astronomy* (March 2011): 24. A nice timeline of the Moon's evolution and the story of how we are finding out more about its internal structure.

Jayawardhana, R. "Deconstructing the Moon." *Astronomy* (September 1998): 40. An update on the giant impact hypothesis for forming the Moon.

Register, B. "The Fate of the Moon Rocks." *Astronomy* (December 1985): 15. What was done with the rocks the astronauts brought back from the Moon.

Schmitt, H. "Exploring Taurus–Littrow: Apollo 17." *National Geographic* (September 1973). First-person account given by the only scientist to walk on the Moon.

Schmitt, H. "From the Moon to Mars." *Scientific American* (July 2009): 36. The only scientist to walk on the Moon reflects on the science from Apollo and future missions to Mars.

Schultz, P. "New Clues to the Moon's Distant Past." *Astronomy* (December 2011): 34. Summary of results and ideas from the LCROSS and LRO missions.

Shirao, M. "Kayuga's High Def Highlights." *Sky & Telescope* (February 2010): 20. Results from the Japanese mission to the Moon, with high definition TV cameras.

Wadhwa, M. "What Are We Learning from the Moon Rocks?" *Astronomy* (June 2013): 54. Very nice discussion of how the rocks tell us about Moon's composition, age, and origin.

Wood, Charles. "The Moon's Far Side: Nearly a New World." *Sky & Telescope* (January 2007): 48. This article compares what we know about the two sides and why they are different.

Zimmerman, R. "How Much Water is on the Moon?" *Astronomy* (January 2014): 50. Results from the LRO's instruments and good overview of issue.

Mercury

Beatty, J. "Mercury Gets a Second Look." *Sky & Telescope* (March 2009): 26. The October 2008 MESSENGER mission flyby.

Beatty, J. "Reunion with Mercury." *Sky & Telescope* (May 2008): 24. The January 2008 MESSENGER encounter with Mercury.

"Mercury: Meet the Planet Nearest the Sun." *Sky & Telescope* (March 2014): 39. Four-page pictorial introduction, including the new MESSENGER probe full map of the planet provided.

Oberg, J. "Torrid Mercury's Icy Poles." *Astronomy* (December 2013): 30. A nice overview of results from MESSENGER mission, including the ice in polar craters.

This OpenStax book is available for free at http://cnx.org/content/col11992/1.8

Sheehan, W., and Dobbins, T. "Mesmerized by Mercury." *Sky & Telescope* (June 2000): 109. History of Mercury observations and how amateur astronomers can contribute.

Talcott, R. "Surprises from MESSENGER's Historic Mercury Fly-by." *Astronomy* (March 2009): 28.

Talcott, R. "Mercury Reveals its Hidden Side." *Astronomy* (May 2008): 26. Results and image from the MESSENGER mission flyby of January 2008.

Websites

The Moon

Apollo Lunar Surface Journal: http://www.hq.nasa.gov/office/pao/History/alsj/. Information, interviews, maps, photos, video and audio clips, and much more on each of the Apollo landing missions.

Lunar & Planetary Institute: http://www.lpi.usra.edu/lunar/missions/. Lunar Science and Exploration web pages.

Lunar Reconnaissance Orbiter Mission Page: http://lro.gsfc.nasa.gov/.

NASA's Guide to Moon Missions and Information: http://nssdc.gsfc.nasa.gov/planetary/planets/moonpage.html.

Origin of the Moon: http://www.psi.edu/projects/moon/moon.html. By William Hartmann, who, with a colleague, first suggested the giant impact hypothesis for how the Moon formed, in 1975.

Sky & Telescope magazine's observing guides and articles about the Moon: http://www.skyandtelescope.com/observing/celestial-objects-to-watch/moon/.

To the Moon: http://www.pbs.org/wgbh/nova/tothemoon/. PBS program on the Apollo landings.

We Choose the Moon: http://wechoosethemoon.org/. A recreation of the Apollo 11 mission.

Mercury

Mercury Unveiled by G. Jeffrey Taylor (summarizing the Mariner 10 Mission): http://www.psrd.hawaii.edu/Jan97/MercuryUnveiled.html.

MESSENGER Mission Website: http://messenger.jhuapl.edu/.

NASA Planetary Data Center Mercury Page: http://nssdc.gsfc.nasa.gov/planetary/planets/mercurypage.html.

Views of the Solar System Mercury Page: http://solarviews.com/eng/mercury.htm.

🛆 COLLABORATIVE GROUP ACTIVITIES

A. We mentioned that no nation on Earth now has the capability to send a human being to the Moon, even though the United States once sent 12 astronauts to land there. What does your group think about this? Should we continue the exploration of space with human beings? Should we put habitats on the Moon? Should we go to Mars? Does humanity have a "destiny in space?" Whatever your answer to these questions, make a list of the arguments and facts that support your position.

B. When they hear about the giant impact hypothesis for the origin of the Moon, many students are intrigued and wonder why we can't cite more evidence for it. In your group, make a list of reasons we cannot find any traces on Earth of the great impact that formed the Moon?

Download for free at http://cnx.org/content/col11992/latest/

C. We discussed that the ice (mixed into the soil) that is found on the Moon was most likely delivered by comets. Have your group make a list of all the reasons the Moon would not have any ice of its own left over from its early days.

D. Can your group make a list of all the things that would be different if Earth had no Moon? Don't restrict your answer to astronomy and geology. Think about our calendars and moonlit romantic strolls, for example. (You may want to review Earth, Moon, and Sky.)

E. If, one day, humanity decides to establish a colony on the Moon, where should we put it? Make a list of the advantages and disadvantages of locating such a human habitat on the near side, the far side, or at the poles. What site would be best for doing visible-light and radio astronomy from observatories on the Moon?

F. A member of the class (but luckily, not a member of your group) suggests that he has always dreamed of building a vacation home on the planet Mercury. Can your group make a list of all reasons such a house would be hard to build and keep in good repair?

G. As you've read in this chapter, craters on the Moon are (mostly) named after scientists. (See the official list at: http://planetarynames.wr.usgs.gov/SearchResults?target=MOON&featureType=Crater,%20craters). The craters on Mercury, on the other hand, are named for writers, artists, composers, and others in the humanities. See the official list at: http://planetarynames.wr.usgs.gov/SearchResults?target=MERCURY&featureType=Crater,%20craters). Living persons are not eligible. Can each person in your group think of a scientist or someone in the arts whom they especially respect? Now check to see if they are listed. Are there scientists or people in the arts who should have their names on the Moon or Mercury and do not?

H. Imagine that a distant relative, hearing you are taking an astronomy course, calls you up and tells you that NASA faked the Moon landings. His most significant argument is that all the photos of the Moon show black skies, but none of them have any stars showing. This proves that the photos were taken against a black backdrop in a studio and not on the Moon. Based on your reading in this chapter, what arguments can your group come up with to rebut this idea?

▣ EXERCISES

Review Questions

1. What is the composition of the Moon, and how does it compare to the composition of Earth? Of Mercury?

2. Why does the Moon not have an atmosphere?

3. What are the principal features of the Moon observable with the unaided eye?

4. Frozen water exists on the lunar surface primarily in which location? Why?

5. Outline the main events in the Moon's geological history.

6. What are the maria composed of? Is this material found elsewhere in the solar system?

7. The mountains on the Moon were formed by what process?

8. With no wind or water erosion of rocks, what is the mechanism for the creation of the lunar "soil?"

This OpenStax book is available for free at http://cnx.org/content/col11992/1.8

9. What differences did Grove K. Gilbert note between volcanic craters on Earth and lunar craters?

10. Explain how high-speed impacts form circular craters. How can this explanation account for the various characteristic features of impact craters?

11. Explain the evidence for a period of heavy bombardment on the Moon about 4 billion years ago.

12. How did our exploration of the Moon differ from that of Mercury (and the other planets)?

13. Summarize the four main hypotheses for the origin of the Moon.

14. What are the difficulties with the capture hypothesis of the Moon's origin?

15. What is the main consequence of Mercury's orbit being so highly eccentric?

16. Describe the basic internal structure of Mercury.

17. How was the rotation rate of Mercury determined?

18. What is the relationship between Mercury's rotational period and orbital period?

19. The features of Mercury are named in honor of famous people in which fields of endeavor?

20. What do our current ideas about the origins of the Moon and Mercury have in common? How do they differ?

Thought Questions

21. One of the primary scientific objectives of the Apollo program was the return of lunar material. Why was this so important? What can be learned from samples? Are they still of value now?

22. Apollo astronaut David Scott dropped a hammer and a feather together on the Moon, and both reached the ground at the same time. What are the two distinct advantages that this experiment on the Moon had over the same kind of experiment as performed by Galileo on Earth?

23. Galileo thought the lunar maria might be seas of water. If you had no better telescope than the one he had, could you demonstrate that they are not composed of water?

24. Why did it take so long for geologists to recognize that the lunar craters had an impact origin rather than a volcanic one?

25. How might a crater made by the impact of a comet with the Moon differ from a crater made by the impact of an asteroid?

26. Why are the lunar mountains smoothly rounded rather than having sharp, pointed peaks (as they were almost always depicted in science-fiction illustrations and films before the first lunar landings)?

27. The lunar highlands have about ten times more craters in a given area than do the maria. Does this mean that the highlands are 10 times older? Explain your reasoning.

28. At the end of the section on the lunar surface, your authors say that lunar night and day each last about two Earth weeks. After looking over the information in Earth, Moon, and Sky and this chapter about the motions of the Moon, can you explain why? (It helps to draw a diagram for yourself.)

29. Give several reasons Mercury would be a particularly unpleasant place to build an astronomical observatory.

Download for free at http://cnx.org/content/col11992/latest/

30. If, in the remote future, we establish a base on Mercury, keeping track of time will be a challenge. Discuss how to define a year on Mercury, and the two ways to define a day. Can you come up with ways that humans raised on Earth might deal with time cycles on Mercury?

31. The Moon has too little iron, Mercury too much. How can both of these anomalies be the result of giant impacts? Explain how the same process can yield such apparently contradictory results.

Figuring For Yourself

32. In the future, astronomers discover a solid moon around a planet orbiting one of the nearest stars. This moon has a diameter of 1948 km and a mass of 1.6×10^{22} kg. What is its density?

33. The Moon was once closer to Earth than it is now. When it was at half its present distance, how long was its period of revolution? (See Orbits and Gravity for the formula to use.)

34. Astronomers believe that the deposit of lava in the giant mare basins did not happen in one flow but in many different eruptions spanning some time. Indeed, in any one mare, we find a variety of rock ages, typically spanning about 100 million years. The individual lava flows as seen in Hadley Rille by the Apollo 15 astronauts were about 4 m thick. Estimate the average time interval between the beginnings of successive lava flows if the total depth of the lava in the mare is 2 km.

35. The Moon requires about 1 month (0.08 year) to orbit Earth. Its distance from us is about 400,000 km (0.0027 AU). Use Kepler's third law, as modified by Newton, to calculate the mass of Earth relative to the Sun.

This OpenStax book is available for free at http://cnx.org/content/col11992/1.8

10

EARTHLIKE PLANETS: VENUS AND MARS

Figure 10.1. Spirit Rover on Mars. This May 2004 image shows the tracks made by the Mars Exploration *Spirit* rover on the surface of the red planet. *Spirit* was active on Mars between 2004 and 2010, twenty times longer than its planners had expected. It "drove" over 7.73 kilometers in the process of examining the martian landscape. (credit: modification of work by NASA/JPL/Cornell)

Chapter Outline

Thinking Ahead

The Moon and Mercury are geologically dead. In contrast, the larger terrestrial planets—Earth, Venus, and Mars—are more active and interesting worlds. We have already discussed Earth, and we now turn to Venus and Mars. These are the nearest planets and the most accessible to spacecraft. Not surprisingly, the greatest effort in planetary exploration has been devoted to these fascinating worlds. In the chapter, we discuss some of the results of more than four decades of scientific exploration of Mars and Venus. Mars is exceptionally interesting, with evidence that points to habitable conditions in the past. Even today, we are discovering things about Mars that make it the most likely place where humans might set up a habitat in the future. However, our robot explorers have clearly shown that neither Venus nor Mars has conditions similar to Earth. How did it happen that these three neighboring terrestrial planets have diverged so dramatically in their evolution?

10.1 THE NEAREST PLANETS: AN OVERVIEW

Learning Objectives

By the end of this section, you will be able to:

Download for free at http://cnx.org/content/col11992/latest/

> Explain why it's difficult to learn about Venus from Earth-based observation alone
> Describe the history of our interest in Mars before the Space Age
> Compare the basic physical properties of Earth, Mars, and Venus, including their orbits

As you might expect from close neighbors, Mars and Venus are among the brightest objects in the night sky. The average distance of Mars from the Sun is 227 million kilometers (1.52 AU), or about half again as far from the Sun as Earth. Venus' orbit is very nearly circular, at a distance of 108 million kilometers (0.72 AU) from the Sun. Like Mercury, Venus sometimes appears as an "evening star" and sometimes as a "morning star." Venus approaches Earth more closely than does any other planet: at its nearest, it is only 40 million kilometers from us. The closest Mars ever gets to Earth is about 56 million kilometers.

Appearance

Venus appears very bright, and even a small telescope reveals that it goes through phases like the Moon. Galileo discovered that Venus displays a full range of phases, and he used this as an argument to show that Venus must circle the Sun and not Earth. The planet's actual surface is not visible because it is shrouded by dense clouds that reflect about 70% of the sunlight that falls on them, frustrating efforts to study the underlying surface, even with cameras in orbit around the planet (Figure 10.2).

Figure 10.2. Venus as Photographed by the Pioneer Venus Orbiter. This ultraviolet image shows an upper-atmosphere cloud structure that would be invisible at visible wavelengths. Note that there is not even a glimpse of the planet's surface. (credit: modification of work by NASA)

In contrast, Mars is more tantalizing as seen through a telescope (Figure 10.3). The planet is distinctly red, due (as we now know) to the presence of iron oxides in its soil. This color may account for its association with war (and blood) in the legends of early cultures. The best resolution obtainable from telescopes on the ground is about 100 kilometers, or about the same as what we can see on the Moon with the unaided eye. At this resolution, no hint of topographic structure can be detected: no mountains, no valleys, not even impact craters. On the other hand, bright polar ice caps can be seen easily, together with dusky surface markings that sometimes change in outline and intensity from season to season.

This OpenStax book is available for free at http://cnx.org/content/col11992/1.8

Download for free at http://cnx.org/content/col11992/latest/

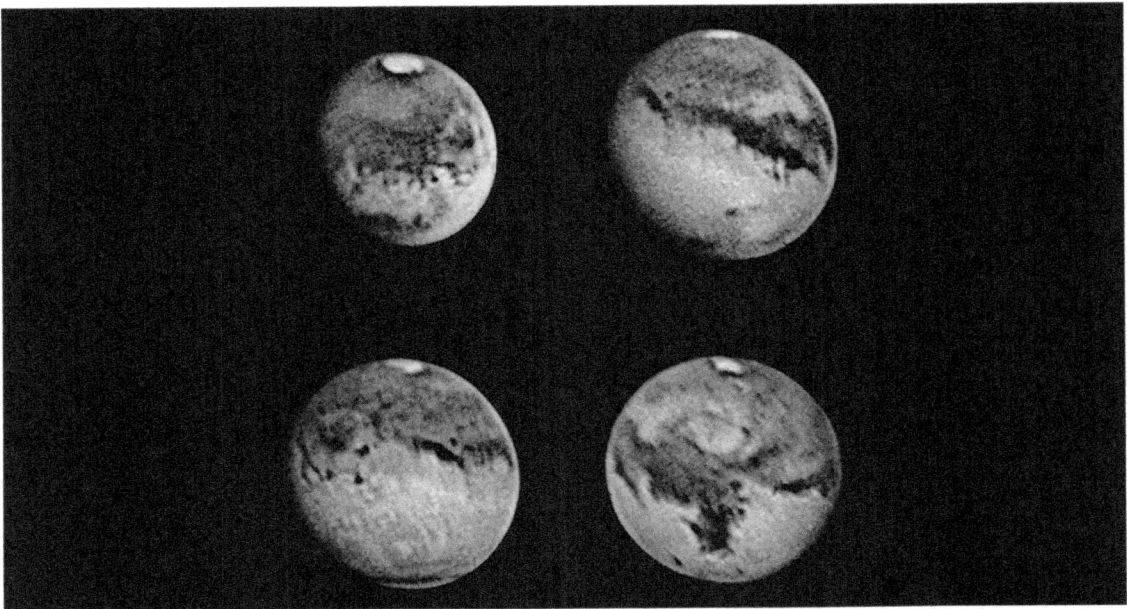

Figure 10.3. Mars as Seen from Earth's Surface. These are among the best Earth-based photos of Mars, taken in 1988 when the planet was exceptionally close to Earth. The polar caps and dark surface markings are evident, but not topographic features. (credit: modification of work by Steve Larson, Lunar and Planetary Laboratory, University of Arizona)

For a few decades around the turn of the twentieth century, some astronomers believed that they saw evidence of an intelligent civilization on Mars. The controversy began in 1877, when Italian astronomer Giovanni Schiaparelli (1835–1910) announced that he could see long, faint, straight lines on Mars that he called *canale*, or channels. In English-speaking countries, the term was mistakenly translated as "canals," implying an artificial origin.

Even before Schiaparelli's observations, astronomers had watched the bright polar caps change size with the seasons and had seen variations in the dark surface features. With a little imagination, it was not difficult to picture the canals as long fields of crops bordering irrigation ditches that brought water from the melting polar ice to the parched deserts of the red planet. (They assumed the polar caps were composed of water ice, which isn't exactly true, as we will see shortly.)

Until has death in 1916, the most effective proponent of intelligent life on Mars was Percival Lowell, a self-made American astronomer and member of the wealthy Lowell family of Boston (see the feature box on Percival Lowell: Dreaming of an Inhabited Mars). A skilled author and speaker, Lowell made what seemed to the public to be a convincing case for intelligent Martians, who had constructed the huge canals to preserve their civilization in the face of a deteriorating climate (Figure 10.4).

Download for free at http://cnx.org/content/col11992/latest/

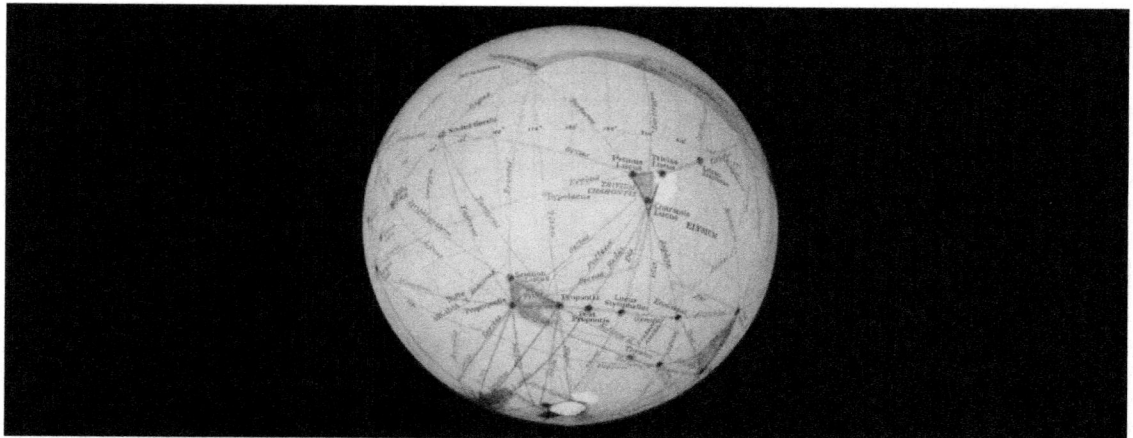

Figure 10.4. Lowell's Mars Globe. One of the remarkable globes of Mars prepared by Percival Lowell, showing a network of dozens of canals, oases, and triangular water reservoirs that he claimed were visible on the red planet.

The argument for a race of intelligent Martians, however, hinged on the reality of the canals, a matter that remained in serious dispute among astronomers. The canal markings were always difficult to study, glimpsed only occasionally because atmospheric conditions caused the tiny image of Mars to shimmer in the telescope. Lowell saw canals everywhere (even a few on Venus), but many other observers could not see them at all and remained unconvinced of their existence. When telescopes larger than Lowell's failed to confirm the presence of canals, the skeptics felt vindicated. Now it is generally accepted that the straight lines were an optical illusion, the result of the human mind's tendency to see order in random features that are glimpsed dimly at the limits of the eye's resolution. When we see small, dim dots of surface markings, our minds tend to connect those dots into straight lines.

VOYAGERS IN ASTRONOMY

Percival Lowell: Dreaming of an Inhabited Mars

Percival Lowell was born into the well-to-do Massachusetts family about whom John Bossidy made the famous toast:

> And this is good old Boston,
> The home of the bean and the cod,
> Where the Lowells talk to the Cabots
> And the Cabots talk only to God.

Percival's brother Lawrence became president of Harvard University, and his sister, Amy, became a distinguished poet. Percival was already interested in astronomy as a boy: he made observations of Mars at age 13. His undergraduate thesis at Harvard dealt with the origin of the solar system, but he did not pursue this interest immediately. Instead, he entered the family business and traveled extensively in Asia. In 1892, however, he decided to dedicate himself to carrying on Schiaparelli's work and solving the mysteries of the martian canals.

This OpenStax book is available for free at http://cnx.org/content/col11992/1.8

Download for free at http://cnx.org/content/col11992/latest/

In 1894, with the help of astronomers at Harvard but using his own funds, Lowell built an observatory on a high plateau in Flagstaff, Arizona, where he hoped the seeing would be clear enough to show him Mars in unprecedented detail. He and his assistants quickly accumulated a tremendous number of drawings and maps, purporting to show a vast network of martian canals (see Figure 10.4). He elaborated his ideas about the inhabitants of the red planet in several books, including *Mars* (1895) and *Mars and Its Canals* (1906), and in hundreds of articles and speeches.

As Lowell put it,

> A mind of no mean order would seem to have presided over the system we see—a mind certainly of considerably more comprehensiveness than that which presides over the various departments of our own public works. Party politics, at all events, have had no part in them; for the system is planet-wide. . . . Certainly what we see hints at the existence of beings who are in advance of, not behind us, in the journey of life.

Lowell's views captured the public imagination and inspired many novels and stories, the most famous of which was H. G. Wells' *War of the Worlds* (1897). In this famous "invasion" novel, the thirsty inhabitants of a dying planet Mars (based entirely on Lowell's ideas) come to conquer Earth with advanced technology.

Although the Lowell Observatory first became famous for its work on the martian canals, both Lowell and the observatory eventually turned to other projects as well. He became interested in the search for a ninth (and then undiscovered) planet in the solar system. In 1930, Pluto was found at the Lowell Observatory, and it is not a coincidence that the name selected for the new planet starts with Lowell's initials. It was also at the Lowell Observatory that the first measurements were made of the great speed at which galaxies are moving away from us, observations that would ultimately lead to our modern view of an expanding universe.

Lowell (Figure 10.5) continued to live at his observatory, marrying at age 53 and publishing extensively. He relished the debate his claims about Mars caused far more than the astronomers on the other side, who often complained that Lowell's work was making planetary astronomy a less respectable field. At the same time, the public fascination with the planets fueled by Lowell's work (and its interpreters) may, several generations later, have helped fan support for the space program and the many missions whose results grace the pages of our text.

Download for free at http://cnx.org/content/col11992/latest/

Figure 10.5. Percival Lowell (1855–1916). This 1914 photograph shows Percival Lowell observing Venus with his 24-inch telescope at Flagstaff, Arizona.

LINK TO LEARNING

In October 1938, the Mercury Theater of the Air on radio dramatized *The War of the Worlds* as a series of radio news reports. This broadcast (https://openstax.org/l/30WarofWorlds) scared many people into thinking that Lowell's Martians were really invading New Jersey, and caused something of a panic. You can listen to the original radio broadcast if you scroll down to "War of the Worlds."

Rotation of the Planets

Astronomers have determined the rotation period of Mars with great accuracy by watching the motion of permanent surface markings; its sidereal day is 24 hours 37 minutes 23 seconds, just a little longer than the rotation period of Earth. This high precision is not obtained by watching Mars for a single rotation, but by noting how many turns it makes over a long period of time. Good observations of Mars date back more than 200 years, a period during which tens of thousands of martian days have passed. As a result, the rotation period can be calculated to within a few hundredths of a second.

The rotational axis of Mars has a tilt of about 25°, similar to the tilt of Earth's axis. Thus, Mars experiences seasons very much like those on Earth. Because of the longer martian year (almost two Earth years), however, each season there lasts about six of our months.

The situation with Venus is different. Since no surface detail can be seen through Venus' clouds, its rotation period can be found only by bouncing radar signals off the planet (as explained for Mercury in the Cratered Worlds chapter). The first radar observations of Venus' rotation were made in the early 1960s. Later, topographical surface features were identified on the planet that showed up in the reflected radar signals. The rotation period of Venus, precisely determined from the motion of such "radar features" across its disk, is 243 days. Even more surprising than how *long* Venus takes to rotate is the fact that it spins in a backward or retrograde direction (east to west).

This OpenStax book is available for free at http://cnx.org/content/col11992/1.8

Stop for a moment and think about how odd this slow rotation makes the calendar on Venus. The planet takes 225 Earth days to orbit the Sun and 243 Earth days to spin on its axis. So the day on Venus (as defined by its spinning once) is longer than the year! As a result, the time the Sun takes to return to the same place in Venus' sky—another way we might define the meaning of a day—turns out to be 117 Earth days. (If you say "See you tomorrow" on Venus, you'll have a long time to wait.) Although we do not know the reason for Venus' slow backward rotation, we can guess that it may have suffered one or more extremely powerful collisions during the formation process of the solar system.

Basic Properties of Venus and Mars

Before discussing each planet individually, let us compare some of their basic properties with each other and with Earth (Table 10.1). Venus is in many ways Earth's twin, with a mass 0.82 times the mass of Earth and an almost identical density. The average amount of geological activity has been also relatively high, almost as high as on Earth. On the other hand, with a surface pressure nearly 100 times greater than ours, Venus' atmosphere is not at all like that of Earth. The surface of Venus is also remarkably hot, with a temperature of 730 K (over 850 °F), hotter than the self-cleaning cycle of your oven. One of the major challenges presented by Venus is to understand why the atmosphere and surface environment of this twin have diverged so sharply from those of our own planet.

Properties of Earth, Venus, and Mars

Property	Earth	Venus	Mars
Semimajor axis (AU)	1.00	0.72	1.52
Period (year)	1.00	0.61	1.88
Mass (Earth = 1)	1.00	0.82	0.11
Diameter (km)	12,756	12,102	6,790
Density (g/cm^3)	5.5	5.3	3.9
Surface gravity (Earth = 1)	1.00	0.91	0.38
Escape velocity (km/s)	11.2	10.4	5.0
Rotation period (hours or days)	23.9 h	243 d	24.6 h
Surface area (Earth = 1)	1.00	0.90	0.28
Atmospheric pressure (bar)	1.00	90	0.007

Table 10.1

Mars, by contrast, is rather small, with a mass only 0.11 times the mass of Earth. It is larger than either the Moon or Mercury, however, and, unlike them, it retains a thin atmosphere. Mars is also large enough to have supported considerable geological activity in the distant past. But the most fascinating thing about Mars is that long ago it probably had a thick atmosphere and seas of liquid water—the conditions we associate with

Download for free at http://cnx.org/content/col11992/latest/

development of life. There is even a chance that some form of life persists today in protected environments below the martian surface.

10.2   THE GEOLOGY OF VENUS

Learning Objectives

By the end of this section, you will be able to:

> Describe the general features of the surface of Venus
> Explain what the study of craters on Venus tells us about the age of its surface
> Compare tectonic activity and volcanoes on Venus with those of Earth
> Explain why the surface of Venus is inhospitable to human life

Since Venus has about the same size and composition as Earth, we might expect its geology to be similar. This is partly true, but Venus does not exhibit the same kind of *plate tectonics* as Earth, and we will see that its lack of erosion results in a very different surface appearance.

Spacecraft Exploration of Venus

Nearly 50 spacecraft have been launched to Venus, but only about half were successful. Although the 1962 US Mariner 2 flyby was the first, the Soviet Union launched most of the subsequent missions to Venus. In 1970, Venera 7 became the first probe to land and broadcast data from the surface of Venus. It operated for 23 minutes before succumbing to the high surface temperature. Additional Venera probes and landers followed, photographing the surface and analyzing the atmosphere and soil.

To understand the geology of Venus, however, we needed to make a global study of its surface, a task made very difficult by the perpetual cloud layers surrounding the planet. The problem resembles the challenge facing air traffic controllers at an airport, when the weather is so cloudy or smoggy that they can't locate the incoming planes visually. The solution is similar in both cases: use a radar instrument to probe through the obscuring layer.

The first global radar map was made by the US Pioneer Venus orbiter in the late 1970s, followed by better maps from the twin Soviet Venera 15 and 16 radar orbiters in the early 1980s. However, most of our information on the geology of Venus is derived from the US *Magellan* spacecraft, which mapped Venus with a powerful *imaging radar*. *Magellan* produced images with a resolution of 100 meters, much better than that of previous missions, yielding our first detailed look at the surface of our sister planet (Figure 10.6). (The *Magellan* spacecraft returned more data to Earth than all previous planetary missions combined; each 100 minutes of data transmission from the spacecraft provided enough information, if translated into characters, to fill two 30-volume encyclopedias.)

This OpenStax book is available for free at http://cnx.org/content/col11992/1.8

Figure 10.6. Radar Map of Venus. This composite image has a resolution of about 3 kilometers. Colors have been added to indicate elevation, with blue meaning low and brown and white high. The large continent Aphrodite stretches around the equator, where the bright (therefore rough) surface has been deformed by tectonic forces in the crust of Venus. (credit: modification of work by NASA/JPL/USGS)

Consider for a moment how good *Magellan*'s resolution of 100 meters really is. It means the radar images from Venus can show anything on the surface larger than a football field. Suddenly, a whole host of topographic features on Venus became accessible to our view. As you look at the radar images throughout this chapter, bear in mind that these are constructed from radar reflections, not from visible-light photographs. For example, bright features on these radar images are an indication of rough terrain, whereas darker regions are smoother.

Probing Through the Clouds of Venus

The radar maps of Venus reveal a planet that looks much the way Earth might look if our planet's surface were not constantly being changed by erosion and deposition of sediment. Because there is no water or ice on Venus and the surface wind speeds are low, almost nothing obscures or erases the complex geological features produced by the movements of Venus' crust, by volcanic eruptions, and by impact craters. Having finally penetrated below the clouds of Venus, we find its surface to be naked, revealing the history of hundreds of millions of years of geological activity.

About 75% of the surface of Venus consists of lowland lava plains. Superficially, these plains resemble the basaltic ocean basins of Earth, but they were not produced in quite the same way. There is no evidence of subduction zones on Venus, indicating that, unlike Earth, this planet never experienced plate tectonics. Although *convection* (the rising of hot materials) in its mantle generated great stresses in the crust of Venus, they did not start large continental plates moving. The formation of the lava plains of Venus more nearly resembles that of the lunar maria. Both were the result of widespread lava eruptions without the crustal spreading associated with plate tectonics.

Rising above the lowland lava plains are two full-scale continents of mountainous terrain. The largest continent on Venus, called Aphrodite, is about the size of Africa (you can see it stand out in Figure 10.6). Aphrodite stretches along the equator for about one-third of the way around the planet. Next in size is the northern highland region Ishtar, which is about the size of Australia. Ishtar contains the highest region on the planet, the Maxwell Mountains, which rise 11 kilometers above the surrounding lowlands. (The Maxwell Mountains are the only feature on Venus named after a man. They commemorate James Clerk Maxwell, whose theory of

Download for free at http://cnx.org/content/col11992/latest/

electromagnetism led to the invention of radar. All other features are named for women, either from history or mythology.)

Craters and the Age of the Venus Surface

One of the first questions astronomers addressed with the high-resolution *Magellan* images was the age of the surface of Venus. Remember that the age of a planetary surface is rarely the age of the world it is on. A young age merely implies an active geology in that location. Such ages can be derived from counting impact craters. Figure 10.7 is an example of what these craters look like on the Venus radar images. The more densely cratered the surface, the greater its age. The largest crater on Venus (called Mead) is 275 kilometers in diameter, slightly larger than the largest known terrestrial crater (Chicxulub), but much smaller than the lunar impact basins.

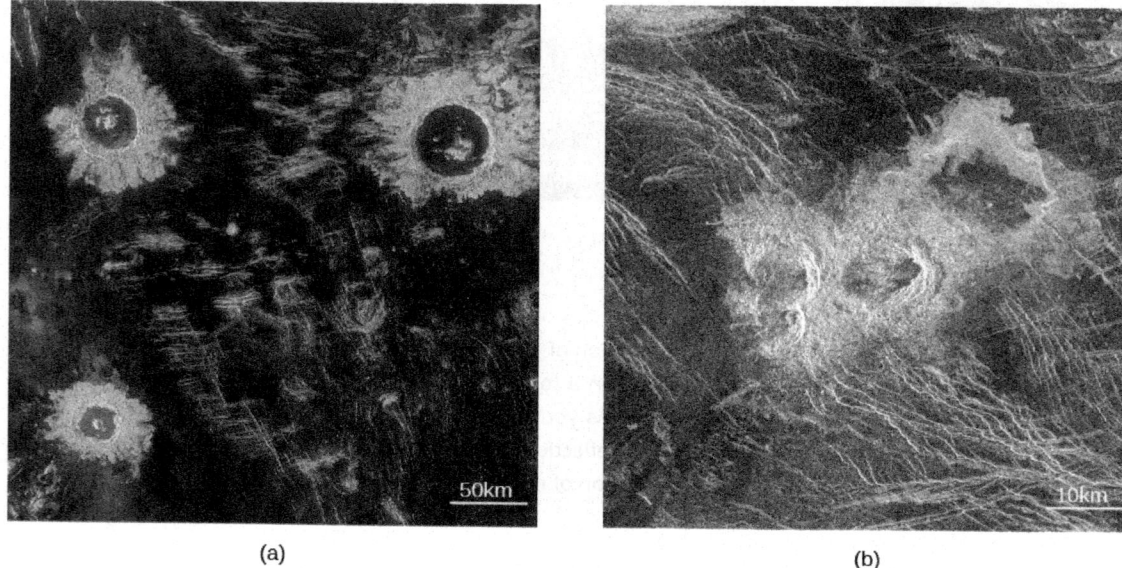

(a)   (b)

Figure 10.7. Impact Craters on Venus. (a) These large impact craters are in the Lavinia region of Venus. Because they are rough, the crater rims and ejecta appear brighter in these radar images than do the smoother surrounding lava plains. The largest of these craters has a diameter of 50 kilometers. (b) This small, complex crater is named after writer Gertrude Stein. The triple impact was caused by the breaking apart of the incoming asteroid during its passage through the thick atmosphere of Venus. The projectile had an initial diameter of between 1 and 2 kilometers. (credit a: modification of work by NASA/JPL; credit b: modification of work by NASA/JPL)

You might think that the thick atmosphere of Venus would protect the surface from impacts, burning up the projectiles long before they could reach the surface. But this is the case for only smaller projectiles. Crater statistics show very few craters less than 10 kilometers in diameter, indicating that projectiles smaller than about 1 kilometer (the size that typically produces a 10-kilometer crater) were stopped by the atmosphere. Those craters with diameters from 10 to 30 kilometers are frequently distorted or multiple, apparently because the incoming projectile broke apart in the atmosphere before it could strike the ground as shown in the Stein crater in Figure 10.7. If we limit ourselves to impacts that produce craters with diameters of 30 kilometers or larger, however, then crater counts are as useful on Venus for measuring surface age as they are on airless bodies such as the Moon.

The large craters in the venusian plains indicate an average surface age that is only between 300 and 600 million years. These results indicate that Venus is indeed a planet with persistent geological activity, intermediate between that of Earth's ocean basins (which are younger and more active) and that of its continents (which are older and less active).

This OpenStax book is available for free at http://cnx.org/content/col11992/1.8

Download for free at http://cnx.org/content/col11992/latest/

Almost all of the large craters on Venus look fresh, with little degradation or filling in by either lava or windblown dust. This is one way we know that the rates of erosion or sediment deposition are very low. We have the impression that relatively little has happened since the venusian plains were last resurfaced by large-scale volcanic activity. Apparently Venus experienced some sort of planet-wide volcanic convulsion between 300 and 600 million years ago, a mysterious event that is unlike anything in terrestrial history.

Volcanoes on Venus

Like Earth, Venus is a planet that has experienced widespread volcanism. In the lowland plains, volcanic eruptions are the principal way the surface is renewed, with large flows of highly fluid lava destroying old craters and generating a fresh surface. In addition, numerous younger volcanic mountains and other structures are associated with surface hot spots—places where convection in the planet's mantle transports the interior heat to the surface.

The largest individual volcano on Venus, called Sif Mons, is about 500 kilometers across and 3 kilometers high—broader but lower than the Hawaiian volcano Mauna Loa. At its top is a volcanic crater, or *caldera*, about 40 kilometers across, and its slopes show individual lava flows up to 500 kilometers long. Thousands of smaller volcanoes dot the surface, down to the limit of visibility of the *Magellan* images, which correspond to cones or domes about the size of a shopping mall parking lot. Most of these seem similar to terrestrial volcanoes. Other volcanoes have unusual shapes, such as the "pancake domes" illustrated in Figure 10.8.

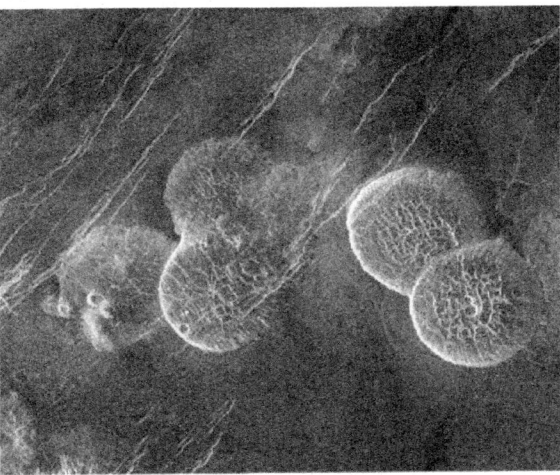

Figure 10.8. Pancake-Shaped Volcanoes on Venus. These remarkable circular domes, each about 25 kilometers across and about 2 kilometers tall, are the result of eruptions of highly viscous (sludgy) lava that spreads out evenly in all directions. (credit: modification of work by NASA/JPL)

All of the volcanism is the result of eruption of lava onto the surface of the planet. But the hot lava rising from the interior of a planet does not always make it to the surface. On both Earth and Venus, this upwelling lava can collect to produce bulges in the crust. Many of the granite mountain ranges on Earth, such as the Sierra Nevada in California, involve such subsurface volcanism. These bulges are common on Venus, where they produce large circular or oval features called *coronae* (singular: corona) (Figure 10.9).

Download for free at http://cnx.org/content/col11992/latest/

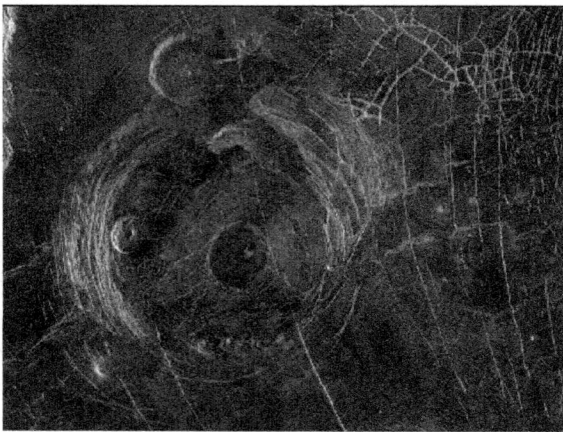

Figure 10.9. The "Miss Piggy" Corona. Fotla Corona is located in the plains to the south of Aphrodite Terra. Curved fracture patterns show where the material beneath has put stress on the surface. A number of pancake and dome volcanoes are also visible. Fotla was a Celtic fertility goddess. Some students see a resemblance between this corona and Miss Piggy of the Muppets (her left ear, at the top of the picture, is the pancake volcano in the upper center of the image). (credit: NASA/JPL)

Tectonic Activity

Convection currents of molten material in the mantle of Venus push and stretch the crust. Such forces are called **tectonic**, and the geological features that result from these forces are called *tectonic features*. On Venus' lowland plains, tectonic forces have broken the lava surface to create remarkable patterns of ridges and cracks (Figure 10.10). In a few places, the crust has even torn apart to generate rift valleys. The circular features associated with coronae are tectonic ridges and cracks, and most of the mountains of Venus also owe their existence to tectonic forces.

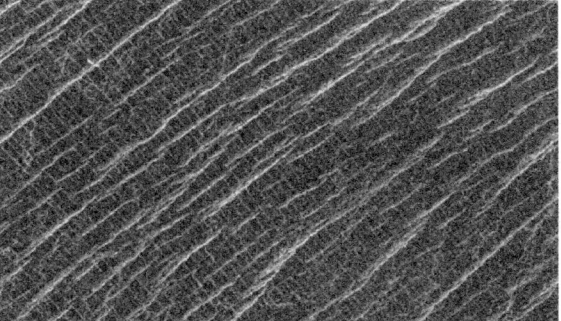

Figure 10.10. Ridges and Cracks. This region of the Lakshmi Plains on Venus has been fractured by tectonic forces to produce a cross-hatched grid of cracks and ridges. Be sure to notice the fainter linear features that run perpendicular to the brighter ones. As this is a radar image, the brightness of the ridges indicates their relative height. This image shows a region about 80 kilometers wide and 37 kilometers high. Lakshmi is a Hindu goddess of prosperity. (credit: modification of work by Magellan Team, JPL, NASA)

The Ishtar continent, which has the highest elevations on Venus, is the most dramatic product of these tectonic forces. Ishtar and its tall Maxwell Mountains resemble the Tibetan Plateau and Himalayan Mountains on Earth. Both are the product of compression of the crust, and both are maintained by the continuing forces of mantle convection.

This OpenStax book is available for free at http://cnx.org/content/col11992/1.8

Download for free at http://cnx.org/content/col11992/latest/

On Venus' Surface

The successful Venera landers of the 1970s found themselves on an extraordinarily inhospitable planet, with a surface pressure of 90 bars and a temperature hot enough to melt lead and zinc. Despite these unpleasant conditions, the spacecraft were able to photograph their surroundings and collect surface samples for chemical analysis before their instruments gave out. The diffuse sunlight striking the surface was tinted red by the clouds, and the illumination level was equivalent to a heavy overcast on Earth.

The probes found that the rock in the landing areas is igneous, primarily basalts. Examples of the Venera photographs are shown in Figure 10.11. Each picture shows a flat, desolate landscape with a variety of rocks, some of which may be ejecta from impacts. Other areas show flat, layered lava flows. There have been no further landings on Venus since the 1970s.

Figure 10.11. Surface of Venus. These views of the surface of Venus are from the Venera 13 spacecraft. Everything is orange because the thick atmosphere of Venus absorbs the bluer colors of light. The horizon is visible in the upper corner of each image. (credit: NASA)

10.3 | THE MASSIVE ATMOSPHERE OF VENUS

Learning Objectives

By the end of this section, you will be able to:

> Describe the general composition and structure of the atmosphere on Venus
> Explain how the greenhouse effect has led to high temperatures on Venus

The thick atmosphere of Venus produces the high surface temperature and shrouds the surface in a perpetual red twilight. Sunlight does not penetrate directly through the heavy clouds, but the surface is fairly well lit by diffuse light (about the same as the light on Earth under a heavy overcast). The weather at the bottom of this deep atmosphere remains perpetually hot and dry, with calm winds. Because of the heavy blanket of clouds and atmosphere, one spot on the surface of Venus is similar to any other as far as weather is concerned.

Composition and Structure of the Atmosphere

The most abundant gas on Venus is carbon dioxide (CO_2), which accounts for 96% of the atmosphere. The second most common gas is nitrogen. The predominance of carbon dioxide over nitrogen is not surprising when you recall that Earth's atmosphere would also be mostly carbon dioxide if this gas were not locked up in marine sediments (see the discussion of Earth's atmosphere in Earth as a Planet).

Table 10.2 compares the compositions of the atmospheres of Venus, Mars, and Earth. Expressed in this way, as percentages, the proportions of the major gases are very similar for Venus and Mars, but in total quantity, their atmospheres are dramatically different. With its surface pressure of 90 bars, the venusian atmosphere is more than 10,000 times more massive than its martian counterpart. Overall, the atmosphere of Venus is very dry; the absence of water is one of the important ways that Venus differs from Earth.

Download for free at http://cnx.org/content/col11992/latest/

Atmospheric Composition of Earth, Venus, and Mars

Gas	Earth	Venus	Mars
Carbon dioxide (CO_2)	0.03%	96%	95.3%
Nitrogen (N_2)	78.1%	3.5%	2.7%
Argon (Ar)	0.93%	0.006%	1.6%
Oxygen (O_2)	21.0%	0.003%	0.15%
Neon (Ne)	0.002%	0.001%	0.0003%

Table 10.2

The atmosphere of Venus has a huge troposphere (region of convection) that extends up to at least 50 kilometers above the surface (Figure 10.12). Within the troposphere, the gas is heated from below and circulates slowly, rising near the equator and descending over the poles. Being at the base of the atmosphere of Venus is something like being a kilometer or more below the ocean surface on Earth. There, the mass of water evens out temperature variations and results in a uniform environment—the same effect the thick atmosphere has on Venus.

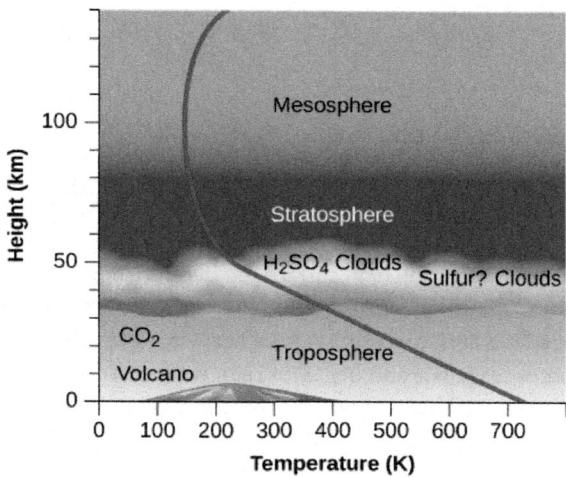

Figure 10.12. Venus' Atmosphere. The layers of the massive atmosphere of Venus shown here are based on data from the Pioneer and Venera entry probes. Height is measured along the left axis, the bottom scale shows temperature, and the red line allows you to read off the temperature at each height. Notice how steeply the temperature rises below the clouds, thanks to the planet's huge greenhouse effect.

In the upper troposphere, between 30 and 60 kilometers above the surface, a thick cloud layer is composed primarily of sulfuric acid droplets. Sulfuric acid (H_2SO_4) is formed from the chemical combination of sulfur dioxide (SO_2) and water (H_2O). In the atmosphere of Earth, sulfur dioxide is one of the primary gases emitted by volcanoes, but it is quickly diluted and washed out by rainfall. In the dry atmosphere of Venus, this unpleasant substance is apparently stable. Below 30 kilometers, the Venus atmosphere is clear of clouds.

This OpenStax book is available for free at http://cnx.org/content/col11992/1.8

Surface Temperature on Venus

The high surface temperature of Venus was discovered by radio astronomers in the late 1950s and confirmed by the Mariner and Venera probes. How can our neighbor planet be so hot? Although Venus is somewhat closer to the Sun than is Earth, its surface is hundreds of degrees hotter than you would expect from the extra sunlight it receives. Scientists wondered what could be heating the surface of Venus to a temperature above 700 K. The answer turned out to be the *greenhouse effect*.

The greenhouse effect works on Venus just as it does on Earth, but since Venus has so much more CO_2—almost a million times more—the effect is much stronger. The thick CO_2 acts as a blanket, making it very difficult for the infrared (heat) radiation from the ground to get back into space. As a result, the surface heats up. The energy balance is only restored when the planet is radiating as much energy as it receives from the Sun, but this can happen only when the temperature of the lower atmosphere is very high. One way of thinking of greenhouse heating is that it must raise the surface temperature of Venus until this energy balance is achieved.

Has Venus always had such a massive atmosphere and high surface temperature, or might it have evolved to such conditions from a climate that was once more nearly earthlike? The answer to this question is of particular interest to us as we look at the increasing levels of CO_2 in Earth's atmosphere. As the greenhouse effect becomes stronger on Earth, are we in any danger of transforming our own planet into a hellish place like Venus?

Let us try to reconstruct the possible evolution of Venus from an earthlike beginning to its present state. Venus may once have had a climate similar to that of Earth, with moderate temperatures, water oceans, and much of its CO_2 dissolved in the ocean or chemically combined with the surface rocks. Then we allow for modest additional heating—by gradual increase in the energy output of the Sun, for example. When we calculate how Venus' atmosphere would respond to such effects, it turns out that even a small amount of extra heat can lead to increased evaporation of water from the oceans and the release of gas from surface rocks.

This in turn means a further increase in the atmospheric CO_2 and H_2O, gases that would amplify the greenhouse effect in Venus' atmosphere. That would lead to still more heat near Venus' surface and the release of further CO_2 and H_2O. Unless some other processes intervene, the temperature thus continues to rise. Such a situation is called the **runaway greenhouse effect**.

We want to emphasize that the runaway greenhouse effect is not just a large greenhouse effect; it is an evolutionary *process*. The atmosphere evolves from having a small greenhouse effect, such as on Earth, to a situation where greenhouse warming is a major factor, as we see today on Venus. Once the large greenhouse conditions develop, the planet establishes a new, much hotter equilibrium near its surface.

Reversing the situation is difficult because of the role water plays. On Earth, most of the CO_2 is either chemically bound in the rocks of our crust or dissolved by the water in our oceans. As Venus got hotter and hotter, its oceans evaporated, eliminating that safety valve. But the water vapor in the planet's atmosphere will not last forever in the presence of ultraviolet light from the Sun. The light element hydrogen can escape from the atmosphere, leaving the oxygen behind to combine chemically with surface rock. The loss of water is therefore an irreversible process: once the water is gone, it cannot be restored. There is evidence that this is just what happened to the water once present on Venus.

We don't know if the same runaway greenhouse effect could one day happen on Earth. Although we are uncertain about the point at which a stable greenhouse effect breaks down and turns into a runaway greenhouse effect, Venus stands as clear testament to the fact that a planet cannot continue heating indefinitely without a major change in its oceans and atmosphere. It is a conclusion that we and our descendants will surely want to pay close attention to.

Download for free at http://cnx.org/content/col11992/latest/

10.4 THE GEOLOGY OF MARS

Learning Objectives

By the end of this section, you will be able to:

> Discuss the main missions that have explored Mars
> Explain what we have learned from examination of meteorites from Mars
> Describe the various features found on the surface of Mars
> Compare the volcanoes and canyons on Mars with those of Earth
> Describe the general conditions on the surface of Mars

Mars is more interesting to most people than Venus because it is more hospitable. Even from the distance of Earth, we can see surface features on Mars and follow the seasonal changes in its polar caps (Figure 10.13). Although the surface today is dry and cold, evidence collected by spacecraft suggests that Mars once had blue skies and lakes of liquid water. Even today, it is the sort of place we can imagine astronauts visiting and perhaps even setting up permanent bases.

Figure 10.13. Mars Photographed by the Hubble Space Telescope. This is one of the best photos of Mars taken from our planet, obtained in June 2001 when Mars was only 68 million kilometers away. The resolution is about 20 kilometers—much better than can be obtained with ground-based telescopes but still insufficient to reveal the underlying geology of Mars. (credit: modification of work by NASA and the Hubble Heritage Team (STScI/AURA))

Spacecraft Exploration of Mars

Mars has been intensively investigated by spacecraft. More than 50 spacecraft have been launched toward Mars, but only about half were fully successful. The first visitor was the US Mariner 4, which flew past Mars in 1965 and transmitted 22 photos to Earth. These pictures showed an apparently bleak planet with abundant impact craters. In those days, craters were unexpected; some people who were romantically inclined still hoped to see canals or something like them. In any case, newspaper headlines sadly announced that Mars was a "dead planet."

This OpenStax book is available for free at http://cnx.org/content/col11992/1.8

Download for free at http://cnx.org/content/col11992/latest/

In 1971, NASA's Mariner 9 became the first spacecraft to orbit another planet, mapping the entire surface of Mars at a resolution of about 1 kilometer and discovering a great variety of geological features, including volcanoes, huge canyons, intricate layers on the polar caps, and channels that appeared to have been cut by running water. Geologically, Mars didn't look so dead after all.

The twin Viking spacecraft of the 1970s were among the most ambitious and successful of all planetary missions. Two *orbiters* surveyed the planet and served to relay communications for two *landers* on the surface. After an exciting and sometimes frustrating search for a safe landing spot, the Viking 1 lander touched down on the surface of Chryse Planitia (the Plains of Gold) on July 20, 1976, exactly 7 years after Neil Armstrong's historic first step on the Moon. Two months later, Viking 2 landed with equal success in another plain farther north, called Utopia. The landers photographed the surface with high resolution and carried out complex experiments searching for evidence of life, while the orbiters provided a global perspective on Mars geology.

Mars languished unvisited for two decades after Viking. Two more spacecraft were launched toward Mars, by NASA and the Russian Space Agency, but both failed before reaching the planet.

The situation changed in the 1990s as NASA began a new exploration program using spacecraft that were smaller and less expensive than Viking. The first of the new missions, appropriately called Pathfinder, landed the first wheeled, solar-powered rover on the martian surface on July 4, 1997 (Figure 10.14). An orbiter called *Mars Global Surveyor* (MGS) arrived a few months later and began high-resolution photography of the entire surface over more than one martian year. The most dramatic discovery by this spacecraft, which is still operating, was evidence of gullies apparently cut by surface water, as we will discuss later. These missions were followed in 2003 by the NASA *Mars Odyssey* orbiter, and the ESA *Mars Express* orbiter, both carrying high-resolution cameras. A gamma-ray spectrometer on *Odyssey* discovered a large amount of subsurface hydrogen (probably in the form of frozen water). Subsequent orbiters included the NASA *Mars Reconnaissance Orbiter* to evaluate future landing sites, MAVEN to study the upper atmosphere, and India's *Mangalyaan*, also focused on study of Mars' thin layers of air. Several of these orbiters are also equipped to communicate with landers and rovers on the surface and serve as data relays to Earth.

Figure 10.14. Surface View from Mars Pathfinder. The scene from the Pathfinder lander shows a windswept plain, sculpted long ago when water flowed out of the martian highlands and into the depression where the spacecraft landed. The *Sojourner* rover, the first wheeled vehicle on Mars, is about the size of a microwave oven. Its flat top contains solar cells that provided electricity to run the vehicle. You can see the ramp from the lander and the path the rover took to the larger rock that the mission team nicknamed "Yogi." (credit: NASA/JPL)

In 2003, NASA began a series of highly successful Mars landers. Twin Mars Exploration Rovers (MER), named *Spirit* and *Opportunity*, have been successful far beyond their planned lifetimes. The design goal for the rovers was 600 meters of travel; in fact, they have traveled jointly more than 50 kilometers. After scouting around its

Download for free at http://cnx.org/content/col11992/latest/

rim, *Opportunity* drove down the steep walls into an impact crater called Victoria, then succeeded with some difficulty in climbing back out to resume its route (Figure 10.15). Dust covering the rovers' solar cells caused a drop in power, but when a seasonal dust storm blew away the dust, the rovers resumed full operation. In order to survive winter, the rovers were positioned on slopes to maximize solar heating and power generation. In 2006, *Spirit* lost power on one of its wheels, and subsequently became stuck in the sand, where it continued operation as a fixed ground station. Meanwhile, in 2008, *Phoenix* (a spacecraft "reborn" of spare parts from a previous Mars mission that had failed) landed near the edge of the north polar cap, at latitude 68°, and directly measured water ice in the soil.

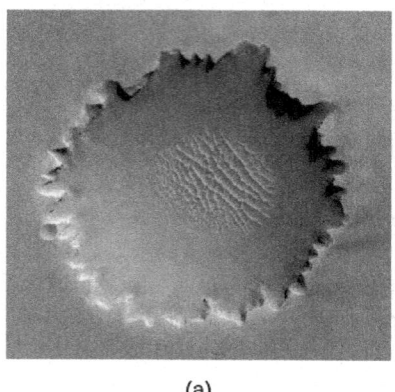
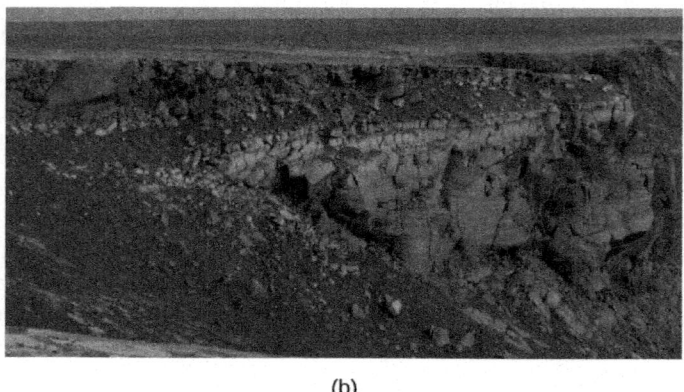

(a)   (b)

Figure 10.15. Victoria Crater. (a) This crater in Meridiani Planum is 800 meters wide, making it slightly smaller than Meteor crater on Earth. Note the dune field in the interior. (b) This image shows the view from the *Opportunity* rover as it scouted the rim of Victoria crater looking for a safe route down into the interior. (credit a: modification of work by NASA/JPL-Caltech/University of Arizona/Cornell/Phio State University; credit b: modification of work by NASA/JPL/Cornell)

In 2011, NASA launched its largest (and most expensive) Mars mission since Viking (see Figure 10.1). The 1-ton rover *Curiosity*, the size of a subcompact car, has plutonium-powered electrical generators, so that it is not dependent on sunlight for power. *Curiosity* made a pinpoint landing on the floor of Gale crater, a site selected for its complex geology and evidence that it had been submerged by water in the past. Previously, Mars landers had been sent to flat terrains with few hazards, as required by their lower targeting accuracy. The scientific goals of *Curiosity* include investigations of climate and geology, and assessment of the habitability of past and present Mars environments. It does not carry a specific life detection instrument, however. So far, scientists have not been able to devise a simple instrument that could distinguish living from nonliving materials on Mars.

LINK TO LEARNING

The *Curiosity* rover required a remarkably complex landing sequence and NASA made a video (https://openstax.org/l/30Curiosityrove) about it called "7 Minutes of Terror" that went viral on the Internet.

A dramatic video summary (https://openstax.org/l/30MarsSurface) of the first two years of *Curiosity*'s exploration of the martian surface can be viewed as well.

This OpenStax book is available for free at http://cnx.org/content/col11992/1.8

Martian Samples

Much of what we know of the Moon, including the circumstances of its origin, comes from studies of lunar samples, but spacecraft have not yet returned martian samples to Earth for laboratory analysis. It is with great interest, therefore, that scientists have discovered that samples of martian material are nevertheless already here on Earth, available for study. These are all members of a rare class of *meteorites* (Figure 10.16)—rocks that have fallen from space.

Figure 10.16. Martian Meteorite. This fragment of basalt, ejected from Mars in a crater-forming impact, eventually arrived on Earth's surface. (credit: NASA)

How would rocks have escaped from Mars? Many impacts have occurred on the red planet, as shown by its heavily cratered surface. Fragments blasted from large impacts can escape from Mars, whose surface gravity is only 38% of Earth's. A long time later (typically a few million years), a very small fraction of these fragments collide with Earth and survive their passage through our atmosphere, just like other meteorites. (We'll discuss meteorites in more detail in the chapter on Cosmic Samples and the Origin of the Solar System.) By the way, rocks from the Moon have also reached our planet as meteorites, although we were able to demonstrate their lunar origin only by comparison with samples returned by the Apollo missions

Most of the martian meteorites are volcanic basalts; most of them are also relatively young—about 1.3 billion years old. We know from details of their composition that they are not from Earth or the Moon. Besides, there was no volcanic activity on the Moon to form them as recently as 1.3 billon years ago. It would be very difficult for ejecta from impacts on Venus to escape through its thick atmosphere. By the process of elimination, the only reasonable origin seems to be Mars, where the Tharsis volcanoes were active at that time.

The martian origin of these meteorites was confirmed by the analysis of tiny gas bubbles trapped inside several of them. These bubbles match the atmospheric properties of Mars as first measured directly by Viking. It appears that some atmospheric gas was trapped in the rock by the shock of the impact that ejected it from Mars and started it on its way toward Earth.

One of the most exciting results from analysis of these martian samples has been the discovery of both water and organic (carbon-based) compounds in them, which suggests that Mars may once have had oceans and perhaps even life on its surface. As we have already hinted, there is other evidence for the presence of flowing water on Mars in the remote past, and even extending to the present.

In this and the following sections, we will summarize the picture of Mars as revealed by all these exploratory missions and by about 40 samples from Mars.

Global Properties of Mars

Mars has a diameter of 6790 kilometers, just over half the diameter of Earth, giving it a total surface area very nearly equal to the continental (land) area of our planet. Its overall density of 3.9 g/cm^3 suggests a composition consisting primarily of silicates but with a small metal core. The planet has no global magnetic field, although

Download for free at http://cnx.org/content/col11992/latest/

there are areas of strong surface magnetization that indicate that there was a global field billions of years ago. Apparently, the red planet has no liquid material in its core today that would conduct electricity.

Thanks to the *Mars Global Surveyor*, we have mapped the entire planet, as shown in Figure 10.17. A laser altimeter on board made millions of separate measurements of the surface topography to a precision of a few meters—good enough to show even the annual deposition and evaporation of the polar caps. Like Earth, the Moon, and Venus, the surface of Mars has continental or highland areas as well as widespread volcanic plains. The total range in elevation from the top of the highest mountain (Olympus Mons) to the bottom of the deepest basin (Hellas) is 31 kilometers.

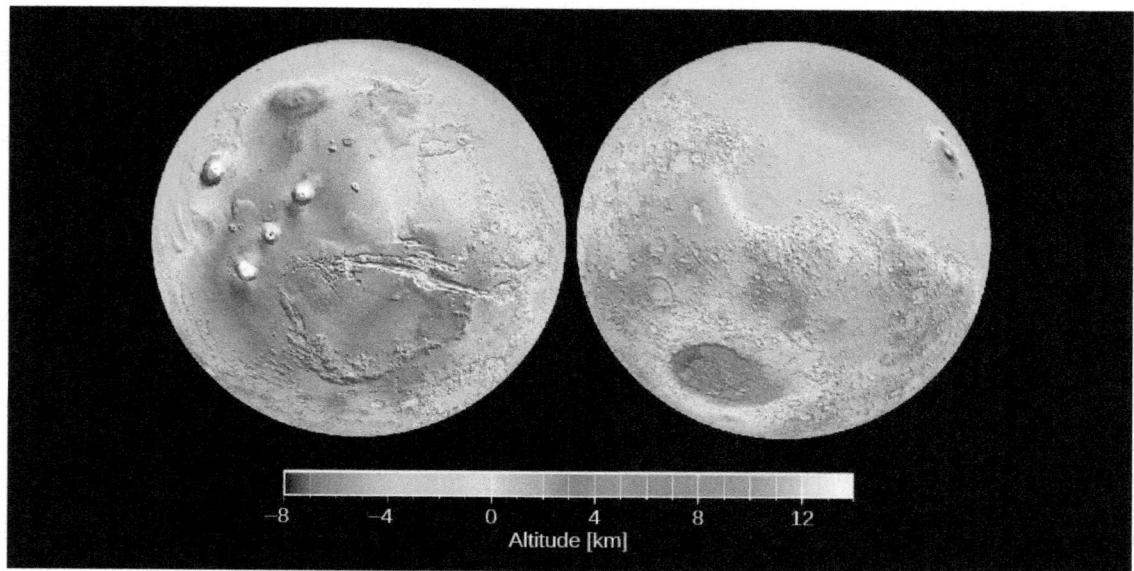

Figure 10.17. Mars Map from Laser Ranging. These globes are highly precise topographic maps, reconstructed from millions of individual elevation measurements made with the *Mars Global Surveyor*. Color is used to indicate elevation. The hemisphere on the left includes the Tharsis bulge and Olympus Mons, the highest mountain on Mars; the hemisphere on the right includes the Hellas basin, which has the lowest elevation on Mars. (credit: modification of work by NASA/JPL)

Approximately half the planet consists of heavily cratered highland terrain, found primarily in the southern hemisphere. The other half, which is mostly in the north, contains younger, lightly cratered volcanic plains at an average elevation about 5 kilometers lower than the highlands. Remember that we saw a similar pattern on Earth, the Moon, and Venus. A geological division into older highlands and younger lowland plains seems to be characteristic of all the terrestrial planets except Mercury.

Lying across the north-south division of Mars is an uplifted continent the size of North America. This is the 10-kilometer-high Tharsis bulge, a volcanic region crowned by four great volcanoes that rise still higher into the martian sky.

Volcanoes on Mars

The lowland plains of Mars look very much like the lunar maria, and they have about the same density of impact craters. Like the lunar maria, they probably formed between 3 and 4 billion years ago. Apparently, Mars experienced extensive volcanic activity at about the same time the Moon did, producing similar basaltic lavas.

The largest volcanic mountains of Mars are found in the Tharsis area (you can see them in Figure 10.17), although smaller volcanoes dot much of the surface. The most dramatic volcano on Mars is Olympus Mons (Mount Olympus), with a diameter larger than 500 kilometers and a summit that towers more than 20

This OpenStax book is available for free at http://cnx.org/content/col11992/1.8

kilometers above the surrounding plains—three times higher than the tallest mountain on Earth (Figure 10.18). The volume of this immense volcano is nearly 100 times greater than that of Mauna Loa in Hawaii. Placed on Earth's surface, Olympus would more than cover the entire state of Missouri.

Figure 10.18. Olympus Mons. The largest volcano on Mars, and probably the largest in the solar system, is Olympus Mons, illustrated in this computer-generated rendering based on data from the *Mars Global Surveyor's* laser altimeter. Placed on Earth, the base of Olympus Mons would completely cover the state of Missouri; the caldera, the circular opening at the top, is 65 kilometers across, about the size of Los Angeles. (credit: NASA/Corbis)

Images taken from orbit allow scientists to search for impact craters on the slopes of these volcanoes in order to estimate their age. Many of the volcanoes show a fair number of such craters, suggesting that they ceased activity a billion years or more ago. However, Olympus Mons has very, very few impact craters. Its present surface cannot be more than about 100 million years old; it may even be much younger. Some of the fresh-looking lava flows might have been formed a hundred years ago, or a thousand, or a million, but geologically speaking, they are quite young. This leads geologists to the conclusion that Olympus Mons possibly remains intermittently active today—something future Mars land developers may want to keep in mind.

Martian Cracks and Canyons

The Tharsis bulge has many interesting geological features in addition to its huge volcanoes. In this part of the planet, the surface itself has bulged upward, forced by great pressures from below, resulting in extensive tectonic cracking of the crust. Among the most spectacular tectonic features on Mars are the canyons called the Valles Marineris (or Mariner Valleys, named after Mariner 9, which first revealed them to us), which are shown in Figure 10.19. They extend for about 5000 kilometers (nearly a quarter of the way around Mars) along the slopes of the Tharsis bulge. If it were on Earth, this canyon system would stretch all the way from Los Angeles to Washington, DC. The main canyon is about 7 kilometers deep and up to 100 kilometers wide, large enough for the Grand Canyon of the Colorado River to fit comfortably into one of its side canyons.

Download for free at http://cnx.org/content/col11992/latest/

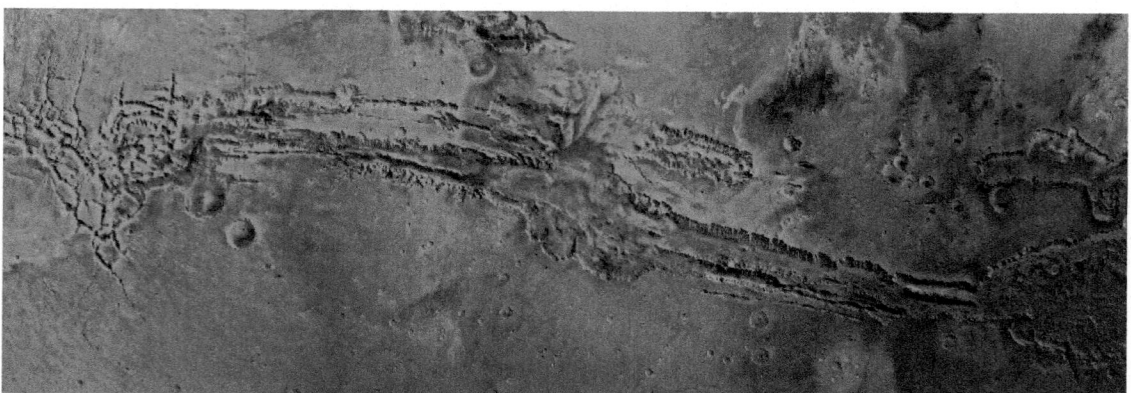

Figure 10.19. Heavily Eroded Canyonlands on Mars. This image shows the Valles Marineris canyon complex, which is 3000 kilometers wide and 8 kilometers deep. (credit: NASA/JPL/USGS)

LINK TO LEARNING

An excellent 4-minute video tour (https://openstax.org/l/30VallesMariner) of Valles Marineris, narrated by planetary scientist Phil Christensen, is available for viewing.

The term "canyon" is somewhat misleading here because the Valles Marineris canyons have no outlets and were not cut by running water. They are basically tectonic cracks, produced by the same crustal tensions that caused the Tharsis uplift. However, water has played a later role in shaping the canyons, primarily by seeping from deep springs and undercutting the cliffs. This undercutting led to landslides that gradually widened the original cracks into the great valleys we see today (Figure 10.20). Today, the primary form of erosion in the canyons is probably wind.

This OpenStax book is available for free at http://cnx.org/content/col11992/1.8

Figure 10.20. Martian Landslides. This Viking orbiter image shows Ophir Chasma, one of the connected valleys of the Valles Marineris canyon system. Look carefully and you can see enormous landslides whose debris is piled up underneath the cliff wall, which tower up to 10 kilometers above the canyon floor. (credit: modification of work by NASA/JPL/USGS)

While the Tharsis bulge and Valles Marineris are impressive, in general, we see fewer tectonic structures on Mars than on Venus. In part, this may reflect a lower general level of geological activity, as would be expected for a smaller planet. But it is also possible that evidence of widespread faulting has been buried by wind-deposited sediment over much of Mars. Like Earth, Mars may have hidden part of its geological history under a cloak of soil.

The View on the Martian Surface

The first spacecraft to land successfully on Mars were Vikings 1 and 2 and Mars Pathfinder. All sent back photos that showed a desolate but strangely beautiful landscape, including numerous angular rocks interspersed with dune like deposits of fine-grained, reddish soil (Figure 10.21).

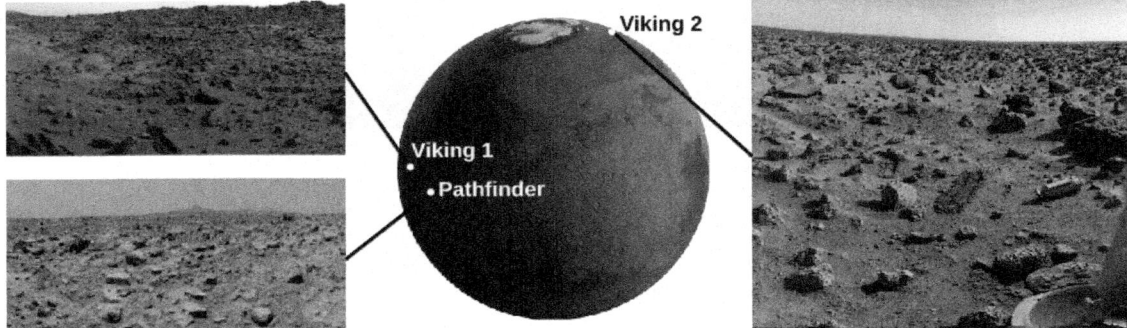

Figure 10.21. Three Martian Landing Sites. The Mars landers Viking 1 in Chryse, Pathfinder in Ares Valley, and Viking 2 in Utopia, all photographed their immediate surroundings. It is apparent from the similarity of these three photos that each spacecraft touched down on a flat, windswept plain littered with rocks ranging from tiny pebbles up to meter-size boulders. It is probable that most of Mars looks like this on the surface. (credit "Viking 1": modification of work by Van der Hoorn/NASA; credit "Pathfinder": modification of work by NASA; credit "Viking 2": modification of work by NASA; credit Mars: modification of work by NASA/Goddard Space Flight Center)

All three of these landers were targeted to relatively flat, lowland terrain. Instruments on the landers found that the soil consisted of clays and iron oxides, as had long been expected from the red color of the planet. All

Download for free at http://cnx.org/content/col11992/latest/

the rocks measured appeared to be of volcanic origin and roughly the same composition. Later landers were targeted to touch down in areas that apparently were flooded sometime in the past, where sedimentary rock layers, formed in the presence of water, are common. (Although we should note that nearly all the planet is blanketed in at least a thin layer of wind-blown dust).

The Viking landers included weather stations that operated for several years, providing a perspective on martian weather. The temperatures they measured varied greatly with the seasons, due to the absence of moderating oceans and clouds. Typically, the summer maximum at Viking 1 was 240 K (–33 °C), dropping to 190 K (–83 °C) at the same location just before dawn. The lowest air temperatures, measured farther north by Viking 2, were about 173 K (–100 °C). During the winter, Viking 2 also photographed water frost deposits on the ground (Figure 10.22). We make a point of saying "water frost" here because at some locations on Mars, it gets cold enough for carbon dioxide (dry ice) to freeze out of the atmosphere as well.

Figure 10.22. Water Frost in Utopia. This image of surface frost was photographed at the Viking 2 landing site during late winter. (credit: NASA/JPL)

Most of the winds measured on Mars are only a few kilometers per hour. However, Mars is capable of great windstorms that can shroud the entire planet with windblown dust. Such high winds can strip the surface of some of its loose, fine dust, leaving the rock exposed. The later rovers found that each sunny afternoon the atmosphere became turbulent as heat rose off the surface. This turbulence generated dust devils, which play an important role in lifting the fine dust into the atmosphere. As the dust devils strip off the top layer of light dust and expose darker material underneath, they can produce fantastic patterns on the ground (Figure 10.23).

Wind on Mars plays an important role in redistributing surface material. Figure 10.23 shows a beautiful area of dark sand dunes on top of lighter material. Much of the material stripped out of the martian canyons has been dumped in extensive dune fields like this, mostly at high latitudes.

This OpenStax book is available for free at http://cnx.org/content/col11992/1.8

Download for free at http://cnx.org/content/col11992/latest/

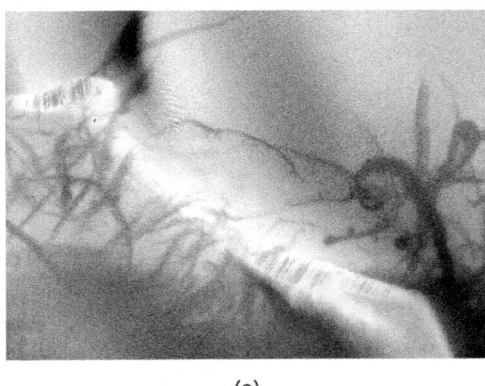
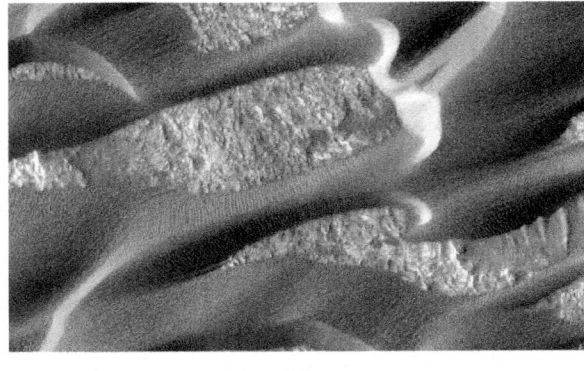

(a)   (b)

Figure 10.23. Dust Devil Tracks and Sand Dunes. (a) This high-resolution photo from the *Mars Global Surveyor* shows the dark tracks of several dust devils that have stripped away a thin coating of light-colored dust. This view is of an area about 3 kilometers across. Dust devils are one of the most important ways that dust gets redistributed by the martian winds. They may also help keep the solar panels of our rovers free of dust. (b) These windblown sand dunes on Mars overlay a lighter sandy surface. Each dune in this high-resolution view is about 1 kilometer across. (credit a: modification of work by NASA/JPL/University of Arizona; credit b: modification of work by NASA/JPL-Caltech/University of Arizona)

10.5 WATER AND LIFE ON MARS

Learning Objectives

By the end of this section, you will be able to:

> Describe the general composition of the atmosphere on Mars
> Explain what we know about the polar ice caps on Mars and how we know it
> Describe the evidence for the presence of water in the past history of Mars
> Summarize the evidence for and against the possibility of life on Mars

Of all the planets and moons in the solar system, Mars seems to be the most promising place to look for life, both fossil microbes and (we hope) some forms of life deeper underground that still survive today. But where (and how) should we look for life? We know that the one requirement shared by all life on Earth is liquid water. Therefore, the guiding principle in assessing habitability on Mars and elsewhere has been to "follow the water." That is the perspective we take in this section, to follow the water on the red planet and hope it will lead us to life.

Atmosphere and Clouds on Mars

The atmosphere of Mars today has an average surface pressure of only 0.007 bar, less than 1% that of Earth. (This is how thin the air is about 30 kilometers above Earth's surface.) Martian air is composed primarily of carbon dioxide (95%), with about 3% nitrogen and 2% argon. The proportions of different gases are similar to those in the atmosphere of Venus (see Table 10.2), but a lot less of each gas is found in the thin air on Mars.

While winds on Mars can reach high speeds, they exert much less force than wind of the same velocity would on Earth because the atmosphere is so thin. The wind is able, however, to loft very fine dust particles, which can sometimes develop planet-wide dust storms. It is this fine dust that coats almost all the surface, giving Mars its distinctive red color. In the absence of surface water, wind erosion plays a major role in sculpting the martian surface (Figure 10.24).

Download for free at http://cnx.org/content/col11992/latest/

Figure 10.24. Wind Erosion on Mars. These long straight ridges, called yardangs, are aligned with the dominant wind direction. This is a high-resolution image from the *Mars Reconnaissance Orbiter* and is about 1 kilometer wide. (credit: NASA/JPL-Caltech/University of Arizona)

LINK TO LEARNING

The issue of how strong the winds on Mars can be plays a big role in the 2015 hit movie The Martian (https://openstax.org/l/30TheMartian) in which the main character is stranded on Mars after being buried in the sand in a windstorm so great that his fellow astronauts have to leave the planet so their ship is not damaged. Astronomers have noted that the martian winds could not possibly be as forceful as depicted in the film. In most ways, however, the depiction of Mars in this movie is remarkably accurate.

Although the atmosphere contains small amounts of water vapor and occasional clouds of water ice, liquid water is not stable under present conditions on Mars. Part of the problem is the low temperatures on the planet. But even if the temperature on a sunny summer day rises above the freezing point, the low pressure means that liquid water still cannot exist on the surface, except at the lowest elevations. At a pressure of less than 0.006 bar, the boiling point is as low or lower than the freezing point, and water changes directly from solid to vapor without an intermediate liquid state (as does "dry ice," carbon dioxide, on Earth). However, salts dissolved in water lower its freezing point, as we know from the way salt is used to thaw roads after snow and ice forms during winter on Earth. Salty water is therefore sometimes able to exist in liquid form on the martian surface, under the right conditions.

Several types of clouds can form in the martian atmosphere. First there are dust clouds, discussed above. Second are water-ice clouds similar to those on Earth. These often form around mountains, just as happens on our planet. Finally, the CO_2 of the atmosphere can itself condense at high altitudes to form hazes of dry ice crystals. The CO_2 clouds have no counterpart on Earth, since on our planet temperatures never drop low enough (down to about 150 K or about 125 °C) for this gas to condense.

The Polar Caps

Through a telescope, the most prominent surface features on Mars are the bright polar caps, which change with the seasons, similar to the seasonal snow cover on Earth. We do not usually think of the winter snow in northern latitudes as a part of our polar caps, but seen from space, the thin winter snow merges with Earth's thick, permanent ice caps to create an impression much like that seen on Mars (Figure 10.25).

This OpenStax book is available for free at http://cnx.org/content/col11992/1.8

Download for free at http://cnx.org/content/col11992/latest/

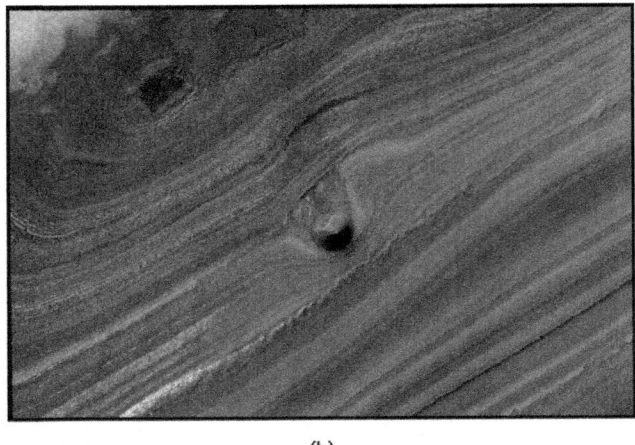

(a)   (b)

Figure 10.25. Martian North Polar Cap. (a) This is a composite image of the north pole in summer, obtained in October 2006 by the *Mars Reconnaissance Orbiter*. It shows the mostly water-ice residual cap sitting atop light, tan-colored, layered sediments. Note that although the border of this photo is circular, it shows only a small part of the planet. (b) Here we see a small section of the layered terrain near the martian north pole. There is a mound about 40 meters high that is sticking out of a trough in the center of the picture. (credit a: modification of work by NASA/JPL/MSSS; credit b: modification of work by NASA/JPL-Caltech/University of Arizona)

The *seasonal caps* on Mars are composed not of ordinary snow but of frozen CO_2 (dry ice). These deposits condense directly from the atmosphere when the surface temperature drops below about 150 K. The caps develop during the cold martian winters and extend down to about 50° latitude by the start of spring.

Quite distinct from these thin seasonal caps of CO_2 are the *permanent* or *residual caps* that are always present near the poles. The southern permanent cap has a diameter of 350 kilometers and is composed of frozen CO_2 deposits together with a great deal of water ice. Throughout the southern summer, it remains at the freezing point of CO_2, 150 K, and this cold reservoir is thick enough to survive the summer heat intact.

The northern permanent cap is different. It is much larger, never shrinking to a diameter less than 1000 kilometers, and is composed of water ice. Summer temperatures in the north are too high for the frozen CO_2 to be retained. Measurements from the *Mars Global Surveyor* have established the exact elevations in the north polar region of Mars, showing that it is a large basin about the size of our own Arctic Ocean basin. The ice cap itself is about 3 kilometers thick, with a total volume of about 10 million km^3 (similar to that of Earth's Mediterranean Sea). If Mars ever had extensive liquid water, this north polar basin would have contained a shallow sea. There is some indication of ancient shorelines visible, but better images will be required to verify this suggestion.

Images taken from orbit also show a distinctive type of terrain surrounding the permanent polar caps, as shown in Figure 10.25. At latitudes above 80° in both hemispheres, the surface consists of recent layered deposits that cover the older cratered ground below. Individual layers are typically ten to a few tens of meters thick, marked by alternating light and dark bands of sediment. Probably the material in the polar deposits includes dust carried by wind from the equatorial regions of Mars.

What do these terraced layers tell us about Mars? Some cyclic process is depositing dust and ice over periods of time. The time scales represented by the polar layers are tens of thousands of years. Apparently the martian climate experiences periodic changes at intervals similar to those between ice ages on Earth. Calculations indicate that the causes are probably also similar: the gravitational pull of the other planets produces variations in Mars' orbit and tilt as the great clockwork of the solar system goes through its paces.

Download for free at http://cnx.org/content/col11992/latest/

The *Phoenix* spacecraft landed near the north polar cap in summer (Figure 10.26). Controllers knew that is would not be able to survive a polar winter, but directly measuring the characteristics of the polar region was deemed important enough to send a dedicated mission. The most exciting discovery came when the spacecraft tried to dig a shallow trench under the spacecraft. When the overlying dust was stripped off, they saw bright white material, apparently some kind of ice. From the way this ice sublimated over the next few days, it was clear that it was frozen water.

1.7 cm

Figure 10.26. Evaporating Ice on Mars. We see a trench dug by the *Phoenix* lander in the north polar region four martian days apart in June 2008. If you look at the shadowed region in the bottom left of the trench, you can see three spots of ice in the left image which have sublimated away in the right image. (credit: modification of work by NASA/JPL-Caltech/University of Arizona/Texas A&M University)

EXAMPLE 10.1

Comparing the Amount of Water on Mars and Earth

It is interesting to estimate the amount of water (in the form of ice) on Mars and to compare this with the amount of water on Earth. In each case, we can find the total volume of a layer on a sphere by multiplying the area of the sphere ($4\pi R^2$) by the thickness of the layer. For Earth, the ocean water is equivalent to a layer 3 km thick spread over the entire planet, and the radius of Earth is 6.378×10^6 m (see Appendix F). For Mars, most of the water we are sure of is in the form of ice near the poles. We can calculate the amount of ice in one of the residual polar caps if it is (for example) 2 km thick and has a radius of 400 km (the area of a circle is πR^2).

Solution

The volume of Earth's water is therefore the area $4\pi R^2$

$$4\pi \left(6.378 \times 10^6 \, \text{m}\right)^2 = 5.1 \times 10^{14} \, \text{m}^2$$

This OpenStax book is available for free at http://cnx.org/content/col11992/1.8

Download for free at http://cnx.org/content/col11992/latest/

multiplied by the thickness of 3000 m:

$$5.1 \times 10^{14} \text{ m}^2 \times 3000 \text{ m} = 1.5 \times 10^{18} \text{ m}^3$$

This gives 1.5×10^{18} m^3 of water. Since water has a density of 1 ton per cubic meter (1000 kg/m^3), we can calculate the mass:

$$1.5 \times 10^{18} \text{ m}^3 \times 1 \text{ ton/m}^3 = 1.5 \times 10^{18} \text{ tons}$$

For Mars, the ice doesn't cover the whole planet, only the caps; the polar cap area is

$$\pi R^2 = \pi (4 \times 10^5 \text{ m})^2 = 5 \times 10^{11} \text{ m}^2$$

(Note that we converted kilometers to meters.)

The volume = area × height, so we have:

$$(2 \times 10^3 \text{ m})(5 \times 10^{11} \text{ m}^2) = 1 \times 10^{15} \text{ m}^3 = 10^{15} \text{ m}^3$$

Therefore, the mass is:

$$10^{15} \text{ m}^3 \times 1 \text{ ton/m}^3 = 10^{15} \text{ tons}$$

This is about 0.1% that of Earth's oceans.

Check Your Learning

A better comparison might be to compare the amount of ice in the Mars polar ice caps to the amount of ice in the Greenland ice sheet on Earth, which has been estimated as 2.85×10^{15} m^3. How does this compare with the ice on Mars?

Answer:

The Greenland ice sheet has about 2.85 times as much ice as in the polar ice caps on Mars. They are about the same to the nearest power of 10.

Channels and Gullies on Mars

Although no bodies of liquid water exist on Mars today, evidence has accumulated that rivers flowed on the red planet long ago. Two kinds of geological features appear to be remnants of ancient watercourses, while a third class—smaller gullies—suggests intermittent outbreaks of liquid water even today. We will examine each of these features in turn.

In the highland equatorial plains, there are multitudes of small, sinuous (twisting) channels—typically a few meters deep, some tens of meters wide, and perhaps 10 or 20 kilometers long (Figure 10.27). They are called runoff channels because they look like what geologists would expect from the surface runoff of ancient rain storms. These runoff channels seem to be telling us that the planet had a very different climate long ago. To estimate the age of these channels, we look at the cratering record. Crater counts show that this part of the planet is more cratered than the lunar maria but less cratered than the lunar highlands. Thus, the runoff channels are probably older than the lunar maria, presumably about 4 billion years old.

The second set of water-related features we see are *outflow channels* (Figure 10.27) are much larger than the runoff channels. The largest of these, which drain into the Chryse basin where Pathfinder landed, are 10 kilometers or more wide and hundreds of kilometers long. Many features of these outflow channels have

Download for free at http://cnx.org/content/col11992/latest/

convinced geologists that they were carved by huge volumes of running water, far too great to be produced by ordinary rainfall. Where could such floodwater have come from on Mars?

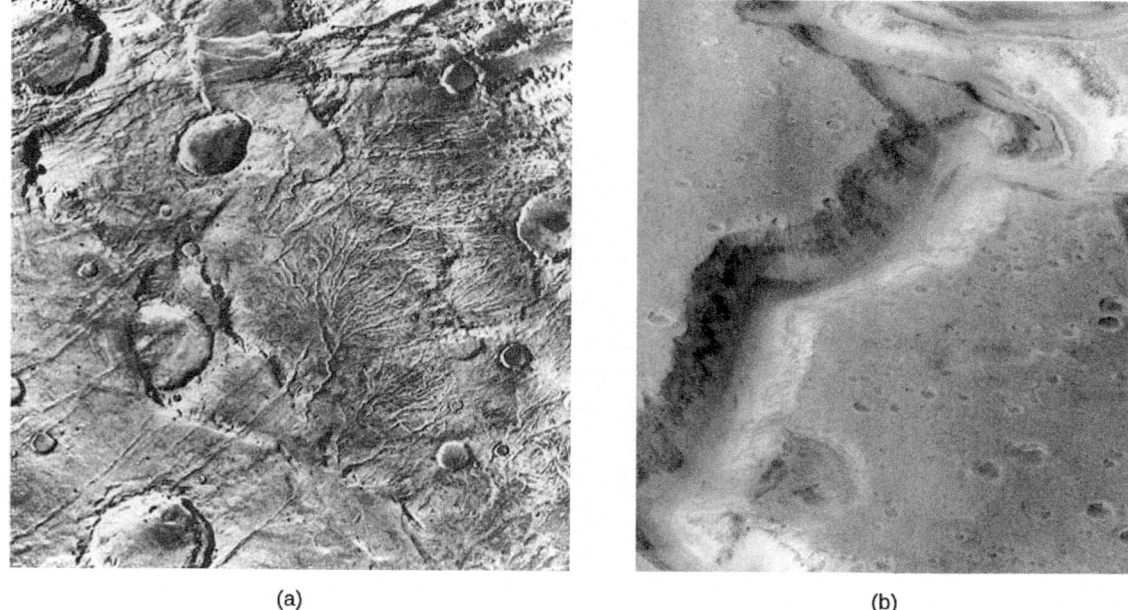

(a)   (b)

Figure 10.27. Runoff and Outflow Channels. (a) These runoff channels in the old martian highlands are interpreted as the valleys of ancient rivers fed by either rain or underground springs. The width of this image is about 200 kilometers. (b) This intriguing channel, called Nanedi Valles, resembles Earth riverbeds in some (but not all) ways. The tight curves and terraces seen in the channel certainly suggest the sustained flow of a fluid like water. The channel is about 2.5 kilometers across. (credit a: modification of work by Jim Secosky/NASA; credit b: modification of work by Jim Secosky/NASA)

As far we can tell, the regions where the outflow channels originate contained abundant water frozen in the soil as permafrost. Some local source of heating must have released this water, leading to a period of rapid and catastrophic flooding. Perhaps this heating was associated with the formation of the volcanic plains on Mars, which date back to roughly the same time as the outflow channels.

Note that neither the runoff channels nor the outflow channels are wide enough to be visible from Earth, nor do they follow straight lines. They could not have been the "canals" Percival Lowell imagined seeing on the red planet.

The third type of water feature, the smaller *gullies*, was discovered by the *Mars Global Surveyor* (Figure 10.28). The *Mars Global Surveyor's* camera images achieved a resolution of a few meters, good enough to see something as small as a truck or bus on the surface. On the steep walls of valleys and craters at high latitudes, there are many erosional features that look like gullies carved by flowing water. These gullies are very young: not only are there no superimposed impact craters, but in some instances, the gullies seem to cut across recent wind-deposited dunes. Perhaps there is liquid water underground that can occasionally break out to produce short-lived surface flows before the water can freeze or evaporate.

This OpenStax book is available for free at http://cnx.org/content/col11992/1.8

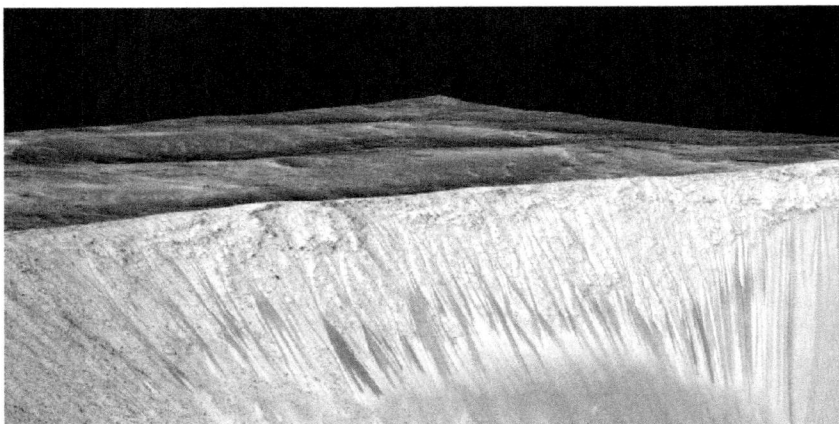

Figure 10.28. Gullies on the Wall of Garni Crater. This high-resolution image is from the *Mars Reconnaissance Orbiter*. The dark streaks, which are each several hundred meters long, change in a seasonal pattern that suggests they are caused by the temporary flow of surface water. (credit: NASA/JPL-Caltech/University of Arizona)

The gullies also have the remarkable property of changing regularly with the martian seasons. Many of the dark streaks (visible in Figure 10.28) elongate within a period of a few days, indicating that something is flowing downhill—either water or dark sediment. If it is water, it requires a continuing source, either from the atmosphere or from springs that tap underground water layers (aquifers.) Underground water would be the most exciting possibility, but this explanation seems inconsistent with the fact that many of the dark streaks start at high elevations on the walls of craters.

Additional evidence that the dark streaks (called by the scientists *recurring slope lineae*) are caused by water was found in 2015 when spectra were obtained of the dark streaks (Figure 10.29). These showed the presence of hydrated salts produced by the evaporation of salty water. If the water is salty, it could remain liquid long enough to flow downstream for distances of a hundred meters or more, before it either evaporates or soaks into the ground. However, this discovery still does not identify the ultimate source of the water.

Download for free at http://cnx.org/content/col11992/latest/

Figure 10.29. Evidence for Liquid Water on Mars. The dark streaks in Horowitz crater, which move downslope, have been called recurring slope lineae. The streaks in the center of the image go down the wall of the crater for about a distance of 100 meters. Spectra taken of this region indicate that these are locations where salty liquid water flows on or just below the surface of Mars. (The vertical dimension is exaggerated by a factor of 1.5 compared to horizontal dimensions.) (credit: NASA/JPL-Caltech/University of Arizona)

Ancient Lakes

The rovers (*Spirit*, *Opportunity*, and *Curiosity*) that have operated on the surface of Mars have been used to hunt for additional evidence of water. They could not reach the most interesting sites, such as the gullies, which are located on steep slopes. Instead, they explored sites that might be dried-out lake beds, dating back to a time when the climate on Mars was warmer and the atmosphere thicker—allowing water to be liquid on the surface.

Spirit was specifically targeted to explore what looked like an ancient lake-bed in Gusev crater, with an outflow channel emptying into it. However, when the spacecraft landed, it found that the former lakebed had been covered by thin lava flows, blocking the rover from access to the sedimentary rocks it had hoped to find. However, *Opportunity* had better luck. Peering at the walls of a small crater, it detected layered sedimentary rock. These rocks contained chemical evidence of evaporation, suggesting there had been a shallow salty lake in that location. In these sedimentary rocks were also small spheres that were rich in the mineral hematite, which forms only in watery environments. Apparently this very large basin had once been underwater.

LINK TO LEARNING

The small spherical rocks were nicknamed "blueberries" by the science team and the discovery of a whole "berry-bowl" of them was announced in this interesting news release (https://openstax.org/l/30berrybowl) from NASA.

The *Curiosity* rover landed inside Gale crater, where photos taken from orbit also suggested past water erosion. It discovered numerous sedimentary rocks, some in the form of mudstones from an ancient lakebed; it also

This OpenStax book is available for free at http://cnx.org/content/col11992/1.8

found indications of rocks formed by the action of shallow water at the time the sediment formed (Figure 10.30).

(a)

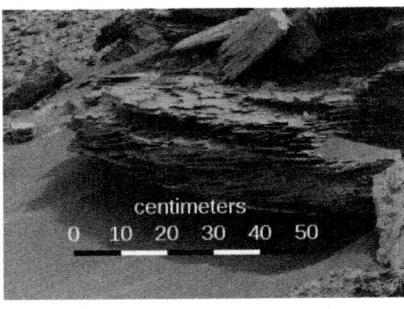
(b)

Figure 10.30. Gale Crater. (a) This scene, photographed by the *Curiosity* rover, shows an ancient lakebed of cracked mudstones. (b) Geologists working with the *Curiosity* rover interpret this image of cross-bedded sandstone in Gale crater as evidence of liquid water passing over a loose bed of sediment at the time this rock formed. (credit a: modification of work by NASA/JPL-Caltech/MSSS; credit b: modification of work by NASA/JPL-Caltech/MSSS)

MAKING CONNECTIONS

Astronomy and Pseudoscience: The "Face on Mars"

People like human faces. We humans have developed great skill in recognizing people and interpreting facial expressions. We also have a tendency to see faces in many natural formations, from clouds to the man in the Moon. One of the curiosities that emerged from the Viking orbiters' global mapping of Mars was the discovery of a strangely shaped mesa in the Cydonia region that resembled a human face. Despite later rumors of a cover-up, the "Face on Mars" was, in fact, recognized by Viking scientists and included in one of the early mission press releases. At the low resolution and oblique lighting under which the Viking image was obtained, the mile-wide mesa had something of a Sphinx-like appearance.

Unfortunately, a small band of individuals decided that this formation was an artificial, carved sculpture of a human face placed on Mars by an ancient civilization that thrived there hundreds of thousands of years ago. A band of "true believers" grew around the face and tried to deduce the nature of the "sculptors" who made it. This group also linked the face to a variety of other pseudoscientific phenomena such as crop circles (patterns in fields of grain, mostly in Britain, now known to be the work of pranksters).

Members of this group accused NASA of covering up evidence of intelligent life on Mars, and they received a great deal of help in publicizing their perspective from tabloid media. Some of the believers picketed the Jet Propulsion Laboratory at the time of the failure of the *Mars Observer* spacecraft, circulating stories that the "failure" of the *Mars Observer* was itself a fake, and that its true (secret) mission was to photograph the face.

The high-resolution *Mars Observer* camera (MOC) was reflown on the *Mars Global Surveyor* mission, which arrived at Mars in 1997. On April 5, 1998, in Orbit 220, the MOC obtained an oblique image of the face at a resolution of 4 meters per pixel, a factor-of-10 improvement in resolution over the Viking image. Another image in 2001 had even higher resolution. Immediately released by NASA, the new images showed a low mesa-like hill cut crossways by several roughly linear ridges and depressions, which were misidentified in

Download for free at http://cnx.org/content/col11992/latest/

the 1976 photo as the eyes and mouth of a face. Only with an enormous dose of imagination can any resemblance to a face be seen in the new images, demonstrating how dramatically our interpretation of geology can change with large improvements in resolution. The original and the higher resolution images can be seen in Figure 10.31.

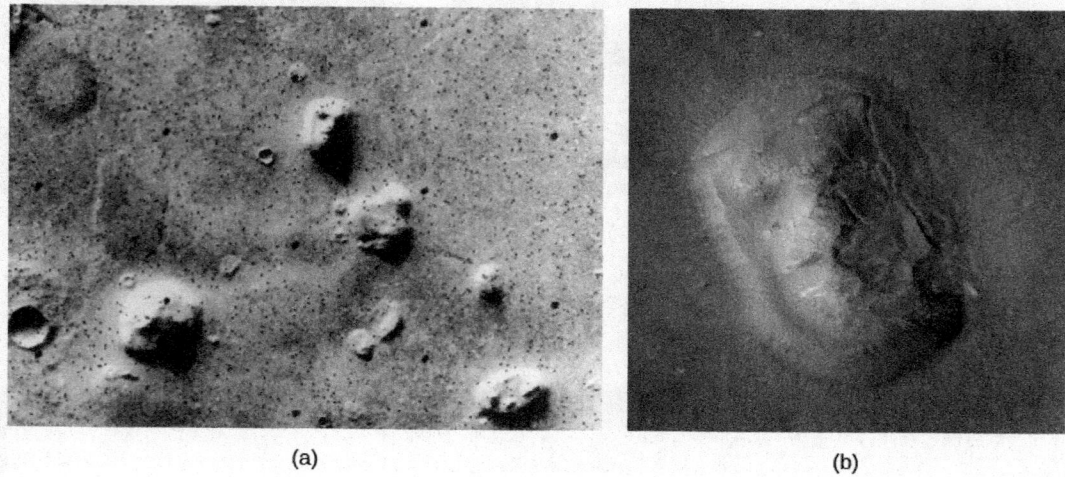

(a)   (b)

Figure 10.31. Face on Mars. The so-called "Face on Mars" is seen (a) in low resolution from Viking (the "face" is in the upper part of the picture) and (b) with 20 times better resolution from the *Mars Global Surveyor*. (credit a: modification of work NASA/JPL; credit b: modification of work by NASA/JPL/MSSS)

After 20 years of promoting pseudoscientific interpretations and various conspiracy theories, can the "Face on Mars" believers now accept reality? Unfortunately, it does not seem so. They have accused NASA of faking the new picture. They also suggest that the secret mission of the *Mars Observer* included a nuclear bomb used to destroy the face before it could be photographed in greater detail by the *Mars Global Surveyor*.

Space scientists find these suggestions incredible. NASA is spending increasing sums for research on life in the universe, and a major objective of current and upcoming Mars missions is to search for evidence of past microbial life on Mars. Conclusive evidence of extraterrestrial life would be one of the great discoveries of science and incidentally might well lead to increased funding for NASA. The idea that NASA or other government agencies would (or could) mount a conspiracy to suppress such welcome evidence is truly bizarre.

Alas, the "Face on Mars" story is only one example of a whole series of conspiracy theories that are kept before the public by dedicated believers, by people out to make a fast buck, and by irresponsible media attention. Others include the "urban legend" that the Air Force has the bodies of extraterrestrials at a secret base, the widely circulated report that UFOs crashed near Roswell, New Mexico (actually it was a balloon carrying scientific instruments to find evidence of Soviet nuclear tests), or the notion that alien astronauts helped build the Egyptian pyramids and many other ancient monuments because our ancestors were too stupid to do it alone.

In response to the increase in publicity given to these "fiction science" ideas, a group of scientists, educators, scholars, and magicians (who know a good hoax when they see one) have formed the Committee for Skeptical Inquiry. Two of the original authors of your book are active on the committee.

This OpenStax book is available for free at http://cnx.org/content/col11992/1.8

For more information about its work delving into the rational explanations for paranormal claims, see their excellent magazine, *The Skeptical Inquirer*, or check out their website at www.csicop.org/.

Climate Change on Mars

The evidence about ancient rivers and lakes of water on Mars discussed so far suggests that, billions of years ago, martian temperatures must have been warmer and the atmosphere must have been more substantial than it is today. But what could have changed the climate on Mars so dramatically?

We presume that, like Earth and Venus, Mars probably formed with a higher surface temperature thanks to the greenhouse effect. But Mars is a smaller planet, and its lower gravity means that atmospheric gases could escape more easily than from Earth and Venus. As more and more of the atmosphere escaped into space, the temperature on the surface gradually fell.

Eventually Mars became so cold that most of the water froze out of the atmosphere, further reducing its ability to retain heat. The planet experienced a sort of *runaway refrigerator effect*, just the opposite of the runaway greenhouse effect that occurred on Venus. Probably, this loss of atmosphere took place within less than a billion years after Mars formed. The result is the cold, dry Mars we see today.

Conditions a few meters below the martian surface, however, may be much different. There, liquid water (especially salty water) might persist, kept warm by the internal heat of Mars or the insulating layers solid and rock. Even on the surface, there may be ways to change the martian atmosphere temporarily.

Mars is likely to experience long-term climate cycles, which may be caused by the changing orbit and tilt of the planet. At times, one or both of the polar caps might melt, releasing a great deal of water vapor into the atmosphere. Perhaps an occasional impact by a comet might produce a temporary atmosphere that is thick enough to permit liquid water on the surface for a few weeks or months. Some have even suggested that future technology might allow us to *terraform* Mars—that is, to engineer its atmosphere and climate in ways that might make the planet more hospitable for long-term human habitation.

The Search for Life on Mars

If there was running water on Mars in the past, perhaps there was life as well. Could life, in some form, remain in the martian soil today? Testing this possibility, however unlikely, was one of the primary objectives of the Viking landers in 1976. These landers carried miniature biological laboratories to test for microorganisms in the martian soil. Martian soil was scooped up by the spacecraft's long arm and placed into the experimental chambers, where it was isolated and incubated in contact with a variety of gases, radioactive isotopes, and nutrients to see what would happen. The experiments looked for evidence of *respiration* by living animals, *absorption* of *nutrients* offered to organisms that might be present, and an *exchange of gases* between the soil and its surroundings for any reason whatsoever. A fourth instrument pulverized the soil and analyzed it carefully to determine what organic (carbon-bearing) material it contained.

The Viking experiments were so sensitive that, had one of the spacecraft landed anywhere on Earth (with the possible exception of Antarctica), it would easily have detected life. But, to the disappointment of many scientists and members of the public, no life was detected on Mars. The soil tests for absorption of nutrients and gas exchange did show some activity, but this was most likely caused by chemical reactions that began as water was added to the soil and had nothing to do with life. In fact, these experiments showed that martian soil seems much more chemically active than terrestrial soils because of its exposure to solar ultraviolet radiation (since Mars has no ozone layer).

Download for free at http://cnx.org/content/col11992/latest/

The organic chemistry experiment showed no trace of organic material, which is apparently destroyed on the martian surface by the sterilizing effect of this ultraviolet light. While the possibility of life on the surface has not been eliminated, most experts consider it negligible. Although Mars has the most earthlike environment of any planet in the solar system, the sad fact is that nobody seems to be home today, at least on the surface.

However, there is no reason to think that life could not have begun on Mars about 4 billion years ago, at the same time it started on Earth. The two planets had very similar surface conditions then. Thus, the attention of scientists has shifted to the search for *fossil* life on Mars. One of the primary questions to be addressed by future spacecraft is whether Mars once supported its own life forms and, if so, how this martian life compared with that on our own planet. Future missions will include the return of martian samples selected from sedimentary rocks at sites that once held water and thus perhaps ancient life. The most powerful searches for martian life (past or present) will thus be carried out in our laboratories here on Earth.

MAKING CONNECTIONS

Planetary Protection

When scientists begin to search for life on another planet, they must make sure that we do not contaminate the other world with life carried from Earth. At the very beginning of spacecraft exploration on Mars, an international agreement specified that all landers were to be carefully sterilized to avoid accidentally transplanting terrestrial microbes to Mars. In the case of Viking, we know the sterilization was successful. Viking's failure to detect martian organisms also implies that these experiments did not detect hitchhiking terrestrial microbes.

As we have learned more about the harsh conditions on the martian surface, the sterilization requirements have been somewhat relaxed. It is evident that no terrestrial microbes could grow on the martian surface, with its low temperature, absence of water, and intense ultraviolet radiation. Microbes from Earth might survive in a dormant, dried state, but they cannot grow and proliferate on Mars.

The problem of contaminating Mars will become more serious, however, as we begin to search for life below the surface, where temperatures are higher and no ultraviolet light penetrates. The situation will be even more daunting if we consider human flights to Mars. Any humans will carry with them a multitude of terrestrial microbes of all kinds, and it is hard to imagine how we can effectively keep the two biospheres isolated from each other if Mars has indigenous life. Perhaps the best situation could be one in which the two life-forms are so different that each is effectively invisible to the other—not recognized on a chemical level as living or as potential food.

The most immediate issue of public concern is not with the contamination of Mars but with any dangers associated with returning Mars samples to Earth. NASA is committed to the complete biological isolation of returned samples until they are demonstrated to be safe. Even though the chances of contamination are extremely low, it is better to be safe than sorry.

Most likely there is no danger, even if there is life on Mars and alien microbes hitch a ride to Earth inside some of the returned samples. In fact, Mars is sending samples to Earth all the time in the form of the Mars meteorites. Since some of these microbes (if they exist) could probably survive the trip to Earth inside their rocky home, we may have been exposed many times over to martian microbes. Either they do not interact with our terrestrial life, or in effect our planet has already been inoculated against such alien bugs.

This OpenStax book is available for free at http://cnx.org/content/col11992/1.8

LINK TO LEARNING

More than any other planet, Mars has inspired science fiction writers over the years. You can find scientifically reasonable stories about Mars in a subject index of such stories online. If you click on Mars (https://openstax.org/l/30MarsStories) as a topic, you will find stories by a number of space scientists, including William Hartmann, Geoffrey Landis, and Ludek Pesek.

10.6 DIVERGENT PLANETARY EVOLUTION

Learning Objectives

By the end of this section, you will be able to:

> Compare the planetary evolution of Venus, Earth, and Mars

Venus, Mars, and our own planet Earth form a remarkably diverse triad of worlds. Although all three orbit in roughly the same inner zone around the Sun and all apparently started with about the same chemical mix of silicates and metals, their evolutionary paths have diverged. As a result, Venus became hot and dry, Mars became cold and dry, and only Earth ended up with what we consider a hospitable climate.

We have discussed the runaway greenhouse effect on Venus and the runaway refrigerator effect on Mars, but we do not understand exactly what started these two planets down these separate evolutionary paths. Was Earth ever in danger of a similar fate? Or might it still be diverted onto one of these paths, perhaps due to stress on the atmosphere generated by human pollutants? One of the reasons for studying Venus and Mars is to seek insight into these questions.

Some people have even suggested that if we understood the evolution of Mars and Venus better, we could possibly reverse their evolution and restore more earthlike environments. While it seems unlikely that humans could ever make either Mars or Venus into a replica of Earth, considering such possibilities is a useful part of our more general quest to understand the delicate environmental balance that distinguishes our planet from its two neighbors. In Cosmic Samples and the Origin of the Solar System, we return to the comparative study of the terrestrial planets and their divergent evolutionary histories.

Download for free at http://cnx.org/content/col11992/latest/

CHAPTER 10 REVIEW

 KEY TERMS

runaway greenhouse effect the process by which the greenhouse effect, rather than remaining stable or being lessened through intervention, continues to grow at an increasing rate

tectonic geological features that result from stresses and pressures in the crust of a planet; tectonic forces can lead to earthquakes and motion of the crust

 SUMMARY

10.1 The Nearest Planets: An Overview

Venus, the nearest planet, is a great disappointment through the telescope because of its impenetrable cloud cover. Mars is more tantalizing, with dark markings and polar caps. Early in the twentieth century, it was widely believed that the "canals" of Mars indicated intelligent life there. Mars has only 11% the mass of Earth, but Venus is nearly our twin in size and mass. Mars rotates in 24 hours and has seasons like Earth; Venus has a retrograde rotation period of 243 days. Both planets have been extensively explored by spacecraft.

10.2 The Geology of Venus

Venus has been mapped by radar, especially with the *Magellan* spacecraft. Its crust consists of 75% lowland lava plains, numerous volcanic features, and many large coronae, which are the expression of subsurface volcanism. The planet has been modified by widespread tectonics driven by mantle convection, forming complex patterns of ridges and cracks and building high continental regions such as Ishtar. The surface is extraordinarily inhospitable, with pressure of 90 bars and temperature of 730 K, but several Russian Venera landers investigated it successfully.

10.3 The Massive Atmosphere of Venus

The atmosphere of Venus is 96% CO_2. Thick clouds at altitudes of 30 to 60 kilometers are made of sulfuric acid, and a CO_2 greenhouse effect maintains the high surface temperature. Venus presumably reached its current state from more earthlike initial conditions as a result of a runaway greenhouse effect, which included the loss of large quantities of water.

10.4 The Geology of Mars

Most of what we know about Mars is derived from spacecraft: highly successful orbiters, landers, and rovers. We have also been able to study a few martian rocks that reached Earth as meteorites. Mars has heavily cratered highlands in its southern hemisphere, but younger, lower volcanic plains over much of its northern half. The Tharsis bulge, as big as North America, includes several huge volcanoes; Olympus Mons is more than 20 kilometers high and 500 kilometers in diameter. The Valles Marineris canyons are tectonic features widened by erosion. Early landers revealed only barren, windswept plains, but later missions have visited places with more geological (and scenic) variety. Landing sites have been selected in part to search for evidence of past water.

10.5 Water and Life on Mars

The martian atmosphere has a surface pressure of less than 0.01 bar and is 95% CO_2. It has dust clouds, water clouds, and carbon dioxide (dry ice) clouds. Liquid water on the surface is not possible today, but there is subsurface permafrost at high latitudes. Seasonal polar caps are made of dry ice; the northern residual cap

This OpenStax book is available for free at http://cnx.org/content/col11992/1.8

is water ice, whereas the southern permanent ice cap is made predominantly of water ice with a covering of carbon dioxide ice. Evidence of a very different climate in the past is found in water erosion features: both runoff channels and outflow channels, the latter carved by catastrophic floods. Our rovers, exploring ancient lakebeds and places where sedimentary rock has formed, have found evidence for extensive surface water in the past. Even more exciting are the gullies that seem to show the presence of flowing salty water on the surface today, hinting at near-surface aquifers. The Viking landers searched for martian life in 1976, with negative results, but life might have flourished long ago. We have found evidence of water on Mars, but following the water has not yet led us to life on that planet.

10.6 Divergent Planetary Evolution

Earth, Venus, and Mars have diverged in their evolution from what may have been similar beginnings. We need to understand why if we are to protect the environment of Earth.

 # FOR FURTHER EXPLORATION

Articles

Venus

Dorminey, B. "Cool Science on a Hot World." *Astronomy* (February 2006): 46. Five-page overview of Venus and the Venus Express mission plans.

Kargel, J. "Rivers of Venus." *Sky & Telescope* (August 1997): 32. On lava channels.

Robertson, D. "Parched Planet." *Sky & Telescope* (April 2008): 26. Overview of our understanding of the planet.

Robinson, C. "Magellan Reveals Venus." *Astronomy* (February 1995): 32.

Stofan, E. "The New Face of Venus." *Sky & Telescope* (August 1993): 22.

Zimmerman, R. "Taking Venus by Storm." *Astronomy* (October 2008): 66. On results from the Venus Express mission.

Mars

Albee, A. "The Unearthly Landscapes of Mars." *Scientific American* (June 2003): 44. Results from the Mars Global Surveyor and Mars Odyssey missions and an overview.

Bell, J. "A Fresh Look at Mars." *Astronomy* (August 2015): 28. Nice summary of recent spacecraft results and how they are revising our understanding of Mars.

Bell, J. "Uncovering Mars' Secret Past." *Sky & Telescope* (July 2009): 22. How rovers and orbiters are helping us to understand Mars history and the role of water.

Bell, J. "The Red Planet's Watery Past." *Scientific American* (December 2006): 62. Rovers are furnishing proof that ancient Mars was wet.

Burnham, R. "Red Planet Rendezvous." *Astronomy* (May 2006): 68. About Mariner Valley and a flyover film constructed from many still images.

Christensen, P. "The Many Faces of Mars." *Scientific American* (July 2005): 32. Results from the Rover mission; evidence that Mars was once wet in places.

Lakdawalla, E. "The History of Water on Mars." *Sky & Telescope* (September 2013): 16. Clear review of our current understanding of the role of water on Mars in different epochs.

Download for free at http://cnx.org/content/col11992/latest/

Malin, M. "Visions of Mars." *Sky & Telescope* (April 1999): 42. A geological tour of the red planet, with new Mars Global Surveyor images.

McEwen, A. "Mars in Motion." *Scientific American* (May 2013): 58. On gullies and other surface changes.

McKay, C. & Garcia, V. "How to Search for Life on Mars." *Scientific American* (June 2014): 44. Experiments future probes could perform.

Naeye, R. "Europe's Eye on Mars." *Sky & Telescope* (December 2005): 30. On the Mars Express mission and the remarkable close-up images it is sending.

Talcott, R. "Seeking Ground Truth on Mars." *Astronomy* (October 2009): 34. How rovers and orbiters are helping scientists understand the red planet's surface.

Websites

European Space Agency Mars Express Page: http://www.esa.int/Our_Activities/Space_Science/Mars_Express.

European Space Agency Venus Express Page: http://www.esa.int/Our_Activities/Space_Science/Venus_Express.

High Resolution Imaging Science Experiment: http://hirise.lpl.arizona.edu/.

Jet Propulsion Lab Mars Exploration Page: http://mars.jpl.nasa.gov/.

Mars Globe HD app: https://itunes.apple.com/us/app/mars-globe-hd/id376020224?mt=8.

Mars Rover 360° Panorama: http://www.360cities.net/image/curiosity-rover-martian-solar-day-2#171.10,26.50,70.0. Interactive.

NASA Center for Mars Exploration: http://www.nasa.gov/mission_pages/mars/main/index.html.

NASA Solar System Exploration Mars Page: http://solarsystem.nasa.gov/planets/mars.

NASA Solar System Exploration Venus Page: http://solarsystem.nasa.gov/planets/venus.

NASA's apps about Mars for phones and tablets can be found at: http://mars.nasa.gov/mobile/info/.

NASA's Magellan Mission to Venus: http://www2.jpl.nasa.gov/magellan/.

Russian (Soviet) Venus Missions and Images: http://mentallandscape.com/C_CatalogVenus.htm.

Venus Atlas app: https://itunes.apple.com/us/app/venus-atlas/id317310503?mt=8.

Venus Express Results Article: http://www.mpg.de/798302/F002_Focus_026-033.pdf.

Videos

50 Years of Mars Exploration: http://www.jpl.nasa.gov/video/details.php?id=1395. NASA's summary of all missions through *MAVEN*; good quick overview (4:08).

Being a Mars Rover: What It's Like to be an Interplanetary Explorer: https://www.youtube.com/watch?v=nRpCOEsPD54. 2013 talk by Dr. Lori Fenton about what it's like on the surface of Mars (1:07:24).

Magellan Maps Venus: http://www.bbc.co.uk/science/space/solarsystem/space_missions/magellan_probe#p005y07s. BBC clip with Dr. Ellen Stofan on the radar images of Venus and what they tell us (3:06).

Our *Curiosity*: https://www.youtube.com/watch?v=XczKXWvokm4. Mars *Curiosity* rover 2-year anniversary video narrated by Neil deGrasse Tyson and Felicia Day (6:01).

Planet Venus: The Deadliest Planet, Venus Surface and Atmosphere: https://www.youtube.com/watch?v=HqFVxWfVtoo. Quick tour of Venus' atmosphere and surface (2:04).

This OpenStax book is available for free at http://cnx.org/content/col11992/1.8

Planetary Protection and Hitchhikers in the Solar System: The Danger of Mingling Microbes: https://www.youtube.com/watch?v=6iGC3uO7jBI. 2009 talk by Dr. Margaret Race on preventing contamination between worlds (1:28:50).

⚎ COLLABORATIVE GROUP ACTIVITIES

A. Your group has been asked by high NASA officials to start planning the first human colony on Mars. Begin by making a list of what sorts of things humans would need to bring along to be able to survive for years on the surface of the red planet.

B. As a publicity stunt, the mayor of Venus, Texas (there really is such a town), proposes that NASA fund a mission to Venus with humans on board. Clearly, the good mayor neglected to take an astronomy course in college. Have your group assemble a list of as many reasons as possible why it is unlikely that humans will soon land on the surface of Venus.

C. Even if humans would have trouble surviving on the surface of Venus, this does not mean we could not learn a lot more about our veiled sister planet. Have your group brainstorm a series of missions (pretend cost is no object) that would provide us with more detailed information about Venus' atmosphere, surface, and interior.

D. Sometime late in the twenty-first century, when travel to Mars has become somewhat routine, a very wealthy couple asks you to plan a honeymoon tour of Mars that includes the most spectacular sights on the red planet. Constitute your group as the Percival Lowell Memorial Tourist Agency, and come up with a list of not-to-be missed tourist stops on Mars.

E. In the popular book and film, called *The Martian*, the drama really begins when our hero is knocked over and loses consciousness as he is half buried by an intense wind storm on Mars. Given what you have learned about Mars' atmosphere in this chapter, have your group discuss how realistic that scenario is. (By the way, the author of the book has himself genially acknowledged in interviews and talks that this is a reasonable question to ask.)

F. Astronomers have been puzzled and annoyed about the extensive media publicity that was given the small group of "true believers" who claimed the "Face on Mars" was not a natural formation (see the Astronomy and Pseudoscience: The "Face on Mars" feature box). Have your group make a list of the reasons many of the media were so enchanted by this story. What do you think astronomers could or should do to get the skeptical, scientific perspective about such issues before the public?

G. Your group is a special committee of scientists set up by the United Nations to specify how any Mars samples should be returned to Earth so that possible martian microbes do not harm Earth life. What precautions would you recommend, starting at Mars and going all the way to the labs that analyze the martian samples back on Earth?

H. Have your group brainstorm about Mars in popular culture. How many movies, songs or other music, and products can you think of connected with Mars? What are some reasons that Mars would be a popular theme for filmmakers, songwriters, and product designers?

Download for free at http://cnx.org/content/col11992/latest/

☐ EXERCISES

Review Questions

1. List several ways that Venus, Earth, and Mars are similar, and several ways they are different.

2. Compare the current atmospheres of Earth, Venus, and Mars in terms of composition, thickness (and pressure at the surface), and the greenhouse effect.

3. How might Venus' atmosphere have evolved to its present state through a runaway greenhouse effect?

4. Describe the current atmosphere on Mars. What evidence suggests that it must have been different in the past?

5. Explain the runaway refrigerator effect and the role it may have played in the evolution of Mars.

6. What evidence do we have that there was running (liquid) water on Mars in the past? What evidence is there for water coming out of the ground even today?

7. What evidence is there that Venus was volcanically active about 300–600 million years ago?

8. Why is Mars red?

9. What is the composition of clouds on Mars?

10. What is the composition of the polar caps on Mars?

11. Describe two anomalous features of the rotation of Venus and what might account for them.

12. How was the *Mars Odyssey* spacecraft able to detect water on Mars without landing on it?

Thought Questions

13. What are the advantages of using radar imaging rather than ordinary cameras to study the topography of Venus? What are the relative advantages of these two approaches to mapping Earth or Mars?

14. Venus and Earth are nearly the same size and distance from the Sun. What are the main differences in the geology of the two planets? What might be some of the reasons for these differences?

15. Why is there so much more carbon dioxide in the atmosphere of Venus than in that of Earth? Why so much more carbon dioxide than on Mars?

16. If the Viking missions were such a rich source of information about Mars, why have we sent the Pathfinder, *Global Surveyor*, and other more recent spacecraft to Mars? Make a list of questions about Mars that still puzzle astronomers.

17. Compare Mars with Mercury and the Moon in terms of overall properties. What are the main similarities and differences?

18. Contrast the mountains on Mars and Venus with those on Earth and the Moon.

19. We believe that all of the terrestrial planets had similar histories when it comes to impacts from space. Explain how this idea can be used to date the formation of the martian highlands, the martian basins, and the Tharsis volcanoes. How certain are the ages derived for these features (in other words, how do we check the ages we derive from this method)?

This OpenStax book is available for free at http://cnx.org/content/col11992/1.8

20. Is it likely that life ever existed on either Venus or Mars? Justify your answer in each case.

21. Suppose that, decades from now, NASA is considering sending astronauts to Mars and Venus. In each case, describe what kind of protective gear they would have to carry, and what their chances for survival would be if their spacesuits ruptured.

22. We believe that Venus, Earth, and Mars all started with a significant supply of water. Explain where that water is now for each planet.

23. One source of information about Mars has been the analysis of meteorites from Mars. Since no samples from Mars have ever been returned to Earth from any of the missions we sent there, how do we know these meteorites are from Mars? What information have they revealed about Mars?

24. The runaway greenhouse effect and its inverse, the runaway refrigerator effect, have led to harsh, uninhabitable conditions on Venus and Mars. Does the greenhouse effect always cause climate changes leading to loss of water and life? Give a reason for your answer.

25. In what way is the high surface temperature of Venus relevant to concerns about global warming on Earth today?

26. What is a dust devil? Would you expect to feel more of a breeze from a dust devil on Mars or on Earth? Explain.

27. Near the martian equator, temperatures at the same spot can vary from an average of –135 °C at night to an average of 30 °C during the day. How can you explain such a wide difference in temperature compared to that on Earth?

Figuring For Yourself

28. Estimate the amount of water there could be in a global (planet-wide) region of subsurface permafrost on Mars (do the calculations for two permafrost thicknesses, 1 and 10 km, and a concentration of ice in the permafrost of 10% by volume). Compare the two results you get with the amount of water in Earth's oceans calculated in Example 10.1.

29. At its nearest, Venus comes within about 41 million km of Earth. How distant is it at its farthest?

30. If you weigh 150 lbs. on the surface of Earth, how much would you weigh on Venus? On Mars?

31. Calculate the relative land area—that is, the amount of the surface not covered by liquids—of Earth, the Moon, Venus, and Mars. (Assume that 70% of Earth is covered with water.)

32. The closest approach distance between Mars and Earth is about 56 million km. Assume you can travel in a spaceship at 58,000 km/h, which is the speed achieved by the New Horizons space probe that went to Pluto and is the fastest speed so far of any space vehicle launched from Earth. How long would it take to get to Mars at the time of closest approach?

Download for free at http://cnx.org/content/col11992/latest/

This OpenStax book is available for free at http://cnx.org/content/col11992/1.8

Download for free at http://cnx.org/content/col11992/latest/

Figure 11.1. Giant Planets. The four giant planets in our solar system all have hydrogen atmospheres, but the warm gas giants, Jupiter and Saturn, have tan, beige, red, and white clouds that are thought to be composed of ammonia ice particles with various colorants called "chromophores." The blue-tinted ice giants, Uranus and Neptune, are much colder and covered in methane ice clouds. (credit: modification of work by Lunar and Planetary Institute, NASA)

Chapter Outline

Thinking Ahead

"What do we learn about the Earth by studying the planets? Humility."—Andrew Ingersoll discussing the results of the Voyager mission in 1986.

Beyond Mars and the asteroid belt, we encounter a new region of the solar system: the realm of the giants. Temperatures here are lower, permitting water and other volatiles to condense as ice. The planets are much larger, distances between them are much greater, and each giant world is accompanied by an extensive system of moons and rings.

From many perspectives, the outer solar system is where the action is, and the giant planets are the most important members of the Sun's family. When compared to these outer giants, the little cinders of rock and metal that orbit closer to the Sun can seem insignificant. These four giant worlds—Jupiter, Saturn, Uranus, Neptune—are the subjects of this chapter. Their rings, moons, and the dwarf planet Pluto are discussed in a later chapter.

11.1 EXPLORING THE OUTER PLANETS

Learning Objectives

By the end of this section, you will be able to:

Download for free at http://cnx.org/content/col11992/latest/

> Provide an overview of the composition of the giant planets
> Chronicle the robotic exploration of the outer solar system
> Summarize the missions sent to orbit the gas giants

The giant planets hold most of the mass in our planetary system. Jupiter alone exceeds the mass of all the other planets combined (Figure 11.2). The material available to build these planets can be divided into three classes by what they are made of: "gases," "ices," and "rocks" (see Table 11.1). The "gases" are primarily hydrogen and helium, the most abundant elements in the universe. The way it is used here, the term "ices" refers to composition only and not whether a substance is actually in a solid state. "Ices" means compounds that form from the next most abundant elements: oxygen, carbon, and nitrogen. Common ices are water, methane, and ammonia, but ices may also include carbon monoxide, carbon dioxide, and others. "Rocks" are even less abundant than ices, and include everything else: magnesium, silicon, iron, and so on.

Figure 11.2. Jupiter. The Cassini spacecraft imaged Jupiter on its way to Saturn in 2012. The giant storm system called the Great Red Spot is visible to the lower right. The dark spot to the lower left is the shadow of Jupiter's moon Europa. (credit: modification of work by NASA/JPL)

Abundances in the Outer Solar System

Type of Material	Name	Approximate % (by Mass)
Gas	Hydrogen (H_2)	75
Gas	Helium (He)	24
Ice	Water (H_2O)	0.6
Ice	Methane (CH_4)	0.4
Ice	Ammonia (NH_3)	0.1
Rock	Magnesium (Mg), iron (Fe), silicon (Si)	0.3

Table 11.1

This OpenStax book is available for free at http://cnx.org/content/col11992/1.8

Download for free at http://cnx.org/content/col11992/latest/

In the outer solar system, gases dominate the two largest planets, Jupiter and Saturn, hence their nickname "gas giants." Uranus and Neptune are called "ice giants" because their interiors contain far more of the "ice" component than their larger cousins. The chemistry for all four giant planet atmospheres is dominated by hydrogen. This hydrogen caused the chemistry of the outer solar system to become *reducing*, meaning that other elements tend to combine with hydrogen first. In the early solar system, most of the oxygen combined with hydrogen to make H_2O and was thus unavailable to form the kinds of oxidized compounds with other elements that are more familiar to us in the inner solar system (such as CO_2). As a result, the compounds detected in the atmosphere of the giant planets are mostly hydrogen-based gases such as methane (CH_4) and ammonia (NH_3), or more complex hydrocarbons (combinations of hydrogen and carbon) such as ethane (C_2H_6) and acetylene (C_2H_2).

Exploration of the Outer Solar System So Far

Eight spacecraft, seven from the United States and one from Europe, have penetrated beyond the asteroid belt into the realm of the giants. Table 11.2 summarizes the spacecraft missions to the outer solar system.

Missions to the Giant Planets

Planet	Spacecraft[1]	Encounter Date	Type
Jupiter	Pioneer 10	December 1973	Flyby
	Pioneer 11	December 1974	Flyby
	Voyager 1	March 1979	Flyby
	Voyager 2	July 1979	Flyby
	Ulysses	February 1992	Flyby during gravity assist
	Galileo	December 1995	Orbiter and probe
	Cassini	December 2002	Flyby
	New Horizons	February 2007	Flyby during gravity assist
	Juno	July 2016	Orbiter
Saturn	Pioneer 11	September 1979	Flyby
	Voyager 1	November 1980	Flyby
	Voyager 2	August 1981	Flyby
	Cassini	July 2004 (Saturn orbit injection 2000)	Orbiter

Table 11.2

---

1   Both the Ulysses and the New Horizons spacecraft (designed to study the Sun and Pluto, respectively) flew past Jupiter for a gravity boost (gaining energy by "stealing" a little bit from the giant planet's rotation).

Download for free at http://cnx.org/content/col11992/latest/

Missions to the Giant Planets

Planet	Spacecraft	Encounter Date	Type
Uranus	Voyager 2	January 1986	Flyby
Neptune	Voyager 2	August 1989	Flyby

Table 11.2

The challenges of exploring so far away from Earth are considerable. Flight times to the giant planets are measured in years to decades, rather than the months required to reach Venus or Mars. Even at the speed of light, messages take hours to pass between Earth and the spacecraft. If a problem develops near Saturn, for example, a wait of hours for the alarm to reach Earth and for instructions to be routed back to the spacecraft could spell disaster. Spacecraft to the outer solar system must therefore be highly reliable and capable of a greater degree of independence and autonomy. Outer solar system missions also must carry their own power sources since the Sun is too far away to provide enough energy. Heaters are required to keep instruments at proper operating temperatures, and spacecraft must have radio transmitters powerful enough to send their data to receivers on distant Earth.

The first spacecraft to investigate the regions past Mars were the NASA Pioneers 10 and 11, launched in 1972 and 1973 as pathfinders to Jupiter. One of their main objectives was simply to determine whether a spacecraft could actually navigate through the belt of asteroids that lies beyond Mars without getting destroyed by collisions with asteroidal dust. Another objective was to measure the radiation hazards in the *magnetosphere* (or zone of magnetic influence) of Jupiter. Both spacecraft passed through the asteroid belt without incident, but the energetic particles in Jupiter's magnetic field nearly wiped out their electronics, providing information necessary for the safe design of subsequent missions.

Pioneer 10 flew past Jupiter in 1973, after which it sped outward toward the limits of the solar system. Pioneer 11 undertook a more ambitious program, using the gravity of Jupiter to aim for Saturn, which it reached in 1979. The twin Voyager spacecraft launched the next wave of outer planet exploration in 1977. Voyagers 1 and 2 each carried 11 scientific instruments, including cameras and spectrometers, as well as devices to measure the characteristics of planetary magnetospheres. Since they kept going outward after their planetary encounters, these are now the most distant spacecraft ever launched by humanity.

Voyager 1 reached Jupiter in 1979 and used a gravity assist from that planet to take it on to Saturn in 1980. Voyager 2 arrived at Jupiter four months later, but then followed a different path to visit all the outer planets, reaching Saturn in 1981, Uranus in 1986, and Neptune in 1989. This trajectory was made possible by the approximate alignment of the four giant planets on the same side of the Sun. About once every 175 years, these planets are in such a position, and it allows a single spacecraft to visit them all by using gravity-assisted flybys to adjust course for each subsequent encounter; such a maneuver has been nicknamed a "Grand Tour" by astronomers.

This OpenStax book is available for free at http://cnx.org/content/col11992/1.8

LINK TO LEARNING

The Jet Propulsion Laboratory has a nice video called Voyager: The Grand Tour (https://openstax.org/l/30JPLGrandT) that describes the Voyager mission and what it found.

MAKING CONNECTIONS

Engineering and Space Science: Teaching an Old Spacecraft New Tricks

By the time Voyager 2 arrived at Neptune in 1989, 12 years after its launch, the spacecraft was beginning to show signs of old age. The arm on which the camera and other instruments were located was "arthritic": it could no longer move easily in all directions. The communications system was "hard of hearing": part of its radio receiver had stopped working. The "brains" had significant "memory loss": some of the onboard computer memory had failed. And the whole spacecraft was beginning to run out of energy: its generators had begun showing serious signs of wear.

To make things even more of a challenge, Voyager's mission at Neptune was in many ways the most difficult of all four flybys. For example, since sunlight at Neptune is 900 times weaker than at Earth, the onboard camera had to take much longer exposures in this light-starved environment. This was a nontrivial requirement, given that the spacecraft was hurtling by Neptune at ten times the speed of a rifle bullet.

The solution was to swivel the camera backward at exactly the rate that would compensate for the forward motion of the spacecraft. Engineers had to preprogram the ship's computer to execute an incredibly complex series of maneuvers for each image. The beautiful Voyager images of Neptune are a testament to the ingenuity of spacecraft engineers.

The sheer distance of the craft from its controllers on Earth was yet another challenge. Voyager 2 received instructions and sent back its data via on-board radio transmitter. The distance from Earth to Neptune is about 4.8 billion kilometers. Over this vast distance, the power that reached us from Voyager 2 at Neptune was approximately10^{-16} watts, or 20 billion times less power than it takes to operate a digital watch. Thirty-eight different antennas on four continents were used by NASA to collect the faint signals from the spacecraft and decode the precious information about Neptune that they contained.

Enter the Orbiters: Galileo and Cassini

The Pioneer and Voyager missions were flybys of the giant planets: they each produced only quick looks before the spacecraft sped onward. For more detailed studies of these worlds, we require spacecraft that can go into orbit around a planet. For Jupiter and Saturn, these orbiters were the Galileo and Cassini spacecraft, respectively. To date, no orbiter missions have been started for Uranus and Neptune, although planetary scientists have expressed keen interest.

The Galileo spacecraft was launched toward Jupiter in 1989 and arrived in 1995. Galileo began its investigations by deploying an entry probe into Jupiter, for the first direct studies of the planet's outer atmospheric layers.

Download for free at http://cnx.org/content/col11992/latest/

The probe plunged at a shallow angle into Jupiter's atmosphere, traveling at a speed of 50 kilometers *per second*—that's fast enough to fly from New York to San Francisco in 100 seconds! This was the highest speed at which any probe has so far entered the atmosphere of a planet, and it put great demands on the heat shield protecting it. The high entry speed was a result of acceleration by the strong gravitational attraction of Jupiter.

Atmospheric friction slowed the probe within 2 minutes, producing temperatures at the front of its heat shield as high as 15,000 °C. As the probe's speed dropped to 2500 kilometers per hour, the remains of the glowing heat shield were jettisoned, and a parachute was deployed to lower the instrumented probe spacecraft more gently into the atmosphere (Figure 11.3). The data from the probe instruments were relayed to Earth via the main Galileo spacecraft.

Figure 11.3. Galileo Probe Falling into Jupiter. This artist's depiction shows the Galileo probe descending into the clouds via parachute just after the protective heat shield separated. The probe made its measurements of Jupiter's atmosphere on December 7, 1995. (credit: modification of work by NASA/Ames Research Center)

The probe continued to operate for an hour, descending 200 kilometers into the atmosphere. A few minutes later the polyester parachute melted, and within a few hours the main aluminum and titanium structure of the probe vaporized to become a part of Jupiter itself. About 2 hours after receipt of the final probe data, the main spacecraft fired its retro-rockets so it could be captured into orbit around the planet, where its primary objectives were to study Jupiter's large and often puzzling moons.

The Cassini mission to Saturn (Figure 11.4), a cooperative venture between NASA and the European Space Agency, was similar to Galileo in its two-fold approach. Launched in 1997, Cassini arrived in 2004 and went into orbit around Saturn, beginning extensive studies of its rings and moons, as well as the planet itself. In January 2005, Cassini deployed an entry probe into the atmosphere of Saturn's large moon, Titan, where it successfully landed on the surface. (We'll discuss the probe and what it found in the chapter on Rings, Moons, and Pluto.)

This OpenStax book is available for free at http://cnx.org/content/col11992/1.8

Download for free at http://cnx.org/content/col11992/latest/

Figure 11.4. Earth as Seen from Saturn. This popular Cassini image shows Earth as a tiny dot (marked with an arrow) seen below Saturn's rings. It was taken in July 2013, when Saturn was 1.4 billion kilometers from Earth. (credit: modification of work by NASA/JPL-Caltech/Space Science Institute)

11.2 | THE GIANT PLANETS

Learning Objectives

By the end of this section, you will be able to:

> Describe the basic physical characteristics, general appearance, and rotation of the giant planets
> Describe the composition and structure of Jupiter, Saturn, Uranus, and Neptune
> Compare and contrast the internal heat sources of the giant planets
> Describe the discovery and characteristics of the giant planets' magnetic fields

Let us now examine the four giant (or *jovian*) planets in more detail. Our approach is not just to catalog their characteristics, but to compare them with each other, noting their similarities and differences and attempting to relate their properties to their differing masses and distances from the Sun.

Basic Characteristics

The giant planets are very far from the Sun. Jupiter is more than five times farther from the Sun than Earth's distance (5 AU), and takes just under 12 years to circle the Sun. Saturn is about twice as far away as Jupiter (almost 10 AU) and takes nearly 30 years to complete one orbit. Uranus orbits at 19 AU with a period of 84 years, while Neptune, at 30 AU, requires 165 years for each circuit of the Sun. These long timescales make it difficult for us short-lived humans to study seasonal change on the outer planets.

Jupiter and Saturn have many similarities in composition and internal structure, although Jupiter is nearly four times more massive. Uranus and Neptune are smaller and differ in composition and internal structure from their large siblings. Some of the main properties of these four planets are summarized in Table 11.3.

Download for free at http://cnx.org/content/col11992/latest/

Basic Properties of the Jovian Planets

Planet	Distance (AU)	Period (years)	Diameter (km)	Mass (Earth = 1)	Density (g/cm^3)	Rotation (hours)
Jupiter	5.2	11.9	142,800	318	1.3	9.9
Saturn	9.5	29.5	120,540	95	0.7	10.7
Uranus	19.2	84.1	51,200	14	1.3	17.2
Neptune	30.0	164.8	49,500	17	1.6	16.1

Table 11.3

Jupiter, the giant among giants, has enough mass to make 318 Earths. Its diameter is about 11 times that of Earth (and about one tenth that of the Sun). Jupiter's average density is 1.3 g/cm^3, much lower than that of any of the terrestrial planets. (Recall that water has a density of 1 g/cm^3.) Jupiter's material is spread out over a volume so large that more than 1400 Earths could fit within it.

Saturn's mass is 95 times that of Earth, and its average density is only 0.7 g/cm^3—the lowest of any planet. Since this is less than the density of water, Saturn would be light enough to float.

Uranus and Neptune each have a mass about 15 times that of Earth and, hence, are only 5% as massive as Jupiter. Their densities of 1.3 g/cm^3 and 1.6 g/cm^3, respectively, are much higher than that of Saturn. This is one piece of evidence that tells us that their composition must differ fundamentally from the gas giants. When astronomers began to discover other planetary systems (exoplanets), we found that planets the size of Uranus and Neptune are common, and that there are even more exoplanets intermediate in size between Earth and these ice giants, a type of planet not found in our solar system.

Appearance and Rotation

When we look at the planets, we see only their atmospheres, composed primarily of hydrogen and helium gas (see Figure 11.1). The uppermost clouds of Jupiter and Saturn, the part we see when looking down at these planets from above, are composed of ammonia crystals. On Neptune, the upper clouds are made of methane. On Uranus, we see no obvious cloud layer at all, but only a deep and featureless haze.

Seen through a telescope, Jupiter is a colorful and dynamic planet. Distinct details in its cloud patterns allow us to determine the rotation rate of its atmosphere at the cloud level, although such atmosphere rotation may have little to do with the spin of the underlying planet. Much more fundamental is the rotation of the mantle and core; these can be determined by periodic variations in radio waves coming from Jupiter, which are controlled by its magnetic field. Since the magnetic field (which we will discuss below) originates deep inside the planet, it shares the rotation of the interior. The rotation period we measure in this way is 9 hours 56 minutes, which gives Jupiter the shortest "day" of any planet. In the same way, we can measure that the underlying rotation period of Saturn is 10 hours 40 minutes. Uranus and Neptune have slightly longer rotation periods of about 17 hours, also determined from the rotation of their magnetic fields.

This OpenStax book is available for free at http://cnx.org/content/col11992/1.8

LINK TO LEARNING

A brief video made from Hubble Space Telescope photos shows the rotation of Jupiter (https://openstax.org/l/30HSTJupRot) with its many atmospheric features.

Remember that Earth and Mars have seasons because their spin axes, instead of "standing up straight," are tilted relative to the orbital plane of the solar system. This means that as Earth revolves around the Sun, sometimes one hemisphere and sometimes the other "leans into" the Sun.

What are the seasons like for the giant planets? The spin axis of Jupiter is tilted by only 3°, so there are no seasons to speak of. Saturn, however, does have seasons, since its spin axis is inclined at 27° to the perpendicular to its orbit. Neptune has about the same tilt as Saturn (29°); therefore, it experiences similar seasons (only more slowly). The strangest seasons of all are on Uranus, which has a spin axis tilted by 98° with respect to the north direction. Practically speaking, we can say that Uranus orbits on its side, and its ring and moon system follow along, orbiting about Uranus' equator (Figure 11.5).

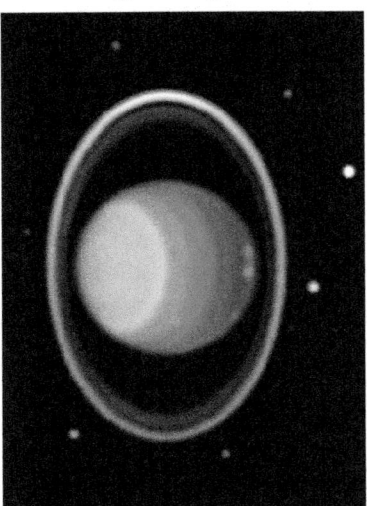 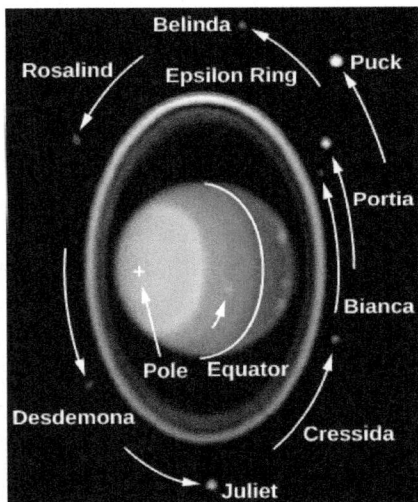

Figure 11.5. Infrared Image of Uranus. The infrared camera on the Hubble Space Telescope took these false-color images of the planet Uranus, its ring system, and moons in 1997. The south pole of the planet (marked with a "+" on the right image) faces the Sun; its green color shows a strong local haze. The two images were taken 90 minutes apart, and during that time the five reddish clouds can be seen to rotate around the parallel to the equator. The rings (which are very faint in the visible light, but prominent in infrared) and eight moons can be seen around the equator. This was the "bull's eye" arrangement that Voyager saw as it approached Uranus in 1986. (credit: modification of work by Erich Karkoschka (University of Arizona), and NASA/ESA)

We don't know what caused Uranus to be tipped over like this, but one possibility is a collision with a large planetary body when our system was first forming. Whatever the cause, this unusual tilt creates dramatic seasons. When Voyager 2 arrived at Uranus, its south pole was facing directly into the Sun. The southern hemisphere was experiencing a 21-year sunlit summer, while during that same period the northern hemisphere was plunged into darkness. For the next 21-year season, the Sun shines on Uranus' equator, and both hemispheres go through cycles of light and dark as the planet rotates (Figure 11.6). Then there are 21 years of an illuminated northern hemisphere and a dark southern hemisphere. After that the pattern of alternating day and night repeats.

Download for free at http://cnx.org/content/col11992/latest/

Just as on Earth, the seasons are even more extreme at the poles. If you were to install a floating platform at the south pole of Uranus, for example, it would experience 42 years of light and 42 years of darkness. Any future astronauts crazy enough to set up camp there could spend most of their lives without ever seeing the Sun.

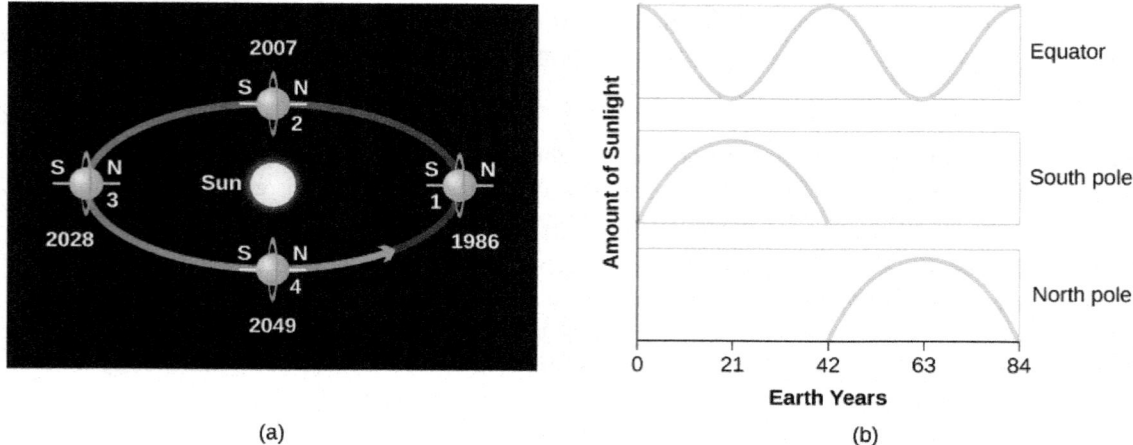

(a)   (b)

Figure 11.6. Strange Seasons on Uranus. (a) This diagram shows the orbit of Uranus as seen from above. At the time Voyager 2 arrived (position 1), the South Pole was facing the Sun. As we move counterclockwise in the diagram, we see the planet 21 years later at each step. (b) This graph compares the amount of sunlight seen at the poles and the equator of Uranus over the course of its 84-year revolution around the Sun.

Composition and Structure

Although we cannot see into these planets, astronomers are confident that the interiors of Jupiter and Saturn are composed primarily of hydrogen and helium. Of course, these gases have been measured only in their atmosphere, but calculations first carried out more than 50 years ago showed that these two light gases are the only possible materials out of which a planet with the observed masses and densities of Jupiter and Saturn could be constructed.

The deep internal structures of these two planets are difficult to predict. This is mainly because these planets are so big that the hydrogen and helium in their centers become tremendously compressed and behave in ways that these gases can never behave on Earth. The best theoretical models we have of Jupiter's structure predict a central pressure greater than 100 million bars and a central density of about 31 g/cm^3. (By contrast, Earth's core has a central pressure of 4 million bars and a central density of 17 g/cm^3.)

At the pressures inside the giant planets, familiar materials can take on strange forms. A few thousand kilometers below the visible clouds of Jupiter and Saturn, pressures become so great that hydrogen changes from a gaseous to a liquid state. Still deeper, this liquid hydrogen is further compressed and begins to act like a metal, something it never does on Earth. (In a metal, electrons are not firmly attached to their parent nuclei but can wander around. This is why metals are such good conductors of electricity.) On Jupiter, the greater part of the interior is liquid metallic hydrogen.

Because Saturn is less massive, it has only a small volume of metallic hydrogen, but most of its interior is liquid. Uranus and Neptune are too small to reach internal pressures sufficient to liquefy hydrogen. We will return to the discussion of the metallic hydrogen layers when we examine the magnetic fields of the giant planets.

Each of these planets has a core composed of heavier materials, as demonstrated by detailed analyses of their gravitational fields. Presumably these cores are the original rock-and-ice bodies that formed before the capture of gas from the surrounding nebula. The cores exist at pressures of tens of millions of bars. While scientists speak of the giant planet cores being composed of rock and ice, we can be sure that neither rock nor ice

This OpenStax book is available for free at http://cnx.org/content/col11992/1.8

assumes any familiar forms at such pressures and temperatures. Remember that what is really meant by "rock" is any material made up primarily of iron, silicon, and oxygen, while the term "ice" in this chapter denotes materials composed primarily of the elements carbon, nitrogen, and oxygen in combination with hydrogen.

Figure 11.7 illustrates the likely interior structures of the four jovian planets. It appears that all four have similar cores of rock and ice. On Jupiter and Saturn, the cores constitute only a few percent of the total mass, consistent with the initial composition of raw materials shown in Table 11.1. However, most of the mass of Uranus and Neptune resides in these cores, demonstrating that the two outer planets were unable to attract massive quantities of hydrogen and helium when they were first forming.

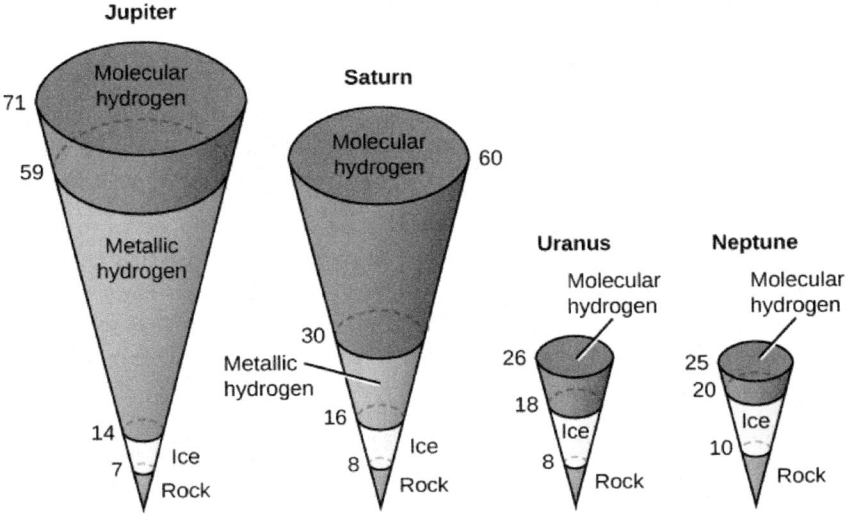

Figure 11.7. Internal Structures of the Jovian Planets. Jupiter and Saturn are composed primarily of hydrogen and helium (but hydrogen dominates), but Uranus and Neptune consist in large part of compounds of carbon, nitrogen, and oxygen. (The diagrams are drawn to scale; numbers show radii in thousands of kilometers.)

Internal Heat Sources

Because of their large sizes, all the giant planets were strongly heated during their formation by the collapse of surrounding material onto their cores. Jupiter, being the largest, was the hottest. Some of this primordial heat can still remain inside such large planets. In addition, it is possible for giant, largely gaseous planets to generate heat after formation by slowly contracting. (With so large a mass, even a minuscule amount of shrinking can generate significant heat.) The effect of these internal energy sources is to raise the temperatures in the interiors and atmospheres of the planets higher than we would expect from the heating effect of the Sun alone.

Jupiter has the largest internal energy source, amounting to 4×10^{17} watts; that is, it is heated from inside with energy equivalent to 4 million billion 100-watt lightbulbs. This energy is about the same as the total solar energy absorbed by Jupiter. The atmosphere of Jupiter is therefore something of a cross between a normal planetary atmosphere (like Earth's), which obtains most of its energy from the Sun, and the atmosphere of a star, which is entirely heated by an internal energy source. Most of the internal energy of Jupiter is primordial heat, left over from the formation of the planet 4.5 billon years ago.

Saturn has an internal energy source about half as large as that of Jupiter, which means (since its mass is only about one quarter as great) that it is producing twice as much energy per kilogram of material as does Jupiter. Since Saturn is expected to have much less primordial heat, there must be another source at work generating most of this 2×10^{17} watts of power. This source is the separation of helium from hydrogen in

Download for free at http://cnx.org/content/col11992/latest/

Saturn's interior. In the liquid hydrogen mantle, the heavier helium forms droplets that sink toward the core, releasing gravitational energy. In effect, Saturn is still differentiating—letting lighter material rise and heavier material fall.

Uranus and Neptune are different. Neptune has a small internal energy source, while Uranus does not emit a measurable amount of internal heat. As a result, these two planets have almost the same atmospheric temperature, in spite of Neptune's greater distance from the Sun. No one knows why these two planets differ in their internal heat, but all this shows how nature can contrive to make each world a little bit different from its neighbors.

Magnetic Fields

Each of the giant planets has a strong magnetic field, generated by electric currents in its rapidly spinning interior. Associated with the magnetic fields are the planets' *magnetospheres*, which are regions around the planet within which the planet's own magnetic field dominates over the general interplanetary magnetic field. The magnetospheres of these planets are their largest features, extending millions of kilometers into space.

In the late 1950s, astronomers discovered that Jupiter was a source of radio waves that got more intense at longer rather than at shorter wavelengths—just the reverse of what is expected from thermal radiation (radiation caused by the normal vibrations of particles within all matter). Such behavior is typical, however, of the radiation emitted when high-speed electrons are accelerated by a magnetic field. We call this **synchrotron radiation** because it was first observed on Earth in particle accelerators, called synchrotrons. This was our first hint that Jupiter must have a strong magnetic field.

Later observations showed that the radio waves are coming from a region surrounding Jupiter with a diameter several times that of the planet itself (Figure 11.8). The evidence suggested that a vast number of charged atomic particles must be circulating around Jupiter, spiraling around the lines of force of a magnetic field associated with the planet. This is just what we observe happening, but on a smaller scale, in the Van Allen belt around Earth. The magnetic fields of Saturn, Uranus, and Neptune, discovered by the spacecraft that first passed close to these planets, work in a similar way, but are not as strong.

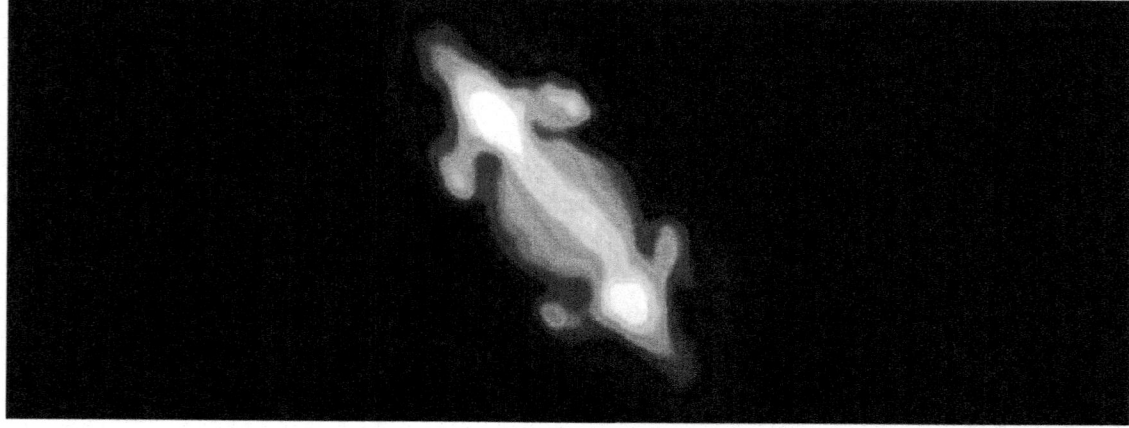

Figure 11.8. Jupiter in Radio Waves. This false-color image of Jupiter was made with the Very Large Array (of radio telescopes) in New Mexico. We see part of the magnetosphere, brightest in the middle because the largest number of charged particles are in the equatorial zone of Jupiter. The planet itself is slightly smaller than the green oval in the center. Different colors are used to indicate different intensities of synchrotron radiation. (credit: modification of work by I. de Pater (UC Berkeley) NRAO, AUI, NSF)

This OpenStax book is available for free at http://cnx.org/content/col11992/1.8

Download for free at http://cnx.org/content/col11992/latest/

LINK TO LEARNING

Learn more about the magnetosphere of Jupiter (https://openstax.org/l/30NASAJupMag) and why we continue to be interested in it from this brief NASA video.

Inside each magnetosphere, charged particles spiral around in the magnetic field; as a result, they can be accelerated to high energies. These charged particles can come from the Sun or from the neighborhood of the planet itself. In Jupiter's case, Io, one of its moons, turns out to have volcanic eruptions that blast charged particles into space and right into the jovian magnetosphere.

The axis of Jupiter's magnetic field (the line that connects the magnetic north pole with the magnetic south pole) is not aligned exactly with the axis of rotation of the planet; rather, it is tipped by about 10°. Uranus and Neptune have even greater magnetic tilts, of 60° and 55°, respectively. Saturn's field, on the other hand, is perfectly aligned with its rotation axis. Why different planets have such different magnetic tilts is not well understood.

The physical processes around the jovian planets turn out to be milder versions of what astronomers find in many distant objects, from the remnants of dead stars to the puzzling distant powerhouses we call quasars. One reason to study the magnetospheres of the giant planets and Earth is that they provide nearby accessible analogues of more energetic and challenging cosmic processes.

11.3 ATMOSPHERES OF THE GIANT PLANETS

Learning Objectives

By the end of this section, you will be able to:

> Discuss the atmospheric composition of the giant planets
> Describe the cloud formation and atmospheric structure of the gas giants
> Characterize the giant planets' wind and weather patterns
> Understand the scale and longevity of storms on the giant planets

The atmospheres of the jovian planets are the parts we can observe or measure directly. Since these planets have no solid surfaces, their atmospheres are more representative of their general compositions than is the case with the terrestrial planets. These atmospheres also present us with some of the most dramatic examples of weather patterns in the solar system. As we will see, storms on these planets can grow bigger than the entire planet Earth.

Atmospheric Composition

When sunlight reflects from the atmospheres of the giant planets, the atmospheric gases leave their "fingerprints" in the spectrum of light. Spectroscopic observations of the jovian planets began in the nineteenth century, but for a long time, astronomers were not able to interpret the spectra they observed. As late as the 1930s, the most prominent features photographed in these spectra remained unidentified. Then better spectra revealed the presence of molecules of methane (CH_4) and ammonia (NH_3) in the atmospheres of Jupiter and Saturn.

Download for free at http://cnx.org/content/col11992/latest/

At first astronomers thought that methane and ammonia might be the main constituents of these atmospheres, but now we know that hydrogen and helium are actually the dominant gases. The confusion arose because neither hydrogen nor helium possesses easily detected spectral features in the visible spectrum. It was not until the Voyager spacecraft measured the far-infrared spectra of Jupiter and Saturn that a reliable abundance for the elusive helium could be found.

The compositions of the two atmospheres are generally similar, except that on Saturn there is less helium as the result of the precipitation of helium that contributes to Saturn's internal energy source. The most precise measurements of composition were made on Jupiter by the Galileo entry probe in 1995; as a result, we know the abundances of some elements in the jovian atmosphere even better than we know those in the Sun.

VOYAGERS IN ASTRONOMY

James Van Allen: Several Planets under His Belt

The career of physicist James Van Allen spanned the birth and growth of the space age, and he played a major role in its development. Born in Iowa in 1914, Van Allen received his PhD from the University of Iowa. He then worked for several research institutions and served in the Navy during World War II.

After the war, Van Allen (Figure 11.9) was appointed Professor of Physics at the University of Iowa. He and his collaborators began using rockets to explore cosmic radiation in Earth's outer atmosphere. To reach extremely high altitudes, Van Allen designed a technique in which a balloon lifts and then launches a small rocket (the rocket is nicknamed "the rockoon").

Figure 11.9. James Van Allen (1914–2006). In this 1950s photograph, Van Allen holds a "rockoon." (credit: modification of work by Frederick W. Kent Collection, University of Iowa Archives)

Over dinner one night in 1950, Van Allen and several colleagues came up with the idea of the International Geophysical Year (IGY), an opportunity for scientists around the world to coordinate their investigations of the physics of Earth, especially research done at high altitudes. In 1955, the United States and the Soviet Union each committed themselves to launching an Earth-orbiting satellite during IGY, a competition that began what came to be known as the space race. The IGY (stretched to 18 months) took place between July 1957 and December 1958.

This OpenStax book is available for free at http://cnx.org/content/col11992/1.8

Download for free at http://cnx.org/content/col11992/latest/

The Soviet Union won the first lap of the race by launching Sputnik 1 in October 1957. The US government spurred its scientists and engineers to even greater efforts to get something into space to maintain the country's prestige. However, the primary US satellite program, Vanguard, ran into difficulties: each of its early launches crashed or exploded. Simultaneously, a second team of rocket engineers and scientists had quietly been working on a military launch vehicle called Jupiter-C. Van Allen spearheaded the design of the instruments aboard a small satellite that this vehicle would carry. On January 31, 1958, Van Allen's Explorer 1 became the first US satellite in space.

Unlike Sputnik, Explorer 1 was equipped to make scientific measurements of high-energy charged particles above the atmosphere. Van Allen and his team discovered a belt of highly charged particles surrounding Earth, and these belts now bear his name. This first scientific discovery of the space program made Van Allen's name known around the world.

Van Allen and his colleagues continued to measure the magnetic and particle environment around planets with increasingly sophisticated spacecraft, including Pioneers 10 and 11, which made exploratory surveys of the environments of Jupiter and Saturn. Some scientists refer to the charged-particle zones around those planets as Van Allen belts as well. (Once, when Van Allen was giving a lecture at the University of Arizona, the graduate students in planetary science asked him if he would leave his belt at the school. It is now proudly displayed as the university's "Van Allen belt.")

Van Allen was a strong supporter of space science and an eloquent senior spokesperson for the American scientific community, warning NASA not to put all its efforts into human spaceflight, but to also use robotic spacecraft as productive tools for space exploration.

Clouds and Atmospheric Structure

The clouds of Jupiter (Figure 11.10) are among the most spectacular sights in the solar system, much beloved by makers of science-fiction films. They range in color from white to orange to red to brown, swirling and twisting in a constantly changing kaleidoscope of patterns. Saturn shows similar but much more subdued cloud activity; instead of vivid colors, its clouds have a nearly uniform butterscotch hue (Figure 11.11).

Download for free at http://cnx.org/content/col11992/latest/

Figure 11.10. Jupiter's Colorful Clouds. The vibrant colors of the clouds on Jupiter present a puzzle to astronomers: given the cool temperatures and the composition of nearly 90% hydrogen, the atmosphere should be colorless. One hypothesis suggests that perhaps colorful hydrogen compounds rise from warm areas. The actual colors are a bit more muted, as shown in Figure 11.2. (credit: modification of work by Voyager Project, JPL, and NASA)

Different gases freeze at different temperatures. At the temperatures and pressures of the upper atmospheres of Jupiter and Saturn, methane remains a gas, but ammonia can condense and freeze. (Similarly, water vapor condenses high in Earth's atmosphere to produce clouds of ice crystals.) The primary clouds that we see around these planets, whether from a spacecraft or through a telescope, are composed of frozen ammonia crystals. The ammonia clouds mark the upper edge of the planets' tropospheres; above that is the stratosphere, the coldest part of the atmosphere. (These layers were initially defined in Earth as a Planet.)

Figure 11.11. Saturn over Five Years. These beautiful images of Saturn were recorded by the Hubble Space Telescope between 1996 and 2000. Since Saturn is tilted by 27°, we see the orientation of Saturn's rings around its equator change as the planet moves along its orbit. Note the horizontal bands in the atmosphere. (credit: modification of work by NASA and The Hubble Heritage Team (STScI/AURA))

The diagrams in Figure 11.12 show the structure and clouds in the atmospheres of all four jovian planets. On both Jupiter and Saturn, the temperature near the cloud tops is about 140 K (only a little cooler than the polar caps of Mars). On Jupiter, this cloud level is at a pressure of about 0.1 bar (one tenth the atmospheric pressure at the surface of Earth), but on Saturn it occurs lower in the atmosphere, at about 1 bar. Because the ammonia

This OpenStax book is available for free at http://cnx.org/content/col11992/1.8

clouds lie so much deeper on Saturn, they are more difficult to see, and the overall appearance of the planet is much blander than is Jupiter's appearance.

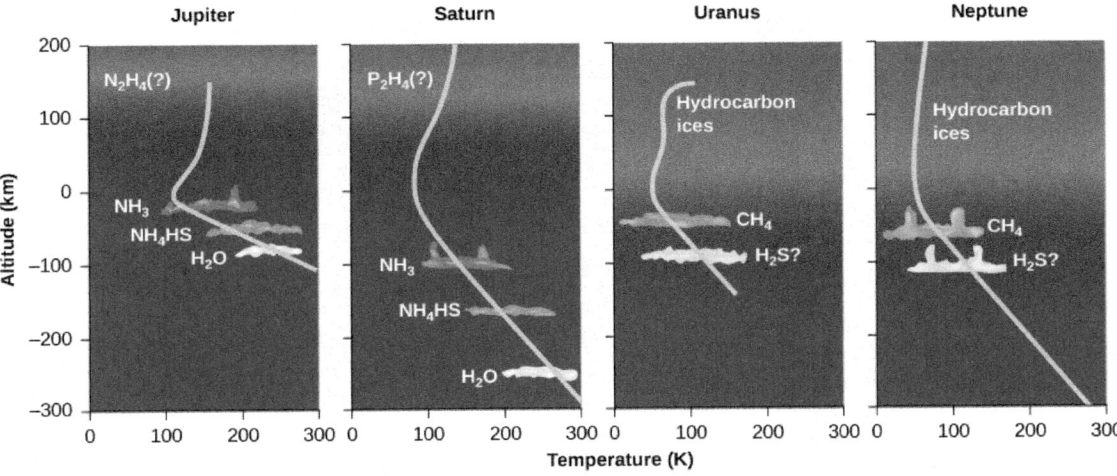

Figure 11.12. Atmospheric Structure of the Jovian Planets. In each diagram, the yellow line shows how the temperature (see the scale on the bottom) changes with altitude (see the scale at the left). The location of the main layers on each planet is also shown.

Within the tropospheres of these planets, the temperature and pressure both increase with depth. Through breaks in the ammonia clouds, we can see tantalizing glimpses of other cloud layers that can form in these deeper regions of the atmosphere—regions that were sampled directly for Jupiter by the Galileo probe that fell into the planet.

As it descended to a pressure of 5 bars, the probe should have passed into a region of frozen water clouds, then below that into clouds of liquid water droplets, perhaps similar to the common clouds of the terrestrial troposphere. At least this is what scientists expected. But the probe saw no water clouds, and it measured a surprisingly low abundance of water vapor in the atmosphere. It soon became clear to the Galileo scientists that the probe happened to descend through an unusually dry, cloud-free region of the atmosphere—a giant downdraft of cool, dry gas. Andrew Ingersoll of Caltech, a member of the Galileo team, called this entry site the "desert" of Jupiter. It's a pity that the probe did not enter a more representative region, but that's the luck of the cosmic draw. The probe continued to make measurements to a pressure of 22 bars but found no other cloud layers before its instruments stopped working. It also detected lightning storms, but only at great distances, further suggesting that the probe itself was in a region of clear weather.

Above the visible ammonia clouds in Jupiter's atmosphere, we find the clear stratosphere, which reaches a minimum temperature near 120 K. At still higher altitudes, temperatures rise again, just as they do in the upper atmosphere of Earth, because here the molecules absorb ultraviolet light from the Sun. The cloud colors are due to impurities, the product of chemical reactions among the atmospheric gases in a process we call **photochemistry**. In Jupiter's upper atmosphere, photochemical reactions create a variety of fairly complex compounds of hydrogen and carbon that form a thin layer of smog far above the visible clouds. We show this smog as a fuzzy orange region in Figure 11.12; however, this thin layer does not block our view of the clouds beneath it.

The visible atmosphere of Saturn is composed of approximately 75% hydrogen and 25% helium, with trace amounts of methane, ethane, propane, and other hydrocarbons. The overall structure is similar to that of Jupiter. Temperatures are somewhat colder, however, and the atmosphere is more extended due to Saturn's

Download for free at http://cnx.org/content/col11992/latest/

lower surface gravity. Thus, the layers are stretched out over a longer distance, as you can see in Figure 11.12. Overall, though, the same atmospheric regions, condensation cloud, and photochemical reactions that we see on Jupiter should be present on Saturn (Figure 11.13).

Figure 11.13. Cloud Structure on Saturn. In this Cassini image, colors have been intensified, so we can see the bands and zones and storms in the atmosphere. The dark band is the shadow of the rings on the planet. (credit: NASA/JPL-Caltech/Space Science Institute)

Saturn has one anomalous cloud structure that has mystified scientists: a hexagonal wave pattern around the north pole, shown in Figure 11.14. The six sides of the hexagon are each longer than the diameter of Earth. Winds are also extremely high on Saturn, with speeds of up to 1800 kilometers per hour measured near the equator.

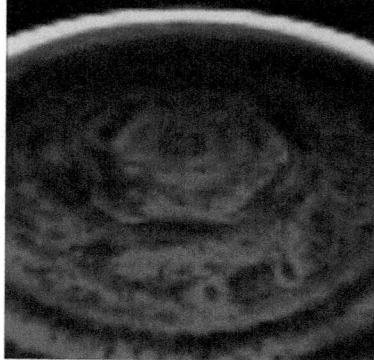

Figure 11.14. Hexagon Pattern on Saturn's North Pole. In this infrared nighttime image from the Cassini mission, the path of Saturn's hexagonal jet stream is visible as the planet's north pole emerges from the darkness of winter. (credit: NASA/JPL/University of Arizona)

This OpenStax book is available for free at http://cnx.org/content/col11992/1.8

Download for free at http://cnx.org/content/col11992/latest/

LINK TO LEARNING

See images of Saturn's hexagon (https://openstax.org/l/30Hexagon) with exaggerated color in this brief NASA video.

Unlike Jupiter and Saturn, Uranus is almost entirely featureless as seen at wavelengths that range from the ultraviolet to the infrared (see its rather boring image in Figure 11.1). Calculations indicate that the basic atmospheric structure of Uranus should resemble that of Jupiter and Saturn, although its upper clouds (at the 1-bar pressure level) are composed of methane rather than ammonia. However, the absence of an internal heat source suppresses up-and-down movement and leads to a very stable atmosphere with little visible structure.

Neptune differs from Uranus in its appearance, although their basic atmospheric temperatures are similar. The upper clouds are composed of methane, which forms a thin cloud layer near the top of the troposphere at a temperature of 70 K and a pressure of 1.5 bars. Most the atmosphere above this level is clear and transparent, with less haze than is found on Uranus. The scattering of sunlight by gas molecules lends Neptune a pale blue color similar to that of Earth's atmosphere (Figure 11.15). Another cloud layer, perhaps composed of hydrogen sulfide ice particles, exists below the methane clouds at a pressure of 3 bars.

Figure 11.15. Neptune. The planet Neptune is seen here as photographed by Voyager in 1989. The blue color, exaggerated with computer processing, is caused by the scattering of sunlight in the planet's upper atmosphere. (credit: modification of work by NASA)

Unlike Uranus, Neptune has an atmosphere in which convection currents—vertical drafts of gas—emanate from the interior, powered by the planet's internal heat source. These currents carry warm gas above the 1.5-bar cloud level, forming additional clouds at elevations about 75 kilometers higher. These high-altitude clouds form bright white patterns against the blue planet beneath. Voyager photographed distinct shadows on the methane cloud tops, permitting the altitudes of the high clouds to be calculated. Figure 11.16 is a remarkable close-up of Neptune's outer layers that could never have been obtained from Earth.

Download for free at http://cnx.org/content/col11992/latest/

Figure 11.16. High Clouds in the Atmosphere of Neptune. These bright, narrow cirrus clouds are made of methane ice crystals. From the shadows they cast on the thicker cloud layer below, we can measure that they are about 75 kilometers higher than the main clouds. (credit: modification of work by NASA/JPL)

Winds and Weather

The atmospheres of the jovian planets have many regions of high pressure (where there is more air) and low pressure (where there is less). Just as it does on Earth, air flows between these regions, setting up wind patterns that are then distorted by the rotation of the planet. By observing the changing cloud patterns on the jovian planets, we can measure wind speeds and track the circulation of their atmospheres.

The atmospheric motions we see on these planets are fundamentally different from those on the terrestrial planets. The giants spin faster, and their rapid rotation tends to smear out of the circulation into horizontal (east-west) patterns parallel to the equator. In addition, there is no solid surface below the atmosphere against which the circulation patterns can rub and lose energy (which is how tropical storms on Earth ultimately die out when they come over land).

As we have seen, on all the giants except Uranus, heat from the inside contributes about as much energy to the atmosphere as sunlight from the outside. This means that deep convection currents of rising hot air and falling cooler air circulate throughout the atmospheres of the planets in the vertical direction.

The main features of Jupiter's visible clouds (see Figure 11.2 and Figure 11.10, for example) are alternating dark and light bands that stretch around the planet parallel to the equator. These bands are semi-permanent features, although they shift in intensity and position from year to year. Consistent with the small tilt of Jupiter's axis, the pattern does not change with the seasons.

More fundamental than these bands are underlying east-west wind patterns in the atmosphere, which do not appear to change at all, even over many decades. These are illustrated in Figure 11.17, which indicates how strong the winds are at each latitude for the giant planets. At Jupiter's equator, a jet stream flows eastward with a speed of about 90 meters per second (300 kilometers per hour), similar to the speed of jet streams in Earth's upper atmosphere. At higher latitudes there are alternating east- and west-moving streams, with each hemisphere an almost perfect mirror image of the other. Saturn shows a similar pattern, but with a much stronger equatorial jet stream, as we noted earlier.

This OpenStax book is available for free at http://cnx.org/content/col11992/1.8

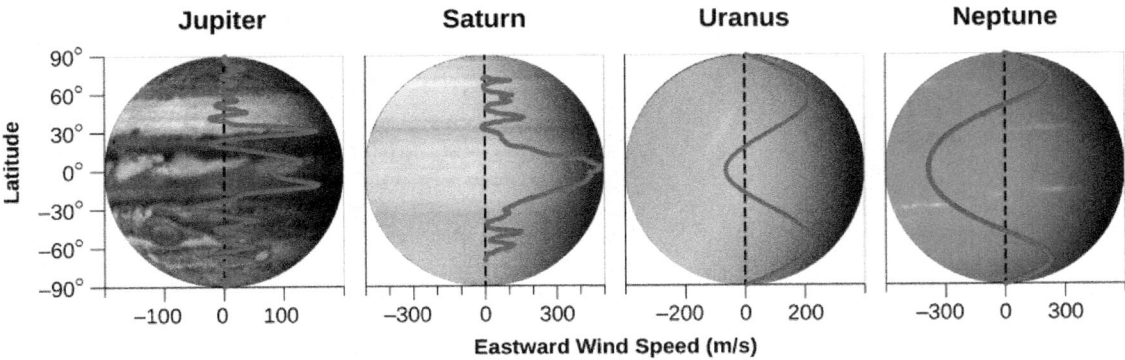

Figure 11.17. Winds on the Giant Planets. This image compares the winds of the giant planets, illustrating that wind speed (shown on the horizontal axis) and wind direction vary with latitude (shown on the vertical axis). Winds are measured relative to a planet's internal rotation speed. A positive velocity means that the winds are blowing in the same direction as, but faster than, the planet's internal rotation. A negative velocity means that the winds are blowing more slowly than the planet's internal rotation. Note that Saturn's winds move faster than those of the other planets.

The light zones on Jupiter are regions of upwelling air capped by white ammonia cirrus clouds. They apparently represent the tops of upward-moving convection currents.[2] The darker belts are regions where the cooler atmosphere moves downward, completing the convection cycle; they are darker because fewer ammonia clouds mean we can see deeper into the atmosphere, perhaps down to a region of ammonium hydrosulfide (NH_4SH) clouds. The Galileo probe sampled one of the clearest of these dry downdrafts.

In spite of the strange seasons induced by the 98° tilt of its axis, Uranus' basic circulation is parallel with its equator, as is the case on Jupiter and Saturn. The mass of the atmosphere and its capacity to store heat are so great that the alternating 42-year periods of sunlight and darkness have little effect. In fact, Voyager measurements show that the atmospheric temperature is even a few degrees higher on the dark winter side than on the hemisphere facing the Sun. This is another indication that the behavior of such giant planet atmospheres is a complex problem that we do not fully understand.

Neptune's weather is characterized by strong east-west winds generally similar to those observed on Jupiter and Saturn. The highest wind speeds near its equator reach 2100 kilometers per hour, even higher than the peak winds on Saturn. The Neptune equatorial jet stream actually approaches supersonic speeds (faster than the speed of sound in Neptune's air).

Giant Storms on Giant Planets

Superimposed on the regular atmospheric circulation patterns we have just described are many local disturbances—weather systems or *storms*, to borrow the term we use on Earth. The most prominent of these are large, oval-shaped, high-pressure regions on both Jupiter (Figure 11.18) and Neptune.

---

2   Recall from earlier chapters that convection is a process in which liquids, heated from underneath, have regions where hot material rises and cooler material descends. You can see convection at work if you heat oatmeal on a stovetop or watch miso soup boil.

Download for free at http://cnx.org/content/col11992/latest/

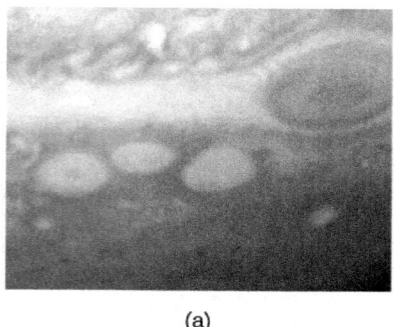 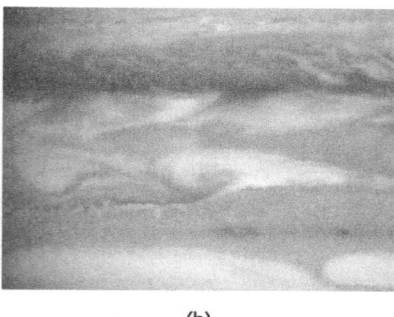

(a)   (b)

Figure 11.18. Storms on Jupiter. Two examples of storms on Jupiter illustrate the use of enhanced color and contrast to bring out faint features. (a) The three oval-shaped white storms below and to the left of Jupiter's Great Red Spot are highly active, and moved closer together over the course of seven months between 1994 and 1995. (b) The clouds of Jupiter are turbulent and ever-changing, as shown in this Hubble Space Telescope image from 2007. (credit a: modification of work by Reta Beebe, Amy Simon (New Mexico State Univ.), and NASA; credit b: modification of work by NASA, ESA, and A. Simon-Miller (NASA Goddard Space Flight Center))

The largest and most famous of Jupiter's storms is the Great Red Spot, a reddish oval in the southern hemisphere that changes slowly; it was 25,000 kilometers long when Voyager arrived in 1979, but it had shrunk to 20,000 kilometers by the end of the Galileo mission in 2000 (Figure 11.19). The giant storm has persisted in Jupiter's atmosphere ever since astronomers were first able to observe it after the invention of the telescope, more than 300 years ago. However, it has continued to shrink, raising speculation that we may see its end within a few decades.

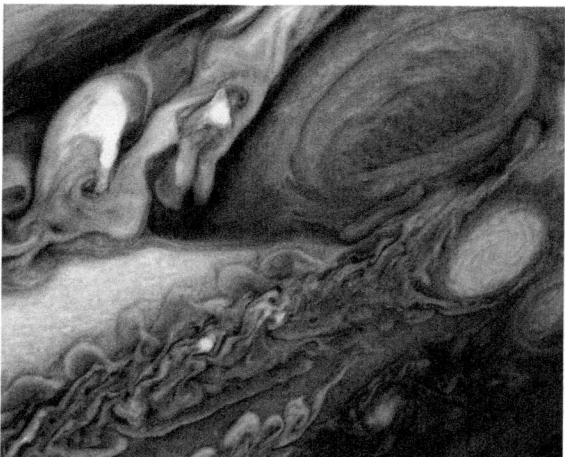

Figure 11.19. Jupiter's Great Red Spot. This is the largest storm system on Jupiter, as seen during the Voyager spacecraft flyby. Below and to the right of the Red Spot is one of the white ovals, which are similar but smaller high-pressure features. The white oval is roughly the size of planet Earth, to give you a sense of the huge scale of the weather patterns we are seeing. The colors on the Jupiter image have been somewhat exaggerated here so astronomers (and astronomy students) can study their differences more effectively. See Figure 11.2 to get a better sense of the colors your eye would actually see near Jupiter. (credit: NASA/JPL)

In addition to its longevity, the Red Spot differs from terrestrial storms in being a high-pressure region; on our planet, such storms are regions of lower pressure. The Red Spot's counterclockwise rotation has a period of six days. Three similar but smaller disturbances (about as big as Earth) formed on Jupiter in the 1930s. They look like white ovals, and one can be seen clearly below and to the right of the Great Red Spot in Figure 11.19. In 1998, the Galileo spacecraft watched as two of these ovals collided and merged into one.

This OpenStax book is available for free at http://cnx.org/content/col11992/1.8

Download for free at http://cnx.org/content/col11992/latest/

We don't know what causes the Great Red Spot or the white ovals, but we do have an idea how they can last so long once they form. On Earth, the lifetime of a large oceanic hurricane or typhoon is typically a few weeks, or even less when it moves over the continents and encounters friction with the land. Jupiter has no solid surface to slow down an atmospheric disturbance; furthermore, the sheer size of the disturbances lends them stability. We can calculate that on a planet with no solid surface, the lifetime of anything as large as the Red Spot should be measured in centuries, while lifetimes for the white ovals should be measured in decades, which is pretty much what we have observed.

Despite Neptune's smaller size and different cloud composition, Voyager showed that it had an atmospheric feature surprisingly similar to Jupiter's Great Red Spot. Neptune's Great Dark Spot was nearly 10,000 kilometers long (Figure 11.15). On both planets, the giant storms formed at latitude 20° S, had the same shape, and took up about the same fraction of the planet's diameter. The Great Dark Spot rotated with a period of 17 days, versus about 6 days for the Great Red Spot. When the Hubble Space Telescope examined Neptune in the mid-1990s, however, astronomers could find no trace of the Great Dark Spot on their images.

Although many of the details of the weather on the jovian planets are not yet understood, it is clear that if you are a fan of dramatic weather, these worlds are the place to look. We study the features in these atmospheres not only for what they have to teach us about conditions in the jovian planets, but also because we hope they can help us understand the weather on Earth just a bit better.

EXAMPLE 11.1

Storms and Winds

The wind speeds in circular storm systems can be formidable on both Earth and the giant planets. Think about our big terrestrial hurricanes. If you watch their behavior in satellite images shown on weather outlets, you will see that they require about one day to rotate. If a storm has a diameter of 400 km and rotates once in 24 h, what is the wind speed?

Solution

Speed equals distance divided by time. The distance in this case is the circumference ($2\pi R$ or πd), or approximately 1250 km, and the time is 24 h, so the speed at the edge of the storm would be about 52 km/h. Toward the center of the storm, the wind speeds can be much higher.

Check Your Learning

Jupiter's Great Red Spot rotates in 6 d and has a circumference equivalent to a circle with radius 10,000 km. Calculate the wind speed at the outer edge of the spot.

Answer:

For the Great Red Spot of Jupiter, the circumference ($2\pi R$) is about 63,000 km. Six d equals 144 h, suggesting a speed of about 436 km/h. This is much faster than wind speeds on Earth.

Download for free at http://cnx.org/content/col11992/latest/

CHAPTER 11 REVIEW

KEY TERMS

photochemistry chemical changes caused by electromagnetic radiation

synchrotron radiation the radiation emitted by charged particles being accelerated in magnetic fields and moving at speeds near that of light

SUMMARY

11.1 Exploring the Outer Planets

The outer solar system contains the four giant planets: Jupiter, Saturn, Uranus, and Neptune. The gas giants Jupiter and Saturn have overall compositions similar to that of the Sun. These planets have been explored by the Pioneer, Voyager, Galileo, and Cassini spacecraft. Voyager 2, perhaps the most successful of all space-science missions, explored Jupiter (1979), Saturn (1981), Uranus (1986), and Neptune (1989)—a grand tour of the giant planets—and these flybys have been the only explorations to date of the ice giants Uranus and Neptune. The Galileo and Cassini missions were long-lived orbiters, and each also deployed an entry probe, one into Jupiter and one into Saturn's moon Titan.

11.2 The Giant Planets

Jupiter is 318 times more massive than Earth. Saturn is about 25% as massive as Jupiter, and Uranus and Neptune are only 5% as massive. All four have deep atmospheres and opaque clouds, and all rotate quickly with periods from 10 to 17 hours. Jupiter and Saturn have extensive mantles of liquid hydrogen. Uranus and Neptune are depleted in hydrogen and helium relative to Jupiter and Saturn (and the Sun). Each giant planet has a core of "ice" and "rock" of about 10 Earth masses. Jupiter, Saturn, and Neptune have major internal heat sources, obtaining as much (or more) energy from their interiors as by radiation from the Sun. Uranus has no measurable internal heat. Jupiter has the strongest magnetic field and largest magnetosphere of any planet, first discovered by radio astronomers from observations of synchrotron radiation.

11.3 Atmospheres of the Giant Planets

The four giant planets have generally similar atmospheres, composed mostly of hydrogen and helium. Their atmospheres contain small quantities of methane and ammonia gas, both of which also condense to form clouds. Deeper (invisible) cloud layers consist of water and possibly ammonium hydrosulfide (Jupiter and Saturn) and hydrogen sulfide (Neptune). In the upper atmospheres, hydrocarbons and other trace compounds are produced by photochemistry. We do not know exactly what causes the colors in the clouds of Jupiter. Atmospheric motions on the giant planets are dominated by east-west circulation. Jupiter displays the most active cloud patterns, with Neptune second. Saturn is generally bland, in spite of its extremely high wind speeds, and Uranus is featureless (perhaps due to its lack of an internal heat source). Large storms (oval-shaped high-pressure systems such as the Great Red Spot on Jupiter and the Great Dark Spot on Neptune) can be found in some of the planet atmospheres.

This OpenStax book is available for free at http://cnx.org/content/col11992/1.8

FOR FURTHER EXPLORATION

Articles

Jupiter

Aguirre, Edwin. "Hubble Zooms in on Jupiter's New Red Spot." *Sky & Telescope* (August 2006): 26.

Beatty, J. "Into the Giant." *Sky & Telescope* (April 1996): 20. On the Galileo probe.

Beebe, R. "Queen of the Giant Storms." *Sky & Telescope* (October 1990): 359. Excellent review of the Red Spot.

Johnson, T. "The Galileo Mission to Jupiter and Its Moons." *Scientific American* (February 2000): 40. Results about Jupiter, Io, Ganymede, and Callisto.

Simon, A. "The Not-So-Great Red Spot." *Sky & Telescope* (March 2016): 18. On how the huge storm on Jupiter is evolving with time.

Smith, B. "Voyage of the Century." *National Geographic* (August 1990): 48. Beautiful summary of the Voyager mission to all four outer planets.

Stern, S. "Jupiter Up Close and Personal." *Astronomy* (August 2007): 28. On the New Horizons mission flyby in February 2007.

Saturn

Gore, R. "The Riddle of the Rings." *National Geographic* (July 1981): 3. Colorful report on the Voyager mission.

McEwen, A. "Cassini Unveils Saturn." *Astronomy* (July 2006): 30. A report on the first two years of discoveries in the Saturn system.

Spilker, L. "Saturn Revolution." *Astronomy* (October 2008): 34. On results from the Cassini mission.

Talcott, R. "Saturn's Sweet Surprises." *Astronomy* (June 2007): 52. On Cassini mission results.

Uranus and Neptune

Cowling, T. "Big Blue: The Twin Worlds of Uranus and Neptune." *Astronomy* (October 1990): 42. Nice, long review of the two planets.

Gore, R. "Neptune: Voyager's Last Picture Show." *National Geographic* (August 1990): 35.

Lunine, J. "Neptune at 150." *Sky & Telescope* (September 1996): 38. Nice review.

Websites

Jupiter

NASA Solar System Exploration: http://Solarsystem.nasa.gov/planets/jupiter

Nine Planets Site: http://nineplanets.org/jupiter.html

Planetary Sciences Site: http://nssdc.gsfc.nasa.gov/planetary/planets/jupiterpage.html

Saturn

NASA Solar System Exploration: http://Solarsystem.nasa.gov/planets/saturn

Nine Planets Site: http://nineplanets.org/saturn.html

Planetary Sciences Site: http://nssdc.gsfc.nasa.gov/planetary/planets/saturnpage.html

Download for free at http://cnx.org/content/col11992/latest/

Uranus

NASA Solar System Exploration: http://Solarsystem.nasa.gov/planets/uranus

Nine Planets Site: http://nineplanets.org/uranus.html

Planetary Sciences Site: http://nssdc.gsfc.nasa.gov/planetary/planets/uranuspage.html

Neptune

NASA Solar System Exploration: http://Solarsystem.nasa.gov/planets/neptune

Nine Planets Site: http://nineplanets.org/neptune.html

Planetary Sciences Site: http://nssdc.gsfc.nasa.gov/planetary/planets/neptunepage.html

Missions

Cassini Mission Site at the Jet Propulsion Lab: http://saturn.jpl.nasa.gov/index.cfm

Cassini-Huygens Mission Site at European Space Agency: http://sci.esa.int/cassini-huygens/

NASA Galileo Mission Site: http://Solarsystem.nasa.gov/galileo/

NASA's Juno Mission to Jupiter: http://www.nasa.gov/mission_pages/juno/main/index.html

Voyager Mission Site at the Jet Propulsion Lab: http://voyager.jpl.nasa.gov/

Videos

Cassini: 15 Years of Exploration: https://www.youtube.com/watch?v=2z8fzz_MBAw. Quick visual summary of mission highlights (2:29).

In the Land of Enchantment: The Epic Story of the Cassini Mission to Saturn: https://www.youtube.com/watch?v=Vx135n8VFxY. An inspiring illustrated lecture by Cassini Mission Imagining Lead Scientist Carolyn Porco (1:37:52).

Jupiter: The Largest Planet: http://www.youtube.com/watch?v=s56pxa9lpvo. Produced by NASA's Goddard Space Flight Center and Science on a Sphere (7:29).

⚎ COLLABORATIVE GROUP ACTIVITIES

A. A new member of Congress has asked your group to investigate why the Galileo probe launched into the Jupiter atmosphere in 1995 survived only 57 minutes and whether this was an example of a terrible scandal. Make a list of all the reasons the probe did not last longer, and why it was not made more durable. (Remember that the probe had to hitch a ride to Jupiter!)

B. Select one of the jovian planets and organize your group to write a script for an evening news weather report for the planet you chose. Be sure you specify roughly how high in the atmosphere the region lies for which you are giving the report.

C. What does your group think should be the next step to learn more about the giant planets? Put cost considerations aside for a moment: What kind of mission would you recommend to NASA to learn more about these giant worlds? Which world or worlds should get the highest priority and why?

D. Suppose that an extremely dedicated (and slightly crazy) astronomer volunteers to become a human probe into Jupiter (and somehow manages to survive the trip through Jupiter's magnetosphere alive). As she

This OpenStax book is available for free at http://cnx.org/content/col11992/1.8

enters the upper atmosphere of Jupiter, would she fall faster or slower than she would fall doing the same suicidal jump into the atmosphere of solid Earth? Groups that have some algebra background could even calculate the force she would feel compared to the force on Earth. (Bonus question: If she were in a capsule, falling into Jupiter feet first, and the floor of the capsule had a scale, what would the scale show as her weight compared to her weight on Earth?)

E. Would you or anyone in your group volunteer for a one-way, life-long mission to a space station orbiting any of the gas giants without ever being able to return to Earth? What are the challenges of such a mission? Should we leave all exploration of the outer solar system to unmanned space probes?

EXERCISES

Review Questions

1. What are the main challenges involved in sending probes to the giant planets?

2. Why is it difficult to drop a probe like Galileo? How did engineers solve this problem?

3. Explain why visual observation of the gas giants is not sufficient to determine their rotation periods, and what evidence was used to deduce the correct periods.

4. What are the seasons like on Jupiter?

5. What is the consequence of Uranus' spin axis being 98° away from perpendicular to its orbital plane?

6. Describe the seasons on the planet Uranus.

7. At the pressures in Jupiter's interior, describe the physical state of the hydrogen found there.

8. Which of the gas giants has the largest icy/rocky core compared to its overall size?

9. In the context of the giant planets and the conditions in their interiors, what is meant by "rock" and "ice"?

10. What is the primary source of Jupiter's internal heat?

11. Describe the interior heat source of Saturn.

12. Which planet has the strongest magnetic field, and hence the largest magnetosphere? What is its source?

13. What are the visible clouds on the four giant planets composed of, and why are they different from each other?

14. Compare the atmospheric circulation (weather) of the four giant planets.

15. What are the main atmospheric heat sources of each of the giant planets?

16. Why do the upper levels of Neptune's atmosphere appear blue?

17. How do storms on Jupiter differ from storm systems on Earth?

Thought Questions

18. Describe the differences in the chemical makeup of the inner and outer parts of the solar system. What is the relationship between what the planets are made of and the temperature where they formed?

19. How did the giant planets grow to be so large?

Download for free at http://cnx.org/content/col11992/latest/

20. Jupiter is denser than water, yet composed for the most part of two light gases, hydrogen and helium. What makes Jupiter as dense as it is?

21. Would you expect to find free oxygen gas in the atmospheres of the giant planets? Why or why not?

22. Why would a tourist brochure (of the future) describing the most dramatic natural sights of the giant planets have to be revised more often than one for the terrestrial planets?

23. The water clouds believed to be present on Jupiter and Saturn exist at temperatures and pressures similar to those in the clouds of the terrestrial atmosphere. What would it be like to visit such a location on Jupiter or Saturn? In what ways would the environment differ from that in the clouds of Earth?

24. Describe the different processes that lead to substantial internal heat sources for Jupiter and Saturn. Since these two objects generate much of their energy internally, should they be called stars instead of planets? Justify your answer.

25. Research the Galileo mission. What technical problems occurred between the mission launch and the arrival of the craft in Jupiter's system, and how did the mission engineers deal with them? (Good sources of information include *Astronomy* and *Sky & Telescope* articles, plus the mission website.)

Figuring For Yourself

26. How many times more pressure exists in the interior of Jupiter compared to that of Earth?

27. Calculate the wind speed at the edge of Neptune's Great Dark Spot, which was 10,000 km in diameter and rotated in 17 d.

28. Calculate how many Earths would fit into the volumes of Saturn, Uranus, and Neptune.

29. As the Voyager spacecraft penetrated into the outer solar system, the illumination from the Sun declined. Relative to the situation at Earth, how bright is the sunlight at each of the jovian planets?

30. The ions in the inner parts of Jupiter's magnetosphere rotate with the same period as Jupiter. Calculate how fast they are moving at the orbit of Jupiter's moon Io (see Appendix G). Will these ions strike Io from behind or in front as it moves about Jupiter?

This OpenStax book is available for free at http://cnx.org/content/col11992/1.8

RINGS, MOONS, AND PLUTO

Figure 12.1. Jupiter Family. This montage, assembled from individual Galileo and Voyager images, shows a "family portrait" of Jupiter (with its giant red spot) and its four large moons. From top to bottom, we see Io, Europa, Ganymede, and Callisto. The colors are exaggerated by image processing to emphasize contrasts. (credit: modification of work by NASA)

Chapter Outline

 Thinking Ahead

"Our imaginations always fall short of anticipating the beauty we find in nature."—Geologist Laurence Soderblom, discussing the 1989 Voyager encounter with Neptune's moons

All four giant planets are accompanied by moons that orbit about them like planets in a miniature solar system. Nearly 200 moons are known in the outer solar system—too many to name individually or discuss in any detail. Astronomers anticipate that additional small moons await future discovery. We have also discovered a fascinating variety of rings around each of the jovian planets.

Before the Voyager missions, even the largest of the outer-planet moons were mere points of light in our telescopes. Then, in less than a decade, we had close-up images, and these moons became individual worlds for us, each with unique features and peculiarities. The Galileo mission added greatly to our knowledge of the moons of Jupiter, and the Cassini mission has done the same for the Saturn system. In 2015, the NASA New Horizons spacecraft completed the initial exploration of the "classical" planets in the solar system with its flyby of Pluto and its moons. We include Pluto here because in some ways it resembles some of the larger moons in the outer solar system. Each new spacecraft mission has revealed many surprises, as the objects in the outer solar system are much more varied and geologically active than scientists had anticipated.

Download for free at http://cnx.org/content/col11992/latest/

12.1 | RING AND MOON SYSTEMS INTRODUCED

Learning Objectives

By the end of this section, you will be able to:

> Name the major moons of each of the jovian planets
> Describe the basic composition of each jovian planet's ring system

The rings and moons (see the moons in Figure 12.2) of the outer solar system are not composed of the same materials as the mostly rocky objects in the inner solar system. We should expect this, since they formed in regions of lower temperature, cool enough so that large quantities of water ice were available as building materials. Most of these objects also contain dark, organic compounds mixed with their ice and rock. Don't be surprised, therefore, to find that many objects in the ring and moon systems are both icy and dark.

Roughly a third of the moons in the outer solar system are in *direct* or regular orbits; that is, they revolve about their parent planet in a west-to-east direction and in the plane of the planet's equator. The majority are irregular moons that orbit in a *retrograde* (east-to-west) direction or else have orbits of high eccentricity (more elliptical than circular) or high inclination (moving in and out of the planet's equatorial plane). These irregular moons are mostly located relatively far from their planet; they were probably formed elsewhere and subsequently captured by the planet they now orbit. (Perhaps the fact that they were not born locally will excuse their ill-mannered behavior.)

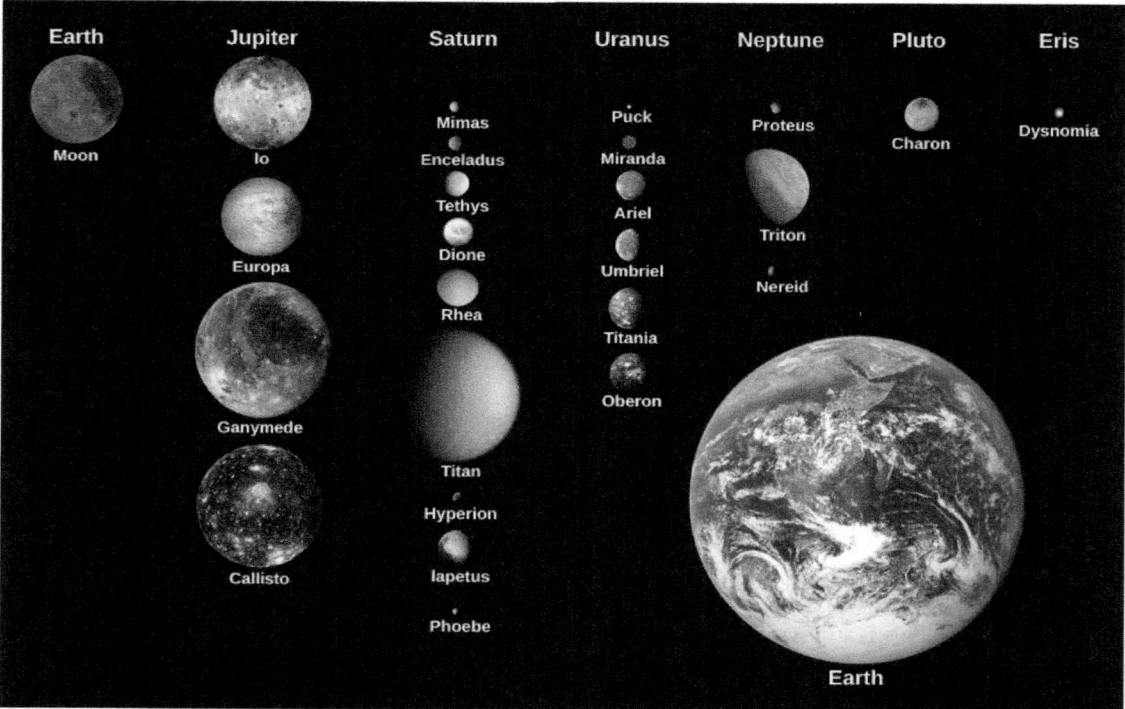

Figure 12.2. Moons of the Solar System. This image shows some selected moons of our solar system and their comparison to the size of Earth's Moon and Earth itself. (credit: modification of work by NASA)

This OpenStax book is available for free at http://cnx.org/content/col11992/1.8

Download for free at http://cnx.org/content/col11992/latest/

The Jupiter System

Jupiter has 67 known moons (that's the number as we write) and a faint ring. These include four large moons—Callisto, Ganymede, Europa, and Io (see Figure 12.1)—discovered in 1610 by Galileo and therefore often called the *Galilean moons*. The smaller of these, Europa and Io, are about the size of our Moon, while the larger, Ganymede and Callisto, are about the same size as the planet Mercury. Most of Jupiter's moons are much smaller. The majority are in retrograde orbits more than 20 million kilometers from Jupiter; these are very likely small captured asteroids.

The Saturn System

Saturn has at least 62 known moons in addition to a magnificent set of rings. The largest of the moons, Titan, is almost as big as Ganymede in Jupiter's system, and it is the only moon with a substantial atmosphere and lakes or seas of liquid hydrocarbons (such as methane and ethane) on the surface. Saturn has six other large regular moons with diameters between 400 and 1600 kilometers, a collection of small moons orbiting in or near the rings, and many captured strays similar to those of Jupiter. Mysteriously, one of Saturn's smaller moons, Enceladus, has active geysers of water being expelled into space.

The rings of Saturn, one of the most impressive sights in the solar system, are broad and flat, with a few major and many minor gaps. They are not solid, but rather a huge collection of icy fragments, all orbiting the equator of Saturn in a traffic pattern that makes rush hour in a big city look simple by comparison. Individual ring particles are composed primarily of water ice and are typically the size of ping-pong balls, tennis balls, and basketballs.

The Uranus System

The ring and moon system of Uranus is tilted at 98°, just like the planet itself. It consists of 11 rings and 27 currently known moons. The five largest moons are similar in size to the six regular moons of Saturn, with diameters of 500 to 1600 kilometers. Discovered in 1977, the rings of Uranus are narrow ribbons of dark material with broad gaps in between. Astronomers suppose that the ring particles are confined to these narrow paths by the gravitational effects of numerous small moons, many of which we have not yet glimpsed.

The Neptune System

Neptune has 14 known moons. The most interesting of these is Triton, a relatively large moon in a retrograde orbit—which is unusual. Triton has a very thin atmosphere, and active eruptions were discovered there by Voyager in its 1989 flyby. To explain its unusual characteristics, astronomers have suggested that Triton may have originated beyond the Neptune system, as a dwarf planet like Pluto. The rings of Neptune are narrow and faint. Like those of Uranus, they are composed of dark materials and are thus not easy to see.

THE GALILEAN MOONS OF JUPITER

Learning Objectives

By the end of this section, you will be able to:

> Describe the major features we can observe about Callisto and what we can deduce from them
> Explain the evidence for tectonic and volcanic activity on Ganymede
> Explain what may be responsible for the unusual features on the icy surface of Europa
> Describe the major distinguishing characteristic of Io
> Explain how tidal forces generate the geological activity we see on Europa and Io

Download for free at http://cnx.org/content/col11992/latest/

From 1996 to 1999, the Galileo spacecraft careered through the jovian system on a complex but carefully planned trajectory that provided repeated close encounters with the large Galilean moons. (Beginning in 2004, we received an even greater bonanza of information about Titan, obtained from the Cassini spacecraft and its Huygens probe, which landed on its surface. We include Titan, Saturn's one big moon, here for comparison.) Table 12.1 summarizes some basic facts about these large moons (plus our own Moon for comparison).

The Largest Moons

Name	Diameter (km)	Mass (Earth's Moon = 1)	Density (g/cm³)	Reflectivity (%)
Moon	3476	1.0	3.3	12
Callisto	4820	1.5	1.8	20
Ganymede	5270	2.0	1.9	40
Europa	3130	0.7	3.0	70
Io	3640	1.2	3.5	60
Titan	5150	1.9	1.9	20

Table 12.1

Callisto: An Ancient, Primitive World

We begin our discussion of the Galilean moons with the outermost one, Callisto, not because it is remarkable but because it is not. This makes it a convenient object with which other, more active, worlds can be compared. Its distance from Jupiter is about 2 million kilometers, and it orbits the planet in 17 days. Like our own Moon, Callisto rotates in the same period as it revolves, so it always keeps the same face toward Jupiter. Callisto's day thus equals its month: 17 days. Its noontime surface temperature is only 130 K (about 140 °C below freezing), so that water ice is stable (it never evaporates) on its surface year round.

Callisto has a diameter of 4820 kilometers, almost the same as the planet Mercury (Figure 12.3). Yet its mass is only one-third as great, which means its density (the mass divided by the volume) must be only one-third as great as well. This tells us that Callisto has far less of the rocky and metallic materials found in the inner planets and must instead be an icy body through much of its interior. Callisto can show us how the geology of an icy object compares with those made primarily of rock.

Unlike the worlds we have studied so far, Callisto has not fully *differentiated* (separated into layers of different density materials). We can tell that it lacks a dense core from the details of its gravitational pull on the Galileo spacecraft. This surprised scientists, who expected that all the big icy moons would be differentiated. It should be easier for an icy body to differentiate than for a rocky one because the melting temperature of ice is so low. Only a little heating will soften the ice and get the process started, allowing the rock and metal to sink to the center while the slushy ice floats to the surface. Yet Callisto seems to have frozen solid before the process of differentiation was complete.

The surface of Callisto is covered with impact craters, like the lunar highlands. The survival of these craters tells us that an icy object can retain impact craters on its surface. Callisto is unique among the planet-sized objects

This OpenStax book is available for free at http://cnx.org/content/col11992/1.8

of the solar system in the apparent absence of interior forces to drive geological change. You might say that this moon was stillborn, and it has remained geologically dead for more than 4 billion years (Figure 12.3).

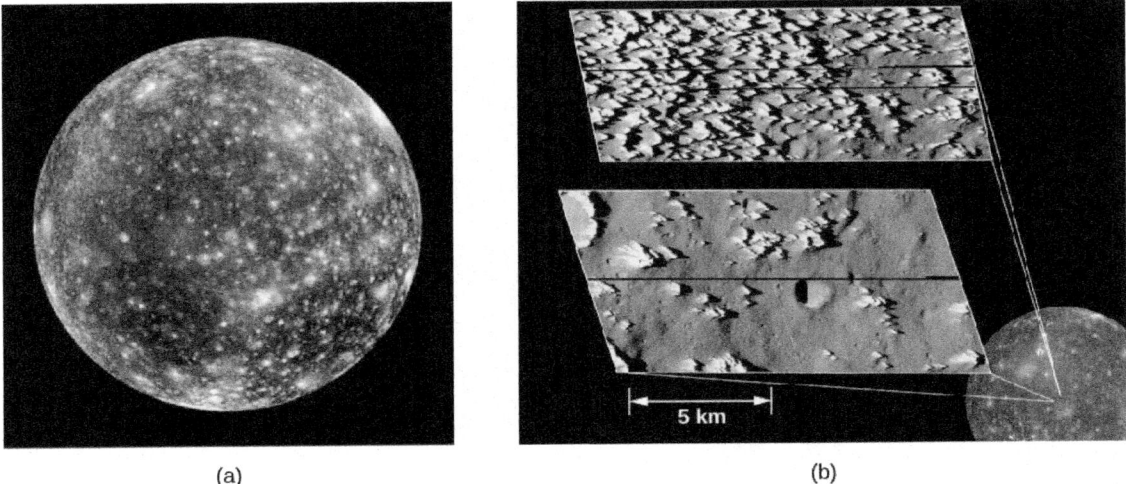

(a)   (b)

Figure 12.3. Callisto. (a) Jupiter's outermost large moon shows a heavily cratered surface. Astronomers believe that the bright areas are mostly ice, while the darker areas are more eroded, ice-poor material. (b) These high-resolution images, taken by NASA's Galileo spacecraft in May 2001, show the icy spires (top) on Callisto's surface, with darker dust that has slid down as the ice erodes, collecting in the low-lying areas. The spires are about 80 to 100 meters tall. As the surface erodes even further, the icy spires eventually disappear, leaving impact craters exposed, as shown in the lower image. (credit a: modification of work by NASA/JPL/DLR; credit b: modification of work by NASA/JPL/Arizona State University, Academic Research Lab)

In thinking about ice so far from the Sun, we must take care not to judge its behavior from the much warmer ice we know and love on Earth. At the temperatures of the outer solar system, ice on the surface is nearly as hard as rock, and it behaves similarly. Ice on Callisto does not deform or flow like ice in glaciers on Earth.

Ganymede, the Largest Moon

Ganymede, the largest moon in the solar system, also shows a great deal of cratering (Figure 12.4). Recall from Other Worlds: An Introduction to the Solar System) that we can use crater counts on solid worlds to estimate the age of the surface. The more craters, the longer the surface has been exposed to battering from space, and the older it must therefore be. About one-quarter of Ganymede's surface seems to be as old and heavily cratered as that of Callisto; the rest formed more recently, as we can tell by the sparse covering of impact craters as well as the relative freshness of those craters. If we judge from crater counts, this fresher terrain on Ganymede is somewhat younger than the lunar maria or the martian volcanic plains, perhaps 2 to 3 billion years old.

The differences between Ganymede and Callisto are more than skin deep. Ganymede is a differentiated world, like the terrestrial planets. Measurements of its gravity field tell us that the rock sank to form a core about the size of our Moon, with a mantle and crust of ice "floating" above it. In addition, the Galileo spacecraft discovered that Ganymede has a magnetic field, the sure signature of a partially molten interior. There is very likely liquid water trapped within the interior. Thus, Ganymede is not a dead world but rather a place of intermittent geological activity powered by an internal heat source. Some surface features could be as young as the surface of Venus (a few hundred million years).

The younger terrain was formed by tectonic and volcanic forces (Figure 12.4). In some places, the crust apparently cracked, flooding many of the craters with water from the interior. Extensive mountain ranges were formed from compression of the crust, forming long ridges with parallel valleys spaced a few kilometers apart.

Download for free at http://cnx.org/content/col11992/latest/

In some areas, older impact craters were split and pulled apart. There are even indications of large-scale crustal movements that are similar to the plate tectonics of Earth.

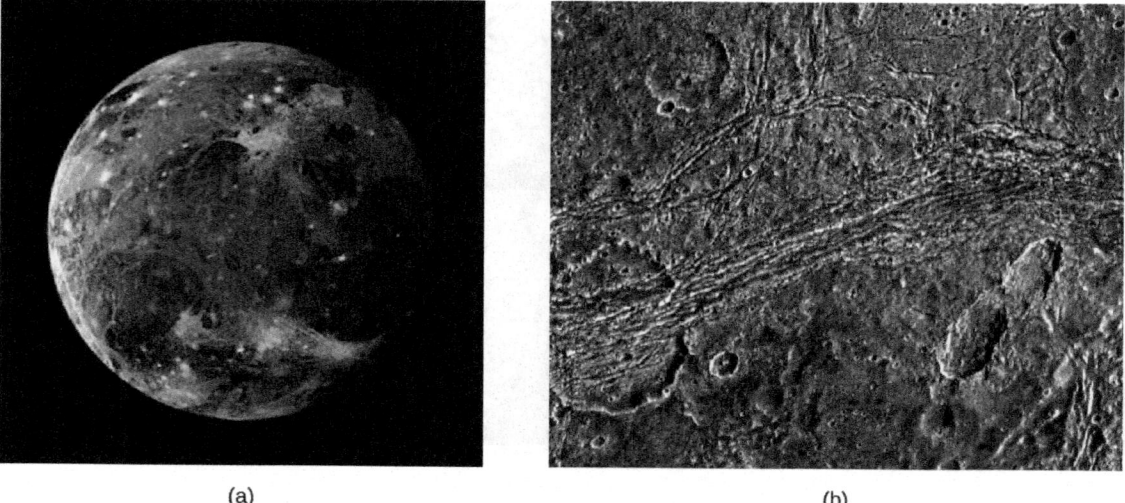

(a)   (b)

Figure 12.4. Ganymede. (a) This global view of Ganymede, the largest moon in the solar system, was taken by Voyager 2. The colors are enhanced to make spotting differences easier. Darker places are older, more heavily cratered regions; the lighter areas are younger (the reverse of our Moon). The brightest spots are sites of geologically recent impacts. (b) This close-up of Nicholson Regio on Ganymede shows an old impact crater (on the lower left-hand side) that has been split and pulled apart by tectonic forces. Against Ganymede's dark terrain, a line of grooves and ridges appears to cut through the crater, deforming its circular shape. (credit a: modification of work by NASA/JPL/DLR; credit b: modification of work by NASA/JPL/Brown University)

Why is Ganymede so different from Callisto? Possibly the small difference in size and internal heating between the two led to this divergence in their evolution. But more likely the gravity of Jupiter is to blame for Ganymede's continuing geological activity. Ganymede is close enough to Jupiter that *tidal forces* from the giant planet may have episodically heated its interior and triggered major convulsions on its crust.

A tidal force results from the unequal gravitational pull on two sides of a body. In a complex kind of modern dance, the large moons of Jupiter are caught in the varying gravity grip of both the giant planet and each other. This leads to gravitational flexing or kneading in their centers, which can heat them—an effect called **tidal heating**. (A fuller explanation is given in the section on Io.) We will see as we move inward to Europa and Io that the role of jovian tides becomes more important for moons close to the planet.

Europa, a Moon with an Ocean

Europa and Io, the inner two Galilean moons, are not icy worlds like most of the moons of the outer planets. With densities and sizes similar to our Moon, they appear to be predominantly rocky objects. How did they fail to acquire a majority share of the ice that must have been plentiful in the outer solar system at the time of their formation?

The most probable cause is Jupiter itself, which was hot enough to radiate a great deal of infrared energy during the first few million years after its formation. This infrared radiation would have heated the disk of material near the planet that would eventually coalesce into the closer moons. Thus, any ice near Jupiter was vaporized, leaving Europa and Io with compositions similar to planets in the inner solar system.

Despite its mainly rocky composition, Europa has an ice-covered surface, as astronomers have long known from examining spectra of sunlight reflected from it. In this it resembles Earth, which has a layer of water on its surface, but in Europa's case the water is capped by a thick crust of ice. There are very few impact craters in this

This OpenStax book is available for free at http://cnx.org/content/col11992/1.8

Download for free at http://cnx.org/content/col11992/latest/

ice, indicating that the surface of Europa is in a continual state of geological self-renewal. Judging from crater counts, the surface must be no more than a few million years old, and perhaps substantially less. In terms of its ability to erase impact craters, Europa is more geologically active than Earth.

When we look at close-up photos of Europa, we see a strange, complicated surface (Figure 12.5). For the most part, the icy crust is extremely smooth, but it is crisscrossed with cracks and low ridges that often stretch for thousands of kilometers. Some of these long lines are single, but most are double or multiple, looking rather like the remnants of a colossal freeway system.

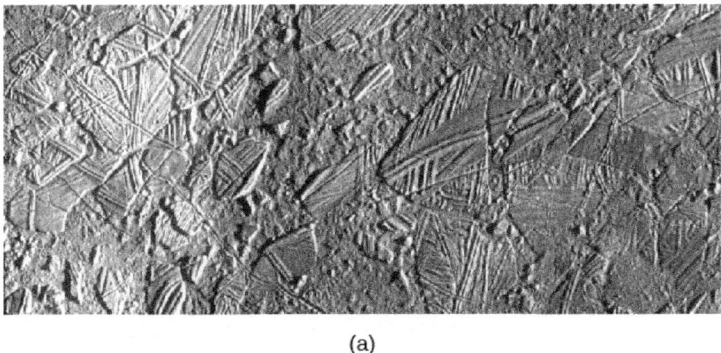 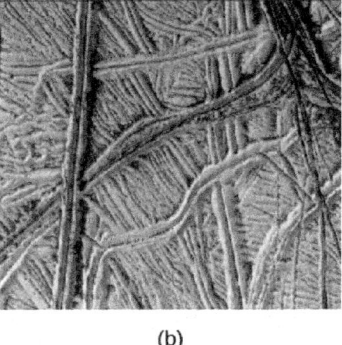

(a)   (b)

Figure 12.5. Evidence for an Ocean on Europa. (a) A close-up of an area called Conamara Chaos is shown here with enhanced color. This view is 70 kilometers wide in its long dimension. It appears that Conamara is a region where Europa's icy crust is (or recently was) relatively thin and there is easier access to the possible liquid or slushy ocean beneath. Not anchored to solid crust underneath, many of the ice blocks here seem to have slid or rotated from their original positions. In fact, the formations seen here look similar to views of floating sea-ice and icebergs in Earth's Arctic Ocean. (b) In this high-resolution view, the ice is *wrinkled* and crisscrossed by long ridges. Where these ridges intersect, we can see which ones are older and which younger; the younger ones cross over the older ones. While superficially this system of ridges resembles a giant freeway system on Europa, the ridges are much wider than our freeways and are a natural result of the flexing of the moon. (credit a: modification of work by NASA/JPL/University of Arizona; credit b: modification of work by NASA/JPL)

It is very difficult to make straight lines on a planetary surface. In discussing Mars, we explained that when Percival Lowell saw what appeared to him to be straight lines (the so-called martian "canals"), he attributed them to the engineering efforts of intelligent beings. We now know the lines on Mars were optical illusions, but the lines on Europa are real. These long cracks can form in the icy crust if it is floating without much friction on an ocean of liquid water (Figure 12.6).

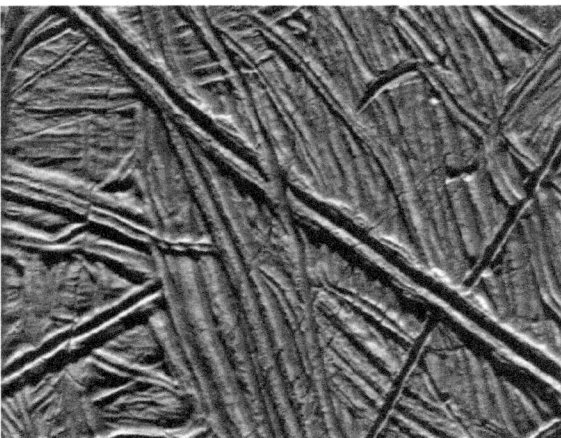

Figure 12.6. Very High-Resolution Galileo Image of One Young Double Ridge on Europa. The area in this picture is only 15 kilometers across. It appears to have formed when viscous icy material was forced up through a long, straight crack in the crust. Note how the young ridge going from top left toward bottom right lies on top of older features, which are themselves on top of even older ones. (credit: modification of work by NASA/JPL)

Download for free at http://cnx.org/content/col11992/latest/

The close-up Galileo images appear to confirm the existence of a global ocean. In many places, the surface of Europa looks just as we would expect for a thick layer of ice that was broken up into giant icebergs and ice floes and then refrozen in place. When the ice breaks, water or slush from below may be able to seep up through the cracks and make the ridges and multiple-line features we observe. Many episodes of ice cracking, shifting, rotating, and refreezing are required to explain the complexity we see. The icy crust might vary in thickness from a kilometer or so up to 20 kilometers. Further confirmation that a liquid ocean exists below the ice comes from measurements of the small magnetic field induced by Europa's interactions with the magnetosphere of Jupiter. The "magnetic signature" of Europa is that of a liquid water ocean, not one of ice or rock.

If Europa really has a large ocean of liquid water under its ice, then it may be the only place in the solar system, other than Earth, with really large amounts of liquid water.[1] To remain liquid, this ocean must be warmed by heat escaping from the interior of Europa. Hot (or at least warm) springs might be active there, analogous to those we have discovered in the deep oceans of Earth. The necessary internal heat is generated by tidal heating (see the discussion later in this chapter).

LINK TO LEARNING

A short film (https://openstax.org/l/30Europa) with planetary scientist Kevin Hand explains why Europa is so interesting for future exploration. Or listen to this more in-depth talk (https://openstax.org/l/30Europa2) on Europa.

What makes the idea of an ocean with warm springs exciting is the discovery in Earth's oceans of large ecosystems clustered around deep ocean hot springs. Such life derives all its energy from the mineral-laden water and thrives independent of the sunlight shining on Earth's surface. Is it possible that similar ecosystems could exist today under the ice of Europa?

Many scientists now think that Europa is the most likely place beyond Earth to find life in the solar system. In response, NASA is designing a Europa mission to characterize its liquid ocean and its ice crust, and to identify locations where material from inside has risen to the surface. Such interior material might reveal direct evidence for microbial life. In planning a future mission, it may be possible to include a small lander craft as well.

Io, a Volcanic Moon

Io, the innermost of Jupiter's Galilean moons, is in many ways a close twin of our Moon, with nearly the same size and density. We might therefore expect it to have experienced a similar history. Its appearance, as photographed from space, tells us another story, however (Figure 12.7). Instead of being a dead cratered world, Io turns out to have the highest level of volcanism in the solar system, greatly exceeding that of Earth.

1　Ganymede and Saturn's moon Enceladus may have smaller amounts of liquid water under their surfaces.

This OpenStax book is available for free at http://cnx.org/content/col11992/1.8

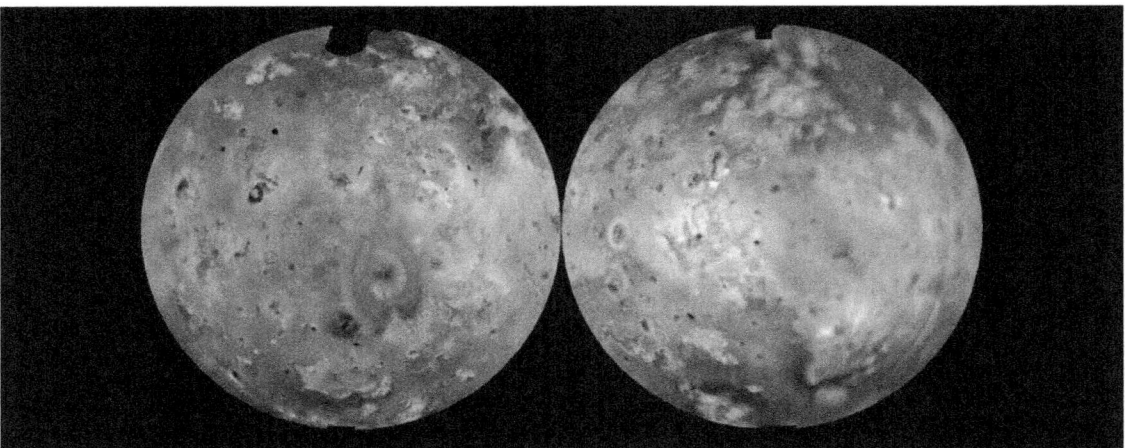

Figure 12.7. Two Sides of Io. This composite image shows both sides of the volcanically active moon Io. The orange deposits are sulfur snow; the white is sulfur dioxide. (Carl Sagan once quipped that Io looks as if it desperately needs a shot of penicillin.) (credit: modification of work by NASA/JPL/USGS)

Io's active volcanism was discovered by the Voyager spacecraft. Eight volcanoes were seen erupting when Voyager 1 passed in March 1979, and six of these were still active four months later when Voyager 2 passed. With the improved instruments carried by the Galileo spacecraft, more than 50 eruptions were found during 1997 alone. Many of the eruptions produce graceful plumes that extend hundreds of kilometers out into space (Figure 12.8).

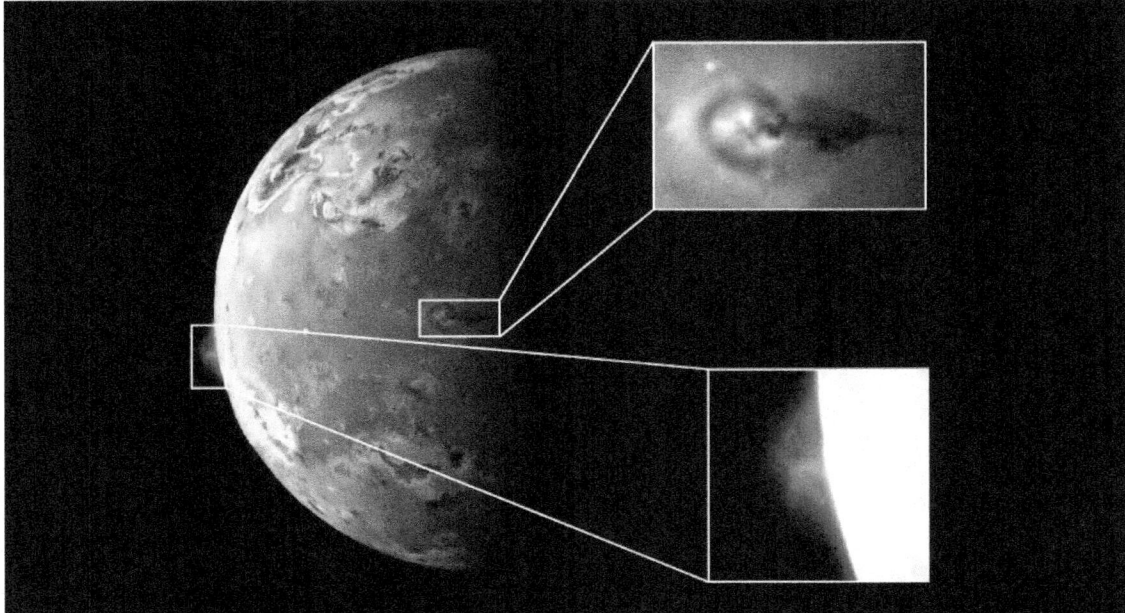

Figure 12.8. Volcanic Eruptions on Io. This composite image from NASA's Galileo spacecraft shows close-ups (the two inset photos) of two separate volcanic eruptions on Jupiter's volcanic moon, Io. In the upper inset image, you can see a close up of a bluish plume rising about 140 kilometers above the surface of the volcano. In the lower inset image is the Prometheus plume, rising about 75 kilometers from Io's surface. The Prometheus plume is named for the Greek god of fire. (credit: modification of work by NASA/JPL)

Download for free at http://cnx.org/content/col11992/latest/

LINK TO LEARNING

Watch a brief movie (https://openstax.org/l/30IoSurf) made from Voyager and Galileo data, showing a rotating Io with its dramatic surface features.

The Galileo data show that most of the volcanism on Io consists of hot silicate lava, like the volcanoes on Earth. Sometimes the hot lava encounters frozen deposits of sulfur and sulfur dioxide. When these icy deposits are suddenly heated, the result is great eruptive plumes far larger than any ejected from terrestrial volcanoes. As the rising plumes cool, the sulfur and sulfur dioxide recondense as solid particles that fall back to the surface in colorful "snowfalls" that extend as much as a thousand kilometers from the vent. Major new surface features were even seen to appear between Galileo orbits, as shown in Figure 12.9.

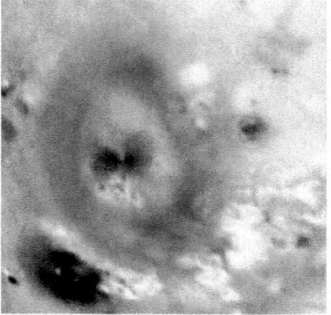 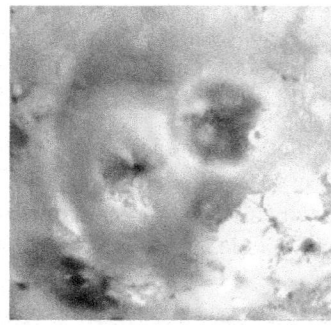 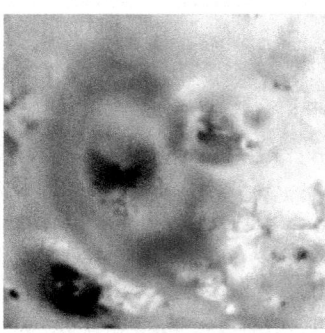

April 1997   September 1997   July 1999

Figure 12.9. Volcanic Changes on Io. These three images were taken of the same 1700-kilometer-square region of Io in April 1997, September 1997, and July 1999. The dark volcanic center called Pillan Patera experienced a huge eruption, producing a dark deposit some 400 kilometers across (seen as the grey area in the upper center of the middle image). In the right image, however, some of the new dark deposit is already being covered by reddish material from the volcano Pele. Also, a small unnamed volcano to the right of Pillan has erupted since 1997, and some of its dark deposit and a yellow ring around it are visible on the right image (to the right of the grey spot). The color range is exaggerated in these images. (credit: modification of work by NASA/JPL/University of Arizona)

As the Galileo mission drew to a close, controllers were willing to take risks in getting close to Io. Approaching this moon is a dangerous maneuver because the belts of atomic particles trapped in Jupiter's magnetic environment are at their most intense near Io's orbit. Indeed, in its very first pass by Io, the spacecraft absorbed damaging radiation beyond its design levels. To keep the system working at all, controllers had to modify or disable various fault-protection software routines in the onboard computers. In spite of these difficulties, the spacecraft achieved four successful Io flybys, obtaining photos and spectra of the surface with unprecedented resolution.

Maps of Io reveal more than 100 recently active volcanoes. Huge flows spread out from many of these vents, covering about 25% of the moon's total surface with still-warm lava. From these measurements, it seems clear that the bright surface colors that first attracted attention to Io are the result of a thin veneer of sulfur compounds. The underlying volcanism is driven by eruptions of molten silicates, just like on Earth (Figure 12.10).

This OpenStax book is available for free at http://cnx.org/content/col11992/1.8

Download for free at http://cnx.org/content/col11992/latest/

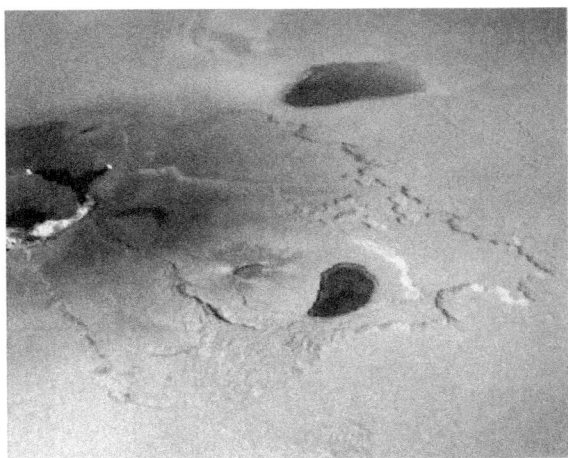

Figure 12.10. Lava Fountains on Io. Galileo captured a number of eruptions along the chain of huge volcanic calderas (or pits) on Io called Tvashtar Catena in this false-color image combining infrared and visible light. The bright orange-yellow areas at left are places where fresh, hot lava is erupting from below ground. (credit: modification of work by NASA/JPL)

Tidal Heating

How can Io remain volcanically active in spite of its small size? The answer, as we hinted earlier, lies in the effect of gravity, through tidal heating. Io is about the same distance from Jupiter as our Moon is from Earth. Yet Jupiter is more than 300 times more massive than Earth, causing forces that pull Io into an elongated shape, with a several-kilometer-high bulge extending toward Jupiter.

If Io always kept exactly the same face turned toward Jupiter, this bulge would not generate heat. However, Io's orbit is not exactly circular due to gravitational perturbations (tugs) from Europa and Ganymede. In its slightly eccentric orbit, Io twists back and forth with respect to Jupiter, at the same time moving nearer and farther from the planet on each revolution. The twisting and flexing heat Io, much as repeated flexing of a wire coat hanger heats the wire.

After billions of years, this constant flexing and heating have taken their toll on Io, driving away water and carbon dioxide and other gases, so that now sulfur and sulfur compounds are the most volatile materials remaining. Its interior is entirely melted, and the crust itself is constantly recycled by volcanic activity.

In moving inward toward Jupiter from Callisto to Io, we have encountered more and more evidence of geological activity and internal heating, culminating in the violent volcanism on Io. Three of these surfaces are compared in Figure 12.11. Just as the character of the planets in our solar system depends in large measure on their distance from the Sun (and on the amount of heat they receive), so it appears that distance from a giant planet like Jupiter can play a large role in the composition and evolution of its moons (at least partly due to differences in internal heating of each moon by Jupiter's unrelenting tidal forces).

Download for free at http://cnx.org/content/col11992/latest/

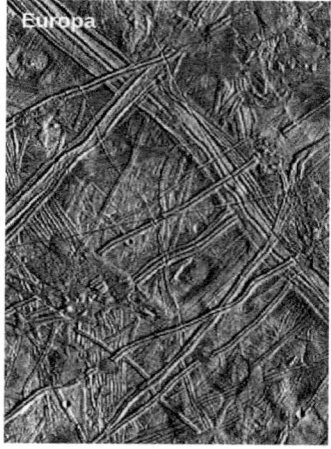
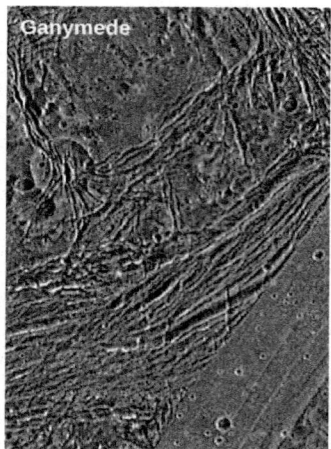
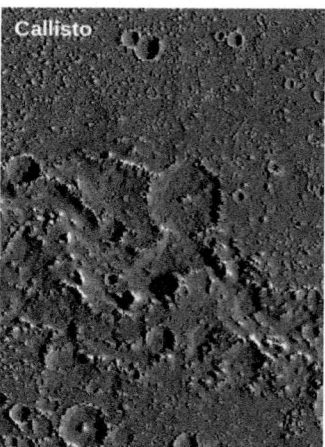

Figure 12.11. Three Icy Moons. These Galileo images compare the surfaces of Europa, Ganymede, and Callisto at the same resolution. Note that the number of craters (and thus the age of the surface we see) increases as we go from Europa to Ganymede to Callisto. The Europa image is one of those where the system of cracks and ridges resembles a freeway system. (credit: modification of work by NASA/JPL/DLR)

12.3 TITAN AND TRITON

Learning Objectives

By the end of this section, you will be able to:

> Explain how the thick atmosphere of Titan makes bodies of liquid on its surface possible
> Describe what we learned from the landing on Titan with the Huygens probe
> Discuss the features we observed on the surface of Triton when Voyager 2 flew by

We shift our attention now to small worlds in the more distant parts of the solar system. Saturn's large moon Titan turns out to be a weird cousin of Earth, with many similarities in spite of frigid temperatures. The Cassini observations of Titan have provided some of the most exciting recent discoveries in planetary science. Neptune's moon Triton also has unusual characteristics and resembles Pluto, which we will discuss in the following section.

Titan, a Moon with Atmosphere and Hydrocarbon Lakes

Titan, first seen in 1655 by the Dutch astronomer Christiaan Huygens, was the first moon discovered after Galileo saw the four large moons of Jupiter. Titan has roughly the same diameter, mass, and density as Callisto or Ganymede. Presumably it also has a similar composition—about half ice and half rock. However, Titan is unique among moons, with a thick atmosphere and lakes and rivers and falling rain (although these are not composed of water but of hydrocarbons such as ethane and methane, which can stay liquid at the frigid temperatures on Titan).

The 1980 Voyager flyby of Titan determined that the surface density of its atmosphere is four times greater than that on Earth. The atmospheric pressure on this moon is 1.6 bars, higher than that on any other moon and, remarkably, even higher than that of the terrestrial planets Mars and Earth. The atmospheric composition is primarily nitrogen, an important way in which Titan's atmosphere resembles Earth's.

Also detected in Titan's atmosphere were carbon monoxide (CO), hydrocarbons (compounds of hydrogen and carbon) such as methane (CH_4), ethane (C_2H_6), and propane (C_3H_8), and nitrogen compounds such as hydrogen cyanide (HCN), cyanogen (C_2N_2), and cyanoacetylene (HC_3N). Their presence indicates an active chemistry in

This OpenStax book is available for free at http://cnx.org/content/col11992/1.8

which sunlight interacts with atmospheric nitrogen and methane to create a rich mix of organic molecules. There are also multiple layers of hydrocarbon haze and clouds in the atmosphere, as illustrated in Figure 12.12.

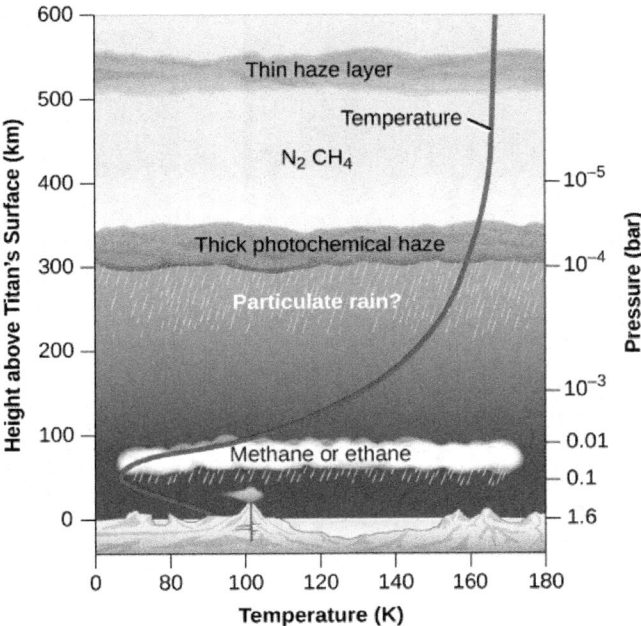

Figure 12.12. Structure of Titan's Atmosphere. Some characteristics of Titan's atmosphere resemble those of Earth's atmosphere, although it is much colder than our planet. The red line indicates the temperature of Titan's atmosphere at different altitudes.

These Voyager discoveries motivated a much more ambitious exploration program using the NASA Cassini Saturn orbiter and a probe to land on Titan called Huygens, built by the European Space Agency. The orbiter, which included several cameras, spectrometers, and a radar imaging system, made dozens of close flybys of Titan between 2004 and 2015, each yielding radar and infrared images of portions of the surface (see Exploring the Outer Planets). The Huygens probe successfully descended by parachute through the atmosphere, photographing the surface from below the clouds, and landing on January 14, 2005. This was the first (and so far the only) spacecraft landing on a moon in the outer solar system.

At the end of its parachute descent, the 319-kilogram Huygens probe safely touched down, slid a short distance, and began sending data back to Earth, including photos and analyses of the atmosphere. It appeared to have landed on a flat, boulder-strewn plain, but both the surface and the boulders were composed of water ice, which is as hard as rock at the temperature of Titan (see Figure 12.13).

The photos taken during descent showed a variety of features, including drainage channels, suggesting that Huygens had landed on the shore of an ancient hydrocarbon lake. The sky was deep orange, and the brightness of the Sun was a thousand times less than sunlight on Earth (but still more than a hundred times brighter than under the full moon on Earth). Titan's surface temperature was 94 K (–179 °C). The warmer spacecraft heated enough of the ice where it landed for its instruments to measure released hydrocarbon gas. Measurements on the surface continued for more than an hour before the probe succumbed to the frigid temperature.

Download for free at http://cnx.org/content/col11992/latest/

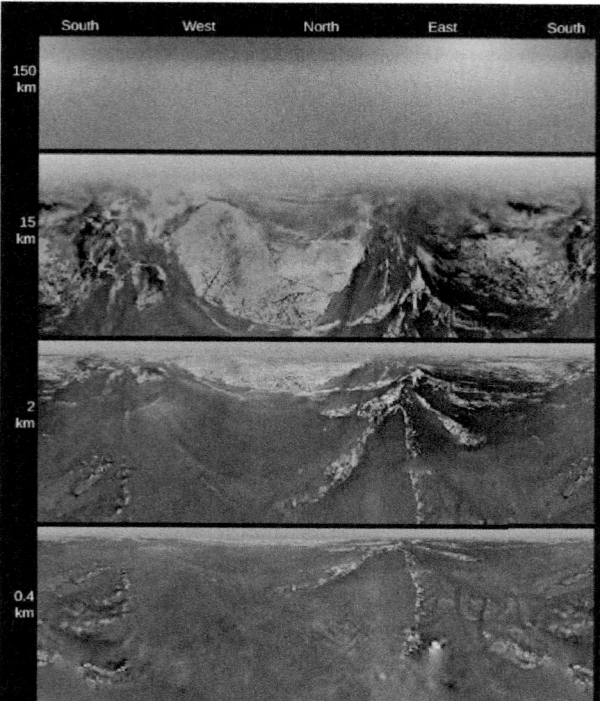

Figure 12.13. Views of the Surface of Titan. The left image shows the views of Titan from the descent camera, in a flattened projection, at different altitudes. The right image, taken after landing, shows a boulder-strewn surface illuminated by faint reddish sunlight. The boulders are composed of water ice. (credit left: modification of work by ESA/NASA/JPL/University of Arizona; credit right: modification of work by ESA/NASA/JPL/University of Arizona; processed by Andrey Pivovarov)

Radar and infrared imaging of Titan from the Cassini orbiter gradually built up a picture of a remarkably active surface on this moon, complex and geologically young (Figure 12.14). There are large methane lakes near the polar regions that interact with the methane in the atmosphere, much as Earth's water oceans interact with the water vapor in our atmosphere. The presence of many erosional features indicates that atmospheric methane can condense and fall as rain, then flow down valleys to the big lakes. Thus, Titan has a low-temperature equivalent of the water cycle on Earth, with liquid on the surface that evaporates, forms clouds, and then condenses to fall as rain—but on Titan the liquid is a combination of methane, ethane, and a trace of other hydrocarbons. It is a weirdly familiar and yet utterly alien landscape.

This OpenStax book is available for free at http://cnx.org/content/col11992/1.8

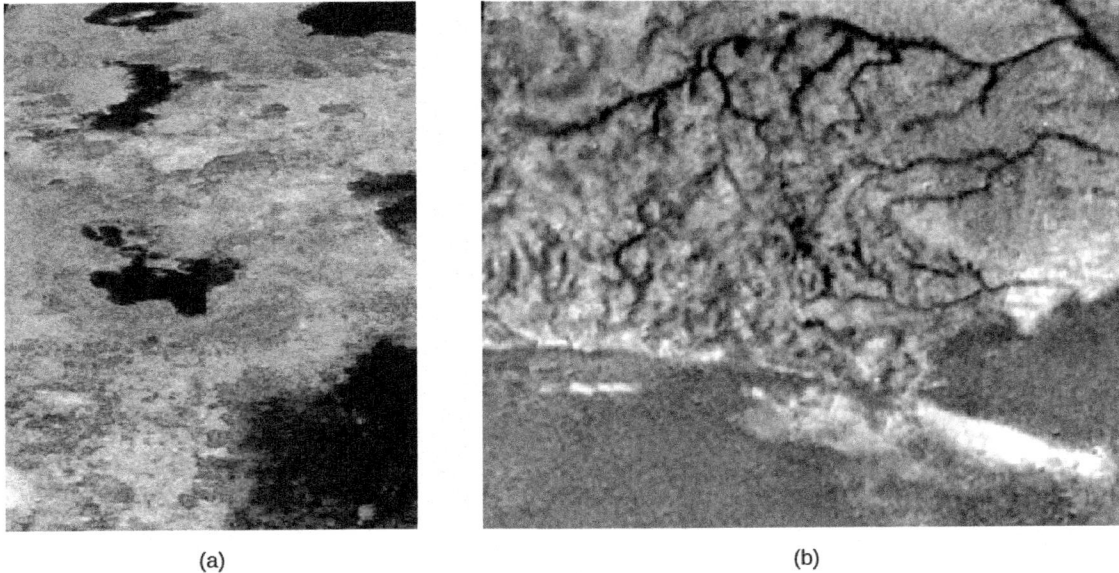

(a)   (b)

Figure 12.14. Titan's Lakes. (a) This Cassini image from a September 2006 flyby shows the liquid lakes on Titan. Their composition is most likely a combination of methane and ethane. (Since this is a radar image, the colors are artificially added. The dark blue areas are the smooth surfaces of the liquid lakes, and yellow is the rougher solid terrain around them.) (b) This mosaic of Titan's surface from the Cassini-Huygens mission shows in detail a high ridge area and many narrow, sinuous erosion channels that appear to be part of a widespread network of "rivers" carved by flowing hydrocarbons. (credit a: modification of work by NASA/JPL-Caltech/USGS; credit b; modification of work by NASA/JPL/ESA/University of Arizona)

These discoveries raise the question of whether there could be life on Titan. Hydrocarbons are fundamental for the formation of the large carbon molecules that are essential to life on our planet. However, the temperature on Titan is far too low for liquid water or for many of the chemical processes that are essential to life as we know it. There remains, though, an intriguing possibility that Titan might have developed a different form of low-temperature carbon-based life that could operate with liquid hydrocarbons playing the role of water. The discovery of such "life as we don't know it" could be even more exciting than finding life like ours on Mars. If such a truly alien life is present on Titan, its existence would greatly expand our understanding of the nature of life and of habitable environments.

LINK TO LEARNING

The Cassini mission scientists and the visual presentation specialists at NASA's Jet Propulsion Laboratory have put together some nice films from the images taken by Cassini and Huygens. See, for example, the Titan approach (https://openstax.org/l/30Titan) and the flyover (https://openstax.org/l/30Titan2) of the Northern lakes district.

Triton and Its Volcanoes

Neptune's largest moon Triton (don't get its name confused with Titan) has a diameter of 2720 kilometers and a density of 2.1 g/cm^3, indicating that it's probably composed of about 75% rock mixed with 25% water ice. Measurements indicate that Triton's surface has the coldest temperature of any of the worlds our robot

Download for free at http://cnx.org/content/col11992/latest/

representatives have visited. Because its reflectivity is so high (about 80%), Triton reflects most of the solar energy that falls on it, resulting in a surface temperature between 35 and 40 K.

The surface material of Triton is made of frozen water, nitrogen, methane, and carbon monoxide. Methane and nitrogen exist as gas in most of the solar system, but they are frozen at Triton's temperatures. Only a small quantity of nitrogen vapor persists to form an atmosphere. Although the surface pressure of this atmosphere is only 16 millionths of a bar, this is sufficient to support thin haze or cloud layers.

Triton's surface, like that of many other moons in the outer solar system, reveals a long history of geological evolution (Figure 12.15). Although some impact craters are found, many regions have been flooded fairly recently by the local version of "lava" (perhaps water or water-ammonia mixtures). There are also mysterious regions of jumbled or mountainous terrain.

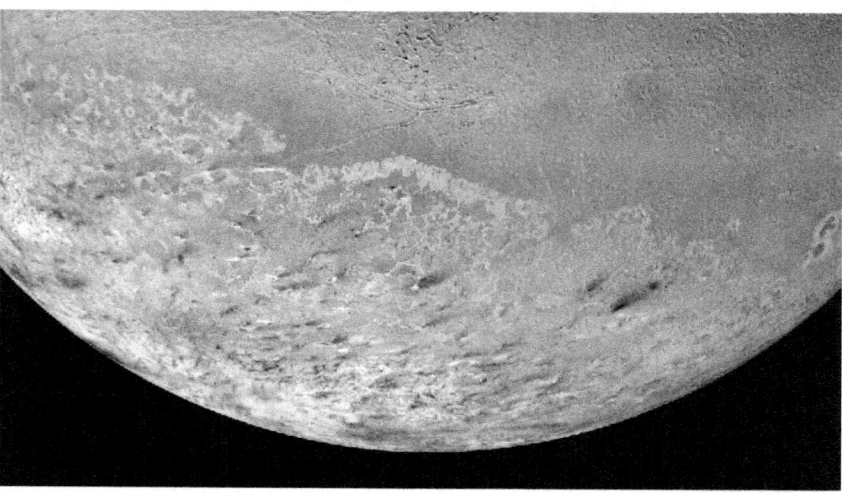

Figure 12.15. Neptune's Moon Triton. This mosaic of Voyager 2 images of Triton shows a wide range of surface features. The pinkish area at the bottom is Triton's large southern polar cap. The south pole of Triton faces the Sun here, and the slight heating effect is driving some of the material northward, where it is colder. (credit: modification of work by NASA/JPL/USGS)

The Voyager flyby of Triton took place at a time when the moon's southern pole was tipped toward the Sun, allowing this part of the surface to enjoy a period of relative warmth. (Remember that "warm" on Triton is still outrageously colder than anything we experience on Earth.) A polar cap covers much of Triton's southern hemisphere, apparently evaporating along the northern edge. This polar cap may consist of frozen nitrogen that was deposited during the previous winter.

Remarkably, the Voyager images showed that the evaporation of Triton's polar cap generates geysers or volcanic plumes of nitrogen gas (see Figure 12.16). (Fountains of such gas rose about 10 kilometers high, visible in the thin atmosphere because dust from the surface rose with them and colored them dark.) These plumes differ from the volcanic plumes of Io in their composition and also in that they derive their energy from sunlight warming the surface rather than from internal heat.

This OpenStax book is available for free at http://cnx.org/content/col11992/1.8

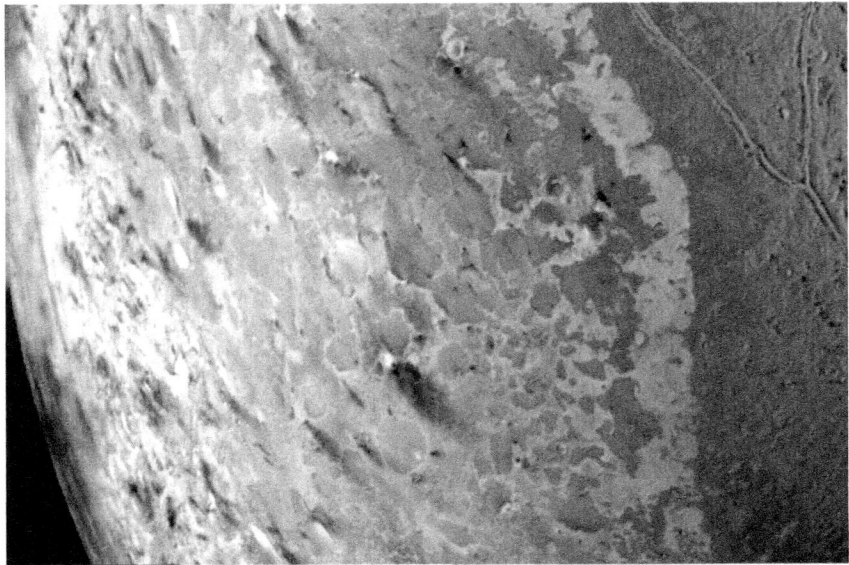

Figure 12.16. Triton's Geysers. This close-up view shows some of the geysers on Neptune's moon Triton, with the long trains of dust pointing to the lower right in this picture. (credit: modification of work by NASA/JPL)

PLUTO AND CHARON

Learning Objectives

By the end of this section, you will be able to:

> Compare the orbital characteristics of Pluto with those of the planets
> Describe information about Pluto's surface deduced from the New Horizons images
> Note some distinguishing characteristics of Pluto's large moon Charon

Pluto is not a moon, but we discuss it here because its size and composition are similar to many moons in the outer solar system. Our understanding of Pluto (and its large moon Charon) have changed dramatically as a result of the New Horizons flyby in 2015.

Is Pluto a Planet?

Pluto was discovered through a careful, systematic search, unlike Neptune, whose position was calculated from gravitational theory. Nevertheless, the history of the search for Pluto began with indications that Uranus had slight departures from its predicted orbit, departures that could be due to the gravitation of an undiscovered "Planet X." Early in the twentieth century, several astronomers, most notably Percival Lowell, then at the peak of his fame as an advocate of intelligent life on Mars, became interested in searching for this ninth planet.

Lowell and his contemporaries based their calculations primarily on tiny unexplained irregularities in the motion of Uranus. Lowell's computations indicated two possible locations for a perturbing Planet X; the more likely of the two was in the constellation Gemini. He predicted a mass for the planet intermediate between the masses of Earth and Neptune (his calculations gave about 6 Earth masses). Other astronomers, however, obtained other solutions from the tiny orbital irregularities, even including one model that indicated two planets beyond Neptune.

Download for free at http://cnx.org/content/col11992/latest/

At his Arizona observatory, Lowell searched without success for the unknown planet from 1906 until his death in 1916, and the search was not renewed until 1929. In February 1930, a young observing assistant named Clyde Tombaugh (see the Clyde Tombaugh: From the Farm to Fame feature box), comparing photographs he made on January 23 and 29 of that year, found a faint object whose motion appeared to be about right for a planet far beyond the orbit of Neptune (Figure 12.17). The new planet was named for Pluto, the Roman god of the underworld, who dwelt in remote darkness, just like the new planet. The choice of this name, among hundreds suggested, was helped by the fact that the first two letters were Percival Lowell's initials.

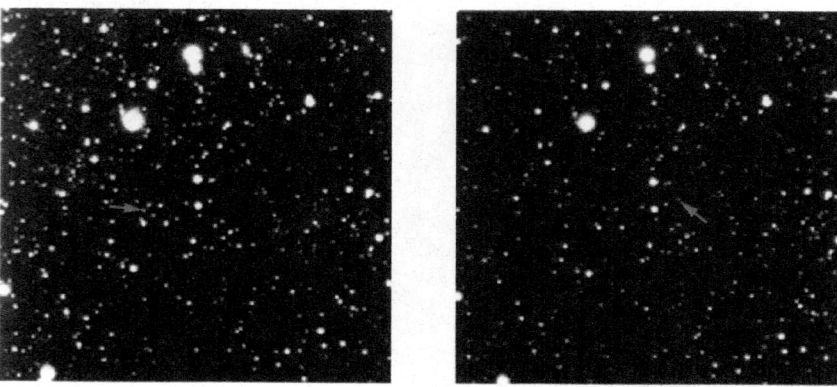

Figure 12.17. Pluto's Motion. Portions of the two photographs by which Clyde Tombaugh discovered Pluto in 1930. The left one was taken on January 23 and the right on January 29. Note that Pluto, indicated by an arrow, has moved among the stars during those six nights. If we hadn't put an arrow next to it, though, you probably would never have spotted the dot that moved. (credit: modification of work by the Lowell Observatory Archives)

Although the discovery of Pluto appeared initially to be a vindication of gravitational theory similar to the earlier triumph of Adams and Le Verrier in predicting the position of Neptune, we now know that Lowell's calculations were wrong. When its mass and size were finally measured, it was found that Pluto could not possibly have exerted any measurable pull on either Uranus or Neptune. Astronomers are now convinced that the reported small anomalies in the motions of Uranus are not, and never were, real.

From the time of its discovery, it was clear that Pluto was not a giant like the other four outer solar system planets. For a long time, it was thought that the mass of Pluto was similar to that of Earth, so that it was classed as a fifth terrestrial planet, somehow misplaced in the far outer reaches of the solar system. There were other anomalies, however, as Pluto's orbit was more eccentric and inclined to the plane of our solar system than that of any other planet. Only after the discovery of its moon Charon in 1978 could the mass of Pluto be measured, and it turned out to be far less than the mass of Earth.

In addition to Charon, Pluto has four small moons. Subsequent observations of Charon showed that this moon is in a retrograde orbit and has a diameter of about 1200 kilometers, more than half the size of Pluto itself (Figure 12.18). This makes Charon the moon whose size is the largest fraction of its parent planet. We could even think of Pluto and Charon as a double world. Seen from Pluto, Charon would be as large as eight full moons on Earth.

This OpenStax book is available for free at http://cnx.org/content/col11992/1.8

Download for free at http://cnx.org/content/col11992/latest/

Figure 12.18. Comparison of the Sizes of Pluto and Its Moon Charon with Earth. This graphic vividly shows how tiny Pluto is relative to a terrestrial planet like Earth. That is the primary justification for putting Pluto in the class of dwarf planets rather than terrestrial planets. (credit: modification of work by NASA)

To many astronomers, Pluto seemed like the odd cousin that everyone hopes will not show up at the next family reunion. Neither its path around the Sun nor its size resembles either the giant planets or the terrestrial planets. In the 1990s, astronomers began to discover additional small objects in the far outer solar system, showing that Pluto was not unique. We will discuss these trans-neptunian objects later with other small bodies, in the chapter on Comets and Asteroids: Debris of the Solar System. One of them (called Eris) is nearly the same size as Pluto, and another (Makemake) is substantially smaller. It became clear to astronomers that Pluto was so different from the other planets that it needed a new classification. Therefore, it was called a *dwarf planet*, meaning a planet much smaller than the terrestrial planets. We now know of many small objects in the vicinity of Pluto and we have classified several as dwarf planets.

A similar history was associated with the discovery of the asteroids. When the first asteroid (Ceres) was discovered at the beginning of the nineteenth century, it was hailed as a new planet. In the following years, however, other objects were found with similar orbits to Ceres. Astronomers decided that these should not all be considered planets, so they invented a new class of objects, called minor planets or asteroids. Today, Ceres is also called a dwarf planet. Both minor planets and dwarf planets are part of a whole belt or zones of similar objects (as we will discuss in Comets and Asteroids: Debris of the Solar System).

So, is Pluto a planet? Our answer is yes, but it is a *dwarf planet*, clearly not in the same league with the eight major planets (four giants and four terrestrials). While some people were upset when Pluto was reclassified, we might point out that a dwarf tree is still a type of tree and (as we shall see) a dwarf galaxy is still a type of galaxy.

Download for free at http://cnx.org/content/col11992/latest/

VOYAGERS IN ASTRONOMY

Clyde Tombaugh: From the Farm to Fame

Clyde Tombaugh discovered Pluto when he was 24 years old, and his position as staff assistant at the Lowell Observatory was his first paying job. Tombaugh had been born on a farm in Illinois, but when he was 16, his family moved to Kansas. There, with his uncle's encouragement, he observed the sky through a telescope the family had ordered from the Sears catalog. Tombaugh later constructed a larger telescope on his own and devoted his nights (when he wasn't too tired from farm work) to making detailed sketches of the planets (Figure 12.19).

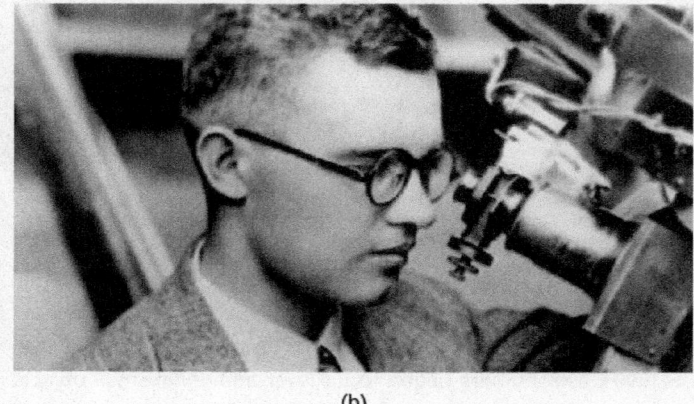

(a)   (b)

Figure 12.19. Clyde Tombaugh (1906–1997). (a) Tombaugh is pictured on his family farm in 1928 with a 9-inch telescope he built. (b) Here Tombaugh is looking through an eyepiece at the Lowell Observatory. (credit b: modification of work by NASA)

In 1928, after a hailstorm ruined the crop, Tombaugh decided he needed a job to help support his family. Although he had only a high school education, he thought of becoming a telescope builder. He sent his planet sketches to the Lowell Observatory, seeking advice about whether such a career choice was realistic. By a wonderful twist of fate, his query arrived just when the Lowell astronomers realized that a renewed search for a ninth planet would require a very patient and dedicated observer.

The large photographic plates (pieces of glass with photographic emulsion on them) that Tombaugh was hired to take at night and search during the day contained an average of about 160,000 star images each. How to find Pluto among them? The technique involved taking two photographs about a week apart. During that week, a planet would move a tiny bit, while the stars remained in the same place relative to each other. A new instrument called a "blink comparator" could quickly alternate the two images in an eyepiece. The stars, being in the same position on the two plates, would not appear to change as the two images were "blinked." But a moving object would appear to wiggle back and forth as the plates were alternated.

After examining more than 2 million stars (and many false alarms), Tombaugh found his planet on February 18, 1930. The astronomers at the observatory checked his results carefully, and the find was announced on March 13, the 149th anniversary of the discovery of Uranus. Congratulations and requests for interviews poured in from around the world. Visitors descended on the observatory in scores, wanting to see the place where the first new planet in almost a century had been discovered, as well as the person who had discovered it.

This OpenStax book is available for free at http://cnx.org/content/col11992/1.8

In 1932, Tombaugh took leave from Lowell, where he had continued to search and blink, to get a college degree. Eventually, he received a master's degree in astronomy and taught navigation for the Navy during World War II. In 1955, after working to develop a rocket-tracking telescope, he became a professor at New Mexico State University, where he helped found the astronomy department. He died in 1997; some of his ashes were placed inside the New Horizons spacecraft to Pluto.

LINK TO LEARNING

Here is a touching video (https://openstax.org/l/30Tbaugh) about Tombaugh's life as described by his children.

The Nature of Pluto

Using data from the New Horizons probe, astronomers have measured the diameter of Pluto as 2370 kilometers, only 60 perent as large as our Moon. From the diameter and mass, we find a density of 1.9 g/cm³, suggesting that Pluto is a mixture of rocky materials and water ice in about the same proportions as many outer-planet moons.

Parts of Pluto's surface are highly reflective, and its spectrum demonstrates the presence on its surface of frozen methane, carbon monoxide, and nitrogen. The maximum surface temperature ranges from about 50 K when Pluto is farthest from the Sun to 60 K when it is closest. Even this small difference is enough to cause a partial sublimation (going from solid to gas) of the methane and nitrogen ice. This generates an atmosphere when Pluto is close to the Sun, and it freezes out when Pluto is farther away. Observations of distant stars seen through this thin atmosphere indicate that the surface pressure is about a ten-thousandth of Earth's. Because Pluto is a few degrees warmer than Triton, its atmospheric pressure is about ten times greater. This atmosphere contains several distinct haze layers, presumably caused by photochemical reactions, like those in Titan's atmosphere (Figure 12.20).

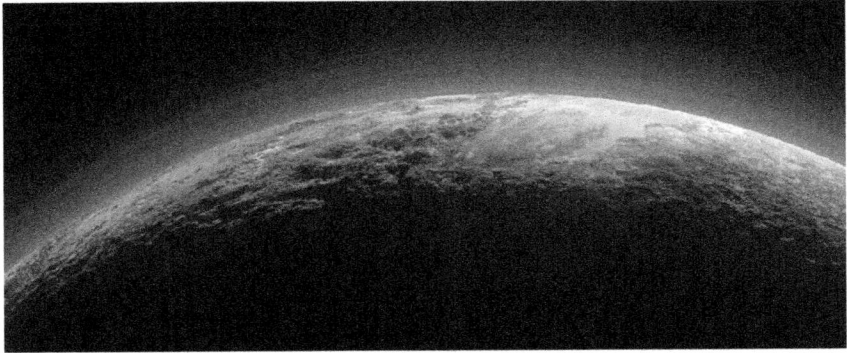

Figure 12.20. Haze Layers in the Atmosphere of Pluto. This is one of the highest-resolution photos of Pluto, taken by the New Horizons spacecraft 15 minutes after its closest approach. It shows 12 layers of haze. Note also the range of mountains with heights up to 3500 meters. (credit: modification of work by NASA/Johns Hopkins University Applied Physics Laboratory/Southwest Research Institute)

Reaching Pluto with a spacecraft was a major challenge, especially in an era when reduced NASA budgets could not support large, expensive missions like Galileo and Cassini. Yet like Galileo and Cassini, a Pluto mission

Download for free at http://cnx.org/content/col11992/latest/

would require a nuclear electric system that used the heat from plutonium to generate the energy to power the instruments and keep them operating far from the warmth of the Sun. NASA made available one of the last of its nuclear generators for such a mission. Assuming an affordable but highly capable spacecraft could be built, there was still the problem of getting to Pluto, nearly 5 billion kilometers from Earth, without waiting decades. The answer was to use Jupiter's gravity to slingshot the spacecraft toward Pluto.

The 2006 launch of New Horizons started the mission with a high speed, and the Jupiter flyby just a year later gave it the required additional boost. The New Horizons spacecraft arrived at Pluto in July 2015, traveling at a relative speed of 14 kilometers per second (or about 50,000 kilometers per hour). With this high speed, the entire flyby sequence was compressed into just one day. Most of the data recorded near closest approach could not be transmitted to Earth until many months later, but when it finally arrived, astronomers were rewarded with a treasure trove of images and data.

First Close-up Views of Pluto

Pluto is not the geologically dead world that many anticipated for such a small object—far from it. The division of the surface into areas with different composition and surface texture is apparent in the global color photo shown in Figure 12.21. The reddish color is enhanced in this image to bring out differences in color more clearly. The darker parts of the surface appear to be cratered, but adjacent to them is a nearly featureless light area in the lower right quadrant of this image. The dark areas show the colors of photochemical haze or smog similar to that in the atmosphere of Titan. The dark material that is staining these old surfaces could come from Pluto's atmospheric haze or from chemical reactions taking place at the surface due to the action of sunlight.

The light areas in the photo are lowland basins. These are apparently seas of frozen nitrogen, perhaps many kilometers deep. Both nitrogen and methane gas are able to escape from Pluto when it is in the part of its orbit close to the Sun, but only very slowly, so there is no reason that a vast bowl of frozen nitrogen could not persist for a long time.

This OpenStax book is available for free at http://cnx.org/content/col11992/1.8

Figure 12.21. Global Color Image of Pluto. This New Horizons image clearly shows the variety of terrains on Pluto. The dark area in the lower left is covered with impact craters, while the large light area in the center and lower right is a flat basin devoid of craters. The colors you see are somewhat enhanced to bring out subtle differences. (credit: modification of work by NASA/Johns Hopkins University Applied Physics Laboratory/Southwest Research Institute)

Figure 12.22 shows some of the remarkable variety of surface features New Horizons revealed. At the right of this image we see the "shoreline" of the vast bowl of nitrogen ice we saw as the smooth region in Figure 12.21. Temporarily nicknamed the "Sputnik Plains," after the first human object to get into space, this round region is roughly a thousand kilometers wide and shows intriguing cells or polygons that have an average width of more than 30 kilometers. The mountains in the middle are great blocks of frozen water ice, some reaching heights of 2 to 3 kilometers.

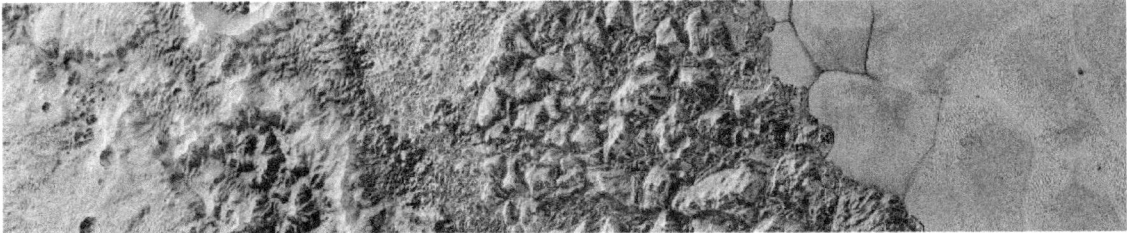

Figure 12.22. Diversity of Terrain on Pluto. This enhanced color view of a strip of Pluto's surface about 80 kilometers long shows a variety of different surface features. From left to right, we first cross a region of "badlands" with some craters showing, and then move across a wide range of mountains made of water ice and coated with the redder material we saw in the previous image. Then, at right, we arrive at the "shoreline" of the great sea of frozen nitrogen that the mission scientists have nicknamed the "Sputnik Plains." This nitrogen sea is divided into mysterious cells or segments that are many kilometers across. (credit: modification of work by NASA/Johns Hopkins University Applied Physics Laboratory/Southwest Research Institute)

Download for free at http://cnx.org/content/col11992/latest/

Figure 12.23 shows another view of the boundary between different types of geology. The width of this image is 250 kilometers, and it shows dark, ancient, heavily cratered terrain; dark, uncratered terrain with a hilly surface; smooth, geologically young terrain; and a small cluster of mountains more than 3000 meters high. In the best images, the light areas of nitrogen ice seem to have flowed much like glaciers on Earth, covering some of the older terrain underneath them.

The isolated mountains in the midst of the smooth nitrogen plains are probably also made of water ice, which is very hard at the temperatures on Pluto and can float on frozen nitrogen. Additional mountains, and some hilly terrain that reminded the mission scientists of snakeskin, are visible in part (b) of Figure 12.23. These are preliminary interpretations from just the first data coming back from New Horizons in 2015 and early 2016. As time goes on, scientists will have a better understanding of the unique geology of Pluto.

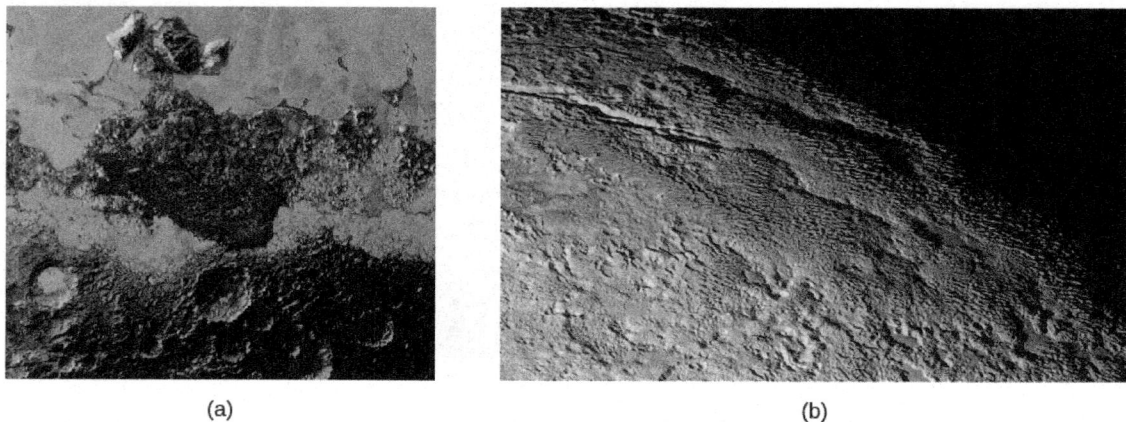

(a)   (b)

Figure 12.23. Diversity of Terrains on Pluto. (a) In this photo, about 250 kilometers across, we can see many different kinds of terrain. At the bottom are older, cratered highlands; a V-shaped region of hills without cratering points toward the bottom of the image. Surrounding the V-shaped dark region is the smooth, brighter frozen nitrogen plain, acting as glaciers on Earth do. Some isolated mountains, made of frozen water ice, are floating in the nitrogen near the top of the picture. (b) This scene is about 390 kilometers across. The rounded mountains, quite different from those we know on Earth, are named Tartarus Dorsa. The patterns, made of repeating ridges with the more reddish terrain between them, are not yet understood. (credit a, b: modification of work by NASA/Johns Hopkins University Applied Physics Laboratory/Southwest Research Institute)

A Quick Look at Charon

To add to the mysteries of Pluto, we show in Figure 12.24 one of the best New Horizons images of Pluto's large moon Charon. Recall from earlier that Charon is roughly half Pluto's size (its diameter is about the size of Texas). Charon keeps the same side toward Pluto, just as our Moon keeps the same side toward Earth. What is unique about the Pluto-Charon system, however, is that Pluto also keeps its same face toward Charon. Like two dancers embracing, these two constantly face each other as they spin across the celestial dance floor. Astronomers call this a double tidal lock.

This OpenStax book is available for free at http://cnx.org/content/col11992/1.8

(a)   (b)

Figure 12.24. Pluto's Large Moon Charon. (a) In this New Horizons image, the color has been enhanced to bring out the color of the moon's strange red polar cap. Charon has a diameter of 1214 kilometers, and the resolution of this image is 3 kilometers. (b) Here we see the moon from a slightly different angle, in true color. The inset shows an area about 390 kilometers from top to bottom. Near the top left is an intriguing feature—what appears to be a mountain in the middle of a depression or moat. (credit a, b: modification of work by NASA/JHUAPL/SwRI)

What New Horizons showed was another complex world. There are scattered craters in the lower part of the image, but much of the rest of the surface appears smooth. Crossing the center of the image is a belt of rough terrain, including what appear to be tectonic valleys, as if some forces had tried to split Charon apart. Topping off this strange image is a distinctly red polar cap, of unknown composition. Many features on Charon are not yet understood, including what appears to be a mountain in the midst of a low-elevation region.

12.5 PLANETARY RINGS

Learning Objectives

By the end of this section, you will be able to:

> Describe the two theories of planetary ring formation
> Compare the major rings of Saturn and explain the role of the moon Enceladus in the formation of the E ring
> Explain how the rings of Uranus and Neptune differ in composition and appearance from the rings of Saturn
> Describe how ring structure is affected by the presence of moons

In addition to their moons, all four of the giant planets have rings, with each ring system consisting of billions of small particles or "moonlets" orbiting close to their planet. Each of these rings displays a complicated structure that is related to interactions between the ring particles and the larger moons. However, the four ring systems are very different from each other in mass, structure, and composition, as outlined in Table 12.2.

Download for free at http://cnx.org/content/col11992/latest/

Properties of the Ring Systems

Planet	Outer Radius (km)	Outer Radius (R_{planet})	Mass (kg)	Reflectivity (%)
Jupiter	128,000	1.8	$10^{10}(?)$	?
Saturn	140,000	2.3	10^{19}	60
Uranus	51,000	2.2	10^{14}	5
Neptune	63,000	2.5	10^{12}	5

Table 12.2

Saturn's large ring system is made up of icy particles spread out into several vast, flat rings containing a great deal of fine structure. The Uranus and Neptune ring systems, on the other hand, are nearly the reverse of Saturn's: they consist of dark particles confined to a few narrow rings with broad empty gaps in between. Jupiter's ring and at least one of Saturn's are merely transient dust bands, constantly renewed by dust grains eroded from small moons. In this section, we focus on the two most massive ring systems, those of Saturn and Uranus.

What Causes Rings?

A ring is a collection of vast numbers of particles, each like a tiny moon obeying Kepler's laws as it follows its own orbit around the planet. Thus, the inner particles revolve faster than those farther out, and the ring as a whole does not rotate as a solid body. In fact, it is better not to think of a ring rotating at all, but rather to consider the revolution (or motion in orbit) of its individual moonlets.

If the ring particles were widely spaced, they would move independently, like separate moonlets. However, in the main rings of Saturn and Uranus the particles are close enough to exert mutual gravitational influence, and occasionally even to rub together or bounce off each other in low-speed collisions. Because of these interactions, we see phenomena such as waves that move across the rings—just the way water waves move over the surface of the ocean.

There are two basic ideas of how such rings come to be. First is the *breakup hypothesis*, which suggests that the rings are the remains of a shattered moon. A passing comet or asteroid might have collided with the moon, breaking it into pieces. Tidal forces then pulled the fragments apart, and they dispersed into a disk. The second hypothesis, which takes the reverse perspective, suggests that the rings are made of particles that were unable to come together to form a moon in the first place.

In either theory, the gravity of the planet plays an important role. Close to the planet (see Figure 12.25), tidal forces can tear bodies apart or inhibit loose particles from coming together. We do not know which explanation holds for any given ring, although many scientists have concluded that at least a few of the rings are relatively young and must therefore be the result of breakup.

This OpenStax book is available for free at http://cnx.org/content/col11992/1.8

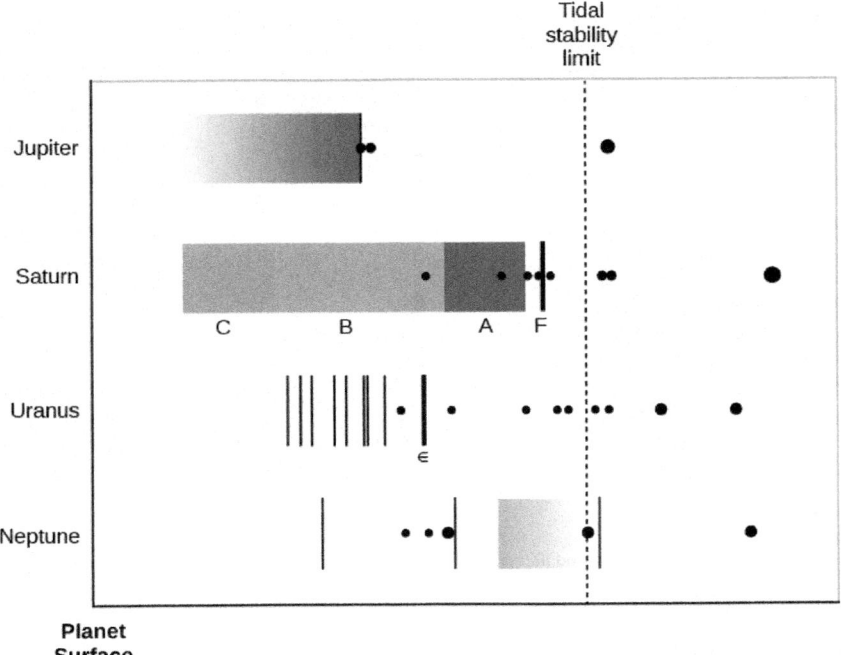

Figure 12.25. Four Ring Systems. This diagram shows the locations of the ring systems of the four giant planets. The left axis represents the planet's surface. The dotted vertical line is the limit inside which gravitational forces can break up moons (each planet's system is drawn to a different scale, so that this stability limit lines up for all four of them). The black dots are the inner moons of each planet on the same scale as its rings. Notice that only really small moons survive inside the stability limit.

Rings of Saturn

Saturn's rings are one of the most beautiful sights in the solar system (Figure 12.26). From outer to inner, the three brightest rings are labeled with the extremely unromantic names of A, B, and C Rings. Table 12.3 gives the dimensions of the rings in both kilometers and units of the radius of Saturn, R_{Saturn}. The B Ring is the brightest and has the most closely packed particles, whereas the A and C Rings are translucent.

The total mass of the B Ring, which is probably close to the mass of the entire ring system, is about equal to that of an icy moon 250 kilometers in diameter (suggesting that the ring could have originated in the breakup of such a moon). Between the A and B Rings is a wide gap named the Cassini Division after Gian Domenico Cassini, who first glimpsed it through a telescope in 1675 and whose name planetary scientists have also given to the Cassini spacecraft exploring the Saturn system.

Download for free at http://cnx.org/content/col11992/latest/

(a)   (b)

Figure 12.26. Saturn's Rings as Seen from Above and Below. (a) The view from above is illuminated by direct sunlight. (b) The illumination seen from below is sunlight that has diffused through gaps in the rings. (credit a, b: modification of work by NASA/JPL-Caltech/Space Science Institute)

Selected Features in the Rings of Saturn

Ring Name[2]	Outer Edge (R_{Saturn})	Outer Edge (km)	Width (km)
F	2.324	140,180	90
A	2.267	136,780	14,600
Cassini Division	2.025	122,170	4590
B	1.949	117,580	25,580
C	1.525	92,000	17,490

Table 12.3

Saturn's rings are very broad and very thin. The width of the main rings is 70,000 kilometers, yet their average thickness is only 20 meters. If we made a scale model of the rings out of paper, we would have to make them 1 kilometer across. On this scale, Saturn itself would loom as high as an 80-story building. The ring particles are composed primarily of water ice, and they range from grains the size of sand up to house-sized boulders. An insider's view of the rings would probably resemble a bright cloud of floating snowflakes and hailstones, with a few snowballs and larger objects, many of them loose aggregates of smaller particles (Figure 12.27).

---

2   The ring letters are assigned in the order of their discovery.

This OpenStax book is available for free at http://cnx.org/content/col11992/1.8

Figure 12.27. Artist's Idealized Impression of the Rings of Saturn as Seen from the Inside. Note that the rings are mostly made of pieces of water ice of different sizes. At the end of its mission, the Cassini spacecraft is planning to cut through one of the gaps in Saturn's rings, but it won't get this close. (credit: modification of work by NASA/JPL/University of Colorado)

In addition to the broad A, B, and C Rings, Saturn has a handful of very narrow rings no more than 100 kilometers wide. The most substantial of these, which lies just outside the A Ring, is called the F Ring; its surprising appearance is discussed below. In general, Saturn's narrow rings resemble the rings of Uranus and Neptune.

There is also a very faint, tenuous ring, called the E Ring, associated with Saturn's small icy moon Enceladus. The particles in the E Ring are very small and composed of water ice. Since such a tenuous cloud of ice crystals will tend to dissipate, the ongoing existence of the E Ring strongly suggests that it is being continually replenished by a source at Enceladus. This icy moon is very small—only 500 kilometers in diameter—but the Voyager images showed that the craters on about half of its surface have been erased, indicating geological activity sometime in the past few million years. It was with great anticipation that the Cassini scientists maneuvered the spacecraft orbit to allow multiple close flybys of Enceladus starting in 2005.

Those awaiting the Cassini flyby results were not disappointed. High-resolution images showed long, dark stripes of smooth ground near its south pole, which were soon nicknamed "tiger stripes" (Figure 12.28). Infrared measurements revealed that these tiger stripes are warmer than their surroundings. Best of all, dozens of cryovolcanic vents on the tiger stripes were seen to be erupting geysers of salty water and ice (Figure 12.29). Estimates suggested that 200 kilograms of material were shooting into space each second—not a lot, but enough for the spacecraft to sample.

Download for free at http://cnx.org/content/col11992/latest/

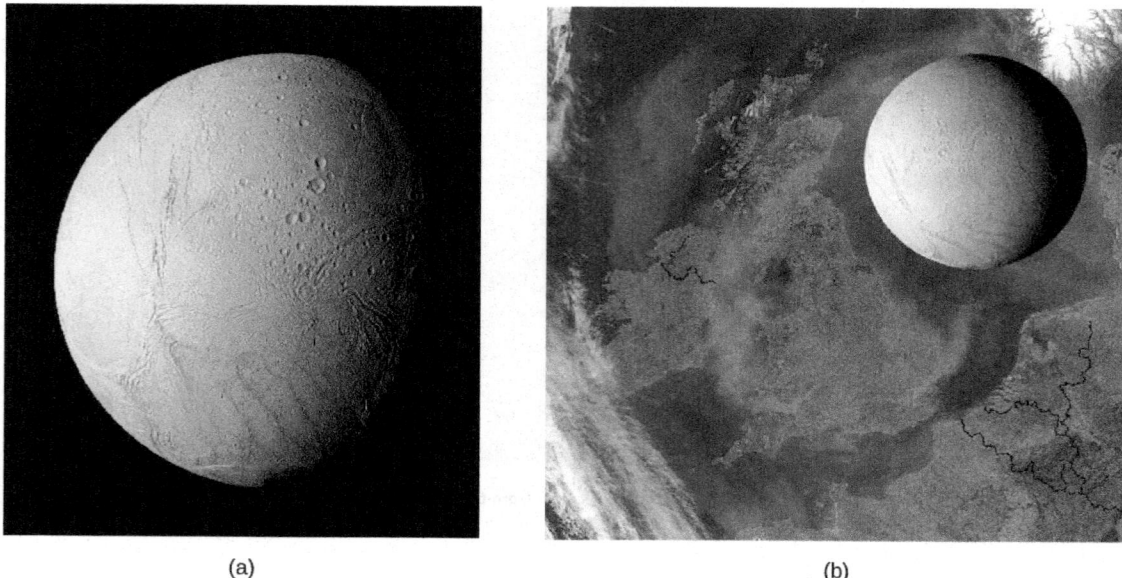

(a)   (b)

Figure 12.28. Enceladus. (a) This image shows both smooth and cratered terrain on Saturn's moon, and also "tiger stripes" in the south polar region (lower part of image). These dark stripes (shown here in exaggerated color) have elevated temperatures and are the source of the many geysers discovered on Enceladus. They are about 130 kilometers long and 40 kilometers apart. (b) Here Enceladus is shown to scale with Great Britain and the coast of Western Europe, to emphasize that it is a small moon, only about 500 kilometers in diameter. (credit a, b: modification of work by NASA/JPL/Space Science Institute)

When Cassini was directed to fly into the plumes, it measured their composition and found them to be similar to material we see liberated from comets (see Comets and Asteroids: Debris of the Solar System). The vapor and ice plumes consisted mostly of water, but with trace amounts of nitrogen, ammonia, methane, and other hydrocarbons. Minerals found in the geysers in trace amounts included ordinary salt, meaning that the geyser plumes were high-pressure sprays of salt water.

Based on the continuing study of Enceladus' bulk properties and the ongoing geysers, in 2015 the Cassini mission scientists tentatively identified a subsurface ocean of water feeding the geysers. These discoveries suggested that in spite of its small size, Enceladus should be added to the list of worlds that we would like to explore for possible life. Since its subsurface ocean is conveniently escaping into space, it might be much easier to sample than the ocean of Europa, which is deeply buried below its thick crust of ice.

This OpenStax book is available for free at http://cnx.org/content/col11992/1.8

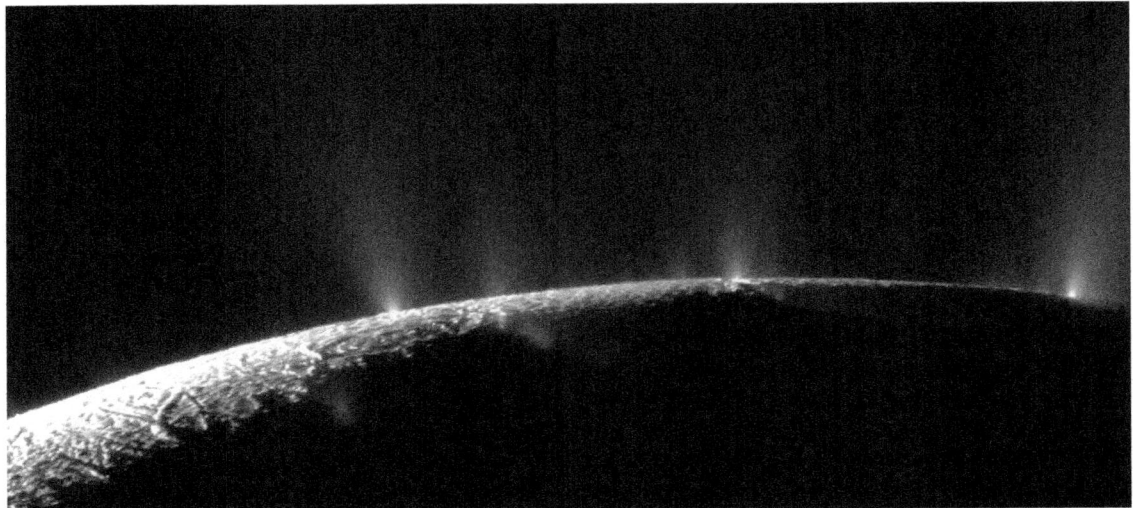

Figure 12.29. Geysers on Enceladus. This Cassini image shows a number of water geysers on Saturn's small moon Enceladus, apparently salty water from a subsurface source escaping through cracks in the surface. You can see curved lines of geysers along the four "tiger stripes" on the surface. (credit: modification of work by NASA/JPL/Space Science Institute)

Rings of Uranus and Neptune

Uranus' rings are narrow and black, making them almost invisible from Earth. The nine main rings were discovered in 1977 from observations made of a star as Uranus passed in front of it. We call such a passage of one astronomical object in front of another an *occultation*. During the 1977 occultation, astronomers expected the star's light to disappear as the planet moved across it. But in addition, the star dimmed briefly several times before Uranus reached it, as each narrow ring passed between the star and the telescope. Thus, the rings were mapped out in detail even though they could not be seen or photographed directly, like counting the number of cars in a train at night by watching the blinking of a light as the cars successively pass in front of it. When Voyager approached Uranus in 1986, it was able to study the rings at close range; the spacecraft also photographed two new rings (Figure 12.30).

Download for free at http://cnx.org/content/col11992/latest/

Figure 12.30. Rings of Uranus. The Voyager team had to expose this image for a long time to get a glimpse of Uranus' narrow dark rings. You can see the grainy structure of "noise" in the electronics of the camera in the picture background. (credit: modification of work by NASA/JPL)

The outermost and most massive of Uranus' rings is called the Epsilon Ring. It is only about 100 kilometers wide and probably no more than 100 meters thick (similar to the F Ring of Saturn). The Epsilon Ring encircles Uranus at a distance of 51,000 kilometers, about twice the radius of Uranus. This ring probably contains as much mass as all of Uranus' other ten rings combined; most of them are narrow ribbons less than 10 kilometers wide, just the reverse of the broad rings of Saturn.

The individual particles in the uranian rings are nearly as black as lumps of coal. While astronomers do not understand the composition of this material in detail, it seems to consist in large part of carbon and hydrocarbon compounds. Organic material of this sort is rather common in the outer solar system. Many of the asteroids and comets are also composed of dark, tarlike materials. In the case of Uranus, its ten small inner moons have a similar composition, suggesting that one or more moons might have broken up to make the rings.

Neptune's rings are generally similar to those of Uranus but even more tenuous (Figure 12.31). There are only four of them, and the particles are not uniformly distributed along their lengths. Because these rings are so difficult to investigate from Earth, it will probably be a long time before we understand them very well.

This OpenStax book is available for free at http://cnx.org/content/col11992/1.8

Figure 12.31. Rings of Neptune. This long exposure of Neptune's rings was photographed by Voyager 2. Note the two denser regions of the outer ring. (credit: modification of work by NASA/JPL)

LINK TO LEARNING

Mark Showalter (of the SETI Institute) and his colleagues maintain the NASA's Planetary Ring Node (https://openstax.org/l/30NASArings) website. It is full of information about the rings and their interactions with moons; check out their press-release images of the Saturn ring system, for example. And Showalter gives an entertaining illustrated talk (https://openstax.org/l/30StrnRngs) about Saturn's ring and moon system.

EXAMPLE 12.1

Resolution of Planetary Rings

Using the occultations of stars by the rings of Saturn, astronomers have been able to measure details in the ring structure to a resolution of 10 km. This is a much higher resolution than can be obtained in a conventional photo of the rings. Let's figure out what angular resolution (in arcsec) a space telescope in Earth orbit would have to achieve to obtain equal resolution.

Solution

To solve this problem, we use the "small-angle formula" to relate angular and linear diameters in the sky. For angles in the sky that are small, the formula is usually written as

$$\frac{\text{angular diameter}}{206{,}265 \text{ arcsec}} = \frac{\text{linear diameter}}{\text{distance}}$$

where angular diameter is expressed in arcsec. The distance of Saturn near opposition is about 9 AU = 1.4×10^9 km. Substituting in the above formula and solving for the angular resolution, we get

Download for free at http://cnx.org/content/col11992/latest/

$$\text{angular resolution} = \frac{206,265 \text{ arcsec} \times 10}{1.4 \times 10^9 \text{ km}}$$

which is about 10^{-3} arcsec, or a milliarcsec. This is not possible for our telescopes to achieve. For comparison, the best resolution from either the Hubble Space Telescope or ground-based telescopes is about 0.1 arcsec, or 100 times worse than what we would need. This is why such occultation measurements are so useful for astronomers.

Check Your Learning

How close to Saturn would a spacecraft have to be to make out detail in its rings as small as 20 km, if its camera has an angular resolution of 5 arcsec?

Answer:

Using our formula,

$$\frac{\text{angular diameter}}{206,265 \text{ arcsec}} = \frac{\text{linear diameter}}{\text{distance}}$$

we get

$$\frac{5 \text{ arcsec}}{206,265 \text{ arcsec}} = \frac{20 \text{ km}}{\text{distance}}.$$

So, the distance is about 825,000 km.

Interactions between Rings and Moons

Much of our fascination with planetary rings is a result of their intricate structures, most of which owe their existence to the gravitational effect of moons, without which the rings would be flat and featureless. Indeed, it is becoming clear that without moons there would probably be no rings at all because, left to themselves, thin disks of small particles gradually spread and dissipate.

Most of the gaps in Saturn's rings, and also the location of the outer edge of the A Ring, result from gravitational resonances with small inner moons. A **resonance** takes place when two objects have orbital periods that are exact ratios of each other, such as 1:2 or 2:3. For example, any particle in the gap at the inner side of the Cassini Division of Saturn's rings would have a period equal to one-half that of Saturn's moon Mimas. Such a particle would be nearest Mimas in the same part of its orbit every second revolution. The repeated gravitational tugs of Mimas, acting always in the same direction, would perturb it, forcing it into a new orbit outside the gap. In this way, the Cassini Division became depleted of ring material over long periods of time.

The Cassini mission revealed a great deal of fine structure in Saturn's rings. Unlike the earlier Voyager flybys, Cassini was able to observe the rings for more than a decade, revealing a remarkable range of changes, on time scales from a few minutes to several years. Many of the features newly seen in Cassini data indicated the presence of condensations or small moons only a few tens of meters across imbedded in the rings. As each small moon moves, it produces waves in the surrounding ring material like the wake left by a moving ship. Even when the moon is too small to be resolved, its characteristic waves could be photographed by Cassini.

One of the most interesting rings of Saturn is the narrow F Ring, which contains several apparent ringlets within its 90-kilometer width. In places, the F Ring breaks up into two or three parallel strands that sometimes show bends or kinks. Most of the rings of Uranus and Neptune are also narrow ribbons like the F Ring of Saturn. Clearly, the gravity of some objects must be keeping the particles in these thin rings from spreading out.

This OpenStax book is available for free at http://cnx.org/content/col11992/1.8

Download for free at http://cnx.org/content/col11992/latest/

As we have seen, the largest features in the rings of Saturn are produced by gravitational resonances with the inner moons, while much of the fine structure is caused by smaller embedded moons. In the case of Saturn's F Ring, close-up images revealed that it is bounded by the orbits of two moons, called Pandora and Prometheus (Figure 12.32). These two small moons (each about 100 kilometers in diameter) are referred to as *shepherd moons*, since their gravitation serves to "shepherd" the ring particles and keep them confined to a narrow ribbon. A similar situation applies to the Epsilon Ring of Uranus, which is shepherded by the moons Cordelia and Ophelia. These two shepherds, each about 50 kilometers in diameter, orbit about 2000 kilometers inside and outside the ring.

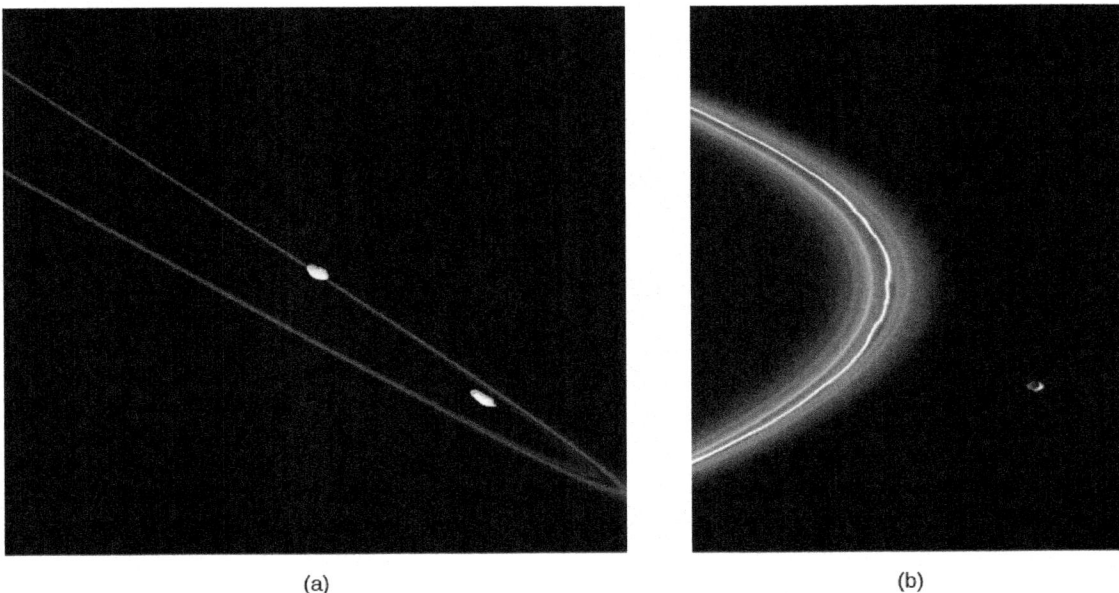

(a)   (b)

Figure 12.32. Saturn's F Ring and Its Shepherd Moons. (a) This Cassini image shows the narrow, complex F Ring of Saturn, with its two small shepherd moons Pandora (left) and Prometheus (right). (b) In this closer view, the shepherd moon Pandora (84 kilometers across) is seen next to the F ring, in which the moon is perturbing the main (brightest) strand of ring particles as it passes. You can see the dark side of Pandora on this image because it is being illuminated by the light reflected from Saturn. (credit a, b: modification of work by NASA/JPL/Space Science Institute)

LINK TO LEARNING

You can download a movie (https://openstax.org/l/30ShprdMns) showing the two shepherd moons on either side of Saturn's F ring.

Theoretical calculations suggest that the other narrow rings in the uranian and neptunian systems should also be controlled by shepherd moons, but none has been located. The calculated diameter for such shepherds (about 10 kilometers) was just at the limit of detectability for the Voyager cameras, so it is impossible to say whether they are present or not. (Given all the narrow rings we see, some scientists still hope to find another more satisfactory mechanism for keeping them confined.)

One of the outstanding problems with understanding the rings is determining their ages. Have the giant planets always had the ring systems we see today, or might these be a recent or transient addition to the solar system? In the case of the main rings of Saturn, their mass is about the same as that of the inner moon

Download for free at http://cnx.org/content/col11992/latest/

Mimas. Thus, they could have been formed by the break-up of a Mimas-sized moon, perhaps very early in solar system history, when there were many interplanetary projectiles left over from planet formation. It is harder to understand how such a catastrophic event could have taken place recently, when the solar system had become a more stable place.

This OpenStax book is available for free at http://cnx.org/content/col11992/1.8

Download for free at http://cnx.org/content/col11992/latest/

CHAPTER 12 REVIEW

 KEY TERMS

resonance an orbital condition in which one object is subject to periodic gravitational perturbations by another, most commonly arising when two objects orbiting a third have periods of revolution that are simple multiples or fractions of each other

tidal heating the heating of a planet or moon's interior by variable tidal forces caused by changing gravitational pull from a nearby planet or moon

 SUMMARY

12.1 Ring and Moon Systems Introduced

The four jovian planets are accompanied by impressive systems of moons and rings. Nearly 200 moons have been discovered in the outer solar system. Of the four ring systems, Saturn's is the largest and is composed primarily of water ice; in contrast, Uranus and Neptune have narrow rings of dark material, and Jupiter has a tenuous ring of dust.

12.2 The Galilean Moons of Jupiter

Jupiter's largest moons are Ganymede and Callisto, both low-density objects that are composed of more than half water ice. Callisto has an ancient cratered surface, while Ganymede shows evidence of extensive tectonic and volcanic activity, persisting until perhaps a billion years ago. Io and Europa are denser and smaller, each about the size of our Moon. Io is the most volcanically active object in the solar system. Various lines of evidence indicate that Europa has a global ocean of liquid water under a thick ice crust. Many scientists think that Europa may offer the most favorable environment in the solar system to search for life.

12.3 Titan and Triton

Saturn's moon Titan has an atmosphere that is thicker than that of Earth. There are lakes and rivers of liquid hydrocarbons, and evidence of a cycle of evaporation, condensation, and return to the surface that is similar to the water cycle on Earth (but with liquid methane and ethane). The Cassini-Huygens lander set down on Titan and showed a scene with boulders, made of water ice, frozen harder than rock. Neptune's cold moon Triton has a very thin atmosphere and nitrogen gas geysers.

12.4 Pluto and Charon

Pluto and Charon have been revealed by the New Horizons spacecraft to be two of the most fascinating objects in the outer solar system. Pluto is small (a dwarf planet) but also surprisingly active, with contrasting areas of dark cratered terrain, light-colored basins of nitrogen ice, and mountains of frozen water that may be floating in the nitrogen ice. Even Pluto's largest moon Charon shows evidence of geological activity. Both Pluto and Charon turn out to be far more dynamic and interesting than could have been imagined before the New Horizons mission.

12.5 Planetary Rings

Rings are composed of vast numbers of individual particles orbiting so close to a planet that its gravitational forces could have broken larger pieces apart or kept small pieces from gathering together. Saturn's rings are broad, flat, and nearly continuous, except for a handful of gaps. The particles are mostly water ice, with typical dimensions of a few centimeters. One Saturn moon, Enceladus, is today erupting geysers of water to maintain

Download for free at http://cnx.org/content/col11992/latest/

the tenuous E Ring, which is composed of very small ice crystals. The rings of Uranus are narrow ribbons separated by wide gaps and contain much less mass. Neptune's rings are similar but contain even less material. Much of the complex structure of the rings is due to waves and resonances induced by moons within the rings or orbiting outside them. The origin and age of each of these ring systems is still a mystery.

FOR FURTHER EXPLORATION

Articles

Moons

Carroll, M. "Titan: What We've Learned about a Strange New World." *Astronomy* (March 2010): 30. Nice review of Cassini mission results.

Elliot, J. "The Warming Wisps of Triton." *Sky & Telescope* (February 1999): 42. About Neptune's intriguing moon.

Hayes, A., "Secrets from Titan's Seas." *Astronomy* (October 2015): 24. Good review of what we now know and what puzzles us about the hydrocarbon lakes of Titan.

Jewitt, D., et al. "The Strangest Satellites in the Solar System." *Scientific American* (August 2006): 40. Small irregular moons in the outer solar system.

Lakdawalla, E. "Ice Worlds of the Ringed Planet." *Sky & Telescope* (June 2009): 27. On the Cassini mission exploration of Enceladus, Iapetus, and other moons.

Mackenzie, D. "Is There Life under the Ice?" *Astronomy* (August 2001): 32. On future exploration of Europa.

Robertson, D. "Where Goes the Rain?" *Sky & Telescope* (March 2013): 26. About the methane weather cycle on Titan and what Cassini experiments are telling us.

Scharf, C. "A Universe of Dark Oceans." *Sky & Telescope* (December 2014): 20. Subsurface oceans on Europa, Ganymede, Enceladus, and Titan.

Showalter, M. "How to Catch a Moon (or Two) of Pluto." *Astronomy Beat* (December 2012): http://www.astrosociety.org/wp-content/uploads/2013/02/ab2012-106.pdf. On the discovery of small moons around Pluto, written by the person who discovered two of them.

Spencer, J. "Galileo's Closest Look at Io." *Sky & Telescope* (May 2001): 40.

Talcott, R. "Cassini Flies through Enceladus' Geysers." *Astronomy* (March 2009): 32.

Zimmerman, R. "Does Methane Flow on Titan?" *Astronomy* (February 2014): 22. Ideas about lakes, channels, and rain.

Pluto

Stern, A. "Pluto: Up Close and Personal." *Astronomy* (July 2015): 22. Good summary of the history of understanding Pluto and our current knowledge on the eve of the New Horizons encounter.

Stern, A. "The Pluto System Explored." *Astronomy* (November 2015): 24. Fine review of what the team learned from the first few data downloads from New Horizons.

Tombaugh, C. "How I Found Pluto" *Astronomy Beat* (May 2009): http://astrosociety.org/wp-content/uploads/2013/02/ab2009-23.pdf.

Rings

Beatty, J. "Saturn's Amazing Rings." *Sky & Telescope* (May 2013): 18. Good 7-page summary of what we know.

This OpenStax book is available for free at http://cnx.org/content/col11992/1.8

Burns, J., et al. "Bejeweled Worlds." *Scientific American* (February 2002): 64. On rings throughout the solar system.

Elliot, J., et al. "Discovering the Rings of Uranus." *Sky & Telescope* (June 1977): 412.

Esposito, L. "The Changing Shape of Planetary Rings." *Astronomy* (September 1987): 6.

Sobel, D. "Secrets of the Rings." *Discover* (April 1994): 86. Discusses the outer planet ring systems.

Tiscareno, M. "Ringworld Revelations." *Sky & Telescope* (February 2007): 32. Cassini results about the rings of Saturn.

Websites

Note: Many of the sites about planets and planetary missions listed for Other Worlds: An Introduction to the Solar System and The Giant Planets also include good information about the moons of the planets.

Cassini Mission to Saturn: http://saturn.jpl.nasa.gov/ and http://www.esa.int/SPECIALS/Cassini-Huygens/index.html and http://ciclops.org

Jupiter's Moons, at JPL: http://solarsystem.nasa.gov/planets/jupiter/moons

Neptune's Moons, at JPL: http://solarsystem.nasa.gov/planets/neptune/moons

New Horizons Mission: http://pluto.jhuapl.edu. Gives the latest news bulletins and images from the Pluto encounter, plus lots of background information.

Pluto, at JPL: http://solarsystem.nasa.gov/planets/pluto

Saturn's Moons, at JPL: http://solarsystem.nasa.gov/planets/saturn/moons

Uranus' Moons, at JPL: http://solarsystem.nasa.gov/planets/uranus/moons

Apps

Two apps you can buy for iPhones or iPads can show you the positions and features of the moons of Jupiter and Saturn for any selected date:

- Jupiter Atlas: https://itunes.apple.com/us/app/ju[iter-atlas/id352033947?mt=8
- Saturn Atlas: https://itunes.apple.com/us/app/saturn-atlas/id352038051?mt=8

Videos

Amazing Moons: https://www.youtube.com/watch?v=CQjZf2bW9XQ. 2016 NASA video on intriguing moons in our solar system (4:16).

Briny Breath of Enceladus: http://www.jpl.nasa.gov/video/details.php?id=846. Brief 2009 JPL film on the geysers of Enceladus (2:36).

Dr. Carolyn Porco's TED Talk on Enceladus: https://www.youtube.com/watch?v=TRQdHrGuVgI (3:26).

Titan: http://www.youtube.com/watch?v=iTrOFefYxFg. Video from Open University, with interviews, animations, and images (8:11).

Europa Mission: http://www.jpl.nasa.gov/events/lectures_archive.php?year=2016&month=2. 2016 talk by two JPL scientists on NASA's plans for a mission to Jupiter's moon, which may have an underground liquid ocean (1:26:22).

Download for free at http://cnx.org/content/col11992/latest/

Great Planet Debate: http://gpd.jhuapl.edu/debate/debateStream.php OR https://www.youtube.com/watch?v=RJ8EErV6-6Q. Neil deGrasse Tyson debates Mark Sykes about how to characterize Pluto, in 2008 (1:14:11).

How I Killed Pluto and Why It Had It Coming: http://www.youtube.com/watch?v=7pbj_IlmiMg. 2011 Silicon Valley Astronomy Lecture by Michael Brown on the "demotion" of Pluto to a dwarf planet (1:27:13).

Seeking Pluto's Frigid Heart: https://www.youtube.com/watch?v=jIxQXGTI_mo. Dramatic 2016 *New York Times* production, narrated by Dennis Overbye (7:43).

Saturn's Restless Rings: https://www.youtube.com/watch?v=X5zcrEze8L4. 2013 talk by Mark Showalter in the Silicon Valley Astronomy Lecture Series (1:30:59).

🏛 COLLABORATIVE GROUP ACTIVITIES

A. Imagine it's the distant future and humans can now travel easily among the planets. Your group is a travel agency, with the task of designing a really challenging tour of the Galilean moons for a group of sports enthusiasts. What kinds of activities are possible on each world? How would rock climbing on Ganymede, for example, differ from rock climbing on Earth? (If you design an activity for Io, you had better bring along very strong radiation shielding. Why?)

B. In the same spirit as Activity A, have your agency design a tour that includes the seven most spectacular sights of any kind on all the moons or rings covered in this chapter. What are the not-to-be-missed destinations that future tourists will want to visit and why? Which of the sights you pick are going to be spectacular if you are on the moon's surface or inside the ring, and which would look interesting only from far away in space?

C. In this chapter we could cover only a few of the dozens of moons in the outer solar system. Using the Internet or your college library, organize your group into a research team and find out more about one of the moons we did not cover in detail. Our favorites include Uranus' Miranda, with its jigsaw puzzle surface; Saturn's Mimas, with a "knockout" crater called Herschel; and Saturn's Iapetus, whose two hemispheres differ significantly. Prepare a report to attract tourists to the world you selected.

D. In a novel entitled *2010*, science fiction writer Arthur C. Clarke, inspired by the information coming back from the Voyager spacecraft, had fun proposing a life form under the ice of Europa that was evolving toward intelligence. Suppose future missions do indeed find some sort of life (not necessarily intelligent but definitely alive) under the ice of Europa—life that evolved completely independently from life on Earth. Have your group discuss what effect such a discovery would have on humanity's view of itself. What should be our attitude toward such a life form? Do we have an obligation to guard it against contamination by our microbes and viruses? Or, to take an extreme position, should we wipe it out before it becomes competitive with Earth life or contaminates our explorers with microorganisms we are not prepared to deal with? Who should be in charge of making such decisions?

E. In the same spirit as Activity D, your group may want to watch the 2013 science fiction film *Europa Report*. The producers tried to include good science in depicting what it would be like for astronauts to visit that jovian moon. How well does your group think they did?

F. A number of modern science fiction writers (especially those with training in science) have written short stories that take place on the moons of Jupiter and Saturn. There is a topical listing of science fiction stories

This OpenStax book is available for free at http://cnx.org/content/col11992/1.8

with good astronomy at http://www.astrosociety.org/scifi. Members of your group can look under "Jupiter" or "Saturn" and find a story that interests you and then report on it to the whole class.

G. Work together to make a list of all the reasons it is hard to send a mission to Pluto. What compromises had to be made so that the New Horizons mission was affordable? How would you design a second mission to learn more about the Pluto system?

H. Your group has been asked by NASA to come up with one or more missions to learn about Europa. Review what we know about this moon so far and then design a robotic mission that would answer some of the questions we have. You can assume that budget is not a factor, but your instruments have to be realistic. (Bear in mind that Europa is cold and far from the Sun.)

I. Imagine your group is the first landing party on Pluto (let's hope you remembered to bring long underwear!). You land in a place where Charon is visible in the sky and you observe Charon for one Earth week. Describe what Charon will look like during that week. Now you move your camp to the opposite hemisphere of Pluto. What will Charon look like there during the course of a week?

J. When, in 2006, the International Astronomical Union (IAU) decided that Pluto should be called a dwarf planet and not a planet, they set up three criteria that a world must meet to be called a planet. Your group should use the Internet to find these criteria. Which of them did Pluto not meet? Read a little bit about the reaction to the IAU's decision among astronomers and the public. How do members of your group feel about Pluto's new classification? (After you have discussed it within the group, you may want to watch *The Great Planet Debate* video recommended in "For Further Exploration.")

🖰 EXERCISES

Review Questions

1. What are the moons of the outer planets made of, and how is their composition different from that of our Moon?

2. Compare the geology of Callisto, Ganymede, and Titan.

3. What is the evidence for a liquid water ocean on Europa, and why is this interesting to scientists searching for extraterrestrial life?

4. Explain the energy source that powers the volcanoes of Io.

5. Compare the properties of Titan's atmosphere with those of Earth's atmosphere.

6. How was Pluto discovered? Why did it take so long to find it?

7. How are Triton and Pluto similar?

8. Describe and compare the rings of Saturn and Uranus, including their possible origins.

9. Why were the rings of Uranus not observed directly from telescopes on the ground on Earth? How were they discovered?

10. List at least three major differences between Pluto and the terrestrial planets.

Download for free at http://cnx.org/content/col11992/latest/

11. The Hubble Space Telescope images of Pluto in 2002 showed a bright spot and some darker areas around it. Now that we have the close-up New Horizons images, what did the large bright region on Pluto turn out to be?

12. Saturn's E ring is broad and thin, and far from Saturn. It requires fresh particles to sustain itself. What is the source of new E-ring particles?

Thought Questions

13. Why do you think the outer planets have such extensive systems of rings and moons, while the inner planets do not?

14. Ganymede and Callisto were the first icy objects to be studied from a geological point of view. Summarize the main differences between their geology and that of the rocky terrestrial planets.

15. Compare the properties of the volcanoes on Io with those of terrestrial volcanoes. Give at least two similarities and two differences.

16. Would you expect to find more impact craters on Io or Callisto? Why?

17. Why is it unlikely that humans will be traveling to Io? (Hint: Review the information about Jupiter's magnetosphere in The Giant Planets.)

18. Why do you suppose the rings of Saturn are made of bright particles, whereas the particles in the rings of Uranus and Neptune are black?

19. Suppose you miraculously removed all of Saturn's moons. What would happen to its rings?

20. We have a lot of good images of the large moons of Jupiter and Saturn from the Galileo and Cassini spacecraft missions (check out NASA's Planetary Photojournal site, at http://photojournal.jpl.nasa.gov, to see the variety). Now that the New Horizons mission has gone to Pluto, why don't we have as many good images of all sides of Pluto and Charon?

21. In the Star Wars movie Star Wars Episode VI: Return of the Jedi, a key battle takes place on the inhabited "forest moon" Endor, which supposedly orbits around a gas giant planet. From what you have learned about planets and moons of the solar system, why would this be an unusual situation?

Figuring For Yourself

22. Which would have the longer orbital period: a moon 1 million km from the center of Jupiter, or a moon 1 million km from the center of Earth? Why?

23. How close to Uranus would a spacecraft have to get to obtain the same resolution as in Example 12.1 with a camera that has an angular resolution of 2 arcsec?

24. Saturn's A, B, and C Rings extend 75,000 to 137,000 km from the center of the planet. Use Kepler's third law to calculate the difference between how long a particle at the inner edge and a particle at the outer edge of the three-ring system would take to revolve about the planet.

25. Use the information in Appendix G to calculate what you would weigh on Titan, Io, and Uranus' moon Miranda.

26. The average distance of Enceladus from Saturn is 238,000 km; the average distance of Titan from Saturn is 1,222,000 km. How much longer does it take Titan to orbit Saturn compared to Enceladus?

This OpenStax book is available for free at http://cnx.org/content/col11992/1.8

13

COMETS AND ASTEROIDS: DEBRIS OF THE SOLAR SYSTEM

Figure 13.1. Hale-Bopp. Comet Hale-Bopp was one of the most attractive and easily visible comets of the twentieth century. It is shown here as it appeared in the sky in March 1997. You can see the comet's long blue ion tail and the shorter white dust tail. You will learn about these two types of comet tails, and how they form, in this chapter. (credit: modification of work by ESO/E. Slawik)

Chapter Outline

Thinking Ahead

Hundreds of smaller members of the solar system—asteroids and comets—are known to have crossed Earth's orbit in the past, and many others will do so in centuries ahead. What could we do if we knew a few years in advance that one of these bodies would hit Earth?

To understand the early history of life on Earth, scientists study ancient fossils. To reconstruct the early history of the solar system, we need cosmic fossils—materials that formed when our system was very young. However, reconstructing the early history of the solar system by looking just at the planets is almost as difficult as determining the circumstances of human birth by merely looking at an adult.

Instead, we turn to the surviving remnants of the creation process—ancient but smaller objects in our cosmic neighborhood. Asteroids are rocky or metallic and contain little *volatile* (easily evaporated) material. Comets are small icy objects that contain frozen water and other volatile materials but with solid grains mixed in. In the deep freeze beyond Neptune, we also have a large reservoir of material unchanged since the formation of the solar system, as well as a number of dwarf planets.

Download for free at http://cnx.org/content/col11992/latest/

13.1 ASTEROIDS

Learning Objectives

By the end of this section, you will be able to:

> Outline the story of the discovery of asteroids and describe their typical orbits
> Describe the composition and classification of the various types of asteroids
> Discuss what was learned from spacecraft missions to several asteroids

The asteroids are mostly found in the broad space between Mars and Jupiter, a region of the solar system called the *asteroid belt*. Asteroids are too small to be seen without a telescope; the first of them was not discovered until the beginning of the nineteenth century.

Discovery and Orbits of the Asteroids

In the late 1700s, many astronomers were hunting for an additional planet they thought should exist in the gap between the orbits of Mars and Jupiter. The Sicilian astronomer Giovanni Piazzi thought he had found this missing planet in 1801, when he discovered the first asteroid (or as it was later called, "minor planet") orbiting at 2.8 AU from the Sun. His discovery, which he named Ceres, was quickly followed by the detection of three other little planets in similar orbits.

Clearly, there was not a single missing planet between Mars and Jupiter but rather a whole group of objects, each much smaller than our Moon. (An analogous discovery history has played out in slow motion in the outer solar system. Pluto was discovered beyond Neptune in 1930 and was initially called a planet, but early in the twenty-first century, several other similar objects were found. We now call all of them dwarf planets.)

By 1890, more than 300 of these minor planets or **asteroids** had been discovered by sharp-eyed observers. In that year, Max Wolf at Heidelberg introduced astronomical photography to the search for asteroids, greatly accelerating the discovery of these dim objects. In the twenty-first century, searchers use computer-driven electronic cameras, another leap in technology. More than half a million asteroids now have well-determined orbits.

Asteroids are given a number (corresponding to the order of discovery) and sometimes also a name. Originally, the names of asteroids were chosen from goddesses in Greek and Roman mythology. After exhausting these and other female names (including, later, those of spouses, friends, flowers, cities, and others), astronomers turned to the names of colleagues (and other people of distinction) whom they wished to honor. For example, asteroids 2410, 4859, and 68448 are named Morrison, Fraknoi, and Sidneywolff, for the three original authors of this textbook.

The largest asteroid is Ceres (numbered 1), with a diameter just less than 1000 kilometers. As we saw, Ceres was considered a planet when it was discovered but later was called an asteroid (the first of many.) Now, it has again been reclassified and is considered one of the dwarf planets, like Pluto (see the chapter on Moons, Rings and Pluto). We still find it convenient, however, to discuss Ceres as the largest of the asteroids. Two other asteroids, Pallas and Vesta, have diameters of about 500 kilometers, and about 15 more are larger than 250 kilometers (see Table 13.1). The number of asteroids increases rapidly with decreasing size; there are about 100 times more objects 10 kilometers across than there are 100 kilometers across. By 2016, nearly a million asteroids have been discovered by astronomers.

This OpenStax book is available for free at http://cnx.org/content/col11992/1.8

LINK TO LEARNING

The Minor Planet Center (https://openstaxcollege.org/l/30minplancen) is a worldwide repository of data on asteroids. Visit it online to find out about the latest discoveries related to the small bodies in our solar system. (Note that some of the material on this site is technical; it's best to click on the menu tab for the "public" for information more at the level of this textbook.)

The Largest Asteroids

#	Name	Year of Discovery	Orbit's Semimajor Axis (AU)	Diameter (km)	Compositional Class
1	Ceres	1801	2.77	940	C (carbonaceous)
2	Pallas	1802	2.77	540	C (carbonaceous)
3	Juno	1804	2.67	265	S (stony)
4	Vesta	1807	2.36	510	basaltic
10	Hygiea	1849	3.14	410	C (carbonaceous)
16	Psyche	1852	2.92	265	M (metallic)
31	Euphrosyne	1854	3.15	250	C (carbonaceous)
52	Europa	1858	3.10	280	C (carbonaceous)
65	Cybele	1861	3.43	280	C (carbonaceous)
87	Sylvia	1866	3.48	275	C (carbonaceous)
451	Patientia	1899	3.06	260	C (carbonaceous)
511	Davida	1903	3.16	310	C (carbonaceous)
704	Interamnia	1910	3.06	310	C (carbonaceous)

Table 13.1

The asteroids all revolve about the Sun in the same direction as the planets, and most of their orbits lie near the plane in which Earth and other planets circle. The majority of asteroids are in the **asteroid belt**, the region between Mars and Jupiter that contains all asteroids with orbital periods between 3.3 to 6 years (Figure 13.2). Although more than 75% of the known asteroids are in the belt, they are not closely spaced (as they are sometimes depicted in science fiction movies). The volume of the belt is actually very large, and the typical spacing between objects (down to 1 kilometer in size) is several million kilometers. (This was fortunate for

Download for free at http://cnx.org/content/col11992/latest/

spacecraft like Galileo, Cassini, *Rosetta*, and New Horizons, which needed to travel through the asteroid belt without a collision.)

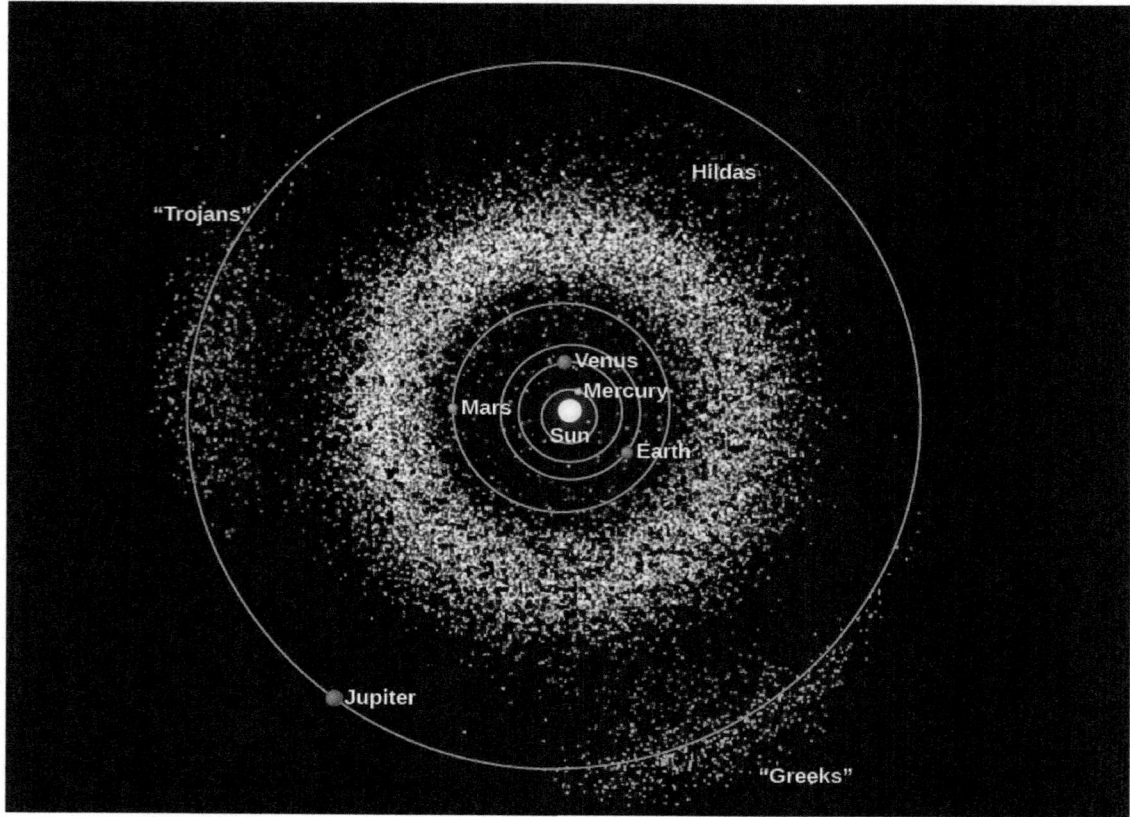

Figure 13.2. Asteroids in the Solar System. This computer-generated diagram shows the positions of the asteroids known in 2006. If the asteroid sizes were drawn to scale, none of the dots representing an asteroid would be visible. Here, the asteroid dots are too big and give a false impression of how crowded the asteroid belt would look if you were in it. Note that in addition to those in the asteroid belt, there are also asteroids in the inner solar system and some along Jupiter's orbit (such as the Trojans and Greeks groups), controlled by the giant planet's gravity.

Still, over the long history of our solar system, there have been a good number of collisions among the asteroids themselves. In 1918, the Japanese astronomer Kiyotsugu Hirayama found that some asteroids fall into *families*, groups with similar orbital characteristics. He hypothesized that each family may have resulted from the breakup of a larger body or, more likely, from the collision of two asteroids. Slight differences in the speeds with which the various fragments left the collision scene account for the small spread in orbits now observed for the different asteroids in a given family. Several dozen such families exist, and observations have shown that individual members of most families have similar compositions, as we would expect if they were fragments of a common parent.

This OpenStax book is available for free at http://cnx.org/content/col11992/1.8

LINK TO LEARNING

You can see a dramatic animated video (https://openstaxcollege.org/l/30anividastorb) showing the orbits of 100,000 asteroids found by one sky survey. As the 3-minute video goes on, you get to see the orbits of the planets and how the asteroids are distributed in the solar system. But note that all such videos are misleading in one sense. The asteroids themselves are really small compared to the distances covered, so they have to be depicted as larger points to be visible. If you were in the asteroid belt, there would be far more empty space than asteroids.

Composition and Classification

Asteroids are as different as black and white. The majority are very dark, with reflectivity of only 3 to 4%, like a lump of coal. However, another large group has a typical reflectivity of 15%. To understand more about these differences and how they are related to chemical composition, astronomers study the spectrum of the light reflected from asteroids for clues about their composition.

The dark asteroids are revealed from spectral studies to be *primitive* bodies (those that have changed little chemically since the beginning of the solar system) composed of silicates mixed with dark, organic carbon compounds. These are known as C-type asteroids ("C" for carbonaceous). Two of the largest asteroids, Ceres and Pallas, are primitive, as are almost all of the asteroids in the outer part of the belt.

The second most populous group is the S-type asteroids, where "S" stands for a stony or silicate composition. Here, the dark carbon compounds are missing, resulting in higher reflectivity and clearer spectral signatures of silicate minerals. The S-type asteroids are also chemically primitive, but their different composition indicates that they were probably formed in a different location in the solar system from the C-type asteroids.

Asteroids of a third class, much less numerous than those of the first two, are composed primarily of metal and are called M-type asteroids ("M" for metallic). Spectroscopically, the identification of metal is difficult, but for at least the largest M-type asteroid, Psyche, this identification has been confirmed by radar. Since a metal asteroid, like an airplane or ship, is a much better reflector of radar than is a stony object, Psyche appears bright when we aim a radar beam at it.

How did such metal asteroids come to be? We suspect that each came from a parent body large enough for its molten interior to settle out or differentiate, and the heavier metals sank to the center. When this parent body shattered in a later collision, the fragments from the core were rich in metals. There is enough metal in even a 1-kilometer M-type asteroid to supply the world with iron and many other industrial metals for the foreseeable future, if we could bring one safely to Earth.

In addition to the M-type asteroids, a few other asteroids show signs of early heating and differentiation. These have basaltic surfaces like the volcanic plains of the Moon and Mars; the large asteroid Vesta (discussed in a moment) is in this last category.

The different classes of asteroids are found at different distances from the Sun (Figure 13.3). By tracing how asteroid compositions vary with distance from the Sun, we can reconstruct some of the properties of the solar nebula from which they originally formed.

Download for free at http://cnx.org/content/col11992/latest/

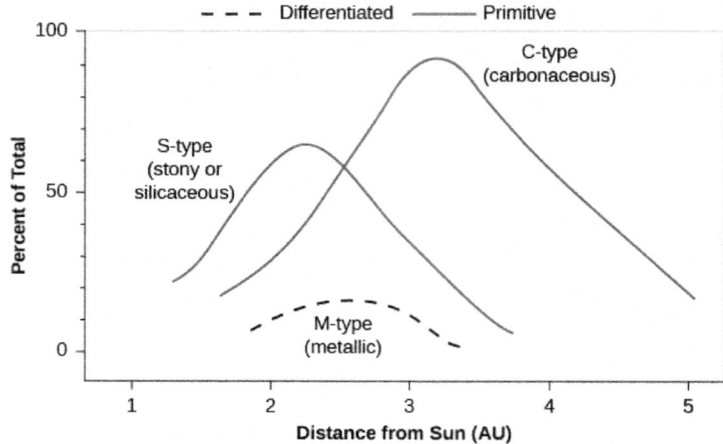

Figure 13.3. Where Different Types of Asteroids Are Found. Asteroids of different composition are distributed at different distances from the Sun. The S-type and C-type are both primitive; the M-type consists of cores of differentiated parent bodies.

Vesta: A Differentiated Asteroid

Vesta is one of the most interesting of the asteroids. It orbits the Sun with a semi-major axis of 2.4 AU in the inner part of the asteroid belt. Its relatively high reflectivity of almost 30% makes it the brightest asteroid, so bright that it is actually visible to the unaided eye if you know just where to look. But its real claim to fame is that its surface is covered with basalt, indicating that Vesta is a differentiated object that must once have been volcanically active, in spite of its small size (about 500 kilometers in diameter).

Meteorites from Vesta's surface (Figure 13.4), identified by comparing their spectra with that of Vesta itself, have landed on Earth and are available for direct study in the laboratory. We thus know a great deal about this asteroid. The age of the lava flows from which these meteorites derived has been measured at 4.4 to 4.5 billion years, very soon after the formation of the solar system. This age is consistent with what we might expect for volcanoes on Vesta; whatever process heated such a small object was probably intense and short-lived. In 2016, a meteorite fell in Turkey that could be identified with a particular lava flow as revealed by the orbiting *Dawn* spacecraft.

Figure 13.4. Piece of Vesta. This meteorite (rock that fell from space) has been identified as a volcanic fragment from the crust of asteroid Vesta. (credit: modification of work by R. Kempton (New England Meteoritical Services))

This OpenStax book is available for free at http://cnx.org/content/col11992/1.8

Download for free at http://cnx.org/content/col11992/latest/

Asteroids Up Close

On the way to its 1995 encounter with Jupiter, the Galileo spacecraft was targeted to fly close to two main-belt S-type asteroids called Gaspra and Ida. The Galileo camera revealed both as long and highly irregular (resembling a battered potato), as befits fragments from a catastrophic collision (Figure 13.5).

Figure 13.5. Mathilde, Gaspra, and Ida. The first three asteroids photographed from spacecraft flybys, printed to the same scale. Gaspra and Ida are S-type and were investigated by the Galileo spacecraft; Mathilde is C-type and was a flyby target for the NEAR-Shoemaker spacecraft. (credit: modification of work by NEAR Project, Galileo Project, NASA)

The detailed images allowed us to count the craters on Gaspra and Ida, and to estimate the length of time their surfaces have been exposed to collisions. The Galileo scientists concluded that these asteroids are only about 200 million years old (that is, the collisions that formed them took place about 200 million years ago). Calculations suggest that an asteroid the size of Gaspra or Ida can expect another catastrophic collision sometime in the next billion years, at which time it will be disrupted to form another generation of still-smaller fragments.

The greatest surprise of the Galileo flyby of Ida was the discovery of a moon (which was then named Dactyl), in orbit about the asteroid (Figure 13.6). Although only 1.5 kilometers in diameter, smaller than many college campuses, Dactyl provides scientists with something otherwise beyond their reach—a measurement of the mass and density of Ida using Kepler's laws. The moon's distance of about 100 kilometers and its orbital period of about 24 hours indicate that Ida has a density of approximately 2.5 g/cm^3, which matches the density of primitive rocks. Subsequently, both large visible-light telescopes and high-powered planetary radar have discovered many other asteroid moons, so that we are now able to accumulate valuable data on asteroid masses and densities.

Download for free at http://cnx.org/content/col11992/latest/

Figure 13.6. Ida and Dactyl. The asteroid Ida and its tiny moon Dactyl (the small body off to its right), were photographed by the Galileo spacecraft in 1993. Irregularly shaped Ida is 56 kilometers in its longest dimension, while Dactyl is about 1.5 kilometers across. The colors have been intensified in this image; to the eye, all asteroids look basically gray. (credit: modification of work by NASA/JPL)

By the way, Phobos and Deimos, the two small moons of Mars, are probably captured asteroids (Figure 13.7). They were first studied at close range by the Viking orbiters in 1977 and later by *Mars Global Surveyor*. Both are irregular, somewhat elongated, and heavily created, resembling other smaller asteroids. Their largest dimensions are about 26 kilometers and 16 kilometers, respectively. The small outer moons of Jupiter and Saturn were probably also captured from passing asteroids, perhaps early in the history of the solar system.

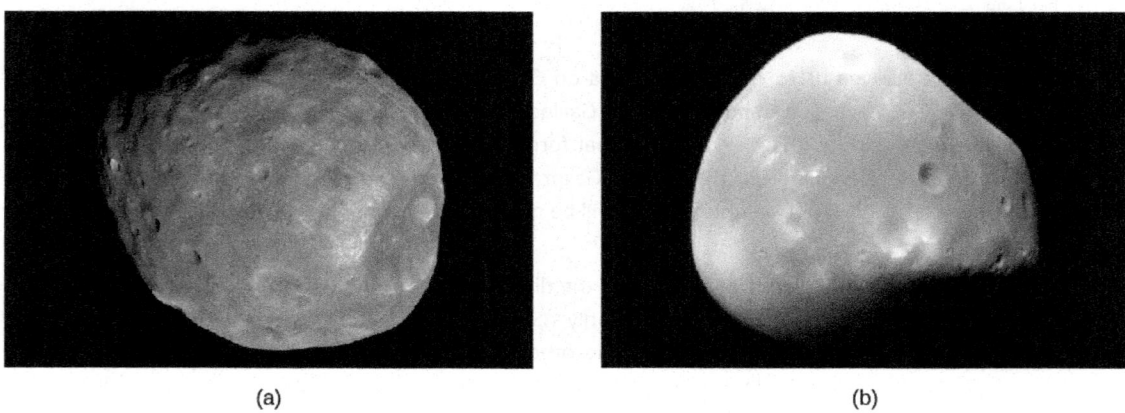

(a)   (b)

Figure 13.7. Moons of Mars. The two small moons of Mars, (a) Phobos and (b) Deimos, were discovered in 1877 by American astronomer Asaph Hall. Their surface materials are similar to many of the asteroids in the outer asteroid belt, leading astronomers to believe that the two moons may be captured asteroids. (credit a: modification of work by NASA; credit b: modification of work by NASA/JPL-Caltech/University of Arizona)

Beginning in the 1990s, spacecraft have provided close looks at several more asteroids. The Near Earth Asteroid Rendezvous (NEAR) spacecraft went into orbit around the S-type asteroid Eros, becoming a temporary moon of this asteroid. On its way to Eros, the *NEAR* spacecraft was renamed after planetary geologist Eugene Shoemaker, a pioneer in our understanding of craters and impacts.

For a year, the NEAR-Shoemaker spacecraft orbited the little asteroid at various altitudes, measuring its surface and interior composition as well as mapping Eros from all sides (Figure 13.8). The data showed that Eros is made of some of the most chemically primitive materials in the solar system. Several other asteroids have been revealed as made of loosely bound rubble throughout, but not Eros. Its uniform density (about the same as that of Earth's crust) and extensive global-scale grooves and ridges show that it is a cracked but solid rock.

This OpenStax book is available for free at http://cnx.org/content/col11992/1.8

Figure 13.8. Looking Down on the North Pole of Eros. This view was constructed from six images of the asteroid taken from an altitude of 200 kilometers. The large crater at the top has been named Psyche (after the maiden who was Eros' lover in classical mythology) and is about 5.3 kilometers wide. A saddle-shaped region can be seen directly below it. Craters of many different sizes are visible. (credit: modification of work by NASA/JHUPL)

Eros has a good deal of loose surface material that appears to have slid down toward lower elevations. In some places, the surface rubble layer is 100 meters deep. The top of loose soil is dotted with scattered, half-buried boulders. There are so many of these boulders that they are more numerous than the craters. Of course, with the gravity so low on this small world, a visiting astronaut would find loose boulders rolling toward her pretty slowly and could easily leap high enough to avoid being hit by one. Although the NEAR-Shoemaker spacecraft was not constructed as a lander, at the end of its orbital mission in 2000, it was allowed to fall gently to the surface, where it continued its chemical analysis for another week.

In 2003, Japan's Hayabusa 1 mission not only visited a small asteroid but also brought back samples to study in laboratories on Earth. The target S-type asteroid, Itokawa (shown in Figure 13.9), is much smaller than Eros, only about 500 meters long. This asteroid is elongated and appears to be the result of the collision of two separate asteroids long ago. There are almost no impact craters, but an abundance of boulders (like a pile of rubble) on the surface.

Download for free at http://cnx.org/content/col11992/latest/

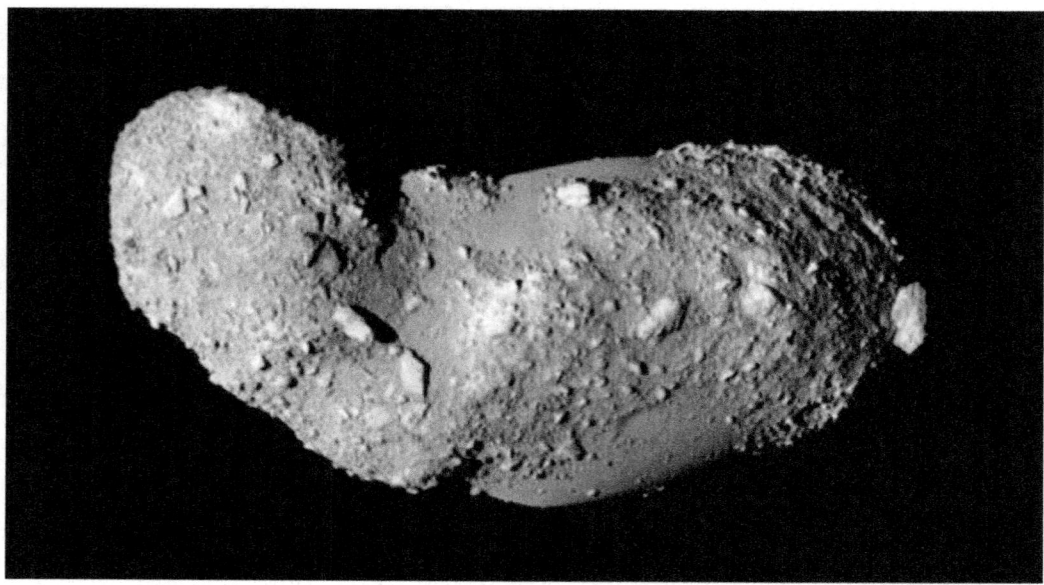

Figure 13.9. Asteroid Itokawa. The surface of asteroid Itokawa appears to have no craters. Astronomers have hypothesized that its surface consists of rocks and ice chunks held together by a small amount of gravity, and its interior is probably also a similar rubble pile. (credit: JAXA)

The *Hayabusa* spacecraft was designed not to land, but to touch the surface just long enough to collect a small sample. This tricky maneuver failed on its first try, with the spacecraft briefly toppling over on its side. Eventually, the controllers were successful in picking up a few grains of surface material and transferring them into the return capsule. The 2010 reentry into Earth's atmosphere over Australia was spectacular (Figure 13.10), with a fiery breakup of the spacecraft, while a small return capsule successfully parachuted to the surface. Months of careful extraction and study of more than a thousand tiny dust particles confirmed that the surface of Itokawa had a composition similar to a well-known class of primitive meteorites. We estimate that the dust grains *Hayabusa* picked up had been exposed on the surface of the asteroid for about 8 million years.

Figure 13.10. Hayabusa Return. This dramatic image shows the *Hayabusa* probe breaking up upon reentry. The return capsule, which separated from the main spacecraft and parachuted to the surface, glows at the bottom right. (credit: modification of work by NASA Ames/Jesse Carpenter/Greg Merkes)

The most ambitious asteroid space mission (called Dawn) has visited the two largest main belt asteroids, Ceres and Vesta, orbiting each for about a year (Figure 13.11). Their large sizes (diameters of about 1000 and 500

This OpenStax book is available for free at http://cnx.org/content/col11992/1.8

kilometers, respectively) make them appropriate for comparison with the planets and large moons. Both turned out to be heavily cratered, implying their surfaces are old. On Vesta, we have now actually located the large impact craters that ejected the basaltic meteorites previously identified as coming from this asteroid. These craters are so large that they sample several layers of Vesta's crustal material.

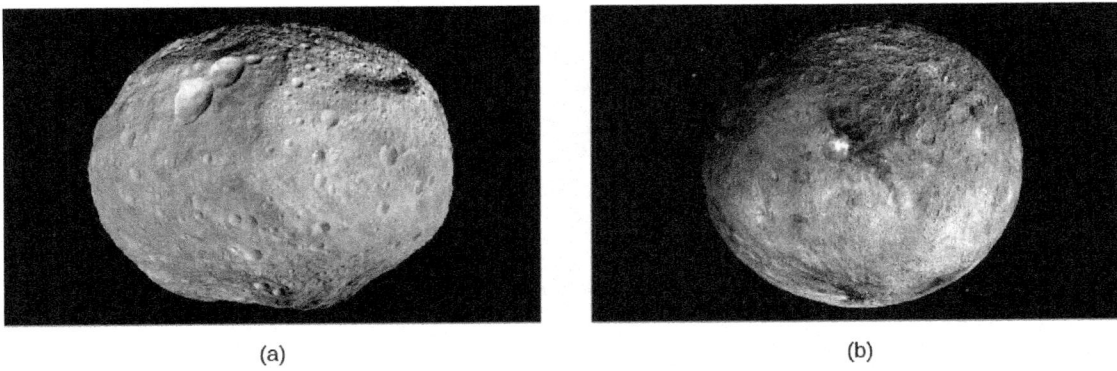

(a)   (b)

Figure 13.11. Vesta and Ceres. The NASA *Dawn* spacecraft took these images of the large asteroids (a) Vesta and (b) Ceres. (a) Note that Vesta is not round, as Ceres (which is considered a dwarf planet) is. A mountain twice the height of Mt. Everest on Earth is visible at the very bottom of the Vesta image. (b) The image of Ceres has its colors exaggerated to bring out differences in composition. You can see a white feature in Occator crater near the center of the image. (credit a, b: modification of work by NASA/JPL-Caltech/UCLA/MPS/DLR/IDA)

Ceres has not had a comparable history of giant impacts, so its surface is covered with craters that look more like those from the lunar highlands. The big surprise at Ceres is the presence of very bright white spots, associated primarily with the central peaks of large craters (Figure 13.12). The light-colored mineral is some kind of salt, either produced when these craters were formed or subsequently released from the interior.

Figure 13.12. White Spots in a Larger Crater on Ceres. These bright features appear to be salt deposits in a Ceres crater called Occator, which is 92 kilometers across. (credit: modification of work by NASA/JPL-Caltech/UCLA/MPS/DLR/IDA)

Download for free at http://cnx.org/content/col11992/latest/

LINK TO LEARNING

The space agencies involved with the Dawn mission have produced nice animated "flyover" videos of Vesta (https://openstaxcollege.org/l/30vestaflyover) and Ceres (https://openstaxcollege.org/l/30ceresflyover) available online.

13.2 ASTEROIDS AND PLANETARY DEFENSE

Learning Objectives

By the end of this section, you will be able to:

> Recognize the threat that near-Earth objects represent for Earth
> Discuss possible defensive strategies to protect our planet

Not all asteroids are in the main asteroid belt. In this section, we consider some special groups of asteroids with orbits that approach or cross the orbit of Earth. These pose the risk of a catastrophic collision with our planet, such as the collision 65 million years ago that killed the dinosaurs.

Earth-Approaching Asteroids

Asteroids that stray far *outside* the main belt are of interest mostly to astronomers. But asteroids that come *inward*, especially those with orbits that come close to or cross the orbit of Earth, are of interest to political leaders, military planners—indeed, everyone alive on Earth. Some of these asteroids briefly become the closest celestial object to us.

In 1994, a 1-kilometer object was picked up passing closer than the Moon, causing a stir of interest in the news media. Today, it is routine to read of small asteroids coming this close to Earth. (They were always there, but only in recent years have astronomers been able to detect such faint objects.)

In 2013, a small asteroid hit our planet, streaking across the sky over the Russian city of Chelyabinsk and exploding with the energy of a nuclear bomb (Figure 13.13). The impactor was a stony object about 20 meters in diameter, exploding about 30 kilometers high with an energy of 500 kilotons (about 30 times larger than the nuclear bombs dropped on Japan in World War II). No one was hurt by the blast itself, although it briefly became as bright as the Sun, drawing many spectators to the windows in their offices and homes. When the blast wave from the explosion then reached the town, it blew out the windows. About 1500 people had to seek medical attention from injuries from the shattered glass.

A much larger atmospheric explosion took place in Russia in 1908, caused by an asteroid about 40 meters in diameter, releasing an energy of 5 megatons, as large the most powerful nuclear weapons of today. Fortunately, the area directly affected, on the Tunguska River in Siberia, was unpopulated, and no one was killed. However, the area of forest destroyed by the blast was large equal to the size of a major city (Figure 13.13).

Together with any comets that come close to our planet, such asteroids are known collectively as **near-Earth objects (NEOs)**. As we will see (and as the dinosaurs found out 65 million years ago,) the collision of a significant-sized NEO could be a catastrophe for life on our planet.

This OpenStax book is available for free at http://cnx.org/content/col11992/1.8

Download for free at http://cnx.org/content/col11992/latest/

(a)   (b)

Figure 13.13. Impacts with Earth. (a) As the Chelyabinsk meteor passed through the atmosphere, it left a trail of smoke and briefly became as bright as the Sun. (b) Hundreds of kilometers of forest trees were knocked down and burned at the Tunguska impact site. (credit a: modification of work by Alex Alishevskikh)

LINK TO LEARNING

Visit the video compilation (https://openstaxcollege.org/l/30vidcomchelmet) of the Chelyabinsk meteor streaking through the sky over the city on February 15, 2013, as taken by people who were in the area when it occurred.

LINK TO LEARNING

View this video of a non-technical talk by David Morrison (https://openstaxcollege.org/l/ 30davmorrison) to watch "The Chelyabinsk Meteor: Can We Survive a Bigger Impact?" Dr. Morrison (SETI Institute and NASA Ames Research Center) discusses the Chelyabinsk impact and how we learn about NEOs and protect ourselves; the talk is from the Silicon Valley Astronomy Lectures series.

Astronomers have urged that the first step in protecting Earth from future impacts by NEOs must be to learn what potential impactors are out there. In 1998, NASA began the Spaceguard Survey, with the goal to discover and track 90% of Earth-approaching asteroids greater than 1 kilometer in diameter. The size of 1 kilometer was selected to include all asteroids capable of causing global damage, not merely local or regional effects. At 1 kilometer or larger, the impact could blast so much dust into the atmosphere that the sunlight would be dimmed for months, causing global crop failures—an event that could threaten the survival of our civilization. The Spaceguard goal of 90% was reached in 2012 when nearly a thousand of these 1-kilometer **near-Earth asteroids (NEAs)** had been found, along with more than 10,000 smaller asteroids. Figure 13.14 shows how the pace of NEA discoveries has been increasing over recent years.

Download for free at http://cnx.org/content/col11992/latest/

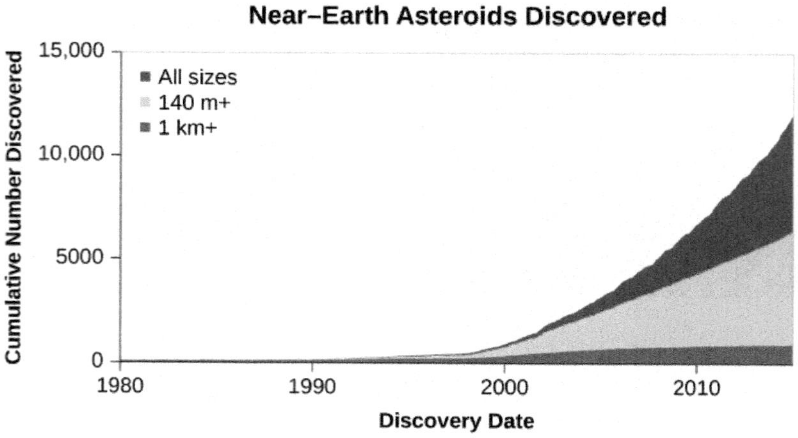

Near–Earth Asteroids Discovered

Figure 13.14. Discovery of Near-Earth Asteroids. The accelerating rate of discovery of NEAs is illustrated in this graph, which shows the total number of known NEAs, the number over 140 kilometers in diameter, and the number over 1 kilometer in diameter, the size that poses the dominant impact risk on Earth.

How did astronomers know when they had discovered 90% of these asteroids? There are several ways to estimate the total number, even before they were individually located. One way is to look at the numbers of large craters on the dark lunar maria. Remember that these craters were made by impacts just like the ones we are considering. They are preserved on the Moon's airless surface, whereas Earth soon erases the imprints of past impacts. Thus, the number of large craters on the Moon allows us to estimate how often impacts have occurred on both the Moon and Earth over the past several billion years. The number of impacts is directly related to the number of asteroids and comets on Earth-crossing orbits.

Another approach is to see how often the surveys (which are automated searches for faint points of light that move among the stars) rediscover a previously known asteroid. At the beginning of a survey, all the NEAs it finds will be new. But as the survey becomes more complete, more and more of the moving points the survey cameras record will be *rediscoveries*. The more rediscoveries each survey experiences, the more complete our inventory of these asteroids must be.

We have been relieved to find that none of the NEAs discovered so far is on a trajectory that will impact Earth within the foreseeable future. However, we can't speak for the handful of asteroids larger than 1 kilometer that have not yet been found, or for the much more numerous smaller ones. It is estimated that there are a million NEAs capable of hitting Earth that are smaller than 1 kilometer but still large enough to destroy a city, and our surveys have found fewer than 10% of them. Researchers who work with asteroid orbits estimate that for smaller (and therefore fainter) asteroids we are not yet tracking, we will have about a 5-second warning that one is going to hit Earth—in other words, we won't see it until it enters the atmosphere. Clearly, this estimate gives us a lot of motivation to continue these surveys to track as many asteroids as possible.

Though entirely predictable over times of a few centuries, the orbits of Earth-approaching asteroids are unstable over long time spans as they are tugged by the gravitational attractions of the planets. These objects will eventually meet one of two fates: either they will impact one of the terrestrial planets or the Sun, or they will be ejected gravitationally from the inner solar system due to a near-encounter with a planet. The probabilities of these two outcomes are about the same. The timescale for impact or ejection is only about a hundred million years, very short compared with the 4-billion-year age of the solar system. Calculations show that only approximately one quarter of the current Earth-approaching asteroids will eventually end up colliding with Earth itself.

This OpenStax book is available for free at http://cnx.org/content/col11992/1.8

If most of the current population of Earth-approaching asteroids will be removed by impact or ejection in a hundred million years, there must be a continuing source of new objects to replenish our supply of NEAs. Most of them come from the asteroid belt between Mars and Jupiter, where collisions between asteroids can eject fragments into Earth-crossing orbits (see Figure 13.15). Others may be "dead" comets that have exhausted their volatile materials (which we'll discuss in the next section).

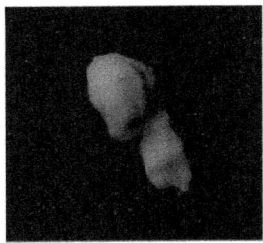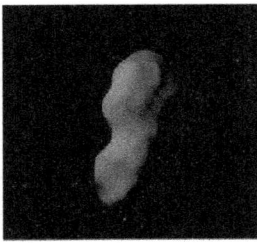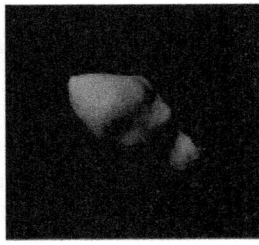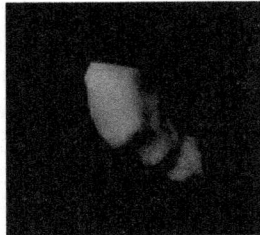

Figure 13.15. Near-Earth Asteroid. Toutatis is a 5-kilometer long NEA that approached within 3 million kilometers of Earth in 1992. This series of images is a reconstruction its size and shape obtained from bouncing radar waves off the asteroid during its close flyby. Toutatis appears to consist of two irregular, lumpy bodies rotating in contact with each other. (Note that the color has been artificially added.) (credit: modification of work by NASA)

One reason scientists are interested in the composition and interior structure of NEAs is that humans will probably need to defend themselves against an asteroid impact someday. If we ever found one of these asteroids on a collision course with us, we would need to deflect it so it would miss Earth. The most straightforward way to deflect it would be to crash a spacecraft into it, either slowing it or speeding it up, slightly changing its orbital period. If this were done several years before the predicted collision, the asteroid would miss the planet entirely—making an asteroid impact the only natural hazard that we could eliminate completely by the application of technology. Alternatively, such deflection could be done by exploding a nuclear bomb near the asteroid to nudge it off course.

To achieve a successful deflection by either technique, we need to know more about the density and interior structure of the asteroid. A spacecraft impact or a nearby explosion would have a greater effect on a solid rocky asteroid such as Eros than on a loose rubble pile. Think of climbing a sand dune compared to climbing a rocky hill with the same slope. On the dune, much of our energy is absorbed in the slipping sand, so the climb is much more difficult and takes more energy.

There is increasing international interest in the problem of asteroid impacts. The United Nations has formed two technical committees on planetary defense, recognizing that the entire planet is at risk from asteroid impacts. However, the fundamental problem remains one of finding NEAs in time for defensive measures to be taken. We must be able to find the next impactor before it finds us. And that's a job for the astronomers.

THE "LONG-HAIRED" COMETS

Learning Objectives

By the end of this section, you will be able to:

> Characterize the general physical appearance of comets
> Explain the range of cometary orbits
> Describe the size and composition of a typical comet's nucleus
> Discuss the atmospheres of comets
> Summarize the discoveries of the Rosetta mission

Download for free at http://cnx.org/content/col11992/latest/

Comets differ from asteroids primarily in their icy composition, a difference that causes them to brighten dramatically as they approach the Sun, forming a temporary atmosphere. In some early cultures, these so-called "hairy stars" were considered omens of disaster. Today, we no longer fear comets, but eagerly anticipate those that come close enough to us to put on a good sky show.

Appearance of Comets

A **comet** is a are relatively small chunk of icy material (typically a few kilometers across) that develops an atmosphere as it approaches the Sun. Later, there may be a very faint, nebulous **tail**, extending several million kilometers away from the main body of the comet. Comets have been observed from the earliest times: accounts of comets are found in the histories of virtually all ancient civilizations. The typical comet, however, is not spectacular in our skies, instead having the appearance of a rather faint, diffuse spot of light somewhat smaller than the Moon and many times less brilliant. (Comets seemed more spectacular to people before the invention of artificial lighting, which compromises our view of the night sky.)

Like the Moon and planets, comets appear to wander among the stars, slowly shifting their positions in the sky from night to night. Unlike the planets, however, most comets appear at unpredictable times, which perhaps explain why they frequently inspired fear and superstition in earlier times. Comets typically remain visible for periods that vary from a couple of weeks to several months. We'll say more about what they are made of and how they become visible after we discuss their motions.

Note that still images of comets give the impression that they are moving rapidly across the sky, like a bright meteor or shooting star. Looking only at such images, it is easy to confuse comets and meteors. But seen in the real sky, they are very different: the meteor burns up in our atmosphere and is gone in a few seconds, whereas the comet may be visible for weeks in nearly the same part of the sky.

Comet Orbits

The study of comets as members of the solar system dates from the time of Isaac Newton, who first suggested that they orbited the Sun on extremely elongated ellipses. Newton's colleague Edmund Halley (see the Edmund Halley: Astronomy's Renaissance Man feature box) developed these ideas, and in 1705, he published calculations of 24 comet orbits. In particular, he noted that the orbits of the bright comets that had appeared in the years 1531, 1607, and 1682 were so similar that the three could well be the same comet, returning to perihelion (closest approach to the Sun) at average intervals of 76 years. If so, he predicted that the object should next return about 1758. Although Halley had died by the time the comet appeared as he predicted, it was given the name Comet Halley (rhymes with "valley") in honor of the astronomer who first recognized it as a permanent member of our solar system, orbiting around the Sun. Its aphelion (furthest point from the Sun) is beyond the orbit of Neptune.

We now know from historical records that Comet Halley has actually been observed and recorded on every passage near the Sun since 239 BCE at intervals ranging from 74 to 79 years. The period of its return varies somewhat because of orbital changes produced by the pull of the giant planets. In 1910, Earth was brushed by the comet's tail, causing much needless public concern. Comet Halley last appeared in our skies in 1986 (Figure 13.16), when it was met by several spacecraft that gave us a wealth of information about its makeup; it will return in 2061.

This OpenStax book is available for free at http://cnx.org/content/col11992/1.8

Figure 13.16. Comet Halley. This composite of three images (one in red, one in green, one in blue) shows Comet Halley as seen with a large telescope in Chile in 1986. During the time the three images were taken in sequence, the comet moved among the stars. The telescope was moved to keep the image of the comet steady, causing the stars to appear in triplicate (once in each color) in the background. (credit: modification of work by ESO)

VOYAGERS IN ASTRONOMY

Edmund Halley: Astronomy's Renaissance Man

Edmund Halley (Figure 13.17), a brilliant astronomer who made contributions in many fields of science and statistics, was by all accounts a generous, warm, and outgoing person. In this, he was quite the opposite of his good friend Isaac Newton, whose great work, the *Principia* (see Orbits and Gravity), Halley encouraged, edited, and helped pay to publish. Halley himself published his first scientific paper at age 20, while still in college. As a result, he was given a royal commission to go to Saint Helena (a remote island off the coast of Africa where Napoleon would later be exiled) to make the first telescopic survey of the southern sky. After returning, he received the equivalent of a master's degree and was elected to the prestigious Royal Society in England, all at the age of 22.

In addition to his work on comets, Halley was the first astronomer to recognize that the so-called "fixed" stars move relative to each other, by noting that several bright stars had changed their positions since Ptolemy's publication of the ancient Greek catalogs. He wrote a paper on the possibility of an infinite universe, proposed that some stars may be variable, and discussed the nature and size of *nebulae* (glowing cloudlike structures visible in telescopes). While in Saint Helena, Halley observed the planet

Download for free at http://cnx.org/content/col11992/latest/

Mercury going across the face of the Sun and developed the mathematics of how such transits could be used to establish the size of the solar system.

In other fields, Halley published the first table of human life expectancies (the precursor of life-insurance statistics); wrote papers on monsoons, trade winds, and tides (charting the tides in the English Channel for the first time); laid the foundations for the systematic study of Earth's magnetic field; studied evaporation and how inland waters become salty; and even designed an underwater diving bell. He served as a British diplomat, advising the emperor of Austria and squiring the future czar of Russia around England (avidly discussing, we are told, both the importance of science and the quality of local brandy).

In 1703, Halley became a professor of geometry at Oxford, and in 1720, he was appointed Astronomer Royal of England. He continued observing Earth and the sky and publishing his ideas for another 20 years, until death claimed him at age 85.

Figure 13.17. Edmund Halley (1656–1742). Halley was a prolific contributor to the sciences. His study of comets at the turn of the eighteenth century helped predict the orbit of the comet that now bears his name.

Only a few comets return in a time measureable in human terms (shorter than a century), like Comet Halley does; these are called *short-period* comets. Many short-period comets have had their orbits changed by coming too close to one of the giant planets—most often Jupiter (and they are thus sometimes called Jupiter-family comets). Most comets have long periods and will take thousands of years to return, if they return at all. As we will see later in this chapter, most Jupiter-family comets come from a different source than the *long-period* comets (those with orbital periods longer than about a century).

Observational records exist for thousands of comets. We were visited by two bright comets in recent decades. First, in March 1996, came Comet Hyakutake, with a very long tail. A year later, Comet Hale-Bopp appeared; it was as bright as the brightest stars and remained visible for several weeks, even in urban areas (see the image that opens this chapter, Figure 13.1).

Table 13.2 lists some well-known comets whose history or appearance is of special interest.

This OpenStax book is available for free at http://cnx.org/content/col11992/1.8

Some Interesting Comets

Name	Period	Significance
Great Comet of 1577	Long	Tycho Brahe showed it was beyond the Moon (a big step in our understanding)
Great Comet of 1843	Long	Brightest recorded comet; visible in daytime
Daylight Comet of 1910	Long	Brightest comet of the twentieth century
West	Long	Nucleus broke into pieces (1976)
Hyakutake	Long	Passed within 15 million km of Earth (1996)
Hale–Bopp	Long	Brightest recent comet (1997)
Swift-Tuttle	133 years	Parent comet of Perseid meteor shower
Halley	76 years	First comet found to be periodic; explored by spacecraft in 1986
Borrelly	6.8 years	Flyby by Deep Space 1 spacecraft (2000)
Biela	6.7 years	Broke up in 1846 and not seen again
Churyumov-Gerasimenko	6.5 years	Target of Rosetta mission (2014–16)
Wild 2	6.4 years	Target of Stardust sample return mission (2004)
Tempel 1	5.7 years	Target of Deep Impact mission (2005)
Encke	3.3 years	Shortest known period

Table 13.2

The Comet's Nucleus

When we look at an active comet, all we normally see is its temporary atmosphere of gas and dust illuminated by sunlight. This atmosphere is called the comet's head or *coma*. Since the gravity of such small bodies is very weak, the atmosphere is rapidly escaping all the time; it must be replenished by new material, which has to come from somewhere. The source is the small, solid **nucleus** inside, just a few kilometers across, usually hidden by the glow from the much-larger atmosphere surrounding it. The nucleus is the real comet, the fragment of ancient icy material responsible for the atmosphere and the tail (Figure 13.18).

Download for free at http://cnx.org/content/col11992/latest/

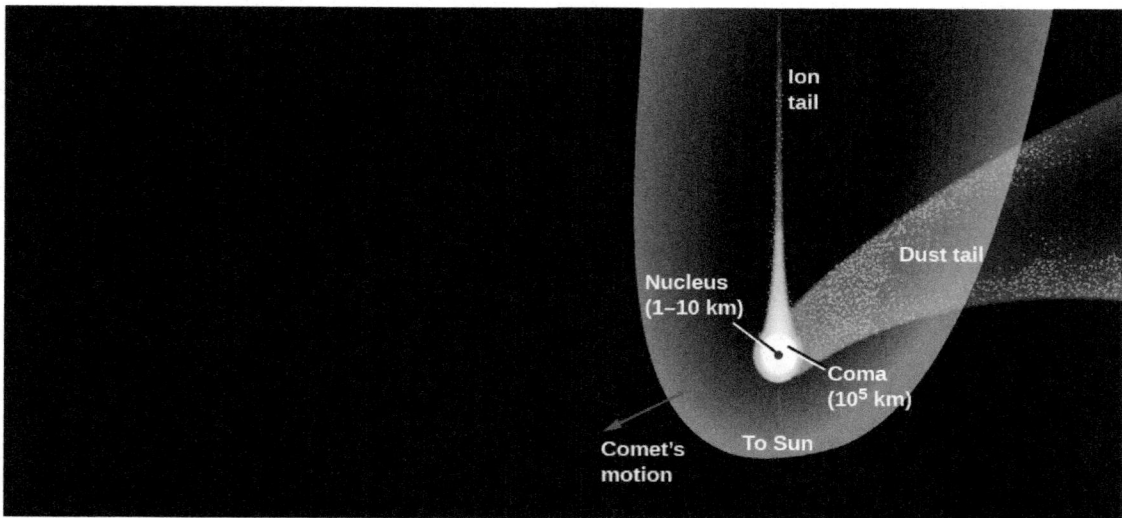

Figure 13.18. Parts of a Comet. This schematic illustration shows the main parts of a comet. Note that the different structures are not to scale.

The modern theory of the physical and chemical nature of comets was first proposed by Harvard astronomer Fred Whipple in 1950. Before Whipple's work, many astronomers thought that a comet's nucleus might be a loose aggregation of solids, sort of an orbiting "gravel bank," Whipple proposed instead that the nucleus is a solid object a few kilometers across, composed in substantial part of water ice (but with other ices as well) mixed with silicate grains and dust. This proposal became known as the "dirty snowball" model.

The water vapor and other volatiles that escape from the nucleus when it is heated can be detected in the comet's head and tail, and therefore, we can use spectra to analyze what atoms and molecules the nucleus ice consists of. However, we are somewhat less certain of the non-icy component. We have never identified a fragment of solid matter from a comet that has survived passage through Earth's atmosphere. However, spacecraft that have approached comets have carried dust detectors, and some comet dust has even been returned to Earth (see Figure 13.19). It seems that much of the "dirt" in the dirty snowball is dark, primitive hydrocarbons and silicates, rather like the material thought to be present on the dark, primitive asteroids.

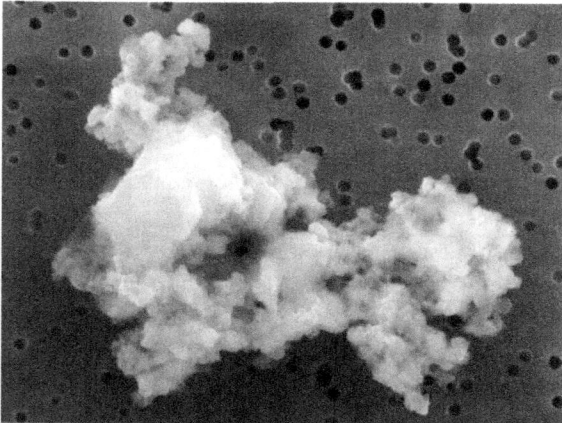

Figure 13.19. Captured Comet Dust. This particle (seen through a microscope) is believed to be a tiny fragment of cometary dust, collected in the upper atmosphere of Earth. It measures about 10 microns, or 1/100 of a millimeter, across. (credit: NASA/JPL)

This OpenStax book is available for free at http://cnx.org/content/col11992/1.8

Since the nuclei of comets are small and dark, they are difficult to study from Earth. Spacecraft did obtain direct measurements of a comet nucleus, however, in 1986, when three spacecraft swept past Comet Halley at close range (see Figure 13.20). Subsequently, other spacecraft have flown close to other comets. In 2005, the NASA *Deep Impact* spacecraft even carried a probe for a high-speed impact with the nucleus of Comet Tempel 1. But by far, the most productive study of a comet has been by the 2015 Rosetta mission, which we will discuss shortly.

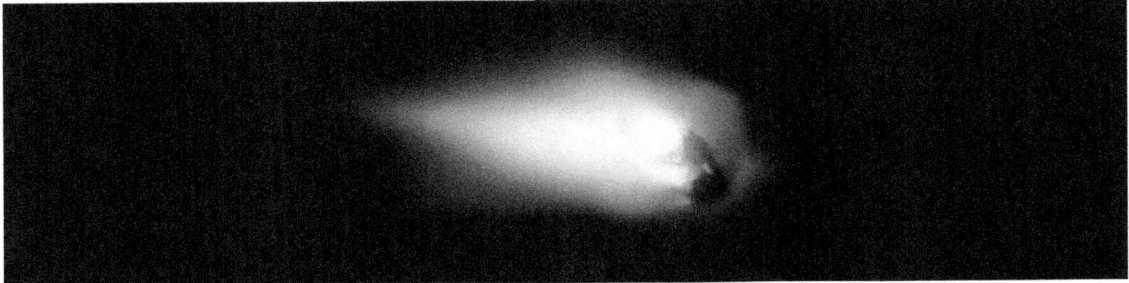

Figure 13.20. Close-up of Comet Halley. This historic photograph of the black, irregularly shaped nucleus of Comet Halley was obtained by the ESA *Giotto* spacecraft from a distance of about 1000 kilometers. The bright areas are jets of material escaping from the surface. The length of the nucleus is 10 kilometers, and details as small as 1 kilometer can be made out. (credit: modification of work by ESA)

The Comet's Atmosphere

The spectacular activity that allows us to see comets is caused by the evaporation of cometary ices heated by sunlight. Beyond the asteroid belt, where comets spend most of their time, these ices are solidly frozen. But as a comet approaches the Sun, it begins to warm up. If water (H_2O) is the dominant ice, significant quantities vaporize as sunlight heats the surface above 200 K. This happens for the typical comet somewhat beyond the orbit of Mars. The evaporating H_2O in turn releases the dust that was mixed with the ice. Since the comet's nucleus is so small, its gravity cannot hold back either the gas or the dust, both of which flow away into space at speeds of about 1 kilometer per second.

The comet continues to absorb energy as it approaches the Sun. A great deal of this energy goes into the evaporation of its ice, as well as into heating the surface. However, recent observations of many comets indicate that the evaporation is not uniform and that most of the gas is released in sudden spurts, perhaps confined to a few areas of the surface. Expanding into space at a speed of about 1 kilometer per second, the comet's atmosphere can reach an enormous size. The diameter of a comet's head is often as large as Jupiter, and it can sometimes approach a diameter of a million kilometers (Figure 13.21).

Figure 13.21. Head of Comet Halley. Here we see the cloud of gas and dust that make up the head, or coma, of Comet Halley in 1986. On this scale, the nucleus (hidden inside the cloud) would be a dot too small to see. (credit: modification of work by NASA/W. Liller)

Download for free at http://cnx.org/content/col11992/latest/

Most comets also develop tails as they approach the Sun. A comet's tail is an extension of its atmosphere, consisting of the same gas and dust that make up its head. As early as the sixteenth century, observers realized that comet tails always point away from the Sun (Figure 13.22), not back along the comet's orbit. Newton proposed that comet tails are formed by a repulsive force of sunlight driving particles away from the head—an idea close to our modern view.

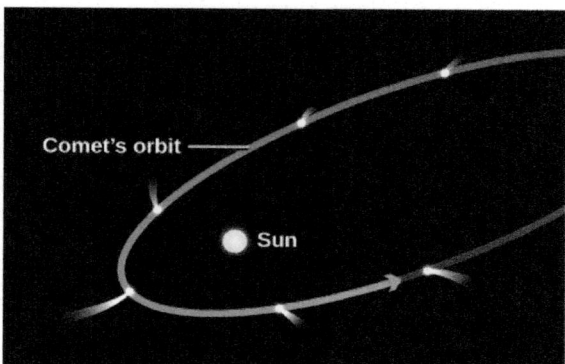

Figure 13.22. Comet Orbit and Tail. The orientation of a typical comet tail changes as the comet passes perihelion. Approaching the Sun, the tail is behind the incoming comet head, but on the way out, the tail precedes the head.

The two different components that make up the tail (the dust and gas) act somewhat differently. The brightest part of the tail is called the *dust tail*, to differentiate it from a fainter, straight tail made of ionized gas, called the ion tail. The ion tail is carried outward by streams of ions (charged particles) emitted by the Sun. As you can see in Figure 13.23, the smoother dust tail curves a bit, as individual dust particles spread out along the comet's orbit, whereas the straight ion is tail pushed more directly outward from the Sun by our star's wind of charged particles

This OpenStax book is available for free at http://cnx.org/content/col11992/1.8

Download for free at http://cnx.org/content/col11992/latest/

(a)

(b)

Figure 13.23. Comet Tails. (a) As a comet nears the Sun, its features become more visible. In this illustration from NASA showing Comet Hale-Bopp, you can see a comet's two tails: the more easily visible dust tail, which can be up to 10 million kilometers long, and the fainter gas tail (or ion tail), which is up to hundreds of millions of kilometers long. The grains that make up the dust tail are the size of smoke particles. (b) Comet Mrkos was photographed in 1957 with a wide-field telescope at Palomar Observatory and also shows a clear distinction between the straight gas tail and the curving dust tail. (credit a: modification of work by ESO/E. Slawik; credit b: modification of work by Charles Kearns, George O. Abell, and Byron Hill)

LINK TO LEARNING

These days, comets close to the Sun can be found with spacecraft designed to observe our star. For example, in early July, 2011, astronomers at the ESA/NASA's Solar and Heliospheric Observatory (SOHO) witnessed a comet (https://openstaxcollege.org/l/30ESANASAcomet) streaking toward the Sun, one of almost 3000 such sightings. You can also watch a brief video by NASA entitled "Why Are We Seeing So Many Sungrazing Comets?"

The Rosetta Comet Mission

In the 1990s, European scientists decided to design a much more ambitious mission that would match orbits with an incoming comet and follow it as it approached the Sun. They also proposed that a smaller spacecraft would actually try to land on the comet. The 2-ton main spacecraft was named *Rosetta*, carrying a dozen scientific instruments, and its 100-kilogram lander with nine more instruments was named *Philae*.

The Rosetta mission was launched in 2004. Delays with the launch rocket caused it to miss its original target comet, so an alternate destination was picked, Comet Churyumov-Gerasimenko (named after the two discoverers, but generally denoted 67P). This comet's period of revolution is 6.45 years, making it a Jupiter-family comet.

Since the European Space Agency did not have access to the plutonium-fueled nuclear power sources used by NASA for deep space missions, *Rosetta* had to be solar powered, requiring especially large solar panels. Even these were not enough to keep the craft operating as it matched orbits with 67P near the comet's aphelion. The only solution was to turn off all the spacecraft systems and let it coast for several years toward the Sun, out of

Download for free at http://cnx.org/content/col11992/latest/

contact with controllers on Earth until solar energy was stronger. The success of the mission depended on an automatic timer to turn the power back on as it neared the Sun. Fortunately, this strategy worked.

In August 2014, *Rosetta* began a gradual approach to the comet nucleus, which is a strangely misshapen object about 5 kilometers across, quite different from the smooth appearance of Halley's nucleus (but equally dark). Its rotation period is 12 hours. On November 12, 2014, the *Philae* lander was dropped, descending slowly for 7 hours before gently hitting the surface. It bounced and rolled, coming to rest under an overhang where there was not enough sunlight to keep its batteries charged. After operating for a few hours and sending data back to the orbiter, *Philae* went silent. The main *Rosetta* spacecraft continued operations, however, as the level of comet activity increased, with steamers of gas jetting from the surface. As the comet approached perihelion in September 2015, the spacecraft backed off to ensure its safety.

The extent of the *Rosetta* images (and data from other instruments) far exceeds anything astronomers had seen before from a comet. The best imaging resolution was nearly a factor of 100 greater than in the best Halley images. At this scale, the comet appears surprisingly rough, with sharp angles, deep pits, and overhangs (Figure 13.24).

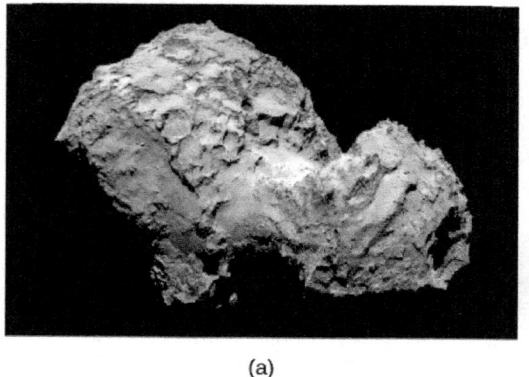

(a)   (b)

Figure 13.24. Comet 67P's Strange Shape and Surface Features. (a) This image from the *Rosetta* camera was taken from a distance of 285 kilometers. The resolution is 5 meters. You can see that the comet consists of two sections with a connecting "neck" between them. (b) This close-up view of Comet Churyumov-Gerasimenko is from the *Philae* lander. One of the lander's three feet is visible in the foreground. The lander itself is mostly in shadow. (credit a: modification of work by ESA/Rosetta/MPS for OSIRIS Team MPS/UPD/LAM/IAA/SSO/INTA/UPM/DASP/IDA; credit b: modification of work by ESA/Rosetta/Philae/CIVA)

The double-lobed shape of 67P's nucleus has been tentatively attributed to the collision and merger of two independent comet nuclei long ago. The spacecraft verified that the comet's dark surface was covered with organic carbon-rich compounds, mixed with sulfides and iron-nickel grains. 67P has an average density of only 0.5 g/cm^3 (recall water in these units has a density of 1 g/cm^3.) This low density indicates that the comet is quite porous, that is, there is a large amount of empty space among its materials.

We already knew that the evaporation of comet ices was sporadic and limited to small jets, but in comet 67P, this was carried to an extreme. At any one time, more than 99% of the surface is inactive. The active vents are only a few meters across, with the material confined to narrow jets that persist for just a few minutes (Figure 13.25). The level of activity is strongly dependent on solar heating, and between July and August 2015, it increased by a factor of 10. Isotopic analysis of deuterium in the water ejected by the comet shows that it is different from the water found on Earth. Thus, apparently comets like 67P did not contribute to the origin of our oceans or the water in our bodies, as some scientists had thought.

This OpenStax book is available for free at http://cnx.org/content/col11992/1.8

  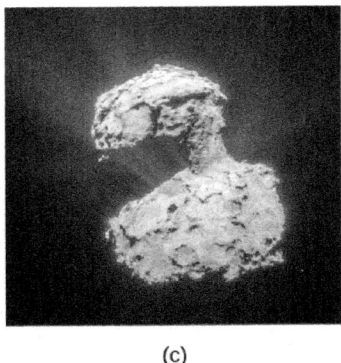

(a)   (b)   (c)

Figure 13.25. Gas Jets on Comet 67P. (a) This activity was photographed by the *Rosetta* spacecraft near perihelion. You can see a jet suddenly appearing; it was active for only a few minutes. (b) This spectacular photo, taken near perihelion, shows the active comet surrounded by multiple jets of gas and dust. (credit a, b: modification of work by ESA/Rosetta/MPS; credit c: modification of work by ESA/Rosetta/NAVCAM)

LINK TO LEARNING

The European Space Agency is continuing to make interesting short videos (https://openstaxcollege.org/l/30ESAvideoros) illustrating the challenges and results of the Rosetta and Philae missions. For example, watch "*Rosetta*'s Moment in the Sun" to see some of the images of the comet generating plumes of gas and dust and hear about some of the dangers an active comet poses for the spacecraft.

13.4 THE ORIGIN AND FATE OF COMETS AND RELATED OBJECTS

Learning Objectives

By the end of this section, you will be able to:

> Describe the traits of the centaur objects
> Chronicle the discovery and describe the composition of the Oort cloud
> Describe trans-Neptunian and Kuiper-belt objects
> Explain the proposed fate of comets that enter the inner solar system

The comets we notice when they come near Earth (especially the ones coming for the first time) are probably the most primitive objects we can study, preserved unchanged for billions of years in the deep freeze of the outer solar system. However, astronomers have discovered many other objects that orbit the Sun beyond the planets.

Centaurs

In the outer solar system, where most objects contain large amounts of water ice, the distinction between asteroids and comets breaks down. Astronomers initially still used the name "asteroids" for new objects discovered going around the Sun with orbits that carry them far beyond Jupiter. The first of these objects is Chiron, found in 1977 on a path that carries it from just inside the orbit of Saturn at its closest approach to the

Download for free at http://cnx.org/content/col11992/latest/

Sun out to almost the distance of Uranus (Figure 13.26). The diameter of Chiron is estimated to be about 200 kilometers, much larger than any known comet.

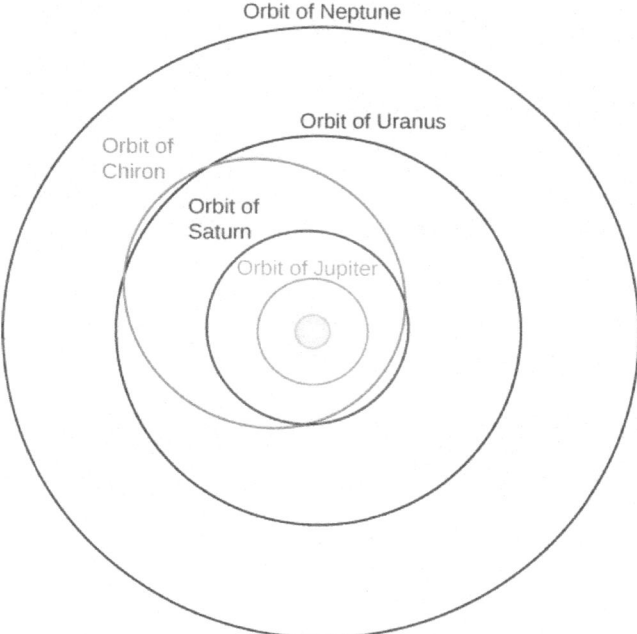

Figure 13.26. Chiron's Orbit. Chiron orbits the Sun every 50 years, with its closest approach being inside the orbit of Saturn and its farthest approach out to the orbit of Uranus.

In 1992, a still-more-distant object named Pholus was discovered with an orbit that takes it 33 AU from the Sun, beyond the orbit of Neptune. Pholus has the reddest surface of any object in the solar system, indicating a strange (and still unknown) surface composition. As more objects are discovered in these distant reaches, astronomers decided that they will be given the names of *centaurs* from classical mythology; this is because the centaurs were half human, half horse, and these new objects display some of the properties of both asteroids and comets.

Beyond the orbit of Neptune lies a cold, dark realm populated by objects called simply trans-Neptunian objects (TNOs). The first discovered, and best known, of these TNOs is the dwarf planet Pluto. We discussed Pluto and the New Horizons spacecraft encounter with it in Rings, Moons, and Pluto. The second TNO was discovered in 1992, and now more than a thousand are known, most of them smaller than Pluto.

The largest ones after Pluto—named Eris, Makemake, and Haumea—are also classed as dwarf planets. Except for their small size, dwarf planets have many properties in common with the larger planets. Pluto has five moons, and two moons have been discovered orbiting Haumea and one each circling Eris and Makemake.

The Kuiper Belt and the Oort Cloud

TNOs are a part of what is called the **Kuiper belt**, a large area of space beyond Neptune that is also the source of many comets. Astronomers study the Kuiper belt in two ways. New, more powerful telescopes allow us to discover many of the larger members of the Kuiper belt directly. We can also measure the composition of comets that come from the Kuiper belt. More than a thousand Kuiper belt objects have been discovered, and astronomers estimate that there are more than 100,000 with diameters large than 100 kilometers, in a disk extending out to about 50 AU from the Sun.

This OpenStax book is available for free at http://cnx.org/content/col11992/1.8

Download for free at http://cnx.org/content/col11992/latest/

The short-period comets (such as Halley) are thought to originate in the Kuiper belt, where small gravitational perturbations from Neptune can gradually shift their orbits until they can penetrate the inner solar system. The long-period comets, however, come from a much more distant reservoir of icy objects, called the Oort cloud.

Careful studies of the orbits of long-period comets revealed that they come initially from very great distances. By following their orbits backward, we can calculate that the *aphelia* (points farthest from the Sun) of newly discovered comets typically have values near 50,000 AU (more than a thousand times farther than Pluto). This clustering of aphelion distances was first noted by Dutch astronomer Jan Oort, who, in 1950, proposed an idea for the origin of those comets that is still accepted today (Figure 13.27).

Figure 13.27. Jan Oort (1900–1992). Jan Oort first suggested that there might be a reservoir of frozen chunks, potential comet nuclei, at the edge of the region of the Sun's gravitational influence. (credit: The Leiden Observatory)

It is possible to calculate that a star's gravitational *sphere of influence*—the distance within which it can exert sufficient gravitation to hold onto orbiting objects—is about one third of its distance to the nearest other stars. Stars in the vicinity of the Sun are spaced in such a way that the Sun's sphere of influence extends a little beyond 50,000 AU, or about 1 light-year. At such great distances, however, objects in orbit about the Sun can be perturbed by the gravity of passing stars. Some of the perturbed objects can then take on orbits that bring them much closer to the Sun (while others might be lost to the solar system forever).

Oort suggested, therefore, that the new comets we were seeing were examples of objects orbiting the Sun near the edge of its sphere of influence, whose orbits had been disturbed by nearby stars, eventually bringing them close to the Sun where we can see them. The reservoir of ancient icy objects from which such comets are derived is now called the **Oort cloud**.

Astronomers estimate that there are about a trillion (10^{12}) comets in the Oort cloud. In addition, we estimate that about 10 times this number of icy objects could be orbiting the Sun in the volume of space between the Kuiper belt (which is gravitationally linked to Neptune) and the Oort cloud. These objects remain undiscovered because they are too faint to be seen directly and their orbits are too stable to permit any of them to be deflected inward close to the Sun. The total number of icy or cometary objects in the outer reaches of our solar system could thus be on the order of 10 trillion (10^{13}), a very large number indeed.

What is the mass represented by 10^{13} comets? We can make an estimate if we assume something about comet sizes and masses. Let us suppose that the nucleus of Comet Halley is typical. Its observed volume is about 600

Download for free at http://cnx.org/content/col11992/latest/

km^3. If the primary constituent is water ice with a density of about 1 g/cm^3, then the total mass of Halley's nucleus must be about 6 × 10^{14} kilograms. This is about one ten billionth (10^{-10}) of the mass of Earth.

If our estimate is reasonable and there are 10^{13} comets with this mass out there, their total mass would be equal to about 1000 Earths—comparable to the mass of all the planets put together. Therefore, icy, cometary material could be the most important constituent of the solar system after the Sun itself.

EXAMPLE 13.1

Mass of the Oort Cloud Comets

Suppose the Oort cloud contains 10^{12} comets with an average diameter of 10 km each. Let's estimate the mass of the total Oort cloud.

Solution

We can start by assuming that typical comets are about the size of Comets Halley and Borrelly, with a diameter of 10 km and a density appropriate to water ice, which is about 1 g/cm^3 or 1000 kg/m^3. We know that density = mass/volume, the volume of a sphere, $V = \frac{4}{3}\pi R^3$, and the radius, $R = \frac{1}{2}D$.

Therefore, for each comet,

$$\text{mass} = \text{density} \times \text{volume}$$
$$= \text{density} \times \frac{4}{3}\pi\left(\frac{1}{2}D\right)^3$$

Given that 10 km = 10^4 m, each comet's mass is

$$\text{mass} = 1000\,\text{kg/m}^3 \times \frac{4}{3} \times 3.14 \times \frac{1}{8} \times \left(10^4\right)^3 \text{m}^3$$
$$\approx 10^{15}\,\text{kg}$$
$$= 10^{12}\,\text{tons}$$

To calculate the total mass of the cloud, we multiply this typical mass for one comet by the number of comets:

$$\text{total mass} = 10^{15}\,\text{kg/comet} \times 10^{12}\,\text{comets}$$
$$= 10^{27}\,\text{kg}$$

Check Your Learning

How does the total mass we calculated above compare to the mass of Jupiter? To the mass of the Sun? (Give a numerical answer.)

Answer:

The mass of Jupiter is about 1.9 × 10^{27} kg. The mass of the Oort cloud calculated above is 10^{27} kg. So the cloud would contain about half a Jupiter of mass. The mass of the Sun is 2 × 10^{30} kg. This means the Oort cloud would be

$$\frac{10^{27}\,\text{kg}}{\left(2 \times 10^{30}\,\text{kg}\right)} = 0.0005 \times \text{the mass of the Sun}$$

This OpenStax book is available for free at http://cnx.org/content/col11992/1.8

Download for free at http://cnx.org/content/col11992/latest/

Early Evolution of the Planetary System

Comets from the Oort cloud help us sample material that formed very far from the Sun, whereas the short-period comets from the Kuiper belt sample materials that were planetesimals in the solar nebula disk but did not form planets. Studies of the Kuiper belt also are influencing our understanding of the early evolution of our planetary system.

The objects in the Oort cloud and the Kuiper belt have different histories, and they may therefore have different compositions. Astronomers are therefore very interested in comparing detailed measurements of the comets derived from these two source regions. Most of the bright comets that have been studied in the past (Halley, Hyakutake, Hale-Bopp) are Oort cloud comets, but P67 and several other comets targeted for spacecraft measurements in the next decade are Jupiter-family comets from the Kuiper belt (see Table 13.2).

The Kuiper belt is made up of ice-and rock planetesimals, a remnant of the building blocks of the planets. Since it is gravitationally linked to Neptune, it can help us understand the formation and history of the solar system. As the giant planets formed, their gravity profoundly influenced the orbits of Kuiper belt objects. Computer simulations of the early evolution of the planetary system suggest that the gravitational interactions between the giant planets and the remaining planetesimals caused the orbit of Jupiter to drift inward, whereas the orbits of Saturn, Uranus, and Neptune all expanded, carrying the Kuiper belt with them.

Another hypotheses involves a fifth giant planet that was expelled from the solar system entirely as the planetary orbits shifted. Neptune's retrograde (backward-orbiting) moon Triton (which is nearly as large as Pluto) may have been a Kuiper belt object captured by Neptune during the period of shifting orbits. It clearly seems that the Kuiper belt may carry important clues to the way our solar system reached its present planetary configuration.

MAKING CONNECTIONS

Comet Hunting as a Hobby

When amateur astronomer David Levy (Figure 13.28), the co-discoverer of Comet Shoemaker-Levy 9, found his first comet, he had already spent 928 fruitless hours searching through the dark night sky. But the discovery of the first comet only whetted his appetite. Since then, he has found 8 others on his own and 13 more working with others. Despite this impressive record, he ranks only third in the record books for number of comet discoveries. But David hopes to break the record someday.

All around the world, dedicated amateur observers spend countless nights scanning the sky for new comets. Astronomy is one of the very few fields of science where amateurs can still make a meaningful contribution, and the discovery of a comet is one of the most exciting ways they can establish their place in astronomical history. Don Machholz, a California amateur (and comet hunter) who has been making a study of comet discoveries, reported that between 1975 and 1995, 38% of all comets discovered were found by amateurs. Those 20 years yielded 67 comets for amateurs, or almost 4 per year. That might sound pretty encouraging to new comet hunters, until they learn that the average number of hours the typical amateur spent searching for a comet before finding one was about 420. Clearly, this is not an activity for impatient personalities.

What do comet hunters do if they think they have found a new comet? First, they must check the object's location in an atlas of the sky to make sure it really is a comet. Since the first sighting of a comet usually occurs when it is still far from the Sun and before it sports a significant tail, it will look like only a small,

Download for free at http://cnx.org/content/col11992/latest/

fuzzy patch. And through most amateur telescopes, so will nebulae (clouds of cosmic gas and dust) and galaxies (distant groupings of stars). Next, they must check that they have not come across a comet that is already known, in which case, they will only get a pat on the back instead of fame and glory. Then they must re-observe or re-image it sometime later to see whether its motion in the sky is appropriate for comets.

Often, comet hunters who think they have made a discovery get another comet hunter elsewhere in the country to confirm it. If everything checks out, the place they contact is the Central Bureau for Astronomical Telegrams at the Harvard-Smithsonian Center for Astrophysics in Cambridge, Massachusetts (http://www.cbat.eps.harvard.edu/). If the discovery is confirmed, the bureau will send the news out to astronomers and observatories around the world. One of the unique rewards of comet hunting is that the discoverer's name becomes associated with the new comet—a bit of cosmic fame that few hobbies can match.

Figure 13.28. David Levy. Amateur astronomer David Levy ranks third in the world for comet discoveries. (credit: Andrew Fraknoi)

The Fate of Comets

Any comet we see today will have spent nearly its entire existence in the Oort cloud or the Kuiper belt at a temperature near absolute zero. But once a comet enters the inner solar system, its previously uneventful life history begins to accelerate. It may, of course, survive its initial passage near the Sun and return to the cold reaches of space where it spent the previous 4.5 billion years. At the other extreme, it may collide with the Sun or come so close that it is destroyed on its first perihelion passage (several such collisions have been observed with space telescopes that monitor the Sun). Sometimes, however, the new comet does not come that close to the Sun but instead interacts with one or more of the planets.

LINK TO LEARNING

SOHO (the Solar and Heliospheric Observatory) has an excellent collection of videos of comets (https://openstaxcollege.org/l/30SOHOcomvid) that come near the Sun. At this site, comet ISON approaches the Sun and is believed to be destroyed in its passage.

This OpenStax book is available for free at http://cnx.org/content/col11992/1.8

A comet that comes within the gravitational influence of a planet has three possible fates. It can (1) impact the planet, ending the story at once; (2) speed up and be ejected, leaving the solar system forever; or (3) be perturbed into an orbit with a shorter period. In the last case, its fate is sealed. Each time it approaches the Sun, it loses part of its material and also has a significant chance of collision with a planet. Once the comet is in this kind of short-period orbit, its lifetime starts being measured in thousands, not billions, of years.

A few comets end their lives catastrophically by breaking apart (sometimes for no apparent reason) (Figure 13.29). Especially spectacular was the fate of the faint Comet Shoemaker-Levy 9, which broke into about 20 pieces when it passed close to Jupiter in July 1992. The fragments of Shoemaker-Levy were actually captured into a very elongated, two-year orbit around Jupiter, more than doubling the number of known jovian moons. This was only a temporary enrichment of Jupiter's family, however, because in July 1994, all the comet fragments crashed unto Jupiter, releasing energy equivalent to millions of megatons of TNT.

 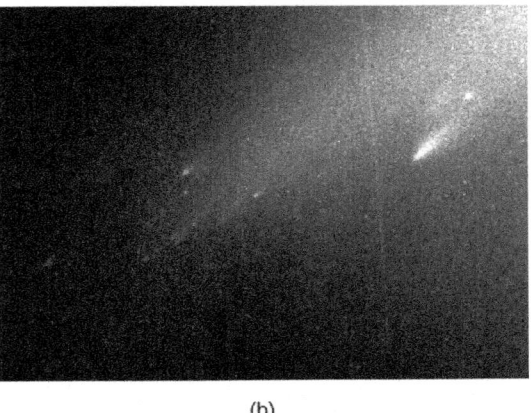

(a)   (b)

Figure 13.29. Breakup of Comet LINEAR. (a) A ground-based view with much less detail and (b) a much more detailed photo with the Hubble Space Telescope, showing the multiple fragments of the nucleus of Comet LINEAR. The comet disintegrated in July 2000 for no apparent reason. (Note in the left view, the fragments all blend their light together, and can't be distinguished. The short diagonal white lines are stars that move in the image, which is keeping track of the moving comet.) (credit a: modification of work by the University of Hawaii; credit b: modification of work by NASA, Harold Weaver (the Johns Hopkins University), and the HST Comet LINEAR Investigation Team)

As each cometary fragment streaked into the jovian atmosphere at a speed of 60 kilometers per second, it disintegrated and exploded, producing a hot fireball that carried the comet dust as well as atmospheric gases to high altitudes. These fireballs were clearly visible in profile, with the actual point of impact just beyond the jovian horizon as viewed from Earth (Figure 13.30). As each explosive plume fell back into Jupiter, a region of the upper atmosphere larger than Earth was heated to incandescence and glowed brilliantly for about 15 minutes, a glow we could detect with infrared-sensitive telescopes.

Download for free at http://cnx.org/content/col11992/latest/

(a)   (b)

Figure 13.30. Comet Impact on Jupiter. (a) The "string" of white objects are fragments of Comet Shoemaker-Levy 9 approaching Jupiter. (b)
The first fragment of the comet impacts Jupiter, with the point of contact on the bottom left side in this image. On the right is Jupiter's moon, Io.
The equally bright spot in the top image is the comet fragment flaring to maximum brightness. The bottom image, taken about 20 minutes
later, shows the lingering flare from the impact. The Great Red Spot is visible near the center of Jupiter. These infrared images were taken with a
German-Spanish telescope on Calar Alto in southern Spain. (credit a: modification of work by ESA; credit b: modification of work by Tom Herbst,
Max-Planck-Institut fuer Astronomie, Heidelberg, Doug Hamilton, Max-Planck-Institut fuer Kernphysik, Heidelberg, Hermann Boehnhardt,
Universitaets-Sternewarte, Muenchen, and Jose Luis Ortiz Moreno, Instituto de Astrofisica de Andalucia, Granada)

After this event, dark clouds of debris settled into the stratosphere of Jupiter, producing long-lived "bruises"
(each still larger than Earth) that could be easily seen through even small telescopes (Figure 13.31). Millions
of people all over the world peered at Jupiter through telescopes or followed the event via television or online.
Another impact feature was seen on Jupiter in summer 2009, indicating that the 1994 events were by no means
unique. Seeing these large, impact explosions on Jupiter helps us to appreciate the disaster that would happen
to our planet if we were hit by a comet or asteroid.

This OpenStax book is available for free at http://cnx.org/content/col11992/1.8

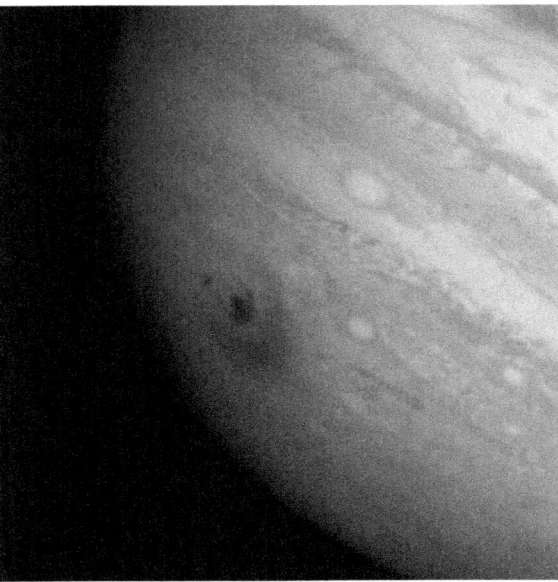

Figure 13.31. Impact Dust Cloud on Jupiter. These features result from the impact of Comet Shoemaker-Levy 9 with Jupiter, seen with the Hubble Space Telescope 105 minutes after the impact that produced the dark rings (the compact back dot came from another fragment). The inner edge of the diffuse, outer ring is about the same size as Earth. Later, the winds on Jupiter blended these features into a broad spot that remained visible for more than a month. (credit: modification of work by H. Hammel, MIT, and NASA/ESA)

For comets that do not meet so dramatic an end, measurements of the amount of gas and dust in their atmospheres permit us to estimate the total losses during one orbit. Typical loss rates are up to a million tons per day from an active comet near the Sun, adding up to some tens of millions of tons per orbit. At that rate, a typical comet will be gone after a few thousand orbits. This will probably be the fate of Comet Halley in the long run.

LINK TO LEARNING

This History Channel video (https://openstaxcollege.org/l/30hischanUNIV) shows a short discussion and animation from the TV documentary series *Universe*, showing the collision of Comet Shoemaker-Levy 9 with Jupiter.

Download for free at http://cnx.org/content/col11992/latest/

CHAPTER 13 REVIEW

 KEY TERMS

asteroid a stony or metallic object orbiting the Sun that is smaller than a major planet but that shows no evidence of an atmosphere or of other types of activity associated with comets

asteroid belt the region of the solar system between the orbits of Mars and Jupiter in which most asteroids are located; the main belt, where the orbits are generally the most stable, extends from 2.2 to 3.3 AU from the Sun

comet a small body of icy and dusty matter that revolves about the Sun; when a comet comes near the Sun, some of its material vaporizes, forming a large head of tenuous gas and often a tail

Kuiper belt a region of space beyond Neptune that is dynamically stable (like the asteroid belt); the source region for most short-period comets

near-Earth asteroid (NEA) an Earth-approaching asteroid, one whose orbit could bring it on a collision course with our planet

near-Earth object (NEO) a comet or asteroid whose path intersects the orbit of Earth

nucleus (of a comet) the solid chunk of ice and dust in the head of a comet

Oort cloud the large spherical region around the Sun from which most "new" comets come; a reservoir of objects with aphelia at about 50,000 AU

tail (of a comet) a tail consisting of two parts: the dust tail is made of dust loosened by the sublimation of ice in a comet that is then pushed by photons from the Sun into a curved stream; the ion tail is a stream of ionized particles evaporated from a comet and then swept away from the Sun by the solar wind

 SUMMARY

13.1 Asteroids

The solar system includes many objects that are much smaller than the planets and their larger moons. The rocky ones are generally called asteroids. Ceres is the largest asteroid; about 15 are larger than 250 kilometers and about 100,000 are larger than 1 kilometer. Most are in the asteroid belt between Mars and Jupiter. The presence of asteroid families in the belt indicates that many asteroids are the remnants of ancient collisions and fragmentation. The asteroids include both primitive and differentiated objects. Most asteroids are classed as C-type, meaning they are composed of carbonaceous materials. Dominating the inner belt are S-type (stony) asteroids, with a few M-type (metallic) ones. We have spacecraft images of several asteroids and returned samples from asteroid Itokawa. Recent observations have detected a number of asteroid moons, making it possible to measure the masses and densities of the asteroids they orbit. The two largest asteroids, Ceres and Vesta, have been extensively studied from orbit by the *Dawn* spacecraft.

13.2 Asteroids and Planetary Defense

Near-Earth asteroids (NEAs), and near-Earth objects (NEOs) in general, are of interest in part because of their potential to hit Earth. They are on unstable orbits, and on timescales of 100 million years, they will either impact one of the terrestrial planets or the Sun, or be ejected. Most of them probably come from the asteroid belt, but some may be dead comets. NASA's Spaceguard Survey has found 90% of the NEAs larger than 1 kilometer, and

This OpenStax book is available for free at http://cnx.org/content/col11992/1.8

none of the ones found so far are on a collision course with Earth. Scientists are actively working on possible technologies for planetary defense in case any NEOs are found on a collision course with Earth years in advance. For now, the most important task is to continue our surveys, so we can find the next Earth impactor before it finds us.

13.3 The "Long-Haired" Comets

Halley first showed that some comets are on closed orbits and return periodically to swing around the Sun. The heart of a comet is its nucleus, a few kilometers in diameter and composed of volatiles (primarily frozen H_2O) and solids (including both silicates and carbonaceous materials). Whipple first suggested this "dirty snowball" model in 1950; it has been confirmed by spacecraft studies of several comets. As the nucleus approaches the Sun, its volatiles evaporate (perhaps in localized jets or explosions) to form the comet's head or atmosphere, which escapes at about 1 kilometer per second. The atmosphere streams away from the Sun to form a long tail. The ESA Rosetta mission to Comet P67 (Churyumov-Gerasimenko) has greatly increased our knowledge of the nature of the nucleus and of the process by which comets release water and other volatiles when heated by sunlight.

13.4 The Origin and Fate of Comets and Related Objects

Oort proposed in 1950 that long-period comets are derived from what we now call the Oort cloud, which surrounds the Sun out to about 50,000 AU (near the limit of the Sun's gravitational sphere of influence) and contains between 10^{12} and 10^{13} comets. Comets also come from the Kuiper belt, a disk-shaped region beyond the orbit of Neptune, extending to 50 AU from the Sun. Comets are primitive bodies left over from the formation of the outer solar system. Once a comet is diverted into the inner solar system, it typically survives no more than a few thousand perihelion passages before losing all its volatiles. Some comets die spectacular deaths: Shoemaker-Levy 9, for example, broke into 20 pieces before colliding with Jupiter in 1994.

FOR FURTHER EXPLORATION

Articles

Asteroids

Asphang, E. "The Small Planets." *Scientific American* (May 2000): 46. On asteroids, including results from the NEAR mission.

Beatty, J. "The Falcon's Wild Flight." *Sky & Telescope* (September 2006): 34. On the Japanese mission to asteroid Itakawa.

Beatty, J. "NEAR Falls for Eros." *Sky & Telescope* (May 2001): 35. On the first landing on an asteroid.

Betz, E. "Dawn Mission Reveals Dwarf Planet Ceres." *Astronomy* (January 2016): 44. First images and discoveries.

Binzel, R. "A New Century for Asteroids." *Sky & Telescope* (July 2001): 44. Nice overview.

Boslaugh, M. "In Search of Death-Plunge Asteroids." *Astronomy* (July 2015): 28. On existing and proposed programs to search for Earth-crossing asteroids.

Cooke, B. "Fatal Attraction." *Astronomy* (May 2006): 46. On near-Earth asteroid Apophis, its orbit, and what we can learn from it.

Durda, D. "Odd Couples." *Astronomy* (December 2005): 54. On binary asteroids.

Durda, D. "All in the Family." *Astronomy* (February 1993): 36. Discusses asteroid families.

Oberg, J. "2013's Historic Russian Meteorite Fall" *Astronomy* (June 2012): 18. On the Chelyabinsk event.

Download for free at http://cnx.org/content/col11992/latest/

Sheppard, S. "Dancing with the Planets." *Sky & Telescope* (June 2016): 16. On Trojan asteroids that "follow" planets like Jupiter.

Talcott, R. "Galileo Views Gaspra." *Astronomy* (February 1992): 52.

Yeomans, D. "Japan Visits an Asteroid." *Astronomy* (March 2006): 32. On the *Hayabusa* probe exploration of asteroid Itakawa.

Zimmerman, R. "Ice Cream Sundaes and Mashed Potatoes." *Astronomy* (February 1999): 54. On the NEAR mission.

Comets

Aguirre, E. "The Great Comet of 1997." *Sky & Telescope* (July 1997): 50. On Comet Hale-Bopp.

Bakich, M. "How to Observe Comets." *Astronomy* (December 2009): 50. A guide for amateur astronomers.

Gore, R. "Halley's Comet '86: Much More Than Met the Eye." *National Geographic* (December 1986): 758. (Also, the March 1987 issue of *Sky & Telescope* was devoted to what we learned from Halley's Comet in 1986.)

Hale, A. "Hale-Bopp Plus Ten." *Astronomy* (July 2005): 76. The co-discoverer of a naked-eye comet tells the story of the discovery and what followed.

Jewett, D. "Mysterious Travelers: Comet Science." *Sky & Telescope* (December. 2013): 18. Nice summary of what we know about comets and questions we have.

Rao, J. "How Often do Bright Comets Appear?" *Sky & Telescope* (November 2013): 30. Nice summary of bright comets in the last century and what factors make a comet spectacular in our skies.

Sekanina, Z. "Sungrazing Comets." *Astronomy* (March 2006): 36.

Sheppard, S. "Beyond the Kuiper Belt." *Sky & Telescope* (March 2015): 26. On Sedna and the Oort cloud.

Stern, S. "Evolution at the Edge." *Astronomy* (September 2005): 46. How comet nuclei evolve with time.

Talcott, R. "Rendezvous with an Evolving Comet [Rosetta at Comet 67P/C-G]." *Astronomy* (September 2015): 44.

Tytell, D. "Deep Impact's Hammer Throw." *Sky & Telescope* (October 2006): 34. On the mission that threw a probe at the nucleus of a comet. See also (June 2005): 40.

Weissman, P. "A Comet Tale." *Sky & Telescope* (February 2006): 36. A nice review of what we know and don't know about the physical nature of comets.

Websites

Asteroids

Dawn Mission: http://dawn.jpl.nasa.gov. Discover more about this mission to the largest asteroids.

NEAR-Shoemaker Mission: http://near.jhuapl.edu/. Review background information and see great images from the mission that went by Mathilde and Eros.

Comets

Deep Impact Mission: http://www.nasa.gov/mission_pages/deepimpact/main/.

Kuiper Belt: http://www2.ess.ucla.edu/~jewitt/kb.html. David Jewitt of the University of Hawaii keeps track of the objects that have been discovered.

Missions to Comets: http://solarsystem.nasa.gov/missions/target/comets. Read about NASA's current and past missions to comets.

This OpenStax book is available for free at http://cnx.org/content/col11992/1.8

Stardust Mission: http://stardust.jpl.nasa.gov/home/index.html. Learn about this mission to collect a sample of a comet and bring it back to Earth.

Videos

Asteroids

Sweating the Small Stuff: The Fear and Fun of Near-Earth Asteroids: https://www.youtube.com/watch?v=5gyAvc5OhII. Harvard Observatory Night Lecture by Jose-Luis Galache (1:18:07).

Unveiling Dwarf Planet Ceres: https://www.youtube.com/watch?v=_G9LudkLWOY. A vonKarman Lecture by Dr. Carol Raymond, Oct. 2015, also includes Vesta results (1:18:38).

Comets

Great Comets, Comets in General, and Comet ISON: https://www.youtube.com/watch?v=DiBkYAnQ_C. Talk by Frank Summers, Space Telescope Science Institute (1:01:10).

Press Conference on the Impact of Comet Shoemaker-Levy 9 with Jupiter: https://www.youtube.com/watch?v=B-tUP8afEIo. Day 2 after impact; July 17, 1994; with the discoverers and Heidi Hammel (1:22:29).

Rosetta: The Story So Far: https://www.ras.org.uk/events-and-meetings/public-lectures/public-lecture-videos/2726-rosetta-the-story-so-far. Royal Astronomical Society Lecture by Dr. Ian Wright (1:00:29).

🧑‍🤝‍🧑 COLLABORATIVE GROUP ACTIVITIES

A. Your group is a congressional committee charged with evaluating the funding for an effort to find all the NEAs (near-Earth asteroids) that are larger than 0.5 kilometers across. Make a list of reasons it would be useful to humanity to find such objects. What should we (could we) do if we found one that will hit Earth in a few years?

B. Many cultures considered comets bad omens. Legends associate comets with the deaths of kings, losses in war, or ends of dynasties. Did any members of your group ever hear about such folktales? Discuss reasons why comets in earlier times may have gotten this bad reputation.

C. Because asteroids have a variety of compositions and a low gravity that makes the removal of materials quite easy, some people have suggested that mining asteroids may be a way to get needed resources in the future. Make a list of materials in asteroids (and comets that come to the inner solar system) that may be valuable to a space-faring civilization. What are the pros and cons of undertaking mining operations on these small worlds?

D. As discussed in the feature box on Comet Hunting as a Hobby, amateur comet hunters typically spend more than 400 hours scanning the skies with their telescopes to find a comet. That's a lot of time to spend (usually alone, usually far from city lights, usually in the cold, and always in the dark). Discuss with members of your group whether you can see yourself being this dedicated. Why do people undertake such quests? Do you envy their dedication?

E. The largest Kuiper belt objects known are also called dwarf planets. All the planets (terrestrial, jovian, and dwarf) in our solar system have so far been named after mythological gods. (The dwarf planet names have moved away from Roman mythology to include the gods of other cultures.) Have your group discuss whether we should continue this naming tradition with newly discovered dwarf planets. Why or why not?

Download for free at http://cnx.org/content/col11992/latest/

F. The total cost of the Rosetta mission to match courses with a comet was about 1.4 billion Euros (about $1.6 billion US). Have your group discuss whether this investment was worth it, giving reasons for whichever side you choose. (On the European Space Agency website, they put this cost in context by saying, "The figure is barely half the price of a modern submarine, or three Airbus 380 jumbo jets, and covers a period of almost 20 years, from the start of the project in 1996 through the end of the mission in 2015.")

G. If an Earth-approaching asteroid were discovered early enough, humanity could take measures to prevent a collision. Discuss possible methods for deflecting or even destroying an asteroid or comet. Go beyond the few methods mentioned in the text and use your creativity. Give pros and cons for each method.

🖱 EXERCISES

Review Questions

1. Why are asteroids and comets important to our understanding of solar system history?

2. Give a brief description of the asteroid belt.

3. Describe the main differences between C-type and S-type asteroids.

4. In addition to the ones mentioned in Exercise 13.3, what is the third, rarer class of asteroids?

5. Vesta is unusual as it contains what mineral on its surface? What does the presence of this material indicate?

6. Compare asteroids of the asteroid belt with Earth-approaching asteroids. What is the main difference between the two groups?

7. Briefly describe NASA's Spaceguard Survey. How many objects have been found in this survey?

8. Who first calculated the orbits of comets based on historical records dating back to antiquity?

9. Describe the nucleus of a typical comet and compare it with an asteroid of similar size.

10. Describe the two types of comet tails and how each are formed.

11. What classification is given to objects such as Pluto and Eris, which are large enough to be round, and whose orbits lie beyond that of Neptune?

12. Describe the origin and eventual fate of the comets we see from Earth.

13. What evidence do we have for the existence of the Kuiper belt? What kind of objects are found there?

14. Give brief descriptions of both the Kuiper belt and the Oort cloud.

Thought Questions

15. Give at least two reasons today's astronomers are so interested in the discovery of additional Earth-approaching asteroids.

16. Suppose you were designing a spacecraft that would match course with an asteroid and follow along its orbit. What sorts of instruments would you put on board to gather data, and what would you like to learn?

17. Suppose you were designing a spacecraft that would match course with a comet and move with it for a while. What sorts of instruments would you put on board to gather data, and what would you like to learn?

This OpenStax book is available for free at http://cnx.org/content/col11992/1.8

Download for free at http://cnx.org/content/col11992/latest/

18. Suppose a comet were discovered approaching the Sun, one whose orbit would cause it to collide with Earth 20 months later, after perihelion passage. (This is approximately the situation described in the science-fiction novel *Lucifer's Hammer* by Larry Niven and Jerry Pournelle.) What could we do? Would there be any way to protect ourselves from a catastrophe?

19. We believe that chains of comet fragments like Comet Shoemaker-Levy 9's have collided not only with the jovian planets, but occasionally with their moons. What sort of features would you look for on the outer planet moons to find evidence of such collisions? (As an extra bonus, can you find any images of such features on a moon like Callisto? You can use an online site of planetary images, such as the *Planetary Photojournal*, at photojournal.jpl.nasa.gov.)

20. Why have we found so many objects in the Kuiper belt in the last two decades and not before then?

21. Why is it hard to give exact diameters for even the larger objects in the Kuiper belt?

Figuring For Yourself

22. Refer to Example 13.1. How would the calculation change if a typical comet in the Oort cloud is only 1 km in diameter?

23. Refer to Example 13.1. How would the calculation change if a typical comet in the Oort cloud is larger—say, 50 km in diameter?

24. The calculation in Example 13.1 refers to the known Oort cloud, the source for most of the comets we see. If, as some astronomers suspect, there are 10 times this many cometary objects in the solar system, how does the total mass of cometary matter compare with the mass of Jupiter?

25. If the Oort cloud contains 10^{12} comets, and ten new comets are discovered coming close to the Sun each year, what percentage of the comets have been "used up" since the beginning of the solar system?

26. The mass of the asteroids is found mostly in the larger asteroids, so to estimate the total mass we need to consider only the larger objects. Suppose the three largest asteroids—Ceres (1000 km in diameter), Pallas (500 km in diameter), and Vesta (500 km in diameter)—account for half the total mass. Assume that each of these three asteroids has a density of 3 g/cm^3 and calculate their total mass. Multiply your result by 2 to obtain an estimate for the mass of the total asteroid belt. How does this compare with the mass of the Oort cloud?

27. Make a similar estimate for the mass of the Kuiper belt. The three largest objects are Pluto, Eris, and Makemake (each roughly 2000 km). In addition, assume there are eight objects (including Haumea, Orcus, Quaoar, Ixion, Varuna, and Charon, and objects that have not been named yet) with diameters of about 1000 km. Assume that all objects have Pluto's density of 2 g/cm^3. Calculate twice the mass of the largest 13 objects and compare it to the mass of the main asteroid belt.

28. What is the period of revolution about the Sun for an asteroid with a semi-major axis of 3 AU in the middle of the asteroid belt?

29. What is the period of revolution for a comet with aphelion at 5 AU and perihelion at the orbit of Earth?

Download for free at http://cnx.org/content/col11992/latest/

This OpenStax book is available for free at http://cnx.org/content/col11992/1.8

14

COSMIC SAMPLES AND THE ORIGIN OF THE SOLAR SYSTEM

Figure 14.1. Planetesimals. This illustration depicts a disk of dust and gas around a new star. Material in this disk comes together to form planetesimals. (credit: modification of work by University of Copenhagen/Lars Buchhave, NASA)

Chapter Outline

✎ Thinking Ahead

Imagine you are a scientist examining a sample of rock that had fallen from space a few days earlier and you find within it some of the chemical building blocks of life. How could you determine whether those "organic" materials came from space or were merely the result of earthly contamination?

We conclude our survey of the solar system with a discussion of its origin and evolution. Some of these ideas were introduced in Other Worlds: An Introduction to the Solar System; we now return to them, using the information we have learned about individual planets and smaller members of the solar system. In addition, astronomers have recently discovered several thousand planets around other stars, including numerous multiplanet systems. This is an important new source of data, providing us a perspective that extends beyond our own particular (and perhaps atypical) solar system.

But first, we want to look at another crucial way that astronomers learn about the ancient history of the solar system: by examining samples of *primitive matter*, the debris of the processes that formed the solar system some 4.5 billion years ago. Unlike the Apollo Moon rocks, these samples of cosmic material come to us free of charge—they literally fall from the sky. We call this material cosmic dust and meteorites.

Download for free at http://cnx.org/content/col11992/latest/

14.1 METEORS

Learning Objectives

By the end of this section, you will be able to:

> Explain what a meteor is and why it is visible in the night sky
> Describe the origins of meteor showers

As we saw in Comets and Asteroids: Debris of the Solar System, the ices in comets evaporate when they get close to the Sun, together spraying millions of tons of rock and dust into the inner solar system. There is also dust from asteroids that have collided and broken up. Earth is surrounded by this material. As each of the larger dust or rock particles enters Earth's atmosphere, it creates a brief fiery trail; this is often called a *shooting star*, but it is properly known as a **meteor**.

Observing Meteors

Meteors are tiny solid particles that enter Earth's atmosphere from interplanetary space. Since the particles move at speeds of many kilometers per second, friction with the air vaporizes them at altitudes between 80 and 130 kilometers. The resulting flashes of light fade out within a few seconds. These "shooting stars" got their name because at night their luminous vapors look like stars moving rapidly across the sky. To be visible, a meteor must be within about 200 kilometers of the observer. On a typical dark, moonless night, an alert observer can see half a dozen meteors per hour. These *sporadic meteors*—those not associated with a meteor shower (explained in the next section)—are random occurrences. Over the entire Earth, the total number of meteors bright enough to be visible totals about 25 million per day.

The typical meteor is produced by a particle with a mass of less than 1 gram—no larger than a pea. How can we see such a small particle? The light you see comes from the much larger region of heated, glowing gas surrounding this little grain of interplanetary material. Because of its high speed, the energy in a pea-sized meteor is as great as that of an artillery shell fired on Earth, but this energy is dispersed high in Earth's atmosphere. (When these tiny projectiles hit an airless body like the Moon, they do make small craters and generally pulverize the surface.)

If a particle the size of a golf ball strikes our atmosphere, it produces a much brighter trail called a fireball (Figure 14.2). A piece as large as a bowling ball has a fair chance of surviving its fiery entry if its approach speed is not too high. The total mass of meteoric material entering Earth's atmosphere is estimated to be about 100 tons per day (which seems like a lot if you imagine it all falling in one place, but remember it is spread out all over our planet's surface).

This OpenStax book is available for free at http://cnx.org/content/col11992/1.8

Figure 14.2. Fireball. When a larger piece of cosmic material strikes Earth's atmosphere, it can make a bright fireball. This time-lapse meteor image was captured in April 2014 at the Atacama Large Millimeter/Submillimeter Array (ALMA). The visible trail results from the burning gas around the particle. (credit: modification of work by ESO/C Malin)

LINK TO LEARNING

While it is difficult to capture images of fireballs and other meteors with still photography, it's easy to capture the movement of these objects on video. The American Meteor Society maintains a website (https://openstaxcollege.org/l/30ammetsocweb) on which their members can share such videos.

Meteor Showers

Many—perhaps most—of the meteors that strike Earth are associated with specific comets. Some of these periodic comets still return to our view; others have long ago fallen apart, leaving only a trail of dust behind them. The dust particles from a given comet retain approximately the orbit of their parent, continuing to move together through space but spreading out over the orbit with time. When Earth, in its travels around the Sun, crosses such a dust stream, we see a sudden burst of meteor activity that usually lasts several hours; such an event is called a **meteor shower**.

The dust particles and pebbles that produce meteor showers are moving together in space before they encounter Earth. Thus, as we look up at the atmosphere, their parallel paths seem to come toward us from a place in the sky called the *radiant*. This is the direction in space from which the meteor stream seems to be diverging, just as long railroad tracks seem to diverge from a single spot on the horizon (Figure 14.3). Meteor showers are often designated by the constellation in which this radiant is located: for example, the Perseid meteor shower has its radiant in the constellation of Perseus. But you are likely to see shower meteors anywhere in the sky, not just in the constellation of the radiant. The characteristics of some of the more famous meteor showers are summarized in Table 14.1.

Download for free at http://cnx.org/content/col11992/latest/

Figure 14.3. Radiant of a Meteor Shower. The tracks of the meteors diverge from a point in the distance, just as long, parallel railroad tracks appear to do. (credit "tracks": Nathan Vaughn)

Major Annual Meteor Showers

Shower Name	Date of Maximum	Associated Parent Object	Comet's Period (years)
Quadrantid	January 3–4	2003EH (asteroid)	—
Lyrid	April 22	Comet Thatcher	415
Eta Aquarid	May 4–5	Comet Halley	76
Delta Aquarid	July 29–30	Comet Machholz	—
Perseid	August 11–12	Comet Swift-Tuttle	133
Orionid	October 20–21	Comet Halley	76
Southern Taurid	October 31	Comet Encke	3
Leonid	November 16–17	Comet Tempel-Tuttle	33
Geminid	December 13	Phaethon (asteroid)	1.4

Table 14.1

The meteoric dust is not always evenly distributed along the orbit of the comet, so during some years more meteors are seen when Earth intersects the dust stream, and in other years fewer. For example, a very clumpy distribution is associated with the Leonid meteors, which in 1833 and again in 1866 (after an interval of 33 years—the period of the comet) yielded the most spectacular showers (sometimes called *meteor storms*) ever recorded (Figure 14.4). During the Leonid storm on November 17, 1866, up to a hundred meteors were observed per second in some locations. The Leonid shower of 2001 was not this intense, but it peaked at nearly a thousand meteors per hour—one every few seconds—observable from any dark viewing site.

This OpenStax book is available for free at http://cnx.org/content/col11992/1.8

Download for free at http://cnx.org/content/col11992/latest/

Figure 14.4. Leonid Meteor Storm. A painting depicts the great meteor shower or storm of 1833, shown with a bit of artistic license.

The most dependable annual meteor display is the Perseid shower, which appears each year for about three nights near August 11. In the absence of bright moonlight, you can see one meteor every few minutes during a typical Perseid shower. Astronomers estimate that the total combined mass of the particles in the Perseid swarm is nearly a billion tons; the comet that gave rise to the particles in that swarm, called Swift-Tuttle, must originally have had at least that much mass. However, if its initial mass were comparable to the mass measured for Comet Halley, then Swift-Tuttle would have contained several hundred billion tons, suggesting that only a very small fraction of the original cometary material survives in the meteor stream.

LINK TO LEARNING

The California Academy of Sciences has a short animated guide (https://openstaxcollege.org/l/30howobsmetsho) on "How to Observe a Meteor Shower."

No shower meteor has ever survived its flight through the atmosphere and been recovered for laboratory analysis. However, there are other ways to investigate the nature of these particles and thereby gain additional insight into the comets from which they are derived. Analysis of the flight paths of meteors shows that most of them are very light or porous, with densities typically less than 1.0 g/cm^3. If you placed a fist-sized lump of meteor material on a table in Earth's gravity, it might well fall apart under its own weight.

Such light particles break up very easily in the atmosphere, accounting for the failure of even relatively large shower meteors to reach the ground. Comet dust is apparently fluffy, rather inconsequential stuff. NASA's Stardust mission used a special substance, called aerogel, to collect these particles. We can also infer this from the tiny comet particles recovered in Earth's atmosphere with high-flying aircraft (see Figure 13.19). This fluff, by its very nature, cannot reach Earth's surface intact. However, more substantial fragments from asteroids do make it into our laboratories, as we will see in the next section.

Download for free at http://cnx.org/content/col11992/latest/

Showering with the Stars

Observing a meteor shower is one of the easiest and most enjoyable astronomy activities for beginners (Figure 14.5). The best thing about it is that you don't need a telescope or binoculars—in fact, they would positively get in your way. What you do need is a site far from city lights, with an unobstructed view of as much sky as possible. While the short bright lines in the sky made by individual meteors could, in theory, be traced back to a radiant point (as shown in Figure 14.3), the quick blips of light that represent the end of the meteor could happen anywhere above you.

Figure 14.5. Perseid Meteor Shower. This twenty-second exposure shows a meteor during the 2015 Perseid meteor shower. (credit: NASA/Bill Ingalls)

The key to observing meteor showers is not to restrict your field of view, but to lie back and scan the sky alertly. Try to select a good shower (see the list in Table 14.1) and a night when the Moon will not be bright at the time you are observing. The Moon, street lights, vehicle headlights, bright flashlights, and cell phone and tablet screens will all get in the way of your seeing the faint meteor streaks.

You will see more meteors after midnight, when you are on the hemisphere of Earth that faces forward—in the direction of Earth's revolution around the Sun. Before midnight, you are observing from the "back side" of Earth, and the only meteors you see will be those that traveled fast enough to catch up with Earth's orbital motion.

When you've gotten away from all the lights, give your eyes about 15 minutes to get "dark adapted"—that is, for the pupils of your eyes to open up as much as possible. (This adaptation is the same thing that happens in a dark movie theater. When you first enter, you can't see a thing, but eventually, as your pupils open wider, you can see pretty clearly by the faint light of the screen—and notice all that spilled popcorn on the floor.)

This OpenStax book is available for free at http://cnx.org/content/col11992/1.8

Download for free at http://cnx.org/content/col11992/latest/

Seasoned meteor observers find a hill or open field and make sure to bring warm clothing, a blanket, and a thermos of hot coffee or chocolate with them. (It's also nice to take along someone with whom you enjoy sitting in the dark.) Don't expect to see fireworks or a laser show: meteor showers are subtle phenomena, best approached with a patience that reflects the fact that some of the dust you are watching burn up may first have been gathered into its parent comet more than 4.5 billion years ago, as the solar system was just forming.

14.2 METEORITES: STONES FROM HEAVEN

Learning Objectives

By the end of this section, you will be able to:

> Explain the origin of meteorites and the difference between a meteor and a meteorite
> Describe how most meteorites have been found
> Explain how primitive stone meteorites are significantly different from other types
> Explain how the study of meteorites informs our understanding of the age of the solar system.

Any fragment of interplanetary debris that survives its fiery plunge through Earth's atmosphere is called a **meteorite**. Meteorites fall only very rarely in any one locality, but over the entire Earth thousands fall each year. Some meteorites are loners, but many are fragments from the breakup in the atmosphere of a single larger object. These rocks from the sky carry a remarkable record of the formation and early history of the solar system.

Extraterrestrial Origin of Meteorites

Occasional meteorites have been found throughout history, but their extraterrestrial origin was not accepted by scientists until the beginning of the nineteenth century. Before that, these strange stones were either ignored or considered to have a supernatural origin.

The falls of the earliest recovered meteorites are lost in the fog of mythology. A number of religious texts speak of stones from heaven, which sometimes arrived at opportune moments to smite the enemies of the authors of those texts. At least one sacred meteorite has apparently survived in the form of the Ka'aba, the holy black stone in Mecca that is revered by Islam as a relic from the time of the Patriarchs—although understandably, no chip from this sacred stone has been subject to detailed chemical analysis.

The modern scientific history of the meteorites begins in the late eighteenth century, when a few scientists suggested that some strange-looking stones had such peculiar composition and structure that they were probably not of terrestrial origin. The idea that indeed "stones fall from the sky" was generally accepted only after a scientific team led by French physicist Jean-Baptiste Biot investigated a well-observed fall in 1803.

Meteorites sometimes fall in groups or showers. Such a fall occurs when a single larger object breaks up during its violent passage through the atmosphere. It is important to remember that such a *shower of meteorites* has nothing to do with a *meteor shower*. No meteorites have ever been recovered in association with meteor showers. Whatever the ultimate source of the meteorites, they do not appear to come from the comets or their associated particle streams.

Download for free at http://cnx.org/content/col11992/latest/

Meteorite Falls and Finds

Meteorites are found in two ways. First, sometimes bright meteors (fireballs) are observed to penetrate the atmosphere to low altitudes. If we search the area beneath the point where the fireball burned out, we may find one or more remnants that reached the ground. Observed *meteorite falls*, in other words, may lead to the recovery of fallen meteorites. (A few meteorites have even hit buildings or, very rarely, people; see Making Connections: Some Striking Meteorites). The 2013 Chelyabinsk fireball, which we discussed in the chapter on Comets and Asteroids: Debris of the Solar System, produced tens of thousands of small meteorites, many of them easy to find because these dark stones fell on snow.

There are, however, many false alarms about meteorite falls. Most observers of a bright fireball conclude that part of it hit the ground, but that is rarely the case. Every few months news outlets report that a meteorite has been implicated in the start of a fire. Such stories have always proved to be wrong. The meteorite is ice-cold in space, and most of its interior remains cold even after its brief fiery plunge through the atmosphere. A freshly fallen meteorite is more likely to acquire a coating of frost than to start a fire.

People sometimes discover unusual-looking rocks that turn out to be meteoritic; these rocks are termed *meteorite finds*. Now that the public has become meteorite-conscious, many unusual fragments, not all of which turn out to be from space, are sent to experts each year. Some scientists divide these objects into two categories: "meteorites" and "meteorwrongs." Outside Antarctica (see the next paragraph), genuine meteorites turn up at an average rate of 25 or so per year. Most of these end up in natural history museums or specialized meteoritical laboratories throughout the world (Figure 14.6).

(a)   (b)

Figure 14.6. Meteorite Find. (a) This early twentieth century photo shows a 15-ton iron meteorite found in the Willamette Valley in Oregon. Although known to Native Americans in the area, it was "discovered" by an enterprising local farmer in 1902, who proceeded to steal it and put it on display. (b) It was eventually purchased for the American Museum of Natural History and is now on display in the museum's Rose Center in New York City as the largest iron meteorite in the United States. In this 1911 photo, two young boys are perched in the meteor's crevices.

Since the 1980s, sources in the Antarctic have dramatically increased our knowledge of meteorites. More than ten thousand meteorites have been recovered from the Antarctic as a result of the motion of the ice in some parts of that continent (Figure 14.7). Meteorites that fall in regions where ice accumulates are buried and then carried slowly to other areas where the ice is gradually worn away. After thousands of years, the rock again finds itself on the surface, along with other meteorites carried to these same locations. The ice thus concentrates the meteorites that have fallen both over a large area and over a long period of time. Once on the surface, the rocks stand out in contrast to the ice and are thus easier to spot than in other places on our rocky planet.

Download for free at http://cnx.org/content/col11992/latest/

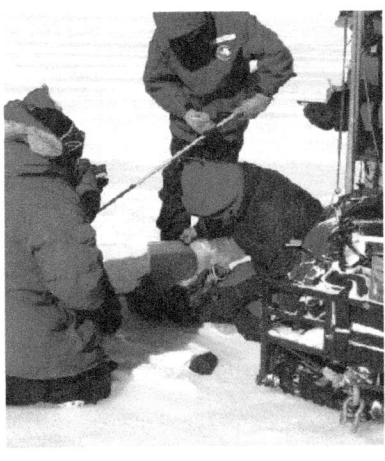

(a)

(b)

Figure 14.7. Antarctic Meteorite. (a) The US Antarctic Search for Meteorites (ANSMET) team recovers a meteorite from the Antarctic ice during a 2001–2002 mission. (b) The team is shown with some of the equipment used in the search. (credit a, b: modification of work by NASA)

MAKING CONNECTIONS

Some Striking Meteorites

Although meteorites fall regularly onto Earth's surface, few of them have much of an impact on human civilization. There is so much water and uninhabited land on our planet that rocks from space typically fall where no one even sees them come down. But given the number of meteorites that land each year, you may not be surprised that a few have struck buildings, cars, and even people. In September 1938, for example, a meteorite plunged through the roof of Edward McCain's garage, where it became embedded in the seat of his Pontiac Coupe (Figure 14.8).

In November 1982, Robert and Wanda Donahue of Wethersfield, Connecticut, were watching *M*A*S*H* on television when a 6-pound meteorite came thundering through their roof, making a hole in the living room ceiling. After bouncing, it finally came to rest under their dining room table.

Eighteen-year-old Michelle Knapp of Peekskill, New York, got quite a surprise one morning in October 1992. She had just purchased her very first car, her grandmother's 1980 Chevy Malibu. But she awoke to find its rear end mangled and a crater in the family driveway—thanks to a 3-pound meteorite. Michelle was not sure whether to be devastated by the loss of her car or thrilled by all the media attention.

In June 1994, Jose Martin and his wife were driving from Madrid, Spain, to a golfing vacation when a fist-sized meteorite crashed through the windshield of their car, bounced off the dashboard, broke Jose's little finger, and then landed in the back seat. Before Martin, the most recent person known to have been struck by a meteorite was Annie Hodges of Sylacauga, Alabama. In November 1954, she was napping on a couch when a meteorite came through the roof, bounced off a large radio set, and hit her first on the arm and then on the leg.

The fireball that exploded at an altitude of about 20 kilometers near the Russian city of Chelyabinsk on February 15, 2013, produced a very large meteorite shower, and quite a few of the small rocks hit

Download for free at http://cnx.org/content/col11992/latest/

buildings. None is known to have hit people, however, and the individual meteorites were so small that they did not do much damage—much less than the shockwave from the exploding fireball, which broke the glass in thousands of windows.

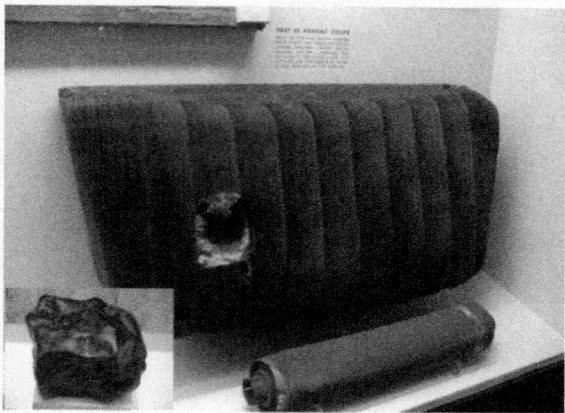

Figure 14.8. Benld Meteorite. A meteorite (inset) left a hole in the seat cushion of Edward McCain's car. (credit: "Shsilver"/Wikimedia Commons)

Meteorite Classification

The meteorites in our collections have a wide range of compositions and histories, but traditionally they have been placed into three broad classes. First are the **irons**, composed of nearly pure metallic nickel-iron. Second are the **stones**, the term used for any silicate or rocky meteorite. Third are the rarer **stony-irons**, made (as the name implies) of mixtures of stone and metallic iron (Figure 14.9).

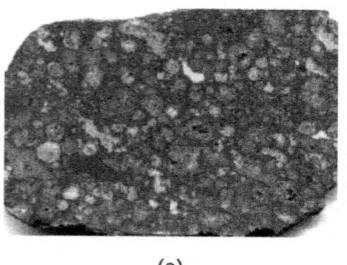
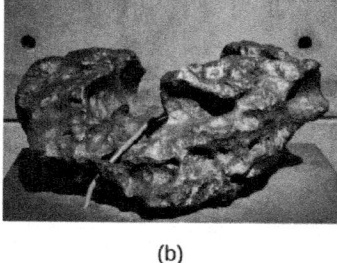
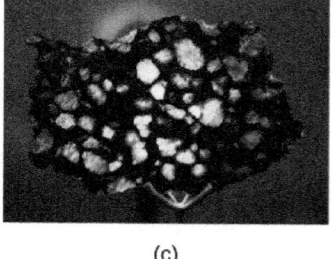

(a)   (b)   (c)

Figure 14.9. Meteorite Types. (a) This piece of the Allende carbonaceous meteorite has white inclusions that may date back to before the formation of the solar nebula. (b) This fragment is from the iron meteorite responsible for the formation of Meteor Crater in Arizona. (c) This piece of the Imilac stony-iron meteorite is a beautiful mixture of green olivine crystals and metallic iron. (credit a: modification of work by James St. John; credit b: modification of work by "Taty2007"/Wikimedia Commons; credit c: modification of work by Juan Manuel Fluxà)

Of these three types, the irons and stony-irons are the most obviously extraterrestrial because of their metallic content. Pure iron almost never occurs naturally on Earth; it is generally found here as an oxide (chemically combined with oxygen) or other mineral ore. Therefore, if you ever come across a chunk of metallic iron, it is sure to be either man-made or a meteorite.

The stones are much more common than the irons but more difficult to recognize. Often laboratory analysis is required to demonstrate that a particular sample is really of extraterrestrial origin, especially if it has lain on

This OpenStax book is available for free at http://cnx.org/content/col11992/1.8

the ground for some time and been subject to weathering. The most scientifically valuable stones are those collected immediately after they fall, or the Antarctic samples preserved in a nearly pristine state by ice.

Table 14.2 summarizes the frequencies of occurrence of the different classes of meteorites among the fall, find, and Antarctic categories.

Frequency of Occurrence of Meteorite Classes

Class	Falls (%)	Finds (%)	Antarctic (%)
Primitive stones	88	51	85
Differentiated stones	8	2	12
Irons	3	42	2
Stony-irons	1	5	1

Table 14.2

Ages and Compositions of Meteorites

It was not until the ages of meteorites were measured and their compositions analyzed in detail that scientists appreciated their true significance. The meteorites include the oldest and most primitive materials available for direct study in the laboratory. The ages of stony meteorites can be determined from the careful measurement of radioactive isotopes and their decay products. Almost all meteorites have radioactive ages between 4.50 and 4.56 billion years, as old as any ages we have measured in the solar system. The few younger exceptions are igneous rocks that have been ejected from cratering events on the Moon or Mars (and have made their way to Earth).

The average age for the most primitive meteorites, calculated using the most accurate values now available for radioactive half-lives, is 4.56 billion years, with an uncertainly of less than 0.01 billion years. This value (which we round off to 4.5 billion years in this book) is taken to represent the *age of the solar system*—the time since the first solids condensed and began to form into larger bodies.

The traditional classification of meteorites into irons, stones, and stony-irons is easy to use because it is obvious from inspection which category a meteorite falls into (although it may be much more difficult to distinguish a meteoritic stone from a terrestrial rock). More scientifically significant, however, is the distinction between *primitive* and *differentiated* meteorites. The differentiated meteorites are fragments of larger parent bodies that were molten before they broke up, allowing the denser materials (such as metals) to sink to their centers. Like many rocks on Earth, they have been subject to a degree of chemical reshuffling, with the different materials sorted according to density. Differentiated meteorites include the irons, which come from the metal cores of their parent bodies; stony-irons, which probably originate in regions between a metal core and a stony mantle; and some stones that are composed of mantle or crust material from the their differentiated parent bodies.

The Most Primitive Meteorites

For information on the *earliest* history of the solar system, we turn to the primitive meteorites—those made of materials that have *not* been subject to great heat or pressure since their formation. We can look at the spectrum of sunlight reflected from asteroids and compare their compositions with those of primitive meteorites. Such analysis indicates that their parent bodies are almost certainly asteroids. Since asteroids are

Download for free at http://cnx.org/content/col11992/latest/

believed to be fragments left over from the formation process of the solar system, it makes sense that they should be the parent bodies of the primitive meteorites.

The great majority of the meteorites that reach Earth are primitive stones. Many of them are composed of light-colored gray silicates with some metallic grains mixed in, but there is also an important group of darker stones called *carbonaceous meteorites*. As their name suggests, these meteorites contain carbon, but we also find various complex organic molecules in them—chemicals based on carbon, which on Earth are the chemical building blocks of life. In addition, some of them contain chemically bound water, and many are depleted in metallic iron. The carbonaceous (or C-type) asteroids are concentrated in the outer part of the asteroid belt.

Among the most useful of these meteorites have been the Allende meteorite that fell in Mexico (see Figure 14.9), the Murchison meteorite that fell in Australia (both in 1969), and the Tagish Lake meteorite that landed in a winter snowdrift on Tagish Lake, Canada, in 2000. (The fragile bits of dark material from the Tagish Lake meteorite were readily visible against the white snow, although at first they were mistaken for wolf droppings.)

The Murchison meteorite (Figure 14.10) is known for the variety of organic chemicals it has yielded. Most of the carbon compounds in carbonaceous meteorites are complex, tarlike substances that defy exact analysis. Murchison also contains 16 amino acids (the building blocks of proteins), 11 of which are rare on Earth. The most remarkable thing about these organic molecules is that they include equal numbers with right-handed and left-handed molecular symmetry. Amino acids can have either kind of symmetry, but all life on Earth has evolved using only the *left-handed* versions to make proteins. The presence of both kinds of amino acids clearly demonstrates that the ones in the meteorites had an extraterrestrial origin.

Figure 14.10. Murchison Meteorite. A fragment of the meteorite that fell near the small town of Murchison, Australia, is shown next to a small sample of its material in a test tube, used for analysis of its chemical makeup.

These naturally occurring amino acids and other complex organic molecules in Murchison—formed without the benefit of the sheltering environment of planet Earth—show that a great deal of interesting chemistry must have taken place when the solar system was forming. If so, then perhaps some of the molecular building blocks of life on Earth were first delivered by primitive meteorites and comets. This is an interesting idea because our planet was probably much too hot for any organic materials to survive its earliest history. But after Earth's surface cooled, the asteroid and comet fragments that pelted it could have refreshed its supply of organic materials.

This OpenStax book is available for free at http://cnx.org/content/col11992/1.8

Download for free at http://cnx.org/content/col11992/latest/

14.3 FORMATION OF THE SOLAR SYSTEM

Learning Objectives

By the end of this section, you will be able to:

> Describe the motion, chemical, and age constraints that must be met by any theory of solar system formation
> Summarize the physical and chemical changes during the solar nebula stage of solar system formation
> Explain the formation process of the terrestrial and giant planets
> Describe the main events of the further evolution of the solar system

As we have seen, the comets, asteroids, and meteorites are surviving remnants from the processes that formed the solar system. The planets, moons, and the Sun, of course, also are the products of the formation process, although the material in them has undergone a wide range of changes. We are now ready to put together the information from all these objects to discuss what is known about the origin of the solar system.

Observational Constraints

There are certain basic properties of the planetary system that any theory of its formation must explain. These may be summarized under three categories: motion constraints, chemical constraints, and age constraints. We call them *constraints* because they place restrictions on our theories; unless a theory can explain the observed facts, it will not survive in the competitive marketplace of ideas that characterizes the endeavor of science. Let's take a look at these constraints one by one.

There are many regularities to the motions in the solar system. We saw that the planets all revolve around the Sun in the same direction and approximately in the plane of the Sun's own rotation. In addition, most of the planets rotate in the same direction as they revolve, and most of the moons also move in counterclockwise orbits (when seen from the north). With the exception of the comets and other trans-neptunian objects, the motions of the system members define a disk or Frisbee shape. Nevertheless, a full theory must also be prepared to deal with the exceptions to these trends, such as the *retrograde rotation* (not revolution) of Venus.

In the realm of chemistry, we saw that Jupiter and Saturn have approximately the same composition—dominated by hydrogen and helium. These are the two largest planets, with sufficient gravity to hold on to any gas present when and where they formed; thus, we might expect them to be representative of the original material out of which the solar system formed. Each of the other members of the planetary system is, to some degree, lacking in the light elements. A careful examination of the composition of solid solar-system objects shows a striking progression from the metal-rich inner planets, through those made predominantly of rocky materials, out to objects with ice-dominated compositions in the outer solar system. The comets in the Oort cloud and the trans-neptunian objects in the Kuiper belt are also icy objects, whereas the asteroids represent a transitional rocky composition with abundant dark, carbon-rich material.

As we saw in Other Worlds: An Introduction to the Solar System, this general chemical pattern can be interpreted as a temperature sequence: hot near the Sun and cooler as we move outward. The inner parts of the system are generally missing those materials that could not condense (form a solid) at the high temperatures found near the Sun. However, there are (again) important exceptions to the general pattern. For example, it is difficult to explain the presence of water on Earth and Mars if these planets formed in a region where the temperature was too hot for ice to condense, unless the ice or water was brought in later from cooler regions. The extreme example is the observation that there are polar deposits of ice on both Mercury and the Moon; these are almost certainly formed and maintained by occasional comet impacts.

Download for free at http://cnx.org/content/col11992/latest/

As far as age is concerned, we discussed that radioactive dating demonstrates that some rocks on the surface of Earth have been present for at least 3.8 billion years, and that certain lunar samples are 4.4 billion years old. The primitive meteorites all have radioactive ages near 4.5 billion years. The age of these unaltered building blocks is considered the age of the planetary system. The similarity of the measured ages tells us that planets formed and their crusts cooled within a few tens of millions of years (at most) of the beginning of the solar system. Further, detailed examination of primitive meteorites indicates that they are made primarily from material that condensed or coagulated out of a hot gas; few identifiable fragments appear to have survived from before this hot-vapor stage 4.5 billion years ago.

The Solar Nebula

All the foregoing constraints are consistent with the general idea, introduced in Other Worlds: An Introduction to the Solar System, that the solar system formed 4.5 billion years ago out of a rotating cloud of vapor and dust—which we call the solar nebula—with an initial composition similar to that of the Sun today. As the solar nebula collapsed under its own gravity, material fell toward the center, where things became more and more concentrated and hot. Increasing temperatures in the shrinking nebula vaporized most of the solid material that was originally present.

At the same time, the collapsing nebula began to rotate faster through the conservation of angular momentum (see the Orbits and Gravity and Earth, Moon, and Sky chapters). Like a figure skater pulling her arms in to spin faster, the shrinking cloud spun more quickly as time went on. Now, think about how a round object spins. Close to the poles, the spin rate is slow, and it gets faster as you get closer to the equator. In the same way, near the poles of the nebula, where orbits were slow, the nebular material fell directly into the center. Faster moving material, on the other hand, collapsed into a flat disk revolving around the central object (Figure 14.11). The existence of this disk-shaped rotating nebula explains the primary motions in the solar system that we discussed in the previous section. And since they formed from a rotating disk, the planets all orbit the same way.

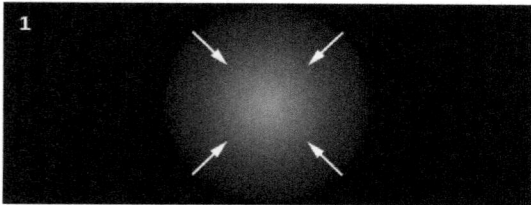

The solar nebula contracts.

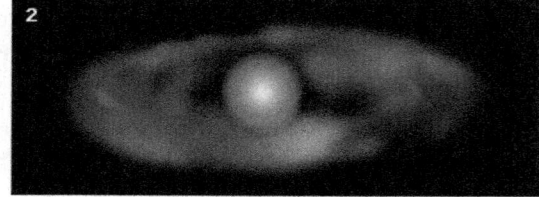

As the nebula shrinks, its motion causes it to flatten.

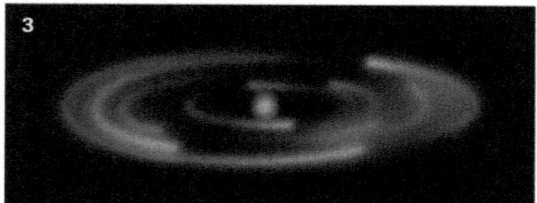

The nebula is a disk of matter with a concentration near the center.

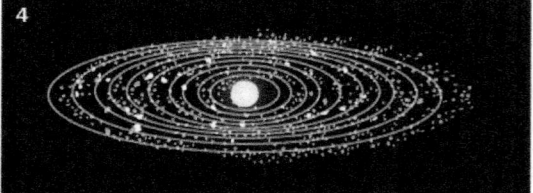

Formation of the protosun. Solid particles condense as the nebula cools, giving rise to the planetesimals, which are the building blocks of the planets.

Figure 14.11. Steps in Forming the Solar System. This illustration shows the steps in the formation of the solar system from the solar nebula. As the nebula shrinks, its rotation causes it to flatten into a disk. Much of the material is concentrated in the hot center, which will ultimately become a star. Away from the center, solid particles can condense as the nebula cools, giving rise to planetesimals, the building blocks of the planets and moons.

This OpenStax book is available for free at http://cnx.org/content/col11992/1.8

Download for free at http://cnx.org/content/col11992/latest/

Picture the solar nebula at the end of the collapse phase, when it was at its hottest. With no more gravitational energy (from material falling in) to heat it, most of the nebula began to cool. The material in the center, however, where it was hottest and most crowded, formed a *star* that maintained high temperatures in its immediate neighborhood by producing its own energy. Turbulent motions and magnetic fields within the disk can drain away angular momentum, robbing the disk material of some of its spin. This allowed some material to continue to fall into the growing star, while the rest of the disk gradually stabilized.

The temperature within the disk decreased with increasing distance from the Sun, much as the planets' temperatures vary with position today. As the disk cooled, the gases interacted chemically to produce compounds; eventually these compounds condensed into liquid droplets or solid grains. This is similar to the process by which raindrops on Earth condense from moist air as it rises over a mountain.

Let's look in more detail at how material condensed at different places in the maturing disk (Figure 14.12). The first materials to form solid grains were the metals and various rock-forming silicates. As the temperature dropped, these were joined throughout much of the solar nebula by sulfur compounds and by carbon- and water-rich silicates, such as those now found abundantly among the asteroids. However, in the inner parts of the disk, the temperature never dropped low enough for such materials as ice or carbonaceous organic compounds to condense, so they were lacking on the innermost planets.

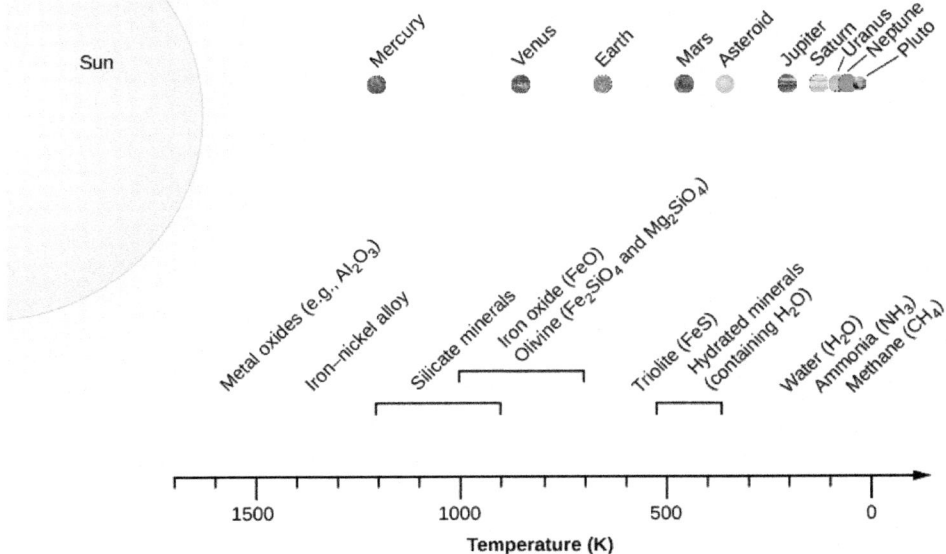

Figure 14.12. Chemical Condensation Sequence in the Solar Nebula. The scale along the bottom shows temperature; above are the materials that would condense out at each temperature under the conditions expected to prevail in the nebula.

Far from the Sun, cooler temperatures allowed the oxygen to combine with hydrogen and condense in the form of water (H_2O) ice. Beyond the orbit of Saturn, carbon and nitrogen combined with hydrogen to make ices such as methane (CH_4) and ammonia (NH_3). This sequence of events explains the basic chemical composition differences among various regions of the solar system.

Download for free at http://cnx.org/content/col11992/latest/

EXAMPLE 14.1

Rotation of the Solar Nebula

We can use the concept of angular momentum to trace the evolution of the collapsing solar nebula. The angular momentum of an object is proportional to the square of its size (diameter) times its period of rotation (D^2/P). If angular momentum is conserved, then any change in the size of a nebula must be compensated for by a proportional change in period, in order to keep D^2/P constant. Suppose the solar nebula began with a diameter of 10,000 AU and a rotation period of 1 million years. What is its rotation period when it has shrunk to the size of Pluto's orbit, which Appendix F tells us has a radius of about 40 AU?

Solution

We are given that the final diameter of the solar nebula is about 80 AU. Noting the initial state before the collapse and the final state at Pluto's orbit, then

$$\frac{P_{\text{fina}}}{P_{\text{initial}}} = \left(\frac{D_{\text{fina}}}{D_{\text{initial}}}\right)^2 = \left(\frac{80}{10,000}\right)^2 = (0.008)^2 = 0.000064$$

With P_{initial} equal to 1,000,000 years, P_{final}, the new rotation period, is 64 years. This is a lot shorter than the actual time Pluto takes to go around the Sun, but it gives you a sense of the kind of speeding up the conservation of angular momentum can produce. As we noted earlier, other mechanisms helped the material in the disk lose angular momentum before the planets fully formed.

Check Your Learning

What would the rotation period of the nebula in our example be when it had shrunk to the size of Jupiter's orbit?

Answer:

The period of the rotating nebula is inversely proportional to D^2. As we have just seen, $\frac{P_{\text{fina}}}{P_{\text{initial}}} = \left(\frac{D_{\text{fina}}}{D_{\text{initial}}}\right)^2$. Initially, we have $P_{\text{initial}} = 10^6$ yr and $D_{\text{initial}} = 10^4$ AU. Then, if D_{final} is in AU, P_{final} (in years) is given by $P_{\text{fina}} = 0.01 D_{\text{fina}}^2$. If Jupiter's orbit has a radius of 5.2 AU, then the diameter is 10.4 AU. The period is then 1.08 years.

Formation of the Terrestrial Planets

The grains that condensed in the solar nebula rather quickly joined into larger and larger chunks, until most of the solid material was in the form of *planetesimals,* chunks a few kilometers to a few tens of kilometers in diameter. Some planetesimals still survive today as comets and asteroids. Others have left their imprint on the cratered surfaces of many of the worlds we studied in earlier chapters. A substantial step up in size is required, however, to go from planetesimal to planet.

Some planetesimals were large enough to attract their neighbors gravitationally and thus to grow by the process called **accretion**. While the intermediate steps are not well understood, ultimately several dozen centers of accretion seem to have grown in the inner solar system. Each of these attracted surrounding

This OpenStax book is available for free at http://cnx.org/content/col11992/1.8

planetesimals until it had acquired a mass similar to that of Mercury or Mars. At this stage, we may think of these objects as *protoplanets*—"not quite ready for prime time" planets.

Each of these protoplanets continued to grow by the accretion of planetesimals. Every incoming planetesimal was accelerated by the gravity of the protoplanet, striking with enough energy to melt both the projectile and a part of the impact area. Soon the entire protoplanet was heated to above the melting temperature of rocks. The result was *planetary differentiation*, with heavier metals sinking toward the core and lighter silicates rising toward the surface. As they were heated, the inner protoplanets lost some of their more volatile constituents (the lighter gases), leaving more of the heavier elements and compounds behind.

Formation of the Giant Planets

In the outer solar system, where the available raw materials included ices as well as rocks, the protoplanets grew to be much larger, with masses ten times greater than Earth. These protoplanets of the outer solar system were so large that they were able to attract and hold the surrounding gas. As the hydrogen and helium rapidly collapsed onto their cores, the giant planets were heated by the energy of contraction. But although these giant planets got hotter than their terrestrial siblings, they were far too small to raise their central temperatures and pressures to the point where nuclear reactions could begin (and it is such reactions that give us our definition of a star). After glowing dull red for a few thousand years, the giant planets gradually cooled to their present state (Figure 14.13).

Figure 14.13. Saturn Seen in Infrared. This image from the Cassini spacecraft is stitched together from 65 individual observations. Sunlight reflected at a wavelength of 2 micrometers is shown as blue, sunlight reflected at 3 micrometers is shown as green, and heat radiated from Saturn's interior at 5 micrometers is red. For example, Saturn's rings reflect sunlight at 2 micrometers, but not at 3 and 5 micrometers, so they appear blue. Saturn's south polar regions are seen glowing with internal heat. (credit: modification of work by NASA/JPL/University of Arizona)

The collapse of gas from the nebula onto the cores of the giant planets explains how these objects acquired nearly the same hydrogen-rich composition as the Sun. The process was most efficient for Jupiter and Saturn; hence, their compositions are most nearly "cosmic." Much less gas was captured by Uranus and Neptune, which is why these two planets have compositions dominated by the icy and rocky building blocks that made up their large cores rather than by hydrogen and helium. The initial formation period ended when much of the available raw material was used up and the solar wind (the flow of atomic particles) from the young Sun blew away the remaining supply of lighter gases.

Further Evolution of the System

All the processes we have just described, from the collapse of the solar nebula to the formation of protoplanets, took place within a few million years. However, the story of the formation of the solar system was not complete

Download for free at http://cnx.org/content/col11992/latest/

at this stage; there were many planetesimals and other debris that did not initially accumulate to form the planets. What was their fate?

The comets visible to us today are merely the tip of the cosmic iceberg (if you'll pardon the pun). Most comets are believed to be in the Oort cloud, far from the region of the planets. Additional comets and icy dwarf planets are in the Kuiper belt, which stretches beyond the orbit of Neptune. These icy pieces probably formed near the present orbits of Uranus and Neptune but were ejected from their initial orbits by the gravitational influence of the giant planets.

In the inner parts of the system, remnant planetesimals and perhaps several dozen protoplanets continued to whiz about. Over the vast span of time we are discussing, collisions among these objects were inevitable. Giant impacts at this stage probably stripped Mercury of part of its mantle and crust, reversed the rotation of Venus, and broke off part of Earth to create the Moon (all events we discussed in other chapters).

Smaller-scale impacts also added mass to the inner protoplanets. Because the gravity of the giant planets could "stir up" the orbits of the planetesimals, the material impacting on the inner protoplanets could have come from almost anywhere within the solar system. In contrast to the previous stage of accretion, therefore, this new material did not represent just a narrow range of compositions.

As a result, much of the debris striking the inner planets was ice-rich material that had condensed in the outer part of the solar nebula. As this comet-like bombardment progressed, Earth accumulated the water and various organic compounds that would later be critical to the formation of life. Mars and Venus probably also acquired abundant water and organic materials from the same source, as Mercury and the Moon are still doing to form their icy polar caps.

Gradually, as the planets swept up or ejected the remaining debris, most of the planetesimals disappeared. In two regions, however, stable orbits are possible where leftover planetesimals could avoid impacting the planets or being ejected from the system. These regions are the asteroid belt between Mars and Jupiter and the Kuiper belt beyond Neptune. The planetesimals (and their fragments) that survive in these special locations are what we now call asteroids, comets, and trans-neptunian objects.

Astronomers used to think that the solar system that emerged from this early evolution was similar to what we see today. Detailed recent studies of the orbits of the planets and asteroids, however, suggest that there were more violent events soon afterward, perhaps involving substantial changes in the orbits of Jupiter and Saturn. These two giant planets control, through their gravity, the distribution of asteroids. Working backward from our present solar system, it appears that orbital changes took place during the first few hundred million years. One consequence may have been scattering of asteroids into the inner solar system, causing the period of "heavy bombardment" recorded in the oldest lunar craters.

14.4 COMPARISON WITH OTHER PLANETARY SYSTEMS

Learning Objectives

By the end of this section, you will be able to:

> Describe how the observations of protoplanetary disks provides evidence for the existence of other planetary systems
> Explain the two primary methods for detection of exoplanets
> Compare the main characteristics of other planetary systems with the features of the solar system

Until the middle 1990s, the practical study of the origin of planets focused on our single known example—the solar system. Although there had been a great deal of speculation about planets circling other stars, none had

This OpenStax book is available for free at http://cnx.org/content/col11992/1.8

actually been detected. Logically enough, in the absence of data, most scientists assumed that our own system was likely to be typical. They were in for a big surprise.

Discovery of Other Planetary Systems

In The Birth of Stars and the Discovery of Planets outside the Solar System, we discuss the formation of stars and planets in some detail. Stars like our Sun are formed when dense regions in a molecular cloud (made of gas and dust) feel an extra gravitational force and begin to collapse. This is a runaway process: as the cloud collapses, the gravitational force gets stronger, concentrating material into a protostar. Roughly half of the time, the protostar will fragment or be gravitationally bound to other protostars, forming a binary or multiple star system—stars that are gravitationally bound and orbit each other. The rest of the time, the protostar collapses in isolation, as was the case for our Sun. In all cases, as we saw, conservation of angular momentum results in a spin-up of the collapsing protostar, with surrounding material flattened into a disk. Today, this kind of structure can actually be observed. The Hubble Space Telescope, as well as powerful new ground-based telescopes, enable astronomers to study directly the nearest of these *circumstellar disks* in regions of space where stars are being born today, such as the Orion Nebula (Figure 14.14) or the Taurus star-forming region.

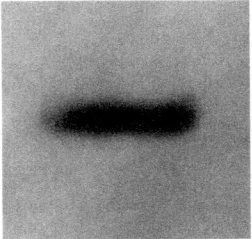

Figure 14.14. Protoplanetary Disk in the Orion Nebula. The Hubble Space Telescope imaged this protoplanetary disk in the Orion Nebula, a region of active star formation, using two different filters. The disk, about 17 times the size of our solar system, is in an edge-on orientation to us, and the newly formed star is shining at the center of the flattened dust cloud. The dark areas indicate absorption, not an absence of material. In the left image we see the light of the nebula and the dark cloud; in the right image, a special filter was used to block the light of the background nebula. You can see gas above and below the disk set to glow by the light of the newborn star hidden by the disk. (credit: modification of work by Mark McCaughrean (Max-Planck-Institute for Astronomy), C. Robert O'Dell (Rice University), and NASA)

Many of the circumstellar disks we have discovered show internal structure. The disks appear to be donut-shaped, with gaps close to the star. Such gaps indicate that the gas and dust in the disk have already collapsed to form large planets (Figure 14.15). The newly born protoplanets are too small and faint to be seen directly, but the depletion of raw materials in the gaps hints at the presence of something invisible in the inner part of the circumstellar disk—and that something is almost certainly one or more planets. Theoretical models of planet formation, like the one seen at right in Figure 14.15, have long supported the idea that planets would clear gaps as they form in disks.

Download for free at http://cnx.org/content/col11992/latest/

(a)   (b)

Figure 14.15. Protoplanetary Disk around HL Tau. (a) This image of a protoplanetary disk around HL Tau was taken with the Atacama Large Millimeter/submillimeter Array (ALMA), which allows astronomers to construct radio images that rival those taken with visible light. (b) Newly formed planets that orbit the central star clear out dust lanes in their paths, just as our theoretical models predict. This computer simulation shows the empty lane and spiral density waves that result as a giant planet is forming within the disk. The planet is not shown to scale. (credit a: modification of work by ALMA (ESO/NAOJ/NRAO); credit b: modification of work by NASA/ESA and A. Feild (STScI))

Our figure shows HL Tau, a one-million-year-old "newborn" star in the Taurus star-forming region. The star is embedded in a shroud of dust and gas that obscures our visible-light view of a circumstellar disk around the star. In 2014 astronomers obtained a dramatic view of the HL Tau circumstellar disk using millimeter waves, which pierce the cocoon of dust around the star, showing dust lanes being carved out by several newly formed protoplanets. As the mass of the protoplanets increases, they travel in their orbits at speeds that are faster than the dust and gas in the circumstellar disk. As the protoplanets plow through the disk, their gravitational reach begins to exceed their cross-sectional area, and they become very efficient at sweeping up material and growing until they clear a gap in the disk. The image of Figure 14.15 shows us that a number of protoplanets are forming in the disk and that they were able to form faster than our earlier ideas had suggested—all in the first million years of star formation.

LINK TO LEARNING

For an explanation of ALMA's ground-breaking observations of HL Tau and what they reveal about plant formation, watch this videocast (https://openstaxcollege.org/l/30eusoobhltavid) from the European Southern Observatory.

Discovering Exoplanets

You might think that with the advanced telescopes and detectors astronomers have today, they could directly image planets around nearby stars (which we call **exoplanets**). This has proved extremely difficult, however, not only because the exoplanets are faint, but also because they are generally lost in the brilliant glare of the star they orbit. As we discuss in more detail in The Birth of Stars and the Discovery of Planets outside the Solar System, the detection techniques that work best are indirect: they observe the effects of the planet on the star it orbits, rather than seeing the planet itself.

The first technique that yielded many planet detections is very high-resolution stellar spectroscopy. The *Doppler effect* lets astronomers measure the star's *radial velocity:* that is, the speed of the star, toward us or away from us, relative to the observer. If there is a massive planet in orbit around the star, the gravity of the planet causes

Download for free at http://cnx.org/content/col11992/latest/

the star to wobble, changing its radial velocity by a small but detectable amount. The distance of the star does not matter, as long as it is bright enough for us to take very high quality spectra.

Measurements of the variation in the star's radial velocity as the planet goes around the star can tell us the mass and orbital period of the planet. If there are several planets present, their effects on the radial velocity can be disentangled, so the entire planetary system can be deciphered—as long as the planets are massive enough to produce a measureable Doppler effect. This detection technique is most sensitive to large planets orbiting close to the star, since these produce the greatest wobble in their stars. It has been used on large ground-based telescopes to detect hundreds of planets, including one around Proxima Centauri, the nearest star to the Sun.

The second indirect technique is based on the slight dimming of a star when one of its planets *transits*, or crosses over the face of the star, as seen from Earth. Astronomers do not see the planet, but only detect its presence from careful measurements of a change in the brightness of the star over long periods of time. If the slight dips in brightness repeat at regular intervals, we can determine the orbital period of the planet. From the amount of starlight obscured, we can measure the planet's size.

While some transits have been measured from Earth, large-scale application of this transit technique requires a telescope in space, above the atmosphere and its distortions of the star images. It has been most successfully applied from the NASA Kepler space observatory, which was built for the sole purpose of "staring" for 5 years at a single part of the sky, continuously monitoring the light from more than 150,000 stars. The primary goal of Kepler was to determine the frequency of occurrence of exoplanets of different sizes around different classes of stars. Like the Doppler technique, the transit observations favor discovery of large planets and short-period orbits.

Recent detection of exoplanets using both the Doppler and transit techniques has been incredibly successful. Within two decades, we went from no knowledge of other planetary systems to a catalog of *thousands* of exoplanets. Most of the exoplanets found so far are more massive than or larger in size than Earth. It is not that Earth analogs do not exist. Rather, the shortage of small rocky planets is an observational bias: smaller planets are more difficult to detect.

Analyses of the data to correct for such biases or selection effects indicate that small planets (like the terrestrial planets in our system) are actually much more common than giant planets. Also relatively common are "super Earths," planets with two to ten times the mass of our planet (Figure 14.16). We don't have any of these in our solar system, but nature seems to have no trouble making them elsewhere. Overall, the Kepler data suggest that approximately one quarter of stars have exoplanet systems, implying the existence of at least 50 billion planets in our Galaxy alone.

Download for free at http://cnx.org/content/col11992/latest/

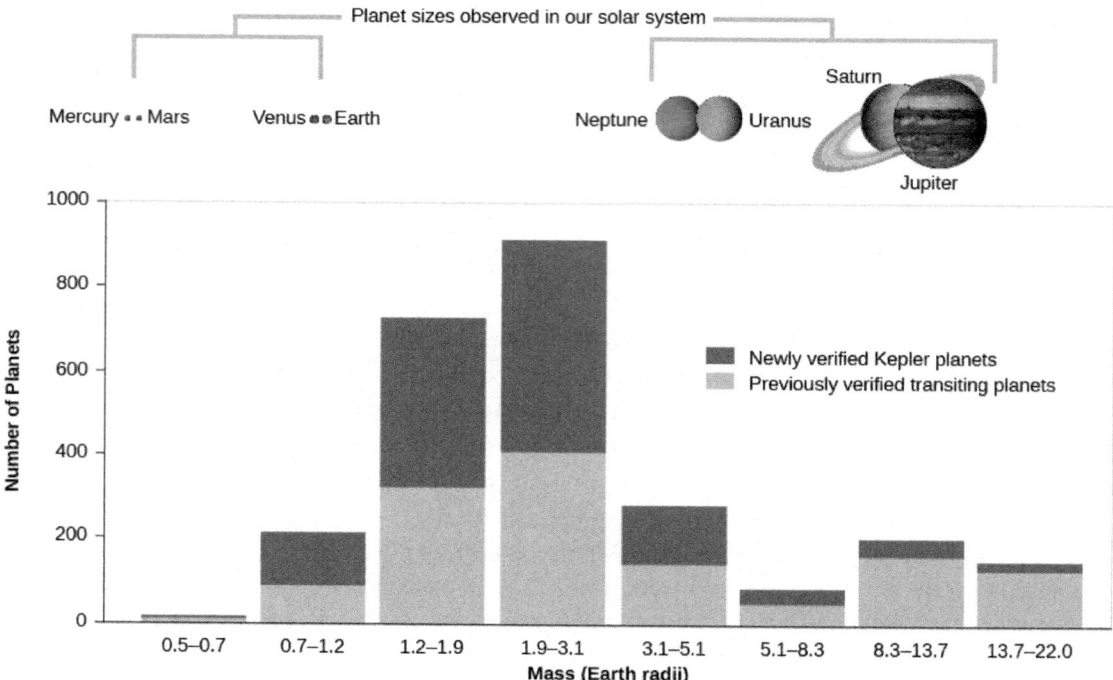

Figure 14.16. Transiting Planets by Size. This bar graph shows the planets found so far using the transit method (the vast majority found by the Kepler mission). The orange parts of each bar indicate the planets announced by the Kepler team in May 2016. Note that the largest number of planets found so far are in two categories that we don't have in our own solar system—planets whose size is between Earth's and Neptune's. (credit: modification of work by NASA)

The Configurations of Other Planetary Systems

Let's look more closely at the progress in the detection of exoplanets. Figure 14.17 shows the planets that were discovered each year by the two techniques we discussed. In the early years of exoplanet discovery, most of the planets were similar in mass to Jupiter. This is because, as mentioned above, the most massive planets were easiest to detect. In more recent years, planets smaller than Neptune and even close to the size of Earth have been detected.

This OpenStax book is available for free at http://cnx.org/content/col11992/1.8

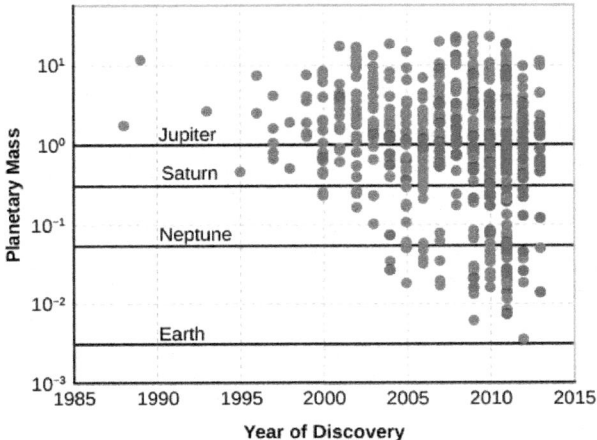

Figure 14.17. Masses of Exoplanets Discovered by Year. Horizontal lines are drawn to reference the masses of Jupiter, Saturn, Neptune, and Earth. The gray dots indicate planets discovered by measuring the radial velocity of the star, and the red dots are for planets that transit their stars. In the early years, the only planets that could be detected were similar in mass to Jupiter. Improvements in technology and observing strategies enabled the detection of lower mass planets as time went on, and now even smaller worlds are being found. (Note that this tally ends in 2014.)

We also know that many exoplanets are in multiplanet systems. This is one characteristic that our solar system shares with exosystems. Looking back at Figure 14.15 and seeing how such large disks can give rise to more than one center of condensation, it is not too surprising that multiplanet systems are a typical outcome of planet formation. Astronomers have tried to measure whether multiple planet systems all lie in the same plane using astrometry. This is a difficult measurement to make with current technology, but it is an important measurement that could help us understand the origin and evolution of planetary systems.

Comparison between Theory and Data

Many of the planetary systems discovered so far do not resemble our own solar system. Consequently, we have had to reassess some aspects of the "standard models" for the formation of planetary systems. Science sometimes works in this way, with new data contradicting our expectations. The press often talks about a scientist making experiments to "confirm" a theory. Indeed, it is comforting when new data support a hypothesis or theory and increase our confidence in an earlier result. But the most exciting and productive moments in science often come when new data *don't* support existing theories, forcing scientists to rethink their position and develop new and deeper insights into the way nature works.

Nothing about the new planetary systems contradicts the basic idea that planets form from the aggregation (clumping) of material within circumstellar disks. However, the existence of " hot Jupiters"—planets of jovian mass that are closer to their stars than the orbit of Mercury—poses the biggest problem. As far as we know, a giant planet cannot be formed without the condensation of water ice, and water ice is not stable so close to the heat of a star. It seems likely that all the giant planets, "hot" or "normal," formed at a distance of several astronomical units from the star, but we now see that they did not necessarily stay there. This discovery has led to a revision in our understanding of planet formation that now includes "planet migrations" within the protoplanetary disk, or later gravitational encounters between sibling planets that scatter one of the planets inward.

Many exoplanets have large orbital eccentricity (recall this means the orbits are not circular). High eccentricities were not expected for planets that form in a disk. This discovery provides further support for the scattering

Download for free at http://cnx.org/content/col11992/latest/

of planets when they interact gravitationally. When planets change each other's motions, their orbits could become much more eccentric than the ones with which they began.

There are several suggestions for ways migration might have occurred. Most involve interactions between the giant planets and the remnant material in the circumstellar disk from which they formed. These interactions would have taken place when the system was very young, while material still remained in the disk. In such cases, the planet travels at a faster velocity than the gas and dust and feels a kind of "headwind" (or friction) that causes it to lose energy and spiral inward. It is still unclear how the spiraling planet stops before it plunges into the star. Our best guess is that this plunge into the star is the fate for many protoplanets; however, clearly some migrating planets can stop their inward motions and escape this destruction, since we find hot Jupiters in many mature planetary systems.

14.5 PLANETARY EVOLUTION

Learning Objectives

By the end of this section, you will be able to:

> Describe the geological activity during the evolution of the planets, particularly on the terrestrial planets
> Describe the factors that affect differences in elevation on the terrestrial planets
> Explain how the differences in atmosphere on Venus, Earth, and Mars evolved from similar starting points in the early history of the solar system

While we await more discoveries and better understanding of other planetary systems, let us look again at the early history of our own solar system, after the dissipation of our dust disk. The era of giant impacts was probably confined to the first 100 million years of solar system history, ending by about 4.4 billion years ago. Shortly thereafter, the planets cooled and began to assume their present aspects. Up until about 4 billion years ago, they continued to acquire volatile materials, and their surfaces were heavily cratered from the remaining debris that hit them. However, as external influences declined, all the terrestrial planets as well as the moons of the outer planets began to follow their own evolutionary courses. The nature of this evolution depended on each object's composition, mass, and distance from the Sun.

Geological Activity

We have seen a wide range in the level of geological activity on the terrestrial planets and icy moons. Internal sources of such activity (as opposed to pummeling from above) require energy, either in the form of primordial heat left over from the formation of a planet or from the decay of radioactive elements in the interior. The larger the planet or moon, the more likely it is to retain its internal heat and the more slowly it cools—this is the "baked potato effect" mentioned in Other Worlds: An Introduction to the Solar System. Therefore, we are more likely to see evidence of continuing geological activity on the surface of larger (solid) worlds (Figure 14.18). Jupiter's moon Io is an interesting exception to this rule; we saw that it has an unusual source of heat from the gravitational flexing of its interior by the tidal pull of Jupiter. Europa is probably also heated by jovian tides. Saturn may be having a similar effect on its moon Enceladus.

This OpenStax book is available for free at http://cnx.org/content/col11992/1.8

Accretion, heating, differentiation

Formation of solid crust, heavy cratering

Widespread mare-like volcanism

Reduced volcanism, possible plate tectonics

Mantle solidification, end of tectonic activity

Cool interior, no activity

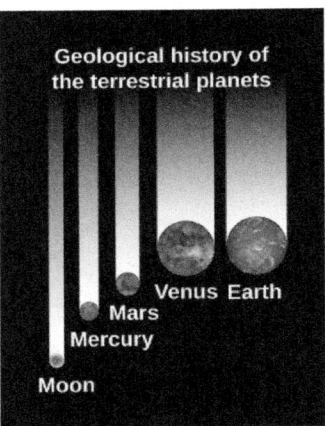

Figure 14.18. Stages in the Geological History of a Terrestrial Planet. In this image, time increases downward along the left side, where the stages are described. Each planet is shown roughly in its present stage. The smaller the planet, the more quickly it passes through these stages.

The Moon, the smallest of the terrestrial worlds, was internally active until about 3.3 billion years ago, when its major volcanism ceased. Since that time, its mantle has cooled and become solid, and today even internal seismic activity has declined to almost zero. The Moon is a geologically dead world. Although we know much less about Mercury, it seems likely that this planet, too, ceased most volcanic activity about the same time the Moon did.

Mars represents an intermediate case, and it has been much more active than the Moon. The southern hemisphere crust had formed by 4 billion years ago, and the northern hemisphere volcanic plains seem to be contemporary with the lunar maria. However, the Tharsis bulge formed somewhat later, and activity in the large Tharsis volcanoes has apparently continued on and off to the present era.

Earth and Venus are the largest and most active terrestrial planets. Our planet experiences global plate tectonics driven by convection in its mantle. As a result, our surface is continually reworked, and most of Earth's surface material is less than 200 million years old. Venus has generally similar levels of volcanic activity, but unlike Earth, it has not experienced plate tectonics. Most of its surface appears to be no more than 500 million years old. We did see that the surface of our sister planet is being modified by a kind of "blob tectonics"—where hot material from below puckers and bursts through the surface, leading to coronae, pancake volcanoes, and other such features. A better understanding of the geological differences between Venus and Earth is a high priority for planetary geologists.

The geological evolution of the icy moons and Pluto has been somewhat different from that of the terrestrial planets. Tidal energy sources have been active, and the materials nature has to work with are not the same. On these outer worlds, we see evidence of low-temperature volcanism, with the silicate lava of the inner planets being supplemented by sulfur compounds on Io, and replaced by water and other ices on Pluto and other outer-planet moons.

Elevation Differences

Let's look at some specific examples of how planets differ. The mountains on the terrestrial planets owe their origins to different processes. On the Moon and Mercury, the major mountains are ejecta thrown up by the large basin-forming impacts that took place billions of years ago. Most large mountains on Mars are volcanoes, produced by repeated eruptions of lava from the same vents. There are similar (but smaller) volcanoes on Earth and Venus. However, the highest mountains on Earth and Venus are the result of compression and uplift of the surface. On Earth, this crustal compression results from collisions of one continental plate with another.

Download for free at http://cnx.org/content/col11992/latest/

It is interesting to compare the maximum heights of the volcanoes on Earth, Venus, and Mars (Figure 14.19). On Venus and Earth, the maximum elevation differences between these mountains and their surroundings are about 10 kilometers. Olympus Mons, in contrast, towers more than 20 kilometers above its surroundings and nearly 30 kilometers above the lowest elevation areas on Mars.

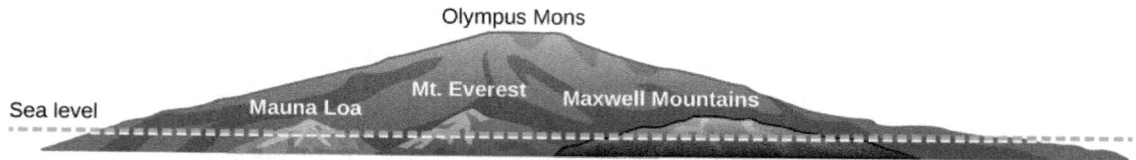

Figure 14.19. Highest Mountains on Mars, Venus, and Earth. Mountains can rise taller on Mars because Mars has less surface gravity and no moving plates. The vertical scale is exaggerated by a factor of three to make comparison easier. The label "sea level" refers only to Earth, of course, since the other two planets don't have oceans. Mauna Loa and Mt. Everest are on Earth, Olympus Mons is on Mars, and the Maxwell Mountains are on Venus.

One reason Olympus Mons (Figure 14.20) is so much higher than its terrestrial counterparts is that the crustal plates on Earth never stop moving long enough to let a really large volcano grow. Instead, the moving plate creates a long row of volcanoes like the Hawaiian Islands. On Mars (and perhaps Venus) the crust remains stationary with respect to the underlying hot spot, and so a single volcano can continue to grow for hundreds of millions of years.

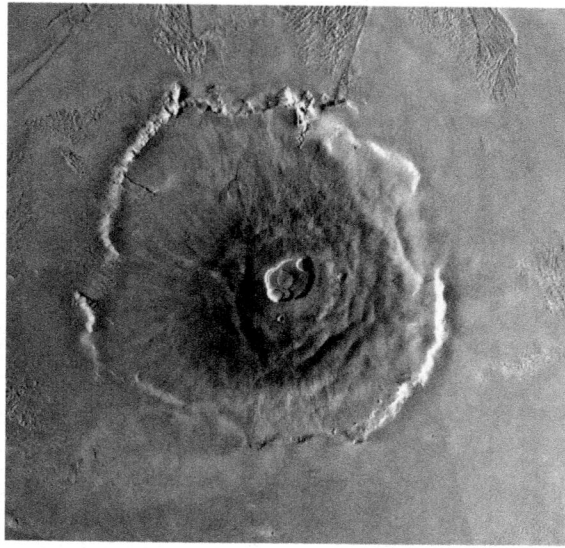

Figure 14.20. Olympus Mons. The largest martian volcano is seen from above in this spectacular composite image created from many Viking orbiter photographs. The volcano is nearly 500 kilometers wide at its base and more than 20 kilometers high. (Its height is almost three times the height of the tallest mountain on Earth.) (credit: modification of work by NASA/USGS)

A second difference relates to the strength of gravity on the three planets. The surface gravity on Venus is nearly the same as that on Earth, but on Mars it is only about one third as great. In order for a mountain to survive, its internal strength must be great enough to support its weight against the force of gravity. Volcanic rocks have known strengths, and we can calculate that on Earth, 10 kilometers is about the limit. For instance, when new lava is added to the top of Mauna Loa in Hawaii, the mountain slumps downward under its own weight. The same height limit applies on Venus, where the force of gravity is the same as Earth's. On Mars, however, with its lesser surface gravity, much greater elevation differences can be supported, which helps explain why Olympus Mons is more than twice as high as the tallest mountains of Venus or Earth.

This OpenStax book is available for free at http://cnx.org/content/col11992/1.8

By the way, the same kind of calculation that determines the limiting height of a mountain can be used to ascertain the largest body that can have an irregular shape. Gravity, if it can, pulls all objects into the most "efficient" shape (where all the outside points are equally distant from the center). All the planets and larger moons are nearly spherical, due to the force of their own gravity pulling them into a sphere. But the smaller the object, the greater the departure from spherical shape that the strength of its rocks can support. For silicate bodies, the limiting diameter is about 400 kilometers; larger objects will always be approximately spherical, while smaller ones can have almost any shape (as we see in photographs of asteroids, such as Figure 14.21).

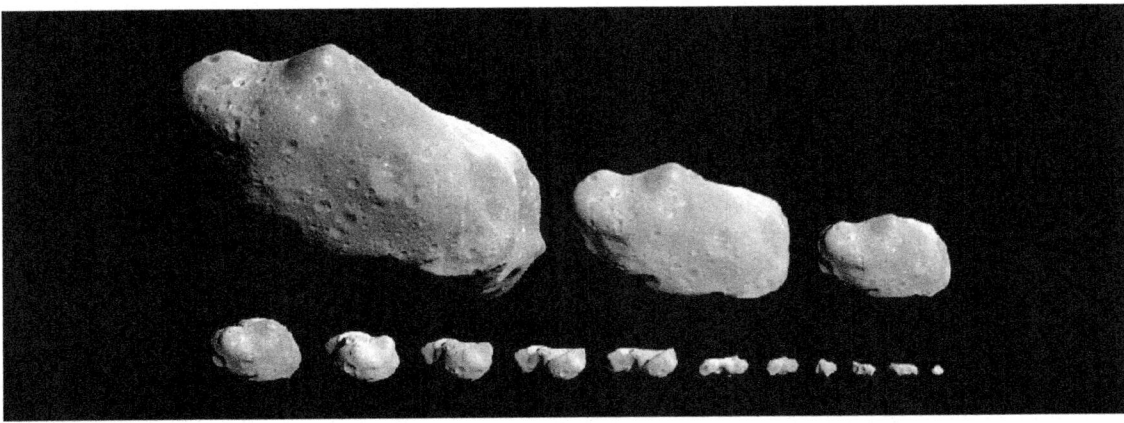

Figure 14.21. Irregular Asteroid. Small objects such as asteroid Ida (shown here in multiple views taken by the Galileo spacecraft camera as it flew past) are generally irregular or elongated; they do not have strong enough gravity to pull them into a spherical shape. Ida is about 60 kilometers long in its longest dimension. (credit: modification of work by NASA/JPL)

Atmospheres

The atmospheres of the planets were formed by a combination of gas escaping from their interiors and the impacts of volatile-rich debris from the outer solar system. Each of the terrestrial planets must have originally had similar atmospheres, but Mercury was too small and too hot to retain its gas. The Moon probably never had an atmosphere since the material composing it was depleted in volatile materials.

The predominant volatile gas on the terrestrial planets is now carbon dioxide (CO_2), but initially there were probably also hydrogen-containing gases. In this more chemically *reduced* (hydrogen-dominated) environment, there should have been large amounts of carbon monoxide (CO) and traces of ammonia (NH_3) and methane (CH_4). Ultraviolet light from the Sun split apart the molecules of reducing gases in the inner solar system, however. Most of the light hydrogen atoms escaped, leaving behind the oxidized (oxygen-dominated) atmospheres we see today on Earth, Venus, and Mars.

The fate of water was different on each of these three planets, depending on its size and distance from the Sun. Early in its history, Mars apparently had a thick atmosphere with abundant liquid water, but it could not retain those conditions. The CO_2 necessary for a substantial greenhouse effect was lost, the temperature dropped, and eventually the remaining water froze. On Venus the reverse process took place, with a runaway greenhouse effect leading to the permanent loss of water. Only Earth managed to maintain the delicate balance that permits liquid water to persist on its surface.

With the water gone, Venus and Mars each ended up with an atmosphere of about 96 percent carbon dioxide and a few percent nitrogen. On Earth, the presence first of water and then of life led to a very different kind of atmosphere. The CO_2 was removed and deposited in marine sediment. The proliferation of life forms that could photosynthesize eventually led to the release of more oxygen than natural chemical reactions can remove from

Download for free at http://cnx.org/content/col11992/latest/

the atmosphere. As a result, thanks to the life on its surface, Earth finds itself with a great deficiency of CO_2, with nitrogen as the most abundant gas, and the only planetary atmosphere that contains free oxygen.

In the outer solar system, Titan is the only moon with a substantial atmosphere. This object must have contained sufficient volatiles—such as ammonia, methane, and nitrogen—to form an atmosphere. Thus, today Titan's atmosphere consists primarily of nitrogen. Compared with those on the inner planets, temperatures on Titan are too low for either carbon dioxide or water to be in vapor form. With these two common volatiles frozen solid, it is perhaps not too surprising that nitrogen has ended up as the primary atmospheric constituent.

We see that nature, starting with one set of chemical constituents, can fashion a wide range of final atmospheres appropriate to the conditions and history of each world. The atmosphere we have on Earth is the result of many eons of evolution and adaptation. And, as we saw, it can be changed by the actions of the life forms that inhabit the planet.

One of the motivations for exploration of our planetary system is the search for life, beginning with a survey for potentially habitable environments. Mercury, Venus, and the Moon are not suitable; neither are most of the moons in the outer solar system. The giant planets, which do not have solid surfaces, also fail the test for habitability.

So far, the search for habitable environments has focused on the presence of liquid water. Earth and Europa both have large oceans, although Europa's ocean is covered with a thick crust of ice. Mars has a long history of liquid water on its surface, although the surface today is mostly dry and cold. However, there is strong evidence for subsurface water on Mars, and even today water flows briefly on the surface under the right conditions. Enceladus may have the most accessible liquid water, which is squirting into space by means of the geysers observed with our Cassini spacecraft. Titan is in many ways the most interesting world we have explored. It is far too cold for liquid water, but with its thick atmosphere and hydrocarbon lakes, it may be the best place to search for "life as we don't know it."

We now come to the end of our study of the planetary system. Although we have learned a great deal about the other planets during the past few decades of spacecraft exploration, much remains unknown. Discoveries in recent years of geological activity on Titan and Enceladus were unexpected, as was the complex surface of Pluto revealed by New Horizons. The study of exoplanetary systems provides a new perspective, teaching us that there is much more variety among planetary systems than scientists had imagined a few decades ago. The exploration of the solar system is one of the greatest human adventures, and, in many ways, it has just begun.

This OpenStax book is available for free at http://cnx.org/content/col11992/1.8

CHAPTER 14 REVIEW

KEY TERMS

accretion the gradual accumulation of mass, as by a planet forming from colliding particles in the solar nebula

exoplanet a planet orbiting a star other than our Sun

iron meteorite a meteorite composed primarily of iron and nickel

meteor a small piece of solid matter that enters Earth's atmosphere and burns up, popularly called a *shooting star* because it is seen as a small flash of light

meteor shower many meteors appearing to radiate from one point in the sky; produced when Earth passes through a cometary dust stream

meteorite a portion of a meteor that survives passage through the atmosphere and strikes the ground

stony meteorite a meteorite composed mostly of stony material, either primitive or differentiated

stony-iron meteorite a type of differentiated meteorite that is a blend of nickel-iron and silicate materials

SUMMARY

14.1 Meteors

When a fragment of interplanetary dust strikes Earth's atmosphere, it burns up to create a meteor. Streams of dust particles traveling through space together produce meteor showers, in which we see meteors diverging from a spot in the sky called the radiant of the shower. Many meteor showers recur each year and are associated with particular comets that have left dust behind as they come close to the Sun and their ices evaporate (or have broken up into smaller pieces).

14.2 Meteorites: Stones from Heaven

Meteorites are the debris from space (mostly asteroid fragments) that survive to reach the surface of Earth. Meteorites are called *finds* or *falls* according to how they are discovered; the most productive source today is the Antarctic ice cap. Meteorites are classified as irons, stony-irons, or stones accordingly to their composition. Most stones are primitive objects, dated to the origin of the solar system 4.5 billion years ago. The most primitive are the carbonaceous meteorites, such as Murchison and Allende. These can contain a number of organic (carbon-rich) molecules.

14.3 Formation of the Solar System

A viable theory of solar system formation must take into account motion constraints, chemical constraints, and age constraints. Meteorites, comets, and asteroids are survivors of the solar nebula out of which the solar system formed. This nebula was the result of the collapse of an interstellar cloud of gas and dust, which contracted (conserving its angular momentum) to form our star, the Sun, surrounded by a thin, spinning disk of dust and vapor. Condensation in the disk led to the formation of planetesimals, which became the building blocks of the planets. Accretion of infalling materials heated the planets, leading to their differentiation. The giant planets were also able to attract and hold gas from the solar nebula. After a few million years of violent impacts, most of the debris was swept up or ejected, leaving only the asteroids and cometary remnants surviving to the present.

Download for free at http://cnx.org/content/col11992/latest/

The first planet circling a distant solar-type star was announced in 1995. Twenty years later, thousands of exoplanets have been identified, including planets with sizes and masses between Earth's and Neptune's, which we don't have in our own solar system. A few percent of exoplanet systems have "hot Jupiters," massive planets that orbit close to their stars, and many exoplanets are also in eccentric orbits. These two characteristics are fundamentally different from the attributes of gas giant planets in our own solar system and suggest that giant planets can migrate inward from their place of formation where it is cold enough for ice to form. Current data indicate that small (terrestrial type) rocky planets are common in our Galaxy; indeed, there must be tens of billions of such earthlike planets.

After their common beginning, each of the planets evolved on its own path. Different possible outcomes are illustrated by comparison of the terrestrial planets (Earth, Venus, Mars, Mercury, and the Moon). All are rocky, differentiated objects. The level of geological activity is proportional to mass: greatest for Earth and Venus, less for Mars, and absent for the Moon and Mercury. However, tides from another nearby world can also generate heat to drive geological activity, as shown by Io, Europa, and Enceladus. Pluto is also active, to the surprise of planetary scientists. On the surfaces of solid worlds, mountains can result from impacts, volcanism, or uplift. Whatever their origin, higher mountains can be supported on smaller planets that have less surface gravity. The atmospheres of the terrestrial planets may have acquired volatile materials from comet impacts. The Moon and Mercury lost their atmospheres; most volatiles on Mars are frozen due to its greater distance from the Sun and its thinner atmosphere; and Venus retained CO_2 but lost H_2O when it developed a massive greenhouse effect. Only Earth still has liquid water on its surface and hence can support life.

 # FOR FURTHER EXPLORATION

Note: Resources about exoplanets are provided in The Birth of Stars and the Discovery of Planets outside the Solar System.

Articles

Meteors and Meteorites

Alper, J. "It Came from Outer Space." *Astronomy* (November 2002): 36. On the analysis of organic materials in meteorites.

Beatty, J. "Catch a Fallen Star." *Sky & Telescope* (August 2009): 22. On the recovery of meteorites from an impact that was seen in the sky.

Durda, D. "The Chelyabinsk Super-Meteor." *Sky & Telescope* (June 2013): 24. A nice summary, with photos and eyewitness reporting.

Garcia, R., & Notkin, G. "Touching the Stars without Leaving Home." *Sky & Telescope* (October 2008): 32. Hunting and collecting meteorites.

Kring, D. "Unlocking the Solar System's Past." *Astronomy* (August 2006): 32. Part of a special issue devoted to meteorites.

Rubin, A. "Secrets of Primitive Meteorites." *Scientific American* (February 2013): 36. What they can teach us about the environment in which the solar system formed.

This OpenStax book is available for free at http://cnx.org/content/col11992/1.8

Evolution of the Solar System and Protoplanetary Disks

Jewitt, D., & Young, E. "Oceans from the Skies." *Scientific American* (March 2015): 36–43. How did Earth and the other inner planets get their water after the initial hot period?

Talcott, R. "How the Solar System Came to Be." *Astronomy* (November 2012): 24. On the formation period of the Sun and the planets.

Young, E. "Cloudy with a Chance of Stars." *Scientific American* (February 2010): 34. On how clouds of interstellar matter turn into star systems.

Websites

Meteors and Meteorites

American Meteor Society: http://www.amsmeteors.org/. For serious observers.

British and Irish Meteorite Society: http://www.bimsociety.org/meteorites1.shtml.

Meteor Showers Online: http://meteorshowersonline.com/. By Gary Kronk.

Meteorite Information: http://www.meteorite-information.com/. A great collection of links for understanding and even collecting meteorites.

Meteorites from Mars: http://www2.jpl.nasa.gov/snc/. A listing and links from the Jet Propulsion Lab.

Meteors and Meteor Showers: http://www.astronomy.com/observing/observe-the-solar-system/2010/04/meteors-and-meteor-showers. From *Astronomy* magazine.

Meteors: http://www.skyandtelescope.com/observing/celestial-objects-to-watch/meteors/. A collection of articles on meteor observing from *Sky & Telescope* magazine.

Nine Planets Meteorites and Meteors Page: http://nineplanets.org/meteorites.html.

Some Interesting Meteorite Falls of the Last Two Centuries: http://www.icq.eps.harvard.edu/meteorites-1.html.

Evolution of the Solar System and Protoplanetary Disks

Circumstellar Disk Learning Site: http://www.disksite.com/. By Dr. Paul Kalas.

Disk Detective Project: http://www.diskdetective.org/. The WISE mission is asking the public to help them find protoplanetary disks in their infrared data.

Videos

Meteors and Meteorites

Meteorites and Meteor-wrongs: https://www.youtube.com/watch?v=VQO335Y3zXo. Video with Dr. Randy Korotev of Washington U. in St. Louis (7:05).

Rare Meteorites from London's Natural History Museum: https://www.youtube.com/watch?v=w-Rsk-ywN44. A tour of the meteorite collection with curator Caroline Smith (18:22). Also see a short news piece about a martian meteorite: https://www.youtube.com/watch?v=1EMR2r53f2s (2:54).

What Is a Meteor Shower (and How to Watch Them): https://www.youtube.com/watch?v=xNmgvlwInCA. Top tips for watching meteor showers from the At-Bristol Science Center (3:18).

Evolution of the Solar System and Protoplanetary Disks

Origins of the Solar System: http://www.pbs.org/wgbh/nova/space/origins-solar-system.html. Video from Nova ScienceNow narrated by Neil deGrasse Tyson (13:02).

Download for free at http://cnx.org/content/col11992/latest/

Where Do Planets Come From?: https://www.youtube.com/watch?v=zdIJUdZWlXo. Public talk by Anjali Tripathi in March 2016 in the Center for Astrophysics Observatory Nights Series (56:14).

⚎ COLLABORATIVE GROUP ACTIVITIES

A. Ever since the true (cosmic) origin of meteorites was understood, people have tried to make money selling them to museums and planetariums. More recently, a growing number of private collectors have been interested in purchasing meteorite fragments, and a network of dealers (some more reputable than others) has sprung up to meet this need. What does your group think of all this? Who should own a meteorite? The person on whose land it falls, the person who finds it, or the local, state, or federal government where it falls? What if it falls on public land? Should there be any limit to what people charge for meteorites? Or should all meteorites be the common property of humanity? (If you can, try to research what the law is now in your area. See, for example, http://www.space.com/18009-meteorite-collectors-public-lands-rules.html.)

B. Your group has been formed to advise a very rich person who wants to buy some meteorites but is afraid of being cheated and sold some Earth rocks. How would you advise your client to make sure that the meteorites she buys are authentic?

C. Your group is a committee set up to give advice to NASA about how to design satellites and telescopes in space to minimize the danger of meteor impacts. Remember that the heavier a satellite is, the harder (more expensive) it is to launch. What would you include in your recommendations?

D. Discuss what you would do if you suddenly found that a small meteorite had crashed in or near your home. Whom would you call first, second, third? What would you do with the sample? (And would any damage to your home be covered by your insurance?)

E. A friend of your group really wants to see a meteor shower. The group becomes a committee to assist her in fulfilling this desire. What time of year would be best? What equipment would you recommend she gets? What advice would you give her?

F. Work with your group to find a table of the phases of the Moon for the next calendar year. Then look at the table of well-known meteor showers in this chapter and report on what phase the Moon will be in during each shower. (The brighter the Moon is in the night sky, the harder it is to see the faint flashes of meteors.)

G. Thinking that all giant planets had to be far from their stars (because the ones in our solar system are) is an example of making theories without having enough data (or examples). Can your group make a list of other instances in science (and human relations) where we have made incorrect judgments without having explored enough examples?

H. Have your group list and then discuss several ways in which the discovery of a diverse group of exoplanets (planets orbiting other stars) has challenged our conventional view of the formation of planetary systems like our solar system.

▣ EXERCISES

This OpenStax book is available for free at http://cnx.org/content/col11992/1.8

Review Questions

1. A friend of yours who has not taken astronomy sees a meteor shower (she calls it a bunch of shooting stars). The next day she confides in you that she was concerned that the stars in the Big Dipper (her favorite star pattern) might be the next ones to go. How would you put her mind at ease?

2. In what ways are meteorites different from meteors? What is the probable origin of each?

3. How are comets related to meteor showers?

4. What do we mean by primitive material? How can we tell if a meteorite is primitive?

5. Describe the solar nebula, and outline the sequence of events within the nebula that gave rise to the planetesimals.

6. Why do the giant planets and their moons have compositions different from those of the terrestrial planets?

7. How do the planets discovered so far around other stars differ from those in our own solar system? List at least two ways.

8. Explain the role of impacts in planetary evolution, including both giant impacts and more modest ones.

9. Why are some planets and moons more geologically active than others?

10. Summarize the origin and evolution of the atmospheres of Venus, Earth, and Mars.

11. Why do meteors in a meteor shower appear to come from just one point in the sky?

Thought Questions

12. What methods do scientists use to distinguish a meteorite from terrestrial material?

13. Why do iron meteorites represent a much higher percentage of finds than of falls?

14. Why is it more useful to classify meteorites according to whether they are primitive or differentiated rather than whether they are stones, irons, or stony-irons?

15. Which meteorites are the most useful for defining the age of the solar system? Why?

16. Suppose a new primitive meteorite is discovered (sometime after it falls in a field of soybeans) and analysis reveals that it contains a trace of amino acids, all of which show the same rotational symmetry (unlike the Murchison meteorite). What might you conclude from this finding?

17. How do we know when the solar system formed? Usually we say that the solar system is 4.5 billion years old. To what does this age correspond?

18. We have seen how Mars can support greater elevation differences than Earth or Venus. According to the same arguments, the Moon should have higher mountains than any of the other terrestrial planets, yet we know it does not. What is wrong with applying the same line of reasoning to the mountains on the Moon?

19. Present theory suggests that giant planets cannot form without condensation of water ice, which becomes vapor at the high temperatures close to a star. So how can we explain the presence of jovian-sized exoplanets closer to their star than Mercury is to our Sun?

20. Why are meteorites of primitive material considered more important than other meteorites? Why have most of them been found in Antarctica?

Download for free at http://cnx.org/content/col11992/latest/

Figuring For Yourself

21. How long would material take to go around if the solar nebula in Example 14.1 became the size of Earth's orbit?

22. Consider the differentiated meteorites. We think the irons are from the cores, the stony-irons are from the interfaces between mantles and cores, and the stones are from the mantles of their differentiated parent bodies. If these parent bodies were like Earth, what fraction of the meteorites would you expect to consist of irons, stony-irons, and stones? Is this consistent with the observed numbers of each? (Hint: You will need to look up what percent of the volume of Earth is taken up by its core, mantle, and crust.)

23. Estimate the maximum height of the mountains on a hypothetical planet similar to Earth but with twice the surface gravity of our planet.

This OpenStax book is available for free at http://cnx.org/content/col11992/1.8

Figure 15.1. Our Star. The Sun—our local star—is quite average in many ways. However, that does not stop it from being a fascinating object to study. From solar flares and coronal mass ejections, like the one seen coming from the Sun in the top right of this image, the Sun is a highly dynamic body at the center of our solar system. This image combines two separate satellite pictures of the Sun—the inner one from the Solar Dynamics Observatory and the outer one from the Solar and Heliospheric Observatory. (credit: modification of work by ESA/NASA)

Chapter Outline

15.1 The Structure and Composition of the Sun

15.2 The Solar Cycle

15.3 Solar Activity above the Photosphere

15.4 Space Weather

✎ Thinking Ahead

"Space weather" may sound like a contradiction. How can there be weather in the vacuum of space? Yet space weather, which refers to changing conditions in space, is an active field of research and can have profound effects on Earth. We are all familiar with the ups and downs of weather on Earth, and how powerful storms can be devastating for people and vegetation. Although we are separated from the Sun by a large distance as well as by the vacuum of space, we now understand that great outbursts on the Sun (solar storms, in effect) can cause changes in the atmosphere and magnetic field of Earth, sometimes even causing serious problems on the ground. In this chapter, we will explore the nature of the Sun's outer layers, the changing conditions and activity there, and the ways that the Sun affects Earth.

By studying the Sun, we also learn much that helps us understand stars in general. The Sun is, in astronomical terms, a rather ordinary star—not unusually hot or cold, old or young, large or small. Indeed, we are lucky that the Sun is typical. Just as studies of Earth help us understand observations of the more distant planets, so too does the Sun serve as a guide to astronomers in interpreting the messages contained in the light we receive from distant stars. As you will learn, the Sun is dynamic, continuously undergoing change, balancing the forces of nature to keep itself in equilibrium. In this chapter, we describe the components of the Sun, how it changes with time, and how those changes affect Earth.

Download for free at http://cnx.org/content/col11992/latest/

15.1 THE STRUCTURE AND COMPOSITION OF THE SUN

Learning Objectives

By the end of this section, you will be able to:

> Explain how the composition of the Sun differs from that of Earth
> Describe the various layers of the Sun and their functions
> Explain what happens in the different parts of the Sun's atmosphere

The Sun, like all stars, is an enormous ball of extremely hot, largely ionized gas, shining under its own power. And we do mean enormous. The Sun could fit 109 Earths side-by-side across its diameter, and it has enough volume (takes up enough space) to hold about 1.3 million Earths.

The Sun does not have a solid surface or continents like Earth, nor does it have a solid core (Figure 15.2). However, it does have a lot of structure and can be discussed as a series of layers, not unlike an onion. In this section, we describe the huge changes that occur in the Sun's extensive interior and atmosphere, and the dynamic and violent eruptions that occur daily in its outer layers.

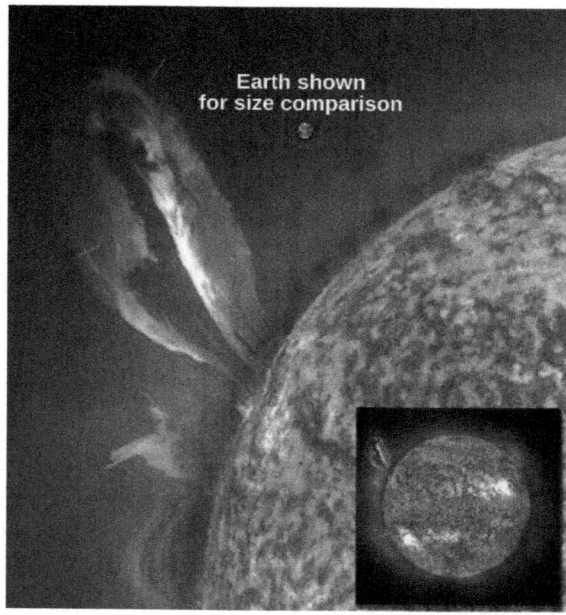

Figure 15.2. Earth and the Sun. Here, Earth is shown to scale with part of the Sun and a giant loop of hot gas erupting from its surface. The inset shows the entire Sun, smaller. (credit: modification of work by SOHO/EIT/ESA)

Some of the basic characteristics of the Sun are listed in Table 15.1. Although some of the terms in that table may be unfamiliar to you right now, you will get to know them as you read further.

This OpenStax book is available for free at http://cnx.org/content/col11992/1.8

Download for free at http://cnx.org/content/col11992/latest/

Characteristics of the Sun

Characteristic	How Found	Value
Mean distance	Radar reflection from planets	1 AU (149,597,892 km)
Maximum distance from Earth		1.521×10^8 km
Minimum distance from Earth		1.471×10^8 km
Mass	Orbit of Earth	333,400 Earth masses (1.99×10^{30} kg)
Mean angular diameter	Direct measure	31´59´´.3
Diameter of photosphere	Angular size and distance	109.3 × Earth diameter (1.39×10^6 km)
Mean density	Mass/volume	1.41 g/cm^3 (1400 kg/m^3)
Gravitational acceleration at photosphere (surface gravity)	GM/R^2	27.9 × Earth surface gravity = 273 m/s^2
Solar constant	Instrument sensitive to radiation at all wavelengths	1370 W/m^2
Luminosity	Solar constant × area of spherical surface 1 AU in radius	3.8×10^{26} W
Spectral class	Spectrum	G2V
Effective temperature	Derived from luminosity and radius of the Sun	5800 K
Rotation period at equator	Sunspots and Doppler shift in spectra taken at the edge of the Sun	24 days 16 hours
Inclination of equator to ecliptic	Motions of sunspots	7°10´.5

Table 15.1

Composition of the Sun's Atmosphere

Let's begin by asking what the solar atmosphere is made of. As explained in Radiation and Spectra, we can use a star's *absorption line spectrum* to determine what elements are present. It turns out that the Sun contains the same elements as Earth but *not* in the same proportions. About 73% of the Sun's mass is hydrogen, and another 25% is helium. All the other chemical elements (including those we know and love in our own bodies, such as carbon, oxygen, and nitrogen) make up only 2% of our star. The 10 most abundant gases in the Sun's visible surface layer are listed in Table 15.2. Examine that table and notice that the composition of the Sun's outer layer is very different from Earth's crust, where we live. (In our planet's crust, the three most abundant

Download for free at http://cnx.org/content/col11992/latest/

elements are oxygen, silicon, and aluminum.) Although not like our planet's, the makeup of the Sun is quite typical of stars in general.

The Abundance of Elements in the Sun

Element	Percentage by Number of Atoms	Percentage By Mass
Hydrogen	92.0	73.4
Helium	7.8	25.0
Carbon	0.02	0.20
Nitrogen	0.008	0.09
Oxygen	0.06	0.80
Neon	0.01	0.16
Magnesium	0.003	0.06
Silicon	0.004	0.09
Sulfur	0.002	0.05
Iron	0.003	0.14

Table 15.2

The fact that our Sun and the stars all have similar compositions and are made up of mostly hydrogen and helium was first shown in a brilliant thesis in 1925 by Cecilia Payne-Gaposchkin, the first woman to get a PhD in astronomy in the United States (Figure 15.3). However, the idea that the simplest light gases—hydrogen and helium—were the most abundant elements in stars was so unexpected and so shocking that she assumed her analysis of the data must be wrong. At the time, she wrote, "The enormous abundance derived for these elements in the stellar atmosphere is almost certainly not real." Even scientists sometimes find it hard to accept new ideas that do not agree with what everyone "knows" to be right.

Figure 15.3. Cecilia Payne-Gaposchkin (1900–1979). Her 1925 doctoral thesis laid the foundations for understanding the composition of the Sun and the stars. Yet, being a woman, she was not given a formal appointment at Harvard, where she worked, until 1938 and was not appointed a professor until 1956. (credit: Smithsonian Institution)

This OpenStax book is available for free at http://cnx.org/content/col11992/1.8

Before Payne-Gaposchkin's work, everyone assumed that the composition of the Sun and stars would be much like that of Earth. It was 3 years after her thesis that other studies proved beyond a doubt that the enormous abundance of hydrogen and helium in the Sun is indeed real. (And, as we will see, the composition of the Sun and the stars is much more typical of the makeup of the universe than the odd concentration of heavier elements that characterizes our planet.)

Most of the elements found in the Sun are in the form of atoms, with a small number of molecules, all in the form of gases: the Sun is so hot that no matter can survive as a liquid or a solid. In fact, the Sun is so hot that many of the atoms in it are *ionized*, that is, stripped of one or more of their electrons. This removal of electrons from their atoms means that there is a large quantity of free electrons and positively charged ions in the Sun, making it an electrically charged environment—quite different from the neutral one in which you are reading this text. (Scientists call such a hot ionized gas a **plasma**.)

In the nineteenth century, scientists observed a spectral line at 530.3 nanometers in the Sun's outer atmosphere, called the corona (a layer we will discuss in a minute.) This line had never been seen before, and so it was assumed that this line was the result of a new element found in the corona, quickly named coronium. It was not until 60 years later that astronomers discovered that this emission was in fact due to highly ionized iron—iron with 13 of its electrons stripped off. This is how we first discovered that the Sun's atmosphere had a temperature of more than a million degrees.

The Layers of the Sun beneath the Visible Surface

Figure 15.4 shows what the Sun would look like if we could see all parts of it from the center to its outer atmosphere; the terms in the figure will become familiar to you as you read on.

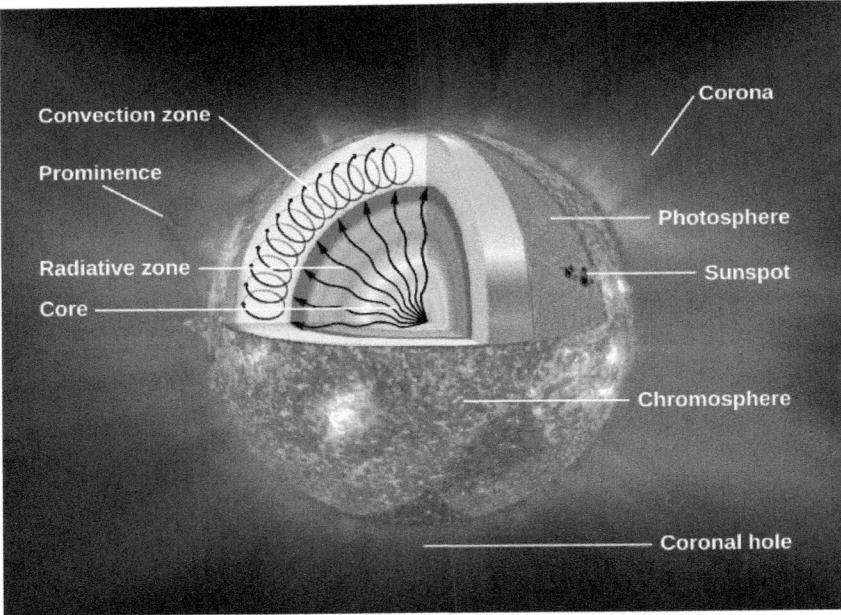

Figure 15.4. Parts of the Sun. This illustration shows the different parts of the Sun, from the hot core where the energy is generated through regions where energy is transported outward, first by radiation, then by convection, and then out through the solar atmosphere. The parts of the atmosphere are also labeled the photosphere, chromosphere, and corona. Some typical features in the atmosphere are shown, such as coronal holes and prominences. (credit: modification of work by NASA/Goddard)

Download for free at http://cnx.org/content/col11992/latest/

The Sun's layers are different from each other, and each plays a part in producing the energy that the Sun ultimately emits. We will begin with the core and work our way out through the layers. The Sun's *core* is extremely dense and is the source of all of its energy. Inside the core, nuclear energy is being released (in ways we will discuss in The Sun: A Nuclear Powerhouse). The core is approximately 20% of the size of the solar interior and is thought to have a temperature of approximately 15 million K, making it the hottest part of the Sun.

Above the core is a region known as the *radiative zone*—named for the primary mode of transporting energy across it. This region starts at about 25% of the distance to the solar surface and extends up to about 70% of the way to the surface. The light generated in the core is transported through the radiative zone very slowly, since the high density of matter in this region means a photon cannot travel too far without encountering a particle, causing it to change direction and lose some energy.

The *convective zone* is the outermost layer of the solar interior. It is a thick layer approximately 200,000 kilometers deep that transports energy from the edge of the radiative zone to the surface through giant convection cells, similar to a pot of boiling oatmeal. The plasma at the bottom of the convective zone is extremely hot, and it bubbles to the surface where it loses its heat to space. Once the plasma cools, it sinks back to the bottom of the convective zone.

Now that we have given a quick overview of the structure of the whole Sun, in this section, we will embark on a journey through the visible layers of the Sun, beginning with the photosphere—the visible surface.

The Solar Photosphere

Earth's air is generally transparent. But on a smoggy day in many cities, it can become opaque, which prevents us from seeing through it past a certain point. Something similar happens in the Sun. Its outer atmosphere is transparent, allowing us to look a short distance through it. But when we try to look through the atmosphere deeper into the Sun, our view is blocked. The **photosphere** is the layer where the Sun becomes opaque and marks the boundary past which we cannot see (Figure 15.5).

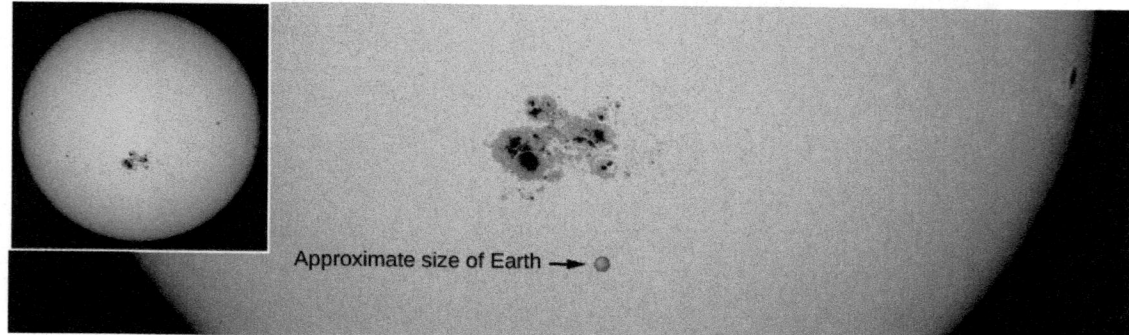

Figure 15.5. Solar Photosphere plus Sunspots. This photograph shows the photosphere—the visible surface of the Sun. Also shown is an enlarged image of a group of sunspots; the size of Earth is shown for comparison. Sunspots appear darker because they are cooler than their surroundings. The typical temperature at the center of a large sunspot is about 3800 K, whereas the photosphere has a temperature of about 5800 K. (credit: modification of work by NASA/SDO)

As we saw, the energy that emerges from the photosphere was originally generated deep inside the Sun (more on this in The Sun: A Nuclear Powerhouse). This energy is in the form of photons, which make their way slowly toward the solar surface. Outside the Sun, we can observe *only* those photons that are emitted into the solar photosphere, where the density of atoms is sufficiently low and the photons can finally escape from the Sun without colliding with another atom or ion.

This OpenStax book is available for free at http://cnx.org/content/col11992/1.8

As an analogy, imagine that you are attending a big campus rally and have found a prime spot near the center of the action. Your friend arrives late and calls you on your cell phone to ask you to join her at the edge of the crowd. You decide that friendship is worth more than a prime spot, and so you work your way out through the dense crowd to meet her. You can move only a short distance before bumping into someone, changing direction, and trying again, making your way slowly to the outside edge of the crowd. All this while, your efforts are not visible to your waiting friend at the edge. Your friend can't see you until you get very close to the edge because of all the bodies in the way. So too photons making their way through the Sun are constantly bumping into atoms, changing direction, working their way slowly outward, and becoming visible only when they reach the atmosphere of the Sun where the density of atoms is too low to block their outward progress.

Astronomers have found that the solar atmosphere changes from almost perfectly transparent to almost completely opaque in a distance of just over 400 kilometers; it is this thin region that we call the *photosphere*, a word that comes from the Greek for "light sphere." When astronomers speak of the "diameter" of the Sun, they mean the size of the region surrounded by the photosphere.

The photosphere looks sharp only from a distance. If you were falling into the Sun, you would not feel any surface but would just sense a gradual increase in the density of the gas surrounding you. It is much the same as falling through a cloud while skydiving. From far away, the cloud looks as if it has a sharp surface, but you do not feel a surface as you fall into it. (One big difference between these two scenarios, however, is temperature. The Sun is so hot that you would be vaporized long before you reached the photosphere. Skydiving in Earth's atmosphere is much safer.)

We might note that the atmosphere of the Sun is not a very dense layer compared to the air in the room where you are reading this text. At a typical point in the photosphere, the pressure is less than 10% of Earth's pressure at sea level, and the density is about one ten-thousandth of Earth's atmospheric density at sea level.

Observations with telescopes show that the photosphere has a mottled appearance, resembling grains of rice spilled on a dark tablecloth or a pot of boiling oatmeal. This structure of the photosphere is called **granulation** (see Figure 15.6). Granules, which are typically 700 to 1000 kilometers in diameter (about the width of Texas), appear as bright areas surrounded by narrow, darker (cooler) regions. The lifetime of an individual granule is only 5 to 10 minutes. Even larger are supergranules, which are about 35,000 kilometers across (about the size of two Earths) and last about 24 hours.

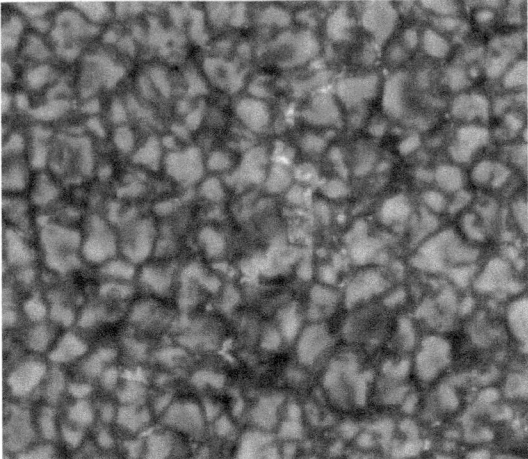 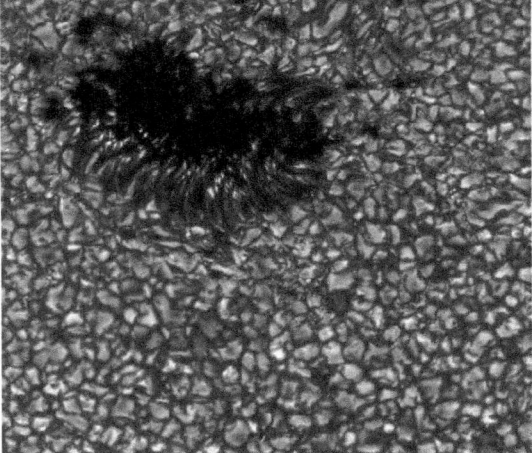

Figure 15.6. Granulation Pattern. The surface markings of the convection cells create a granulation pattern on this dramatic image (left) taken from the Japanese Hinode spacecraft. You can see the same pattern when you heat up miso soup. The right image shows an irregular-shaped sunspot and granules on the Sun's surface, seen with the Swedish Solar Telescope on August 22, 2003. (credit left: modification of work by Hinode JAXA/NASA/PPARC; credit right: ISP/SST/Oddbjorn Engvold, Jun Elin Wiik, Luc Rouppe van der Voort)

Download for free at http://cnx.org/content/col11992/latest/

The motions of the granules can be studied by examining the Doppler shifts in the spectra of gases just above them (see The Doppler Effect). The bright granules are columns of hotter gases rising at speeds of 2 to 3 kilometers per second from below the photosphere. As this rising gas reaches the photosphere, it spreads out, cools, and sinks down again into the darker regions between the granules. Measurements show that the centers of the granules are hotter than the intergranular regions by 50 to 100 K.

LINK TO LEARNING

See the "boiling" action of granulation in this 30-second time-lapse video (https://openstax.org/l/30SolarGran) from the Swedish Institute for Solar Physics.

The Chromosphere

The Sun's outer gases extend far beyond the photosphere (Figure 15.7). Because they are transparent to most visible radiation and emit only a small amount of light, these outer layers are difficult to observe. The region of the Sun's atmosphere that lies immediately above the photosphere is called the **chromosphere**. Until this century, the chromosphere was visible only when the photosphere was concealed by the Moon during a total solar eclipse (see the chapter on Earth, Moon, and Sky). In the seventeenth century, several observers described what appeared to them as a narrow red "streak" or "fringe" around the edge of the Moon during a brief instant after the Sun's photosphere had been covered. The name *chromosphere*, from the Greek for "colored sphere," was given to this red streak.

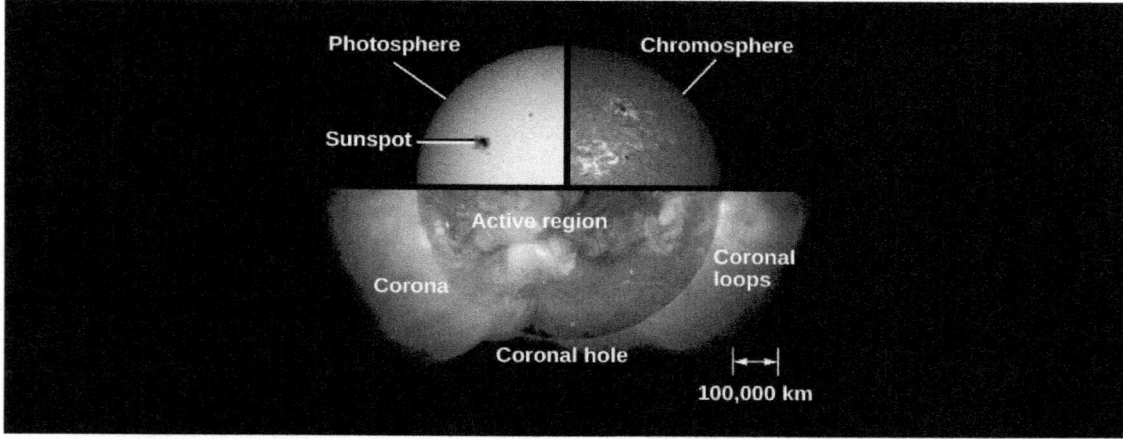

Figure 15.7. The Sun's Atmosphere. Composite image showing the three components of the solar atmosphere: the photosphere or surface of the Sun taken in ordinary light; the chromosphere, imaged in the light of the strong red spectral line of hydrogen (H-alpha); and the corona as seen with X-rays. (credit: modification of work by NASA)

Observations made during eclipses show that the chromosphere is about 2000 to 3000 kilometers thick, and its spectrum consists of bright emission lines, indicating that this layer is composed of hot gases emitting light at discrete wavelengths. The reddish color of the chromosphere arises from one of the strongest emission lines in the visible part of its spectrum—the bright red line caused by hydrogen, the element that, as we have already seen, dominates the composition of the Sun.

This OpenStax book is available for free at http://cnx.org/content/col11992/1.8

In 1868, observations of the chromospheric spectrum revealed a yellow emission line that did not correspond to any previously known element on Earth. Scientists quickly realized they had found a new element and named it *helium* (after *helios*, the Greek word for "Sun"). It took until 1895 for helium to be discovered on our planet. Today, students are probably most familiar with it as the light gas used to inflate balloons, although it turns out to be the second-most abundant element in the universe.

The temperature of the chromosphere is about 10,000 K. This means that the chromosphere is hotter than the photosphere, which should seem surprising. In all the situations we are familiar with, temperatures fall as one moves away from the source of heat, and the chromosphere is farther from the center of the Sun than the photosphere is.

The Transition Region

The increase in temperature does not stop with the chromosphere. Above it is a region in the solar atmosphere where the temperature changes from 10,000 K (typical of the chromosphere) to nearly a million degrees. The hottest part of the solar atmosphere, which has a temperature of a million degrees or more, is called the **corona**. Appropriately, the part of the Sun where the rapid temperature rise occurs is called the **transition region**. It is probably only a few tens of kilometers thick. Figure 15.8 summarizes how the temperature of the solar atmosphere changes from the photosphere outward.

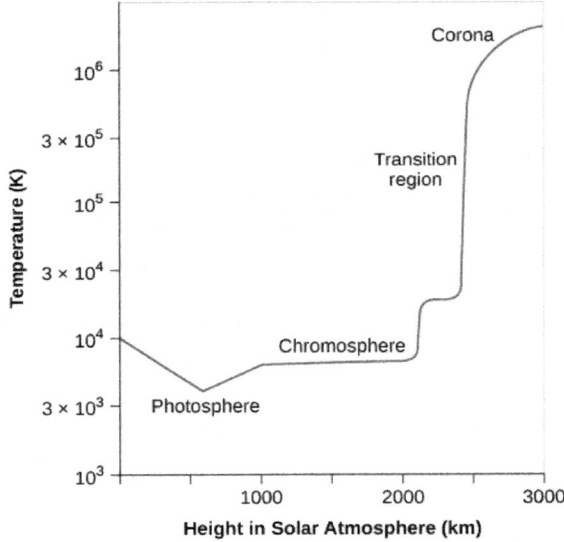

Figure 15.8. Temperatures in the Solar Atmosphere. On this graph, temperature is shown increasing upward, and height above the photosphere is shown increasing to the right. Note the very rapid increase in temperature over a very short distance in the transition region between the chromosphere and the corona.

In 2013, NASA launched the Interface Region Imaging Spectrograph (IRIS) to study the transition region to understand better how and why this sharp temperature increase occurs. IRIS is the first space mission that is able to obtain high spatial resolution images of the different features produced over this wide temperature range and to see how they change with time and location (Figure 15.9).

Download for free at http://cnx.org/content/col11992/latest/

Figure 15.9. Portion of the Transition Region. This image shows a giant ribbon of relatively cool gas threading through the lower portion of the hot corona. This ribbon (the technical term is filament) is made up of many individual threads. Time-lapse movies of this filament showed that it gradually heated as it moved through the corona. Scientists study events like this in order to try to understand what heats the chromosphere and corona to high temperatures. The "whiskers" at the edge of the Sun are spicules, jets of gas that shoot material up from the Sun's surface and disappear after only a few minutes. This single image gives a hint of just how complicated it is to construct a model of the all the different structures and heating mechanisms in the solar atmosphere. (credit: JAXA/NASA/Hinode)

Figure 15.4 and the red graph in Figure 15.8 make the Sun seem rather like an onion, with smooth spherical shells, each one with a different temperature. For a long time, astronomers did indeed think of the Sun this way. However, we now know that while this idea of layers—photosphere, chromosphere, transition region, corona—describes the big picture fairly well, the Sun's atmosphere is really more complicated, with hot and cool regions intermixed. For example, clouds of carbon monoxide gas with temperatures colder than 4000 K have now been found at the same height above the photosphere as the much hotter gas of the chromosphere.

The Corona

The outermost part of the Sun's atmosphere is called the *corona*. Like the chromosphere, the corona was first observed during total eclipses (Figure 15.10). Unlike the chromosphere, the corona has been known for many centuries: it was referred to by the Roman historian Plutarch and was discussed in some detail by Kepler.

The corona extends millions of kilometers above the photosphere and emits about half as much light as the full moon. The reason we don't see this light until an eclipse occurs is the overpowering brilliance of the photosphere. Just as bright city lights make it difficult to see faint starlight, so too does the intense light from the photosphere hide the faint light from the corona. While the best time to see the corona from Earth is during a total solar eclipse, it can be observed easily from orbiting spacecraft. Its brighter parts can now be photographed with a special instrument—a coronagraph—that removes the Sun's glare from the image with an occulting disk (a circular piece of material held so it is just in front of the Sun).

This OpenStax book is available for free at http://cnx.org/content/col11992/1.8

Figure 15.10. Coronagraph. This image of the Sun was taken March 2, 2016. The larger dark circle in the center is the disk the blocks the Sun's glare, allowing us to see the corona. The smaller inner circle is where the Sun would be if it were visible in this image. (credit: modification of work by NASA/SOHO)

Studies of its spectrum show the corona to be very low in density. At the bottom of the corona, there are only about 10^9 atoms per cubic centimeter, compared with about 10^{16} atoms per cubic centimeter in the upper photosphere and 10^{19} molecules per cubic centimeter at sea level in Earth's atmosphere. The corona thins out very rapidly at greater heights, where it corresponds to a high vacuum by Earth laboratory standards. The corona extends so far into space—far past Earth—that here on our planet, we are technically living in the Sun's atmosphere.

The Solar Wind

One of the most remarkable discoveries about the Sun's atmosphere is that it produces a stream of charged particles (mainly protons and electrons) that we call the **solar wind**. These particles flow outward from the Sun into the solar system at a speed of about 400 kilometers per second (almost 1 million miles per hour)! The solar wind exists because the gases in the corona are so hot and moving so rapidly that they cannot be held back by solar gravity. (This wind was actually discovered by its effects on the charged tails of comets; in a sense, we can see the comet tails blow in the solar breeze the way wind socks at an airport or curtains in an open window flutter on Earth.)

Although the solar wind material is very, very rarified (i.e., *extremely* low density), the Sun has an enormous surface area. Astronomers estimate that the Sun is losing about 10 million tons of material each year through this wind. While this amount of lost mass seems large by Earth standards, it is completely insignificant for the Sun.

From where in the Sun does the solar wind emerge? In visible photographs, the solar corona appears fairly uniform and smooth. X-ray and extreme ultraviolet pictures, however, show that the corona has loops, plumes, and both bright and dark regions. Large dark regions of the corona that are relatively cool and quiet are called **coronal holes** (Figure 15.11). In these regions, magnetic field lines stretch far out into space away from the Sun, rather than looping back to the surface. The solar wind comes predominantly from coronal holes, where gas can stream away from the Sun into space unhindered by magnetic fields. Hot coronal gas, on the other hand, is present mainly where magnetic fields have trapped and concentrated it.

Download for free at http://cnx.org/content/col11992/latest/

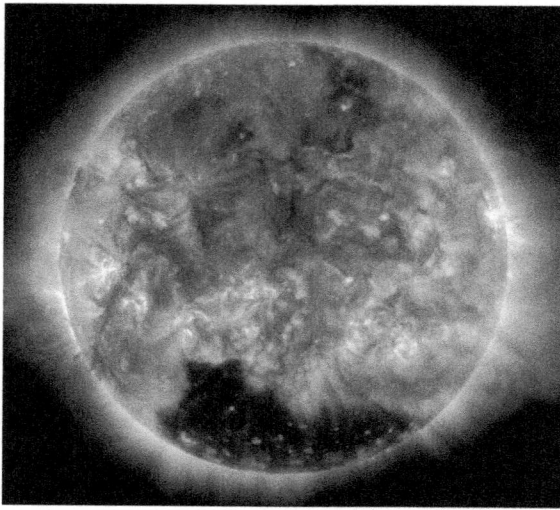

Figure 15.11. Coronal Hole. The dark area visible near the Sun's south pole on this Solar Dynamics Observer spacecraft image is a coronal hole. (credit: modification of work by NASA/SDO)

At the surface of Earth, we are protected to some degree from the solar wind by our atmosphere and Earth's magnetic field (see Earth as a Planet). However, the magnetic field lines come into Earth at the north and south magnetic poles. Here, charged particles accelerated by the solar wind can follow the field down into our atmosphere. As the particles strike molecules of air, they cause them to glow, producing beautiful curtains of light called the **auroras**, or the northern and southern lights (Figure 15.12).

Figure 15.12. Aurora. The colorful glow in the sky results from charged particles in a solar wind interacting with Earth's magnetic fields. The stunning display captured here occurred over Jokulsarlon Lake in Iceland in 2013. (credit: Moyan Brenn)

This OpenStax book is available for free at http://cnx.org/content/col11992/1.8

LINK TO LEARNING

This NASA video (https://openstax.org/l/30Aurora) explains and demonstrates the nature of the auroras and their relationship to Earth's magnetic field.

15.2 THE SOLAR CYCLE

Learning Objectives

By the end of this section, you will be able to:

> Describe the sunspot cycle and, more generally, the solar cycle
> Explain how magnetism is the source of solar activity

Before the invention of the telescope, the Sun was thought to be an unchanging and perfect sphere. We now know that the Sun is in a perpetual state of change: its surface is a seething, bubbling cauldron of hot gas. Areas that are darker and cooler than the rest of the surface come and go. Vast plumes of gas erupt into the chromosphere and corona. Occasionally, there are even giant explosions on the Sun that send enormous streamers of charged particles and energy hurtling toward Earth. When they arrive, these can cause power outages and other serious effects on our planet.

Sunspots

The first evidence that the Sun changes came from studies of **sunspots**, which are large, dark features seen on the surface of the Sun caused by increased magnetic activity. They look darker because the spots are typically at a temperature of about 3800 K, whereas the bright regions that surround them are at about 5800 K (Figure 15.13). Occasionally, these spots are large enough to be visible to the unaided eye, and we have records going back over a thousand years from observers who noticed them when haze or mist reduced the Sun's intensity. (We emphasize what your parents have surely told you: looking at the Sun for even a brief time can cause permanent eye damage. This is the one area of astronomy where we don't encourage you to do your own observing without getting careful instructions or filters from your instructor.)

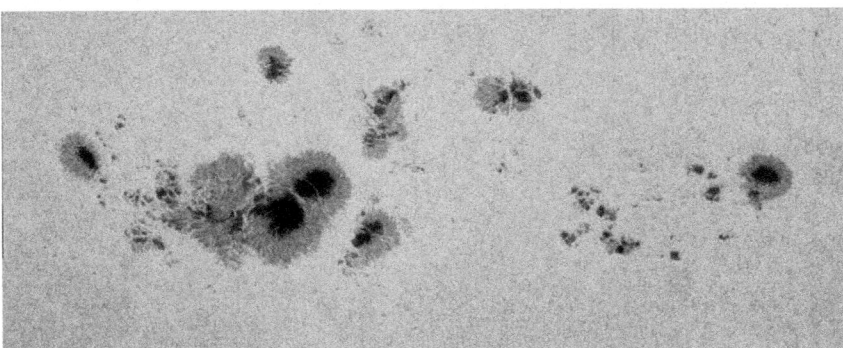

Figure 15.13. Sunspots. This image of sunspots, cooler and thus darker regions on the Sun, was taken in July 2012. You can see the dark, central region of each sunspot (called the umbra) surrounded by a less dark region (the penumbra). The largest spot shown here is about 11 Earths wide. Although sunspots appear dark when seen next to the hotter gases of the photosphere, an average sunspot, cut out of the solar surface and left standing in the night sky, would be about as bright as the full moon. The mottled appearance of the Sun's surface is granulation. (credit: NASA Goddard Space Flight Center, Alan Friedman)

Download for free at http://cnx.org/content/col11992/latest/

While we understand that sunspots look darker because they are cooler, they are nevertheless hotter than the surfaces of many stars. If they could be removed from the Sun, they would shine brightly. They appear dark only in contrast with the hotter, brighter photosphere around them.

Individual sunspots come and go, with lifetimes that range from a few hours to a few months. If a spot lasts and develops, it usually consists of two parts: an inner darker core, the *umbra*, and a surrounding less dark region, the *penumbra*. Many spots become much larger than Earth, and a few, like the largest one shown in Figure 15.13, have reached diameters over 140,000 kilometers. Frequently, spots occur in groups of 2 to 20 or more. The largest groups are very complex and may have over 100 spots. Like storms on Earth, sunspots are not fixed in position, but they drift slowly compared with the Sun's rotation.

By recording the apparent motions of the sunspots as the turning Sun carried them across its disk (Figure 15.14), Galileo, in 1612, demonstrated that the Sun rotates on its axis with a rotation period of approximately 1 month. Our star turns in a west-to-east direction, like the orbital motions of the planets. The Sun, however, is a gas and does not have to rotate rigidly, the way a solid body like Earth does. Modern observations show that the speed of rotation of the Sun varies according to latitude, that is, it's different as you go north or south of the Sun's equator. The rotation period is about 25 days at the equator, 28 days at latitude 40°, and 36 days at latitude 80°. We call this behavior **differential rotation**.

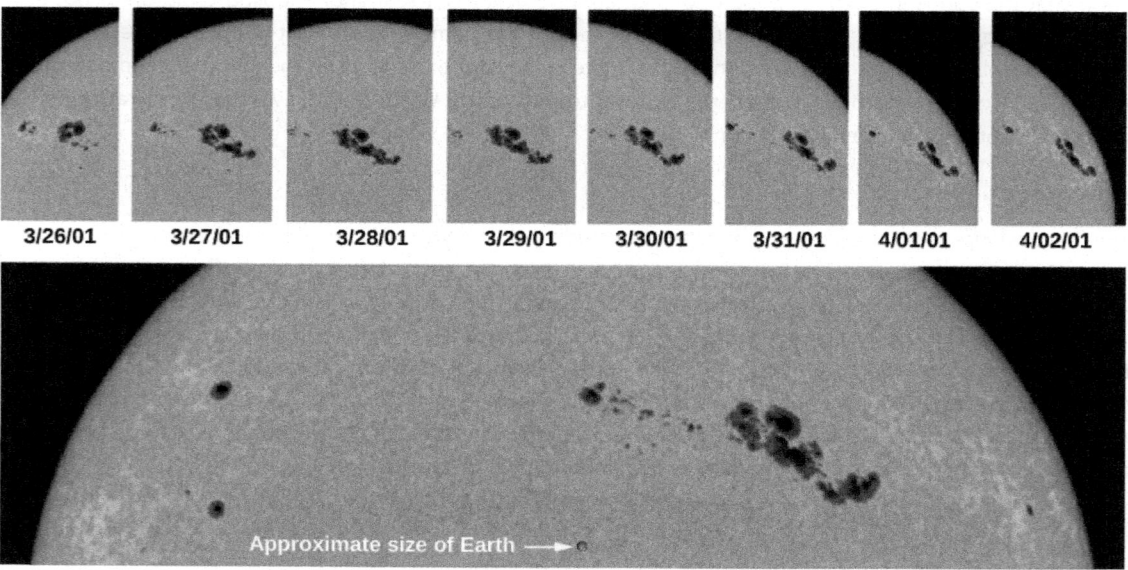

Figure 15.14. Sunspots Rotate Across Sun's Surface. This sequence of photographs of the Sun's surface tracks the movement of sunspots across the visible hemisphere of the Sun. On March 30, 2001, this group of sunspots extended across an area about 13 times the diameter of Earth. This region produced many flares and coronal mass ejections. (credit: modification of work by SOHO/NASA/ESA)

The Sunspot Cycle

Between 1826 and 1850, Heinrich Schwabe, a German pharmacist and amateur astronomer, kept daily records of the number of sunspots. What he was really looking for was a planet inside the orbit of Mercury, which he hoped to find by observing its dark silhouette as it passed between the Sun and Earth. He failed to find the hoped-for planet, but his diligence paid off with an even-more important discovery: the **sunspot cycle**. He found that the number of sunspots varied systematically, in cycles about a decade long.

This OpenStax book is available for free at http://cnx.org/content/col11992/1.8

What Schwabe observed was that, although individual spots are short lived, the total number visible on the Sun at any one time was likely to be very much greater at certain times—the periods of *sunspot maximum*—than at other times—the periods of *sunspot minimum*. We now know that sunspot maxima occur at an *average* interval of 11 years, but the intervals between successive maxima have ranged from as short as 9 years to as long as 14 years. During sunspot maxima, more than 100 spots can often be seen at once. Even then, less than one-half of one percent of the Sun's surface is covered by spots (Figure 15.22). During sunspot minima, sometimes no spots are visible. The Sun's activity reached its most recent maximum in 2014.

LINK TO LEARNING

Watch this brief video (https://openstax.org/l/30SolarCyc) from NASA's Goddard Space Flight Center that explains the sunspot cycle.

Magnetism and the Solar Cycle

Now that we have discussed the Sun's activity cycle, you might be asking, "Why does the Sun change in such a regular way?" Astronomers now understand that it is the Sun's changing magnetic field that drives solar activity.

The solar magnetic field is measured using a property of atoms called the *Zeeman effect*. Recall from Radiation and Spectra that an atom has many energy levels and that spectral lines are formed when electrons shift from one level to another. If each energy level is precisely defined, then the difference between them is also quite precise. As an electron changes levels, the result is a sharp, narrow spectral line (either an absorption or emission line, depending on whether the electron's energy increases or decreases in the transition).

In the presence of a strong magnetic field, however, each energy level is separated into several levels very close to one another. The separation of the levels is proportional to the strength of the field. As a result, spectral lines formed in the presence of a magnetic field are not single lines but a series of very closely spaced lines corresponding to the subdivisions of the atomic energy levels. This splitting of lines in the presence of a magnetic field is what we call the Zeeman effect (after the Dutch scientist who first discovered it in 1896).

Measurements of the Zeeman effect in the spectra of the light from sunspot regions show them to have strong magnetic fields (Figure 15.15). Bear in mind that magnets always have a north pole and a south pole. Whenever sunspots are observed in pairs, or in groups containing two principal spots, one of the spots usually has the magnetic polarity of a north-seeking magnetic pole and the other has the opposite polarity. Moreover, during a given cycle, the leading spots of pairs (or leading principle spots of groups) in the Northern Hemisphere all tend to have the same polarity, whereas those in the Southern Hemisphere all tend to have the opposite polarity.

Download for free at http://cnx.org/content/col11992/latest/

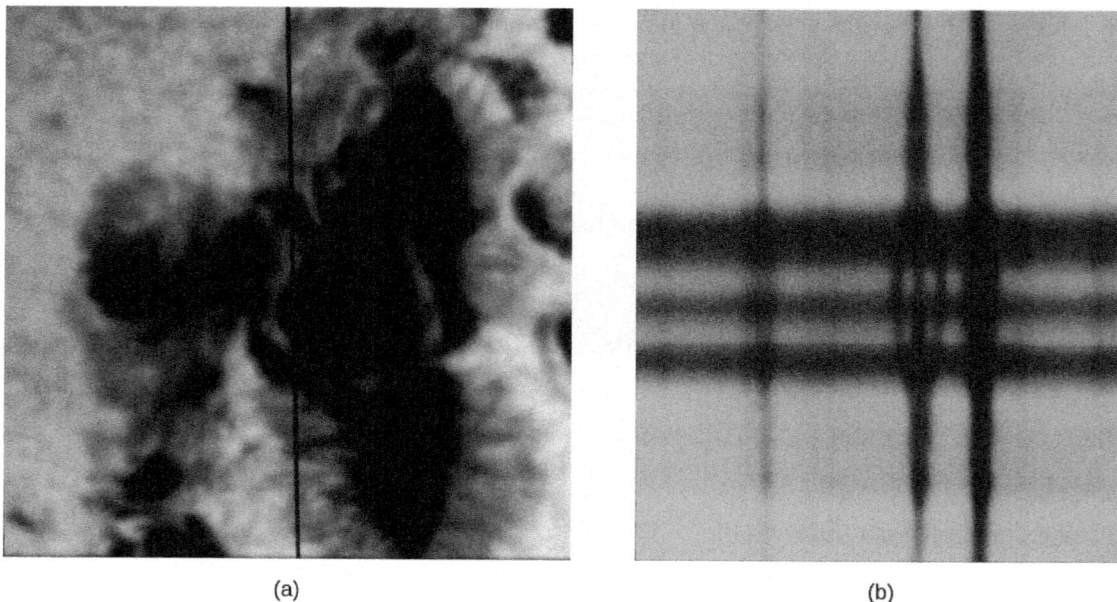

(a)   (b)

Figure 15.15. Zeeman Effect. These photographs show how magnetic fields in sunspots are measured by means of the Zeeman effect. (left) The vertical black line indicates the position of the spectrograph slit through which light is passed to obtain the spectrum in (right). (credit: modification of work by NSO/AURA/NSF)

During the next sunspot cycle, however, the polarity of the leading spots is reversed in each hemisphere. For example, if during one cycle, the leading spots in the Northern Hemisphere all had the polarity of a north-seeking pole, then the leading spots in the Southern Hemisphere would have the polarity of a south-seeking pole. During the next cycle, the leading spots in the Northern Hemisphere would have south-seeking polarity, whereas those in the Southern Hemisphere would have north-seeking polarity. Therefore, strictly speaking, the sunspot cycle does not repeat itself in regard to magnetic polarity until two 11-year cycles have passed. A visual representation of the Sun's magnetic fields, called a *magnetogram*, can be used to see the relationship between sunspots and the Sun's magnetic field (Figure 15.16).

This OpenStax book is available for free at http://cnx.org/content/col11992/1.8

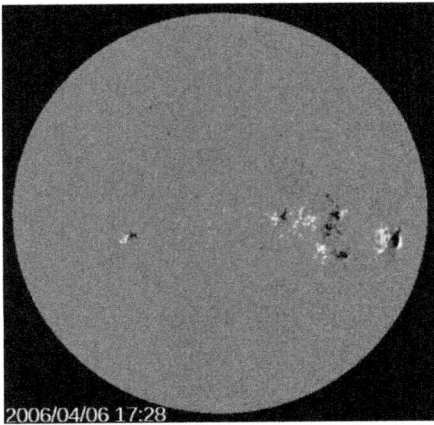
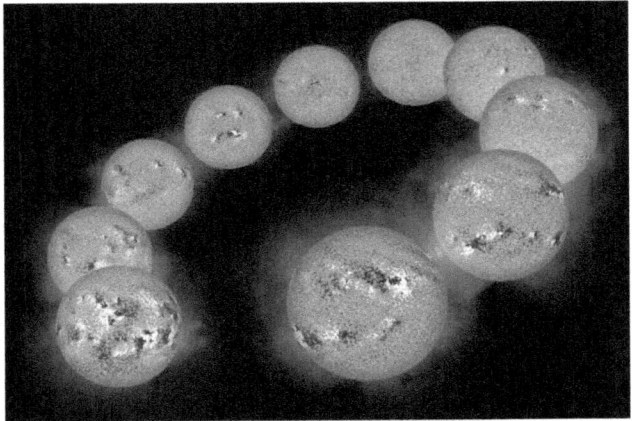

2006/04/06 17:28

Figure 15.16. Magnetogram and Solar Cycle. In the image on the left, called a magnetogram, we see the magnetic polarity of sunspots. The black areas are where the magnetism is pointing toward the Sun's core, whereas the white regions are where it is pointing away from the core, toward us. This dramatic sequence on the right shows the activity cycle of the Sun. The 10 maps of the magnetic field on the surface of the Sun span a period of 7.5 years. The two magnetic polarities (N and S) of the magnetic field are shown against a blue disk as dark blue to black (N) and as light blue to white (S). The earliest image, taken on January 8, 1992, is at the lower left and was taken just after solar maximum. Each image, from left to right around the arc, was taken one-half to one year after the preceding one. The last image was taken on July 25, 1999, as the Sun was approaching the next solar maximum. Note a few striking patterns in the magnetic maps: the direction from white to black polarity in the Southern Hemisphere is opposite from that in the Northern Hemisphere. (credit left: modification of work by NASA/SDO; credit right: modification of work by NASA/SOHO)

Why is the Sun such a strong and complicated magnet? Astronomers have found that it is the Sun's *dynamo* that generates the magnetic field. A dynamo is a machine that converts kinetic energy (i.e., the energy of motion) into electricity. On Earth, dynamos are found in power plants where, for example, the energy from wind or flowing water is used to cause turbines to rotate. In the Sun, the source of kinetic energy is the churning of turbulent layers of ionized gas within the Sun's interior that we mentioned earlier. These generate electric currents—moving electrons—which in turn generate magnetic fields.

Most solar researchers agree that the solar dynamo is located in the convection zone or in the interface layer between the convection zone and the radiative zone below it. As the magnetic fields from the Sun's dynamo interact, they break, reconnect, and rise through the Sun's surface.

We should say that, although we have good observations that show us *how* the Sun changes during each solar cycle, it is still very difficult to build physical models of something as complicated as the Sun that can account satisfactorily for *why* it changes. Researchers have not yet developed a generally accepted model that describes in detail the physical processes that control the solar cycle. Calculations do show that differential rotation (the idea that the Sun rotates at different rates at different latitudes) and convection just below the solar surface can twist and distort the magnetic fields. This causes them to grow and then decay, regenerating with opposite polarity approximately every 11 years. The calculations also show that as the fields grow stronger near solar maximum, they flow from the interior of the Sun toward its surface in the form of loops. When a large loop emerges from the solar surface, it creates regions of sunspot activity (Figure 15.17).

Download for free at http://cnx.org/content/col11992/latest/

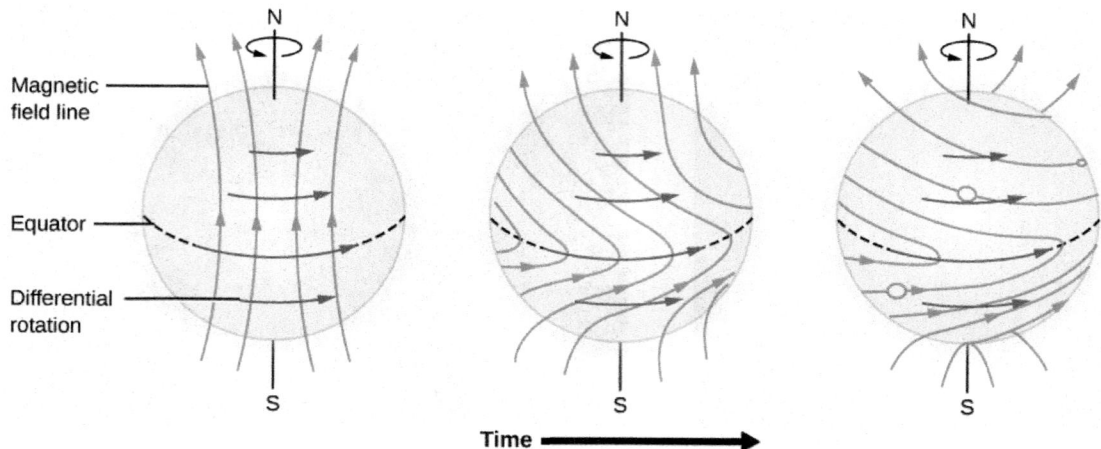

Figure 15.17. Magnetic Field Lines Wind Up. Because the Sun spins faster at the equator than near the poles, the magnetic fields in the Sun tend to wind up as shown, and after a while make loops. This is an idealized diagram; the real situation is much more complex.

This idea of magnetic loops offers a natural explanation of why the leading and trailing sunspots in an active region have opposite polarity. The leading sunspot coincides with one end of the loop and the trailing spot with the other end. Magnetic fields also hold the key to explaining why sunspots are cooler and darker than the regions without strong magnetic fields. The forces produced by the magnetic field resist the motions of the bubbling columns of rising hot gases. Since these columns carry most of the heat from inside the Sun to the surface by means of convection, and strong magnetic fields inhibit this convection, the surface of the Sun is allowed to cool. As a result, these regions are seen as darker, cooler sunspots.

Beyond this general picture, researchers are still trying to determine why the magnetic fields are as large as they are, why the polarity of the field in each hemisphere flips from one cycle to the next, why the length of the solar cycle can vary from one cycle to the next, and why events like the Maunder Minimum occur.

LINK TO LEARNING

In this video (https://openstax.org/l/30MagField) solar scientist Holly Gilbert discusses the Sun's magnetic field.

15.3 | SOLAR ACTIVITY ABOVE THE PHOTOSPHERE

Learning Objectives

By the end of this section, you will be able to:

> Describe the various ways in which the solar activity cycle manifests itself, including flares, coronal mass ejections, prominences, and plages

Sunspots are not the only features that vary during a solar cycle. There are dramatic changes in the chromosphere and corona as well. To see what happens in the chromosphere, we must observe the emission

This OpenStax book is available for free at http://cnx.org/content/col11992/1.8

lines from elements such as hydrogen and calcium, which emit useful spectral lines at the temperatures in that layer. The hot corona, on the other hand, can be studied by observations of X-rays and of extreme ultraviolet and other wavelengths at high energies.

Plages and Prominences

As we saw, emission lines of hydrogen and calcium are produced in the hot gases of the chromosphere. Astronomers routinely photograph the Sun through filters that transmit light only at the wavelengths that correspond to these emission lines. Pictures taken through these special filters show bright "clouds" in the chromosphere around sunspots; these bright regions are known as **plages** (Figure 15.18). These are regions within the chromosphere that have higher temperature and density than their surroundings. The plages actually contain all of the elements in the Sun, not just hydrogen and calcium. It just happens that the spectral lines of hydrogen and calcium produced by these clouds are bright and easy to observe.

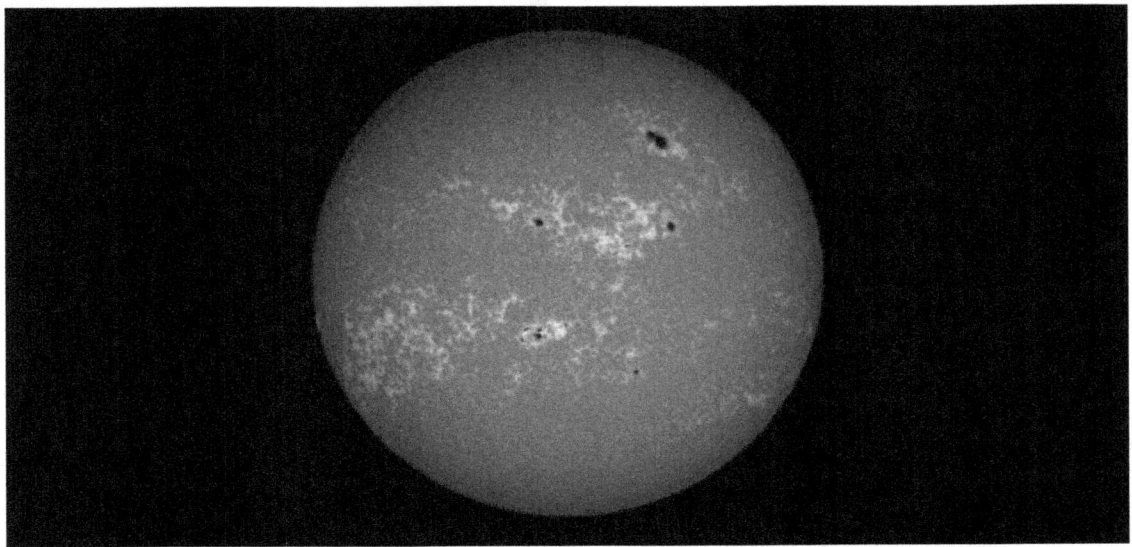

Figure 15.18. Plages on the Sun. This image of the Sun was taken with a filter that transmits only the light of the spectral line produced by singly ionized calcium. The bright cloud-like regions are the plages. (credit: modification of work by NASA)

Moving higher into the Sun's atmosphere, we come to the spectacular phenomena called **prominences** (Figure 15.19), which usually originate near sunspots. Eclipse observers often see prominences as red features rising above the eclipsed Sun and reaching high into the corona. Some, the *quiescent* prominences, are graceful loops of plasma (ionized gas) that can remain nearly stable for many hours or even days. The relatively rare *eruptive* prominences appear to send matter upward into the corona at high speeds, and the most active *surge* prominences may move as fast as 1300 kilometers per second (almost 3 million miles per hour). Some eruptive prominences have reached heights of more than 1 million kilometers above the photosphere; Earth would be completely lost inside one of those awesome displays (Figure 15.19).

Download for free at http://cnx.org/content/col11992/latest/

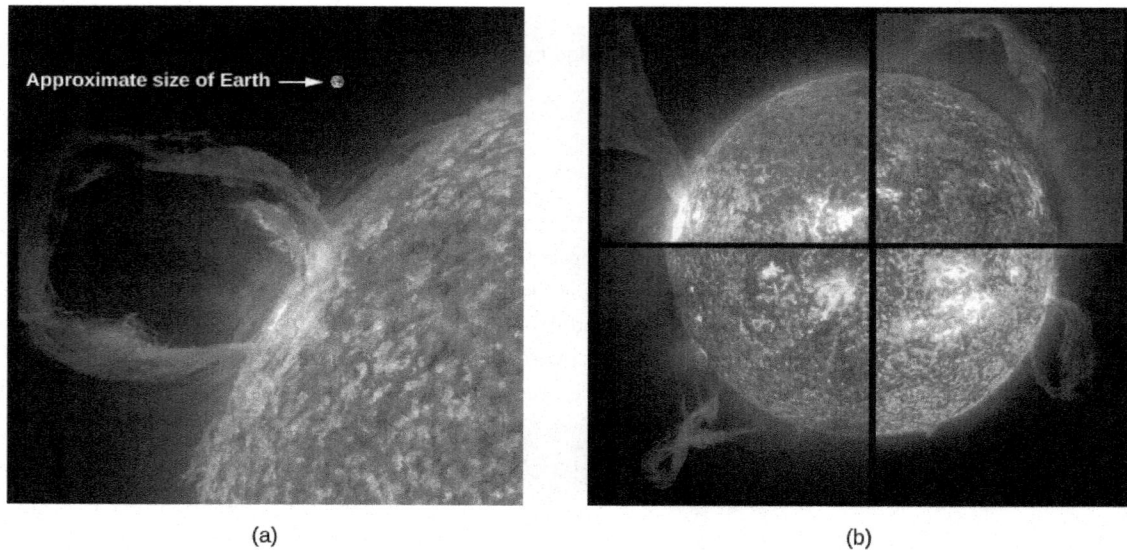

(a)   (b)

Figure 15.19. Prominences. (a) This image of an eruptive prominence was taken in the light of singly ionized helium in the extreme ultraviolet part of the spectrum. The prominence is a particularly large one. An image of Earth is shown at the same scale for comparison. (b) A prominence is a huge cloud of relatively cool (about 60,000 K in this case), fairly dense gas suspended in the much hotter corona. These pictures, taken in ultraviolet, are color coded so that white corresponds to the hottest temperatures and dark red to cooler ones. The four images were taken, moving clockwise from the upper left, on May 15, 2001; March 28, 2000; January 18, 2000; and February 2, 2001. (credit a: modification of work by NASA/SOHO; credit b: modification of work by NASA/SDO)

Flares and Coronal Mass Ejections

The most violent event on the surface of the Sun is a rapid eruption called a **solar flare** (Figure 15.20). A typical flare lasts for 5 to 10 minutes and releases a total amount of energy equivalent to that of perhaps a million hydrogen bombs. The largest flares last for several hours and emit enough energy to power the entire United States at its current rate of electrical consumption for 100,000 years. Near sunspot maximum, small flares occur several times per day, and major ones may occur every few weeks.

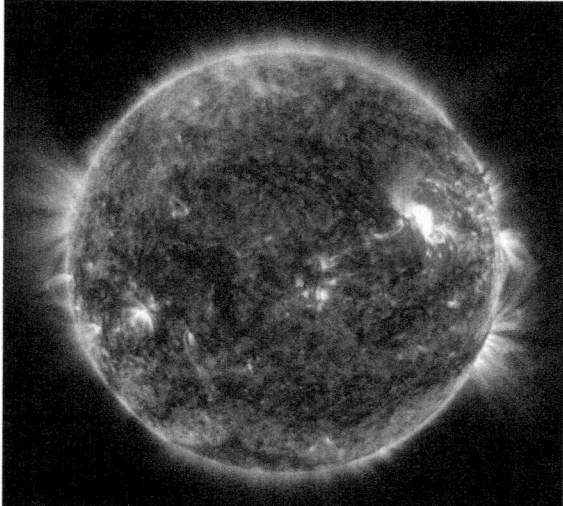

Figure 15.20. Solar Flare. The bright white area seen on the right side of the Sun in this image from the Solar Dynamics Observer spacecraft is a solar flare that was observed on June 25, 2015. (credit: NASA/SDO)

This OpenStax book is available for free at http://cnx.org/content/col11992/1.8

Flares, like the one shown in Figure 15.21, are often observed in the red light of hydrogen, but the visible emission is only a tiny fraction of the energy released when a solar flare explodes. At the moment of the explosion, the matter associated with the flare is heated to temperatures as high as 10 million K. At such high temperatures, a flood of X-ray and ultraviolet radiation is emitted.

Flares seem to occur when magnetic fields pointing in opposite directions release energy by interacting with and destroying each other—much as a stretched rubber band releases energy when it breaks.

What is different about flares is that their magnetic interactions cover a large volume in the solar corona and release a tremendous amount of electromagnetic radiation. In some cases, immense quantities of coronal material—mainly protons and electrons—may also be ejected at high speeds (500–1000 kilometers per second) into interplanetary space. Such a **coronal mass ejection (CME)** can affect Earth in a number of ways (which we will discuss in the section on space weather).

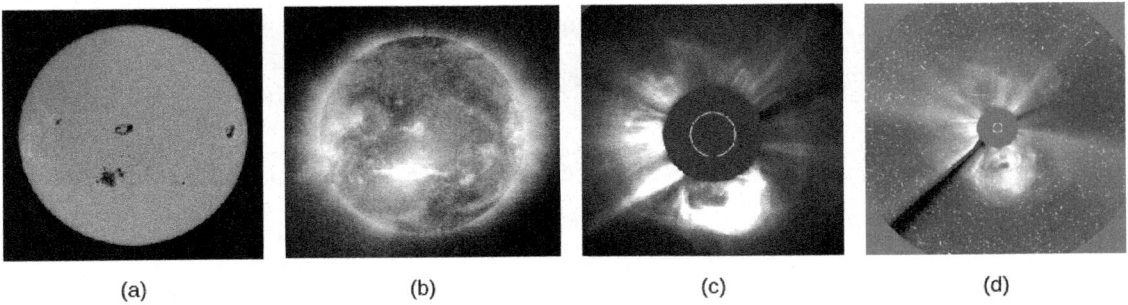

(a)　　　　(b)　　　　(c)　　　　(d)

Figure 15.21. Flare and Coronal Mass Ejection. This sequence of four images shows the evolution over time of a giant eruption on the Sun. (a) The event began at the location of a sunspot group, and (b) a flare is seen in far-ultraviolet light. (c) Fourteen hours later, a CME is seen blasting out into space. (d) Three hours later, this CME has expanded to form a giant cloud of particles escaping from the Sun and is beginning the journey out into the solar system. The white circle in (c) and (d) shows the diameter of the solar photosphere. The larger dark area shows where light from the Sun has been blocked out by a specially designed instrument to make it possible to see the faint emission from the corona. (credit a, b, c, d: modification of work by SOHO/EIT, SOHO/LASCO, SOHO/MDI (ESA & NASA))

LINK TO LEARNING

See a coronal mass ejection (https://openstax.org/l/30CorMaEj) recorded by the Solar Dynamics Observatory.

Active Regions

To bring the discussion of the last two sections together, astronomers now realize that sunspots, flares, and bright regions in the chromosphere and corona tend to occur together on the Sun in time and space. That is, they all tend to have similar longitudes and latitudes, but they are located at different heights in the atmosphere. Because they all occur together, they vary with the sunspot cycle.

Download for free at http://cnx.org/content/col11992/latest/

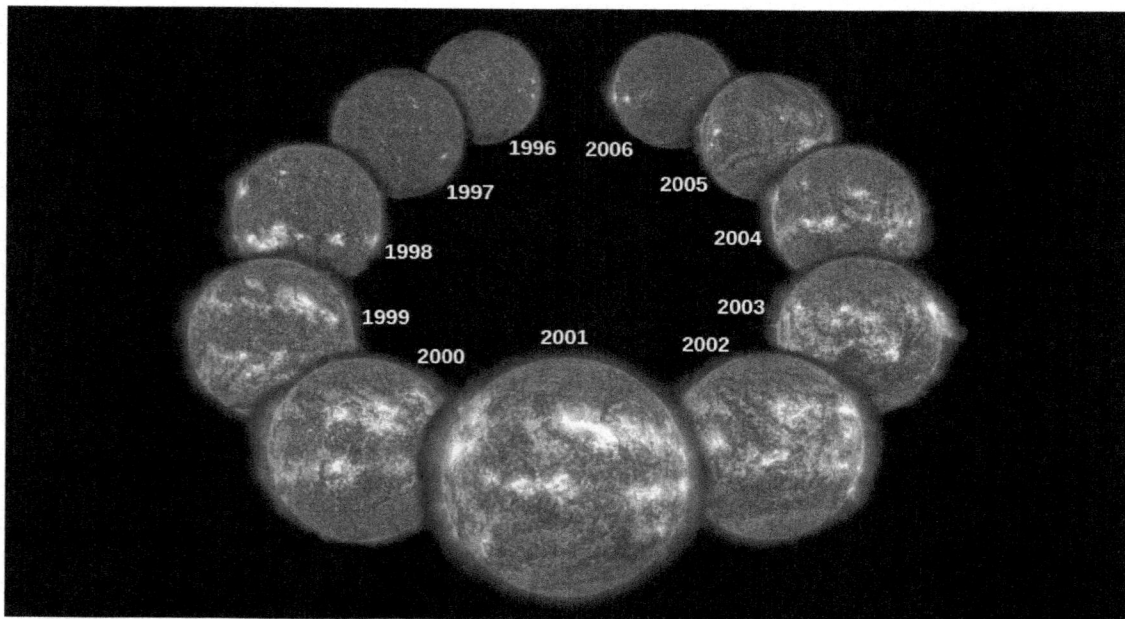

Figure 15.22. Solar Cycle. This dramatic sequence of images taken from the SOHO satellite over a period of 11 years shows how active regions change during the solar cycle. The images were taken in the ultraviolet region of the spectrum and show that active regions on the Sun increase and decrease during the cycle. Sunspots are located in the cooler photosphere, beneath the hot gases shown in this image, and vary in phase with the emission from these hot gases—more sunspots and more emission from hot gases occur together. (credit: modification of work by ESA/NASA/SOHO)

For example, flares are more likely to occur near sunspot maximum, and the corona is much more conspicuous at that time (see Figure 15.22). A place on the Sun where a number of these phenomena are seen is called an **active region** (Figure 15.23). As you might deduce from our earlier discussion, active regions are always associated with strong magnetic fields.

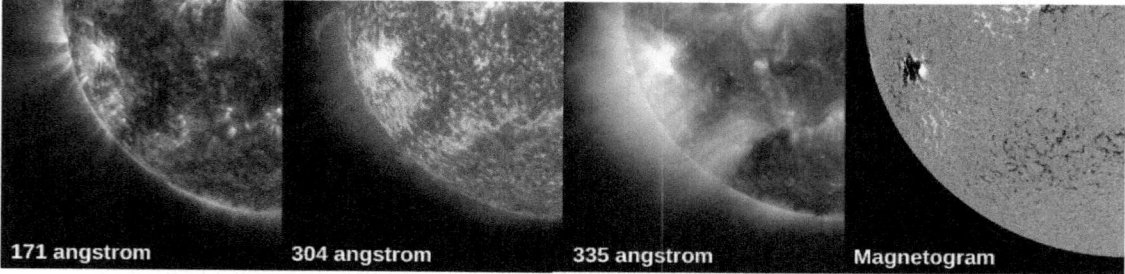

Figure 15.23. Solar Active Region Observed at Different Heights in the Sun's Atmosphere. These four images of a solar flare on October 22, 2012, show from the left: light from the Sun at a wavelength of 171 angstroms, which shows the structure of loops of solar material in the corona; ultraviolet at 304 angstroms, which shows light from the region of the Sun's atmosphere where flares originate; light at 335 angstroms, which highlights radiation from active regions in the corona; a magnetogram, which shows magnetically active regions on the Sun. Note how these different types of activity all occur above a sunspot region with a strong magnetic field. (credit: modification of work by NASA/SDO/Goddard)

15.4 SPACE WEATHER

Learning Objectives

By the end of this section, you will be able to:

This OpenStax book is available for free at http://cnx.org/content/col11992/1.8

Download for free at http://cnx.org/content/col11992/latest/

> » Explain what space weather is and how it affects Earth

In the previous sections, we have seen that some of the particles coming off the Sun—either steadily as in the solar wind or in great bursts like CMEs—will reach Earth and its *magnetosphere* (the zone of magnetic influence that surrounds our planet). As if scientists did not have enough trouble trying to predict weather on Earth, this means that they are now facing the challenge of predicting the effects of solar storms on Earth. This field of research is called *space weather*; when that weather turns stormy, our technology turns out to be at risk.

With thousands of satellites in orbit, astronauts taking up long-term residence in the International Space Station, millions of people using cell phones, GPS, and wireless communication, and nearly everyone relying on the availability of dependable electrical power, governments are now making major investments in trying to learn how to predict when solar storms will occur and how strongly they will affect Earth.

Some History

What we now study as space weather was first recognized (though not yet understood) in 1859, in what is now known as the Carrington Event. In early September of that year, two amateur astronomers, including Richard Carrington in England, independently observed a solar flare. This was followed a day or two later by a significant solar storm reaching the region of Earth's magnetic field, which was soon overloaded with charged particles (see Earth as a Planet).

As a result, aurora activity was intense and the northern lights were visible well beyond their normal locations near the poles—as far south as Hawaii and the Caribbean. The glowing lights in the sky were so intense that some people reported getting up in the middle of the night, thinking it must be daylight.

The 1859 solar storm happened at a time when a new technology was beginning to tie people in the United States and some other countries together: the telegraph system. This was a machine and network for sending messages in code through overhead electrical wires (a bit like a very early version of the internet). The charged particles that overwhelmed Earth's magnetic field descended toward our planet's surface and affected the wires of the telegraph system. Sparks were seen coming out of exposed wires and out of the telegraph machines in the system's offices.

The observation of the bright flare that preceded these effects on Earth led to scientific speculation that a connection existed between solar activity and impacts on Earth—this was the beginning of our understanding of what today we call space weather.

LINK TO LEARNING

Watch NASA scientists answer some questions (https://openstax.org/l/30SpcWeath) about space weather, and discuss (https://openstax.org/l/30SpcWeath2) some effects it can have in space and on Earth.

Sources of Space Weather

Three solar phenomena— coronal holes, solar flares, and CMEs—account for most of the space weather we experience. Coronal holes allow the solar wind to flow freely away from the Sun, unhindered by solar magnetic fields. When the solar wind reaches Earth, as we saw, it causes Earth's magnetosphere to contract and then expand after the solar wind passes by. These changes can cause (usually mild) electromagnetic disturbances on Earth.

Download for free at http://cnx.org/content/col11992/latest/

More serious are solar flares, which shower the upper atmosphere of Earth with X-rays, energetic particles, and intense ultraviolet radiation. The X-rays and ultraviolet radiation can ionize atoms in Earth's upper atmosphere, and the freed electrons can build up a charge on the surface of a spacecraft. When this static charge discharges, it can damage the electronics in the spacecraft—just as you can receive a shock when you walk across a carpet in your stocking feet in a dry climate and then touch a light switch or some other metal object.

Most disruptive are coronal mass ejections. A CME is an erupting bubble of tens of millions of tons of gas blown away from the Sun into space. When this bubble reaches Earth a few days after leaving the Sun, it heats the ionosphere, which expands and reaches farther into space. As a consequence, friction between the atmosphere and spacecraft increases, dragging satellites to lower altitudes.

At the time of a particularly strong flare and CME in March 1989, the system responsible for tracking some 19,000 objects orbiting Earth temporarily lost track of 11,000 of them because their orbits were changed by the expansion of Earth's atmosphere. During solar maximum, a number of satellites are brought to such a low altitude that they are destroyed by friction with the atmosphere. Both the Hubble Space Telescope and the International Space Station (Figure 15.24) require reboosts to higher altitude so that they can remain in orbit.

Figure 15.24. International Space Station. The International Space Station is see above Earth, as photographed in 2010 by the departing crew of the Space Shuttle Atlantis. (credit: NASA)

Solar Storm Damage on Earth

When a CME reaches Earth, it distorts Earth's magnetic field. Since a changing magnetic field induces electrical current, the CME accelerates electrons, sometimes to very high speeds. These "killer electrons" can penetrate deep into satellites, sometimes destroying their electronics and permanently disabling operation. This has happened with some communications satellites.

Disturbances in Earth's magnetic field can cause disruptions in communications, especially cell phone and wireless systems. In fact, disruptions can be expected to occur several times a year during solar maximum. Changes in Earth's magnetic field due to CMEs can also cause surges in power lines large enough to burn out transformers and cause major power outages. For example, in 1989, parts of Montreal and Quebec Province in Canada were without power for up to 9 hours as a result of a major solar storm. Electrical outages due to CMEs are more likely to occur in North America than in Europe because North America is closer to Earth's magnetic pole, where the currents induced by CMEs are strongest.

This OpenStax book is available for free at http://cnx.org/content/col11992/1.8

Besides changing the orbits of satellites, CMEs can also distort the signals sent by them. These effects can be large enough to reduce the accuracy of GPS-derived positions so that they cannot meet the limits required for airplane systems, which must know their positions to within 160 feet. Such disruptions caused by CMEs have occasionally forced the Federal Aviation Administration to restrict flights for minutes or, in a few cases, even days.

Solar storms also expose astronauts, passengers in high-flying airplanes, and even people on the surface of Earth to increased amounts of radiation. Astronauts, for example, are limited in the total amount of radiation to which they can be exposed during their careers. A single ill-timed solar outburst could end an astronaut's career. This problem becomes increasingly serious as astronauts spend more time in space. For example, the typical *daily* dose of radiation aboard the Russian Mir space station was equivalent to about eight chest X-rays. One of the major challenges in planning the human exploration of Mars is devising a way to protect astronauts from high-energy solar radiation.

Advance warning of solar storms would help us minimize their disruptive effects. Power networks could be run at less than their full capacity so that they could absorb the effects of power surges. Communications networks could be prepared for malfunctions and have backup plans in place. Spacewalks could be timed to avoid major solar outbursts. Scientists are now trying to find ways to predict where and when flares and CMEs will occur, and whether they will be big, fast events or small, slow ones with little consequence for Earth.

The strategy is to relate changes in the appearance of small, active regions and changes in local magnetic fields on the Sun to subsequent eruptions. However, right now, our predictive capability is still poor, and so the only real warning we have is from actually seeing CMEs and flares occur. Since a CME travels outward at about 500 kilometers per second, an observation of an eruption provides several days warning at the distance of Earth. However, the severity of the impact on Earth depends on how the magnetic field associated with the CME is oriented relative to Earth's magnetic field. The orientation can be measured only when the CME flows past a satellite we have put up for this purpose. However, it is located only about an hour upstream from Earth.

Space weather predictions are now available online to scientists and the public. Outlooks are given a week ahead, bulletins are issued when there is an event that is likely to be of interest to the public, and warnings and alerts are posted when an event is imminent or already under way (Figure 15.25).

Figure 15.25. NOAA Space Weather Prediction Operations Center. Bill Murtagh, a space weather forecaster, leads a workshop on preparedness for events like geomagnetic storms. (credit: modification of work by FEMA/Jerry DeFelice)

Download for free at http://cnx.org/content/col11992/latest/

LINK TO LEARNING

To find public information and alerts about space weather, you can turn to the National Space Weather Prediction Center (https://openstax.org/l/30NSWPC) or SpaceWeather (https://openstax.org/l/30SpcWeath3) for consolidated information from many sources.

Fortunately, we can expect calmer space weather for the next few years, since the most recent solar maximum, which was relatively weak, occurred in 2014, and scientists believe the current solar cycle to be one of the least active in recent history. We expect more satellites to be launched that will allow us to determine whether CMEs are headed toward Earth and how big they are. Models are being developed that will then allow scientists to use early information about the CME to predict its likely impact on Earth.

The hope is that by the time of the next maximum, solar weather forecasting will have some of the predictive capability that meteorologists have achieved for terrestrial weather at Earth's surface. However, the most difficult events to predict are the largest and most damaging storms—hurricanes on Earth and extreme, rare storm events on the Sun. Thus, it is inevitable that the Sun will continue to surprise us.

EXAMPLE 15.1

The Timing of Solar Events

A basic equation is useful in figuring out when events on the Sun will impact Earth:

$$\text{distance} = \text{velocity} \times \text{time, or } D = v \times t$$

Dividing both sides by v, we get

$$T = D/v$$

Suppose you observe a major solar flare while astronauts are orbiting Earth. If the average speed of solar wind is 400 km/s and the distance to the Sun as 1.496×10^8 km, how long it will before the charged particles ejected from the Sun during the flare reach the space station?

Solution

The time required for solar wind particles to reach Earth is $T = D/v$.

$$\frac{1.496 \times 10^8 \text{ km}}{400 \text{ km/s}} = 3.74 \times 10^5 \text{ s, or } \frac{3.74 \times 10^5 \text{ s}}{60 \text{ s/min} \times 60 \text{ min/h} \times 24 \text{ h/d}} = 4.3 \text{ d}$$

Check Your Learning

How many days would it take for the particles to reach Earth if the solar wind speed increased to 500 km/s?

Answer:

$$\frac{1.496 \times 10^8 \text{ km}}{500 \text{ km/s}} = 2.99 \times 10^5 \text{ s, or } \frac{2.99 \times 10^5 \text{ s}}{60 \text{ s/min} \times 60 \text{ min/h} \times 24 \text{ h/d}} = 3.46 \text{ d}$$

Download for free at http://cnx.org/content/col11992/latest/

Earth's Climate and the Sunspot Cycle: Is There a Connection?

While the Sun rises faithfully every day at a time that can be calculated precisely, scientists have determined that the Sun's energy output is not truly constant but varies over the centuries by a small amount—probably less than 1%. We've seen that the number of sunspots varies, with the time between sunspot maxima of about 11 years, and that the number of sunspots at maximum is not always the same. Considerable evidence shows that between the years 1645 and 1715, the number of sunspots, even at sunspot maximum, was much lower than it is now. This interval of significantly low sunspot numbers was first noted by Gustav Spörer in 1887 and then by E. W. Maunder in 1890; it is now called the **Maunder Minimum**. The variation in the number of sunspots over the past three centuries is shown in Figure 15.26. Besides the Maunder Minimum in the seventeenth century, sunspot numbers were somewhat lower during the first part of the nineteenth century than they are now; this period is called the Little Maunder Minimum.

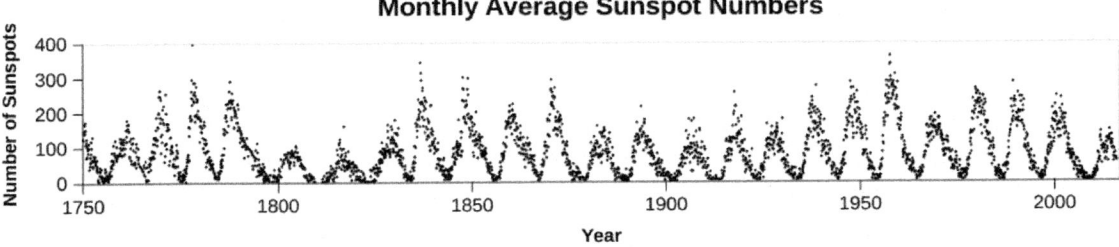

Figure 15.26. Numbers of Sunspots over Time. This diagram shows how the number of sunspots has changed with time since counts of the numbers of spots began to be recorded on a consistent scale. Note the low number of spots during the early years of the nineteenth century, the Little Maunder Minimum. (credit: modification of work by NASA/ARC)

When the number of sunspots is high, the Sun is active in various other ways as well, and, as we will see in several sections below, some of this activity affects Earth directly. For example, there are more auroral displays when the sunspot number is high. Auroras are caused when energetically charged particles from the Sun interact with Earth's magnetosphere, and the Sun is more likely to spew out particles when it is active and the sunspot number is high. Historical accounts also indicate that auroral activity was abnormally low throughout the several decades of the Maunder Minimum.

The Maunder Minimum was a time of exceptionally low temperatures in Europe—so low that this period is described as the Little Ice Age. This coincidence in time caused scientists to try to understand whether small changes in the Sun could affect the climate on Earth. There is clear evidence that it was unusually cold in Europe during part of the seventeenth century. The River Thames in London froze at least 11 times, ice appeared in the oceans off the coasts of southeast England, and low summer temperatures led to short growing seasons and poor harvests. However, whether and how changes on the Sun on this timescale influence Earth's climate is still a matter of debate among scientists.

Other small changes in climate like the Little Ice Age have occurred and have had their impacts on human history. For example, explorers from Norway first colonized Iceland and then reached Greenland by 986. From there, they were able to make repeated visits to the northeastern coasts of North America, including Newfoundland, between about 1000 and 1350. (The ships of the time did not allow the Norse explorers to travel all the way to North America directly, but only from Greenland, which served as a station for further exploration.)

Most of Greenland is covered by ice, and the Greenland station was never self-sufficient; rather, it depended on imports of food and other goods from Norway for its survival. When a little ice age began in the thirteenth century, voyaging became very difficult, and support of the Greenland colony was no longer possible. The last-

Download for free at http://cnx.org/content/col11992/latest/

known contact with it was made by a ship from Iceland blown off course in 1410. When European ships again began to visit Greenland in 1577, the entire colony there had disappeared.

The estimated dates for these patterns of migration follow what we know about solar activity. Solar activity was unusually high between 1100 and 1250, which includes the time when the first European contacts were made with North America. Activity was low from 1280 to 1340 and there was a little ice age, which was about the time regular contact with North America and between Greenland and Europe stopped.

One must be cautious, however, about assuming that low sunspot numbers or variations in the Sun's output of energy *caused* the Little Ice Age. There is no satisfactory model that can explain how a reduction in solar activity might cause cooler temperatures on Earth. An alternative possibility is that the cold weather during the Little Ice Age was related to volcanic activity. Volcanoes can eject aerosols (tiny droplets or particles) into the atmosphere that efficiently reflect sunlight. Observations show, for example, that the Pinatubo eruption in 1991 ejected SO_2 aerosols into the atmosphere, which reduced the amount of sunlight reaching Earth's surface enough to lower global temperatures by 0.4 °C.

Satellite data show that the energy output from the Sun during a solar cycle varies by only about 0.1%. We know of no physical process that would explain how such a small variation could cause global temperature changes. The level of solar activity may, however, have other effects. For example, although the Sun's total energy output varies by only 0.1% during a solar cycle, its extreme ultraviolet radiation is 10 times higher at times of solar maximum than at solar minimum. This large variation can affect the chemistry and temperature structure of the upper atmosphere. One effect might be a reduction in the ozone layer and a cooling of the stratosphere near Earth's poles. This, in turn, could change the circulation patterns of winds aloft and, hence, the tracks of storms. There is some recent evidence that variations in regional rainfall correlate better with solar activity than does the global temperature of Earth. But, as you can see, the relationship between what happens on the Sun and what happens to Earth's climate over the short term is still an area that scientists are investigating and debating.

Whatever the effects of solar activity may be on local rainfall or temperature patterns, we want to emphasize one important idea: Our climate change data and the models developed to account for the data consistently show that solar variability is *not* the cause of the global warming that has occurred during the past 50 years.

This OpenStax book is available for free at http://cnx.org/content/col11992/1.8

CHAPTER 15 REVIEW

 KEY TERMS

active region an area on the Sun where magnetic fields are concentrated; sunspots, prominences, flares, and CMEs all tend to occur in active regions

aurora light radiated by atoms and ions in the ionosphere excited by charged particles from the Sun, mostly seen in the magnetic polar regions

chromosphere the part of the solar atmosphere that lies immediately above the photospheric layers

corona (of the Sun) the outer (hot) atmosphere of the Sun

coronal hole a region in the Sun's outer atmosphere that appears darker because there is less hot gas there

coronal mass ejection (CME) a solar flare in which immense quantities of coronal material—mainly protons and electrons—is ejected at high speeds (500–1000 kilometers per second) into interplanetary space

differential rotation the phenomenon that occurs when different parts of a rotating object rotate at different rates at different latitudes

granulation the rice-grain-like structure of the solar photosphere; granulation is produced by upwelling currents of gas that are slightly hotter, and therefore brighter, than the surrounding regions, which are flowing downward into the Sun

Maunder Minimum a period during the eighteenth century when the number of sunspots seen throughout the solar cycle was unusually low

photosphere the region of the solar (or stellar) atmosphere from which continuous radiation escapes into space

plage a bright region of the solar surface observed in the light of some spectral line

plasma a hot ionized gas

prominence a large, bright, gaseous feature that appears above the surface of the Sun and extends into the corona

solar flare a sudden and temporary outburst of electromagnetic radiation from an extended region of the Sun's surface

solar wind a flow of hot, charged particles leaving the Sun

sunspot large, dark features seen on the surface of the Sun caused by increased magnetic activity

sunspot cycle the semiregular 11-year period with which the frequency of sunspots fluctuates

transition region the region in the Sun's atmosphere where the temperature rises very rapidly from the relatively low temperatures that characterize the chromosphere to the high temperatures of the corona

Download for free at http://cnx.org/content/col11992/latest/

📄 SUMMARY

15.1 The Structure and Composition of the Sun

The Sun, our star, has several layers beneath the visible surface: the core, radiative zone, and convective zone. These, in turn, are surrounded by a number of layers that make up the solar atmosphere. In order of increasing distance from the center of the Sun, they are the photosphere, with a temperature that ranges from 4500 K to about 6800 K; the chromosphere, with a typical temperature of 10^4 K; the transition region, a zone that may be only a few kilometers thick, where the temperature increases rapidly from 10^4 K to 10^6 K; and the corona, with temperatures of a few million K. The Sun's surface is mottled with upwelling convection currents seen as hot, bright granules. Solar wind particles stream out into the solar system through coronal holes. When such particles reach the vicinity of Earth, they produce auroras, which are strongest near Earth's magnetic poles. Hydrogen and helium together make up 98% of the mass of the Sun, whose composition is much more characteristic of the universe at large than is the composition of Earth.

15.2 The Solar Cycle

Sunspots are dark regions where the temperature is up to 2000 K cooler than the surrounding photosphere. Their motion across the Sun's disk allows us to calculate how fast the Sun turns on its axis. The Sun rotates more rapidly at its equator, where the rotation period is about 25 days, than near the poles, where the period is slightly longer than 36 days. The number of visible sunspots varies according to a sunspot cycle that averages 11 years in length. Spots frequently occur in pairs. During a given 11-year cycle, all leading spots in the Northern Hemisphere have the same magnetic polarity, whereas all leading sports in the Southern Hemisphere have the opposite polarity. In the subsequent 11-year cycle, the polarity reverses. For this reason, the magnetic activity cycle of the Sun is understood to last for 22 years. This activity cycle is connected with the behavior of the Sun's magnetic field, but the exact mechanism is not yet understood.

15.3 Solar Activity above the Photosphere

Signs of more intense solar activity, an increase in the number of sunspots, as well as prominences, plages, solar flares, and coronal mass ejections, all tend to occur in active regions—that is, in places on the Sun with the same latitude and longitude but at different heights in the atmosphere. Active regions vary with the solar cycle, just like sunspots do.

15.4 Space Weather

Space weather is the effect of solar activity on our own planet, both in our magnetosphere and on Earth's surface. Coronal holes allow more of the Sun's material to flow out into space. Solar flares and coronal mass ejections can cause auroras, disrupt communications, damage satellites, and cause power outages on Earth.

FOR FURTHER EXPLORATION

Articles

Berman, B. "How Solar Storms Could Shut Down Earth." *Astronomy* (September 2013): 22. Up-to-date review of how events on the Sun can hurt our civilization.

Frank, A. "Blowin' in the Solar Wind." *Astronomy* (October 1998): 60. On results from the SOHO spacecraft.

Holman, G. "The Mysterious Origins of Solar Flares." *Scientific American* (April 2006): 38. New ideas involving magnetic reconnection and new observations of flares.

This OpenStax book is available for free at http://cnx.org/content/col11992/1.8

Download for free at http://cnx.org/content/col11992/latest/

James, C. "Solar Forecast: Storm Ahead." *Sky & Telescope* (July 2007): 24. Nice review on the effects of the Sun's outbursts and on Earth and how we monitor "space weather."

Schaefer, B. "Sunspots That Changed the World." *Sky & Telescope* (April 1997): 34. Historical events connected with sunspots and solar activity.

Schrijver, C. and Title, A. "Today's Science of the Sun." *Sky & Telescope* (February 2001): 34; (March 2001): 34. Excellent reviews of recent results about the solar atmosphere.

Wadhwa, M. "Order from Chaos: Genesis Samples the Solar Wind." *Astronomy* (October 2013): 54. On a satellite that returned samples of the Sun's wind.

Websites

Dr. Sten Odenwald's "Solar Storms" site: http://www.solarstorms.org/.

ESA/NASA's Solar & Heliospheric Observatory: http://sohowww.nascom.nasa.gov. A satellite mission with a rich website to explore.

High Altitude Observatory Introduction to the Sun: http://www.hao.ucar.edu/education/basic.php. For beginners.

NASA's Solar Missions: https://www.nasa.gov/mission_pages/sunearth/missions/index.html. Good summary of the many satellites and missions NASA has.

NOAA Profile of Space Weather: http://www.swpc.noaa.gov/sites/default/files/images/u33/primer_2010_new.pdf. A primer.

NOAA Space Weather Prediction Center Information Pages: http://www.swpc.noaa.gov/content/education-and-outreach. Includes primers, videos, a curriculum and training modules.

Nova Sun Lab: http://www.pbs.org/wgbh/nova/labs/lab/sun/. Videos, scientist profiles, a research challenge related to the active Sun from the PBS science program.

Space Weather: Storms on the Sun: http://www.swpc.noaa.gov/sites/default/files/images/u33/swx_booklet.pdf. An illustrated booklet from NOAA.

Stanford Solar Center: http://solar-center.stanford.edu/. An excellent site with information for students and teachers.

Apps

These can tell you and your students more about what's happening on the Sun in real time.

NASA's 3-D Sun: http://3dsun.org/.

NASA Space Weather: https://itunes.apple.com/us/app/nasa-space-weather/id422621403?mt=8.

Solaris Alpha: https://play.google.com/store/apps/details?id=com.tomoreilly.solarisalpha.

Solar Monitor Pro: http://www.solarmonitor.eu/.

Videos

Journey into the Sun: https://www.youtube.com/watch?v=fqKFQ7z0Nuk. 2010 KQED Quest TV Program mostly about the Solar Dynamics Observatory spacecraft, its launch and capabilities, but with good general information on how the Sun works (12:24).

Download for free at http://cnx.org/content/col11992/latest/

NASA | SDO: Three Years in Three Minutes--With Expert Commentary: https://www.youtube.com/watch?v=QaCG0wAjJSY&src. Video of 3 years of observations of the Sun by the Solar Dynamics Observatory made into a speeded up movie, with commentary by solar physicist Alex Young (5:03).

Our Explosive Sun: http://www.youtube.com/watch?v=kI6YGSIJqrE. Video of a 2011 public lecture in the Silicon Valley Astronomy Lecture Series by Dr. Thomas Berger about solar activity and recent satellite missions to observe and understand it (1:20:22).

Out There Raining Fire: http://www.nytimes.com/video/science/100000003489464/out-there-raining-fire.html?emc=eta1. Nice overview and introduction to the Sun by science reporter Dennis Overbye of the NY Times (2:28)

Space Weather Impacts: http://www.swpc.noaa.gov/content/education-and-outreach. Video from NOAA (2:47); https://www.youtube.com/playlist?list=PLBdd8cMH5jFmvVR2sZubIUzBO6JI0Pvx0. Videos from the National Weather Service (four short videos) (14:41).

Space Weather: Storms on the Sun: http://www.youtube.com/watch?v=vWsmp4o-qVg. Science bulletin from the American Museum of Natural History, giving the background to what happens on the Sun to cause space weather (6:10).

Sun Storms: http://www.livescience.com/11754-sun-storms-havoc-electronic-world.html. From the Starry Night company about storms from the Sun now and in the past (4:49).

Sunspot Group AR 2339 Crosses the Sun: http://apod.nasa.gov/apod/ap150629.html. Short video (with music) animates Solar Dynamics Observatory images of an especially large sunspot group going across the Sun's face (1:15).

What Happens on the Sun Doesn't Stay on the Sun: https://www.youtube.com/watch?v=bg_gD2-ujCk. From the National Oceanic and Atmospheric Administration: introduction to the Sun, space weather, its effects, and how we monitor it (4:56).

🏛 COLLABORATIVE GROUP ACTIVITIES

A. Have your group make a list of all the ways the Sun personally affects your life on Earth. (Consider the everyday effects as well as the unusual effects due to high solar activity.)

B. Long before the nature of the Sun was fully understood, astronomer (and planet discoverer) William Herschel (1738–1822) proposed that the hot Sun may have a cool interior and may be inhabited. Have your group discuss this proposal and come up with modern arguments against it.

C. We discussed how the migration of Europeans to North America was apparently affected by short-term climate change. If Earth were to become significantly hotter, either because of changes in the Sun or because of greenhouse warming, one effect would be an increase in the rate of melting of the polar ice caps. How would this affect modern civilization?

D. Suppose we experience another Maunder Minimum on Earth, and it is accompanied by a drop in the average temperature like the Little Ice Age in Europe. Have your group discuss how this would affect civilization and international politics. Make a list of the most serious effects that you can think of.

E. Watching sunspots move across the disk of the Sun is one way to show that our star rotates on its axis. Can your group come up with other ways to show the Sun's rotation?

This OpenStax book is available for free at http://cnx.org/content/col11992/1.8

Download for free at http://cnx.org/content/col11992/latest/

F. Suppose in the future, we are able to forecast space weather as well as we forecast weather on Earth. And suppose we have a few days of warning that a big solar storm is coming that will overload Earth's magnetosphere with charged particles and send more ultraviolet and X-rays toward our planet. Have your group discuss what steps we might take to protect our civilization?

G. Have your group members research online to find out what satellites are in space to help astronomers study the Sun. In addition to searching for NASA satellites, you might also check for satellites launched by the European Space Agency and the Japanese Space Agency.

H. Some scientists and engineers are thinking about building a "solar sail"—something that can use the Sun's wind or energy to propel a spacecraft away from the Sun. The Planetary Society is a nonprofit organization that is trying to get solar sails launched, for example. Have your group do a report on the current state of solar-sail projects and what people are dreaming about for the future.

⎙ EXERCISES

Review Questions

1. Describe the main differences between the composition of Earth and that of the Sun.

2. Describe how energy makes its way from the nuclear core of the Sun to the atmosphere. Include the name of each layer and how energy moves through the layer.

3. Make a sketch of the Sun's atmosphere showing the locations of the photosphere, chromosphere, and corona. What is the approximate temperature of each of these regions?

4. Why do sunspots look dark?

5. Which aspects of the Sun's activity cycle have a period of about 11 years? Which vary during intervals of about 22 years?

6. Summarize the evidence indicating that over several hundreds of years or more there have been variations in the level of the solar activity.

7. What it the Zeeman effect and what does it tell us about the Sun?

8. Explain how the theory of the Sun's dynamo results in an average 22-year solar activity cycle. Include the location and mechanism for the dynamo.

9. Compare and contrast the four different types of solar activity above the photosphere.

10. What are the two sources of particles coming from the Sun that cause space weather? How are they different?

11. How does activity on the Sun affect human technology on Earth and in the rest of the solar system?

12. How does activity on the Sun affect natural phenomena on Earth?

Thought Questions

13. Table 15.1 indicates that the density of the Sun is 1.41 g/cm³. Since other materials, such as ice, have similar densities, how do you know that the Sun is not made of ice?

Download for free at http://cnx.org/content/col11992/latest/

14. Starting from the core of the Sun and going outward, the temperature decreases. Yet, above the photosphere, the temperature increases. How can this be?

15. Since the rotation period of the Sun can be determined by observing the apparent motions of sunspots, a correction must be made for the orbital motion of Earth. Explain what the correction is and how it arises. Making some sketches may help answer this question.

16. Suppose an (extremely hypothetical) elongated sunspot forms that extends from a latitude of 30° to a latitude of 40° along a fixed of longitude on the Sun. How will the appearance of that sunspot change as the Sun rotates? (Figure 15.17 should help you figure this out.)

17. The text explains that plages are found near sunspots, but Figure 15.18 shows that they appear even in areas without sunspots. What might be the explanation for this?

18. Why would a flare be observed in visible light, when they are so much brighter in X-ray and ultraviolet light?

19. How can the prominences, which are so big and 'float' in the corona, stay gravitationally attached to the Sun while flares can escape?

20. If you were concerned about space weather and wanted to avoid it, where would be the safest place on Earth for you to live?

21. Suppose you live in northern Canada and an extremely strong flare is reported on the Sun. What precautions might you take? What might be a positive result?

Figuring For Yourself

22. The edge of the Sun doesn't have to be absolutely sharp in order to look that way to us. It just has to go from being transparent to being completely opaque in a distance that is smaller than your eye can resolve. Remember from Astronomical Instruments that the ability to resolve detail depends on the size of the telescope's aperture. The pupil of your eye is very small relative to the size of a telescope and therefore is very limited in the amount of detail you can see. In fact, your eye cannot see details that are smaller than 1/30 of the diameter of the Sun (about 1 arcminute). Nearly all the light from the Sun emerges from a layer that is only about 400 km thick. What fraction is this of the diameter of the Sun? How does this compare with the ability of the human eye to resolve detail? Suppose we could see light emerging directly from a layer that was 300,000 km thick. Would the Sun appear to have a sharp edge?

23. Show that the statement that 92% of the Sun's atoms are hydrogen is consistent with the statement that 73% of the Sun's mass is made up of hydrogen, as found in Table 15.2. (Hint: Make the simplifying assumption, which is nearly correct, that the Sun is made up entirely of hydrogen and helium.)

24. From Doppler shifts of the spectral lines in the light coming from the east and west edges of the Sun, astronomers find that the radial velocities of the two edges differ by about 4 km/s, meaning that the Sun's rotation rate is 2 km/s. Find the approximate period of rotation of the Sun in days. The circumference of a sphere is given by $2\pi R$, where R is the radius of the sphere.

25. Assuming an average sunspot cycle of 11 years, how many revolutions does the equator of the Sun make during that one cycle? Do higher latitudes make more or fewer revolutions compared to the equator?

26. This chapter gives the average sunspot cycle as 11 years. Verify this using Figure 15.26.

27. The escape velocity from any astronomical object can be calculated as $v_{escape} = \sqrt{2GM/R}$. Using the data in Appendix E, calculate the escape velocity from the photosphere of the Sun. Since coronal mass ejections escape from the corona, would the escape velocity from there be more or less than from the photosphere?

This OpenStax book is available for free at http://cnx.org/content/col11992/1.8

28. Suppose you observe a major solar flare while astronauts are orbiting Earth. Use the data in the text to calculate how long it will before the charged particles ejected from the Sun during the flare reach them.

29. Suppose an eruptive prominence rises at a speed of 150 km/s. If it does not change speed, how far from the photosphere will it extend after 3 hours? How does this distance compare with the diameter of Earth?

30. From the information in Figure 15.21, estimate the speed with which the particles in the CME in parts (c) and (d) are moving away from the Sun.

Download for free at http://cnx.org/content/col11992/latest/

This OpenStax book is available for free at http://cnx.org/content/col11992/1.8

Figure 16.1. The Sun. It takes an incredible amount of energy for the Sun to shine, as it has and will continue to do for billions of years. (credit: modification of work by Ed Dunens)

Chapter Outline

16.1 Sources of Sunshine: Thermal and Gravitational Energy
16.2 Mass, Energy, and the Theory of Relativity
16.3 The Solar Interior: Theory
16.4 The Solar Interior: Observations

Thinking Ahead

The Sun puts out an incomprehensible amount of energy—so much that its ultraviolet radiation can cause sunburns from 93 million miles away. It is also very old. As you learned earlier, evidence shows that the Sun formed about 4.5 billion years ago and has been shining ever since. How can the Sun produce so much energy for so long?

The Sun's energy output is about 4×10^{26} watts. This is unimaginably bright: brighter than a trillion cities together each with a trillion 100-watt light bulbs. Most known methods of generating energy fall far short of the capacity of the Sun. The total amount of energy produced over the entire life of the Sun is staggering, since the Sun has been shining for billions of years. Scientists were unable to explain the seemingly unlimited energy of stars like the Sun prior to the twentieth century.

16.1 SOURCES OF SUNSHINE: THERMAL AND GRAVITATIONAL ENERGY

Learning Objectives

By the end of this section, you will be able to:

> Identify different forms of energy
> Understand the law of conservation of energy

Download for free at http://cnx.org/content/col11992/latest/

> Explain ways that energy can be transformed

Energy is a challenging concept to grasp because it exists in so many different forms that it defies any single simple explanation. In many ways, comprehending energy is like comprehending wealth: There are very different forms of wealth and they follow different rules, depending on if they are the stock market, real estate, a collection of old comic books, great piles of cash, or one of the many other ways to make and lose money. It is easier to discuss one or two forms of wealth—or energy—than to discuss that concept in general.

When striving to understand how the Sun can put out so much energy for so long, scientists considered many different types of energy. Nineteenth-century scientists knew of two possible sources for the Sun's energy: chemical and gravitational energy. The source of chemical energy most familiar to them was the burning (the chemical term is *oxidation*) of wood, coal, gasoline, or other fuel. We know exactly how much energy the burning of these materials can produce. We can thus calculate that even if the immense mass of the Sun consisted of a burnable material like coal or wood, our star could not produce energy at its present rate for more than few thousand years. However, we know from geologic evidence that water was present on Earth's surface nearly 4 billion years ago, so the Sun must have been shining brightly (and making Earth warm) at least as long as that. Today, we also know that at the temperatures found in the Sun, nothing like solid wood or coal could survive.

ASTRONOMY BASICS

What's Watt?

Just a word about the units we are using. A watt (W) is a unit of *power*, which is energy used or given off per unit time. It is measured in joules per second (J/s). You know from your everyday experience that it is not just *how much* energy you expend, but *how long* you take to do it. (Burning 10 Calories in 10 minutes requires a very different kind of exercise than burning those 10 Calories in an hour.) Watts tell you the *rate* at which energy is being used; for example, a 100-watt bulb uses 100 joules (J) of energy every second.

And how big is a joule? A 73-kilogram (160-pound) astronomy instructor running at about 4.4 meters per second (10 miles per hour) because he is late for class has a motion energy of about 700 joules.

Conservation of Energy

Other nineteenth-century attempts to determine what makes the Sun shine used the law of conservation of energy. Simply stated, this law says that energy cannot be created or destroyed, but can be transformed from one type to another, such as from heat to mechanical energy. The steam engine, which was key to the Industrial Revolution, provides a good example. In this type of engine, the hot steam from a boiler drives the movement of a piston, converting heat energy into motion energy.

Conversely, motion can be transformed into heat. If you clap your hands vigorously at the end of an especially good astronomy lecture, your palms become hotter. If you rub ice on the surface of a table, the heat produced by friction melts the ice. The brakes on cars use friction to reduce speed, and in the process, transform motion energy into heat energy. That is why after bringing a car to a stop, the brakes can be very hot; this also explains why brakes can overheat when used carelessly while descending long mountain roads.

In the nineteenth century, scientists thought that the source of the Sun's heat might be the mechanical motion of meteorites falling into it. Their calculations showed, however, that in order to produce the total amount of

This OpenStax book is available for free at http://cnx.org/content/col11992/1.8

Download for free at http://cnx.org/content/col11992/latest/

energy emitted by the Sun, the mass in meteorites that would have to fall into the Sun every 100 years would equal the mass of Earth. The resulting increase in the Sun's mass would, according to Kepler's third law, change the period of Earth's orbit by 2 seconds per year. Such a change would be easily measurable and was not, in fact, occurring. Scientists could then disprove this as the source of the Sun's energy.

Gravitational Contraction as a Source of Energy

Proposing an alternative explanation, British physicist Lord Kelvin and German scientist Hermann von Helmholtz (Figure 16.2), in about the middle of the nineteenth century, proposed that the Sun might produce energy by the conversion of gravitational energy into heat. They suggested that the outer layers of the Sun might be "falling" inward because of the force of gravity. In other words, they proposed that the Sun could be shrinking in size, staying hot and bright as a result.

(a)   (b)

Figure 16.2. Kelvin (1824–1907) and Helmholtz (1821–1894). (a) British physicist William Thomson (Lord Kelvin) and (b) German scientist Hermann von Helmholtz proposed that the contraction of the Sun under its own gravity might account for its energy. (credit a: modification of work by Wellcome Library, London; credit b: modification of work by Wellcome Library, London)

To imagine what would happen if this hypothesis were true, picture the outer layer of the Sun starting to fall inward. This outer layer is a gas made up of individual atoms, all moving about in random directions. If a layer falls inward, the atoms acquire an additional speed because of falling motion. As the outer layer falls inward, it also contracts, moving the atoms closer together. Collisions become more likely, and some of them transfer the extra speed associated with the falling motion to other atoms. This, in turn, increases the speeds of those atoms. The temperature of a gas is a measure of the kinetic energy (motion) of the atoms within it; hence, the temperature of this layer of the Sun increases. Collisions also excite electrons within the atoms to higher-energy orbits. When these electrons return to their normal orbits, they emit photons, which can then escape from the Sun (see Radiation and Spectra).

Kelvin and Helmholtz calculated that a contraction of the Sun at a rate of only about 40 meters per year would be enough to produce the amount of energy that it is now radiating. Over the span of human history, the decrease in the Sun's size from such a slow contraction would be undetectable.

If we assume that the Sun began its life as a large, diffuse cloud of gas, then we can calculate how much energy has been radiated by the Sun during its entire lifetime as it has contracted from a very large diameter to its present size. The amount of energy is on the order of 10^{42} joules. Since the solar luminosity is 4×10^{26} watts

Download for free at http://cnx.org/content/col11992/latest/

(joules/second) or about 10^{34} joules per year, contraction could keep the Sun shining at its present rate for roughly 100 million years.

In the nineteenth century, 100 million years at first seemed plenty long enough, since Earth was then widely thought to be much younger than this. But toward the end of that century and into the twentieth, geologists and physicists showed that Earth (and, hence, the Sun) is actually much older. Contraction therefore cannot be the primary source of solar energy (although, as we shall see in The Birth of Stars and the Discovery of Planets Outside the Solar System, contraction is an important source of energy for a while in stars that are just being born). Scientists were thus confronted with a puzzle of enormous proportions. Either an unknown type of energy was responsible for the most important energy source known to humanity, or estimates of the age of the solar system (and life on Earth) had to be seriously modified. Charles Darwin, whose theory of evolution required a longer time span than the theories of the Sun seemed to permit, was discouraged by these results and continued to worry about them until his death in 1882.

It was only in the twentieth century that the true source of the Sun's energy was identified. The two key pieces of information required to solve the puzzle were the structure of the nucleus of the atom and the fact that mass can be converted into energy.

16.2 MASS, ENERGY, AND THE THEORY OF RELATIVITY

Learning Objectives

By the end of this section, you will be able to:

> Explain how matter can be converted into energy
> Describe the particles that make up atoms
> Describe the nucleus of an atom
> Understand the nuclear forces that hold atoms together
> Trace the nuclear reactions in the solar interior

As we have seen, energy cannot be created or destroyed, but only converted from one form to another. One of the remarkable conclusions derived by Albert Einstein (see Albert Einstein) when he developed his theory of relativity is that matter can be considered a form of energy too and can be converted into energy. Furthermore, energy can also be converted into matter. This seemed to contradict what humans had learned over thousands of years by studying nature. Matter is something we can see and touch, whereas energy is something objects have when they do things like move or heat up. The idea that matter or energy can be converted into each other seemed as outrageous as saying you could accelerate a car by turning the bumper into more speed, or that you could create a bigger front seat by slowing down your car. That would be pretty difficult to believe; yet, the universe actually works somewhat like that.

Converting Matter into Energy

The remarkable equivalence between matter and energy is given in one of the most famous equations:

$$E = mc^2$$

In this equation, E stands for energy, m stands for mass, and c, the constant that relates the two, is the speed of light (3×10^8 meters per second). Note that mass is a measure of the quantity of matter, so the significance of this equation is that matter can be converted into energy and energy can be converted into matter. Let's compare this equation of converting matter and energy to some common conversion equations that have the same form:

This OpenStax book is available for free at http://cnx.org/content/col11992/1.8

$$\text{inches} = \text{feet} \times 12, \text{ or cents} = \text{dollars} \times 100$$

Just as each conversion formula allows you to calculate the conversion of one thing into another, when we convert matter into energy, we consider how much mass the matter has. The conversion factor in this case turns out not to be either 12 or 100, as in our examples, but another constant quantity: the speed of light squared. Note that matter does not have to travel at the speed of light (or the speed of light squared) for this conversion to occur. The factor of c^2 is just the number that Einstein showed must be used to relate mass and energy.

Notice that this formula does not tell us *how* to convert mass into energy, just as the formula for cents does not tell us where to exchange coins for a dollar bill. The formulas merely tell us what the equivalent values are if we succeed in making the conversion. When Einstein first derived his formula in 1905, no one had the faintest idea how to convert mass into energy in any practical way. Einstein himself tried to discourage speculation that the large-scale conversion of atomic mass into energy would be feasible in the near future. Today, as a result of developments in nuclear physics, we regularly convert mass into energy in power plants, nuclear weapons, and high-energy physics experiments in particle accelerators.

Because the speed of light squared (c^2) is a very large quantity, the conversion of even a small amount of mass results in a very large amount of energy. For example, the complete conversion of 1 gram of matter (about 1/28 ounce, or approximately 1 paperclip) would produce as much energy as the burning of 15,000 barrels of oil.

Scientists soon realized that the conversion of mass into energy is the source of the Sun's heat and light. With Einstein's equation of $E = mc^2$, we can calculate that the amount of energy radiated by the Sun could be produced by the complete conversion of about 4 million tons of matter into energy inside the Sun each second. Destroying 4 million tons per second sounds like a lot when compared to earthly things, but bear in mind that the Sun is a very big reservoir of matter. In fact, we will see that the Sun contains more than enough mass to destroy such huge amounts of matter and still continue shining at its present rate for billions of years.

But knowing all that still does not tell us *how* mass can be converted into energy. To understand the process that actually occurs in the Sun, we need to explore the structure of the atom a bit further.

VOYAGERS IN ASTRONOMY

Albert Einstein

For a large part of his life, Albert Einstein (Figure 16.3) was one of the most recognized celebrities of his day. Strangers stopped him on the street, and people all over the world asked him for endorsements, advice, and assistance. In fact, when Einstein and the great film star Charlie Chaplin met in California, they found they shared similar feelings about the loss of privacy that came with fame. Einstein's name was a household word despite the fact that most people did not understand the ideas that had made him famous.

Einstein was born in 1879 in Ulm, Germany. Legend has it that he did not do well in school (even in arithmetic), and thousands of students have since attempted to justify a bad grade by referring to this story. Alas, like many legends, this one is not true. Records indicate that although he tended to rebel against the authoritarian teaching style in vogue in Germany at that time, Einstein was a good student.

After graduating from the Federal Polytechnic Institute in Zurich, Switzerland, Einstein at first had trouble getting a job (even as a high school teacher), but he eventually became an examiner in the Swiss Patent Office. Working in his spare time, without the benefit of a university environment but using his superb

Download for free at http://cnx.org/content/col11992/latest/

physical intuition, he wrote four papers in 1905 that would ultimately transform the way physicists looked at the world.

One of these, which earned Einstein the Nobel Prize in 1921, set part of the foundation of *quantum mechanics*—the rich, puzzling, and remarkable theory of the subatomic realm. But his most important paper presented the *special theory of relativity*, a reexamination of space, time, and motion that added a whole new level of sophistication to our understanding of those concepts. The famed equation $E = mc^2$ was actually a relatively minor part of this theory, added in a later paper.

In 1916, Einstein published his *general theory of relativity*, which was, among other things, a fundamentally new description of gravity (see Black Holes and Curved Spacetime). When this theory was confirmed by measurements of the "bending of starlight" during a 1919 eclipse (*The New York Times* headline read, "Lights All Askew in the Heavens"), Einstein became world famous.

In 1933, to escape Nazi persecution, Einstein left his professorship in Berlin and settled in the United States at the newly created Institute for Advanced Studies at Princeton. He remained there until his death in 1955, writing, lecturing, and espousing a variety of intellectual and political causes. For example, he agreed to sign a letter written by Leo Szilard and other scientists in 1939, alerting President Roosevelt to the dangers of allowing Nazi Germany to develop the atomic bomb first. And in 1952, Einstein was offered the second presidency of Israel. In declining the position, he said, "I know a little about nature and hardly anything about men."

Figure 16.3. Albert Einstein (1879–1955). This portrait of Einstein was taken in 1912. (credit: modification of work by J. F. Langhans)

Elementary Particles

The fundamental components of atoms are the proton, neutron, and electron (see The Structure of the Atom).

Protons, neutrons, and electrons are by no means all the particles that exist. First, for each kind of particle, there is a corresponding but opposite *antiparticle*. If the particle carries a charge, its antiparticle has the opposite charge. The antielectron is the *positron*, which has the same mass as the electron but is positively charged. Similarly, the antiproton has a negative charge. The remarkable thing about such *antimatter* is that when a particle comes into contact with its antiparticle, the original particles are annihilated, and substantial amounts of energy in the form of photons are produced.

This OpenStax book is available for free at http://cnx.org/content/col11992/1.8

Since our world is made exclusively of ordinary particles of matter, antimatter cannot survive for very long. But individual antiparticles are found in cosmic rays (particles that arrive at the top of Earth's atmosphere from space) and can be created in particle accelerators. And, as we will see in a moment, antimatter is created in the core of the Sun and other stars.

Science fiction fans may be familiar with antimatter from the *Star Trek* television series and films. The Starship Enterprise is propelled by the careful combining of matter and antimatter in the ship's engine room. According to $E = mc^2$, the annihilation of matter and antimatter can produce a huge amount of energy, but keeping the antimatter fuel from touching the ship before it is needed must be a big problem. No wonder Scotty, the chief engineer in the original TV show, always looked worried!

In 1933, physicist Wolfgang Pauli (Figure 16.4) suggested that there might be another type of elementary particle. Energy seemed to disappear when certain types of nuclear reactions took place, violating the law of conservation of energy. Pauli was reluctant to accept the idea that one of the basic laws of physics was wrong, and he suggested a "desperate remedy." Perhaps a so-far-undetected particle, which was given the name **neutrino** ("little neutral one"), carried away the "missing" energy. He suggested that neutrinos were particles with zero mass, and that like photons, they moved with the speed of light.

Figure 16.4. Wolfgang Pauli in 1945. Pauli is considered the "father" of the neutrino, having conceived of it in 1933.

The elusive neutrino was not detected until 1956. The reason it was so hard to find is that neutrinos interact very weakly with other matter and therefore are very difficult to detect. Earth is more transparent to a neutrino than the thinnest and cleanest pane of glass is to a photon of light. In fact, most neutrinos can pass completely through a star or planet without being absorbed. As we shall see, this behavior of neutrinos makes them a very important tool for studying the Sun. Since Pauli's prediction, scientists have learned a lot more about the neutrino. We now know that there are three different types of neutrinos, and in 1998, neutrinos were discovered to have a tiny amount of mass. Indeed, it is so tiny that electrons are at least 500,000 times more massive. Ongoing research is focused on determining the mass of neutrinos more precisely, and it may still turn out that one of the three types is massless. We will return to the subject of neutrinos later in this chapter.

Some of the properties of the proton, electron, neutron, and neutrino are summarized in Table 16.1. (Other subatomic particles have been produced by experiments with particle accelerators, but they do not play a role in the generation of solar energy.)

Download for free at http://cnx.org/content/col11992/latest/

Properties of Some Common Particles

Particle	Mass (kg)	Charge
Proton	1.67265×10^{-27}	+1
Neutron	1.67495×10^{-27}	0
Electron	9.11×10^{-31}	−1
Neutrino	$<2 \times 10^{-36}$ (uncertain)	0

Table 16.1

The Atomic Nucleus

The nucleus of an atom is not just a loose collection of elementary particles. Inside the nucleus, particles are held together by a very powerful force called the strong nuclear force. This is short-range force, only capable of acting over distances about the size of the atomic nucleus. A quick thought experiment shows how important this force is. Take a look at your finger and consider the atoms composing it. Among them is carbon, one of the basic elements of life. Focus your imagination on the nucleus of one of your carbon atoms. It contains six protons, which have a positive charge, and six neutrons, which are neutral. Thus, the nucleus has a net charge of six positives. If only the electrical force were acting, the protons in this and every carbon atom would find each other very repulsive and fly apart.

The strong nuclear force is an attractive force, stronger than the electrical force, and it keeps the particles of the nucleus tightly bound together. We saw earlier that if under the force of gravity a star "shrinks"—bringing its atoms closer together—gravitational energy is released. In the same way, if particles come together under the strong nuclear force and unite to form an atomic nucleus, some of the nuclear energy is released. The energy given up in such a process is called the *binding energy* of the nucleus.

When such binding energy is released, the resulting nucleus has slightly less mass than the sum of the masses of the particles that came together to form it. In other words, the energy comes from the loss of mass. This slight deficit in mass is only a small fraction of the mass of one proton. But because each bit of lost mass can provide a lot of energy (remember, $E = mc^2$), this nuclear energy release can be quite substantial.

Measurements show that the binding energy is greatest for atoms with a mass near that of the iron nucleus (with a combined number of protons and neutrons equal to 56) and less for both the lighter and the heavier nuclei. Iron, therefore, is the most stable element: since it gives up the most energy when it forms, it would require the most energy to break it back down into its component particles.

What this means is that, in general, when light atomic nuclei come together to form a heavier one (up to iron), mass is lost and energy is released. This joining together of atomic nuclei is called nuclear **fusion**.

Energy can also be produced by breaking up heavy atomic nuclei into lighter ones (down to iron); this process is called nuclear **fission**. Nuclear fission was the process we learned to use first—in atomic bombs and in nuclear reactors used to generate electrical power—and it may therefore be more familiar to you. Fission also sometimes occurs spontaneously in some unstable nuclei through the process of natural radioactivity. But fission requires big, complex nuclei, whereas we know that the stars are made up predominantly of small, simple nuclei. So we must look to fusion first to explain the energy of the Sun and the stars (Figure 16.5).

This OpenStax book is available for free at http://cnx.org/content/col11992/1.8

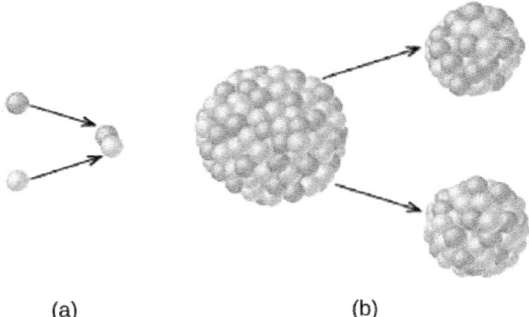

Figure 16.5. Fusion and Fission. (a) In fusion, light atomic nuclei join together to form a heavier nuclei, releasing energy in the process. (b) In fission, energy is produced by the breaking up of heavy, complex nuclei into lighter ones.

Nuclear Attraction versus Electrical Repulsion

So far, we seem to have a very attractive prescription for producing the energy emitted by the Sun: "roll" some nuclei together and join them via nuclear fusion. This will cause them to lose some of their mass, which then turns into energy. However, every nucleus, even simple hydrogen, has protons—and protons all have positive charges. Since like charges repel via the electrical force, the closer we get two nuclei to each other, the more they repel. It's true that if we can get them within "striking distance" of the nuclear force, they will then come together with a much stronger attraction. But that striking distance is very tiny, about the size of a nucleus. How can we get nuclei close enough to participate in fusion?

The answer turns out to be heat—tremendous heat—which speeds the protons up enough to overcome the electrical forces that try to keep protons apart. Inside the Sun, as we saw, the most common element is hydrogen, whose nucleus contains only a single proton. Two protons can fuse only in regions where the temperature is greater than about 12 million K, and the speed of the protons average around 1000 kilometers per second or more. (In old-fashioned units, that's over 2 million miles per hour!)

In our Sun, such extreme temperatures are reached only in the regions near its center, which has a temperature of 15 million K. Calculations show that nearly all of the Sun's energy is generated within about 150,000 kilometers of its core, or within less than 10% of its total volume.

Even at these high temperatures, it is exceedingly difficult to force two protons to combine. On average, a proton will rebound from other protons in the Sun's crowded core for about 14 billion years, at the rate of *100 million collisions per second*, before it fuses with a second proton. This is, however, only the *average* waiting time. Some of the enormous numbers of protons in the Sun's inner region are "lucky" and take only a few collisions to achieve a fusion reaction: they are the protons responsible for producing the energy radiated by the Sun. Since the Sun is about 4.5 billion years old, most of its protons have not yet been involved in fusion reactions.

Nuclear Reactions in the Sun's Interior

The Sun, then, taps the energy contained in the nuclei of atoms through nuclear fusion. Let's look at what happens in more detail. Deep inside the Sun, a three-step process takes four hydrogen nuclei and fuses them together to form a single helium nucleus. The helium nucleus is slightly less massive than the four hydrogen nuclei that combine to form it, and that mass is converted into energy.

The initial step required to form one helium nucleus from four hydrogen nuclei is shown in Figure 16.6. At the high temperatures inside the Sun's core, two protons combine to make a *deuterium* nucleus, which is an isotope (or version) of hydrogen that contains one proton and one neutron. In effect, one of the original protons has been converted into a neutron in the fusion reaction. Electric charge has to be conserved in nuclear reactions,

Download for free at http://cnx.org/content/col11992/latest/

and it is conserved in this one. A **positron** (antimatter electron) emerges from the reaction and carries away the positive charge originally associated with one of the protons.

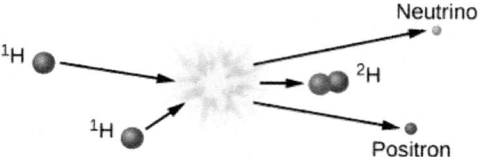

Figure 16.6. Proton-Proton Chain, Step 1. This is the first step in the process of fusing hydrogen into helium in the Sun. High temperatures are required because this reaction starts with two hydrogen nuclei, which are protons (shown in blue at left) that must overcome electrical repulsion to combine, forming a hydrogen nucleus with a proton and a neutron (shown in red). Note that hydrogen containing one proton and one neutron is given its own name: deuterium. Also produced in this reaction are a positron, which is an antielectron, and an elusive particle named the neutrino.

Since it is antimatter, this positron will instantly collide with a nearby electron, and both will be annihilated, producing electromagnetic energy in the form of gamma-ray photons. This gamma ray, which has been created in the center of the Sun, finds itself in a world crammed full of fast-moving nuclei and electrons. The gamma ray collides with particles of matter and transfers its energy to one of them. The particle later emits another gamma-ray photon, but often the emitted photon has a bit less energy than the one that was absorbed.

Such interactions happen to gamma rays again and again and again as they make their way slowly toward the outer layers of the Sun, until their energy becomes so reduced that they are no longer gamma rays but X-rays (recall what you learned in The Electromagnetic Spectrum). Later, as the photons lose still more energy through collisions in the crowded center of the Sun, they become ultraviolet photons.

By the time they reach the Sun's surface, most of the photons have given up enough energy to be ordinary light—and they are the sunlight we see coming from our star. (To be precise, each gamma-ray photon is ultimately converted into many separate lower-energy photons of sunlight.) So, the sunlight given off by the Sun today had its origin as a gamma ray produced by nuclear reactions deep in the Sun's core. The length of time that photons require to reach the surface depends on how far a photon on average travels between collisions, and the travel time depends on what model of the complicated solar interior we accept. Estimates are somewhat uncertain but indicate that the emission of energy from the surface of the Sun can lag its production in the interior by 100,000 years to as much as 1,000,000 years.

In addition to the positron, the fusion of two hydrogen atoms to form deuterium results in the emission of a neutrino. Because neutrinos interact so little with ordinary matter, those produced by fusion reactions near the center of the Sun travel directly to the Sun's surface and then out into space, in all directions. Neutrinos move at nearly the speed of light, and they escape the Sun about two seconds after they are created.

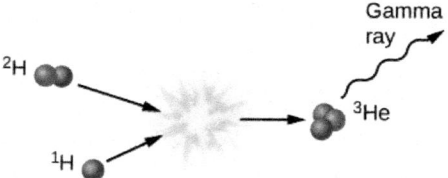

Figure 16.7. Proton-Proton Chain, Step 2. This is the second step of the proton-proton chain, the fusion reaction that converts hydrogen into helium in the Sun. This step combines one hydrogen nucleus, which is a proton (shown in blue), with the deuterium nucleus from the previous step (shown as a red and blue particle). The product of this is an isotope of helium with two protons (blue) and one neutron (red) and energy in the form of gamma-ray radiation.

This OpenStax book is available for free at http://cnx.org/content/col11992/1.8

Download for free at http://cnx.org/content/col11992/latest/

The second step in forming helium from hydrogen is to add another proton to the deuterium nucleus to create a helium nucleus that contains two protons and one neutron (Figure 16.7). In the process, some mass is again lost and more gamma radiation is emitted. Such a nucleus is helium because an element is defined by its number of protons; any nucleus with two protons is called helium. But this form of helium, which we call helium-3 (and write in shorthand as ^{3}He) is not the isotope we see in the Sun's atmosphere or on Earth. That helium has two neutrons and two protons and hence is called helium-4 (^{4}He).

To get to helium-4 in the Sun, helium-3 must combine with another helium-3 in the third step of fusion (illustrated in Figure 16.8). Note that two energetic protons are left over from this step; each of them comes out of the reaction ready to collide with other protons and to start step 1 in the chain of reactions all over again.

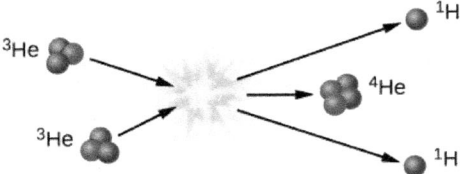

Figure 16.8. Proton-Proton Chain, Step 3. This is the third step in the fusion of hydrogen into helium in the Sun. Note that the two helium-3 nuclei from the second step (see Figure 16.7) must combine before the third step becomes possible. The two protons that come out of this step have the energy to collide with other protons in the Sun and start step one again.

LINK TO LEARNING

These animations of proton-proton reactions (https://openstaxcollege.org/l/30proproreactio) show the steps required for fusion of hydrogen into helium in the Sun.

LINK TO LEARNING

Visit the Tokamak Fusion Reactor (https://openstaxcollege.org/l/30tokamakfusrea) at the General Atomics Lab in San Diego, CA, for an 8-minute tour.

The Proton-Proton Chain

The nuclear reactions in the Sun that we have been discussing can be described succinctly through the following nuclear formulas:

$$^1\text{H} + {}^1\text{H} \longrightarrow {}^2\text{H} + e^+ + \nu$$
$$^2\text{H} + {}^1\text{H} \longrightarrow {}^3\text{He} + \gamma$$
$$^3\text{He} + {}^3\text{He} \longrightarrow {}^4\text{He} + {}^1\text{H} + {}^1\text{H}$$

Here, the superscripts indicate the total number of neutrons plus protons in the nucleus, e^+ is the positron, ν is the neutrino, and γ indicates that gamma rays are emitted. Note that the third step requires two helium-3 nuclei to start; the first two steps must happen twice before the third step can occur.

Download for free at http://cnx.org/content/col11992/latest/

Although, as we discussed, the first step in this chain of reactions is very difficult and generally takes a long time, the other steps happen more quickly. After the deuterium nucleus is formed, it survives an average of only about 6 seconds before being converted into ^{3}He. About a million years after that (on average), the ^{3}He nucleus will combine with another to form ^{4}He.

We can compute the amount of energy these reactions generate by calculating the difference in the initial and final masses. The masses of hydrogen and helium atoms in the units normally used by scientists are 1.007825 u and 4.00268 u, respectively. (The unit of mass, u, is defined to be 1/12 the mass of an atom of carbon, or approximately the mass of a proton.) Here, we include the mass of the entire atom, not just the nucleus, because electrons are involved as well. When hydrogen is converted into helium, two positrons are created (remember, the first step happens twice), and these are annihilated with two free electrons, adding to the energy produced.

$$4 \times 1.007825 \quad = 4.03130 \, u \text{ (mass of initial hydrogen atoms)}$$
$$- 4.00268 \, u \text{ (mass of final helium a oms)}$$
$$= 0.02862 \, u \text{ (mass lost in the transformation)}$$

The mass lost, 0.02862 u, is 0.71% of the mass of the initial hydrogen. Thus, if 1 kilogram of hydrogen is converted into helium, then the mass of the helium is only 0.9929 kilogram, and 0.0071 kilogram of material is converted into energy. The speed of light (c) is 3 × 10^8 meters per second, so the energy released by the conversion of just 1 kilogram of hydrogen into helium is:

$$E = mc^2$$
$$E = 0.0071 \, \text{kg} \times \left(3 \times 10^8 \, \text{m/s}\right)^2 = 6.4 \times 10^{14} \, \text{J}$$

This amount, the energy released when a single kilogram (2.2 pounds) of hydrogen undergoes fusion, would supply all of the electricity used in the United States for about 2 weeks.

To produce the Sun's luminosity of 4 × 10^{26} watts, some 600 million tons of hydrogen must be converted into helium *each second*, of which about 4 million tons are converted from matter into energy. As large as these numbers are, the store of hydrogen (and thus of nuclear energy) in the Sun is still *more* enormous, and can last a long time—billions of years, in fact.

At the temperatures inside the stars with masses smaller than about 1.2 times the mass of our Sun (a category that includes the Sun itself), most of the energy is produced by the reactions we have just described, and this set of reactions is called the **proton-proton chain** (or sometimes, the p-p chain). In the proton-proton chain, protons collide directly with other protons to form helium nuclei.

In hotter stars, another set of reactions, called the carbon-nitrogen-oxygen (CNO) cycle, accomplishes the same net result. In the CNO cycle, carbon and hydrogen nuclei collide to initiate a series of reactions that form nitrogen, oxygen, and ultimately, helium. The nitrogen and oxygen nuclei do not survive but interact to form carbon again. Therefore, the outcome is the same as in the proton-proton chain: four hydrogen atoms disappear, and in their place, a single helium atom is created. The CNO cycle plays only a minor role in the Sun but is the main source of energy for stars with masses greater than about the mass of the Sun.

So you can see that we have solved the puzzle that so worried scientists at the end of the nineteenth century. The Sun can maintain its high temperature and energy output for billions of years through the fusion of the simplest element in the universe, hydrogen. Because most of the Sun (and the other stars) is made of hydrogen, it is an ideal "fuel" for powering a star. As will be discussed in the following chapters, we can define a star as a ball of gas capable of getting its core hot enough to initiate the fusion of hydrogen. There are balls of gas that

This OpenStax book is available for free at http://cnx.org/content/col11992/1.8

lack the mass required to do this (Jupiter is a local example); like so many hopefuls in Hollywood, they will never be stars.

MAKING CONNECTIONS

Fusion on Earth

Wouldn't it be wonderful if we could duplicate the Sun's energy mechanism in a controlled way on Earth? (We have already duplicated it in an uncontrolled way in hydrogen bombs, but we hope our storehouses of these will never be used.) Fusion energy would have many advantages: it would use hydrogen (or deuterium, which is heavy hydrogen) as fuel, and there is abundant hydrogen in Earth's lakes and oceans. Water is much more evenly distributed around the world than oil or uranium, meaning that a few countries would no longer hold an energy advantage over the others. And unlike fission, which leaves dangerous byproducts, the nuclei that result from fusion are perfectly safe.

The problem is that, as we saw, it takes extremely high temperatures for nuclei to overcome their electrical repulsion and undergo fusion. When the first hydrogen bombs were exploded in tests in the 1950s, the "fuses" to get them hot enough were fission bombs. Interactions at such temperatures are difficult to sustain and control. To make fusion power on Earth, after all, we have to do what the Sun does: produce temperatures and pressures high enough to get hydrogen nuclei on intimate terms with one another.

The European Union, the United States, South Korea, Japan, China, Russia, Switzerland, and India are collaborating on the International Thermonuclear Experimental Reactor (ITER), a project to demonstrate the feasibility of controlled fusion (Figure 16.9). The facility is being built in France. Construction will require over 10,000,000 components and 2000 workers for assembly. The date for the start of operations is yet to be determined.

ITER is based on the Tokamak design, in which a large doughnut-shaped container is surrounded by superconducting magnets to confine and control the hydrogen nuclei in a strong magnetic field. Previous fusion experiments have produced about 15 million watts of energy, but only for a second or two, and they have required 100 million watts to produce the conditions necessary to achieve fusion. The goal of ITER is to build the first fusion device capable of producing 500 million watts of fusion energy for up to 1000 seconds. The challenge is keeping the deuterium and tritium—which will participate in fusion reactions—hot enough and dense enough, for a long enough time to produce energy.

Download for free at http://cnx.org/content/col11992/latest/

Figure 16.9. ITER Design. The bright yellow areas in this model show where the superconducting magnets will circle the chamber within which fusion will take place. A huge magnet will keep the charged nuclei of heavy hydrogen confined. The goal is to produce 500 megawatts of energy. (credit: modification of work by Stephan Mosel)

16.3 | THE SOLAR INTERIOR: THEORY

Learning Objectives

By the end of this section, you will be able to:

> Describe the state of equilibrium of the Sun
> Understand the energy balance of the Sun
> Explain how energy moves outward through the Sun
> Describe the structure of the solar interior

Fusion of protons can occur in the center of the Sun only if the temperature exceeds 12 million K. How do we know that the Sun is actually this hot? To determine what the interior of the Sun might be like, it is necessary to resort to complex calculations. Since we can't see the interior of the Sun, we have to use our understanding of physics, combined with what we see at the surface, to construct a mathematical model of what must be happening in the interior. Astronomers use observations to build a computer program containing everything they think they know about the physical processes going on in the Sun's interior. The computer then calculates the temperature and pressure at every point inside the Sun and determines what nuclear reactions, if any, are taking place. For some calculations, we can use observations to determine whether the computer program is producing results that match what we see. In this way, the program evolves with ever-improving observations.

The computer program can also calculate how the Sun will change with time. After all, the Sun must change. In its center, the Sun is slowly depleting its supply of hydrogen and creating helium instead. Will the Sun get hotter? Cooler? Larger? Smaller? Brighter? Fainter? Ultimately, the changes in the center could be catastrophic,

This OpenStax book is available for free at http://cnx.org/content/col11992/1.8

since eventually all the hydrogen fuel hot enough for fusion will be exhausted. Either a new source of energy must be found, or the Sun will cease to shine. We will describe the ultimate fate of the Sun in later chapters. For now, let's look at some of the things we must teach the computer about the Sun in order to carry out such calculations.

The Sun Is a Plasma

The Sun is so hot that all of the material in it is in the form of an ionized gas, called a plasma. Plasma acts much like a hot gas, which is easier to describe mathematically than either liquids or solids. The particles that constitute a gas are in rapid motion, frequently colliding with one another. This constant bombardment is the *pressure* of the gas (Figure 16.10).

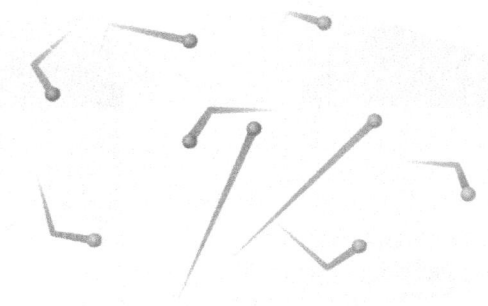

Figure 16.10. Gas Pressure. The particles in a gas are in rapid motion and produce pressure through collisions with the surrounding material. Here, particles are shown bombarding the sides of an imaginary container.

More particles within a given volume of gas produce more pressure because the combined impact of the moving particles increases with their number. The pressure is also greater when the molecules or atoms are moving faster. Since the molecules move faster when the temperature is hotter, higher temperatures produce higher pressure.

The Sun Is Stable

The Sun, like the majority of other stars, is stable; it is neither expanding nor contracting. Such a star is said to be in a condition of *equilibrium*. All the forces within it are balanced, so that at each point within the star, the temperature, pressure, density, and so on are maintained at constant values. We will see in later chapters that even these stable stars, including the Sun, are changing as they evolve, but such evolutionary changes are so gradual that, for all intents and purposes, the stars are still in a state of equilibrium at any given time.

The mutual gravitational attraction between the masses of various regions within the Sun produces tremendous forces that tend to collapse the Sun toward its center. Yet we know from the history of Earth that the Sun has been emitting roughly the same amount of energy for billions of years, so clearly it has managed to resist collapse for a very long time. The gravitational forces must therefore be counterbalanced by some other force. That force is due to the pressure of gases within the Sun (Figure 16.11). Calculations show that, in order to exert enough pressure to prevent the Sun from collapsing due to the force of gravity, the gases at its center must be maintained at a temperature of 15 million K. Think about what this tells us. Just from the fact that the Sun is not contracting, we can conclude that its temperature must indeed be high enough at the center for protons to undergo fusion.

Download for free at http://cnx.org/content/col11992/latest/

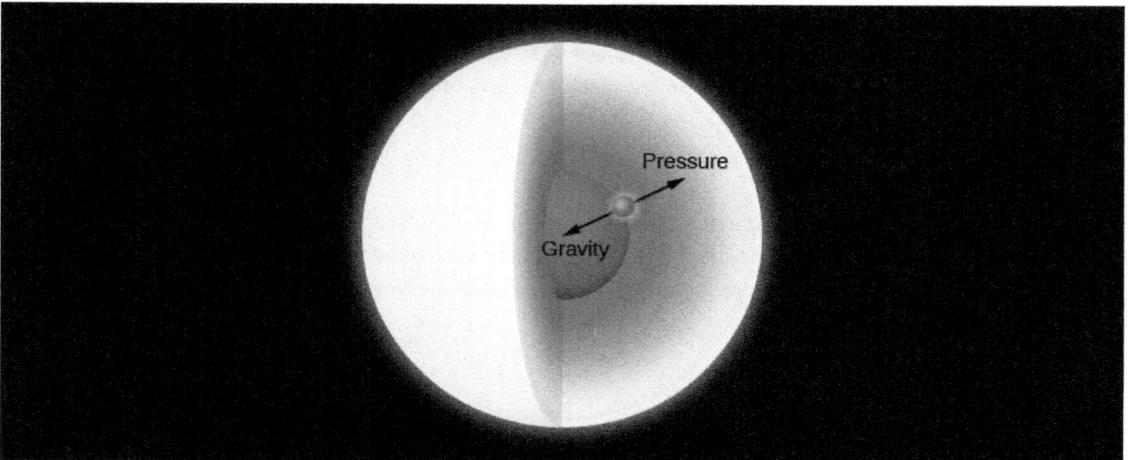

Figure 16.11. Hydrostatic Equilibrium. In the interior of a star, the inward force of gravity is exactly balanced at each point by the outward force of gas pressure.

The Sun maintains its stability in the following way. If the internal pressure in such a star were not great enough to balance the weight of its outer parts, the star would collapse somewhat, contracting and building up the pressure inside. On the other hand, if the pressure were greater than the weight of the overlying layers, the star would expand, thus decreasing the internal pressure. Expansion would stop, and equilibrium would again be reached when the pressure at every internal point equaled the weight of the stellar layers above that point. An analogy is an inflated balloon, which will expand or contract until an equilibrium is reached between the pressure of the air inside and outside. The technical term for this condition is **hydrostatic equilibrium**. Stable stars are all in hydrostatic equilibrium; so are the oceans of Earth as well as Earth's atmosphere. The air's own pressure keeps it from falling to the ground.

The Sun Is Not Cooling Down

As everyone who has ever left a window open on a cold winter night knows, heat always flows from hotter to cooler regions. As energy filters outward toward the surface of a star, it must be flowing from inner, hotter regions. The temperature cannot ordinarily get cooler as we go inward in a star, or energy would flow in and heat up those regions until they were at least as hot as the outer ones. Scientists conclude that the temperature is highest at the center of a star, dropping to lower and lower values toward the stellar surface. (The high temperature of the Sun's chromosphere and corona may therefore appear to be a paradox. But remember from The Sun: A Garden-Variety Star that these high temperatures are maintained by magnetic effects, which occur in the Sun's atmosphere.)

The outward flow of energy through a star robs it of its internal heat, and the star would cool down if that energy were not replaced. Similarly, a hot iron begins to cool as soon as it is unplugged from its source of electric energy. Therefore, a source of fresh energy must exist within each star. In the Sun's case, we have seen that this energy source is the ongoing fusion of hydrogen to form helium.

Heat Transfer in a Star

Since the nuclear reactions that generate the Sun's energy occur deep within it, the energy must be transported from the center of the Sun to its surface—where we see it in the form of both heat and light. There are three ways in which energy can be transferred from one place to another. In **conduction**, atoms or molecules pass on their energy by colliding with others nearby. This happens, for example, when the handle of a metal spoon

This OpenStax book is available for free at http://cnx.org/content/col11992/1.8

heats up as you stir a cup of hot coffee. In **convection**, currents of warm material rise, carrying their energy with them to cooler layers. A good example is hot air rising from a fireplace. In **radiation**, energetic photons move away from hot material and are absorbed by some material to which they convey some or all of their energy. You can feel this when you put your hand close to the coils of an electric heater, allowing infrared photons to heat up your hand. Conduction and convection are both important in the interiors of planets. In stars, which are much more transparent, radiation and convection are important, whereas conduction can usually be ignored.

Stellar *convection* occurs as currents of hot gas flow up and down through the star (Figure 16.12). Such currents travel at moderate speeds and do not upset the overall stability of the star. They don't even result in a net transfer of mass either inward or outward because, as hot material rises, cool material falls and replaces it. This results in a convective circulation of rising and falling cells as seen in Figure 16.12. In much the same way, heat from a fireplace can stir up air currents in a room, some rising and some falling, without driving any air into or out the room. Convection currents carry heat very efficiently outward through a star. In the Sun, convection turns out to be important in the central regions and near the surface.

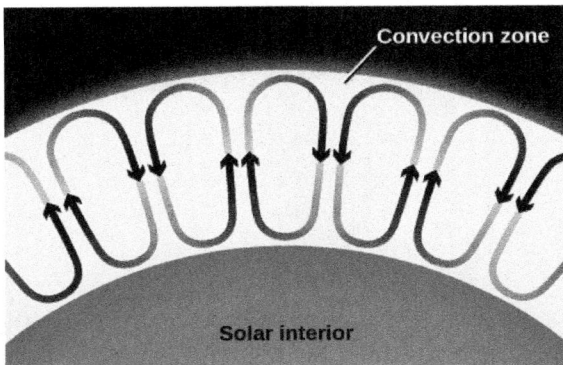

Figure 16.12. Convection. Rising convection currents carry heat from the Sun's interior to its surface, whereas cooler material sinks downward. Of course, nothing in a real star is as simple as diagrams in textbooks suggest.

Unless convection occurs, the only significant mode of energy transport through a star is by electromagnetic radiation. Radiation is not an efficient means of energy transport in stars because gases in stellar interiors are very opaque, that is, a photon does not go far (in the Sun, typically about 0.01 meter) before it is absorbed. (The processes by which atoms and ions can interrupt the outward flow of photons—such as becoming ionized—were discussed in the section on the Formation of Spectral Lines.) The absorbed energy is always reemitted, but it can be reemitted in any direction. A photon absorbed when traveling outward in a star has almost as good a chance of being radiated back toward the center of the star as toward its surface.

A particular quantity of energy, therefore, zigzags around in an almost random manner and takes a long time to work its way from the center of a star to its surface (Figure 16.13). Estimates are somewhat uncertain, but in the Sun, as we saw, the time required is probably between 100,000 and 1,000,000 years. If the photons were not absorbed and reemitted along the way, they would travel at the speed of light and could reach the surface in a little over 2 seconds, just as neutrinos do (Figure 16.14).

Download for free at http://cnx.org/content/col11992/latest/

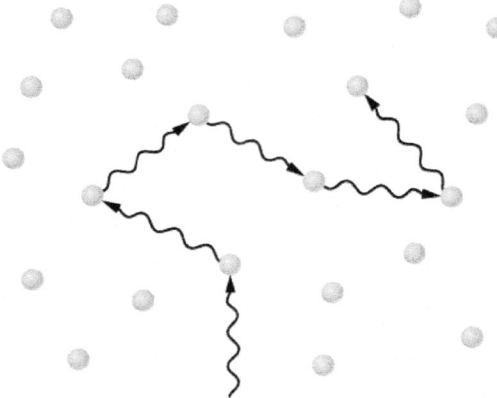

Figure 16.13. Photons Deep in the Sun. A photon moving through the dense gases in the solar interior travels only a short distance before it interacts with one of the surrounding atoms. The photon usually has a lower energy after each interaction and may then travel in any random direction.

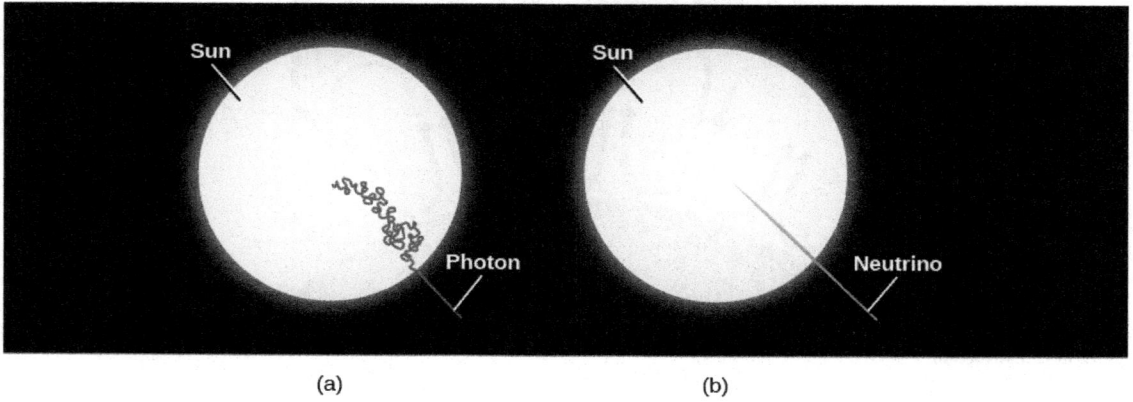

Figure 16.14. Photon and Neutrino Paths in the Sun. (a) Because photons generated by fusion reactions in the solar interior travel only a short distance before being absorbed or scattered by atoms and sent off in random directions, estimates are that it takes between 100,000 and 1,000,000 years for energy to make its way from the center of the Sun to its surface. (b) In contrast, neutrinos do not interact with matter but traverse straight through the Sun at the speed of light, reaching the surface in only a little more than 2 seconds.

MAKING CONNECTIONS

Heat Transfer and Cooking

The three ways that heat energy moves from higher-temperature regions to cooler regions are all used in cooking, and this is important to all of us who enjoy making or eating food.

Conduction is heat transfer by physical contact during which the energetic motion of particles in one region spread to other regions and even to adjacent objects in close contact. A tasty example of this is cooking a steak on a hot iron skillet. When a flame makes the bottom of a skillet hot, the particles in it vibrate actively and collide with neighboring particles, spreading the heat energy throughout the skillet (the ability to spread heat uniformly is a key criterion for selecting materials for cookware). A steak sitting on the surface of the skillet picks up heat energy by the particles in the surface of the skillet colliding with particles on the surface of the steak. Many cooks will put a little oil on the pan, and this layer of oil,

This OpenStax book is available for free at http://cnx.org/content/col11992/1.8

besides preventing sticking, increases heat transfer by filling in gaps and increasing the contact surface area.

Convection is heat transfer by the motion of matter that rises because it is hot and less dense. Heating a fluid makes it expand, which makes it less dense, so it rises. An oven is a great example of this: the fire is at the bottom of the oven and heats the air down there, causing it to expand (becoming less dense), so it rises up to where the food is. The rising hot air carries the heat from the fire to the food by convection. This is how conventional ovens work. You may also be familiar with convection ovens that use a fan to circulate hot air for more even cooking. A scientist would object to that name because normal non-fan ovens that rely on hot air rising to circulate the heat are convection ovens; technically, the ovens that use fans to help move heat are "advection" ovens. (You may not have heard about this because the scientists who complain loudly about misusing the terms convection and advection don't get out much.)

Radiation is the transfer of heat energy by electromagnetic radiation. Although microwave ovens are an obvious example of using radiation to heat food, a simpler example is a toy oven. Toy ovens are powered by a very bright light bulb. The child-chefs prepare a mix for brownies or cookies, put it into a tray, and place it in the toy oven under the bright light bulb. The light and heat from the bulb hit the brownie mix and cook it. If you have ever put your hand near a bright light, you have undoubtedly noticed your hand getting warmed by the light.

Model Stars

Scientists use the principles we have just described to calculate what the Sun's interior is like. These physical ideas are expressed as mathematical equations that are solved to determine the values of temperature, pressure, density, the efficiency with which photons are absorbed, and other physical quantities throughout the Sun. The solutions obtained, based on a specific set of physical assumptions, provide a theoretical model for the interior of the Sun.

Figure 16.15 schematically illustrates the predictions of a theoretical model for the Sun's interior. Energy is generated through fusion in the core of the Sun, which extends only about one-quarter of the way to the surface but contains about one-third of the total mass of the Sun. At the center, the temperature reaches a maximum of approximately 15 million K, and the density is nearly 150 times that of water. The energy generated in the core is transported toward the surface by radiation until it reaches a point about 70% of the distance from the center to the surface. At this point, convection begins, and energy is transported the rest of the way, primarily by rising columns of hot gas.

Download for free at http://cnx.org/content/col11992/latest/

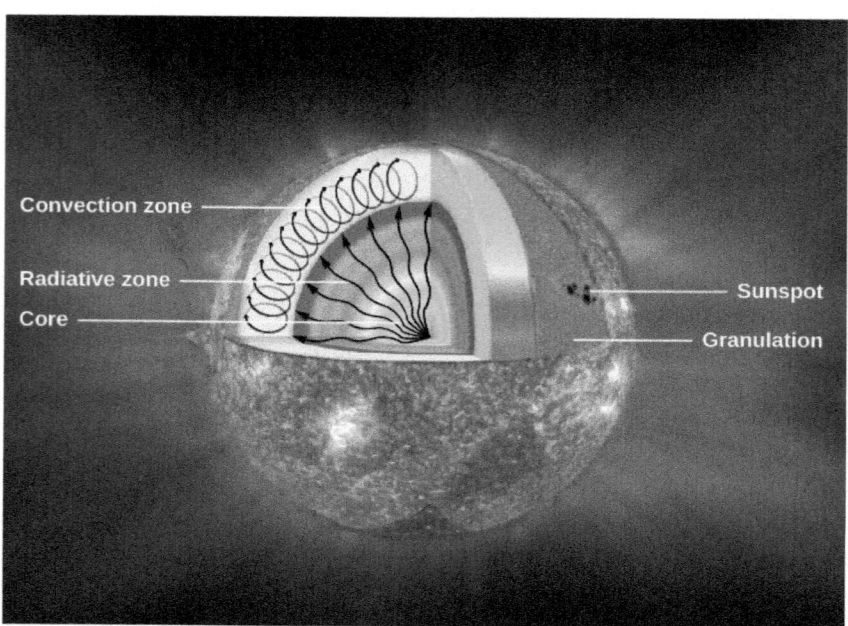

Figure 16.15. Interior Structure of the Sun. Energy is generated in the core by the fusion of hydrogen to form helium. This energy is transmitted outward by radiation—that is, by the absorption and reemission of photons. In the outermost layers, energy is transported mainly by convection. (credit: modification of work by NASA/Goddard)

Figure 16.16 shows how the temperature, density, rate of energy generation, and composition vary from the center of the Sun to its surface.

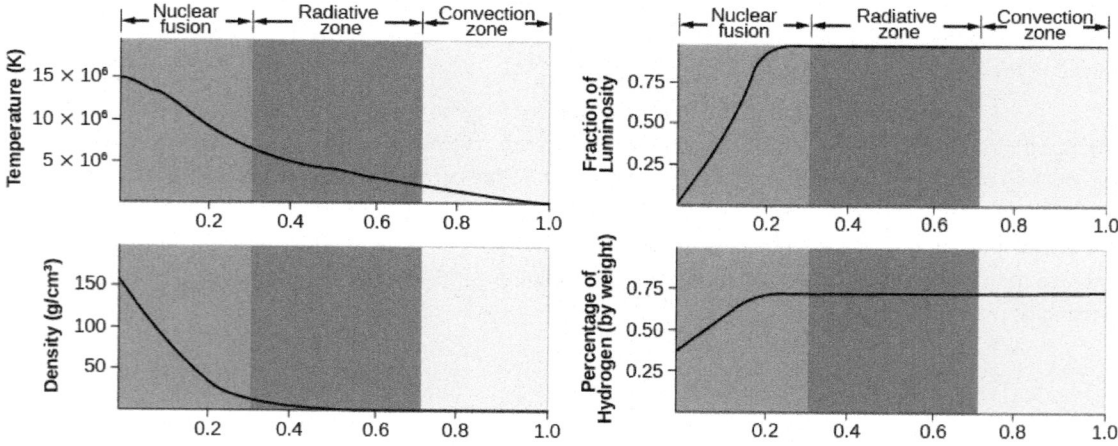

Figure 16.16. Interior of the Sun. Diagrams showing how temperature, density, rate of energy generation, and the percentage (by mass) abundance of hydrogen vary inside the Sun. The horizontal scale shows the fraction of the Sun's radius: the left edge is the very center, and the right edge is the visible surface of the Sun, which is called the photosphere.

16.4 THE SOLAR INTERIOR: OBSERVATIONS

Learning Objectives

By the end of this section, you will be able to:

This OpenStax book is available for free at http://cnx.org/content/col11992/1.8

Download for free at http://cnx.org/content/col11992/latest/

> > Explain how the Sun pulsates
> > Explain what helioseismology is and what it can tell us about the solar interior
> > Discuss how studying neutrinos from the Sun has helped understand neutrinos

Recall that when we observe the Sun's photosphere (the surface layer we see from the outside), we are not seeing very deeply into our star, certainly not into the regions where energy is generated. That's why the title of this section—observations of the solar interior—should seem very surprising. However, astronomers have indeed devised two types of measurements that can be used to obtain information about the inner parts of the Sun. One technique involves the analysis of tiny changes in the motion of small regions at the Sun's surface. The other relies on the measurement of the neutrinos emitted by the Sun.

Solar Pulsations

Astronomers discovered that the Sun pulsates—that is, it alternately expands and contracts—just as your chest expands and contracts as you breathe. This pulsation is very slight, but it can be detected by measuring the *radial velocity* of the solar surface—the speed with which it moves toward or away from us. The velocities of small regions on the Sun are observed to change in a regular way, first toward Earth, then away, then toward, and so on. It is as if the Sun were "breathing" through thousands of individual lungs, each having a size in the range of 4000 to 15,000 kilometers, each fluctuating back and forth (Figure 16.17).

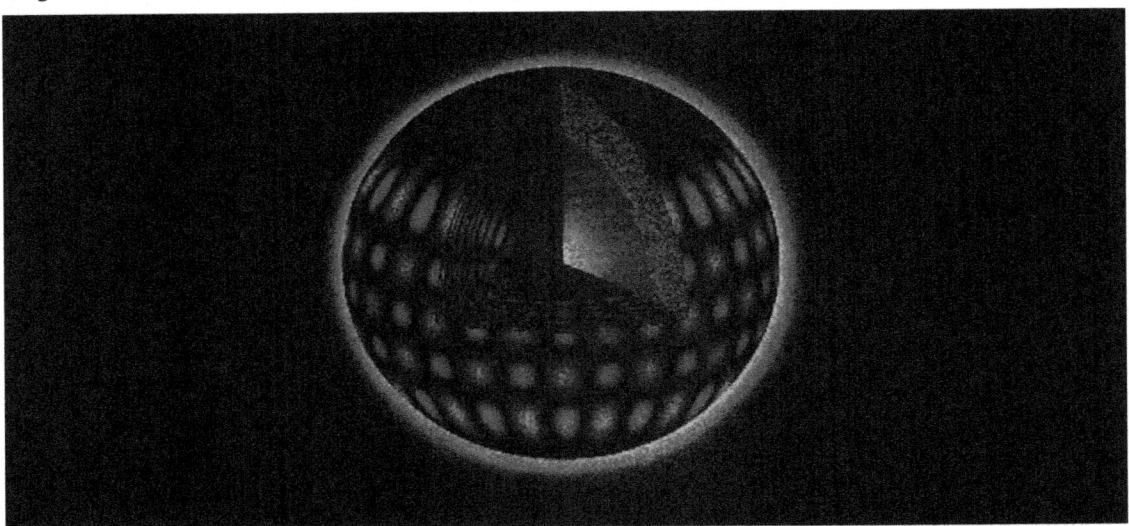

Figure 16.17. Oscillations in the Sun. New observational techniques permit astronomers to measure small differences in velocity at the Sun's surface to infer what the deep solar interior is like. In this computer simulation, red shows surface regions that are moving away from the observer (inward motion); blue marks regions moving toward the observer (outward motion). Note that the velocity changes penetrate deep into the Sun's interior. (credit: modification of work by GONG, NOAO)

The typical velocity of one of the oscillating regions on the Sun is only a few hundred meters per second, and it takes about 5 minutes to complete a full cycle from maximum to minimum velocity and back again. The change in the size of the Sun measured at any given point is no more than a few kilometers.

The remarkable thing is that these small velocity variations can be used to determine what the interior of the Sun is like. The motion of the Sun's surface is caused by waves that reach it from deep in the interior. Study of the amplitude and cycle length of velocity changes provides information about the temperature, density, and composition of the layers through which the waves passed before they reached the surface. The situation is somewhat analogous to the use of seismic waves generated by earthquakes to infer the properties

Download for free at http://cnx.org/content/col11992/latest/

of Earth's interior. For this reason, studies of solar oscillations (back-and-forth motions) are referred to as **helioseismology**.

It takes a little over an hour for waves to traverse the Sun from center to surface, so the waves, like neutrinos, provide information about what the solar interior is like at the present time. In contrast, remember that the sunlight we see today emerging from the Sun was actually generated in the core several hundred thousand years ago.

Helioseismology has shown that convection extends inward from the surface 30% of the way toward the center; we have used this information in drawing Figure 16.15. Pulsation measurements also show that the *differential rotation* that we see at the Sun's surface, with the fastest rotation occurring at the equator, persists down through the convection zone. Below the convection zone, however, the Sun, even though it is gaseous throughout, rotates as if it were a solid body like a bowling ball. Another finding from helioseismology is that the abundance of helium inside the Sun, except in the center where nuclear reactions have converted hydrogen into helium, is about the same as at its surface. That result is important to astronomers because it means we are correct when we use the abundance of the elements measured in the solar atmosphere to construct models of the solar interior.

Helioseismology also allows scientists to look beneath a sunspot and see how it works. In The Sun: A Garden-Variety Star, we said that sunspots are cool because strong magnetic fields block the outward flow of energy. Figure 16.18 shows how gas moves around underneath a sunspot. Cool material from the sunspot flows downward, and material surrounding the sunspot is pulled inward, carrying magnetic field with it and thus maintaining the strong field that is necessary to form a sunspot. As the new material enters the sunspot region, it too cools, becomes denser, and sinks, thus setting up a self-perpetuating cycle that can last for weeks.

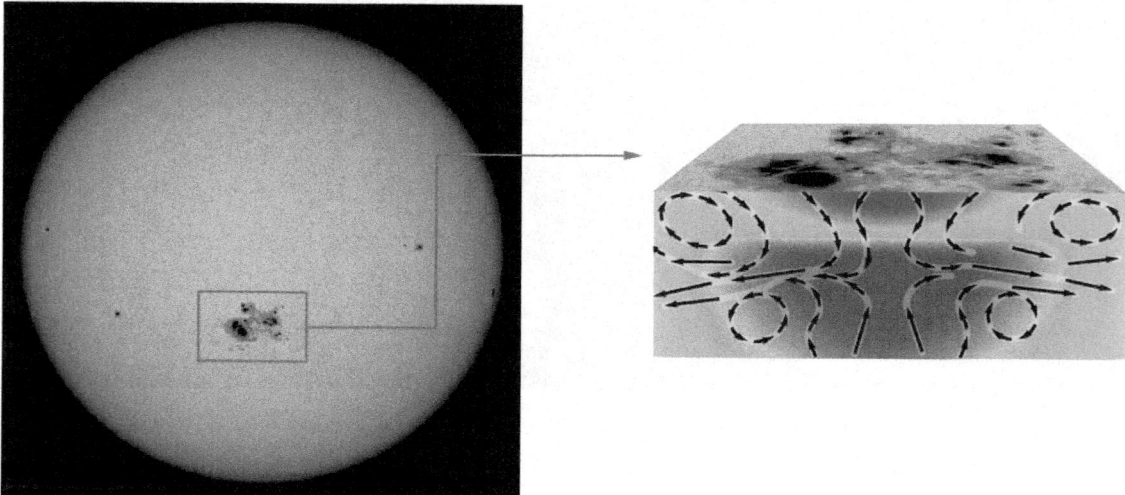

Figure 16.18. Sunspot Structure. This drawing shows our new understanding, from helioseismology, of what lies beneath a sunspot. The black arrows show the direction of the flow of material. The intense magnetic field associated with the sunspot stops the upward flow of hot material and creates a kind of plug that blocks the hot gas. As the material above the plug cools (shown in blue), it becomes denser and plunges inward, drawing more gas and more magnetic field behind it into the spot. The concentrated magnetic field causes more cooling, thereby setting up a self-perpetuating cycle that allows a spot to survive for several weeks. Since the plug keeps hot material from flowing up into the sunspot, the region below the plug, represented by red in this picture, becomes hotter. This material flows sideways and then upward, eventually reaching the solar surface in the area surrounding the sunspot. (credit: modification of work by NASA, SDO)

The downward-flowing cool material acts as a kind of plug that block the upward flow of hot material, which is then diverted sideways and eventually reaches the solar surface in the region around the sunspot. This outward

This OpenStax book is available for free at http://cnx.org/content/col11992/1.8

flow of hot material accounts for the paradox that we described in The Sun: A Garden-Variety Star—namely, that the Sun emits slightly more energy when more of its surface is covered by cool sunspots.

Helioseismology has become an important tool for predicting solar storms that might impact Earth. Active regions can appear and grow large in only a few days. The solar rotation period is about 28 days. Therefore, regions capable of producing solar flares and coronal mass ejections can develop on the far side of the Sun, where, for a long time, we couldn't see them directly.

Fortunately, we now have space telescopes monitoring the Sun from all angles, so we know if there are sunspots forming on the opposite side of the Sun. Moreover, sound waves travel slightly faster in regions of high magnetic field, and waves generated in active regions traverse the Sun about 6 seconds faster than waves generated in quiet regions. By detecting this subtle difference, scientists can provide warnings of a week or more to operators of electric utilities and satellites about when a potentially dangerous active region might rotate into view. With this warning, it is possible to plan for disruptions, put key instruments into safe mode, or reschedule spacewalks in order to protect astronauts.

Solar Neutrinos

The second technique for obtaining information about the Sun's interior involves the detection of a few of those elusive neutrinos created during nuclear fusion. Recall from our earlier discussion that neutrinos created in the center of the Sun make their way directly out of the Sun and travel to Earth at nearly the speed of light. As far as neutrinos are concerned, the Sun is transparent.

About 3% of the total energy generated by nuclear fusion in the Sun is carried away by neutrinos. So many protons react and form neutrinos inside the Sun's core that, scientists calculate, 35 million billion (3.5×10^{16}) solar neutrinos pass through each square meter of Earth's surface every second. If we can devise a way to detect even a few of these solar neutrinos, then we can obtain information directly about what is going on in the center of the Sun. Unfortunately for those trying to "catch" some neutrinos, Earth and everything on it are also nearly transparent to passing neutrinos, just like the Sun.

On very, very rare occasions, however, one of the billions and billions of solar neutrinos will interact with another atom. The first successful detection of solar neutrinos made use of cleaning fluid (C_2Cl_4), which is the least expensive way to get a lot of chlorine atoms together. The nucleus of a chlorine (Cl) atom in the cleaning fluid can be turned into a radioactive argon nucleus by an interaction with a neutrino. Because the argon is radioactive, its presence can be detected. However, since the interaction of a neutrino with chlorine happens so rarely, a huge amount of chlorine is needed.

Raymond Davis, Jr. (Figure 16.19) and his colleagues at Brookhaven National Laboratory, placed a tank containing nearly 400,000 liters of cleaning fluid 1.5 kilometers beneath Earth's surface in a gold mine at Lead, South Dakota. A mine was chosen so that the surrounding material of Earth would keep cosmic rays (high-energy particles from space) from reaching the cleaning fluid and creating false signals. (Cosmic-ray particles are stopped by thick layers of Earth, but neutrinos find them of no significance.) Calculations show that solar neutrinos should produce about one atom of radioactive argon in the tank each day.

Download for free at http://cnx.org/content/col11992/latest/

(a)   (b)

Figure 16.19. Davis Experiment. (a) Raymond Davis received the Nobel Prize in physics in 2002. (b) Davis' experiment at the bottom of an abandoned gold mine first revealed problems with our understanding of neutrinos. (credit a: modification of work by Brookhaven National Laboratory; credit b: modification of work by the United States Department of Energy)

This was an amazing project: they counted argon atoms about once per month—and remember, they were looking for a tiny handful of argon atoms in a massive tank of chlorine atoms. When all was said and done, Davis' experiment, begun in 1970, detected only about one-third as many neutrinos as predicted by solar models! This was a shocking result because astronomers thought they had a pretty good understanding of both neutrinos and the Sun's interior. For many years, astronomers and physicists wrestled with Davis' results, trying to find a way out of the dilemma of the "missing" neutrinos.

Eventually Davis' result was explained by the surprising discovery that there are actually three types of neutrinos. Solar fusion produces only one type of neutrino, the so-called electron neutrino, and the initial experiments to detect solar neutrinos were designed to detect this one type. Subsequent experiments showed that these neutrinos change to a different type during their journey from the center of the Sun through space to Earth in a process called *neutrino oscillation*.

An experiment, conducted at the Sudbury Neutrino Observatory in Canada, was the first one designed to capture all three types of neutrinos (Figure 16.20). The experiment was located in a mine 2 kilometers underground. The neutrino detector consisted of a 12-meter-diameter transparent acrylic plastic sphere, which contained 1000 metric tons of heavy water. Remember that an ordinary water nucleus contains two hydrogen atoms and one oxygen atom. Heavy water instead contains two deuterium atoms and one oxygen atom, and incoming neutrinos can occasionally break up the loosely bound proton and neutron that make up the deuterium nucleus. The sphere of heavy water was surrounded by a shield of 1700 metric tons of very pure water, which in turn was surrounded by 9600 photomultipliers, devices that detect flashes of light produced after neutrons interact with the heavy water.

This OpenStax book is available for free at http://cnx.org/content/col11992/1.8

Figure 16.20. Sudbury Neutrino Detector. The 12-meter sphere of the Sudbury Neutrino Detector lies more than 2 kilometers underground and holds 1000 metric tons of heavy water. (credit: A.B. McDonald (Queen's University) et al., The Sudbury Neutrino Observatory Institute)

To the enormous relief of astronomers who make models of the Sun, the Sudbury experiment detected about 1 neutrino per hour and has shown that the *total* number of neutrinos reaching the heavy water is just what solar models predict. Only one-third of these, however, are electron neutrinos. It appears that two-thirds of the electron neutrinos produced by the Sun transform themselves into one of the other types of neutrinos as they make their way from the core of the Sun to Earth. This is why the earlier experiments saw only one-third the number of neutrinos expected.

Although it is not intuitively obvious, such neutrino oscillations can happen only if the mass of the electron neutrino is not zero. Other experiments indicate that its mass is tiny (even compared to the electron). The 2015 Nobel Prize in physics was awarded to researchers Takaaki Kajita and Arthur B. McDonald for their work establishing the changeable nature of neutrinos. (Raymond Davis shared the 2002 Nobel Prize with Japan's Masatoshi Koshiba for the experiments that led to our understanding of the neutrino problem in the first place.) But the fact that the neutrino has mass at all has deep implications for both physics and astronomy. For example, we will look at the role that neutrinos play in the inventory of the mass of the universe in The Big Bang.

The Borexino experiment, an international experiment conducted in Italy, detected neutrinos coming from the Sun that were identified as coming from different reactions. Whereas the p-p chain is the reaction producing most of the Sun's energy, it is not the only nuclear reaction occurring in the Sun's core. There are side reactions involving nuclei of such elements as beryllium and boron. By probing the number of neutrinos that come from each reaction, the Borexino experiment has helped us confirm in detail our understanding of nuclear fusion in the Sun. In 2014, the Borexino experiment also identified neutrinos that were produced by the first step in the p-p chain, confirming the models of solar astronomers.

It's amazing that a series of experiments that began with enough cleaning fluid to fill a swimming pool brought down the shafts of an old gold mine is now teaching us about the energy source of the Sun and the properties of matter! This is a good example of how experiments in astronomy and physics, coupled with the best theoretical models we can devise, continue to lead to fundamental changes in our understanding of nature.

Download for free at http://cnx.org/content/col11992/latest/

CHAPTER 16 REVIEW

 KEY TERMS

conduction process by which heat is directly transmitted through a substance when there is a difference of temperature between adjoining regions caused by atomic or molecular collisions

convection movement caused within a gas or liquid by the tendency of hotter, and therefore less dense material, to rise and colder, denser material to sink under the influence of gravity, which consequently results in transfer of heat

fission breaking up of heavier atomic nuclei into lighter ones

fusion building up of heavier atomic nuclei from lighter ones

helioseismology study of pulsations or oscillations of the Sun in order to determine the characteristics of the solar interior

hydrostatic equilibrium balance between the weights of various layers, as in a star or Earth's atmosphere, and the pressures that support them

neutrino fundamental particle that has no charge and a mass that is tiny relative to an electron; it rarely interacts with ordinary matter and comes in three different types

positron particle with the same mass as an electron, but positively charged

proton-proton chain series of thermonuclear reactions by which nuclei of hydrogen are built up into nuclei of helium

radiation emission of energy as electromagnetic waves or photons also the transmitted energy itself

 SUMMARY

16.1 Sources of Sunshine: Thermal and Gravitational Energy

The Sun produces an enormous amount of energy every second. Since Earth and the solar system are roughly 4.5 billion years old, this means that the Sun has been producing vast amounts for energy for a very, very long time. Neither chemical burning nor gravitational contraction can account for the total amount of energy radiated by the Sun during all this time.

16.2 Mass, Energy, and the Theory of Relativity

Solar energy is produced by interactions of particles—that is, protons, neutrons, electrons, positrons, and neutrinos. Specifically, the source of the Sun's energy is the fusion of hydrogen to form helium. The series of reactions required to convert hydrogen to helium is called the proton-proton chain. A helium atom is about 0.71% less massive than the four hydrogen atoms that combine to form it, and that lost mass is converted to energy (with the amount of energy given by the formula $E = mc^2$).

16.3 The Solar Interior: Theory

Even though we cannot see inside the Sun, it is possible to calculate what its interior must be like. As input for these calculations, we use what we know about the Sun. It is made entirely of hot gas. Apart from some very tiny changes, the Sun is neither expanding nor contracting (it is in hydrostatic equilibrium) and puts out energy

This OpenStax book is available for free at http://cnx.org/content/col11992/1.8

Download for free at http://cnx.org/content/col11992/latest/

at a constant rate. Fusion of hydrogen occurs in the center of the Sun, and the energy generated is carried to the surface by radiation and then convection. A solar model describes the structure of the Sun's interior. Specifically, it describes how pressure, temperature, mass, and luminosity depend on the distance from the center of the Sun.

16.4 The Solar Interior: Observations

Studies of solar oscillations (helioseismology) and neutrinos can provide observational data about the Sun's interior. The technique of helioseismology has so far shown that the composition of the interior is much like that of the surface (except in the core, where some of the original hydrogen has been converted into helium), and that the convection zone extends about 30% of the way from the Sun's surface to its center. Helioseismology can also detect active regions on the far side of the Sun and provide better predictions of solar storms that may affect Earth. Neutrinos from the Sun call tell us about what is happening in the solar interior. A recent experiment has shown that solar models do predict accurately the number of electron neutrinos produced by nuclear reactions in the core of the Sun. However, two-thirds of these neutrinos are converted into different types of neutrinos during their long journey from the Sun to Earth, a result that also indicates that neutrinos are not massless particles.

 FOR FURTHER EXPLORATION

Articles

Harvey, J. et al. "GONG: To See Inside Our Sun." *Sky & Telescope* (November 1987): 470.

Hathaway, D. "Journey to the Heart of the Sun." *Astronomy* (January 1995): 38.

Kennedy, J. "GONG: Probing the Sun's Hidden Heart." *Sky & Telescope* (October 1996): 20. A discussion on hydroseismology.

LoPresto, J. "Looking Inside the Sun." *Astronomy* (March 1989): 20. A discussion on hydroseismology.

McDonald, A. et al. "Solving the Solar Neutrino Problem." *Scientific American* (April 2003): 40. A discussion on how underground experiments with neutrino detectors helped explain the seeming absence of neutrinos from the Sun.

Trefil, J. "How Stars Shine." *Astronomy* (January 1998): 56.

Websites

Albert Einstein Online: http://www.westegg.com/einstein/.

Ghost Particle: http://www.pbs.org/wgbh/nova/neutrino/.

GONG Project Site: http://gong.nso.edu/.

Helioseismology: http://solar-center.stanford.edu/about/helioseismology.html.

Princeton Plasma Physics Lab: http://www.pppl.gov/.

Solving the Mystery of the Solar Neutrinos: http://www.nobelprize.org/nobel_prizes/themes/physics/bahcall/.

Super Kamiokande Neutrino Mass Page: http://www.ps.uci.edu/~superk/.

Download for free at http://cnx.org/content/col11992/latest/

Videos

Deep Secrets of the Neutrino: Physics Underground: https://www.youtube.com/watch?v=Ar9ydagYkYg. 2010 Public Lecture by Peter Rowson at the Stanford Linear Accelerator Center (1:22:00).

The Elusive Neutrino and the Nature of Physics: https://www.youtube.com/watch?v=CBfUHzkcaHQ. Panel at the 2014 World Science Festival (1:30:00).

The Ghost Particle: http://www.dailymotion.com/video/x20rn7s_nova-the-ghost-particle-discovery-science-universe-documentary_tv. 2006 NOVA episode (52:49).

COLLABORATIVE GROUP ACTIVITIES

A. In this chapter, we learned that meteorites falling into the Sun could not be the source of the Sun's energy because the necessary increase in the mass of the Sun would lengthen Earth's orbital period by 2 seconds per year. Have your group discuss what effects this would cause for our planet and for us as the centuries went on.

B. Solar astronomers can learn more about the Sun's interior if they can observe the Sun's oscillations 24 hours each day. This means that they cannot have their observations interrupted by the day/night cycle. Such an experiment, called the GONG (Global Oscillation Network Group) project, was first set up in the 1990s. To save money, this experiment was designed to make use of the minimum possible number of telescopes. It turns out that if the sites are selected carefully, the Sun can be observed all but about 10% of the time with only six observing stations. What factors do you think have to be taken into consideration in selecting the observing sites? Can your group suggest six general geographic locations that would optimize the amount of time that the Sun can be observed? Check your answer by looking at the GONG website.

C. What would it be like if we actually manage to get controlled fusion on Earth to be economically feasible? If the hydrogen in *water* becomes the fuel for releasing enormous amounts of energy (instead of fossil fuels), have your group discuss how this affects the world economy and international politics. (Think of the role that oil and natural gas deposits now play on the world scene and in international politics.)

D. Your group is a delegation sent to the city council of a small mining town to explain why the government is putting a swimming-pool-sized vat of commercial cleaning fluid down one of the shafts of an old gold mine. How would you approach this meeting? Assuming that the members of the city council do not have much science background, how would you explain the importance of the project to them? Suggest some visual aids you could use.

E. When Raymond Davis first suggested his experiment in the underground gold mine, which had significant costs associated with it, some people said it wasn't worth the expense since we already understood the conditions and reactions in the core of the Sun. Yet his experiment led to a major change in our understanding of neutrinos and the physics of subatomic particles. Can your group think of other "expensive" experiments in astronomy that led to fundamental improvements in our understanding of nature?

EXERCISES

This OpenStax book is available for free at http://cnx.org/content/col11992/1.8

Review Questions

1. How do we know the age of the Sun?

2. Explain how we know that the Sun's energy is not supplied either by chemical burning, as in fires here on Earth, or by gravitational contraction (shrinking).

3. What is the ultimate source of energy that makes the Sun shine?

4. What are the formulas for the three steps in the proton-proton chain?

5. How is a neutrino different from a neutron? List all the ways you can think of.

6. Describe in your own words what is meant by the statement that the Sun is in hydrostatic equilibrium.

7. Two astronomy students travel to South Dakota. One stands on Earth's surface and enjoys some sunshine. At the same time, the other descends into a gold mine where neutrinos are detected, arriving in time to detect the creation of a new radioactive argon nucleus. Although the photon at the surface and the neutrinos in the mine arrive at the same time, they have had very different histories. Describe the differences.

8. What do measurements of the number of neutrinos emitted by the Sun tell us about conditions deep in the solar interior?

9. Do neutrinos have mass? Describe how the answer to this question has changed over time and why.

10. Neutrinos produced in the core of the Sun carry energy to its exterior. Is the mechanism for this energy transport conduction, convection, or radiation?

11. What conditions are required before proton-proton chain fusion can start in the Sun?

12. Describe the two main ways that energy travels through the Sun.

Thought Questions

13. Someone suggests that astronomers build a special gamma-ray detector to detect gamma rays produced during the proton-proton chain in the core of the Sun, just like they built a neutrino detector. Explain why this would be a fruitless effort.

14. Earth contains radioactive elements whose decay produces neutrinos. How might we use neutrinos to determine how these elements are distributed in Earth's interior?

15. The Sun is much larger and more massive than Earth. Do you think the average density of the Sun is larger or smaller than that of Earth? Write down your answer before you look up the densities. Now find the values of the densities elsewhere in this text. Were you right? Explain clearly the meanings of density and mass.

16. A friend who has not had the benefit of an astronomy course suggests that the Sun must be full of burning coal to shine as brightly as it does. List as many arguments as you can against this hypothesis.

17. Which of the following transformations is (are) fusion and which is (are) fission: helium to carbon, carbon to iron, uranium to lead, boron to carbon, oxygen to neon? (See Appendix K for a list of the elements.)

18. Why is a higher temperature required to fuse hydrogen to helium by means of the CNO cycle than is required by the process that occurs in the Sun, which involves only isotopes of hydrogen and helium?

Download for free at http://cnx.org/content/col11992/latest/

19. Earth's atmosphere is in hydrostatic equilibrium. What this means is that the pressure at any point in the atmosphere must be high enough to support the weight of air above it. How would you expect the pressure on Mt. Everest to differ from the pressure in your classroom? Explain why.

20. Explain what it means when we say that Earth's oceans are in hydrostatic equilibrium. Now suppose you are a scuba diver. Would you expect the pressure to increase or decrease as you dive below the surface to a depth of 200 feet? Why?

21. What mechanism transfers heat away from the surface of the Moon? If the Moon is losing energy in this way, why does it not simply become colder and colder?

22. Suppose you are standing a few feet away from a bonfire on a cold fall evening. Your face begins to feel hot. What is the mechanism that transfers heat from the fire to your face? (Hint: Is the air between you and the fire hotter or cooler than your face?)

23. Give some everyday examples of the transport of heat by convection and by radiation.

24. Suppose the proton-proton cycle in the Sun were to slow down suddenly and generate energy at only 95% of its current rate. Would an observer on Earth see an immediate decrease in the Sun's brightness? Would she immediately see a decrease in the number of neutrinos emitted by the Sun?

25. Do you think that nuclear fusion takes place in the atmospheres of stars? Why or why not?

26. Why is fission not an important energy source in the Sun?

27. Why do you suppose so great a fraction of the Sun's energy comes from its central regions? Within what fraction of the Sun's radius does practically all of the Sun's luminosity originate (see Figure 16.16)? Within what radius of the Sun has its original hydrogen been partially used up? Discuss what relationship the answers to these questions bear to one another.

28. Explain how mathematical computer models allow us to understand what is going on inside of the Sun.

Figuring For Yourself

29. Estimate the amount of mass that is converted to energy when a proton combines with a deuterium nucleus to form tritium, ^{3}He.

30. How much energy is released when a proton combines with a deuterium nucleus to produce tritium, ^{3}He?

31. The Sun converts 4×10^9 kg of mass to energy every second. How many years would it take the Sun to convert a mass equal to the mass of Earth to energy?

32. Assume that the mass of the Sun is 75% hydrogen and that all of this mass could be converted to energy according to Einstein's equation $E = mc^2$. How much total energy could the Sun generate? If m is in kg and c is in m/s, then E will be expressed in J. (The mass of the Sun is given in Appendix E.)

33. In fact, the conversion of mass to energy in the Sun is not 100% efficient. As we have seen in the text, the conversion of four hydrogen atoms to one helium atom results in the conversion of about 0.02862 times the mass of a proton to energy. How much energy in joules does one such reaction produce? (See Appendix E for the mass of the hydrogen atom, which, for all practical purposes, is the mass of a proton.)

34. Now suppose that all of the hydrogen atoms in the Sun were converted into helium. How much total energy would be produced? (To calculate the answer, you will have to estimate how many hydrogen atoms are in the Sun. This will give you good practice with scientific notation, since the numbers involved are very large! See Appendix C for a review of scientific notation.)

This OpenStax book is available for free at http://cnx.org/content/col11992/1.8

35. Models of the Sun indicate that only about 10% of the total hydrogen in the Sun will participate in nuclear reactions, since it is only the hydrogen in the central regions that is at a high enough temperature. Use the total energy radiated per second by the Sun, 3.8×10^{26} watts, alongside the exercises and information given here to estimate the lifetime of the Sun. (Hint: Make sure you keep track of the units: if the luminosity is the energy radiated per second, your answer will also be in seconds. You should convert the answer to something more meaningful, such as years.)

36. Show that the statement in the text is correct: namely, that roughly 600 million tons of hydrogen must be converted to helium in the Sun each second to explain its energy output. (Hint: Recall Einstein's most famous formula, and remember that for each kg of hydrogen, 0.0071 kg of mass is converted into energy.) How long will it be before 10% of the hydrogen is converted into helium? Does this answer agree with the lifetime you calculated in Exercise 16.35?

37. Every second, the Sun converts 4 million tons of matter to energy. How long will it take the Sun to reduce its mass by 1% (the mass of the Sun is 2×10^{30})? Compare your answer with the lifetime of the Sun so far.

38. Raymond Davis Jr.'s neutrino detector contained approximately 10^{30} chlorine atoms. During his experiment, he found that one neutrino reacted with a chlorine atom to produce one argon atom each day.

    A. How many days would he have to run the experiment for 1% of his tank to be filled with argon atoms?

    B. Convert your answer from A. into years.

    C. Compare this answer to the age of the universe, which is approximately 14 billion years (1.4×10^{10} y).

    D. What does this tell you about how frequently neutrinos interact with matter?

Download for free at http://cnx.org/content/col11992/latest/

This OpenStax book is available for free at http://cnx.org/content/col11992/1.8

17

ANALYZING STARLIGHT

Figure 17.1. Star Colors. This long time exposure shows the colors of the stars. The circular motion of the stars across the image is provided by Earth's rotation. The various colors of the stars are caused by their different temperatures. (credit: modification of work by ESO/A.Santerne)

Chapter Outline

✎ Thinking Ahead

Everything we know about stars—how they are born, what they are made of, how far away they are, how long they live, and how they will die—we learn by decoding the messages contained in the light and radiation that reaches Earth. What questions should we ask, and how do we find the answers?

We can begin our voyage to the stars by looking at the night sky. It is obvious that stars do not all appear equally bright, nor are they all the same color. To understand the stars, we must first determine their basic properties, such as what their temperatures are, how much material they contain (their masses), and how much energy they produce. Since our Sun is a star, of course the same techniques, including spectroscopy, used to study the Sun can be used to find out what stars are like. As we learn more about the stars, we will use these characteristics to begin assembling clues to the main problems we are interested in solving: How do stars form? How long do they survive? What is their ultimate fate?

17.1 THE BRIGHTNESS OF STARS

Learning Objectives

By the end of this section, you will be able to:

> Explain the difference between luminosity and apparent brightness

Download for free at http://cnx.org/content/col11992/latest/

> Understand how astronomers specify brightness with magnitudes

Luminosity

Perhaps the most important characteristic of a star is its **luminosity**—the total amount of energy at all wavelengths that it emits per second. Earlier, we saw that the Sun puts out a tremendous amount of energy every second. (And there are stars far more luminous than the Sun out there.) To make the comparison among stars easy, astronomers express the luminosity of other stars in terms of the Sun's luminosity. For example, the luminosity of Sirius is about 25 times that of the Sun. We use the symbol L_{Sun} to denote the Sun's luminosity; hence, that of Sirius can be written as 25 L_{Sun}. In a later chapter, we will see that if we can measure how much energy a star emits and we also know its mass, then we can calculate how long it can continue to shine before it exhausts its nuclear energy and begins to die.

Apparent Brightness

Astronomers are careful to distinguish between the luminosity of the star (the total energy output) and the amount of energy that happens to reach our eyes or a telescope on Earth. Stars are democratic in how they produce radiation; they emit the same amount of energy in every direction in space. Consequently, only a minuscule fraction of the energy given off by a star actually reaches an observer on Earth. We call the amount of a star's energy that reaches a given area (say, one square meter) each second here on Earth its **apparent brightness**. If you look at the night sky, you see a wide range of apparent brightnesses among the stars. Most stars, in fact, are so dim that you need a telescope to detect them.

If all stars were the same luminosity—if they were like standard bulbs with the same light output—we could use the difference in their apparent brightnesses to tell us something we very much want to know: how far away they are. Imagine you are in a big concert hall or ballroom that is dark except for a few dozen 25-watt bulbs placed in fixtures around the walls. Since they are all 25-watt bulbs, their luminosity (energy output) is the same. But from where you are standing in one corner, they do *not* have the same apparent brightness. Those close to you appear brighter (more of their light reaches your eye), whereas those far away appear dimmer (their light has spread out more before reaching you). In this way, you can tell which bulbs are closest to you. In the same way, if all the stars had the same luminosity, we could immediately infer that the brightest-appearing stars were close by and the dimmest-appearing ones were far away.

To pin down this idea more precisely, recall from the Radiation and Spectra chapter that we know exactly how light fades with increasing distance. The energy we receive is inversely proportional to the square of the distance. If, for example, we have two stars of the same luminosity and one is twice as far away as the other, it will look four times dimmer than the closer one. If it is three times farther away, it will look nine (three squared) times dimmer, and so forth.

Alas, the stars do not all have the same luminosity. (Actually, we are pretty glad about that because having many different types of stars makes the universe a much more interesting place.) But this means that if a star looks dim in the sky, we cannot tell whether it appears dim because it has a low luminosity but is relatively nearby, or because it has a high luminosity but is very far away. To measure the luminosities of stars, we must first compensate for the dimming effects of distance on light, and to do that, we must know how far away they are. Distance is among the most difficult of all astronomical measurements. We will return to how it is determined after we have learned more about the stars. For now, we will describe how astronomers specify the apparent brightness of stars.

This OpenStax book is available for free at http://cnx.org/content/col11992/1.8

The Magnitude Scale

The process of measuring the apparent brightness of stars is called *photometry* (from the Greek *photo* meaning "light" and *–metry* meaning "to measure"). As we saw Observing the Sky: The Birth of Astronomy, astronomical photometry began with Hipparchus. Around 150 B.C.E., he erected an observatory on the island of Rhodes in the Mediterranean. There he prepared a catalog of nearly 1000 stars that included not only their positions but also estimates of their apparent brightnesses.

Hipparchus did not have a telescope or any instrument that could measure apparent brightness accurately, so he simply made estimates with his eyes. He sorted the stars into six brightness categories, each of which he called a **magnitude**. He referred to the brightest stars in his catalog as first-magnitudes stars, whereas those so faint he could barely see them were sixth-magnitude stars. During the nineteenth century, astronomers attempted to make the scale more precise by establishing exactly how much the apparent brightness of a sixth-magnitude star differs from that of a first-magnitude star. Measurements showed that we receive about 100 times more light from a first-magnitude star than from a sixth-magnitude star. Based on this measurement, astronomers then defined an accurate magnitude system in which a difference of five magnitudes corresponds exactly to a brightness ratio of 100:1. In addition, the magnitudes of stars are decimalized; for example, a star isn't just a "second-magnitude star," it has a magnitude of 2.0 (or 2.1, 2.3, and so forth). So what number is it that, when multiplied together five times, gives you this factor of 100? Play on your calculator and see if you can get it. The answer turns out to be about 2.5, which is the fifth root of 100. This means that a magnitude 1.0 star and a magnitude 2.0 star differ in brightness by a factor of about 2.5. Likewise, we receive about 2.5 times as much light from a magnitude 2.0 star as from a magnitude 3.0 star. What about the difference between a magnitude 1.0 star and a magnitude 3.0 star? Since the difference is 2.5 times for each "step" of magnitude, the total difference in brightness is 2.5 × 2.5 = 6.25 times.

Here are a few rules of thumb that might help those new to this system. If two stars differ by 0.75 magnitudes, they differ by a factor of about 2 in brightness. If they are 2.5 magnitudes apart, they differ in brightness by a factor of 10, and a 4-magnitude difference corresponds to a difference in brightness of a factor of 40. You might be saying to yourself at this point, "Why do astronomers continue to use this complicated system from more than 2000 years ago?" That's an excellent question and, as we shall discuss, astronomers today can use other ways of expressing how bright a star looks. But because this system is still used in many books, star charts, and computer apps, we felt we had to introduce students to it (even though we were very tempted to leave it out.)

The brightest stars, those that were traditionally referred to as first-magnitude stars, actually turned out (when measured accurately) not to be identical in brightness. For example, the brightest star in the sky, Sirius, sends us about 10 times as much light as the average first-magnitude star. On the modern magnitude scale, Sirius, the star with the brightest apparent magnitude, has been assigned a magnitude of –1.5. Other objects in the sky can appear even brighter. Venus at its brightest is of magnitude –4.4, while the Sun has a magnitude of –26.8. Figure 17.2 shows the range of observed magnitudes from the brightest to the faintest, along with the actual magnitudes of several well-known objects. The important fact to remember when using magnitude is that the system goes backward: the *larger* the magnitude, the *fainter* the object you are observing.

Download for free at http://cnx.org/content/col11992/latest/

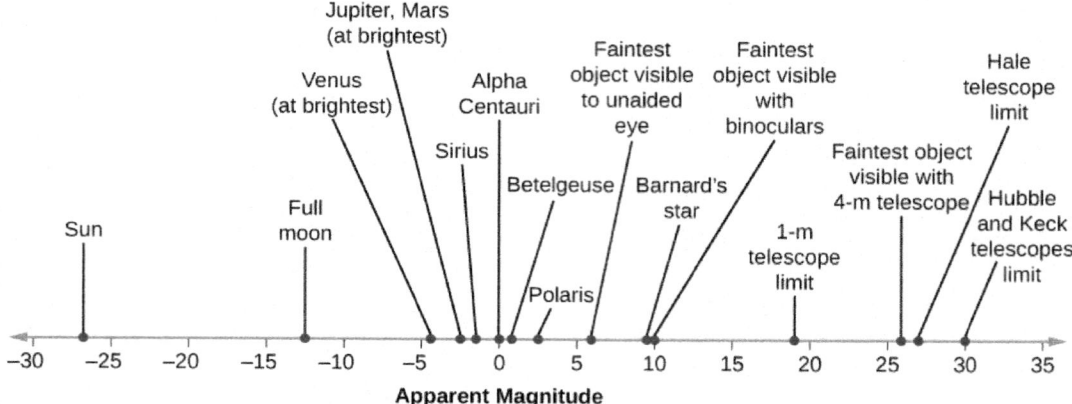

Figure 17.2. Apparent Magnitudes of Well-Known Objects. The faintest magnitudes that can be detected by the unaided eye, binoculars, and large telescopes are also shown.

EXAMPLE 17.1

The Magnitude Equation

Even scientists can't calculate fifth roots in their heads, so astronomers have summarized the above discussion in an equation to help calculate the difference in brightness for stars with different magnitudes. If m_1 and m_2 are the magnitudes of two stars, then we can calculate the ratio of their brightness $\left(\frac{b_2}{b_1}\right)$ using this equation:

$$m_1 - m_2 = 2.5 \log\left(\frac{b_2}{b_1}\right) \qquad \text{or} \qquad \frac{b_2}{b_1} = 2.5^{\,m_1 - m_2}$$

Here is another way to write this equation:

$$\frac{b_2}{b_1} = \left(100^{0.2}\right)^{m_1 - m_2}$$

Let's do a real example, just to show how this works. Imagine that an astronomer has discovered something special about a dim star (magnitude 8.5), and she wants to tell her students how much dimmer the star is than Sirius. Star 1 in the equation will be our dim star and star 2 will be Sirius.

Solution

Remember, Sirius has a magnitude of –1.5. In that case:

$$\frac{b_2}{b_1} = \left(100^{0.2}\right)^{8.5 - (-1.5)} = \left(100^{0.2}\right)^{10}$$

$$= (100)^2 = 100 \times 100 = 10,000$$

Check Your Learning

This OpenStax book is available for free at http://cnx.org/content/col11992/1.8

Download for free at http://cnx.org/content/col11992/latest/

It is a common misconception that Polaris (magnitude 2.0) is the brightest star in the sky, but, as we saw, that distinction actually belongs to Sirius (magnitude −1.5). How does Sirius' apparent brightness compare to that of Polaris?

Answer:

$$\frac{b_{\text{Sirius}}}{b_{\text{Polaris}}} = \left(100^{0.2}\right)^{2.0 - (-1.5)} = \left(100^{0.2}\right)^{3.5} = 100^{0.7} = 25$$

(Hint: If you only have a basic calculator, you may wonder how to take 100 to the 0.7th power. But this is something you can ask Google to do. Google now accepts mathematical questions and will answer them. So try it for yourself. Ask Google, "What is 100 to the 0.7th power?")

Our calculation shows that Sirius' apparent brightness is 25 times greater than Polaris' apparent brightness.

Other Units of Brightness

Although the magnitude scale is still used for visual astronomy, it is not used at all in newer branches of the field. In radio astronomy, for example, no equivalent of the magnitude system has been defined. Rather, radio astronomers measure the amount of energy being collected each second by each square meter of a radio telescope and express the brightness of each source in terms of, for example, watts per square meter.

Similarly, most researchers in the fields of infrared, X-ray, and gamma-ray astronomy use energy per area per second rather than magnitudes to express the results of their measurements. Nevertheless, astronomers in all fields are careful to distinguish between the *luminosity* of the source (even when that luminosity is all in X-rays) and the amount of energy that happens to reach us on Earth. After all, the luminosity is a really important characteristic that tells us a lot about the object in question, whereas the energy that reaches Earth is an accident of cosmic geography.

To make the comparison among stars easy, in this text, we avoid the use of magnitudes as much as possible and will express the luminosity of other stars in terms of the Sun's luminosity. For example, the luminosity of Sirius is 25 times that of the Sun. We use the symbol L_{Sun} to denote the Sun's luminosity; hence, that of Sirius can be written as 25 L_{Sun}.

17.2 COLORS OF STARS

Learning Objectives

By the end of this section, you will be able to:

> Compare the relative temperatures of stars based on their colors
> Understand how astronomers use color indexes to measure the temperatures of stars

Look at the beautiful picture of the stars in the Sagittarius Star Cloud shown in Figure 17.3. The stars show a multitude of colors, including red, orange, yellow, white, and blue. As we have seen, stars are not all the same color because they do not all have identical temperatures. To define *color* precisely, astronomers have devised quantitative methods for characterizing the color of a star and then using those colors to determine stellar

Download for free at http://cnx.org/content/col11992/latest/

temperatures. In the chapters that follow, we will provide the temperature of the stars we are describing, and this section tells you how those temperatures are determined from the colors of light the stars give off.

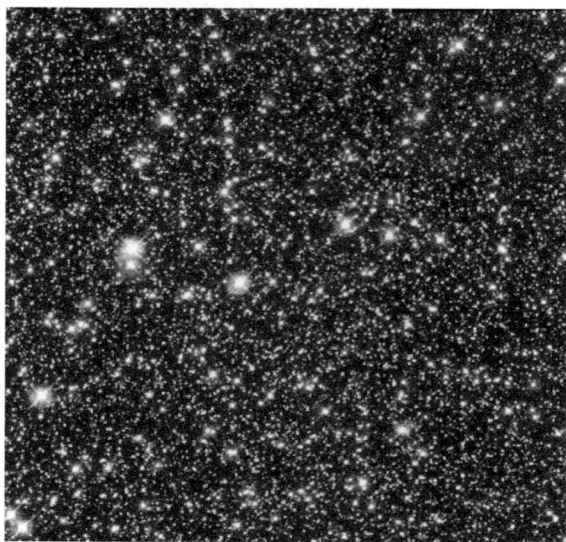

Figure 17.3. Sagittarius Star Cloud. This image, which was taken by the Hubble Space Telescope, shows stars in the direction toward the center of the Milky Way Galaxy. The bright stars glitter like colored jewels on a black velvet background. The color of a star indicates its temperature. Blue-white stars are much hotter than the Sun, whereas red stars are cooler. On average, the stars in this field are at a distance of about 25,000 light-years (which means it takes light 25,000 years to traverse the distance from them to us) and the width of the field is about 13.3 light-years. (credit: Hubble Heritage Team (AURA/STScI/NASA))

Color and Temperature

As we learned in The Electromagnetic Spectrum section, Wien's law relates stellar color to stellar temperature. Blue colors dominate the visible light output of very hot stars (with much additional radiation in the ultraviolet). On the other hand, cool stars emit most of their visible light energy at red wavelengths (with more radiation coming off in the infrared) (Table 17.1). The color of a star therefore provides a measure of its intrinsic or true surface temperature (apart from the effects of reddening by interstellar dust, which will be discussed in Between the Stars: Gas and Dust in Space). Color does not depend on the distance to the object. This should be familiar to you from everyday experience. The color of a traffic signal, for example, appears the same no matter how far away it is. If we could somehow take a star, observe it, and then move it much farther away, its apparent brightness (magnitude) would change. But this change in brightness is the same for all wavelengths, and so its color would remain the same.

Example Star Colors and Corresponding Approximate Temperatures

Star Color	Approximate Temperature	Example
Blue	25,000 K	Spica
White	10,000 K	Vega
Yellow	6000 K	Sun

Table 17.1

This OpenStax book is available for free at http://cnx.org/content/col11992/1.8

Example Star Colors and Corresponding Approximate Temperatures

Star Color	Approximate Temperature	Example
Orange	4000 K	Aldebaran
Red	3000 K	Betelgeuse

Table 17.1

LINK TO LEARNING

Go to this interactive simulation from the University of Colorado (https://openstax.org/l/30UofCsimstar) to see the color of a star changing as the temperature is changed.

The hottest stars have temperatures of over 40,000 K, and the coolest stars have temperatures of about 2000 K. Our Sun's surface temperature is about 6000 K; its peak wavelength color is a slightly greenish-yellow. In space, the Sun would look white, shining with about equal amounts of reddish and bluish wavelengths of light. It looks somewhat yellow as seen from Earth's surface because our planet's nitrogen molecules scatter some of the shorter (i.e., blue) wavelengths out of the beams of sunlight that reach us, leaving more long wavelength light behind. This also explains why the sky is blue: the blue sky is sunlight scattered by Earth's atmosphere.

Color Indices

In order to specify the exact color of a star, astronomers normally measure a star's apparent brightness through filters, each of which transmits only the light from a particular narrow band of wavelengths (colors). A crude example of a filter in everyday life is a green-colored, plastic, soft drink bottle, which, when held in front of your eyes, lets only the green colors of light through.

One commonly used set of filters in astronomy measures stellar brightness at three wavelengths corresponding to ultraviolet, blue, and yellow light. The filters are named: U (ultraviolet), B (blue), and V (visual, for yellow). These filters transmit light near the wavelengths of 360 nanometers (nm), 420 nm, and 540 nm, respectively. The brightness measured through each filter is usually expressed in magnitudes. The difference between any two of these magnitudes—say, between the blue and the visual magnitudes (B–V)—is called a **color index**.

LINK TO LEARNING

Go to this light and filters simulator (https://openstax.org/l/30lightfiltsim) for a demonstration of how different light sources and filters can combine to determine the observed spectrum. You can also see how the perceived colors are associated with the spectrum.

By agreement among astronomers, the ultraviolet, blue, and visual magnitudes of the UBV system are adjusted to give a color index of 0 to a star with a surface temperature of about 10,000 K, such as Vega. The B–V color

Download for free at http://cnx.org/content/col11992/latest/

indexes of stars range from –0.4 for the bluest stars, with temperatures of about 40,000 K, to +2.0 for the reddest stars, with temperatures of about 2000 K. The B–V index for the Sun is about +0.65. Note that, by convention, the B–V index is always the "bluer" minus the "redder" color.

Why use a color index if it ultimately implies temperature? Because the brightness of a star through a filter is what astronomers actually measure, and we are always more comfortable when our statements have to do with measurable quantities.

17.3 THE SPECTRA OF STARS (AND BROWN DWARFS)

Learning Objectives

By the end of this section, you will be able to:

> Describe how astronomers use spectral classes to characterize stars
> Explain the difference between a star and a brown dwarf

Measuring colors is only one way of analyzing starlight. Another way is to use a spectrograph to spread out the light into a spectrum (see the Radiation and Spectra and the Astronomical Instruments chapters). In 1814, the German physicist Joseph Fraunhofer observed that the spectrum of the Sun shows dark lines crossing a continuous band of colors. In the 1860s, English astronomers Sir William Huggins and Lady Margaret Huggins (Figure 17.4) succeeded in identifying some of the lines in stellar spectra as those of known elements on Earth, showing that the same chemical elements found in the Sun and planets exist in the stars. Since then, astronomers have worked hard to perfect experimental techniques for obtaining and measuring spectra, and they have developed a theoretical understanding of what can be learned from spectra. Today, spectroscopic analysis is one of the cornerstones of astronomical research.

Figure 17.4. William Huggins (1824–1910) and Margaret Huggins (1848–1915). William and Margaret Huggins were the first to identify the lines in the spectrum of a star other than the Sun; they also took the first spectrogram, or photograph of a stellar spectrum.

Formation of Stellar Spectra

When the spectra of different stars were first observed, astronomers found that they were not all identical. Since the dark lines are produced by the chemical elements present in the stars, astronomers first thought that the spectra differ from one another because stars are not all made of the same chemical elements. This hypothesis turned out to be wrong. *The primary reason that stellar spectra look different is because the stars have different temperatures.* Most stars have nearly the same composition as the Sun, with only a few exceptions.

Hydrogen, for example, is by far the most abundant element in most stars. However, lines of hydrogen are not seen in the spectra of the hottest and the coolest stars. In the atmospheres of the hottest stars, hydrogen atoms

This OpenStax book is available for free at http://cnx.org/content/col11992/1.8

are completely ionized. Because the electron and the proton are separated, ionized hydrogen cannot produce absorption lines. (Recall from the Formation of Spectral Lines section, the lines are the result of electrons in orbit around a nucleus changing energy levels.)

In the atmospheres of the coolest stars, hydrogen atoms have their electrons attached and can switch energy levels to produce lines. However, practically all of the hydrogen atoms are in the lowest energy state (unexcited) in these stars and thus can absorb only those photons able to lift an electron from that first energy level to a higher level. Photons with enough energy to do this lie in the ultraviolet part of the electromagnetic spectrum, and there are very few ultraviolet photons in the radiation from a cool star. What this means is that if you observe the spectrum of a very hot or very cool star with a typical telescope on the surface of Earth, the most common element in that star, hydrogen, will show very weak spectral lines or none at all.

The hydrogen lines in the visible part of the spectrum (called *Balmer lines*) are strongest in stars with intermediate temperatures—not too hot and not too cold. Calculations show that the optimum temperature for producing visible hydrogen lines is about 10,000 K. At this temperature, an appreciable number of hydrogen atoms are excited to the second energy level. They can then absorb additional photons, rise to still-higher levels of excitation, and produce a dark absorption line. Similarly, every other chemical element, in each of its possible stages of ionization, has a characteristic temperature at which it is most effective in producing absorption lines in any particular part of the spectrum.

Classification of Stellar Spectra

Astronomers use the patterns of lines observed in stellar spectra to sort stars into a **spectral class**. Because a star's temperature determines which absorption lines are present in its spectrum, these spectral classes are a measure of its surface temperature. There are seven standard spectral classes. From hottest to coldest, these seven spectral classes are designated O, B, A, F, G, K, and M. Recently, astronomers have added three additional classes for even cooler objects—L, T, and Y.

At this point, you may be looking at these letters with wonder and asking yourself why astronomers didn't call the spectral types A, B, C, and so on. You will see, as we tell you the history, that it's an instance where tradition won out over common sense.

In the 1880s, Williamina Fleming devised a system to classify stars based on the strength of hydrogen absorption lines. Spectra with the strongest lines were classified as "A" stars, the next strongest "B," and so on down the alphabet to "O" stars, in which the hydrogen lines were very weak. But we saw above that hydrogen lines alone are not a good indicator for classifying stars, since their lines disappear from the visible light spectrum when the stars get too hot or too cold.

In the 1890s, Annie Jump Cannon revised this classification system, focusing on just a few letters from the original system: A, B, F, G, K, M, and O. Instead of starting over, Cannon also rearranged the existing classes—in order of decreasing temperature—into the sequence we have learned: O, B, A, F, G, K, M. As you can read in the feature on Annie Cannon: Classifier of the Stars in this chapter, she classified around 500,000 stars over her lifetime, classifying up to three stars per minute by looking at the stellar spectra.

LINK TO LEARNING

For a deep dive into spectral types, explore the interactive project at the Sloan Digital Sky Survey (https://openstax.org/l/30sloandigsky) in which you can practice classifying stars yourself.

Download for free at http://cnx.org/content/col11992/latest/

To help astronomers remember this crazy order of letters, Cannon created a mnemonic, "Oh Be A Fine Girl, Kiss Me." (If you prefer, you can easily substitute "Guy" for "Girl.") Other mnemonics, which we hope will not be relevant for you, include "Oh Brother, Astronomers Frequently Give Killer Midterms" and "Oh Boy, An F Grade Kills Me!" With the new L, T, and Y spectral classes, the mnemonic might be expanded to "Oh Be A Fine Girl (Guy), Kiss Me Like That, Yo!"

Each of these spectral classes, except possibly for the Y class which is still being defined, is further subdivided into 10 subclasses designated by the numbers 0 through 9. A B0 star is the hottest type of B star; a B9 star is the coolest type of B star and is only slightly hotter than an A0 star.

And just one more item of vocabulary: for historical reasons, astronomers call all the elements heavier than helium *metals*, even though most of them do not show metallic properties. (If you are getting annoyed at the peculiar jargon that astronomers use, just bear in mind that every field of human activity tends to develop its own specialized vocabulary. Just try reading a credit card or social media agreement form these days without training in law!)

Let's take a look at some of the details of how the spectra of the stars change with temperature. (It is these details that allowed Annie Cannon to identify the spectral types of stars as quickly as three per minute!) As Figure 17.5 shows, in the hottest O stars (those with temperatures over 28,000 K), only lines of ionized helium and highly ionized atoms of other elements are conspicuous. Hydrogen lines are strongest in A stars with atmospheric temperatures of about 10,000 K. Ionized metals provide the most conspicuous lines in stars with temperatures from 6000 to 7500 K (spectral type F). In the coolest M stars (below 3500 K), absorption bands of titanium oxide and other molecules are very strong. By the way, the spectral class assigned to the Sun is G2. The sequence of spectral classes is summarized in Table 17.2.

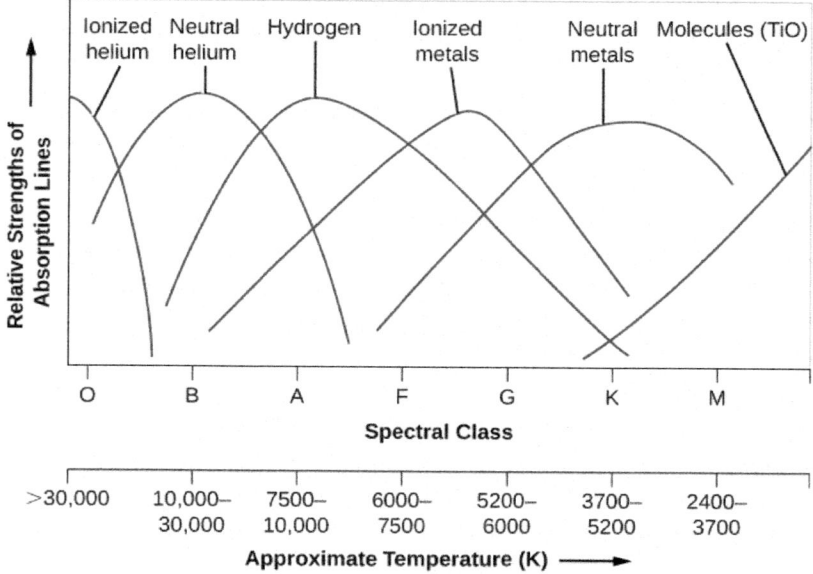

Figure 17.5. Absorption Lines in Stars of Different Temperatures. This graph shows the strengths of absorption lines of different chemical species (atoms, ions, molecules) as we move from hot (left) to cool (right) stars. The sequence of spectral types is also shown.

This OpenStax book is available for free at http://cnx.org/content/col11992/1.8

Download for free at http://cnx.org/content/col11992/latest/

Spectral Classes for Stars

Spectral Class	Color	Approximate Temperature (K)	Principal Features	Examples
O	Blue	> 30,000	Neutral and ionized helium lines, weak hydrogen lines	10 Lacertae
B	Blue-white	10,000–30,000	Neutral helium lines, strong hydrogen lines	Rigel, Spica
A	White	7500–10,000	Strongest hydrogen lines, weak ionized calcium lines, weak ionized metal (e.g., iron, magnesium) lines	Sirius, Vega
F	Yellow-white	6000–7500	Strong hydrogen lines, strong ionized calcium lines, weak sodium lines, many ionized metal lines	Canopus, Procyon
G	Yellow	5200–6000	Weaker hydrogen lines, strong ionized calcium lines, strong sodium lines, many lines of ionized and neutral metals	Sun, Capella
K	Orange	3700–5200	Very weak hydrogen lines, strong ionized calcium lines, strong sodium lines, many lines of neutral metals	Arcturus, Aldebaran
M	Red	2400–3700	Strong lines of neutral metals and molecular bands of titanium oxide dominate	Betelgeuse, Antares
L	Red	1300–2400	Metal hydride lines, alkali metal lines (e.g., sodium, potassium, rubidium)	Teide 1
T	Magenta	700–1300	Methane lines	Gliese 229B
Y	Infrared[1]	< 700	Ammonia lines	WISE 1828+2650

Table 17.2

To see how spectral classification works, let's use Figure 17.5. Suppose you have a spectrum in which the hydrogen lines are about half as strong as those seen in an A star. Looking at the lines in our figure, you see that the star could be either a B star or a G star. But if the spectrum also contains helium lines, then it is a B star, whereas if it contains lines of ionized iron and other metals, it must be a G star.

If you look at Figure 17.6, you can see that you, too, could assign a spectral class to a star whose type was not already known. All you have to do is match the pattern of spectral lines to a standard star (like the ones shown in the figure) whose type has already been determined.

---

1   Absorption by sodium and potassium atoms makes Y dwarfs appear a bit less red than L dwarfs.

Download for free at http://cnx.org/content/col11992/latest/

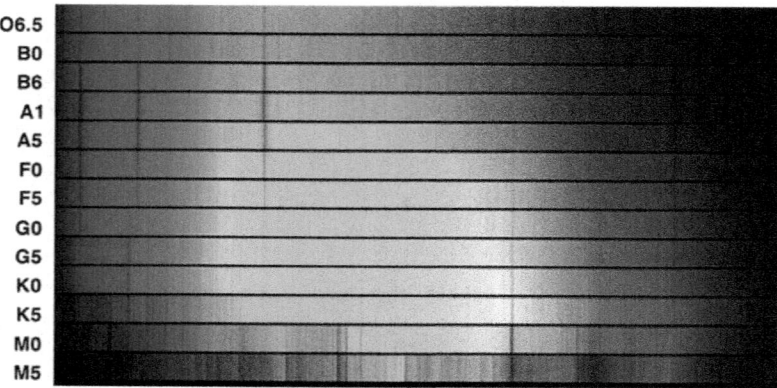

Figure 17.6. Spectra of Stars with Different Spectral Classes. This image compares the spectra of the different spectral classes. The spectral class assigned to each of these stellar spectra is listed at the left of the picture. The strongest four lines seen at spectral type A1 (one in the red, one in the blue-green, and two in the blue) are Balmer lines of hydrogen. Note how these lines weaken at both higher and lower temperatures, as Figure 17.5 also indicates. The strong pair of closely spaced lines in the yellow in the cool stars is due to neutral sodium (one of the neutral metals in Figure 17.5). (Credit: modification of work by NOAO/AURA/NSF)

Both colors and spectral classes can be used to estimate the temperature of a star. Spectra are harder to measure because the light has to be bright enough to be spread out into all colors of the rainbow, and detectors must be sensitive enough to respond to individual wavelengths. In order to measure colors, the detectors need only respond to the many wavelengths that pass simultaneously through the colored filters that have been chosen—that is, to *all* the blue light or *all* the yellow-green light.

VOYAGERS IN ASTRONOMY

Annie Cannon: Classifier of the Stars

Annie Jump Cannon was born in Delaware in 1863 (Figure 17.7). In 1880, she went to Wellesley College, one of the new breed of US colleges opening up to educate young women. Wellesley, only 5 years old at the time, had the second student physics lab in the country and provided excellent training in basic science. After college, Cannon spent a decade with her parents but was very dissatisfied, longing to do scientific work. After her mother's death in 1893, she returned to Wellesley as a teaching assistant and also to take courses at Radcliffe, the women's college associated with Harvard.

Figure 17.7. Annie Jump Cannon (1863–1941). Cannon is well-known for her classifications of stellar spectra. (credit: modification of work by Smithsonian Institution)

In the late 1800s, the director of the Harvard Observatory, Edward C. Pickering, needed lots of help with his ambitious program of classifying stellar spectra. The basis for these studies was a monumental collection of nearly a million photographic spectra of stars, obtained from many years of observations

This OpenStax book is available for free at http://cnx.org/content/col11992/1.8

Download for free at http://cnx.org/content/col11992/latest/

made at Harvard College Observatory in Massachusetts as well as at its remote observing stations in South America and South Africa. Pickering quickly discovered that educated young women could be hired as assistants for one-third or one-fourth the salary paid to men, and they would often put up with working conditions and repetitive tasks that men with the same education would not tolerate. These women became known as the Harvard Computers. (We should emphasize that astronomers were not alone in reaching such conclusions about the relatively new idea of upper-class, educated women working outside the home: women were exploited and undervalued in many fields. This is a legacy from which our society is just beginning to emerge.)

Cannon was hired by Pickering as one of the "computers" to help with the classification of spectra. She became so good at it that she could visually examine and determine the spectral types of several hundred stars per hour (dictating her conclusions to an assistant). She made many discoveries while investigating the Harvard photographic plates, including 300 variable stars (stars whose luminosity changes periodically). But her main legacy is a marvelous catalog of spectral types for hundreds of thousands of stars, which served as a foundation for much of twentieth-century astronomy.

In 1911, a visiting committee of astronomers reported that "she is the one person in the world who can do this work quickly and accurately" and urged Harvard to give Cannon an official appointment in keeping with her skill and renown. Not until 1938, however, did Harvard appoint her an astronomer at the university; she was then 75 years old.

Cannon received the first honorary degree Oxford awarded to a woman, and she became the first woman to be elected an officer of the American Astronomical Society, the main professional organization of astronomers in the US. She generously donated the money from one of the major prizes she had won to found a special award for women in astronomy, now known as the Annie Jump Cannon Prize. True to form, she continued classifying stellar spectra almost to the very end of her life in 1941.

Spectral Classes L, T, and Y

The scheme devised by Cannon worked well until 1988, when astronomers began to discover objects even cooler than M9-type stars. We use the word *object* because many of the new discoveries are not true stars. A star is defined as an object that during some part of its lifetime derives 100% of its energy from the same process that makes the Sun shine—the fusion of hydrogen nuclei (protons) into helium. Objects with masses less than about 7.5% of the mass of our Sun (about 0.075 M_{Sun}) do not become hot enough for hydrogen fusion to take place. Even before the first such "failed star" was found, this class of objects, with masses intermediate between stars and planets, was given the name **brown dwarfs**.

Brown dwarfs are very difficult to observe because they are extremely faint and cool, and they put out most of their light in the infrared part of the spectrum. It was only after the construction of very large telescopes, like the Keck telescopes in Hawaii, and the development of very sensitive infrared detectors, that the search for brown dwarfs succeeded. The first brown dwarf was discovered in 1988, and, as of the summer of 2015, there are more than 2200 known brown dwarfs.

Initially, brown dwarfs were given spectral classes like M10[+] or "much cooler than M9," but so many are now known that it is possible to begin assigning spectral types. The hottest brown dwarfs are given types L0–L9 (temperatures in the range 2400–1300 K), whereas still cooler (1300–700 K) objects are given types T0–T9 (see Figure 17.8). In class L brown dwarfs, the lines of titanium oxide, which are strong in M stars, have disappeared. This is because the L dwarfs are so cool that atoms and molecules can gather together into dust particles in their atmospheres; the titanium is locked up in the dust grains rather than being available to form molecules of

Download for free at http://cnx.org/content/col11992/latest/

titanium oxide. Lines of steam (hot water vapor) are present, along with lines of carbon monoxide and neutral sodium, potassium, cesium, and rubidium. Methane (CH_4) lines are strong in class-T brown dwarfs, as methane exists in the atmosphere of the giant planets in our own solar system.

In 2009, astronomers discovered ultra-cool brown dwarfs with temperatures of 500–600 K. These objects exhibited absorption lines due to ammonia (NH_3), which are not seen in T dwarfs. A new spectral class, Y, was created for these objects. As of 2015, over two dozen brown dwarfs belonging to spectral class Y have been discovered, some with temperatures comparable to that of the human body (about 300 K).

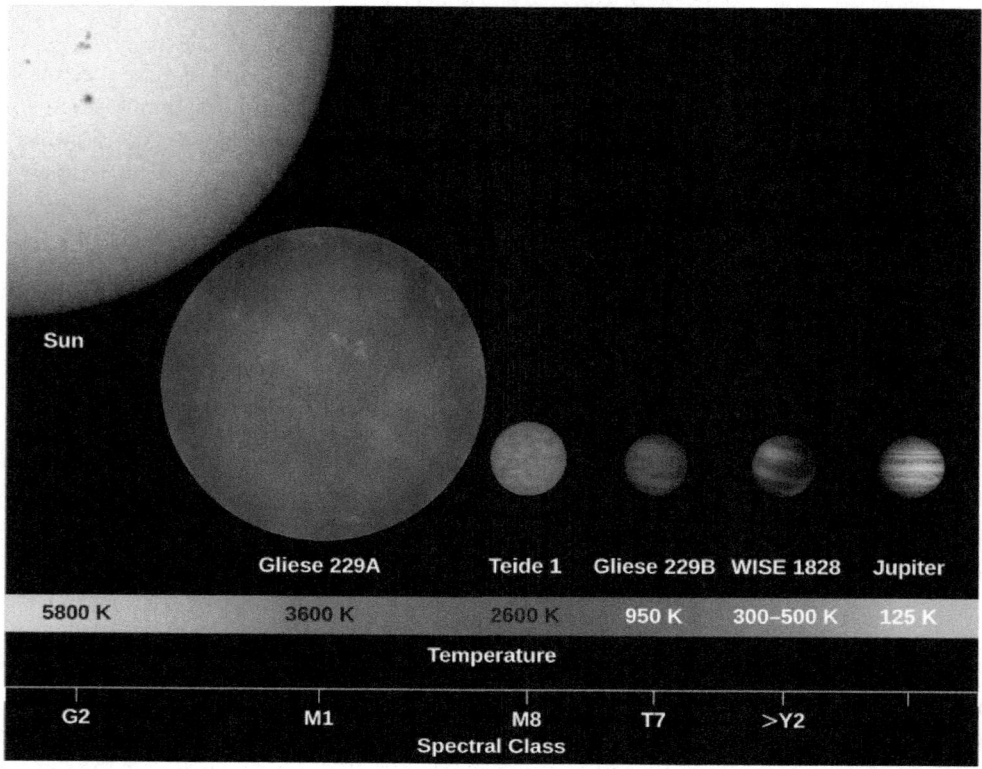

Figure 17.8. Brown Dwarfs. This illustration shows the sizes and surface temperatures of brown dwarfs Teide 1, Gliese 229B, and WISE1828 in relation to the Sun, a red dwarf star (Gliese 229A), and Jupiter. (credit: modification of work by MPIA/V. Joergens)

Most brown dwarfs start out with atmospheric temperatures and spectra like those of true stars with spectral classes of M6.5 and later, even though the brown dwarfs are not hot and dense enough in their interiors to fuse hydrogen. In fact, the spectra of brown dwarfs and true stars are so similar from spectral types late M through L that it is not possible to distinguish the two types of objects based on spectra alone. An independent measure of mass is required to determine whether a specific object is a brown dwarf or a very low mass star. Since brown dwarfs cool steadily throughout their lifetimes, the spectral type of a given brown dwarf changes with time over a billion years or more from late M through L, T, and Y spectral types.

Low-Mass Brown Dwarfs vs. High-Mass Planets

An interesting property of brown dwarfs is that they are all about the same radius as Jupiter, regardless of their masses. Amazingly, this covers a range of masses from about 13 to 80 times the mass of Jupiter (M_J). This can make distinguishing a low-mass brown dwarf from a high-mass planet very difficult.

This OpenStax book is available for free at http://cnx.org/content/col11992/1.8

So, what is the difference between a low-mass brown dwarf and a high-mass planet? The International Astronomical Union considers the distinctive feature to be *deuterium fusion*. Although brown dwarfs do not sustain regular (proton-proton) hydrogen fusion, they are capable of fusing deuterium (a rare form of hydrogen with one proton and one neutron in its nucleus). The fusion of deuterium can happen at a lower temperature than the fusion of hydrogen. If an object has enough mass to fuse deuterium (about 13 M_J or 0.012 M_{Sun}), it is a brown dwarf. Objects with less than 13 M_J do not fuse deuterium and are usually considered planets.

17.4 USING SPECTRA TO MEASURE STELLAR RADIUS, COMPOSITION, AND MOTION

Learning Objectives

By the end of this section, you will be able to:

> Understand how astronomers can learn about a star's radius and composition by studying its spectrum
> Explain how astronomers can measure the motion and rotation of a star using the Doppler effect
> Describe the proper motion of a star and how it relates to a star's space velocity

Analyzing the spectrum of a star can teach us all kinds of things in addition to its temperature. We can measure its detailed chemical composition as well as the pressure in its atmosphere. From the pressure, we get clues about its size. We can also measure its motion toward or away from us and estimate its rotation.

Clues to the Size of a Star

As we shall see in The Stars: A Celestial Census, stars come in a wide variety of sizes. At some periods in their lives, stars can expand to enormous dimensions. Stars of such exaggerated size are called **giants**. Luckily for the astronomer, stellar spectra can be used to distinguish giants from run-of-the-mill stars (such as our Sun).

Suppose you want to determine whether a star is a giant. A giant star has a large, extended photosphere. Because it is so large, a giant star's atoms are spread over a great volume, which means that the density of particles in the star's photosphere is low. As a result, the pressure in a giant star's photosphere is also low. This low pressure affects the spectrum in two ways. First, a star with a lower-pressure photosphere shows narrower spectral lines than a star of the same temperature with a higher-pressure photosphere (Figure 17.9). The difference is large enough that careful study of spectra can tell which of two stars at the same temperature has a higher pressure (and is thus more compressed) and which has a lower pressure (and thus must be extended). This effect is due to collisions between particles in the star's photosphere—more collisions lead to broader spectral lines. Collisions will, of course, be more frequent in a higher-density environment. Think about it like traffic—collisions are much more likely during rush hour, when the density of cars is high.

Second, more atoms are ionized in a giant star than in a star like the Sun with the same temperature. The ionization of atoms in a star's outer layers is caused mainly by photons, and the amount of energy carried by photons is determined by temperature. But how long atoms *stay* ionized depends in part on pressure. Compared with what happens in the Sun (with its relatively dense photosphere), ionized atoms in a giant star's photosphere are less likely to pass close enough to electrons to interact and combine with one or more of them, thereby becoming neutral again. Ionized atoms, as we discussed earlier, have different spectra from atoms that are neutral.

Download for free at http://cnx.org/content/col11992/latest/

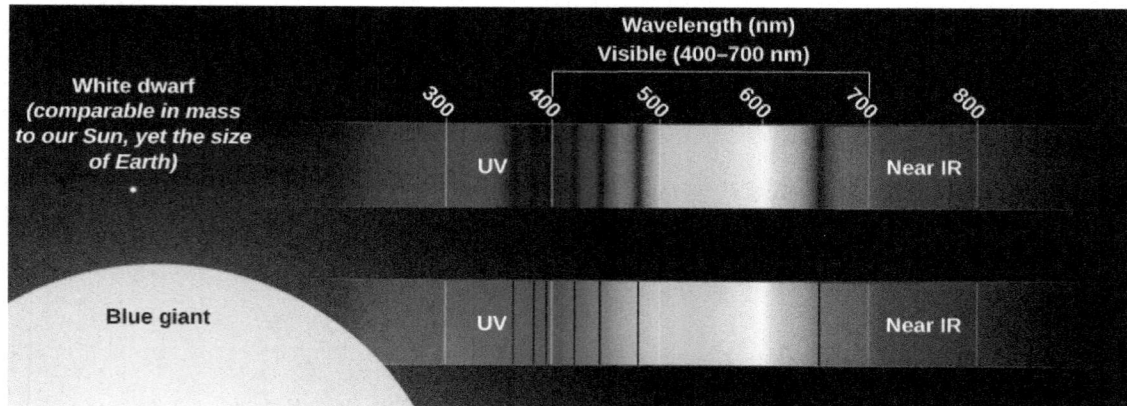

Figure 17.9. Spectral Lines. This figure illustrates one difference in the spectral lines from stars of the same temperature but different pressures. A giant star with a very-low-pressure photosphere shows very narrow spectral lines (bottom), whereas a smaller star with a higher-pressure photosphere shows much broader spectral lines (top). (credit: modification of work by NASA, ESA, A. Field, and J. Kalirai (STScI))

Abundances of the Elements

Absorption lines of a majority of the known chemical elements have now been identified in the spectra of the Sun and stars. If we see lines of iron in a star's spectrum, for example, then we know immediately that the star must contain iron.

Note that the *absence* of an element's spectral lines does not necessarily mean that the element itself is absent. As we saw, the temperature and pressure in a star's atmosphere will determine what types of atoms are able to produce absorption lines. Only if the physical conditions in a star's photosphere are such that lines of an element *should* (according to calculations) be there can we conclude that the absence of observable spectral lines implies low abundance of the element.

Suppose two stars have identical temperatures and pressures, but the lines of, say, sodium are stronger in one than in the other. Stronger lines mean that there are more atoms in the stellar photosphere absorbing light. Therefore, we know immediately that the star with stronger sodium lines contains more sodium. Complex calculations are required to determine exactly how much more, but those calculations can be done for any element observed in any star with any temperature and pressure.

Of course, astronomy textbooks such as ours always make these things sound a bit easier than they really are. If you look at the stellar spectra such as those in Figure 17.6, you may get some feeling for how hard it is to decode all of the information contained in the thousands of absorption lines. First of all, it has taken many years of careful laboratory work on Earth to determine the precise wavelengths at which hot gases of each element have their spectral lines. Long books and computer databases have been compiled to show the lines of each element that can be seen at each temperature. Second, stellar spectra usually have many lines from a number of elements, and we must be careful to sort them out correctly. Sometimes nature is unhelpful, and lines of different elements have identical wavelengths, thereby adding to the confusion. And third, as we saw in the chapter on Radiation and Spectra, the motion of the star can change the observed wavelength of each of the lines. So, the observed wavelengths may not match laboratory measurements exactly. In practice, analyzing stellar spectra is a demanding, sometimes frustrating task that requires both training and skill.

Studies of stellar spectra have shown that hydrogen makes up about three-quarters of the mass of most stars. Helium is the second-most abundant element, making up almost a quarter of a star's mass. Together, hydrogen and helium make up from 96 to 99% of the mass; in some stars, they amount to more than 99.9%. Among the 4% or less of "heavy elements," oxygen, carbon, neon, iron, nitrogen, silicon, magnesium, and sulfur are among

This OpenStax book is available for free at http://cnx.org/content/col11992/1.8

the most abundant. Generally, but not invariably, the elements of lower atomic weight are more abundant than those of higher atomic weight.

Take a careful look at the list of elements in the preceding paragraph. Two of the most abundant are hydrogen and oxygen (which make up water); add carbon and nitrogen and you are starting to write the prescription for the chemistry of an astronomy student. We are made of elements that are common in the universe—just mixed together in a far more sophisticated form (and a much cooler environment) than in a star.

As we mentioned in The Spectra of Stars (and Brown Dwarfs) section, astronomers use the term "metals" to refer to all elements heavier than hydrogen and helium. The fraction of a star's mass that is composed of these elements is referred to as the star's *metallicity*. The metallicity of the Sun, for example, is 0.02, since 2% of the Sun's mass is made of elements heavier than helium.

Appendix K lists how common each element is in the universe (compared to hydrogen); these estimates are based primarily on investigation of the Sun, which is a typical star. Some very rare elements, however, have not been detected in the Sun. Estimates of the amounts of these elements in the universe are based on laboratory measurements of their abundance in primitive meteorites, which are considered representative of unaltered material condensed from the solar nebula (see the Cosmic Samples and the Origin of the Solar System chapter).

Radial Velocity

When we measure the spectrum of a star, we determine the wavelength of each of its lines. If the star is not moving with respect to the Sun, then the wavelength corresponding to each element will be the same as those we measure in a laboratory here on Earth. But if stars are moving toward or away from us, we must consider the *Doppler effect* (see The Doppler Effect section). We should see all the spectral lines of moving stars shifted toward the red end of the spectrum if the star is moving away from us, or toward the blue (violet) end if it is moving toward us (Figure 17.10). The greater the shift, the faster the star is moving. Such motion, along the line of sight between the star and the observer, is called **radial velocity** and is usually measured in kilometers per second.

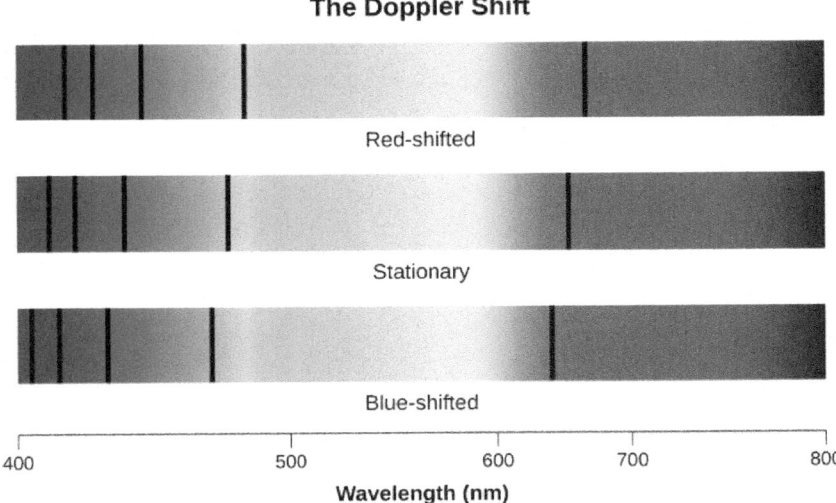

Figure 17.10. Doppler-Shifted Stars. When the spectral lines of a moving star shift toward the red end of the spectrum, we know that the star is moving away from us. If they shift toward the blue end, the star is moving toward us.

Download for free at http://cnx.org/content/col11992/latest/

William Huggins, pioneering yet again, in 1868 made the first radial velocity determination of a star. He observed the Doppler shift in one of the hydrogen lines in the spectrum of Sirius and found that this star is moving toward the solar system. Today, radial velocity can be measured for any star bright enough for its spectrum to be observed. As we will see in The Stars: A Celestial Census, radial velocity measurements of double stars are crucial in deriving stellar masses.

Proper Motion

There is another type of motion stars can have that cannot be detected with stellar spectra. Unlike radial motion, which is along our line of sight (i.e., toward or away from Earth), this motion, called **proper motion**, is *transverse*: that is, across our line of sight. We see it as a change in the relative positions of the stars on the celestial sphere (Figure 17.11). These changes are very slow. Even the star with the largest proper motion takes 200 years to change its position in the sky by an amount equal to the width of the full Moon, and the motions of other stars are smaller yet.

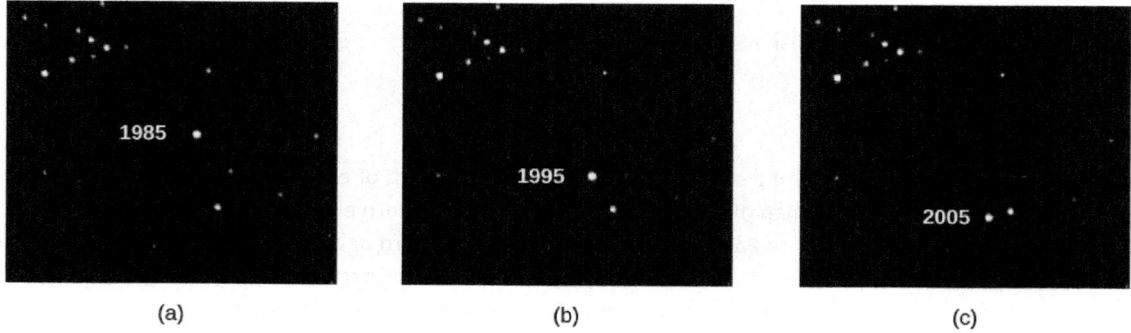

Figure 17.11. Large Proper Motion. Three photographs of Barnard's star, the star with the largest known proper motion, show how this faint star has moved over a period of 20 years. (modification of work by Steve Quirk)

For this reason, with our naked eyes, we do not notice any change in the positions of the bright stars during the course of a human lifetime. If we could live long enough, however, the changes would become obvious. For example, some 50,000 years from now, terrestrial observers will find the handle of the Big Dipper unmistakably more bent than it is now (Figure 17.12).

This OpenStax book is available for free at http://cnx.org/content/col11992/1.8

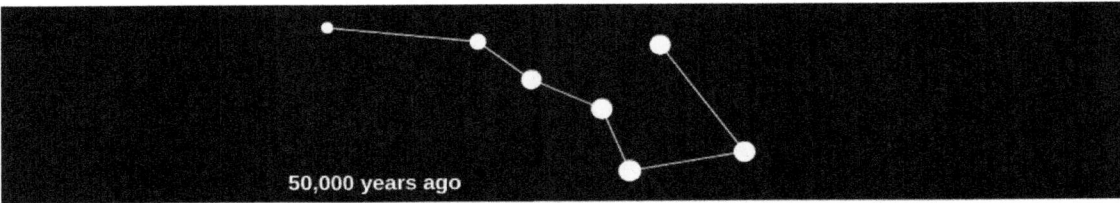

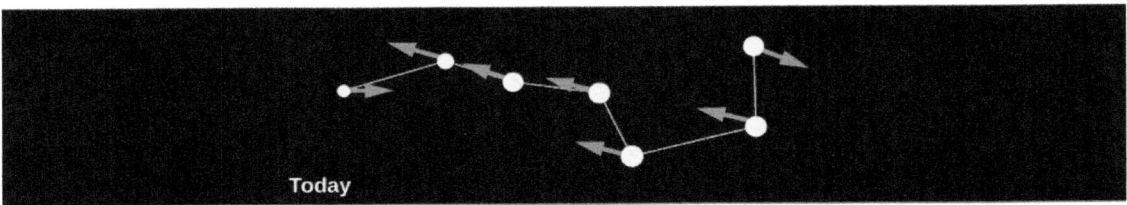

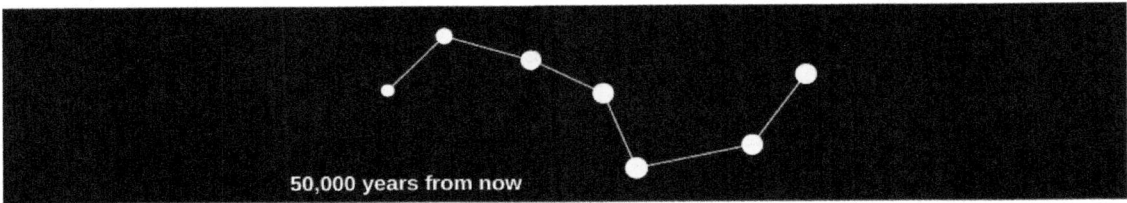

Figure 17.12. Changes in the Big Dipper. This figure shows changes in the appearance of the Big Dipper due to proper motion of the stars over 100,000 years.

We measure the proper motion of a star in arcseconds (1/3600 of a degree) per year. That is, the measurement of proper motion tells us only by how much of an angle a star has changed its position on the celestial sphere. If two stars at different distances are moving at the same velocity perpendicular to our line of sight, the closer one will show a larger shift in its position on the celestial sphere in a year's time. As an analogy, imagine you are standing at the side of a freeway. Cars will appear to whiz past you. If you then watch the traffic from a vantage point half a mile away, the cars will move much more slowly across your field of vision. In order to convert this angular motion to a velocity, we need to know how far away the star is.

To know the true **space velocity** of a star—that is, its total speed and the direction in which it is moving through space relative to the Sun—we must know its radial velocity, proper motion, and distance (Figure 17.13). A star's space velocity can also, over time, cause its distance from the Sun to change significantly. Over several hundred thousand years, these changes can be large enough to affect the apparent brightnesses of nearby stars. Today, Sirius, in the constellation Canis Major (the Big Dog) is the brightest star in the sky, but 100,000 years ago, the star Canopus in the constellation Carina (the Keel) was the brightest one. A little over 200,000 years from now, Sirius will have moved away and faded somewhat, and Vega, the bright blue star in Lyra, will take over its place of honor as the brightest star in Earth's skies.

Download for free at http://cnx.org/content/col11992/latest/

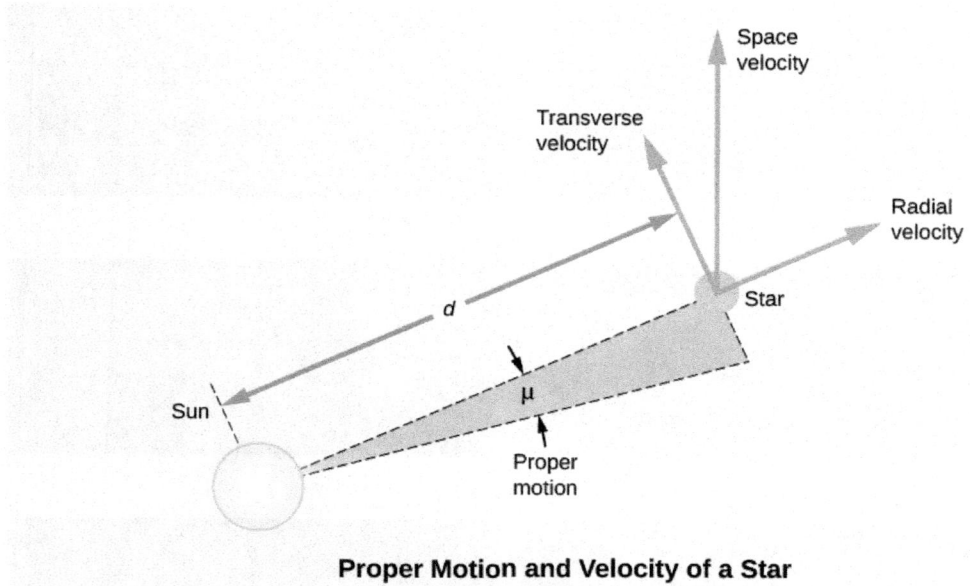

Proper Motion and Velocity of a Star

Figure 17.13. Space Velocity and Proper Motion. This figure shows the true space velocity of a star. The radial velocity is the component of the space velocity projected along the line of sight from the Sun to a star. The transverse velocity is a component of the space velocity projected on the sky. What astronomers measure is proper motion (μ), which is the change in the apparent direction on the sky measured in fractions of a degree. To convert this change in direction to a speed in, say, kilometers per second, it is necessary to also know the distance (*d*) from the Sun to the star.

Rotation

We can also use the Doppler effect to measure how fast a star rotates. If an object is rotating, then one of its sides is approaching us while the other is receding (unless its axis of rotation happens to be pointed exactly toward us). This is clearly the case for the Sun or a planet; we can observe the light from either the approaching or receding edge of these nearby objects and directly measure the Doppler shifts that arise from the rotation.

Stars, however, are so far away that they all appear as unresolved points. The best we can do is to analyze the light from the entire star at once. Due to the Doppler effect, the lines in the light that come from the side of the star rotating toward us are shifted to shorter wavelengths and the lines in the light from the opposite edge of the star are shifted to longer wavelengths. You can think of each spectral line that we observe as the sum or composite of spectral lines originating from different speeds with respect to us. Each point on the star has its own Doppler shift, so the absorption line we see from the whole star is actually much wider than it would be if the star were not rotating. If a star is rotating rapidly, there will be a greater spread of Doppler shifts and all its spectral lines should be quite broad. In fact, astronomers call this effect *line broadening*, and the amount of broadening can tell us the speed at which the star rotates (Figure 17.14).

This OpenStax book is available for free at http://cnx.org/content/col11992/1.8

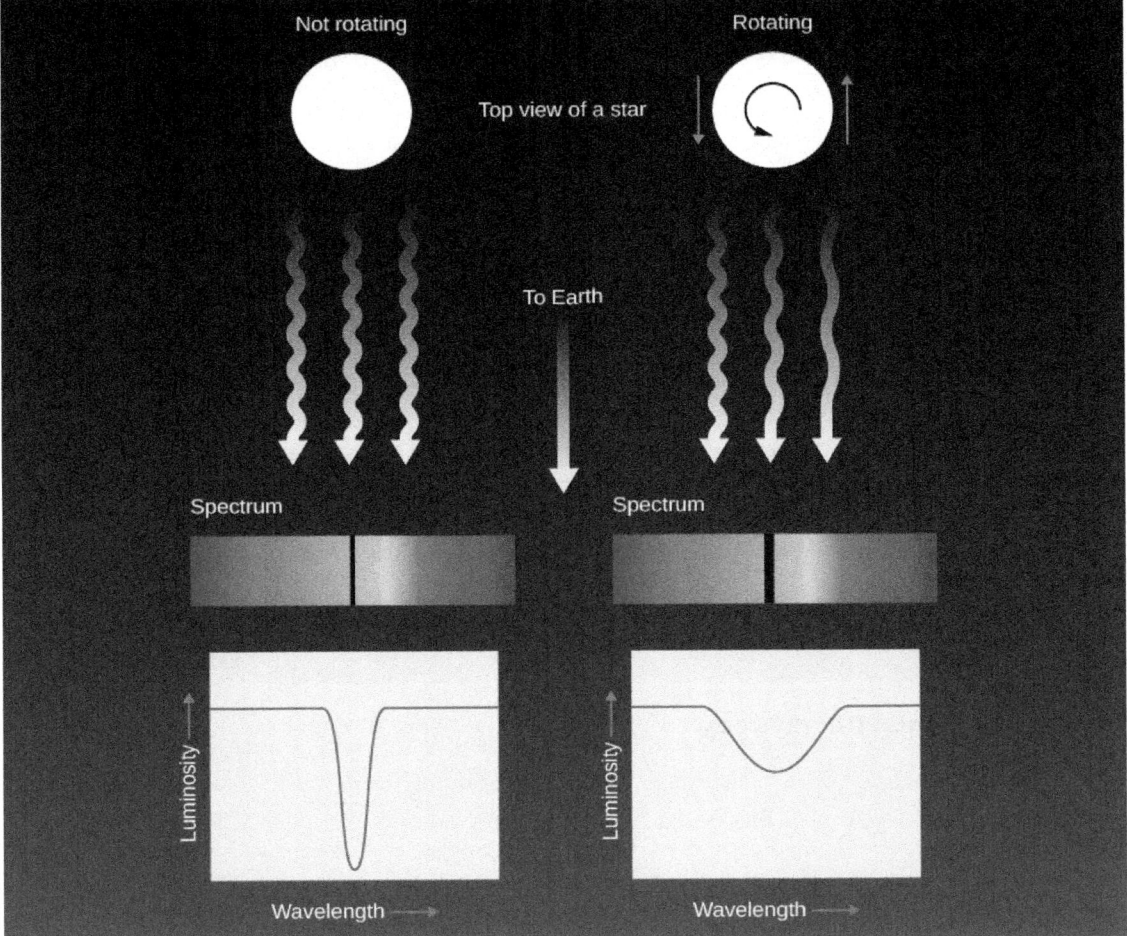

Figure 17.14. Using a Spectrum to Determine Stellar Rotation. A rotating star will show broader spectral lines than a nonrotating star.

Measurements of the widths of spectral lines show that many stars rotate faster than the Sun, some with periods of less than a day! These rapid rotators spin so fast that their shapes are "flattened" into what we call *oblate spheroids*. An example of this is the star Vega, which rotates once every 12.5 hours. Vega's rotation flattens its shape so much that its diameter at the equator is 23% wider than its diameter at the poles (Figure 17.15). The Sun, with its rotation period of about a month, rotates rather slowly. Studies have shown that stars decrease their rotational speed as they age. Young stars rotate very quickly, with rotational periods of days or less. Very old stars can have rotation periods of several months.

Download for free at http://cnx.org/content/col11992/latest/

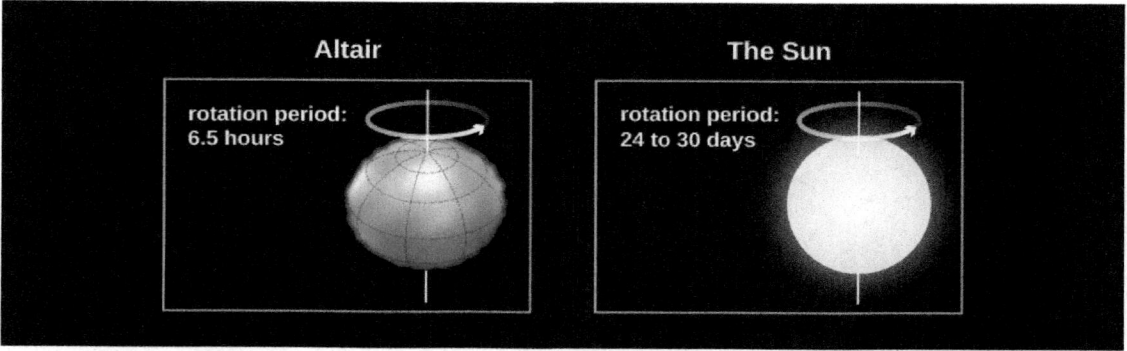

Figure 17.15. Comparison of Rotating Stars. This illustration compares the more rapidly rotating star Altair to the slower rotating Sun.

As you can see, spectroscopy is an extremely powerful technique that helps us learn all kinds of information about stars that we simply could not gather any other way. We will see in later chapters that these same techniques can also teach us about galaxies, which are the most distant objects that can we observe. Without spectroscopy, we would know next to nothing about the universe beyond the solar system.

MAKING CONNECTIONS

Astronomy and Philanthropy

Throughout the history of astronomy, contributions from wealthy patrons of the science have made an enormous difference in building new instruments and carrying out long-term research projects. Edward Pickering's stellar classification project, which was to stretch over several decades, was made possible by major donations from Anna Draper. She was the widow of Henry Draper, a physician who was one of the most accomplished amateur astronomers of the nineteenth century and the first person to successfully photograph the spectrum of a star. Anna Draper gave several hundred thousand dollars to Harvard Observatory. As a result, the great spectroscopic survey is still known as the Henry Draper Memorial, and many stars are still referred to by their "HD" numbers in that catalog (such as HD 209458).

In the 1870s, the eccentric piano builder and real estate magnate James Lick (Figure 17.16) decided to leave some of his fortune to build the world's largest telescope. When, in 1887, the pier to house the telescope was finished, Lick's body was entombed in it. Atop the foundation rose a 36-inch refractor, which for many years was the main instrument at the Lick Observatory near San Jose.

This OpenStax book is available for free at http://cnx.org/content/col11992/1.8

Figure 17.16. Henry Draper (1837–1882) and James Lick (1796–1876). (a) Draper stands next to a telescope used for photography. After his death, his widow funded further astronomy work in his name. (b) Lick was a philanthropist who provided funds to build a 36-inch refractor not only as a memorial to himself but also to aid in further astronomical research.

The Lick telescope remained the largest in the world until 1897, when George Ellery Hale persuaded railroad millionaire Charles Yerkes to finance the construction of a 40-inch telescope near Chicago. More recently, Howard Keck, whose family made its fortune in the oil industry, gave $70 million from his family foundation to the California Institute of Technology to help build the world's largest telescope atop the 14,000-foot peak of Mauna Kea in Hawaii (see the chapter on Astronomical Instruments to learn more about these telescopes). The Keck Foundation was so pleased with what is now called the Keck telescope that they gave $74 million more to build Keck II, another 10-meter reflector on the same volcanic peak.

Now, if any of you become millionaires or billionaires, and astronomy has sparked your interest, do keep an astronomical instrument or project in mind as you plan your estate. But frankly, private philanthropy could not possibly support the full enterprise of scientific research in astronomy. Much of our exploration of the universe is financed by federal agencies such as the National Science Foundation and NASA in the United States, and by similar government agencies in the other countries. In this way, all of us, through a very small share of our tax dollars, are philanthropists for astronomy.

Download for free at http://cnx.org/content/col11992/latest/

CHAPTER 17 REVIEW

 KEY TERMS

apparent brightness a measure of the amount of light received by Earth from a star or other object—that is, how bright an object appears in the sky, as contrasted with its luminosity

brown dwarf an object intermediate in size between a planet and a star; the approximate mass range is from about 1/100 of the mass of the Sun up to the lower mass limit for self-sustaining nuclear reactions, which is about 0.075 the mass of the Sun; brown dwarfs are capable of deuterium fusion, but not hydrogen fusion

color index difference between the magnitudes of a star or other object measured in light of two different spectral regions—for example, blue minus visual (B–V) magnitudes

giant a star of exaggerated size with a large, extended photosphere

luminosity the rate at which a star or other object emits electromagnetic energy into space; the total power output of an object

magnitude an older system of measuring the amount of light we receive from a star or other luminous object; the larger the magnitude, the less radiation we receive from the object

proper motion the angular change per year in the direction of a star as seen from the Sun

radial velocity motion toward or away from the observer; the component of relative velocity that lies in the line of sight

space velocity the total (three-dimensional) speed and direction with which an object is moving through space relative to the Sun

spectral class (or spectral type) the classification of stars according to their temperatures using the characteristics of their spectra; the types are O, B, A, F, G, K, and M with L, T, and Y added recently for cooler star-like objects that recent survey have revealed

 SUMMARY

17.1 The Brightness of Stars

The total energy emitted per second by a star is called its luminosity. How bright a star looks from the perspective of Earth is its apparent brightness. The apparent brightness of a star depends on both its luminosity and its distance from Earth. Thus, the determination of apparent brightness and measurement of the distance to a star provide enough information to calculate its luminosity. The apparent brightnesses of stars are often expressed in terms of magnitudes, which is an old system based on how human vision interprets relative light intensity.

17.2 Colors of Stars

Stars have different colors, which are indicators of temperature. The hottest stars tend to appear blue or blue-white, whereas the coolest stars are red. A color index of a star is the difference in the magnitudes measured at any two wavelengths and is one way that astronomers measure and express the temperature of stars.

This OpenStax book is available for free at http://cnx.org/content/col11992/1.8

17.3 The Spectra of Stars (and Brown Dwarfs)

The differences in the spectra of stars are principally due to differences in temperature, not composition. The spectra of stars are described in terms of spectral classes. In order of decreasing temperature, these spectral classes are O, B, A, F, G, K, M, L, T, and Y. These are further divided into subclasses numbered from 0 to 9. The classes L, T, and Y have been added recently to describe newly discovered star-like objects—mainly brown dwarfs—that are cooler than M9. Our Sun has spectral type G2.

17.4 Using Spectra to Measure Stellar Radius, Composition, and Motion

Spectra of stars of the same temperature but different atmospheric pressures have subtle differences, so spectra can be used to determine whether a star has a large radius and low atmospheric pressure (a giant star) or a small radius and high atmospheric pressure. Stellar spectra can also be used to determine the chemical composition of stars; hydrogen and helium make up most of the mass of all stars. Measurements of line shifts produced by the Doppler effect indicate the radial velocity of a star. Broadening of spectral lines by the Doppler effect is a measure of rotational velocity. A star can also show proper motion, due to the component of a star's space velocity across the line of sight.

 FOR FURTHER EXPLORATION

Articles

Berman, B. "Magnitude Cum Laude." *Astronomy* (December 1998): 92. How we measure the apparent brightnesses of stars is discussed.

Dvorak, J. "The Women Who Created Modern Astronomy [including Annie Cannon]." *Sky & Telescope* (August 2013): 28.

Hearnshaw, J. "Origins of the Stellar Magnitude Scale." *Sky & Telescope* (November 1992): 494. A good history of how we have come to have this cumbersome system is discussed.

Hirshfeld, A. "The Absolute Magnitude of Stars." *Sky & Telescope* (September 1994): 35.

Kaler, J. "Stars in the Cellar: Classes Lost and Found." *Sky & Telescope* (September 2000): 39. An introduction is provided for spectral types and the new classes L and T.

Kaler, J. "Origins of the Spectral Sequence." *Sky & Telescope* (February 1986): 129.

Skrutskie, M. "2MASS: Unveiling the Infrared Universe." *Sky & Telescope* (July 2001): 34. This article focuses on an all-sky survey at 2 microns.

Sneden, C. "Reading the Colors of the Stars." *Astronomy* (April 1989): 36. This article includes a discussion of what we learn from spectroscopy.

Steffey, P. "The Truth about Star Colors." *Sky & Telescope* (September 1992): 266. The color index and how the eye and film "see" colors are discussed.

Tomkins, J. "Once and Future Celestial Kings." *Sky & Telescope* (April 1989): 59. Calculating the motion of stars and determining which stars were, are, and will be brightest in the sky are discussed.

Websites

Discovery of Brown Dwarfs: http://w.astro.berkeley.edu/~basri/bdwarfs/SciAm-book.pdf.

Listing of Nearby Brown Dwarfs: http://www.solstation.com/stars/pc10bd.htm.

Download for free at http://cnx.org/content/col11992/latest/

Spectral Types of Stars: http://www.skyandtelescope.com/astronomy-equipment/the-spectral-types-of-stars/.

Stellar Velocities https://www.e-education.psu.edu/astro801/content/l4_p7.html.

Unheard Voices! The Contributions of Women to Astronomy: A Resource Guide: http://multiverse.ssl.berkeley.edu/women and http://www.astrosociety.org/education/astronomy-resource-guides/women-in-astronomy-an-introductory-resource-guide/.

Videos

When You Are Just Too Small to be a Star: https://www.youtube.com/watch?v=zXCDsb4n4KU. 2013 Public Talk on Brown Dwarfs and Planets by Dr. Gibor Basri of the University of California–Berkeley (1:32:52).

COLLABORATIVE GROUP ACTIVITIES

A. The Voyagers in Astronomy feature on Annie Cannon: Classifier of the Stars discusses some of the difficulties women who wanted to do astronomy faced in the first half of the twentieth century. What does your group think about the situation for women today? Do men and women have an equal chance to become scientists? Discuss with your group whether, in your experience, boys and girls were equally encouraged to do science and math where you went to school.

B. In the section on magnitudes in The Brightness of Stars, we discussed how this old system of classifying how bright different stars appear to the eye first developed. Your authors complained about the fact that this old system still has to be taught to every generation of new students. Can your group think of any other traditional systems of doing things in science and measurement where tradition rules even though common sense says a better system could certainly be found. Explain. (Hint: Try Daylight Savings Time, or metric versus English units.)

C. Suppose you could observe a star that has only one spectral line. Could you tell what element that spectral line comes from? Make a list of reasons with your group about why you answered yes or no.

D. A wealthy alumnus of your college decides to give $50 million to the astronomy department to build a world-class observatory for learning more about the characteristics of stars. Have your group discuss what kind of equipment they would put in the observatory. Where should this observatory be located? Justify your answers. (You may want to refer back to the Astronomical Instruments chapter and to revisit this question as you learn more about the stars and equipment for observing them in future chapters.)

E. For some astronomers, introducing a new spectral type for the stars (like the types L, T, and Y discussed in the text) is similar to introducing a new area code for telephone calls. No one likes to disrupt the old system, but sometimes it is simply necessary. Have your group make a list of steps an astronomer would have to go through to persuade colleagues that a new spectral class is needed.

EXERCISES

Review Questions

1. What two factors determine how bright a star appears to be in the sky?

This OpenStax book is available for free at http://cnx.org/content/col11992/1.8

2. Explain why color is a measure of a star's temperature.

3. What is the main reason that the spectra of all stars are not identical? Explain.

4. What elements are stars mostly made of? How do we know this?

5. What did Annie Cannon contribute to the understanding of stellar spectra?

6. Name five characteristics of a star that can be determined by measuring its spectrum. Explain how you would use a spectrum to determine these characteristics.

7. How do objects of spectral types L, T, and Y differ from those of the other spectral types?

8. Do stars that look brighter in the sky have larger or smaller magnitudes than fainter stars?

9. The star Antares has an apparent magnitude of 1.0, whereas the star Procyon has an apparent magnitude of 0.4. Which star appears brighter in the sky?

10. Based on their colors, which of the following stars is hottest? Which is coolest? Archenar (blue), Betelgeuse (red), Capella (yellow).

11. Order the seven basic spectral types from hottest to coldest.

12. What is the defining difference between a brown dwarf and a true star?

Thought Questions

13. If the star Sirius emits 23 times more energy than the Sun, why does the Sun appear brighter in the sky?

14. How would two stars of equal luminosity—one blue and the other red—appear in an image taken through a filter that passes mainly blue light? How would their appearance change in an image taken through a filter that transmits mainly red light?

15. Table 17.2 lists the temperature ranges that correspond to the different spectral types. What part of the star do these temperatures refer to? Why?

16. Suppose you are given the task of measuring the colors of the brightest stars, listed in Appendix J, through three filters: the first transmits blue light, the second transmits yellow light, and the third transmits red light. If you observe the star Vega, it will appear equally bright through each of the three filters. Which stars will appear brighter through the blue filter than through the red filter? Which stars will appear brighter through the red filter? Which star is likely to have colors most nearly like those of Vega?

17. Star X has lines of ionized helium in its spectrum, and star Y has bands of titanium oxide. Which is hotter? Why? The spectrum of star Z shows lines of ionized helium and also molecular bands of titanium oxide. What is strange about this spectrum? Can you suggest an explanation?

18. The spectrum of the Sun has hundreds of strong lines of nonionized iron but only a few, very weak lines of helium. A star of spectral type B has very strong lines of helium but very weak iron lines. Do these differences mean that the Sun contains more iron and less helium than the B star? Explain.

Download for free at http://cnx.org/content/col11992/latest/

19. What are the approximate spectral classes of stars with the following characteristics?

  A. Balmer lines of hydrogen are very strong; some lines of ionized metals are present.

  B. The strongest lines are those of ionized helium.

  C. Lines of ionized calcium are the strongest in the spectrum; hydrogen lines show only moderate strength; lines of neutral and metals are present.

  D. The strongest lines are those of neutral metals and bands of titanium oxide.

20. Look at the chemical elements in Appendix K. Can you identify any relationship between the abundance of an element and its atomic weight? Are there any obvious exceptions to this relationship?

21. Appendix I lists some of the nearest stars. Are most of these stars hotter or cooler than the Sun? Do any of them emit more energy than the Sun? If so, which ones?

22. Appendix J lists the stars that appear brightest in our sky. Are most of these hotter or cooler than the Sun? Can you suggest a reason for the difference between this answer and the answer to the previous question? (Hint: Look at the luminosities.) Is there any tendency for a correlation between temperature and luminosity? Are there exceptions to the correlation?

23. What star appears the brightest in the sky (other than the Sun)? The second brightest? What color is Betelgeuse? Use Appendix J to find the answers.

24. Suppose hominids one million years ago had left behind maps of the night sky. Would these maps represent accurately the sky that we see today? Why or why not?

25. Why can only a lower limit to the rate of stellar rotation be determined from line broadening rather than the actual rotation rate? (Refer to Figure 17.14.)

26. Why do you think astronomers have suggested three different spectral types (L, T, and Y) for the brown dwarfs instead of M? Why was one not enough?

27. Sam, a college student, just bought a new car. Sam's friend Adam, a graduate student in astronomy, asks Sam for a ride. In the car, Adam remarks that the colors on the temperature control are wrong. Why did he say that?

Figure 17.16. (credit: modification of work by Michael Sheehan)

28. Would a red star have a smaller or larger magnitude in a red filter than in a blue filter?

29. Two stars have proper motions of one arcsecond per year. Star A is 20 light-years from Earth, and Star B is 10 light-years away from Earth. Which one has the faster velocity in space?

This OpenStax book is available for free at http://cnx.org/content/col11992/1.8

30. Suppose there are three stars in space, each moving at 100 km/s. Star A is moving across (i.e., perpendicular to) our line of sight, Star B is moving directly away from Earth, and Star C is moving away from Earth, but at a 30° angle to the line of sight. From which star will you observe the greatest Doppler shift? From which star will you observe the smallest Doppler shift?

31. What would you say to a friend who made this statement, "The visible-light spectrum of the Sun shows weak hydrogen lines and strong calcium lines. The Sun must therefore contain more calcium than hydrogen."?

Figuring For Yourself

32. In Appendix J, how much more luminous is the most luminous of the stars than the least luminous?

For Exercise 17.33 through Exercise 17.38, use the equations relating magnitude and apparent brightness given in the section on the magnitude scale in The Brightness of Stars and Example 17.1.

33. Verify that if two stars have a difference of five magnitudes, this corresponds to a factor of 100 in the ratio $\left(\frac{b_2}{b_1}\right)$; that 2.5 magnitudes corresponds to a factor of 10; and that 0.75 magnitudes corresponds to a factor of 2.

34. As seen from Earth, the Sun has an apparent magnitude of about –26.7. What is the apparent magnitude of the Sun as seen from Saturn, about 10 AU away? (Remember that one AU is the distance from Earth to the Sun and that the brightness decreases as the inverse square of the distance.) Would the Sun still be the brightest star in the sky?

35. An astronomer is investigating a faint star that has recently been discovered in very sensitive surveys of the sky. The star has a magnitude of 16. How much less bright is it than Antares, a star with magnitude roughly equal to 1?

36. The center of a faint but active galaxy has magnitude 26. How much less bright does it look than the very faintest star that our eyes can see, roughly magnitude 6?

37. You have enough information from this chapter to estimate the distance to Alpha Centauri, the second nearest star, which has an apparent magnitude of 0. Since it is a G2 star, like the Sun, assume it has the same luminosity as the Sun and the difference in magnitudes is a result only of the difference in distance. Estimate how far away Alpha Centauri is. Describe the necessary steps in words and then do the calculation. (As we will learn in the Celestial Distances chapter, this method—namely, assuming that stars with identical spectral types emit the same amount of energy—is actually used to estimate distances to stars.) If you assume the distance to the Sun is in AU, your answer will come out in AU.

38. Do the previous problem again, this time using the information that the Sun is 150,000,000 km away. You will get a very large number of km as your answer. To get a better feeling for how the distances compare, try calculating the time it takes light at a speed of 299,338 km/s to travel from the Sun to Earth and from Alpha Centauri to Earth. For Alpha Centauri, figure out how long the trip will take in years as well as in seconds.

39. Star A and Star B have different apparent brightnesses but identical luminosities. If Star A is 20 light-years away from Earth and Star B is 40 light-years away from Earth, which star appears brighter and by what factor?

40. Star A and Star B have different apparent brightnesses but identical luminosities. Star A is 10 light-years away from Earth and appears 36 times brighter than Star B. How far away is Star B?

Download for free at http://cnx.org/content/col11992/latest/

41. The star Sirius A has an apparent magnitude of −1.5. Sirius A has a dim companion, Sirius B, which is 10,000 times less bright than Sirius A. What is the apparent magnitude of Sirius B? Can Sirius B be seen with the naked eye?

42. Our Sun, a type G star, has a surface temperature of 5800 K. We know, therefore, that it is cooler than a type O star and hotter than a type M star. Given what you learned about the temperature ranges of these types of stars, how many times hotter than our Sun is the hottest type O star? How many times cooler than our Sun is the coolest type M star?

This OpenStax book is available for free at http://cnx.org/content/col11992/1.8

Figure 18.1. Variety of Stars. Stars come in a variety of sizes, masses, temperatures, and luminosities. This image shows part of a cluster of stars in the Small Magellanic Cloud (catalog number NGC 290). Located about 200,000 light-years away, NGC 290 is about 65 light-years across. Because the stars in this cluster are all at about the same distance from us, the differences in apparent brightness correspond to differences in luminosity; differences in temperature account for the differences in color. The various colors and luminosities of these stars provide clues about their life stories. (credit: modification of work by E. Olszewski (University of Arizona), European Space Agency, NASA)

Chapter Outline

✎ Thinking Ahead

How do stars form? How long do they live? And how do they die? Stop and think how hard it is to answer these questions.

Stars live such a long time that nothing much can be gained from staring at one for a human lifetime. To discover how stars evolve from birth to death, it was necessary to measure the characteristics of many stars (to take a celestial census, in effect) and then determine which characteristics help us understand the stars' life stories. Astronomers tried a variety of hypotheses about stars until they came up with the right approach to understanding their development. But the key was first making a thorough census of the stars around us.

18.1 A STELLAR CENSUS

Learning Objectives

By the end of this section, you will be able to:

> Explain why the stars visible to the unaided eye are not typical
> Describe the distribution of stellar masses found close to the Sun

Download for free at http://cnx.org/content/col11992/latest/

Before we can make our own survey, we need to agree on a unit of distance appropriate to the objects we are studying. The stars are all so far away that kilometers (and even astronomical units) would be very cumbersome to use; so—as discussed in Science and the Universe: A Brief Tour—astronomers use a much larger "measuring stick" called the *light-year*. A light-year is the distance that light (the fastest signal we know) travels in 1 year. Since light covers an astounding 300,000 kilometers per second, and since there are a lot of seconds in 1 year, a light-year is a very large quantity: 9.5 trillion (9.5×10^{12}) kilometers to be exact. (Bear in mind that the light-year is a unit of *distance* even though the term *year* appears in it.) If you drove at the legal US speed limit without stopping for food or rest, you would not arrive at the end of a light-year in space until roughly 12 million years had passed. And the closest star is more than 4 light-years away.

Notice that we have not yet said much about how such enormous distances can be measured. That is a complicated question, to which we will return in Celestial Distances. For now, let us assume that distances have been measured for stars in our cosmic vicinity so that we can proceed with our census.

Small Is Beautiful—Or at Least More Common

When we do a census of people in the United States, we count the inhabitants by neighborhood. We can try the same approach for our stellar census and begin with our own immediate neighborhood. As we shall see, we run into two problems—just as we do with a census of human beings. First, it is hard to be sure we have counted *all* the inhabitants; second, our local neighborhood may not contain all possible types of people.

Table 18.1 shows an estimate of the number of stars of each spectral type[1] in our own local neighborhood—within 21 light-years of the Sun. (The Milky Way Galaxy, in which we live, is about 100,000 light-years in diameter, so this figure really applies to a *very* local neighborhood, one that contains a *tiny* fraction of all the billions of stars in the Milky Way.) You can see that there are many more low-luminosity (and hence low mass) stars than high-luminosity ones. Only three of the stars in our local neighborhood (one F type and two A types) are significantly more luminous and more massive than the Sun. This is truly a case where small triumphs over large—at least in terms of numbers. The Sun is more massive than the vast majority of stars in our vicinity.

Stars within 21 Light-Years of the Sun

Spectral Type	Number of Stars
A	2
F	1
G	7
K	17
M	94
White dwarfs	8
Brown dwarfs	33

Table 18.1

1　The spectral types of stars were defined and discussed in **Analyzing Starlight**.

This OpenStax book is available for free at http://cnx.org/content/col11992/1.8

Download for free at http://cnx.org/content/col11992/latest/

This table is based on data published through 2015, and it is likely that more faint objects remain to be discovered (see Figure 18.2). Along with the L and T brown dwarfs already observed in our neighborhood, astronomers expect to find perhaps hundreds of additional T dwarfs. Many of these are likely to be even cooler than the coolest currently known T dwarf. The reason the lowest-mass dwarfs are so hard to find is that they put out very little light—ten thousand to a million times less light than the Sun. Only recently has our technology progressed to the point that we can detect these dim, cool objects.

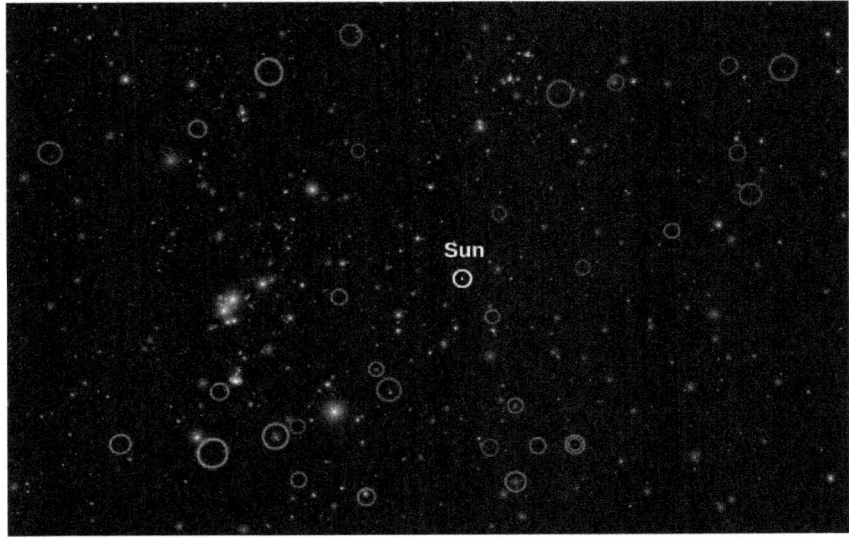

Figure 18.2. Dwarf Simulation. This computer simulation shows the stars in our neighborhood as they would be seen from a distance of 30 light-years away. The Sun is in the center. All the brown dwarfs are circled; those found earlier are circled in blue, the ones found recently with the WISE infrared telescope in space (whose scientists put this diagram together) are circled in red. The common M stars, which are red and faint, are made to look brighter than they really would be so that you can see them in the simulation. Note that luminous hot stars like our Sun are very rare. (credit: modification of work by NASA/ JPL-Caltech)

To put all this in perspective, we note that even though the stars counted in the table are our closest neighbors, you can't just look up at the night sky and see them without a telescope; stars fainter than the Sun cannot be seen with the unaided eye unless they are *very* nearby. For example, stars with luminosities ranging from 1/100 to 1/10,000 the luminosity of the Sun (L_{Sun}) are very common, but a star with a luminosity of 1/100 L_{Sun} would have to be within 5 light-years to be visible to the naked eye—and only three stars (all in one system) are this close to us. The nearest of these three stars, Proxima Centauri, still cannot be seen without a telescope because it has such a low luminosity.

Astronomers are working hard these days to complete the census of our local neighborhood by finding our faintest neighbors. Recent discoveries of nearby stars have relied heavily upon infrared telescopes that are able to find these many cool, low-mass stars. You should expect the number of known stars within 26 light-years of the Sun to keep increasing as more and better surveys are undertaken.

Bright Does Not Necessarily Mean Close

If we confine our census to the local neighborhood, we will miss many of the most interesting kinds of stars. After all, the neighborhood in which you live does not contain all the types of people—distinguished according to age, education, income, race, and so on—that live in the entire country. For example, a few people do live to be over 100 years old, but there may be no such individual within several miles of where you live. In order to sample the full range of the human population, you would have to extend your census to a much larger area. Similarly, some types of stars simply are not found nearby.

Download for free at http://cnx.org/content/col11992/latest/

A clue that we are missing something in our stellar census comes from the fact that only six of the 20 stars that appear brightest in our sky— Sirius, Vega, Altair, Alpha Centauri, Fomalhaut, and Procyon—are found within 26 light-years of the Sun (Figure 18.3). Why are we missing most of the brightest stars when we take our census of the local neighborhood?

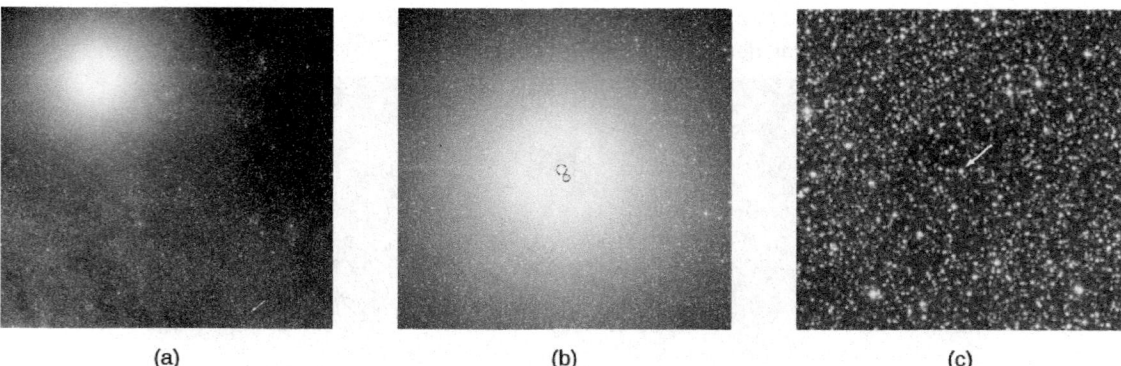

(a)   (b)   (c)

Figure 18.3. The Closest Stars. (a) This image, taken with a wide-angle telescope at the European Southern Observatory in Chile, shows the system of three stars that is our nearest neighbor. (b) Two bright stars that are close to each other (Alpha Centauri A and B) blend their light together. (c) Indicated with an arrow (since you'd hardly notice it otherwise) is the much fainter Proxima Centauri star, which is spectral type M. (credit: modification of work by ESO)

The answer, interestingly enough, is that the stars that appear brightest are *not* the ones closest to us. The brightest stars look the way they do because they emit a very large amount of energy—so much, in fact, that they do not have to be nearby to look brilliant. You can confirm this by looking at Appendix J, which gives distances for the 20 stars that appear brightest from Earth. The most distant of these stars is more than *1000 light-years* from us. In fact, it turns out that most of the stars visible without a telescope are hundreds of light-years away and many times more luminous than the Sun. Among the 6000 stars visible to the unaided eye, only about 50 are intrinsically fainter than the Sun. Note also that several of the stars in Appendix J are spectral type B, a type that is completely missing from Table 18.1.

The most luminous of the bright stars listed in Appendix J emit more than 50,000 times more energy than does the Sun. These highly luminous stars are missing from the solar neighborhood because they are very rare. None of them happens to be in the tiny volume of space immediately surrounding the Sun, and only this small volume was surveyed to get the data shown in Table 18.1.

For example, let's consider the most luminous stars—those 100 or more times as luminous as the Sun. Although such stars are rare, they are visible to the unaided eye, even when hundreds to thousands of light-years away. A star with a luminosity 10,000 times greater than that of the Sun can be seen without a telescope out to a distance of 5000 light-years. The volume of space included within a distance of 5000 light-years, however, is enormous; so even though highly luminous stars are intrinsically rare, many of them are readily visible to our unaided eye.

The contrast between these two samples of stars, those that are close to us and those that can be seen with the unaided eye, is an example of a **selection effect**. When a population of objects (stars in this example) includes a great variety of different types, we must be careful what conclusions we draw from an examination of any particular subgroup. Certainly we would be fooling ourselves if we assumed that the stars visible to the unaided eye are characteristic of the general stellar population; this subgroup is heavily weighted to the most luminous stars. It requires much more effort to assemble a complete data set for the nearest stars, since most are so faint that they can be observed only with a telescope. However, it is only by doing so that astronomers are able to

This OpenStax book is available for free at http://cnx.org/content/col11992/1.8

know about the properties of the vast majority of the stars, which are actually much smaller and fainter than our own Sun. In the next section, we will look at how we measure some of these properties.

18.2 MEASURING STELLAR MASSES

Learning Objectives

By the end of this section, you will be able to:

> Distinguish the different types of binary star systems
> Understand how we can apply Newton's version of Kepler's third law to derive the sum of star masses in a binary star system
> Apply the relationship between stellar mass and stellar luminosity to determine the physical characteristics of a star

The mass of a star—how much material it contains—is one of its most important characteristics. If we know a star's mass, as we shall see, we can estimate how long it will shine and what its ultimate fate will be. Yet the mass of a star is very difficult to measure directly. Somehow, we need to put a star on the cosmic equivalent of a scale.

Luckily, not all stars live like the Sun, in isolation from other stars. About half the stars are **binary stars**—two stars that orbit each other, bound together by gravity. Masses of binary stars can be calculated from measurements of their orbits, just as the mass of the Sun can be derived by measuring the orbits of the planets around it (see Orbits and Gravity).

Binary Stars

Before we discuss in more detail how mass can be measured, we will take a closer look at stars that come in pairs. The first binary star was discovered in 1650, less than half a century after Galileo began to observe the sky with a telescope. John Baptiste Riccioli (1598–1671), an Italian astronomer, noted that the star Mizar, in the middle of the Big Dipper's handle, appeared through his telescope as two stars. Since that discovery, thousands of binary stars have been cataloged. (Astronomers call any pair of stars that appear to be close to each other in the sky *double stars*, but not all of these form a true binary, that is, not all of them are physically associated. Some are just chance alignments of stars that are actually at different distances from us.) Although stars most commonly come in pairs, there are also triple and quadruple systems.

One well-known binary star is Castor, located in the constellation of Gemini. By 1804, astronomer William Herschel, who also discovered the planet Uranus, had noted that the fainter component of Castor had slightly changed its position relative to the brighter component. (We use the term "component" to mean a member of a star system.) Here was evidence that one star was moving around another. It was actually the first evidence that gravitational influences exist outside the solar system. The orbital motion of a binary star is shown in Figure 18.4. A binary star system in which both of the stars can be seen with a telescope is called a **visual binary**.

Download for free at http://cnx.org/content/col11992/latest/

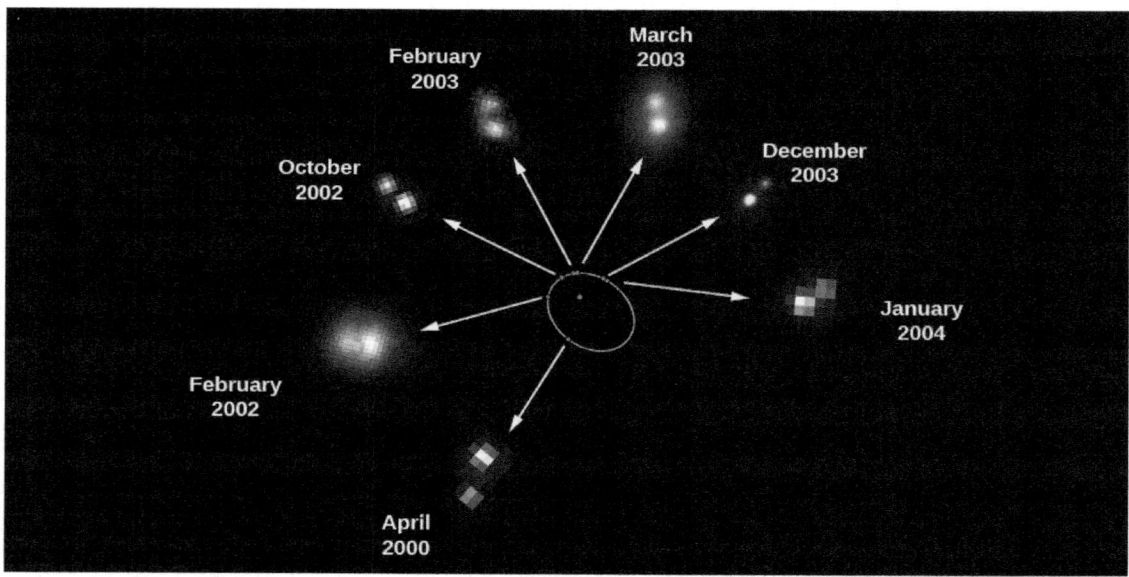

Figure 18.4. Revolution of a Binary Star. This figure shows seven observations of the mutual revolution of two stars, one a brown dwarf and one an ultra-cool L dwarf. Each red dot on the orbit, which is shown by the blue ellipse, corresponds to the position of one of the dwarfs relative to the other. The reason that the pair of stars looks different on the different dates is that some images were taken with the Hubble Space Telescope and others were taken from the ground. The arrows point to the actual observations that correspond to the positions of each red dot. From these observations, an international team of astronomers directly measured the mass of an ultra-cool brown dwarf star for the first time. Barely the size of the planet Jupiter, the dwarf star weighs in at just 8.5% of the mass of our Sun. (credit: modification of work by ESA/NASA and Herve Bouy (Max-Planck-Institut für Extraterrestrische Physik/ESO, Germany))

Edward C. Pickering (1846–1919), at Harvard, discovered a second class of binary stars in 1889—a class in which only one of the stars is actually seen directly. He was examining the spectrum of Mizar and found that the dark absorption lines in the brighter star's spectrum were usually double. Not only were there two lines where astronomers normally saw only one, but the spacing of the lines was constantly changing. At times, the lines even became single. Pickering correctly deduced that the brighter component of Mizar, called Mizar A, is itself really two stars that revolve about each other in a period of 104 days. A star like Mizar A, which appears as a single star when photographed or observed visually through the telescope, but which spectroscopy shows really to be a double star, is called a **spectroscopic binary**.

Mizar, by the way, is a good example of just how complex such star systems can be. Mizar has been known for centuries to have a faint companion called Alcor, which can be seen without a telescope. Mizar and Alcor form an *optical double*—a pair of stars that appear close together in the sky but do not orbit each other. Through a telescope, as Riccioli discovered in 1650, Mizar can be seen to have another, closer companion that does orbit it; Mizar is thus a visual binary. The two components that make up this visual binary, known as Mizar A and Mizar B, are both spectroscopic binaries. So, Mizar is really a quadruple system of stars.

Strictly speaking, it is not correct to describe the motion of a binary star system by saying that one star orbits the other. Gravity is a *mutual* attraction. Each star exerts a gravitational force on the other, with the result that both stars orbit a point between them called the *center of mass*. Imagine that the two stars are seated at either end of a seesaw. The point at which the fulcrum would have to be located in order for the seesaw to balance is the center of mass, and it is always closer to the more massive star (Figure 18.5).

This OpenStax book is available for free at http://cnx.org/content/col11992/1.8

Download for free at http://cnx.org/content/col11992/latest/

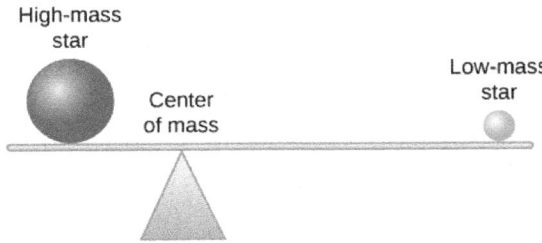

Figure 18.5. Binary Star System. In a binary star system, both stars orbit their center of mass. The image shows the relative positions of two, different-mass stars from their center of mass, similar to how two masses would have to be located on a seesaw in order to keep it level. The star with the higher mass will be found closer to the center of mass, while the star with the lower mass will be farther from it.

Figure 18.6 shows two stars (A and B) moving around their center of mass, along with one line in the spectrum of each star that we observe from the system at different times. When one star is approaching us relative to the center of mass, the other star is receding from us. In the top left illustration, star A is moving toward us, so the line in its spectrum is Doppler-shifted toward the blue end of the spectrum. Star B is moving away from us, so its line shows a redshift. When we observe the composite spectrum of the two stars, the line appears double. When the two stars are both moving across our line of sight (neither away from nor toward us), they both have the same radial velocity (that of the pair's center of mass); hence, the spectral lines of the two stars come together. This is shown in the two bottom illustrations in Figure 18.6.

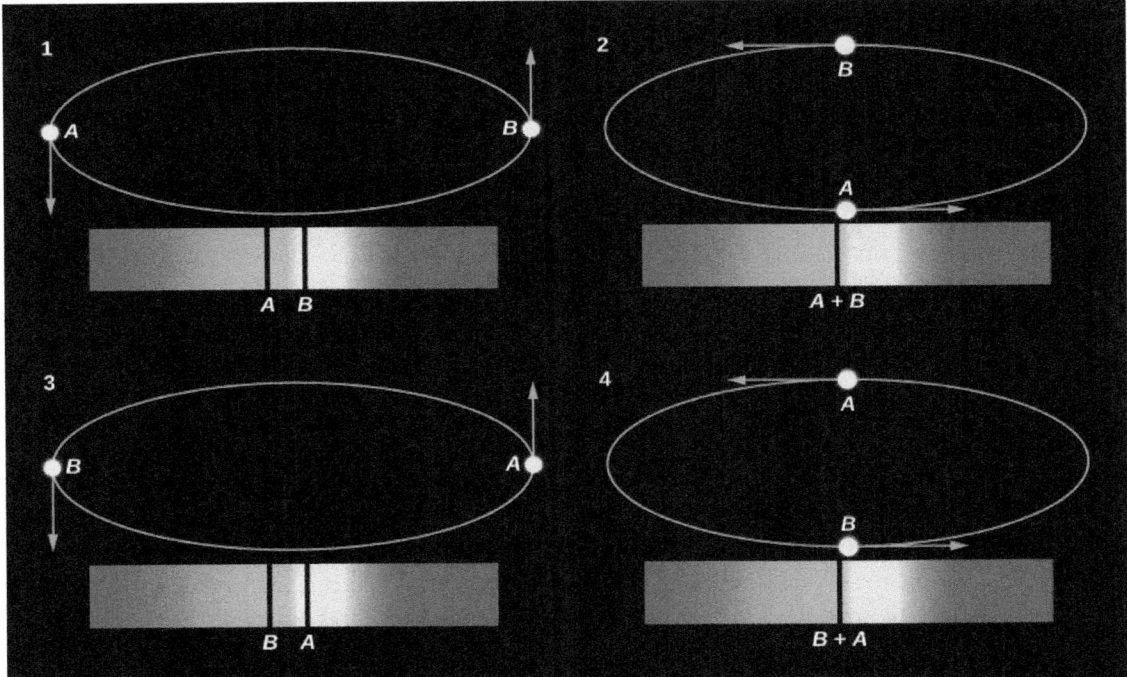

Figure 18.6. Motions of Two Stars Orbiting Each Other and What the Spectrum Shows. We see changes in velocity because when one star is moving toward Earth, the other is moving away; half a cycle later, the situation is reversed. Doppler shifts cause the spectral lines to move back and forth. In diagrams 1 and 3, lines from both stars can be seen well separated from each other. When the two stars are moving perpendicular to our line of sight (that is, they are not moving either toward or away from us), the two lines are exactly superimposed, and so in diagrams 2 and 4, we see only a single spectral line. Note that in the diagrams, the orbit of the star pair is tipped slightly with respect to the viewer (or if the viewer were looking at it in the sky, the orbit would be tilted with respect to the viewer's line of sight). If the orbit were exactly in the plane of the page or screen (or the sky), then it would look nearly circular, but we would see no change in radial velocity (no part of the motion would be toward us or away from us.) If the orbit were perpendicular to the plane of the page or screen, then the stars would appear to move back and forth in a straight line, and we would see the largest-possible radial velocity variations.

Download for free at http://cnx.org/content/col11992/latest/

A plot showing how the velocities of the stars change with time is called a *radial velocity curve*; the curve for the binary system in Figure 18.6 is shown in Figure 18.7.

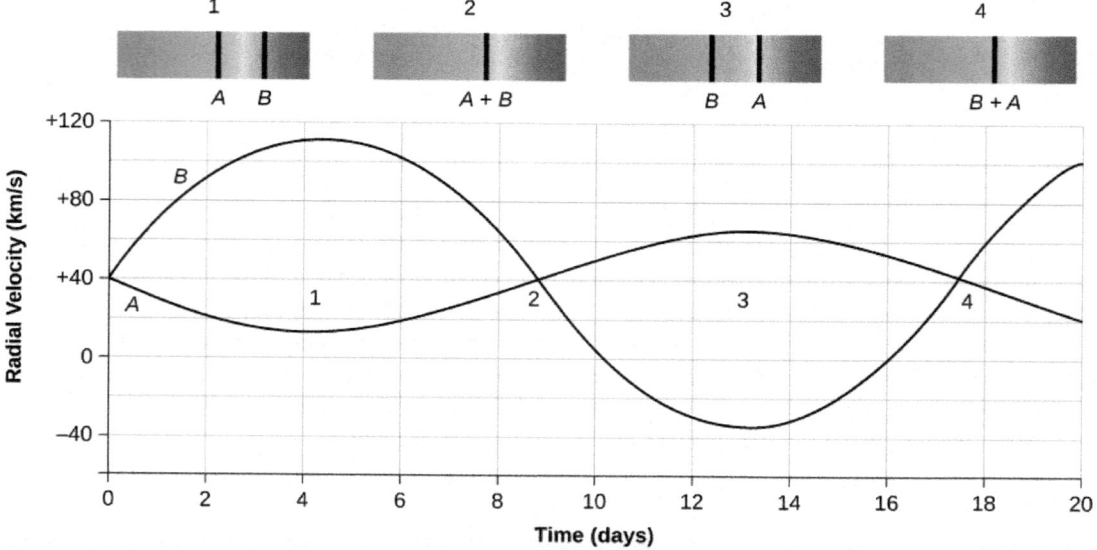

Figure 18.7. Radial Velocities in a Spectroscopic Binary System. These curves plot the radial velocities of two stars in a spectroscopic binary system, showing how the stars alternately approach and recede from Earth. Note that positive velocity means the star is moving away from us relative to the center of mass of the system, which in this case is 40 kilometers per second. Negative velocity means the star is moving toward us relative to the center of mass. The positions on the curve corresponding to the illustrations in Figure 18.6 are marked with the diagram number (1–4).

LINK TO LEARNING

This animation (https://openstax.org/l/30binstaranim) lets you follow the orbits of a binary star system in various combinations of the masses of the two stars.

Masses from the Orbits of Binary Stars

We can estimate the masses of binary star systems using Newton's reformulation of Kepler's third law (discussed in Newton's Universal Law of Gravitation). Kepler found that the time a planet takes to go around the Sun is related by a specific mathematical formula to its distance from the Sun. In our binary star situation, if two objects are in mutual revolution, then the period (*P*) with which they go around each other is related to the semimajor axis (*D*) of the orbit of one with respect to the other, according to this equation

$$D^3 = (M_1 + M_2)P^2$$

where *D* is in astronomical units, *P* is measured in years, and $M_1 + M_2$ is the sum of the masses of the two stars in units of the Sun's mass. This is a very useful formula for astronomers; it says that if we can observe the size of the orbit and the period of mutual revolution of the stars in a binary system, we can calculate the sum of their masses.

Most spectroscopic binaries have periods ranging from a few days to a few months, with separations of usually less than 1 AU between their member stars. Recall that an AU is the distance from Earth to the Sun, so this is a

This OpenStax book is available for free at http://cnx.org/content/col11992/1.8

small separation and very hard to see at the distances of stars. This is why many of these systems are known to be double only through careful study of their spectra.

We can analyze a radial velocity curve (such as the one in Figure 18.7) to determine the masses of the stars in a spectroscopic binary. This is complex in practice but not hard in principle. We measure the speeds of the stars from the Doppler effect. We then determine the period—how long the stars take to go through an orbital cycle—from the velocity curve. Knowing how fast the stars are moving and how long they take to go around tells us the circumference of the orbit and, hence, the separation of the stars in kilometers or astronomical units. From Kepler's law, the period and the separation allow us to calculate the sum of the stars' masses.

Of course, knowing the sum of the masses is not as useful as knowing the mass of each star separately. But the relative orbital speeds of the two stars can tell us how much of the total mass each star has. As we saw in our seesaw analogy, the more massive star is closer to the center of mass and therefore has a smaller orbit. Therefore, it moves more slowly to get around in the same time compared to the more distant, lower-mass star. If we sort out the speeds relative to each other, we can sort out the masses relative to each other. In practice, we also need to know how the binary system is oriented in the sky to our line of sight, but if we do, and the just-described steps are carried out carefully, the result is a calculation of the masses of each of the two stars in the system.

To summarize, a good measurement of the motion of two stars around a common center of mass, combined with the laws of gravity, allows us to determine the masses of stars in such systems. These mass measurements are absolutely crucial to developing a theory of how stars evolve. One of the best things about this method is that it is independent of the location of the binary system. It works as well for stars 100 light-years away from us as for those in our immediate neighborhood.

To take a specific example, Sirius is one of the few binary stars in Appendix J for which we have enough information to apply Kepler's third law:

$$D^3 = (M_1 + M_2)P^2$$

In this case, the two stars, the one we usually call Sirius and its very faint companion, are separated by about 20 AU and have an orbital period of about 50 years. If we place these values in the formula we would have

$$(20)^3 = (M_1 + M_2)(50)^2$$
$$8000 = (M_1 + M_2)(2500)$$

This can be solved for the sum of the masses:

$$M_1 + M_2 = \frac{8000}{2500} = 3.2$$

Therefore, the sum of masses of the two stars in the Sirius binary system is 3.2 times the Sun's mass. In order to determine the individual mass of each star, we would need the velocities of the two stars and the orientation of the orbit relative to our line of sight.

The Range of Stellar Masses

How large can the mass of a star be? Stars more massive than the Sun are rare. None of the stars within 30 light-years of the Sun has a mass greater than four times that of the Sun. Searches at large distances from the Sun have led to the discovery of a few stars with masses up to about 100 times that of the Sun, and a handful of stars (a few out of several billion) may have masses as large as 250 solar masses. However, most stars have less mass than the Sun.

According to theoretical calculations, the smallest mass that a true star can have is about 1/12 that of the Sun. By a "true" star, astronomers mean one that becomes hot enough to fuse protons to form helium (as discussed

Download for free at http://cnx.org/content/col11992/latest/

in The Sun: A Nuclear Powerhouse). Objects with masses between roughly 1/100 and 1/12 that of the Sun may produce energy for a brief time by means of nuclear reactions involving deuterium, but they do not become hot enough to fuse protons. Such objects are intermediate in mass between stars and planets and have been given the name **brown dwarfs** (Figure 18.8). Brown dwarfs are similar to Jupiter in radius but have masses from approximately 13 to 80 times larger than the mass of Jupiter.[2]

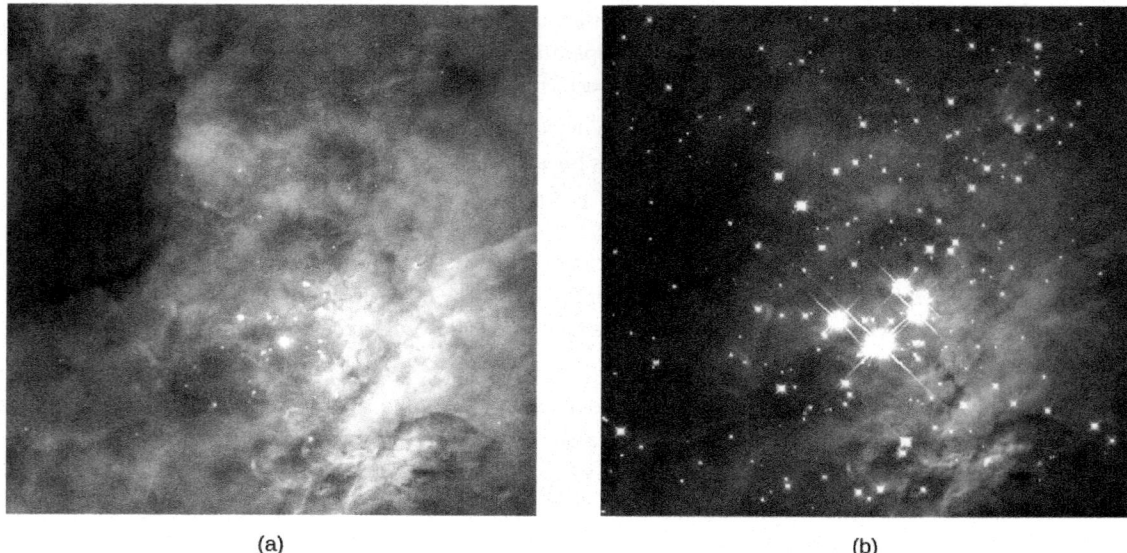

(a)　　　　　　　　　　　　　　　　　　　　　　　　　　　　　　　　(b)

Figure 18.8. Brown Dwarfs in Orion. These images, taken with the Hubble Space Telescope, show the region surrounding the Trapezium star cluster inside the star-forming region called the Orion Nebula. (a) No brown dwarfs are seen in the visible light image, both because they put out very little light in the visible and because they are hidden within the clouds of dust in this region. (b) This image was taken in infrared light, which can make its way to us through the dust. The faintest objects in this image are brown dwarfs with masses between 13 and 80 times the mass of Jupiter. (credit a: NASA, C.R. O'Dell and S.K. Wong (Rice University); credit b: NASA; K.L. Luhman (Harvard-Smithsonian Center for Astrophysics) and G. Schneider, E. Young, G. Rieke, A. Cotera, H. Chen, M. Rieke, R. Thompson (Steward Observatory))

Still-smaller objects with masses less than about 1/100 the mass of the Sun (or 10 Jupiter masses) are called planets. They may radiate energy produced by the radioactive elements that they contain, and they may also radiate heat generated by slowly compressing under their own weight (a process called gravitational contraction). However, their interiors will never reach temperatures high enough for any nuclear reactions, to take place. Jupiter, whose mass is about 1/1000 the mass of the Sun, is unquestionably a planet, for example. Until the 1990s, we could only detect planets in our own solar system, but now we have thousands of them elsewhere as well. (We will discuss these exciting observations in The Birth of Stars and the Discovery of Planets outside the Solar System.)

The Mass-Luminosity Relation

Now that we have measurements of the characteristics of many different types of stars, we can search for relationships among the characteristics. For example, we can ask whether the mass and luminosity of a star are related. It turns out that for most stars, they are: The more massive stars are generally also the more luminous. This relationship, known as the **mass-luminosity relation**, is shown graphically in Figure 18.9. Each point represents a star whose mass and luminosity are both known. The horizontal position on the graph shows

---

2　Exactly where to put the dividing line between planets and brown dwarfs is a subject of some debate among astronomers as we write this book (as is, in fact, the exact definition of each of these objects). Even those who accept deuterium fusion (see **The Birth of Stars and the Discovery of Planets outside the Solar System**) as the crucial issue for brown dwarfs concede that, depending on the composition of the star and other factors, the lowest mass for such a dwarf could be anywhere from 11 to 16 Jupiter masses.

This OpenStax book is available for free at http://cnx.org/content/col11992/1.8

the star's mass, given in units of the Sun's mass, and the vertical position shows its luminosity in units of the Sun's luminosity.

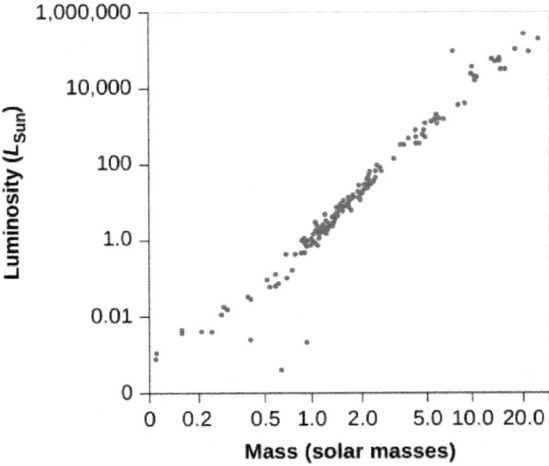

Figure 18.9. Mass-Luminosity Relation. The plotted points show the masses and luminosities of stars. The three points lying below the sequence of points are all white dwarf stars.

We can also say this in mathematical terms.

$$L \sim M^{3.9}$$

It's a reasonably good approximation to say that luminosity (expressed in units of the Sun's luminosity) varies as the fourth power of the mass (in units of the Sun's mass). (The symbol ~ means the two quantities are proportional.) If two stars differ in mass by a factor of 2, then the more massive one will be 2^4, or about 16 times brighter; if one star is 1/3 the mass of another, it will be approximately 81 times less luminous.

EXAMPLE 18.1

Calculating the Mass from the Luminosity of a Star

The mass-luminosity formula can be rewritten so that a value of mass can be determined if the luminosity is known.

Solution

First, we must get our units right by expressing both the mass and the luminosity of a star in units of the Sun's mass and luminosity:

$$L/L_{Sun} = (M/M_{Sun})^4$$

Now we can take the 4th root of both sides, which is equivalent to taking both sides to the 1/4 = 0.25 power. The formula in this case would be:

$$M/M_{Sun} = (L/L_{Sun})^{0.25} = (L/L_{Sun})^{0.25}$$

Check Your Learning

Download for free at http://cnx.org/content/col11992/latest/

In the previous section, we determined the sum of the masses of the two stars in the Sirius binary system (Sirius and its faint companion) using Kepler's third law to be 3.2 solar masses. Using the mass-luminosity relationship, calculate the mass of each individual star.

Answer:

In Appendix J, Sirius is listed with a luminosity 23 times that of the Sun. This value can be inserted into the mass-luminosity relationship to get the mass of Sirius:

$M/M_{Sun} = 23^{0.25} = 2.2$

The mass of the companion star to Sirius is then 3.2 – 2.2 = 1.0 solar mass.

Notice how good this mass-luminosity relationship is. Most stars (see Figure 18.9) fall along a line running from the lower-left (low mass, low luminosity) corner of the diagram to the upper-right (high mass, high luminosity) corner. About 90% of all stars obey the mass-luminosity relation. Later, we will explore why such a relationship exists and what we can learn from the roughly 10% of stars that "disobey" it.

18.3 DIAMETERS OF STARS

Learning Objectives

By the end of this section, you will be able to:

> Describe the methods used to determine star diameters
> Identify the parts of an eclipsing binary star light curve that correspond to the diameters of the individual components

It is easy to measure the diameter of the Sun. Its angular diameter—that is, its apparent size on the sky—is about 1/2°. If we know the angle the Sun takes up in the sky and how far away it is, we can calculate its true (linear) diameter, which is 1.39 million kilometers, or about 109 times the diameter of Earth.

Unfortunately, the Sun is the only star whose angular diameter is easily measured. All the other stars are so far away that they look like pinpoints of light through even the largest ground-based telescopes. (They often seem to be bigger, but that is merely distortion introduced by turbulence in Earth's atmosphere.) Luckily, there are several techniques that astronomers can use to estimate the sizes of stars.

Stars Blocked by the Moon

One technique, which gives very precise diameters but can be used for only a few stars, is to observe the dimming of light that occurs when the Moon passes in front of a star. What astronomers measure (with great precision) is the time required for the star's brightness to drop to zero as the edge of the Moon moves across the star's disk. Since we know how rapidly the Moon moves in its orbit around Earth, it is possible to calculate the angular diameter of the star. If the distance to the star is also known, we can calculate its diameter in kilometers. This method works only for fairly bright stars that happen to lie along the zodiac, where the Moon (or, much more rarely, a planet) can pass in front of them as seen from Earth.

This OpenStax book is available for free at http://cnx.org/content/col11992/1.8

Eclipsing Binary Stars

Accurate sizes for a large number of stars come from measurements of **eclipsing binary** star systems, and so we must make a brief detour from our main story to examine this type of star system. Some binary stars are lined up in such a way that, when viewed from Earth, each star passes in front of the other during every revolution (Figure 18.10). When one star blocks the light of the other, preventing it from reaching Earth, the luminosity of the system decreases, and astronomers say that an eclipse has occurred.

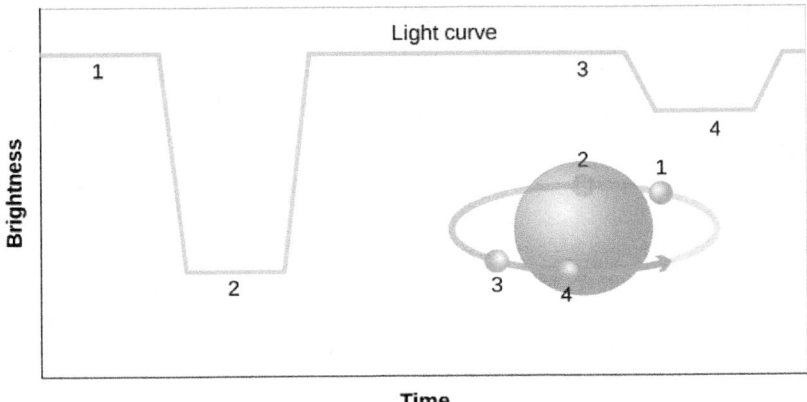

Figure 18.10. Light Curve of an Eclipsing Binary. The light curve of an eclipsing binary star system shows how the combined light from both stars changes due to eclipses over the time span of an orbit. This light curve shows the behavior of a hypothetical eclipsing binary star with total eclipses (one star passes directly in front of and behind the other). The numbers indicate parts of the light curve corresponding to various positions of the smaller star in its orbit. In this diagram, we have assumed that the smaller star is also the hotter one so that it emits more flux (energy per second per square meter) than the larger one. When the smaller, hotter star goes behind the larger one, its light is completely blocked, and so there is a strong dip in the light curve. When the smaller star goes in front of the bigger one, a small amount of light from the bigger star is blocked, so there is a smaller dip in the light curve.

The discovery of the first eclipsing binary helped solve a long-standing puzzle in astronomy. The star Algol, in the constellation of Perseus, changes its brightness in an odd but regular way. Normally, Algol is a fairly bright star, but at intervals of 2 days, 20 hours, 49 minutes, it fades to one-third of its regular brightness. After a few hours, it brightens to normal again. This effect is easily seen, even without a telescope, if you know what to look for.

In 1783, a young English astronomer named John Goodricke (1764–1786) made a careful study of Algol (see the feature on John Goodricke for a discussion of his life and work). Even though Goodricke could neither hear nor speak, he made a number of major discoveries in the 21 years of his brief life. He suggested that Algol's unusual brightness variations might be due to an invisible companion that regularly passes in front of the brighter star and blocks its light. Unfortunately, Goodricke had no way to test this idea, since it was not until about a century later that equipment became good enough to measure Algol's spectrum.

In 1889, the German astronomer Hermann Vogel (1841–1907) demonstrated that, like Mizar, Algol is a spectroscopic binary. The spectral lines of Algol were not observed to be double because the fainter star of the pair gives off too-little light compared with the brighter star for its lines to be conspicuous in the composite spectrum. Nevertheless, the periodic shifting back and forth of the brighter star's lines gave evidence that it was revolving about an unseen companion. (The lines of both components need not be visible for a star to be recognized as a spectroscopic binary.)

The discovery that Algol is a spectroscopic binary verified Goodricke's hypothesis. The plane in which the stars revolve is turned nearly edgewise to our line of sight, and each star passes in front of the other during every revolution. (The eclipse of the fainter star in the Algol system is not very noticeable because the part of it that

Download for free at http://cnx.org/content/col11992/latest/

is covered contributes little to the total light of the system. This second eclipse can, however, be detected by careful measurements.)

Any binary star produces eclipses if viewed from the proper direction, near the plane of its orbit, so that one star passes in front of the other (see Figure 18.10). But from our vantage point on Earth, only a few binary star systems are oriented in this way.

MAKING CONNECTIONS

Astronomy and Mythology: Algol the Demon Star and Perseus the Hero

The name Algol comes from the Arabic *Ras al Ghul*, meaning "the demon's head."[3] The word "ghoul" in English has the same derivation. As discussed in Observing the Sky: The Birth of Astronomy, many of the bright stars have Arabic names because during the long dark ages in medieval Europe, it was Arabic astronomers who preserved and expanded the Greek and Roman knowledge of the skies. The reference to the demon is part of the ancient Greek legend of the hero Perseus, who is commemorated by the constellation in which we find Algol and whose adventures involve many of the characters associated with the northern constellations.

Perseus was one of the many half-god heroes fathered by Zeus (Jupiter in the Roman version), the king of the gods in Greek mythology. Zeus had, to put it delicately, a roving eye and was always fathering somebody or other with a human maiden who caught his fancy. (Perseus derives from *Per Zeus*, meaning "fathered by Zeus.") Set adrift with his mother by an (understandably) upset stepfather, Perseus grew up on an island in the Aegean Sea. The king there, taking an interest in Perseus' mother, tried to get rid of the young man by assigning him an extremely difficult task.

In a moment of overarching pride, a beautiful young woman named Medusa had compared her golden hair to that of the goddess Athena (Minerva for the Romans). The Greek gods did not take kindly to being compared to mere mortals, and Athena turned Medusa into a gorgon: a hideous, evil creature with writhing snakes for hair and a face that turned anyone who looked at it into stone. Perseus was given the task of slaying this demon, which seemed like a pretty sure way to get him out of the way forever.

But because Perseus had a god for a father, some of the other gods gave him tools for the job, including Athena's reflective shield and the winged sandals of Hermes (Mercury in the Roman story). By flying over her and looking only at her reflection, Perseus was able to cut off Medusa's head without ever looking at her directly. Taking her head (which, conveniently, could still turn onlookers to stone even without being attached to her body) with him, Perseus continued on to other adventures.

He next came to a rocky seashore, where boasting had gotten another family into serious trouble with the gods. Queen Cassiopeia had dared to compare her own beauty to that of the Nereids, sea nymphs who were daughters of Poseidon (Neptune in Roman mythology), the god of the sea. Poseidon was so offended that he created a sea-monster named Cetus to devastate the kingdom. King Cepheus, Cassiopeia's beleaguered husband, consulted the oracle, who told him that he must sacrifice his beautiful daughter Andromeda to the monster.

When Perseus came along and found Andromeda chained to a rock near the sea, awaiting her fate, he rescued her by turning the monster to stone. (Scholars of mythology actually trace the essence of this story back to far-older legends from ancient Mesopotamia, in which the god-hero Marduk vanquishes a monster named Tiamat. Symbolically, a hero like Perseus or Marduk is usually associated with the Sun,

This OpenStax book is available for free at http://cnx.org/content/col11992/1.8

the monster with the power of night, and the beautiful maiden with the fragile beauty of dawn, which the Sun releases after its nightly struggle with darkness.)

Many of the characters in these Greek legends can be found as constellations in the sky, not necessarily resembling their namesakes but serving as reminders of the story. For example, vain Cassiopeia is sentenced to be very close to the celestial pole, rotating perpetually around the sky and hanging upside down every winter. The ancients imagined Andromeda still chained to her rock (it is much easier to see the chain of stars than to recognize the beautiful maiden in this star grouping). Perseus is next to her with the head of Medusa swinging from his belt. Algol represents this gorgon head and has long been associated with evil and bad fortune in such tales. Some commentators have speculated that the star's change in brightness (which can be observed with the unaided eye) may have contributed to its unpleasant reputation, with the ancients regarding such a change as a sort of evil "wink."

Diameters of Eclipsing Binary Stars

We now turn back to the main thread of our story to discuss how all this can be used to measure the sizes of stars. The technique involves making a light curve of an eclipsing binary, a graph that plots how the brightness changes with time. Let us consider a hypothetical binary system in which the stars are very different in size, like those illustrated in Figure 18.11. To make life easy, we will assume that the orbit is viewed exactly edge-on.

Even though we cannot see the two stars separately in such a system, the light curve can tell us what is happening. When the smaller star just starts to pass behind the larger star (a point we call *first contact*), the brightness begins to drop. The eclipse becomes total (the smaller star is completely hidden) at the point called *second contact*. At the end of the total eclipse (*third contact*), the smaller star begins to emerge. When the smaller star has reached *last contact*, the eclipse is completely over.

To see how this allows us to measure diameters, look carefully at Figure 18.11. During the time interval between the first and second contacts, the smaller star has moved a distance equal to its own diameter. During the time interval from the first to third contacts, the smaller star has moved a distance equal to the diameter of the larger star. If the spectral lines of both stars are visible in the spectrum of the binary, then the speed of the smaller star with respect to the larger one can be measured from the Doppler shift. But knowing the speed with which the smaller star is moving and how long it took to cover some distance can tell the span of that distance—in this case, the diameters of the stars. The speed multiplied by the time interval from the first to second contact gives the diameter of the smaller star. We multiply the speed by the time between the first and third contacts to get the diameter of the larger star.

---

3   Fans of Batman comic books and movies will recognize that this name was given to an archvillain in the series.

Download for free at http://cnx.org/content/col11992/latest/

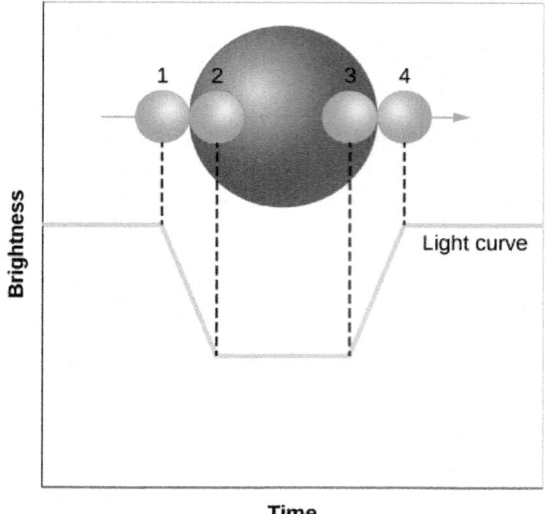

Figure 18.11. Light Curve of an Edge-On Eclipsing Binary. Here we see the light curve of a hypothetical eclipsing binary star whose orbit we view exactly edge-on, in which the two stars fully eclipse each other. From the time intervals between contacts, it is possible to estimate the diameters of the two stars.

In actuality, the situation with eclipsing binaries is often a bit more complicated: orbits are generally not seen exactly edge-on, and the light from each star may be only partially blocked by the other. Furthermore, binary star orbits, just like the orbits of the planets, are ellipses, not circles. However, all these effects can be sorted out from very careful measurements of the light curve.

Using the Radiation Law to Get the Diameter

Another method for measuring star diameters makes use of the Stefan-Boltzmann law for the relationship between energy radiated and temperature (see Radiation and Spectra). In this method, the *energy flux* (energy emitted per second per square meter by a blackbody, like the Sun) is given by

$$F = \sigma T^4$$

where σ is a constant and *T* is the temperature. The surface area of a sphere (like a star) is given by

$$A = 4\pi R^2$$

The luminosity (*L*) of a star is then given by its surface area in square meters times the energy flux:

$$L = (A \times F)$$

Previously, we determined the masses of the two stars in the Sirius binary system. Sirius gives off 8200 times more energy than its fainter companion star, although both stars have nearly identical temperatures. The extremely large difference in luminosity is due to the difference in radius, since the temperatures and hence the energy fluxes for the two stars are nearly the same. To determine the relative sizes of the two stars, we take the ratio of the corresponding luminosities:

This OpenStax book is available for free at http://cnx.org/content/col11992/1.8

Download for free at http://cnx.org/content/col11992/latest/

$$\frac{L_{Sirius}}{L_{companion}} = \frac{\left(A_{Sirius} \times F_{Sirius}\right)}{\left(A_{companion} \times F_{companion}\right)}$$

$$= \frac{A_{Sirius}}{A_{companion}} = \frac{4\pi R^2{}_{Sirius}}{4\pi R^2{}_{companion}} = \frac{R^2{}_{Sirius}}{R^2{}_{companion}}$$

$$\frac{L_{Sirius}}{L_{companion}} = 8200 = \frac{R^2{}_{Sirius}}{R^2{}_{companion}}$$

Therefore, the relative sizes of the two stars can be found by taking the square root of the relative luminosity. Since $\sqrt{8200} = 91$, the radius of Sirius is 91 times larger than the radium of its faint companion.

The method for determining the radius shown here requires both stars be visible, which is not always the case.

Stellar Diameters

The results of many stellar size measurements over the years have shown that most nearby stars are roughly the size of the Sun, with typical diameters of a million kilometers or so. Faint stars, as we might have expected, are generally smaller than more luminous stars. However, there are some dramatic exceptions to this simple generalization.

A few of the very luminous stars, those that are also red (indicating relatively low surface temperatures), turn out to be truly enormous. These stars are called, appropriately enough, giant stars or supergiant stars. An example is Betelgeuse, the second brightest star in the constellation of Orion and one of the dozen brightest stars in our sky. Its diameter, remarkably, is greater than 10 AU (1.5 *billion* kilometers!), large enough to fill the entire inner solar system almost as far out as Jupiter. In Stars from Adolescence to Old Age, we will look in detail at the evolutionary process that leads to the formation of such giant and supergiant stars.

LINK TO LEARNING

Watch this star size comparison video (https://openstax.org/l/30starsizecomp) for a striking visual that highlights the size of stars versus planets and the range of sizes among stars.

18.4 THE H–R DIAGRAM

Learning Objectives

By the end of this section, you will be able to:

> Identify the physical characteristics of stars that are used to create an H–R diagram, and describe how those characteristics vary among groups of stars
> Discuss the physical properties of most stars found at different locations on the H–R diagram, such as radius, and for main sequence stars, mass

In this chapter and Analyzing Starlight, we described some of the characteristics by which we might classify stars and how those are measured. These ideas are summarized in Table 18.2. We have also given an example of a relationship between two of these characteristics in the mass-luminosity relation. When the characteristics

Download for free at http://cnx.org/content/col11992/latest/

of large numbers of stars were measured at the beginning of the twentieth century, astronomers were able to begin a deeper search for patterns and relationships in these data.

Measuring the Characteristics of Stars

Characteristic	Technique
Surface temperature	1. Determine the color (very rough). 2. Measure the spectrum and get the spectral type.
Chemical composition	Determine which lines are present in the spectrum.
Luminosity	Measure the apparent brightness and compensate for distance.
Radial velocity	Measure the Doppler shift in the spectrum.
Rotation	Measure the width of spectral lines.
Mass	Measure the period and radial velocity curves of spectroscopic binary stars.
Diameter	1. Measure the way a star's light is blocked by the Moon. 2. Measure the light curves and Doppler shifts for eclipsing binary stars.

Table 18.2

To help understand what sorts of relationships might be found, let's look briefly at a range of data about human beings. If you want to understand humans by comparing and contrasting their characteristics—without assuming any previous knowledge of these strange creatures—you could try to determine which characteristics lead you in a fruitful direction. For example, you might plot the heights of a large sample of humans against their weights (which is a measure of their mass). Such a plot is shown in Figure 18.12 and it has some interesting features. In the way we have chosen to present our data, height increases upward, whereas weight increases to the left. Notice that humans are not randomly distributed in the graph. Most points fall along a sequence that goes from the upper left to the lower right.

This OpenStax book is available for free at http://cnx.org/content/col11992/1.8

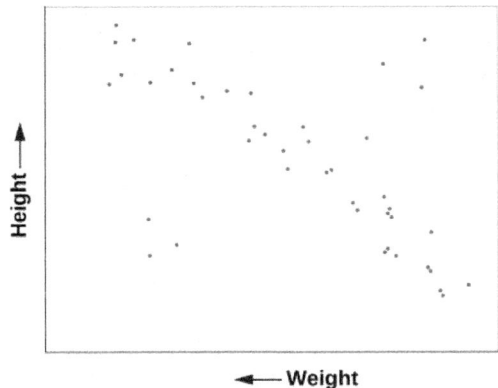

Figure 18.12. Height versus Weight. The plot of the heights and weights of a representative group of human beings. Most points lie along a "main sequence" representing most people, but there are a few exceptions.

We can conclude from this graph that human height and weight are related. Generally speaking, taller human beings weigh more, whereas shorter ones weigh less. This makes sense if you are familiar with the structure of human beings. Typically, if we have bigger bones, we have more flesh to fill out our larger frame. It's not mathematically exact—there is a wide range of variation—but it's not a bad overall rule. And, of course, there are some dramatic exceptions. You occasionally see a short human who is very overweight and would thus be more to the bottom left of our diagram than the average sequence of people. Or you might have a very tall, skinny fashion model with great height but relatively small weight, who would be found near the upper right.

A similar diagram has been found extremely useful for understanding the lives of stars. In 1913, American astronomer Henry Norris Russell plotted the luminosities of stars against their spectral classes (a way of denoting their surface temperatures). This investigation, and a similar independent study in 1911 by Danish astronomer Ejnar Hertzsprung, led to the extremely important discovery that the temperature and luminosity of stars are related (Figure 18.13).

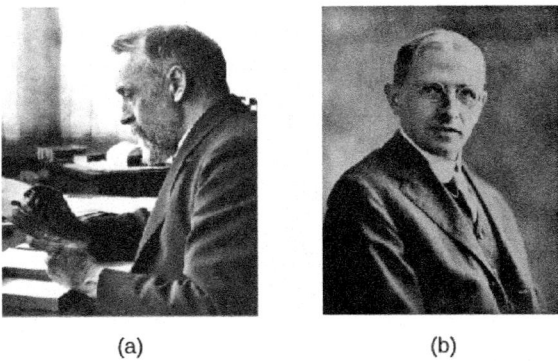

(a)   (b)

Figure 18.13. Hertzsprung (1873–1967) and Russell (1877–1957). (a) Ejnar Hertzsprung and (b) Henry Norris Russell independently discovered the relationship between the luminosity and surface temperature of stars that is summarized in what is now called the H–R diagram.

Download for free at http://cnx.org/content/col11992/latest/

VOYAGERS IN ASTRONOMY

Henry Norris Russell

When Henry Norris Russell graduated from Princeton University, his work had been so brilliant that the faculty decided to create a new level of honors degree beyond "summa cum laude" for him. His students later remembered him as a man whose thinking was three times faster than just about anybody else's. His memory was so phenomenal, he could correctly quote an enormous number of poems and limericks, the entire Bible, tables of mathematical functions, and almost anything he had learned about astronomy. He was nervous, active, competitive, critical, and very articulate; he tended to dominate every meeting he attended. In outward appearance, he was an old-fashioned product of the nineteenth century who wore high-top black shoes and high starched collars, and carried an umbrella every day of his life. His 264 papers were enormously influential in many areas of astronomy.

Born in 1877, the son of a Presbyterian minister, Russell showed early promise. When he was 12, his family sent him to live with an aunt in Princeton so he could attend a top preparatory school. He lived in the same house in that town until his death in 1957 (interrupted only by a brief stay in Europe for graduate work). He was fond of recounting that both his mother and his maternal grandmother had won prizes in mathematics, and that he probably inherited his talents in that field from their side of the family.

Before Russell, American astronomers devoted themselves mainly to surveying the stars and making impressive catalogs of their properties, especially their spectra (as described in Analyzing Starlight. Russell began to see that interpreting the spectra of stars required a much more sophisticated understanding of the physics of the atom, a subject that was being developed by European physicists in the 1910s and 1920s. Russell embarked on a lifelong quest to ascertain the physical conditions inside stars from the clues in their spectra; his work inspired, and was continued by, a generation of astronomers, many trained by Russell and his collaborators.

Russell also made important contributions in the study of binary stars and the measurement of star masses, the origin of the solar system, the atmospheres of planets, and the measurement of distances in astronomy, among other fields. He was an influential teacher and popularizer of astronomy, writing a column on astronomical topics for *Scientific American* magazine for more than 40 years. He and two colleagues wrote a textbook for college astronomy classes that helped train astronomers and astronomy enthusiasts over several decades. That book set the scene for the kind of textbook you are now reading, which not only lays out the facts of astronomy but also explains how they fit together. Russell gave lectures around the country, often emphasizing the importance of understanding modern physics in order to grasp what was happening in astronomy.

Harlow Shapley, director of the Harvard College Observatory, called Russell "the dean of American astronomers." Russell was certainly regarded as the leader of the field for many years and was consulted on many astronomical problems by colleagues from around the world. Today, one of the highest recognitions that an astronomer can receive is an award from the American Astronomical Society called the Russell Prize, set up in his memory.

Features of the H–R Diagram

Following Hertzsprung and Russell, let us plot the temperature (or spectral class) of a selected group of nearby stars against their luminosity and see what we find (Figure 18.14). Such a plot is frequently called

This OpenStax book is available for free at http://cnx.org/content/col11992/1.8

Download for free at http://cnx.org/content/col11992/latest/

the *Hertzsprung-Russell diagram*, abbreviated **H-R diagram**. It is one of the most important and widely used diagrams in astronomy, with applications that extend far beyond the purposes for which it was originally developed more than a century ago.

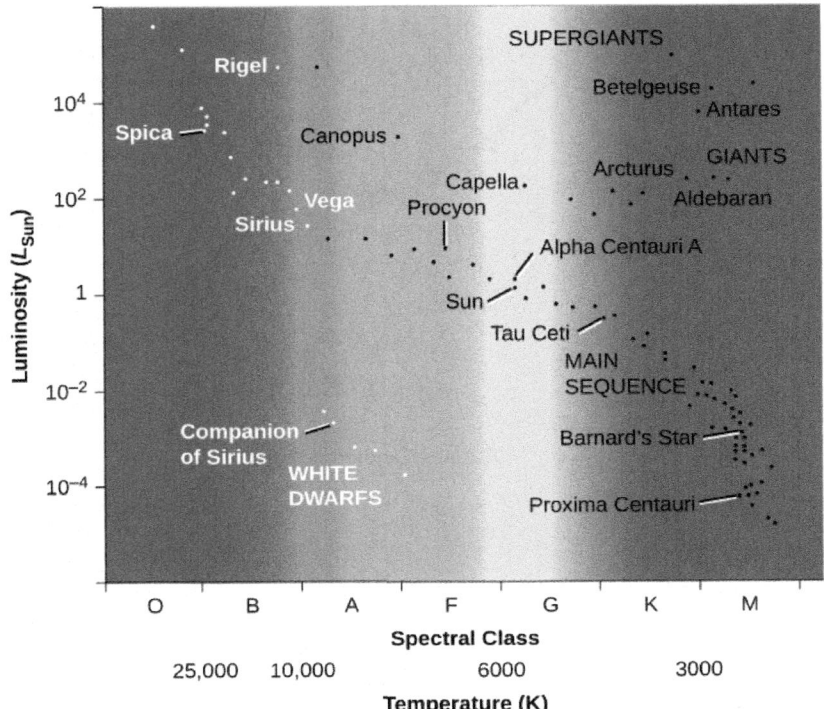

Figure 18.14. H-R Diagram for a Selected Sample of Stars. In such diagrams, luminosity is plotted along the vertical axis. Along the horizontal axis, we can plot either temperature or spectral type (also sometimes called spectral class). Several of the brightest stars are identified by name. Most stars fall on the main sequence.

It is customary to plot H-R diagrams in such a way that temperature increases toward the left and luminosity toward the top. Notice the similarity to our plot of height and weight for people (Figure 18.12). Stars, like people, are not distributed over the diagram at random, as they would be if they exhibited all combinations of luminosity and temperature. Instead, we see that the stars cluster into certain parts of the H-R diagram. The great majority are aligned along a narrow sequence running from the upper left (hot, highly luminous) to the lower right (cool, less luminous). This band of points is called the **main sequence**. It represents a relationship between *temperature* and *luminosity* that is followed by most stars. We can summarize this relationship by saying that hotter stars are more luminous than cooler ones.

A number of stars, however, lie above the main sequence on the H-R diagram, in the upper-right region, where stars have low temperature and high luminosity. How can a star be at once cool, meaning each square meter on the star does not put out all that much energy, and yet very luminous? The only way is for the star to be enormous—to have so many square meters on its surface that the *total* energy output is still large. These stars must be *giants* or *supergiants*, the stars of huge diameter we discussed earlier.

There are also some stars in the lower-left corner of the diagram, which have high temperature and low luminosity. If they have high surface temperatures, each square meter on that star puts out a lot of energy. How then can the overall star be dim? It must be that it has a very small total surface area; such stars are known as **white dwarfs** (white because, at these high temperatures, the colors of the electromagnetic radiation that

Download for free at http://cnx.org/content/col11992/latest/

they emit blend together to make them look bluish-white). We will say more about these puzzling objects in a moment. Figure 18.15 is a schematic H–R diagram for a large sample of stars, drawn to make the different types more apparent.

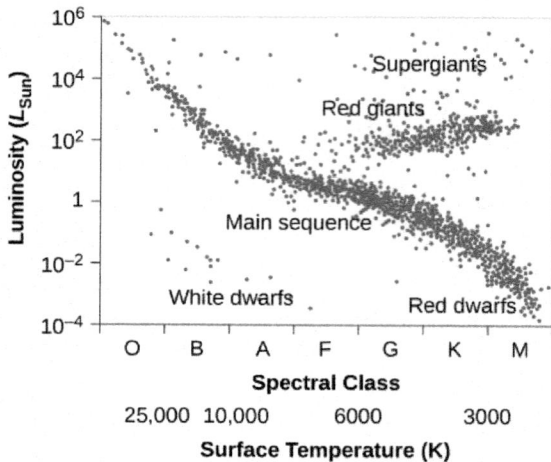

Figure 18.15. Schematic H–R Diagram for Many Stars. Ninety percent of all stars on such a diagram fall along a narrow band called the main sequence. A minority of stars are found in the upper right; they are both cool (and hence red) and bright, and must be giants. Some stars fall in the lower left of the diagram; they are both hot and dim, and must be white dwarfs.

Now, think back to our discussion of star surveys. It is difficult to plot an H–R diagram that is truly representative of all stars because most stars are so faint that we cannot see those outside our immediate neighborhood. The stars plotted in Figure 18.14 were selected because their distances are known. This sample omits many intrinsically faint stars that are nearby but have not had their distances measured, so it shows fewer faint main-sequence stars than a "fair" diagram would. To be truly representative of the stellar population, an H–R diagram should be plotted for all stars within a certain distance. Unfortunately, our knowledge is reasonably complete only for stars within 10 to 20 light-years of the Sun, among which there are no giants or supergiants. Still, from many surveys (and more can now be done with new, more powerful telescopes), we estimate that about 90% of the true stars overall (excluding brown dwarfs) in our part of space are main-sequence stars, about 10% are white dwarfs, and fewer than 1% are giants or supergiants.

These estimates can be used directly to understand the lives of stars. Permit us another quick analogy with people. Suppose we survey people just like astronomers survey stars, but we want to focus our attention on the location of young people, ages 6 to 18 years. Survey teams fan out and take data about where such youngsters are found at all times during a 24-hour day. Some are found in the local pizza parlor, others are asleep at home, some are at the movies, and many are in school. After surveying a very large number of young people, one of the things that the teams determine is that, averaged over the course of the 24 hours, one-third of all youngsters are found in school.

How can they interpret this result? Does it mean that two-thirds of students are truants and the remaining one-third spend all their time in school? No, we must bear in mind that the survey teams counted youngsters throughout the full 24-hour day. Some survey teams worked at night, when most youngsters were at home asleep, and others worked in the late afternoon, when most youngsters were on their way home from school (and more likely to be enjoying a pizza). If the survey was truly representative, we *can* conclude, however, that if an average of one-third of all youngsters are found in school, then humans ages 6 to 18 years must spend about one-third of *their time* in school.

This OpenStax book is available for free at http://cnx.org/content/col11992/1.8

Download for free at http://cnx.org/content/col11992/latest/

We can do something similar for stars. We find that, on average, 90% of all stars are located on the main sequence of the H–R diagram. If we can identify some activity or life stage with the main sequence, then it follows that stars must spend 90% of their lives in that activity or life stage.

Understanding the Main Sequence

In The Sun: A Nuclear Powerhouse, we discussed the Sun as a representative star. We saw that what stars such as the Sun "do for a living" is to convert protons into helium deep in their interiors via the process of nuclear fusion, thus producing energy. The fusion of protons to helium is an excellent, long-lasting source of energy for a star because the bulk of every star consists of hydrogen atoms, whose nuclei are protons.

Our computer models of how stars evolve over time show us that a typical star will spend about 90% of its life fusing the abundant hydrogen in its core into helium. This then is a good explanation of why 90% of all stars are found on the main sequence in the H–R diagram. But if all the stars on the main sequence are doing the same thing (fusing hydrogen), why are they distributed along a sequence of points? That is, why do they differ in luminosity and surface temperature (which is what we are plotting on the H–R diagram)?

To help us understand how main-sequence stars differ, we can use one of the most important results from our studies of model stars. Astrophysicists have been able to show that the structure of stars that are in equilibrium and derive all their energy from nuclear fusion is completely and uniquely determined by just two quantities: the *total mass* and the *composition* of the star. This fact provides an interpretation of many features of the H–R diagram.

Imagine a cluster of stars forming from a cloud of interstellar "raw material" whose chemical composition is similar to the Sun's. (We'll describe this process in more detail in The Birth of Stars and Discovery of Planets outside the Solar System, but for now, the details will not concern us.) In such a cloud, all the clumps of gas and dust that become stars begin with the same chemical composition and differ from one another only in mass. Now suppose that we compute a model of each of these stars for the time at which it becomes stable and derives its energy from nuclear reactions, but before it has time to alter its composition appreciably as a result of these reactions.

The models calculated for these stars allow us to determine their luminosities, temperatures, and sizes. If we plot the results from the models—one point for each model star—on the H–R diagram, we get something that looks just like the main sequence we saw for real stars.

And here is what we find when we do this. The model stars with the largest masses are the hottest and most luminous, and they are located at the upper left of the diagram.

The least-massive model stars are the coolest and least luminous, and they are placed at the lower right of the plot. The other model stars all lie along a line running diagonally across the diagram. In other words, *the main sequence turns out to be a sequence of stellar masses*.

This makes sense if you think about it. The most massive stars have the most gravity and can thus compress their centers to the greatest degree. This means they are the hottest inside and the best at generating energy from nuclear reactions deep within. As a result, they shine with the greatest luminosity and have the hottest surface temperatures. The stars with lowest mass, in turn, are the coolest inside and least effective in generating energy. Thus, they are the least luminous and wind up being the coolest on the surface. Our Sun lies somewhere in the middle of these extremes (as you can see in Figure 18.14). The characteristics of representative main-sequence stars (excluding brown dwarfs, which are not true stars) are listed in Table 18.3.

Download for free at http://cnx.org/content/col11992/latest/

Characteristics of Main-Sequence Stars

Spectral Type	Mass (Sun = 1)	Luminosity (Sun = 1)	Temperature	Radius (Sun = 1)
O5	40	7×10^5	40,000 K	18
B0	16	2.7×10^5	28,000 K	7
A0	3.3	55	10,000 K	2.5
F0	1.7	5	7500 K	1.4
G0	1.1	1.4	6000 K	1.1
K0	0.8	0.35	5000 K	0.8
M0	0.4	0.05	3500 K	0.6

Table 18.3

Note that we've seen this 90% figure come up before. This is exactly what we found earlier when we examined the mass-luminosity relation (Figure 18.9). We observed that 90% of all stars seem to follow the relationship; these are the 90% of all stars that lie on the main sequence in our H–R diagram. Our models and our observations agree.

What about the other stars on the H–R diagram—the giants and supergiants, and the white dwarfs? As we will see in the next few chapters, these are what main-sequence stars turn into as they age: They are the later stages in a star's life. As a star consumes its nuclear fuel, its source of energy changes, as do its chemical composition and interior structure. These changes cause the star to alter its luminosity and surface temperature so that it no longer lies on the main sequence on our diagram. Because stars spend much less time in these later stages of their lives, we see fewer stars in those regions of the H–R diagram.

Extremes of Stellar Luminosities, Diameters, and Densities

We can use the H–R diagram to explore the extremes in size, luminosity, and density found among the stars. Such extreme stars are not only interesting to fans of the *Guinness Book of World Records*; they can teach us a lot about how stars work. For example, we saw that the most massive main-sequence stars are the most luminous ones. We know of a few extreme stars that are a million times more luminous than the Sun, with masses that exceed 100 times the Sun's mass. These superluminous stars, which are at the upper left of the H–R diagram, are exceedingly hot, very blue stars of spectral type O. These are the stars that would be the most conspicuous at vast distances in space.

The cool supergiants in the upper corner of the H–R diagram are as much as 10,000 times as luminous as the Sun. In addition, these stars have diameters very much larger than that of the Sun. As discussed above, some supergiants are so large that if the solar system could be centered in one, the star's surface would lie beyond the orbit of Mars (see Figure 18.16). We will have to ask, in coming chapters, what process can make a star swell up to such an enormous size, and how long these "swollen" stars can last in their distended state.

This OpenStax book is available for free at http://cnx.org/content/col11992/1.8

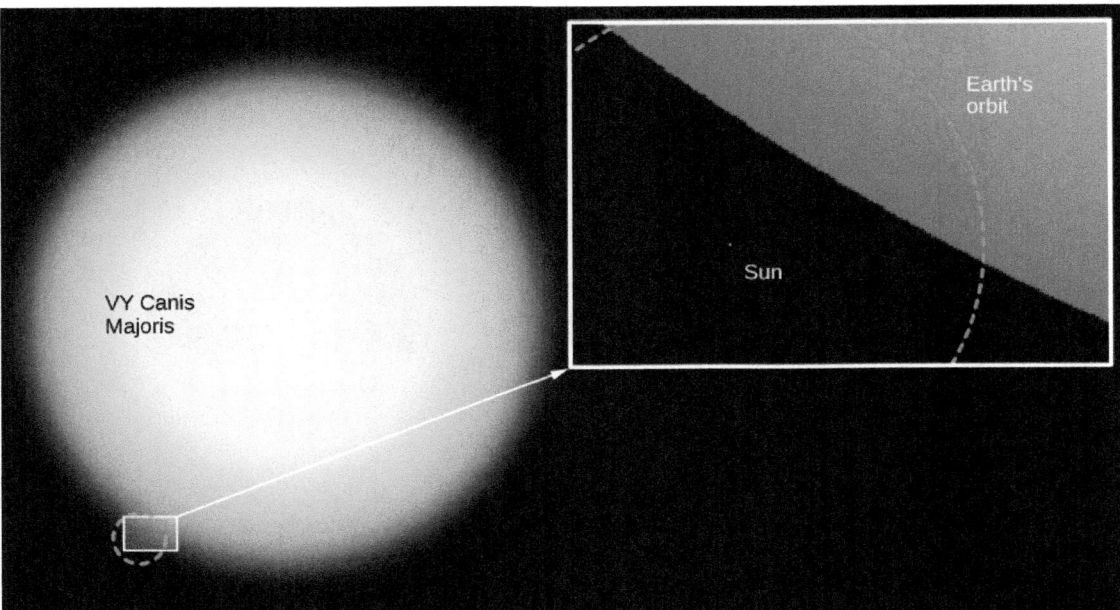

Figure 18.16. The Sun and a Supergiant. Here you see how small the Sun looks in comparison to one of the largest known stars: VY Canis Majoris, a supergiant.

In contrast, the very common red, cool, low-luminosity stars at the lower end of the main sequence are much smaller and more compact than the Sun. An example of such a red dwarf is Ross 614B, with a surface temperature of 2700 K and only 1/2000 of the Sun's luminosity. We call such a star a dwarf because its diameter is only 1/10 that of the Sun. A star with such a low luminosity also has a low mass (about 1/12 that of the Sun). This combination of mass and diameter means that it is so compressed that the star has an average density about 80 times that of the Sun. Its density must be higher, in fact, than that of any known solid found on the surface of Earth. (Despite this, the star is made of gas throughout because its center is so hot.)

The faint, red, main-sequence stars are not the stars of the most extreme densities, however. The white dwarfs, at the lower-left corner of the H–R diagram, have densities many times greater still.

The White Dwarfs

The first white dwarf star was detected in 1862. Called Sirius B, it forms a binary system with Sirius A, the brightest-appearing star in the sky. It eluded discovery and analysis for a long time because its faint light tends to be lost in the glare of nearby Sirius A (Figure 18.17). (Since Sirius is often called the Dog Star—being the brightest star in the constellation of Canis Major, the big dog—Sirius B is sometimes nicknamed the Pup.)

Download for free at http://cnx.org/content/col11992/latest/

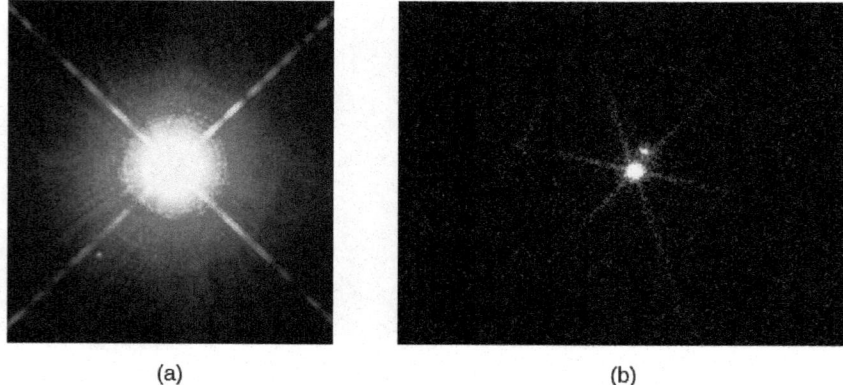

(a)   (b)

Figure 18.17. Two Views of Sirius and Its White Dwarf Companion. (a) The (visible light) image, taken with the Hubble Space Telescope, shows bright Sirius A, and, below it and off to its left, faint Sirius B. (b) This image of the Sirius star system was taken with the Chandra X-Ray Telescope. Now, the bright object is the white dwarf companion, Sirius B. Sirius A is the faint object above it; what we are seeing from Sirius is probably not actually X-ray radiation but rather ultraviolet light that has leaked into the detector. Note that the ultraviolet intensities of these two objects are completely reversed from the situation in visible light because Sirius B is hotter and emits more higher-frequency radiation. (credit a: modification of work by NASA, H.E. Bond and E. Nelan (Space Telescope Science Institute), M. Barstow and M. Burleigh (University of Leicester) and J.B. Holberg (University of Arizona); credit b: modification of work by NASA/SAO/CXC)

We have now found thousands of white dwarfs. Table 18.1 shows that about 7% of the true stars (spectral types O–M) in our local neighborhood are white dwarfs. A good example of a typical white dwarf is the nearby star 40 Eridani B. Its surface temperature is a relatively hot 12,000 K, but its luminosity is only 1/275 L_{Sun}. Calculations show that its radius is only 1.4% of the Sun's, or about the same as that of Earth, and its volume is 2.5 × 10^{-6} that of the Sun. Its mass, however, is 0.43 times the Sun's mass, just a little less than half. To fit such a substantial mass into so tiny a volume, the star's density must be about 170,000 times the density of the Sun, or more than 200,000 g/cm^3. A teaspoonful of this material would have a mass of some 50 tons! At such enormous densities, matter cannot exist in its usual state; we will examine the particular behavior of this type of matter in The Death of Stars. For now, we just note that white dwarfs are dying stars, reaching the end of their productive lives and ready for their stories to be over.

The British astrophysicist (and science popularizer) Arthur Eddington (1882–1944) described the first known white dwarf this way:

> The message of the companion of Sirius, when decoded, ran: "I am composed of material three thousand times denser than anything you've ever come across. A ton of my material would be a little nugget you could put in a matchbox." What reply could one make to something like that? Well, the reply most of us made in 1914 was, "Shut up; don't talk nonsense."

Today, however, astronomers not only accept that stars as dense as white dwarfs exist but (as we will see) have found even denser and stranger objects in their quest to understand the evolution of different types of stars.

This OpenStax book is available for free at http://cnx.org/content/col11992/1.8

CHAPTER 18 REVIEW

 KEY TERMS

binary stars two stars that revolve about each other

brown dwarf an object intermediate in size between a planet and a star; the approximate mass range is from about 1/100 of the mass of the Sun up to the lower mass limit for self-sustaining nuclear reactions, which is about 1/12 the mass of the Sun

eclipsing binary a binary star in which the plane of revolution of the two stars is nearly edge-on to our line of sight, so that the light of one star is periodically diminished by the other passing in front of it

H–R diagram (Hertzsprung–Russell diagram) a plot of luminosity against surface temperature (or spectral type) for a group of stars

main sequence a sequence of stars on the Hertzsprung–Russell diagram, containing the majority of stars, that runs diagonally from the upper left to the lower right

mass-luminosity relation the observed relation between the masses and luminosities of many (90% of all) stars

selection effect the selection of sample data in a nonrandom way, causing the sample data to be unrepresentative of the entire data set

spectroscopic binary a binary star in which the components are not resolved but whose binary nature is indicated by periodic variations in radial velocity, indicating orbital motion

visual binary a binary star in which the two components are telescopically resolved

white dwarf a low-mass star that has exhausted most or all of its nuclear fuel and has collapsed to a very small size; such a star is near its final state of life

 SUMMARY

18.1 A Stellar Census

To understand the properties of stars, we must make wide-ranging surveys. We find the stars that appear brightest to our eyes are bright primarily because they are intrinsically very luminous, not because they are the closest to us. Most of the nearest stars are intrinsically so faint that they can be seen only with the aid of a telescope. Stars with low mass and low luminosity are much more common than stars with high mass and high luminosity. Most of the brown dwarfs in the local neighborhood have not yet been discovered.

18.2 Measuring Stellar Masses

The masses of stars can be determined by analysis of the orbit of binary stars—two stars that orbit a common center of mass. In visual binaries, the two stars can be seen separately in a telescope, whereas in a spectroscopic binary, only the spectrum reveals the presence of two stars. Stellar masses range from about 1/12 to more than 100 times the mass of the Sun (in rare cases, going to 250 times the Sun's mass). Objects with masses between 1/12 and 1/100 that of the Sun are called brown dwarfs. Objects in which no nuclear reactions can take place are planets. The most massive stars are, in most cases, also the most luminous, and this correlation is known as the mass-luminosity relation.

Download for free at http://cnx.org/content/col11992/latest/

18.3 Diameters of Stars

The diameters of stars can be determined by measuring the time it takes an object (the Moon, a planet, or a companion star) to pass in front of it and block its light. Diameters of members of eclipsing binary systems (where the stars pass in front of each other) can be determined through analysis of their orbital motions.

18.4 The H–R Diagram

The Hertzsprung–Russell diagram, or H–R diagram, is a plot of stellar luminosity against surface temperature. Most stars lie on the main sequence, which extends diagonally across the H–R diagram from high temperature and high luminosity to low temperature and low luminosity. The position of a star along the main sequence is determined by its mass. High-mass stars emit more energy and are hotter than low-mass stars on the main sequence. Main-sequence stars derive their energy from the fusion of protons to helium. About 90% of the stars lie on the main sequence. Only about 10% of the stars are white dwarfs, and fewer than 1% are giants or supergiants.

 FOR FURTHER EXPLORATION

Articles

Croswell, K. "The Periodic Table of the Cosmos." *Scientific American* (July 2011):45–49. A brief introduction to the history and uses of the H–R diagram.

Davis, J. "Measuring the Stars." *Sky & Telescope* (October 1991): 361. The article explains direct measurements of stellar diameters.

DeVorkin, D. "Henry Norris Russell." *Scientific American* (May 1989): 126.

Kaler, J. "Journeys on the H–R Diagram." *Sky & Telescope* (May 1988): 483.

McAllister, H. "Twenty Years of Seeing Double." *Sky & Telescope* (November 1996): 28. An update on modern studies of binary stars.

Parker, B. "Those Amazing White Dwarfs." *Astronomy* (July 1984): 15. The article focuses on the history of their discovery.

Pasachoff, J. "The H–R Diagram's 100th Anniversary." *Sky & Telescope* (June 2014): 32.

Roth, J., and Sinnott, R. "Our Studies of Celestial Neighbors." *Sky & Telescope* (October 1996): 32. A discussion is provided on finding the nearest stars.

Websites

Eclipsing Binary Stars: http://www.midnightkite.com/index.aspx?URL=Binary. Dan Bruton at Austin State University has created this collection of animations, articles, and links showing how astronomers use eclipsing binary light curves.

Henry Norris Russell: http://www.nasonline.org/publications/biographical-memoirs/memoir-pdfs/russell-henry-n.pdf. A biographic memoir by Harlow Shapley.

Henry Norris Russell: http://www.phys-astro.sonoma.edu/brucemedalists/russell/RussellBio.pdf. A Bruce Medal profile of Russell.

Hertzsprung–Russell Diagram: http://skyserver.sdss.org/dr1/en/proj/advanced/hr/. This site from the Sloan Digital Sky Survey introduces the H–R diagram and gives you information for making your own. You can go step

This OpenStax book is available for free at http://cnx.org/content/col11992/1.8

by step by using the menu at the left. Note that in the project instructions, the word "here" is a link and takes you to the data you need.

Stars of the Week: http://stars.astro.illinois.edu/sow/sowlist.html. Astronomer James Kaler does "biographical summaries" of famous stars—not the Hollywood type, but ones in the real sky.

Videos

WISE Mission Surveys Nearby Stars: http://www.jpl.nasa.gov/video/details.php?id=1089. Short video about the WISE telescope survey of brown dwarfs and M dwarfs in our immediate neighborhood (1:21).

COLLABORATIVE GROUP ACTIVITIES

A. Two stars are seen close together in the sky, and your group is given the task of determining whether they are a visual binary or whether they just happen to be seen in nearly the same direction. You have access to a good observatory. Make a list of the types of measurements you would make to determine whether they orbit each other.

B. Your group is given information about five main sequence stars that are among the brightest-appearing stars in the sky and yet are pretty far away. Where would these stars be on the H–R diagram and why? Next, your group is given information about five main-sequence stars that are typical of the stars closest to us. Where would these stars be on the H–R diagram and why?

C. A very wealthy (but eccentric) alumnus of your college donates a lot of money for a fund that will help in the search for more brown dwarfs. Your group is the committee in charge of this fund. How would you spend the money? (Be as specific as you can, listing instruments and observing programs.)

D. Use the internet to search for information about the stars with the largest known diameter. What star is considered the record holder (this changes as new measurements are made)? Read about some of the largest stars on the web. Can your group list some reasons why it might be hard to know which star is the largest?

E. Use the internet to search for information about stars with the largest mass. What star is the current "mass champion" among stars? Try to research how the mass of one or more of the most massive stars was measured, and report to the group or the whole class.

EXERCISES

Review Questions

1. How does the mass of the Sun compare with that of other stars in our local neighborhood?

2. Name and describe the three types of binary systems.

3. Describe two ways of determining the diameter of a star.

4. What are the largest- and smallest-known values of the mass, luminosity, surface temperature, and diameter of stars (roughly)?

Download for free at http://cnx.org/content/col11992/latest/

5. You are able to take spectra of both stars in an eclipsing binary system. List all properties of the stars that can be measured from their spectra and light curves.

6. Sketch an H–R diagram. Label the axes. Show where cool supergiants, white dwarfs, the Sun, and main-sequence stars are found.

7. Describe what a typical star in the Galaxy would be like compared to the Sun.

8. How do we distinguish stars from brown dwarfs? How do we distinguish brown dwarfs from planets?

9. Describe how the mass, luminosity, surface temperature, and radius of main-sequence stars change in value going from the "bottom" to the "top" of the main sequence.

10. One method to measure the diameter of a star is to use an object like the Moon or a planet to block out its light and to measure the time it takes to cover up the object. Why is this method used more often with the Moon rather than the planets, even though there are more planets?

11. We discussed in the chapter that about half of stars come in pairs, or multiple star systems, yet the first eclipsing binary was not discovered until the eighteenth century. Why?

Thought Questions

12. Is the Sun an average star? Why or why not?

13. Suppose you want to determine the average educational level of people throughout the nation. Since it would be a great deal of work to survey every citizen, you decide to make your task easier by asking only the people on your campus. Will you get an accurate answer? Will your survey be distorted by a selection effect? Explain.

14. Why do most known visual binaries have relatively long periods and most spectroscopic binaries have relatively short periods?

15. Figure 18.11 shows the light curve of a hypothetical eclipsing binary star in which the light of one star is completely blocked by another. What would the light curve look like for a system in which the light of the smaller star is only partially blocked by the larger one? Assume the smaller star is the hotter one. Sketch the relative positions of the two stars that correspond to various portions of the light curve.

16. There are fewer eclipsing binaries than spectroscopic binaries. Explain why.

17. Within 50 light-years of the Sun, visual binaries outnumber eclipsing binaries. Why?

18. Which is easier to observe at large distances—a spectroscopic binary or a visual binary?

19. The eclipsing binary Algol drops from maximum to minimum brightness in about 4 hours, remains at minimum brightness for 20 minutes, and then takes another 4 hours to return to maximum brightness. Assume that we view this system exactly edge-on, so that one star crosses directly in front of the other. Is one star much larger than the other, or are they fairly similar in size? (Hint: Refer to the diagrams of eclipsing binary light curves.)

This OpenStax book is available for free at http://cnx.org/content/col11992/1.8

20. Review this spectral data for five stars.

Table A

Star	Spectrum
1	G, main sequence
2	K, giant
3	K, main sequence
4	O, main sequence
5	M, main sequence

Which is the hottest? Coolest? Most luminous? Least luminous? In each case, give your reasoning.

21. Which changes by the largest factor along the main sequence from spectral types O to M—mass or luminosity?

22. Suppose you want to search for brown dwarfs using a space telescope. Will you design your telescope to detect light in the ultraviolet or the infrared part of the spectrum? Why?

23. An astronomer discovers a type-M star with a large luminosity. How is this possible? What kind of star is it?

24. Approximately 6000 stars are bright enough to be seen without a telescope. Are any of these white dwarfs? Use the information given in this chapter to explain your reasoning.

25. Use the data in Appendix J to plot an H–R diagram for the brightest stars. Use the data from Table 18.3 to show where the main sequence lies. Do 90% of the brightest stars lie on or near the main sequence? Explain why or why not.

26. Use the diagram you have drawn for Exercise 18.25 to answer the following questions: Which star is more massive—Sirius or Alpha Centauri? Rigel and Regulus have nearly the same spectral type. Which is larger? Rigel and Betelgeuse have nearly the same luminosity. Which is larger? Which is redder?

27. Use the data in Appendix I (http://cnx.org/content/The Nearest Stars/latest/) to plot an H–R diagram for this sample of nearby stars. How does this plot differ from the one for the brightest stars in Exercise 18.25? Why?

28. If a visual binary system were to have two equal-mass stars, how would they be located relative to the center of the mass of the system? What would you observe as you watched these stars as they orbited the center of mass, assuming very circular orbits, and assuming the orbit was face on to your view?

29. Two stars are in a visual binary star system that we see face on. One star is very massive whereas the other is much less massive. Assuming circular orbits, describe their relative orbits in terms of orbit size, period, and orbital velocity.

30. Describe the spectra for a spectroscopic binary for a system comprised of an F-type and L-type star. Assume that the system is too far away to be able to easily observe the L-type star.

31. Figure 18.7 shows the velocity of two stars in a spectroscopic binary system. Which star is the most massive? Explain your reasoning.

Download for free at http://cnx.org/content/col11992/latest/

32. You go out stargazing one night, and someone asks you how far away the brightest stars we see in the sky without a telescope are. What would be a good, general response? (Use Appendix J for more information.)

33. If you were to compare three stars with the same surface temperature, with one star being a giant, another a supergiant, and the third a main-sequence star, how would their radii compare to one another?

34. Are supergiant stars also extremely massive? Explain the reasoning behind your answer.

35. Consider the following data on four stars:

Table B

Star	Luminosity (in L_{Sun})	Type
1	100	B, main sequence
2	1/100	B, white dwarf
3	1/100	M, main sequence
4	100	M, giant

Which star would have the largest radius? Which star would have the smallest radius? Which star is the most common in our area of the Galaxy? Which star is the least common?

Figuring For Yourself

36. If two stars are in a binary system with a combined mass of 5.5 solar masses and an orbital period of 12 years, what is the average distance between the two stars?

37. It is possible that stars as much as 200 times the Sun's mass or more exist. What is the luminosity of such a star based upon the mass-luminosity relation?

38. The lowest mass for a true star is 1/12 the mass of the Sun. What is the luminosity of such a star based upon the mass-luminosity relationship?

39. Spectral types are an indicator of temperature. For the first 10 stars in Appendix J, the list of the brightest stars in our skies, estimate their temperatures from their spectral types. Use information in the figures and/or tables in this chapter and describe how you made the estimates.

40. We can estimate the masses of most of the stars in Appendix J from the mass-luminosity relationship in Figure 18.9. However, remember this relationship works only for main sequence stars. Determine which of the first 10 stars in Appendix J are main sequence stars. Use one of the figures in this chapter. Make a table of stars' masses.

41. In Diameters of Stars, the relative diameters of the two stars in the Sirius system were determined. Let's use this value to explore other aspects of this system. This will be done through several steps, each in its own exercise. Assume the temperature of the Sun is 5800 K, and the temperature of Sirius A, the larger star of the binary, is
10,000 K. The luminosity of Sirius A can be found in Appendix J, and is given as about 23 times that of the Sun. Using the values provided, calculate the radius of Sirius A relative to that of the Sun.

42. Now calculate the radius of Sirius' white dwarf companion, Sirius B, to the Sun.

43. How does this radius of Sirius B compare with that of Earth?

This OpenStax book is available for free at http://cnx.org/content/col11992/1.8

Download for free at http://cnx.org/content/col11992/latest/

44. From the previous calculations and the results from Diameters of Stars, it is possible to calculate the density of Sirius B relative to the Sun. It is worth noting that the radius of the companion is very similar to that of Earth, whereas the mass is very similar to the Sun's. How does the companion's density compare to that of the Sun? Recall that density = mass/volume, and the volume of a sphere = $(4/3)\pi R^3$. How does this density compare with that of water and other materials discussed in this text? Can you see why astronomers were so surprised and puzzled when they first determined the orbit of the companion to Sirius?

45. How much would you weigh if you were suddenly transported to the white dwarf Sirius B? You may use your own weight (or if don't want to own up to what it is, assume you weigh 70 kg or 150 lb). In this case, assume that the companion to Sirius has a mass equal to that of the Sun and a radius equal to that of Earth. Remember Newton's law of gravity:

$$F = GM_1 M_2 / R^2$$

and that your weight is proportional to the force that you feel. What kind of star should you travel to if you want to *lose* weight (and not gain it)?

46. The star Betelgeuse has a temperature of 3400 K and a luminosity of 13,200 L_{Sun}. Calculate the radius of Betelgeuse relative to the Sun.

47. Using the information provided in Table 18.1, what is the average stellar density in our part of the Galaxy? Use only the true stars (types O–M) and assume a spherical distribution with radius of 26 light-years.

48. Confirm that the angular diameter of the Sun of 1/2° corresponds to a linear diameter of 1.39 million km. Use the average distance of the Sun and Earth to derive the answer. (Hint: This can be solved using a trigonometric function.)

49. An eclipsing binary star system is observed with the following contact times for the main eclipse:

Table C

Contact	Time	Date
First contact	12:00 p.m.	March 12
Second contact	4:00 p.m.	March 13
Third contact	9:00 a.m.	March 18
Fourth contact	1:00 p.m.	March 19

The orbital velocity of the smaller star relative to the larger is 62,000 km/h. Determine the diameters for each star in the system.

50. If a 100 solar mass star were to have a luminosity of 10^7 times the Sun's luminosity, how would such a star's density compare when it is on the main sequence as an O-type star, and when it is a cool supergiant (M-type)? Use values of temperature from Figure 18.14 or Figure 18.15 and the relationship between luminosity, radius, and temperature as given in Exercise 18.47.

51. If Betelgeuse had a mass that was 25 times that of the Sun, how would its average density compare to that of the Sun? Use the definition of $density = \frac{mass}{volume}$, where the volume is that of a sphere.

Download for free at http://cnx.org/content/col11992/latest/

This OpenStax book is available for free at http://cnx.org/content/col11992/1.8

19

CELESTIAL DISTANCES

Figure 19.1. Globular Cluster M80. This beautiful image shows a giant cluster of stars called Messier 80, located about 28,000 light-years from Earth. Such crowded groups, which astronomers call globular clusters, contain hundreds of thousands of stars, including some of the RR Lyrae variables discussed in this chapter. Especially obvious in this picture are the bright red giants, which are stars similar to the Sun in mass that are nearing the ends of their lives. (credit: modification of work by The Hubble Heritage Team (AURA/ STScI/ NASA))

Chapter Outline

Thinking Ahead

How large is the universe? What is the most distant object we can see? These are among the most fundamental questions astronomers can ask. But just as babies must crawl before they can take their first halting steps, so too must we start with a more modest question: How far away are the stars? And even this question proves to be very hard to answer. After all, stars are mere points of light. Suppose you see a point of light in the darkness when you are driving on a country road late at night. How can you tell whether it is a nearby firefly, an oncoming motorcycle some distance away, or the porchlight of a house much farther down the road? It's not so easy, is it? Astronomers faced an even more difficult problem when they tried to estimate how far away the stars are.

In this chapter, we begin with the fundamental definitions of distances on Earth and then extend our reach outward to the stars. We will also examine the newest satellites that are surveying the night sky and discuss the special types of stars that can be used as trail markers to distant galaxies.

19.1 FUNDAMENTAL UNITS OF DISTANCE

Learning Objectives

By the end of this section, you will be able to:

Download for free at http://cnx.org/content/col11992/latest/

> Understand the importance of defining a standard distance unit
> Explain how the meter was originally defined and how it has changed over time
> Discuss how radar is used to measure distances to the other members of the solar system

The first measures of distances were based on human dimensions—the inch as the distance between knuckles on the finger, or the yard as the span from the extended index finger to the nose of the British king. Later, the requirements of commerce led to some standardization of such units, but each nation tended to set up its own definitions. It was not until the middle of the eighteenth century that any real efforts were made to establish a uniform, international set of standards.

The Metric System

One of the enduring legacies of the era of the French emperor Napoleon is the establishment of the *metric system* of units, officially adopted in France in 1799 and now used in most countries around the world. The fundamental metric unit of length is the *meter*, originally defined as one ten-millionth of the distance along Earth's surface from the equator to the pole. French astronomers of the seventeenth and eighteenth centuries were pioneers in determining the dimensions of Earth, so it was logical to use their information as the foundation of the new system.

Practical problems exist with a definition expressed in terms of the size of Earth, since anyone wishing to determine the distance from one place to another can hardly be expected to go out and re-measure the planet. Therefore, an intermediate standard meter consisting of a bar of platinum-iridium metal was set up in Paris. In 1889, by international agreement, this bar was defined to be exactly one meter in length, and precise copies of the original meter bar were made to serve as standards for other nations.

Other units of length are derived from the meter. Thus, 1 kilometer (km) equals 1000 meters, 1 centimeter (cm) equals 1/100 meter, and so on. Even the old British and American units, such as the inch and the mile, are now defined in terms of the metric system.

Modern Redefinitions of the Meter

In 1960, the official definition of the meter was changed again. As a result of improved technology for generating spectral lines of precisely known wavelengths (see the chapter on Radiation and Spectra), the meter was redefined to equal 1,650,763.73 wavelengths of a particular atomic transition in the element krypton-86. The advantage of this redefinition is that anyone with a suitably equipped laboratory can reproduce a standard meter, without reference to any particular metal bar.

In 1983, the meter was defined once more, this time in terms of the velocity of light. Light in a vacuum can travel a distance of one meter in 1/299,792,458.6 second. Today, therefore, light travel time provides our basic unit of length. Put another way, a distance of *one light-second* (the amount of space light covers in one second) is defined to be 299,792,458.6 meters. That's almost 300 million meters that light covers in just one second; light really is *very* fast! We could just as well use the light-second as the fundamental unit of length, but for practical reasons (and to respect tradition), we have defined the meter as a small fraction of the light-second.

Distance within the Solar System

The work of Copernicus and Kepler established the *relative* distances of the planets—that is, how far from the Sun one planet is compared to another (see Observing the Sky: The Birth of Astronomy and Orbits and Gravity). But their work could not establish the *absolute* distances (in light-seconds or meters or other standard units of length). This is like knowing the height of all the students in your class only as compared to the height of your astronomy instructor, but not in inches or centimeters. Somebody's height has to be measured directly.

This OpenStax book is available for free at http://cnx.org/content/col11992/1.8

Download for free at http://cnx.org/content/col11992/latest/

Similarly, to establish absolute distances, astronomers had to measure one distance in the solar system directly. Generally, the closer to us the object is, the easier such a measurement would be. Estimates of the distance to Venus were made as Venus crossed the face of the Sun in 1761 and 1769, and an international campaign was organized to estimate the distance to the asteroid Eros in the early 1930s, when its orbit brought it close to Earth. More recently, Venus crossed (or *transited*) the surface of the Sun in 2004 and 2012, and allowed us to make a modern distance estimate, although, as we will see below, by then it wasn't needed (Figure 19.2).

LINK TO LEARNING

If you would like more information on just how the motion of Venus across the Sun helped us pin down distances in the solar system, you can turn to a nice explanation (https://openstaxcollege.org/l/30VenusandSun) by a NASA astronomer.

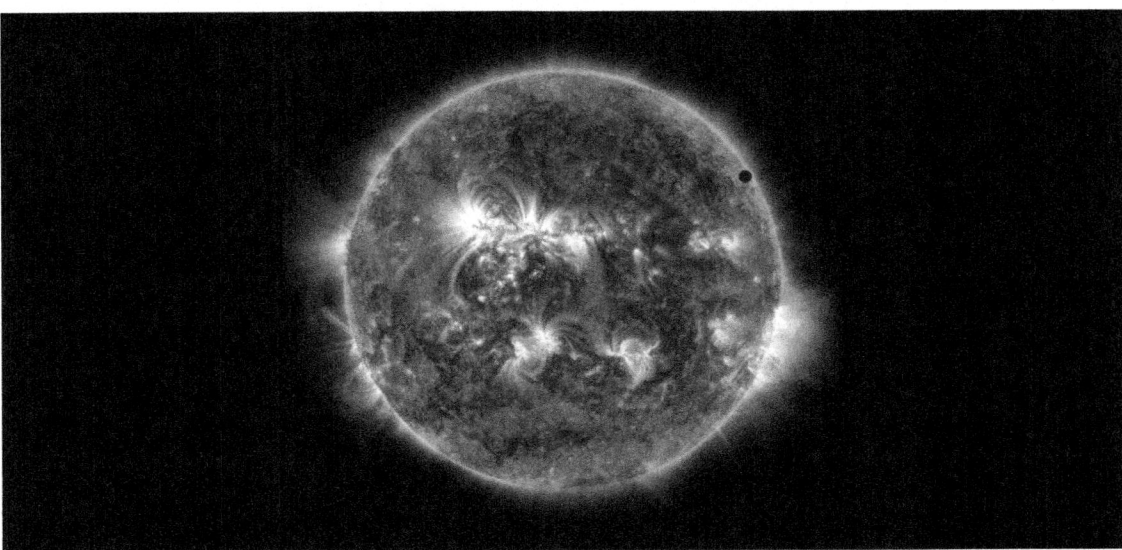

Figure 19.2. Venus Transits the Sun, 2012. This striking "picture" of Venus crossing the face of the Sun (it's the black dot at about 2 o'clock) is more than just an impressive image. Taken with the Solar Dynamics Observatory spacecraft and special filters, it shows a modern transit of Venus. Such events allowed astronomers in the 1800s to estimate the distance to Venus. They measured the time it took Venus to cross the face of the Sun from different latitudes on Earth. The differences in times can be used to estimate the distance to the planet. Today, radar is used for much more precise distance estimates. (credit: modification of work by NASA/SDO, AIA)

The key to our modern determination of solar system dimensions is radar, a type of radio wave that can bounce off solid objects (Figure 19.3). As discussed in several earlier chapters, by timing how long a radar beam (traveling at the speed of light) takes to reach another world and return, we can measure the distance involved very accurately. In 1961, radar signals were bounced off Venus for the first time, providing a direct measurement of the distance from Earth to Venus in terms of light-seconds (from the roundtrip travel time of the radar signal).

Subsequently, radar has been used to determine the distances to Mercury, Mars, the satellites of Jupiter, the rings of Saturn, and several asteroids. Note, by the way, that it is not possible to use radar to measure the distance to the Sun directly because the Sun does not reflect radar very efficiently. But we can measure the distance to many other solar system objects and use Kepler's laws to give us the distance to the Sun.

Download for free at http://cnx.org/content/col11992/latest/

Figure 19.3. Radar Telescope. This dish-shaped antenna, part of the NASA Deep Space Network in California's Mojave Desert, is 70 meters wide. Nicknamed the "Mars antenna," this radar telescope can send and receive radar waves, and thus measure the distances to planets, satellites, and asteroids. (credit: NASA/JPL-Caltech)

From the various (related) solar system distances, astronomers selected the average distance from Earth to the Sun as our standard "measuring stick" within the solar system. When Earth and the Sun are closest, they are about 147.1 million kilometers apart; when Earth and the Sun are farthest, they are about 152.1 million kilometers apart. The average of these two distances is called the astronomical unit (AU). We then express all the other distances in the solar system in terms of the AU. Years of painstaking analyses of radar measurements have led to a determination of the length of the AU to a precision of about one part in a billion. The length of 1 AU can be expressed in light travel time as 499.004854 light-seconds, or about 8.3 light-minutes. If we use the definition of the meter given previously, this is equivalent to 1 AU = 149,597,870,700 meters.

These distances are, of course, given here to a much higher level of precision than is normally needed. In this text, we are usually content to express numbers to a couple of significant places and leave it at that. For our purposes, it will be sufficient to round off these numbers:

$$\text{speed of light: } c = 3 \times 10^8 \text{ m/s} = 3 \times 10^5 \text{ km/s}$$

$$\text{length of light-second: } \text{ls} = 3 \times 10^8 \text{ m} = 3 \times 10^5 \text{ km}$$

$$\text{astronomical unit: } \text{AU} = 1.50 \times 10^{11} \text{ m} = 1.50 \times 10^8 \text{ km} = 500 \text{ light-seconds}$$

We now know the absolute distance scale within our own solar system with fantastic accuracy. This is the first link in the chain of cosmic distances.

LINK TO LEARNING

The distances between the celestial bodies in our solar system are sometimes difficult to grasp or put into perspective. This interactive website (https://openstaxcollege.org/l/30DistanceScale) provides a "map" that shows the distances by using a scale at the bottom of the screen and allows you to scroll

This OpenStax book is available for free at http://cnx.org/content/col11992/1.8

Download for free at http://cnx.org/content/col11992/latest/

(using your arrow keys) through screens of "empty space" to get to the next planet—all while your current distance from the Sun is visible on the scale.

SURVEYING THE STARS

Learning Objectives

By the end of this section, you will be able to:

> Understand the concept of triangulating distances to distant objects, including stars
> Explain why space-based satellites deliver more precise distances than ground-based methods
> Discuss astronomers' efforts to study the stars closest to the Sun

It is an enormous step to go from the planets to the stars. For example, our Voyager 1 probe, which was launched in 1977, has now traveled farther from Earth than any other spacecraft. As this is written in 2016, Voyager 1 is 134 AU from the Sun.[1] The nearest star, however, is hundreds of thousands of AU from Earth. Even so, we can, in principle, survey distances to the stars using the same technique that a civil engineer employs to survey the distance to an inaccessible mountain or tree—the method of *triangulation*.

Triangulation in Space

A practical example of triangulation is your own depth perception. As you are pleased to discover every morning when you look in the mirror, your two eyes are located some distance apart. You therefore view the world from two different vantage points, and it is this dual perspective that allows you to get a general sense of how far away objects are.

To see what we mean, take a pen and hold it a few inches in front of your face. Look at it first with one eye (closing the other) and then switch eyes. Note how the pen seems to shift relative to objects across the room. Now hold the pen at arm's length: the shift is less. If you play with moving the pen for a while, you will notice that the farther away you hold it, the less it seems to shift. Your brain automatically performs such comparisons and gives you a pretty good sense of how far away things in your immediate neighborhood are.

If your arms were made of rubber, you could stretch the pen far enough away from your eyes that the shift would become imperceptible. This is because our depth perception fails for objects more than a few tens of meters away. In order to see the shift of an object a city block or more from you, your eyes would need to be spread apart a lot farther.

Let's see how surveyors take advantage of the same idea. Suppose you are trying to measure the distance to a tree across a deep river (Figure 19.4). You set up two observing stations some distance apart. That distance (line AB in Figure 19.4) is called the *baseline*. Now the direction to the tree (C in the figure) in relation to the baseline is observed from each station. Note that C appears in different directions from the two stations. This apparent change in direction of the remote object due to a change in vantage point of the observer is called **parallax**.

---

1   To have some basis for comparison, the dwarf planet Pluto orbits at an average distance of 40 AU from the Sun, and the dwarf planet Eris is currently roughly 96 AU from the Sun.

Download for free at http://cnx.org/content/col11992/latest/

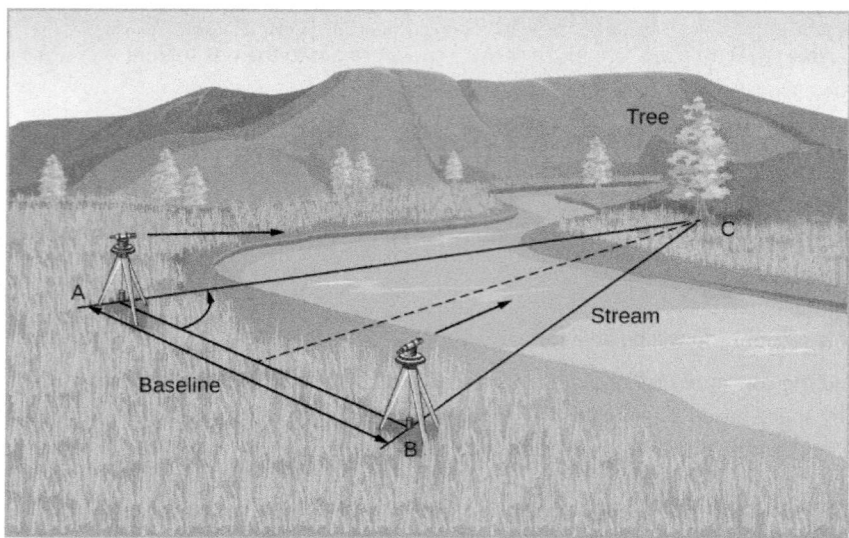

Figure 19.4. Triangulation. Triangulation allows us to measure distances to inaccessible objects. By getting the angle to a tree from two different vantage points, we can calculate the properties of the triangle they make and thus the distance to the tree.

The parallax is also the angle that lines AC and BC make—in mathematical terms, the angle subtended by the baseline. A knowledge of the angles at A and B and the length of the baseline, AB, allows the triangle ABC to be solved for any of its dimensions—say, the distance AC or BC. The solution could be reached by constructing a scale drawing or by using trigonometry to make a numerical calculation. If the tree were farther away, the whole triangle would be longer and skinnier, and the parallax angle would be smaller. Thus, we have the general rule that the smaller the parallax, the more distant the object we are measuring must be.

In practice, the kinds of baselines surveyors use for measuring distances on Earth are completely useless when we try to gauge distances in space. The farther away an astronomical object lies, the longer the baseline has to be to give us a reasonable chance of making a measurement. Unfortunately, nearly all astronomical objects are very far away. To measure their distances requires a very large baseline and highly precise angular measurements. The Moon is the only object near enough that its distance can be found fairly accurately with measurements made without a telescope. Ptolemy determined the distance to the Moon correctly to within a few percent. He used the turning Earth itself as a baseline, measuring the position of the Moon relative to the stars at two different times of night.

With the aid of telescopes, later astronomers were able to measure the distances to the nearer planets and asteroids using Earth's diameter as a baseline. This is how the AU was first established. To reach for the stars, however, requires a much longer baseline for triangulation and extremely sensitive measurements. Such a baseline is provided by Earth's annual trip around the Sun.

Distances to Stars

As Earth travels from one side of its orbit to the other, it graciously provides us with a baseline of 2 AU, or about 300 million kilometers. Although this is a much bigger baseline than the diameter of Earth, the stars are *so far away* that the resulting parallax shift is *still* not visible to the naked eye—not even for the closest stars.

In the chapter on Observing the Sky: The Birth of Astronomy, we discussed how this dilemma perplexed the ancient Greeks, some of whom had actually suggested that the Sun might be the center of the solar system, with Earth in motion around it. Aristotle and others argued, however, that Earth could not be revolving about

This OpenStax book is available for free at http://cnx.org/content/col11992/1.8

Download for free at http://cnx.org/content/col11992/latest/

the Sun. If it were, they said, we would surely observe the parallax of the nearer stars against the background of more distant objects as we viewed the sky from different parts of Earth's orbit (Figure 19.6). Tycho Brahe (1546–1601) advanced the same faulty argument nearly 2000 years later, when his careful measurements of stellar positions with the unaided eye revealed no such shift.

These early observers did not realize how truly distant the stars were and how small the change in their positions therefore was, even with the entire orbit of Earth as a baseline. The problem was that they did not have tools to measure parallax shifts too small to be seen with the human eye. By the eighteenth century, when there was no longer serious doubt about Earth's revolution, it became clear that the stars must be extremely distant. Astronomers equipped with telescopes began to devise instruments capable of measuring the tiny shifts of nearby stars relative to the background of more distant (and thus unshifting) celestial objects.

This was a significant technical challenge, since, even for the nearest stars, parallax angles are usually only a fraction of a second of arc. Recall that one second of arc (arcsec) is an angle of only 1/3600 of a degree. A coin the size of a US quarter would appear to have a diameter of 1 arcsecond if you were viewing it from a distance of about 5 kilometers (3 miles). Think about how small an angle that is. No wonder it took astronomers a long time before they could measure such tiny shifts.

The first successful detections of stellar parallax were in the year 1838, when Friedrich Bessel in Germany (Figure 19.5), Thomas Henderson, a Scottish astronomer working at the Cape of Good Hope, and Friedrich Struve in Russia independently measured the parallaxes of the stars 61 Cygni, Alpha Centauri, and Vega, respectively. Even the closest star, Alpha Centauri, showed a total displacement of only about 1.5 arcseconds during the course of a year.

(a)　　　　　　　　　　　(b)　　　　　　　　　　　(c)

Figure 19.5. Friedrich Wilhelm Bessel (1784–1846), Thomas J. Henderson (1798–1844), and Friedrich Struve (1793–1864). (a) Bessel made the first authenticated measurement of the distance to a star (61 Cygni) in 1838, a feat that had eluded many dedicated astronomers for almost a century. But two others, (b) Scottish astronomer Thomas J. Henderson and (c) Friedrich Struve, in Russia, were close on his heels.

Figure 19.6 shows how such measurements work. Seen from opposite sides of Earth's orbit, a nearby star shifts position when compared to a pattern of more distant stars. Astronomers actually define parallax to be *one-half* the angle that a star shifts when seen from opposite sides of Earth's orbit (the angle labeled *P* in Figure 19.6). The reason for this definition is just that they prefer to deal with a baseline of 1 AU instead of 2 AU.

Download for free at http://cnx.org/content/col11992/latest/

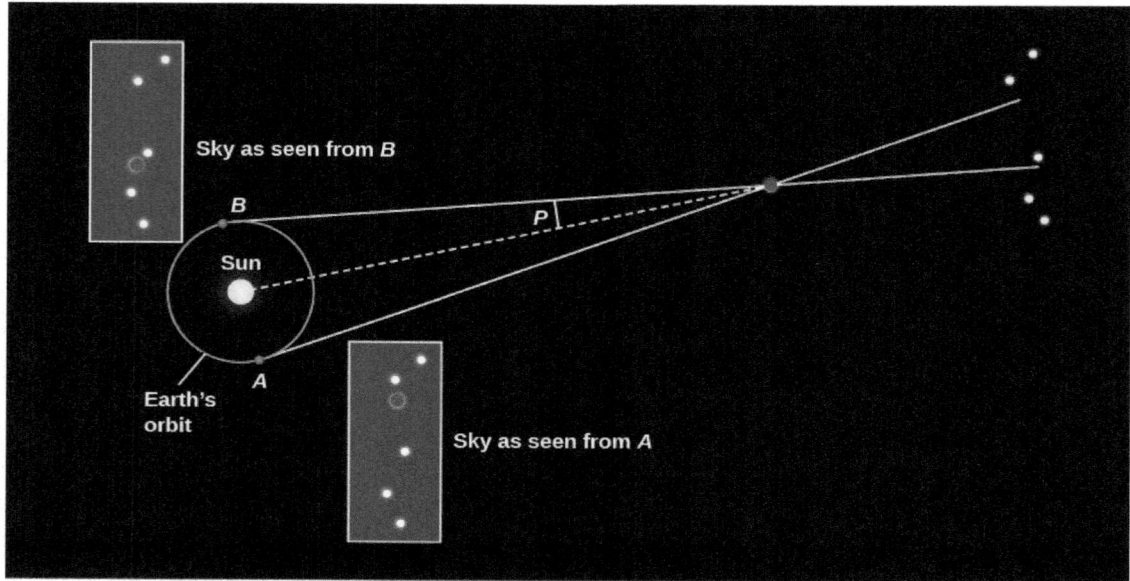

Figure 19.6. Parallax. As Earth revolves around the Sun, the direction in which we see a nearby star varies with respect to distant stars. We define the parallax of the nearby star to be one half of the total change in direction, and we usually measure it in arcseconds.

Units of Stellar Distance

With a baseline of one AU, how far away would a star have to be to have a parallax of 1 arcsecond? The answer turns out to be 206,265 AU, or 3.26 light-years. This is equal to 3.1×10^{13} kilometers (in other words, 31 trillion kilometers). We give this unit a special name, the **parsec** (pc)—derived from "the distance at which we have a *par*allax of one *sec*ond." The distance (D) of a star in parsecs is just the reciprocal of its parallax (p) in arcseconds; that is,

$$D = \frac{1}{p}$$

Thus, a star with a parallax of 0.1 arcsecond would be found at a distance of 10 parsecs, and one with a parallax of 0.05 arcsecond would be 20 parsecs away.

Back in the days when most of our distances came from parallax measurements, a parsec was a useful unit of distance, but it is not as intuitive as the light-year. One advantage of the light-year as a unit is that it emphasizes the fact that, as we look out into space, we are also looking back into time. The light that we see from a star 100 light-years away left that star 100 years ago. What we study is not the star as it is now, but rather as it was in the past. The light that reaches our telescopes today from distant galaxies left them before Earth even existed.

In this text, we will use light-years as our unit of distance, but many astronomers still use parsecs when they write technical papers or talk with each other at meetings. To convert between the two distance units, just bear in mind: 1 parsec = 3.26 light-year, and 1 light-year = 0.31 parsec.

EXAMPLE 19.1

How Far Is a Light-Year?

This OpenStax book is available for free at http://cnx.org/content/col11992/1.8

A light-year is the distance light travels in 1 year. Given that light travels at a speed of 300,000 km/s, how many kilometers are there in a light-year?

Solution

We learned earlier that speed = distance/time. We can rearrange this equation so that distance = velocity × time. Now, we need to determine the number of seconds in a year.

There are approximately 365 days in 1 year. To determine the number of seconds, we must estimate the number of seconds in 1 day.

We can change units as follows (notice how the units of time cancel out):

$$1 \text{ day} \times 24 \text{ hr/day} \times 60 \text{ min/hr} \times 60 \text{ s/min} = 86,400 \text{ s/day}$$

Next, to get the number of seconds per year:

$$365 \text{ days/year} \times 86,400 \text{ s/day} = 31,536,000 \text{ s/year}$$

Now we can multiply the speed of light by the number of seconds per year to get the distance traveled by light in 1 year:

$$
\begin{aligned}
\text{distance} \ &= \text{velocity} \times \text{time} \\
&= 300,000 \text{ km/s} \times 31,536,000 \text{ s} \\
&= 9.46 \times 10^{12} \text{ km}
\end{aligned}
$$

That's almost 10,000,000,000,000 km that light covers in a year. To help you imagine how long this distance is, we'll mention that a string 1 light-year long could fit around the circumference of Earth 236 million times.

Check Your Learning

The number above is really large. What happens if we put it in terms that might be a little more understandable, like the diameter of Earth? Earth's diameter is about 12,700 km.

Answer:

$$
\begin{aligned}
\text{1 light-year} \ &= 9.46 \times 10^{12} \text{ km} \\
&= 9.46 \times 10^{12} \text{ km} \times \frac{1 \text{ Earth diameter}}{12,700 \text{ km}} \\
&= 7.45 \times 10^{8} \text{ Earth diameters}
\end{aligned}
$$

That means that 1 light-year is about 745 million times the diameter of Earth.

ASTRONOMY BASICS

Naming Stars

You may be wondering why stars have such a confusing assortment of names. Just look at the first three stars to have their parallaxes measured: 61 Cygni, Alpha Centauri, and Vega. Each of these names comes from a different tradition of designating stars.

Download for free at http://cnx.org/content/col11992/latest/

The brightest stars have names that derive from the ancients. Some are from the Greek, such as Sirius, which means "the scorched one"—a reference to its brilliance. A few are from Latin, but many of the best-known names are from Arabic because, as discussed in Observing the Sky: The Birth of Astronomy, much of Greek and Roman astronomy was "rediscovered" in Europe after the Dark Ages by means of Arabic translations. Vega, for example, means "swooping Eagle," and Betelgeuse (pronounced "Beetle-juice") means "right hand of the central one."

In 1603, German astronomer Johann Bayer (1572–1625) introduced a more systematic approach to naming stars. For each constellation, he assigned a Greek letter to the brightest stars, roughly in order of brightness. In the constellation of Orion, for example, Betelgeuse is the brightest star, so it got the first letter in the Greek alphabet—alpha—and is known as Alpha Orionis. ("Orionis" is the possessive form of Orion, so Alpha Orionis means "the first of Orion.") A star called Rigel, being the second brightest in that constellation, is called Beta Orionis (Figure 19.7). Since there are 24 letters in the Greek alphabet, this system allows the labeling of 24 stars in each constellation, but constellations have many more stars than that.

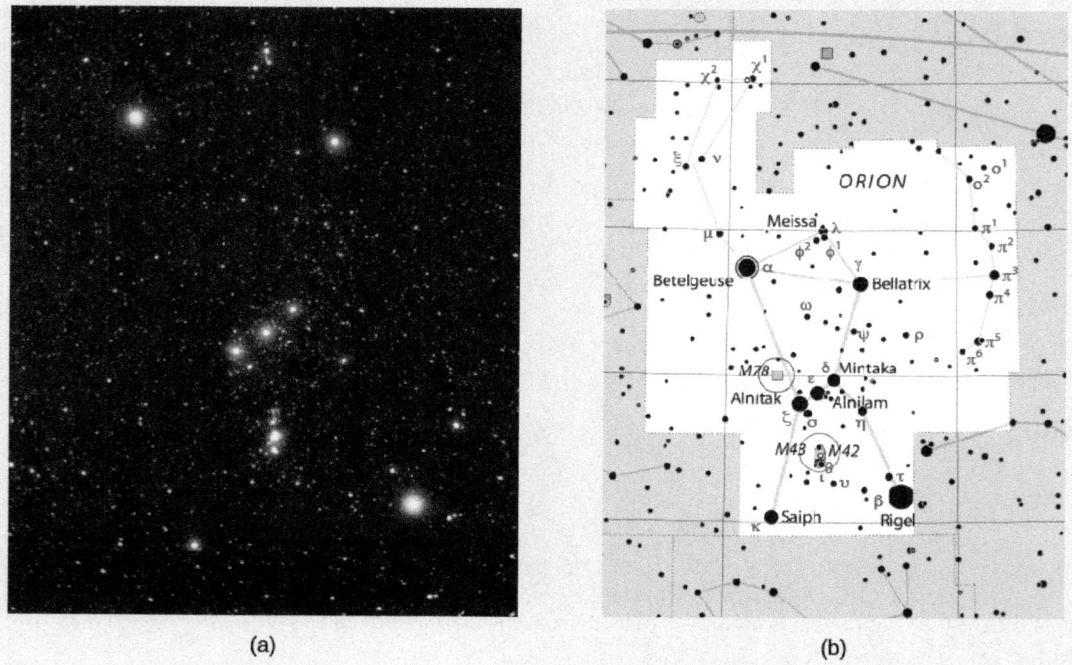

(a)   (b)

Figure 19.7. Objects in Orion. (a) This image shows the brightest objects in or near the star pattern of Orion, the hunter (of Greek mythology), in the constellation of Orion. (b) Note the Greek letters of Bayer's system in this diagram of the Orion constellation. The objects denoted M42, M43, and M78 are not stars but nebulae—clouds of gas and dust; these numbers come from a list of "fuzzy objects" made by Charles Messier in 1781. (credit a: modification of work by Matthew Spinelli; credit b: modification of work by ESO, IAU and *Sky & Telescope*)

In 1725, the English Astronomer Royal John Flamsteed introduced yet another system, in which the brighter stars eventually got a number in each constellation in order of their location in the sky or, more precisely, their right ascension. (The system of sky coordinates that includes right ascension was discussed in Earth, Moon, and Sky.) In this system, Betelgeuse is called 58 Orionis and 61 Cygni is the 61st star in the constellation of Cygnus, the swan.

This OpenStax book is available for free at http://cnx.org/content/col11992/1.8

Download for free at http://cnx.org/content/col11992/latest/

It gets worse. As astronomers began to understand more and more about stars, they drew up a series of specialized star catalogs, and fans of those catalogs began calling stars by their catalog numbers. If you look at Appendix I—our list of the nearest stars (many of which are much too faint to get an ancient name, Bayer letter, or Flamsteed number)—you will see references to some of these catalogs. An example is a set of stars labeled with a BD number, for "Bonner Durchmusterung." This was a mammoth catalog of over 324,000 stars in a series of zones in the sky, organized at the Bonn Observatory in the 1850s and 1860s. Keep in mind that this catalog was made before photography or computers came into use, so the position of each star had to be measured (at least twice) by eye, a daunting undertaking.

There is also a completely different system for keeping track of stars whose luminosity varies, and another for stars that brighten explosively at unpredictable times. Astronomers have gotten used to the many different star-naming systems, but students often find them bewildering and wish astronomers would settle on one. Don't hold your breath: in astronomy, as in many fields of human thought, tradition holds a powerful attraction. Still, with high-speed computer databases to aid human memory, names may become less and less necessary. Today's astronomers often refer to stars by their precise locations in the sky rather than by their names or various catalog numbers.

The Nearest Stars

No known star (other than the Sun) is within 1 light-year or even 1 parsec of Earth. The stellar neighbors nearest the Sun are three stars in the constellation of Centaurus. To the unaided eye, the brightest of these three stars is Alpha Centauri, which is only 30° from the south celestial pole and hence not visible from the mainland United States. Alpha Centauri itself is a binary star—two stars in mutual revolution—too close together to be distinguished without a telescope. These two stars are 4.4 light-years from us. Nearby is a third faint star, known as Proxima Centauri. Proxima, with a distance of 4.3 light-years, is slightly closer to us than the other two stars. If Proxima Centauri is part of a triple star system with the binary Alpha Centauri, as seems likely, then its orbital period may be longer than 500,000 years.

Proxima Centauri is an example of the most common type of star, and our most common type of stellar neighbor (as we saw in Stars: A Celestial Census.) Low-mass red M dwarfs make up about 70% of all stars and dominate the census of stars within 10 parsecs of the Sun. The latest survey of the solar neighborhood has counted 357 stars and brown dwarfs within 10 parsecs, and 248 of these are red dwarfs. Yet, if you wanted to see an M dwarf with your naked eye, you would be out of luck. These stars only produce a fraction of the Sun's light, and nearly all of them require a telescope to be detected.

The nearest star visible without a telescope from most of the United States is the brightest appearing of all the stars, Sirius, which has a distance of a little more than 8 light-years. It too is a binary system, composed of a faint white dwarf orbiting a bluish-white, main-sequence star. It is an interesting coincidence of numbers that light reaches us from the Sun in about 8 minutes and from the next brightest star in the sky in about 8 years.

EXAMPLE 19.2

Calculating the Diameter of the Sun

Download for free at http://cnx.org/content/col11992/latest/

For nearby stars, we can measure the apparent shift in their positions as Earth orbits the Sun. We wrote earlier that an object must be 206,265 AU distant to have a parallax of one second of arc. This must seem like a very strange number, but you can figure out why this is the right value. We will start by estimating the diameter of the Sun and then apply the same idea to a star with a parallax of 1 arcsecond. Make a sketch that has a round circle to represent the Sun, place Earth some distance away, and put an observer on it. Draw two lines from the point where the observer is standing, one to each side of the Sun. Sketch a circle centered at Earth with its circumference passing through the center of the Sun. Now think about proportions. The Sun spans about half a degree on the sky. A full circle has 360°. The circumference of the circle centered on Earth and passing through the Sun is given by:

$$\text{circumference} = 2\pi \times 93{,}000{,}000 \text{ miles}$$

Then, the following two ratios are equal:

$$\frac{0.5°}{360°} = \frac{\text{diameter of Sun}}{2\pi \times 93{,}000{,}000}$$

Calculate the diameter of the Sun. How does your answer compare to the actual diameter?

Solution

To solve for the diameter of the Sun, we can evaluate the expression above.

$$\text{diameter of the sun} = \frac{0.5°}{360°} \times 2\pi \times 93{,}000{,}000 \text{ miles}$$
$$= 811{,}577 \text{ miles}$$

This is very close to the true value of about 848,000 miles.

Check Your Learning

Now apply this idea to calculating the distance to a star that has a parallax of 1 arcsec. Draw a picture similar to the one we suggested above and calculate the distance in AU. (Hint: Remember that the parallax angle is defined by 1 AU, not 2 AU, and that 3600 arcseconds = 1 degree.)

Answer:

206,265 AU

Measuring Parallaxes in Space

The measurements of stellar parallax were revolutionized by the launch of the spacecraft Hipparcos in 1989, which measured distances for thousands of stars out to about 300 light-years with an accuracy of 10 to 20% (see Figure 19.8 and the feature on Parallax and Space Astronomy). However, even 300 light-years are less than 1% the size of our Galaxy's main disk.

In December 2013, the successor to Hipparcos, named *Gaia*, was launched by the European Space Agency. *Gaia* is expected to measure the position and distances to almost one billion stars with an accuracy of a few ten-millionths of an arcsecond. *Gaia's* distance limit will extend well beyond Hipparcos, studying stars out to 30,000 light-years (100 times farther than Hipparcos, covering nearly 1/3 of the galactic disk). *Gaia* will also be able to measure proper motions[2] for thousands of stars in the halo of the Milky Way—something that can only be done for the brightest stars right now. At the end of *Gaia's* mission, we will not only have a three-dimensional

2   Proper motion (as discussed in **Analyzing Starlight**, is the motion of a star across the sky (perpendicular to our line of sight.)

This OpenStax book is available for free at http://cnx.org/content/col11992/1.8

Download for free at http://cnx.org/content/col11992/latest/

map of a large fraction of our own Milky Way Galaxy, but we will also have a strong link in the chain of cosmic distances that we are discussing in this chapter. Yet, to extend this chain beyond *Gaia's* reach and explore distances to nearby galaxies, we need some completely new techniques.

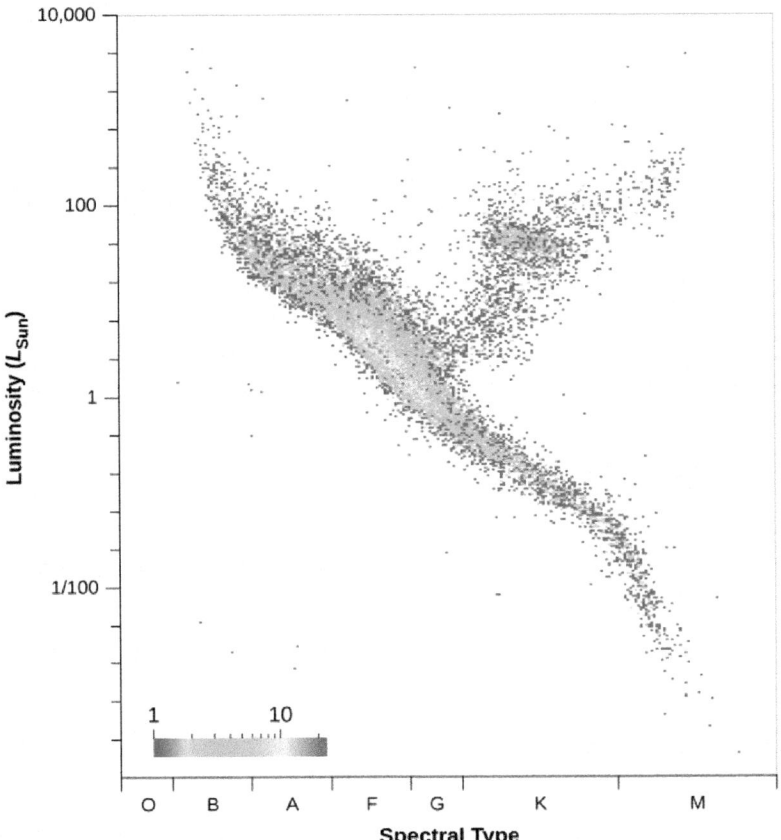

Figure 19.8. H–R Diagram of Stars Measured by Gaia and Hipparcos. This plot includes 16,631 stars for which the parallaxes have an accuracy of 10% or better. The colors indicate the numbers of stars at each point of the diagram, with red corresponding to the largest number and blue to the lowest. Luminosity is plotted along the vertical axis, with luminosity increasing upward. An infrared color is plotted as a proxy for temperature, with temperature decreasing to the right. Most of the data points are distributed along the diagonal running from the top left corner (high luminosity, high temperature) to the bottom right (low temperature, low luminosity). These are main sequence stars. The large clump of data points above the main sequence on the right side of the diagram is composed of red giant stars. (credit: modification of work by the European Space Agency)

MAKING CONNECTIONS

Parallax and Space Astronomy

One of the most difficult things about precisely measuring the tiny angles of parallax shifts from Earth is that you have to observe the stars through our planet's atmosphere. As we saw in Astronomical Instruments, the effect of the atmosphere is to spread out the points of starlight into fuzzy disks, making exact measurements of their positions more difficult. Astronomers had long dreamed of being

Download for free at http://cnx.org/content/col11992/latest/

able to measure parallaxes from space, and two orbiting observatories have now turned this dream into reality.

The name of the Hipparcos satellite, launched in 1989 by the European Space Agency, is both an abbreviation for High Precision Parallax Collecting Satellite and a tribute to Hipparchus, the pioneering Greek astronomer whose work we discussed in the Observing the Sky: The Birth of Astronomy. The satellite was designed to make the most accurate parallax measurements in history, from 36,000 kilometers above Earth. However, its onboard rocket motor failed to fire, which meant it did not get the needed boost to reach the desired altitude. Hipparcos ended up spending its 4-year life in an elliptical orbit that varied from 500 to 36,000 kilometers high. In this orbit, the satellite plunged into Earth's radiation belts every 5 hours or so, which finally took its toll on the solar panels that provided energy to power the instruments.

Nevertheless, the mission was successful, resulting in two catalogs. One gives positions of 120,000 stars to an accuracy of one-thousandth of an arcsecond—about the diameter of a golf ball in New York as viewed from Europe. The second catalog contains information for more than a million stars, whose positions have been measured to thirty-thousandths of an arcsecond. We now have accurate parallax measurements of stars out to distances of about 300 light-years. (With ground-based telescopes, accurate measurements were feasible out to only about 60 light-years.)

In order to build on the success of Hipparcos, in 2013, the European Space Agency launched a new satellite called *Gaia*. The Gaia mission is scheduled to last for 5 years. Because *Gaia* carries larger telescopes than Hipparcos, it can observe fainter stars and measure their positions 200 times more accurately. The main goal of the Gaia mission is to make an accurate three-dimensional map of that portion of the Galaxy within about 30,000 light-years by observing 1 billion stars 70 times each, measuring their positions and hence their parallaxes as well as their brightnesses.

For a long time, the measurement of parallaxes and accurate stellar positions was a backwater of astronomical research—mainly because the accuracy of measurements did not improve much for about 100 years. However, the ability to make measurements from space has revolutionized this field of astronomy and will continue to provide a critical link in our chain of cosmic distances.

LINK TO LEARNING

The European Space Agency (ESA) maintains a Gaia mission website (https://openstaxcollege.org/l/30GaiaMission) where you can learn more about the Gaia mission and to get the latest news on *Gaia* observations.

To learn more about Hipparcos, explore this European Space Agency webpage (https://openstaxcollege.org/l/30Hipparcos) with an ESA vodcast *Charting the Galaxy—from Hipparcos to Gaia*.

This OpenStax book is available for free at http://cnx.org/content/col11992/1.8

Download for free at http://cnx.org/content/col11992/latest/

19.3 VARIABLE STARS: ONE KEY TO COSMIC DISTANCES

Learning Objectives

By the end of this section, you will be able to:

> Describe how some stars vary their light output and why such stars are important
> Explain the importance of pulsating variable stars, such as cepheids and RR Lyrae-type stars, to our study of the universe

Let's briefly review the key reasons that measuring distances to the stars is such a struggle. As discussed in The Brightness of Stars, our problem is that stars come in a bewildering variety of intrinsic luminosities. (If stars were light bulbs, we'd say they come in a wide range of wattages.) Suppose, instead, that all stars had the same "wattage" or luminosity. In that case, the more distant ones would always look dimmer, and we could tell how far away a star is simply by how dim it appeared. In the real universe, however, when we look at a star in our sky (with eye or telescope) and measure its apparent brightness, we cannot know whether it looks dim because it's a low-wattage bulb or because it is far away, or perhaps some of each.

Astronomers need to discover something else about the star that allows us to "read off" its intrinsic luminosity—in effect, to know what the star's true wattage is. With this information, we can then attribute how dim it looks from Earth to its distance. Recall that the apparent brightness of an object decreases with the square of the distance to that object. If two objects have the same luminosity but one is three times farther than the other, the more distant one will look nine times fainter. Therefore, if we know the luminosity of a star and its apparent brightness, we can calculate how far away it is. Astronomers have long searched for techniques that would somehow allow us to determine the luminosity of a star—and it is to these techniques that we turn next.

Variable Stars

The breakthrough in measuring distances to remote parts of our Galaxy, and to other galaxies as well, came from the study of variable stars. Most stars are constant in their luminosity, at least to within a percent or two. Like the Sun, they generate a steady flow of energy from their interiors. However, some stars are seen to vary in brightness and, for this reason, are called *variable stars*. Many such stars vary on a regular cycle, like the flashing bulbs that decorate stores and homes during the winter holidays.

Let's define some tools to help us keep track of how a star varies. A graph that shows how the brightness of a variable star changes with time is called a **light curve** (Figure 19.9). The *maximum* is the point of the light curve where the star has its greatest brightness; the *minimum* is the point where it is faintest. If the light variations repeat themselves periodically, the interval between the two maxima is called the *period* of the star. (If this kind of graph looks familiar, it is because we introduced it in Diameters of Stars.)

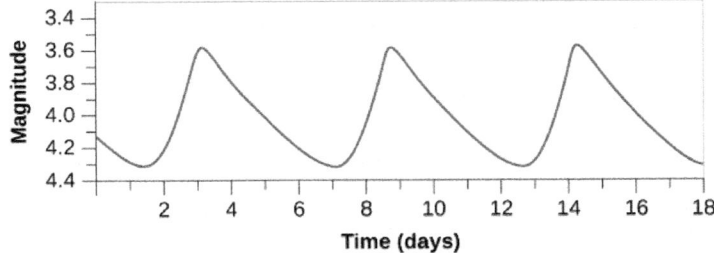

Figure 19.9. Cepheid Light Curve. This light curve shows how the brightness changes with time for a typical cepheid variable, with a period of about 6 days.

Download for free at http://cnx.org/content/col11992/latest/

Pulsating Variables

There are two special types of variable stars for which—as we will see—measurements of the light curve give us accurate distances. These are called **cepheid** and **RR Lyrae** variables, both of which are **pulsating variable stars**. Such a star actually changes its diameter with time—periodically expanding and contracting, as your chest does when you breathe. We now understand that these stars are going through a brief unstable stage late in their lives.

The expansion and contraction of pulsating variables can be measured by using the Doppler effect. The lines in the spectrum shift toward the blue as the surface of the star moves toward us and then shift to the red as the surface shrinks back. As the star pulsates, it also changes its overall color, indicating that its temperature is also varying. And, most important for our purposes, the luminosity of the pulsating variable also changes in a regular way as it expands and contracts.

Cepheid Variables

Cepheids are large, yellow, pulsating stars named for the first-known star of the group, Delta Cephei. This, by the way, is another example of how confusing naming conventions get in astronomy; here, a whole class of stars in named after the constellation in which the first one happened to be found. (We textbook authors can only apologize to our students for the whole mess!)

The variability of Delta Cephei was discovered in 1784 by the young English astronomer John Goodricke (see John Goodricke). The star rises rather rapidly to maximum light and then falls more slowly to minimum light, taking a total of 5.4 days for one cycle. The curve in Figure 19.9 represents a simplified version of the light curve of Delta Cephei.

Several hundred cepheid variables are known in our Galaxy. Most cepheids have periods in the range of 3 to 50 days and luminosities that are about 1000 to 10,000 times greater than that of the Sun. Their variations in luminosity range from a few percent to a factor of 10.

Polaris, the North Star, is a cepheid variable that, for a long time, varied by one tenth of a magnitude, or by about 10% in visual luminosity, in a period of just under 4 days. Recent measurements indicate that the amount by which the brightness of Polaris changes is decreasing and that, sometime in the future, this star will no longer be a pulsating variable. This is just one more piece of evidence that stars really do evolve and change in fundamental ways as they age, and that being a cepheid variable represents a stage in the life of the star.

The Period-Luminosity Relation

The importance of cepheid variables lies in the fact that their periods and average luminosities turn out to be directly related. The longer the period (the longer the star takes to vary), the greater the luminosity. This **period-luminosity relation** was a remarkable discovery, one for which astronomers still (pardon the expression) thank their lucky stars. The period of such a star is easy to measure: a good telescope and a good clock are all you need. Once you have the period, the relationship (which can be put into precise mathematical terms) will give you the luminosity of the star.

Let's be clear on what that means. The relation allows you to essentially "read off" how bright the star really is (how much energy it puts out). Astronomers can then compare this intrinsic brightness with the apparent brightness of the star. As we saw, the difference between the two allows them to calculate the distance.

The relation between period and luminosity was discovered in 1908 by Henrietta Leavitt (Figure 19.10), a staff member at the Harvard College Observatory (and one of a number of women working for low wages assisting Edward Pickering, the observatory's director; see Annie Cannon: Classifier of the Stars). Leavitt discovered hundreds of variable stars in the Large Magellanic Cloud and Small Magellanic Cloud, two great star systems

This OpenStax book is available for free at http://cnx.org/content/col11992/1.8

that are actually neighboring galaxies (although they were not known to be galaxies then). A small fraction of these variables were cepheids (Figure 19.11).

Figure 19.10. Henrietta Swan Leavitt (1868–1921). Leavitt worked as an astronomer at the Harvard College Observatory. While studying photographs of the Magellanic Clouds, she found over 1700 variable stars, including 20 cepheids. Since all the cepheids in these systems were at roughly the same distance, she was able to compare their luminosities and periods of variation. She thus discovered a fundamental relationship between these characteristics that led to a new and much better way of estimating cosmic distances. (credit: modification of work by AIP)

These systems presented a wonderful opportunity to study the behavior of variable stars independent of their distance. For all practical purposes, the Magellanic Clouds are so far away that astronomers can assume that all the stars in them are at roughly the same distance from us. (In the same way, all the suburbs of Los Angeles are roughly the same distance from New York City. Of course, if you are *in* Los Angeles, you will notice annoying distances between the suburbs, but compared to how far away New York City is, the differences seem small.) If all the variable stars in the Magellanic Clouds are at roughly the same distance, then any difference in their apparent brightnesses must be caused by differences in their intrinsic luminosities.

Download for free at http://cnx.org/content/col11992/latest/

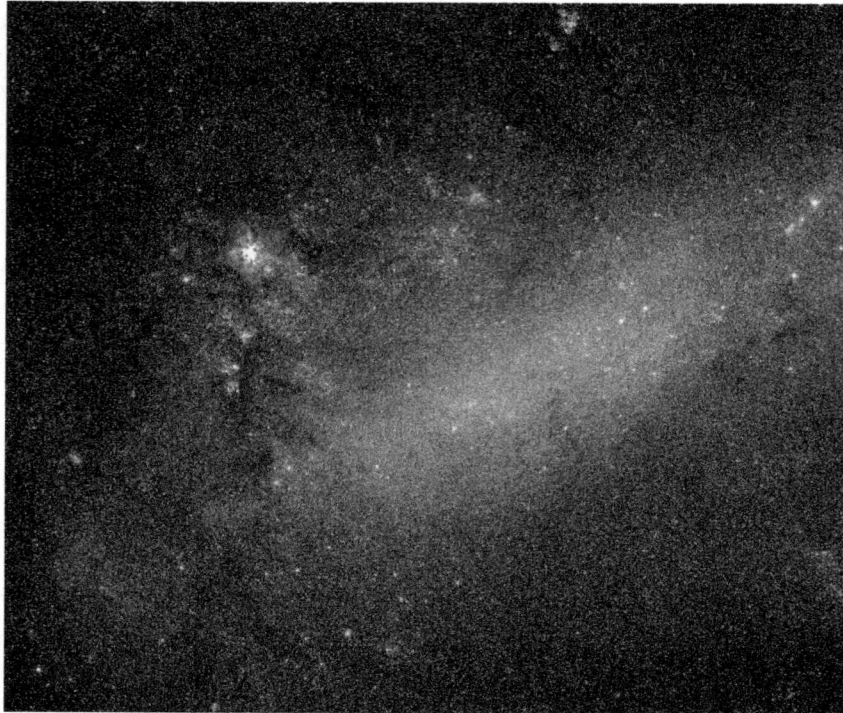

Figure 19.11. Large Magellanic Cloud. The Large Magellanic Cloud (so named because Magellan's crew were the first Europeans to record it) is a small, irregularly shaped galaxy near our own Milky Way. It was in this galaxy that Henrietta Leavitt discovered the cepheid period-luminosity relation. (credit: ESO)

Leavitt found that the brighter-appearing cepheids always have the longer periods of light variation. Thus, she reasoned, the period must be related to the luminosity of the stars. When Leavitt did this work, the distance to the Magellanic Clouds was not known, so she was only able to show that luminosity was related to period. She could not determine exactly what the relationship is.

To define the period-luminosity relation with actual numbers (to *calibrate* it), astronomers first had to measure the actual distances to a few nearby cepheids in another way. (This was accomplished by finding cepheids associated in clusters with other stars whose distances could be estimated from their spectra, as discussed in the next section of this chapter.) But once the relation was thus defined, it could give us the distance to any cepheid, wherever it might be located (Figure 19.12).

This OpenStax book is available for free at http://cnx.org/content/col11992/1.8

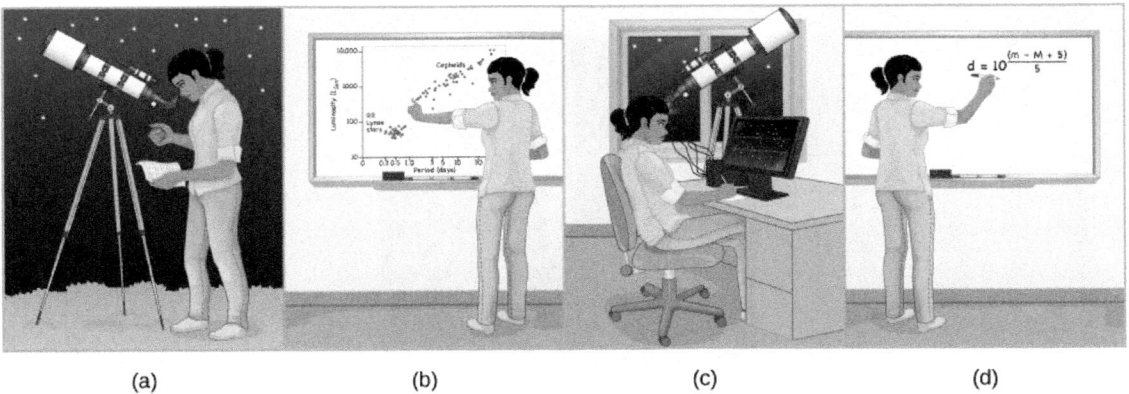

Figure 19.12. How to Use a Cepheid to Measure Distance. (a) Find a cepheid variable star and measure its period. (b) Use the period-luminosity relation to calculate the star's luminosity. (c) Measure the star's apparent brightness. (d) Compare the luminosity with the apparent brightness to calculate the distance.

Here at last was the technique astronomers had been searching for to break the confines of distance that parallax imposed on them. Cepheids can be observed and monitored, it turns out, in many parts of our own Galaxy and in other nearby galaxies as well. Astronomers, including Ejnar Hertzsprung and Harvard's Harlow Shapley, immediately saw the potential of the new technique; they and many others set to work exploring more distant reaches of space using cepheids as signposts. In the 1920s, Edwin Hubble made one of the most significant astronomical discoveries of all time using cepheids, when he observed them in nearby galaxies and discovered the expansion of the universe. As we will see, this work still continues, as the Hubble Space Telescope and other modern instruments try to identify and measure individual cepheids in galaxies farther and farther away. The most distant known variable stars are all cepheids, with some about 60 million light-years away.

VOYAGERS IN ASTRONOMY

John Goodricke

The brief life of John Goodricke (Figure 19.13) is a testament to the human spirit under adversity. Born deaf and unable to speak, Goodricke nevertheless made a number of pioneering discoveries in astronomy through patient and careful observations of the heavens.

Download for free at http://cnx.org/content/col11992/latest/

Figure 19.13. John Goodricke (1764–1786). This portrait of Goodricke by artist J. Scouler hangs in the Royal Astronomical Society in London. There is some controversy about whether this is actually what Goodricke looked like or whether the painting was much retouched to please his family. (credit: James Scouler)

Born in Holland, where his father was on a diplomatic mission, Goodricke was sent back to England at age eight to study at a special school for the deaf. He did sufficiently well to enter Warrington Academy, a secondary school that offered no special assistance for students with handicaps. His mathematics teacher there inspired an interest in astronomy, and in 1781, at age 17, Goodricke began observing the sky at his family home in York, England. Within a year, he had discovered the brightness variations of the star Algol (discussed in The Stars: A Celestial Census) and suggested that an unseen companion star was causing the changes, a theory that waited over 100 years for proof. His paper on the subject was read before the Royal Society (the main British group of scientists) in 1783 and won him a medal from that distinguished group.

In the meantime, Goodricke had discovered two other stars that varied regularly, Beta Lyrae and Delta Cephei, both of which continued to interest astronomers for years to come. Goodricke shared his interest in observing with his older cousin, Edward Pigott, who went on to discover other variable stars during his much longer life. But Goodricke's time was quickly drawing to a close; at age 21, only 2 weeks after he was elected to the Royal Society, he caught a cold while making astronomical observations and never recovered.

Today, the University of York has a building named Goodricke Hall and a plaque that honors his contributions to science. Yet if you go to the churchyard cemetery where he is buried, an overgrown tombstone has only the initials "J. G." to show where he lies. Astronomer Zdenek Kopal, who looked carefully into Goodricke's life, speculated on why the marker is so modest: perhaps the rather staid Goodricke relatives were ashamed of having a "deaf-mute" in the family and could not sufficiently appreciate how much a man who could not hear could nevertheless see.

RR Lyrae Stars

A related group of stars, whose nature was understood somewhat later than that of the cepheids, are called RR Lyrae variables, named for the star RR Lyrae, the best-known member of the group. More common than the cepheids, but less luminous, thousands of these pulsating variables are known in our Galaxy. The periods of RR Lyrae stars are always less than 1 day, and their changes in brightness are typically less than about a factor of two.

This OpenStax book is available for free at http://cnx.org/content/col11992/1.8

Download for free at http://cnx.org/content/col11992/latest/

Astronomers have observed that the RR Lyrae stars occurring in any particular cluster all have about the same apparent brightness. Since stars in a cluster are all at approximately the same distance, it follows that RR Lyrae variables must all have nearly the same intrinsic luminosity, which turns out to be about 50 L_{Sun}. In this sense, RR Lyrae stars are a little bit like standard light bulbs and can also be used to obtain distances, particularly within our Galaxy. Figure 19.14 displays the ranges of periods and luminosities for both the cepheids and the RR Lyrae stars.

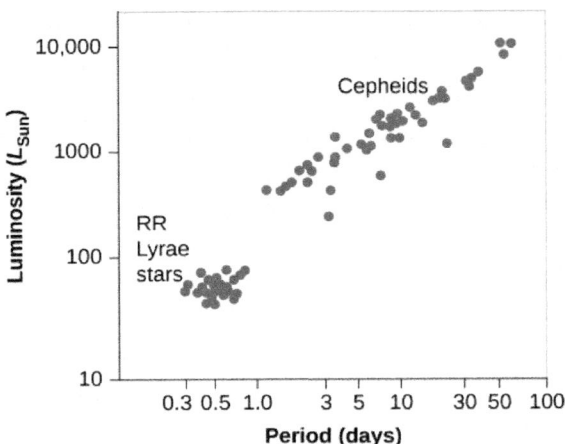

Figure 19.14. Period-Luminosity Relation for Cepheid Variables. In this class of variable stars, the time the star takes to go through a cycle of luminosity changes is related to the average luminosity of the star. Also shown are the period and luminosity for RR Lyrae stars.

19.4 THE H–R DIAGRAM AND COSMIC DISTANCES

Learning Objectives

By the end of this section, you will be able to:

> Understand how spectral types are used to estimate stellar luminosities
> Examine how these techniques are used by astronomers today

Variable stars are not the only way that we can estimate the luminosity of stars. Another way involves the H–R diagram, which shows that the intrinsic brightness of a star can be estimated if we know its spectral type.

Distances from Spectral Types

As satisfying and productive as variable stars have been for distance measurement, these stars are rare and are not found near all the objects to which we wish to measure distances. Suppose, for example, we need the distance to a star that is not varying, or to a group of stars, none of which is a variable. In this case, it turns out the H–R diagram can come to our rescue.

If we can observe the spectrum of a star, we can estimate its distance from our understanding of the H–R diagram. As discussed in Analyzing Starlight, a detailed examination of a stellar spectrum allows astronomers to classify the star into one of the *spectral types* indicating surface temperature. (The types are O, B, A, F, G, K, M, L, T, and Y; each of these can be divided into numbered subgroups.) In general, however, the spectral type alone is not enough to allow us to estimate luminosity. Look again at Figure 18.15. A G2 star could be a main-sequence star with a luminosity of 1 L_{Sun}, or it could be a giant with a luminosity of 100 L_{Sun}, or even a supergiant with a still higher luminosity.

Download for free at http://cnx.org/content/col11992/latest/

We can learn more from a star's spectrum, however, than just its temperature. Remember, for example, that we can detect pressure differences in stars from the details of the spectrum. This knowledge is very useful because giant stars are larger (and have lower pressures) than main-sequence stars, and supergiants are still larger than giants. If we look in detail at the spectrum of a star, we can determine whether it is a main-sequence star, a giant, or a supergiant.

Suppose, to start with the simplest example, that the spectrum, color, and other properties of a distant G2 star match those of the Sun exactly. It is then reasonable to conclude that this distant star is likely to be a main-sequence star just like the Sun and to have the same luminosity as the Sun. But if there are subtle differences between the solar spectrum and the spectrum of the distant star, then the distant star may be a giant or even a supergiant.

The most widely used system of star classification divides stars of a given spectral class into six categories called **luminosity classes**. These luminosity classes are denoted by Roman numbers as follows:

- Ia: Brightest supergiants
- Ib: Less luminous supergiants
- II: Bright giants
- III: Giants
- IV: Subgiants (intermediate between giants and main-sequence stars)
- V: Main-sequence stars

The full spectral specification of a star includes its luminosity class. For example, a main-sequence star with spectral class F3 is written as F3 V. The specification for an M2 giant is M2 III. Figure 19.15 illustrates the approximate position of stars of various luminosity classes on the H–R diagram. The dashed portions of the lines represent regions with very few or no stars.

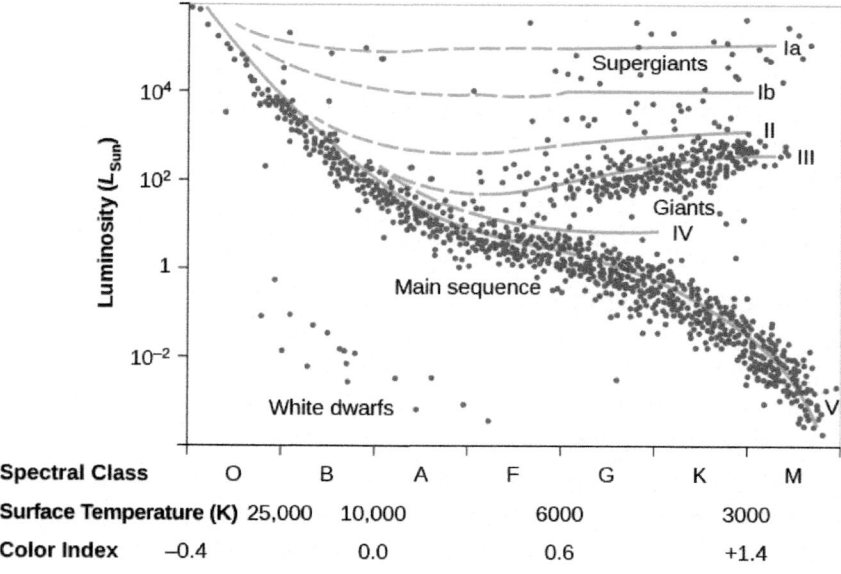

Figure 19.15. Luminosity Classes. Stars of the same temperature (or spectral class) can fall into different luminosity classes on the Hertzsprung-Russell diagram. By studying details of the spectrum for each star, astronomers can determine which luminosity class they fall in (whether they are main-sequence stars, giant stars, or supergiant stars).

This OpenStax book is available for free at http://cnx.org/content/col11992/1.8

With both its spectral and luminosity classes known, a star's position on the H–R diagram is uniquely determined. Since the diagram plots luminosity versus temperature, this means we can now read off the star's luminosity (once its spectrum has helped us place it on the diagram). As before, if we know how luminous the star really is and see how dim it looks, the difference allows us to calculate its distance. (For historical reasons, astronomers sometimes call this method of distance determination *spectroscopic parallax*, even though the method has nothing to do with parallax.)

The H–R diagram method allows astronomers to estimate distances to nearby stars, as well as some of the most distant stars in our Galaxy, but it is anchored by measurements of parallax. The distances measured using parallax are the gold standard for distances: they rely on no assumptions, only geometry. Once astronomers take a spectrum of a nearby star for which we also know the parallax, we know the luminosity that corresponds to that spectral type. Nearby stars thus serve as benchmarks for more distant stars because we can assume that two stars with identical spectra have the same intrinsic luminosity.

A Few Words about the Real World

Introductory textbooks such as ours work hard to present the material in a straightforward and simplified way. In doing so, we sometimes do our students a disservice by making scientific techniques seem too clean and painless. In the real world, the techniques we have just described turn out to be messy and difficult, and often give astronomers headaches that last long into the day.

For example, the relationships we have described such as the period-luminosity relation for certain variable stars aren't exactly straight lines on a graph. The points representing many stars scatter widely when plotted, and thus, the distances derived from them also have a certain built-in scatter or uncertainty.

The distances we measure with the methods we have discussed are therefore only accurate to within a certain percentage of error—sometimes 10%, sometimes 25%, sometimes as much as 50% or more. A 25% error for a star estimated to be 10,000 light-years away means it could be anywhere from 7500 to 12,500 light-years away. This would be an unacceptable uncertainty if you were loading fuel into a spaceship for a trip to the star, but it is not a bad first figure to work with if you are an astronomer stuck on planet Earth.

Nor is the construction of H–R diagrams as easy as you might think at first. To make a good diagram, one needs to measure the characteristics and distances of many stars, which can be a time-consuming task. Since our own solar neighborhood is already well mapped, the stars astronomers most want to study to advance our knowledge are likely to be far away and faint. It may take hours of observing to obtain a single spectrum. Observers may have to spend many nights at the telescope (and many days back home working with their data) before they get their distance measurement. Fortunately, this is changing because surveys like Gaia will study billions of stars, producing public datasets that all astronomers can use.

Despite these difficulties, the tools we have been discussing allow us to measure a remarkable range of distances—parallaxes for the nearest stars, RR Lyrae variable stars; the H–R diagram for clusters of stars in our own and nearby galaxies; and cepheids out to distances of 60 million light-years. Table 19.1 describes the distance limits and overlap of each method.

Each technique described in this chapter builds on at least one other method, forming what many call the *cosmic distance ladder*. Parallaxes are the foundation of all stellar distance estimates, spectroscopic methods use nearby stars to calibrate their H–R diagrams, and RR Lyrae and cepheid distance estimates are grounded in H–R diagram distance estimates (and even in a parallax measurement to a nearby cepheid, Delta Cephei).

This chain of methods allows astronomers to push the limits when looking for even more distant stars. Recent work, for example, has used RR Lyrae stars to identify dim companion galaxies to our own Milky Way out at

Download for free at http://cnx.org/content/col11992/latest/

distances of 300,000 light-years. The H–R diagram method was recently used to identify the two most distant stars in the Galaxy: red giant stars way out in the halo of the Milky Way with distances of almost 1 million light-years.

We can combine the distances we find for stars with measurements of their composition, luminosity, and temperature—made with the techniques described in Analyzing Starlight and The Stars: A Celestial Census. Together, these make up the arsenal of information we need to trace the evolution of stars from birth to death, the subject to which we turn in the chapters that follow.

Distance Range of Celestial Measurement Methods

Method	Distance Range
Trigonometric parallax	4–30,000 light-years when the Gaia mission is complete
RR Lyrae stars	Out to 300,000 light-years
H–R diagram and spectroscopic distances	Out to 1,200,000 light-years
Cepheid stars	Out to 60,000,000 light-years

Table 19.1

This OpenStax book is available for free at http://cnx.org/content/col11992/1.8

CHAPTER 19 REVIEW

KEY TERMS

cepheid a star that belongs to a class of yellow supergiant pulsating stars; these stars vary periodically in brightness, and the relationship between their periods and luminosities is useful in deriving distances to them

light curve a graph that displays the time variation of the light from a variable or eclipsing binary star or, more generally, from any other object whose radiation output changes with time

luminosity class a classification of a star according to its luminosity within a given spectral class; our Sun, a G2V star, has luminosity class V, for example

parallax an apparent displacement of a nearby star that results from the motion of Earth around the Sun

parsec a unit of distance in astronomy, equal to 3.26 light-years; at a distance of 1 parsec, a star has a parallax of 1 arcsecond

period-luminosity relation an empirical relation between the periods and luminosities of certain variable stars

pulsating variable star a variable star that pulsates in size and luminosity

RR Lyrae one of a class of giant pulsating stars with periods shorter than 1 day, useful for finding distances

SUMMARY

19.1 Fundamental Units of Distance

Early measurements of length were based on human dimensions, but today, we use worldwide standards that specify lengths in units such as the meter. Distances within the solar system are now determined by timing how long it takes radar signals to travel from Earth to the surface of a planet or other body and then return.

19.2 Surveying the Stars

For stars that are relatively nearby, we can "triangulate" the distances from a baseline created by Earth's annual motion around the Sun. Half the shift in a nearby star's position relative to very distant background stars, as viewed from opposite sides of Earth's orbit, is called the parallax of that star and is a measure of its distance. The units used to measure stellar distance are the light-year, the distance light travels in 1 year, and the parsec (pc), the distance of a star with a parallax of 1 arcsecond (1 parsec = 3.26 light-years). The closest star, a red dwarf, is over 1 parsec away. The first successful measurements of stellar parallaxes were reported in 1838. Parallax measurements are a fundamental link in the chain of cosmic distances. The Hipparcos satellite has allowed us to measure accurate parallaxes for stars out to about 300 light-years, and the Gaia mission will result in parallaxes out to 30,000 light-years.

19.3 Variable Stars: One Key to Cosmic Distances

Cepheids and RR Lyrae stars are two types of pulsating variable stars. Light curves of these stars show that their luminosities vary with a regularly repeating period. RR Lyrae stars can be used as standard bulbs, and cepheid variables obey a period-luminosity relation, so measuring their periods can tell us their luminosities. Then, we can calculate their distances by comparing their luminosities with their apparent brightnesses, and this can allow us to measure distances to these stars out to over 60 million light-years.

Download for free at http://cnx.org/content/col11992/latest/

Stars with identical temperatures but different pressures (and diameters) have somewhat different spectra. Spectral classification can therefore be used to estimate the luminosity class of a star as well as its temperature. As a result, a spectrum can allow us to pinpoint where the star is located on an H–R diagram and establish its luminosity. This, with the star's apparent brightness, again yields its distance. The various distance methods can be used to check one against another and thus make a kind of distance ladder which allows us to find even larger distances.

 FOR FURTHER EXPLORATION

Articles

Adams, A. "The Triumph of Hipparcos." *Astronomy* (December 1997): 60. Brief introduction.

Dambeck, T. "Gaia's Mission to the Milky Way." *Sky & Telescope* (March 2008): 36–39. An introduction to the mission to measure distances and positions of stars with unprecedented accuracy.

Hirshfeld, A. "The Absolute Magnitude of Stars." *Sky & Telescope* (September 1994): 35. Good review of how we measure luminosity, with charts.

Hirshfeld, A. "The Race to Measure the Cosmos." *Sky & Telescope* (November 2001): 38. On parallax.

Trefil, J. Puzzling Out Parallax." *Astronomy* (September 1998): 46. On the concept and history of parallax.

Turon, C. "Measuring the Universe." *Sky & Telescope* (July 1997): 28. On the Hipparcos mission and its results.

Zimmerman, R. "Polaris: The Code-Blue Star." *Astronomy* (March 1995): 45. On the famous cepheid variable and how it is changing.

Websites

ABCs of Distance: http://www.astro.ucla.edu/~wright/distance.htm. Astronomer Ned Wright (UCLA) gives a concise primer on many different methods of obtaining distances. This site is at a higher level than our textbook, but is an excellent review for those with some background in astronomy.

American Association of Variable Star Observers (AAVSO): https://www.aavso.org/. This organization of amateur astronomers helps to keep track of variable stars; its site has some background material, observing instructions, and links.

Friedrich Wilhelm Bessel: http://messier.seds.org/xtra/Bios/bessel.html. A brief site about the first person to detect stellar parallax, with references and links.

Gaia: http://sci.esa.int/gaia/. News from the Gaia mission, including images and a blog of the latest findings.

Hipparchos: http://sci.esa.int/hipparcos/. Background, results, catalogs of data, and educational resources from the Hipparchos mission to observe parallaxes from space. Some sections are technical, but others are accessible to students.

John Goodricke: The Deaf Astronomer: http://www.bbc.com/news/magazine-20725639. A biographical article from the BBC.

Women in Astronomy: http://www.astrosociety.org/education/astronomy-resource-guides/women-in-astronomy-an-introductory-resource-guide/. More about Henrietta Leavitt's and other women's contributions to astronomy and the obstacles they faced.

This OpenStax book is available for free at http://cnx.org/content/col11992/1.8

Videos

Gaia's Mission: Solving the Celestial Puzzle: https://www.youtube.com/watch?v=oGri4YNggoc. Describes the Gaia mission and what scientists hope to learn, from Cambridge University (19:58).

Hipparcos: Route Map to the Stars: https://www.youtube.com/watch?v=4d8a75fs7KI. This ESA video describes the mission to measure parallax and its results (14:32)

How Big Is the Universe: https://www.youtube.com/watch?v=K_xZuopg4Sk. Astronomer Pete Edwards from the British Institute of Physics discusses the size of the universe and gives a step-by-step introduction to the concepts of distances (6:22)

Search for Miss Leavitt: http://perimeterinstitute.ca/videos/search-miss-leavitt., Video of talk by George Johnson on his search for Miss Leavitt (55:09).

Women in Astronomy: http://www.youtube.com/watch?v=5vMR7su4fi8. Emily Rice (CUNY) gives a talk on the contributions of women to astronomy, with many historical and contemporary examples, and an analysis of modern trends (52:54).

🕮 COLLABORATIVE GROUP ACTIVITIES

A. In this chapter, we explain the various measurements that have been used to establish the size of a standard meter. Your group should discuss why we have changed the definitions of our standard unit of measurement in science from time to time. What factors in our modern society contribute to the growth of technology? Does technology "drive" science, or does science "drive" technology? Or do you think the two are so intertwined that it's impossible to say which is the driver?

B. Cepheids are scattered throughout our own Milky Way Galaxy, but the period-luminosity relation was discovered from observations of the Magellanic Clouds, a satellite galaxy now known to be about 160,000 light-years away. What reasons can you give to explain why the relation was not discovered from observations of cepheids in our own Galaxy? Would your answer change if there were a small cluster in our own Galaxy that contained 20 cepheids? Why or why not?

C. You want to write a proposal to use the Hubble Space Telescope to look for the brightest cepheids in galaxy M100 and estimate their luminosities. What observations would you need to make? Make a list of all the reasons such observations are harder than it first might appear.

D. Why does your group think so many different ways of naming stars developed through history? (Think back to the days before everyone connected online.) Are there other fields where things are named confusingly and arbitrarily? How do stars differ from other phenomena that science and other professions tend to catalog?

E. Although cepheids and RR Lyrae variable stars tend to change their brightness pretty regularly (while they are in that stage of their lives), some variable stars are unpredictable or change their their behavior even during the course of a single human lifetime. Amateur astronomers all over the world follow such variable stars patiently and persistently, sending their nightly observations to huge databases that are being kept on the behavior of many thousands of stars. None of the hobbyists who do this get paid for making such painstaking observations. Have your group discuss why they do it. Would you ever consider a hobby that involves so much work, long into the night, often on work nights? If observing variable stars doesn't pique

Download for free at http://cnx.org/content/col11992/latest/

your interest, is there something you think you could do as a volunteer after college that does excite you? Why?

F. In Figure 19.8, the highest concentration of stars occurs in the middle of the main sequence. Can your group give reasons why this might be so? Why are there fewer very hot stars and fewer very cool stars on this diagram?

G. In this chapter, we discuss two astronomers who were differently abled than their colleagues. John Goodricke could neither hear nor speak, and Henrietta Leavitt struggled with hearing impairment for all of her adult life. Yet they each made fundamental contributions to our understanding of the universe. Does your group know people who are handling a disability? What obstacles would people with different disabilities face in trying to do astronomy and what could be done to ease their way? For a set of resources in this area, see http://astronomerswithoutborders.org/gam2013/programs/1319-people-with-disabilities-astronomy-resources.html.

EXERCISES

Review Questions

1. Explain how parallax measurements can be used to determine distances to stars. Why can we not make accurate measurements of parallax beyond a certain distance?

2. Suppose you have discovered a new cepheid variable star. What steps would you take to determine its distance?

3. Explain how you would use the spectrum of a star to estimate its distance.

4. Which method would you use to obtain the distance to each of the following?
    A. An asteroid crossing Earth's orbit

    B. A star astronomers believe to be no more than 50 light-years from the Sun

    C. A tight group of stars in the Milky Way Galaxy that includes a significant number of variable stars

    D. A star that is not variable but for which you can obtain a clearly defined spectrum

5. What are the luminosity class and spectral type of a star with an effective temperature of 5000 K and a luminosity of 100 L_{Sun}?

Thought Questions

6. The meter was redefined as a reference to Earth, then to krypton, and finally to the speed of light. Why do you think the reference point for a meter continued to change?

7. While a meter is the fundamental unit of length, most distances traveled by humans are measured in miles or kilometers. Why do you think this is?

8. Most distances in the Galaxy are measured in light-years instead of meters. Why do you think this is the case?

9. The AU is defined as the *average* distance between Earth and the Sun, not the distance between Earth and the Sun. Why does this need to be the case?

This OpenStax book is available for free at http://cnx.org/content/col11992/1.8

10. What would be the advantage of making parallax measurements from Pluto rather than from Earth? Would there be a disadvantage?

11. Parallaxes are measured in fractions of an arcsecond. One arcsecond equals 1/60 arcmin; an arcminute is, in turn, 1/60th of a degree (°). To get some idea of how big 1° is, go outside at night and find the Big Dipper. The two pointer stars at the ends of the bowl are 5.5° apart. The two stars across the top of the bowl are 10° apart. (Ten degrees is also about the width of your fist when held at arm's length and projected against the sky.) Mizar, the second star from the end of the Big Dipper's handle, appears double. The fainter star, Alcor, is about 12 arcmin from Mizar. For comparison, the diameter of the full moon is about 30 arcmin. The belt of Orion is about 3° long. Keeping all this in mind, why did it take until 1838 to make parallax measurements for even the nearest stars?

12. For centuries, astronomers wondered whether comets were true celestial objects, like the planets and stars, or a phenomenon that occurred in the atmosphere of Earth. Describe an experiment to determine which of these two possibilities is correct.

13. The Sun is much closer to Earth than are the nearest stars, yet it is not possible to measure accurately the diurnal parallax of the Sun relative to the stars by measuring its position relative to background objects in the sky directly. Explain why.

14. Parallaxes of stars are sometimes measured relative to the positions of galaxies or distant objects called quasars. Why is this a good technique?

15. Estimating the luminosity class of an M star is much more important than measuring it for an O star if you are determining the distance to that star. Why is that the case?

16. Figure 19.9 is the light curve for the prototype cepheid variable Delta Cephei. How does the luminosity of this star compare with that of the Sun?

17. Which of the following can you determine about a star without knowing its distance, and which can you not determine: radial velocity, temperature, apparent brightness, or luminosity? Explain.

18. A G2 star has a luminosity 100 times that of the Sun. What kind of star is it? How does its radius compare with that of the Sun?

19. A star has a temperature of 10,000 K and a luminosity of 10^{-2} L_{Sun}. What kind of star is it?

20. What is the advantage of measuring a parallax distance to a star as compared to our other distance measuring methods?

21. What is the disadvantage of the parallax method, especially for studying distant parts of the Galaxy?

22. Luhman 16 and WISE 0720 are brown dwarfs, also known as failed stars, and are some of the new closest neighbors to Earth, but were only discovered in the last decade. Why do you think they took so long to be discovered?

23. Most stars close to the Sun are red dwarfs. What does this tell us about the average star formation event in our Galaxy?

24. Why would it be easier to measure the characteristics of intrinsically less luminous cepheids than more luminous ones?

25. When Henrietta Leavitt discovered the period-luminosity relationship, she used cepheid stars that were all located in the Large Magellanic Cloud. Why did she need to use stars in another galaxy and not cepheids located in the Milky Way?

Download for free at http://cnx.org/content/col11992/latest/

Figuring For Yourself

26. A radar astronomer who is new at the job claims she beamed radio waves to Jupiter and received an echo exactly 48 min later. Should you believe her? Why or why not?

27. The New Horizons probe flew past Pluto in July 2015. At the time, Pluto was about 32 AU from Earth. How long did it take for communication from the probe to reach Earth, given that the speed of light in km/hr is 1.08×10^9?

28. Estimate the maximum and minimum time it takes a radar signal to make the round trip between Earth and Venus, which has a semimajor axis of 0.72 AU.

29. The Apollo program (not the lunar missions with astronauts) being conducted at the Apache Point Observatory uses a 3.5-m telescope to direct lasers at retro-reflectors left on the Moon by the Apollo astronauts. If the Moon is 384,472 km away, approximately how long do the operators need to wait to see the laser light return to Earth?

30. In 1974, the Arecibo Radio telescope in Puerto Rico was used to transmit a signal to M13, a star cluster about 25,000 light-years away. How long will it take the message to reach M13, and how far has the message travelled so far (in light-years)?

31. Demonstrate that 1 pc equals 3.09×10^{13} km and that it also equals 3.26 light-years. Show your calculations.

32. The best parallaxes obtained with Hipparcos have an accuracy of 0.001 arcsec. If you want to measure the distance to a star with an accuracy of 10%, its parallax must be 10 times larger than the typical error. How far away can you obtain a distance that is accurate to 10% with Hipparcos data? The disk of our Galaxy is 100,000 light-years in diameter. What fraction of the diameter of the Galaxy's disk is the distance for which we can measure accurate parallaxes?

33. Astronomers are always making comparisons between measurements in astronomy and something that might be more familiar. For example, the Hipparcos web pages tell us that the measurement accuracy of 0.001 arcsec is equivalent to the angle made by a golf ball viewed from across the Atlantic Ocean, or to the angle made by the height of a person on the Moon as viewed from Earth, or to the length of growth of a human hair in 10 sec as seen from 10 meters away. Use the ideas in Example 19.2 to verify one of the first two comparisons.

34. *Gaia* will have greatly improved precision over the measurements of Hipparcos. The average uncertainty for most *Gaia* parallaxes will be about 50 microarcsec, or 0.00005 arcsec. How many times better than Hipparcos (see Exercise 19.32) is this precision?

35. Using the same techniques as used in Exercise 19.32, how far away can *Gaia* be used to measure distances with an uncertainty of 10%? What fraction of the Galactic disk does this correspond to?

36. The human eye is capable of an angular resolution of about one arcminute, and the average distance between eyes is approximately 2 in. If you blinked and saw something move about one arcmin across, how far away from you is it? (Hint: You can use the setup in Example 19.2 as a guide.)

37. How much better is the resolution of the *Gaia* spacecraft compared to the human eye (which can resolve about 1 arcmin)?

38. The most recently discovered system close to Earth is a pair of brown dwarfs known as Luhman 16. It has a distance of 6.5 light-years. How many parsecs is this?

39. What would the parallax of Luhman 16 (see Exercise 19.38) be as measured from Earth?

This OpenStax book is available for free at http://cnx.org/content/col11992/1.8

40. The New Horizons probe that passed by Pluto during July 2015 is one of the fastest spacecraft ever assembled. It was moving at about 14 km/s when it went by Pluto. If it maintained this speed, how long would it take New Horizons to reach the nearest star, Proxima Centauri, which is about 4.3 light-years away? (Note: It isn't headed in that direction, but you can pretend that it is.)

41. What physical properties are different for an M giant with a luminosity of 1000 L_{Sun} and an M dwarf with a luminosity of 0.5 L_{Sun}? What physical properties are the same?

Download for free at http://cnx.org/content/col11992/latest/

This OpenStax book is available for free at http://cnx.org/content/col11992/1.8

20

BETWEEN THE STARS: GAS AND DUST IN SPACE

Figure 20.1. NGC 3603 and Its Parent Cloud. This image, taken by the Hubble Space Telescope, shows the young star cluster NGC 3603 interacting with the cloud of gas from which it recently formed. The bright blue stars of the cluster have blown a bubble in the gas cloud. The remains of this cloud can be seen in the lower right part of the frame, glowing in response to the starlight illuminating it. In its darker parts, shielded from the harsh light of NGC 3603, new stars continue to form. Although the stars of NGC 3603 formed only recently, the most massive of them are already dying and ejecting their mass, producing the blue ring and streak features visible in the upper left part of the image. Thus, this image shows the full life cycle of stars, from formation out of interstellar gas, through life on the main sequence, to death and the return of stellar matter to interstellar space. (credit: modification of work by NASA, Wolfgang Brandner (JPL/IPAC), Eva K. Grebel (University of Washington), You-Hua Chu (University of Illinois Urbana-Champaign))

Chapter Outline

✏ Thinking Ahead

Where do stars come from? We already know from earlier chapters that stars must die because ultimately they exhaust their nuclear fuel. We might hypothesize that new stars come into existence to replace the ones that die. In order to form new stars, however, we need the raw material to make them. It also turns out that stars eject mass throughout their lives (a kind of wind blows from their surface layers) and that material must go somewhere. What does this "raw material" of stars look like? How would you detect it, especially if it is not yet in the form of stars and cannot generate its own energy?

One of the most exciting discoveries of twentieth-century astronomy was that our Galaxy contains vast quantities of this "raw material"—atoms or molecules of gas and tiny solid dust particles found between the stars. Studying this diffuse matter between the stars helps us understand how new stars form and gives us important clues about our own origins billions of years ago.

Download for free at http://cnx.org/content/col11992/latest/

20.1 THE INTERSTELLAR MEDIUM

Learning Objectives

By the end of this section, you will be able to:

> Explain how much interstellar matter there is in the Milky Way, and what its typical density is
> Describe how the interstellar medium is divided into gaseous and solid components

Astronomers refer to all the material between stars as *interstellar* matter; the entire collection of interstellar matter is called the **interstellar medium (ISM)**. Some interstellar material is concentrated into giant clouds, each of which is known as a **nebula** (plural "nebulae," Latin for "clouds"). The best-known nebulae are the ones that we can see glowing or reflecting visible light; there are many pictures of these in this chapter.

Interstellar clouds do not last for the lifetime of the universe. Instead, they are like clouds on Earth, constantly shifting, merging with each other, growing, or dispersing. Some become dense and massive enough to collapse under their own gravity, forming new stars. When stars die, they, in turn, eject some of their material into interstellar space. This material can then form new clouds and begin the cycle over again.

About 99% of the material between the stars is in the form of a *gas*—that is, it consists of individual atoms or molecules. The most abundant elements in this gas are hydrogen and helium (which we saw are also the most abundant elements in the stars), but the gas also includes other elements. Some of the gas is in the form of molecules—combinations of atoms. The remaining 1% of the interstellar material is solid—frozen particles consisting of many atoms and molecules that are called *interstellar grains* or **interstellar dust** (Figure 20.2). A typical dust grain consists of a core of rocklike material (silicates) or graphite surrounded by a mantle of ices; water, methane, and ammonia are probably the most abundant ices.

This OpenStax book is available for free at http://cnx.org/content/col11992/1.8

Download for free at http://cnx.org/content/col11992/latest/

Figure 20.2. Various Types of Interstellar Matter. The reddish nebulae in this spectacular photograph glow with light emitted by hydrogen atoms. The darkest areas are clouds of dust that block the light from stars behind them. The upper part of the picture is filled with the bluish glow of light reflected from hot stars embedded in the outskirts of a huge, cool cloud of dust and gas. The cool supergiant star Antares can be seen as a big, reddish patch in the lower-left part of the picture. The star is shedding some of its outer atmosphere and is surrounded by a cloud of its own making that reflects the red light of the star. The red nebula in the middle right partially surrounds the star Sigma Scorpii. (To the right of Antares, you can see M4, a much more distant cluster of extremely old stars.) (credit: modification of work by ESO/Digitized Sky Survey 2)

If all the interstellar gas within the Galaxy were spread out smoothly, there would be only about one atom of gas per cm^3 in interstellar space. (In contrast, the air in the room where you are reading this book has roughly 10^{19} atoms per cm^3.) The dust grains are even scarcer. A km^3 of space would contain only a few hundred to a few thousand tiny grains, each typically less than one ten-thousandth of a millimeter in diameter. These numbers are just averages, however, because the gas and dust are distributed in a patchy and irregular way, much as water vapor in Earth's atmosphere is often concentrated into clouds.

In some interstellar clouds, the density of gas and dust may exceed the average by as much as a thousand times or more, but even this density is more nearly a vacuum than any we can make on Earth. To show what we mean, let's imagine a vertical tube of air reaching from the ground to the top of Earth's atmosphere with a cross-section of 1 square meter. Now let us extend the same-size tube from the top of the atmosphere all the way to the edge of the observable universe—over 10 billion light-years away. Long though it is, the second tube would still contain fewer atoms than the one in our planet's atmosphere.

While the *density* of interstellar matter is very low, the volume of space in which such matter is found is huge, and so its *total mass* is substantial. To see why, we must bear in mind that stars occupy only a tiny fraction of the volume of the Milky Way Galaxy. For example, it takes light only about four seconds to travel a distance equal to the diameter of the Sun, but more than four *years* to travel from the Sun to the nearest star. Even though the spaces among the stars are sparsely populated, there's a lot of space out there!

Astronomers estimate that the total mass of gas and dust in the Milky Way Galaxy is equal to about 15% of the mass contained in stars. This means that the mass of the interstellar matter in our Galaxy amounts to about 10

Download for free at http://cnx.org/content/col11992/latest/

billion times the mass of the Sun. There is plenty of raw material in the Galaxy to make generations of new stars and planets (and perhaps even astronomy students).

EXAMPLE 20.1

Estimating Interstellar Mass

You can make a rough estimate of how much interstellar mass our Galaxy contains and also how many new stars could be made from this interstellar matter. All you need to know is how big the Galaxy is and the average density using this formula:

$$\text{total mass} = \text{volume} \times \text{density of atoms} \times \text{mass per atom}$$

You have to remember to use consistent units—such as meters and kilograms. We will assume that our Galaxy is shaped like a cylinder; the volume of a cylinder equals the area of its base times its height

$$V = \pi R^2 \, h$$

where R is the radius of the cylinder and h is its height.

Suppose that the average density of hydrogen gas in our Galaxy is one atom per cm^3. Each hydrogen atom has a mass of 1.7×10^{-27} kg. If the Galaxy is a cylinder with a diameter of 100,000 light-years and a height of 300 light-years, what is the mass of this gas? How many solar-mass stars (2.0×10^{30} kg) could be produced from this mass of gas if it were all turned into stars?

Solution

Recall that 1 light-year = 9.5×10^{12} km = 9.5×10^{17} cm, so the volume of the Galaxy is

$$V = \pi R^2 \, h = \pi (100{,}000 \times 9.5 \times 10^{17} \text{ cm})^2 \, (300 \times 9.5 \times 10^{17} \text{ cm}) = 8.0 \times 10^{66} \text{ cm}^3$$

The total mass is therefore

$$M = V \times \text{density of atoms} \times \text{mass per atom}$$

$$8.0 \times 10^{66} \text{ cm}^3 \times (1 \text{ atom/cm}^3) \times 1.7 \times 10^{-27} \text{ kg} = 1.4 \times 10^{40} \text{ kg}$$

This is sufficient to make

$$N = \frac{M}{(2.0 \times 10^{30} \text{ kg})} = 6.9 \times 10^9$$

stars equal in mass to the Sun. That's roughly 7 billion stars.

Check Your Learning

You can use the same method to estimate the mass of interstellar gas around the Sun. The distance from the Sun to the nearest other star, Proxima Centauri, is 4.2 light-years. We will see in Interstellar Matter around the Sun that the gas in the immediate vicinity of the Sun is less dense than average, about 0.1 atoms per cm^3. What is the total mass of interstellar hydrogen in a sphere centered on the Sun and extending out to Proxima Centauri? How does this compare to the mass of the Sun? It is helpful to remember that the volume of a sphere is related to its radius:

$$V = (4/3)\pi R^3$$

This OpenStax book is available for free at http://cnx.org/content/col11992/1.8

Download for free at http://cnx.org/content/col11992/latest/

Answer:

The volume of a sphere stretching from the Sun to Proxima Centauri is:

$$V = (4/3)\pi R^3 = (4/3)\pi \left(4.2 \times 9.5 \times 10^{17} \text{ cm}\right)^3 = 2.7 \times 10^{56} \text{ cm}^3$$

Therefore, the mass of hydrogen in this sphere is:

$$M = V \times \left(0.1 \text{ atom/cm}^3\right) \times 1.7 \times 10^{-27} \text{ kg} = 4.5 \times 10^{28} \text{ kg}$$

This is only $(4.5 \times 10^{28}$ kg$)/(2.0 \times 10^{30}$ kg$) = 2.2\%$ the mass of the Sun.

ASTRONOMY BASICS

Naming the Nebulae

As you look at the captions for some of the spectacular photographs in this chapter and The Birth of Stars and the Discovery of Planets outside the Solar System, you will notice the variety of names given to the nebulae. A few, which in small telescopes look like something recognizable, are sometimes named after the creatures or objects they resemble. Examples include the Crab, Tarantula, and Keyhole Nebulae. But most have only numbers that are entries in a catalog of astronomical objects.

Perhaps the best-known catalog of nebulae (as well as star clusters and galaxies) was compiled by the French astronomer Charles Messier (1730–1817). Messier's passion was discovering comets, and his devotion to this cause earned him the nickname "The Comet Ferret" from King Louis XV. When comets are first seen coming toward the Sun, they look like little fuzzy patches of light; in small telescopes, they are easy to confuse with nebulae or with groupings of many stars so far away that their light is all blended together. Time and again, Messier's heart leapt as he thought he had discovered one of his treasured comets, only to find that he had "merely" observed a nebula or cluster.

In frustration, Messier set out to catalog the position and appearance of over 100 objects that could be mistaken for comets. For him, this list was merely a tool in the far more important work of comet hunting. He would be very surprised if he returned today to discover that no one recalls his comets anymore, but his catalog of "fuzzy things that are not comets" is still widely used. When Figure 20.2 refers to M4, it denotes the fourth entry in Messier's list.

A far more extensive listing was compiled under the title of the *New General Catalog* (*NGC*) *of Nebulae and Star Clusters* in 1888 by John Dreyer, working at the observatory in Armagh, Ireland. He based his compilation on the work of William Herschel and his son John, plus many other observers who followed them. With the addition of two further listings (called the *Index Catalogs*), Dreyer's compilation eventually included 13,000 objects. Astronomers today still use his NGC numbers when referring to most nebulae and star groups.

Download for free at http://cnx.org/content/col11992/latest/

20.2 INTERSTELLAR GAS

Learning Objectives

By the end of this section, you will be able to:

> Name the major types of interstellar gas
> Discuss how we can observe each type
> Describe the temperature and other major properties of each type

Interstellar gas, depending on where it is located, can be as cold as a few degrees above absolute zero or as hot as a million degrees or more. We will begin our voyage through the interstellar medium by exploring the different conditions under which we find gas.

Ionized Hydrogen (H II) Regions—Gas Near Hot Stars

Some of the most spectacular astronomical photographs show interstellar gas located near hot stars (Figure 20.3). The strongest line in the visible region of the hydrogen spectrum is the red line in the Balmer series[1] (as explained in the chapter on Radiation and Spectra); this emission line accounts for the characteristic red glow in images like Figure 20.3.

Figure 20.3. Orion Nebula. The red glow that pervades the great Orion Nebula is produced by the first line in the Balmer series of hydrogen. Hydrogen emission indicates that there are hot young stars nearby that ionize these clouds of gas. When electrons then recombine with protons and move back down into lower energy orbits, emission lines are produced. The blue color seen at the edges of some of the clouds is produced by small particles of dust that scatter the light from the hot stars. Dust can also be seen silhouetted against the glowing gas. (credit: NASA,ESA, M. Robberto (Space Telescope Science Institute/ESA) and the Hubble Space Telescope Orion Treasury Project Team)

Hot stars are able to heat nearby gas to temperatures close to 10,000 K. The ultraviolet radiation from the stars also ionizes the hydrogen (remember that during ionization, the electron is stripped completely away from the proton). Such a detached proton won't remain alone forever when attractive electrons are around; it will capture a free electron, becoming a neutral hydrogen once more. However, such a neutral atom can then

---

1   Scientists also call this red Balmer line the H-alpha line, with alpha meaning it is the first spectral line in the Balmer series.

This OpenStax book is available for free at http://cnx.org/content/col11992/1.8

Download for free at http://cnx.org/content/col11992/latest/

absorb ultraviolet radiation again and start the cycle over. At a typical moment, most of the atoms near a hot star are in the ionized state.

Since hydrogen is the main constituent of interstellar gas, we often characterize a region of space according to whether its hydrogen is neutral or ionized. A cloud of ionized hydrogen is called an **H II region**. (Scientists who work with spectra use the Roman numeral I to indicate that an atom is neutral; successively higher Roman numerals are used for each higher stage of ionization. H II thus refers to hydrogen that has lost its one electron; Fe III is iron with two electrons missing.)

The electrons that are captured by the hydrogen nuclei cascade down through the various energy levels of the hydrogen atoms on their way to the lowest level, or ground state. During each transition downward, they give up energy in the form of light. The process of converting ultraviolet radiation into visible light is called *fluorescence*. Interstellar gas contains other elements besides hydrogen. Many of them are also ionized in the vicinity of hot stars; they then capture electrons and emit light, just as hydrogen does, allowing them to be observed by astronomers. But generally, the red hydrogen line is the strongest, and that is why H II regions look red.

A fluorescent light on Earth works using the same principles as a fluorescent H II region. When you turn on the current, electrons collide with atoms of mercury vapor in the tube. The mercury is excited to a high-energy state because of these collisions. When the electrons in the mercury atoms return to lower-energy levels, some of the energy they emit is in the form of ultraviolet photons. These, in turn, strike a phosphor-coated screen on the inner wall of the light tube. The atoms in the screen absorb the ultraviolet photons and emit visible light as they cascade downward among the energy levels. (The difference is that these atoms give off a wider range of light colors, which mix to give the characteristic white glow of fluorescent lights, whereas the hydrogen atoms in an H II region give off a more limited set of colors.)

Neutral Hydrogen Clouds

The very hot stars required to produce H II regions are rare, and only a small fraction of interstellar matter is close enough to such hot stars to be ionized by them. Most of the volume of the interstellar medium is filled with neutral (nonionized) hydrogen. How do we go about looking for it?

Unfortunately, neutral hydrogen atoms at temperatures typical of the gas in interstellar space neither emit nor absorb light in the visible part of the spectrum. Nor, for the most part, do the other trace elements that are mixed with the interstellar hydrogen. However, some of these other elements can *absorb* visible light even at typical interstellar temperatures. This means that when we observe a bright source such as a hot star or a galaxy, we can sometimes see additional lines in its spectrum produced when interstellar gas absorbs light at particular frequencies (see Figure 20.4). Some of the strongest interstellar absorption lines are produced by calcium and sodium, but many other elements can be detected as well in sufficiently sensitive observations (as discussed in Radiation and Spectra).

Download for free at http://cnx.org/content/col11992/latest/

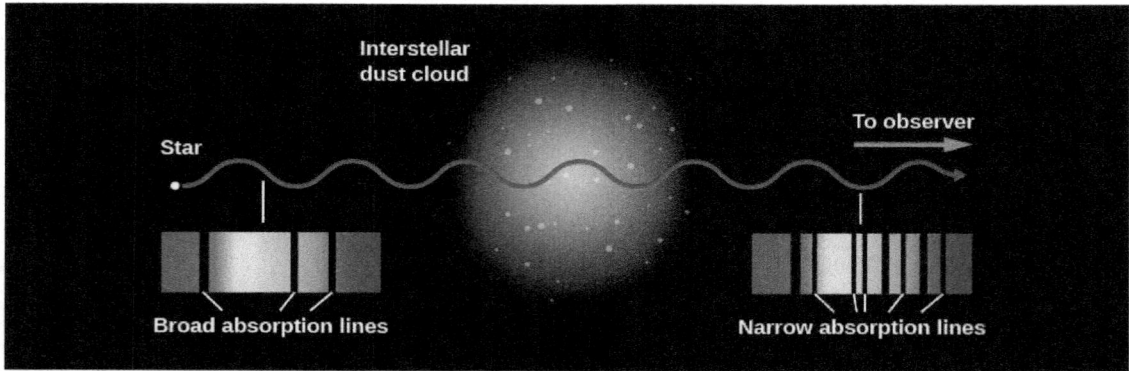

Figure 20.4. Absorption Lines though an Interstellar Dust Cloud. When there is a significant amount of cool interstellar matter between us and a star, we can see the absorption lines of the gas in the star's spectrum. We can distinguish the two kinds of lines because, whereas the star's lines are broad, the lines from the gas are narrower.

The first evidence for absorption by interstellar clouds came from the analysis of a spectroscopic binary star (see The Stars: A Celestial Census), published in 1904. While most of the lines in the spectrum of this binary shifted alternately from longer to shorter wavelengths and back again, as we would expect from the Doppler effect for stars in orbit around each other, a few lines in the spectrum remained fixed in wavelength. Since both stars are moving in a binary system, lines that showed no motion puzzled astronomers. The lines were also peculiar in that they were much, much narrower than the rest of the lines, indicating that the gas producing them was at a very low pressure. Subsequent work demonstrated that these lines were not formed in the star's atmosphere at all, but rather in a cold cloud of gas located between Earth and the binary star.

While these and similar observations proved there was interstellar gas, they could not yet detect hydrogen, the most common element, due to its lack of spectral features in the visible part of the spectrum. (The Balmer line of hydrogen is in the visible range, but only excited hydrogen atoms produce it. In the cold interstellar medium, the hydrogen atoms are all in the ground state and no electrons are in the higher-energy levels required to produce either emission or absorption lines in the Balmer series.) Direct detection of hydrogen had to await the development of telescopes capable of seeing very-low-energy changes in hydrogen atoms in other parts of the spectrum. The first such observations were made using radio telescopes, and radio emission and absorption by interstellar hydrogen remains one of our main tools for studying the vast amounts of cold hydrogen in the universe to this day.

In 1944, while he was still a student, the Dutch astronomer Hendrik van de Hulst predicted that hydrogen would produce a strong line at a wavelength of 21 centimeters. That's quite a long wavelength, implying that the wave has such a low frequency and low energy that it cannot come from electrons jumping between energy levels (as we discussed in Radiation and Spectra). Instead, energy is emitted when the electron does a flip, something like an acrobat in a circus flipping upright after standing on his head.

The flip works like this: a hydrogen atom consists of a proton and an electron bound together. Both the proton and the electron act is if they were spinning like tops, and spin axes of the two tops can either be pointed in the same direction (aligned) or in opposite directions (anti-aligned). If the proton and electron were spinning in opposite directions, the atom as a whole would have a very slightly lower energy than if the two spins were aligned (Figure 20.5). If an atom in the lower-energy state (spins opposed) acquired a small amount of energy, then the spins of the proton and electron could be aligned, leaving the atom in a slightly *excited state*. If the atom then lost that same amount of energy again, it would return to its ground state. The amount of energy involved corresponds to a wave with a wavelength of 21 centimeters; hence, it is known as the *21-centimeter line*.

This OpenStax book is available for free at http://cnx.org/content/col11992/1.8

Download for free at http://cnx.org/content/col11992/latest/

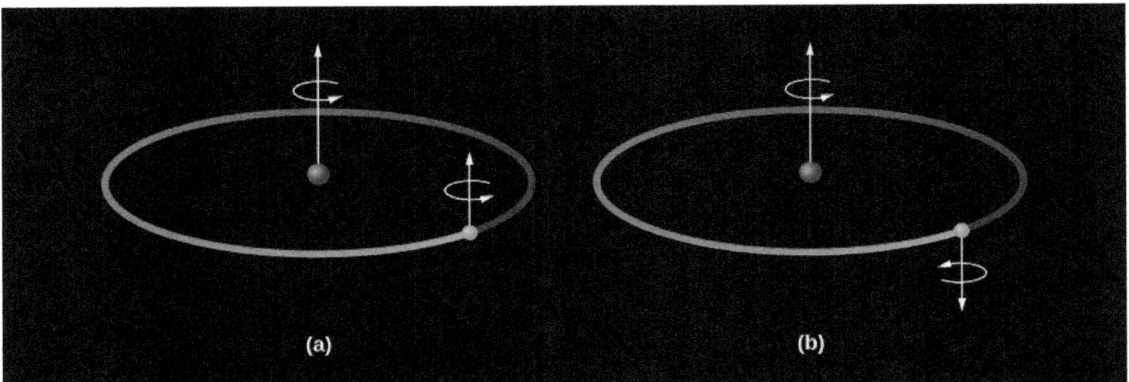

Figure 20.5. Formation of the 21-Centimeter Line. When the electron in a hydrogen atom is in the orbit closest to the nucleus, the proton and the electron may be spinning either (a) in the same direction or (b) in opposite directions. When the electron flips over, the atom gains or loses a tiny bit of energy by either absorbing or emitting electromagnetic energy with a wavelength of 21 centimeters.

Neutral hydrogen atoms can acquire small amounts of energy through collisions with other hydrogen atoms or with free electrons. Such collisions are extremely rare in the sparse gases of interstellar space. An individual atom may wait centuries before such an encounter aligns the spins of its proton and electron. Nevertheless, over many millions of years, a significant fraction of the hydrogen atoms are excited by a collision. (Out there in cold space, that's about as much excitement as an atom typically experiences.)

An excited atom can later lose its excess energy either by colliding with another particle or by giving off a radio wave with a wavelength of 21 centimeters. If there are no collisions, an excited hydrogen atom will wait an average of about 10 million years before emitting a photon and returning to its state of lowest energy. Even though the probability that any single atom will emit a photon is low, there are so many hydrogen atoms in a typical gas cloud that collectively they will produce an observable line at 21 centimeters.

Equipment sensitive enough to detect the 21-cm line of neutral hydrogen became available in 1951. Dutch astronomers had built an instrument to detect the 21-cm waves that they had predicted, but a fire destroyed it. As a result, two Harvard physicists, Harold Ewen and Edward Purcell, made the first detection (Figure 20.6), soon followed by confirmations from the Dutch and a group in Australia. Since the detection of the 21-cm line, many other radio lines produced by both atoms and molecules have been discovered (as we will discuss in a moment), and these have allowed astronomers to map out the neutral gas throughout our home Galaxy. Astronomers have also detected neutral interstellar gas, including hydrogen, at many other wavelengths from the infrared to the ultraviolet.

Download for free at http://cnx.org/content/col11992/latest/

Figure 20.6. Harold Ewen (1922–2015) and Edward Purcell (1912–1997). We see Harold Ewen in 1952 working with the horn antenna (atop the physics laboratory at Harvard) that made the first detection of interstellar 21-cm radiation. The inset shows Edward Purcell, the winner of the 1952 Nobel Prize in physics, a few years later. (credit: modification of work by NRAO)

Modern radio observations show that most of the neutral hydrogen in our Galaxy is confined to an extremely flat layer, less than 300 light-years thick, that extends throughout the disk of the Milky Way Galaxy. This gas has densities ranging from about 0.1 to about 100 atoms per cm^3, and it exists at a wide range of temperatures, from as low as about 100 K (–173 °C) to as high as about 8000 K. These regions of warm and cold gas are interspersed with each other, and the density and temperature at any particular point in space are constantly changing.

Ultra-Hot Interstellar Gas

While the temperatures of 10,000 K found in H II regions might seem warm, they are not the hottest phase of the interstellar medium. Some of the interstellar gas is at a temperature of a *million* degrees, even though there is no visible source of heat nearby. The discovery of this ultra-hot interstellar gas was a big surprise. Before the launch of astronomical observatories into space, which could see radiation in the ultraviolet and X-ray parts of the spectrum, astronomers assumed that most of the region between stars was filled with hydrogen at temperatures no warmer than those found in H II regions. But telescopes launched above Earth's atmosphere obtained ultraviolet spectra that contained interstellar lines produced by oxygen atoms that have been ionized five times. To strip five electrons from their orbits around an oxygen nucleus requires a lot of energy. Subsequent observations with orbiting X-ray telescopes revealed that the Galaxy is filled with numerous bubbles of X-ray-emitting gas. To emit X-rays, and to contain oxygen atoms that have been ionized five times, gas must be heated to temperatures of a million degrees or more.

Theorists have now shown that the source of energy producing these remarkable temperatures is the explosion of massive stars at the ends of their lives (Figure 20.7). Such explosions, called *supernovae*, will be discussed in detail in the chapter on The Death of Stars. For now, we'll just say that some stars, nearing the ends of their lives, become unstable and literally explode. These explosions launch gas into interstellar space at velocities of tens of thousands of kilometers per second (up to about 30% the speed of light). When this ejected gas collides with interstellar gas, it produces shocks that heat the gas to millions or tens of millions of degrees.

This OpenStax book is available for free at http://cnx.org/content/col11992/1.8

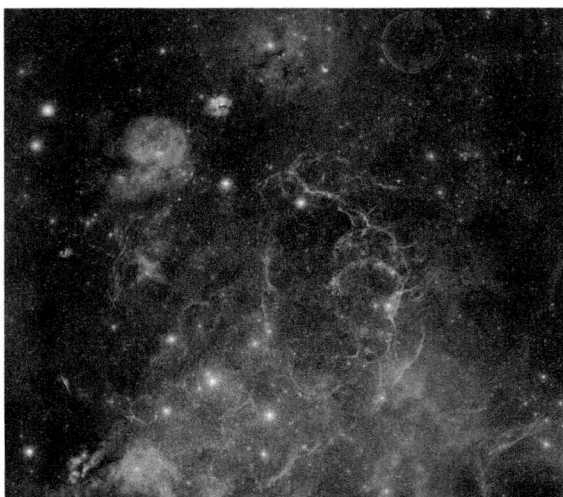

Figure 20.7. Vela Supernova Remnant. About 11,000 years ago, a dying star in the constellation of Vela exploded, becoming as bright as the full moon in Earth's skies. You can see the faint rounded filaments from that explosion in the center of this colorful image. The edges of the remnant are colliding with the interstellar medium, heating the gas they plow through to temperatures of millions of K. Telescopes in space also reveal a glowing sphere of X-ray radiation from the remnant. (credit: Digitized Sky Survey, ESA/ESO/NASA FITS Liberator, Davide De Martin)

Astronomers estimate that one supernova explodes roughly every 100 years somewhere in the Galaxy. On average, shocks launched by supernovae sweep through any given point in the Galaxy about once every few million years. These shocks keep some interstellar space filled with gas at temperatures of millions of degrees, and they continually disturb the colder gas, keeping it in constant, turbulent motion.

Molecular Clouds

A few simple molecules out in space, such as CN and CH, were discovered decades ago because they produce absorption lines in the visible-light spectra of stars behind them. When more sophisticated equipment for obtaining spectra in radio and infrared wavelengths became available, astronomers—to their surprise—found much more complex molecules in interstellar clouds as well.

Just as atoms leave their "fingerprints" in the spectrum of visible light, so the vibration and rotation of atoms within molecules can leave spectral fingerprints in radio and infrared waves. If we spread out the radiation at such longer wavelengths, we can detect emission or absorption lines in the spectra that are characteristic of specific molecules. Over the years, experiments in our laboratories have shown us the exact wavelengths associated with changes in the rotation and vibration of many common molecules, giving us a template of possible lines against which we can now compare our observations of interstellar matter.

The discovery of complex molecules in space came as a surprise because most of interstellar space is filled with ultraviolet light from stars, and this light is capable of *dissociating* molecules (breaking them apart into individual atoms). In retrospect, however, the presence of molecules is not surprising. As we will discuss further in the next section, and have already seen above, interstellar space also contains significant amounts of dust capable of blocking out starlight. When this dust accumulates in a single location, the result is a dark cloud where ultraviolet starlight is blocked and molecules can survive. The largest of these structures are created where gravity pulls interstellar gas together to form giant **molecular clouds**, structures as massive as a million times the mass of the Sun. Within these, most of the interstellar hydrogen has formed the molecule H_2 (molecular hydrogen). Other, more complex molecules are also present in much smaller quantities.

Download for free at http://cnx.org/content/col11992/latest/

Giant molecular clouds have densities of hundreds to thousands of atoms per cm^3, much denser than interstellar space is on average. As a result, though they account for a very small fraction of the volume of interstellar space, they contain a significant fraction—20-30%—of the total mass of the Milky Way's gas. Because of their high density, molecular clouds block ultraviolet starlight, the main agent for heating most interstellar gas. As a result, they tend to be extremely cold, with typical temperatures near 10 K (–263 °C). Giant molecular clouds are also the sites where new stars form, as we will discuss below.

It is in these dark regions of space, protected from starlight, that molecules can form. Chemical reactions occurring both in the gas and on the surface of dust grains lead to much more complex compounds, hundreds of which have been identified in interstellar space. Among the simplest of these are water (H_2O), carbon monoxide (CO), which is produced by fires on Earth, and ammonia (NH_3), whose smell you recognize in strong home cleaning products. Carbon monoxide is particularly abundant in interstellar space and is the primary tool that astronomers use to study giant molecular clouds. Unfortunately, the most abundant molecule, H_2, is particularly difficult to observe directly because in most giant molecular clouds, it is too cold to emit even at radio wavelengths. CO, which tends to be present wherever H_2 is found, is a much better emitter and is often used by astronomers to trace molecular hydrogen.

The more complex molecules astronomers have found are mostly combinations of hydrogen, oxygen, carbon, nitrogen, and sulfur atoms. Many of these molecules are *organic* (those that contain carbon and are associated with the carbon chemistry of life on Earth.) They include formaldehyde (used to preserve living tissues), alcohol (see the feature box on Cocktails in Space), and antifreeze.

In 1996, astronomers discovered acetic acid (the prime ingredient of vinegar) in a cloud lying in the direction of the constellation of Sagittarius. To balance the sour with the sweet, a simple sugar (glycolaldehyde) has also been found. The largest compounds yet discovered in interstellar space are *fullerenes*, molecules in which 60 or 70 carbon atoms are arranged in a cage-like configuration (see Figure 20.8). See Table 20.1 for a list of a few of the more interesting interstellar molecules that have been found so far.

Figure 20.8. Fullerene C60. This three-dimensional perspective shows the characteristic cage-like arrangement of the 60 carbon atoms in a molecule of fullerene C60. Fullerene C60 is also known as a "buckyball," or as its full name, buckminsterfullerene, because of its similarity to the multisided architectural domes designed by American inventor R. Buckminster Fuller.

This OpenStax book is available for free at http://cnx.org/content/col11992/1.8

Download for free at http://cnx.org/content/col11992/latest/

Some Interesting Interstellar Molecules

Name	Chemical Formula	Use on Earth
Ammonia	NH_3	Household cleansers
Formaldehyde	H_2CO	Embalming fluid
Acetylene	HC_2H	Fuel for a welding torch
Acetic acid	$C_2H_2O_4$	The essence of vinegar
Ethyl alcohol	CH_3CH_2OH	End-of-semester parties
Ethylene glycol	$HOCH_2CH_2OH$	Antifreeze ingredient
Benzene	C_6H_6	Carbon ring, ingredient in varnishes and dyes

Table 20.1

The cold interstellar clouds also contain cyanoacetylene (HC_3N) and acetaldehyde (CH_3CHO), generally regarded as starting points for *amino acid* formation. These are building blocks of proteins, which are among the fundamental chemicals from which living organisms on Earth are constructed. The presence of these organic molecules does not imply that life exists in space, but it does show that the chemical building blocks of life can form under a wide range of conditions in the universe. As we learn more about how complex molecules are produced in interstellar clouds, we gain an increased understanding of the kinds of processes that preceded the beginnings of life on Earth billions of years ago.

LINK TO LEARNING

Interested in learning more about fullerenes, buckyballs, or buckminsterfullerenes (as they're called)? Watch a brief video from NASA's Jet Propulsion Laboratory (https://openstax.org/l/30NASAjetprop) that explains what they are and illustrates how they were discovered in space.

MAKING CONNECTIONS

Cocktails in Space

Among the molecules astronomers have identified in interstellar clouds is alcohol, which comes in two varieties: methyl (or wood) alcohol and ethyl alcohol (the kind you find in cocktails). Ethyl alcohol is a pretty complex molecule, written by chemists as C_2H_5OH. It is quite plentiful in space (relatively speaking). In clouds where it has been identified, we detect up to one molecule for every m^3. The largest

Download for free at http://cnx.org/content/col11992/latest/

of the clouds (which can be several hundred light-years across) have enough ethyl alcohol to make 10^{28} fifths of liquor.

We need not fear, however, that future interstellar astronauts will become interstellar alcoholics. Even if a spaceship were equipped with a giant funnel 1 kilometer across and could scoop it through such a cloud at the speed of light, it would take about a thousand years to gather up enough alcohol for one standard martini.

Furthermore, the very same clouds also contain water (H_2O) molecules. Your scoop would gather them up as well, and there are a lot more of them because they are simpler and thus easier to form. For the fun of it, one astronomical paper actually calculated the proof of a typical cloud. *Proof* is the ratio of alcohol to water in a drink, where 0 proof means all water, 100 proof means half alcohol and half water, and 200 proof means all alcohol. The proof of the interstellar cloud was only 0.2, not enough to qualify as a stiff drink

20.3 COSMIC DUST

Learning Objectives

By the end of this section, you will be able to:

> Describe how we can detect interstellar dust
> Understand the role and importance of infrared observations in studying dust
> Explain the terms extinction and interstellar reddening

Figure 20.9 shows a striking example of what is actually a common sight through large telescopes: a dark region on the sky that appears to be nearly empty of stars. For a long time, astronomers debated whether these dark regions were empty "tunnels" through which we looked beyond the stars of the Milky Way Galaxy into intergalactic space, or clouds of some dark material that blocked the light of the stars beyond. The astronomer William Herschel (discoverer of the planet Uranus) thought it was the former, once remarking after seeing one, "Here truly is a hole in heaven!" However, American astronomer E. E. Barnard is generally credited with showing from his extensive series of nebula photographs that the latter interpretation is the correct one (see the feature box on Edward Emerson Barnard).

This OpenStax book is available for free at http://cnx.org/content/col11992/1.8

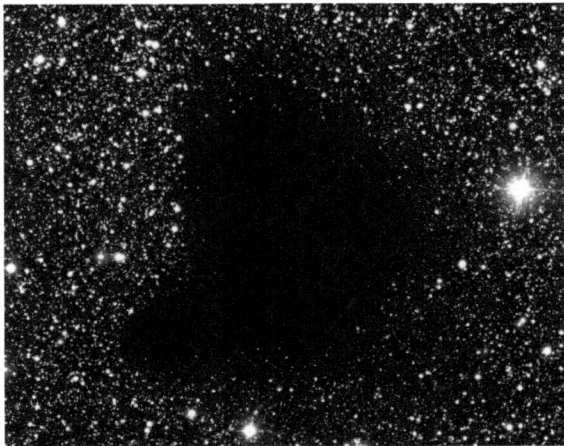

Figure 20.9. Barnard 68. This object, first catalogued by E. E. Barnard, is a dark interstellar cloud. Its striking appearance is due to the fact that, since it is relatively close to Earth, there are no bright stars between us and it, and its dust obscures the light from the stars behind it. (It looks a little bit like a sideways heart; one astronomers sent a photo of this object to his sweetheart as a valentine.) (credit: modification of work by ESO)

Dusty clouds in space betray their presence in several ways: by blocking the light from distant stars, by emitting energy in the infrared part of the spectrum, by reflecting the light from nearby stars, and by making distant stars look redder than they really are.

VOYAGERS IN ASTRONOMY

Edward Emerson Barnard

Born in 1857 in Nashville, Tennessee, two months after his father died, Edward Barnard (Figure 20.10) grew up in such poor circumstances that he had to drop out of school at age nine to help support his ailing mother. He soon became an assistant to a local photographer, where he learned to love both photography and astronomy, destined to become the dual passions of his life. He worked as a photographer's aide for 17 years, studying astronomy on his own. In 1883, he obtained a job as an assistant at the Vanderbilt University Observatory, which enabled him at last to take some astronomy courses.

Married in 1881, Barnard built a house for his family that he could ill afford. But as it happened, a patent medicine manufacturer offered a $200 prize (a lot of money in those days) for the discovery of any new comet. With the determination that became characteristic of him, Barnard spent every clear night searching for comets. He discovered seven of them between 1881 and 1887, earning enough money to make the payments on his home; this "Comet House" later became a local attraction. (By the end of his life, Barnard had found 17 comets through diligent observation.)

In 1887, Barnard got a position at the newly founded Lick Observatory, where he soon locked horns with the director, Edward Holden, a blustering administrator who made Barnard's life miserable. (To be fair, Barnard soon tried to do the same for him.) Despite being denied the telescope time that he needed for his photographic work, in 1892, Barnard managed to discover the first new moon found around Jupiter since Galileo's day, a stunning observational feat that earned him world renown. Now in a position to

Download for free at http://cnx.org/content/col11992/latest/

demand more telescope time, he perfected his photographic techniques and soon began to publish the best images of the Milky Way taken up to that time. It was during the course of this work that he began to examine the dark regions among the crowded star lanes of the Galaxy and to realize that they must be vast clouds of obscuring material (rather than "holes" in the distribution of stars).

Astronomer-historian Donald Osterbrock has called Barnard an "observaholic:" his daily mood seemed to depend entirely on how clear the sky promised to be for his night of observing. He was a driven, neurotic man, concerned about his lack of formal training, fearful of being scorned, and afraid that he might somehow slip back into the poverty of his younger days. He had difficulty taking vacations and lived for his work: only serious illness could deter him from making astronomical observations.

In 1895, Barnard, having had enough of the political battles at Lick, accepted a job at the Yerkes Observatory near Chicago, where he remained until his death in 1923. He continued his photographic work, publishing compilations of his images that became classic photographic atlases, and investigating the varieties of nebulae revealed in his photographs. He also made measurements of the sizes and features of planets, participated in observations of solar eclipses, and carefully cataloged dark nebulae (see Figure 20.9). In 1916, he discovered the star with the largest proper motion, the second-closest star system to our own (see Analyzing Starlight). It is now called Barnard's Star in his honor.

Figure 20.10. Edward Emerson Barnard (1857–1923). Barnard's observations provided information that furthered many astronomical explorations. (credit: The Lick Observatory)

Detecting Dust

The dark cloud seen in Figure 20.9 blocks the light of the many stars that lie behind it; note how the regions in other parts of the photograph are crowded with stars. Barnard 68 is an example of a relatively dense cloud or *dark nebula* containing tiny, solid dust grains. Such opaque clouds are conspicuous on any photograph of the Milky Way, the galaxy in which the Sun is located (see the figures in The Milky Way Galaxy). The "dark rift," which runs lengthwise down a long part of the Milky Way in our sky and appears to split it in two, is produced by a collection of such obscuring clouds.

While dust clouds are too cold to radiate a measurable amount of energy in the visible part of the spectrum, they glow brightly in the infrared (Figure 20.11). The reason is that small dust grains absorb visible light and

This OpenStax book is available for free at http://cnx.org/content/col11992/1.8

ultraviolet radiation very efficiently. The grains are heated by the absorbed radiation, typically to temperatures from 10 to about 500 K, and re-radiate this heat at infrared wavelengths.

(a)   (b)

Figure 20.11. Visible and Infrared Images of the Horsehead Nebula in Orion. This dark cloud is one of the best-known images in astronomy, probably because it really does resemble a horse's head. The horse-head shape is an extension of a large cloud of dust that fills the lower part of the picture. (a) Seen in visible light, the dust clouds are especially easy to see against the bright background. (b) This infrared radiation image from the region of the horse head was recorded by NASA's Wide-Field Infrared Survey Explorer. Note how the regions that appear dark in visible light appear bright in the infrared. The dust is heated by nearby stars and re-radiates this heat in the infrared. Only the top of the horse's head is visible in the infrared image. Bright dots seen in the nebula below and to the left and at the top of the horse head are young, newly formed stars. The insets show the horse head and the bright nebula in more detail. (credit a: modification of work by ESO and Digitized Sky Survey; credit b: modification of work by NASA/JPL-Caltech)

Thanks to their small sizes and low temperatures, interstellar grains radiate most of their energy at infrared to microwave frequencies, with wavelengths of tens to hundreds of microns. Earth's atmosphere is opaque to radiation at these wavelengths, so emission by interstellar dust is best measured from space. Observations from above Earth's atmosphere show that dust clouds are present throughout the plane of the Milky Way (Figure 20.12).

Figure 20.12. Infrared Emission from the Plane of the Milky Way. This infrared image taken by the Spitzer Space Telescope shows a field in the plane of the Milky Way Galaxy. (Our Galaxy is in the shape of a frisbee; the plane of the Milky Way is the flat disk of that frisbee. Since the Sun, Earth, and solar system are located in the plane of the Milky Way and at a large distance from its center, we view the Galaxy edge on, much as we might look at a glass plate from its edge.) This emission is produced by tiny dust grains, which emit at 3.6 microns (blue in this image), 8.0 microns (green), and 24 microns (red). The densest regions of dust are so cold and opaque that they appear as dark clouds even at these infrared wavelengths. The red bubbles visible throughout indicate regions where the dust has been warmed up by young stars. This heating increases the emission at 24 microns, leading to the redder color in this image. (credit: modification of work by NASA/JPL-Caltech/University of Wisconsin)

Download for free at http://cnx.org/content/col11992/latest/

Some dense clouds of dust are close to luminous stars and scatter enough starlight to become visible. Such a cloud of dust, illuminated by starlight, is called a *reflection nebula*, since the light we see is starlight reflected off the grains of dust. One of the best-known examples is the nebulosity around each of the brightest stars in the Pleiades cluster (see Figure 20.1). The dust grains are small, and such small particles turn out to scatter light with blue wavelengths more efficiently than light at red wavelengths. A reflection nebula, therefore, usually appears bluer than its illuminating star (Figure 20.13).

Figure 20.13. Pleiades Star Cluster. The bluish light surrounding the stars in this image is an example of a reflection nebula. Like fog around a street lamp, a reflection nebula shines only because the dust within it scatters light from a nearby bright source. The Pleiades cluster is currently passing through an interstellar cloud that contains dust grains, which scatter the light from the hot blue stars in the cluster. The Pleiades cluster is about 400 light-years from the Sun. (credit: NASA, ESA and AURA/Caltech)

Gas and dust are generally intermixed in space, although the proportions are not exactly the same everywhere. The presence of dust is apparent in many photographs of *emission nebulae* in the constellation of Sagittarius, where we see an H II region surrounded by a blue reflection nebula. Which type of nebula appears brighter depends on the kinds of stars that cause the gas and dust to glow. Stars cooler than about 25,000 K have so little ultraviolet radiation of wavelengths shorter than 91.2 nanometers—which is the wavelength required to ionize hydrogen—that the reflection nebulae around such stars outshine the emission nebulae. Stars hotter than 25,000 K emit enough ultraviolet energy that the emission nebulae produced around them generally outshine the reflection nebulae.

Interstellar Reddening

The tiny interstellar dust grains absorb some of the starlight they intercept. But at least half of the starlight that interacts with a grain is merely scattered, that is, it is redirected rather than absorbed. Since neither the absorbed nor the scattered starlight reaches us directly, both absorption and scattering make stars look dimmer. The effects of both processes are called **interstellar extinction** (Figure 20.14).

Astronomers first came to understand interstellar extinction around the early 1930s, as the explanation of a puzzling observation. In the early part of the twentieth century, astronomers discovered that some stars look red even though their spectral lines indicate that they must be extremely hot (and thus should look blue). The solution to this seeming contradiction turned out to be that the light from these hot stars is not only dimmed but also reddened by interstellar dust, a phenomenon known as interstellar **reddening**.

This OpenStax book is available for free at http://cnx.org/content/col11992/1.8

Figure 20.14. Barnard 68 in Infrared. In this image, we see Barnard 68, the same object shown in Figure 20.9. The difference is that, in the previous image, the blue, green, and red channels showed light in the visible (or very nearly visible) part of the spectrum. In this image, the red color shows radiation emitted in the infrared at a wavelength of 2.2 microns. Interstellar extinction is much smaller at infrared than at visible wavelengths, so the stars behind the cloud become visible in the infrared channel. (credit: ESO)

Dust does not interact with all the colors of light the same way. Much of the violet, blue, and green light from these stars has been scattered or absorbed by dust, so it does not reach Earth. Some of their orange and red light, with longer wavelengths, on the other hand, more easily penetrates the intervening dust and completes its long journey through space to enter Earth-based telescopes (Figure 20.15). Thus, the star looks redder from Earth than it would if you could see it from nearby. (Strictly speaking, *reddening* is not the most accurate term for this process, since no red color is added; instead, blues and related colors are subtracted, so it should more properly be called "deblueing.") In the most extreme cases, stars can be so reddened that they are entirely undetectable at visible wavelengths and can be seen only at infrared or longer wavelengths (Figure 20.14).

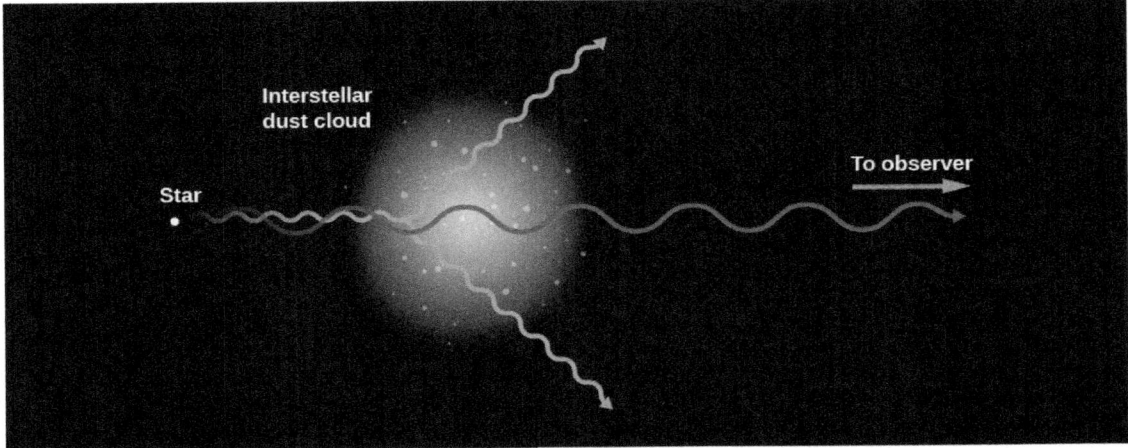

Figure 20.15. Scattering of Light by Dust. Interstellar dust scatters blue light more efficiently than red light, thereby making distant stars appear redder and giving clouds of dust near stars a bluish hue. Here, a red ray of light from a star comes straight through to the observer, whereas a blue ray is shown scattering. A similar scattering process makes Earth's sky look blue.

Download for free at http://cnx.org/content/col11992/latest/

We have all seen an example of reddening on Earth. The Sun appears much redder at sunset than it does at noon. The lower the Sun is in the sky, the longer the path its light must travel through the atmosphere. Over this greater distance, there is a greater chance that sunlight will be scattered. Since red light is less likely to be scattered than blue light, the Sun appears more and more red as it approaches the horizon.

By the way, scattering of sunlight is also what causes our sky to look blue, even though the gases that make up Earth's atmosphere are transparent. As sunlight comes in, it scatters from the molecules of air. The small size of the molecules means that the blue colors scatter much more efficiently than the greens, yellows, and reds. Thus, the blue in sunlight is scattered out of the beam and all over the sky. The light from the Sun that comes to your eye, on the other hand, is missing some of its blue, so the Sun looks a bit yellower, even when it is high in the sky, than it would from space.

The fact that starlight is reddened by interstellar dust means that long-wavelength radiation is transmitted through the Galaxy more efficiently than short-wavelength radiation. Consequently, if we wish to see farther in a direction with considerable interstellar material, we should look at long wavelengths. This simple fact provides one of the motivations for the development of infrared astronomy. In the infrared region at 2 microns (2000 nanometers), for example, the obscuration is only one-sixth as great as in the visible region (500 nanometers), and we can therefore study stars that are more than twice as distant before their light is blocked by interstellar dust. This ability to see farther by observing in the infrared portion of the spectrum represents a major gain for astronomers trying to understand the structure of our Galaxy or probing its puzzling, but distant, center (see The Milky Way Galaxy).

Interstellar Grains

Before we get to the details about interstellar dust, we should perhaps get one concern out of the way. Why couldn't it be the interstellar *gas* that reddens distant stars and not the dust? We already know from everyday experience that atomic or molecular gas is almost transparent. Consider Earth's atmosphere. Despite its very high density compared with that of interstellar gas, it is so transparent as to be practically invisible. (Gas does have a few specific spectral lines, but they absorb only a tiny fraction of the light as it passes through.)The quantity of *gas* required to produce the observed absorption of light in interstellar space would have to be enormous. The gravitational attraction of so great a mass of gas would affect the motions of stars in ways that could easily be detected. Such motions are not observed, and thus, the interstellar absorption cannot be the result of gases.

Although gas does not absorb much light, we know from everyday experience that tiny solid or liquid particles can be very efficient absorbers. Water vapor in the air is quite invisible. When some of that vapor condenses into tiny water droplets, however, the resulting cloud is opaque. Dust storms, smoke, and smog offer familiar examples of the efficiency with which solid particles absorb light. On the basis of arguments like these, astronomers have concluded that widely scattered *solid* particles in interstellar space must be responsible for the observed dimming of starlight. What are these particles made of? And how did they form?

Observations like the pictures in this chapter show that a great deal of this dust exists; hence, it must be primarily composed of elements that are abundant in the universe (and in interstellar matter). After hydrogen and helium, the most abundant elements are oxygen, carbon, and nitrogen. These three elements, along with magnesium, silicon, iron—and perhaps hydrogen itself—turn out to be the most important components of interstellar dust.

Many of the dust particles can be characterized as sootlike (rich in carbon) or sandlike (containing silicon and oxygen). Grains of interstellar dust are found in meteorites and can be identified because the abundances of certain isotopes are different from what we see in other solar system material. Several different interstellar

This OpenStax book is available for free at http://cnx.org/content/col11992/1.8

dust substances have been identified in this way in the laboratory, including graphite and diamonds. (Don't get excited; these diamonds are only a billionth of a meter in size and would hardly make an impressive engagement ring!)

The most widely accepted model pictures the grains with rocky cores that are either like soot (rich in carbon) or like sand (rich in silicates). In the dark clouds where molecules can form, these cores are covered by icy mantles (Figure 20.16). The most common ices in the grains are water (H_2O), methane (CH_4), and ammonia (NH_3)—all built out of atoms that are especially abundant in the realm of the stars. The ice mantles, in turn, are sites for some of the chemical reactions that produce complex organic molecules.

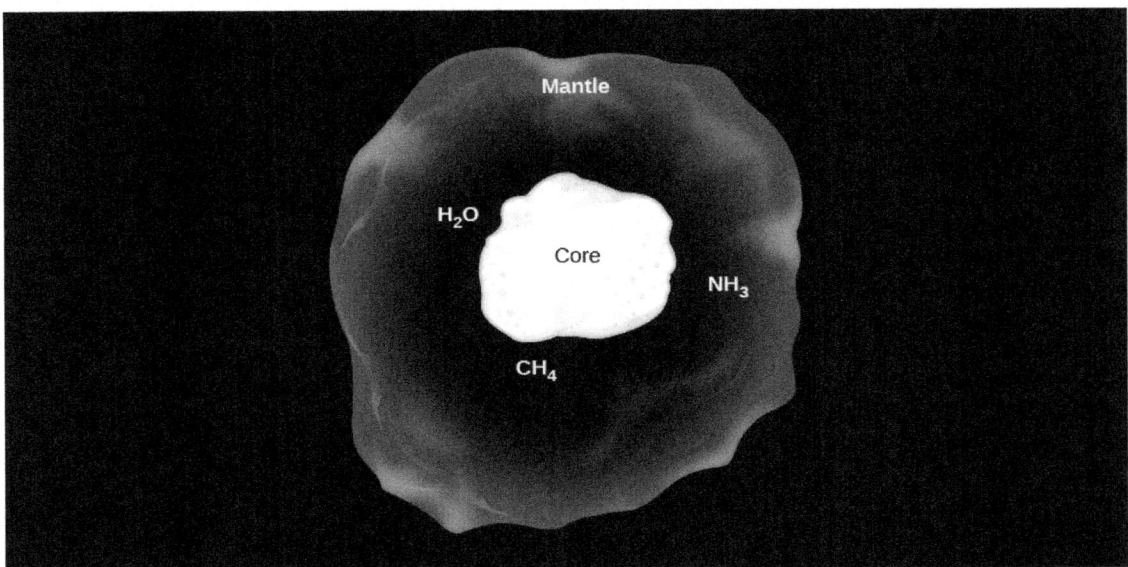

Figure 20.16. Model of an Interstellar Dust Grain. A typical interstellar grain is thought to consist of a core of rocky material (silicates) or graphite, surrounded by a mantle of ices. Typical grain sizes are 10^{-8} to 10^{-7} meters. (This is from 1/100 to 1/10 of a micron; by contrast, human hair is about 10–200 microns wide.)

Typical individual grains must be just slightly smaller than the wavelength of visible light. If the grains were a lot smaller, they would not block the light efficiently, as Figure 20.13 and other images in this chapter show that it does.

On the other hand, if the dust grains were much larger than the wavelength of light, then starlight would not be reddened. Things that are much larger than the wavelength of light would block both blue and red light with equal efficiency. In this way we can deduce that a characteristic interstellar dust grain contains 10^6 to 10^9 atoms and has a diameter of 10^{-8} to 10^{-7} meters (10 to 100 nanometers). This is actually more like the specks of solid matter in cigarette smoke than the larger grains of dust you might find under your desk when you are too busy studying astronomy to clean properly.

20.4 COSMIC RAYS

Learning Objectives

By the end of this section, you will be able to:

> Define cosmic rays and describe their composition

Download for free at http://cnx.org/content/col11992/latest/

> Explain why it is hard to study the origin of cosmic rays, and the current leading hypotheses about where they might come from

In addition to gas and dust, a third class of particles, noteworthy for the high speeds with which they travel, is found in interstellar space. **Cosmic rays** were discovered in 1911 by an Austrian physicist, Victor Hess, who flew simple instruments aboard balloons and showed that high-speed particles arrive at Earth from space (Figure 20.17). The term "cosmic ray" is misleading, implying it might be like a ray of light, but we are stuck with the name. They are definitely particles and have nearly the same composition as ordinary interstellar gas. Their behavior, however, is radically different from the gas we have discussed so far.

Figure 20.17. Victor Hess (1883–1964). Cosmic-ray pioneer Victor Hess returns from a 1912 balloon flight that reached an altitude of 5.3 kilometers. It was on such balloon flights that Hess discovered cosmic rays.

The Nature of Cosmic Rays

Cosmic rays are mostly high-speed atomic nuclei and electrons. Speeds equal to 90% of the speed of light are typical. Almost 90% of the cosmic rays are hydrogen nuclei (protons) stripped of their accompanying electron. Helium and heavier nuclei constitute about 9% more. About 1% of cosmic rays have masses equal to the mass of the electron, and 10–20% of these carry positive charge rather than the negative charge that characterizes electrons. A positively charged particle with the mass of an electron is called a *positron* and is a form of *antimatter* (we discussed antimatter in The Sun: A Nuclear Powerhouse).

The abundances of various atomic nuclei in cosmic rays mirror the abundances in stars and interstellar gas, with one important exception. The light elements lithium, beryllium, and boron are far more abundant in cosmic rays than in the Sun and stars. These light elements are formed when high-speed, cosmic-ray nuclei of carbon, nitrogen, and oxygen collide with protons in interstellar space and break apart. (By the way, if you, like most readers, have not memorized all the elements and want to see how any of those we mention fit into the sequence of elements, you will find them all listed in Appendix K in order of the number of protons they contain.)

Cosmic rays reach Earth in substantial numbers, and we can determine their properties either by capturing them directly or by observing the reactions that occur when they collide with atoms in our atmosphere. The total energy deposited by cosmic rays in Earth's atmosphere is only about one-billionth the energy received from the Sun, but it is comparable to the energy received in the form of starlight. Some of the cosmic rays come to Earth from the surface of the Sun, but most come from outside the solar system.

This OpenStax book is available for free at http://cnx.org/content/col11992/1.8

Where Do They Come From?

There is a serious problem in identifying the source of cosmic rays. Since light travels in straight lines, we can tell where it comes from simply by looking. Cosmic rays are charged particles, and their direction of motion can be changed by magnetic fields. The paths of cosmic rays are curved both by magnetic fields in interstellar space and by Earth's own field. Calculations show that low-energy cosmic rays may spiral many times around Earth before entering the atmosphere where we can detect them. If an airplane circles an airport many times before landing, it is difficult for an observer to determine the direction from which it originated. So, too, after a cosmic ray circles Earth several times, it is impossible to know where its journey began.

There are a few clues, however, about where cosmic rays might be generated. We know, for example, that magnetic fields in interstellar space are strong enough to keep all but the most energetic cosmic rays from escaping the Galaxy. It therefore seems likely that they are produced somewhere inside the Galaxy. The only likely exceptions are those with the very highest energy. Such cosmic rays move so rapidly that they are not significantly influenced by interstellar magnetic fields, and thus, they could escape our Galaxy. By analogy, they could escape other galaxies as well, so some of the highest-energy cosmic rays that we detect may have been created in some distant galaxy. Still, most cosmic rays must have their source inside the Milky Way Galaxy.

We can also estimate how far typical cosmic rays travel before striking Earth. The light elements lithium, beryllium, and boron hold the key. Since these elements are formed when carbon, nitrogen, and oxygen strike interstellar protons, we can calculate how long, on average, cosmic rays must travel through space in order to experience enough collisions to account for the amount of lithium and the other light elements that they contain. It turns out that the required distance is about 30 times around the Galaxy. At speeds near the speed of light, it takes perhaps 3–10 million years for the average cosmic ray to travel this distance. This is only a small fraction of the age of the Galaxy or the universe, so cosmic rays must have been created fairly recently on a cosmic timescale.

The best candidates for a source of cosmic rays are the supernova explosions, which mark the violent deaths of some stars (and which we will discuss in The Death of Stars). The material ejected by the explosion produces a shock wave, which travels through the interstellar medium. Charged particles can become trapped, bouncing back and forth across the front of the shock wave many times. With each pass through the shock, the magnetic fields inside it accelerate the particles more and more. Eventually, they are traveling at close to the speed of light and can escape from the shock to become cosmic rays. Some collapsed stars (including star remnants left over from supernova explosions) may, under the right circumstances, also serve as accelerators of particles. In any case, we again find that the raw material of the Galaxy is enriched by the life cycle of stars. In the next section, we will look at this enrichment process in more detail.

LINK TO LEARNING

You can watch a brief video about the Calorimetric Electron Telescope (CALET) (https://openstax.org/l/30CALETvid) mission, a cosmic ray detector at the International Space Station. The link takes you to NASA Johnson's "Space Station Live: Cosmic Ray Detector for ISS."

Download for free at http://cnx.org/content/col11992/latest/

20.5 THE LIFE CYCLE OF COSMIC MATERIAL

Learning Objectives

By the end of this section, you will be able to:

> Explain how interstellar matter flows into and out of our Galaxy and transforms from one phase to another, and understand how star formation and evolution affects the properties of the interstellar medium
> Explain how the heavy elements and dust grains found in interstellar space got there and describe how dust grains help produce molecules that eventually find their way into planetary systems

Flows of Interstellar Gas

The most important thing to understand about the interstellar medium is that it is not static. Interstellar gas orbits through the Galaxy, and as it does so, it can become more or less dense, hotter and colder, and change its state of ionization. A particular parcel of gas may be neutral hydrogen at some point, then find itself near a young, hot star and become part of an H II region. The star may then explode as a supernova, heating the nearby gas up to temperatures of millions of degrees. Over millions of years, the gas may cool back down and become neutral again, before it collects into a dense region that gravity gathers into a giant molecular cloud (Figure 20.18)

1 kpc (3200 LY)

Figure 20.18. Large-Scale Distribution of Interstellar Matter. This image is from a computer simulation of the Milky Way Galaxy's interstellar medium as a whole. The majority of gas, visible in greenish colors, is neutral hydrogen. In the densest regions in the spiral arms, shown in yellow, the gas is collected into giant molecular clouds. Low-density holes in the spiral arms, shown in blue, are the result of supernova explosions. (credit: modification of work by Mark Krumholz)

At any given time in the Milky Way, the majority of the interstellar gas by mass and volume is in the form of atomic hydrogen. The much-denser molecular clouds occupy a tiny fraction of the volume of interstellar space but add roughly 30% to the total mass of gas between the stars. Conversely, the hot gas produced by supernova explosions contributes a negligible mass but occupies a significant fraction of the volume of interstellar space.

This OpenStax book is available for free at http://cnx.org/content/col11992/1.8

Download for free at http://cnx.org/content/col11992/latest/

H II regions, though they are visually spectacular, constitute only a very small fraction of either the mass or volume of interstellar material.

However, the interstellar medium is not a closed system. Gas from intergalactic space constantly falls onto the Milky Way due to its gravity, adding new gas to the interstellar medium. Conversely, in giant molecular clouds where gas collects together due to gravity, the gas can collapse to form new stars, as discussed in The Birth of Stars and the Discovery of Planets outside the Solar System. This process locks interstellar matter into stars. As the stars age, evolve, and eventually die, massive stars lose a large fraction of their mass, and low-mass stars lose very little. On average, roughly one-third of the matter incorporated into stars goes back into interstellar space. Supernova explosions have so much energy that they can drive interstellar mass out of the Galaxy and back into intergalactic space. Thus, the total amount of mass of the interstellar medium is set by a competition between the gain of mass from intergalactic space, the conversion of interstellar mass into stars, and the loss of interstellar mass back into intergalactic space due to supernovae. This entire process is known as the **baryon cycle**—baryon is from the Latin word for "heavy," and the cycle has this name because it is the repeating process that the heavier components of the universe—the atoms—undergo.

The Cycle of Dust and Heavy Elements

While much of the mass of the interstellar medium is material accreted during the last few billion years from intergalactic space, this is not true of the elements heavier than hydrogen and helium, or of the dust. Instead, these components of the interstellar medium were made inside stars in the Milky Way, which returned them to the interstellar medium at the end of their lives. We will talk more about this process in later chapters, but for now just bear in mind what we learned in The Sun: A Nuclear Powerhouse. What stars "do for a living" is fuse heavier elements from lighter ones, producing energy in the process. As stars mature, they begin to lose some of the newly made elements to the reservoir of interstellar matter.

The same is true of dust grains. Dust forms when grains can condense in regions where gas is dense and cool. One place where the right conditions are found is the winds from luminous cool stars (the red giants and supergiants we discussed in The Stars: A Celestial Census). Grains can also condense in the matter thrown off by a supernova explosion as the ejected gases begin to cool.

The dust grains produced by stars may grow even further when they spend time in the dense parts of the interstellar medium, inside molecular clouds. In these environments, grains can stick together or gather additional atoms from the gas around them, growing larger. They also facilitate the production of other compounds, including some of the more complex molecules we discussed earlier.

The surfaces of the dust grains (see Cosmic Dust)—which would seem very large if you were an atom—provide "nooks and crannies" where these atoms can stick long enough to find partners and form molecules. (Think of the dust grains as "interstellar social clubs" where lonely atoms can meet and form meaningful relationships.) Eventually, the dust grains become coated with ices. The presence of the dust shields the molecules inside the clouds from ultraviolet radiation and cosmic rays that would break them up.

When stars finally begin to form within the cloud, they heat the grains and evaporate the ices. The gravitational attraction of the newly forming stars also increases the density of the surrounding cloud material. Many more chemical reactions take place on the surfaces of grains in the gas surrounding the newly forming stars, and these areas are where organic molecules are formed. These molecules can be incorporated into newly formed planetary systems, and the early Earth may have been seeded in just such a way.

Indeed, scientists speculate that some of the water on Earth may have come from interstellar grains. Recent observations from space have shown that water is abundant in dense interstellar clouds. Since stars are formed from this material, water must be present when solar systems, including our own, come into existence. The

Download for free at http://cnx.org/content/col11992/latest/

water in our oceans and lakes may have come initially from water locked into the rocky material that accreted to form Earth. Alternatively, the water may have been brought to Earth when asteroids and comets (formed from the same cloud that made the planets) later impacted it. Scientists estimate that one comet impact every thousand years during Earth's first billion years would have been enough to account for the water we see today. Of course, both sources may have contributed to the water we now enjoy drinking and swimming in.

Any interstellar grains that are incorporated into newly forming stars (instead of the colder planets and smaller bodies around them) will be destroyed by their high temperatures. But eventually, each new generation of stars will evolve to become red giants, with stellar winds of their own. Some of these stars will also become supernovae and explode. Thus, the process of recycling cosmic material can start all over again.

20.6 INTERSTELLAR MATTER AROUND THE SUN

Learning Objectives

By the end of this section, you will be able to:

> Describe how interstellar matter is arranged around our solar system
> Explain why scientists think that the Sun is located in a hot bubble

We want to conclude our discussion of interstellar matter by asking how this material is organized in our immediate neighborhood. As we discussed above, orbiting X-ray observatories have shown that the Galaxy is full of bubbles of hot, X-ray-emitting gas. They also revealed a diffuse background of X-rays that appears to fill the entire sky from our perspective (Figure 20.19). While some of this emission comes from the interaction of the solar wind with the interstellar medium, a majority of it comes from beyond the solar system. The natural explanation for why there is X-ray-emitting gas all around us is that the Sun is itself inside one of the bubbles. We therefore call our "neighborhood" the Local Hot Bubble, or **Local Bubble** for short. The Local Bubble is much less dense—an average of approximately 0.01 atoms per cm^3—than the average interstellar density of about 1 atom per cm^3. This local gas has a temperature of about a million degrees, just like the gas in the other superbubbles that spread throughout our Galaxy, but because there is so little hot material, this high temperature does not affect the stars or planets in the area in any way.

What caused the Local Bubble to form? Scientists are not entirely sure, but the leading candidate is winds from stars and supernova explosions. In a nearby region in the direction of the constellations Scorpius and Centaurus, a lot of star formation took place about 15 million years ago. The most massive of these stars evolved very quickly until they produced strong winds, and some ended their lives by exploding. These processes filled the region around the Sun with hot gas, driving away cooler, denser gas. The rim of this expanding superbubble reached the Sun about 7.6 million years ago and now lies more than 200 light-years past the Sun in the general direction of the constellations of Orion, Perseus, and Auriga.

This OpenStax book is available for free at http://cnx.org/content/col11992/1.8

Download for free at http://cnx.org/content/col11992/latest/

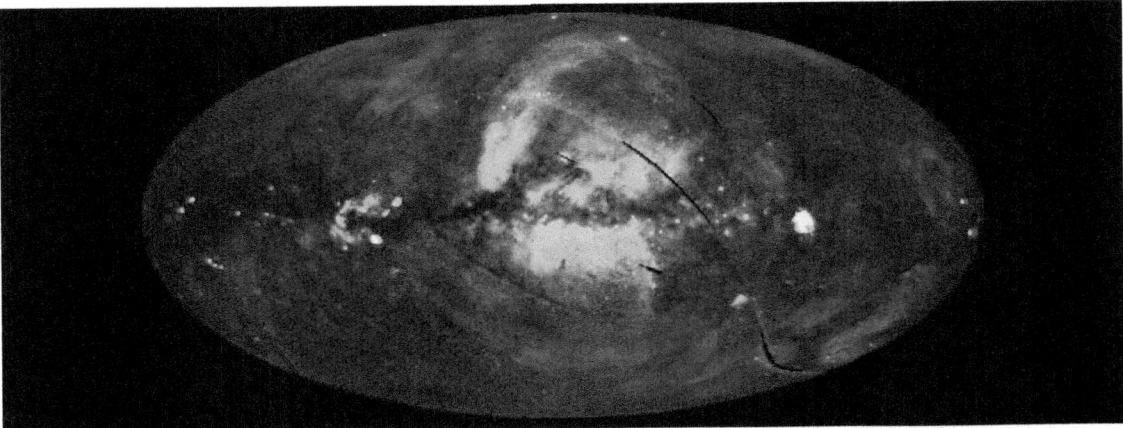

Figure 20.19. Sky in X-Rays. This image, made by the ROSAT satellite, shows the whole sky in X-rays as seen from Earth. Different colors indicate different X-ray energies: red is 0.25 kiloelectron volts, green is 0.75 kiloelectron volts, and blue is 1.5 kiloelectron volts. The image is oriented so the plane of the Galaxy runs across the middle of the image. The ubiquitous red color, which does not disappear completely even in the galactic plane, is evidence for a source of X-rays all around the Sun. (credit: modification of work by NASA)

A few clouds of interstellar matter do exist within the Local Bubble. The Sun itself seems to have entered a cloud about 10,000 years ago. This cloud is warm (with a temperature of about 7000 K) and has a density of 0.3 hydrogen atom per cm^3—higher than most of the Local Bubble but still so tenuous that it is also referred to as **Local Fluff** (Figure 20.20). (Aren't these astronomical names fun sometimes?)

While this is a pretty thin cloud, we estimate that it contributes 50 to 100 times more particles than the solar wind to the diffuse material between the planets in our solar system. These interstellar particles have been detected and their numbers counted by the spacecraft traveling between the planets. Perhaps someday, scientists will devise a way to collect them without destroying them and to return them to Earth, so that we can touch—or at least study in our laboratories—these messengers from distant stars.

Download for free at http://cnx.org/content/col11992/latest/

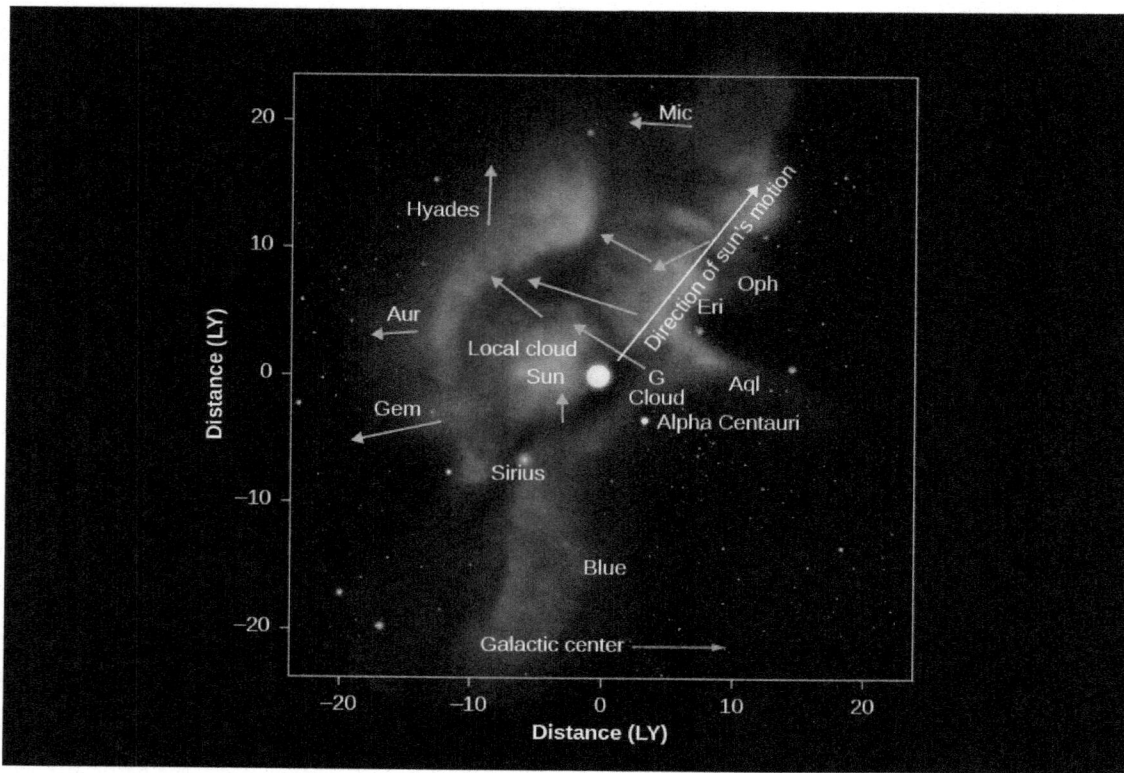

Figure 20.20. Local Fluff. The Sun and planets are currently moving through the Local Interstellar Cloud, which is also called the Local Fluff. Fluff is an appropriate description because the density of this cloud is only about 0.3 atom per cm3. In comparison, Earth's atmosphere at the edge of space has around 1.2 × 1013 molecules per cm3. This image shows the patches of interstellar matter (mostly hydrogen gas) within about 20 light-years of the Sun. The temperature of the Local Interstellar Cloud is about 7,000 K. The arrows point toward the directions that different parts of the cloud are moving. The names associated with each arrow indicate the constellations located on the sky toward which the parts of the cloud are headed. The solar system is thought to have entered the Local Interstellar Cloud, which is a small cloud located within a much larger superbubble that is expanding outward from the Scorpius-Centaurus region of the sky, at some point between 44,000 and 150,000 years ago and is expected to remain within it for another 10,000 to 20,000 years. (credit: modification of work by NASA/Goddard/Adler/ University Chicago/Wesleyan)

This OpenStax book is available for free at http://cnx.org/content/col11992/1.8

Download for free at http://cnx.org/content/col11992/latest/

CHAPTER 20 REVIEW

KEY TERMS

baryon cycle the cycling of mass in and out of the interstellar medium, including accretion of gas from intergalactic space, loss of gas back into intergalactic space, and conversion of interstellar gas into stars

cosmic rays atomic nuclei (mostly protons) and electrons that are observed to strike Earth's atmosphere with exceedingly high energies.

H II region the region of ionized hydrogen in interstellar space

interstellar dust tiny solid grains in interstellar space thought to consist of a core of rocklike material (silicates) or graphite surrounded by a mantle of ices; water, methane, and ammonia are probably the most abundant ices

interstellar extinction the attenuation or absorption of light by dust in the interstellar medium

interstellar medium (ISM) (or interstellar matter) the gas and dust between the stars in a galaxy

Local Bubble (or Local Hot Bubble) a region of low-density, million degree gas in which the Sun and solar system are currently located

Local Fluff a slightly denser cloud inside the Local Bubble, inside which the Sun also lies

molecular cloud a large, dense, cold interstellar cloud; because of its size and density, this type of cloud can keep ultraviolet radiation from reaching its interior, where molecules are able to form

nebula a cloud of interstellar gas or dust; the term is most often used for clouds that are seen to glow with visible light or infrared

reddening (interstellar) the reddening of starlight passing through interstellar dust because dust scatters blue light more effectively than red

SUMMARY

20.1 The Interstellar Medium

About 15% of the visible matter in the Galaxy is in the form of gas and dust, serving as the raw material for new stars. About 99% of this interstellar matter is in the form of gas—individual atoms or molecules. The most abundant elements in the interstellar gas are hydrogen and helium. About 1% of the interstellar matter is in the form of solid interstellar dust grains.

20.2 Interstellar Gas

Interstellar gas may be hot or cold. Gas found near hot stars emits light by fluorescence, that is, light is emitted when an electron is captured by an ion and cascades down to lower-energy levels. Glowing clouds (nebulae) of ionized hydrogen are called H II regions and have temperatures of about 10,000 K. Most hydrogen in interstellar space is not ionized and can best be studied by radio measurements of the 21-centimeter line. Some of the gas in interstellar space is at a temperature of a million degrees, even though it is far away in hot stars; this ultra-hot gas is probably heated when rapidly moving gas ejected in supernova explosions sweeps through space. In some places, gravity gathers interstellar gas into giant clouds, within which the gas is protected from

Download for free at http://cnx.org/content/col11992/latest/

starlight and can form molecules; more than 200 different molecules have been found in space, including the basic building blocks of proteins, which are fundamental to life as we know it here on Earth.

20.3 Cosmic Dust

Interstellar dust can be detected: (1) when it blocks the light of stars behind it, (2) when it scatters the light from nearby stars, and (3) because it makes distant stars look both redder and fainter. These effects are called reddening and interstellar extinction, respectively. Dust can also be detected in the infrared because it emits heat radiation. Dust is found throughout the plane of the Milky Way. The dust particles are about the same size as the wavelength of light and consist of rocky cores that are either sootlike (carbon-rich) or sandlike (silicates) with mantles made of ices such as water, ammonia, and methane.

20.4 Cosmic Rays

Cosmic rays are particles that travel through interstellar space at a typical speed of 90% of the speed of light. The most abundant elements in cosmic rays are the nuclei of hydrogen and helium, but electrons and positrons are also found. It is likely that many cosmic rays are produced in supernova shocks.

20.5 The Life Cycle of Cosmic Material

Interstellar matter is constantly flowing through the Galaxy and changing from one phase to another. At the same time, gas is constantly being added to the Galaxy by accretion from extragalactic space, while mass is removed from the interstellar medium by being locked in stars. Some of the mass in stars is, in turn, returned to the interstellar medium when those stars evolve and die. In particular, the heavy elements in interstellar space were all produced inside stars, while the dust grains are made in the outer regions of stars that have swelled to be giants. These elements and grains, in turn, can then be incorporated into new stars and planetary systems that form out of the interstellar medium.

20.6 Interstellar Matter around the Sun

The Sun is located at the edge of a low-density cloud called the Local Fluff. The Sun and this cloud are located within the Local Bubble, a region extending to at least 300 light-years from the Sun, within which the density of interstellar material is extremely low. Astronomers think this bubble was blown by some nearby stars that experienced a strong wind and some supernova explosions.

 FOR FURTHER EXPLORATION

Articles

Goodman, A. "Recycling the Universe." *Sky & Telescope* November (2000): 44. Review of how stellar evolution, the interstellar medium, and supernovae all work together to recycle cosmic material.

Greenberg, J. "The Secrets of Stardust." *Scientific American* December (2000): 70. The makeup and evolutionary role of solid particles between the stars.

Knapp, G. "The Stuff between the Stars." *Sky & Telescope* May (1995): 20. An introduction to the interstellar medium.

Nadis, S. "Searching for the Molecules of Life in Space." *Sky & Telescope* January (2002): 32. Recent observations of water in the interstellar medium by satellite telescopes.

Olinto, A. "Solving the Mystery of Cosmic Rays." *Astronomy* April (2014): 30. What accelerates them to such high energies.

Reynolds, R. "The Gas between the Stars." *Scientific American* January (2002): 34. On the interstellar medium.

This OpenStax book is available for free at http://cnx.org/content/col11992/1.8

Websites And Apps

Barnard, E. E., Biographical Memoir: http://www.nasonline.org/publications/biographical-memoirs/memoir-pdfs/barnard-edward.pdf.

Cosmicopia: http://helios.gsfc.nasa.gov/cosmic.html. NASA's learning site explains about the history and modern understanding of cosmic rays.

DECO: https://wipac.wisc.edu/deco. A smart-phone app for turning your phone into a cosmic-ray detector.

Hubble Space Telescope Images of Nebulae: http://hubblesite.org/gallery/album/nebula/. Click on any of the beautiful images in this collection, and you are taken to a page with more information; while looking at these images, you may also want to browse through the slide sequence on the meaning of colors in the Hubble pictures (http://hubblesite.org/gallery/behind_the_pictures/meaning_of_color/).

Interstellar Medium Online Tutorial: http://www-ssg.sr.unh.edu/ism/intro.htm. Nontechnical introduction to the interstellar medium (ISM) and how we study it; by the University of New Hampshire astronomy department.

Messier Catalog of Nebulae, Clusters, and Galaxies: http://astropixels.com/messier/messiercat.html. Astronomer Fred Espenak provides the full catalog, with information and images. (The Wikipedia list does something similar: https://en.wikipedia.org/wiki/List_of_Messier_objects.)

Nebulae: What Are They?: http://www.universetoday.com/61103/what-is-a-nebula/. Concise introduction by Matt Williams.

Videos

Barnard 68: The Hole in the Sky: https://www.youtube.com/watch?v=8No6I0Uc3No. About this dark cloud and dark clouds in interstellar space in general (02:08).

Horsehead Nebula in New Light: http://www.esa.int/spaceinvideos/Videos/2013/04/The_Horsehead_Nebula_in_new_light. Tour of the dark nebula in different wavelengths; no audio narration, just music, but explanatory material appears on the screen (03:03).

Hubblecast 65: A Whole New View of the Horsehead Nebula: http://www.spacetelescope.org/videos/heic1307a/. Report on nebulae in general and about the Horsehead specifically, with ESO astronomer Joe Liske (06:03).

Interstellar Reddening: https://www.youtube.com/watch?v=H2M80RAQB6k. Video demonstrating how reddening works, with Scott Miller of Penn State; a bit nerdy but useful (03:45).

♟ COLLABORATIVE GROUP ACTIVITIES

A. The Sun is located in a region where the density of interstellar matter is low. Suppose that instead it were located in a dense cloud 20 light-years in diameter that dimmed the visible light from stars lying outside it by a factor of 100. Have your group discuss how this would have affected the development of civilization on Earth. For example, would it have presented a problem for early navigators?

B. Your group members should look through the pictures in this chapter. How big are the nebulae you see in the images? Are there any clues either in the images or in the captions? Are the clouds they are part of significantly bigger than the nebulae we can see? Why? Suggest some ways that we can determine the sizes of nebulae.

Download for free at http://cnx.org/content/col11992/latest/

C. How do the members of your group think astronomers are able to estimate the distances of such nebulae in our own Galaxy? (Hint: Look at the images. Can you see anything between us and the nebula in some cases. Review Celestial Distances, if you need to remind yourself about methods of measuring distances.)

D. The text suggests that a tube of air extending from the surface of Earth to the top of the atmosphere contains more atoms than a tube of the same diameter extending from the top of the atmosphere to the edge of the observable universe. Scientists often do what they call "back of the envelope calculations," in which they make very rough approximations just to see whether statements or ideas are true. Try doing such a "quick and dirty" estimate for this statement with your group. What are the steps in comparing the numbers of atoms contained in the two different tubes? What information do you need to make the approximations? Can you find it in this text? And is the statement true?

E. If your astronomy course has involved learning about the solar system before you got to this chapter, have your group discuss where else besides interstellar clouds astronomers have been discovering organic molecules (the chemical building blocks of life). How might the discoveries of such molecules in our own solar system be related to the molecules in the clouds discussed in this chapter?

F. Two stars both have a reddish appearance in telescopes. One star is actually red; the other's light has been reddened by interstellar dust on its way to us. Have your group make a list of the observations you could perform to determine which star is which.

G. You have been asked to give a talk to your little brother's middle school class on astronomy, and you decide to talk about how nature recycles gas and dust. Have your group discuss what images from this book you would use in your talk. In what order? What is the one big idea you would like the students to remember when the class is over?

H. This chapter and the next (on The Birth of Stars) include some of the most beautiful images of nebulae that glow with the light produced when starlight interacts with gas and dust. Have your group select one to four of your favorite such nebulae and prepare a report on them to share with the rest of the class. (Include such things as their location, distance, size, way they are glowing, and what is happening within them.)

📖 EXERCISES

Review Questions

1. Identify several dark nebulae in photographs in this chapter. Give the figure numbers of the photographs, and specify where the dark nebulae are to be found on them.

2. Why do nebulae near hot stars look red? Why do dust clouds near stars usually look blue?

3. Describe the characteristics of the various kinds of interstellar gas (HII regions, neutral hydrogen clouds, ultra-hot gas clouds, and molecular clouds).

4. Prepare a table listing the different ways in which dust and gas can be detected in interstellar space.

5. Describe how the 21-cm line of hydrogen is formed. Why is this line such an important tool for understanding the interstellar medium?

6. Describe the properties of the dust grains found in the space between stars.

7. Why is it difficult to determine where cosmic rays come from?

This OpenStax book is available for free at http://cnx.org/content/col11992/1.8

Download for free at http://cnx.org/content/col11992/latest/

8. What causes reddening of starlight? Explain how the reddish color of the Sun's disk at sunset is caused by the same process.

9. Why do molecules, including H_2 and more complex organic molecules, only form inside dark clouds? Why don't they fill all interstellar space?

10. Why can't we use visible light telescopes to study molecular clouds where stars and planets form? Why do infrared or radio telescopes work better?

11. The mass of the interstellar medium is determined by a balance between sources (which add mass) and sinks (which remove it). Make a table listing the major sources and sinks, and briefly explain each one.

12. Where does interstellar dust come from? How does it form?

Thought Questions

13. Figure 20.2 shows a reddish glow around the star Antares, and yet the caption says that is a dust cloud. What observations would you make to determine whether the red glow is actually produced by dust or whether it is produced by an H II region?

14. If the red glow around Antares is indeed produced by reflection of the light from Antares by dust, what does its red appearance tell you about the likely temperature of Antares? Look up the spectral type of Antares in Appendix J. Was your estimate of the temperature about right? In most of the images in this chapter, a red glow is associated with ionized hydrogen. Would you expect to find an H II region around Antares? Explain your answer.

15. Even though neutral hydrogen is the most abundant element in interstellar matter, it was detected first with a radio telescope, not a visible light telescope. Explain why. (The explanation given in Analyzing Starlight for the fact that hydrogen lines are not strong in stars of all temperatures may be helpful.)

16. The terms H II and H_2 are both pronounced "H two." What is the difference in meaning of those two terms? Can there be such a thing as H III?

17. Suppose someone told you that she had discovered H II around the star Aldebaran. Would you believe her? Why or why not?

18. Describe the spectrum of each of the following:
    A. starlight reflected by dust,

    B. a star behind invisible interstellar gas, and

    C. an emission nebula.

19. According to the text, a star must be hotter than about 25,000 K to produce an H II region. Both the hottest white dwarfs and main-sequence O stars have temperatures hotter than 25,000 K. Which type of star can ionize more hydrogen? Why?

20. From the comments in the text about which kinds of stars produce emission nebulae and which kinds are associated with reflection nebulae, what can you say about the temperatures of the stars that produce NGC 1999 (Figure 20.13)?

21. One way to calculate the size and shape of the Galaxy is to estimate the distances to faint stars just from their observed apparent brightnesses and to note the distance at which stars are no longer observable. The first astronomers to try this experiment did not know that starlight is dimmed by interstellar dust. Their estimates of the size of the Galaxy were much too small. Explain why.

22. New stars form in regions where the density of gas and dust is relatively high. Suppose you wanted to search for some recently formed stars. Would you more likely be successful if you observed at visible wavelengths or at infrared wavelengths? Why?

23. Thinking about the topics in this chapter, here is an Earth analogy. In big cities, you can see much farther on days without smog. Why?

24. Stars form in the Milky Way at a rate of about 1 solar mass per year. At this rate, how long would it take for all the interstellar gas in the Milky Way to be turned into stars if there were no fresh gas coming in from outside? How does this compare to the estimated age of the universe, 14 billion years? What do you conclude from this?

25. The 21-cm line can be used not just to find out where hydrogen is located in the sky, but also to determine how fast it is moving toward or away from us. Describe how this might work.

26. Astronomers recently detected light emitted by a supernova that was originally observed in 1572, just reaching Earth now. This light was reflected off a dust cloud; astronomers call such a reflected light a "light echo" (just like reflected sound is called an echo). How would you expect the spectrum of the light echo to compare to that of the original supernova?

27. We can detect 21-cm emission from other galaxies as well as from our own Galaxy. However, 21-cm emission from our own Galaxy fills most of the sky, so we usually see both at once. How can we distinguish the extragalactic 21-cm emission from that arising in our own Galaxy? (Hint: Other galaxies are generally moving relative to the Milky Way.)

28. We have said repeatedly that blue light undergoes more extinction than red light, which is true for visible and shorter wavelengths. Is the same true for X-rays? Look at Figure 20.19. The most dust is in the galactic plane in the middle of the image, and the red color in the image corresponds to the reddest (lowest-energy) light. Based on what you see in the galactic plane, are X-rays experiencing more extinction at redder or bluer colors? You might consider comparing Figure 20.19 to Figure 20.14.

29. Suppose that, instead of being inside the Local Bubble, the Sun were deep inside a giant molecular cloud. What would the night sky look like as seen from Earth at various wavelengths?

30. Suppose that, instead of being inside the Local Bubble, the Sun were inside an H II region. What would the night sky look like at various wavelengths?

Figuring For Yourself

31. A molecular cloud is about 1000 times denser than the average of the interstellar medium. Let's compare this difference in densities to something more familiar. Air has a density of about 1 kg/m^3, so something 1000 times denser than air would have a density of about 1000 kg/m^3. How does this compare to the typical density of water? Of granite? (You can find figures for these densities on the internet.) Is the density difference between a molecular cloud and the interstellar medium larger or smaller than the density difference between air and water or granite?

32. Would you expect to be able to detect an H II region in X-ray emission? Why or why not? (Hint: You might apply Wien's law)

33. Suppose that you gathered a ball of interstellar gas that was equal to the size of Earth (a radius of about 6000 km). If this gas has a density of 1 hydrogen atom per cm^3, typical of the interstellar medium, how would its mass compare to the mass of a bowling ball (5 or 6 kg)? How about if it had the typical density of the Local Bubble, about 0.01 atoms per cm^3? The volume of a sphere is $V = (4/3)\pi R^3$.

This OpenStax book is available for free at http://cnx.org/content/col11992/1.8

Download for free at http://cnx.org/content/col11992/latest/

34. At the average density of the interstellar medium, 1 atom per cm³, how big a volume of material must be used to make a star with the mass of the Sun? What is the radius of a sphere this size? Express your answer in light-years.

35. Consider a grain of sand that contains 1 mg of oxygen (a typical amount for a medium-sized sand grain, since sand is mostly SiO_2). How many oxygen atoms does the grain contain? What is the radius of the sphere you would have to spread them out over if you wanted them to have the same density as the interstellar medium, about 1 atom per cm³? You can look up the mass of an oxygen atom.

36. H II regions can exist only if there is a nearby star hot enough to ionize hydrogen. Hydrogen is ionized only by radiation with wavelengths shorter than 91.2 nm. What is the temperature of a star that emits its maximum energy at 91.2 nm? (Use Wien's law from Radiation and Spectra.) Based on this result, what are the spectral types of those stars likely to provide enough energy to produce H II regions?

37. In the text, we said that the five-times ionized oxygen (OVI) seen in hot gas must have been produced by supernova shocks that heated the gas to millions of degrees, and not by starlight, the way H II is produced. Producing OVI by light requires wavelengths shorter than 10.9 nm. The hottest observed stars have surface temperatures of about 50,000 K. Could they produce OVI?

38. Dust was originally discovered because the stars in certain clusters seemed to be fainter than expected. Suppose a star is behind a cloud of dust that dims its brightness by a factor of 100. Suppose you do not realize the dust is there. How much in error will your distance estimate be? Can you think of any measurement you might make to detect the dust?

39. How would the density inside a cold cloud ($T = 10$ K) compare with the density of the ultra-hot interstellar gas ($T = 10^6$ K) if they were in pressure equilibrium? (It takes a large cloud to be able to shield its interior from heating so that it can be at such a low temperature.) (Hint: In pressure equilibrium, the two regions must have nT equal, where n is the number of particles per unit volume and T is the temperature.) Which region do you think is more suitable for the creation of new stars? Why?

40. The text says that the Local Fluff, which surrounds the Sun, has a temperature of 7500 K and a density 0.1 atom per cm³. The Local Fluff is embedded in hot gas with a temperature of 10^6 K and a density of about 0.01 atom per cm³. Are they in equilibrium? (Hint: In pressure equilibrium, the two regions must have nT equal, where n is the number of particles per unit volume and T is the temperature.) What is likely to happen to the Local Fluff?

Download for free at http://cnx.org/content/col11992/latest/

This OpenStax book is available for free at http://cnx.org/content/col11992/1.8

Download for free at http://cnx.org/content/col11992/latest/

Figure 21.1. Where Stars Are Born. We see a close-up of part of the Carina Nebula taken with the Hubble Space Telescope. This image reveals jets powered by newly forming stars embedded in a great cloud of gas and dust. Parts of the clouds are glowing from the energy of very young stars recently formed within them. (credit: modification of work by NASA, ESA, and M. Livio and the Hubble 20th Anniversary Team (STScI))

Chapter Outline

Thinking Ahead

"There are countless suns and countless earths all rotating round their suns in exactly the same way as the planets of our system. We see only the suns because they are the largest bodies and are luminous, but their planets remain invisible to us because they are smaller and non-luminous. . . . The unnumbered worlds in the universe are all similar in form and rank and subject to the same forces and the same laws."—Giordano Bruno in *On the Infinite Universe and Worlds* (1584). Bruno was tried for heresy by the Roman Inquisition and burned at the stake in 1600.

We've discussed stars as nuclear furnaces that convert light elements into heavier ones. A star's nuclear evolution begins when hydrogen is fused into helium, but that can only occur when the core temperature exceeds 10 to 12 million K. Since stars form from cold interstellar material, we must understand how they collapse and eventually reach this "ignition temperature" to explain the birth of stars. Star formation is a continuous process, from the birth of our Galaxy right up to today. We estimate that every year in our Galaxy, on average, three solar masses of interstellar matter are converted into stars. This may sound like a small amount of mass for an object as large as a galaxy, but only three new stars (out of billions in the *Galaxy*) are formed each year.

Download for free at http://cnx.org/content/col11992/latest/

Do planets orbit other stars or is ours the only planetary system? In the past few decades, new technology has enabled us to answer that question by revealing nearly 3500 exoplanets in over 2600 planetary systems. Even before planets were detected, astronomers had predicted that planetary systems were likely to be byproducts of the star-formation process. In this chapter, we look at how interstellar matter is transformed into stars and planets.

21.1 STAR FORMATION

Learning Objectives

By the end of this section, you will be able to:

> Identify the sometimes-violent processes by which parts of a molecular cloud collapse to produce stars
> Recognize some of the structures seen in images of molecular clouds like the one in Orion
> Explain how the environment of a molecular cloud enables the formation of stars
> Describe how advancing waves of star formation cause a molecular cloud to evolve

As we begin our exploration of how stars are formed, let's review some basics about stars discussed in earlier chapters:

- Stable (main-sequence) stars such as our Sun maintain equilibrium by producing energy through nuclear fusion in their cores. The ability to generate energy by fusion defines a star.

- Each second in the Sun, approximately 600 million tons of hydrogen undergo fusion into helium, with about 4 million tons turning into energy in the process. This rate of hydrogen use means that eventually the Sun (and all other stars) will run out of central fuel.

- Stars come with many different masses, ranging from 1/12 solar masses (M_{Sun}) to roughly 100–200 M_{Sun}. There are far more low-mass than high-mass stars.

- The most massive main-sequence stars (spectral type O) are also the most luminous and have the highest surface temperature. The lowest-mass stars on the main sequence (spectral type M or L) are the least luminous and the coolest.

- A galaxy of stars such as the Milky Way contains enormous amounts of gas and dust—enough to make billions of stars like the Sun.

If we want to find stars still in the process of formation, we must look in places that have plenty of the raw material from which stars are assembled. Since stars are made of gas, we focus our attention (and our telescopes) on the dense and cold clouds of gas and dust that dot the Milky Way (see Figure 21.1 and Figure 21.2).

This OpenStax book is available for free at http://cnx.org/content/col11992/1.8

Download for free at http://cnx.org/content/col11992/latest/

(a)   (b)

Figure 21.2. Pillars of Dust and Dense Globules in M16. (a) This Hubble Space Telescope image of the central regions of M16 (also known as the Eagle Nebula) shows huge columns of cool gas, (including molecular hydrogen, H2) and dust. These columns are of higher density than the surrounding regions and have resisted evaporation by the ultraviolet radiation from a cluster of hot stars just beyond the upper-right corner of this image. The tallest pillar is about 1 light-year long, and the M16 region is about 7000 light-years away from us. (b) This close-up view of one of the pillars shows some very dense globules, many of which harbor embryonic stars. Astronomers coined the term *evaporating gas globules* (EGGs) for these structures, in part so that they could say we found EGGs inside the Eagle Nebula. It is possible that because these EGGs are exposed to the relentless action of the radiation from nearby hot stars, some may not yet have collected enough material to form a star. (credit a : modification of work by NASA, ESA, and the Hubble Heritage Team (STScI/AURA); credit b: modification of work by NASA, ESA, STScI, J. Hester and P. Scowen (Arizona State University))

Molecular Clouds: Stellar Nurseries

As we saw in Between the Stars: Gas and Dust in Space, the most massive reservoirs of interstellar matter—and some of the most massive objects in the Milky Way Galaxy—are the **giant molecular clouds**. These clouds have cold interiors with characteristic temperatures of only 10–20 K; most of their gas atoms are bound into molecules. These clouds turn out to be the birthplaces of most stars in our Galaxy.

The masses of molecular clouds range from a thousand times the mass of the Sun to about 3 million solar masses. Molecular clouds have a complex filamentary structure, similar to cirrus clouds in Earth's atmosphere, but much less dense. The molecular cloud filaments can be up to 1000 light-years long. Within the clouds are cold, dense regions with typical masses of 50 to 500 times the mass of the Sun; we give these regions the highly technical name *clumps*. Within these clumps, there are even denser, smaller regions called cores. The cores are the embryos of stars. The conditions in these cores—low temperature and high density—are just what is required to make stars. Remember that the essence of the life story of any star is the ongoing competition between two forces: *gravity* and *pressure*. The force of gravity, pulling inward, tries to make a star collapse. Internal pressure produced by the motions of the gas atoms, pushing outward, tries to force the star to expand. When a star is first forming, low temperature (and hence, low pressure) and high density (hence, greater gravitational attraction) both work to give gravity the advantage. In order to form a star—that is, a dense, hot ball of matter capable of starting nuclear reactions deep within—we need a typical core of interstellar atoms and molecules to shrink in radius and increase in density by a factor of nearly 10^{20}. It is the force of gravity that produces this drastic collapse.

Download for free at http://cnx.org/content/col11992/latest/

The Orion Molecular Cloud

Let's discuss what happens in regions of star formation by considering a nearby site where stars are forming right now. One of the best-studied stellar nurseries is in the constellation of Orion, The Hunter, about 1500 light-years away (Figure 21.3). The pattern of the hunter is easy to recognize by the conspicuous "belt" of three stars that mark his waist. The Orion molecular cloud is much larger than the star pattern and is truly an impressive structure. In its long dimension, it stretches over a distance of about 100 light-years. The total quantity of molecular gas is about 200,000 times the mass of the Sun. Most of the cloud does not glow with visible light but betrays its presence by the radiation that the dusty gas gives off at infrared and radio wavelengths.

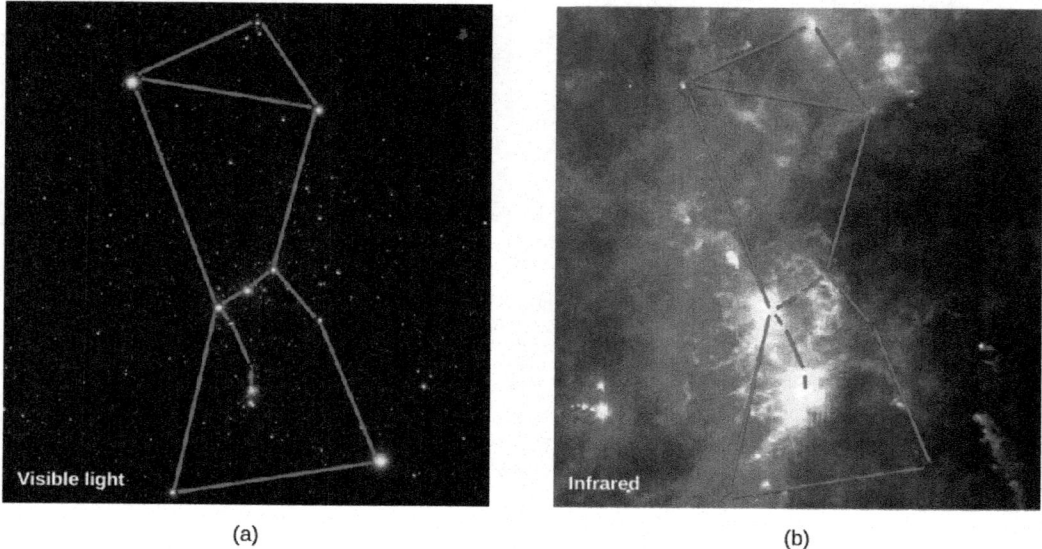

(a)   (b)

Figure 21.3. Orion in Visible and Infrared. (a) The Orion star group was named after the legendary hunter in Greek mythology. Three stars close together in a link mark Orion's belt. The ancients imagined a sword hanging from the belt; the object at the end of the blue line in this sword is the Orion Nebula. (b) This wide-angle, infrared view of the same area was taken with the Infrared Astronomical Satellite. Heated dust clouds dominate in this false-color image, and many of the stars that stood out on part (a) are now invisible. An exception is the cool, red-giant star Betelgeuse, which can be seen as a yellowish point at the left vertex of the blue triangle (at Orion's left armpit). The large, yellow ring to the right of Betelgeuse is the remnant of an exploded star. The infrared image lets us see how large and full of cooler material the Orion molecular cloud really is. On the visible-light image at left, you see only two colorful regions of interstellar matter—the two, bright yellow splotches at the left end of and below Orion's belt. The lower one is the Orion Nebula and the higher one is the region of the Horsehead Nebula. (credit: modification of work by NASA, visible light: Akira Fujii; infrared: Infrared Astronomical Satellite)

The stars in Orion's belt are typically about 5 million years old, whereas the stars near the middle of the "sword" hanging from Orion's belt are only 300,000 to 1 million years old. The region about halfway down the sword where star formation is still taking place is called the Orion Nebula. About 2200 young stars are found in this region, which is only slightly larger than a dozen light-years in diameter. The Orion Nebula also contains a tight cluster of stars called the Trapezium (Figure 21.4). The brightest Trapezium stars can be seen easily with a small telescope.

This OpenStax book is available for free at http://cnx.org/content/col11992/1.8

Download for free at http://cnx.org/content/col11992/latest/

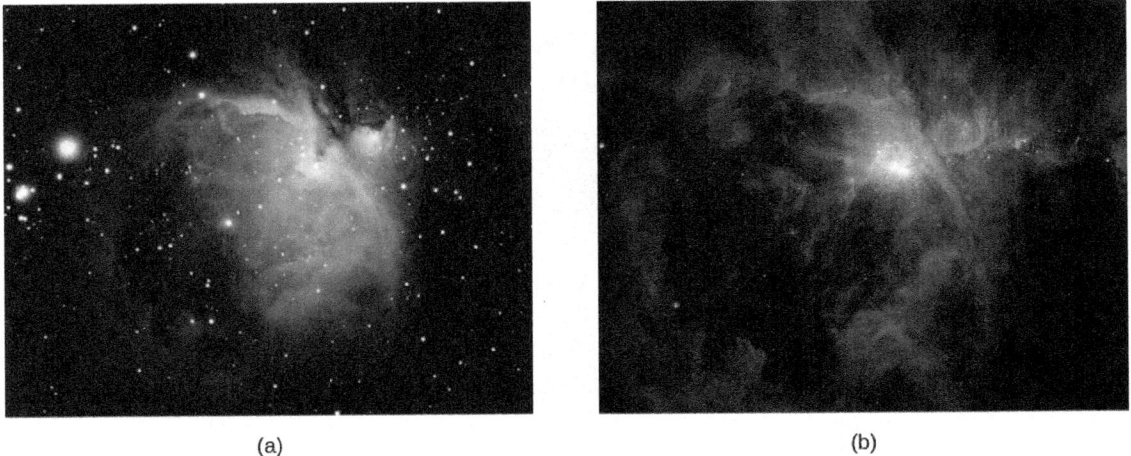

(a)   (b)

Figure 21.4. Orion Nebula. (a) The Orion Nebula is shown in visible light. (b) With near-infrared radiation, we can see more detail within the dusty nebula since infrared can penetrate dust more easily than can visible light. (credit a: modification of work by Filip Lolić; credit b: modification of work by NASA/JPL-Caltech/T. Megeath (University of Toledo, Ohio))

Compare this with our own solar neighborhood, where the typical spacing between stars is about 3 light-years. Only a small number of stars in the Orion cluster can be seen with visible light, but infrared images—which penetrate the dust better—detect the more than 2000 stars that are part of the group (Figure 21.5).

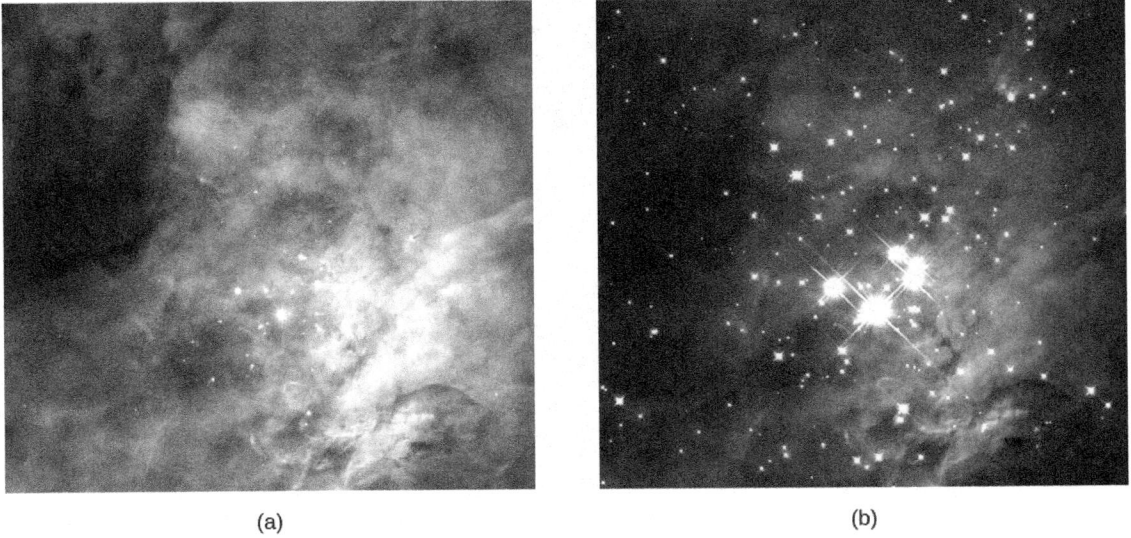

(a)   (b)

Figure 21.5. Central Region of the Orion Nebula. The Orion Nebula harbors some of the youngest stars in the solar neighborhood. At the heart of the nebula is the Trapezium cluster, which includes four very bright stars that provide much of the energy that causes the nebula to glow so brightly. In these images, we see a section of the nebula in (a) visible light and (b) infrared. The four bright stars in the center of the visible-light image are the Trapezium stars. Notice that most of the stars seen in the infrared are completely hidden by dust in the visible-light image. (credit a: modification of work by NASA, C.R. O'Dell and S.K. Wong (Rice University); credit b: modification of work by NASA; K.L. Luhman (Harvard-Smithsonian Center for Astrophysics); and G. Schneider, E. Young, G. Rieke, A. Cotera, H. Chen, M. Rieke, R. Thompson (Steward Observatory, University of Arizona))

Studies of Orion and other star-forming regions show that star formation is not a very efficient process. In the region of the Orion Nebula, about 1% of the material in the cloud has been turned into stars. That is why we still see a substantial amount of gas and dust near the Trapezium stars. The leftover material is eventually heated,

Download for free at http://cnx.org/content/col11992/latest/

either by the radiation and winds from the hot stars that form or by explosions of the most massive stars. (We will see in later chapters that the most massive stars go through their lives very quickly and end by exploding.)

LINK TO LEARNING

Take a journey through the Orion Nebula (https://openstaxcollege.org/l/30OriNebula) to view a nice narrated video tour of this region.

Whether gently or explosively, the material in the neighborhood of the new stars is blown away into interstellar space. Older groups or clusters of stars can now be easily observed in visible light because they are no longer shrouded in dust and gas (Figure 21.6).

Figure 21.6. Westerlund 2. This young cluster of stars known as Westerlund 2 formed within the Carina star-forming region about 2 million years ago. Stellar winds and pressure produced by the radiation from the hot stars within the cluster are blowing and sculpting the surrounding gas and dust. The nebula still contains many globules of dust. Stars are continuing to form within the denser globules and pillars of the nebula. This Hubble Space Telescope image includes near-infrared exposures of the star cluster and visible-light observations of the surrounding nebula. Colors in the nebula are dominated by the red glow of hydrogen gas, and blue-green emissions from glowing oxygen. (credit: NASA, ESA, the Hubble Heritage Team (STScI/AURA), A. Nota (ESA/STScI), and the Westerlund 2 Science Team)

Although we do not know what initially caused stars to begin forming in Orion, there is good evidence that the first generation of stars triggered the formation of additional stars, which in turn led to the formation of still more stars (Figure 21.7).

This OpenStax book is available for free at http://cnx.org/content/col11992/1.8

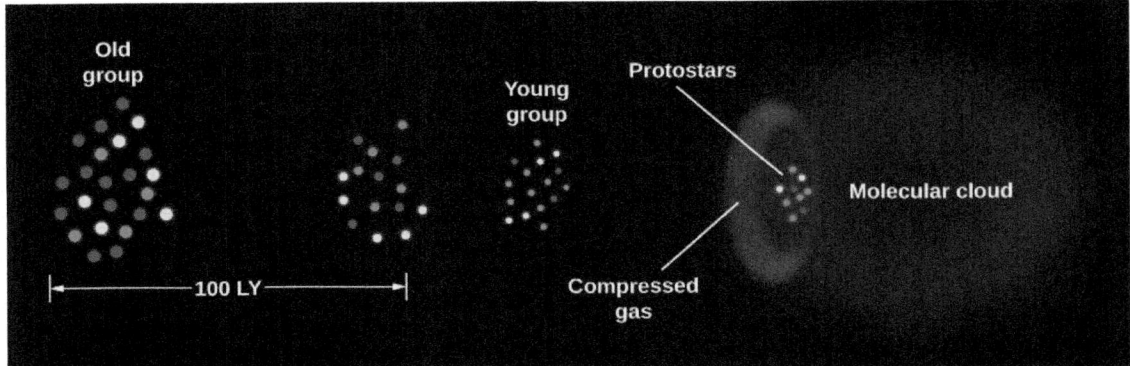

Figure 21.7. Propagating Star Formation. Star formation can move progressively through a molecular cloud. The oldest group of stars lies to the left of the diagram and has expanded because of the motions of individual stars. Eventually, the stars in the group will disperse and no longer be recognizable as a cluster. The youngest group of stars lies to the right, next to the molecular cloud. This group of stars is only 1 to 2 million years old. The pressure of the hot, ionized gas surrounding these stars compresses the material in the nearby edge of the molecular cloud and initiates the gravitational collapse that will lead to the formation of more stars.

The basic idea of triggered star formation is this: when a massive star is formed, it emits a large amount of ultraviolet radiation and ejects high-speed gas in the form of a stellar wind. This injection of energy heats the gas around the stars and causes it to expand. When massive stars exhaust their supply of fuel, they explode, and the energy of the explosion also heats the gas. The hot gases pile into the surrounding cold molecular cloud, compressing the material in it and increasing its density. If this increase in density is large enough, gravity will overcome pressure, and stars will begin to form in the compressed gas. Such a chain reaction—where the brightest and hottest stars of one area become the cause of star formation "next door"—seems to have occurred not only in Orion but also in many other molecular clouds.

There are many molecular clouds that form only (or mainly) low-mass stars. Because low-mass stars do not have strong winds and do not die by exploding, triggered star formation cannot occur in these clouds. There are also stars that form in relative isolation in small cores. Therefore, not all star formation is originally triggered by the death of massive stars. However, there are likely to be other possible triggers, such as spiral density waves and other processes we do not yet understand.

The Birth of a Star

Although regions such as Orion give us clues about how star formation begins, the subsequent stages are still shrouded in mystery (and a lot of dust). There is an enormous difference between the density of a molecular cloud core and the density of the youngest stars that can be detected. Direct observations of this collapse to higher density are nearly impossible for two reasons. First, the dust-shrouded interiors of molecular clouds where stellar births take place cannot be observed with visible light. Second, the timescale for the initial collapse—thousands of years—is very short, astronomically speaking. Since each star spends such a tiny fraction of its life in this stage, relatively few stars are going through the collapse process at any given time. Nevertheless, through a combination of theoretical calculations and the limited observations available, astronomers have pieced together a picture of what the earliest stages of stellar evolution are likely to be.

The first step in the process of creating stars is the formation of dense cores within a clump of gas and dust (Figure 21.8(a)). It is generally thought that all the material for the star comes from the core, the larger structure surrounding the forming star. Eventually, the gravitational force of the infalling gas becomes strong enough to overwhelm the pressure exerted by the cold material that forms the dense cores. The material then undergoes a rapid collapse, and the density of the core increases greatly as a result. During the time a dense

Download for free at http://cnx.org/content/col11992/latest/

core is contracting to become a true star, but before the fusion of protons to produce helium begins, we call the object a **protostar**.

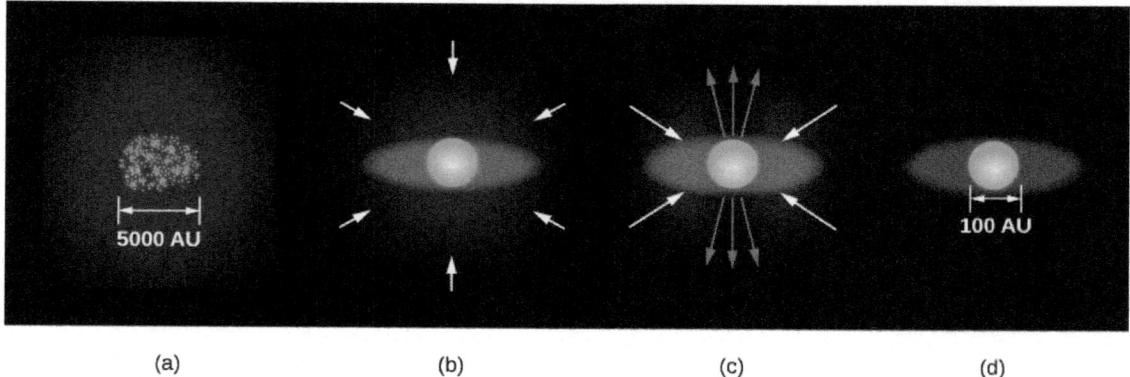

Figure 21.8. Formation of a Star. (a) Dense cores form within a molecular cloud. (b) A protostar with a surrounding disk of material forms at the center of a dense core, accumulating additional material from the molecular cloud through gravitational attraction. (c) A stellar wind breaks out but is confined by the disk to flow out along the two poles of the star. (d) Eventually, this wind sweeps away the cloud material and halts the accumulation of additional material, and a newly formed star, surrounded by a disk, becomes observable. These sketches are not drawn to the same scale. The diameter of a typical envelope that is supplying gas to the newly forming star is about 5000 AU. The typical diameter of the disk is about 100 AU or slightly larger than the diameter of the orbit of Pluto.

The natural turbulence inside a clump tends to give any portion of it some initial spinning motion (even if it is very slow). As a result, each collapsing core is expected to spin. According to the law of conservation of angular momentum (discussed in the chapter on Orbits and Gravity), a rotating body spins more rapidly as it decreases in size. In other words, if the object can turn its material around a smaller circle, it can move that material more quickly—like a figure skater spinning more rapidly as she brings her arms in tight to her body. This is exactly what happens when a core contracts to form a protostar: as it shrinks, its rate of spin increases.

But all directions on a spinning sphere are not created equal. As the protostar rotates, it is much easier for material to fall right onto the poles (which spin most slowly) than onto the equator (where material moves around most rapidly). Therefore, gas and dust falling in toward the protostar's equator are "held back" by the rotation and form a whirling extended disk around the equator (part b in Figure 21.8). You may have observed this same "equator effect" on the amusement park ride in which you stand with your back to a cylinder that is spun faster and faster. As you spin really fast, you are pushed against the wall so strongly that you cannot possibly fall toward the center of the cylinder. Gas can, however, fall onto the protostar easily from directions away from the star's equator.

The protostar and disk at this stage are embedded in an envelope of dust and gas from which material is still falling onto the protostar. This dusty envelope blocks visible light, but infrared radiation can get through. As a result, in this phase of its evolution, the protostar itself is emitting infrared radiation and so is observable only in the infrared region of the spectrum. Once almost all of the available material has been accreted and the central protostar has reached nearly its final mass, it is given a special name: it is called a *T Tauri star*, named after one of the best studied and brightest members of this class of stars, which was discovered in the constellation of Taurus. (Astronomers have a tendency to name types of stars after the first example they discover or come to understand. It's not an elegant system, but it works.) Only stars with masses less than or similar to the mass of the Sun become T Tauri stars. Massive stars do not go through this stage, although they do appear to follow the formation scenario illustrated in Figure 21.8.

This OpenStax book is available for free at http://cnx.org/content/col11992/1.8

Winds and Jets

Recent observations suggest that T Tauri stars may actually be stars in a middle stage between protostars and hydrogen-fusing stars such as the Sun. High-resolution infrared images have revealed jets of material as well as *stellar winds* coming from some T Tauri stars, proof of interaction with their environment. A **stellar wind** consists mainly of protons (hydrogen nuclei) and electrons streaming away from the star at speeds of a few hundred kilometers per second (several hundred thousand miles per hour). When the wind first starts up, the disk of material around the star's equator blocks the wind in its direction. Where the wind particles *can* escape most effectively is in the direction of the star's poles.

Astronomers have actually seen evidence of these beams of particles shooting out in opposite directions from the popular regions of newly formed stars. In many cases, these beams point back to the location of a protostar that is still so completely shrouded in dust that we cannot yet see it (Figure 21.9).

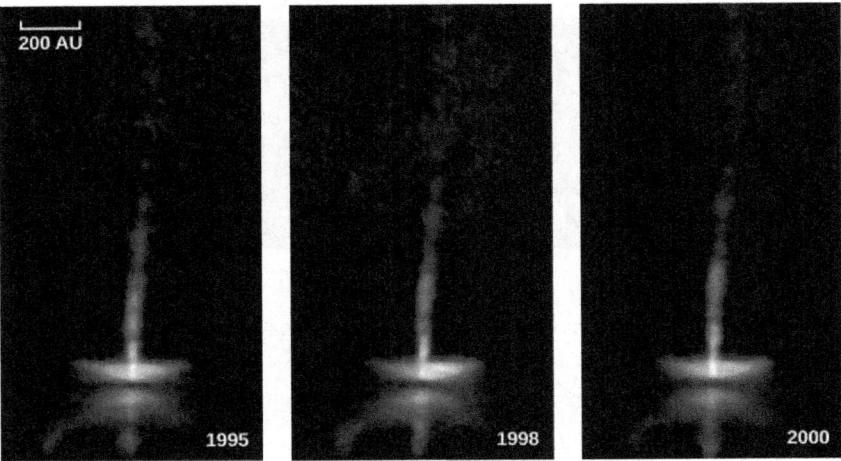

Figure 21.9. Gas Jets Flowing away from a Protostar. Here we see the neighborhood of a protostar, known to us as HH 34 because it is a Herbig-Haro object. The star is about 450 light-years away and only about 1 million years old. Light from the star itself is blocked by a disk, which is larger than 60 billion kilometers in diameter and is seen almost edge-on. Jets are seen emerging perpendicular to the disk. The material in these jets is flowing outward at speeds up to 580,000 kilometers per hour. The series of three images shows changes during a period of 5 years. Every few months, a compact clump of gas is ejected, and its motion outward can be followed. The changes in the brightness of the disk may be due to motions of clouds within the disk that alternately block some of the light and then let it through. This image corresponds to the stage in the life of a protostar shown in part (c) of Figure 21.8. (credit: modification of work by Hubble Space Telescope, NASA, ESA)

On occasion, the jets of high-speed particles streaming away from the protostar collide with a somewhat-denser lump of gas nearby, excite its atoms, and cause them to emit light. These glowing regions, each of which is known as a **Herbig-Haro (HH) object** after the two astronomers who first identified them, allow us to trace the progress of the jet to a distance of a light-year or more from the star that produced it. Figure 21.10 shows two spectacular images of HH objects.

Download for free at http://cnx.org/content/col11992/latest/

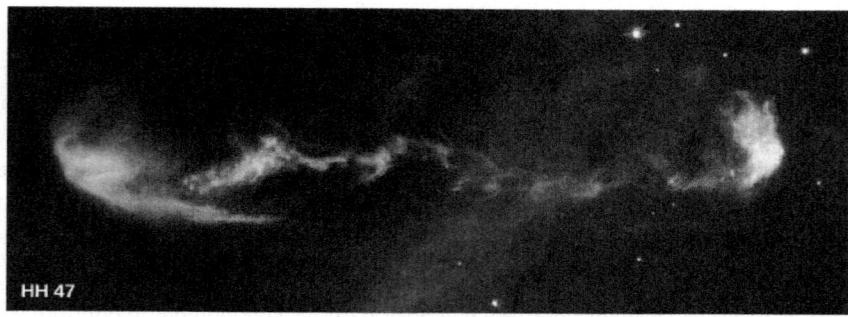

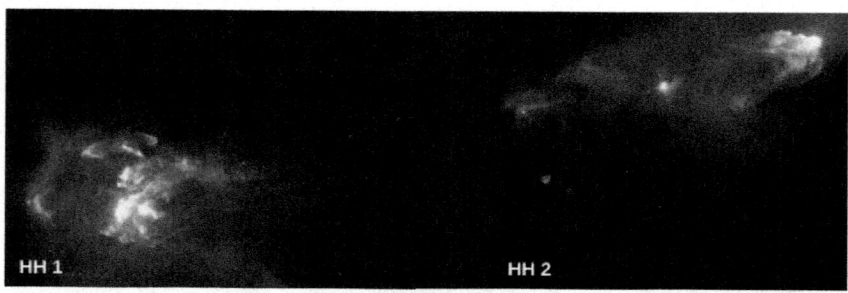

Figure 21.10. Outflows from Protostars. These images were taken with the Hubble Space Telescope and show jets flowing outward from newly formed stars. In the HH47 image, a protostar 1500 light-years away (invisible inside a dust disk at the left edge of the image) produces a very complicated jet. The star may actually be wobbling, perhaps because it has a companion. Light from the star illuminates the white region at the left because light can emerge perpendicular to the disk (just as the jet does). At right, the jet is plowing into existing clumps of interstellar gas, producing a shock wave that resembles an arrowhead. The HH1/2 image shows a double-beam jet emanating from a protostar (hidden in a dust disk in the center) in the constellation of Orion. Tip to tip, these jets are more than 1 light-year long. The bright regions (first identified by Herbig and Haro) are places where the jet is a slamming into a clump of interstellar gas and causing it to glow. (credit "HH 47": modification of work by NASA, ESA, and P. Hartigan (Rice University); credit "HH 1 and HH 2: modification of work by J. Hester, WFPC2 Team, NASA)

The wind from a forming star will ultimately sweep away the material that remains in the obscuring envelope of dust and gas, leaving behind the naked disk and protostar, which can then be seen with visible light. We should note that at this point, the protostar itself is still contracting slowly and has not yet reached the main-sequence stage on the H–R diagram (a concept introduced in the chapter The Stars: A Celestial Census). The disk can be detected directly when observed at infrared wavelengths or when it is seen silhouetted against a bright background (Figure 21.11).

This OpenStax book is available for free at http://cnx.org/content/col11992/1.8

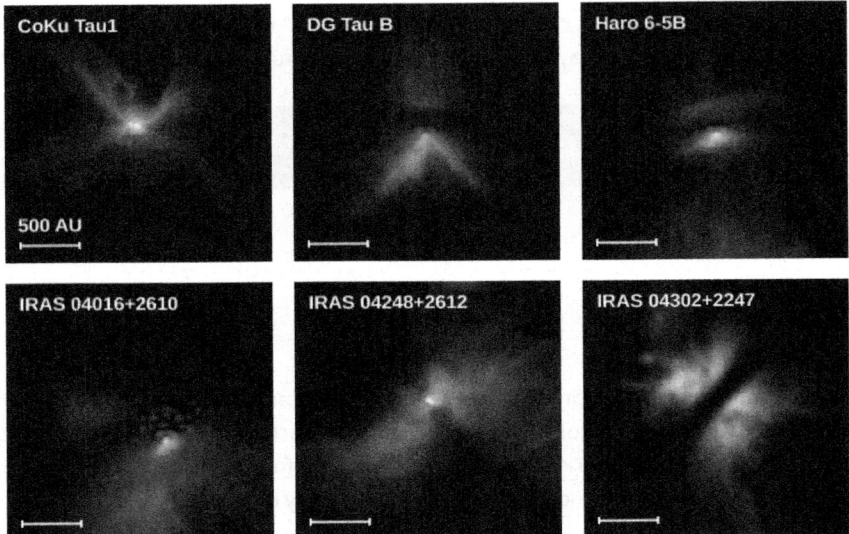

Figure 21.11. Disks around Protostars. These Hubble Space Telescope infrared images show disks around young stars in the constellation of Taurus, in a region about 450 light-years away. In some cases, we can see the central star (or stars—some are binaries). In other cases, the dark, horizontal bands indicate regions where the dust disk is so thick that even infrared radiation from the star embedded within it cannot make its way through. The brightly glowing regions are starlight reflected from the upper and lower surfaces of the disk, which are less dense than the central, dark regions. (Credit: modification of work by D. Padgett (IPAC/Caltech), W. Brandner (IPAC), K. Stapelfeldt (JPL) and NASA)

This description of a protostar surrounded by a rotating disk of gas and dust sounds very much like what happened in our solar system when the Sun and planets formed. Indeed, one of the most important discoveries from the study of star formation in the last decade of the twentieth century was that disks are an inevitable byproduct of the process of creating stars. The next questions that astronomers set out to answer was: will the disks around protostars also form planets? And if so, how often? We will return to these questions later in this chapter.

To keep things simple, we have described the formation of single stars. Many stars, however, are members of binary or triple systems, where several stars are born together. In this case, the stars form in nearly the same way. Widely separated binaries may each have their own disk; close binaries may share a single disk.

21.2 THE H–R DIAGRAM AND THE STUDY OF STELLAR EVOLUTION

Learning Objectives

By the end of this section, you will be able to:

> Determine the age of a protostar using an H–R diagram and the protostar's luminosity and temperature
> Explain the interplay between gravity and pressure, and how the contracting protostar changes its position in the H–R diagram as a result

One of the best ways to summarize all of these details about how a star or protostar changes with time is to use a Hertzsprung-Russell (H–R) diagram. Recall from The Stars: A Celestial Census that, when looking at an H–R diagram, the temperature (the horizontal axis) is plotted increasing toward the left. As a star goes through the stages of its life, its luminosity and temperature change. Thus, its position on the H–R diagram, in which luminosity is plotted against temperature, also changes. As a star ages, we must replot it in different places on the diagram. Therefore, astronomers often speak of a star *moving* on the H–R diagram, or of its evolution

Download for free at http://cnx.org/content/col11992/latest/

tracing out a path on the diagram. In this context, "tracing out a path" has nothing to do with the star's motion through space; this is just a shorthand way of saying that its temperature and luminosity change as it evolves.

LINK TO LEARNING

Watch an animation (https://openstaxcollege.org/l/30aniomelen) of the stars in the Omega Centauri cluster as they rearrange according to luminosity and temperature, forming a Hertzsprung-Russell (H–R) diagram.

To estimate just how much the luminosity and temperature of a star change as it ages, we must resort to calculations. Theorists compute a series of models for a star, with each successive model representing a later point in time. Stars may change for a variety of reasons. Protostars, for example, change in size because they are contracting, and their temperature and luminosity change as they do so. After nuclear fusion begins in the star's core (see Stars from Adolescence to Old Age), main-sequence stars change because they are using up their nuclear fuel.

Given a model that represents a star at one stage of its evolution, we can calculate what it will be like at a slightly later time. At each step, the model predicts the luminosity and size of the star, and from these values, we can figure out its surface temperature. A series of points on an H–R diagram, calculated in this way, allows us to follow the life changes of a star and hence is called its *evolutionary track*.

Evolutionary Tracks

Let's now use these ideas to follow the evolution of protostars that are on their way to becoming main-sequence stars. The evolutionary tracks of newly forming stars with a range of stellar masses are shown in Figure 21.12. These young stellar objects are not yet producing energy by nuclear reactions, but they derive energy from gravitational contraction—through the sort of process proposed for the Sun by Helmhotz and Kelvin in this last century (see the chapter on The Sun: A Nuclear Powerhouse).

This OpenStax book is available for free at http://cnx.org/content/col11992/1.8

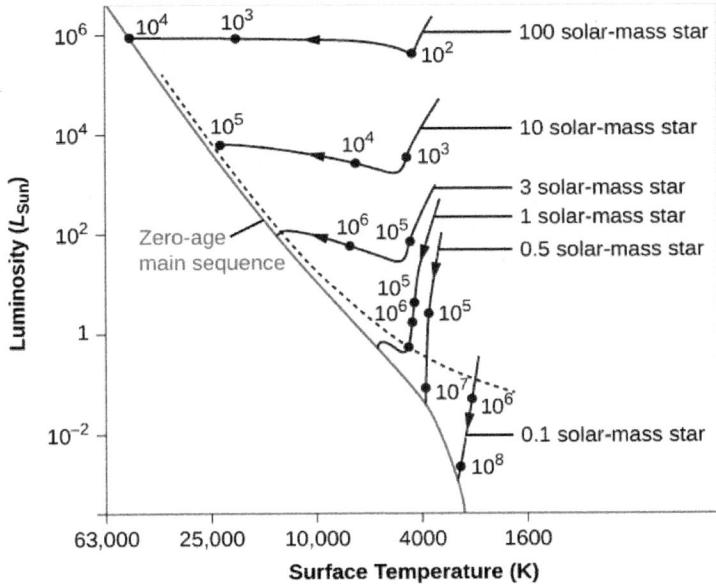

Figure 21.12. Evolutionary Tracks for Contracting Protostars. Tracks are plotted on the H–R diagram to show how stars of different masses change during the early parts of their lives. The number next to each dark point on a track is the rough number of years it takes an embryo star to reach that stage (the numbers are the result of computer models and are therefore not well known). Note that the surface temperature (K) on the horizontal axis increases toward the left. You can see that the more mass a star has, the shorter time it takes to go through each stage. Stars above the dashed line are typically still surrounded by infalling material and are hidden by it.

Initially, a protostar remains fairly cool with a very large radius and a very low density. It is transparent to infrared radiation, and the heat generated by gravitational contraction can be radiated away freely into space. Because heat builds up slowly inside the protostar, the gas pressure remains low, and the outer layers fall almost unhindered toward the center. Thus, the protostar undergoes very rapid collapse, a stage that corresponds to the roughly vertical lines at the right of Figure 21.12. As the star shrinks, its surface area gets smaller, and so its total luminosity decreases. The rapid contraction stops only when the protostar becomes dense and opaque enough to trap the heat released by gravitational contraction.

When the star begins to retain its heat, the contraction becomes much slower, and changes inside the contracting star keep the luminosity of stars like our Sun roughly constant. The surface temperatures start to build up, and the star "moves" to the left in the H–R diagram. Stars first become visible only after the stellar wind described earlier clears away the surrounding dust and gas. This can happen during the rapid-contraction phase for low-mass stars, but high-mass stars remain shrouded in dust until they end their early phase of gravitational contraction (see the dashed line in Figure 21.12).

To help you keep track of the various stages that stars go through in their lives, it can be useful to compare the development of a star to that of a human being. (Clearly, you will not find an exact correspondence, but thinking through the stages in human terms may help you remember some of the ideas we are trying to emphasize.) Protostars might be compared to human embryos—as yet unable to sustain themselves but drawing resources from their environment as they grow. Just as the birth of a child is the moment it is called upon to produce its own energy (through eating and breathing), so astronomers say that a star is born when it is able to sustain itself through nuclear reactions (by making its own energy.)

When the star's central temperature becomes high enough (about 10 million K) to fuse hydrogen into helium, we say that the star has reached the main sequence (a concept introduced in The Stars: A Celestial Census). It

Download for free at http://cnx.org/content/col11992/latest/

is now a full-fledged star, more or less in equilibrium, and its rate of change slows dramatically. Only the gradual depletion of hydrogen as it is transformed into helium in the core slowly changes the star's properties.

The mass of a star determines exactly where it falls on the main sequence. As Figure 21.12 shows, massive stars on the main sequence have high temperatures and high luminosities. Low-mass stars have low temperatures and low luminosities.

Objects of extremely low mass never achieve high-enough central temperatures to ignite nuclear reactions. The lower end of the main sequence stops where stars have a mass just barely great enough to sustain nuclear reactions at a sufficient rate to stop gravitational contraction. This critical mass is calculated to be about 0.075 times the mass of the Sun. As we discussed in the chapter on Analyzing Starlight, objects below this critical mass are called either brown dwarfs or planets. At the other extreme, the upper end of the main sequence terminates at the point where the energy radiated by the newly forming massive star becomes so great that it halts the accretion of additional matter. The upper limit of stellar mass is between 100 and 200 solar masses.

Evolutionary Timescales

How long it takes a star to form depends on its mass. The numbers that label the points on each track in Figure 21.12 are the times, in years, required for the embryo stars to reach the stages we have been discussing. Stars of masses much higher than the Sun's reach the main sequence in a few thousand to a million years. The Sun required millions of years before it was born. Tens of millions of years are required for stars of lower mass to evolve to the lower main sequence. (We will see that this turns out to be a general principle: massive stars go through *all* stages of evolution faster than low-mass stars do.)

We will take up the subsequent stages in the life of a star in Stars from Adolescence to Old Age, examining what happens after stars arrive in the main sequence and begin a "prolonged adolescence" and "adulthood" of fusing hydrogen to form helium. But now we want to examine the connection between the formation of stars and planets.

21.3 EVIDENCE THAT PLANETS FORM AROUND OTHER STARS

Learning Objectives

By the end of this section, you will be able to:

> Trace the evolution of dust surrounding a protostar, leading to the development of rocky planets and gas giants
> Estimate the timescale for growth of planets using observations of the disks surrounding young stars
> Evaluate evidence for planets around forming stars based on the structures seen in images of the circumstellar dust disks

Having developed on a planet and finding it essential to our existence, we have a special interest in how planets fit into the story of star formation. Yet planets outside the solar system are extremely difficult to detect. Recall that we see planets in our own system only because they reflect sunlight and are close by. When we look to the other stars, we find that the amount of light a planet reflects is a depressingly tiny fraction of the light its star gives off. Furthermore, from a distance, planets are lost in the glare of their much-brighter parent stars.

Disks around Protostars: Planetary Systems in Formation

It is a lot easier to detect the spread-out raw material from which planets might be assembled than to detect planets after they are fully formed. From our study of the solar system, we understand that planets form by the

This OpenStax book is available for free at http://cnx.org/content/col11992/1.8

gathering together of gas and dust particles in orbit around a newly created star. Each dust particle is heated by the young protostar and radiates in the infrared region of the spectrum. Before any planets form, we can detect such radiation from all of the spread-out individual dust particles that are destined to become parts of planets. We can also detect the silhouette of the disk if it blocks bright light coming from a source behind it (Figure 21.13).

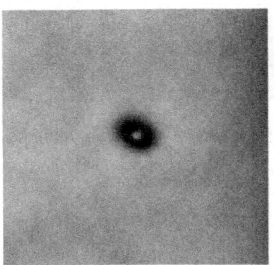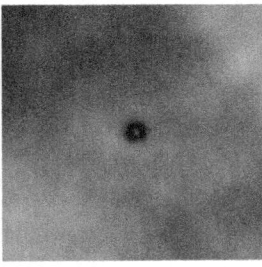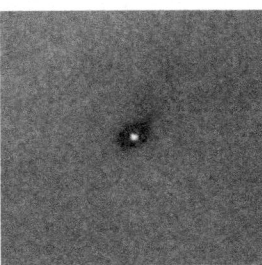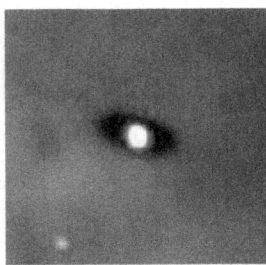

Figure 21.13. Disks around Protostars. These Hubble Space Telescope images show four disks around young stars in the Orion Nebula. The dark, dusty disks are seen silhouetted against the bright backdrop of the glowing gas in the nebula. The size of each image is about 30 times the diameter of our planetary system; this means the disks we see here range in size from two to eight times the orbit of Pluto. The red glow at the center of each disk is a young star, no more than a million years old. These images correspond to the stage in the life of a protostar shown in part (d) of Figure 21.8. (credit: modification of work by Mark McCaughrean (Max-Planck-Institute for Astronomy), C. Robert O'Dell (Rice University), and NASA)

Once the dust particles gather together and form a few planets (and maybe some moons), the overwhelming majority of the dust is hidden in the interiors of the planets where we cannot see it. All we can now detect is the radiation from the outside surfaces, which cover a drastically smaller area than the huge, dusty disk from which they formed. The amount of infrared radiation is therefore greatest before the dust particles combine into planets. For this reason, our search for planets begins with a search for infrared radiation from the material required to make them.

A disk of gas and dust appears to be an essential part of star formation. Observations show that nearly all very young protostars have disks and that the disks range in size from 10 to 1000 AU. (For comparison, the average diameter of the orbit of Pluto, which can be considered the rough size of our own planetary system, is 80 AU, whereas the outer diameter of the Kuiper belt of smaller icy bodies is about 100 AU.) The mass contained in these disks is typically 1–10% of the mass of our own Sun, which is more than the mass of all the planets in our solar system put together. Such observations already demonstrate that a large fraction of stars begin their lives with enough material in the right place to form a planetary system.

The Timing of Planet Formation and Growth

We can use observations of how the disks change with time to estimate how long it takes for planets to form. If we measure the temperature and luminosity of a protostar, then, as we saw, we can place it in an H–R diagram like the one shown in Figure 21.12. By comparing the real star with our models of how protostars should evolve with time, we can estimate its age. We can then look at how the disks we observe change with the ages of the stars that they surround.

What such observations show is that if a protostar is less than about 1 to 3 million years old, its disk extends all the way from very close to the surface of the star out to tens or hundreds of AU away. In older stars, we find disks with outer parts that still contain large amounts of dust, but the inner regions have lost most of their dust. In these objects, the disk looks like a donut, with the protostar centered in its hole. The inner, dense parts of most disks have disappeared by the time the stars are 10 million years old (Figure 21.14).

Download for free at http://cnx.org/content/col11992/latest/

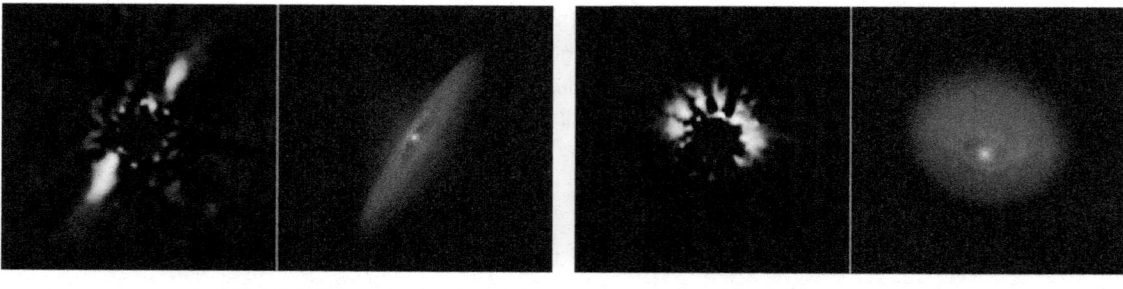

HD 141943   HD 191089

Figure 21.14. Protoplanetary Disks around Two Stars. The left view of each star shows infrared observations by the Hubble Space Telescope of their protoplanetary disks. The central star is much brighter than the surrounding disk, so the instrument includes a coronograph, which has a small shield that blocks the light of the central star but allows the surrounding disk to be imaged. The right image of each star shows models of the disks based on the observations. The star HD 141943 has an age of about 17 million years, while HD 191089 is about 12 million years old. (credit: modification of work by NASA, ESA, R. Soummer and M. Perrin (STScI), L. Pueyo (STScI/Johns Hopkins University), C. Chen and D. Golimowski (STScI), J.B. Hagan (STScI/Purdue University), T. Mittal (University of California, Berkeley/Johns Hopkins University), E. Choquet, M. Moerchen, and M. N'Diaye (STScI), A. Rajan (Arizona State University), S. Wolff (STScI/Purdue University), J. Debes and D. Hines (STScI), and G. Schneider (Steward Observatory/University of Arizona))

Calculations show that the formation of one or more planets could produce such a donut-like distribution of dust. Suppose a planet forms a few AU away from the protostar, presumably due to the gathering together of matter from the disk. As the planet grows in mass, the process clears out a dust-free region in its immediate neighborhood. Calculations also show that any small dust particles and gas that were initially located in the region between the protostar and the planet, and that are not swept up by the planet, will then fall onto the star very quickly in about 50,000 years.

Matter lying outside the planet's orbit, in contrast, is prevented from moving into the hole by the gravitational forces exerted by the planet. (We saw something similar in Saturn's rings, where the action of the shepherd moons keeps the material near the edge of the rings from spreading out.) If the formation of a planet is indeed what produces and sustains holes in the disks that surround very young stars, then planets must form in 3 to 30 million years. This is a short period compared with the lifetimes of most stars and shows that the formation of planets may be a quick byproduct of the birth of stars.

Calculations show that *accretion* can drive the rapid growth of planets—small, dust-grain-size particles orbiting in the disk collide and stick together, with the larger collections growing more rapidly as they attract and capture smaller ones. Once these clumps grow to about 10 centimeters in size or so, they enter a perilous stage in their development. At that size, unless they can grow to larger than about 100 meters in diameter, they are subject to drag forces produced by friction with the gas in the disk—and their orbits can rapidly decay, plunging them into the host star. Therefore, these bodies must rapidly grow to nearly 1 kilometer in size in diameter to avoid a fiery fate. At this stage, they are considered planetesimals (the small chunks of solid matter—ice and dust particles—that you learned about in Other Worlds: An Introduction to the Solar System). Once they survive to those sizes, the largest survivors will continue to grow by accreting smaller planetesimals; ultimately, this process results in a few large planets.

If the growing planets reach a mass bigger than about 10 times the mass of Earth, their gravity is strong enough to capture and hold on to hydrogen gas that remains in the disk. At that point, they will grow in mass and radius rapidly, reaching giant planet dimensions. However, to do so requires that the rapidly evolving central star hasn't yet driven away the gas in the disk with its increasingly vigorous wind (see the earlier section on Star Formation). From observations, we see that the disk can be blown away within 10 million years, so growth of a giant planet must also be a very fast process, astronomically speaking.

This OpenStax book is available for free at http://cnx.org/content/col11992/1.8

Download for free at http://cnx.org/content/col11992/latest/

Debris Disks and Shepherd Planets

The dust around newly formed stars is gradually either incorporated into the growing planets in the newly forming planetary system or ejected through gravitational interactions with the planets into space. The dust will disappear after about 30 million years unless the disk is continually supplied with new material. Local comets and asteroids are the most likely sources of new dust. As the planet-size bodies grow, they stir up the orbits of smaller objects in the area. These small bodies collide at high speeds, shatter, and produce tiny particles of silicate dust and ices that can keep the disk supplied with the debris from these collisions.

Over several hundred million years, the comets and asteroids will gradually be reduced in number, the frequency of collisions will go down, and the supply of fresh dust will diminish. Remember that the heavy bombardment in the early solar system ended when the Sun was only about 500 million years old. Observations show that the dusty "debris disks" around stars also become largely undetectable by the time the stars reach an age of 400 to 500 million years. It is likely, however, that some small amount of cometary material will remain in orbit, much like our Kuiper belt, a flattened disk of comets outside the orbit of Neptune.

In a young planetary system, even if we cannot see the planets directly, the planets can concentrate the dust particles into clumps and arcs that are much larger than the planets themselves and more easily imaged. This is similar to how the tiny moons of Saturn shepherd the particles in the rings and produce large arcs and structures in Saturn's rings.

Debris disks—many with just such clumps and arcs—have now been found around many stars, such as HL Tau, located about 450 light-years from Earth in the constellation Taurus (Figure 21.15). In some stars, the brightness of the rings varies with position; around other stars, there are bright arcs and gaps in the rings. The brightness indicates the relative concentration of dust, since what we are seeing is infrared (heat radiation) from the dust particles in the rings. More dust means more radiation.

Figure 21.15. Dust Ring around a Young Star. This image was made by ALMA (the Atacama Large Millimeter/Submillimeter Array) at a wavelength of 1.3 millimeters and shows the young star HL Tau and its protoplanetary disk. It reveals multiple rings and gaps that indicate the presence of emerging planets, which are sweeping their orbits clear of dust and gas. (credit: modification of work by ALMA (ESO/NAOJ/NRAO))

Download for free at http://cnx.org/content/col11992/latest/

LINK TO LEARNING

Watch a short video clip (https://openstaxcollege.org/l/30vidNRAOdir) of the director of NRAO (National Radio Astronomy Observatory) describing the high-resolution observations of the young star HL Tau. While you're there, watch an artist's animation of a protoplanetary disk to see newly formed planets traveling around a host (parent) star.

21.4 PLANETS BEYOND THE SOLAR SYSTEM: SEARCH AND DISCOVERY

Learning Objectives

By the end of this section, you will be able to:

> Describe the orbital motion of planets in our solar system using Kepler's laws
> Compare the indirect and direct observational techniques for exoplanet detection

For centuries, astronomers have dreamed of finding planets around other stars, including other planets like Earth. Direct observations of such distant planets are very difficult, however. You might compare a planet orbiting a star to a mosquito flying around one of those giant spotlights at a shopping center opening. From close up, you might spot the mosquito. But imagine viewing the scene from some distance away—say, from an airplane. You could see the spotlight just fine, but what are your chances of catching the mosquito in that light? Instead of making direct images, astronomers have relied on indirect observations and have now succeeded in detecting a multitude of planets around other stars.

In 1995, after decades of effort, we found the first such **exoplanet** (a planet outside our solar system) orbiting a main-sequence star, and today we know that most stars form with planets. This is an example of how persistence and new methods of observation advance the knowledge of humanity. By studying exoplanets, astronomers hope to better understand our solar system in context of the rest of the universe. For instance, how does the arrangement of our solar system compare to planetary systems in the rest of the universe? What do exoplanets tell us about the process of planet formation? And how does knowing the frequency of exoplanets influence our estimates of whether there is life elsewhere?

Searching for Orbital Motion

Most exoplanet detections are made using techniques where we observe the *effect* that the planet exerts on the host star. For example, the gravitational tug of an unseen planet will cause a small wobble in the host star. Or, if its orbit is properly aligned, a planet will periodically cross in front of the star, causing the brightness of the star to dim.

To understand how a planet can move its host star, consider a single Jupiter-like planet. Both the planet and the star actually revolve about their *common center of mass*. Remember that gravity is a mutual attraction. The star and the planet each exert a force on the other, and we can find a stable point, the center of mass, between them about which both objects move. The smaller the mass of a body in such a system, the larger its orbit. A massive star barely swings around the center of mass, while a low-mass planet makes a much larger "tour."

Suppose the planet is like Jupiter and has a mass about one-thousandth that of its star; in this case, the size of the star's orbit is one-thousandth the size of the planet's. To get a sense of how difficult observing such motion

Download for free at http://cnx.org/content/col11992/latest/

might be, let's see how hard Jupiter would be to detect in this way from the distance of a nearby star. Consider an alien astronomer trying to observe our own system from Alpha Centauri, the closest star system to our own (about 4.3 light-years away). There are two ways this astronomer could try to detect the orbital motion of the Sun. One way would be to look for changes in the Sun's position on the sky. The second would be to use the Doppler effect to look for changes in its velocity. Let's discuss each of these in turn.

The diameter of Jupiter's apparent orbit viewed from Alpha Centauri is 10 seconds of arc, and that of the Sun's orbit is 0.010 seconds of arc. (Remember, 1 second of arc is 1/3600 degree.) If they could measure the apparent position of the Sun (which is bright and easy to detect) to sufficient precision, they would describe an orbit of diameter 0.010 seconds of arc with a period equal to that of Jupiter, which is 12 years.

In other words, if they watched the Sun for 12 years, they would see it wiggle back and forth in the sky by this minuscule fraction of a degree. From the observed motion and the period of the "wiggle," they could deduce the mass of Jupiter and its distance using Kepler's laws. (To refresh your memory about these laws, see the chapter on Orbits and Gravity.)

Measuring positions in the sky this accurately is extremely difficult, and so far, astronomers have not made any confirmed detections of planets using this technique. However, we have been successful in using spectrometers to measure the changing velocity of stars with planets around them.

As the star and planet orbit each other, part of their motion will be in our line of sight (toward us or away from us). Such motion can be measured using the *Doppler effect* and the star's spectrum. As the star moves back and forth in orbit around the system's center of mass in response to the gravitational tug of an orbiting planet, the lines in its spectrum will shift back and forth.

Let's again consider the example of the Sun. Its *radial velocity* (motion toward or away from us) changes by about 13 meters per second with a period of 12 years because of the gravitational pull of Jupiter. This corresponds to about 30 miles per hour, roughly the speed at which many of us drive around town. Detecting motion at this level in a star's spectrum presents an enormous technical challenge, but several groups of astronomers around the world, using specialized spectrographs designed for this purpose, have succeeded. Note that the change in speed does not depend on the distance of the star from the observer. Using the Doppler effect to detect planets will work at any distance, as long as the star is bright enough to provide a good spectrum and a large telescope is available to make the observations (Figure 21.16).

Download for free at http://cnx.org/content/col11992/latest/

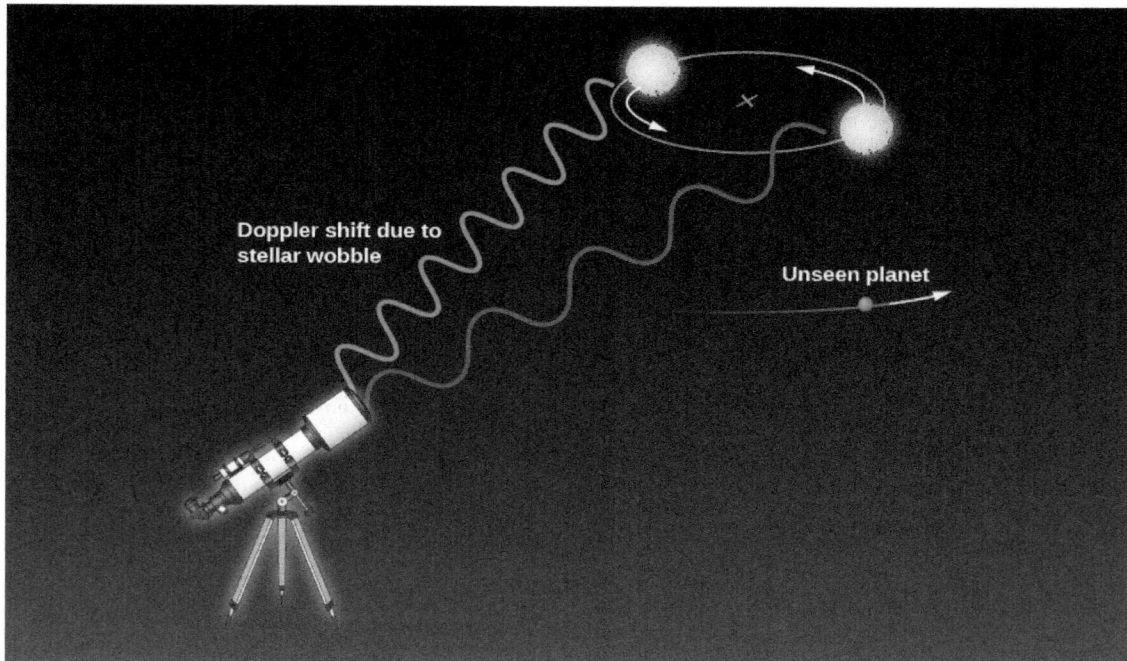

Figure 21.16. Doppler Method of Detecting Planets. The motion of a star around a common center of mass with an orbiting planet can be detected by measuring the changing speed of the star. When the star is moving away from us, the lines in its spectrum show a tiny redshift; when it is moving toward us, they show a tiny blueshift. The change in color (wavelength) has been exaggerated here for illustrative purposes. In reality, the Doppler shifts we measure are extremely small and require sophisticated equipment to be detected.

The first successful use of the Doppler effect to find a planet around another star was in 1995. Michel Mayor and Didier Queloz of the Geneva Observatory (Figure 21.17) used this technique to find a planet orbiting a star resembling our Sun called 51 Pegasi, about 40 light-years away. (The star can be found in the sky near the great square of Pegasus, the flying horse of Greek mythology, one of the easiest-to-find star patterns.) To everyone's surprise, the planet takes a mere 4.2 days to orbit around the star. (Remember that Mercury, the innermost planet in our solar system, takes 88 days to go once around the Sun, so 4.2 days seems fantastically short.)

Figure 21.17. Planet Discoverers. In 1995, Didier Queloz and Michel Mayor of the Geneva Observatory were the first to discover a planet around a regular star (51 Pegasi). They are seen here at an observatory in Chile where they are continuing their planet hunting. (credit: Weinstein/Ciel et Espace Photos)

This OpenStax book is available for free at http://cnx.org/content/col11992/1.8

Download for free at http://cnx.org/content/col11992/latest/

Mayor and Queloz's findings mean the planet must be very close to 51 Pegasi, circling it about 7 million kilometers away (Figure 21.18). At that distance, the energy of the star should heat the planet's surface to a temperature of a few thousand degrees Celsius (a bit hot for future tourism). From its motion, astronomers calculate that it has at least half the mass of Jupiter[1], making it clearly a jovian and not a terrestrial-type planet.

Figure 21.18. Hot Jupiter. Artist Greg Bacon painted this impression of a hot, Jupiter-type planet orbiting close to a sunlike star. The artist shows bands on the planet like Jupiter, but we only estimate the mass of most hot, Jupiter-type planets from the Doppler method and don't know what conditions on the planet are like. (credit: ESO)

Since that initial planet discovery, the rate of progress has been breathtaking. Hundreds of giant planets have been discovered using the Doppler technique. Many of these giant planets are orbiting close to their stars—astronomers have called these *hot Jupiters*.

The existence of giant planets so close to their stars was a surprise, and these discoveries have forced us to rethink our ideas about how planetary systems form. But for now, bear in mind that the Doppler-shift method—which relies on the pull of a planet making its star "wiggle" back and forth around the center of mass—is most effective at finding planets that are both close to their stars and massive. These planets cause the biggest "wiggles" in the motion of their stars and the biggest Doppler shifts in the spectrum. Plus, they will be found sooner, since astronomers like to monitor the star for at least one full orbit (and perhaps more) and hot Jupiters take the shortest time to complete their orbit.

So if such planets exist, we would expect to be finding this type first. Scientists call this a *selection effect*—where our technique of discovery selects certain kinds of objects as "easy finds." As an example of a selection effect in everyday life, imagine you decide you are ready for a new romantic relationship in your life. To begin with, you only attend social events on campus, all of which require a student ID to get in. Your selection of possible partners will then be limited to students at your college. That may not give you as diverse a group to choose from as you want. In the same way, when we first used the Doppler technique, it selected massive planets close

---

1   The Doppler method only allows us to find the minimum mass of a planet. To determine the exact mass using the Doppler shift and Kepler's laws, we must also have the angle at which the planet's orbit is oriented to our view—something we don't have any independent way of knowing in most cases. Still, if the minimum mass is half of Jupiter's, the actual mass can only be larger than that, and we are sure that we are dealing with a jovian planet.

Download for free at http://cnx.org/content/col11992/latest/

to their stars as the most likely discoveries. As we spend longer times watching target stars and as our ability to measure smaller Doppler shifts improves, this technique can reveal more distant and less massive planets too.

LINK TO LEARNING

View a series of animations (https://openstaxcollege.org/l/30keplawsolarani) demonstrating solar system motion and Kepler's laws, and select animation 1 (Kepler's laws) from the dropdown playlist. To view an animation demonstrating the radial velocity curve for an exoplanet, select animation 29 (radial velocity curve for an exoplanet) and animation 30 (radial velocity curve for an exoplanet—elliptical orbit) from the dropdown playlist.

Transiting Planets

The second method for indirect detection of exoplanets is based not on the motion of the star but on its brightness. When the orbital plane of the planet is tilted or inclined so that it is viewed edge-on, we will see the planet cross in front of the star once per orbit, causing the star to dim slightly; this event is known as **transit**. Figure 21.19 shows a sketch of the transit at three time steps: (1) out of transit, (2) the start of transit, and (3) full transit, along with a sketch of the light curve, which shows the drop in the brightness of the host star. The amount of light blocked—the depth of the transit—depends on the area of the planet (its size) compared to the star. If we can determine the size of the star, the transit method tells us the size of the planet.

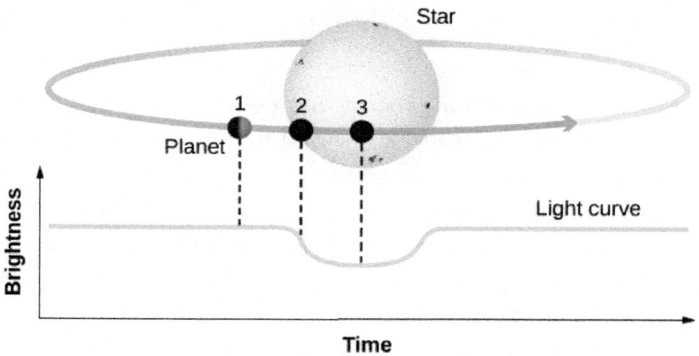

Figure 21.19. Planet Transits. As the planet transits, it blocks out some of the light from the star, causing a temporary dimming in the brightness of the star. The top figure shows three moments during the transit event and the bottom panel shows the corresponding light curve: (1) out of transit, (2) transit ingress, and (3) the full drop in brightness.

The interval between successive transits is the length of the year for that planet, which can be used (again using Kepler's laws) to find its distance from the star. Larger planets like Jupiter block out more starlight than small earthlike planets, making transits by giant planets easier to detect, even from ground-based observatories. But by going into space, above the distorting effects of Earth's atmosphere, the transit technique has been extended to exoplanets as small as Mars.

This OpenStax book is available for free at http://cnx.org/content/col11992/1.8

Download for free at http://cnx.org/content/col11992/latest/

EXAMPLE 21.1

Transit Depth

In a transit, the planet's circular disk blocks the light of the star's circular disk. The area of a circle is πR^2. The amount of light the planet blocks, called the transit depth, is then given by

$$\frac{\pi R^2_{\text{planet}}}{\pi R^2_{\text{star}}} = \frac{R^2_{\text{planet}}}{R^2_{\text{star}}} = \left(\frac{R_{\text{planet}}}{R_{\text{star}}}\right)^2$$

Now calculate the transit depth for a star the size of the Sun with a gas giant planet the size of Jupiter.

Solution

The radius of Jupiter is 71,400 km, while the radius of the Sun is 695,700 km. Substituting into the equation, we get $\left(\frac{R_{\text{planet}}}{R_{\text{star}}}\right)^2 = \left(\frac{71,400 \text{ km}}{695,700 \text{ km}}\right)^2 = 0.01$ or 1%, which can easily be detected with the instruments on board the Kepler spacecraft.

Check Your Learning

What is the transit depth for a star half the size of the Sun with a much smaller planet, like the size of Earth?

Answer:

The radius of Earth is 6371 km. Therefore,

$$\left(\frac{R_{\text{planet}}}{R_{\text{star}}}\right)^2 = \left(\frac{6371 \text{ km}}{695,700/2 \text{ km}}\right)^2 = \left(\frac{6371 \text{ km}}{347,850 \text{ km}}\right)^2 = 0.0003\text{, or significantly less than 1%.}$$

The Doppler method allows us to estimate the mass of a planet. If the same object can be studied by both the Doppler and transit techniques, we can measure both the mass and the size of the exoplanet. This is a powerful combination that can be used to derive the average density (mass/volume) of the planet. In 1999, using measurements from ground-based telescopes, the first transiting planet was detected orbiting the star HD 209458. The planet transits its parent star for about 3 hours every 3.5 days as we view it from Earth. Doppler measurements showed that the planet around HD 209458 has about 70% the mass of Jupiter, but its radius is about 35% larger than Jupiter's. This was the first case where we could determine what an exoplanet was made of—with that mass and radius, HD 209458 must be a gas and liquid world like Jupiter or Saturn.

It is even possible to learn something about the planet's atmosphere. When the planet passes in front of HD 209458, the atoms in the planet's atmosphere absorb starlight. Observations of this absorption were first made at the wavelengths of yellow sodium lines and showed that the atmosphere of the planet contains sodium; now, other elements can be measured as well.

Download for free at http://cnx.org/content/col11992/latest/

LINK TO LEARNING

Try a transit simulator (https://openstaxcollege.org/l/30transimul) that demonstrates how a planet passing in front of its parent star can lead to the planet's detection. Follow the instructions to run the animation on your computer.

Transiting planets reveal such a wealth of information that the French Space Agency (CNES) and the European Space Agency (ESA) launched the CoRoT space telescope in 2007 to detect transiting exoplanets. CoRoT discovered 32 transiting exoplanets, including the first transiting planet with a size and density similar to Earth. In 2012, the spacecraft suffered an onboard computer failure, ending the mission. Meanwhile, NASA built a much more powerful transit observatory called Kepler.

In 2009, NASA launched the Kepler space telescope, dedicated to the discovery of transiting exoplanets. This spacecraft stared continuously at more than 150,000 stars in a small patch of sky near the constellation of Cygnus—just above the plane of our Milky Way Galaxy (Figure 21.20). Kepler's cameras and ability to measure small changes in brightness very precisely enabled the discovery of thousands of exoplanets, including many multi-planet systems. The spacecraft required three reaction wheels—a type of wheel used to help control slight rotation of the spacecraft—to stabilize the pointing of the telescope and monitor the brightness of the same group of stars over and over again. Kepler was launched with four reaction wheels (one a spare), but by May 2013, two wheels had failed and the telescope could no longer be accurately pointed toward the target area. Kepler had been designed to operate for 4 years, and ironically, the pointing failure occurred exactly 4 years and 1 day after it began observing.

This OpenStax book is available for free at http://cnx.org/content/col11992/1.8

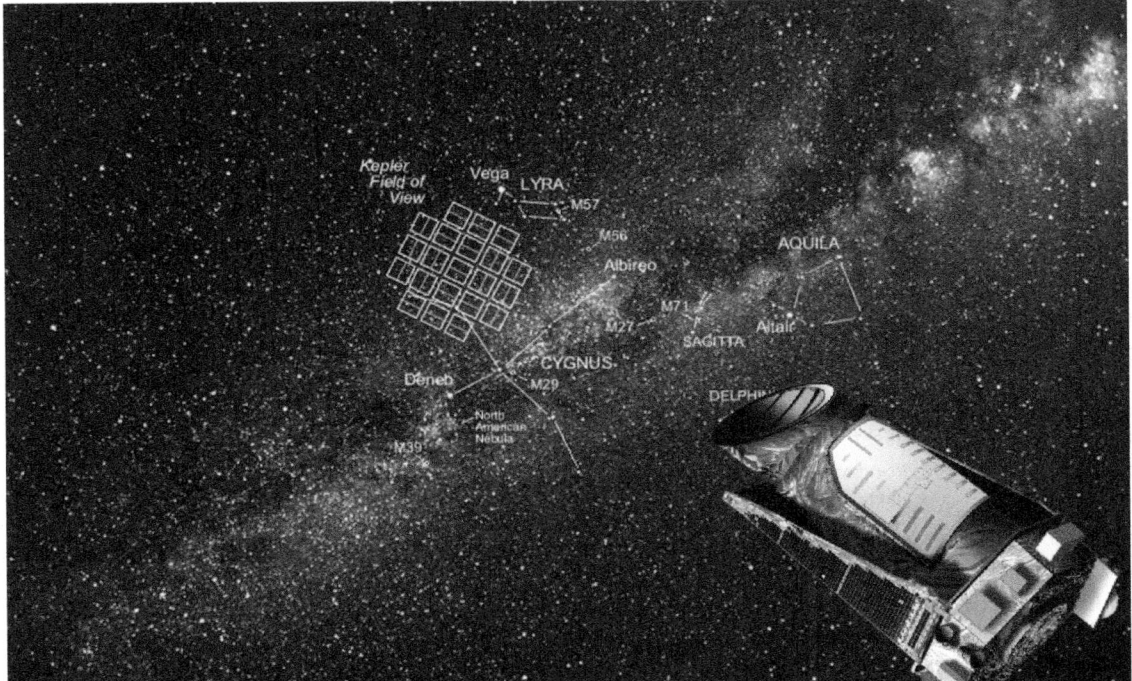

Figure 21.20. Kepler's Field of View. The boxes show the region where the Kepler spacecraft cameras took images of over 150,000 stars regularly, to find transiting planets. (credit "field of view": modification of work by NASA/Kepler mission; credit "spacecraft": modification of work by NASA/Kepler mission/Wendy Stenzel)

What do we mean, exactly, by "discovery" of transiting exoplanets? A single transit shows up as a very slight drop in the brightness of the star, lasting several hours. However, astronomers must be on guard against other factors that might produce a false transit, especially when working at the limit of precision of the telescope. We must wait for a second transit of similar depth. But when another transit is observed, we don't initially know whether it might be due to another planet in a different orbit. The "discovery" occurs only when a third transit is found with similar depth and the same spacing in time as the first pair.

Computers normally conduct the analysis, which involves searching for tiny, periodic dips in the light from each star, extending over 4 years of observation. But the Kepler mission also has a program in which non-astronomers—citizen scientists—can examine the data. These dedicated volunteers have found several transits that were missed by the computer analyses, showing that the human eye and brain sometimes recognize unusual events that a computer was not programmed to look for.

Measuring three or four evenly spaced transits is normally enough to "discover" an exoplanet. But in a new field like exoplanet research, we would like to find further independent verification. The strongest confirmation happens when ground-based telescopes are also able to detect a Doppler shift with the same period as the transits. However, this is generally not possible for Earth-size planets. One of the most convincing ways to verify that a dip in brightness is due to a planet is to find more planets orbiting the same star—a *planetary system*. Multi-planet systems also provide alternative ways to estimate the masses of the planets, as we will discuss in the next section.

The selection effects (or biases) in the Kepler data are similar to those in Doppler observations. Large planets are easier to find than small ones, and short-period planets are easier than long-period planets. If we require three transits to establish the presence of a planet, we are of course limited to discovering planets with orbital

Download for free at http://cnx.org/content/col11992/latest/

periods less than one-third of the observing interval. Thus, it was only in its fourth and final year of operation that Kepler was able to find planets with orbits like Earth's that require 1 year to go around their star.

Direct Detection

The best possible evidence for an earthlike planet elsewhere would be an image. After all, "seeing is believing" is a very human prejudice. But imaging a distant planet is a formidable challenge indeed. Suppose, for example, you were a great distance away and wished to detect reflected light from Earth. Earth intercepts and reflects less than one billionth of the Sun's radiation, so its apparent brightness in visible light is less than one billionth that of the Sun. Compounding the challenge of detecting such a faint speck of light, the planet is swamped by the blaze of radiation from its parent star.

Even today, the best telescope mirrors' optics have slight imperfections that prevent the star's light from coming into focus in a completely sharp point.

Direct imaging works best for young gas giant planets that emit infrared light and reside at large separations from their host stars. Young giant planets emit more infrared light because they have more internal energy, stored from the process of planet formation. Even then, clever techniques must be employed to subtract out the light from the host star. In 2008, three such young planets were discovered orbiting HR 8799, a star in the constellation of Pegasus (Figure 21.21). Two years later, a fourth planet was detected closer to the star. Additional planets may reside even closer to HR 8799, but if they exist, they are currently lost in the glare of the star.

Since then, a number of planets around other stars have been found using direct imaging. However, one challenge is to tell whether the objects we are seeing are indeed planets or if they are brown dwarfs (failed stars) in orbit around a star.

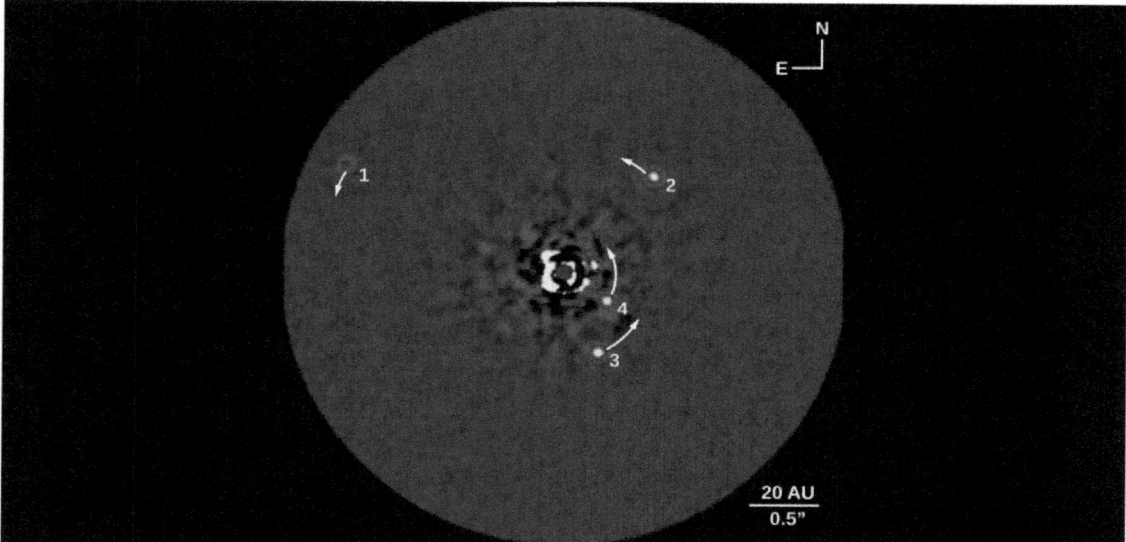

Figure 21.21. Exoplanets around HR 8799. This image shows Keck telescope observations of four directly imaged planets orbiting HR 8799. A size scale for the system gives the distance in AU (remember that one astronomical unit is the distance between Earth and the Sun.) (credit: modification of work by Ben Zuckerman)

Direct imaging is an important technique for characterizing an exoplanet. The brightness of the planet can be measured at different wavelengths. These observations provide an estimate for the temperature of the planet's atmosphere; in the case of HR 8799 planet 1, the color suggests the presence of thick clouds. Spectra can also be

This OpenStax book is available for free at http://cnx.org/content/col11992/1.8

obtained from the faint light to analyze the atmospheric constituents. A spectrum of HR 8799 planet 1 indicates a hydrogen-rich atmosphere, while the closer planet 4 shows evidence for methane in the atmosphere.

Another way to overcome the blurring effect of Earth's atmosphere is to observe from space. Infrared may be the optimal wavelength range in which to observe because planets get brighter in the infrared while stars like our Sun get fainter, thereby making it easier to detect a planet against the glare of its star. Special optical techniques can be used to suppress the light from the central star and make it easier to see the planet itself. However, even if we go into space, it will be difficult to obtain images of Earth-size planets.

21.5 | EXOPLANETS EVERYWHERE: WHAT WE ARE LEARNING

Learning Objectives

By the end of this section, you will be able to:

> Explain what we have learned from our discovery of exoplanets
> Identify which kind of exoplanets appear to be the most common in the Galaxy
> Discuss the kinds of planetary systems we are finding around other stars

Before the discovery of exoplanets, most astronomers expected that other planetary systems would be much like our own—planets following roughly circular orbits, with the most massive planets several AU from their parent star. Such systems do exist in large numbers, but many exoplanets and planetary systems are very different from those in our solar system. Another surprise is the existence of whole classes of exoplanets that we simply don't have in our solar system: planets with masses between the mass of Earth and Neptune, and planets that are several times more massive than Jupiter.

Kepler Results

The Kepler telescope has been responsible for the discovery of most exoplanets, especially at smaller sizes, as illustrated in Figure 21.22, where the Kepler discoveries are plotted in yellow. You can see the wide range of sizes, including planets substantially larger than Jupiter and smaller than Earth. The absence of Kepler-discovered exoplanets with orbital periods longer than a few hundred days is a consequence of the 4-year lifetime of the mission. (Remember that three evenly spaced transits must be observed to register a discovery.) At the smaller sizes, the absence of planets much smaller than one earth radius is due to the difficulty of detecting transits by very small planets. In effect, the "discovery space" for Kepler was limited to planets with orbital periods less than 400 days and sizes larger than Mars.

Download for free at http://cnx.org/content/col11992/latest/

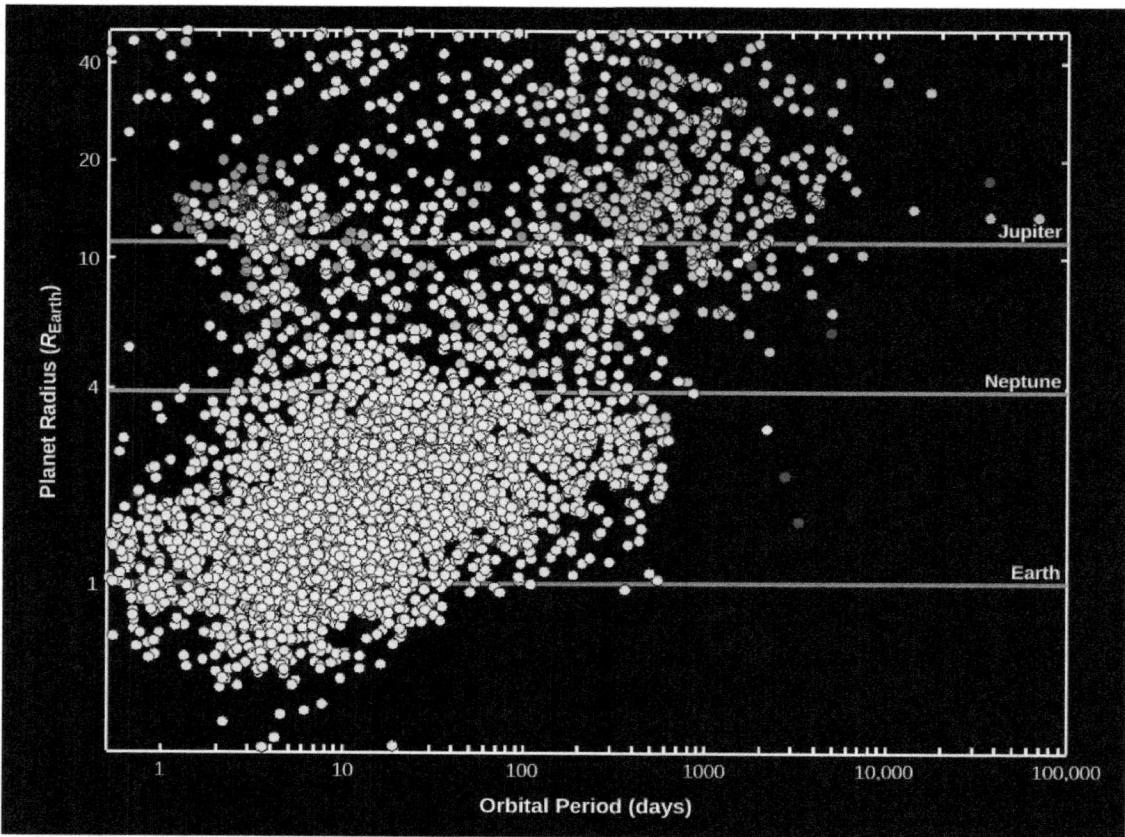

Figure 21.22. Exoplanet Discoveries through 2015. The vertical axis shows the radius of each planet compared to Earth. Horizontal lines show the size of Earth, Neptune, and Jupiter. The horizontal axis shows the time each planet takes to make one orbit (and is given in Earth days). Recall that Mercury takes 88 days and Earth takes a little more than 365 days to orbit the Sun. The yellow and red dots show planets discovered by transits, and the blue dots are the discoveries by the radial velocity (Doppler) technique. (credit: modification of work by NASA/Kepler mission)

One of the primary objectives of the Kepler mission was to find out how many stars hosted planets and especially to estimate the frequency of earthlike planets. Although Kepler looked at only a very tiny fraction of the stars in the Galaxy, the sample size was large enough to draw some interesting conclusions. While the observations apply only to the stars observed by Kepler, those stars are reasonably representative, and so astronomers can extrapolate to the entire Galaxy.

Figure 21.23 shows that the Kepler discoveries include many rocky, Earth-size planets, far more than Jupiter-size gas planets. This immediately tells us that the initial Doppler discovery of many hot Jupiters was a biased sample, in effect, finding the odd planetary systems because they were the easiest to detect. However, there is one huge difference between this observed size distribution and that of planets in our solar system. The most common planets have radii between 1.4 and 2.8 that of Earth, sizes for which we have no examples in the solar system. These have been nicknamed **super-Earths**, while the other large group with sizes between 2.8 and 4 that of Earth are often called **mini-Neptunes**.

This OpenStax book is available for free at http://cnx.org/content/col11992/1.8

Download for free at http://cnx.org/content/col11992/latest/

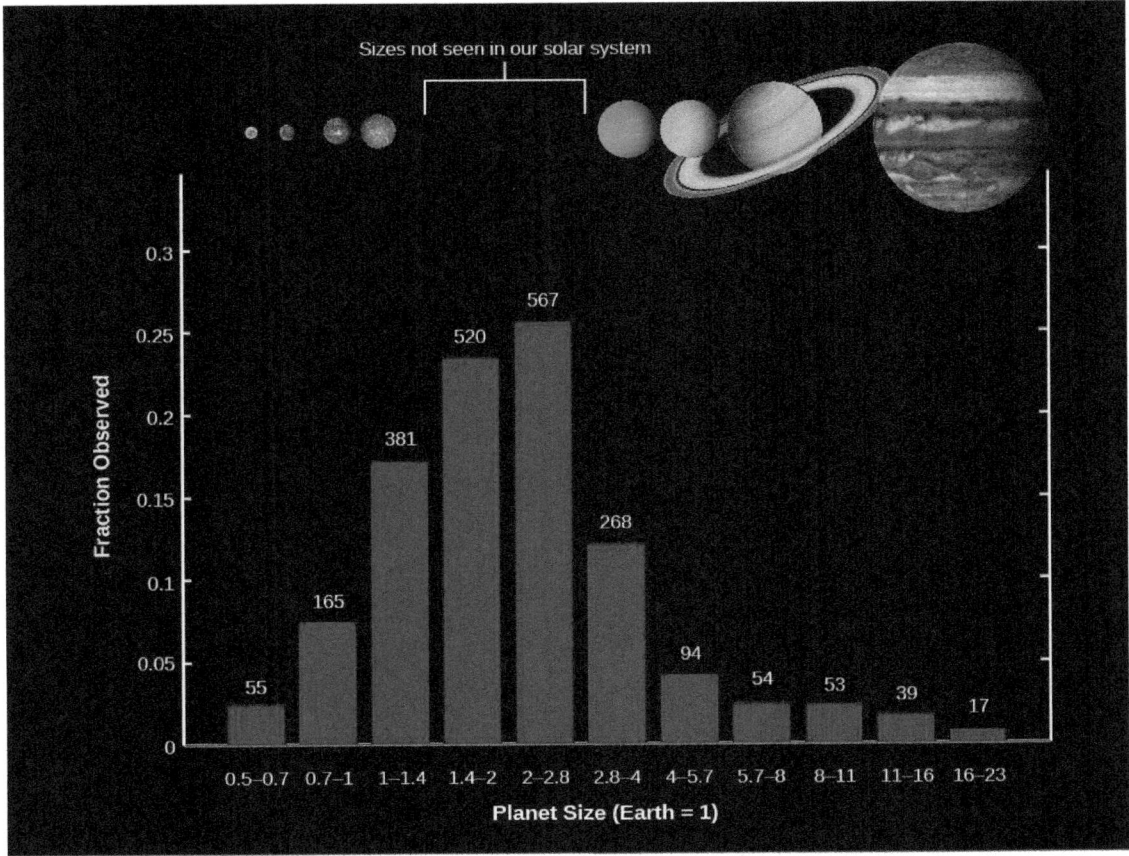

Figure 21.23. Kepler Discoveries. This bar graph shows the number of planets of each size range found among the first 2213 Kepler planet discoveries. Sizes range from half the size of Earth to 20 times that of Earth. On the vertical axis, you can see the fraction that each size range makes up of the total. Note that planets that are between 1.4 and 4 times the size of Earth make up the largest fractions, yet this size range is not represented among the planets in our solar system. (credit: modification of work by NASA/Kepler mission)

What a remarkable discovery it is that the most common types of planets in the Galaxy are completely absent from our solar system and were unknown until Kepler's survey. However, recall that really small planets were difficult for the Kepler instruments to find. So, to estimate the frequency of Earth-size exoplanets, we need to correct for this sampling bias. The result is the corrected size distribution shown in Figure 21.24. Notice that in this graph, we have also taken the step of showing not the number of Kepler detections but the average number of planets per star for solar-type stars (spectral types F, G, and K).

Download for free at http://cnx.org/content/col11992/latest/

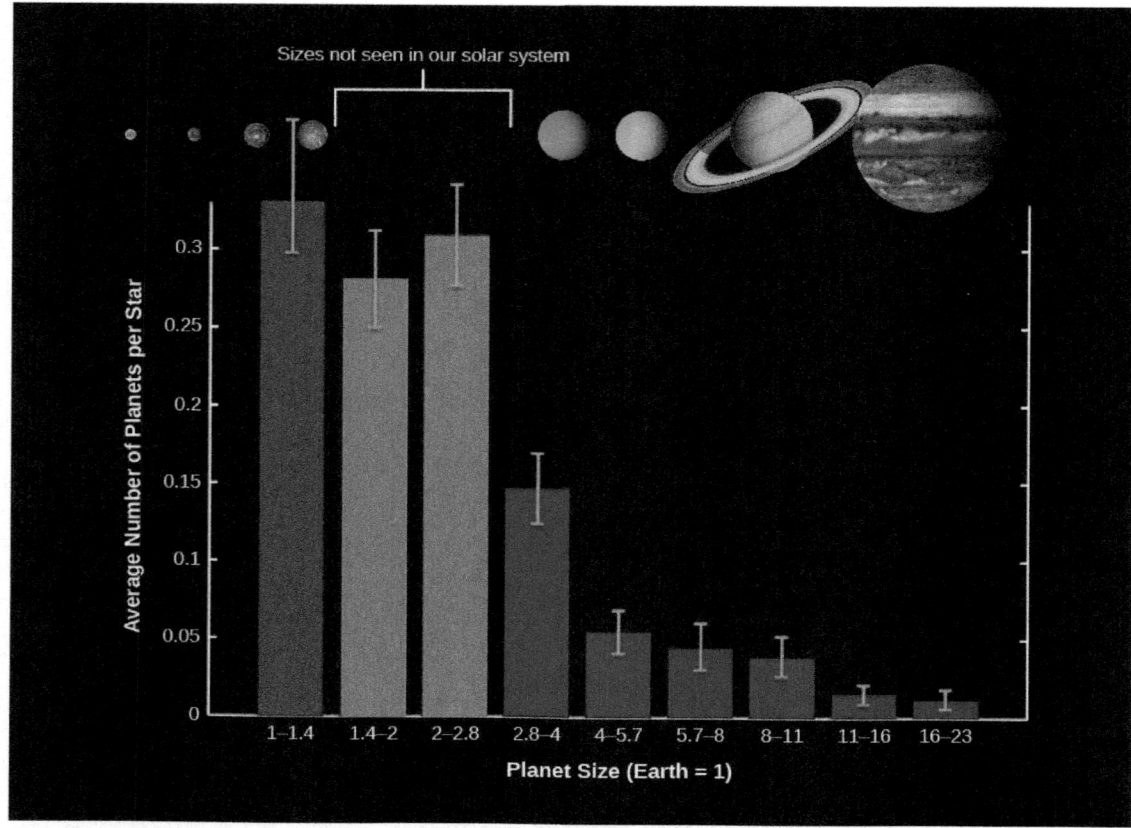

Figure 21.24. Size Distribution of Planets for Stars Similar to the Sun. We show the average number of planets per star in each planet size range. (The average is less than one because some stars will have zero planets of that size range.) This distribution, corrected for biases in the Kepler data, shows that Earth-size planets may actually be the most common type of exoplanets. (credit: modification of work by NASA/Kepler mission)

We see that the most common planet sizes of are those with radii from 1 to 3 times that of Earth—what we have called "Earths" and "super-Earths." Each group occurs in about one-third to one-quarter of stars. In other words, if we group these sizes together, we can conclude there is nearly one such planet per star! And remember, this census includes primarily planets with orbital periods less than 2 years. We do not yet know how many undiscovered planets might exist at larger distances from their star.

To estimate the number of Earth-size planets in our Galaxy, we need to remember that there are approximately 100 billion stars of spectral types F, G, and K. Therefore, we estimate that there are about 30 billion Earth-size planets in our Galaxy. If we include the super-Earths too, then there could be one hundred billion in the whole Galaxy. This idea—that planets of roughly Earth's size are so numerous—is surely one of the most important discoveries of modern astronomy.

Planets with Known Densities

For several hundred exoplanets, we have been able to measure both the size of the planet from transit data and its mass from Doppler data, yielding an estimate of its density. Comparing the average density of exoplanets to the density of planets in our solar system helps us understand whether they are rocky or gaseous in nature. This has been particularly important for understanding the structure of the new categories of super-Earths and mini-Neptunes with masses between 3–10 times the mass of Earth. A key observation so far is that planets

This OpenStax book is available for free at http://cnx.org/content/col11992/1.8

Download for free at http://cnx.org/content/col11992/latest/

that are more than 10 times the mass of Earth have substantial gaseous envelopes (like Uranus and Neptune) whereas lower-mass planets are predominately rocky in nature (like the terrestrial planets).

Figure 21.25 compares all the exoplanets that have both mass and radius measurements. The dependence of the radius on planet mass is also shown for a few illustrative cases—hypothetical planets made of pure iron, rock, water, or hydrogen.

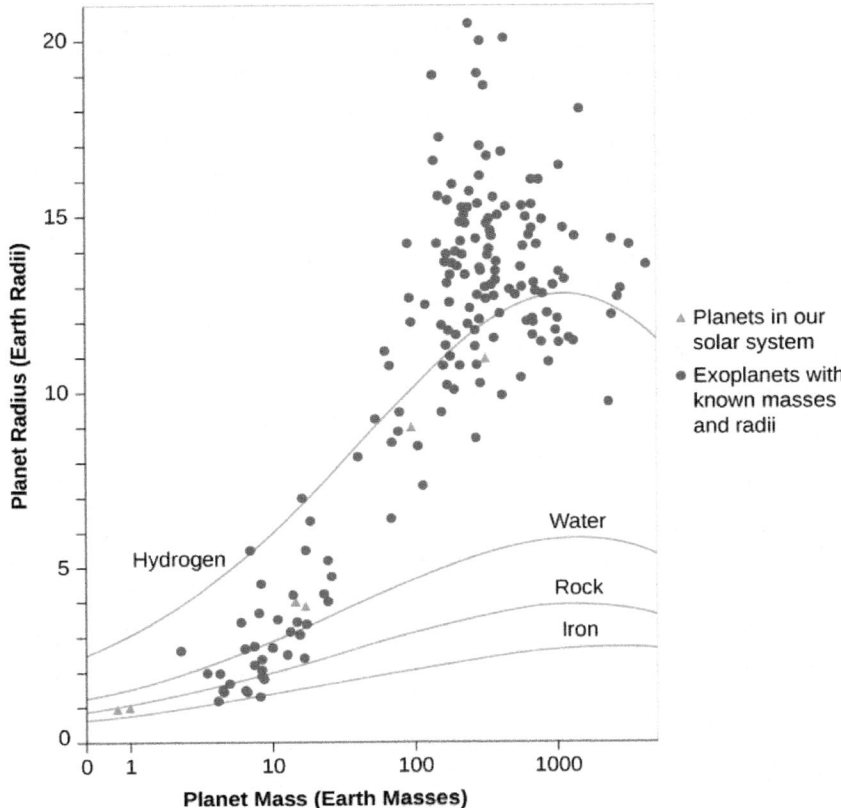

Figure 21.25. Exoplanets with Known Densities. Exoplanets with known masses and radii (red circles) are plotted along with solid lines that show the theoretical size of pure iron, rock, water, and hydrogen planets with increasing mass. Masses are given in multiples of Earth's mass. (For comparison, Jupiter contains enough mass to make 320 Earths.) The green triangles indicate planets in our solar system.

At lower masses, notice that as the mass of these hypothetical planets increases, the radius also increases. That makes sense—if you were building a model of a planet out of clay, your toy planet would increase in size as you added more clay. However, for the highest mass planets ($M > 1000\ M_{Earth}$) in Figure 21.25, notice that the radius stops increasing and the planets with greater mass are actually smaller. This occurs because increasing the mass also increases the gravity of the planet, so that compressible materials (even rock is compressible) will become more tightly packed, shrinking the size of the more massive planet.

In reality, planets are not pure compositions like the hypothetical water or iron planet. Earth is composed of a solid iron core, an outer liquid-iron core, a rocky mantle and crust, and a relatively thin atmospheric layer. Exoplanets are similarly likely to be differentiated into compositional layers. The theoretical lines in Figure 21.25 are simply guides that suggest a range of possible compositions.

Astronomers who work on the complex modeling of the interiors of rocky planets make the simplifying assumption that the planet consists of two or three layers. This is not perfect, but it is a reasonable

Download for free at http://cnx.org/content/col11992/latest/

approximation and another good example of how science works. Often, the first step in understanding something new is to narrow down the range of possibilities. This sets the stage for refining and deepening our knowledge. In Figure 21.25, the two green triangles with roughly 1 M_{Earth} and 1 R_{Earth} represent Venus and Earth. Notice that these planets fall between the models for a pure iron and a pure rock planet, consistent with what we would expect for the known mixed-chemical composition of Venus and Earth.

In the case of gaseous planets, the situation is more complex. Hydrogen is the lightest element in the periodic table, yet many of the detected exoplanets in Figure 21.25 with masses greater than 100 M_{Earth} have radii that suggest they are lower in density than a pure hydrogen planet. Hydrogen is the lightest element, so what is happening here? Why do some gas giant planets have inflated radii that are larger than the fictitious pure hydrogen planet? Many of these planets reside in short-period orbits close to the host star where they intercept a significant amount of radiated energy. If this energy is trapped deep in the planet atmosphere, it can cause the planet to expand.

Planets that orbit close to their host stars in slightly eccentric orbits have another source of energy: the star will raise tides in these planets that tend to circularize the orbits. This process also results in tidal dissipation of energy that can inflate the atmosphere. It would be interesting to measure the size of gas giant planets in wider orbits where the planets should be cooler—the expectation is that unless they are very young, these cooler gas giant exoplanets (sometimes called "cold Jupiters") should not be inflated. But we don't yet have data on these more distant exoplanets.

Exoplanetary Systems

As we search for exoplanets, we don't expect to find only one planet per star. Our solar system has eight major planets, half a dozen dwarf planets, and millions of smaller objects orbiting the Sun. The evidence we have of planetary systems in formation also suggest that they are likely to produce multi-planet systems.

The first planetary system was found around the star Upsilon Andromedae in 1999 using the Doppler method, and many others have been found since then (about 2600 as of 2016). If such exoplanetary system are common, let's consider which systems we expect to find in the Kepler transit data.

A planet will transit its star only if Earth lies in the plane of the planet's orbit. If the planets in other systems do not have orbits in the same plane, we are unlikely to see multiple transiting objects. Also, as we have noted before, Kepler was sensitive only to planets with orbital periods less than about 4 years. What we expect from Kepler data, then, is evidence of coplanar planetary systems confined to what would be the realm of the terrestrial planets in our solar system.

In fact, today we have data on about 2600 such exoplanet systems. Many have only two known planets, but a few have as many as five. For the most part, these are very compact systems with most of their planets closer to their star than Mercury is to the Sun. The figure below shows one of the largest exoplanet systems: that of the star called Kepler-62 (Figure 21.26). Our solar system is shown to the same scale, for comparison.

This OpenStax book is available for free at http://cnx.org/content/col11992/1.8

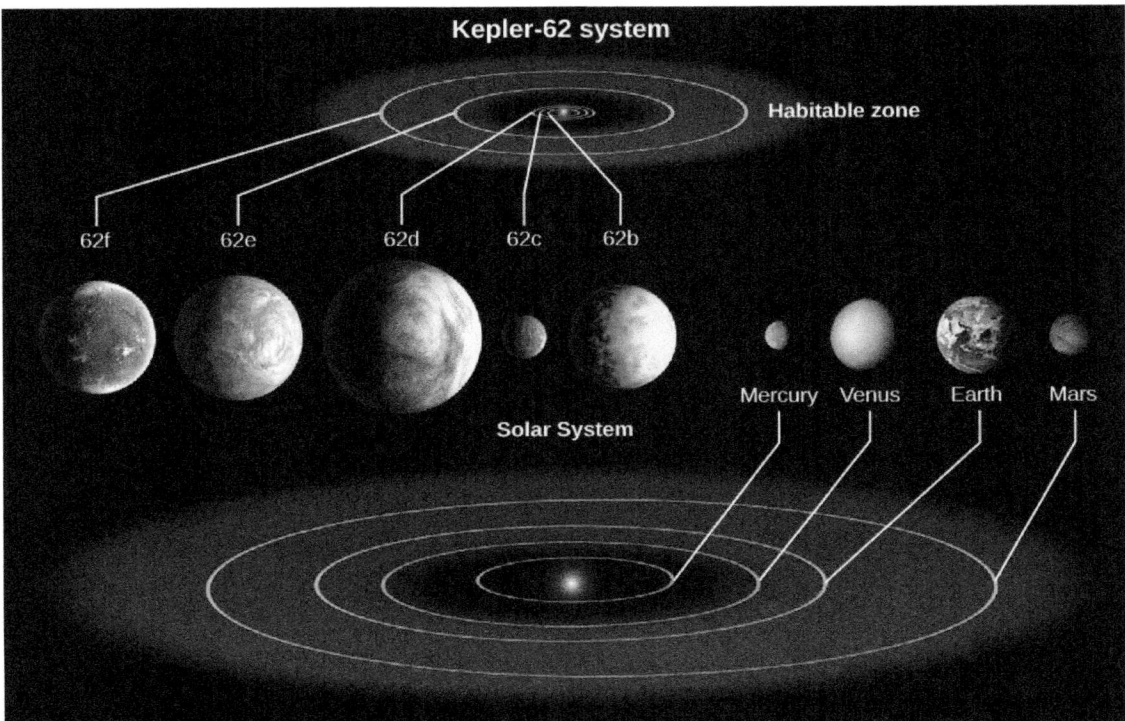

Figure 21.26. Exoplanet System Kepler-62, with the Solar System Shown to the Same Scale. The green areas are the "habitable zones," the range of distance from the star where surface temperatures are likely to be consistent with liquid water. (credit: modification of work by NASA/ Ames/JPL-Caltech)

All but one of the planets in the K-62 system are larger than Earth. These are super-Earths, and one of them (62d) is in the size range of a mini-Neptune, where it is likely to be largely gaseous. The smallest planet in this system is about the size of Mars. The three inner planets orbit very close to their star, and only the outer two have orbits larger than Mercury in our system. The green areas represent each star's "habitable zone," which is the distance from the star where we calculate that surface temperatures would be consistent with liquid water. The Kepler-62 habitable zone is much smaller than that of the Sun because the star is intrinsically fainter.

With closely spaced systems like this, the planets can interact gravitationally with each other. The result is that the observed transits occur a few minutes earlier or later than would be predicted from simple orbits. These gravitational interactions have allowed the Kepler scientists to calculate masses for the planets, providing another way to learn about exoplanets.

Kepler has discovered some interesting and unusual planetary systems. For example, most astronomers expected planets to be limited to single stars. But we have found planets orbiting close double stars, so that the planet would see two suns in its sky, like those of the fictional planet Tatooine in the *Star Wars* films. At the opposite extreme, planets can orbit one star of a wide, double-star system without major interference from the second star.

21.6 | NEW PERSPECTIVES ON PLANET FORMATION

Learning Objectives

By the end of this section, you will be able to:

Download for free at http://cnx.org/content/col11992/latest/

> Explain how exoplanet discoveries have revised our understanding of planet formation
> Discuss how planetary systems quite different from our solar system might have come about

Traditionally, astronomers have assumed that the planets in our solar system formed at about their current distances from the Sun and have remained there ever since. The first step in the formation of a giant planet is to build up a solid core, which happens when planetesimals collide and stick. Eventually, this core becomes massive enough to begin sweeping up gaseous material in the disk, thereby building the gas giants Jupiter and Saturn.

How to Make a Hot Jupiter

The traditional model for the formation of planets works only if the giant planets are formed far from the central star (about 5–10 AU), where the disk is cold enough to have a fairly high density of solid matter. It cannot explain the hot Jupiters, which are located very close to their stars where any rocky raw material would be completely vaporized. It also cannot explain the elliptical orbits we observe for some exoplanets because the orbit of a protoplanet, whatever its initial shape, will quickly become circular through interactions with the surrounding disk of material and will remain that way as the planet grows by sweeping up additional matter.

So we have two options: either we find a new model for forming planets close to the searing heat of the parent star, or we find a way to change the orbits of planets so that cold Jupiters can travel inward *after* they form. Most research now supports the latter explanation.

Calculations show that if a planet forms while a substantial amount of gas remains in the disk, then some of the planet's orbital angular momentum can be transferred to the disk. As it loses momentum (through a process that reminds us of the effects of friction), the planet will spiral inward. This process can transport giant planets, initially formed in cold regions of the disk, closer to the central star—thereby producing hot Jupiters. Gravitational interactions between planets in the chaotic early solar system can also cause planets to slingshot inward from large distances. But for this to work, the other planet has to carry away the angular momentum and move to a more distant orbit.

In some cases, we can use the combination of transit plus Doppler measurements to determine whether the planets orbit in the same plane and in the same direction as the star. For the first few cases, things seemed to work just as we anticipated: like the solar system, the gas giant planets orbited in their star's equatorial plane and in the same direction as the spinning star.

Then, some startling discoveries were made of gas giant planets that orbited at right angles or even in the opposite sense as the spin of the star. How could this happen? Again, there must have been interactions between planets. It's possible that before the system settled down, two planets came close together, so that one was kicked into an usual orbit. Or perhaps a passing star perturbed the system after the planets were newly formed.

Forming Planetary Systems

When the Milky Way Galaxy was young, the stars that formed did not contain many heavy elements like iron. Several generations of star formation and star death were required to enrich the interstellar medium for subsequent generations of stars. Since planets seem to form "inside out," starting with the accretion of the materials that can make the rocky cores with which planets start, astronomers wondered when in the history of the Galaxy, planet formation would turn on.

The star Kepler-444 has shed some light on this question. This is a tightly packed system of five planets—the smallest comparable in size to Mercury and the largest similar in size to Venus. All five planets were detected with the Kepler spacecraft as they transited their parent star. All five planets orbit their host star in less than the

This OpenStax book is available for free at http://cnx.org/content/col11992/1.8

time it takes Mercury to complete one orbit about the Sun. Remarkably, the host star Kepler-444 is more than 11 billion years old and formed when the Milky Way was only 2 billion years old. So the heavier elements needed to make rocky planets must have already been available then. This ancient planetary system sets the clock on the beginning of rocky planet formation to be relatively soon after the formation of our Galaxy.

Kepler data demonstrate that while rocky planets inside Mercury's orbit are missing from our solar system, they are common around other stars, like Kepler-444. When the first systems packed with close-in rocky planets were discovered, we wondered why they were so different from our solar system. When many such systems were discovered, we began to wonder if it was our solar system that was different. This led to speculation that additional rocky planets might once have existed close to the Sun in our solar system.

There is some evidence from the motions in the outer solar system that Jupiter may have migrated inward long ago. If correct, then gravitational perturbations from Jupiter could have dislodged the orbits of close-in rocky planets, causing them to fall into the Sun. Consistent with this picture, astronomers now think that Uranus and Neptune probably did not form at their present distances from the Sun but rather closer to where Jupiter and Saturn are now. The reason for this idea is that density in the disk of matter surrounding the Sun at the time the planets formed was so low outside the orbit of Saturn that it would take several billion years to build up Uranus and Neptune. Yet we saw earlier in the chapter that the disks around protostars survive only a few million years.

Therefore, scientists have developed computer models demonstrating that Uranus and Neptune could have formed near the current locations of Jupiter and Saturn, and then been kicked out to larger distances through gravitational interactions with their neighbors. All these wonderful new observations illustrate how dangerous it can be to draw conclusions about a phenomenon in science (in this case, how planetary systems form and arrange themselves) when you are only working with a single example.

Exoplanets have given rise to a new picture of planetary system formation—one that is much more chaotic than we originally thought. If we think of the planets as being like skaters in a rink, our original model (with only our own solar system as a guide) assumed that the planets behaved like polite skaters, all obeying the rules of the rink and all moving in nearly the same direction, following roughly circular paths. The new picture corresponds more to a roller derby, where the skaters crash into one another, change directions, and sometimes are thrown entirely out of the rink.

Habitable Exoplanets

While thousands of exoplanets have been discovered in the past two decades, every observational technique has fallen short of finding more than a few candidates that resemble Earth (Figure 21.27). Astronomers are not sure exactly what properties would define another Earth. Do we need to find a planet that is *exactly* the same size and mass as Earth? That may be difficult and may not be important from the perspective of habitability. After all, we have no reason to think that life could not have arisen on Earth if our planet had been a little bit smaller or larger. And, remember that how habitable a planet is depends on both its distance from its star and the nature of its atmosphere. The greenhouse effect can make some planets warmer (as it did for Venus and is doing more and more for Earth).

Download for free at http://cnx.org/content/col11992/latest/

Figure 21.27. Many Earthlike Planets. This painting, commissioned by NASA, conveys the idea that there may be many planets resembling Earth out there as our methods for finding them improve. (credit: NASA/JPL-Caltech/R. Hurt (SSC-Caltech))

We can ask other questions to which we don't yet know the answers. Does this "twin" of Earth need to orbit a solar-type star, or can we consider as candidates the numerous exoplanets orbiting K- and M-class stars? (In the summer of 2016, astronomers reported the discovery of a planet with at least 1.3 times the mass of Earth around the nearest star, Proxima Centauri, which is spectral type M and located 4.2 light years from us.) We have a special interest in finding planets that could support life like ours, in which case, we need to find exoplanets within their star's habitable zone, where surface temperatures are consistent with liquid water on the surface. This is probably the most important characteristic defining an Earth-analog exoplanet.

The search for potentially habitable worlds is one of the prime drivers for exoplanet research in the next decade. Astronomers are beginning to develop realistic plans for new instruments that can even look for signs of life on distant worlds (examining their atmospheres for gases associated with life, for example). If we require telescopes in space to find such worlds, we need to recognize that years are required to plan, build, and launch such space observatories. The discovery of exoplanets and the knowledge that most stars have planetary systems are transforming our thinking about life beyond Earth. We are closer than ever to knowing whether habitable (and inhabited) planets are common. This work lends a new spirit of optimism to the search for life elsewhere, a subject to which we will return in Life in the Universe.

LINK TO LEARNING

Check out the habitability of various stars and planets by trying out the interactive Circumstellar Habitable Zone Simulator (https://openstaxcollege.org/l/30cirhabzonsim) and select a star system to investigate.

This OpenStax book is available for free at http://cnx.org/content/col11992/1.8

CHAPTER 21 REVIEW

KEY TERMS

exoplanet a planet orbiting a star other than our Sun

giant molecular clouds large, cold interstellar clouds with diameters of dozens of light-years and typical masses of 10^5 solar masses; found in the spiral arms of galaxies, these clouds are where stars form

Herbig-Haro (HH) object luminous knots of gas in an area of star formation that are set to glow by jets of material from a protostar

mini-Neptune a planet that is intermediate between the largest terrestrial planet in our solar system (Earth) and the smallest jovian planet (Neptune); generally, mini-Neptunes have sizes between 2.8 and 4 times Earth's size

protostar a very young star still in the process of formation, before nuclear fusion begins

stellar wind the outflow of gas, sometimes at speeds as high as hundreds of kilometers per second, from a star

super-Earth a planet larger than Earth, generally between 1.4 and 2.8 times the size of our planet

transit when one astronomical object moves in front of another

SUMMARY

21.1 Star Formation

Most stars form in giant molecular clouds with masses as large as 3×10^6 solar masses. The most well-studied molecular cloud is Orion, where star formation is currently taking place. Molecular clouds typically contain regions of higher density called clumps, which in turn contain several even-denser cores of gas and dust, each of which may become a star. A star can form inside a core if its density is high enough that gravity can overwhelm the internal pressure and cause the gas and dust to collapse. The accumulation of material halts when a protostar develops a strong stellar wind, leading to jets of material being observed coming from the star. These jets of material can collide with the material around the star and produce regions that emit light that are known as Herbig-Haro objects.

21.2 The H–R Diagram and the Study of Stellar Evolution

The evolution of a star can be described in terms of changes in its temperature and luminosity, which can best be followed by plotting them on an H–R diagram. Protostars generate energy (and internal heat) through gravitational contraction that typically continues for millions of years, until the star reaches the main sequence.

21.3 Evidence That Planets Form around Other Stars

Observational evidence shows that most protostars are surrounded by disks with large-enough diameters and enough mass (as much as 10% that of the Sun) to form planets. After a few million years, the inner part of the disk is cleared of dust, and the disk is then shaped like a donut with the protostar centered in the hole—something that can be explained by the formation of planets in that inner zone. Around a few older stars, we see disks formed from the debris produced when small bodies (comets and asteroids) collide with each other. The distribution of material in the rings of debris disks is probably determined by shepherd planets, just as Saturn's shepherd moons affect the orbits of the material in its rings. Protoplanets that grow to be 10 times

Download for free at http://cnx.org/content/col11992/latest/

the mass of Earth or bigger while there is still considerable gas in their disk can then capture more of that gas and become giant planets like Jupiter in the solar system.

21.4 Planets beyond the Solar System: Search and Discovery

Several observational techniques have successfully detected planets orbiting other stars. These techniques fall into two general categories—direct and indirect detection. The Doppler and transit techniques are our most powerful indirect tools for finding exoplanets. Some planets are also being found by direct imaging.

21.5 Exoplanets Everywhere: What We Are Learning

Although the Kepler mission is finding thousands of new exoplanets, these are limited to orbital periods of less than 400 days and sizes larger than Mars. Still, we can use the Kepler discoveries to extrapolate the distribution of planets in our Galaxy. The data so far imply that planets like Earth are the most common type of planet, and that there may be 100 billion Earth-size planets around Sun-like stars in the Galaxy. About 2600 planetary systems have been discovered around other stars. In many of them, planets are arranged differently than in our solar system.

21.6 New Perspectives on Planet Formation

The ensemble of exoplanets is incredibly diverse and has led to a revision in our understanding of planet formation that includes the possibility of vigorous, chaotic interactions, with planet migration and scattering. It is possible that the solar system is unusual (and not representative) in how its planets are arranged. Many systems seem to have rocky planets farther inward than we do, for example, and some even have "hot Jupiters" very close to their star. Ambitious space experiments should make it possible to image earthlike planets outside the solar system and even to obtain information about their habitability as we search for life elsewhere.

FOR FURTHER EXPLORATION

Articles

Star Formation

Blaes, O. "A Universe of Disks." *Scientific American* (October 2004): 48. On accretion disks and jets around young stars and black holes.

Croswell, K. "The Dust Belt Next Door [Tau Ceti]." *Scientific American* (January 2015): 24. Short intro to recent observations of planets and a wide dust belt.

Frank, A. "Starmaker: The New Story of Stellar Birth." *Astronomy* (July 1996): 52.

Jayawardhana, R. "Spying on Stellar Nurseries." *Astronomy* (November 1998): 62. On protoplanetary disks.

O'Dell, C. R. "Exploring the Orion Nebula." *Sky & Telescope* (December 1994): 20. Good review with Hubble results.

Ray, T. "Fountains of Youth: Early Days in the Life of a Star." *Scientific American* (August 2000): 42. On outflows from young stars.

Young, E. "Cloudy with a Chance of Stars." *Scientific American* (February 2010): 34. On how clouds of interstellar matter turn into star systems.

Young, Monica "Making Massive Stars." *Sky & Telescope* (October 2015): 24. Models and observations on how the most massive stars form.

This OpenStax book is available for free at http://cnx.org/content/col11992/1.8

Exoplanets

Billings, L. "In Search of Alien Jupiters." *Scientific American* (August 2015): 40–47. The race to image jovian planets with current instruments and why a direct image of a terrestrial planet is still in the future.

Heller, R. "Better Than Earth." *Scientific American* (January 2015): 32–39. What kinds of planets may be habitable; super-Earths and jovian planet moons should also be considered.

Laughlin, G. "How Worlds Get Out of Whack." *Sky & Telescope* (May 2013): 26. On how planets can migrate from the places they form in a star system.

Marcy, G. "The New Search for Distant Planets." *Astronomy* (October 2006): 30. Fine brief overview. (The same issue has a dramatic fold-out visual atlas of extrasolar planets, from that era.)

Redd, N. "Why Haven't We Found Another Earth?" *Astronomy* (February 2016): 25. Looking for terrestrial planets in the habitable zone with evidence of life.

Seager, S. "Exoplanets Everywhere." *Sky & Telescope* (August 2013): 18. An excellent discussion of some of the frequently asked questions about the nature and arrangement of planets out there.

Seager, S. "The Hunt for Super-Earths." *Sky & Telescope* (October 2010): 30. The search for planets that are up to 10 times the mass of Earth and what they can teach us.

Villard, R. "Hunting for Earthlike Planets." *Astronomy* (April 2011): 28. How we expect to find and characterize super-Earth (planets somewhat bigger than ours) using new instruments and techniques that could show us what their atmospheres are made of.

Websites

Exoplanet Exploration: http://planetquest.jpl.nasa.gov/. PlanetQuest (from the Navigator Program at the Jet Propulsion Lab) is probably the best site for students and beginners, with introductory materials and nice illustrations; it focuses mostly on NASA work and missions.

Exoplanets: http://www.planetary.org/exoplanets/. Planetary Society's exoplanets pages with a dynamic catalog of planets found and good explanations.

Exoplanets: The Search for Planets beyond Our Solar System: http://www.iop.org/publications/iop/2010/page_42551.html. From the British Institute of Physics in 2010.

Extrasolar Planets Encyclopedia: http://exoplanet.eu/. Maintained by Jean Schneider of the Paris Observatory, has the largest catalog of planet discoveries and useful background material (some of it more technical).

Formation of Stars: https://www.spacetelescope.org/science/formation_of_stars/. Star Formation page from the Hubble Space Telescope, with links to images and information.

Kepler Mission: http://kepler.nasa.gov/. The public website for the remarkable telescope in space that is searching planets using the transit technique and is our best hope for finding earthlike planets.

Proxima Centauri Planet Discovery: http://www.eso.org/public/news/eso1629/.

Apps

Exoplanet: http://itunes.apple.com/us/app/exoplanet/id327702034?mt=8. Allows you to browse through a regularly updated visual catalog of exoplanets that have been found so far.

Journey to the Exoplanets: http://itunes.apple.com/us/app/journey-to-the-exoplanets/id463532472?mt=8. Produced by the staff of *Scientific American*, with input from scientists and space artists; gives background information and visual tours of the nearer star systems with planets.

Download for free at http://cnx.org/content/col11992/latest/

Videos

A Star Is Born: http://www.discovery.com/tv-shows/other-shows/videos/how-the-universe-works-a-star-is-born/. Discovery Channel video with astronomer Michelle Thaller (2:25).

Are We Alone: An Evening Dialogue with the Kepler Mission Leaders: http://www.youtube.com/watch?v=O7ItAXfl0Lw. A non-technical panel discussion on Kepler results and ideas about planet formation with Bill Borucki, Natalie Batalha, and Gibor Basri (moderated by Andrew Fraknoi) at the University of California, Berkeley (2:07:01).

Finding the Next Earth: The Latest Results from Kepler: https://www.youtube.com/watch?v=ZbijeR_AALo. Natalie Batalha (San Jose State University & NASA Ames) public talk in the Silicon Valley Astronomy Lecture Series (1:28:38).

From Hot Jupiters to Habitable Worlds: https://vimeo.com/37696087 (Part 1) and https://vimeo.com/37700700 (Part 2). Debra Fischer (Yale University) public talk in Hawaii sponsored by the Keck Observatory (15:20 Part 1, 21:32 Part 2).

Search for Habitable Exoplanets: http://www.youtube.com/watch?v=RLWb_T9yaDU. Sara Seeger (MIT) public talk at the SETI Institute, with Kepler results (1:10:35).

Strange Planetary Vistas: http://www.youtube.com/watch?v=_8ww9eLRSCg. Josh Carter (CfA) public talk at Harvard's Center for Astrophysics with a friendly introduction to exoplanets for non-specialists (46:35).

⚎ COLLABORATIVE GROUP ACTIVITIES

A. Your group is a subcommittee of scientists examining whether any of the "hot Jupiters" (giant planets closer to their stars than Mercury is to the Sun) could have life on or near them. Can you come up with places on, in, or near such planets where life could develop or where some forms of life might survive?

B. A wealthy couple (who are alumni of your college or university and love babies) leaves the astronomy program several million dollars in their will, to spend in the best way possible to search for "infant stars in our section of the Galaxy." Your group has been assigned the task of advising the dean on how best to spend the money. What kind of instruments and search programs would you recommend, and why?

C. Some people consider the discovery of any planets (even hot Jupiters) around other stars one of the most important events in the history of astronomical research. Some astronomers have been surprised that the public is not more excited about the planet discoveries. One reason that has been suggested for this lack of public surprise and excitement is that science fiction stories have long prepared us for there being planets around other stars. (The Starship Enterprise on the 1960s *Star Trek* TV series found some in just about every weekly episode.) What does your group think? Did you know about the discovery of planets around other stars before taking this course? Do you consider it exciting? Were you surprised to hear about it? Are science fiction movies and books good or bad tools for astronomy education in general, do you think?

D. What if future space instruments reveal an earthlike exoplanet with significant amounts of oxygen and methane in its atmosphere? Suppose the planet and its star are 50 light-years away. What does your group suggest astronomers do next? How much effort and money would you recommend be put into finding out more about this planet and why?

This OpenStax book is available for free at http://cnx.org/content/col11992/1.8

E. Discuss with your group the following question: which is easier to find orbiting a star with instruments we have today: a jovian planet or a proto-planetary disk? Make a list of arguments for each side of this question.

F. (This activity should be done when your group has access to the internet.) Go to the page which indexes all the publicly released Hubble Space Telescope images by subject: http://hubblesite.org/newscenter/ archive/browse/image/. Under "Star," go to "Protoplanetary Disk" and find a system—not mentioned in this chapter—that your group likes, and prepare a short report to the class about why you find it interesting. Then, under "Nebula," go to "Emission" and find a region of star formation not mentioned in this chapter, and prepare a short report to the class about what you find interesting about it.

G. There is a "citizen science" website called Planet Hunters (http://www.planethunters.org/) where you can participate in identifying exoplanets from the data that Kepler provided. Your group should access the site, work together to use it, and classify two light curves. Report back to the class on what you have done.

H. Yuri Milner, a Russian-American billionaire, recently pledged $100 million to develop the technology to send many miniaturized probes to a star in the Alpha Centauri triple star system (which includes Proxima Centauri, the nearest star to us, now known to have at least one planet.) Each tiny probe will be propelled by powerful lasers at 20% the speed of light, in the hope that one or more might arrive safely and be able to send back information about what it's like there. Your group should search online for more information about this project (called "Breakthrough: Starshot") and discuss your reactions to this project. Give specific reasons for your arguments.

🗐 EXERCISES

Review Questions

1. Give several reasons the Orion molecular cloud is such a useful "laboratory" for studying the stages of star formation.

2. Why is star formation more likely to occur in cold molecular clouds than in regions where the temperature of the interstellar medium is several hundred thousand degrees?

3. Why have we learned a lot about star formation since the invention of detectors sensitive to infrared radiation?

4. Describe what happens when a star forms. Begin with a dense core of material in a molecular cloud and trace the evolution up to the time the newly formed star reaches the main sequence.

5. Describe how the T Tauri star stage in the life of a low-mass star can lead to the formation of a Herbig-Haro (H-H) object.

6. Look at the four stages shown in Figure 21.8. In which stage(s) can we see the star in visible light? In infrared radiation?

7. The evolutionary track for a star of 1 solar mass remains nearly vertical in the H–R diagram for a while (see Figure 21.12). How is its luminosity changing during this time? Its temperature? Its radius?

8. Two protostars, one 10 times the mass of the Sun and one half the mass of the Sun are born at the same time in a molecular cloud. Which one will be first to reach the main sequence stage, where it is stable and getting energy from fusion?

Download for free at http://cnx.org/content/col11992/latest/

9. Compare the scale (size) of a typical dusty disk around a forming star with the scale of our solar system.

10. Why is it so hard to see planets around other stars and so easy to see them around our own?

11. Why did it take astronomers until 1995 to discover the first exoplanet orbiting another star like the Sun?

12. Which types of planets are most easily detected by Doppler measurements? By transits?

13. List three ways in which the exoplanets we have detected have been found to be different from planets in our solar system.

14. List any similarities between discovered exoplanets and planets in our solar system.

15. What revisions to the theory of planet formation have astronomers had to make as a result of the discovery of exoplanets?

16. Why are young Jupiters easier to see with direct imaging than old Jupiters?

Thought Questions

17. A friend of yours who did not do well in her astronomy class tells you that she believes all stars are old and none could possibly be born today. What arguments would you use to persuade her that stars are being born somewhere in the Galaxy during your lifetime?

18. Observations suggest that it takes more than 3 million years for the dust to begin clearing out of the inner regions of the disks surrounding protostars. Suppose this is the minimum time required to form a planet. Would you expect to find a planet around a 10-M_{Sun} star? (Refer to Figure 21.12.)

19. Suppose you wanted to observe a planet around another star with direct imaging. Would you try to observe in visible light or in the infrared? Why? Would the planet be easier to see if it were at 1 AU or 5 AU from its star?

20. Why were giant planets close to their stars the first ones to be discovered? Why has the same technique not been used yet to discover giant planets at the distance of Saturn?

21. Exoplanets in eccentric orbits experience large temperature swings during their orbits. Suppose you had to plan for a mission to such a planet. Based on Kepler's second law, does the planet spend more time closer or farther from the star? Explain.

Figuring For Yourself

22. When astronomers found the first giant planets with orbits of only a few days, they did not know whether those planets were gaseous and liquid like Jupiter or rocky like Mercury. The observations of HD 209458 settled this question because observations of the transit of the star by this planet made it possible to determine the radius of the planet. Use the data given in the text to estimate the density of this planet, and then use that information to explain why it must be a gas giant.

23. An exoplanetary system has two known planets. Planet X orbits in 290 days and Planet Y orbits in 145 days. Which planet is closest to its host star? If the star has the same mass as the Sun, what is the semi-major axis of the orbits for Planets X and Y?

24. Kepler's third law says that the orbital period (in years) is proportional to the square root of the cube of the mean distance (in AU) from the Sun ($P \propto a^{1.5}$). For mean distances from 0.1 to 32 AU, calculate and plot a curve showing the expected Keplerian period. For each planet in our solar system, look up the mean distance from the Sun in AU and the orbital period in years and overplot these data on the theoretical Keplerian curve.

This OpenStax book is available for free at http://cnx.org/content/col11992/1.8

25. Calculate the transit depth for an M dwarf star that is 0.3 times the radius of the Sun with a gas giant planet the size of Jupiter.

26. If a transit depth of 0.00001 can be detected with the Kepler spacecraft, what is the smallest planet that could be detected around a 0.3 R_{sun} M dwarf star?

27. What fraction of gas giant planets seems to have inflated radii?

Download for free at http://cnx.org/content/col11992/latest/

This OpenStax book is available for free at http://cnx.org/content/col11992/1.8

Download for free at http://cnx.org/content/col11992/latest/

Figure 22.1. Ant Nebula. During the later phases of stellar evolution, stars expel some of their mass, which returns to the interstellar medium to form new stars. This Hubble Space Telescope image shows a star losing mass. Known as Menzel 3, or the Ant Nebula, this beautiful region of expelled gas is about 3000 light-years away from the Sun. We see a central star that has ejected mass preferentially in two opposite directions. The object is about 1.6 light-years long. The image is color coded—red corresponds to an emission line of sulfur, green to nitrogen, blue to hydrogen, and blue/violet to oxygen. (credit: modification of work by NASA, ESA and The Hubble Heritage Team (STScI/AURA))

Chapter Outline

Thinking Ahead

The Sun and other stars cannot last forever. Eventually they will exhaust their nuclear fuel and cease to shine. But how do they change during their long lifetimes? And what do these changes mean for the future of Earth?

We now turn from the birth of stars to the rest of their life stories. This is not an easy task since stars live much longer than astronomers. Thus, we cannot hope to see the life story of any single star unfold before our eyes or telescopes. To learn about their lives, we must survey as many of the stellar inhabitants of the Galaxy as possible. With thoroughness and a little luck, we can catch at least a few of them in each stage of their lives. As you've learned, stars have many different characteristics, with the differences sometimes resulting from their different masses, temperatures, and luminosities, and at other times derived from changes that occur as they age. Through a combination of observation and theory, we can use these differences to piece together the life story of a star.

Download for free at http://cnx.org/content/col11992/latest/

22.1 | EVOLUTION FROM THE MAIN SEQUENCE TO RED GIANTS

Learning Objectives

By the end of this section, you will be able to:

> Explain the zero-age main sequence
> Describe what happens to main-sequence stars of various masses as they exhaust their hydrogen supply

One of the best ways to get a "snapshot" of a group of stars is by plotting their properties on an H–R diagram. We have already used the H–R diagram to follow the evolution of protostars up to the time they reach the main sequence. Now we'll see what happens next.

Once a star has reached the main-sequence stage of its life, it derives its energy almost entirely from the conversion of hydrogen to helium via the process of nuclear fusion in its core (see The Sun: A Nuclear Powerhouse). Since hydrogen is the most abundant element in stars, this process can maintain the star's equilibrium for a long time. Thus, all stars remain on the main sequence for most of their lives. Some astronomers like to call the main-sequence phase the star's "prolonged adolescence" or "adulthood" (continuing our analogy to the stages in a human life).

The left-hand edge of the main-sequence band in the H–R diagram is called the **zero-age main sequence** (see Figure 18.15). We use the term *zero-age* to mark the time when a star stops contracting, settles onto the main sequence, and begins to fuse hydrogen in its core. The zero-age main sequence is a continuous line in the H–R diagram that shows where stars of different masses but similar chemical composition can be found when they begin to fuse hydrogen.

Since only 0.7% of the hydrogen used in fusion reactions is converted into energy, fusion does not change the *total* mass of the star appreciably during this long period. It does, however, change the chemical composition in its central regions where nuclear reactions occur: hydrogen is gradually depleted, and helium accumulates. This change of composition changes the luminosity, temperature, size, and interior structure of the star. When a star's luminosity and temperature begin to change, the point that represents the star on the H–R diagram moves away from the zero-age main sequence.

Calculations show that the temperature and density in the inner region slowly increase as helium accumulates in the center of a star. As the temperature gets hotter, each proton acquires more energy of motion on average; this means it is more likely to interact with other protons, and as a result, the rate of fusion also increases. For the proton-proton cycle described in The Sun: A Nuclear Powerhouse, the rate of fusion goes up roughly as the temperature to the fourth power.

If the rate of fusion goes up, the rate at which energy is being generated also increases, and the luminosity of the star gradually rises. Initially, however, these changes are small, and stars remain within the main-sequence band on the H–R diagram for most of their lifetimes.

EXAMPLE 22.1

Star Temperature and Rate of Fusion

If a star's temperature were to double, by what factor would its rate of fusion increase?

This OpenStax book is available for free at http://cnx.org/content/col11992/1.8

Solution

Since the rate of fusion (like temperature) goes up to the fourth power, it would increase by a factor of 2^4, or 16 times.

Check Your Learning

If the rate of fusion of a star increased 256 times, by what factor would the temperature increase?

Answer:

The temperature would increase by a factor of $256^{0.25}$ (that is, the 4th root of 256), or 4 times.

Lifetimes on the Main Sequence

How many years a star remains in the main-sequence band depends on its mass. You might think that a more massive star, having more fuel, would last longer, but it's not that simple. The lifetime of a star in a particular stage of evolution depends on how much nuclear fuel it has and on *how quickly* it uses up that fuel. (In the same way, how long people can keep spending money depends not only on how much money they have but also on how quickly they spend it. This is why many lottery winners who go on spending sprees quickly wind up poor again.) In the case of stars, more massive ones use up their fuel much more quickly than stars of low mass.

The reason massive stars are such spendthrifts is that, as we saw above, the rate of fusion depends *very* strongly on the star's core temperature. And what determines how hot a star's central regions get? It is the *mass* of the star—the weight of the overlying layers determines how high the pressure in the core must be: higher mass requires higher pressure to balance it. Higher pressure, in turn, is produced by higher temperature. The higher the temperature in the central regions, the faster the star races through its storehouse of central hydrogen. Although massive stars have more fuel, they burn it so prodigiously that their lifetimes are much shorter than those of their low-mass counterparts. You can also understand now why the most massive main-sequence stars are also the most luminous. Like new rock stars with their first platinum album, they spend their resources at an astounding rate.

The main-sequence lifetimes of stars of different masses are listed in Table 22.1. This table shows that the most massive stars spend only a few million years on the main sequence. A star of 1 solar mass remains there for roughly 10 billion years, while a star of about 0.4 solar mass has a main-sequence lifetime of some 200 billion years, which is longer than the current age of the universe. (Bear in mind, however, that every star spends *most* of its total lifetime on the main sequence. Stars devote an average of 90% of their lives to peacefully fusing hydrogen into helium.)

Lifetimes of Main-Sequence Stars

Spectral Type	Surface Temperature (K)	Mass (Mass of Sun = 1)	Lifetime on Main Sequence (years)
O5	54,000	40	1 million
B0	29,200	16	10 million

Table 22.1

Download for free at http://cnx.org/content/col11992/latest/

Lifetimes of Main-Sequence Stars

Spectral Type	Surface Temperature (K)	Mass (Mass of Sun = 1)	Lifetime on Main Sequence (years)
A0	9600	3.3	500 million
F0	7350	1.7	2.7 billion
G0	6050	1.1	9 billion
K0	5240	0.8	14 billion
M0	3750	0.4	200 billion

Table 22.1

These results are not merely of academic interest. Human beings developed on a planet around a G-type star. This means that the Sun's stable main-sequence lifetime is so long that it afforded life on Earth plenty of time to evolve. When searching for intelligent life like our own on planets around other stars, it would be a pretty big waste of time to search around O- or B-type stars. These stars remain stable for such a short time that the development of creatures complicated enough to take astronomy courses is very unlikely.

From Main-Sequence Star to Red Giant

Eventually, all the hydrogen in a star's core, where it is hot enough for fusion reactions, is used up. The core then contains only helium, "contaminated" by whatever small percentage of heavier elements the star had to begin with. The helium in the core can be thought of as the accumulated "ash" from the nuclear "burning" of hydrogen during the main-sequence stage.

Energy can no longer be generated by hydrogen fusion in the stellar core because the hydrogen is all gone and, as we will see, the fusion of helium requires much higher temperatures. Since the central temperature is not yet high enough to fuse helium, there is no nuclear energy source to supply heat to the central region of the star. The long period of stability now ends, gravity again takes over, and the core begins to contract. Once more, the star's energy is partially supplied by gravitational energy, in the way described by Kelvin and Helmholtz (see Sources of Sunshine: Thermal and Gravitational Energy). As the star's core shrinks, the energy of the inward-falling material is converted to heat.

The heat generated in this way, like all heat, flows outward to where it is a bit cooler. In the process, the heat raises the temperature of a layer of hydrogen that spent the whole long main-sequence time just outside the core. Like an understudy waiting in the wings of a hit Broadway show for a chance at fame and glory, this hydrogen was almost (but not quite) hot enough to undergo fusion and take part in the main action that sustains the star. Now, the additional heat produced by the shrinking core puts this hydrogen "over the limit," and a shell of hydrogen nuclei just outside the core becomes hot enough for hydrogen fusion to begin.

New energy produced by fusion of this hydrogen now pours outward from this shell and begins to heat up layers of the star farther out, causing them to expand. Meanwhile, the helium core continues to contract, producing more heat right around it. This leads to more fusion in the shell of fresh hydrogen outside the core (Figure 22.2). The additional fusion produces still more energy, which also flows out into the upper layer of the star.

This OpenStax book is available for free at http://cnx.org/content/col11992/1.8

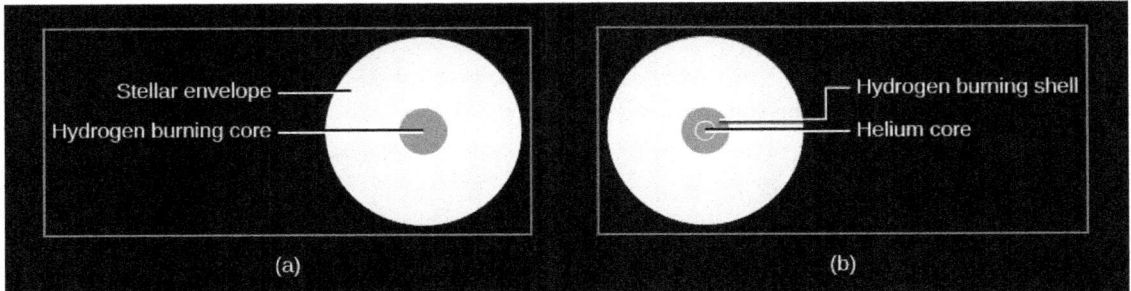

Figure 22.2. Star Layers during and after the Main Sequence. (a) During the main sequence, a star has a core where fusion takes place and a much larger envelope that is too cold for fusion. (b) When the hydrogen in the core is exhausted (made of helium, not hydrogen), the core is compressed by gravity and heats up. The additional heat starts hydrogen fusion in a layer just outside the core. Note that these parts of the Sun are not drawn to scale.

Most stars actually generate more energy each second when they are fusing hydrogen in the shell surrounding the helium core than they did when hydrogen fusion was confined to the central part of the star; thus, they increase in luminosity. With all the new energy pouring outward, the outer layers of the star begin to expand, and the star eventually grows and grows until it reaches enormous proportions (Figure 22.3).

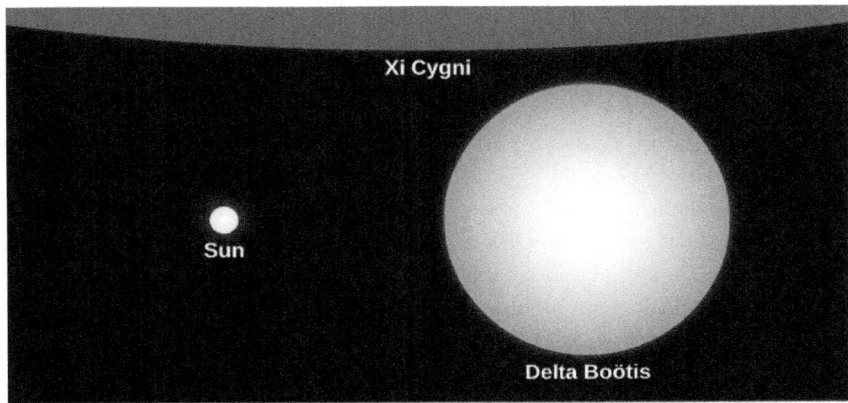

Figure 22.3. Relative Sizes of Stars. This image compares the size of the Sun to that of Delta Boötis, a giant star, and Xi Cygni, a supergiant. Note that Xi Cygni is so large in comparison to the other two stars that only a small portion of it is visible at the top of the frame.

When you take the lid off a pot of boiling water, the steam can expand and it cools down. In the same way, the expansion of a star's outer layers causes the temperature at the surface to decrease. As it cools, the star's overall color becomes redder. (We saw in Radiation and Spectra that a red color corresponds to cooler temperature.)

So the star becomes simultaneously more luminous and cooler. On the H–R diagram, the star therefore leaves the main-sequence band and moves upward (brighter) and to the right (cooler surface temperature). Over time, massive stars become red supergiants, and lower-mass stars like the Sun become red giants. (We first discussed such giant stars in The Stars: A Celestial Census; here we see how such "swollen" stars originate.) You might also say that these stars have "split personalities": their cores are contracting while their outer layers are expanding. (Note that red giant stars do not actually look deep red; their colors are more like orange or orange-red.)

Just how different are these red giants and supergiants from a main-sequence star? Table 22.2 compares the Sun with the red supergiant Betelgeuse, which is visible above Orion's belt as the bright red star that marks the

Download for free at http://cnx.org/content/col11992/latest/

hunter's armpit. Relative to the Sun, this supergiant has a much larger radius, a much lower average density, a cooler surface, and a much hotter core.

Comparing a Supergiant with the Sun

Property	Sun	Betelgeuse
Mass (2×10^{33} g)	1	16
Radius (km)	700,000	500,000,000
Surface temperature (K)	5,800	3,600
Core temperature (K)	15,000,000	160,000,000
Luminosity (4×10^{26} W)	1	46,000
Average density (g/cm^3)	1.4	1.3×10^{-7}
Age (millions of years)	4,500	10

Table 22.2

Red giants can become so large that if we were to replace the Sun with one of them, its outer atmosphere would extend to the orbit of Mars or even beyond (Figure 22.4). This is the next stage in the life of a star as it moves (to continue our analogy to human lives) from its long period of "youth" and "adulthood" to "old age." (After all, many human beings today also see their outer layers expand a bit as they get older.) By considering the relative ages of the Sun and Betelgeuse, we can also see that the idea that "bigger stars die faster" is indeed true here. Betelgeuse is a mere 10 million years old, which is relatively young compared with our Sun's 4.5 billion years, but it is already nearing its death throes as a red supergiant.

This OpenStax book is available for free at http://cnx.org/content/col11992/1.8

Figure 22.4. Betelgeuse. Betelgeuse is in the constellation Orion, the hunter; in the right image, it is marked with a yellow "X" near the top left. In the left image, we see it in ultraviolet with the Hubble Space Telescope, in the first direct image ever made of the surface of another star. As shown by the scale at the bottom, Betelgeuse has an extended atmosphere so large that, if it were at the center of our solar system, it would stretch past the orbit of Jupiter. (credit: Modification of work by Andrea Dupree (Harvard-Smithsonian CfA), Ronald Gilliland (STScI), NASA and ESA)

Models for Evolution to the Giant Stage

As we discussed earlier, astronomers can construct computer models of stars with different masses and compositions to see how stars change throughout their lives. Figure 22.5, which is based on theoretical calculations by University of Illinois astronomer Icko Iben, shows an H–R diagram with several tracks of evolution from the main sequence to the giant stage. Tracks are shown for stars with different masses (from 0.5 to 15 times the mass of our Sun) and with chemical compositions similar to that of the Sun. The red line is the initial or zero-age main sequence. The numbers along the tracks indicate the time, in years, required for each star to reach those points in their evolution after leaving the main sequence. Once again, you can see that the more massive a star is, the more quickly it goes through each stage in its life.

Download for free at http://cnx.org/content/col11992/latest/

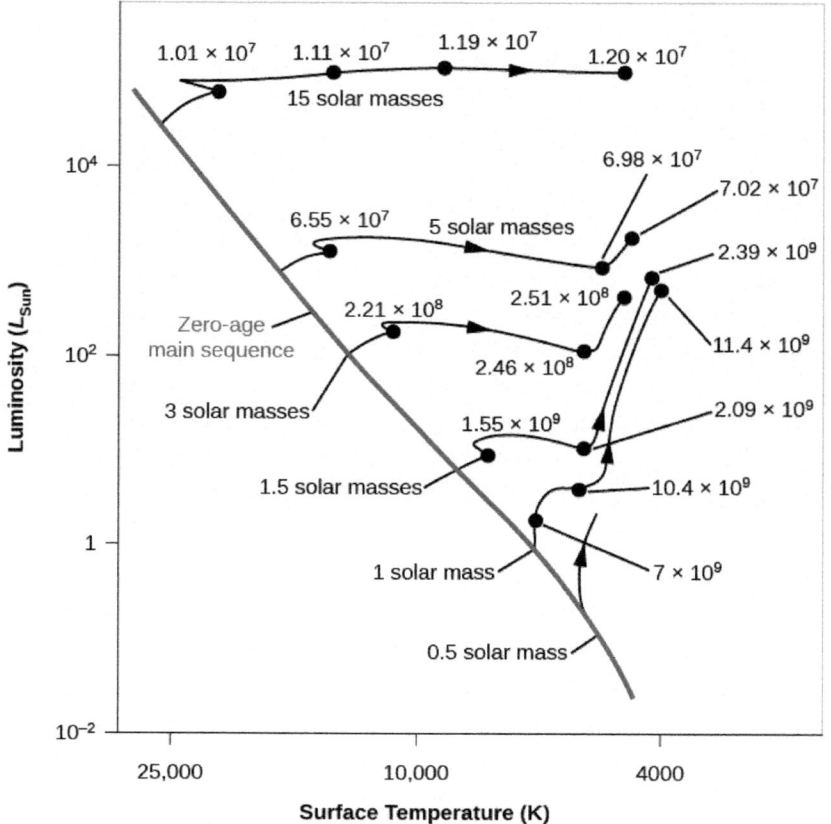

Figure 22.5. Evolutionary Tracks of Stars of Different Masses. The solid black lines show the predicted evolution from the main sequence through the red giant or supergiant stage on the H–R diagram. Each track is labeled with the mass of the star it is describing. The numbers show how many years each star takes to become a giant after leaving the main sequence. The red line is the zero-age main sequence.

Note that the most massive star in this diagram has a mass similar to that of Betelgeuse, and so its evolutionary track shows approximately the history of Betelgeuse. The track for a 1-solar-mass star shows that the Sun is still in the main-sequence phase of evolution, since it is only about 4.5 billion years old. It will be billions of years before the Sun begins its own "climb" away from the main sequence—the expansion of its outer layers that will make it a red giant.

22.2 | STAR CLUSTERS

Learning Objectives

By the end of this section, you will be able to:

> Explain how star clusters help us understand the stages of stellar evolution
> List the different types of star clusters and describe how they differ in number of stars, structure, and age
> Explain why the chemical composition of globular clusters is different from that of open clusters

The preceding description of stellar evolution is based on calculations. However, no star completes its main-sequence lifetime or its evolution to a red giant quickly enough for us to observe these structural changes as they happen. Fortunately, nature has provided us with an indirect way to test our calculations.

This OpenStax book is available for free at http://cnx.org/content/col11992/1.8

Download for free at http://cnx.org/content/col11992/latest/

Instead of observing the evolution of a single star, we can look at a group or *cluster* of stars. We look for a group of stars that is very close together in space, held together by gravity, often moving around a common center. Then it is reasonable to assume that the individual stars in the group all formed at nearly the same time, from the same cloud, and with the same composition. We expect that these stars will differ only in mass. And their masses determine how quickly they go through each stage of their lives.

Since stars with higher masses evolve more quickly, we can find clusters in which massive stars have already completed their main-sequence phase of evolution and become red giants, while stars of lower mass in the same cluster are still on the main sequence, or even—if the cluster is very young—undergoing pre-main-sequence gravitational contraction. We can see many stages of stellar evolution among the members of a single cluster, and we can see whether our models can explain why the H–R diagrams of clusters of different ages look the way they do.

The three basic types of clusters astronomers have discovered are globular clusters, open clusters, and stellar associations. Their properties are summarized in Table 22.3. As we will see in the next section of this chapter, globular clusters contain only very old stars, whereas open clusters and associations contain young stars.

Characteristics of Star Clusters

Characteristic	Globular Clusters	Open Clusters	Associations
Number in the Galaxy	150	Thousands	Thousands
Location in the Galaxy	Halo and central bulge	Disk (and spiral arms)	Spiral arms
Diameter (in light-years)	50–450	<30	100–500
Mass M_{Sun}	10^4–10^6	10^2–10^3	10^2–10^3
Number of stars	10^4–10^6	50–1000	10^2–10^4
Color of brightest stars	Red	Red or blue	Blue
Luminosity of cluster (L_{Sun})	10^4–10^6	10^2–10^6	10^4–10^7
Typical ages	Billions of years	A few hundred million years to, in the case of unusually large clusters, more than a billion years	Up to about 10^7 years

Table 22.3

Globular Clusters

Globular clusters were given this name because they are nearly symmetrical round systems of, typically, hundreds of thousands of stars. The most massive globular cluster in our own Galaxy is Omega Centauri, which

Download for free at http://cnx.org/content/col11992/latest/

is about 16,000 light-years away and contains several million stars (Figure 22.6). Note that the brightest stars in this cluster, which are red giants that have already completed the main-sequence phase of their evolution, are red-orange in color. These stars have typical surface temperatures around 4000 K. As we will see, globular clusters are among the oldest parts of our Milky Way Galaxy.

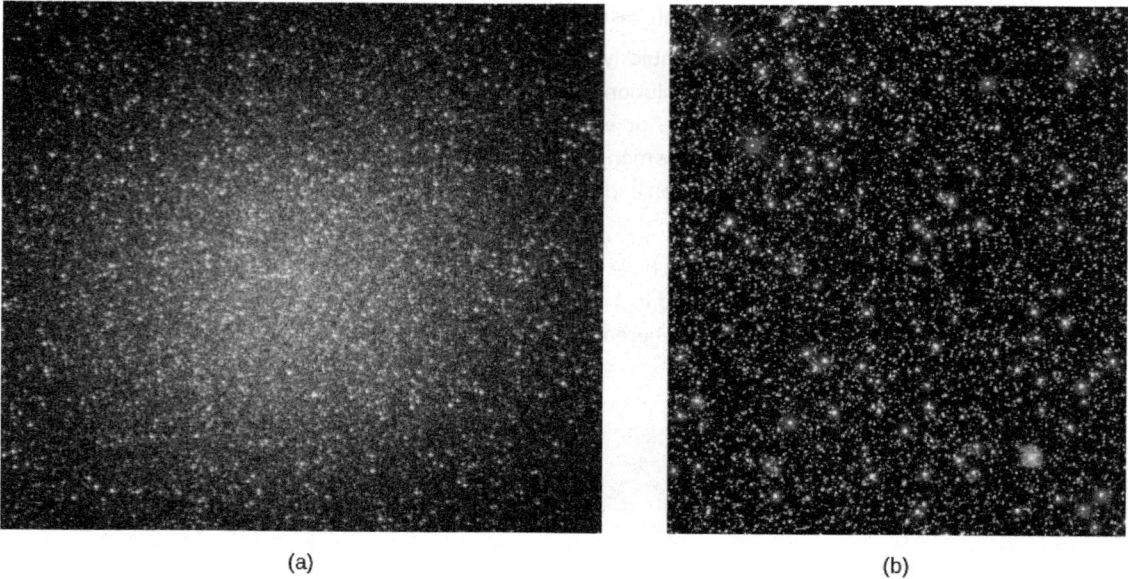

(a)   (b)

Figure 22.6. Omega Centauri. (a) Located at about 16,000 light-years away, Omega Centauri is the most massive globular cluster in our Galaxy. It contains several million stars. (b) This image, taken with the Hubble Space Telescope, zooms in near the center of Omega Centauri. The image is about 6.3 light-years wide. The most numerous stars in the image, which are yellow-white in color, are main-sequence stars similar to our Sun. The brightest stars are red giants that have begun to exhaust their hydrogen fuel and have expanded to about 100 times the diameter of our Sun. The blue stars have started helium fusion. (credit a: modification of work by NASA, ESA and the Hubble Heritage Team (STScI/AURA); credit b: modification of work by NASA, ESA, and the Hubble SM4 ERO Team)

What would it be like to live inside a globular cluster? In the dense central regions, the stars would be roughly a million times closer together than in our own neighborhood. If Earth orbited one of the inner stars in a globular cluster, the nearest stars would be light-months, not light-years, away. They would still appear as points of light, but would be brighter than any of the stars we see in our own sky. The Milky Way would probably be difficult to see through the bright haze of starlight produced by the cluster.

About 150 globular clusters are known in our Galaxy. Most of them are in a spherical halo (or cloud) surrounding the flat disk formed by the majority of our Galaxy's stars. All the globular clusters are very far from the Sun, and some are found at distances of 60,000 light-years or more from the main disk of the Milky Way. The diameters of globular star clusters range from 50 light-years to more than 450 light-years.

Open Clusters

Open clusters are found in the disk of the Galaxy. They have a range of ages, some as old as, or even older than, our Sun. The youngest open clusters are still associated with the interstellar matter from which they formed. Open clusters are smaller than globular clusters, usually having diameters of less than 30 light-years, and they typically contain only several dozen to several hundreds of stars (Figure 22.7). The stars in open clusters usually appear well separated from one another, even in the central regions, which explains why they are called "open." Our Galaxy contains thousands of open clusters, but we can see only a small fraction of them. Interstellar dust, which is also concentrated in the disk, dims the light of more distant clusters so much that they are undetectable.

This OpenStax book is available for free at http://cnx.org/content/col11992/1.8

Download for free at http://cnx.org/content/col11992/latest/

Figure 22.7. Jewel Box (NGC 4755). This open cluster of young, bright stars is about 6400 light-years away from the Sun. Note the contrast in color between the bright yellow supergiant and the hot blue main-sequence stars. The name comes from John Herschel's nineteenth-century description of it as "a casket of variously colored precious stones." (credit: ESO/Y. Beletsky)

Although the individual stars in an open cluster can survive for billions of years, they typically remain together as a cluster for only a few million years, or at most, a few hundred million years. There are several reasons for this. In small open clusters, the average speed of the member stars within the cluster may be higher than the cluster's escape velocity,[1] and the stars will gradually "evaporate" from the cluster. Close encounters of member stars may also increase the velocity of one of the members beyond the escape velocity. Every few hundred million years or so, the cluster may have a close encounter with a giant molecular cloud, and the gravitational force exerted by the cloud may tear the cluster apart.

Several open clusters are visible to the unaided eye. Most famous among them is the Pleiades (Figure 20.13), which appears as a tiny group of six stars (some people can see even more than six, and the Pleiades is sometimes called the Seven Sisters). This cluster is arranged like a small dipping spoon and is seen in the constellation of Taurus, the bull. A good pair of binoculars shows dozens of stars in the cluster, and a telescope reveals hundreds. (A car company, Subaru, takes its name from the Japanese term for this cluster; you can see the star group on the Subaru logo.)

The Hyades is another famous open cluster in Taurus. To the naked eye, it appears as a V-shaped group of faint stars marking the face of the bull. Telescopes show that Hyades actually contains more than 200 stars.

Stellar Associations

An **association** is a group of extremely young stars, typically containing 5 to 50 hot, bright O and B stars scattered over a region of space some 100–500 light-years in diameter. As an example, most of the stars in the constellation Orion form one of the nearest stellar associations. Associations also contain hundreds to thousands of low-mass stars, but these are much fainter and less conspicuous. The presence of really hot, luminous stars indicates that star formation in the association has occurred in the last million years or so. Since O stars go through their entire lives in only about a million years, they would not still be around unless star

---

1   Escape velocity is the speed needed to overcome the gravity of some object or group of objects. The rockets we send up from Earth, for example, must travel faster than the escape velocity of our planet to be able to get to other worlds.

Download for free at http://cnx.org/content/col11992/latest/

formation has occurred recently. It is therefore not surprising that associations are found in regions rich in the gas and dust required to form new stars. It's like a brand new building still surrounded by some of the construction materials used to build it and with the landscape still showing signs of construction. On the other hand, because associations, like ordinary open clusters, lie in regions occupied by dusty interstellar matter, many are hidden from our view.

22.3 CHECKING OUT THE THEORY

Learning Objectives

By the end of this section, you will be able to:

> Explain how the H–R diagram of a star cluster can be related to the cluster's age and the stages of evolution of its stellar members
> Describe how the main-sequence turnoff of a cluster reveals its age

In the previous section, we indicated that that open clusters are younger than globular clusters, and associations are typically even younger. In this section, we will show how we determine the ages of these star clusters. The key observation is that the stars in these different types of clusters are found in different places in the H–R diagram, and we can use their locations in the diagram in combination with theoretical calculations to estimate how long they have lived.

H–R Diagrams of Young Clusters

What does theory predict for the H–R diagram of a cluster whose stars have recently condensed from an interstellar cloud? Remember that at every stage of evolution, massive stars evolve more quickly than their lower-mass counterparts. After a few million years ("recently" for astronomers), the most massive stars should have completed their contraction phase and be on the main sequence, while the less massive ones should be off to the right, still on their way to the main sequence. These ideas are illustrated in Figure 22.8, which shows the H–R diagram calculated by R. Kippenhahn and his associates at Munich University for a hypothetical cluster with an age of 3 million years.

This OpenStax book is available for free at http://cnx.org/content/col11992/1.8

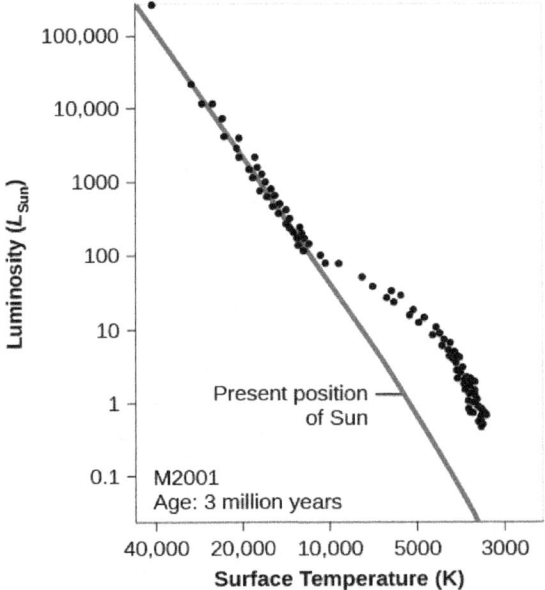

Figure 22.8. Young Cluster H–R Diagram. We see an H–R diagram for a hypothetical young cluster with an age of 3 million years. Note that the high-mass (high-luminosity) stars have already arrived at the main-sequence stage of their lives, while the lower-mass (lower-luminosity) stars are still contracting toward the zero-age main sequence (the red line) and are not yet hot enough to derive all of their energy from the fusion of hydrogen.

There are real star clusters that fit this description. The first to be studied (in about 1950) was NGC 2264, which is still associated with the region of gas and dust from which it was born (Figure 22.9).

Figure 22.9. Young Cluster NGC 2264. Located about 2600 light-years from us, this region of newly formed stars, known as the Christmas Tree Cluster, is a complex mixture of hydrogen gas (which is ionized by hot embedded stars and shown in red), dark obscuring dust lanes, and brilliant young stars. The image shows a scene about 30 light-years across. (credit: ESO)

Download for free at http://cnx.org/content/col11992/latest/

The NGC 2264 cluster's H–R diagram is shown in Figure 22.10. The cluster in the middle of the Orion Nebula (shown in Figure 21.4 and Figure 21.5) is in a similar stage of evolution.

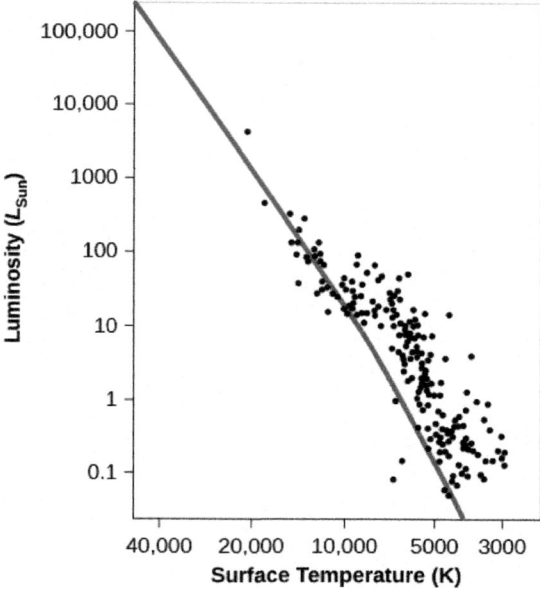

Figure 22.10. NGC 2264 H–R Diagram. Compare this H–R diagram to that in Figure 22.8; although the points scatter a bit more here, the theoretical and observational diagrams are remarkably, and satisfyingly, similar.

As clusters get older, their H–R diagrams begin to change. After a short time (less than a million years after they reach the main sequence), the most massive stars use up the hydrogen in their cores and evolve off the main sequence to become red giants and supergiants. As more time passes, stars of lower mass begin to leave the main sequence and make their way to the upper right of the H–R diagram.

LINK TO LEARNING

To see the evolution of a star cluster in a dwarf galaxy, you can watch this brief animation (https://openstax.org/l/30StarCluster) of how its H–R diagram changes.

Figure 22.11 is a photograph of NGC 3293, a cluster that is about 10 million years old. The dense clouds of gas and dust are gone. One massive star has evolved to become a red giant and stands out as an especially bright orange member of the cluster.

This OpenStax book is available for free at http://cnx.org/content/col11992/1.8

Download for free at http://cnx.org/content/col11992/latest/

Figure 22.11. NGC 3293. All the stars in an open star cluster like NGC 3293 form at about the same time. The most massive stars, however, exhaust their nuclear fuel more rapidly and hence evolve more quickly than stars of low mass. As stars evolve, they become redder. The bright orange star in NGC 3293 is the member of the cluster that has evolved most rapidly. (credit: ESO/G. Beccari)

Figure 22.12 shows the H–R diagram of the open cluster M41, which is roughly 100 million years old; by this time, a significant number of stars have moved off to the right and become red giants. Note the gap that appears in this H–R diagram between the stars near the main sequence and the red giants. A gap does not necessarily imply that stars avoid a region of certain temperatures and luminosities. In this case, it simply represents a domain of temperature and luminosity through which stars evolve very quickly. We see a gap for M41 because at this particular moment, we have not caught a star in the process of scurrying across this part of the diagram.

Download for free at http://cnx.org/content/col11992/latest/

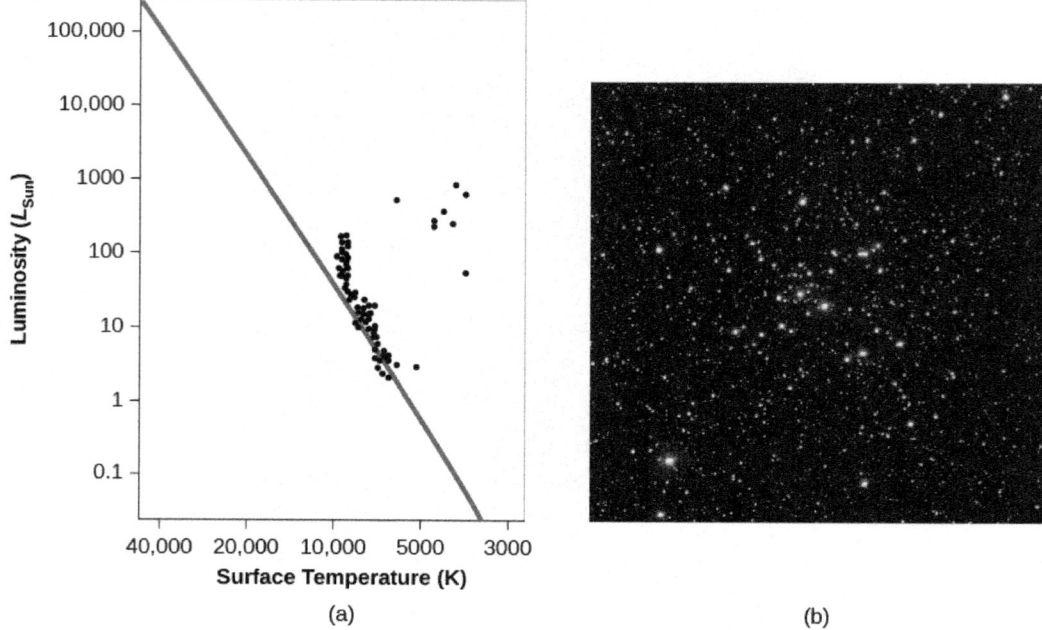

Figure 22.12. Cluster M41. (a) Cluster M41 is older than NGC 2264 (see Figure 22.10) and contains several red giants. Some of its more massive stars are no longer close to the zero-age main sequence (red line). (b) This ground-based photograph shows the open cluster M41. Note that it contains several orange-color stars. These are stars that have exhausted hydrogen in their centers, and have swelled up to become red giants. (credit b: modification of work by NOAO/AURA/NSF)

H–R Diagrams of Older Clusters

After 4 billion years have passed, many more stars, including stars that are only a few times more massive than the Sun, have left the main sequence (Figure 22.13). This means that no stars are left near the top of the main sequence; only the low-mass stars near the bottom remain. The older the cluster, the lower the point on the main sequence (and the lower the mass of the stars) where stars begin to move toward the red giant region. The location in the H–R diagram where the stars have begun to leave the main sequence is called the **main-sequence turnoff**.

This OpenStax book is available for free at http://cnx.org/content/col11992/1.8

Download for free at http://cnx.org/content/col11992/latest/

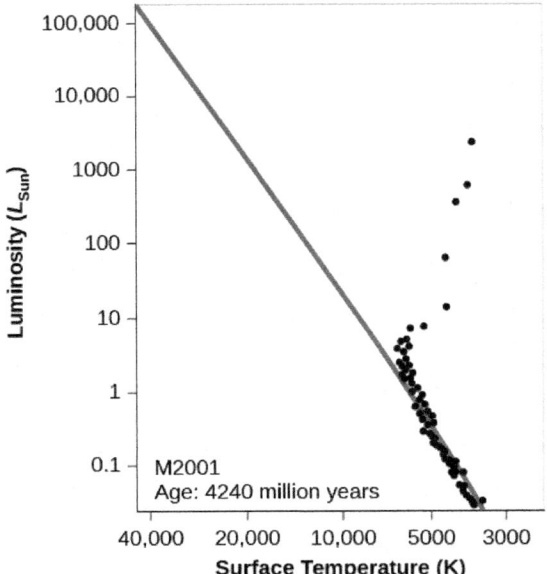

Figure 22.13. H–R Diagram for an Older Cluster. We see the H–R diagram for a hypothetical older cluster at an age of 4.24 billion years. Note that most of the stars on the upper part of the main sequence have turned off toward the red-giant region. And the most massive stars in the cluster have already died and are no longer on the diagram.

The oldest clusters of all are the globular clusters. Figure 22.14 shows the H–R diagram of globular cluster 47 Tucanae. Notice that the luminosity and temperature scales are different from those of the other H–R diagrams in this chapter. In Figure 22.13, for example, the luminosity scale on the left side of the diagram goes from 0.1 to 100,000 times the Sun's luminosity. But in Figure 22.14, the luminosity scale has been significantly reduced in extent. So many stars in this old cluster have had time to turn off the main sequence that only the very bottom of the main sequence remains.

Download for free at http://cnx.org/content/col11992/latest/

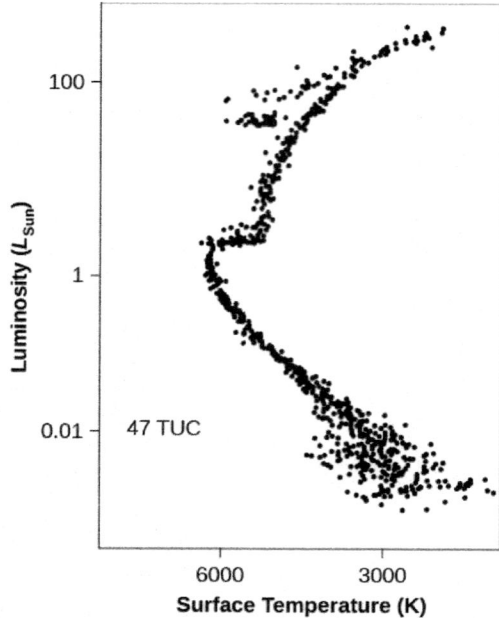

Figure 22.14. Cluster 47 Tucanae. This H–R diagram is for the globular cluster 47. Note that the scale of luminosity differs from that of the other H–R diagrams in this chapter. We are only focusing on the lower portion of the main sequence, the only part where stars still remain in this old cluster.

LINK TO LEARNING

Check out this brief NASA video with a 3-D visualization (https://openstax.org/l/30HRDiagram) of how an H–R diagram is created for the globular cluster Omega Centauri.

Just how old are the different clusters we have been discussing? To get their actual ages (in years), we must compare the appearances of our *calculated* H–R diagrams of different ages to *observed* H–R diagrams of real clusters. In practice, astronomers use the position at the top of the main sequence (that is, the luminosity at which stars begin to move off the main sequence to become red giants) as a measure of the age of a cluster (the main-sequence turnoff we discussed previously). For example, we can compare the luminosities of the brightest stars that are still on the main sequence in Figure 22.10 and Figure 22.13.

Using this method, some associations and open clusters turn out to be as young as 1 million years old, while others are several hundred million years old. Once all of the interstellar matter surrounding a cluster has been used to form stars or has dispersed and moved away from the cluster, star formation ceases, and stars of progressively lower mass move off the main sequence, as shown in Figure 22.10, Figure 22.12, and Figure 22.13.

To our surprise, even the youngest of the globular clusters in our Galaxy are found to be older than the oldest open cluster. All of the globular clusters have main sequences that turn off at a luminosity less than that of the Sun. Star formation in these crowded systems ceased billions of years ago, and no new stars are coming on to the main sequence to replace the ones that have turned off (see Figure 22.15).

This OpenStax book is available for free at http://cnx.org/content/col11992/1.8

Download for free at http://cnx.org/content/col11992/latest/

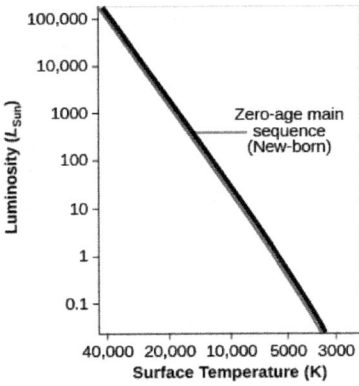 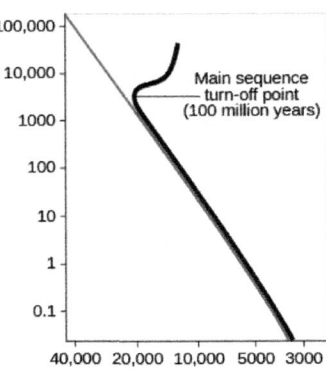 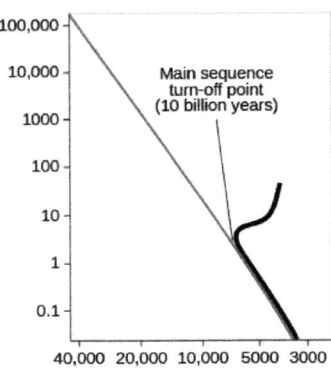

Figure 22.15. H–R Diagrams for Clusters of Different Ages. This sketch shows how the turn-off point from the main sequence gets lower as we make H–R diagrams for clusters that are older and older.

Indeed, the globular clusters are the oldest structures in our Galaxy (and in other galaxies as well). The youngest have ages of about 11 billion years and some appear to be even older. Since these are the oldest objects we know of, this estimate is one of the best limits we have on the age of the universe itself—it must be at least 11 billion years old. We will return to the fascinating question of determining the age of the entire universe in the chapter on The Big Bang.

22.4 | FURTHER EVOLUTION OF STARS

Learning Objectives

By the end of this section, you will be able to:

> Explain what happens in a star's core when all of the hydrogen has been used up
> Define "planetary nebulae" and discuss their origin
> Discuss the creation of new chemical elements during the late stages of stellar evolution

The "life story" we have related so far applies to almost all stars: each starts as a contracting protostar, then lives most of its life as a stable main-sequence star, and eventually moves off the main sequence toward the red-giant region.

As we have seen, the pace at which each star goes through these stages depends on its mass, with more massive stars evolving more quickly. But after this point, the life stories of stars of different masses diverge, with a wider range of possible behavior according to their masses, their compositions, and the presence of any nearby companion stars.

Because we have written this book for students taking their first astronomy course, we will recount a simplified version of what happens to stars as they move toward the final stages in their lives. We will (perhaps to your heartfelt relief) not delve into all the possible ways aging stars can behave and the strange things that happen when a star is orbited by a second star in a binary system. Instead, we will focus only on the key stages in the evolution of single stars and show how the evolution of high-mass stars differs from that of low-mass stars (such as our Sun).

Download for free at http://cnx.org/content/col11992/latest/

Helium Fusion

Let's begin by considering stars with composition like that of the Sun and whose *initial* masses are comparatively low—no more than about twice the mass of our Sun. (Such mass may not seem too low, but stars with masses less than this all behave in a fairly similar fashion. We will see what happens to more massive stars in the next section.) Because there are much more low-mass stars than high-mass stars in the Milky Way, the vast majority of stars—including our Sun—follow the scenario we are about to relate. By the way, we carefully used the term *initial masses* of stars because, as we will see, stars can lose quite a bit of mass in the process of aging and dying.

Remember that red giants start out with a helium core where no energy generation is taking place, surrounded by a shell where hydrogen is undergoing fusion. The core, having no source of energy to oppose the inward pull of gravity, is shrinking and growing hotter. As time goes on, the temperature in the core can rise to much hotter values than it had in its main-sequence days. Once it reaches a temperature of 100 million K (but not before such point), three helium atoms can begin to fuse to form a single carbon nucleus. This process is called the **triple-alpha process**, so named because physicists call the nucleus of the helium atom an alpha particle.

When the triple-alpha process begins in low-mass (about 0.8 to 2.0 solar masses) stars, calculations show that the entire core is ignited in a quick burst of fusion called a **helium flash**. (More massive stars also ignite helium but more gradually and not with a flash.) As soon as the temperature at the center of the star becomes high enough to start the triple-alpha process, the extra energy released is transmitted quickly through the entire helium core, producing very rapid heating. The heating speeds up the nuclear reactions, which provide more heating, and which accelerates the nuclear reactions even more. We have runaway generation of energy, which reignites the entire helium core in a flash.

You might wonder why the next major step in nuclear fusion in stars involves three helium nuclei and not just two. Although it is a lot easier to get two helium nuclei to collide, the product of this collision is not stable and falls apart very quickly. It takes three helium nuclei coming together *simultaneously* to make a stable nuclear structure. Given that each helium nucleus has two positive protons and that such protons repel one another, you can begin to see the problem. It takes a temperature of 100 million K to slam three helium nuclei (six protons) together and make them stick. But when that happens, the star produces a carbon nucleus.

ASTRONOMY BASICS

Stars in Your Little Finger

Stop reading for a moment and look at your little finger. It's full of carbon atoms because carbon is a fundamental chemical building block for life on Earth. Each of those carbon atoms was once inside a red giant star and was fused from helium nuclei in the triple-alpha process. All the carbon on Earth—in you, in the charcoal you use for barbecuing, and in the diamonds you might exchange with a loved one—was "cooked up" by previous generations of stars. How the carbon atoms (and other elements) made their way from inside some of those stars to become part of Earth is something we will discuss in the next chapter. For now, we want to emphasize that our description of stellar evolution is, in a very real sense, the story of our own cosmic "roots"—the history of how our own atoms originated among the stars. We are made of "star-stuff."

This OpenStax book is available for free at http://cnx.org/content/col11992/1.8

Download for free at http://cnx.org/content/col11992/latest/

Becoming a Giant Again

After the helium flash, the star, having survived the "energy crisis" that followed the end of the main-sequence stage and the exhaustion of the hydrogen fuel at its center, finds its balance again. As the star readjusts to the release of energy from the triple-alpha process in its core, its internal structure changes once more: its surface temperature increases and its overall luminosity decreases. The point that represents the star on the H–R diagram thus moves to a new position to the left of and somewhat below its place as a red giant (Figure 22.16). The star then continues to fuse the helium in its core for a while, returning to the kind of equilibrium between pressure and gravity that characterized the main-sequence stage. During this time, a newly formed carbon nucleus at the center of the star can sometimes be joined by another helium nucleus to produce a nucleus of oxygen—another building block of life.

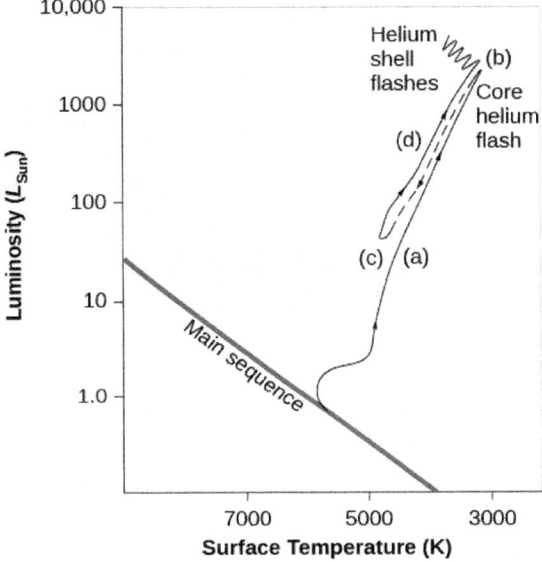

Figure 22.16. Evolution of a Star Like the Sun on an H–R Diagram. Each stage in the star's life is labeled. (a) The star evolves from the main sequence to be a red giant, decreasing in surface temperature and increasing in luminosity. (b) A helium flash occurs, leading to a readjustment of the star's internal structure and to (c) a brief period of stability during which helium is fused to carbon and oxygen in the core (in the process the star becomes hotter and less luminous than it was as a red giant). (d) After the central helium is exhausted, the star becomes a giant again and moves to higher luminosity and lower temperature. By this time, however, the star has exhausted its inner resources and will soon begin to die. Where the evolutionary track becomes a dashed line, the changes are so rapid that they are difficult to model.

However, at a temperature of 100 million K, the inner core is converting its helium fuel or carbon (and a bit of oxygen) at a rapid rate. Thus, the new period of stability cannot last very long: it is far shorter than the main-sequence stage. Soon, all the helium hot enough for fusion will be used up, just like the hot hydrogen that was used up earlier in the star's evolution. Once again, the inner core will not be able to generate energy via fusion. Once more, gravity will take over, and the core will start to shrink again. We can think of stellar evolution as a story of a constant struggle against gravitational collapse. A star can avoid collapsing as long as it can tap energy sources, but once any particular fuel is used up, it starts to collapse again.

The star's situation is analogous to the end of the main-sequence stage (when the central hydrogen got used up), but the star now has a somewhat more complicated structure. Again, the star's core begins to collapse under its own weight. Heat released by the shrinking of the carbon and oxygen core flows into a shell of helium just above the core. This helium, which had not been hot enough for fusion into carbon earlier, is heated just enough for fusion to begin and to generate a new flow of energy.

Download for free at http://cnx.org/content/col11992/latest/

Farther out in the star, there is also a shell where fresh hydrogen has been heated enough to fuse helium. The star now has a multi-layered structure like an onion: a carbon-oxygen core, surrounded by a shell of helium fusion, a layer of helium, a shell of hydrogen fusion, and finally, the extended outer layers of the star (see Figure 22.17). As energy flows outward from the two fusion shells, once again the outer regions of the star begin to expand. Its brief period of stability is over; the star moves back to the red-giant domain on the H–R diagram for a short time (see Figure 22.16). But this is a brief and final burst of glory.

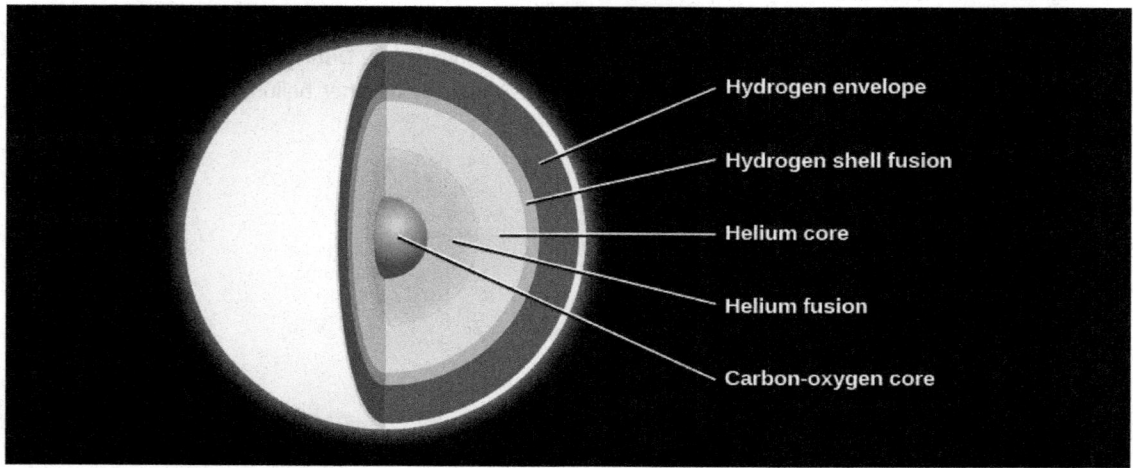

Figure 22.17. Layers inside a Low-Mass Star before Death. Here we see the layers inside a star with an initial mass that is less than twice the mass of the Sun. These include, from the center outward, the carbon-oxygen core, a layer of helium hot enough to fuse, a layer of cooler helium, a layer of hydrogen hot enough to fuse, and then cooler hydrogen beyond.

Recall that the last time the star was in this predicament, helium fusion came to its rescue. The temperature at the star's center eventually became hot enough for the *product* of the previous step of fusion (helium) to become the *fuel* for the next step (helium fusing into carbon). But the step after the fusion of helium nuclei requires a temperature so hot that the kinds of lower-mass stars (less than 2 solar masses) we are discussing simply cannot compress their cores to reach it. No further types of fusion are possible for such a star.

In a star with a mass similar to that of the Sun, the formation of a carbon-oxygen core thus marks the end of the generation of nuclear energy at the center of the star. The star must now confront the fact that its death is near. We will discuss how stars like this end their lives in The Death of Stars, but in the meantime, Table 22.4 summarizes the stages discussed so far in the life of a star with the same mass as that of the Sun. One thing that gives us confidence in our calculations of stellar evolution is that when we make H–R diagrams of older clusters, we actually see stars in each of the stages that we have been discussing.

The Evolution of a Star with the Sun's Mass

Stage	Time in This Stage (years)	Surface Temperature (K)	Luminosity (L_{Sun})	Diameter (Sun = 1)
Main sequence	11 billion	6000	1	1
Becomes red giant	1.3 billion	3100 at minimum	2300 at maximum	165

Table 22.4

This OpenStax book is available for free at http://cnx.org/content/col11992/1.8

Download for free at http://cnx.org/content/col11992/latest/

The Evolution of a Star with the Sun's Mass

Stage	Time in This Stage (years)	Surface Temperature (K)	Luminosity (L_{Sun})	Diameter (Sun = 1)
Helium fusion	100 million	4800	50	10
Giant again	20 million	3100	5200	180

Table 22.4

Mass Loss from Red-Giant Stars and the Formation of Planetary Nebulae

When stars swell up to become red giants, they have very large radii and therefore a low escape velocity.[2] Radiation pressure, stellar pulsations, and violent events like the helium flash can all drive atoms in the outer atmosphere away from the star, and cause it to lose a substantial fraction of its mass into space. Astronomers estimate that by the time a star like the Sun reaches the point of the helium flash, for example, it will have lost as much as 25% of its mass. And it can lose still more mass when it ascends the red-giant branch for the second time. As a result, aging stars are surrounded by one or more expanding shells of gas, each containing as much as 10–20% of the Sun's mass (or 0.1–0.2 M_{Sun}).

When nuclear energy generation in the carbon-oxygen core ceases, the star's core begins to shrink again and to heat up as it gets more and more compressed. (Remember that this compression will not be halted by another type of fusion in these low-mass stars.) The whole star follows along, shrinking and also becoming very hot—reaching surface temperatures as high as 100,000 K. Such hot stars are very strong sources of stellar winds and ultraviolet radiation, which sweep outward into the shells of material ejected when the star was a red giant. The winds and the ultraviolet radiation heat the shells, ionize them, and set them aglow (just as ultraviolet radiation from hot, young stars produces H II regions; see Between the Stars: Gas and Dust in Space).

The result is the creation of some of the most beautiful objects in the cosmos (see the gallery in Figure 22.18 and Figure 22.1). These objects were given an extremely misleading name when first found in the eighteenth century: **planetary nebulae**. The name is derived from the fact that a few planetary nebulae, when viewed through a small telescope, have a round shape bearing a superficial resemblance to planets. Actually, they have nothing to do with planets, but once names are put into regular use in astronomy, it is extremely difficult to change them. There are tens of thousands of planetary nebulae in our own Galaxy, although many are hidden from view because their light is absorbed by interstellar dust.

---

2   Recall that the force of gravity depends not only on the mass doing the pulling, but also on our distance from the center of gravity. As a red giant star gets a lot bigger, a point on the surface of the star is now farther from the center, and thus has less gravity. That's why the speed needed to escape the star goes down.

Download for free at http://cnx.org/content/col11992/latest/

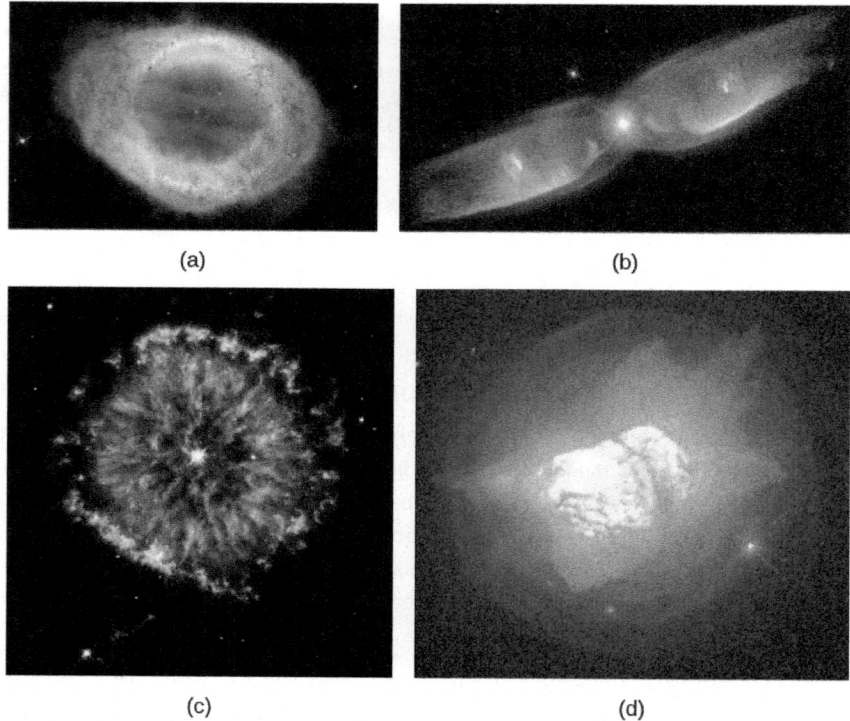

Figure 22.18. Gallery of Planetary Nebulae. This series of beautiful images depicting some intriguing planetary nebulae highlights the capabilities of the Hubble Space Telescope. (a) Perhaps the best known planetary nebula is the Ring Nebula (M57), located about 2000 light-years away in the constellation of Lyra. The ring is about 1 light-year in diameter, and the central star has a temperature of about 120,000 °C. Careful study of this image has shown scientists that, instead of looking at a spherical shell around this dying star, we may be looking down the barrel of a tube or cone. The blue region shows emission from very hot helium, which is located very close to the star; the red region isolates emission from ionized nitrogen, which is radiated by the coolest gas farthest from the star; and the green region represents oxygen emission, which is produced at intermediate temperatures and is at an intermediate distance from the star. (b) This planetary nebula, M2-9, is an example of a butterfly nebula. The central star (which is part of a binary system) has ejected mass preferentially in two opposite directions. In other images, a disk, perpendicular to the two long streams of gas, can be seen around the two stars in the middle. The stellar outburst that resulted in the expulsion of matter occurred about 1200 years ago. Neutral oxygen is shown in red, once-ionized nitrogen in green, and twice-ionized oxygen in blue. The planetary nebula is about 2100 light-years away in the constellation of Ophiuchus. (c) In this image of the planetary nebula NGC 6751, the blue regions mark the hottest gas, which forms a ring around the central star. The orange and red regions show the locations of cooler gas. The origin of these cool streamers is not known, but their shapes indicate that they are affected by radiation and stellar winds from the hot star at the center. The temperature of the star is about 140,000 °C. The diameter of the nebula is about 600 times larger than the diameter of our solar system. The nebula is about 6500 light-years away in the constellation of Aquila. (d) This image of the planetary nebula NGC 7027 shows several stages of mass loss. The faint blue concentric shells surrounding the central region identify the mass that was shed slowly from the surface of the star when it became a red giant. Somewhat later, the remaining outer layers were ejected but not in a spherically symmetric way. The dense clouds formed by this late ejection produce the bright inner regions. The hot central star can be seen faintly near the center of the nebulosity. NGC 7027 is about 3000 light-years away in the direction of the constellation of Cygnus. (credit a: modification of work by NASA, ESA, and the Hubble Heritage (STScI/AURA)-ESA/Hubble Collaboration; credit b: modification of work by Bruce Balick (University of Washington), Vincent Icke (Leiden University, The Netherlands), Garrelt Mellema (Stockholm University), and NASA; credit c: modification of work by NASA, The Hubble Heritage Team (STScI/AURA); credit d: modification of work by H. Bond (STScI) and NASA)

As Figure 22.18 shows, sometimes a planetary nebula appears to be a simple ring. Others have faint shells surrounding the bright ring, which is evidence that there were multiple episodes of mass loss when the star was a red giant (see image (d) in Figure 22.18). In a few cases, we see two lobes of matter flowing in opposite directions. Many astronomers think that a considerable number of planetary nebulae basically consist of the same structure, but that the shape we see depends on the viewing angle (Figure 22.19). According to this idea, the dying star is surrounded by a very dense, doughnut-shaped disk of gas. (Theorists do not yet have a definite explanation for why the dying star should produce this ring, but many believe that binary stars, which are common, are involved.)

This OpenStax book is available for free at http://cnx.org/content/col11992/1.8

Download for free at http://cnx.org/content/col11992/latest/

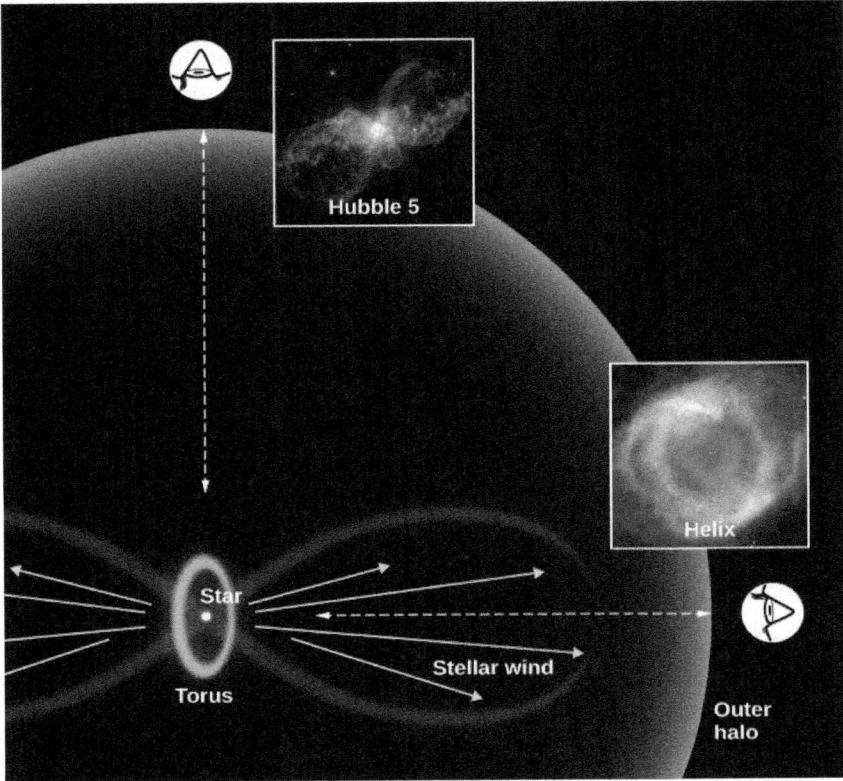

Figure 22.19. Model to Explain the Different Shapes of Planetary Nebulae. The range of different shapes that we see among planetary nebulae may, in many cases, arise from the same geometric shape, but seen from a variety of viewing directions. The basic shape is a hot central star surrounded by a thick torus (or doughnut-shaped disk) of gas. The star's wind cannot flow out into space very easily in the direction of the torus, but can escape more freely in the two directions perpendicular to it. If we view the nebula along the direction of the flow (Helix Nebula), it will appear nearly circular (like looking directly down into an empty ice-cream cone). If we look along the equator of the torus, we see both outflows and a very elongated shape (Hubble 5). Current research on planetary nebulae focuses on the reasons for having a torus around the star in the first place. Many astronomers suggest that the basic cause may be that many of the central stars are actually close binary stars, rather than single stars. (credit "Hubble 5": modification of work by Bruce Balick (University of Washington), Vincent Icke (Leiden University, The Netherlands), Garrelt Mellema (Stockholm University), and NASA/ESA; credit "Helix": modification of work by NASA, ESA, C.R. O'Dell (Vanderbilt University), and M. Meixner, P. McCullough)

As the star continues to lose mass, any less dense gas that leaves the star cannot penetrate the torus, but the gas *can* flow outward in directions perpendicular to the disk. If we look perpendicular to the direction of outflow, we see the disk and both of the outward flows. If we look "down the barrel" and into the flows, we see a ring. At intermediate angles, we may see wonderfully complex structures. Compare the viewpoints in Figure 22.19 with the images in Figure 22.18.

Planetary nebula shells usually expand at speeds of 20–30 km/s, and a typical planetary nebula has a diameter of about 1 light-year. If we assume that the gas shell has expanded at a constant speed, we can calculate that the shells of all the planetary nebulae visible to us were ejected within the past 50,000 years at most. After this amount of time, the shells have expanded so much that they are too thin and tenuous to be seen. That's a pretty short time that each planetary nebula can be observed (when compared to the whole lifetime of the star). Given the number of such nebulae we nevertheless see, we must conclude that a large fraction of all stars evolve through the planetary nebula phase. Since we saw that low-mass stars are much more common than high-mass stars, this confirms our view of planetary nebulae as sort of "last gasp" of low-mass star evolution.

Download for free at http://cnx.org/content/col11992/latest/

Cosmic Recycling

The loss of mass by dying stars is a key step in the gigantic cosmic recycling scheme we discussed in Between the Stars: Gas and Dust in Space. Remember that stars form from vast clouds of gas and dust. As they end their lives, stars return part of their gas to the galactic reservoirs of raw material. Eventually, some of the expelled material from aging stars will participate in the formation of new star systems.

However, the atoms returned to the Galaxy by an aging star are not necessarily the same ones it received initially. The star, after all, has fused hydrogen and helium to form new elements over the course of its life. And during the red-giant stage, material from the star's central regions is dredged up and mixed with its outer layers, which can cause further nuclear reactions and the creation of still more new elements. As a result, the winds that blow outward from such stars include atoms that were "newly minted" inside the stars' cores. (As we will see, this mechanism is even more effective for high-mass stars, but it does work for stars with masses like that of the Sun.) In this way, the raw material of the Galaxy is not only resupplied but also receives infusions of new elements. You might say this cosmic recycling plan allows the universe to get more "interesting" all the time.

MAKING CONNECTIONS

The Red Giant Sun and the Fate of Earth

How will the evolution of the Sun affect conditions on Earth in the future? Although the Sun has appeared reasonably steady in size and luminosity over recorded human history, that brief span means nothing compared with the timescales we have been discussing. Let's examine the long-term prospects for our planet.

The Sun took its place on the zero-age main sequence approximately 4.5 billion years ago. At that time, it emitted only about 70% of the energy that it radiates today. One might expect that Earth would have been a lot colder than it is now, with the oceans frozen solid. But if this were the case, it would be hard to explain why simple life forms existed when Earth was less than a billion years old. Scientists now think that the explanation may be that much more carbon dioxide was present in Earth's atmosphere when it was young, and that a much stronger greenhouse effect kept Earth warm. (In the greenhouse effect, gases like carbon dioxide or water vapor allow the Sun's light to come in but do not allow the infrared radiation from the ground to escape back into space, so the temperature near Earth's surface increases.)

Carbon dioxide in Earth's atmosphere has steadily declined as the Sun has increased in luminosity. As the brighter Sun increases the temperature of Earth, rocks weather faster and react with carbon dioxide, removing it from the atmosphere. The warmer Sun and the weaker greenhouse effect have kept Earth at a nearly constant temperature for most of its life. This remarkable coincidence, which has resulted in fairly stable climatic conditions, has been the key in the development of complex life-forms on our planet.

As a result of changes caused by the buildup of helium in its core, the Sun will continue to increase in luminosity as it grows older, and more and more radiation will reach Earth. For a while, the amount of carbon dioxide will continue to decrease. (Note that this effect counteracts increases in carbon dioxide from human activities, but on a much-too-slow timescale to undo the changes in climate that are likely to occur in the next 100 years.)

Eventually, the heating of Earth will melt the polar caps and increase the evaporation of the oceans. Water vapor is also an efficient greenhouse gas and will more than compensate for the decrease in

This OpenStax book is available for free at http://cnx.org/content/col11992/1.8

carbon dioxide. Sooner or later (atmospheric models are not yet good enough to say exactly when, but estimates range from 500 million to 2 billion years), the increased water vapor will cause a runaway greenhouse effect.

About 1 billion years from now, Earth will lose its water vapor. In the upper atmosphere, sunlight will break down water vapor into hydrogen, and the fast-moving hydrogen atoms will escape into outer space. Like Humpty Dumpty, the water molecules cannot be put back together again. Earth will start to resemble the Venus of today, and temperatures will become much too high for life as we know it.

All of this will happen before the Sun even becomes a red giant. Then the bad news really starts. The Sun, as it expands, will swallow Mercury and Venus, and friction with our star's outer atmosphere will make these planets spiral inward until they are completely vaporized. It is not completely clear whether Earth will escape a similar fate. As described in this chapter, the Sun will lose some of its mass as it becomes a red giant. The gravitational pull of the Sun decreases when it loses mass. The result would be that the diameter of Earth's orbit would increase (remember Kepler's third law). However, recent calculations also show that forces due to the tides raised on the Sun by Earth will act in the opposite direction, causing Earth's orbit to shrink. Thus, many astrophysicists conclude that Earth will be vaporized along with Mercury and Venus. Whether or not this dire prediction is true, there is little doubt that all life on Earth will surely be incinerated. But don't lose any sleep over this—we are talking about events that will occur billions of years from now.

What then are the prospects for preserving Earth life as we know it? The first strategy you might think of would be to move humanity to a more distant and cooler planet. However, calculations indicate that there are long periods of time (several hundred million years) when no planet is habitable. For example, Earth becomes far too warm for life long before Mars warms up enough.

A better alternative may be to move the entire Earth progressively farther from the Sun. The idea is to use gravity in the same way NASA has used it to send spacecraft to distant planets. When a spacecraft flies near a planet, the planet's motion can be used to speed up the spacecraft, slow it down, or redirect it. Calculations show that if we were to redirect an asteroid so that it follows just the right orbit between Earth and Jupiter, it could transfer orbital energy from Jupiter to Earth and move Earth slowly outward, pulling us away from the expanding Sun on each flyby. Since we have hundreds of millions of years to change Earth's orbit, the effect of each flyby need not be large. (Of course, the people directing the asteroid had better get the orbit exactly right and not cause the asteroid to hit Earth.)

It may seem crazy to think about projects to move an entire planet to a different orbit. But remember that we are talking about the distant future. If, by some miracle, human beings are able to get along for all that time and don't blow ourselves to bits, our technology is likely to be far more sophisticated than it is today. It may also be that if humans survive for hundreds of millions of years, we may spread to planets or habitats around other stars. Indeed, Earth, by then, might be a museum world to which youngsters from other planets return to learn about the origin of our species. It is also possible that evolution will by then have changed us in ways that allow us to survive in very different environments. Wouldn't it be exciting to see how the story of the story of the human race turns out after all those billions of years?

Download for free at http://cnx.org/content/col11992/latest/

22.5 THE EVOLUTION OF MORE MASSIVE STARS

Learning Objectives

By the end of this section, you will be able to:

> Explain how and why massive stars evolve much more rapidly than lower-mass stars like our Sun
> Discuss the origin of the elements heavier than carbon within stars

If what we have described so far were the whole story of the evolution of stars and elements, we would have a big problem on our hands. We will see in later chapters that in our best models of the first few minutes of the universe, everything starts with the two simplest elements—hydrogen and helium (plus a tiny bit of lithium). All the predictions of the models imply that no heavier elements were produced at the beginning of the universe. Yet when we look around us on Earth, we see lots of other elements besides hydrogen and helium. These elements must have been made (fused) somewhere in the universe, *and the only place hot enough to make them is inside stars.* One of the fundamental discoveries of twentieth-century astronomy is that the stars are the source of most of the chemical richness that characterizes our world and our lives.

We have already seen that carbon and some oxygen are manufactured inside the lower-mass stars that become red giants. But where do the heavier elements we know and love (such as the silicon and iron inside Earth, and the gold and silver in our jewelry) come from? The kinds of stars we have been discussing so far never get hot enough at their centers to make these elements. It turns out that such heavier elements can be formed only late in the lives of *more massive* stars.

Making New Elements in Massive Stars

Massive stars evolve in much the same way that the Sun does (but always more quickly)—up to the formation of a carbon-oxygen core. One difference is that for stars with more than about twice the mass of the Sun, helium begins fusion more gradually, rather than with a sudden flash. Also, when more massive stars become red giants, they become so bright and large that we call them *supergiants*. Such stars can expand until their outer regions become as large as the orbit of Jupiter, which is precisely what the Hubble Space Telescope has shown for the star Betelgeuse (see Figure 22.4). They also lose mass very effectively, producing dramatic winds and outbursts as they age. Figure 22.20 shows a wonderful image of the very massive star Eta Carinae, with a great deal of ejected material clearly visible.

This OpenStax book is available for free at http://cnx.org/content/col11992/1.8

Download for free at http://cnx.org/content/col11992/latest/

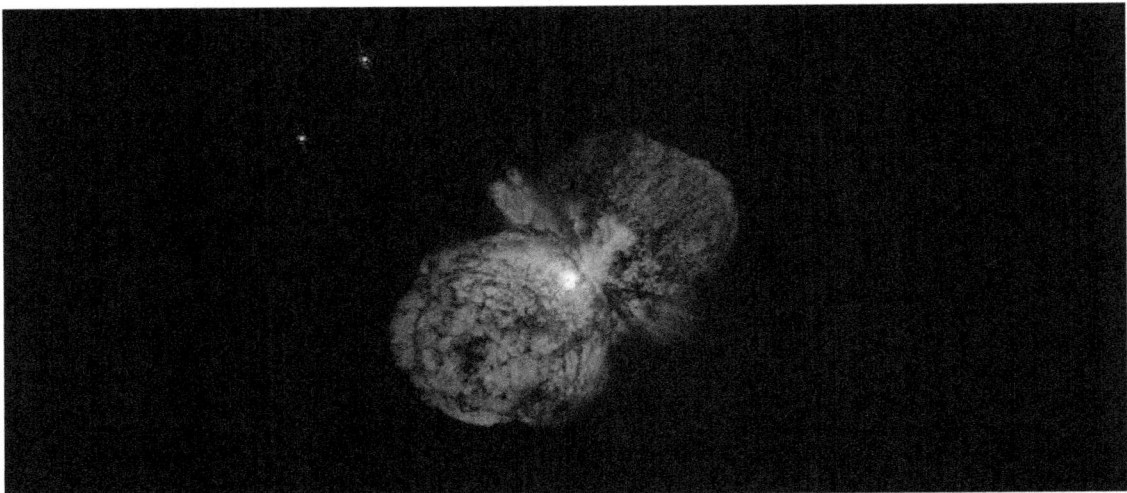

Figure 22.20. Eta Carinae. With a mass at least 100 times that of the Sun, the hot supergiant Eta Carinae is one of the most massive stars known. This Hubble Space Telescope image records the two giant lobes and equatorial disk of material it has ejected in the course of its evolution. The pink outer region is material ejected in an outburst seen in 1843, the largest of such mass loss event that any star is known to have survived. Moving away from the star at a speed of about 1000 km/s, the material is rich in nitrogen and other elements formed in the interior of the star. The inner blue-white region is the material ejected at lower speeds and is thus still closer to the star. It appears blue-white because it contains dust and reflects the light of Eta Carinae, whose luminosity is 4 million times that of our Sun. (credit: modification of work by Jon Morse (University of Colorado) & NASA)

But the crucial way that massive stars diverge from the story we have outlined is that they can start additional kinds of fusion in their centers and in the shells surrounding their central regions. The outer layers of a star with a mass greater than about 8 solar masses have a weight that is enough to compress the carbon-oxygen core until it becomes hot enough to ignite fusion of carbon nuclei. Carbon can fuse into still more oxygen, and at still higher temperatures, oxygen and then neon, magnesium, and finally silicon can build even heavier elements. Iron is, however, the endpoint of this process. The fusion of iron atoms produces products that are *more* massive than the nuclei that are being fused and therefore the process *requires* energy, as opposed to releasing energy, which all fusion reactions up to this point have done. This required energy comes at the expense of the star itself, which is now on the brink of death (Figure 22.21). What happens next will be described in the chapter on The Death of Stars.

Download for free at http://cnx.org/content/col11992/latest/

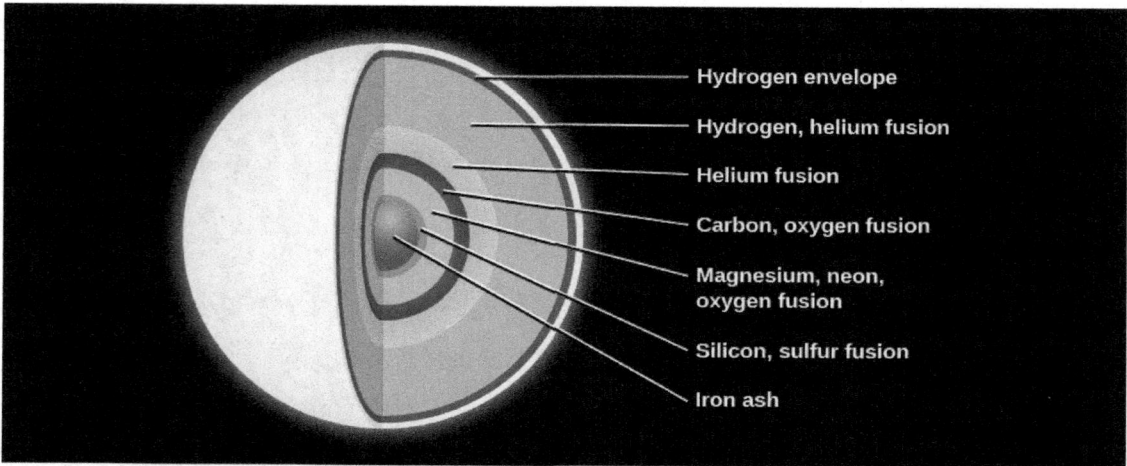

Figure 22.21. Interior Structure of a Massive Star Just before It Exhausts Its Nuclear Fuel. High-mass stars can fuse elements heavier than carbon. As a massive star nears the end of its evolution, its interior resembles an onion. Hydrogen fusion is taking place in an outer shell, and progressively heavier elements are undergoing fusion in the higher-temperature layers closer to the center. All of these fusion reactions generate energy and enable the star to continue shining. Iron is different. The fusion of iron requires energy, and when iron is finally created in the core, the star has only minutes to live.

Physicists have now found nuclear pathways whereby virtually all chemical elements of atomic weights up to that of iron can be built up by this **nucleosynthesis** (the making of new atomic nuclei) in the centers of the more massive red giant stars. This still leaves the question of where elements *heavier* than iron come from. We will see in the next chapter that when massive stars finally exhaust their nuclear fuel, they most often die in a spectacular explosion—a supernova. Heavier elements can be synthesized in the stunning violence of such explosions.

Not only can we explain in this way where the elements that make up our world and others come from, but our theories of nucleosynthesis inside stars are even able to predict the *relative abundances* with which the elements occur in nature. The way stars build up elements during various nuclear reactions really can explain why some elements (oxygen, carbon, and iron) are common and others are quite rare (gold, silver, and uranium).

Elements in Globular Clusters and Open Clusters Are Not the Same

The fact that the elements are made in stars over time explains an important difference between globular and open clusters. Hydrogen and helium, which are the most abundant elements in stars in the solar neighborhood, are also the most abundant constituents of stars in both kinds of clusters. However, the abundances of the elements *heavier* than helium are very different.

In the Sun and most of its neighboring stars, the combined abundance (by mass) of the elements heavier than hydrogen and helium is 1–4% of the star's mass. Spectra show that most open-cluster stars also have 1–4% of their matter in the form of heavy elements. Globular clusters, however, are a different story. The heavy-element abundance of stars in typical globular clusters is found to be only 1/10 to 1/100 that of the Sun. A few very old stars not in clusters have been discovered with even lower abundances of heavy elements.

The differences in chemical composition are a direct consequence of the formation of a cluster of stars. The very first generation of stars initially contained only hydrogen and helium. We have seen that these stars, in order to generate energy, created heavier elements in their interiors. In the last stages of their lives, they ejected matter, now enriched in heavy elements, into the reservoirs of raw material between the stars. Such matter was then incorporated into a new generation of stars.

This OpenStax book is available for free at http://cnx.org/content/col11992/1.8

This means that the relative abundance of the heavy elements must be less and less as we look further into the past. We saw that the globular clusters are much older than the open clusters. Since globular-cluster stars formed much earlier (that is, they are an earlier generation of stars) than those in open clusters, they have only a relatively small abundance of elements heavier than hydrogen and helium.

As time passes, the proportion of heavier elements in the "raw material" that makes new stars and planets increases. This means that the first generation of stars that formed in our Galaxy would not have been accompanied by a planet like Earth, full of silicon, iron, and many other heavy elements. Earth (and the astronomy students who live on it) was possible only after generations of stars had a chance to make and recycle their heavier elements.

Now the search is on for true *first*-generation stars, made only of hydrogen and helium. Theories predict that such stars should be very massive, live fast, and die quickly. They should have lived and died long ago. The place to look for them is in very distant galaxies that formed when the universe was only a few hundred million years old, but whose light is only arriving at Earth now.

Approaching Death

Compared with the main-sequence lifetimes of stars, the events that characterize the last stages of stellar evolution pass very quickly (especially for massive stars). As the star's luminosity increases, its rate of nuclear fuel consumption goes up rapidly—just at that point in its life when its fuel supply is beginning to run down.

After the prime fuel—hydrogen—is exhausted in a star's core, we saw that other sources of nuclear energy are available to the star in the fusion of, first, helium, and then of other more complex elements. But the energy yield of these reactions is much less than that of the fusion of hydrogen to helium. And to trigger these reactions, the central temperature must be higher than that required for the fusion of hydrogen to helium, leading to even more rapid consumption of fuel. Clearly this is a losing game, and very quickly the star reaches its end. As it does so, however, some remarkable things can happen, as we will see in The Death of Stars.

Download for free at http://cnx.org/content/col11992/latest/

CHAPTER 22 REVIEW

 KEY TERMS

association a loose group of young stars whose spectral types, motions, and positions in the sky indicate a common origin

globular cluster one of about 150 large, spherical star clusters (each with hundreds of thousands of stars) that form a system of clusters in the center of our Galaxy

helium flash a nearly explosive ignition of helium in the triple-alpha process in the dense core of a red giant star

main-sequence turnoff location in the H–R diagram where stars begin to leave the main sequence

nucleosynthesis the building up of heavy elements from lighter ones by nuclear fusion

open cluster a comparatively loose cluster of stars, containing from a few dozen to a few thousand members, located in the spiral arms or disk of our Galaxy; sometimes referred to as a galactic cluster

planetary nebula a shell of gas ejected by and expanding away from an extremely hot low-mass star that is nearing the end of its life (the nebulae glow because of the ultra-violet energy of the central star)

triple-alpha process a nuclear reaction by which three helium nuclei are built up (fused) into one carbon nucleus

zero-age main sequence a line denoting the main sequence on the H–R diagram for a system of stars that have completed their contraction from interstellar matter and are now deriving all their energy from nuclear reactions, but whose chemical composition has not yet been altered substantially by nuclear reactions

 SUMMARY

22.1 Evolution from the Main Sequence to Red Giants

When stars first begin to fuse hydrogen to helium, they lie on the zero-age main sequence. The amount of time a star spends in the main-sequence stage depends on its mass. More massive stars complete each stage of evolution more quickly than lower-mass stars. The fusion of hydrogen to form helium changes the interior composition of a star, which in turn results in changes in its temperature, luminosity, and radius. Eventually, as stars age, they evolve away from the main sequence to become red giants or supergiants. The core of a red giant is contracting, but the outer layers are expanding as a result of hydrogen fusion in a shell outside the core. The star gets larger, redder, and more luminous as it expands and cools.

22.2 Star Clusters

Star clusters provide one of the best tests of our calculations of what happens as stars age. The stars in a given cluster were formed at about the same time and have the same composition, so they differ mainly in mass, and thus, in their life stage. There are three types of star clusters: globular, open, and associations. Globular clusters have diameters of 50–450 light-years, contain hundreds of thousands of stars, and are distributed in a halo around the Galaxy. Open clusters typically contain hundreds of stars, are located in the plane of the Galaxy, and have diameters less than 30 light-years. Associations are found in regions of gas and dust and contain extremely young stars.

This OpenStax book is available for free at http://cnx.org/content/col11992/1.8

22.3 Checking Out the Theory

The H–R diagram of stars in a cluster changes systematically as the cluster grows older. The most massive stars evolve most rapidly. In the youngest clusters and associations, highly luminous blue stars are on the main sequence; the stars with the lowest masses lie to the right of the main sequence and are still contracting toward it. With passing time, stars of progressively lower masses evolve away from (or turn off) the main sequence. In globular clusters, which are all at least 11 billion years old, there are no luminous blue stars at all. Astronomers can use the turnoff point from the main sequence to determine the age of a cluster.

22.4 Further Evolution of Stars

After stars become red giants, their cores eventually become hot enough to produce energy by fusing helium to form carbon (and sometimes a bit of oxygen.) The fusion of three helium nuclei produces carbon through the triple-alpha process. The rapid onset of helium fusion in the core of a low-mass star is called the helium flash. After this, the star becomes stable and reduces its luminosity and size briefly. In stars with masses about twice the mass of the Sun or less, fusion stops after the helium in the core has been exhausted. Fusion of hydrogen and helium in shells around the contracting core makes the star a bright red giant again, but only temporarily. When the star is a red giant, it can shed its outer layers and thereby expose hot inner layers. Planetary nebulae (which have nothing to do with planets) are shells of gas ejected by such stars, set glowing by the ultraviolet radiation of the dying central star.

22.5 The Evolution of More Massive Stars

In stars with masses higher than about 8 solar masses, nuclear reactions involving carbon, oxygen, and still heavier elements can build up nuclei as heavy as iron. The creation of new chemical elements is called nucleosynthesis. The late stages of evolution occur very quickly. Ultimately, all stars must use up all of their available energy supplies. In the process of dying, most stars eject some matter, enriched in heavy elements, into interstellar space where it can be used to form new stars. Each succeeding generation of stars therefore contains a larger proportion of elements heavier than hydrogen and helium. This progressive enrichment explains why the stars in open clusters (which formed more recently) contain more heavy elements than do those in ancient globular clusters, and it tells us where most of the atoms on Earth and in our bodies come from.

FOR FURTHER EXPLORATION

Articles

Balick, B. & Frank, A. "The Extraordinary Deaths of Ordinary Stars." *Scientific American* (July 2004): 50. About planetary nebulae, the last gasps of low-mass stars, and the future of our own Sun.

Djorgovsky, G. "The Dynamic Lives of Globular Clusters." *Sky & Telescope* (October 1998): 38. Cluster evolution and blue straggler stars.

Frank, A. "Angry Giants of the Universe." *Astronomy* (October 1997): 32. On luminous blue variables like Eta Carinae.

Garlick, M. "The Fate of the Earth." *Sky & Telescope* (October 2002): 30. What will happen when our Sun becomes a red giant.

Harris, W. & Webb, J. "Life Inside a Globular Cluster." *Astronomy* (July 2014): 18. What would night sky be like there?

Download for free at http://cnx.org/content/col11992/latest/

Iben, I. & Tutokov, A. "The Lives of the Stars: From Birth to Death and Beyond." *Sky & Telescope* (December 1997): 36.

Kaler, J. "The Largest Stars in the Galaxy." *Astronomy* (October 1990): 30. On red supergiants.

Kalirai, J. "New Light on Our Sun's Fate." *Astronomy* (February 2014): 44. What will happen to stars like our Sun between the main sequence and the white dwarf stages.

Kwok, S. "What Is the Real Shape of the Ring Nebula?" *Sky & Telescope* (July 2000): 33. On seeing planetary nebulae from different angles.

Kwok, S. "Stellar Metamorphosis." *Sky & Telescope* (October 1998): 30. How planetary nebulae form.

Stahler, S. "The Inner Life of Star Clusters." *Scientific American* (March 2013): 44–49. How all stars are born in clusters, but different clusters evolve differently.

Subinsky, R. "All About 47 Tucanae." *Astronomy* (September 2014): 66. What we know about this globular cluster and how to see it.

Websites

BBC Page on Giant Stars: http://www.bbc.co.uk/science/space/universe/sights/giant_stars. Includes basic information and links to brief video excerpts.

Encylopedia Brittanica Article on Star Clusters: http://www.britannica.com/topic/star-cluster. Written by astronomer Helen Sawyer Hogg-Priestley.

Hubble Image Gallery: Planetary Nebulae: http://hubblesite.org/gallery/album/nebula/planetary/. Click on each image to go to a page with more information available. (See also a similar gallery at the National Optical Astronomy Observatories: https://www.noao.edu/image_gallery/planetary_nebulae.html).

Hubble Image Gallery: Star Clusters: http://hubblesite.org/gallery/album/star/star_cluster/. Each image comes with an explanatory caption when you click on it. (See also a similar European Southern Observatory Gallery at: https://www.eso.org/public/images/archive/category/starclusters/).

Measuring the Age of a Star Cluster: https://www.e-education.psu.edu/astro801/content/l7_p6.html. From Penn State.

Videos

Life Cycle of Stars: https://www.youtube.com/watch?v=PM9CQDIQI0A. Short summary of stellar evolution from the Institute of Physics in Great Britain, with astronomer Tim O'Brien (4:58).

Missions Take an Unparalleled Look into Superstar Eta Carinae: https://www.youtube.com/watch?v=0rJQi6oaZf0. NASA Goddard video about observations in 2014 and what we know about the pair of stars in this complicated system (4:00).

Star Clusters: Open and Globular Clusters: https://www.youtube.com/watch?v=rGPRLxrYbYA. Three Short Hubblecast Videos from 2007–2008 on discoveries involving star clusters (12:24).

Tour of Planetary Nebula NGC 5189: https://www.youtube.com/watch?v=1D2cwiZld0o. Brief Hubblecast episode with Joe Liske, explaining planetary nebulae in general and one example in particular (5:22).

This OpenStax book is available for free at http://cnx.org/content/col11992/1.8

COLLABORATIVE GROUP ACTIVITIES

A. Have your group take a look at the list of the brightest stars in the sky in Appendix J. What fraction of them are past the main-sequence phase of evolution? The text says that stars spend 90% of their lifetimes in the main-sequence phase of evolution. This suggests that if we have a fair (or representative) sample of stars, 90% of them should be main-sequence stars. Your group should brainstorm why 90% of the brightest stars are not in the main-sequence phase of evolution.

B. Reading an H–R diagram can be tricky. Suppose your group is given the H–R diagram of a star cluster. Stars above and to the right of the main sequence could be either red giants that had evolved away from the main sequence or very young stars that are still evolving toward the main sequence. Discuss how you would decide which they are.

C. In the chapter on Life in the Universe, we discuss some of the efforts now underway to search for radio signals from possible intelligent civilizations around other stars. Our present resources for carrying out such searches are very limited and there are many stars in our Galaxy. Your group is a committee set up by the International Astronomical Union to come up with a list of the best possible stars with which such a search should begin. Make a list of criteria for choosing the stars on the list, and explain the reasons behind each entry (keeping in mind some of the ideas about the life story of stars and timescales that we discuss in the present chapter.)

D. Have your group make a list of the reasons why a star that formed at the very beginning of the universe (soon after the Big Bang) could not have a planet with astronomy students reading astronomy textbooks (even if the star has the same mass as that of our Sun).

E. Since we are pretty sure that when the Sun becomes a giant star, all life on Earth will be wiped out, does your group think that we should start making preparations of any kind? Let's suppose that a political leader who fell asleep during large parts of his astronomy class suddenly hears about this problem from a large donor and appoints your group as a task force to make suggestions on how to prepare for the end of Earth. Make a list of arguments for why such a task force is not really necessary.

F. Use star charts to identify at least one open cluster visible at this time of the year. (Such charts can be found in *Sky & Telescope* and *Astronomy* magazines each month and their websites; see Appendix B.) The Pleiades and Hyades are good autumn subjects, and Praesepe is good for springtime viewing. Go out and look at these clusters with binoculars and describe what you see.

G. Many astronomers think that planetary nebulae are among the most attractive and interesting objects we can see in the Galaxy. In this chapter, we could only show you a few examples of the pictures of these objects taken with the Hubble or large telescopes on the ground. Have members of your group search further for planetary nebula images online, and make a "top ten" list of your favorite ones (do not include more than three that were featured in this chapter.) Make a report (with images) for the whole class and explain why you found your top five especially interesting. (You may want to check Figure 22.19 in the process.)

🖱 EXERCISES

Download for free at http://cnx.org/content/col11992/latest/

Review Questions

1. Compare the following stages in the lives of a human being and a star: prenatal, birth, adolescence/adulthood, middle age, old age, and death. What does a star with the mass of our Sun do in each of these stages?

2. What is the first event that happens to a star with roughly the mass of our Sun that exhausts the hydrogen in its core and stops the generation of energy by the nuclear fusion of hydrogen to helium? Describe the sequence of events that the star undergoes.

3. Astronomers find that 90% of the stars observed in the sky are on the main sequence of an H–R diagram; why does this make sense? Why are there far fewer stars in the giant and supergiant region?

4. Describe the evolution of a star with a mass similar to that of the Sun, from the protostar stage to the time it first becomes a red giant. Give the description in words and then sketch the evolution on an H–R diagram.

5. Describe the evolution of a star with a mass similar to that of the Sun, from just after it first becomes a red giant to the time it exhausts the last type of fuel its core is capable of fusing.

6. A star is often described as "moving" on an H–R diagram; why is this description used and what is actually happening with the star?

7. On which edge of the main sequence band on an H–R diagram would the zero-age main sequence be?

8. How do stars typically "move" through the main sequence band on an H–R diagram? Why?

9. Certain stars, like Betelgeuse, have a lower surface temperature than the Sun and yet are more luminous. How do these stars produce so much more energy than the Sun?

10. Gravity always tries to collapse the mass of a star toward its center. What mechanism can oppose this gravitational collapse for a star? During what stages of a star's life would there be a "balance" between them?

11. Why are star clusters so useful for astronomers who want to study the evolution of stars?

12. Would the Sun more likely have been a member of a globular cluster or open cluster in the past?

13. Suppose you were handed two H–R diagrams for two different clusters: diagram A has a majority of its stars plotted on the upper left part of the main sequence with the rest of the stars off the main sequence; and diagram B has a majority of its stars plotted on the lower right part of the main sequence with the rest of the stars off the main sequence. Which diagram would be for the older cluster? Why?

14. Referring to the H–R diagrams in Exercise 22.13, which diagram would more likely be the H–R diagram for an association?

15. The nuclear process for fusing helium into carbon is often called the "triple-alpha process." Why is it called as such, and why must it occur at a much higher temperature than the nuclear process for fusing hydrogen into helium?

16. Pictures of various planetary nebulae show a variety of shapes, but astronomers believe a majority of planetary nebulae have the same basic shape. How can this paradox be explained?

17. Describe the two "recycling" mechanisms that are associated with stars (one during each star's life and the other connecting generations of stars).

18. In which of these star groups would you mostly likely find the least heavy-element abundance for the stars within them: open clusters, globular clusters, or associations?

This OpenStax book is available for free at http://cnx.org/content/col11992/1.8

19. Explain how an H–R diagram of the stars in a cluster can be used to determine the age of the cluster.

20. Where did the carbon atoms in the trunk of a tree on your college campus come from originally? Where did the neon in the fabled "neon lights of Broadway" come from originally?

21. What is a planetary nebula? Will we have one around the Sun?

Thought Questions

22. Is the Sun on the zero-age main sequence? Explain your answer.

23. How are planetary nebulae comparable to a fluorescent light bulb in your classroom?

24. Which of the planets in our solar system have orbits that are smaller than the photospheric radius of Betelgeuse listed in in Table 22.2?

25. Would you expect to find an earthlike planet (with a solid surface) around a very low-mass star that formed right at the beginning of a globular cluster's life? Explain.

26. In the H–R diagrams for some young clusters, stars of both very low and very high luminosity are off to the right of the main sequence, whereas those of intermediate luminosity are on the main sequence. Can you offer an explanation for that? Sketch an H–R diagram for such a cluster.

27. If the Sun were a member of the cluster NGC 2264, would it be on the main sequence yet? Why or why not?

28. If all the stars in a cluster have nearly the same age, why are clusters useful in studying evolutionary effects (different stages in the lives of stars)?

29. Suppose a star cluster were at such a large distance that it appeared as an unresolved spot of light through the telescope. What would you expect the overall color of the spot to be if it were the image of the cluster immediately after it was formed? How would the color differ after 10^{10} years? Why?

30. Suppose an astronomer known for joking around told you she had found a type-O main-sequence star in our Milky Way Galaxy that contained no elements heavier than helium. Would you believe her? Why?

31. Stars that have masses approximately 0.8 times the mass of the Sun take about 18 billion years to turn into red giants. How does this compare to the current age of the universe? Would you expect to find a globular cluster with a main-sequence turnoff for stars of 0.8 solar mass or less? Why or why not?

32. Automobiles are often used as an analogy to help people better understand how more massive stars have much shorter main-sequence lifetimes compared to less massive stars. Can you explain such an analogy using automobiles?

Figuring For Yourself

33. The text says a star does not change its mass very much during the course of its main-sequence lifetime. While it is on the main sequence, a star converts about 10% of the hydrogen initially present into helium (remember it's only the core of the star that is hot enough for fusion). Look in earlier chapters to find out what percentage of the hydrogen mass involved in fusion is lost because it is converted to energy. By how much does the mass of the whole star change as a result of fusion? Were we correct to say that the mass of a star does not change significantly while it is on the main sequence?

Download for free at http://cnx.org/content/col11992/latest/

34. The text explains that massive stars have shorter lifetimes than low-mass stars. Even though massive stars have more fuel to burn, they use it up faster than low-mass stars. You can check and see whether this statement is true. The lifetime of a star is directly proportional to the amount of mass (fuel) it contains and inversely proportional to the rate at which it uses up that fuel (i.e., to its luminosity). Since the lifetime of the Sun is about 10^{10} y, we have the following relationship:

$$T = 10^{10}\frac{M}{L}\text{ y}$$

where T is the lifetime of a main-sequence star, M is its mass measured in terms of the mass of the Sun, and L is its luminosity measured in terms of the Sun's luminosity.

    A. Explain in words why this equation works.

    B. Use the data in Table 18.3 to calculate the ages of the main-sequence stars listed.

    C. Do low-mass stars have longer main-sequence lifetimes?

    D. Do you get the same answers as those in Table 22.1?

35. You can use the equation in Exercise 22.34 to estimate the approximate ages of the clusters in Figure 22.10, Figure 22.12, and Figure 22.13. Use the information in the figures to determine the luminosity of the most massive star still on the main sequence. Now use the data in Table 18.3 to estimate the mass of this star. Then calculate the age of the cluster. This method is similar to the procedure used by astronomers to obtain the ages of clusters, except that they use actual data and model calculations rather than simply making estimates from a drawing. How do your ages compare with the ages in the text?

36. You can estimate the age of the planetary nebula in image (c) in Figure 22.18. The diameter of the nebula is 600 times the diameter of our own solar system, or about 0.8 light-year. The gas is expanding away from the star at a rate of about 25 mi/s. Considering that distance = velocity × time, calculate how long ago the gas left the star if its speed has been constant the whole time. Make sure you use consistent units for time, speed, and distance.

37. If star A has a core temperature T, and star B has a core temperature $3T$, how does the rate of fusion of star A compare to the rate of fusion of star B?

This OpenStax book is available for free at http://cnx.org/content/col11992/1.8

Figure 23.1. Stellar Life Cycle. This remarkable picture of NGC 3603, a nebula in the Milky Way Galaxy, was taken with the Hubble Space Telescope. This image illustrates the life cycle of stars. In the bottom half of the image, we see clouds of dust and gas, where it is likely that star formation will take place in the near future. Near the center, there is a cluster of massive, hot young stars that are only a few million years old. Above and to the right of the cluster, there is an isolated star surrounded by a ring of gas. Perpendicular to the ring and on either side of it, there are two bluish blobs of gas. The ring and the blobs were ejected by the star, which is nearing the end of its life. (credit: modification of work by NASA, Wolfgang Brandner (JPL/IPAC), Eva K. Grebel (University of Washington), You-Hua Chu (University of Illinois Urbana-Champaign))

Chapter Outline

Thinking Ahead

Do stars die with a bang or a whimper? In the preceding two chapters, we followed the life story of stars, from the process of birth to the brink of death. Now we are ready to explore the ways that stars end their lives. Sooner or later, each star exhausts its store of nuclear energy. Without a source of internal pressure to balance the weight of the overlying layers, every star eventually gives way to the inexorable pull of gravity and collapses under its own weight.

Following the rough distinction made in the last chapter, we will discuss the end-of-life evolution of stars of lower and higher mass separately. What determines the outcome—bang or whimper—is the mass of the star *when it is ready to die*, not the mass it was born with. As we noted in the last chapter, stars can lose a significant amount of mass in their middle and old age.

Download for free at http://cnx.org/content/col11992/latest/

23.1 | THE DEATH OF LOW-MASS STARS

Learning Objectives

By the end of this section, you will be able to:

> Describe the physical characteristics of degenerate matter and explain how the mass and radius of degenerate stars are related
> Plot the future evolution of a white dwarf and show how its observable features will change over time
> Distinguish which stars will become white dwarfs

Let's begin with those stars whose final mass just before death is less than about 1.4 times the mass of the Sun (M_{Sun}). (We will explain why this mass is the crucial dividing line in a moment.) Note that most stars in the universe fall into this category. The number of stars decreases as mass increases; really massive stars are rare (see The Stars: A Celestial Census). This is similar to the music business where only a few musicians ever become superstars. Furthermore, many stars with an initial mass much greater than 1.4 M_{Sun} will be reduced to that level by the time they die. For example, we now know that stars that start out with masses of at least 8.0 M_{Sun} (and possibly as much as 10 M_{Sun}) manage to lose enough mass during their lives to fit into this category (an accomplishment anyone who has ever attempted to lose weight would surely envy).

A Star in Crisis

In the last chapter, we left the life story of a star with a mass like the Sun's just after it had climbed up to the red-giant region of the H–R diagram for a second time and had shed some of its outer layers to form a planetary nebula. Recall that during this time, the *core* of the star was undergoing an "energy crisis." Earlier in its life, during a brief stable period, helium in the core had gotten hot enough to fuse into carbon (and oxygen). But after this helium was exhausted, the star's core had once more found itself without a source of pressure to balance gravity and so had begun to contract.

This collapse is the final event in the life of the core. Because the star's mass is relatively low, it cannot push its core temperature high enough to begin another round of fusion (in the same way larger-mass stars can). The core continues to shrink until it reaches a density equal to nearly a million times the density of water! That is 200,000 times greater than the average density of Earth. At this extreme density, a new and different way for matter to behave kicks in and helps the star achieve a final state of equilibrium. In the process, what remains of the star becomes one of the strange *white dwarfs* that we met in The Stars: A Celestial Census.

Degenerate Stars

Because white dwarfs are far denser than any substance on Earth, the matter inside them behaves in a very unusual way—unlike anything we know from everyday experience. At this high density, gravity is incredibly strong and tries to shrink the star still further, but all the *electrons* resist being pushed closer together and set up a powerful pressure inside the core. This pressure is the result of the fundamental rules that govern the behavior of electrons (the quantum physics you were introduced to in The Sun: A Nuclear Powerhouse). According to these rules (known to physicists as the Pauli exclusion principle), which have been verified in studies of atoms in the laboratory, no two electrons can be in the same place at the same time doing the same thing. We specify the *place* of an electron by its position in space, and we specify what it is doing by its motion and the way it is spinning.

The temperature in the interior of a star is always so high that the atoms are stripped of virtually all their electrons. For most of a star's life, the density of matter is also relatively low, and the electrons in the star are

This OpenStax book is available for free at http://cnx.org/content/col11992/1.8

moving rapidly. This means that no two of them will be in the same place moving in exactly the same way at the same time. But this all changes when a star exhausts its store of nuclear energy and begins its final collapse.

As the star's core contracts, electrons are squeezed closer and closer together. Eventually, a star like the Sun becomes so dense that further contraction would in fact require two or more electrons to violate the rule against occupying the same place and moving in the same way. Such a dense gas is said to be degenerate (a term coined by physicists and not related to the electron's moral character). The electrons in a **degenerate gas** resist further crowding with tremendous pressure. (It's as if the electrons said, "You can press inward all you want, but there is simply no room for any other electrons to squeeze in here without violating the rules of our existence.")

The degenerate electrons do not require an input of heat to maintain the pressure they exert, and so a star with this kind of structure, if nothing disturbs it, can last essentially forever. (Note that the repulsive force between degenerate electrons is different from, and much stronger than, the normal electrical repulsion between charges that have the same sign.)

The electrons in a degenerate gas do move about, as do particles in any gas, but not with a lot of freedom. A particular electron cannot change position or momentum until another electron in an adjacent stage gets out of the way. The situation is much like that in the parking lot after a big football game. Vehicles are closely packed, and a given car cannot move until the one in front of it moves, leaving an empty space to be filled.

Of course, the dying star also has atomic nuclei in it, not just electrons, but it turns out that the nuclei must be squeezed to much higher densities before their quantum nature becomes apparent. As a result, in white dwarfs, the nuclei do not exhibit degeneracy pressure. Hence, in the white dwarf stage of stellar evolution, it is the degeneracy pressure of the electrons, and not of the nuclei, that halts the collapse of the core.

White Dwarfs

White dwarfs, then, are stable, compact objects with electron-degenerate cores that cannot contract any further. Calculations showing that white dwarfs are the likely end state of low-mass stars were first carried out by the Indian-American astrophysicist Subrahmanyan Chandrasekhar. He was able to show how much a star will shrink before the degenerate electrons halt its further contraction and hence what its final diameter will be (Figure 23.2).

When Chandrasekhar made his calculation about white dwarfs, he found something very surprising: the radius of a white dwarf shrinks as the mass in the star increases (the larger the mass, the more tightly packed the electrons can become, resulting in a smaller radius). According to the best theoretical models, a white dwarf with a mass of about 1.4 M_{Sun} or larger would have a radius of zero. What the calculations are telling us is that even the force of degenerate electrons cannot stop the collapse of a star with more mass than this. The maximum mass that a star can end its life with and still become a white dwarf—1.4 M_{Sun}—is called the **Chandrasekhar limit**. Stars with end-of-life masses that exceed this limit have a different kind of end in store—one that we will explore in the next section.

Download for free at http://cnx.org/content/col11992/latest/

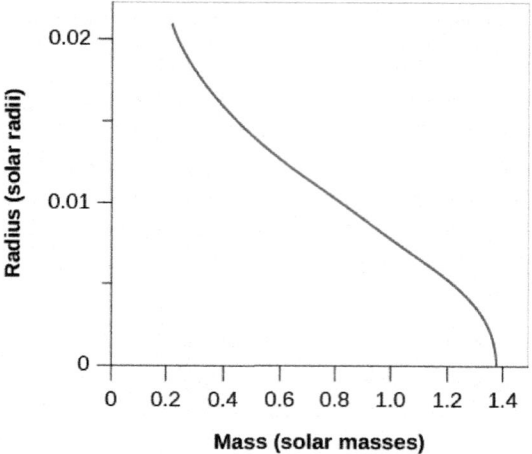

Figure 23.2. Relating Masses and Radii of White Dwarfs. Models of white-dwarf structure predict that as the mass of the star increases (toward the right), its radius gets smaller and smaller.

VOYAGERS IN ASTRONOMY

Subrahmanyan Chandrasekhar

Born in 1910 in Lahore, India, Subrahmanyan Chandrasekhar (known as Chandra to his friends and colleagues) grew up in a home that encouraged scholarship and an interest in science (Figure 23.3). His uncle, C. V. Raman, was a physicist who won the 1930 Nobel Prize. A precocious student, Chandra tried to read as much as he could about the latest ideas in physics and astronomy, although obtaining technical books was not easy in India at the time. He finished college at age 19 and won a scholarship to study in England. It was during the long boat voyage to get to graduate school that he first began doing calculations about the structure of white dwarf stars.

Chandra developed his ideas during and after his studies as a graduate student, showing—as we have discussed—that white dwarfs with masses greater than 1.4 times the mass of the Sun cannot exist and that the theory predicts the existence of other kinds of stellar corpses. He wrote later that he felt very shy and lonely during this period, isolated from students, afraid to assert himself, and sometimes waiting for hours to speak with some of the famous professors he had read about in India. His calculations soon brought him into conflict with certain distinguished astronomers, including Sir Arthur Eddington, who publicly ridiculed Chandra's ideas. At a number of meetings of astronomers, such leaders in the field as Henry Norris Russell refused to give Chandra the opportunity to defend his ideas, while allowing his more senior critics lots of time to criticize them.

Yet Chandra persevered, writing books and articles elucidating his theories, which turned out not only to be correct, but to lay the foundation for much of our modern understanding of the death of stars. In 1983, he received the Nobel Prize in physics for this early work.

In 1937, Chandra came to the United States and joined the faculty at the University of Chicago, where he remained for the rest of his life. There he devoted himself to research and teaching, making major contributions to many fields of astronomy, from our understanding of the motions of stars through the

This OpenStax book is available for free at http://cnx.org/content/col11992/1.8

Galaxy to the behavior of the bizarre objects called black holes (see Black Holes and Curved Spacetime). In 1999, NASA named its sophisticated orbiting X-ray telescope (designed in part to explore such stellar corpses) the Chandra X-ray Observatory.

Figure 23.3. S. Chandrasekhar (1910–1995). Chandra's research provided the basis for much of what we now know about stellar corpses. (credit: modification of work by American Institute of Physics)

Chandra spent a great deal of time with his graduate students, supervising the research of more than 50 PhDs during his life. He took his teaching responsibilities very seriously: during the 1940s, while based at the Yerkes Observatory, he willingly drove the more than 100-mile trip to the university each week to teach a class of only a few students.

Chandra also had a deep devotion to music, art, and philosophy, writing articles and books about the relationship between the humanities and science. He once wrote that "one can learn science the way one enjoys music or art. . . . Heisenberg had a marvelous phrase 'shuddering before the beautiful'. . . that is the kind of feeling I have."

LINK TO LEARNING

Using the Hubble Space Telescope, astronomers were able to detect images (https://openstaxcollege.org/l/30hubimgwhidwa) of faint white dwarf stars and other "stellar corpses" in the M4 star cluster, located about 7200 light-years away.

The Ultimate Fate of White Dwarfs

If the birth of a main-sequence star is defined by the onset of fusion reactions, then we must consider the end of all fusion reactions to be the time of a star's death. As the core is stabilized by degeneracy pressure, a last shudder of fusion passes through the outside of the star, consuming the little hydrogen still remaining. Now the star is a true white dwarf: nuclear fusion in its interior has ceased. Figure 23.4 shows the path of a star like the Sun on the H–R diagram during its final stages.

Download for free at http://cnx.org/content/col11992/latest/

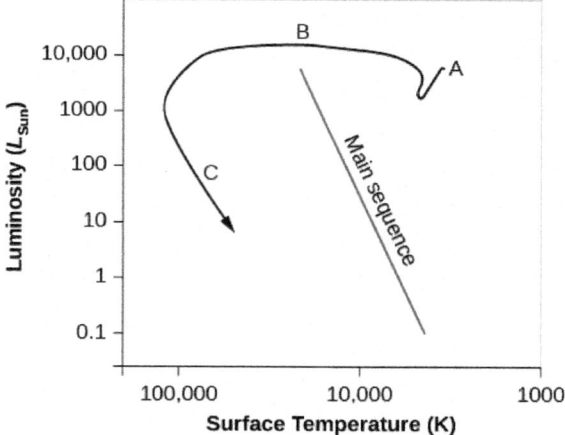

Figure 23.4. Evolutionary Track for a Star Like the Sun. This diagram shows the changes in luminosity and surface temperature for a star with a mass like the Sun's as it nears the end of its life. After the star becomes a giant again (point A on the diagram), it will lose more and more mass as its core begins to collapse. The mass loss will expose the hot inner core, which will appear at the center of a planetary nebula. In this stage, the star moves across the diagram to the left as it becomes hotter and hotter during its collapse (point B). At first, the luminosity remains nearly constant, but as the star begins to cool off, it becomes less and less bright (point C). It is now a white dwarf and will continue to cool slowly for billions of years until all of its remaining store of energy is radiated away. (This assumes the Sun will lose between 46–50% of its mass during the giant stages, based upon various theoretical models).

Since a stable white dwarf can no longer contract or produce energy through fusion, its only energy source is the heat represented by the motions of the atomic nuclei in its interior. The light it emits comes from this internal stored heat, which is substantial. Gradually, however, the white dwarf radiates away all its heat into space. After many billions of years, the nuclei will be moving much more slowly, and the white dwarf will no longer shine (Figure 23.5). It will then be a black dwarf—a cold stellar corpse with the mass of a star and the size of a planet. It will be composed mostly of carbon, oxygen, and neon, the products of the most advanced fusion reactions of which the star was capable.

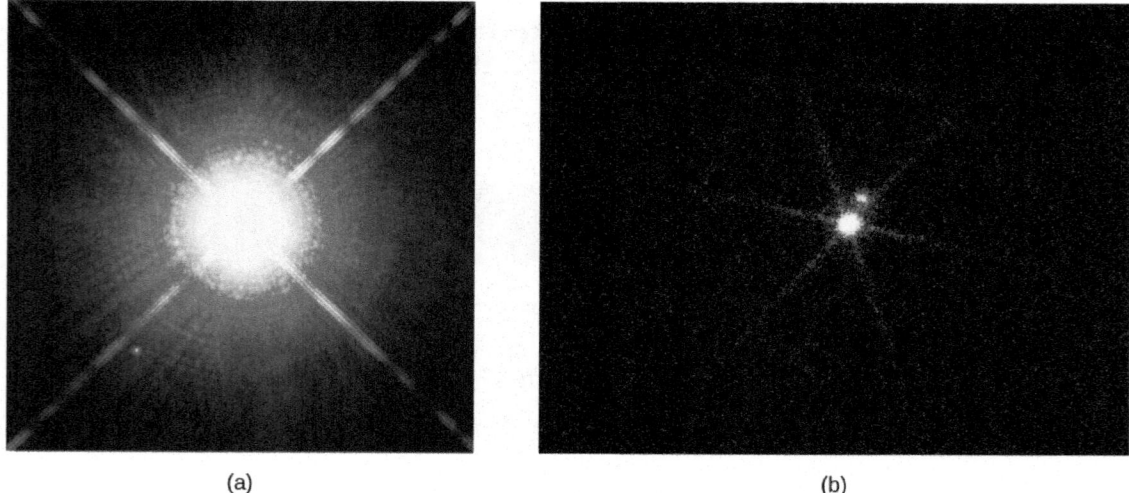

(a)   (b)

Figure 23.5. Visible Light and X-Ray Images of the Sirius Star System. (a) This image taken by the Hubble Space Telescope shows Sirius A (the large bright star), and its companion star, the white dwarf known as Sirius B (the tiny, faint star at the lower left). Sirius A and B are 8.6 light-years from Earth and are our fifth-closest star system. Note that the image has intentionally been overexposed to allow us to see Sirius B. (b) The same system is shown in X-ray taken with the Chandra Space Telescope. Note that Sirius A is fainter in X-rays than the hot white dwarf that is Sirius B. (credit a: modification of work by NASA, ESA, H. Bond, M. Barstow(University of Leicester); credit b: modification of work by NASA/SAO/CXC)

This OpenStax book is available for free at http://cnx.org/content/col11992/1.8

Download for free at http://cnx.org/content/col11992/latest/

We have one final surprise as we leave our low-mass star in the stellar graveyard. Calculations show that as a degenerate star cools, the atoms inside it in essence "solidify" into a giant, highly compact lattice (organized rows of atoms, just like in a crystal). When carbon is compressed and crystallized in this way, it becomes a giant *diamond-like* star. A white dwarf star might make the most impressive engagement present you could ever see, although any attempt to mine the diamond-like material inside would crush an ardent lover instantly!

LINK TO LEARNING

Learn about a recent "diamond star" find, a cold, white dwarf star (https://openstaxcollege.org/l/30diamondstar) detected in 2014, which is considered the coldest and dimmest found to date, at the website of the National Radio Astronomy Observatory.

Evidence That Stars Can Shed a Lot of Mass as They Evolve

Whether or not a star will become a white dwarf depends on how much mass is lost in the red-giant and earlier phases of evolution. All stars that have masses below the Chandrasekhar limit when they run out of fuel will become white dwarfs, no matter what mass they were born with. But which stars shed enough mass to reach this limit?

One strategy for answering this question is to look in young, open clusters (which were discussed in Star Clusters). The basic idea is to search for young clusters that contain one or more white dwarf stars. Remember that more massive stars go through all stages of their evolution more rapidly than less massive ones. Suppose we find a cluster that has a white dwarf member and also contains stars on the main sequence that have 6 times the mass of the Sun. This means that only those stars with masses greater than 6 M_{Sun} have had time to exhaust their supply of nuclear energy and complete their evolution to the white dwarf stage. The star that turned into the white dwarf must therefore have had a main-sequence mass of more than 6 M_{Sun}, since stars with lower masses have not yet had time to use up their stores of nuclear energy. The star that became the white dwarf must, therefore, have gotten rid of at least 4.6 M_{Sun} so that its mass at the time nuclear energy generation ceased could be less than 1.4 M_{Sun}.

Astronomers continue to search for suitable clusters to make this test, and the evidence so far suggests that stars with masses up to about 8 M_{Sun} can shed enough mass to end their lives as white dwarfs. Stars like the Sun will probably lose about 45% of their initial mass and become white dwarfs with masses less than 1.4 M_{Sun}.

23.2 EVOLUTION OF MASSIVE STARS: AN EXPLOSIVE FINISH

Learning Objectives

By the end of this section, you will be able to:

> Describe the interior of a massive star before a supernova
> Explain the steps of a core collapse and explosion
> List the hazards associated with nearby supernovae

Download for free at http://cnx.org/content/col11992/latest/

Thanks to mass loss, then, stars with starting masses up to at least 8 M_{Sun} (and perhaps even more) probably end their lives as white dwarfs. But we know stars can have masses as large as 150 (or more) M_{Sun}. They have a different kind of death in store for them. As we will see, these stars die with a bang.

Nuclear Fusion of Heavy Elements

After the helium in its core is exhausted (see The Evolution of More Massive Stars), the evolution of a massive star takes a significantly different course from that of lower-mass stars. In a massive star, the weight of the outer layers is sufficient to force the carbon core to contract until it becomes hot enough to fuse carbon into oxygen, neon, and magnesium. This cycle of contraction, heating, and the ignition of another nuclear fuel repeats several more times. After each of the possible nuclear fuels is exhausted, the core contracts again until it reaches a new temperature high enough to fuse still-heavier nuclei. The products of carbon fusion can be further converted into silicon, sulfur, calcium, and argon. And these elements, when heated to a still-higher temperature, can combine to produce iron. Massive stars go through these stages very, very quickly. In really massive stars, some fusion stages toward the very end can take only months or even days! This is a far cry from the millions of years they spend in the main-sequence stage.

At this stage of its evolution, a massive star resembles an onion with an iron core. As we get farther from the center, we find shells of decreasing temperature in which nuclear reactions involve nuclei of progressively lower mass—silicon and sulfur, oxygen, neon, carbon, helium, and finally, hydrogen (Figure 23.6).

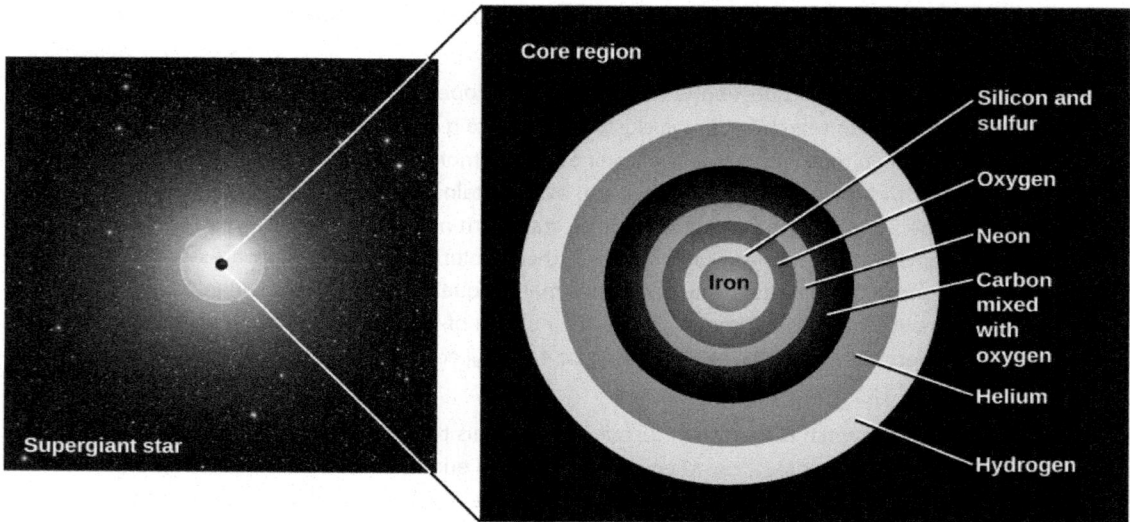

Figure 23.6. Structure of an Old Massive Star. Just before its final gravitational collapse, the core of a massive star resembles an onion. The iron core is surrounded by layers of silicon and sulfur, oxygen, neon, carbon mixed with some oxygen, helium, and finally hydrogen. Outside the core, the composition is mainly hydrogen and helium. (Note that this diagram is not precisely to scale but is just meant to convey the general idea of what such a star would be like.) (credit: modification of work by ESO, Digitized Sky Survey)

But there is a limit to how long this process of building up elements by fusion can go on. The fusion of silicon into iron turns out to be the last step in the sequence of nonexplosive element production. Up to this point, each fusion reaction has *produced* energy because the nucleus of each fusion product has been a bit more stable than the nuclei that formed it. As discussed in The Sun: A Nuclear Powerhouse, light nuclei give up some of their binding energy in the process of fusing into more tightly bound, heavier nuclei. It is this released energy that maintains the outward pressure in the core so that the star does not collapse. But of all the nuclei known, iron is the most tightly bound and thus the most stable.

This OpenStax book is available for free at http://cnx.org/content/col11992/1.8

Download for free at http://cnx.org/content/col11992/latest/

You might think of the situation like this: all smaller nuclei want to "grow up" to be like iron, and they are willing to pay (*produce* energy) to move toward that goal. But iron is a mature nucleus with good self-esteem, perfectly content being iron; it requires payment (must *absorb* energy) to change its stable nuclear structure. This is the exact opposite of what has happened in each nuclear reaction so far: instead of *providing* energy to balance the inward pull of gravity, any nuclear reactions involving iron would *remove* some energy from the core of the star.

Unable to generate energy, the star now faces catastrophe.

Collapse into a Ball of Neutrons

When nuclear reactions stop, the core of a massive star is supported by degenerate electrons, just as a white dwarf is. For stars that begin their evolution with masses of at least 10 M_{Sun}, this core is likely made mainly of iron. (For stars with initial masses in the range 8 to 10 M_{Sun}, the core is likely made of oxygen, neon, and magnesium, because the star never gets hot enough to form elements as heavy as iron. The exact composition of the cores of stars in this mass range is very difficult to determine because of the complex physical characteristics in the cores, particularly at the very high densities and temperatures involved.) We will focus on the more massive iron cores in our discussion.

While no energy is being generated within the white dwarf core of the star, fusion still occurs in the shells that surround the core. As the shells finish their fusion reactions and stop producing energy, the ashes of the last reaction fall onto the white dwarf core, increasing its mass. As Figure 23.2 shows, a higher mass means a smaller core. The core can contract because even a degenerate gas is still mostly empty space. Electrons and atomic nuclei are, after all, extremely small. The electrons and nuclei in a stellar core may be crowded compared to the air in your room, but there is still lots of space between them.

The electrons at first resist being crowded closer together, and so the core shrinks only a small amount. Ultimately, however, the iron core reaches a mass so large that even degenerate electrons can no longer support it. When the density reaches 4×10^{11} g/cm^3 (400 billion times the density of water), some electrons are actually squeezed into the atomic nuclei, where they combine with protons to form neutrons and neutrinos. This transformation is not something that is familiar from everyday life, but becomes very important as such a massive star core collapses.

Some of the electrons are now gone, so the core can no longer resist the crushing mass of the star's overlying layers. The core begins to shrink rapidly. More and more electrons are now pushed into the atomic nuclei, which ultimately become so saturated with neutrons that they cannot hold onto them.

At this point, the neutrons are squeezed out of the nuclei and can exert a new force. As is true for electrons, it turns out that the neutrons strongly resist being in the same place and moving in the same way. The force that can be exerted by such *degenerate neutrons* is much greater than that produced by degenerate electrons, so unless the core is too massive, they can ultimately stop the collapse.

This means the collapsing core can reach a stable state as a crushed ball made mainly of neutrons, which astronomers call a **neutron star**. We don't have an exact number (a "Chandrasekhar limit") for the maximum mass of a neutron star, but calculations tell us that the upper mass limit of a body made of neutrons might only be about 3 M_{Sun}. So if the mass of the core were greater than this, then even neutron degeneracy would not be able to stop the core from collapsing further. The dying star must end up as something even more extremely compressed, which until recently was believed to be only one possible type of object—the state of ultimate compaction known as a black hole (which is the subject of our next chapter). This is because no force was believed to exist that could stop a collapse beyond the neutron star stage.

Download for free at http://cnx.org/content/col11992/latest/

Collapse and Explosion

When the collapse of a high-mass star's core is stopped by degenerate neutrons, the core is saved from further destruction, but it turns out that the rest of the star is literally blown apart. Here's how it happens.

The collapse that takes place when electrons are absorbed into the nuclei is very rapid. In less than a second, a core with a mass of about 1 M_{Sun}, which originally was approximately the size of Earth, collapses to a diameter of less than 20 kilometers. The speed with which material falls inward reaches one-fourth the speed of light. The collapse halts only when the density of the core exceeds the density of an atomic nucleus (which is the densest form of matter we know). A typical neutron star is so compressed that to duplicate its density, we would have to squeeze all the people in the world into a single sugar cube! This would give us one sugar cube's worth (one cubic centimeter's worth) of a neutron star.

The neutron degenerate core strongly resists further compression, abruptly halting the collapse. The shock of the sudden jolt initiates a shock wave that starts to propagate outward. However, this shock alone is not enough to create a star explosion. The energy produced by the outflowing matter is quickly absorbed by atomic nuclei in the dense, overlying layers of gas, where it breaks up the nuclei into individual neutrons and protons.

Our understanding of nuclear processes indicates (as we mentioned above) that each time an electron and a proton in the star's core merge to make a neutron, the merger releases a *neutrino*. These ghostly subatomic particles, introduced in The Sun: A Nuclear Powerhouse, carry away some of the nuclear energy. It is their presence that launches the final disastrous explosion of the star. The total energy contained in the neutrinos is huge. In the initial second of the star's explosion, the power carried by the neutrinos (10^{46} watts) is greater than the power put out by all the stars in over a billion galaxies.

While neutrinos ordinarily do not interact very much with ordinary matter (we earlier accused them of being downright antisocial), matter near the center of a collapsing star is so dense that the neutrinos do interact with it to some degree. They deposit some of this energy in the layers of the star just outside the core. This huge, sudden input of energy reverses the infall of these layers and drives them explosively outward. Most of the mass of the star (apart from that which went into the neutron star in the core) is then ejected outward into space. As we saw earlier, such an explosion requires a star of at least 8 M_{Sun}, and the neutron star can have a mass of at most 3 M_{Sun}. Consequently, at least five times the mass of our Sun is ejected into space in each such explosive event!

The resulting explosion is called a supernova (Figure 23.7). When these explosions happen close by, they can be among the most spectacular celestial events, as we will discuss in the next section. (Actually, there are at least two different types of supernova explosions: the kind we have been describing, which is the collapse of a massive star, is called, for historical reasons, a **type II supernova**. We will describe how the types differ later in this chapter).

This OpenStax book is available for free at http://cnx.org/content/col11992/1.8

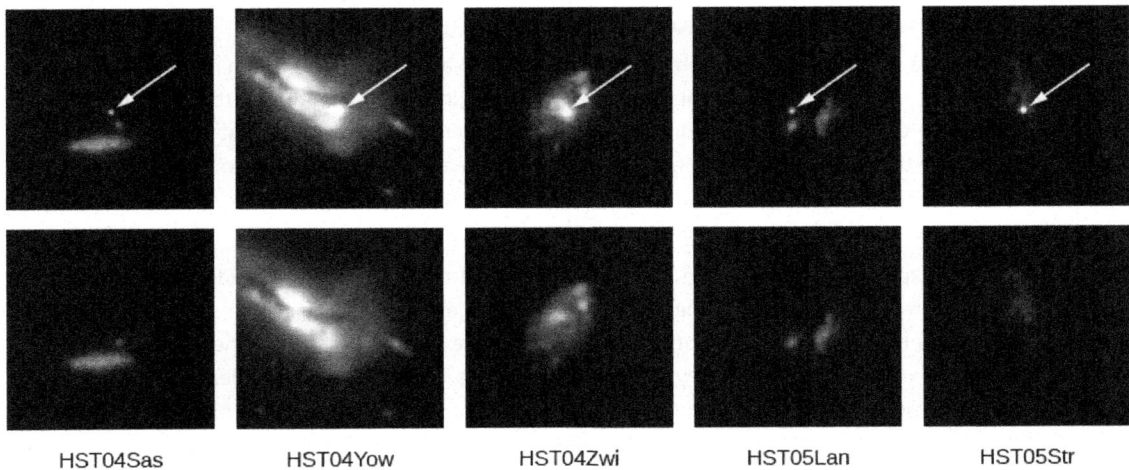

HST04Sas HST04Yow HST04Zwi HST05Lan HST05Str

Figure 23.7. Five Supernova Explosions in Other Galaxies. The arrows in the top row of images point to the supernovae. The bottom row shows the host galaxies before or after the stars exploded. Each of these supernovae exploded between 3.5 and 10 billion years ago. Note that the supernovae when they first explode can be as bright as an entire galaxy. (credit: modification of work by NASA, ESA, and A. Riess (STScI))

Table 23.1 summarizes the discussion so far about what happens to stars and substellar objects of different initial masses at the ends of their lives. Like so much of our scientific understanding, this list represents a progress report: it is the best we can do with our present models and observations. The mass limits corresponding to various outcomes may change somewhat as models are improved. There is much we do not yet understand about the details of what happens when stars die.

The Ultimate Fate of Stars and Substellar Objects with Different Masses

Initial Mass (Mass of Sun = 1)[1]	Final State at the End of Its Life
< 0.01	Planet
0.01 to 0.08	Brown dwarf
0.08 to 0.25	White dwarf made mostly of helium
0.25 to 8	White dwarf made mostly of carbon and oxygen
8 to 10	White dwarf made of oxygen, neon, and magnesium
10 to 40	Supernova explosion that leaves a neutron star
> 40	Supernova explosion that leaves a black hole

Table 23.1

---

1 Stars in the mass ranges 0.25–8 and 8–10 may later produce a type of supernova different from the one we have discussed so far. These are discussed in **The Evolution of Binary Star Systems**.

Download for free at http://cnx.org/content/col11992/latest/

The Supernova Giveth and the Supernova Taketh Away

After the supernova explosion, the life of a massive star comes to an end. But the death of each massive star is an important event in the history of its galaxy. The elements built up by fusion during the star's life are now "recycled" into space by the explosion, making them available to enrich the gas and dust that form new stars and planets. Because these heavy elements ejected by supernovae are critical for the formation of planets and the origin of life, it's fair to say that without mass loss from supernovae and planetary nebulae, neither the authors nor the readers of this book would exist.

But the supernova explosion has one more creative contribution to make, one we alluded to in Stars from Adolescence to Old Age when we asked where the atoms in your jewelry came from. The supernova explosion produces a flood of energetic neutrons that barrel through the expanding material. These neutrons can be absorbed by iron and other nuclei where they can turn into protons. Thus, they build up elements that are more massive than iron, including such terrestrial favorites as gold and silver. This is the only place we know where such heavier atoms as lead or uranium can be made. Next time you wear some gold jewelry (or give some to your sweetheart), bear in mind that those gold atoms were once part of an exploding star!

When supernovae explode, these elements (as well as the ones the star made during more stable times) are ejected into the existing gas between the stars and mixed with it. Thus, supernovae play a crucial role in enriching their galaxy with heavier elements, allowing, among other things, the chemical elements that make up earthlike planets and the building blocks of life to become more common as time goes on (Figure 23.8).

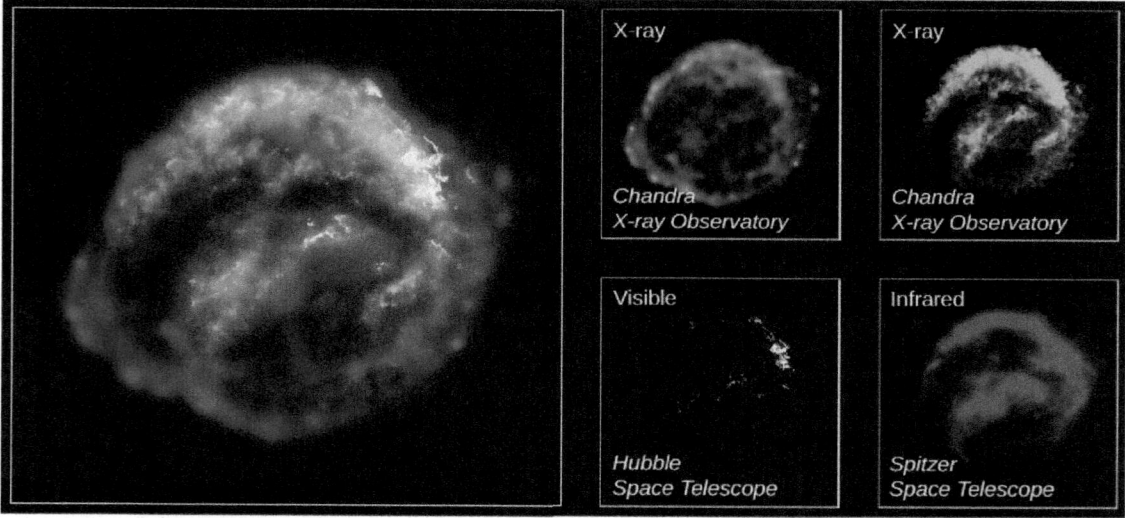

Figure 23.8. Kepler Supernova Remant. This image shows the expanding remains of a supernova explosion, which was first seen about 400 years ago by sky watchers, including the famous astronomer Johannes Kepler. The bubble-shaped shroud of gas and dust is now 14 light-years wide and is expanding at 2,000 kilometers per second (4 million miles per hour). The remnant emits energy at wavelengths from X-rays (shown in blue and green) to visible light (yellow) and into the infrared (red). The expanding shell is rich in iron, which was produced in the star that exploded. The main image combines the individual single-color images seen at the bottom into one multi-wavelength picture. (credit: modification of work by NASA, ESA, R. Sankrit and W. Blair (Johns Hopkins University))

Supernovae are also thought to be the source of many of the high-energy *cosmic ray* particles discussed in Cosmic Rays. Trapped by the magnetic field of the Galaxy, the particles from exploded stars continue to circulate around the vast spiral of the Milky Way. Scientists speculate that high-speed cosmic rays hitting the genetic material of Earth organisms over billions of years may have contributed to the steady *mutations*—subtle changes in the genetic code—that drive the evolution of life on our planet. In all the ways we have mentioned, supernovae have played a part in the development of new generations of stars, planets, and life.

This OpenStax book is available for free at http://cnx.org/content/col11992/1.8

Download for free at http://cnx.org/content/col11992/latest/

But supernovae also have a dark side. Suppose a life form has the misfortune to develop around a star that happens to lie near a massive star destined to become a supernova. Such life forms may find themselves snuffed out when the harsh radiation and high-energy particles from the neighboring star's explosion reach their world. If, as some astronomers speculate, life can develop on many planets around long-lived (lower-mass) stars, then the suitability of that life's *own star* and planet may not be all that matters for its long-term evolution and survival. Life may well have formed around a number of pleasantly stable stars only to be wiped out because a massive nearby star suddenly went supernova. Just as children born in a war zone may find themselves the unjust victims of their violent neighborhood, life too close to a star that goes supernova may fall prey to having been born in the wrong place at the wrong time.

What is a safe distance to be from a supernova explosion? A lot depends on the violence of the particular explosion, what type of supernova it is (see The Evolution of Binary Star Systems), and what level of destruction we are willing to accept. Calculations suggest that a supernova less than 50 light-years away from us would certainly end all life on Earth, and that even one 100 light-years away would have drastic consequences for the radiation levels here. One minor extinction of sea creatures about 2 million years ago on Earth may actually have been caused by a supernova at a distance of about 120 light-years.

The good news is that there are at present no massive stars that promise to become supernovae within 50 light-years of the Sun. (This is in part because the kinds of massive stars that become supernovae are overall quite rare.) The massive star closest to us, Spica (in the constellation of Virgo), is about 260 light-years away, probably a safe distance, even if it were to explode as a supernova in the near future.

EXAMPLE 23.1

Extreme Gravity

In this section, you were introduced to some very dense objects. How would those objects' gravity affect you? Recall that the force of gravity, F, between two bodies is calculated as

$$F = \frac{GM_1 M_2}{R^2}$$

where G is the gravitational constant, 6.67×10^{-11} Nm^2/kg^2, M_1 and M_2 are the masses of the two bodies, and R is their separation. Also, from Newton's second law,

$$F = M \times a$$

where a is the acceleration of a body with mass M.

So let's consider the situation of a mass—say, you—standing on a body, such as Earth or a white dwarf (where we assume you will be wearing a heat-proof space suit). You are M_1 and the body you are standing on is M_2. The distance between you and the center of gravity of the body on which you stand is its radius, R. The force exerted on you is

$$F = M_1 \times a = GM_1 M_2 \Big/ R^2$$

Solving for a, the acceleration of gravity on that world, we get

$$g = \frac{(G \times M)}{R^2}$$

Download for free at http://cnx.org/content/col11992/latest/

Note that we have replaced the general symbol for acceleration, a, with the symbol scientists use for the acceleration of gravity, g.

Say that a particular white dwarf has the mass of the Sun (2×10^{30} kg) but the radius of Earth (6.4×10^6 m). What is the acceleration of gravity at the surface of the white dwarf?

Solution

The acceleration of gravity at the surface of the white dwarf is

$$g(\text{white dwarf}) = \frac{(G \times M_{\text{Sun}})}{R_{\text{Earth}}{}^2} = \frac{\left(6.67 \times 10^{-11} \text{ m}^2/\text{kg s}^2 \times 2 \times 10^{30} \text{ kg}\right)}{\left(6.4 \times 10^6 \text{ m}\right)^2} = 3.26 \times 10^6 \text{ m/s}^2$$

Compare this to g on the surface of Earth, which is 9.8 m/s².

Check Your Learning

What is the acceleration of gravity at the surface if the white dwarf has the twice the mass of the Sun and is only half the radius of Earth?

Answer:

$$g(\text{white dwarf}) = \frac{(G \times 2M_{\text{Sun}})}{(0.5R_{\text{Earth}})^2} = \frac{\left(6.67 \times 10^{-11} \text{ m}^2/\text{kg s}^2 \times 4 \times 10^{30} \text{ kg}\right)}{\left(3.2 \times 10^6\right)^2} = 2.61 \times 10^7 \text{ m/s}^2$$

23.3 SUPERNOVA OBSERVATIONS

Learning Objectives

By the end of this section, you will be able to:

> Describe the observed features of SN 1987A both before and after the supernova
> Explain how observations of various parts of the SN 1987A event helped confirm theories about supernovae

Supernovae were discovered long before astronomers realized that these spectacular cataclysms mark the death of stars (see Making Connections: Supernovae in History). The word *nova* means "new" in Latin; before telescopes, when a star too dim to be seen with the unaided eye suddenly flared up in a brilliant explosion, observers concluded it must be a brand-new star. Twentieth-century astronomers reclassified the explosions with the greatest luminosity as *super*novae.

From historical records of such explosions, from studies of the remnants of supernovae in our Galaxy, and from analyses of supernovae in other galaxies, we estimate that, on average, one supernova explosion occurs somewhere in the Milky Way Galaxy every 25 to 100 years. Unfortunately, however, no supernova explosion has been observable in our Galaxy since the invention of the telescope. Either we have been exceptionally unlucky or, more likely, recent explosions have taken place in parts of the Galaxy where interstellar dust blocks light from reaching us.

This OpenStax book is available for free at http://cnx.org/content/col11992/1.8

Download for free at http://cnx.org/content/col11992/latest/

Supernovae in History

Although many supernova explosions in our own Galaxy have gone unnoticed, a few were so spectacular that they were clearly seen and recorded by sky watchers and historians at the time. We can use these records, going back two millennia, to help us pinpoint where the exploding stars were and thus where to look for their remnants today.

The most dramatic supernova was observed in the year 1006. It appeared in May as a brilliant point of light visible during the daytime, perhaps 100 times brighter than the planet Venus. It was bright enough to cast shadows on the ground during the night and was recorded with awe and fear by observers all over Europe and Asia. No one had seen anything like it before; Chinese astronomers, noting that it was a temporary spectacle, called it a "guest star."

Astronomers David Clark and Richard Stephenson have scoured records from around the world to find more than 20 reports of the 1006 supernova (SN 1006) (Figure 23.9). This has allowed them to determine with some accuracy where in the sky the explosion occurred. They place it in the modern constellation of Lupus; at roughly the position they have determined, we find a supernova remnant, now quite faint. From the way its filaments are expanding, it indeed appears to be about 1000 years old.

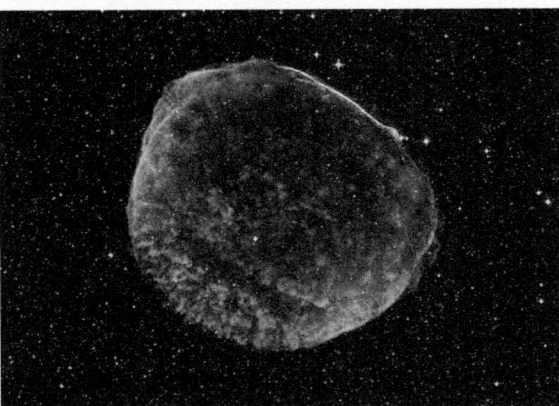

Figure 23.9. Supernova 1006 Remnant. This composite view of SN 1006 from the Chandra X-Ray Observatory shows the X-rays coming from the remnant in blue, visible light in white-yellow, and radio emission in red. (credit: modification of work by NASA, ESA, Zolt Levay(STScI))

Another guest star, now known as SN 1054, was clearly recorded in Chinese records in July 1054. The remnant of that star is one of the most famous and best-studied objects in the sky, called the Crab Nebula (Figure 23.14). It is a marvelously complex object, which has been key to understanding the death of massive stars. When its explosion was first seen, we estimate that it was about as bright as the planet Jupiter: nowhere near as dazzling as the 1006 event but still quite dramatic to anyone who kept track of objects in the sky. Another fainter supernova was seen in 1181.

The next supernova became visible in November 1572 and, being brighter than the planet Venus, was quickly spotted by a number of observers, including the young Tycho Brahe (see Orbits and Gravity). His careful measurements of the star over a year and a half showed that it was not a comet or something in Earth's atmosphere since it did not move relative to the stars. He correctly deduced that it must be a

Download for free at http://cnx.org/content/col11992/latest/

phenomenon belonging to the realm of the stars, not of the solar system. The remnant of Tycho's Supernova (as it is now called) can still be detected in many different bands of the electromagnetic spectrum.

Not to be outdone, Johannes Kepler, Tycho Brahe's scientific heir, found his own supernova in 1604, now known as Kepler's Supernova (Figure 23.8). Fainter than Tycho's, it nevertheless remained visible for about a year. Kepler wrote a book about his observations that was read by many with an interest in the heavens, including Galileo.

No supernova has been spotted in our Galaxy for the past 300 years. Since the explosion of a visible supernova is a chance event, there is no way to say when the next one might occur. Around the world, dozens of professional and amateur astronomers keep a sharp lookout for "new" stars that appear overnight, hoping to be the first to spot the next guest star in our sky and make a little history themselves.

At their maximum brightness, the most luminous supernovae have about 10 billion times the luminosity of the Sun. For a brief time, a supernova may outshine the entire galaxy in which it appears. After maximum brightness, the star's light fades and disappears from telescopic visibility within a few months or years. At the time of their outbursts, supernovae eject material at typical velocities of 10,000 kilometers per second (and speeds twice that have been observed). A speed of 20,000 kilometers per second corresponds to about 45 million miles per hour, truly an indication of great cosmic violence.

Supernovae are classified according to the appearance of their spectra, but in this chapter, we will focus on the two main causes of supernovae. Type Ia supernovae are ignited when a lot of material is dumped on degenerate white dwarfs (Figure 23.10); these supernovae will be discussed later in this chapter. For now, we will continue our story about the death of massive stars and focus on type II supernovae, which are produced when the core of a massive star collapses.

This OpenStax book is available for free at http://cnx.org/content/col11992/1.8

Download for free at http://cnx.org/content/col11992/latest/

SN 2014J January 31, 2014

Figure 23.10. Supernova 2014J. This image of supernova 2014J, located in Messier 82 (M82), which is also known as the Cigar galaxy, was taken by the Hubble Space Telescope and is superposed on a mosaic image of the galaxy also taken with Hubble. The supernova event is indicated by the box and the inset. This explosion was produced by a type Ia supernova, which is theorized to be triggered in binary systems consisting of a white dwarf and another star—and could be a second white dwarf, a star like our Sun, or a giant star. This type of supernova will be discussed later in this chapter. At a distance of approximately 11.5 million light-years from Earth, this is the closest supernova of type Ia discovered in the past few decades. In the image, you can see reddish plumes of hydrogen coming from the central region of the galaxy, where a considerable number of young stars are being born. (credit: modification of work by NASA, ESA, A. Goobar (Stockholm University), and the Hubble Heritage Team (STScI/AURA))

Supernova 1987A

Our most detailed information about what happens when a type II supernova occurs comes from an event that was observed in 1987. Before dawn on February 24, Ian Shelton, a Canadian astronomer working at an observatory in Chile, pulled a photographic plate from the developer. Two nights earlier, he had begun a survey of the Large Magellanic Cloud, a small galaxy that is one of the Milky Way's nearest neighbors in space. Where he expected to see only faint stars, he saw a large bright spot. Concerned that his photograph was flawed, Shelton went outside to look at the Large Magellanic Cloud . . . and saw that a new object had indeed appeared in the sky (see Figure 23.11). He soon realized that he had discovered a supernova, one that could be seen with the unaided eye even though it was about 160,000 light-years away.

Download for free at http://cnx.org/content/col11992/latest/

Figure 23.11. Hubble Space Telescope Image of SN 1987A. The supernova remnant with its inner and outer red rings of material is located in the Large Magellanic Cloud. This image is a composite of several images taken in 1994, 1996, and 1997—about a decade after supernova 1987A was first observed. (credit: modification of work by the Hubble Heritage Team (AURA/STScI/NASA/ESA))

Now known as SN 1987A, since it was the first supernova discovered in 1987, this brilliant newcomer to the southern sky gave astronomers their first opportunity to study the death of a relatively nearby star with modern instruments. It was also the first time astronomers had observed a star *before* it became a supernova. The star that blew up had been included in earlier surveys of the Large Magellanic Cloud, and as a result, we know the star was a blue supergiant just before the explosion.

By combining theory and observations at many different wavelengths, astronomers have reconstructed the life story of the star that became SN 1987A. Formed about 10 million years ago, it originally had a mass of about $20\ M_{Sun}$. For 90% of its life, it lived quietly on the main sequence, converting hydrogen into helium. At this time, its luminosity was about 60,000 times that of the Sun (L_{Sun}), and its spectral type was O. When the hydrogen in the center of the star was exhausted, the core contracted and ultimately became hot enough to fuse helium. By this time, the star was a red supergiant, emitting about 100,000 times more energy than the Sun. While in this stage, the star lost some of its mass.

This lost material has actually been detected by observations with the Hubble Space Telescope (Figure 23.12). The gas driven out into space by the subsequent supernova explosion is currently colliding with the material the star left behind when it was a red giant. As the two collide, we see a glowing ring.

This OpenStax book is available for free at http://cnx.org/content/col11992/1.8

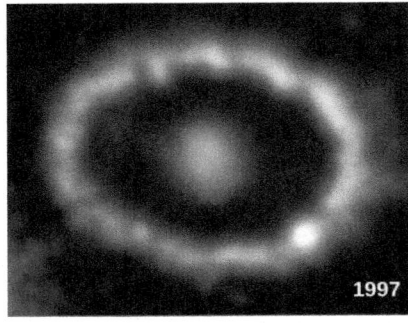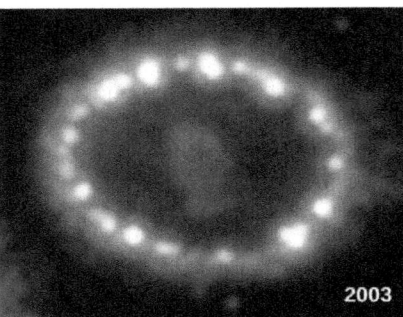

Figure 23.12. Ring around Supernova 1987A. These two images show a ring of gas expelled about 30,000 years ago when the star that exploded in 1987 was a red giant. The supernova, which has been artificially dimmed, is located at the center of the ring. The left-hand image was taken in 1997 and the right-hand image in 2003. Note that the number of bright spots has increased from 1 to more than 15 over this time interval. These spots occur where high-speed gas ejected by the supernova and moving at millions of miles per hour has reached the ring and blasted into it. The collision has heated the gas in the ring and caused it to glow more brightly. The fact that we see individual spots suggests that material ejected by the supernova is first hitting narrow, inward-projecting columns of gas in the clumpy ring. The hot spots are the first signs of a dramatic and violent collision between the new and old material that will continue over the next few years. By studying these bright spots, astronomers can determine the composition of the ring and hence learn about the nuclear processes that build heavy elements inside massive stars. (credit: modification of work by NASA, P. Challis, R. Kirshner (Harvard-Smithsonian Center for Astrophysics) and B. Sugerman (STScI))

Helium fusion lasted only about 1 million years. When the helium was exhausted at the center of the star, the core contracted again, the radius of the surface also decreased, and the star became a blue supergiant with a luminosity still about equal to 100,000 L_{Sun}. This is what it still looked like on the outside when, after brief periods of further fusion, it reached the iron crisis we discussed earlier and exploded.

Some key stages of evolution of the star that became SN 1987A, including the ones following helium exhaustion, are listed in Table 23.2. While we don't expect you to remember these numbers, note the patterns in the table: each stage of evolution happens more quickly than the preceding one, the temperature and pressure in the core increase, and progressively heavier elements are the source of fusion energy. Once iron was created, the collapse began. It was a catastrophic collapse, lasting only a few tenths of a second; the speed of infall in the outer portion of the iron core reached 70,000 kilometers per second, about one-fourth the speed of light.

Evolution of the Star That Exploded as SN 1987A

Phase	Central Temperature (K)	Central Density (g/cm³)	Time Spent in This Phase
Hydrogen fusion	40×10^6	5	8×10^6 years
Helium fusion	190×10^6	970	10^6 years
Carbon fusion	870×10^6	170,000	2000 years
Neon fusion	1.6×10^9	3.0×10^6	6 months
Oxygen fusion	2.0×10^9	5.6×10^6	1 year
Silicon fusion	3.3×10^9	4.3×10^7	Days
Core collapse	200×10^9	2×10^{14}	Tenths of a second

Table 23.2

Download for free at http://cnx.org/content/col11992/latest/

In the meantime, as the core was experiencing its last catastrophe, the outer shells of neon, oxygen, carbon, helium, and hydrogen in the star did not yet know about the collapse. Information about the physical movement of different layers travels through a star at the speed of sound and cannot reach the surface in the few tenths of a second required for the core collapse to occur. Thus, the surface layers of our star hung briefly suspended, much like a cartoon character who dashes off the edge of a cliff and hangs momentarily in space before realizing that he is no longer held up by anything.

The collapse of the core continued until the densities rose to several times that of an atomic nucleus. The resistance to further collapse then became so great that the core rebounded. Infalling material ran into the "brick wall" of the rebounding core and was thrown outward with a great shock wave. Neutrinos poured out of the core, helping the shock wave blow the star apart. The shock reached the surface of the star a few hours later, and the star began to brighten into the supernova Ian Shelton observed in 1987.

The Synthesis of Heavy Elements

The variations in the brightness of SN 1987A in the days and months after its discovery, which are shown in Figure 23.13, helped confirm our ideas about heavy element production. In a single day, the star soared in brightness by a factor of about 1000 and became just visible without a telescope. The star then continued to increase slowly in brightness until it was about the same apparent magnitude as the stars in the Little Dipper. Up until about day 40 after the outburst, the energy being radiated away was produced by the explosion itself. But then SN 1987A did not continue to fade away, as we might have expected the light from the explosion to do. Instead, SN 1987A remained bright as energy from newly created radioactive elements came into play.

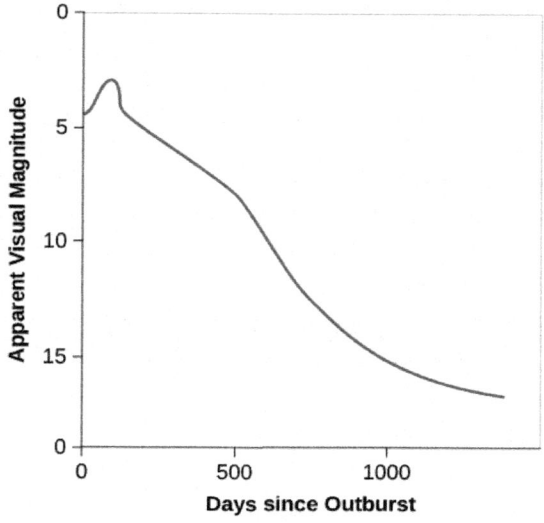

Figure 23.13. Change in the Brightness of SN 1987A over Time. Note how the rate of decline of the supernova's light slowed between days 40 and 500. During this time, the brightness was mainly due to the energy emitted by newly formed (and quickly decaying) radioactive elements. Remember that magnitudes are a backward measure of brightness: the larger the magnitude, the dimmer the object looks.

One of the elements formed in a supernova explosion is radioactive nickel, with an atomic mass of 56 (that is, the total number of protons plus neutrons in its nucleus is 56). Nickel-56 is unstable and changes spontaneously (with a half-life of about 6 days) to cobalt-56. (Recall that a half-life is the time it takes for half the nuclei in a sample to undergo radioactive decay.) Cobalt-56 in turn decays with a half-life of about 77 days to iron-56, which is stable. Energetic gamma rays are emitted when these radioactive nuclei decay. Those gamma rays then serve

This OpenStax book is available for free at http://cnx.org/content/col11992/1.8

Download for free at http://cnx.org/content/col11992/latest/

as a new source of energy for the expanding layers of the supernova. The gamma rays are absorbed in the overlying gas and re-emitted at visible wavelengths, keeping the remains of the star bright.

As you can see in Figure 23.13, astronomers did observe brightening due to radioactive nuclei in the first few months following the supernova's outburst and then saw the extra light die away as more and more of the radioactive nuclei decayed to stable iron. The gamma-ray heating was responsible for virtually all of the radiation detected from SN 1987A after day 40. Some gamma rays also escaped directly without being absorbed. These were detected by Earth-orbiting telescopes at the wavelengths expected for the decay of radioactive nickel and cobalt, clearly confirming our understanding that new elements were indeed formed in the crucible of the supernova.

Neutrinos from SN 1987A

If there had been any human observers in the Large Magellanic Cloud about 160,000 years ago, the explosion we call SN 1987A would have been a brilliant spectacle in their skies. Yet we know that less than 1/10 of 1% of the energy of the explosion appeared as visible light. About 1% of the energy was required to destroy the star, and the rest was carried away by neutrinos. The overall energy in these neutrinos was truly astounding. In the initial second of the event, as we noted earlier in our general discussion of supernovae, their total luminosity exceeded the luminosity of all the stars in over a billion galaxies. And the supernova generated this energy in a volume less than 50 kilometers in diameter! Supernovae are one of the most violent events in the universe, and their *light* turns out to be only the tip of the iceberg in revealing how much energy they produce.

In 1987, the neutrinos from SN 1987A were detected by two instruments—which might be called "neutrino telescopes"—almost a full day before Shelton's observations. (This is because the neutrinos get out of the exploding star more easily than light does, and also because you don't need to wait until nightfall to catch a "glimpse" of them.) Both neutrino telescopes, one in a deep mine in Japan and the other under Lake Erie, consist of several thousand tons of purified water surrounded by several hundred light-sensitive detectors. Incoming neutrinos interact with the water to produce positrons and electrons, which move rapidly through the water and emit deep blue light.

Altogether, 19 neutrinos were detected. Since the neutrino telescopes were in the Northern Hemisphere and the supernova occurred in the Southern Hemisphere, the detected neutrinos had already passed through Earth and were on their way back out into space when they were captured.

Only a few neutrinos were detected because the probability that they will interact with ordinary matter is very, very low. It is estimated that the supernova actually released 10^{58} neutrinos. A tiny fraction of these, about 30 billion, eventually passed through each square centimeter of Earth's surface. About a million people actually experienced a neutrino interaction within their bodies as a result of the supernova. This interaction happened to only a single nucleus in each person and thus had absolutely no biological effect; it went completely unnoticed by everyone concerned.

Since the neutrinos come directly from the heart of the supernova, their energies provided a measure of the temperature of the core as the star was exploding. The central temperature was about 200 billion K, a stunning figure to which no earthly analog can bring much meaning. With neutrino telescopes, we are peering into the final moment in the life stories of massive stars and observing conditions beyond all human experience. Yet we are also seeing the unmistakable hints of our own origins.

Download for free at http://cnx.org/content/col11992/latest/

23.4 PULSARS AND THE DISCOVERY OF NEUTRON STARS

Learning Objectives

By the end of this section, you will be able to:

> Explain the research method that led to the discovery of neutron stars, located hundreds or thousands of light-years away
> Describe the features of a neutron star that allow it to be detected as a pulsar
> List the observational evidence that links pulsars and neutron stars to supernovae

After a type II supernova explosion fades away, all that is left behind is either a neutron star or something even more strange, a black hole. We will describe the properties of black holes in Black Holes and Curved Spacetime, but for now, we want to examine how the neutron stars we discussed earlier might become observable.

Neutron stars are the densest objects in the universe; the force of gravity at their surface is 10^{11} times greater than what we experience at Earth's surface. The interior of a neutron star is composed of about 95% neutrons, with a small number of protons and electrons mixed in. In effect, a neutron star is a giant atomic nucleus, with a mass about 10^{57} times the mass of a proton. Its diameter is more like the size of a small town or an asteroid than a star. (Table 23.3 compares the properties of neutron stars and white dwarfs.) Because it is so small, a neutron star probably strikes you as the object least likely to be observed from thousands of light-years away. Yet neutron stars do manage to signal their presence across vast gulfs of space.

Properties of a Typical White Dwarf and a Neutron Star

Property	White Dwarf	Neutron Star
Mass (Sun = 1)	0.6 (always <1.4)	Always >1.4 and <3
Radius	7000 km	10 km
Density	8×10^5 g/cm^3	10^{14} g/cm^3

Table 23.3

The Discovery of Neutron Stars

In 1967, Jocelyn Bell, a research student at Cambridge University, was studying distant radio sources with a special detector that had been designed and built by her advisor Antony Hewish to find rapid variations in radio signals. The project computers spewed out reams of paper showing where the telescope had surveyed the sky, and it was the job of Hewish's graduate students to go through it all, searching for interesting phenomena. In September 1967, Bell discovered what she called "a bit of scruff"—a strange radio signal unlike anything seen before.

What Bell had found, in the constellation of Vulpecula, was a source of rapid, sharp, intense, and extremely regular pulses of radio radiation. Like the regular ticking of a clock, the pulses arrived precisely every 1.33728 seconds. Such exactness first led the scientists to speculate that perhaps they had found signals from an

This OpenStax book is available for free at http://cnx.org/content/col11992/1.8

intelligent civilization. Radio astronomers even half-jokingly dubbed the source "LGM" for "little green men." Soon, however, three similar sources were discovered in widely separated directions in the sky.

When it became apparent that this type of radio source was fairly common, astronomers concluded that they were highly unlikely to be signals from other civilizations. By today, more than 2500 such sources have been discovered; they are now called **pulsars**, short for "pulsating radio sources."

The pulse periods of different pulsars range from a little longer than 1/1000 of a second to nearly 10 seconds. At first, the pulsars seemed particularly mysterious because nothing could be seen at their location on visible-light photographs. But then a pulsar was discovered right in the center of the Crab Nebula, a cloud of gas produced by SN 1054, a supernova that was recorded by the Chinese in 1054 (Figure 23.14). The energy from the Crab Nebula pulsar arrives in sharp bursts that occur 30 times each second—with a regularity that would be the envy of a Swiss watchmaker. In addition to pulses of radio energy, we can observe pulses of visible light and X-rays from the Crab Nebula. The fact that the pulsar was just in the region of the supernova remnant where we expect the leftover neutron star to be immediately alerted astronomers that pulsars might be connected with these elusive "corpses" of massive stars.

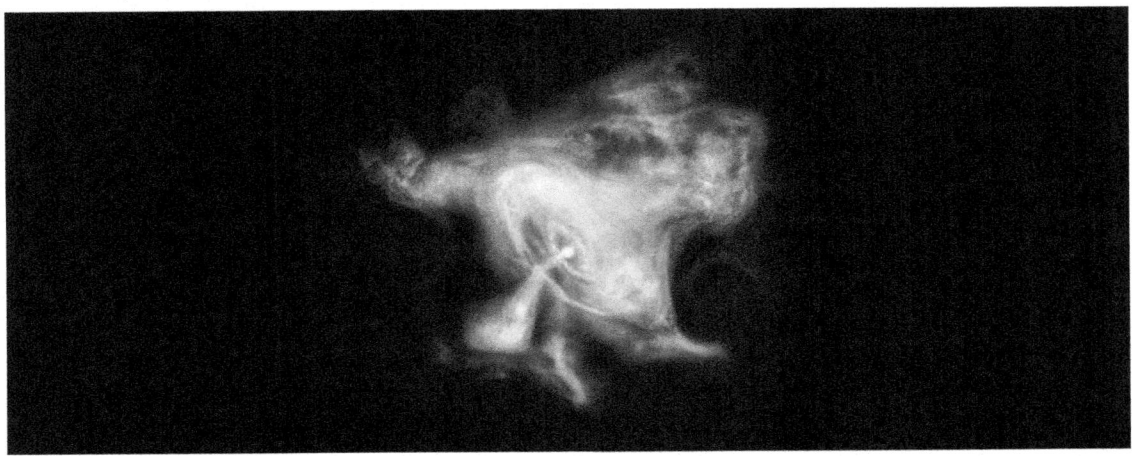

Figure 23.14. Crab Nebula. This image shows X-ray emmisions from the Crab Nebula, which is about 6500 light-years away. The pulsar is the bright spot at the center of the concentric rings. Data taken over about a year show that particles stream away from the inner ring at about half the speed of light. The jet that is perpendicular to this ring is a stream of matter and antimatter electrons also moving at half the speed of light. (credit: modification of work by NASA/CXC/SAO)

The Crab Nebula is a fascinating object. The whole nebula glows with radiation at many wavelengths, and its overall energy output is more than 100,000 times that of the Sun—not a bad trick for the remnant of a supernova that exploded almost a thousand years ago. Astronomers soon began to look for a connection between the pulsar and the large energy output of the surrounding nebula.

LINK TO LEARNING

View an interesting interview (https://openstaxcollege.org/l/30jocbellint) with Jocelyn Bell (Burnell) to learn about her life and work (this is part of a project at the American Institute of Physics to record interviews with pathbreaking scientists while they are still alive).

Download for free at http://cnx.org/content/col11992/latest/

A Spinning Lighthouse Model

By applying a combination of theory and observation, astronomers eventually concluded that pulsars must be *spinning neutron stars*. According to this model, a neutron star is something like a lighthouse on a rocky coast (Figure 23.15). To warn ships in all directions and yet not cost too much to operate, the light in a modern lighthouse turns, sweeping its beam across the dark sea. From the vantage point of a ship, you see a pulse of light each time the beam points in your direction. In the same way, radiation from a small region on a neutron star sweeps across the oceans of space, giving us a pulse of radiation each time the beam points toward Earth.

Figure 23.15. Lighthouse. A lighthouse in California warns ships on the ocean not to approach too close to the dangerous shoreline. The lighted section at the top rotates so that its beam can cover all directions. (credit: Anita Ritenour)

Neutron stars are ideal candidates for such a job because the collapse has made them so small that they can turn very rapidly. Recall the principle of the *conservation of angular momentum* from Newton's Great Synthesis: if an object gets smaller, it can spin more rapidly. Even if the parent star was rotating very slowly when it was on the main sequence, its rotation had to speed up as it collapsed to form a neutron star. With a diameter of only 10 to 20 kilometers, a neutron star can complete one full spin in only a fraction of a second. This is just the sort of time period we observe between pulsar pulses.

Any magnetic field that existed in the original star will be highly compressed when the core collapses to a neutron star. At the surface of the neutron star, in the outer layer consisting of ordinary matter (and not just pure neutrons), protons and electrons are caught up in this spinning field and accelerated nearly to the speed of light. In only two places—the north and south magnetic poles—can the trapped particles escape the strong hold of the magnetic field (Figure 23.16). The same effect can be seen (in reverse) on Earth, where charged particles from space are *kept out* by our planet's magnetic field everywhere except near the poles. As a result, Earth's auroras (caused when charged particles hit the atmosphere at high speed) are seen mainly near the poles.

This OpenStax book is available for free at http://cnx.org/content/col11992/1.8

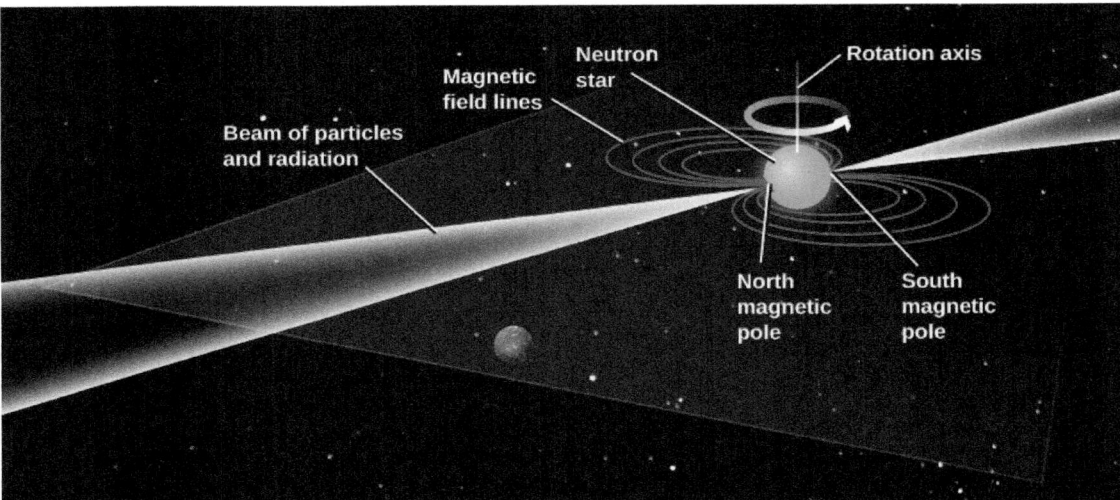

Figure 23.16. Model of a Pulsar. A diagram showing how beams of radiation at the magnetic poles of a neutron star can give rise to pulses of emission as the star rotates. As each beam sweeps over Earth, like a lighthouse beam sweeping over a distant ship, we see a short pulse of radiation. This model requires that the magnetic poles be located in different places from the rotation poles. (credit "stars": modification of work by Tony Hisgett)

Note that in a neutron star, the magnetic north and south poles do not have to be anywhere close to the north and south poles defined by the star's rotation. In the same way, we discussed in the chapter on The Giant Planets that the magnetic poles on the planets Uranus and Neptune are not lined up with the poles of the planet's spin. Figure 23.16 shows the poles of the magnetic field perpendicular to the poles of rotation, but the two kinds of poles could make any angle.

In fact, the misalignment of the rotational axis with the magnetic axis plays a crucial role in the generation of the observed pulses in this model. At the two magnetic poles, the particles from the neutron star are focused into a narrow beam and come streaming out of the whirling magnetic region at enormous speeds. They emit energy over a broad range of the electromagnetic spectrum. The radiation itself is also confined to a narrow beam, which explains why the pulsar acts like a lighthouse. As the rotation carries first one and then the other magnetic pole of the star into our view, we see a pulse of radiation each time.

Tests of the Model

This explanation of pulsars in terms of beams of radiation from highly magnetic and rapidly spinning neutron stars is a very clever idea. But what evidence do we have that it is the correct model? First, we can measure the masses of some pulsars, and they do turn out be in the range of 1.4 to 1.8 times that of the Sun—just what theorists predict for neutron stars. The masses are found using Kepler's law for those few pulsars that are members of binary star systems.

But there is an even-better confirming argument, which brings us back to the Crab Nebula and its vast energy output. When the high-energy charged particles from the neutron star pulsar hit the slower-moving material from the supernova, they energize this material and cause it to "glow" at many different wavelengths—just what we observe from the Crab Nebula. The pulsar beams are a power source that "light up" the nebula long after the initial explosion of the star that made it.

Who "pays the bills" for all the energy we see coming out of a remnant like the Crab Nebula? After all, when energy emerges from one place, it must be depleted in another. The ultimate energy source in our model is the rotation of the neutron star, which propels charged particles outward and spins its magnetic field at enormous

Download for free at http://cnx.org/content/col11992/latest/

speeds. As its rotational energy is used to excite the Crab Nebula year after year, the pulsar inside the nebula slows down. As it slows, the pulses come a little less often; more time elapses before the slower neutron star brings its beam back around.

Several decades of careful observations have now shown that the Crab Nebula pulsar is not a perfectly regular clock as we originally thought: instead, it is gradually slowing down. Having measured how much the pulsar is slowing down, we can calculate how much rotation energy the neutron star is losing. Remember that it is very densely packed and spins amazingly quickly. Even a tiny slowing down can mean an immense loss of energy.

To the satisfaction of astronomers, the rotational energy lost by the pulsar turns out to be the same as the amount of energy emerging from the nebula surrounding it. In other words, the slowing down of a rotating neutron star can explain precisely why the Crab Nebula is glowing with the amount of energy we observe.

The Evolution of Pulsars

From observations of the pulsars discovered so far, astronomers have concluded that one new pulsar is born somewhere in the Galaxy every 25 to 100 years, the same rate at which supernovae are estimated to occur. Calculations suggest that the typical lifetime of a pulsar is about 10 million years; after that, the neutron star no longer rotates fast enough to produce significant beams of particles and energy, and is no longer observable. We estimate that there are about 100 million neutron stars in our Galaxy, most of them rotating too slowly to come to our notice.

The Crab pulsar is rather young (only about 960 years old) and has a short period, whereas other, older pulsars have already slowed to longer periods. Pulsars thousands of years old have lost too much energy to emit appreciably in the visible and X-ray wavelengths, and they are observed only as radio pulsars; their periods are a second or longer.

There is one other reason we can see only a fraction of the pulsars in the Galaxy. Consider our lighthouse model again. On Earth, all ships approach on the same plane—the surface of the ocean—so the lighthouse can be built to sweep its beam over that surface. But in space, objects can be anywhere in three dimensions. As a given pulsar's beam sweeps over a circle in space, there is absolutely no guarantee that this circle will include the direction of Earth. In fact, if you think about it, many more circles in space will *not* include Earth than will include it. Thus, we estimate that we are unable to observe a large number of neutron stars because their pulsar beams miss us entirely.

At the same time, it turns out that only a few of the pulsars discovered so far are embedded in the visible clouds of gas that mark the remnant of a supernova. This might at first seem mysterious, since we know that supernovae give rise to neutron stars and we should expect each pulsar to have begun its life in a supernova explosion. But the lifetime of a pulsar turns out to be about 100 times longer than the length of time required for the expanding gas of a supernova remnant to disperse into interstellar space. Thus, most pulsars are found with no other trace left of the explosion that produced them.

In addition, some pulsars are ejected by a supernova explosion that is not the same in all directions. If the supernova explosion is stronger on one side, it can kick the pulsar entirely out of the supernova remnant (some astronomers call this "getting a birth kick"). We know such kicks happen because we see a number of young supernova remnants in nearby galaxies where the pulsar is to one side of the remnant and racing away at several hundred miles per second (Figure 23.17).

This OpenStax book is available for free at http://cnx.org/content/col11992/1.8

Figure 23.17. Speeding Pulsar. This intriguing image (which combines X-ray, visible, and radio observations) shows the jet trailing behind a pulsar (at bottom right, lined up between the two bright stars). With a length of 37 light-years, the jet trail (seen in purple) is the longest ever observed from an object in the Milky Way. (There is also a mysterious shorter, comet-like tail that is almost perpendicular to the purple jet.) Moving at a speed between 2.5 and 5 million miles per hour, the pulsar is traveling away from the core of the supernova remnant where it originated. (credit: X-ray: NASA/CXC/ISDC/L.Pavan et al, Radio: CSIRO/ATNF/ATCA Optical: 2MASS/UMass/IPAC-Caltech/NASA/NSF)

MAKING CONNECTIONS

Touched by a Neutron Star

On December 27, 2004, Earth was bathed with a stream of X-ray and gamma-ray radiation from a neutron star known as SGR 1806-20. What made this event so remarkable was that, despite the distance of the source, its tidal wave of radiation had measurable effects on Earth's atmosphere. The apparent brightness of this gamma-ray flare was greater than any historical star explosion.

The primary effect of the radiation was on a layer high in Earth's atmosphere called the *ionosphere*. At night, the ionosphere is normally at a height of about 85 kilometers, but during the day, energy from the Sun ionizes more molecules and lowers the boundary of the ionosphere to a height of about 60 kilometers. The pulse of X-ray and gamma-ray radiation produced about the same level of ionization as the daytime Sun. It also caused some sensitive satellites above the atmosphere to shut down their electronics.

Measurements by telescopes in space indicate that SGR 1806-20 was a special type of fast-spinning neutron star called a *magnetar*. Astronomers Robert Duncan and Christopher Thomson gave them this name because their magnetic fields are stronger than that of any other type of astronomical source—in this case, about 800 trillion times stronger than the magnetic field of Earth.

A magnetar is thought to consist of a superdense core of neutrons surrounded by a rigid crust of atoms about a mile deep with a surface made of iron. The magnetar's field is so strong that it creates huge stresses inside that can sometimes crack open the hard crust, causing a starquarke. The vibrating crust produces an enormous blast of radiation. An astronaut 0.1 light-year from this particular magnetar would have received a fatal does from the blast in less than a second.

Download for free at http://cnx.org/content/col11992/latest/

Fortunately, we were far enough away from magnetar SGR 1806-20 to be safe. Could a magnetar ever present a real danger to Earth? To produce enough energy to disrupt the ozone layer, a magnetar would have to be located within the cloud of comets that surround the solar system, and we know no magnetars are that close. Nevertheless, it is a fascinating discovery that events on distant star corpses can have measurable effects on Earth.

23.5 THE EVOLUTION OF BINARY STAR SYSTEMS

Learning Objectives

By the end of this section, you will be able to:

> Describe the kind of binary star system that leads to a nova event
> Describe the type of binary star system that leads to a type Ia supernovae event
> Indicate how type Ia supernovae differ from type II supernovae

The discussion of the life stories of stars presented so far has suffered from a bias—what we might call "single-star chauvinism." Because the human race developed around a star that goes through life alone, we tend to think of most stars in isolation. But as we saw in The Stars: A Celestial Census, it now appears that as many as half of all stars may develop in *binary* systems—those in which two stars are born in each other's gravitational embrace and go through life orbiting a common center of mass.

For these stars, the presence of a close-by companion can have a profound influence on their evolution. Under the right circumstances, stars can exchange material, especially during the stages when one of them swells up into a giant or supergiant, or has a strong wind. When this happens and the companion stars are sufficiently close, material can flow from one star to another, decreasing the mass of the donor and increasing the mass of the recipient. Such *mass transfer* can be especially dramatic when the recipient is a stellar remnant such as a white dwarf or a neutron star. While the detailed story of how such binary stars evolve is beyond the scope of our book, we do want to mention a few examples of how the stages of evolution described in this chapter may change when there are two stars in a system.

White Dwarf Explosions: The Mild Kind

Let's consider the following system of two stars: one has become a white dwarf and the other is gradually transferring material onto it. As fresh hydrogen from the outer layers of its companion accumulates on the surface of the hot white dwarf, it begins to build up a layer of hydrogen. As more and more hydrogen accumulates and heats up on the surface of the degenerate star, the new layer eventually reaches a temperature that causes fusion to begin in a sudden, explosive way, blasting much of the new material away.

In this way, the white dwarf quickly (but only briefly) becomes quite bright, hundreds or thousands of times its previous luminosity. To observers before the invention of the telescope, it seemed that a new star suddenly appeared, and they called it a **nova**.[2] Novae fade away in a few months to a few years.

---

2   We now know that this historical terminology is quite misleading since novae do not originate from new stars. In fact, quite to the contrary, novae originate from white dwarfs, which are actually the endpoint of stellar evolution for low-mass stars. But since the system of two stars was too faint to be visible to the naked eye, it did seem to people, before telescopes were invented, that a star had appeared where nothing had been visible.

This OpenStax book is available for free at http://cnx.org/content/col11992/1.8

Download for free at http://cnx.org/content/col11992/latest/

Hundreds of novae have been observed, each occurring in a binary star system and each later showing a shell of expelled material. A number of stars have more than one nova episode, as more material from its neighboring star accumulates on the white dwarf and the whole process repeats. As long as the episodes do not increase the mass of the white dwarf beyond the Chandrasekhar limit (by transferring too much mass too quickly), the dense white dwarf itself remains pretty much unaffected by the explosions on its surface.

White Dwarf Explosions: The Violent Kind

If a white dwarf accumulates matter from a companion star at a much faster rate, it can be pushed over the Chandrasekhar limit. The evolution of such a binary system is shown in Figure 23.18. When its mass approaches the Chandrasekhar mass limit (exceeds 1.4 M_{Sun}), such an object can no longer support itself as a white dwarf, and it begins to contract. As it does so, it heats up, and new nuclear reactions can begin in the degenerate core. The star "simmers" for the next century or so, building up internal temperature. This simmering phase ends in less than a second, when an enormous amount of fusion (especially of carbon) takes place all at once, resulting in an explosion. The fusion energy produced during the final explosion is so great that it completely destroys the white dwarf. Gases are blown out into space at velocities of about 10,000 kilometers per second, and afterward, no trace of the white dwarf remains.

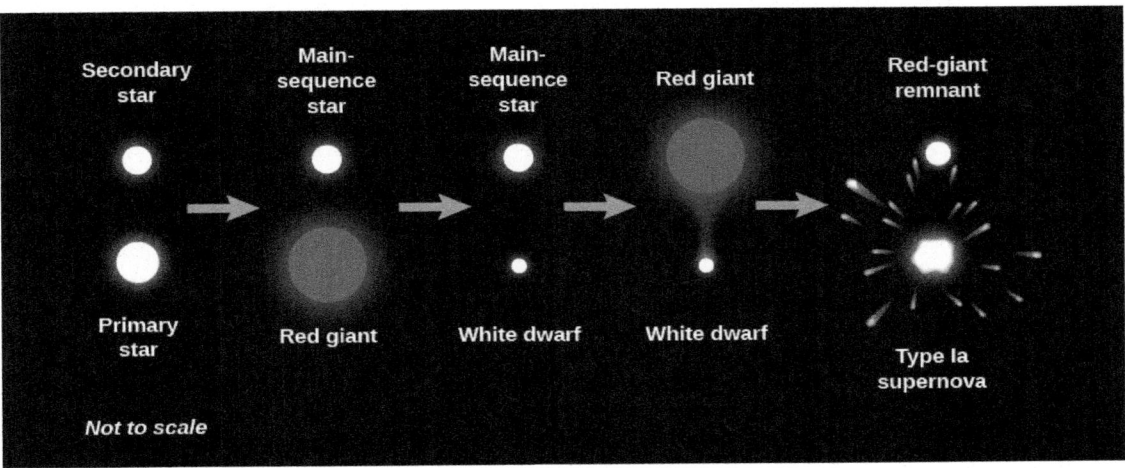

Figure 23.18. Evolution of a Binary System. The more massive star evolves first to become a red giant and then a white dwarf. The white dwarf then begins to attract material from its companion, which in turn evolves to become a red giant. Eventually, the white dwarf acquires so much mass that it is pushed over the Chandrasekhar limit and becomes a type Ia supernova.

Such an explosion is also called a supernova, since, like the destruction of a high-mass star, it produces a huge amount of energy in a very short time. However, unlike the explosion of a high-mass star, which can leave behind a neutron star or black hole remnant, the white dwarf is completely destroyed in the process, leaving behind no remnant. We call these white dwarf explosions type Ia supernovae.

We distinguish type I supernovae from those of supernovae of type II originating from the death of massive stars discussed earlier by the absence of hydrogen in their observed spectra. Hydrogen is the most common element in the universe and is a major component of massive, evolved stars. However, as we learned earlier, hydrogen is absent from the white dwarf remnant, which is primarily composed of carbon and oxygen for masses comparable to the Chandrasekhar mass limit.

The "a" subdesignation of type Ia supernovae further refers to the presence of strong silicon absorption lines, which are absent from supernovae originating from the collapse of massive stars. Silicon is one of the products

Download for free at http://cnx.org/content/col11992/latest/

that results from the fusion of carbon and oxygen, which bears out the scenario we described above—that there is a sudden onset of the fusion of the carbon (and oxygen) of which the white dwarf was made.

Observational evidence now strongly indicates that SN 1006, Tycho's Supernova, and Kepler's Supernova (see Supernovae in History) were all type Ia supernovae. For instance, in contrast to the case of SN 1054, which yielded the spinning pulsar in the Crab Nebula, none of these historical supernovae shows any evidence of stellar remnants that have survived their explosions. Perhaps even more puzzling is that, so far, astronomers have not been able to identify the companion star feeding the white dwarf in any of these historical supernovae.

Consequently, in order to address the mystery of the absent companion stars and other outstanding puzzles, astronomers have recently begun to investigate alternative mechanisms of generating type Ia supernovae. All proposed mechanisms rely upon white dwarfs composed of carbon and oxygen, which are needed to meet the observed absence of hydrogen in the type Ia spectrum. And because any isolated white dwarf below the Chandrasekhar mass is stable, all proposed mechanisms invoke a binary companion to explode the white dwarf. The leading alternative mechanism scientists believe creates a type Ia supernova is the merger of two white dwarf stars in a binary system. The two white dwarfs may have unstable orbits, such that over time, they would slowly move closer together until they merge. If their combined mass is greater than the Chandrasekhar limit, the result could also be a type Ia supernova explosion.

LINK TO LEARNING

You can watch a short video (https://openstaxcollege.org/l/30supernovavid) about Supernova SN 2014J, a type Ia supernova discovered in the Messier 82 (M82) galaxy on January 21, 2014, as well as see brief animations of the two mechanisms by which such a supernova could form.

Type Ia supernovae are of great interest to astronomers in other areas of research. This type of supernova is brighter than supernovae produced by the collapse of a massive star. Thus, type Ia supernovae can be seen at very large distances, and they are found in all types of galaxies. The energy output from most type Ia supernovae is consistent, with little variation in their maximum luminosities, or in how their light output initially increases and then slowly decreases over time. These properties make type Ia supernovae extremely valuable "standard bulbs" for astronomers looking out at great distances—well beyond the limits of our own Galaxy. You'll learn more about their use in measuring distances to other galaxies in The Big Bang.

In contrast, type II supernovae are about 5 times less luminous than type Ia supernovae and are only seen in galaxies that have recent, massive star formation. Type II supernovae are also less consistent in their energy output during the explosion and can have a range a peak luminosity values.

Neutron Stars with Companions

Now let's look at an even-more mismatched pair of stars in action. It is possible that, under the right circumstances, a binary system can even survive the explosion of one of its members as a type II supernova. In that case, an ordinary star can eventually share a system with a neutron star. If material is then transferred from the "living" star to its "dead" (and highly compressed) companion, this material will be pulled in by the strong gravity of the neutron star. Such infalling gas will be compressed and heated to incredible temperatures. It will quickly become so hot that it will experience an explosive burst of fusion. The energies involved are so great that we would expect much of the radiation from the burst to emerge as X-rays. And indeed, high-energy

This OpenStax book is available for free at http://cnx.org/content/col11992/1.8

observatories above Earth's atmosphere (see Astronomical Instruments) have recorded many objects that undergo just these types of X-ray *bursts*.

If the neutron star and its companion are positioned the right way, a significant amount of material can be transferred to the neutron star and can set it spinning faster (as spin energy is also transferred). The radius of the neutron star would also decrease as more mass was added. Astronomers have found pulsars in binary systems that are spinning at a rate of more than 500 times per second! (These are sometimes called **millisecond pulsars** since the pulses are separated by a few thousandths of a second.)

Such a rapid spin could not have come from the birth of the neutron star; it must have been externally caused. (Recall that the Crab Nebula pulsar, one of the youngest pulsars known, was spinning "only" 30 times per second.) Indeed, some of the fast pulsars are observed to be part of binary systems, while others may be alone only because they have "fully consumed" their former partner stars through the mass transfer process. (These have sometimes been called " black widow pulsars.")

LINK TO LEARNING

View this short video (https://openstaxcollege.org/l/30scotronvid) to see Dr. Scott Ransom, of the National Radio Astronomy Observatory, explain how millisecond pulsars come about, with some nice animations.

And if you thought that a neutron star interacting with a "normal" star was unusual, there are also binary systems that consist of two neutron stars. One such system has the stars in very close orbits to one another, so much that they continually alter each other's orbit. Another binary neutron star system includes two pulsars that are orbiting each other every 2 hours and 25 minutes. As we discussed earlier, pulsars radiate away their energy, and these two pulsars are slowly moving toward one another, such that in about 85 million years, they will actually merge.

We have now reached the end of our description of the final stages of stars, yet one piece of the story remains to be filled in. We saw that stars whose core masses are less than 1.4 M_{Sun} at the time they run out of fuel end their lives as white dwarfs. Dying stars with core masses between 1.4 and about 3 M_{Sun} become neutron stars. But there are stars whose core masses are greater than 3 M_{Sun} when they exhaust their fuel supplies. What becomes of them? The truly bizarre result of the death of such massive stellar cores (called a black hole) is the subject of our next chapter. But first, we will look at an astronomical mystery that turned out to be related to the deaths of stars and was solved through clever sleuthing and a combination of observation and theory.

23.6 | THE MYSTERY OF THE GAMMA-RAY BURSTS

Learning Objectives

By the end of this section, you will be able to:

> Give a brief history of how gamma-ray bursts were discovered and what instruments made the discovery possible
> Explain why astronomers think that gamma-ray bursts beam their energy rather than it radiating uniformly in all directions
> Describe how the radiation from a gamma-ray burst and its afterglow is produced

Download for free at http://cnx.org/content/col11992/latest/

> Explain how short-duration gamma-ray bursts differ from longer ones, and describe the process that makes short-duration gamma-ray bursts
> Explain why gamma-ray bursts may help us understand the early universe

Everybody loves a good mystery, and astronomers are no exception. The mystery we will discuss in this section was first discovered in the mid-1960s, not via astronomical research, but as a result of a search for the tell-tale signs of nuclear weapon explosions. The US Defense Department launched a series of *Vela* satellites to make sure that no country was violating a treaty that banned the detonation of nuclear weapons in space.

Since nuclear explosions produce the most energetic form of electromagnetic waves called *gamma rays* (see Radiation and Spectra), the *Vela* satellites contained detectors to search for this type of radiation. The satellites did not detect any confirmed events from human activities, but they did—to everyone's surprise—detect short bursts of gamma rays coming from random directions in the sky. News of the discovery was first published in 1973; however, the origin of the bursts remained a mystery. No one knew what produced the brief flashes of gamma rays or how far away the sources were.

From a Few Bursts to Thousands

With the launch of the Compton Gamma-Ray Observatory by NASA in 1991, astronomers began to identify many more bursts and to learn more about them (Figure 23.19). Approximately once per day, the NASA satellite detected a flash of gamma rays somewhere in the sky that lasted from a fraction of a second to several hundred seconds. Before the Compton measurements, astronomers had expected that the most likely place for the bursts to come from was the main disk of our own (pancake-shaped) Galaxy. If this had been the case, however, more bursts would have been seen in the crowded plane of the Milky Way than above or below it. Instead, the sources of the bursts were distributed *isotropically*; that is, they could appear anywhere in the sky with no preference for one region over another. Almost never did a second burst come from the same location.

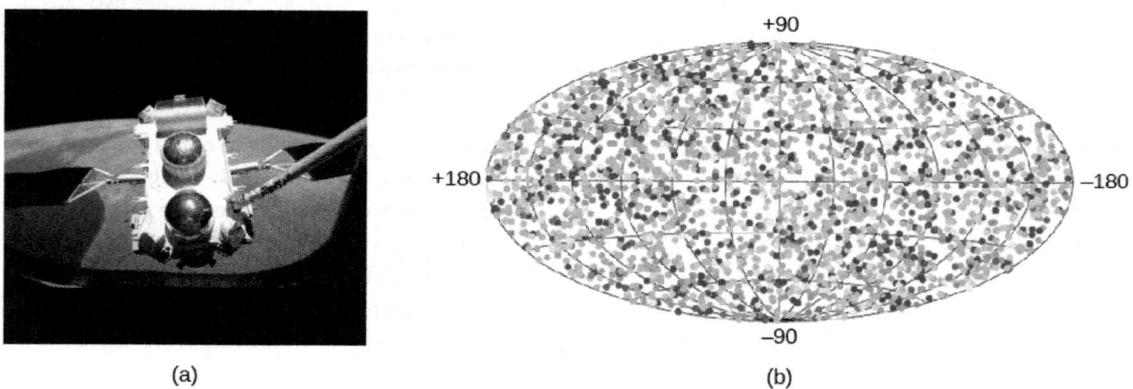

(a)   (b)

Figure 23.19. Compton Detects Gamma-Ray Bursts. (a) In 1991, the Compton Gamma-Ray Observatory was deployed by the Space Shuttle Atlantis. Weighing more than 16 tons, it was one of the largest scientific payloads ever launched into space. (b) This map of gamma-ray burst positions measured by the Compton Gamma-Ray Observatory shows the isotropic (same in all directions), uniform distribution of bursts on the sky. The map is oriented so that the disk of the Milky Way would stretch across the center line (or equator) of the oval. Note that the bursts show no preference at all for the plane of the Milky Way, as many other types of objects in the sky do. Colors indicate the total energy in the burst: red dots indicate long-duration, bright bursts; blue and purple dots show short, weaker bursts. (credit a: modification of work by NASA; credit b: modification of work by NASA/GSFC)

This OpenStax book is available for free at http://cnx.org/content/col11992/1.8

LINK TO LEARNING

To get a good visual sense of the degree to which the bursts come from all over the sky, watch this short animated NASA video (https://openstaxcollege.org/l/30swiftnasavid) showing the location of the first 500 bursts found by the later *Swift* satellite.

For several years, astronomers actively debated whether the burst sources were relatively nearby or very far away—the two possibilities for bursts that are isotropically distributed. Nearby locations might include the cloud of comets that surrounds the solar system or the halo of our Galaxy, which is large and spherical, and also surrounds us in all directions. If, on the other hand, the bursts occurred at very large distances, they could come from faraway galaxies, which are also distributed uniformly in all directions.

Both the very local and the very distant hypotheses required something strange to be going on. If the bursts were coming from the cold outer reaches of our own solar system or from the halo of our Galaxy, then astronomers had to hypothesize some new kind of physical process that could produce unpredictable flashes of high-energy gamma rays in these otherwise-quiet regions of space. And if the bursts came from galaxies millions or billions of light-years away, then they must be extremely powerful to be observable at such large distances; indeed they had to be the among the biggest explosions in the universe.

The First Afterglows

The problem with trying to figure out the source of the gamma-ray bursts was that our instruments for detecting gamma rays could not pinpoint the exact place in the sky where the burst was happening. Early gamma-ray telescopes did not have sufficient *resolution.* This was frustrating because astronomers suspected that if they *could* pinpoint the exact position of one of these rapid bursts, then they would be able to identify a counterpart (such as a star or galaxy) at other wavelengths and learn much more about the burst, including where it came from. This would, however, require either major improvements in gamma-ray detector technology to provide better resolution or detection of the burst at some other wavelength. In the end, both techniques played a role.

The breakthrough came with the launch of the Italian Dutch *BeppoSAX* satellite in 1996. *BeppoSAX* included a new type of gamma-ray telescope capable of identifying the position of a source much more accurately than previous instruments, to within a few minutes of arc on the sky. By itself, however, it was still not sophisticated enough to determine the exact source of the gamma-ray burst. After all, a box a few minutes of arc on a side could still contain many stars or other celestial objects.

However, the angular resolution of *BeppoSAX* was good enough to tell astronomers where to point other, more precise telescopes in the hopes of detecting longer-lived electromagnetic emission from the bursts at other wavelengths. Detection of a burst at visible-light or radio wavelengths could provide a position accurate to a few *seconds* of arc and allow the position to be pinpointed to an individual star or galaxy. *BeppoSAX* carried its own X-ray telescope onboard the spacecraft to look for such a counterpart, and astronomers using visible-light and radio facilities on the ground were eager to search those wavelengths as well.

Two crucial *BeppoSAX* burst observations in 1997 helped to resolve the mystery of the gamma-ray bursts. The first burst came in February from the direction of the constellation Orion. Within 8 hours, astronomers working with the satellite had identified the position of the burst, and reoriented the spacecraft to focus *BeppoSAX*'s X-ray detector on the source. To their excitement, they detected a slowly fading X-ray source 8 hours after the event—the first successful detection of an afterglow from a gamma-ray burst. This provided an even-better

Download for free at http://cnx.org/content/col11992/latest/

location of the burst (accurate to about 40 seconds of arc), which was then distributed to astronomers across the world to try to detect it at even longer wavelengths.

That very night, the 4.2-meter William Herschel Telescope on the Canary Islands found a fading visible-light source at the same position as the X-ray afterglow, confirming that such an afterglow could be detected in visible light as well. Eventually, the afterglow faded away, but left behind at the location of the original gamma-ray burst was a faint, fuzzy source right where the fading point of light had been—a distant galaxy (Figure 23.20). This was the first piece of evidence that gamma-ray bursts were indeed very energetic objects from very far away. However, it also remained possible that the burst source was much closer to us and just happened to align with a more distant galaxy, so this one observation alone was not a conclusive demonstration of the extragalactic origin of gamma-ray bursts.

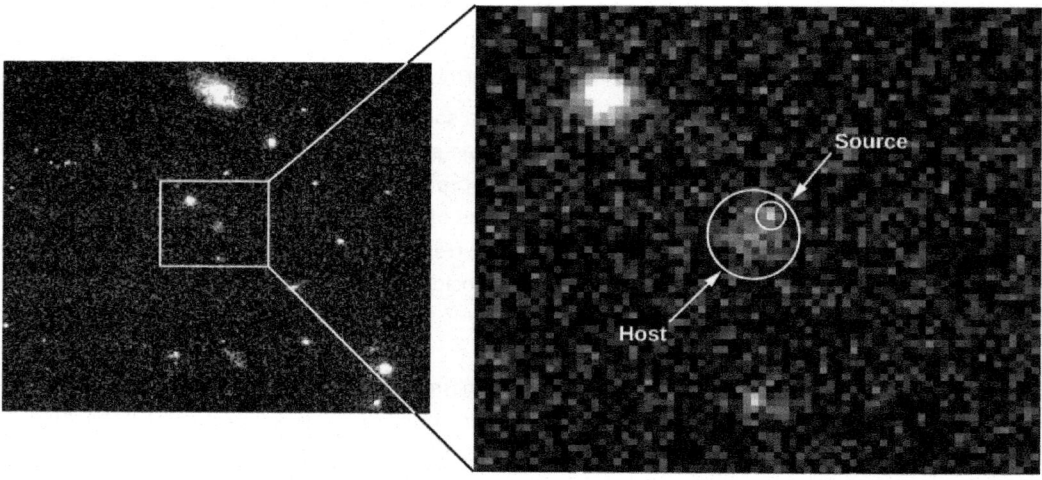

Figure 23.20. Gamma-Ray Burst. This false-color Hubble Space Telescope image, taken in September 1997, shows the fading afterglow of the gamma-ray burst of February 28, 1997 and the host galaxy in which the burst originated. The left view shows the region of the burst. The enlargement shows the burst source and what appears to be its host galaxy. Note that the gamma-ray source is not in the center of the galaxy. (credit: modification of work by Andrew Fruchter (STScI), Elena Pian (ITSRE-CNR), and NASA, ESA)

On May 8 of the same year, a burst came from the direction of the constellation Camelopardalis. In a coordinated international effort, *BeppoSAX* again fixed a reasonably precise position, and almost immediately a telescope on Kitt Peak in Arizona was able to catch the visible-light afterglow. Within 2 days, the largest telescope in the world (the Keck in Hawaii) collected enough light to record a spectrum of the burst. The May gamma-ray burst afterglow spectrum showed absorption features from a fuzzy object that was 4 billion light-years from the Sun, meaning that the location of the burst had to be at least this far away—and possibly even farther. (How astronomers can get the distance of such an object from the Doppler shift in the spectrum is something we will discuss in Galaxies.) What that spectrum showed was clear evidence that the gamma-ray burst had taken place in a distant galaxy.

Networking to Catch More Bursts

After initial observations showed that the precise locations and afterglows of gamma-ray bursts could be found, astronomers set up a system to catch and pinpoint bursts on a regular basis. But to respond as quickly as needed to obtain usable results, astronomers realized that they needed to rely on automated systems rather than human observers happening to be in the right place at the right time.

This OpenStax book is available for free at http://cnx.org/content/col11992/1.8

Now, when an orbiting high-energy telescope discovers a burst, its rough location is immediately transmitted to a *Gamma-Ray Coordinates Network* based at NASA's Goddard Space Flight Center, alerting observers on the ground within a few seconds to look for the visible-light afterglow.

The first major success with this system was achieved by a team of astronomers from the University of Michigan, Lawrence Livermore National Laboratory, and Los Alamos National Laboratories, who designed an automated device they called the *Robotic Optical Transient Search Experiment* (*ROTSE*), which detected a very bright visible-light counterpart in 1999. At peak, the burst was almost as bright as Neptune—despite a distance (measured later by spectra from larger telescopes) of 9 billion light-years.

More recently, astronomers have been able to take this a step further, using wide-field-of-view telescopes to stare at large fractions of the sky in the hope that a gamma-ray burst will occur at the right place and time, and be recorded by the telescope's camera. These wide-field telescopes are not sensitive to faint sources, but ROTSE showed that gamma-ray burst afterglows could sometimes be very bright.

Astronomers' hopes were vindicated in March 2008, when an extremely bright gamma-ray burst occurred and its light was captured by two wide-field camera systems in Chile: the Polish "Pi of the Sky" and the Russian-Italian TORTORA [Telescopio Ottimizzato per la Ricerca dei Transienti Ottici Rapidi (Italian for Telescope Optimized for the Research of Rapid Optical Transients)] (see Figure 23.21). According to the data taken by these telescopes, for a period of about 30 seconds, the light from the gamma-ray burst was bright enough that it could have been seen by the unaided eye had a person been looking in the right place at the right time. Adding to our amazement, later observations by larger telescopes demonstrated that the burst occurred at a distance of 8 billion light-years from Earth!

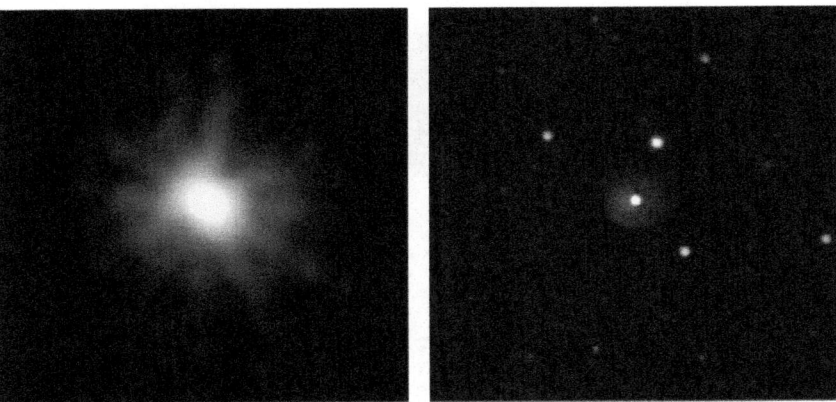

Figure 23.21. Gamma-Ray Burst Observed in March 2008. The extremely luminous afterglow of GRB 080319B was imaged by the Swift Observatory in X-rays (left) and visible light/ultraviolet (right). (credit: modification of work by NASA/Swift/Stefan Immler, et al.)

To Beam or Not to Beam

The enormous distances to these events meant they had to have been astoundingly energetic to appear as bright as they were across such an enormous distance. In fact, they required so much energy that it posed a problem for gamma-ray burst models: if the source was radiating energy in all directions, then the energy released in gamma rays alone during a bright burst (such as the 1999 or 2008 events) would have been equivalent to the energy produced if the entire mass of a Sun-like star were suddenly converted into pure radiation.

For a source to produce this much energy this quickly (in a burst) is a real challenge. Even if the star producing the gamma-ray burst was much more massive than the Sun (as is probably the case), there is no known means

Download for free at http://cnx.org/content/col11992/latest/

of converting so much mass into radiation within a matter of seconds. However, there is one way to reduce the power required of the "mechanism" that makes gamma-ray bursts. So far, our discussion has assumed that the source of the gamma rays gives off the same amount of energy in all directions, like an incandescent light bulb.

But as we discuss in Pulsars and the Discovery of Neutron Stars, not all sources of radiation in the universe are like this. Some produce thin beams of radiation that are concentrated into only one or two directions. A laser pointer and a lighthouse on the ocean are examples of such beamed sources on Earth (Figure 23.22). If, when a burst occurs, the gamma rays come out in only one or two narrow beams, then our estimates of the luminosity of the source can be reduced, and the bursts may be easier to explain. In that case, however, the beam has to point toward Earth for us to be able to see the burst. This, in turn, would imply that for every burst we see from Earth, there are probably many others that we never detect because their beams point in other directions.

Figure 23.22. Burst That Is Beamed. This artist's conception shows an illustration of one kind of gamma-ray burst. The collapse of the core of a massive star into a black hole has produced two bright beams of light originating from the star's poles, which an observer pointed along one of these axes would see as a gamma-ray burst. The hot blue stars and gas clouds in the vicinity are meant to show that the event happened in an active star-forming region. (credit: NASA/Swift/Mary Pat Hrybyk-Keith and John Jones)

Long-Duration Gamma-Ray Bursts: Exploding Stars

After identifying and following large numbers of gamma-ray bursts, astronomers began to piece together clues about what kind of event is thought to be responsible for producing the gamma-ray burst. Or, rather, what kind of *events*, because there are at least two distinct types of gamma-ray bursts. The two—like the different types of supernovae—are produced in completely different ways.

Observationally, the crucial distinction is how long the burst lasts. Astronomers now divide gamma-ray bursts into two categories: short-duration ones (defined as lasting less than 2 seconds, but typically a fraction of a second) and long-duration ones (defined as lasting more than 2 seconds, but typically about a minute).

All of the examples we have discussed so far concern the long-duration gamma-ray bursts. These constitute most of the gamma-ray bursts that our satellites detect, and they are also brighter and easier to pinpoint. Many hundreds of long-duration gamma-ray bursts, and the properties of the galaxies in which they occurred, have now been studied in detail. Long-duration gamma-ray bursts are universally observed to come from distant galaxies that are still actively making stars. They are usually found to be located in regions of the galaxy with strong star-formation activity (such as spiral arms). Recall that the more massive a star is, the less time it spends

This OpenStax book is available for free at http://cnx.org/content/col11992/1.8

Download for free at http://cnx.org/content/col11992/latest/

in each stage of its life. This suggests that the bursts come from a young and short-lived, and therefore massive type of star.

Furthermore, in several cases when a burst has occurred in a galaxy relatively close to Earth (within a few billion light-years), it has been possible to search for a supernova at the same position—and in nearly all of these cases, astronomers have found evidence of a supernova of type Ic going off. A type Ic is a particular type of supernova, which we did not discuss in the earlier parts of this chapter; these are produced by a massive star that has been stripped of its outer hydrogen layer. However, only a tiny fraction of type Ic supernovae produce gamma-ray bursts.

Why would a massive star with its outer layers missing sometimes produce a gamma-ray burst at the same time that it explodes as a supernova? The explanation astronomers have in mind for the extra energy is the collapse of the star's core to form a spinning, magnetic black hole or neutron star. Because the star corpse is both magnetic and spinning rapidly, its sudden collapse is complex and can produce swirling jets of particles and powerful beams of radiation—just like in a quasar or active galactic nucleus (objects you will learn about Active Galaxies, Quasars, and Supermassive Black Holes), but on a much faster timescale. A small amount of the infalling mass is ejected in a narrow beam, moving at speeds close to that of light. Collisions among the particles in the beam can produce intense bursts of energy that we see as a gamma-ray burst.

Within a few minutes, the expanding blast from the fireball plows into the interstellar matter in the dying star's neighborhood. This matter might have been ejected from the star itself at earlier stages in its evolution. Alternatively, it could be the gas out of which the massive star and its neighbors formed.

As the high-speed particles from the blast are slowed, they transfer their energy to the surrounding matter in the form of a shock wave. That shocked material emits radiation at longer wavelengths. This accounts for the afterglow of X-rays, visible light, and radio waves—the glow comes at longer and longer wavelengths as the blast continues to lose energy.

Short-Duration Gamma-Ray Bursts: Colliding Stellar Corpses

What about the shorter gamma-ray bursts? The gamma-ray emission from these events lasts less than 2 seconds, and in some cases may last only milliseconds—an amazingly short time. Such a timescale is difficult to achieve if they are produced in the same way as long-duration gamma-ray bursts, since the collapse of the stellar interior onto the black hole should take at least a few seconds.

Astronomers looked fruitlessly for afterglows from short-duration gamma-ray bursts found by *BeppoSAX* and other satellites. Evidently, the afterglows fade away too quickly. Fast-responding visible-light telescopes like ROTSE were not helpful either: no matter how fast these telescopes responded, the bursts were not bright enough at visible wavelengths to be detected by these small telescopes.

Once again, it took a new satellite to clear up the mystery. In this case, it was the *Swift Gamma-Ray Burst Satellite*, launched in 2004 by a collaboration between NASA and the Italian and UK space agencies (Figure 23.23). The design of *Swift* is similar to that of *BeppoSAX*. However, *Swift* is much more agile and flexible: after a gamma-ray burst occurs, the X-ray and UV telescopes can be repointed automatically within a few minutes (rather than a few hours). Thus, astronomers can observe the afterglow much earlier, when it is expected to be much brighter. Furthermore, the X-ray telescope is far more sensitive and can provide positions that are 30 times more precise than those provided by *BeppoSAX*, allowing bursts to be identified even without visible-light or radio observations.

Download for free at http://cnx.org/content/col11992/latest/

Figure 23.23. Artist's Illustration of Swift. The US/UK/Italian spacecraft *Swift* contains on-board gamma-ray, X-ray, and ultraviolet detectors, and has the ability to automatically reorient itself to a gamma-ray burst detected by the gamma-ray instrument. Since its launch in 2005, *Swift* has detected and observed over a thousand bursts, including dozens of short-duration bursts. (credit: NASA, Spectrum Astro)

On May 9, 2005, *Swift* detected a flash of gamma rays lasting 0.13 seconds in duration, originating from the constellation Coma Berenices. Remarkably, the galaxy at the X-ray position looked completely different from any galaxy in which a long-duration burst had been seen to occur. The afterglow originated from the halo of a giant elliptical galaxy 2.7 billion light-years away, with no signs of any young, massive stars in its spectrum. Furthermore, no supernova was ever detected after the burst, despite extensive searching.

What could produce a burst less than a second long, originating from a region with no star formation? The leading model involves the *merger* of two compact stellar corpses: two neutron stars, or perhaps a neutron star and a black hole. Since many stars come in binary or multiple systems, it's possible to have systems where two such star corpses orbit one another. According to general relativity (which will be discussed in Black Holes and Curved Spacetime), the orbits of a binary star system composed of such objects should slowly decay with time, eventually (after millions or billions of years) causing the two objects to slam together in a violent but brief explosion. Because the decay of the binary orbit is so slow, we would expect more of these mergers to occur in old galaxies in which star formation has long since stopped.

LINK TO LEARNING

To learn more about the merger of two neutron stars and how they can produce a burst that lasts less than a second, check out this computer simulation (https://openstaxcollege.org/l/30comsimneustr) by NASA.

While it was impossible to be sure of this model based on only a single event (it is possible this burst actually came from a background galaxy and lined up with the giant elliptical only by chance), several dozen more short-

This OpenStax book is available for free at http://cnx.org/content/col11992/1.8

Download for free at http://cnx.org/content/col11992/latest/

duration gamma-ray bursts have since been located by *Swift*, many of which also originate from galaxies with very low star-formation rates. This has given astronomers greater confidence that this model is the correct one. Still, to be fully convinced, astronomers are searching for a "smoking gun" signature for the merger of two ultra-dense stellar remnants.

There are two examples we can think of that would provide more direct evidence. One is a very special kind of explosion, produced when neutrons stripped from the neutron stars during the violent final phase of the merger fuse together into heavy elements and then release heat due to radioactivity, producing a short-lived but red supernova sometimes called a *kilonova*. (The term is used because it is about a thousand times brighter than an ordinary nova, but not quite as "super" as a traditional supernova.) Hubble observations of one short-duration gamma-ray burst in 2013 show suggestive evidence of such a signature, but need to be confirmed by future observations.

The second "smoking gun" has been even more exciting to see: the detection of *gravitational waves.* As will be discussed in Black Holes and Curved Spacetime, gravitational waves are ripples in the fabric of spacetime that general relativity predicts should be produced by the acceleration of extremely massive and dense objects—such as two neutron stars or black holes spiraling toward each other and colliding. The first example of gravitational waves has been observed recently from the merger of two large black holes. If a gravitational wave is observed one day to be coincident in time and space with a gamma-ray burst, this will not only confirm our theories of the origin of short gamma-ray bursts but would also be among the most spectacular demonstrations yet of Einstein's theory of general relativity.

Probing the Universe with Gamma-Ray Bursts

The story of how astronomers came to explain the origin of the different kinds of bursts is a good example of how the scientific process sometimes resembles good detective work. While the mystery of short-duration gamma-ray bursts is still being unraveled, the focus of studies for long-duration gamma-ray bursts has begun to change from understanding the origin of the bursts themselves (which is now fairly well-established) to using them as tools to understand the broader universe.

The reason that long-duration gamma-ray bursts are useful has to do with their extreme luminosities, if only for a short time. In fact, long-duration gamma-ray bursts are so bright that they could easily be seen at distances that correspond to a few hundred million years after the expansion of the universe began, which is when theorists think that the first generation of stars formed. Some theories predict that the first stars are likely to be massive and complete their evolution in only a million years or so. If this turns out to be the case, then gamma-ray bursts (which signal the death of some of these stars) may provide us with the best way of probing the universe when stars and galaxies first began to form.

So far, the most distant gamma-ray burst found (on April 29, 2009) originated a remarkable 13.2 billion light-years away—meaning it happened only 600 million years after the Big Bang itself. This is comparable to the earliest and most distant galaxies found by the Hubble Space Telescope. It is not quite old enough to expect that it formed from the first generation of stars, but its appearance at this distance still gives us useful information about the production of stars in the early universe. Astronomers continue to scan the skies, looking for even more distant events signaling the deaths of stars from even further back in time.

Download for free at http://cnx.org/content/col11992/latest/

CHAPTER 23 REVIEW

 KEY TERMS

Chandrasekhar limit the upper limit to the mass of a white dwarf (equals 1.4 times the mass of the Sun)

degenerate gas a gas that resists further compression because no two electrons can be in the same place at the same time doing the same thing (Pauli exclusion principle)

millisecond pulsar a pulsar that rotates so quickly that it can give off hundreds of pulses per second (and its period is therefore measured in milliseconds)

neutron star a compact object of extremely high density composed almost entirely of neutrons

nova the cataclysmic explosion produced in a binary system, temporarily increasing its luminosity by hundreds to thousands of times

pulsar a variable radio source of small physical size that emits very rapid radio pulses in very regular periods that range from fractions of a second to several seconds; now understood to be a rotating, magnetic neutron star that is energetic enough to produce a detectable beam of radiation and particles

type II supernova a stellar explosion produced at the endpoint of the evolution of stars whose mass exceeds roughly 10 times the mass of the Sun

 SUMMARY

23.1 The Death of Low-Mass Stars

During the course of their evolution, stars shed their outer layers and lose a significant fraction of their initial mass. Stars with masses of 8 M_{Sun} or less can lose enough mass to become white dwarfs, which have masses less than the Chandrasekhar limit (about 1.4 M_{Sun}). The pressure exerted by degenerate electrons keeps white dwarfs from contracting to still-smaller diameters. Eventually, white dwarfs cool off to become black dwarfs, stellar remnants made mainly of carbon, oxygen, and neon.

23.2 Evolution of Massive Stars: An Explosive Finish

In a massive star, hydrogen fusion in the core is followed by several other fusion reactions involving heavier elements. Just before it exhausts all sources of energy, a massive star has an iron core surrounded by shells of silicon, sulfur, oxygen, neon, carbon, helium, and hydrogen. The fusion of iron requires energy (rather than releasing it). If the mass of a star's iron core exceeds the Chandrasekhar limit (but is less than 3 M_{Sun}), the core collapses until its density exceeds that of an atomic nucleus, forming a neutron star with a typical diameter of 20 kilometers. The core rebounds and transfers energy outward, blowing off the outer layers of the star in a type II supernova explosion.

23.3 Supernova Observations

A supernova occurs on average once every 25 to 100 years in the Milky Way Galaxy. Despite the odds, no supernova in our Galaxy has been observed from Earth since the invention of the telescope. However, one nearby supernova (SN 1987A) has been observed in a neighboring galaxy, the Large Magellanic Cloud. The star that evolved to become SN 1987A began its life as a blue supergiant, evolved to become a red supergiant, and returned to being a blue supergiant at the time it exploded. Studies of SN 1987A have detected neutrinos from the core collapse and confirmed theoretical calculations of what happens during such explosions, including

This OpenStax book is available for free at http://cnx.org/content/col11992/1.8

the formation of elements beyond iron. Supernovae are a main source of high-energy cosmic rays and can be dangerous for any living organisms in nearby star systems.

23.4 Pulsars and the Discovery of Neutron Stars

At least some supernovae leave behind a highly magnetic, rapidly rotating neutron star, which can be observed as a pulsar if its beam of escaping particles and focused radiation is pointing toward us. Pulsars emit rapid pulses of radiation at regular intervals; their periods are in the range of 0.001 to 10 seconds. The rotating neutron star acts like a lighthouse, sweeping its beam in a circle and giving us a pulse of radiation when the beam sweeps over Earth. As pulsars age, they lose energy, their rotations slow, and their periods increase.

23.5 The Evolution of Binary Star Systems

When a white dwarf or neutron star is a member of a close binary star system, its companion star can transfer mass to it. Material falling *gradually* onto a white dwarf can explode in a sudden burst of fusion and make a nova. If material falls *rapidly* onto a white dwarf, it can push it over the Chandrasekhar limit and cause it to explode completely as a type Ia supernova. Another possible mechanism for a type Ia supernova is the merger of two white dwarfs. Material falling onto a neutron star can cause powerful bursts of X-ray radiation. Transfer of material and angular momentum can speed up the rotation of pulsars until their periods are just a few thousandths of a second.

23.6 The Mystery of the Gamma-Ray Bursts

Gamma-ray bursts last from a fraction of a second to a few minutes. They come from all directions and are now known to be associated with very distant objects. The energy is most likely beamed, and, for the ones we can detect, Earth lies in the direction of the beam. Long-duration bursts (lasting more than a few seconds) come from massive stars with their outer hydrogen layers missing that explode as supernovae. Short-duration bursts are believed to be mergers of stellar corpses (neutron stars or black holes).

FOR FURTHER EXPLORATION

Articles

Death of Stars

Hillebrandt, W., et al. "How To Blow Up a Star." *Scientific American* (October 2006): 42. On supernova mechanisms.

Irion, R. "Pursuing the Most Extreme Stars." *Astronomy* (January 1999): 48. On pulsars.

Kalirai, J. "New Light on Our Sun's Fate." *Astronomy* (February 2014): 44. What will happen to stars like our Sun between the main sequence and the white dwarf stages.

Kirshner, R. "Supernova 1987A: The First Ten Years." *Sky & Telescope* (February 1997): 35.

Maurer, S. "Taking the Pulse of Neutron Stars." *Sky & Telescope* (August 2001): 32. Review of recent ideas and observations of pulsars.

Zimmerman, R. "Into the Maelstrom." *Astronomy* (November 1998): 44. About the Crab Nebula.

Gamma-Ray Bursts

Fox, D. & Racusin, J. "The Brightest Burst." *Sky & Telescope* (January 2009): 34. Nice summary of the brightest burst observed so far, and what we have learned from it.

Download for free at http://cnx.org/content/col11992/latest/

Nadis, S. "Do Cosmic Flashes Reveal Secrets of the Infant Universe?" *Astronomy* (June 2008): 34. On different types of gamma-ray bursts and what we can learn from them.

Naeye, R. "Dissecting the Bursts of Doom." *Sky & Telescope* (August 2006): 30. Excellent review of gamma-ray bursts—how we discovered them, what they might be, and what they can be used for in probing the universe.

Zimmerman, R. "Speed Matters." *Astronomy* (May 2000): 36. On the quick-alert networks for finding afterglows.

Zimmerman, R. "Witness to Cosmic Collisions." *Astronomy* (July 2006): 44. On the Swift mission and what it is teaching astronomers about gamma-ray bursts.

Websites

Death of Stars

Crab Nebula: http://chandra.harvard.edu/xray_sources/crab/crab.html. A short, colorfully written introduction to the history and science involving the best-known supernova remant.

Introduction to Neutron Stars: https://www.astro.umd.edu/~miller/nstar.html. Coleman Miller of the University of Maryland maintains this site, which goes from easy to hard as you get into it, but it has lots of good information about corpses of massive stars.

Introduction to Pulsars (by Maryam Hobbs at the Australia National Telescope Facility): http://www.atnf.csiro.au/outreach/education/everyone/pulsars/index.html.

Magnetars, Soft Gamma Repeaters, and Very Strong Magnetic Fields: http://solomon.as.utexas.edu/magnetar.html. Robert Duncan, one of the originators of the idea of magnetars, assembled this site some years ago.

Gamma-Ray Bursts

Brief Intro to Gamma-Ray Bursts (from PBS' *Seeing in the Dark*): http://www.pbs.org/seeinginthedark/astronomy-topics/gamma-ray-bursts.html.

Discovery of Gamma-ray Bursts: http://science.nasa.gov/science-news/science-at-nasa/1997/ast19sep97_2/.

Gamma-Ray Bursts: Introduction to a Mystery (at NASA's Imagine the Universe site): http://imagine.gsfc.nasa.gov/docs/science/know_l1/bursts.html.

Introduction from the *Swift* Satellite Site: http://swift.sonoma.edu/about_swift/grbs.html.

Missions to Detect and Learn More about Gamma-ray Bursts:

- Fermi Space Telescope: http://fermi.gsfc.nasa.gov/public/.
- *INTEGRAL* Spacecraft: http://www.esa.int/science/integral.
- *SWIFT* Spacecraft: http://swift.sonoma.edu/.

Videos

Death of Stars

BBC interview with Antony Hewish: http://www.bbc.co.uk/archive/scientists/10608.shtml. (40:54).

Black Widow Pulsars: The Vengeful Corpses of Stars: https://www.youtube.com/watch?v=Fn-3G_N0hy4. A public talk in the Silicon Valley Astronomy Lecture Series by Dr. Roger Romani (Stanford University) (1:01:47).

Hubblecast 64: It all ends with a bang!: http://www.spacetelescope.org/videos/hubblecast64a/. HubbleCast Program introducing Supernovae with Dr. Joe Liske (9:48).

This OpenStax book is available for free at http://cnx.org/content/col11992/1.8

Space Movie Reveals Shocking Secrets of the Crab Pulsar: http://hubblesite.org/newscenter/archive/releases/2002/24/video/c/. A sequence of Hubble and Chandra Space Telescope images of the central regions of the Crab Nebula have been assembled into a very brief movie accompanied by animation showing how the pulsar affects its environment; it comes with some useful background material (40:06).

Gamma-Ray Bursts

Gamma-Ray Bursts: The Biggest Explosions Since the Big Bang!: https://www.youtube.com/watch?v=ePo_EdgV764. Edo Berge in a popular-level lecture at Harvard (58:50).

Gamma-Ray Bursts: Flashes in the Sky: https://www.youtube.com/watch?v=23EhcAP3O8Q. American Museum of Natural History Science Bulletin on the *Swift* satellite (5:59).

Overview Animation of Gamma-Ray Burst: http://news.psu.edu/video/296729/2013/11/27/overview-animation-gamma-ray-burst. Brief Animation of what causes a long-duration gamma-ray burst (0:55).

🏛 COLLABORATIVE GROUP ACTIVITIES

A. Someone in your group uses a large telescope to observe an expanding shell of gas. Discuss what measurements you could make to determine whether you have discovered a planetary nebula or the remnant of a supernova explosion.

B. The star Sirius (the brightest star in our northern skies) has a white-dwarf companion. Sirius has a mass of about 2 M_{Sun} and is still on the main sequence, while its companion is already a star corpse. Remember that a white dwarf can't have a mass greater than 1.4 M_{Sun}. Assuming that the two stars formed at the same time, your group should discuss how Sirius could have a white-dwarf companion. Hint: Was the initial mass of the white-dwarf star larger or smaller than that of Sirius?

C. Discuss with your group what people today would do if a brilliant star suddenly became visible during the daytime? What kind of fear and superstition might result from a supernova that was really bright in our skies? Have your group invent some headlines that the tabloid newspapers and the less responsible web news outlets would feature.

D. Suppose a supernova exploded only 40 light-years from Earth. Have your group discuss what effects there may be on Earth when the radiation reaches us and later when the particles reach us. Would there be any way to protect people from the supernova effects?

E. When pulsars were discovered, the astronomers involved with the discovery talked about finding "little green men." If you had been in their shoes, what tests would you have performed to see whether such a pulsating source of radio waves was natural or the result of an alien intelligence? Today, several groups around the world are actively searching for possible radio signals from intelligent civilizations. How might you expect such signals to differ from pulsar signals?

F. Your little brother, who has not had the benefit of an astronomy course, reads about white dwarfs and neutron stars in a magazine and decides it would be fun to go near them or even try to land on them. Is this a good idea for future tourism? Have your group make a list of reasons it would not be safe for children (or adults) to go near a white dwarf and a neutron star.

G. A lot of astronomers' time and many instruments have been devoted to figuring out the nature of gamma-ray bursts. Does your group share the excitement that astronomers feel about these mysterious high-energy events? What are some reasons that people outside of astronomy might care about learning about gamma-ray bursts?

EXERCISES

Review Questions

1. How does a white dwarf differ from a neutron star? How does each form? What keeps each from collapsing under its own weight?

2. Describe the evolution of a star with a mass like that of the Sun, from the main-sequence phase of its evolution until it becomes a white dwarf.

3. Describe the evolution of a massive star (say, 20 times the mass of the Sun) up to the point at which it becomes a supernova. How does the evolution of a massive star differ from that of the Sun? Why?

4. How do the two types of supernovae discussed in this chapter differ? What kind of star gives rise to each type?

5. A star begins its life with a mass of 5 M_{Sun} but ends its life as a white dwarf with a mass of 0.8 M_{Sun}. List the stages in the star's life during which it most likely lost some of the mass it started with. How did mass loss occur in each stage?

6. If the formation of a neutron star leads to a supernova explosion, explain why only three of the hundreds of known pulsars are found in supernova remnants.

7. How can the Crab Nebula shine with the energy of something like 100,000 Suns when the star that formed the nebula exploded almost 1000 years ago? Who "pays the bills" for much of the radiation we see coming from the nebula?

8. How is a nova different from a type Ia supernova? How does it differ from a type II supernova?

9. Apart from the masses, how are binary systems with a neutron star different from binary systems with a white dwarf?

10. What observations from SN 1987A helped confirm theories about supernovae?

11. Describe the evolution of a white dwarf over time, in particular how the luminosity, temperature, and radius change.

12. Describe the evolution of a pulsar over time, in particular how the rotation and pulse signal changes over time.

13. How would a white dwarf that formed from a star that had an initial mass of 1 M_{Sun} be different from a white dwarf that formed from a star that had an initial mass of 9 M_{Sun}?

14. What do astronomers think are the causes of longer-duration gamma-ray bursts and shorter-duration gamma-ray bursts?

15. How did astronomers finally solve the mystery of what gamma-ray bursts were? What instruments were required to find the solution?

This OpenStax book is available for free at http://cnx.org/content/col11992/1.8

Thought Questions

16. Arrange the following stars in order of their evolution:

    A. A star with no nuclear reactions going on in the core, which is made primarily of carbon and oxygen.

    B. A star of uniform composition from center to surface; it contains hydrogen but has no nuclear reactions going on in the core.

    C. A star that is fusing hydrogen to form helium in its core.

    D. A star that is fusing helium to carbon in the core and hydrogen to helium in a shell around the core.

    E. A star that has no nuclear reactions going on in the core but is fusing hydrogen to form helium in a shell around the core.

17. Would you expect to find any white dwarfs in the Orion Nebula? (See The Birth of Stars and the Discovery of Planets outside the Solar System to remind yourself of its characteristics.) Why or why not?

18. Suppose no stars more massive than about 2 M_{Sun} had ever formed. Would life as we know it have been able to develop? Why or why not?

19. Would you be more likely to observe a type II supernova (the explosion of a massive star) in a globular cluster or in an open cluster? Why?

20. Astronomers believe there are something like 100 million neutron stars in the Galaxy, yet we have only found about 2000 pulsars in the Milky Way. Give several reasons these numbers are so different. Explain each reason.

21. Would you expect to observe every supernova in our own Galaxy? Why or why not?

22. The Large Magellanic Cloud has about one-tenth the number of stars found in our own Galaxy. Suppose the mix of high- and low-mass stars is exactly the same in both galaxies. Approximately how often does a supernova occur in the Large Magellanic Cloud?

23. Look at the list of the nearest stars in Appendix I. Would you expect any of these to become supernovae? Why or why not?

24. If most stars become white dwarfs at the ends of their lives and the formation of white dwarfs is accompanied by the production of a planetary nebula, why are there more white dwarfs than planetary nebulae in the Galaxy?

25. If a 3 and 8 M_{Sun} star formed together in a binary system, which star would:

    A. Evolve off the main sequence first?

    B. Form a carbon- and oxygen-rich white dwarf?

    C. Be the location for a nova explosion?

26. You have discovered two star clusters. The first cluster contains mainly main-sequence stars, along with some red giant stars and a few white dwarfs. The second cluster also contains mainly main-sequence stars, along with some red giant stars, and a few neutron stars—but no white dwarf stars. What are the relative ages of the clusters? How did you determine your answer?

27. A supernova remnant was recently discovered and found to be approximately 150 years old. Provide possible reasons that this supernova explosion escaped detection.

Download for free at http://cnx.org/content/col11992/latest/

28. Based upon the evolution of stars, place the following elements in order of least to most common in the Galaxy: gold, carbon, neon. What aspects of stellar evolution formed the basis for how you ordered the elements?

29. What observations or types of telescopes would you use to distinguish a binary system that includes a main-sequence star and a white dwarf star from one containing a main-sequence star and a neutron star?

30. How would the spectra of a type II supernova be different from a type Ia supernova? Hint: Consider the characteristics of the objects that are their source.

Figuring For Yourself

31. The ring around SN 1987A (Figure 23.12) initially became illuminated when energetic photons from the supernova interacted with the material in the ring. The radius of the ring is approximately 0.75 light-year from the supernova location. How long after the supernova did the ring become illuminated?

32. What is the acceleration of gravity (g) at the surface of the Sun? (See Appendix E for the Sun's key characteristics.) How much greater is this than g at the surface of Earth? Calculate what you would weigh on the surface of the Sun. Your weight would be your Earth weight multiplied by the ratio of the acceleration of gravity on the Sun to the acceleration of gravity on Earth. (Okay, we know that the Sun does not have a solid surface to stand on and that you would be vaporized if you were at the Sun's photosphere. Humor us for the sake of doing these calculations.)

33. What is the escape velocity from the Sun? How much greater is it than the escape velocity from Earth?

34. What is the average density of the Sun? How does it compare to the average density of Earth?

35. Say that a particular white dwarf has the mass of the Sun but the radius of Earth. What is the acceleration of gravity at the surface of the white dwarf? How much greater is this than g at the surface of Earth? What would you weigh at the surface of the white dwarf (again granting us the dubious notion that you could survive there)?

36. What is the escape velocity from the white dwarf in Exercise 23.35? How much greater is it than the escape velocity from Earth?

37. What is the average density of the white dwarf in Exercise 23.35? How does it compare to the average density of Earth?

38. Now take a neutron star that has twice the mass of the Sun but a radius of 10 km. What is the acceleration of gravity at the surface of the neutron star? How much greater is this than g at the surface of Earth? What would you weigh at the surface of the neutron star (provided you could somehow not become a puddle of protoplasm)?

39. What is the escape velocity from the neutron star in Exercise 23.38? How much greater is it than the escape velocity from Earth?

40. What is the average density of the neutron star in Exercise 23.38? How does it compare to the average density of Earth?

41. One way to calculate the radius of a star is to use its luminosity and temperature and assume that the star radiates approximately like a blackbody. Astronomers have measured the characteristics of central stars of planetary nebulae and have found that a typical central star is 16 times as luminous and 20 times as hot (about 110,000 K) as the Sun. Find the radius in terms of the Sun's. How does this radius compare with that of a typical white dwarf?

This OpenStax book is available for free at http://cnx.org/content/col11992/1.8

Download for free at http://cnx.org/content/col11992/latest/

42. According to a model described in the text, a neutron star has a radius of about 10 km. Assume that the pulses occur once per rotation. According to Einstein's theory of relatively, nothing can move faster than the speed of light. Check to make sure that this pulsar model does not violate relativity. Calculate the rotation speed of the Crab Nebula pulsar at its equator, given its period of 0.033 s. (Remember that distance equals velocity × time and that the circumference of a circle is given by 2πR).

43. Do the same calculations as in Exercise 23.42 but for a pulsar that rotates 1000 times per second.

44. If the Sun were replaced by a white dwarf with a surface temperature of 10,000 K and a radius equal to Earth's, how would its luminosity compare to that of the Sun?

45. A supernova can eject material at a velocity of 10,000 km/s. How long would it take a supernova remnant to expand to a radius of 1 AU? How long would it take to expand to a radius of 1 light-years? Assume that the expansion velocity remains constant and use the relationship: $\text{expansion time} = \frac{\text{distance}}{\text{expansion velocity}}$.

46. A supernova remnant was observed in 2007 to be expanding at a velocity of 14,000 km/s and had a radius of 6.5 light-years. Assuming a constant expansion velocity, in what year did this supernova occur?

47. The ring around SN 1987A (Figure 23.12) started interacting with material propelled by the shockwave from the supernova beginning in 1997 (10 years after the explosion). The radius of the ring is approximately 0.75 light-year from the supernova location. How fast is the supernova material moving, assume a constant rate of motion in km/s?

48. Before the star that became SN 1987A exploded, it evolved from a red supergiant to a blue supergiant while remaining at the same luminosity. As a red supergiant, its surface temperature would have been approximately 4000 K, while as a blue supergiant, its surface temperature was 16,000 K. How much did the radius change as it evolved from a red to a blue supergiant?

49. What is the radius of the progenitor star that became SN 1987A? Its luminosity was 100,000 times that of the Sun, and it had a surface temperature of 16,000 K.

50. What is the acceleration of gravity at the surface of the star that became SN 1987A? How does this g compare to that at the surface of Earth? The mass was 20 times that of the Sun and the radius was 41 times that of the Sun.

51. What was the escape velocity from the surface of the SN 1987A progenitor star? How much greater is it than the escape velocity from Earth? The mass was 20 times that of the Sun and the radius was 41 times that of the Sun.

52. What was the average density of the star that became SN 1987A? How does it compare to the average density of Earth? The mass was 20 times that of the Sun and the radius was 41 times that of the Sun.

53. If the pulsar shown in Figure 23.16 is rotating 100 times per second, how many pulses would be detected in one minute? The two beams are located along the pulsar's equator, which is aligned with Earth.

Download for free at http://cnx.org/content/col11992/latest/

This OpenStax book is available for free at http://cnx.org/content/col11992/1.8

Download for free at http://cnx.org/content/col11992/latest/

OPTICAL

ILLUSTRATION

24

BLACK HOLES AND CURVED SPACETIME

Figure 24.1. Stellar Mass Black Hole. On the left, a visible-light image shows a region of the sky in the constellation of Cygnus; the red box marks the position of the X-ray source Cygnus X-1. It is an example of a black hole created when a massive star collapses at the end of its life. Cygnus X-1 is in a binary star system, and the artist's illustration on the right shows the black hole pulling material away from a massive blue companion star. This material forms a disk (shown in red and orange) that rotates around the black hole before falling into it or being redirected away from the black hole in the form of powerful jets. The material in the disk (before it falls into the black hole) is so hot that it glows with X-rays, explaining why this object is an X-ray source. (credit left: modification of work by DSS; credit right: modification of work by NASA/CXC/M.Weiss)

Chapter Outline

Thinking Ahead

For most of the twentieth century, black holes seemed the stuff of science fiction, portrayed either as monster vacuum cleaners consuming all the matter around them or as tunnels from one universe to another. But the truth about black holes is almost stranger than fiction. As we continue our voyage into the universe, we will discover that black holes are the key to explaining many mysterious and remarkable objects—including collapsed stars and the active centers of giant galaxies.

24.1 INTRODUCING GENERAL RELATIVITY

Learning Objectives

By the end of this section, you will be able to:

Download for free at http://cnx.org/content/col11992/latest/

> Discuss some of the key ideas of the theory of general relativity
> Recognize that one's experiences of gravity and acceleration are interchangeable and indistinguishable
> Distinguish between Newtonian ideas of gravity and Einsteinian ideas of gravity
> Recognize why the theory of general relativity is necessary for understanding the nature of black holes

Most stars end their lives as white dwarfs or neutron stars. When a *very* massive star collapses at the end of its life, however, not even the mutual repulsion between densely packed neutrons can support the core against its own weight. If the remaining mass of the star's core is more than about three times that of the Sun (M_{Sun}), our theories predict that *no known force can stop it from collapsing forever!* Gravity simply overwhelms all other forces and crushes the core until it occupies an infinitely small volume. A star in which this occurs may become one of the strangest objects ever predicted by theory—a black hole.

To understand what a black hole is like and how it influences its surroundings, we need a theory that can describe the action of gravity under such extreme circumstances. To date, our best theory of gravity is the **general theory of relativity**, which was put forward in 1916 by Albert Einstein.

General relativity was one of the major intellectual achievements of the twentieth century; if it were music, we would compare it to the great symphonies of Beethoven or Mahler. Until recently, however, scientists had little need for a better theory of gravity; Isaac Newton's ideas that led to his law of universal gravitation (see Orbits and Gravity) are perfectly sufficient for most of the objects we deal with in everyday life. In the past half century, however, general relativity has become more than just a beautiful idea; it is now essential in understanding pulsars, quasars (which will be discussed in Active Galaxies, Quasars, and Supermassive Black Holes), and many other astronomical objects and events, including the black holes we will discuss here.

We should perhaps mention that this is the point in an astronomy course when many students start to feel a little nervous (and perhaps wish they had taken botany or some other earthbound course to satisfy the science requirement). This is because in popular culture, Einstein has become a symbol for mathematical brilliance that is simply beyond the reach of most people (Figure 24.2).

Figure 24.2. Albert Einstein (1879–1955). This famous scientist, seen here younger than in the usual photos, has become a symbol for high intellect in popular culture. (credit: NASA)

So, when we wrote that the theory of general relativity was Einstein's work, you may have worried just a bit, convinced that anything Einstein did must be beyond your understanding. This popular view is unfortunate and mistaken. Although the detailed calculations of general relativity do involve a good deal of higher mathematics, the basic ideas are not difficult to understand (and are, in fact, almost poetic in the way they give us a new perspective on the world). Moreover, general relativity goes beyond Newton's famous "inverse-square" law of

This OpenStax book is available for free at http://cnx.org/content/col11992/1.8

gravity; it helps *explain* how matter interacts with other matter in space and time. This explanatory power is one of the requirements that any successful scientific theory must meet.

The Principle of Equivalence

The fundamental insight that led to the formulation of the general theory of relativity starts with a very simple thought: if you were able to jump off a high building and fall freely, you would not feel your own weight. In this chapter, we will describe how Einstein built on this idea to reach sweeping conclusions about the very fabric of space and time itself. He called it the "happiest thought of my life."

Einstein himself pointed out an everyday example that illustrates this effect (see Figure 24.3). Notice how your weight seems to be reduced in a high-speed elevator when it accelerates from a stop to a rapid descent. Similarly, your weight seems to increase in an elevator that starts to move quickly upward. This effect is not just a feeling you have: if you stood on a scale in such an elevator, you could measure your weight changing (you can actually perform this experiment in some science museums).

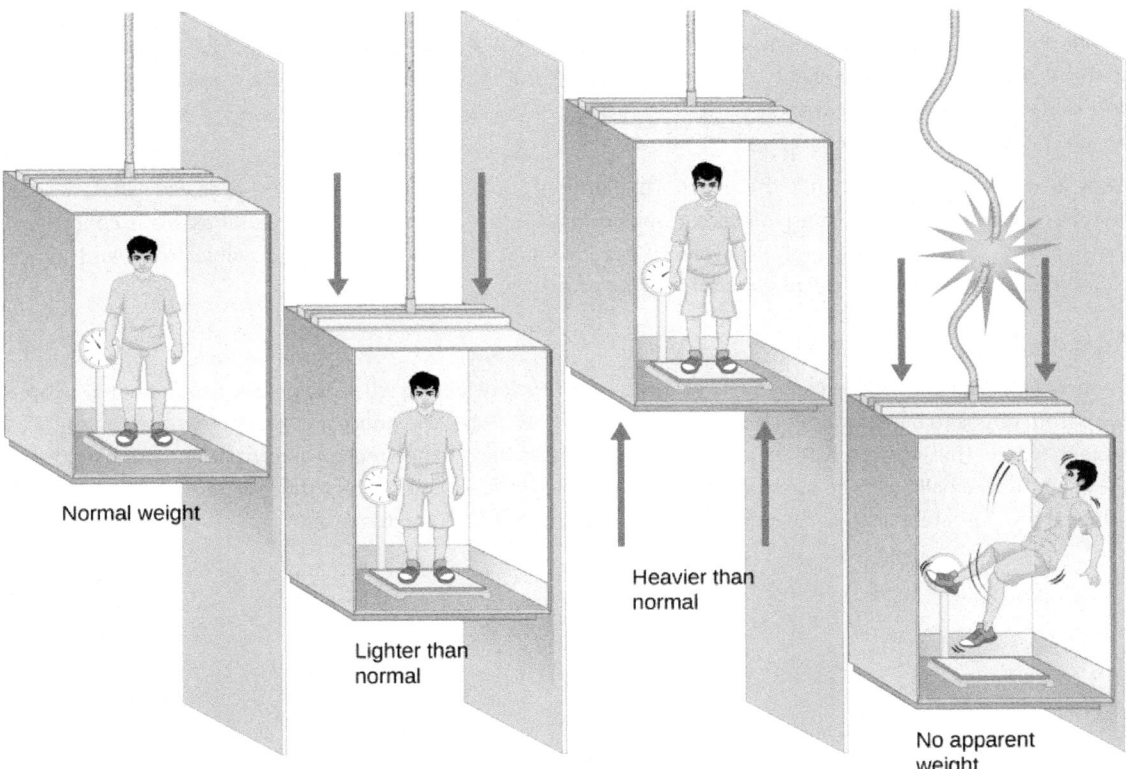

Normal weight

Lighter than normal

Heavier than normal

No apparent weight

Figure 24.3. Your Weight in an Elevator. In an elevator at rest, you feel your normal weight. In an elevator that accelerates as it descends, you would feel lighter than normal. In an elevator that accelerates as it ascends, you would feel heavier than normal. If an evil villain cut the elevator cable, you would feel weightless as you fell to your doom.

In a *freely falling* elevator, with no air friction, you would lose your weight altogether. We generally don't like to cut the cables holding elevators to try this experiment, but near-weightlessness can be achieved by taking an airplane to high altitude and then dropping rapidly for a while. This is how NASA trains its astronauts for the experience of free fall in space; the scenes of weightlessness in the 1995 movie *Apollo 13* were filmed in

Download for free at http://cnx.org/content/col11992/latest/

the same way. (Moviemakers have since devised other methods using underwater filming, wire stunts, and computer graphics to create the appearance of weightlessness seen in such movies as *Gravity* and *The Martian*.)

LINK TO LEARNING

Watch how NASA uses a "weightless" environment (https://openstax.org/l/30NASAweightra) to help train astronauts.

Another way to state Einstein's idea is this: suppose we have a spaceship that contains a windowless laboratory equipped with all the tools needed to perform scientific experiments. Now, imagine that an astronomer wakes up after a long night celebrating some scientific breakthrough and finds herself sealed into this laboratory. She has no idea how it happened but notices that she is weightless. This could be because she and the laboratory are far away from any source of gravity, and both are either at rest or moving at some steady speed through space (in which case she has plenty of time to wake up). But it could also be because she and the laboratory are falling freely toward a planet like Earth (in which case she might first want to check her distance from the surface before making coffee).

What Einstein postulated is that there is no experiment she can perform inside the sealed laboratory to determine whether she is floating in space or falling freely in a gravitational field.[1] As far as she is concerned, the two situations are completely *equivalent*. This idea that free fall is indistinguishable from, and hence equivalent to, zero gravity is called the **equivalence principle**.

Gravity or Acceleration?

Einstein's simple idea has big consequences. Let's begin by considering what happens if two foolhardy people jump from opposite banks into a bottomless chasm (Figure 24.4). If we ignore air friction, then we can say that while they freely fall, they both accelerate downward at the same rate and feel no external force acting on them. They can throw a ball back and forth, always aiming it straight at each other, as if there were no gravity. The ball falls at the same rate that they do, so it always remains in a line between them.

---

1   Strictly speaking, this is true only if the laboratory is infinitesimally small. Different locations in a real laboratory that is falling freely due to gravity cannot all be at identical distances from the object(s) responsible for producing the gravitational force. In this case, objects in different locations will experience slightly different accelerations. But this point does not invalidate the principle of equivalence that Einstein derived from this line of thinking.

This OpenStax book is available for free at http://cnx.org/content/col11992/1.8

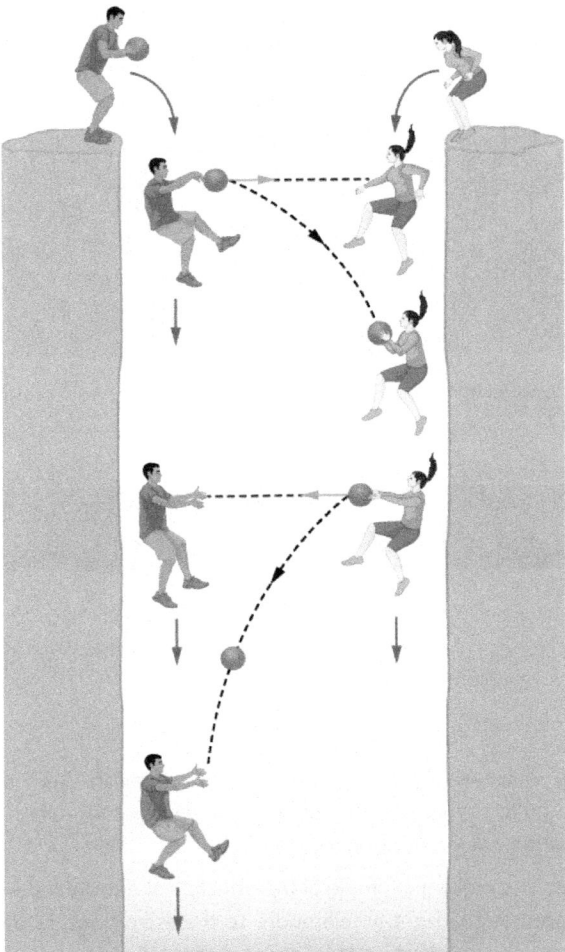

Figure 24.4. Free Fall. Two people play catch as they descend into a bottomless abyss. Since the people and ball all fall at the same speed, it appears to them that they can play catch by throwing the ball in a straight line between them. Within their frame of reference, there appears to be no gravity.

Such a game of catch is very different on the surface of Earth. Everyone who grows up feeling gravity knows that a ball, once thrown, falls to the ground. Thus, in order to play catch with someone, you must aim the ball upward so that it follows an arc—rising and then falling as it moves forward—until it is caught at the other end.

Now suppose we isolate our falling people and ball inside a large box that is falling with them. No one inside the box is aware of any gravitational force. If they let go of the ball, it doesn't fall to the bottom of the box or anywhere else but merely stays there or moves in a straight line, depending on whether it is given any motion.

Astronauts in the International Space Station (ISS) that is orbiting Earth live in an environment just like that of the people sealed in a freely falling box (Figure 24.5). The orbiting ISS is actually "falling" freely around Earth. While in free fall, the astronauts live in a strange world where there seems to be no gravitational force. One can give a wrench a shove, and it moves at constant speed across the orbiting laboratory. A pencil set in midair remains there as if no force were acting on it.

Download for free at http://cnx.org/content/col11992/latest/

Figure 24.5. Astronauts aboard the Space Shuttle. Shane Kimbrough and Sandra Magnus are shown aboard the Endeavour in 2008 with various fruit floating freely. Because the shuttle is in free fall as it orbits Earth, everything—including astronauts—stays put or moves uniformly relative to the walls of the spacecraft. This free-falling state produces a lack of apparent gravity inside the spacecraft. (credit: NASA)

LINK TO LEARNING

In the "weightless" environment of the International Space Station, moving takes very little effort. Watch astronaut Karen Nyberg (https://openstax.org/l/30ISSzerogravid) demonstrate how she can propel herself with the force of a single human hair.

Appearances are misleading, however. There is a force in this situation. Both the ISS and the astronauts continually fall around Earth, pulled by its gravity. But since all fall together—shuttle, astronauts, wrench, and pencil—inside the ISS all gravitational forces appear to be absent.

Thus, the orbiting ISS provides an excellent example of the principle of equivalence—how local effects of gravity can be completely compensated by the right acceleration. To the astronauts, falling around Earth creates the same effects as being far off in space, remote from all gravitational influences.

The Paths of Light and Matter

Einstein postulated that the equivalence principle is a fundamental fact of nature, and that there is *no* experiment inside any spacecraft by which an astronaut can ever distinguish between being weightless in remote space and being in free fall near a planet like Earth. This would apply to experiments done with beams of light as well. But the minute we use light in our experiments, we are led to some very disturbing conclusions—and it is these conclusions that lead us to general relativity and a new view of gravity.

It seems apparent to us, from everyday observations, that beams of light travel in straight lines. Imagine that a spaceship is moving through empty space far from any gravity. Send a laser beam from the back of the ship to the front, and it will travel in a nice straight line and land on the front wall exactly opposite the point from which it left the rear wall. If the equivalence principle really applies universally, then this same experiment performed in free fall around Earth should give us the same result.

Now imagine that the astronauts again shine a beam of light along the length of their ship. But, as shown in Figure 24.6, this time the orbiting space station falls a bit between the time the light leaves the back wall and the time it hits the front wall. (The amount of the fall is grossly exaggerated in Figure 24.6 to illustrate the effect.)

Download for free at http://cnx.org/content/col11992/latest/

Therefore, if the beam of light follows a straight line but the ship's path curves downward, then the light should strike the front wall at a point higher than the point from which it left.

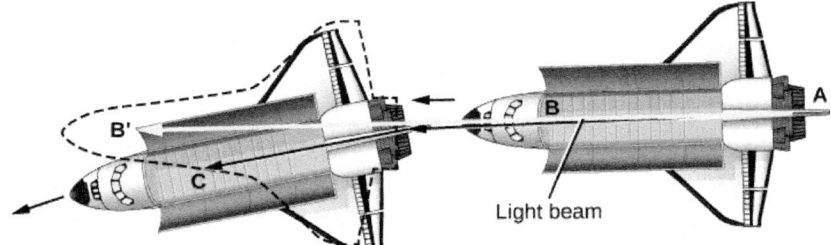

Figure 24.6. Curved Light Path. In a spaceship moving to the left (in this figure) in its orbit about a planet, light is beamed from the rear, A, toward the front, B. Meanwhile, the ship is falling out of its straight path (exaggerated here). We might therefore expect the light to strike at B', above the target in the ship. Instead, the light follows a curved path and strikes at C. In order for the principle of equivalence to be correct, gravity must be able to curve the path of a light beam just as it curves the path of the spaceship.

However, this would violate the principle of equivalence—the two experiments would give different results. We are thus faced with giving up one of our two assumptions. Either the principle of equivalence is not correct, or light does not always travel in straight lines. Instead of dropping what probably seemed at the time like a ridiculous idea, Einstein worked out what happens if light sometimes does *not* follow a straight path.

Let's suppose the principle of equivalence is right. Then the light beam must arrive directly opposite the point from which it started in the ship. The light, like the ball thrown back and forth, *must fall with the ship* that is in orbit around Earth (see Figure 24.6). This would make its path curve downward, like the path of the ball, and thus the light would hit the front wall exactly opposite the spot from which it came.

Thinking this over, you might well conclude that it doesn't seem like such a big problem: why *can't* light fall the way balls do? But, as discussed in Radiation and Spectra, light is profoundly different from balls. Balls have mass, while light does not.

Here is where Einstein's intuition and genius allowed him to make a profound leap. He gave physical meaning to the strange result of our thought experiment. Einstein suggested that the light curves down to meet the front of the shuttle because Earth's gravity actually bends the *fabric of space and time*. This radical idea—which we will explain next—keeps the behavior of light the same in both empty space and free fall, but it changes some of our most basic and cherished ideas about space and time. The reason we take Einstein's suggestion seriously is that, as we will see, experiments now clearly show his intuitive leap was correct.

24.2 SPACETIME AND GRAVITY

Learning Objectives

By the end of this section, you will be able to:

> Describe Einstein's view of gravity as the warping of spacetime in the presence of massive objects
> Understand that Newton's concept of the gravitational force between two massive objects and Einstein's concept of warped spacetime are different explanations for the same observed accelerations of one massive object in the presence of another massive object

Is light actually bent from its straight-line path by the mass of Earth? How can light, which has no mass, be affected by gravity? Einstein preferred to think that it is *space and time* that are affected by the presence of a large mass; light beams, and everything else that travels through space and time, then find their paths affected.

Download for free at http://cnx.org/content/col11992/latest/

Light always follows the shortest path—but that path may not always be straight. This idea is true for human travel on the curved surface of planet Earth, as well. Say you want to fly from Chicago to Rome. Since an airplane can't go through the solid body of the Earth, the shortest distance is not a straight line but the arc of a *great circle*.

Linkages: Mass, Space, and Time

To show what Einstein's insight really means, let's first consider how we locate an event in space and time. For example, imagine you have to describe to worried school officials the fire that broke out in your room when your roommate tried cooking shish kebabs in the fireplace. You explain that your dorm is at 6400 College Avenue, a street that runs in the left-right direction on a map of your town; you are on the fifth floor, which tells where you are in the up-down direction; and you are the sixth room back from the elevator, which tells where you are in the forward-backward direction. Then you explain that the fire broke out at 6:23 p.m. (but was soon brought under control), which specifies the event in time. *Any* event in the universe, whether nearby or far away, can be pinpointed using the three dimensions of space and the one dimension of time.

Newton considered space and time to be completely independent, and that continued to be the accepted view until the beginning of the twentieth century. But Einstein showed that there is an intimate connection between space and time, and that only by considering the two together—in what we call **spacetime**—can we build up a correct picture of the physical world. We examine spacetime a bit more closely in the next subsection.

The gist of Einstein's general theory is that the presence of matter curves or warps the fabric of spacetime. This curving of spacetime is identified with gravity. When something else—a beam of light, an electron, or the starship *Enterprise*—enters such a region of distorted spacetime, its path will be different from what it would have been in the absence of the matter. As American physicist John Wheeler summarized it: "Matter tells spacetime how to curve; spacetime tells matter how to move."

The amount of distortion in spacetime depends on the mass of material that is involved and on how concentrated and compact it is. Terrestrial objects, such as the book you are reading, have far too little mass to introduce any significant distortion. Newton's view of gravity is just fine for building bridges, skyscrapers, or amusement park rides. General relativity does, however, have some practical applications. The GPS (Global Positioning System) in every smartphone can tell you where you are within 5 to 10 meters only because the effects of general and special relativity on the GPS satellites in orbit around the Earth are taken into account.

Unlike a book or your roommate, stars produce measurable distortions in spacetime. A white dwarf, with its stronger surface gravity, produces more distortion just above its surface than does a red giant with the same mass. So, you see, we *are* eventually going to talk about collapsing stars again, but not before discussing Einstein's ideas (and the evidence for them) in more detail.

Spacetime Examples

How can we understand the distortion of spacetime by the presence of some (significant) amount of mass? Let's try the following analogy. You may have seen maps of New York City that squeeze the full three dimensions of this towering metropolis onto a flat sheet of paper and still have enough information so tourists will not get lost. Let's do something similar with diagrams of spacetime.

Figure 24.7, for example, shows the progress of a motorist driving east on a stretch of road in Kansas where the countryside is absolutely flat. Since our motorist is traveling only in the east-west direction and the terrain is flat, we can ignore the other two dimensions of space. The amount of time elapsed since he left home is shown on the *y*-axis, and the distance traveled eastward is shown on the *x*-axis. From A to B he drove at a uniform speed; unfortunately, it was too fast a uniform speed and a police car spotted him. From B to C he stopped to

This OpenStax book is available for free at http://cnx.org/content/col11992/1.8

receive his ticket and made no progress through space, only through time. From C to D he drove more slowly because the police car was behind him.

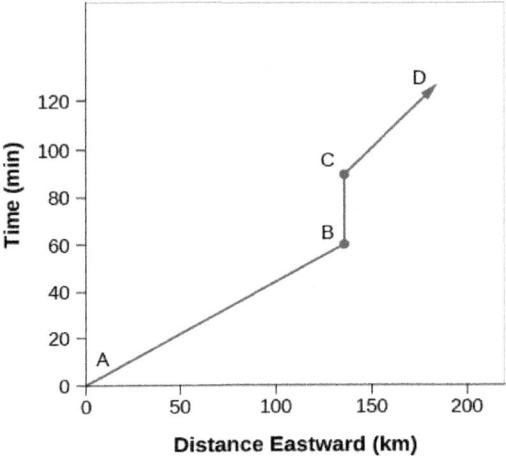

Figure 24.7. Spacetime Diagram. This diagram shows the progress of a motorist traveling east across the flat Kansas landscape. Distance traveled is plotted along the horizontal axis. The time elapsed since the motorist left the starting point is plotted along the vertical axis.

Now let's try illustrating the distortions of spacetime in two dimensions. In this case, we will (in our imaginations) use a rubber sheet that can stretch or warp if we put objects on it.

Let's imagine stretching our rubber sheet taut on four posts. To complete the analogy, we need something that normally travels in a straight line (as light does). Suppose we have an extremely intelligent ant—a friend of the comic book superhero Ant-Man, perhaps—that has been trained to walk in a straight line.

We begin with just the rubber sheet and the ant, simulating empty space with no mass in it. We put the ant on one side of the sheet and it walks in a beautiful straight line over to the other side (Figure 24.8). We next put a small grain of sand on the rubber sheet. The sand does distort the sheet a tiny bit, but this is not a distortion that we or the ant can measure. If we send the ant so it goes close to, but not on top of, the sand grain, it has little trouble continuing to walk in a straight line.

Now we grab something with a little more mass—say, a small pebble. It bends or distorts the sheet just a bit around its position. If we send the ant into this region, it finds its path slightly altered by the distortion of the sheet. The distortion is not large, but if we follow the ant's path carefully, we notice it deviating slightly from a straight line.

The effect gets more noticeable as we increase the mass of the object that we put on the sheet. Let's say we now use a massive paperweight. Such a heavy object distorts or warps the rubber sheet very effectively, putting a good sag in it. From our point of view, we can see that the sheet near the paperweight is no longer straight.

Download for free at http://cnx.org/content/col11992/latest/

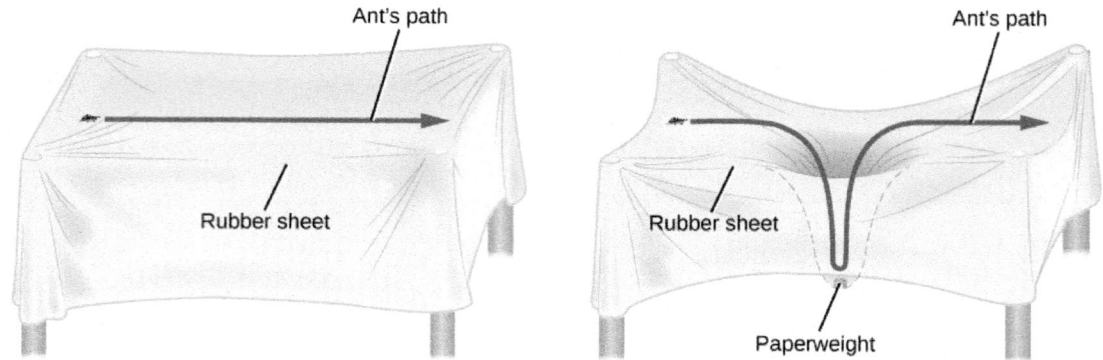

Figure 24.8. Three-Dimensional Analogy for Spacetime. On a flat rubber sheet, a trained ant has no trouble walking in a straight line. When a massive object creates a big depression in the sheet, the ant, which must walk where the sheet takes it, finds its path changed (warped) dramatically.

Now let's again send the ant on a journey that takes it close to, but not on top of, the paperweight. Far away from the paperweight, the ant has no trouble doing its walk, which looks straight to us. As it nears the paperweight, however, the ant is forced down into the sag. It must then climb up the other side before it can return to walking on an undistorted part of the sheet. All this while, the ant is following the shortest path it can, but through no fault of its own (after all, ants can't fly, so it has to stay on the sheet) this path is curved by the distortion of the sheet itself.

In the same way, according to Einstein's theory, light always follows the shortest path through spacetime. But the mass associated with large concentrations of matter distorts spacetime, and the shortest, most direct paths are no longer straight lines, but curves.

How large does a mass have to be before we can measure a change in the path followed by light? In 1916, when Einstein first proposed his theory, no distortion had been detected at the surface of Earth (so Earth might have played the role of the grain of sand in our analogy). Something with a mass like our Sun's was necessary to detect the effect Einstein was describing (we will discuss how this effect was measured using the Sun in the next section).

The paperweight in our analogy might be a white dwarf or a neutron star. The distortion of spacetime is greater near the surfaces of these compact, massive objects than near the surface of the Sun. And when, to return to the situation described at the beginning of the chapter, a star core with more than three times the mass of the Sun collapses forever, the distortions of spacetime very close to it can become truly mind-boggling.

24.3 | TESTS OF GENERAL RELATIVITY

Learning Objectives

By the end of this section, you will be able to:

> Describe unusual motion of Mercury around the Sun and explain how general relativity explains the observed behavior
> Provide examples of evidence for light rays being bent by massive objects, as predicted by general relativity's theory about the warping of spacetime

What Einstein proposed was nothing less than a major revolution in our understanding of space and time. It was a new theory of gravity, in which mass determines the curvature of spacetime and that curvature, in

This OpenStax book is available for free at http://cnx.org/content/col11992/1.8

turn, controls how objects move. Like all new ideas in science, no matter who advances them, Einstein's theory had to be tested by comparing its predictions against the experimental evidence. This was quite a challenge because the effects of the new theory were apparent only when the mass was quite large. (For smaller masses, it required measuring techniques that would not become available until decades later.)

When the distorting mass is small, the predictions of general relativity must agree with those resulting from Newton's law of universal gravitation, which, after all, has served us admirably in our technology and in guiding space probes to the other planets. In familiar territory, therefore, the differences between the predictions of the two models are subtle and difficult to detect. Nevertheless, Einstein was able to demonstrate one proof of his theory that could be found in existing data and to suggest another one that would be tested just a few years later.

The Motion of Mercury

Of the planets in our solar system, Mercury orbits closest to the Sun and is thus most affected by the distortion of spacetime produced by the Sun's mass. Einstein wondered if the distortion might produce a noticeable difference in the motion of Mercury that was not predicted by Newton's law. It turned out that the difference was subtle, but it was definitely there. Most importantly, it had already been measured.

Mercury has a highly elliptical orbit, so that it is only about two-thirds as far from the Sun at perihelion as it is at aphelion. (These terms were defined in the chapter on Orbits and Gravity.) The gravitational effects (perturbations) of the other planets on Mercury produce a calculable advance of Mercury's perihelion. What this means is that each successive perihelion occurs in a slightly different direction as seen from the Sun (Figure 24.9).

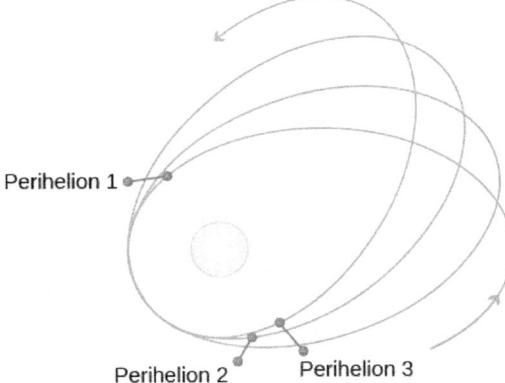

Figure 24.9. Mercury's Wobble. The major axis of the orbit of a planet, such as Mercury, rotates in space slightly because of various perturbations. In Mercury's case, the amount of rotation (or orbital precession) is a bit larger than can be accounted for by the gravitational forces exerted by other planets; this difference is precisely explained by the general theory of relativity. Mercury, being the planet closest to the Sun, has its orbit most affected by the warping of spacetime near the Sun. The change from orbit to orbit has been significantly exaggerated on this diagram.

According to Newtonian gravitation, the gravitational forces exerted by the planets will cause Mercury's perihelion to advance by about 531 seconds of arc (arcsec) per century. In the nineteenth century, however, it was observed that the actual advance is 574 arcsec per century. The discrepancy was first pointed out in 1859 by Urbain Le Verrier, the codiscoverer of Neptune. Just as discrepancies in the motion of Uranus allowed astronomers to discover the presence of Neptune, so it was thought that the discrepancy in the motion of Mercury could mean the presence of an undiscovered inner planet. Astronomers searched for this planet near the Sun, even giving it a name: Vulcan, after the Roman god of fire. (The name would later be used for the home planet of a famous character on a popular television show about future space travel.)

Download for free at http://cnx.org/content/col11992/latest/

But no planet has ever been found nearer to the Sun than Mercury, and the discrepancy was still bothering astronomers when Einstein was doing his calculations. General relativity, however, predicts that due to the curvature of spacetime around the Sun, the perihelion of Mercury should advance slightly more than is predicted by Newtonian gravity. The result is to make the major axis of Mercury's orbit rotate slowly in space because of the Sun's gravity alone. The prediction of general relativity is that the direction of perihelion should change by an additional 43 arcsec per century. This is remarkably close to the observed discrepancy, and it gave Einstein a lot of confidence as he advanced his theory. The relativistic advance of perihelion was later also observed in the orbits of several asteroids that come close to the Sun.

Deflection of Starlight

Einstein's second test was something that had not been observed before and would thus provide an excellent confirmation of his theory. Since spacetime is more curved in regions where the gravitational field is strong, we would expect light passing very near the Sun to appear to follow a curved path (Figure 24.10), just like that of the ant in our analogy. Einstein calculated from general relativity theory that starlight just grazing the Sun's surface should be deflected by an angle of 1.75 arcsec. Could such a deflection be observed?

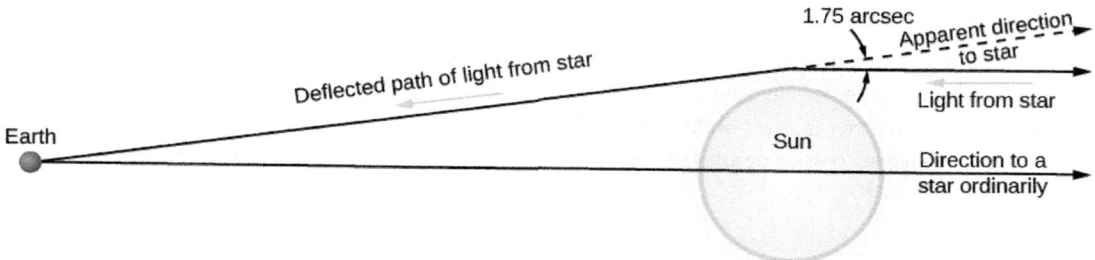

Figure 24.10. Curvature of Light Paths near the Sun. Starlight passing near the Sun is deflected slightly by the "warping" of spacetime. (This deflection of starlight is one small example of a phenomenon called gravitational lensing, which we'll discuss in more detail in The Evolution and Distribution of Galaxies.) Before passing by the Sun, the light from the star was traveling parallel to the bottom edge of the figure. When it passed near the Sun, the path was altered slightly. When we see the light, we assume the light beam has been traveling in a straight path throughout its journey, and so we measure the position of the star to be slightly different from its true position. If we were to observe the star at another time, when the Sun is not in the way, we would measure its true position.

We encounter a small "technical problem" when we try to photograph starlight coming very close to the Sun: the Sun is an outrageously bright source of starlight itself. But during a total solar eclipse, much of the Sun's light is blocked out, allowing the stars near the Sun to be photographed. In a paper published during World War I, Einstein (writing in a German journal) suggested that photographic observations during an eclipse could reveal the deflection of light passing near the Sun.

The technique involves taking a photograph of the stars six months prior to the eclipse and measuring the position of all the stars accurately. Then the same stars are photographed during the eclipse. This is when the starlight has to travel to us by skirting the Sun and moving through measurably warped spacetime. As seen from Earth, the stars closest to the Sun will seem to be "out of place"—slightly away from their regular positions as measured when the Sun is not nearby.

A single copy of that paper, passed through neutral Holland, reached the British astronomer Arthur S. Eddington, who noted that the next suitable eclipse was on May 29, 1919. The British organized two expeditions to observe it: one on the island of Príncipe, off the coast of West Africa, and the other in Sobral, in northern Brazil. Despite some problems with the weather, both expeditions obtained successful photographs. The stars seen near the Sun were indeed displaced, and to the accuracy of the measurements, which was about 20%, the shifts were consistent with the predictions of general relativity. More modern experiments with radio waves

This OpenStax book is available for free at http://cnx.org/content/col11992/1.8

traveling close to the Sun have confirmed that the actual displacements are within 1% of what general relativity predicts.

The confirmation of the theory by the eclipse expeditions in 1919 was a triumph that made Einstein a world celebrity.

 # TIME IN GENERAL RELATIVITY

Learning Objectives

By the end of this section, you will be able to:

> Describe how Einsteinian gravity slows clocks and can decrease a light wave's frequency of oscillation
> Recognize that the gravitational decrease in a light wave's frequency is compensated by an increase in the light wave's wavelength—the so-called gravitational redshift—so that the light continues to travel at constant speed

General relativity theory makes various predictions about the behavior of space and time. One of these predictions, put in everyday terms, is that *the stronger the gravity, the slower the pace of time*. Such a statement goes very much counter to our intuitive sense of time as a flow that we all share. Time has always seemed the most democratic of concepts: all of us, regardless of wealth or status, appear to move together from the cradle to the grave in the great current of time.

But Einstein argued that it only seems this way to us because all humans so far have lived and died in the gravitational environment of Earth. We have had no chance to test the idea that the pace of time might depend on the strength of gravity, because we have not experienced radically different gravities. Moreover, the differences in the flow of time are extremely small until truly large masses are involved. Nevertheless, Einstein's prediction has now been tested, both on Earth and in space.

The Tests of Time

An ingenious experiment in 1959 used the most accurate atomic clock known to compare time measurements on the ground floor and the top floor of the physics building at Harvard University. For a clock, the experimenters used the frequency (the number of cycles per second) of gamma rays emitted by radioactive cobalt. Einstein's theory predicts that such a cobalt clock on the ground floor, being a bit closer to Earth's center of gravity, should run very slightly slower than the same clock on the top floor. This is precisely what the experiments observed. Later, atomic clocks were taken up in high-flying aircraft and even on one of the Gemini space flights. In each case, the clocks farther from Earth ran a bit faster. While in 1959 it didn't matter much if the clock at the top of the building ran faster than the clock in the basement, today that effect is highly relevant. Every smartphone or device that synchronizes with a GPS must correct for this (as we will see in the next section) since the clocks on satellites will run faster than clocks on Earth.

The effect is more pronounced if the gravity involved is the Sun's and not Earth's. If stronger gravity slows the pace of time, then it will take longer for a light or radio wave that passes very near the edge of the Sun to reach Earth than we would expect on the basis of Newton's law of gravity. (It takes longer because spacetime is curved in the vicinity of the Sun.) The smaller the distance between the ray of light and the edge of the Sun at closest approach, the longer will be the delay in the arrival time.

In November 1976, when the two Viking spacecraft were operating on the surface of Mars, the planet went behind the Sun as seen from Earth (Figure 24.11). Scientists had preprogrammed Viking to send a radio wave toward Earth that would go extremely close to the outer regions of the Sun. According to general relativity,

Download for free at http://cnx.org/content/col11992/latest/

there would be a delay because the radio wave would be passing through a region where time ran more slowly. The experiment was able to confirm Einstein's theory to within 0.1%.

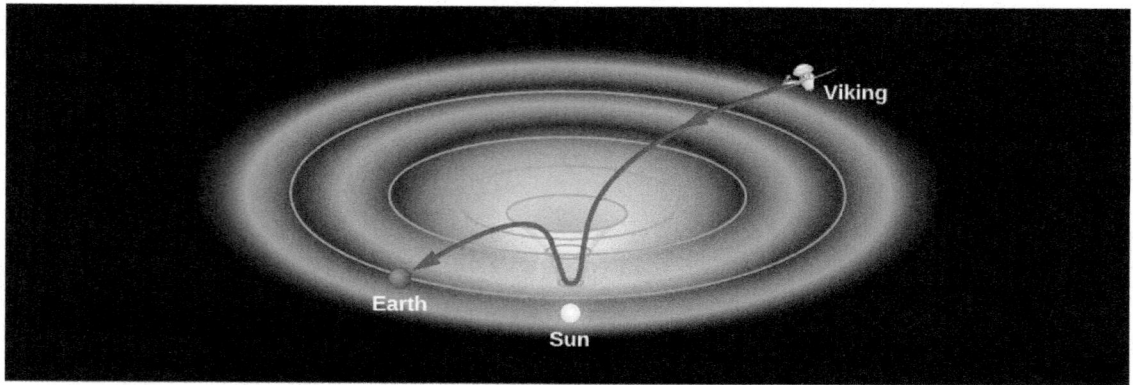

Figure 24.11. Time Delays for Radio Waves near the Sun. Radio signals from the Viking lander on Mars were delayed when they passed near the Sun, where spacetime is curved relatively strongly. In this picture, spacetime is pictured as a two-dimensional rubber sheet.

Gravitational Redshift

What does it mean to say that time runs more slowly? When light emerges from a region of strong gravity where time slows down, the light experiences a change in its frequency and wavelength. To understand what happens, let's recall that a wave of light is a repeating phenomenon—crest follows crest with great regularity. In this sense, each light wave is a little clock, keeping time with its wave cycle. If stronger gravity slows down the pace of time (relative to an outside observer), then the rate at which crest follows crest must be correspondingly slower—that is, the waves become less *frequent*.

To maintain constant light speed (the key postulate in Einstein's theories of special and general relativity), the lower frequency must be compensated by a longer wavelength. This kind of increase in wavelength (when caused by the motion of the source) is what we called a *redshift* in Radiation and Spectra. Here, because it is gravity and not motion that produces the longer wavelengths, we call the effect a **gravitational redshift**.

The advent of space-age technology made it possible to measure gravitational redshift with very high accuracy. In the mid-1970s, a hydrogen *maser*, a device akin to a laser that produces a microwave radio signal at a particular wavelength, was carried by a rocket to an altitude of 10,000 kilometers. Instruments on the ground were used to compare the frequency of the signal emitted by the rocket-borne maser with that from a similar maser on Earth. The experiment showed that the stronger gravitational field at Earth's surface really did slow the flow of time relative to that measured by the maser in the rocket. The observed effect matched the predictions of general relativity to within a few parts in 100,000.

These are only a few examples of tests that have confirmed the predictions of general relativity. Today, general relativity is accepted as our best description of gravity and is used by astronomers and physicists to understand the behavior of the centers of galaxies, the beginning of the universe, and the subject with which we began this chapter—the death of truly massive stars.

Relativity: A Practical Application

By now you may be asking: why should I be bothered with relativity? Can't I live my life perfectly well without it? The answer is you can't. Every time a pilot lands an airplane or you use a GPS to determine where you are on a drive or hike in the back country, you (or at least your GPS-enabled device) must take the effects of both general and special relativity into account.

This OpenStax book is available for free at http://cnx.org/content/col11992/1.8

GPS relies on an array of 24 satellites orbiting the Earth, and at least 4 of them are visible from any spot on Earth. Each satellite carries a precise atomic clock. Your GPS receiver detects the signals from those satellites that are overhead and calculates your position based on the time that it has taken those signals to reach you. Suppose you want to know where you are within 50 feet (GPS devices can actually do much better than this). Since it takes only 50 billionths of a second for light to travel 50 feet, the clocks on the satellites must be synchronized to at least this accuracy—and relativistic effects must therefore be taken into account.

The clocks on the satellites are orbiting Earth at a speed of 14,000 kilometers per hour and are moving much faster than clocks on the surface of Earth. According to Einstein's theory of relativity, the clocks on the satellites are ticking more slowly than Earth-based clocks by about 7 millionths of a second per day. (We have not discussed the *special* theory of relativity, which deals with changes when objects move very fast, so you'll have to take our word for this part.)

The orbits of the satellites are 20,000 kilometers above Earth, where gravity is about four times weaker than at Earth's surface. General relativity says that the orbiting clocks should tick about 45 millionths of a second faster than they would on Earth. The net effect is that the time on a satellite clock advances by about 38 microseconds per day. If these relativistic effects were not taken into account, navigational errors would start to add up and positions would be off by about 7 miles in only a single day.

BLACK HOLES

Learning Objectives

By the end of this section, you will be able to:

> Explain the event horizon surrounding a black hole
> Discuss why the popular notion of black holes as great sucking monsters that can ingest material at great distances from them is erroneous
> Use the concept of warped spacetime near a black hole to track what happens to any object that might fall into a black hole
> Recognize why the concept of a singularity—with its infinite density and zero volume—presents major challenges to our understanding of matter

Let's now apply what we have learned about gravity and spacetime curvature to the issue we started with: the collapsing core in a very massive star. We saw that if the core's mass is greater than about 3 M_{Sun}, theory says that nothing can stop the core from collapsing forever. We will examine this situation from two perspectives: first from a pre-Einstein point of view, and then with the aid of general relativity.

Classical Collapse

Let's begin with a thought experiment. We want to know what speeds are required to escape from the gravitational pull of different objects. A rocket must be launched from the surface of Earth at a very high speed if it is to escape the pull of Earth's gravity. In fact, any object—rocket, ball, astronomy book—that is thrown into the air with a velocity less than 11 kilometers per second will soon fall back to Earth's surface. Only those objects launched with a speed greater than this *escape velocity* can get away from Earth.

The escape velocity from the surface of the Sun is higher yet—618 kilometers per second. Now imagine that we begin to compress the Sun, forcing it to shrink in diameter. Recall that the pull of gravity depends on both the mass that is pulling you and your distance from the center of gravity of that mass. If the Sun is compressed, its *mass* will remain the same, but the *distance* between a point on the Sun's surface and the center will get smaller

Download for free at http://cnx.org/content/col11992/latest/

and smaller. Thus, as we compress the star, the pull of gravity for an object on the shrinking surface will get stronger and stronger (Figure 24.12).

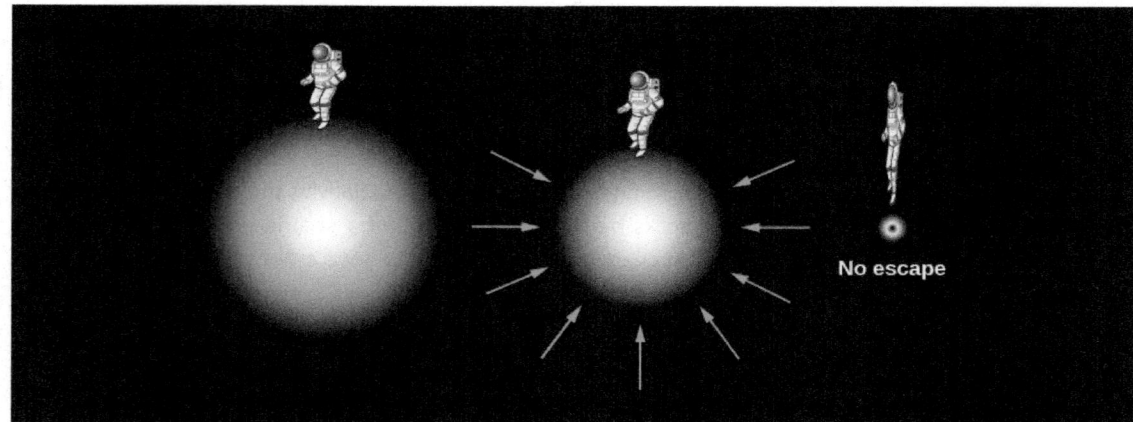

Figure 24.12. Formation of a Black Hole. At left, an imaginary astronaut floats near the surface of a massive star-core about to collapse. As the same mass falls into a smaller sphere, the gravity at its surface goes up, making it harder for anything to escape from the stellar surface. Eventually the mass collapses into so small a sphere that the escape velocity exceeds the speed of light and nothing can get away. Note that the size of the astronaut has been exaggerated. In the last picture, the astronaut is just outside the sphere we will call the event horizon and is stretched and squeezed by the strong gravity.

When the shrinking Sun reaches the diameter of a neutron star (about 20 kilometers), the velocity required to escape its gravitational pull will be about half the speed of light. Suppose we continue to compress the Sun to a smaller and smaller diameter. (We saw this can't happen to a star like our Sun in the real world because of electron degeneracy, i.e., the mutual repulsion between tightly packed electrons; this is just a quick "thought experiment" to get our bearings).

Ultimately, as the Sun shrinks, the escape velocity near the surface would exceed the speed of light. If the speed you need to get away is faster than the fastest possible speed in the universe, then nothing, not even light, is able to escape. An object with such large escape velocity emits no light, and anything that falls into it can never return.

In modern terminology, we call an object from which light cannot escape a **black hole**, a name popularized by the America scientist John Wheeler starting in the late 1960s (Figure 24.13). The idea that such objects might exist is, however, not a new one. Cambridge professor and amateur astronomer John Michell wrote a paper in 1783 about the possibility that stars with escape velocities exceeding that of light might exist. And in 1796, the French mathematician Pierre-Simon, marquis de Laplace, made similar calculations using Newton's theory of gravity; he called the resulting objects "dark bodies."

This OpenStax book is available for free at http://cnx.org/content/col11992/1.8

Download for free at http://cnx.org/content/col11992/latest/

Figure 24.13. John Wheeler (1911–2008). This brilliant physicist did much pioneering work in general relativity theory and popularized the term *black hole* starting in the late 1960s. (credit: modification of work by Roy Bishop)

While these early calculations provided strong hints that something strange should be expected if very massive objects collapse under their own gravity, we really need general relativity theory to give an adequate description of what happens in such a situation.

Collapse with Relativity

General relativity tells us that gravity is really a curvature of spacetime. As gravity increases (as in the collapsing Sun of our thought experiment), the curvature gets larger and larger. Eventually, if the Sun could shrink down to a diameter of about 6 kilometers, only light beams sent out perpendicular to the surface would escape. All others would fall back onto the star (Figure 24.14). If the Sun could then shrink just a little more, even that one remaining light beam would no longer be able to escape.

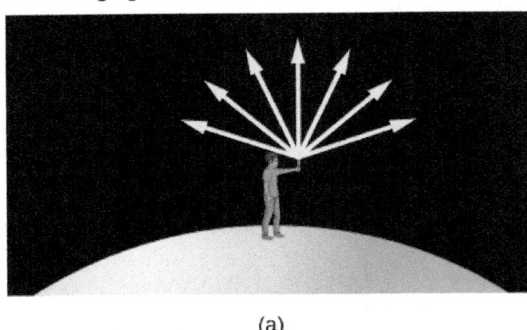
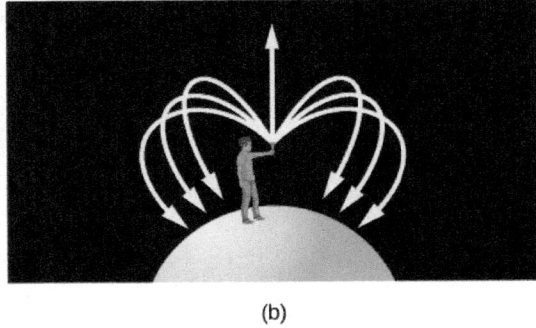

(a)   (b)

Figure 24.14. Light Paths near a Massive Object. Suppose a person could stand on the surface of a normal star with a flashlight. The light leaving the flashlight travels in a straight line no matter where the flashlight is pointed. Now consider what happens if the star collapses so that it is just a little larger than a black hole. All the light paths, except the one straight up, curve back to the surface. When the star shrinks inside the event horizon and becomes a black hole, even a beam directed straight up returns.

Keep in mind that gravity is not pulling on the light. The concentration of matter has curved spacetime, and light (like the trained ant of our earlier example) is "doing its best" to go in a straight line, yet is now confronted with a world in which straight lines that used to go outward have become curved paths that lead back in. The

Download for free at http://cnx.org/content/col11992/latest/

collapsing star is a *black hole* in this view, because the very concept of "out" has no geometrical meaning. The star has become trapped in its own little pocket of spacetime, from which there is no escape.

The star's geometry cuts off communication with the rest of the universe at precisely the moment when, in our earlier picture, the escape velocity becomes equal to the speed of light. The size of the star at this moment defines a surface that we call the **event horizon**. It's a wonderfully descriptive name: just as objects that sink below our horizon cannot be seen on Earth, so anything happening inside the event horizon can no longer interact with the rest of the universe.

Imagine a future spacecraft foolish enough to land on the surface of a massive star just as it begins to collapse in the way we have been describing. Perhaps the captain is asleep at the gravity meter, and before the crew can say "Albert Einstein," they have collapsed with the star inside the event horizon. Frantically, they send an escape pod straight outward. But paths outward twist around to become paths inward, and the pod turns around and falls toward the center of the black hole. They send a radio message to their loved ones, bidding good-bye. But radio waves, like light, must travel through spacetime, and curved spacetime allows nothing to get out. Their final message remains unheard. Events inside the event horizon can never again affect events outside it.

The characteristics of an event horizon were first worked out by astronomer and mathematician Karl Schwarzschild (Figure 24.15). A member of the German army in World War I, he died in 1916 of an illness he contracted while doing artillery shell calculations on the Russian front. His paper on the theory of event horizons was among the last things he finished as he was dying; it was the first exact solution to Einstein's equations of general relativity. The radius of the event horizon is called the *Schwarzschild radius* in his memory.

Figure 24.15. Karl Schwarzschild (1873–1916). This German scientist was the first to demonstrate mathematically that a black hole is possible and to determine the size of a nonrotating black hole's event horizon.

The event horizon is the boundary of the black hole; calculations show that it does not get smaller once the whole star has collapsed inside it. It is the region that separates the things trapped inside it from the rest of the universe. Anything coming from the outside is also trapped once it comes inside the event horizon. The horizon's size turns out to depend only on the mass inside it. If the Sun, with its mass of 1 M_{Sun}, were to become a black hole (fortunately, it can't—this is just a thought experiment), the Schwarzschild radius would be about 3 kilometers; thus, the entire black hole would be about one-third the size of a neutron star of that same mass. Feed the black hole some mass, and the horizon will grow—but not very much. Doubling the mass will make the black hole 6 kilometers in radius, still very tiny on the cosmic scale.

This OpenStax book is available for free at http://cnx.org/content/col11992/1.8

The event horizons of more massive black holes have larger radii. For example, if a globular cluster of 100,000 stars (solar masses) could collapse to a black hole, it would be 300,000 kilometers in radius, a little less than half the radius of the Sun. If the entire Galaxy could collapse to a black hole, it would be only about 10^{12} kilometers in radius—about a tenth of a light year. Smaller masses have correspondingly smaller horizons: for Earth to become a black hole, it would have to be compressed to a radius of only 1 centimeter—less than the size of a grape. A typical asteroid, if crushed to a small enough size to be a black hole, would have the dimensions of an atomic nucleus.

EXAMPLE 24.1

The Milky Way's Black Hole

The size of the event horizon of a black hole depends on the mass of the black hole. The greater the mass, the larger the radius of the event horizon. General relativity calculations show that the formula for the Schwarzschild radius (R_S) of the event horizon is

$$R_S = \frac{2GM}{c^2}$$

where c is the speed of light, G is the gravitational constant, and M is the mass of the black hole. Note that in this formula, 2, G, and c are all constant; only the mass changes from black hole to black hole.

As we will see in the chapter on The Milky Way Galaxy, astronomers have traced the paths of several stars near the center of our Galaxy and found that they seem to be orbiting an unseen object—dubbed Sgr A* (pronounced "Sagittarius A-star")—with a mass of about 4 million solar masses. What is the size of its Schwarzschild radius?

Solution

We can substitute data for G, M, and c (from Appendix E) directly into the equation:

$$R_S = \frac{2GM}{c^2} = \frac{2(6.67 \times 10^{-11} \, \text{N} \cdot \text{m}^2/\text{kg}^2)(4 \times 10^6)(1.99 \times 10^{30} \, \text{kg})}{(3.00 \times 10^8 \, \text{m/s})^2}$$
$$= 1.18 \times 10^{10} \, \text{m}$$

This distance is about one-fifth of the radius of Mercury's orbit around the Sun, yet the object contains 4 million solar masses and cannot be seen with our largest telescopes. You can see why astronomers are convinced this object is a black hole.

Check Your Learning

What would be the size of a black hole that contained only as much mass as a typical pickup truck (about 3000 kg)? (Note that something with so little mass could never actually form a black hole, but it's interesting to think about the result.)

Answer:

Substituting the data into our equation gives

$$R_S = \frac{2GM}{c^2} = \frac{2(6.67 \times 10^{-11} \, \text{N} \cdot \text{m}^2/\text{kg}^2)(3000 \, \text{kg})}{(3.00 \times 10^8 \, \text{m/s})^2} = 1.33 \times 10^{-23} \, \text{m}.$$

Download for free at http://cnx.org/content/col11992/latest/

For comparison, the size of a proton is usually considered to be about 8×10^{-16} m, which would be about ten million times larger.

A Black Hole Myth

Much of the modern folklore about black holes is misleading. One idea you may have heard is that black holes go about sucking things up with their gravity. Actually, it is only very close to a black hole that the strange effects we have been discussing come into play. The gravitational attraction far away from a black hole is the same as that of the star that collapsed to form it.

Remember that the gravity of any star some distance away acts as if all its mass were concentrated at a point in the center, which we call the center of gravity. For real stars, we merely *imagine* that all mass is concentrated there; for black holes, all the mass *really is* concentrated at a point in the center.

So, if you are a star or distant planet orbiting around a star that becomes a black hole, your orbit may not be significantly affected by the collapse of the star (although it may be affected by any mass loss that precedes the collapse). If, on the other hand, you venture close to the event horizon, it would be very hard for you to resist the "pull" of the warped spacetime near the black hole. You have to get really close to the black hole to experience any significant effect.

If another star or a spaceship were to pass one or two solar radii from a black hole, Newton's laws would be adequate to describe what would happen to it. Only very near the event horizon of a black hole is the gravitation so strong that Newton's laws break down. The black hole remnant of a massive star coming into our neighborhood would be far, far safer to us than its earlier incarnation as a brilliant, hot star.

MAKING CONNECTIONS

Gravity and Time Machines

Time machines are one of the favorite devices of science fiction. Such a device would allow you to move through time at a different pace or in a different direction from everyone else. General relativity suggests that it is possible, in theory, to construct a time machine using gravity that could take you into the future.

Let's imagine a place where gravity is terribly strong, such as near a black hole. General relativity predicts that the stronger the gravity, the slower the pace of time (as seen by a distant observer). So, imagine a future astronaut, with a fast and strongly built spaceship, who volunteers to go on a mission to such a high-gravity environment. The astronaut leaves in the year 2222, just after graduating from college at age 22. She takes, let's say, exactly 10 years to get to the black hole. Once there, she orbits some distance from it, taking care not to get pulled in.

She is now in a high-gravity realm where time passes much more slowly than it does on Earth. This isn't just an effect on the mechanism of her clocks—*time itself* is running slowly. That means that every way she has of measuring time will give the same slowed-down reading when compared to time passing on Earth. Her heart will beat more slowly, her hair will grow more slowly, her antique wristwatch will tick more slowly, and so on. She is not aware of this slowing down because all her readings of time, whether made by her own bodily functions or with mechanical equipment, are measuring the same—slower—time. Meanwhile, back on Earth, time passes as it always does.

This OpenStax book is available for free at http://cnx.org/content/col11992/1.8

Our astronaut now emerges from the region of the black hole, her mission of exploration finished, and returns to Earth. Before leaving, she carefully notes that (according to her timepieces) she spent about 2 weeks around the black hole. She then takes exactly 10 years to return to Earth. Her calculations tell her that since she was 22 when she left the Earth, she will be 42 plus 2 weeks when she returns. So, the year on Earth, she figures, should be 2242, and her classmates should now be approaching their midlife crises.

But our astronaut should have paid more attention in her astronomy class! Because time slowed down near the black hole, much less time passed for her than for the people on Earth. While her clocks measured 2 weeks spent near the black hole, more than 2000 weeks (depending on how close she got) could well have passed on Earth. That's equal to 40 years, meaning her classmates will be senior citizens in their 80s when she (a mere 42-year-old) returns. On Earth it will be not 2242, but 2282—and she will say that she has arrived *in the future*.

Is this scenario real? Well, it has a few practical challenges: we don't think any black holes are close enough for us to reach in 10 years, and we don't think any spaceship or human can survive near a black hole. But the key point about the slowing down of time is a natural consequence of Einstein's general theory of relativity, and we saw that its predictions have been confirmed by experiment after experiment.

Such developments in the understanding of science also become inspiration for science fiction writers. Recently, the film *Interstellar* featured the protagonist traveling close to a massive black hole; the resulting delay in his aging relative to his earthbound family is a key part of the plot.

Science fiction novels, such as *Gateway* by Frederik Pohl and *A World out of Time* by Larry Niven, also make use of the slowing down of time near black holes as major turning points in the story. For a list of science fiction stories based on good astronomy, you can go to www.astrosociety.org/scifi.

A Trip into a Black Hole

The fact that scientists cannot see inside black holes has not kept them from trying to calculate what they are like. One of the first things these calculations showed was that the formation of a black hole obliterates nearly all information about the star that collapsed to form it. Physicists like to say "black holes have no hair," meaning that nothing sticks out of a black hole to give us clues about what kind of star produced it or what material has fallen inside. The only information a black hole can reveal about itself is its mass, its spin (rotation), and whether it has any electrical charge.

What happens to the collapsing star-core that made the black hole? Our best calculations predict that the material will continue to collapse under its own weight, forming an infinitely *squozen* point—a place of zero volume and infinite density—to which we give the name **singularity**. At the singularity, spacetime ceases to exist. The laws of physics as we know them break down. We do not yet have the physical understanding or the mathematical tools to describe the singularity itself, or even if singularities actually occur. From the outside, however, the entire structure of a basic black hole (one that is not rotating) can be described as a singularity surrounded by an event horizon. Compared to humans, black holes are really very simple objects.

Scientists have also calculated what would happen if an astronaut were to fall into a black hole. Let's take up an observing position a long, safe distance away from the event horizon and watch this astronaut fall toward it. At first he falls away from us, moving ever faster, just as though he were approaching any massive star. However, as he nears the event horizon of the black hole, things change. The strong gravitational field around the black hole will make his clocks run more slowly, when seen from our outside perspective.

Download for free at http://cnx.org/content/col11992/latest/

If, as he approaches the event horizon, he sends out a signal once per second according to his clock, we will see the spacing between his signals grow longer and longer until it becomes infinitely long when he reaches the event horizon. (Recalling our discussion of gravitational redshift, we could say that if the infalling astronaut uses a blue light to send his signals every second, we will see the light get redder and redder until its wavelength is nearly infinite.) As the spacing between clock ticks approaches infinity, it will appear to us that the astronaut is slowly coming to a stop, frozen in time at the event horizon.

In the same way, all matter falling into a black hole will also appear to an outside observer to stop at the event horizon, frozen in place and taking an infinite time to fall through it. But don't think that matter falling into a black hole will therefore be easily visible at the event horizon. The tremendous redshift will make it very difficult to observe any radiation from the "frozen" victims of the black hole.

This, however, is only how we, located far away from the black hole, see things. To the astronaut, his time goes at its normal rate and he falls right on through the event horizon into the black hole. (Remember, this horizon is not a physical barrier, but only a region in space where the curvature of spacetime makes escape impossible.)

You may have trouble with the idea that you (watching from far away) and the astronaut (falling in) have such different ideas about what has happened. This is the reason Einstein's ideas about space and time are called theories of *relativity*. What each observer measures about the world depends on (is relative to) his or her frame of reference. The observer in strong gravity measures time and space differently from the one sitting in weaker gravity. When Einstein proposed these ideas, many scientists also had difficulty with the idea that two such different views of the same event could be correct, each in its own "world," and they tried to find a mistake in the calculations. There were no mistakes: we and the astronaut really would see him fall into a black hole very differently.

For the astronaut, there is no turning back. Once inside the event horizon, the astronaut, along with any signals from his radio transmitter, will remain hidden forever from the universe outside. He will, however, not have a long time (from his perspective) to feel sorry for himself as he approaches the black hole. Suppose he is falling feet first. The force of gravity that the singularity exerts on his feet is greater than on his head, so he will be stretched slightly. Because the singularity is a point, the left side of his body will be pulled slightly toward the right, and the right slightly toward the left, bringing each side closer to the singularity. The astronaut will therefore be slightly squeezed in one direction and stretched in the other. Some scientists like to call this process of stretching and narrowing *spaghettification*. The point at which the astronaut becomes so stretched that he perishes depends on the size of the black hole. For black holes with masses billions of times the mass of the Sun, such as those found at the centers of galaxies, the spaghettification becomes significant only after the astronaut passes through the event horizon. For black holes with masses of a few solar masses, the astronaut will be stretched and ripped apart even before he reaches the event horizon.

Earth exerts similar *tidal forces* on an astronaut performing a spacewalk. In the case of Earth, the tidal forces are so small that they pose no threat to the health and safety of the astronaut. Not so in the case of a black hole. Sooner or later, as the astronaut approaches the black hole, the tidal forces will become so great that the astronaut will be ripped apart, eventually reduced to a collection of individual atoms that will continue their inexorable fall into the singularity.

LINK TO LEARNING

From the previous discussion, you will probably agree that jumping into a black hole is definitely a once-in-a-lifetime experience! You can see an engaging explanation (https://openstax.org/l/

This OpenStax book is available for free at http://cnx.org/content/col11992/1.8

Download for free at http://cnx.org/content/col11992/latest/

30ndegtystidfor) of death by black hole by Neil deGrasse Tyson, where he explains the effect of tidal forces on the human body until it dies by spaghettification.

A similar explanation can be seen in this Discovery Channel video (https://openstax.org/l/30dischatidfor) excerpt.

24.6 EVIDENCE FOR BLACK HOLES

Learning Objectives

By the end of this section, you will be able to:

> Describe what to look for when seeking and confirming the presence of a stellar black hole
> Explain how a black hole is inherently black yet can be associated with luminous matter
> Differentiate between stellar black holes and the black holes in the centers of galaxies

Theory tells us what black holes are like. But do they actually exist? And how do we go about looking for something that is many light years away, only about a few dozen kilometers across (if a stellar black hole), and completely black? It turns out that the trick is not to look for the black hole itself but instead to look for what it does to a nearby companion star.

As we saw, when very massive stars collapse, they leave behind their gravitational influence. What if a member of a double-star system becomes a black hole, and its companion manages to survive the death of the massive star? While the black hole disappears from our view, we may be able to deduce its presence from the things it does to its companion.

Requirements for a Black Hole

So, here is a prescription for finding a black hole: start by looking for a star whose motion (determined from the Doppler shift of its spectral lines) shows it to be a member of a binary star system. If both stars are visible, neither can be a black hole, so focus your attention on just those systems where only one star of the pair is visible, even with our most sensitive telescopes.

Being invisible is not enough, however, because a relatively faint star might be hard to see next to the glare of a brilliant companion or if it is shrouded by dust. And even if the star really is invisible, it could be a neutron star. Therefore, we must also have evidence that the unseen star has a mass too high to be a neutron star and that it is a collapsed object—an extremely small stellar remnant.

We can use Kepler's law (see Orbits and Gravity) and our knowledge of the visible star to measure the mass of the invisible member of the pair. If the mass is greater than about 3 M_{Sun}, then we are likely seeing (or, more precisely, not seeing) a black hole—as long as we can make sure the object really is a collapsed star.

If matter falls toward a compact object of high gravity, the material is accelerated to high speed. Near the event horizon of a black hole, matter is moving at velocities that approach the speed of light. As the atoms whirl chaotically toward the event horizon, they rub against each other; internal friction can heat them to temperatures of 100 million K or more. Such hot matter emits radiation in the form of flickering X-rays. The last part of our prescription, then, is to look for a source of X-rays associated with the binary system. Since X-rays do not penetrate Earth's atmosphere, such sources must be found using X-ray telescopes in space.

Download for free at http://cnx.org/content/col11992/latest/

In our example, the infalling gas that produces the X-ray emission comes from the black hole's companion star. As we saw in The Death of Stars, stars in close binary systems can exchange mass, especially as one of the members expands into a red giant. Suppose that one star in a double-star system has evolved to a black hole and that the second star begins to expand. If the two stars are not too far apart, the outer layers of the expanding star may reach the point where the black hole exerts more gravitational force on them than do the inner layers of the red giant to which the atmosphere belongs. The outer atmosphere then passes through the point of no return between the stars and falls toward the black hole.

The mutual revolution of the giant star and the black hole causes the material falling toward the black hole to spiral around it rather than flow directly into it. The infalling gas whirls around the black hole in a pancake of matter called an **accretion disk**. It is within the inner part of this disk that matter is revolving about the black hole so fast that internal friction heats it up to X-ray–emitting temperatures (see Figure 24.1).

Another way to form an accretion disk in a binary star system is to have a powerful stellar wind come from the black hole's companion. Such winds are a characteristic of several stages in a star's life. Some of the ejected gas in the wind will then flow close enough to the black hole to be captured by it into the disk (Figure 24.16).

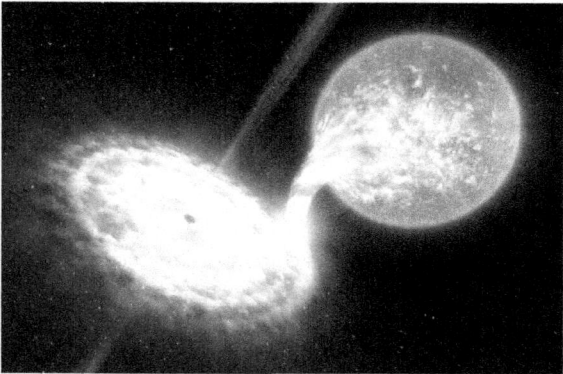

Figure 24.16. Binary Black Hole. This artist's rendition shows a black hole and star (red). As matter streams from the star, it forms a disk around the black hole. Some of the swirling material close to the black hole is pushed outward perpendicular to the disk in two narrow jets. (credit: modification of work by ESO/L. Calçada)

We should point out that, as often happens, the measurements we have been discussing are not quite as simple as they are described in introductory textbooks. In real life, Kepler's law allows us to calculate only the combined mass of the two stars in the binary system. We must learn more about the visible star of the pair and its history to ascertain the distance to the binary pair, the true size of the visible star's orbit, and how the orbit of the two stars is tilted toward Earth, something we can rarely measure. And neutron stars can also have accretion disks that produce X-rays, so astronomers must study the properties of these X-rays carefully when trying to determine what kind of object is at the center of the disk. Nevertheless, a number of systems that clearly contain black holes have now been found.

The Discovery of Stellar-Mass Black Holes

Because X-rays are such important tracers of black holes that are having some of their stellar companions for lunch, the search for black holes had to await the launch of sophisticated X-ray telescopes into space. These instruments must have the resolution to locate the X-ray sources accurately and thereby enable us to match them to the positions of binary star systems.

The first black hole binary system to be discovered is called Cygnus X-1 (see Figure 24.1). The visible star in this binary system is spectral type O. Measurements of the Doppler shifts of the O star's spectral lines show that it

This OpenStax book is available for free at http://cnx.org/content/col11992/1.8

has an unseen companion. The X-rays flickering from it strongly indicate that the companion is a small collapsed object. The mass of the invisible collapsed companion is about 15 times that of the Sun. The companion is therefore too massive to be either a white dwarf or a neutron star.

A number of other binary systems also meet all the conditions for containing a black hole. Table 24.1 lists the characteristics of some of the best examples.

Some Black Hole Candidates in Binary Star Systems

Name/Catalog Designation[2]	Companion Star Spectral Type	Orbital Period (days)	Black Hole Mass Estimates (M_{Sun})
LMC X-1	O giant	3.9	10.9
Cygnus X-1	O supergiant	5.6	15
XTE J1819.3-254 (V4641 Sgr)	B giant	2.8	6–7
LMC X-3	B main sequence	1.7	7
4U1543-475 (IL Lup)	A main sequence	1.1	9
GRO J1655-40 (V1033 Sco)	F subgiant	2.6	7
GRS 1915+105	K giant	33.5	14
GS202+1338 (V404 Cyg)	K giant	6.5	12
XTE J1550-564	K giant	1.5	11
A0620-00 (V616 Mon)	K main sequence	0.33	9–13
H1705-250 (Nova Oph 1977)	K main sequence	0.52	5–7
GRS1124-683 (Nova Mus 1991)	K main sequence	0.43	7
GS2000+25 (QZ Vul)	K main sequence	0.35	5–10
GRS1009-45 (Nova Vel 1993)	K dwarf	0.29	8–9
XTE J1118+480	K dwarf	0.17	7
XTE J1859+226	K dwarf	0.38	5.4

Table 24.1

---

2   As you can tell, there is no standard way of naming these candidates. The chain of numbers is the location of the source in right ascension and declination (the longitude and latitude system of the sky); some of the letters preceding the numbers refer to objects (e.g., LMC) and constellations (e.g., Cygnus), while other letters refer to the satellite that discovered the candidate—A for Ariel, G for Ginga, and so on. The notations in parentheses are those used by astronomers who study binary star system or novae.

Download for free at http://cnx.org/content/col11992/latest/

Some Black Hole Candidates in Binary Star Systems

Name/Catalog Designation	Companion Star Spectral Type	Orbital Period (days)	Black Hole Mass Estimates (M_{Sun})
GRO J0422+32	M dwarf	0.21	4

Table 24.1

Feeding a Black Hole

After an isolated star, or even one in a binary star system, becomes a black hole, it probably won't be able to grow much larger. Out in the suburban regions of the Milky Way Galaxy where we live (see The Milky Way Galaxy), stars and star systems are much too far apart for other stars to provide "food" to a hungry black hole. After all, material must approach very close to the event horizon before the gravity is any different from that of the star before it became the black hole.

But, as will see, the central regions of galaxies are quite different from their outer parts. Here, stars and raw material can be quite crowded together, and they can interact much more frequently with each other. Therefore, black holes in the centers of galaxies may have a much better opportunity to find mass close enough to their event horizons to pull in. Black holes are not particular about what they "eat": they are happy to consume other stars, asteroids, gas, dust, and even other black holes. (If two black holes merge, you just get a black hole with more mass and a larger event horizon.)

As a result, black holes in crowded regions can grow, eventually swallowing thousands or even millions of times the mass of the Sun. Ground-based observations have provided compelling evidence that there is a black hole in the center of our own Galaxy with a mass of about 4 million times the mass of the Sun (we'll discuss this further in the chapter on The Milky Way Galaxy). Observations with the Hubble Space Telescope have shown dramatic evidence for the existence of black holes in the centers of many other galaxies. These black holes can contain more than a billion solar masses. The feeding frenzy of such supermassive black holes may be responsible for some of the most energetic phenomena in the universe (see Active Galaxies, Quasars, and Supermassive Black Holes). And evidence from more recent X-ray observations is also starting to indicate the existence of "middle-weight" black holes, whose masses are dozens to thousands of times the mass of the Sun. The crowded inner regions of the globular clusters we described in Stars from Adolescence to Old Age may be just the right breeding grounds for such intermediate-mass black holes.

Over the past decades, many observations, especially with the Hubble Space Telescope and with X-ray satellites, have been made that can be explained only if black holes really do exist. Furthermore, the observational tests of Einstein's general theory of relativity have convinced even the most skeptical scientists that his picture of warped or curved spacetime is indeed our best description of the effects of gravity near these black holes.

GRAVITATIONAL WAVE ASTRONOMY

Learning Objectives

By the end of this section, you will be able to:

> Describe what a gravitational wave is, what can produce it, and how fast it propagates
> Understand the basic mechanisms used to detect gravitational waves

This OpenStax book is available for free at http://cnx.org/content/col11992/1.8

Download for free at http://cnx.org/content/col11992/latest/

Another part of Einstein's ideas about gravity can be tested as a way of checking the theory that underlies black holes. According to general relativity, the geometry of spacetime depends on where matter is located. Any rearrangement of matter—say, from a sphere to a sausage shape—creates a disturbance in spacetime. This disturbance is called a **gravitational wave**, and relativity predicts that it should spread outward at the speed of light. The big problem with trying to study such waves is that they are tremendously weaker than electromagnetic waves and correspondingly difficult to detect.

Proof from a Pulsar

We've had indirect evidence for some time that gravitational waves exist. In 1974, astronomers Joseph Taylor and Russell Hulse discovered a pulsar (with the designation PSR1913+16) orbiting another neutron star. Pulled by the powerful gravity of its companion, the pulsar is moving at about one-tenth the speed of light in its orbit.

According to general relativity, this system of stellar corpses should be radiating energy in the form of gravitational waves at a high enough rate to cause the pulsar and its companion to spiral closer together. If this is correct, then the orbital period should decrease (according to Kepler's third law) by one ten-millionth of a second per orbit. Continuing observations showed that the period is decreasing by precisely this amount. Such a loss of energy in the system can be due only to the radiation of gravitational waves, thus confirming their existence. Taylor and Hulse shared the 1993 Nobel Prize in physics for this work.

Direct Observations

Although such an indirect proof convinced physicists that gravitational waves exist, it is even more satisfying to detect the waves directly. What we need are phenomena that are powerful enough to produce gravitational waves with amplitudes large enough that we can measure them. Theoretical calculations suggest some of the most likely events that would give a burst of gravitational waves strong enough that our equipment on Earth could measure it:

- the coalescence of two neutron stars in a binary system that spiral together until they merge
- the swallowing of a neutron star by a black hole
- the coalescence (merger) of two black holes
- the implosion of a really massive star to form a neutron star or a black hole
- the first "shudder" when space and time came into existence and the universe began

For the last four decades, scientists have been developing an audacious experiment to try to detect gravitational waves from a source on this list. The US experiment, which was built with collaborators from the UK, Germany, Australia and other countries, is named LIGO (Laser Interferometer Gravitational-Wave Observatory). LIGO currently has two observing stations, one in Louisiana and the other in the state of Washington. The effects of gravitational waves are so small that confirmation of their detection will require simultaneous measurements by two widely separated facilities. Local events that might cause small motions within the observing stations and mimic gravitational waves—such as small earthquakes, ocean tides, and even traffic—should affect the two sites differently.

Each of the LIGO stations consists of two 4-kilometer-long, 1.2-meter-diameter vacuum pipes arranged in an L-shape. A test mass with a mirror on it is suspended by wire at each of the four ends of the pipes. Ultra-stable laser light is reflected from the mirrors and travels back and forth along the vacuum pipes (Figure 24.17). If gravitational waves pass through the LIGO instrument, then, according to Einstein's theory, the waves will affect local spacetime—they will alternately stretch and shrink the distance the laser light must travel between the mirrors ever so slightly. When one arm of the instrument gets longer, the other will get shorter, and vice versa.

Download for free at http://cnx.org/content/col11992/latest/

Figure 24.17. Gravitational Wave Telescope. An aerial view of the LIGO facility at Livingston, Louisiana. Extending to the upper left and far right of the image are the 4-kilometer-long detectors. (credit: modification of work by Caltech/MIT/LIGO Laboratory)

The challenge of this experiment lies in that phrase "ever so slightly." In fact, to detect a gravitational wave, the change in the distance to the mirror must be measured with an accuracy of *one ten-thousandth the diameter of a proton*. In 1972, Rainer Weiss of MIT wrote a paper suggesting how this seemingly impossible task might be accomplished.

A great deal of new technology had to be developed, and work on the laboratory, with funding from the National Science Foundation, began in 1979. A full-scale prototype to demonstrate the technology was built and operated from 2002 to 2010, but the prototype was not expected to have the sensitivity required to actually detect gravitational waves from an astronomical source. Advanced LIGO, built to be more precise with the improved technology developed in the prototype, went into operation in 2015—and almost immediately detected gravitational waves.

What LIGO found was gravitational waves produced in the final fraction of a second of the merger of two black holes (Figure 24.18). The black holes had masses of 20 and 36 times the mass of the Sun, and the merger took place 1.3 billion years ago—the gravitational waves occurred so far away that it has taken that long for them, traveling at the speed of light, to reach us.

In the cataclysm of the merger, about three times the mass of the Sun was converted to energy (recall $E = mc^2$). During the tiny fraction of a second for the merger to take place, this event produced power about 50 times the power produced by all the stars in the entire visible universe—but the power was all in the form of gravitational waves and hence was invisible to our instruments, except to LIGO. The event was recorded in Louisiana about 7 milliseconds before the detection in Washington—just the right distance given the speed at which gravitational waves travel—and indicates that the source was located somewhere in the southern hemisphere sky. Unfortunately, the merger of two black holes is not expected to produce any light, so this is the only observation we have of the event.

This OpenStax book is available for free at http://cnx.org/content/col11992/1.8

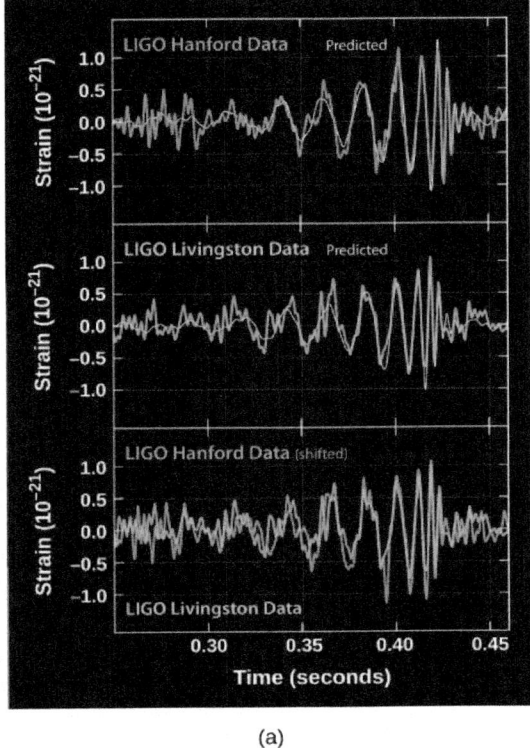

(a)

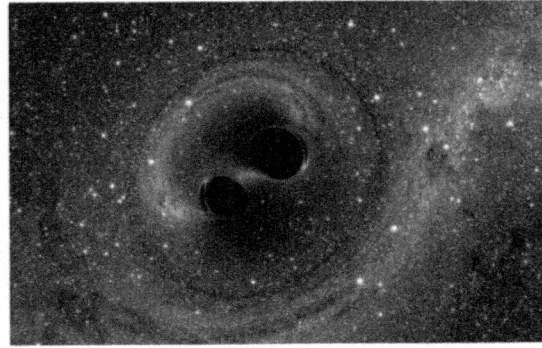

(b)

Figure 24.18. Signal Produced by a Gravitational Wave. (a) The top panel shows the signal measured at Hanford, Washington; the middle panel shows the signal measured at Livingston, Louisiana. The smoother thin curve in each panel shows the predicted signal, based on Einstein's general theory of relativity, produced by the merger of two black holes. The bottom panel shows a superposition of the waves detected at the two LIGO observatories. Note the remarkable agreement of the two independent observations and of the observations with theory. (b) The painting shows an artist's impression of two massive black holes spiraling inward toward an eventual merger. (credit a, b: modification of work by SXS)

This detection by LIGO (and another one of a different black hole merger a few months later) opens a whole new window on the universe. One of the experimenters compared the beginning of gravitational wave astronomy to the era when silent films were replaced by movies with sound (comparing the vibration of spacetime during the passing of a gravitational wave to the vibrations that sound makes).

We can now learn about events, such as the merger of black holes, that can be studied in no other way. For example, this first detected merger involved black holes with masses greater than previously observed for stellar-mass black holes. Such a discovery suggests that we may need to make changes to existing models of the evolution of massive stars.

Observing the merger of black holes via gravitational waves also means that we can now make tests of Einstein's general theory of relativity where its effects are very strong—close to black holes—and not weak, as they are near Earth. One remarkable result from this first detection is that the signal measured matched so closely the theoretical predictions made using Einstein's theory. Once again, Einstein's revolutionary idea is found to be the correct description of nature.

Several facilities similar to LIGO are under construction in other countries to contribute to gravitational wave astronomy and help us pinpoint more precisely where in the sky the signals we detect come from. The European Space Agency (ESA) is also exploring the possibility of building an even larger detector for gravitational waves in space. The goal is to launch a facility called eLISA sometime in the mid 2030s. The design calls for three arms

Download for free at http://cnx.org/content/col11992/latest/

or paths, each a million kilometers in length, for the laser light to travel. This facility could detect the distant merger of supermassive black holes such as might have occurred when the first generation of stars formed only a few hundred million years after the Big Bang. In December 2015, ESA launched LISA Pathfinder to test the technology required to hold two gold-platinum cubes in a state of weightless, perfect rest relative one another. While LISA Pathfinder cannot itself detect gravitational waves, such stability will be required if eLISA is to be able to detect the small changes in path length produced by passing gravitational waves.

We should end by acknowledging that the ideas discussed in this chapter may seem strange and overwhelming, especially the first time you read them. The consequences of the general theory of relatively take some getting used to. But you have to admit that they make the universe more interesting and bizarre than you probably thought before you took this course.

This OpenStax book is available for free at http://cnx.org/content/col11992/1.8

CHAPTER 24 REVIEW

 KEY TERMS

accretion disk the disk of gas and dust found orbiting newborn stars, as well as compact stellar remnants such as white dwarfs, neutron stars, and black holes when they are in binary systems and are sufficiently close to their binary companions to draw off material

black hole a region in spacetime where gravity is so strong that nothing—not even light—can escape

equivalence principle concept that a gravitational force and a suitable acceleration are indistinguishable within a sufficiently local environment

event horizon a boundary in spacetime such that events inside the boundary can have no effect on the world outside it—that is, the boundary of the region around a black hole where the curvature of spacetime no longer provides any way out

general theory of relativity Einstein's theory relating gravity and the structure (geometry) of space and time

gravitational redshift an increase in wavelength of an electromagnetic wave (light) when propagating from or near a massive object

gravitational wave a disturbance in the curvature of spacetime caused by changes in how matter is distributed; gravitational waves propagate at (or near) the speed of light.

singularity the point of zero volume and infinite density to which any object that becomes a black hole must collapse, according to the theory of general relativity

spacetime system of one time and three space coordinates, with respect to which the time and place of an event can be specified

 SUMMARY

24.1 Introducing General Relativity

Einstein proposed the equivalence principle as the foundation of the theory of general relativity. According to this principle, there is no way that anyone or any experiment in a sealed environment can distinguish between free fall and the absence of gravity.

24.2 Spacetime and Gravity

By considering the consequences of the equivalence principle, Einstein concluded that we live in a curved spacetime. The distribution of matter determines the curvature of spacetime; other objects (and even light) entering a region of spacetime must follow its curvature. Light must change its path near a massive object not because light is bent by gravity, but because spacetime is.

24.3 Tests of General Relativity

In weak gravitational fields, the predictions of general relativity agree with the predictions of Newton's law of gravity. However, in the stronger gravity of the Sun, general relativity makes predictions that differ from Newtonian physics and can be tested. For example, general relativity predicts that light or radio waves will be deflected when they pass near the Sun, and that the position where Mercury is at perihelion would change by 43

Download for free at http://cnx.org/content/col11992/latest/

arcsec per century even if there were no other planets in the solar system to perturb its orbit. These predictions have been verified by observation.

24.4 Time in General Relativity

General relativity predicts that the stronger the gravity, the more slowly time must run. Experiments on Earth and with spacecraft have confirmed this prediction with remarkable accuracy. When light or other radiation emerges from a compact smaller remnant, such as a white dwarf or neutron star, it shows a gravitational redshift due to the slowing of time.

24.5 Black Holes

Theory suggests that stars with stellar cores more massive than three times the mass of the Sun at the time they exhaust their nuclear fuel will collapse to become black holes. The surface surrounding a black hole, where the escape velocity equals the speed of light, is called the event horizon, and the radius of the surface is called the Schwarzschild radius. Nothing, not even light, can escape through the event horizon from the black hole. At its center, each black hole is thought to have a singularity, a point of infinite density and zero volume. Matter falling into a black hole appears, as viewed by an outside observer, to freeze in position at the event horizon. However, if we were riding on the infalling matter, we would pass through the event horizon. As we approach the singularity, the tidal forces would tear our bodies apart even before we reach the singularity.

24.6 Evidence for Black Holes

The best evidence of stellar-mass black holes comes from binary star systems in which (1) one star of the pair is not visible, (2) the flickering X-ray emission is characteristic of an accretion disk around a compact object, and (3) the orbit and characteristics of the visible star indicate that the mass of its invisible companion is greater than $3\ M_{Sun}$. A number of systems with these characteristics have been found. Black holes with masses of millions to billions of solar masses are found in the centers of large galaxies.

24.7 Gravitational Wave Astronomy

General relativity predicts that the rearrangement of matter in space should produce gravitational waves. The existence of such waves was first confirmed in observations of a pulsar in orbit around another neutron star whose orbits were spiraling closer and losing energy in the form of gravitational waves. In 2015, LIGO found gravitational waves directly by detecting the signal produced by the merger of two stellar-mass black holes, opening a new window on the universe.

 FOR FURTHER EXPLORATION

Articles

Black Holes

Charles, P. & Wagner, R. "Black Holes in Binary Stars: Weighing the Evidence." *Sky & Telescope* (May 1996): 38. Excellent review of how we find stellar-mass black holes.

Gezari, S. "Star-Shredding Black Holes." *Sky & Telescope* (June 2013): 16. When black holes and stars collide.

Jayawardhana, R. "Beyond Black." *Astronomy* (June 2002): 28. On finding evidence of the existence of event horizons and thus black holes.

Nadis, S. "Black Holes: Seeing the Unseeable." *Astronomy* (April 2007): 26. A brief history of the black hole idea and an introduction to potential new ways to observe them.

This OpenStax book is available for free at http://cnx.org/content/col11992/1.8

Psallis, D. & Sheperd, D. "The Black Hole Test." *Scientific American* (September 2015): 74–79. The Event Horizon Telescope (a network of radio telescopes) will test some of the stranger predictions of general relativity for the regions near black holes. The September 2015 issue of *Scientific American* was devoted to a celebration of the 100th anniversary of the general theory of relativity.

Rees, M. "To the Edge of Space and Time." *Astronomy* (July 1998): 48. Good, quick overview.

Talcott, R. "Black Holes in our Backyard." *Astronomy* (September 2012): 44. Discussion of different kinds of black holes in the Milky Way and the 19 objects known to be black holes.

Gravitational Waves

Bartusiak, M. "Catch a Gravity Wave." *Astronomy* (October 2000): 54.

Gibbs, W. "Ripples in Spacetime." *Scientific American* (April 2002): 62.

Haynes, K., & Betz, E. "A Wrinkle in Spacetime Confirms Einstein's Gravitation." *Astronomy* (May 2016): 22. On the direct detection of gravity waves.

Sanders, G., and Beckett, D. "LIGO: An Antenna Tuned to the Songs of Gravity." *Sky & Telescope* (October 2000): 41.

Websites

Black Holes

Black Hole Encyclopedia: http://blackholes.stardate.org. From StarDate at the University of Texas McDonald Observatory.

Black Holes: http://science.nasa.gov/astrophysics/focus-areas/black-holes. NASA overview of black holes, along with links to the most recent news and discoveries.

Black Holes FAQ: http://cfpa.berkeley.edu/Education/BHfaq.html. Frequently asked questions about black holes, answered by Ted Bunn of UC–Berkeley's Center for Particle Astrophysics.

Black Holes: Gravity's Relentless Pull: http://hubblesite.org/explore_astronomy/black_holes/home.html. The Hubble Space Telescope's Journey to a Black Hole and Black Hole Encyclopedia (a good introduction for beginners).

Introduction to Black Holes: http://www.damtp.cam.ac.uk/research/gr/public/bh_intro.html. The Cambridge University Relativity Group's pages on black holes and related calculations.

March 1918: Testing Einstein: http://www.nature.com/nature/podcast/index-pastcast-2014-03-20.html. Nature Podcast about the 1919 eclipse expedition that proved Einstein's General Theory of Relativity.

Movies from the Edge of Spacetime: http://archive.ncsa.illinois.edu/Cyberia/NumRel/MoviesEdge.html. Physicists simulate the behavior of various black holes.

Virtual Trips into Black Holes and Neutron Stars: http://antwrp.gsfc.nasa.gov/htmltest/rjn_bht.html. By Robert Nemiroff at Michigan Technological University.

Gravitational Waves

Advanced LIGO: https://www.advancedligo.mit.edu. The full story on this gravitational wave observatory.

eLISA: https://www.elisascience.org.

Gravitational Waves Detected, Confirming Einstein's Theory: http://www.nytimes.com/2016/02/12/science/ligo-gravitational-waves-black-holes-einstein.html. *New York Times* article and videos on the discovery of gravitational waves.

Download for free at http://cnx.org/content/col11992/latest/

Gravitational Waves Discovered from Colliding Black Holes: http://www.scientificamerican.com/article/gravitational-waves-discovered-from-colliding-black-holes1. *Scientific American* coverage of the discovery of gravitational waves (note the additional materials available in the menu at the right).

LIGO Caltech: https://www.ligo.caltech.edu.

Videos

Black Holes

Black Holes: The End of Time or a New Beginning?: https://www.youtube.com/watch?v=mgtJRsdKe6Q. 2012 Silicon Valley Astronomy Lecture by Roger Blandford (1:29:52).

Death by Black Hole: http://www.openculture.com/2009/02/death_by_black_hole_and_its_kind_of_funny.htm. Neil deGrasse Tyson explains spaghettification with only his hands (5:34).

Hearts of Darkness: Black Holes in Space: https://www.youtube.com/watch?v=4tiAOldypLk. 2010 Silicon Valley Astronomy Lecture by Alex Filippenko (1:56:11).

Gravitational Waves

Journey of a Gravitational Wave: https://www.youtube.com/watch?v=FlDtXIBrAYE. Introduction from LIGO Caltech (2:55).

LIGO's First Detection of Gravitational Waves: https://www.youtube.com/watch?v=gw-i_VKd6Wo. Explanation and animations from PBS Digital Studio (9:31).

Two Black Holes Merge into One: https://www.youtube.com/watch?v=I_88S8DWbcU. Simulation from LIGO Caltech (0:35).

What the Discovery of Gravitational Waves Means: https://www.youtube.com/watch?v=jMVAgCPYYHY. TED Talk by Allan Adams (10:58).

🏛 COLLABORATIVE GROUP ACTIVITIES

A. A computer science major takes an astronomy course like the one you are taking and becomes fascinated with black holes. Later in life, he founds his own internet company and becomes very wealthy when it goes public. He sets up a foundation to support the search for black holes in our Galaxy. Your group is the allocation committee of this foundation. How would you distribute money each year to increase the chances that more black holes will be found?

B. Suppose for a minute that stars evolve *without* losing any mass at any stage of their lives. Your group is given a list of binary star systems. Each binary contains one main-sequence star and one invisible companion. The spectral types of the main-sequence stars range from spectral type O to M. Your job is to determine whether any of the invisible companions might be black holes. Which ones are worth observing? Why? (Hint: Remember that in a binary star system, the two stars form at the same time, but the pace of their evolution depends on the mass of each star.)

C. You live in the far future, and the members of your group have been convicted (falsely) of high treason. The method of execution is to send everyone into a black hole, but you get to pick which one. Since you are doomed to die, you would at least like to see what the inside of a black hole is like—even if you can't tell anyone outside about it. Would you choose a black hole with a mass equal to that of Jupiter or one with

This OpenStax book is available for free at http://cnx.org/content/col11992/1.8

a mass equal to that of an entire galaxy? Why? What would happen to you as you approached the event horizon in each case? (Hint: Consider the difference in force on your feet and your head as you cross over the event horizon.)

D. General relativity is one of the areas of modern astrophysics where we can clearly see the frontiers of human knowledge. We have begun to learn about black holes and warped spacetime recently and are humbled by how much we still don't know. Research in this field is supported mostly by grants from government agencies. Have your group discuss what reasons there are for our tax dollars to support such "far out" (seemingly impractical) work. Can you make a list of "far out" areas of research in past centuries that later led to practical applications? What if general relativity does not have many practical applications? Do you think a small part of society's funds should still go to exploring theories about the nature of space and time?

E. Once you all have read this chapter, work with your group to come up with a plot for a science fiction story that uses the properties of black holes.

F. Black holes seem to be fascinating not just to astronomers but to the public, and they have become part of popular culture. Searching online, have group members research examples of black holes in music, advertising, cartoons, and the movies, and then make a presentation to share the examples you found with the whole class.

G. As mentioned in the Gravity and Time Machines feature box in this chapter, the film *Interstellar* has a lot of black hole science in its plot and scenery. That's because astrophysicist Kip Thorne at Caltech had a big hand in writing the initial treatment for the movie, and later producing it. Get your group members together (be sure you have popcorn) for a viewing of the movie and then try to use your knowledge of black holes from this chapter to explain the plot. (Note that the film also uses the concept of a *wormhole,* which we don't discuss in this chapter. A wormhole is a theoretically possible way to use a large, spinning black hole to find a way to travel from one place in the universe to another without having to go through regular spacetime to get there.)

🖰 EXERCISES

Review Questions

1. How does the equivalence principle lead us to suspect that spacetime might be curved?

2. If general relativity offers the best description of what happens in the presence of gravity, why do physicists still make use of Newton's equations in describing gravitational forces on Earth (when building a bridge, for example)?

3. Einstein's general theory of relativity made or allowed us to make predictions about the outcome of several experiments that had not yet been carried out at the time the theory was first published. Describe three experiments that verified the predictions of the theory after Einstein proposed it.

4. If a black hole itself emits no radiation, what evidence do astronomers and physicists today have that the theory of black holes is correct?

5. What characteristics must a binary star have to be a good candidate for a black hole? Why is each of these characteristics important?

Download for free at http://cnx.org/content/col11992/latest/

6. A student becomes so excited by the whole idea of black holes that he decides to jump into one. It has a mass 10 times the mass of our Sun. What is the trip like for him? What is it like for the rest of the class, watching from afar?

7. What is an event horizon? Does our Sun have an event horizon around it?

8. What is a gravitational wave and why was it so hard to detect?

9. What are some strong sources of gravitational waves that astronomers hope to detect in the future?

10. Suppose the amount of mass in a black hole doubles. Does the event horizon change? If so, how does it change?

Thought Questions

11. Imagine that you have built a large room around the people in Figure 24.4 and that this room is falling at exactly the same rate as they are. Galileo showed that if there is no air friction, light and heavy objects that are dropping due to gravity will fall at the same rate. Suppose that this were not true and that instead heavy objects fall faster. Also suppose that the man in Figure 24.4 is twice as massive as the woman. What would happen? Would this violate the equivalence principle?

12. A monkey hanging from a tree branch sees a hunter aiming a rifle directly at him. The monkey then sees a flash and knows that the rifle has been fired. Reacting quickly, the monkey lets go of the branch and drops so that the bullet can pass harmlessly over his head. Does this act save the monkey's life? Why or why not? (Hint: Consider the similarities between this situation and that of Exercise 24.11.)

13. Why would we not expect to detect X-rays from a disk of matter about an ordinary star?

14. Look elsewhere in this book for necessary data, and indicate what the final stage of evolution—white dwarf, neutron star, or black hole—will be for each of these kinds of stars.

    A. Spectral type-O main-sequence star

    B. Spectral type-B main-sequence star

    C. Spectral type-A main-sequence star

    D. Spectral type-G main-sequence star

    E. Spectral type-M main-sequence star

15. Which is likely to be more common in our Galaxy: white dwarfs or black holes? Why?

16. If the Sun could suddenly collapse to a black hole, how would the period of Earth's revolution about it differ from what it is now?

17. Suppose the people in Figure 24.4 are in an elevator moving upward with an acceleration equal to *g*, but in the opposite direction. The woman throws the ball to the man with a horizontal force. What happens to the ball?

18. You arrange to meet a friend at 5:00 p.m. on Valentine's Day on the observation deck of the Empire State Building in New York City. You arrive right on time, but your friend is not there. She arrives 5 minutes late and says the reason is that time runs faster at the top of a tall building, so she is on time but you were early. Is your friend right? Does time run slower or faster at the top of a building, as compared with its base? Is this a reasonable excuse for your friend arriving 5 minutes late?

This OpenStax book is available for free at http://cnx.org/content/col11992/1.8

19. You are standing on a scale in an elevator when the cable snaps, sending the elevator car into free fall. Before the automatic brakes stop your fall, you glance at the scale reading. Does the scale show your real weight? An apparent weight? Something else?

Figuring For Yourself

20. Look up *G*, *c*, and the mass of the Sun in Appendix E and calculate the radius of a black hole that has the same mass as the Sun. (Note that this is only a theoretical calculation. The Sun does not have enough mass to become a black hole.)

21. Suppose you wanted to know the size of black holes with masses that are larger or smaller than the Sun. You could go through all the steps in Exercise 24.20, wrestling with a lot of large numbers with large exponents. You could be clever, however, and evaluate all the constants in the equation once and then simply vary the mass. You could even express the mass in terms of the Sun's mass and make future calculations really easy. Show that the event horizon equation is equivalent to saying that the radius of the event horizon is equal to 3 km times the mass of the black hole in units of the Sun's mass.

22. Use the result from Exercise 24.21 to calculate the radius of a black hole with a mass equal to: the Earth, a B0-type main-sequence star, a globular cluster, and the Milky Way Galaxy. Look elsewhere in this text and the appendixes for tables that provide data on the mass of these four objects.

23. Since the force of gravity a significant distance away from the event horizon of a black hole is the same as that of an ordinary object of the same mass, Kepler's third law is valid. Suppose that Earth collapsed to the size of a golf ball. What would be the period of revolution of the Moon, orbiting at its current distance of 400,000 km? Use Kepler's third law to calculate the period of revolution of a spacecraft orbiting at a distance of 6000 km.

Download for free at http://cnx.org/content/col11992/latest/

This OpenStax book is available for free at http://cnx.org/content/col11992/1.8

25

THE MILKY WAY GALAXY

Figure 25.1. Milky Way Galaxy. The Milky Way rises over Square Tower, an ancestral pueblo building at Hovenweep National Monument in Utah. Many stars and dark clouds of dust combine to make a spectacular celestial sight of our home Galaxy. The location has been designated an International Dark Sky Park by the International Dark Sky Association.

Chapter Outline

✎ Thinking Ahead

Today, we know that our Sun is just one of the many billions of stars that make up the huge cosmic island we call the Milky Way Galaxy. How can we "weigh" such an enormous system of stars and measure its total mass?

One of the most striking features you can see in a truly dark sky—one without light pollution—is the band of faint white light called the Milky Way, which stretches from one horizon to the other. The name comes from an ancient Greek legend that compared its faint white splash of light to a stream of spilled milk. But folktales differ from culture to culture: one East African tribe thought of the hazy band as the smoke of ancient campfires, several Native American stories tell of a path across the sky traveled by sacred animals, and in Siberia, the diffuse arc was known as the seam of the tent of the sky.

In 1610, Galileo made the first telescopic survey of the Milky Way and discovered that it is composed of a multitude of individual stars. Today, we know that the Milky Way comprises our view inward of the huge cosmic pinwheel that we call the Milky Way Galaxy and that is our home. Moreover, our Galaxy is now recognized as just one galaxy among many billions of other galaxies in the cosmos.

Download for free at http://cnx.org/content/col11992/latest/

25.1 | THE ARCHITECTURE OF THE GALAXY

Learning Objectives

By the end of this section, you will be able to:

> Explain why William and Caroline Herschel concluded that the Milky Way has a flattened structure centered on the Sun and solar system
> Describe the challenges of determining the Galaxy's structure from our vantage point within it
> Identify the main components of the Galaxy

The **Milky Way Galaxy** surrounds us, and you might think it is easy to study because it is so close. However, the very fact that we are embedded within it presents a difficult challenge. Suppose you were given the task of mapping New York City. You could do a much better job from a helicopter flying over the city than you could if you were standing in Times Square. Similarly, it would be easier to map our Galaxy if we could only get a little way outside it, but instead we are trapped inside and way out in its suburbs—far from the galactic equivalent of Times Square.

Herschel Measures the Galaxy

In 1785, William Herschel (Figure 25.2) made the first important discovery about the architecture of the Milky Way Galaxy. Using a large reflecting telescope that he had built, William and his sister Caroline counted stars in different directions of the sky. They found that most of the stars they could see lay in a flattened structure encircling the sky, and that the numbers of stars were about the same in any direction around this structure. Herschel therefore concluded that the stellar system to which the Sun belongs has the shape of a disk or wheel (he might have called it a Frisbee except Frisbees hadn't been invented yet), and that the Sun must be near the hub of the wheel (Figure 25.3).

Figure 25.2. William Herschel (1738–1822) and Caroline Herschel (1750–1848). William Herschel was a German musician who emigrated to England and took up astronomy in his spare time. He discovered the planet Uranus, built several large telescopes, and made measurements of the Sun's place in the Galaxy, the Sun's motion through space, and the comparative brightnesses of stars. This painting shows William and his sister Caroline polishing a telescope lens. (credit: modification of work by the Wellcome Library)

This OpenStax book is available for free at http://cnx.org/content/col11992/1.8

To understand why Herschel reached this conclusion, imagine that you are a member of a band standing in formation during halftime at a football game. If you count the band members you see in different directions and get about the same number each time, you can conclude that the band has arranged itself in a circular pattern with you at the center. Since you see no band members above you or underground, you know that the circle made by the band is much flatter than it is wide.

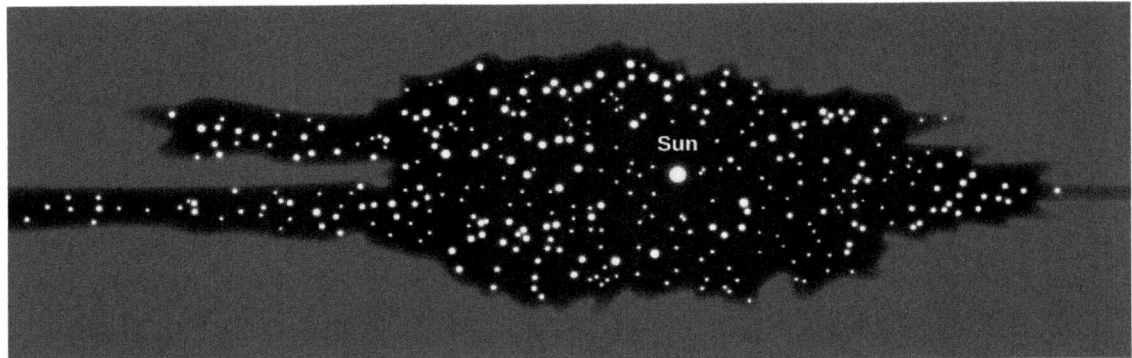

Figure 25.3. Herschel's Diagram of the Milky Way. Herschel constructed this cross section of the Galaxy by counting stars in various directions.

We now know that Herschel was right about the shape of our system, but wrong about where the Sun lies within the disk. As we saw in Between the Stars: Gas and Dust in Space, we live in a dusty Galaxy. Because interstellar dust absorbs the light from stars, Herschel could see only those stars within about 6000 light-years of the Sun. Today we know that this is a very small section of the entire 100,000-light-year-diameter disk of stars that makes up the Galaxy.

VOYAGERS IN ASTRONOMY

Harlow Shapley: Mapmaker to the Stars

Until the early 1900s, astronomers generally accepted Herschel's conclusion that the Sun is near the center of the Galaxy. The discovery of the Galaxy's true size and our actual location came about largely through the efforts of Harlow Shapley. In 1917, he was studying RR Lyrae variable stars in globular clusters. By comparing the known intrinsic luminosity of these stars to how bright they appeared, Shapley could calculate how far away they are. (Recall that it is distance that makes the stars look dimmer than they would be "up close," and that the brightness fades as the distance squared.) Knowing the distance to any star in a cluster then tells us the distance to the cluster itself.

Globular clusters can be found in regions that are free of interstellar dust and so can be seen at very large distances. When Shapley used the distances and directions of 93 globular clusters to map out their positions in space, he found that the clusters are distributed in a spherical volume, which has its center not at the Sun but at a distant point along the Milky Way in the direction of Sagittarius. Shapley then made the bold assumption, verified by many other observations since then, that the point on which the system of globular clusters is centered is also the center of the entire Galaxy (Figure 25.4).

Download for free at http://cnx.org/content/col11992/latest/

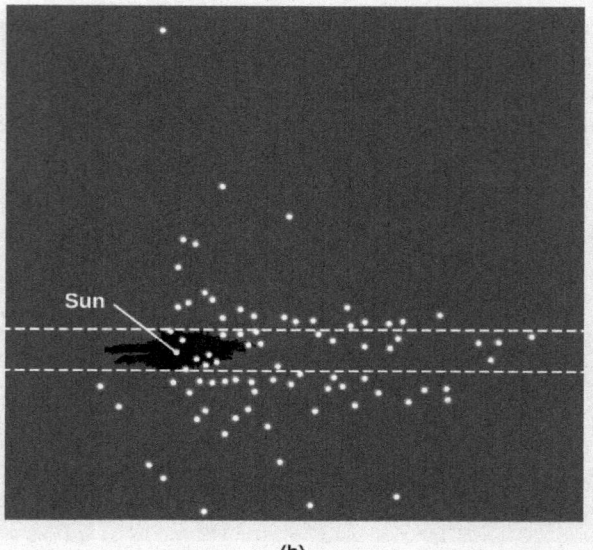

(a)   (b)

Figure 25.4. Harlow Shapley and His Diagram of the Milky Way. (a) Shapley poses for a formal portrait. (b) His diagram shows the location of globular clusters, with the position of the Sun also marked. The black area shows Herschel's old diagram, centered on the Sun, approximately to scale.

Shapley's work showed once and for all that our star has no special place in the Galaxy. We are in a nondescript region of the Milky Way, only one of 200 to 400 billion stars that circle the distant center of our Galaxy.

Born in 1885 on a farm in Missouri, Harlow Shapley at first dropped out of school with the equivalent of only a fifth-grade education. He studied at home and at age 16 got a job as a newspaper reporter covering crime stories. Frustrated by the lack of opportunities for someone who had not finished high school, Shapley went back and completed a six-year high-school program in only two years, graduating as class valedictorian.

In 1907, at age 22, he went to the University of Missouri, intent on studying journalism, but found that the school of journalism would not open for a year. Leafing through the college catalog (or so he told the story later), he chanced to see "Astronomy" among the subjects beginning with "A." Recalling his boyhood interest in the stars, he decided to study astronomy for the next year (and the rest, as the saying goes, is history).

Upon graduation Shapley received a fellowship for graduate study at Princeton and began to work with the brilliant Henry Norris Russell (see the Henry Norris Russell feature box). For his PhD thesis, Shapley made major contributions to the methods of analyzing the behavior of eclipsing binary stars. He was also able to show that cepheid variable stars are not binary systems, as some people thought at the time, but individual stars that pulsate with striking regularity.

Impressed with Shapley's work, George Ellery Hale offered him a position at the Mount Wilson Observatory, where the young man took advantage of the clear mountain air and the 60-inch reflector to do his pioneering study of variable stars in globular clusters.

Shapley subsequently accepted the directorship of the Harvard College Observatory, and over the next 30 years, he and his collaborators made contributions to many fields of astronomy, including the study of

This OpenStax book is available for free at http://cnx.org/content/col11992/1.8

neighboring galaxies, the discovery of dwarf galaxies, a survey of the distribution of galaxies in the universe, and much more. He wrote a series of nontechnical books and articles and became known as one of the most effective popularizers of astronomy. Shapley enjoyed giving lectures around the country, including at many smaller colleges where students and faculty rarely got to interact with scientists of his caliber.

During World War II, Shapley helped rescue many scientists and their families from Eastern Europe; later, he helped found UNESCO, the United Nations Educational, Scientific, and Cultural Organization. He wrote a pamphlet called *Science from Shipboard* for men and women in the armed services who had to spend many weeks on board transport ships to Europe. And during the difficult period of the 1950s, when congressional committees began their "witch hunts" for communist sympathizers (including such liberal leaders as Shapley), he spoke out forcefully and fearlessly in defense of the freedom of thought and expression. A man of many interests, he was fascinated by the behavior of ants, and wrote scientific papers about them as well as about galaxies.

By the time he died in 1972, Shapley was acknowledged as one of the pivotal figures of modern astronomy, a "twentieth-century Copernicus" who mapped the Milky Way and showed us our place in the Galaxy.

LINK TO LEARNING

To find more information about Shapley's life and work (https://openstaxcollege.org/l/30shapbrumed) , see the entry for him on the Bruce Medalists website. (This site features the winners of the Bruce Medal of the Astronomical Society of the Pacific, one of the highest honors in astronomy; the list is a who's who of some of the greatest astronomers of the last twelve decades.)

Disks and Haloes

With modern instruments, astronomers can now penetrate the "smog" of the Milky Way by studying radio and infrared emissions from distant parts of the Galaxy. Measurements at these wavelengths (as well as observations of other galaxies like ours) have given us a good idea of what the Milky Way would look like if we *could* observe it from a distance.

Figure 25.5 sketches what we would see if we could view the Galaxy face-on and edge-on. The brightest part of the Galaxy consists of a thin, circular, rotating disk of stars distributed across a region about 100,000 light-years in diameter and about 1000 light-years thick. (Given how thin the disk is, perhaps a CD is a more appropriate analogy than a wheel.) In addition to stars, the dust and gas from which stars form are also found mostly in the thin disk of the Galaxy. The mass of the interstellar matter is about 15% of the mass of the stars in this disk.

Download for free at http://cnx.org/content/col11992/latest/

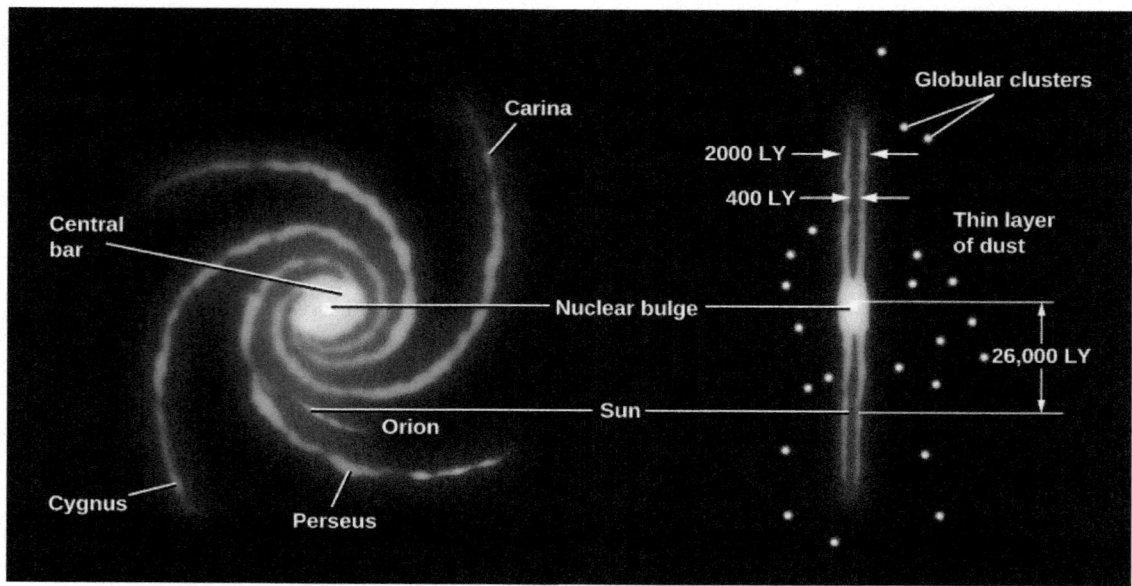

Figure 25.5. Schematic Representation of the Galaxy. The left image shows the face-on view of the spiral disk; the right image shows the view looking edge-on along the disk. The major spiral arms are labeled. The Sun is located on the inside edge of the short Orion spur.

As the diagram in Figure 25.5 shows, the stars, gas, and dust are not spread evenly throughout the disk but are concentrated into a central bar and a series of spiral arms. Recent infrared observations have confirmed that the central bar is composed mostly of old yellow-red stars. The two main spiral arms appear to connect with the ends of the bar. They are highlighted by the blue light from young hot stars. We know many other spiral galaxies that also have bar-shaped concentrations of stars in their central regions; for that reason they are called *barred spirals*. Figure 25.6 shows two other galaxies—one without a bar and one with a strong bar—to give you a basis for comparison to our own. We will describe our spiral structure in more detail shortly. The Sun is located about halfway between the center of the Galaxy and the edge of the disk and only about 70 light-years above its central plane.

This OpenStax book is available for free at http://cnx.org/content/col11992/1.8

Download for free at http://cnx.org/content/col11992/latest/

(a)   (b)

Figure 25.6. Unbarred and Barred Spiral Galaxies. (a) This image shows the unbarred spiral galaxy M74. It contains a small central bulge of mostly old yellow-red stars, along with spiral arms that are highlighted with the blue light from young hot stars. (b) This image shows the strongly barred spiral galaxy NGC 1365. The bulge and the fainter bar both appear yellowish because the brightest stars in them are mostly old yellow and red giants. Two main spiral arms project from the ends of the bar. As in M74, these spiral arms are populated with blue stars and red patches of glowing gas—hallmarks of recent star formation. The Milky Way Galaxy is thought to have a barred spiral structure that is intermediate between these two examples. (credit a: modification of work by ESO/PESSTO/S. Smartt; credit b: modification of work by ESO)

Our thin disk of young stars, gas, and dust is embedded in a thicker but more diffuse disk of older stars; this thicker disk extends about 3000 light-years above and below the midplane of the thin disk and contains only about 5% as much mass as the thin disk.

Close in to the galactic center (within about 10,000 light-years), the stars are no longer confined to the disk but form a **central bulge** (or nuclear bulge). When we observe with visible light, we can glimpse the stars in the bulge only in those rare directions where there happens to be relatively little interstellar dust. The first picture that actually succeeded in showing the bulge as a whole was taken at infrared wavelengths (Figure 25.7).

Figure 25.7. Inner Part of the Milky Way Galaxy. This beautiful infrared map, showing half a billion stars, was obtained as part of the Two Micron All Sky Survey (2MASS). Because interstellar dust does not absorb infrared as strongly as visible light, this view reveals the previously hidden bulge of old stars that surrounds the center of our Galaxy, along with the Galaxy's thin disk component. (credit: modification of work by 2MASS/J. Carpenter, T. H. Jarrett, and R. Hurt)

The fact that much of the bulge is obscured by dust makes its shape difficult to determine. For a long time, astronomers assumed it was spherical. However, infrared images and other data indicate that the bulge is about two times longer than it is wide, and shaped rather like a peanut. The relationship between this elongated inner bulge and the larger bar of stars remains uncertain. At the very center of the nuclear bulge is a tremendous concentration of matter, which we will discuss later in this chapter.

Download for free at http://cnx.org/content/col11992/latest/

In our Galaxy, the thin and thick disks and the nuclear bulge are embedded in a spherical **halo** of very old, faint stars that extends to a distance of at least 150,000 light-years from the galactic center. Most of the globular clusters are also found in this halo.

The mass in the Milky Way extends even farther out, well beyond the boundary of the luminous stars to a distance of at least 200,000 light-years from the center of the Galaxy. This invisible mass has been give the name *dark matter* because it emits no light and cannot be seen with any telescope. Its composition is unknown, and it can be detected only because of its gravitational effects on the motions of luminous matter that we can see. We know that this extensive **dark matter halo** exists because of its effects on the orbits of distant star clusters and other dwarf galaxies that are associated with the Galaxy. This mysterious halo will be a subject of the section on The Mass of the Galaxy, and the properties of dark matter will be discussed more in the chapter on The Big Bang.

Some vital statistics of the thin and thick disks and the stellar halo are given in Table 25.1, with an illustration in Figure 25.8. Note particularly how the ages of stars correlate with where they are found. As we shall see, this information holds important clues to how the Milky Way Galaxy formed.

Characteristics of the Milky Way Galaxy

Property	Thin Disk	Thick Disk	Stellar Halo (Excludes Dark Matter)
Stellar mass	$4 \times 10^{10}\ M_{Sun}$	A few percent of the thin disk mass	$10^{10}\ M_{Sun}$
Luminosity	$3 \times 10^{10}\ L_{Sun}$	A few percent of the thin disk luminosity	$8 \times 10^8\ L_{Sun}$
Typical age of stars	1 million to 10 billion years	11 billion years	13 billion years
Heavier-element abundance	High	Intermediate	Very low
Rotation	High	Intermediate	Very low

Table 25.1

This OpenStax book is available for free at http://cnx.org/content/col11992/1.8

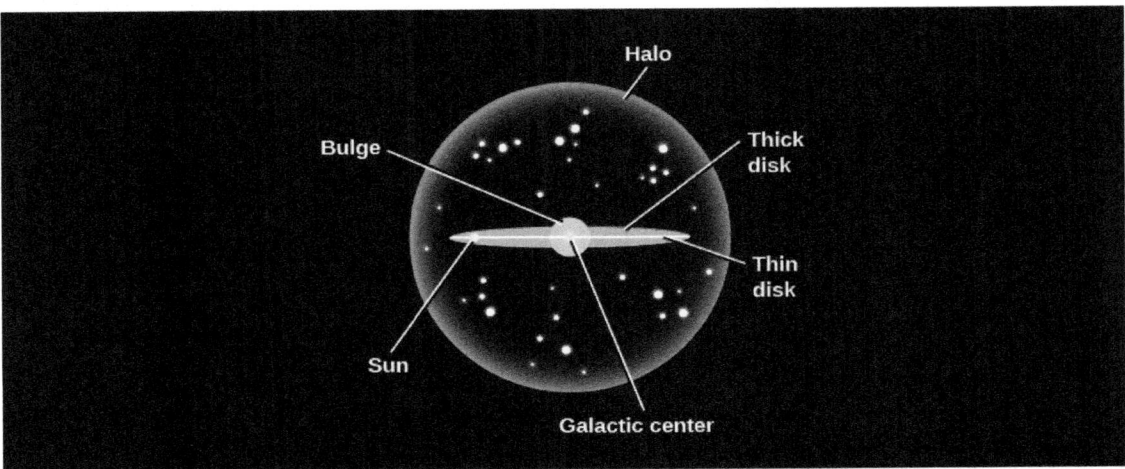

Figure 25.8. Major Parts of the Milky Way Galaxy. This schematic shows the major components of our Galaxy.

Establishing this overall picture of the Galaxy from our dust-shrouded viewpoint inside the thin disk has been one of the great achievements of modern astronomy (and one that took decades of effort by astronomers working with a wide range of telescopes). One thing that helped enormously was the discovery that our Glaxy is not unique in its characteristics. There are many other flat, spiral-shaped islands of stars, gas, and dust in the universe. For example, the Milky Way somewhat resembles the Andromeda galaxy, which, at a distance of about 2.3 million light-years, is our nearest neighboring giant spiral galaxy. Just as you can get a much better picture of yourself if someone else takes the photo from a distance away, pictures and other diagnostic observations of nearby galaxies that resemble ours have been vital to our understanding of the properties of the Milky Way.

MAKING CONNECTIONS

The Milky Way Galaxy in Myth and Legend

To most of us living in the twenty-first century, the Milky Way Galaxy is an elusive sight. We must make an effort to leave our well-lit homes and streets and venture beyond our cities and suburbs into less populated environments. Once the light pollution subsides to negligible levels, the Milky Way can be readily spotted arching over the sky on clear, moonless nights. The Milky Way is especially bright in late summer and early fall in the Northern Hemisphere. Some of the best places to view the Milky Way are in our national and state parks, where residential and industrial developments have been kept to a minimum. Some of these parks host special sky-gazing events that are definitely worth checking out—especially during the two weeks surrounding the new moon, when the faint stars and Milky Way don't have to compete with the Moon's brilliance.

Go back a few centuries, and these starlit sights would have been the norm rather than the exception. Before the advent of electric or even gas lighting, people relied on short-lived fires to illuminate their homes and byways. Consequently, their night skies were typically much darker. Confronted by myriad stellar patterns and the Milky Way's gauzy band of diffuse light, people of all cultures developed myths to make sense of it all.

Download for free at http://cnx.org/content/col11992/latest/

Some of the oldest myths relating to the Milky Way are maintained by the aboriginal Australians through their rock painting and storytelling. These legacies are thought to go back tens of thousands of years, to when the aboriginal people were being "dreamed" along with the rest of the cosmos. The Milky Way played a central role as an arbiter of the Creation. Taking the form of a great serpent, it joined with the Earth serpent to dream and thus create all the creatures on Earth.

The ancient Greeks viewed the Milky Way as a spray of milk that spilled from the breast of the goddess Hera. In this legend, Zeus had secretly placed his infant son Heracles at Hera's breast while she was asleep in order to give his half-human son immortal powers. When Hera awoke and found Heracles suckling, she pushed him away, causing her milk to spray forth into the cosmos (Figure 25.9).

The dynastic Chinese regarded the Milky Way as a "silver river" that was made to separate two star-crossed lovers. To the east of the Milky Way, Zhi Nu, the weaving maiden, was identified with the bright star Vega in the constellation of Lyra the Harp. To the west of the Milky Way, her lover Niu Lang, the cowherd, was associated with the star Altair in the constellation of Aquila the Eagle. They had been exiled on opposite sides of the Milky Way by Zhi Nu's mother, the Queen of Heaven, after she heard of their secret marriage and the birth of their two children. However, once a year, they are permitted to reunite. On the seventh day of the seventh lunar month (which typically occurs in our month of August), they would meet on a bridge over the Milky Way that thousands of magpies had made (Figure 25.9). This romantic time continues to be celebrated today as Qi Xi, meaning "Double Seventh," with couples reenacting the cosmic reunion of Zhi Nu and Niu Lang.

(a)

(b)

Figure 25.9. The Milky Way in Myth. (a) *Origin of the Milky Way* by Jacopo Tintoretto (circa 1575) illustrates the Greek myth that explains the formation of the Milky Way. (b) *The Moon of the Milky Way* by Japanese painter Tsukioka Yoshitoshi depicts the Chinese legend of Zhi Nu and Niu Lang.

To the Quechua Indians of Andean Peru, the Milky Way was seen as the celestial abode for all sorts of cosmic creatures. Arrayed along the Milky Way are myriad dark patches that they identified with partridges, llamas, a toad, a snake, a fox, and other animals. The Quechua's orientation toward the dark

This OpenStax book is available for free at http://cnx.org/content/col11992/1.8

Download for free at http://cnx.org/content/col11992/latest/

regions rather than the glowing band of starlight appears to be unique among all the myth makers. Likely, their access to the richly structured southern Milky Way had something to do with it.

Among Finns, Estonians, and related northern European cultures, the Milky Way is regarded as the "pathway of birds" across the night sky. Having noted that birds seasonally migrate along a north-south route, they identified this byway with the Milky Way. Recent scientific studies have shown that this myth is rooted in fact: the birds of this region use the Milky Way as a guide for their annual migrations.

Today, we regard the Milky Way as our galactic abode, where the foment of star birth and star death plays out on a grand stage, and where sundry planets have been found to be orbiting all sorts of stars. Although our perspective on the Milky Way is based on scientific investigations, we share with our forebears an affinity for telling stories of origin and transformation. In these regards, the Milky Way continues to fascinate and inspire us.

25.2 SPIRAL STRUCTURE

Learning Objectives

By the end of this section, you will be able to:

> Describe the structure of the Milky Way Galaxy and how astronomers discovered it
> Compare theoretical models for the formation of spiral arms in disk galaxies

Astronomers were able to make tremendous progress in mapping the spiral structure of the Milky Way after the discovery of the 21-cm line that comes from cool hydrogen (see Between the Stars: Gas and Dust in Space). Remember that the obscuring effect of interstellar dust prevents us from seeing stars at large distances in the disk at visible wavelengths. However, radio waves of 21-cm wavelength pass right through the dust, enabling astronomers to detect hydrogen atoms throughout the Galaxy. More recent surveys of the infrared emission from stars in the disk have provided a similar dust-free perspective of our Galaxy's stellar distribution. Despite all this progress over the past fifty years, we are still just beginning to pin down the precise structure of our Galaxy.

The Arms of the Milky Way

Our radio observations of the disk's gaseous component indicate that the Galaxy has two major spiral arms that emerge from the bar and several fainter arms and shorter spurs. You can see a recently assembled map of our Galaxy's arm structure—derived from studies in the infrared—in Figure 25.10.

Download for free at http://cnx.org/content/col11992/latest/

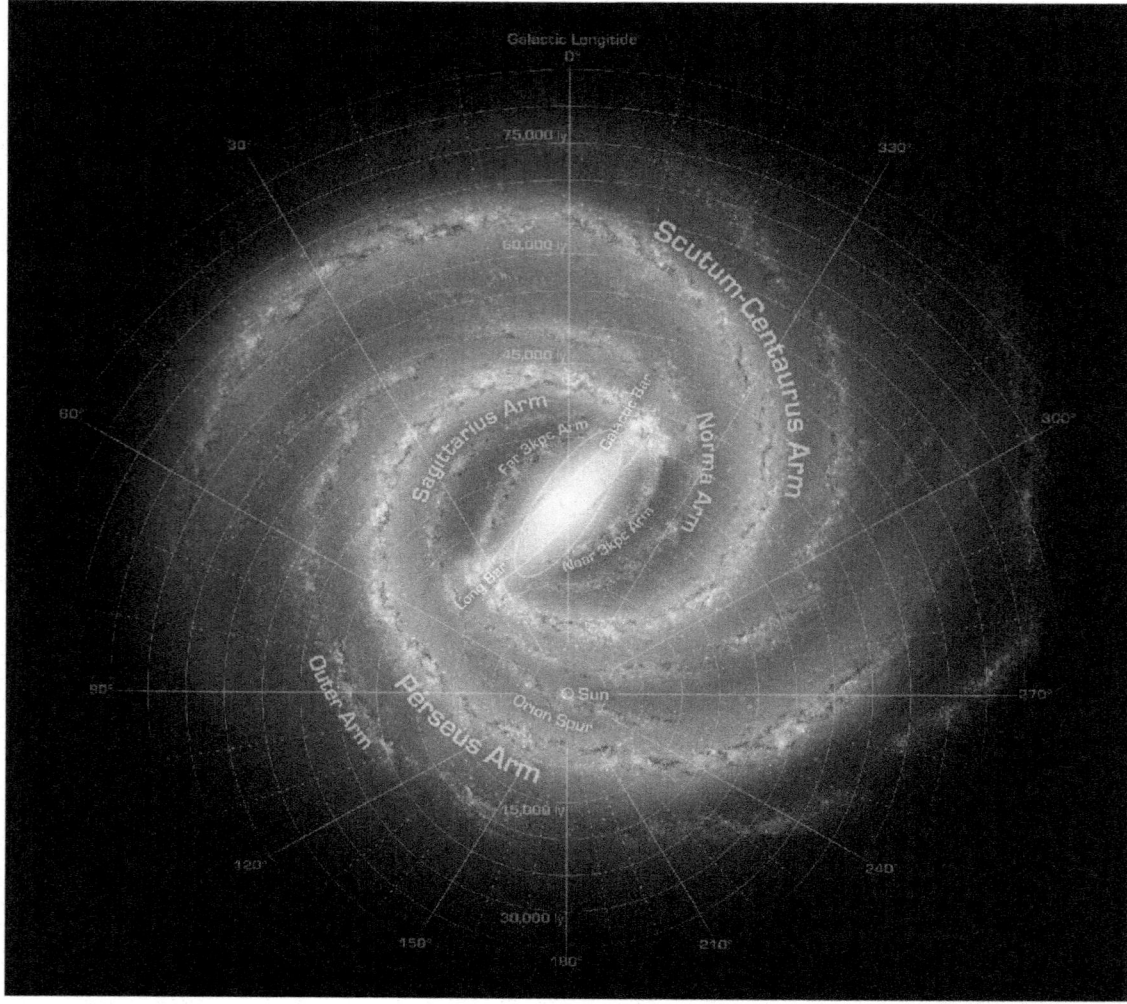

Figure 25.10. Milky Way Bar and Arms. Here, we see the Milky Way Galaxy as it would look from above. This image, assembled from data from NASA's WISE mission, shows that the Milky Way Galaxy has a modest bar in its central regions. Two spiral arms, Scutum-Centaurus and Perseus, emerge from the ends of the bar and wrap around the bulge. The Sagittarius and Outer arms have fewer stars than the other two arms. (credit: modification of work by NASA/JPL-Caltech/R. Hurt (SSC/Caltech))

The Sun is near the inner edge of a short arm called the Orion Spur, which is about 10,000 light-years long and contains such conspicuous features as the Cygnus Rift (the great dark nebula in the summer Milky Way) and the bright Orion Nebula. Figure 25.11 shows a few other objects that share this small section of the Galaxy with us and are easy to see. Remember, the farther away we try to look from our own arm, the more the dust in the Galaxy builds up and makes it hard to see with visible light.

This OpenStax book is available for free at http://cnx.org/content/col11992/1.8

Download for free at http://cnx.org/content/col11992/latest/

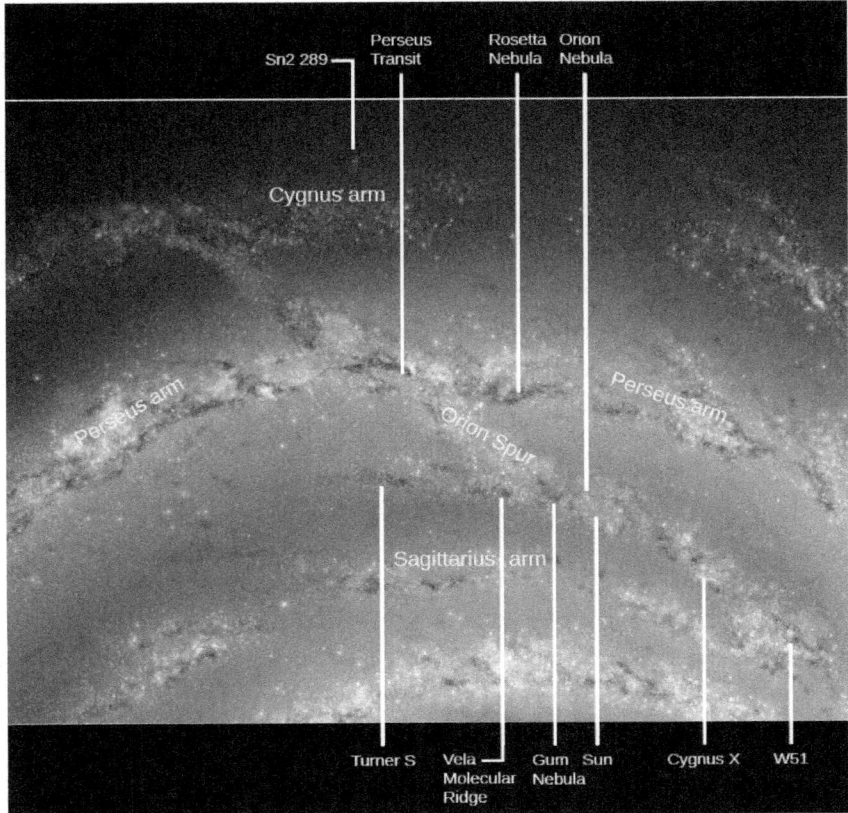

Figure 25.11. Orion Spur. The Sun is located in the Orion Spur, which is a minor spiral arm located between two other arms. In this diagram, the white lines point to some other noteworthy objects that share this feature of the Milky Way Galaxy with the Sun. (credit: modification of work by NASA/JPL-Caltech)

Formation of Spiral Structure

At the Sun's distance from its center, the Galaxy does not rotate like a solid wheel or a CD inside your player. Instead, the way individual objects turn around the center of the Galaxy is more like the solar system. Stars, as well as the clouds of gas and dust, obey Kepler's third law. Objects farther from the center take longer to complete an orbit around the Galaxy than do those closer to the center. In other words, stars (and interstellar matter) in larger orbits in the Galaxy trail behind those in smaller ones. This effect is called **differential galactic rotation**.

Differential rotation would appear to explain why so much of the material in the disk of the Milky Way is concentrated into elongated features that resemble **spiral arms**. No matter what the original distribution of the material might be, the differential rotation of the Galaxy can stretch it out into spiral features. Figure 25.12 shows the development of spiral arms from two irregular blobs of interstellar matter. Notice that as the portions of the blobs closest to the galactic center move faster, those farther out trail behind.

Download for free at http://cnx.org/content/col11992/latest/

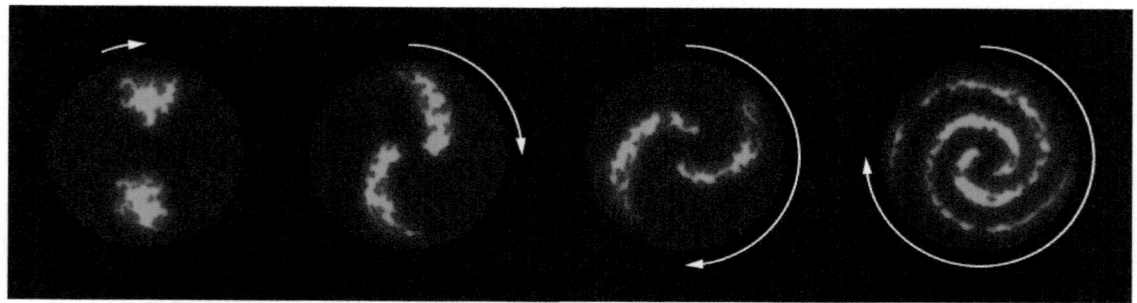

Figure 25.12. Simplified Model for the Formation of Spiral Arms. This sketch shows how spiral arms might form from irregular clouds of interstellar material stretched out by the different rotation rates throughout the Galaxy. The regions farthest from the galactic center take longer to complete their orbits and thus lag behind the inner regions. If this were the only mechanism for creating spiral arms, then over time the spiral arms would completely wind up and disappear. Since many galaxies have spiral arms, they must be long-lived, and there must be other processes at work to maintain them.

But this picture of spiral arms presents astronomers with an immediate problem. If that's all there were to the story, differential rotation—over the roughly 13-billion-year history of the Galaxy—would have wound the Galaxy's arms tighter and tighter until all semblance of spiral structure had disappeared. But did the Milky Way actually have spiral arms when it formed 13 billion years ago? And do spiral arms, once formed, last for that long a time?

With the advent of the Hubble Space Telescope, it has become possible to observe the structure of very distant galaxies and to see what they were like shortly after they began to form more than 13 billion years ago. What the observations show is that galaxies in their infancy had bright, clumpy star-forming regions, but no regular spiral structure.

Over the next few billion years, the galaxies began to "settle down." The galaxies that were to become spirals lost their massive clumps and developed a central bulge. The turbulence in these galaxies decreased, rotation began to dominate the motions of the stars and gas, and stars began to form in a much quieter disk. Smaller star-forming clumps began to form fuzzy, not-very-distinct spiral arms. Bright, well-defined spiral arms began to appear only when the galaxies were about 3.6 billion years old. Initially, there were two well-defined arms. Multi-armed structures in galaxies like we see in the Milky Way appeared only when the universe was about 8 billion years old.

We will discuss the history of galaxies in more detail in The Evolution and Distribution of Galaxies. But, even from our brief discussion, you can get the sense that the spiral structures we now observe in mature galaxies have come along later in the full story of how things develop in the universe.

Scientists have used supercomputer calculations to model the formation and evolution of the arms. These calculations follow the motions of up to 100 million "star particles" to see whether gravitational forces can cause them to form spiral structure. What these calculations show is that giant molecular clouds (which we discussed in Between the Stars: Gas and Dust in Space) have enough gravitational influence over their surroundings to initiate the formation of structures that look like spiral arms. These arms then become self-perpetuating and can survive for at least several billion years. The arms may change their brightness over time as star formation comes and goes, but they are not temporary features. The concentration of matter in the arms exerts sufficient gravitational force to keep the arms together over long periods of time.

This OpenStax book is available for free at http://cnx.org/content/col11992/1.8

Download for free at http://cnx.org/content/col11992/latest/

25.3 THE MASS OF THE GALAXY

Learning Objectives

By the end of this section, you will be able to:

> Describe historical attempts to determine the mass of the Galaxy
> Interpret the observed rotation curve of our Galaxy to suggest the presence of dark matter whose distribution extends well beyond the Sun's orbit

When we described the sections of the Milky Way, we said that the stars are now known to be surrounded by a much larger halo of invisible matter. Let's see how this surprising discovery was made.

Kepler Helps Weigh the Galaxy

The Sun, like all the other stars in the Galaxy, orbits the center of the Milky Way. Our star's orbit is nearly circular and lies in the Galaxy's disk. The speed of the Sun in its orbit is about 200 kilometers per second, which means it takes us approximately 225 million years to go once around the center of the Galaxy. We call the period of the Sun's revolution the *galactic year*. It is a long time compared to human time scales; during the entire lifetime of Earth, only about 20 galactic years have passed. This means that we have gone only a tiny fraction of the way around the Galaxy in all the time that humans have gazed into the sky.

We can use the information about the Sun's orbit to estimate the mass of the Galaxy (just as we could "weigh" the Sun by monitoring the orbit of a planet around it—see Orbits and Gravity). Let's assume that the Sun's orbit is circular and that the Galaxy is roughly spherical, (we know the Galaxy is shaped more like a disk, but to simplify the calculation we will make this assumption, which illustrates the basic approach). Long ago, Newton showed that if you have matter distributed in the shape of a sphere, then it is simple to calculate the pull of gravity on some object just outside that sphere: you can assume that gravity acts as if all the matter were concentrated at a point in the center of the sphere. For our calculation, then, we can assume that all the mass that lies inward of the Sun's position is concentrated at the center of the Galaxy, and that the Sun orbits that point from a distance of about 26,000 light-years.

This is the sort of situation to which Kepler's third law (as modified by Newton) can be directly applied. Plugging numbers into Kepler's formula, we can calculate the sum of the masses of the Galaxy and the Sun. However, the mass of the Sun is completely trivial compared to the mass of the Galaxy. Thus, for all practical purposes, the result (about 100 billion times the mass of the Sun) is the mass of the Milky Way. More sophisticated calculations based on more sophisticated models give a similar result.

Our estimate tells us how much mass is contained in the volume inside the Sun's orbit. This is a good estimate for the total mass of the Galaxy only if hardly any mass lies outside the Sun's orbit. For many years astronomers thought this assumption was reasonable. The number of bright stars and the amount of *luminous matter* (meaning any material from which we can detect electromagnetic radiation) both drop off dramatically at distances of more than about 30,000 light-years from the galactic center. Little did we suspect how wrong our assumption was.

A Galaxy of Mostly Invisible Matter

In science, what seems to be a reasonable assumption can later turn out to be wrong (which is why we continue to do observations and experiments every chance we get). There is a lot more to the Milky Way than meets the eye (or our instruments). While there is relatively little luminous matter beyond 30,000 light-years, we now know that a lot of *invisible matter* exists at great distances from the galactic center.

Download for free at http://cnx.org/content/col11992/latest/

We can understand how astronomers detected this invisible matter by remembering that according to Kepler's third law, objects orbiting at large distances from a massive object will move more slowly than objects that are closer to that central mass. In the case of the solar system, for example, the outer planets move more slowly in their orbits than the planets close to the Sun.

There are a few objects, including globular clusters and some nearby small satellite galaxies, that lie well outside the luminous boundary of the Milky Way. If most of the mass of our Galaxy were concentrated within the luminous region, then these very distant objects should travel around their galactic orbits at lower speeds than, for example, the Sun does.

It turns out, however, that the few objects seen at large distances from the luminous boundary of the Milky Way Galaxy are *not* moving more slowly than the Sun. There are some globular clusters and RR Lyrae stars between 30,000 and 150,000 light-years from the center of the Galaxy, and their orbital velocities are even greater than the Sun's (Figure 25.13).

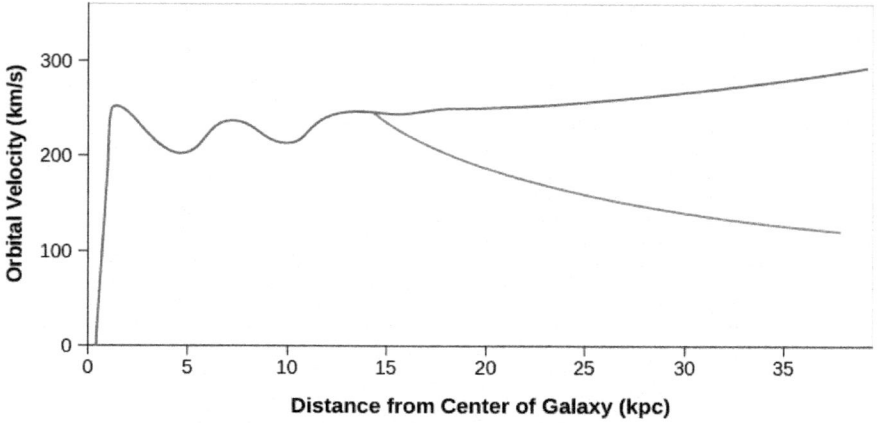

Figure 25.13. Rotation Curve of the Galaxy. The orbital speed of carbon monoxide (CO) and hydrogen (H) gas at different distances from the center of the Milky Way Galaxy is shown in red. The blue curve shows what the rotation curve would look like if all the matter in the Galaxy were located inside a radius of 30,000 light-years. Instead of going down, the speed of gas clouds farther out remains high, indicating a great deal of mass beyond the Sun's orbit. The horizontal axis shows the distance from the galactic center in kiloparsecs (where a kiloparsec equals 3,260 light-years).

What do these higher speeds mean? Kepler's third law tells us how fast objects must orbit a source of gravity if they are neither to fall in (because they move too slowly) nor to escape (because they move too fast). If the Galaxy had only the mass calculated by Kepler, then the high-speed outer objects should long ago have escaped the grip of the Milky Way. The fact that they have not done so means that our Galaxy must have more gravity than can be supplied by the luminous matter—in fact, a *lot* more gravity. The high speed of these outer objects tells us that the source of this extra gravity must extend outward from the center far beyond the Sun's orbit.

If the gravity were supplied by stars or by something else that gives off radiation, we should have spotted this additional outer material long ago. We are therefore forced to the reluctant conclusion that this matter is invisible and has, except for its gravitational pull, gone entirely undetected.

Studies of the motions of the most remote globular clusters and the small galaxies that orbit our own show that the total mass of the Galaxy is at least 2×10^{12} M_{Sun}, which is about twenty times greater than the amount of luminous matter. Moreover, the **dark matter** (as astronomers have come to call the invisible material) extends to a distance of at least 200,000 light-years from the center of the Galaxy. Observations indicate that this dark matter halo is almost but not quite spherical.

This OpenStax book is available for free at http://cnx.org/content/col11992/1.8

The obvious question is: what is the dark matter made of? Let's look at a list of "suspects" taken from our study of astronomy so far. Since this matter is invisible, it clearly cannot be in the form of ordinary stars. And it cannot be gas in any form (remember that there has to be a lot of it). If it were neutral hydrogen gas, its 21-cm wavelength spectral-line emission would have been detected as radio waves. If it were ionized hydrogen, it should be hot enough to emit visible radiation. If a lot of hydrogen atoms out there had combined into hydrogen molecules, these should produce dark features in the ultraviolet spectra of objects lying beyond the Galaxy, but such features have not been seen. Nor can the dark matter consist of interstellar dust, since in the required quantities, the dust would significantly obscure the light from distant galaxies.

What are our other possibilities? The dark matter cannot be a huge number of black holes (of stellar mass) or old neutron stars, since interstellar matter falling onto such objects would produce more X-rays than are observed. Also, recall that the formation of black holes and neutron stars is preceded by a substantial amount of mass loss, which scatters heavy elements into space to be incorporated into subsequent generations of stars. If the dark matter consisted of an enormous number of any of those objects, they would have blown off and recycled a lot of heavier elements over the history of the Galaxy. In that case, the young stars we observe in our Galaxy today would contain much greater abundances of heavy elements than they actually do.

Brown dwarfs and lone Jupiter-like planets have also been ruled out. First of all, there would have to be an awful lot of them to make up so much dark matter. But we have a more direct test of whether so many low-mass objects could actually be lurking out there. As we learned in Black Holes and Curved Spacetime, the general theory of relativity predicts that the path traveled by light is changed when it passes near a concentration of mass. It turns out that when the two objects appear close enough together in the sky, the mass closer to us can bend the light from farther away. With just the right alignment, the image of the more distant object also becomes significantly brighter. By looking for the temporary brightening that occurs when a dark matter object in our own Galaxy moves across the path traveled by light from stars in the Magellanic Clouds, astronomers have now shown that the dark matter cannot be made up of a lot of small objects with masses between one-millionth and one-tenth the mass of the Sun.

What's left? One possibility is that the dark matter is composed of exotic subatomic particles of a type not yet detected on Earth. Very sophisticated (and difficult) experiments are now under way to look for such particles. Stay tuned to see whether anything like that turns up.

We should add that the problem of dark matter is by no means confined to the Milky Way. Observations show that dark matter must also be present in other galaxies (whose outer regions also orbit too fast "for their own good"—they also have flat rotation curves). As we will see, dark matter even exists in great clusters of galaxies whose members are now known to move around under the influence of far more gravity than can be accounted for by luminous matter alone.

Stop a moment and consider how astounding the conclusion we have reached really is. Perhaps as much as 95% of the mass in our Galaxy (and many other galaxies) is not only invisible, but we do not even know what it is made of. The stars and raw material we can observe may be merely the tip of the cosmic iceberg; underlying it all may be other matter, perhaps familiar, perhaps startlingly new. Understanding the nature of this dark matter is one of the great challenges of astronomy today; you will learn more about this in A Universe of (Mostly) Dark Matter and Dark Energy.

25.4 | THE CENTER OF THE GALAXY

Learning Objectives

By the end of this section, you will be able to:

Download for free at http://cnx.org/content/col11992/latest/

> Describe the radio and X-ray observations that indicate energetic phenomena are occurring at the galactic center
> Explain what has been revealed by high-resolution near-infrared imaging of the galactic center
> Discuss how these near-infrared images, when combined with Kepler's third law of motion, can be used to derive the mass of the central gravitating object

At the beginning of this chapter, we hinted that the core of our Galaxy contains a large concentration of mass. In fact, we now have evidence that the very center contains a black hole with a mass equivalent to 4.6 million Suns and that all this mass fits within a sphere that has less than the diameter of Mercury's orbit. Such monster black holes are called **supermassive black holes** by astronomers, to indicate that the mass they contain is far greater than that of the typical black hole created by the death of a single star. It is amazing that we have very convincing evidence that this black hole really does exist. After all, recall from the chapter on Black Holes and Curved Spacetime that we cannot see a black hole directly because by definition it radiates no energy. And we cannot even see into the center of the Galaxy in visible light because of absorption by the interstellar dust that lies between us and the galactic center. Light from the central region of the Galaxy is dimmed by a factor of a trillion (10^{12}) by all this dust.

Fortunately, we are not so blind at other wavelengths. Infrared and radio radiation, which have long wavelengths compared to the sizes of the interstellar dust grains, flow unimpeded past the dust particles and so reach our telescopes with hardly any dimming. In fact, the very bright radio source in the nucleus of the Galaxy, now known as Sagittarius A* (pronounced "Sagittarius A-star" and abbreviated Sgr A*), was the first cosmic radio source astronomers discovered.

A Journey toward the Center

Let's take a voyage to the mysterious heart of our Galaxy and see what's there. Figure 25.14 is a radio image of a region about 1500 light-years across, centered on Sagittarius A, a bright radio source that contains the smaller Sagittarius A*. Much of the radio emission comes from hot gas heated either by clusters of hot stars (the stars themselves do not produce radio emission and can't be seen in the image) or by supernova blast waves. Most of the hollow circles visible on the radio image are supernova remnants. The other main source of radio emission is from electrons moving at high speed in regions with strong magnetic fields. The bright thin arcs and "threads" on the figure show us where this type of emission is produced.

This OpenStax book is available for free at http://cnx.org/content/col11992/1.8

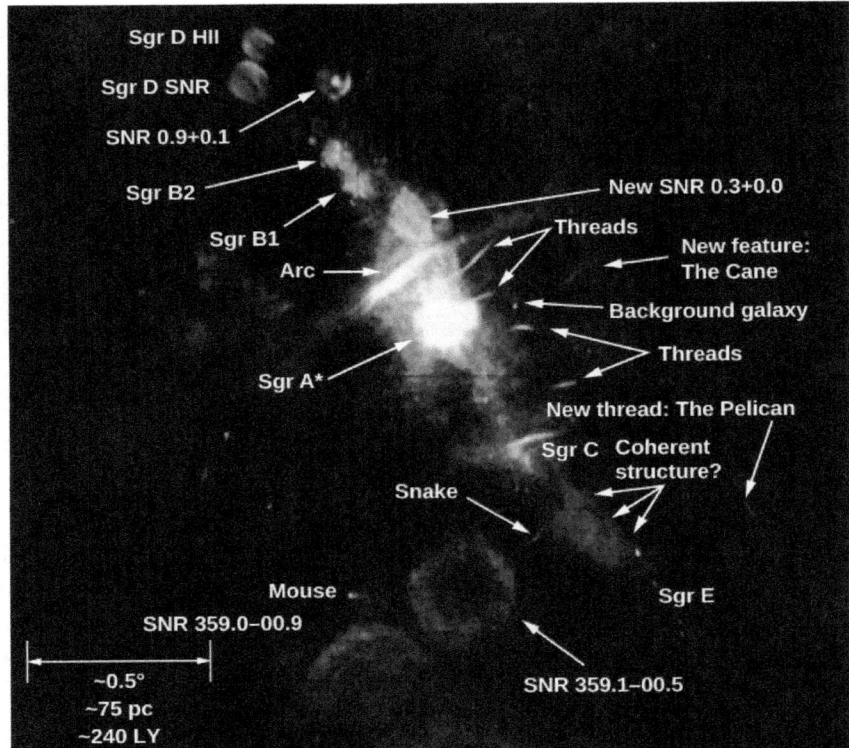

Figure 25.14. Radio Image of Galactic Center Region. This radio map of the center of the Galaxy (at a wavelength of 90 centimeters) was constructed from data obtained with the Very Large Array (VLA) of radio telescopes in Socorro, New Mexico. Brighter regions are more intense in radio waves. The galactic center is inside the region labeled Sagittarius A. Sagittarius B1 and B2 are regions of active star formation. Many filaments or threadlike features are seen, as well as a number of shells (labeled SNR), which are supernova remnants. The scale bar at the bottom left is about 240 light-years long. Notice that radio astronomers also give fanciful animal names to some of the structures, much as visible-light nebulae are sometimes given the names of animals they resemble. (credit: modification of work by N. E. Kassim, D. S. Briggs, T. J. W. Lazio, T. N. LaRosa, and J. Imamura (NRL/RSD))

Now let's focus in on the central region using a more energetic form of electromagnetic radiation. Figure 25.15 shows the X-ray emission from a smaller region 400 light-years wide and 900 light-years across centered in Sagittarius A*. Seen in this picture are hundreds of hot white dwarfs, neutron stars, and stellar black holes with accretion disks glowing with X-rays. The diffuse haze in the picture is emission from gas that lies among the stars and is at a temperature of 10 million K.

Download for free at http://cnx.org/content/col11992/latest/

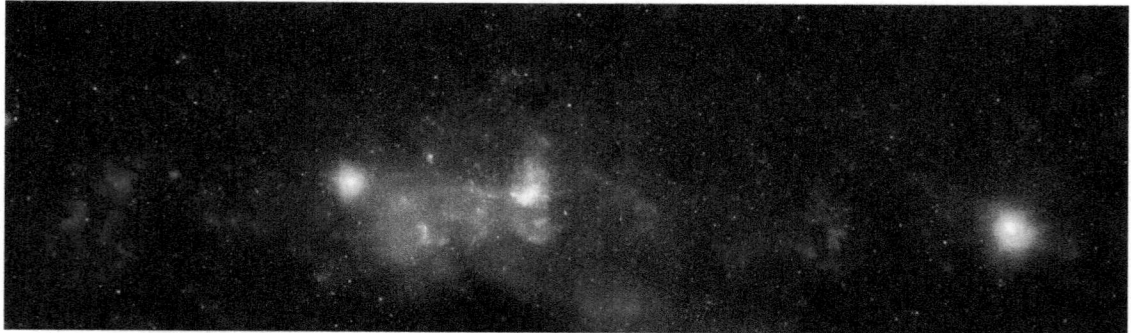

Figure 25.15. Galactic Center in X-Rays. This artificial-color mosaic of 30 images taken with the Chandra X-ray satellite shows a region 400 × 900 light-years in extent and centered on Sagittarius A*, the bright white source in the center of the picture. The X-ray-emitting point sources are white dwarfs, neutron stars, and stellar black holes. The diffuse "haze" is emission from gas at a temperature of 10 million K. This hot gas is flowing away from the center out into the rest of the Galaxy. The colors indicate X-ray energy bands: red (low energy), green (medium energy), and blue (high energy). (credit: modification of work by NASA/CXC/ UMass/D. Wang et al.)

As we approach the center of the Galaxy, we find the supermassive black hole Sagittarius A*. There are also thousands of stars within a parsec of Sagittarius A*. Most of these are old, reddish main-sequence stars. But there are also about a hundred hot OB stars that must have formed within the last few million years. There is as yet no good explanation for how stars could have formed recently so close to a supermassive black hole. Perhaps they formed in a dense cluster of stars that was originally at a larger distance from the black hole and subsequently migrated closer.

There is currently no star formation at the galactic center, but there is lots of dust and molecular gas that is revolving around the black hole, along with some ionized gas streamers that are heated by the hot stars. Figure 25.16 is a radio map that shows these gas streamers.

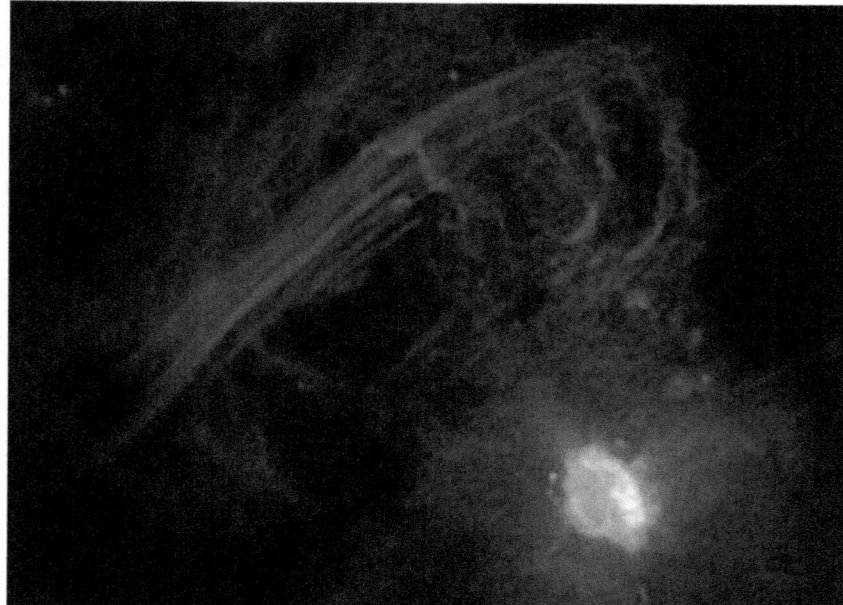

Figure 25.16. Sagittarius A. This image, taken with the Very Large Array of radio telescopes, shows the radio emission from hot, ionized gas in the center of the Milky Way. The lines slanting across the top of the image are gas streamers. Sagittarius A* is the bright spot in the lower right. (credit: modification of work by Farhad Zadeh et al. (Northwestern), VLA, NRAO)

This OpenStax book is available for free at http://cnx.org/content/col11992/1.8

Finding the Heart of the Galaxy

Just what is Sagittarius A*, which lies right at the center our Galaxy? To establish that there really is a black hole there, we must show that there is a very large amount of mass crammed into a very tiny volume. As we saw in Black Holes and Curved Spacetime, proving that a black hole exists is a challenge because the black hole itself emits no radiation. What astronomers must do is prove that a black hole is the only possible explanation for our observations—that a small region contains far more mass than could be accounted for by a very dense cluster of stars or something else made of ordinary matter.

To put some numbers with this discussion, the radius of the event horizon of a *galactic black hole* with a mass of about 4 million M_{Sun} would be only about 17 times the size of the Sun—the equivalent of a single red giant star. The corresponding density within this region of space would be much higher than that of any star cluster or any other ordinary astronomical object. Therefore, we must measure both the diameter of Sagittarius A* and its mass. Both radio and infrared observations are required to give us the necessary evidence.

First, let's look at how the mass can be measured. If we zero in on the inner few light-days of the Galaxy with an infrared telescope equipped with adaptive optics, we see a region crowded with individual stars (Figure 25.17). These stars have now been observed for almost two decades, and astronomers have detected their rapid orbital motions around the very center of the Galaxy.

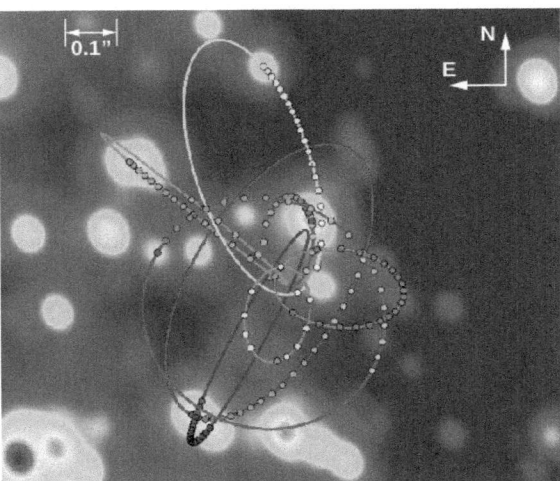

Figure 25.17. Near-Infrared View of the Galactic Center. This image shows the inner 1 arcsecond, or 0.13 light-year, at the center of the Galaxy, as observed with the giant Keck Telescope. Tracks of the orbiting stars measured from 1995 to 2014 have been added to this "snapshot." The stars are moving around the center very fast, and their tracks are all consistent with a single massive "gravitator" that resides in the very center of this image. (credit: modification of work by Andrea Ghez, UCLA Galactic Center Group, W.M. Keck Observatory Laser Team)

LINK TO LEARNING

Check out an animated version (https://openstaxcollege.org/l/30anifiginfgal) of Figure 25.17, showing the motion of the stars over the years.

If we combine observations of their periods and the size of their orbits with Kepler's third law, we can estimate the mass of the object that keeps them in their orbits. One of the stars has been observed for its full orbit of 15.6 years. Its closest approach takes it to a distance of only 124 AU or about 17 light-hours from the black hole.

Download for free at http://cnx.org/content/col11992/latest/

This orbit, when combined with observations of other stars close to the galactic center, indicates that a mass of 4.6 million M_{Sun} must be concentrated inside the orbit—that is, within 17 light-hours of the center of the Galaxy.

Even tighter limits on the size of the concentration of mass at the center of the Galaxy come from radio astronomy, which provided the first clue that a black hole might lie at the center of the Galaxy. As matter spirals inward toward the event horizon of a black hole, it is heated in a whirling *accretion disk* and produces radio radiation. (Such accretion disks were explained in Black Holes and Curved Spacetime.) Measurements of the size of the accretion disk with the Very Long Baseline Array, which provides very high spatial resolution, show that the diameter of the radio source Sagittarius A* is no larger than about 0.3 AU, or about the size of Mercury's orbit. (In light units, that's only 2.5 light-*minutes*!)

The observations thus show that 4.6 million solar masses are crammed into a volume that has a diameter that is no larger than the orbit of Mercury. If this were anything other than a supermassive black hole—low-mass stars that emit very little light or neutron stars or a very large number of small black holes— calculations show that these objects would be so densely packed that they would collapse to a single black hole within a hundred thousand years. That is a very short time compared with the age of the Galaxy, which probably began forming more than 13 billion years ago. Since it seems very unlikely that that we would have caught such a complex cluster of objects just before it collapsed, the evidence for a supermassive black hole at the center of the Galaxy is convincing indeed.

Finding the Source

Where did our galactic black hole come from? The origin of supermassive black holes in galaxies like ours is currently an active field of research. One possibility is that a large cloud of gas near the center of the Milky Way collapsed directly to form a black hole. Since we find large black holes at the centers of most other large galaxies (see Active Galaxies, Quasars, and Supermassive Black Holes)—even ones that are very young—this collapse probably would have taken place when the Milky Way was just beginning to take shape. The initial mass of this black hole might have been only a few tens of solar masses. Another way it could have started is that a massive star might have exploded to leave behind a seed black hole, or a dense cluster of stars might have collapsed into a black hole.

Once a black hole exists at the center of a galaxy, it can grow over the next several billion years by devouring nearby stars and gas clouds in the crowded central regions. It can also grow by merging with other black holes.

It appears that the monster black hole at the center of our Galaxy is not finished "eating." At the present time, we observe clouds of gas and dust falling into the galactic center at the rate of about 1 M_{Sun} per thousand years. Stars are also on the black hole's menu. The density of stars near the galactic center is high enough that we would expect a star to pass near the black hole and be swallowed by it every ten thousand years or so. As this happens, some of the energy of infall is released as radiation. As a result, the center of the Galaxy might flare up and even briefly outshine all the stars in the Milky Way. Other objects might also venture too close to the black hole and be pulled in. How great a flare we observe would depend on the mass of the object falling in.

In 2013, the Chandra X-ray satellite detected a flare from the center of our Galaxy that was 400 times brighter than the usual output from Sagittarius A*. A year later, a second flare, only half as bright, was also detected. This is much less energy than swallowing a whole star would produce. There are two theories to account for the flares. First, an asteroid might have ventured too close to the black hole and been heated to a very high temperature before being swallowed up. Alternatively, the flares might have involved interactions of the magnetic fields near the galactic center in a process similar to the one described for solar flares (see The Sun: A Garden-Variety Star). Astronomers continue to monitor the galactic center area for flares or other activity. Although the monster in the center of the Galaxy is not close enough to us to represent any danger, we still want to keep our eyes on it.

This OpenStax book is available for free at http://cnx.org/content/col11992/1.8

VOYAGERS IN ASTRONOMY

Andrea Ghez

A lover of puzzles, Andrea Ghez has been pursuing one of the greatest mysteries in astronomy: what strange entity lurks within the center of our Milky Way Galaxy?

Figure 25.18. Andrea Ghez. Research by Ghez and her team has helped shape our understanding of supermassive black holes. (credit: modification of work by John D. and Catherine T. MacArthur Foundation)

As a child living in Chicago during the late 1960s, Andrea Ghez (Figure 25.18) was fascinated by the Apollo Moon landings. But she was also drawn to ballet and to solving all sorts of puzzles. By high school, she had lost the ballet bug in favor of competing in field hockey, playing the flute, and digging deeper into academics. Her undergraduate years at MIT were punctuated by a number of changes in her major—from mathematics to chemistry, mechanical engineering, aerospace engineering, and finally physics—where she felt her options were most open. As a physics major, she became involved in astronomical research under the guidance of one of her instructors. Once she got to do some actual observing at Kitt Peak National Observatory in Arizona, and later at Cerro Tololo Inter-American Observatory in Chile, Ghez had found her calling.

Pursuing her graduate studies at Caltech, she stuck with physics but oriented her efforts toward observational astrophysics, an area where Caltech had access to cutting-edge facilities. Though initially attracted to studying the black holes that were suspected of dwelling inside most massive galaxies, Ghez ended up spending most of her graduate study and later postdoctoral research at the University of Arizona studying stars in formation. By taking very high-resolution (detailed) imaging of regions where new stars are born, she discovered that most stars form as members of binary systems. As technologies advanced, she was able to track the orbits danced by these stellar pairings and thereby could ascertain their respective masses.

Now an astronomy professor at UCLA, Ghez has since used similar high-resolution imaging techniques to study the orbits of stars in the innermost core of the Milky Way. These orbits take years to delineate, so Ghez and her science team have logged more than 20 years of taking super-resolution infrared images with the giant Keck telescopes in Hawaii. Based on the resulting stellar orbits, the UCLA Galactic Center Group has settled (as we saw) on a gravitational solution that requires the presence of a supermassive black hole with a mass equivalent to 4.6 million Suns—all nestled within a space smaller than that

Download for free at http://cnx.org/content/col11992/latest/

occupied by our solar system. Ghez's achievements have been recognized with one of the "genius" awards given by the MacArthur Foundation. More recently, her team discovered glowing clouds of warm ionized gas that co-orbit with the stars but may be more vulnerable to the disruptive effects of the central black hole. By monitoring these clouds, the team hopes to better understand the evolution of supermassive black holes and their immediate environs. They also hope to test Einstein's theory of general relativity by carefully scrutinizing the orbits of stars that career closest to the intensely gravitating black hole.

Besides her pioneering work as an astronomer, Ghez competes as a master swimmer, enjoys family life as a mother of two children, and actively encourages other women to pursue scientific careers.

25.5 STELLAR POPULATIONS IN THE GALAXY

Learning Objectives

By the end of this section, you will be able to:

> Distinguish between population I and population II stars according to their locations, motions, heavy-element abundances, and ages
> Explain why the oldest stars in the Galaxy are poor in elements heavier than hydrogen and helium, while stars like the Sun and even younger stars are typically richer in these heavy elements

In the first section of his chapter, we described the thin disk, thick disk, and stellar halo. Look back at Table 25.1 and note some of the patterns. Young stars lie in the thin disk, are rich in metals, and orbit the Galaxy's center at high speed. The stars in the halo are old, have low abundances of elements heavier than hydrogen and helium, and have highly elliptical orbits randomly oriented in direction (see Figure 25.19). Halo stars can plunge through the disk and central bulge, but they spend most of their time far above or below the plane of the Galaxy. The stars in the thick disk are intermediate between these two extremes. Let's first see why age and heavier-element abundance are correlated and then see what these correlations tell us about the origin of our Galaxy.

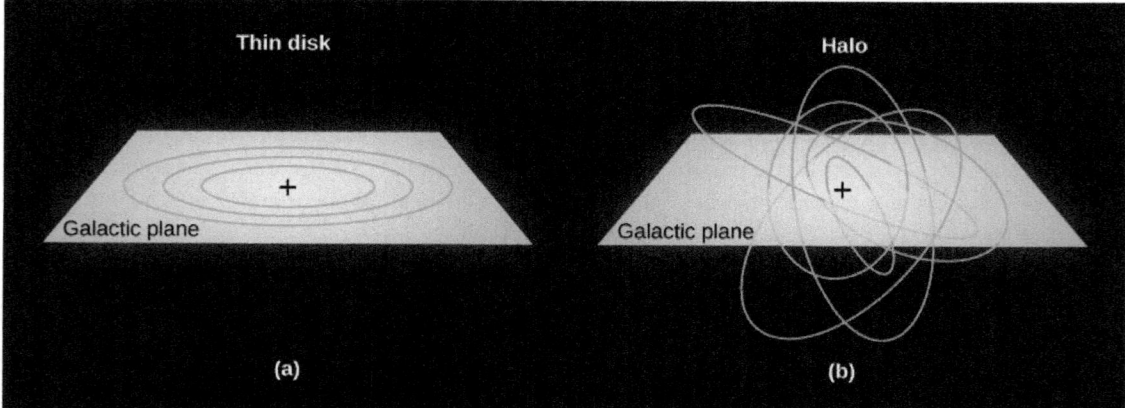

Figure 25.19. How Objects Orbit the Galaxy. (a) In this image, you see stars in the thin disk of our Galaxy in nearly circular orbits. (b) In this image, you see the motion of stars in the Galaxy's halo in randomly oriented and elliptical orbits.

This OpenStax book is available for free at http://cnx.org/content/col11992/1.8

Download for free at http://cnx.org/content/col11992/latest/

Two Kinds of Stars

The discovery that there are two different kinds of stars was first made by Walter Baade during World War II. As a German national, Baade was not allowed to do war research as many other U.S.-based scientists were doing, so he was able to make regular use of the Mount Wilson telescopes in southern California. His observations were aided by the darker skies that resulted from the wartime blackout of Los Angeles.

Among the things a large telescope and dark skies enabled Baade to examine carefully were *other* galaxies—neighbors of our Milky Way Galaxy. We will discuss other galaxies in the next chapter (Galaxies), but for now we will just mention that the nearest Galaxy that resembles our own (with a similar disk and spiral structure) is often called the Andromeda galaxy, after the constellation in which we find it.

Baade was impressed by the similarity of the mainly reddish stars in the Andromeda galaxy's nuclear bulge to those in our Galaxy's globular clusters and the halo. He also noted the difference in color between all these and the bluer stars found in the spiral arms near the Sun (Figure 25.20). On this basis, he called the bright blue stars in the spiral arms **population I** and all the stars in the halo and globular clusters **population II**.

Figure 25.20. Andromeda Galaxy (M31). This neighboring spiral looks similar to our own Galaxy in that it is a disk galaxy with a central bulge. Note the bulge of older, yellowish stars in the center, the bluer and younger stars in the outer regions, and the dust in the disk that blocks some of the light from the bulge. (credit: Adam Evans)

We now know that the populations differ not only in their locations in the Galaxy, but also in their chemical composition, age, and orbital motions around the center of the Galaxy. Population I stars are found only in the disk and follow nearly circular orbits around the galactic center. Examples are bright supergiant stars, main-sequence stars of high luminosity (spectral classes O and B), which are concentrated in the spiral arms, and members of young open star clusters. Interstellar matter and molecular clouds are found in the same places as population I stars.

Population II stars show no correlation with the location of the spiral arms. These objects are found throughout the Galaxy. Some are in the disk, but many others follow eccentric elliptical orbits that carry them high above the galactic disk into the halo. Examples include stars surrounded by planetary nebulae and RR Lyrae variable stars. The stars in globular clusters, found almost entirely in the Galaxy's halo, are also classified as population II.

Download for free at http://cnx.org/content/col11992/latest/

Today, we know much more about stellar evolution than astronomers did in the 1940s, and we can determine the ages of stars. Population I includes stars with a wide range of ages. While some are as old as 10 billion years, others are still forming today. For example, the Sun, which is about 5 billion years old, is a population I star. But so are the massive young stars in the Orion Nebula that have formed in the last few million years. Population II, on the other hand, consists entirely of old stars that formed very early in the history of the Galaxy; typical ages are 11 to 13 billion years.

We also now have good determinations of the compositions of stars. These are based on analyses of the stars' detailed spectra. Nearly all stars appear to be composed mostly of hydrogen and helium, but their abundances of the heavier elements differ. In the Sun and other population I stars, the heavy elements (those heavier than hydrogen and helium) account for 1–4% of the total stellar mass. Population II stars in the outer galactic halo and in globular clusters have much lower abundances of the heavy elements—often less than one-hundredth the concentrations found in the Sun and in rare cases even lower. The oldest population II star discovered to date has less than one ten-millionth as much iron as the Sun, for example.

As we discussed in earlier chapters, heavy elements are created deep within the interiors of stars. They are added to the Galaxy's reserves of raw material when stars die, and their material is recycled into new generations of stars. Thus, as time goes on, stars are born with larger and larger supplies of heavy elements. Population II stars formed when the abundance of elements heavier than hydrogen and helium was low. Population I stars formed later, after mass lost by dying members of the first generations of stars had seeded the interstellar medium with elements heavier than hydrogen and helium. Some are still forming now, when further generations have added to the supply of heavier elements available to new stars.

The Real World

With rare exceptions, we should never trust any theory that divides the world into just two categories. While they can provide a starting point for hypotheses and experiments, they are often oversimplifications that need refinement a research continue. The idea of two populations helped organize our initial thoughts about the Galaxy, but we now know it cannot explain everything we observe. Even the different structures of the Galaxy—disk, halo, central bulge—are not so cleanly separated in terms of their locations, ages, and the heavy element content of the stars within them.

The exact definition of the Galaxy's disk depends on what objects we use to define it, and it has no sharp boundary. The hottest young stars and their associated gas and dust clouds are mostly in a region about 300 light-years thick. Older stars define a thicker disk that is about 3000 light-years thick. Halo stars spend most of their time high above or below the disk but pass through it on their highly elliptical orbits and so are sometimes found relatively near the Sun.

The highest density of stars is found in the central bulge, that bar-shaped inner region of the Galaxy. There are a few hot, young stars in the bulge, but most of the bulge stars are more than 10 billion years old. Yet unlike the halo stars of similar age, the abundance of heavy elements in the bulge stars is about the same as in the Sun. Why would that be?

Astronomers think that star formation in the crowded nuclear bulge occurred very rapidly just after the Milky Way Galaxy formed. After a few million years, the first generation of massive and short-lived stars then expelled heavy elements in supernova explosions and thereby enriched subsequent generations of stars. Thus, even stars that formed in the bulge more than 10 billion years ago started with a good supply of heavy elements.

Exactly the opposite occurred in the Small Magellanic Cloud, a small galaxy near the Milky Way, visible from Earth's Southern Hemisphere. Even the youngest stars in this galaxy are deficient in heavy elements. We think this is because the little galaxy is not especially crowded, and star formation has occurred quite slowly. As a result there have been, so far, relatively few supernova explosions. Smaller galaxies also have more trouble

This OpenStax book is available for free at http://cnx.org/content/col11992/1.8

holding onto the gas expelled by supernova explosions in order to recycle it. Low-mass galaxies exert only a modest gravitational force, and the high-speed gas ejected by supernovae can easily escape from them.

Which elements a star is endowed with thus depends not only on when the star formed in the history of its galaxy, but also on how many stars in its part of the galaxy had already completed their lives by the time the star is ready to form.

 ## THE FORMATION OF THE GALAXY

Learning Objectives

By the end of this section, you will be able to:

> Describe the roles played by the collapse of a single cloud and mergers with other galaxies in building the Milky Way Galaxy we see today
> Provide examples of globular clusters and satellite galaxies affected by the Milky Way's strong gravity.

Information about stellar populations holds vital clues to how our Galaxy was built up over time. The flattened disk shape of the Galaxy suggests that it formed through a process similar to the one that leads to the formation of a protostar (see The Birth of Stars and the Discovery of Planets outside the Solar System). Building on this idea, astronomers first developed models that assumed the Galaxy formed from a single rotating cloud. But, as we shall see, this turns out to be only part of the story.

The Protogalactic Cloud and the Monolithic Collapse Model

Because the oldest stars—those in the halo and in globular clusters—are distributed in a sphere centered on the nucleus of the Galaxy, it makes sense to assume that the *protogalactic* cloud that gave birth to our Galaxy was roughly spherical. The oldest stars in the halo have ages of 12 to 13 billion years, so we estimate that the formation of the Galaxy began about that long ago. (See the chapter on The Big Bang for other evidence that galaxies in general began forming a little more than 13 billion years ago.) Then, just as in the case of star formation, the protogalactic cloud collapsed and formed a thin rotating disk. Stars born before the cloud collapsed did not participate in the collapse, but have continued to orbit in the halo to the present day (Figure 25.21).

Download for free at http://cnx.org/content/col11992/latest/

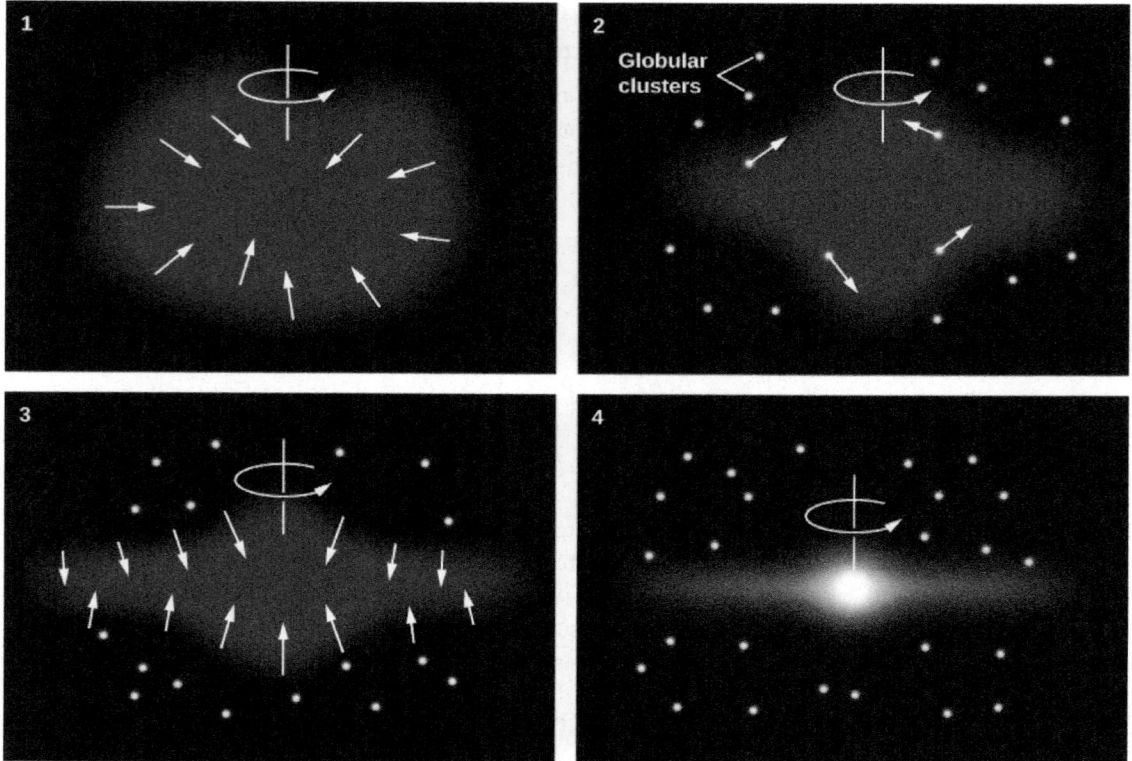

Figure 25.21. Monolithic Collapse Model for the Formation of the Galaxy. According to this model, the Milky Way Galaxy initially formed from a rotating cloud of gas that collapsed due to gravity. Halo stars and globular clusters either formed prior to the collapse or were formed elsewhere. Stars in the disk formed later, when the gas from which they were made was already "contaminated" with heavy elements produced in earlier generations of stars.

Gravitational forces caused the gas in the thin disk to fragment into clouds or clumps with masses like those of star clusters. These individual clouds then fragmented further to form stars. Since the oldest stars in the disk are nearly as old as the youngest stars in the halo, the collapse must have been rapid (astronomically speaking), requiring perhaps no more than a few hundred million years.

Collision Victims and the Multiple Merger Model

In past decades, astronomers have learned that the evolution of the Galaxy has not been quite as peaceful as this monolithic collapse model suggests. In 1994, astronomers discovered a small new galaxy in the direction of the constellation of Sagittarius. The Sagittarius dwarf galaxy is currently about 70,000 light-years away from Earth and 50,000 light-years from the center of the Galaxy. It is the closest galaxy known (Figure 25.22). It is very elongated, and its shape indicates that it is being torn apart by our Galaxy's gravitational tides—just as Comet Shoemaker-Levy 9 was torn apart when it passed too close to Jupiter in 1992.

The Sagittarius galaxy is much smaller than the Milky Way, with only about 150,000 stars, all of which seem destined to end up in the bulge and halo of our own Galaxy. But don't sound the funeral bells for the little galaxy quite yet; the ingestion of the Sagittarius dwarf will take another 100 million years or so, and the stars themselves will survive.

This OpenStax book is available for free at http://cnx.org/content/col11992/1.8

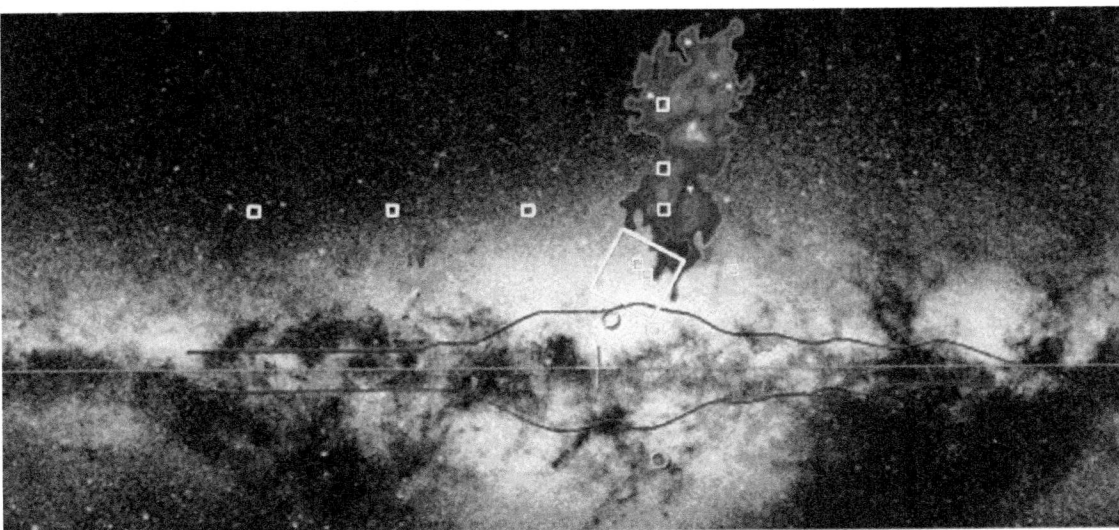

Figure 25.22. Sagittarius Dwarf. In 1994, British astronomers discovered a galaxy in the constellation of Sagittarius, located only about 50,000 light-years from the center of the Milky Way and falling into our Galaxy. This image covers a region approximately 70° × 50° and combines a black-and-white view of the disk of our Galaxy with a red contour map showing the brightness of the dwarf galaxy. The dwarf galaxy lies on the other side of the galactic center from us. The white stars in the red region mark the locations of several globular clusters contained within the Sagittarius dwarf galaxy. The cross marks the galactic center. The horizontal line corresponds to the galactic plane. The blue outline on either side of the galactic plane corresponds to the infrared image in Figure 25.7. The boxes mark regions where detailed studies of individual stars led to the discovery of this galaxy. (credit: modification of work by R. Ibata (UBC), R. Wyse (JHU), R. Sword (IoA))

Since that discovery, evidence has been found for many more close encounters between our Galaxy and other neighbor galaxies. When a small galaxy ventures too close, the force of gravity exerted by our Galaxy tugs harder on the near side than on the far side. The net effect is that the stars that originally belonged to the small galaxy are spread out into a long stream that orbits through the halo of the Milky Way (Figure 25.23).

Download for free at http://cnx.org/content/col11992/latest/

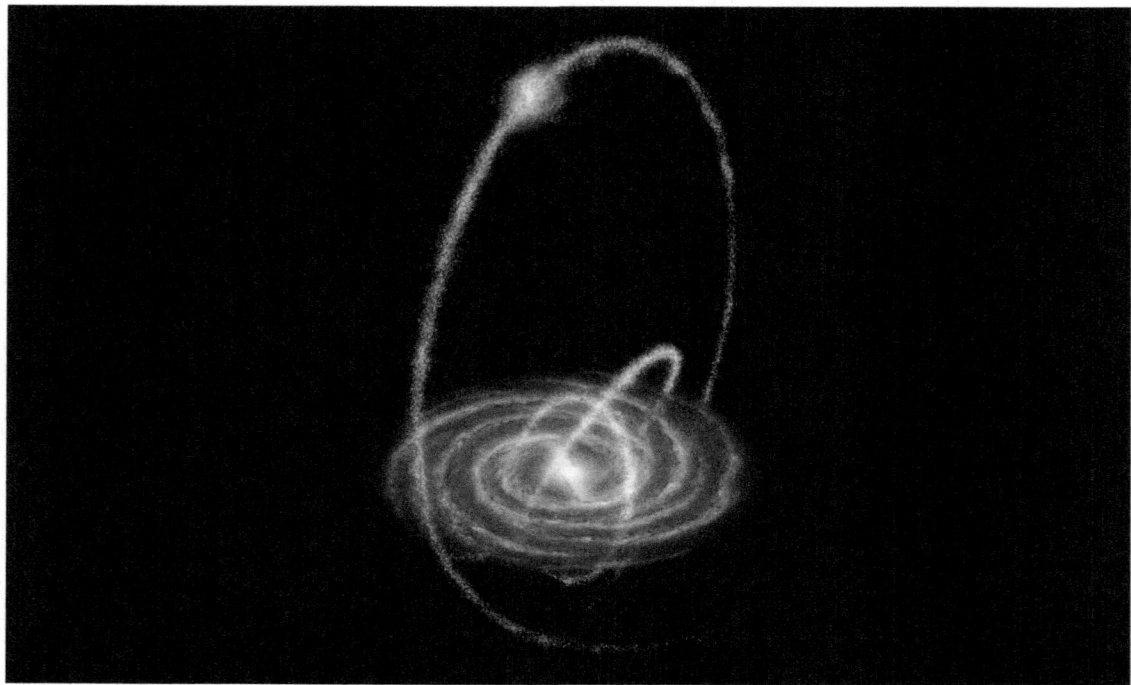

Figure 25.23. Streams in the Galactic Halo. When a small galaxy is swallowed by the Milky Way, its member stars are stripped away and form streams of stars in the galactic halo. This image is based on calculations of what some of these tidal streams might look like if the Milky Way swallowed 50 dwarf galaxies over the past 10 billion years. (credit: modification of work by NASA/JPL-Caltech/R. Hurt (SSC/Caltech))

Such a tidal stream can maintain its identity for billions of years. To date, astronomers have now identified streams originating from 12 small galaxies that ventured too close to the much larger Milky Way. Six more streams are associated with globular clusters. It has been suggested that large globular clusters, like Omega Centauri, are actually dense nuclei of cannibalized dwarf galaxies. The globular cluster M54 is now thought to be the nucleus of the Sagittarius dwarf we discussed earlier, which is currently merging with the Milky Way (Figure 25.24). The stars in the outer regions of such galaxies are stripped off by the gravitational pull of the Milky Way, but the central dense regions may survive.

This OpenStax book is available for free at http://cnx.org/content/col11992/1.8

Figure 25.24. Globular Cluster M54. This beautiful Hubble Space Telescope image shows the globular cluster that is now believed to be the nucleus of the Sagittarius Dwarf Galaxy. (credit: ESA/Hubble & NASA)

Calculations indicate that the Galaxy's thick disk may be a product of one or more such collisions with other galaxies. Accretion of a satellite galaxy would stir up the orbits of the stars and gas clouds originally in the thin disk and cause them to move higher above and below the mid-plane of the Galaxy. Meanwhile, the Galaxy's stars would add to the fluffed-up mix. If such a collision happened about 10 billion years ago, then any gas in the two galaxies that had not yet formed into stars would have had plenty of time to settle back down into the thin disk. The gas could then have begun forming subsequent generations of population I stars. This timing is also consistent with the typical ages of stars in the thick disk.

The Milky Way has more collisions in store. An example is the Canis Major dwarf galaxy, which has a mass of about 1% of the mass of the Milky Way. Already long tidal tails have been stripped from this galaxy, which have wrapped themselves around the Milky Way three times. Several of the globular clusters found in the Milky Way may also have come from the Canis Major dwarf, which is expected to merge gradually with the Milky Way over about the next billion years.

In about 3 billion years, the Milky Way itself will be swallowed up, since it and the Andromeda galaxy are on a collision course. Our computer models show that after a complex interaction, the two will merge to form a larger, more rounded galaxy (Figure 25.25).

Download for free at http://cnx.org/content/col11992/latest/

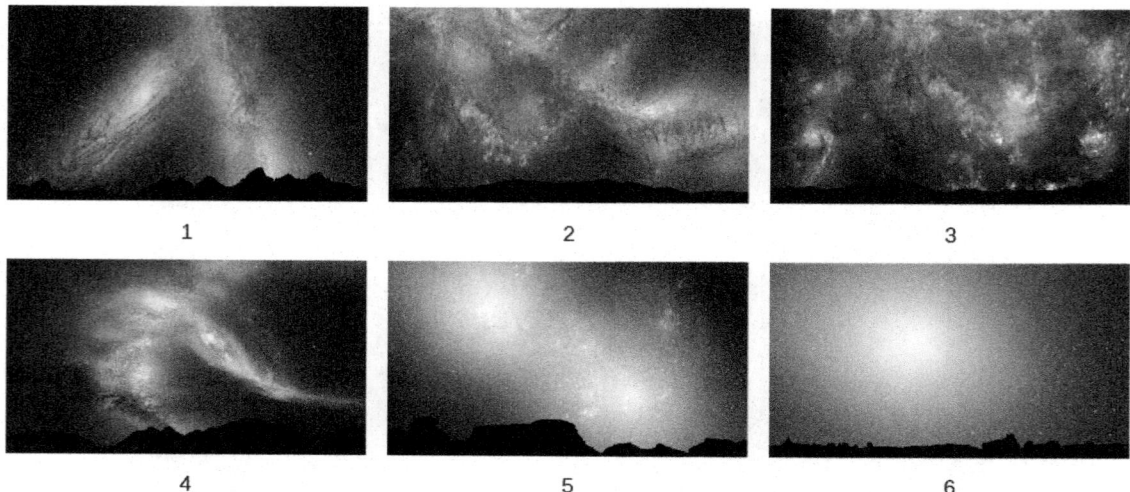

Figure 25.25. Collision of the Milky Way with Andromeda. In about 3 billion years, the Milky Way Galaxy and Andromeda Galaxy will begin a long process of colliding, separating, and then coming back together to form an elliptical galaxy. The whole interaction will take 3 to 4 billion years. These images show the following sequence: (1) In 3.75 billion years, Andromeda has approached the Milky Way. (2) New star formation fills the sky 3.85 billion years from now. (3) Star formation continues at 3.9 billion years. (4) The galaxy shapes change as they interact, with Andromeda being stretched and our Galaxy becoming warped, about 4 billion years from now. (5) In 5.1 billion years, the cores of the two galaxies are bright lobes. (6) In 7 billion years, the merged galaxies form a huge elliptical galaxy whose brightness fills the night sky. This artist's illustrations show events from a vantage point 25,000 light-years from the center of the Milky Way. However, we should mention that the Sun may not be at that distance throughout the sequence of events, as the collision readjusts the orbits of many stars within each galaxy. (credit: NASA; ESA; Z. Levay, R. van der Marel, STScI; T. Hallas, and A. Mellinger)

We are thus coming to realize that "environmental influences" (and not just a galaxy's original characteristics) play an important role in determining the properties and development of our Galaxy. In future chapters we will see that collisions and mergers are a major factor in the evolution of many other galaxies as well.

This OpenStax book is available for free at http://cnx.org/content/col11992/1.8

Download for free at http://cnx.org/content/col11992/latest/

CHAPTER 25 REVIEW

 KEY TERMS

central bulge (or nuclear bulge) the central (round) part of the Milky Way or a similar galaxy

dark matter nonluminous mass, whose presence can be inferred only because of its gravitational influence on luminous matter; the composition of the dark matter is not known

dark matter halo the mass in the Milky Way that extends well beyond the boundary of the luminous stars to a distance of at least 200,000 light-years from the center of the Galaxy; although we deduce its existence from its gravity, the composition of this matter remains a mystery

differential galactic rotation the idea that different parts of the Galaxy turn at different rates, since the parts of the Galaxy follow Kepler's third law: more distant objects take longer to complete one full orbit around the center of the Galaxy

halo the outermost extent of our Galaxy (or another galaxy), containing a sparse distribution of stars and globular clusters in a more or less spherical distribution

Milky Way Galaxy the band of light encircling the sky, which is due to the many stars and diffuse nebulae lying near the plane of the Milky Way Galaxy

population I star a star containing heavy elements; typically young and found in the disk

population II star a star with very low abundance of heavy elements; found throughout the Galaxy

spiral arm a spiral-shaped region, characterized by relatively dense interstellar material and young stars, that is observed in the disks of spiral galaxies

supermassive black hole the object in the center of most large galaxies that is so massive and compact that light cannot escape from it; the Milky Way's supermassive black hole contains 4.6 millions of Suns' worth of mass

 SUMMARY

25.1 The Architecture of the Galaxy

The Milky Way Galaxy consists of a thin disk containing dust, gas, and young and old stars; a spherical halo containing populations of very old stars, including RR Lyrae variable stars and globular star clusters; a thick, more diffuse disk with stars that have properties intermediate between those in the thin disk and the halo; a peanut-shaped nuclear bulge of mostly old stars around the center; and a supermassive black hole at the very center. The Sun is located roughly halfway out of the Milky Way, about 26,000 light-years from the center.

25.2 Spiral Structure

The gaseous distribution in the Galaxy's disk has two main spiral arms that emerge from the ends of the central bar, along with several fainter arms and short spurs; the Sun is located in one of those spurs. Measurements show that the Galaxy does not rotate as a solid body, but instead its stars and gas follow differential rotation, such that the material closer to the galactic center completes its orbit more quickly. Observations show that galaxies like the Milky Way take several billion years after they began to form to develop spiral structure.

Download for free at http://cnx.org/content/col11992/latest/

25.3 The Mass of the Galaxy

The Sun revolves completely around the galactic center in about 225 million years (a galactic year). The mass of the Galaxy can be determined by measuring the orbital velocities of stars and interstellar matter. The total mass of the Galaxy is about 2×10^{12} M_{Sun}. As much as 95% of this mass consists of dark matter that emits no electromagnetic radiation and can be detected only because of the gravitational force it exerts on visible stars and interstellar matter. This dark matter is located mostly in the Galaxy's halo; its nature is not well understood at present.

25.4 The Center of the Galaxy

A supermassive black hole is located at the center of the Galaxy. Measurements of the velocities of stars located within a few light-days of the center show that the mass inside their orbits around the center is about 4.6 million M_{Sun}. Radio observations show that this mass is concentrated in a volume with a diameter similar to that of Mercury's orbit. The density of this matter concentration exceeds that of the densest known star clusters by a factor of nearly a million. The only known object with such a high density and total mass is a black hole.

25.5 Stellar Populations in the Galaxy

We can roughly divide the stars in the Galaxy into two categories. Old stars with few heavy elements are referred to as population II stars and are found in the halo and in globular clusters. Population I stars contain more heavy elements than globular cluster and halo stars, are typically younger and found in the disk, and are especially concentrated in the spiral arms. The Sun is a member of population I. Population I stars formed after previous generations of stars had produced heavy elements and ejected them into the interstellar medium. The bulge stars, most of which are more than 10 billion years old, have unusually high amounts of heavy elements, presumably because there were many massive first-generation stars in this dense region, and these quickly seeded the next generations of stars with heavier elements.

25.6 The Formation of the Galaxy

The Galaxy began forming a little more than 13 billion years ago. Models suggest that the stars in the halo and globular clusters formed first, while the Galaxy was spherical. The gas, somewhat enriched in heavy elements by the first generation of stars, then collapsed from a spherical distribution to a rotating disk-shaped distribution. Stars are still forming today from the gas and dust that remain in the disk. Star formation occurs most rapidly in the spiral arms, where the density of interstellar matter is highest. The Galaxy captured (and still is capturing) additional stars and globular clusters from small galaxies that ventured too close to the Milky Way. In 3 to 4 billion years, the Galaxy will begin to collide with the Andromeda galaxy, and after about 7 billion years, the two galaxies will merge to form a giant elliptical galaxy.

 FOR FURTHER EXPLORATION

Articles

Blitz, L. "The Dark Side of the Milky Way." *Scientific American* (October 2011): 36–43. How we find dark matter and what it tells us about our Galaxy, its warped disk, and its satellite galaxies.

Dvorak, J. "Journey to the Heart of the Milky Way." *Astronomy* (February 2008): 28. Measuring nearby stars to determine the properties of the black hole at the center.

Gallagher, J., Wyse, R., & Benjamin, R. "The New Milky Way." *Astronomy* (September 2011): 26. Highlights all aspects of the Milky Way based on recent observations.

This OpenStax book is available for free at http://cnx.org/content/col11992/1.8

Goldstein, A. "Finding our Place in the Milky Way." *Astronomy* (August 2015): 50. On the history of observations that pinpointed the Sun's location in the Galaxy.

Haggard, D., & Bower, G. "In the Heart of the Milky Way." *Sky & Telescope* (February 2016): 16. On observations of the Galaxy's nucleus and the supermassive black hole and magnetar there.

Ibata, R., & Gibson, B. "The Ghosts of Galaxies Past." *Scientific American* (April 2007): 40. About star streams in the Galaxy that are evidence of past mergers and collisions.

Irion, R. "A Crushing End for Our Galaxy." *Science* (January 7, 2000): 62. On the role of mergers in the evolution of the Milky Way.

Irion, R. "Homing in on Black Holes." *Smithsonian* (April 2008). On how astronomers probe the large black hole at the center of the Milky Way Galaxy.

Kruesi, L. "How We Mapped the Milky Way." *Astronomy* (October 2009): 28.

Kruesi, L. "What Lurks in the Monstrous Heart of the Milky Way?" *Astronomy* (October 2015): 30. On the center of the Galaxy and the black hole there.

Laughlin, G., & Adams, F. "Celebrating the Galactic Millennium." *Astronomy* (November 2001): 39. The long-term future of the Milky Way in the next 90 billion years.

Loeb, A., & Cox, T.J. "Our Galaxy's Date with Destruction." *Astronomy* (June 2008): 28. Describes the upcoming merger of Milky Way and Andromeda.

Szpir, M. "Passing the Bar Exam." *Astronomy* (March 1999): 46. On evidence that our Galaxy is a barred spiral.

Tanner, A. "A Trip to the Galactic Center." *Sky & Telescope* (April 2003): 44. Nice introduction, with observations pointing to the presence of a black hole.

Trimble, V., & Parker, S. "Meet the Milky Way." *Sky & Telescope* (January 1995): 26. Overview of our Galaxy.

Wakker, B., & Richter, P. "Our Growing, Breathing Galaxy." *Scientific American* (January 2004): 38. Evidence that our Galaxy is still being built up by the addition of gas and smaller neighbors.

Waller, W. "Redesigning the Milky Way." *Sky & Telescope* (September 2004): 50. On recent multi-wavelength surveys of the Galaxy.

Whitt, K. "The Milky Way from the Inside." *Astronomy* (November 2001): 58. Fantastic panorama image of the Galaxy, with finder charts and explanations.

Websites

International Dark Sky Sanctuaries: http://darksky.org/idsp/sanctuaries/. A listing of dark-sky sanctuaries, parks, and reserves.

Multiwavelength Milky Way: http://mwmw.gsfc.nasa.gov/mmw_sci.html. This NASA site shows the plane of our Galaxy in a variety of wavelength bands, and includes background material and other resources.

Shapley-Curtis Debate in 1920: http://apod.nasa.gov/diamond_jubilee/debate_1920.html. In 1920, astronomers Harlow Shapley and Heber Curtis engaged in a historic debate about how large our Galaxy was and whether other galaxies existed. Here you can find historical and educational material about the debate.

UCLA Galactic Center Group: http://www.galacticcenter.astro.ucla.edu/. Learn more about the work of Andrea Ghez and colleagues on the central region of the Milky Way Galaxy.

Videos

Crash of the Titans: http://www.spacetelescope.org/videos/hubblecast55a/. This Hubblecast from 2012 features Jay Anderson and Roeland van der Marel explaining how Andromeda will collide with the Milky Way in the distant future (5:07).

Diner at the Center of the Galaxy: https://www.youtube.com/watch?v=UP7ig8Gxftw. A short discussion from NASA ScienceCast of NuSTAR observations of flares from our Galaxy's central black hole (3:23).

Hunt for a Supermassive Black Hole: https://www.ted.com/talks/andrea_ghez_the_hunt_for_a_supermassive_black_hole. 2009 TED talk by Andrea Ghez on searching for supermassive black holes, particularly the one at the center of the Milky Way (16:19).

Journey to the Galactic Center: https://www.youtube.com/watch?v=36xZsgZ0oSo. A brief silent trip into the cluster of stars near the galactic center showing their motions around the center (3:00).

🏛 COLLABORATIVE GROUP ACTIVITIES

A. You are captured by space aliens, who take you inside a complex cloud of interstellar gas, dust, and a few newly formed stars. To escape, you need to make a map of the cloud. Luckily, the aliens have a complete astronomical observatory with equipment for measuring all the bands of the electromagnetic spectrum. Using what you have learned in this chapter, have your group discuss what kinds of maps you would make of the cloud to plot your most effective escape route.

B. The diagram that Herschel made of the Milky Way has a very irregular outer boundary (see Figure 25.3). Can your group think of a reason for this? How did Herschel construct his map?

C. Suppose that for your final exam in this course, your group is assigned telescope time to observe a star selected for you by your professor. The professor tells you the position of the star in the sky (its right ascension and declination) but nothing else. You can make any observations you wish. How would you go about determining whether the star is a member of population I or population II?

D. The existence of dark matter comes as a great surprise, and its nature remains a mystery today. Someday astronomers will know a lot more about it (you can learn more about current findings in The Evolution and Distribution of Galaxies). Can your group make a list of earlier astronomical observations that began as a surprise and mystery, but wound up (with more observations) as well-understood parts of introductory textbooks?

E. Physicist Gregory Benford has written a series of science fiction novels that take place near the center of the Milky Way Galaxy in the far future. Suppose your group were writing such a story. Make a list of ways that the environment near the galactic center differs from the environment in the "galactic suburbs," where the Sun is located. Would life as we know it have an easier or harder time surviving on planets that orbit stars near the center (and why)?

F. These days, in most urban areas, city lights completely swamp the faint light of the Milky Way in our skies. Have each member of your group survey 5 to 10 friends or relatives (you could spread out on campus to investigate or use social media or the phone), explaining what the Milky Way is and then asking if they have seen it. Also ask their age. Report back to your group and discuss your reactions to the survey. Is there any

This OpenStax book is available for free at http://cnx.org/content/col11992/1.8

relationship between a person's age and whether they have seen the Milky Way? How important is it that many kids growing up on Earth today never (or rarely) get to see our home Galaxy in the sky?

EXERCISES

Review Questions

1. Explain why we see the Milky Way as a faint band of light stretching across the sky.

2. Explain where in a spiral galaxy you would expect to find globular clusters, molecular clouds, and atomic hydrogen.

3. Describe several characteristics that distinguish population I stars from population II stars.

4. Briefly describe the main parts of our Galaxy.

5. Describe the evidence indicating that a black hole may be at the center of our Galaxy.

6. Explain why the abundances of heavy elements in stars correlate with their positions in the Galaxy.

7. What will be the long-term future of our Galaxy?

Thought Questions

8. Suppose the Milky Way was a band of light extending only halfway around the sky (that is, in a semicircle). What, then, would you conclude about the Sun's location in the Galaxy? Give your reasoning.

9. Suppose somebody proposed that rather than invoking dark matter to explain the increased orbital velocities of stars beyond the Sun's orbit, the problem could be solved by assuming that the Milky Way's central black hole was much more massive. Does simply increasing the assumed mass of the Milky Way's central supermassive black hole correctly resolve the issue of unexpectedly high orbital velocities in the Galaxy? Why or why not?

10. The globular clusters revolve around the Galaxy in highly elliptical orbits. Where would you expect the clusters to spend most of their time? (Think of Kepler's laws.) At any given time, would you expect most globular clusters to be moving at high or low speeds with respect to the center of the Galaxy? Why?

11. Shapley used the positions of globular clusters to determine the location of the galactic center. Could he have used open clusters? Why or why not?

12. Consider the following five kinds of objects: open cluster, giant molecular cloud, globular cluster, group of O and B stars, and planetary nebulae.
    A. Which occur only in spiral arms?

    B. Which occur only in the parts of the Galaxy other than the spiral arms?

    C. Which are thought to be very young?

    D. Which are thought to be very old?

    E. Which have the hottest stars?

13. The dwarf galaxy in Sagittarius is the one closest to the Milky Way, yet it was discovered only in 1994. Can you think of a reason it was not discovered earlier? (Hint: Think about what else is in its constellation.)

Download for free at http://cnx.org/content/col11992/latest/

14. Suppose three stars lie in the disk of the Galaxy at distances of 20,000 light-years, 25,000 light-years, and 30,000 light-years from the galactic center, and suppose that right now all three are lined up in such a way that it is possible to draw a straight line through them and on to the center of the Galaxy. How will the relative positions of these three stars change with time? Assume that their orbits are all circular and lie in the plane of the disk.

15. Why does star formation occur primarily in the disk of the Galaxy?

16. Where in the Galaxy would you expect to find Type II supernovae, which are the explosions of massive stars that go through their lives very quickly? Where would you expect to find Type I supernovae, which involve the explosions of white dwarfs?

17. Suppose that stars evolved without losing mass—that once matter was incorporated into a star, it remained there forever. How would the appearance of the Galaxy be different from what it is now? Would there be population I and population II stars? What other differences would there be?

Figuring For Yourself

18. Assume that the Sun orbits the center of the Galaxy at a speed of 220 km/s and a distance of 26,000 light-years from the center.

    A. Calculate the circumference of the Sun's orbit, assuming it to be approximately circular. (Remember that the circumference of a circle is given by $2\pi R$, where R is the radius of the circle. Be sure to use consistent units. The conversion from light-years to km/s can be found in an online calculator or appendix, or you can calculate it for yourself: the speed of light is 300,000 km/s, and you can determine the number of seconds in a year.)

    B. Calculate the Sun's period, the "galactic year." Again, be careful with the units. Does it agree with the number we gave above?

19. The Sun orbits the center of the Galaxy in 225 million years at a distance of 26,000 light-years. Given that $a^3 = (M_1 + M_2) \times P^2$, where a is the semimajor axis and P is the orbital period, what is the mass of the Galaxy within the Sun's orbit?

20. Suppose the Sun orbited a little farther out, but the mass of the Galaxy inside its orbit remained the same as we calculated in Exercise 25.19. What would be its period at a distance of 30,000 light-years?

21. We have said that the Galaxy rotates differentially; that is, stars in the inner parts complete a full 360° orbit around the center of the Galaxy more rapidly than stars farther out. Use Kepler's third law and the mass we derived in Exercise 25.19 to calculate the period of a star that is only 5000 light-years from the center. Now do the same calculation for a globular cluster at a distance of 50,000 light-years. Suppose the Sun, this star, and the globular cluster all fall on a straight line through the center of the Galaxy. Where will they be relative to each other after the Sun completes one full journey around the center of the Galaxy? (Assume that all the mass in the Galaxy is concentrated at its center.)

22. If our solar system is 4.6 billion years old, how many galactic years has planet Earth been around?

23. Suppose the average mass of a star in the Galaxy is one-third of a solar mass. Use the value for the mass of the Galaxy that we calculated in Exercise 25.19, and estimate how many stars are in the Milky Way. Give some reasons it is reasonable to assume that the mass of an average star is less than the mass of the Sun.

24. The first clue that the Galaxy contains a lot of dark matter was the observation that the orbital velocities of stars did not decreases with increasing distance from the center of the Galaxy. Construct a rotation curve for the solar system by using the orbital velocities of the planets, which can be found in Appendix F. How does this curve differ from the rotation curve for the Galaxy? What does it tell you about where most of the mass in the solar system is concentrated?

This OpenStax book is available for free at http://cnx.org/content/col11992/1.8

25. The best evidence for a black hole at the center of the Galaxy also comes from the application of Kepler's third law. Suppose a star at a distance of 20 light-hours from the center of the Galaxy has an orbital speed of 6200 km/s. How much mass must be located inside its orbit?

26. The next step in deciding whether the object in Exercise 25.25 is a black hole is to estimate the density of this mass. Assume that all of the mass is spread uniformly throughout a sphere with a radius of 20 light-hours. What is the density in kg/km^3? (Remember that the volume of a sphere is given by $V = \frac{4}{3}\pi R^3$.)

    Explain why the density might be even higher than the value you have calculated. How does this density compare with that of the Sun or other objects we have talked about in this book?

27. Suppose the Sagittarius dwarf galaxy merges completely with the Milky Way and adds 150,000 stars to it. Estimate the percentage change in the mass of the Milky Way. Will this be enough mass to affect the orbit of the Sun around the galactic center? Assume that all of the Sagittarius galaxy's stars end up in the nuclear bulge of the Milky Way Galaxy and explain your answer.

Download for free at http://cnx.org/content/col11992/latest/

This OpenStax book is available for free at http://cnx.org/content/col11992/1.8

Download for free at http://cnx.org/content/col11992/latest/

Figure 26.1. Spiral Galaxy. NGC 6946 is a spiral galaxy also known as the "Fireworks galaxy." It is at a distance of about 18 million light-years, in the direction of the constellations Cepheus and Cygnus. It was discovered by William Herschel in 1798. This galaxy is about one-third the size of the Milky Way. Note on the left how the colors of the galaxy change from the yellowish light of old stars in the center to the blue color of hot, young stars and the reddish glow of hydrogen clouds in the spiral arms. As the image shows, this galaxy is rich in dust and gas, and new stars are still being born here. (credit left: modification of work by NASA, ESA, STScI, R. Gendler, and the Subaru Telescope (NAOJ); credit right: modification of work by X-ray: NASA/CXC/MSSL/R.Soria et al, Optical: AURA/Gemini OBs)

Chapter Outline

Thinking Ahead

In the last chapter, we explored our own Galaxy. But is it the only one? If there are others, are they like the Milky Way? How far away are they? Can we see them? As we shall learn, some galaxies turn out to be so far away that it has taken billions of years for their light to reach us. These remote galaxies can tell us what the universe was like when it was young.

In this chapter, we start our exploration of the vast realm of galaxies. Like tourists from a small town making their first visit to the great cities of the world, we will be awed by the beauty and variety of the galaxies. And yet, we will recognize that much of what we see is not so different from our experiences at home, and we will be impressed by how much we can learn by looking at structures built long ago.

We begin our voyage with a guide to the properties of galaxies, much as a tourist begins with a guidebook to the main features of the cities on the itinerary. In later chapters, we will look more carefully at the past history of galaxies, how they have changed over time, and how they acquired their many different forms. First, we'll begin our voyage through the galaxies with the question: is our Galaxy the only one?

Download for free at http://cnx.org/content/col11992/latest/

26.1 THE DISCOVERY OF GALAXIES

Learning Objectives

By the end of this section, you will be able to:

> Describe the discoveries that confirmed the existence of galaxies that lie far beyond the Milky Way Galaxy
> Explain why galaxies used to be called nebulae and why we don't include them in that category any more

Growing up at a time when the Hubble Space Telescope orbits above our heads and giant telescopes are springing up on the great mountaintops of the world, you may be surprised to learn that we were not sure about the existence of other galaxies for a very long time. The very idea that other galaxies exist used to be controversial. Even into the 1920s, many astronomers thought the Milky Way encompassed *all* that exists in the universe. The evidence found in 1924 that meant our Galaxy is not alone was one of the great scientific discoveries of the twentieth century.

It was not that scientists weren't asking questions. They questioned the composition and structure of the universe as early as the eighteenth century. However, with the telescopes available in earlier centuries, galaxies looked like small fuzzy patches of light that were difficult to distinguish from the star clusters and gas-and-dust clouds that are part of our own Galaxy. All objects that were not sharp points of light were given the same name, *nebulae*, the Latin word for "clouds." Because their precise shapes were often hard to make out and no techniques had yet been devised for measuring their distances, the nature of the nebulae was the subject of much debate.

As early as the eighteenth century, the philosopher Immanuel Kant (1724–1804) suggested that some of the nebulae might be distant systems of stars (other Milky Ways), but the evidence to support this suggestion was beyond the capabilities of the telescopes of that time.

Other Galaxies

By the early twentieth century, some nebulae had been correctly identified as star clusters, and others (such as the Orion Nebula) as gaseous nebulae. Most nebulae, however, looked faint and indistinct, even with the best telescopes, and their distances remained unknown. (For more on how such nebulae are named, by the way, see the feature box on Naming the Nebulae in the chapter on interstellar matter.) If these nebulae were nearby, with distances comparable to those of observable stars, they were most likely clouds of gas or groups of stars within our Galaxy. If, on the other hand, they were remote, far beyond the edge of the Galaxy, they could be other star systems containing billions of stars.

To determine what the nebulae are, astronomers had to find a way of measuring the distances to at least some of them. When the 2.5-meter (100-inch) telescope on Mount Wilson in Southern California went into operation, astronomers finally had the large telescope they needed to settle the controversy.

Working with the 2.5-meter telescope, Edwin Hubble was able to resolve individual stars in several of the brighter spiral-shaped nebulae, including M31, the great spiral in Andromeda (Figure 26.2). Among these stars, he discovered some faint variable stars that—when he analyzed their light curves—turned out to be cepheids. Here were reliable indicators that Hubble could use to measure the distances to the nebulae using the technique pioneered by Henrietta Leavitt (see the chapter on Celestial Distances). After painstaking work, he estimated that the Andromeda galaxy was about 900,000 light-years away from us. At that enormous distance, it had to be a separate galaxy of stars located well outside the boundaries of the Milky Way. Today, we know the

This OpenStax book is available for free at http://cnx.org/content/col11992/1.8

Download for free at http://cnx.org/content/col11992/latest/

Andromeda galaxy is actually slightly more than twice as distant as Hubble's first estimate, but his conclusion about its true nature remains unchanged.

Figure 26.2. Andromeda Galaxy. Also known by its catalog number M31, the Andromeda galaxy is a large spiral galaxy very similar in appearance to, and slightly larger than, our own Galaxy. At a distance of about 2.5 million light-years, Andromeda is the spiral galaxy that is nearest to our own in space. Here, it is seen with two of its satellite galaxies, M32 (top) and M110 (bottom). (credit: Adam Evans)

No one in human history had ever measured a distance so great. When Hubble's paper on the distances to nebulae was read before a meeting of the American Astronomical Society on the first day of 1925, the entire room erupted in a standing ovation. A new era had begun in the study of the universe, and a new scientific field—extragalactic astronomy—had just been born.

VOYAGERS IN ASTRONOMY

Edwin Hubble: Expanding the Universe

The son of a Missouri insurance agent, Edwin Hubble (Figure 26.3) graduated from high school at age 16. He excelled in sports, winning letters in track and basketball at the University of Chicago, where he studied both science and languages. Both his father and grandfather wanted him to study law, however, and he gave in to family pressure. He received a prestigious Rhodes scholarship to Oxford University in England, where he studied law with only middling enthusiasm. Returning to be the United States, he spent a year teaching high school physics and Spanish as well as coaching basketball, while trying to determine his life's direction.

Download for free at http://cnx.org/content/col11992/latest/

Figure 26.3. Edwin Hubble (1889–1953). Edwin Hubble established some of the most important ideas in the study of galaxies.

The pull of astronomy eventually proved too strong to resist, and so Hubble went back to the University of Chicago for graduate work. Just as he was about to finish his degree and accept an offer to work at the soon-to be completed 2.5-meter telescope, the United States entered World War I, and Hubble enlisted as an officer. Although the war had ended by the time he arrived in Europe, he received more officer's training abroad and enjoyed a brief time of further astronomical study at Cambridge before being sent home.

In 1919, at age 30, he joined the staff at Mount Wilson and began working with the world's largest telescope. Ripened by experience, energetic, disciplined, and a skillful observer, Hubble soon established some of the most important ideas in modern astronomy. He showed that other galaxies existed, classified them on the basis of their shapes, found a pattern to their motion (and thus put the notion of an expanding universe on a firm observational footing), and began a lifelong program to study the distribution of galaxies in the universe. Although a few others had glimpsed pieces of the puzzle, it was Hubble who put it all together and showed that an understanding of the large-scale structure of the universe was feasible.

His work brought Hubble much renown and many medals, awards, and honorary degrees. As he became better known (he was the first astronomer to appear on the cover of *Time* magazine), he and his wife enjoyed and cultivated friendships with movie stars and writers in Southern California. Hubble was instrumental (if you'll pardon the pun) in the planning and building of the 2.5-meter telescope on Palomar Mountain, and he had begun to use it for studying galaxies when he passed away from a stroke in 1953.

When astronomers built a space telescope that would allow them to extend Hubble's work to distances he could only dream about, it seemed natural to name it in his honor. It was fitting that observations with the Hubble Space Telescope (and his foundational work on expansion of the universe) contributed to the 2011 Nobel Prize in Physics, given for the discovery that the expansion of the universe is accelerating (a topic we will expand upon in the chapter on The Big Bang).

This OpenStax book is available for free at http://cnx.org/content/col11992/1.8

Download for free at http://cnx.org/content/col11992/latest/

26.2 TYPES OF GALAXIES

Learning Objectives

By the end of this section, you will be able to:

> Describe the properties and features of elliptical, spiral, and irregular galaxies
> Explain what may cause a galaxy's appearance to change over time

Having established the existence of other galaxies, Hubble and others began to observe them more closely—noting their shapes, their contents, and as many other properties as they could measure. This was a daunting task in the 1920s when obtaining a single photograph or spectrum of a galaxy could take a full night of tireless observing. Today, larger telescopes and electronic detectors have made this task less difficult, although observing the most distant galaxies (those that show us the universe in its earliest phases) still requires enormous effort.

The first step in trying to understand a new type of object is often simply to describe it. Remember, the first step in understanding stellar spectra was simply to sort them according to appearance (see Analyzing Starlight). As it turns out, the biggest and most luminous galaxies come in one of two basic shapes: either they are flatter and have spiral arms, like our own Galaxy, or they appear to be elliptical (blimp- or cigar-shaped). Many smaller galaxies, in contrast, have an irregular shape.

Spiral Galaxies

Our own Galaxy and the Andromeda galaxy are typical, large **spiral galaxies** (see Figure 26.2). They consist of a central bulge, a halo, a disk, and spiral arms. Interstellar material is usually spread throughout the disks of spiral galaxies. Bright emission nebulae and hot, young stars are present, especially in the spiral arms, showing that new star formation is still occurring. The disks are often dusty, which is especially noticeable in those systems that we view almost edge on (Figure 26.4).

(a)   (b)

Figure 26.4. Spiral Galaxies. (a) The spiral arms of M100, shown here, are bluer than the rest of the galaxy, indicating young, high-mass stars and star-forming regions. (b) We view this spiral galaxy, NGC 4565, almost exactly edge on, and from this angle, we can see the dust in the plane of the galaxy; it appears dark because it absorbs the light from the stars in the galaxy. (credit a: modification of work by Hubble Legacy Archive, NASA, ESA, and Judy Schmidt; credit b: modification of work by "Jschulman555"/ Wikimedia)

Download for free at http://cnx.org/content/col11992/latest/

In galaxies that we see face on, the bright stars and emission nebulae make the arms of spirals stand out like those of a pinwheel on the fourth of July. Open star clusters can be seen in the arms of nearer spirals, and globular clusters are often visible in their halos. Spiral galaxies contain a mixture of young and old stars, just as the Milky Way does. All spirals rotate, and the direction of their spin is such that the arms appear to trail much like the wake of a boat.

About two-thirds of the nearby spiral galaxies have boxy or peanut-shaped bars of stars running through their centers (Figure 26.5). Showing great originality, astronomers call these galaxies barred spirals.

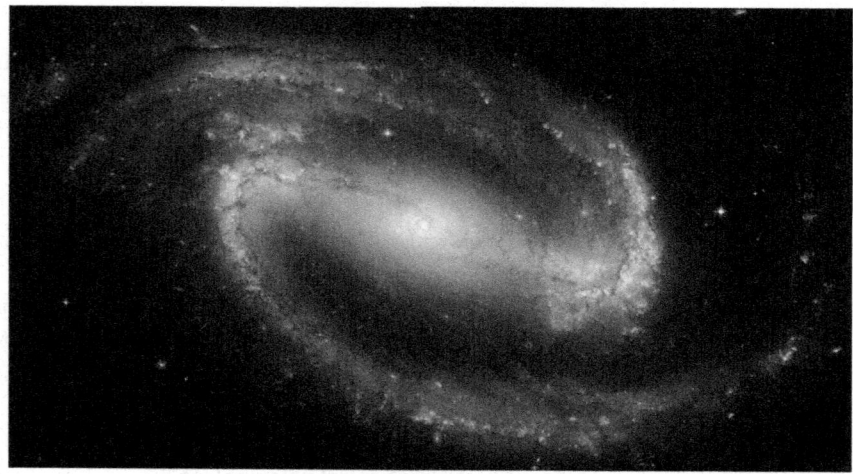

Figure 26.5. Barred Spiral Galaxy. NGC 1300, shown here, is a barred spiral galaxy. Note that the spiral arms begin at the ends of the bar. (credit: NASA, ESA, and the Hubble Heritage Team(STScI/AURA))

As we noted in The Milky Way Galaxy chapter, our Galaxy has a modest bar too (see Figure 25.10). The spiral arms usually begin from the ends of the bar. The fact that bars are so common suggests that they are long lived; it may be that most spiral galaxies form a bar at some point during their evolution.

In both barred and unbarred spiral galaxies, we observe a range of different shapes. At one extreme, the central bulge is large and luminous, the arms are faint and tightly coiled, and bright emission nebulae and supergiant stars are inconspicuous. Hubble, who developed a system of classifying galaxies by shape, gave these galaxies the designation Sa. Galaxies at this extreme may have no clear spiral arm structure, resulting in a lens-like appearance (they are sometimes referred to as lenticular galaxies). These galaxies seem to share as many properties with elliptical galaxies as they do with spiral galaxies

At the other extreme, the central bulge is small and the arms are loosely wound. In these Sc galaxies, luminous stars and emission nebulae are very prominent. Our Galaxy and the Andromeda galaxy are both intermediate between the two extremes. Photographs of spiral galaxies, illustrating the different types, are shown in Figure 26.6, along with elliptical galaxies for comparison.

This OpenStax book is available for free at http://cnx.org/content/col11992/1.8

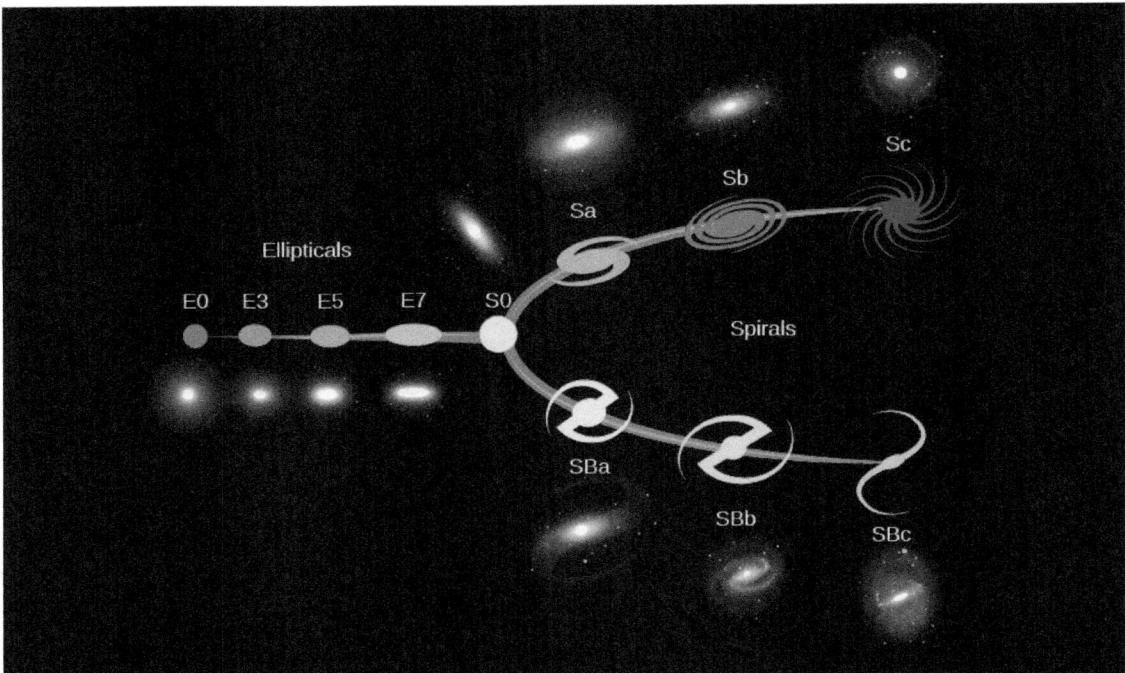

Figure 26.6. Hubble Classification of Galaxies. This figure shows Edwin Hubble's original classification of galaxies. Elliptical galaxies are on the left. On the right, you can see the basic spiral shapes illustrated, alongside images of actual barred and unbarred spirals. (credit: modification of work by NASA, ESA)

The luminous parts of spiral galaxies appear to range in diameter from about 20,000 to more than 100,000 light-years. Recent studies have found that there is probably a large amount of galactic material that extends well beyond the apparent edge of galaxies. This material appears to be thin, cold gas that is difficult to detect in most observations.

From the observational data available, the masses of the visible portions of spiral galaxies are estimated to range from 1 billion to 1 trillion Suns (10^9 to 10^{12} M_{Sun}). The total luminosities of most spirals fall in the range of 100 million to 100 billion times the luminosity of our Sun (10^8 to 10^{11} L_{Sun}). Our Galaxy and M31 are relatively large and massive, as spirals go. There is also considerable dark matter in and around the galaxies, just as there is in the Milky Way; we deduce its presence from how fast stars in the outer parts of the Galaxy are moving in their orbits.

Elliptical Galaxies

Elliptical galaxies consist almost entirely of old stars and have shapes that are spheres or ellipsoids (somewhat squashed spheres) (Figure 26.7). They contain no trace of spiral arms. Their light is dominated by older reddish stars (the population II stars discussed in The Milky Way Galaxy). In the larger nearby ellipticals, many globular clusters can be identified. Dust and emission nebulae are not conspicuous in elliptical galaxies, but many do contain a small amount of interstellar matter.

Download for free at http://cnx.org/content/col11992/latest/

(a)   (b)

Figure 26.7. Elliptical Galaxies. (a) ESO 325-G004 is a giant elliptical galaxy. Other elliptical galaxies can be seen around the edges of this image. (b) This elliptical galaxy probably originated from the collision of two spiral galaxies. (credit a: modification of work by NASA, ESA, and The Hubble Heritage Team (STScI/AURA); credit b: modification of work by ESA/Hubble, NASA)

Elliptical galaxies show various degrees of flattening, ranging from systems that are approximately spherical to those that approach the flatness of spirals. The rare giant ellipticals (for example, ESO 325-G004 in Figure 26.7) reach luminosities of 10^{11} L_{Sun}. The mass in a giant elliptical can be as large as 10^{13} M_{Sun}. The diameters of these large galaxies extend over several hundred thousand light-years and are considerably larger than the largest spirals. Although individual stars orbit the center of an elliptical galaxy, the orbits are not all in the same direction, as occurs in spirals. Therefore, ellipticals don't appear to rotate in a systematic way, making it difficult to estimate how much dark matter they contain.

We find that elliptical galaxies range all the way from the giants, just described, to dwarfs, which may be the most common kind of galaxy. *Dwarf ellipticals* (sometimes called dwarf spheroidals) escaped our notice for a long time because they are very faint and difficult to see. An example of a dwarf elliptical is the Leo I Dwarf Spheroidal galaxy shown in Figure 26.8. The luminosity of this typical dwarf is about equal to that of the brightest globular clusters.

Intermediate between the giant and dwarf elliptical galaxies are systems such as M32 and M110, the two companions of the Andromeda galaxy. While they are often referred to as dwarf ellipticals, these galaxies are significantly larger than galaxies such as Leo I.

This OpenStax book is available for free at http://cnx.org/content/col11992/1.8

Download for free at http://cnx.org/content/col11992/latest/

Figure 26.8. Dwarf Elliptical Galaxy. M32, a dwarf elliptical galaxy and one of the companions to the giant Andromeda galaxy M31. M32 is a dwarf by galactic standards, as it is only 2400 light-years across. (credit: NOAO/AURA/NSF)

Irregular Galaxies

Hubble classified galaxies that do not have the regular shapes associated with the categories we just described into the catchall bin of an **irregular galaxy**, and we continue to use his term. Typically, irregular galaxies have lower masses and luminosities than spiral galaxies. Irregular galaxies often appear disorganized, and many are undergoing relatively intense star formation activity. They contain both young population I stars and old population II stars.

The two best-known irregular galaxies are the Large Magellanic Cloud and Small Magellanic Cloud (Figure 26.9), which are at a distance of a little more than 160,000 light-years away and are among our nearest extragalactic neighbors. Their names reflect the fact that Ferdinand Magellan and his crew, making their round-the-world journey, were the first European travelers to notice them. Although not visible from the United States and Europe, these two systems are prominent from the Southern Hemisphere, where they look like wispy clouds in the night sky. Since they are only about one-tenth as distant as the Andromeda galaxy, they present an excellent opportunity for astronomers to study nebulae, star clusters, variable stars, and other key objects in the setting of another galaxy. For example, the Large Magellanic Cloud contains the 30 Doradus complex (also known as the Tarantula Nebula), one of the largest and most luminous groups of supergiant stars known in any galaxy.

Download for free at http://cnx.org/content/col11992/latest/

Figure 26.9. 4-Meter Telescope at Cerro Tololo Inter-American Observatory Silhouetted against the Southern Sky. The Milky Way is seen to the right of the dome, and the Large and Small Magellanic Clouds are seen to the left. (credit: Roger Smith/NOAO/AURA/NSF)

The Small Magellanic Cloud is considerably less massive than the Large Magellanic Cloud, and it is six times longer than it is wide. This narrow wisp of material points directly toward our Galaxy like an arrow. The Small Magellanic Cloud was most likely contorted into its current shape through gravitational interactions with the Milky Way. A large trail of debris from this interaction between the Milky Way and the Small Magellanic Cloud has been strewn across the sky and is seen as a series of gas clouds moving at abnormally high velocity, known as the Magellanic Stream. We will see that this kind of interaction between galaxies will help explain the irregular shapes of this whole category of small galaxies,

LINK TO LEARNING

View this beautiful album showcasing the different types of galaxies (https://openstaxcollege.org/l/30galaxphohubb) that have been photographed by the Hubble Space Telescope.

Galaxy Evolution

Encouraged by the success of the H-R diagram for stars (see Analyzing Starlight), astronomers studying galaxies hoped to find some sort of comparable scheme, where differences in appearance could be tied to different evolutionary stages in the life of galaxies. Wouldn't it be nice if every elliptical galaxy evolved into a

This OpenStax book is available for free at http://cnx.org/content/col11992/1.8

spiral, for example, just as every main-sequence star evolves into a red giant? Several simple ideas of this kind were tried, some by Hubble himself, but none stood the test of time (and observation).

Because no simple scheme for evolving one type of galaxy into another could be found, astronomers then tended to the opposite point of view. For a while, most astronomers thought that all galaxies formed very early in the history of the universe and that the differences between them had to do with the rate of star formation. Ellipticals were those galaxies in which all the interstellar matter was converted rapidly into stars. Spirals were galaxies in which star formation occurred slowly over the entire lifetime of the galaxy. This idea turned out to be too simple as well.

Today, we understand that at least some galaxies have changed types over the billions of years since the universe began. As we shall see in later chapters, collisions and mergers between galaxies may dramatically change spiral galaxies into elliptical galaxies. Even isolated spirals (with no neighbor galaxies in sight) can change their appearance over time. As they consume their gas, the rate of star formation will slow down, and the spiral arms will gradually become less conspicuous. Over long periods, spirals therefore begin to look more like the galaxies at the middle of Figure 26.6 (which astronomers refer to as S0 types).

Over the past several decades, the study of how galaxies evolve over the lifetime of the universe has become one of the most active fields of astronomical research. We will discuss the evolution of galaxies in more detail in The Evolution and Distribution of Galaxies, but let's first see in a little more detail just what different galaxies are like.

26.3 PROPERTIES OF GALAXIES

Learning Objectives

By the end of this section, you will be able to:

> Describe the methods through which astronomers can estimate the mass of a galaxy
> Characterize each type of galaxy by its mass-to-light ratio

The technique for deriving the masses of galaxies is basically the same as that used to estimate the mass of the Sun, the stars, and our own Galaxy. We measure how fast objects in the outer regions of the galaxy are orbiting the center, and then we use this information along with Kepler's third law to calculate how much mass is inside that orbit.

Masses of Galaxies

Astronomers can measure the rotation speed in spiral galaxies by obtaining spectra of either stars or gas, and looking for wavelength shifts produced by the Doppler effect. Remember that the faster something is moving toward or away from us, the greater the shift of the lines in its spectrum. Kepler's law, together with such observations of the part of the Andromeda galaxy that is bright in visible light, for example, show it to have a galactic mass of about 4×10^{11} M_{Sun} (enough material to make 400 billion stars like the Sun).

The total mass of the Andromeda galaxy is greater than this, however, because we have not included the mass of the material that lies beyond its visible edge. Fortunately, there is a handful of objects—such as isolated stars, star clusters, and satellite galaxies—beyond the visible edge that allows astronomers to estimate how much additional matter is hidden out there. Recent studies show that the amount of dark matter beyond the visible edge of Andromeda may be as large as the mass of the bright portion of the galaxy. Indeed, using Kepler's third law and the velocities of its satellite galaxies, the Andromeda galaxy is estimated to have a mass closer to $1.4 \times$

Download for free at http://cnx.org/content/col11992/latest/

10^{12} M_{Sun}. The mass of the Milky Way Galaxy is estimated to be 8.5×10^{11} M_{Sun}, and so our Milky Way is turning out to be somewhat smaller than Andromeda.

Elliptical galaxies do not rotate in a systematic way, so we cannot determine a rotational velocity; therefore, we must use a slightly different technique to measure their mass. Their stars are still orbiting the galactic center, but not in the organized way that characterizes spirals. Since elliptical galaxies contain stars that are billions of years old, we can assume that the galaxies themselves are not flying apart. Therefore, if we can measure the various speeds with which the stars are moving in their orbits around the center of the galaxy, we can calculate how much mass the galaxy must contain in order to hold the stars within it.

In practice, the spectrum of a galaxy is a composite of the spectra of its many stars, whose different motions produce different Doppler shifts (some red, some blue). The result is that the lines we observe from the entire galaxy contain the combination of many Doppler shifts. When some stars provide blueshifts and others provide redshifts, they create a wider or broader absorption or emission feature than would the same lines in a hypothetical galaxy in which the stars had no orbital motion. Astronomers call this phenomenon line broadening. The amount by which each line broadens indicates the range of speeds at which the stars are moving with respect to the center of the galaxy. The range of speeds depends, in turn, on the force of gravity that holds the stars within the galaxies. With information about the speeds, it is possible to calculate the mass of an elliptical galaxy.

Table 26.1 summarizes the range of masses (and other properties) of the various types of galaxies. Interestingly enough, the most and least massive galaxies are ellipticals. On average, irregular galaxies have less mass than spirals.

Characteristics of the Different Types of Galaxies

Characteristic	Spirals	Ellipticals	Irregulars
Mass (M_{Sun})	10^9 to 10^{12}	10^5 to 10^{13}	10^8 to 10^{11}
Diameter (thousands of light-years)	15 to 150	3 to >700	3 to 30
Luminosity (L_{Sun})	10^8 to 10^{11}	10^6 to 10^{11}	10^7 to 2×10^9
Populations of stars	Old and young	Old	Old and young
Interstellar matter	Gas and dust	Almost no dust; little gas	Much gas; some have little dust, some much dust
Mass-to-light ratio in the visible part	2 to 10	10 to 20	1 to 10
Mass-to-light ratio for total galaxy	100	100	?

Table 26.1

This OpenStax book is available for free at http://cnx.org/content/col11992/1.8

Download for free at http://cnx.org/content/col11992/latest/

Mass-to-Light Ratio

A useful way of characterizing a galaxy is by noting the ratio of its mass (in units of the Sun's mass) to its light output (in units of the Sun's luminosity). This single number tells us roughly what kind of stars make up most of the luminous population of the galaxy, and it also tells us whether a lot of dark matter is present. For stars like the Sun, the **mass-to-light ratio** is 1 by our definition.

Galaxies are not, of course, composed entirely of stars that are identical to the Sun. The overwhelming majority of stars are less massive and less luminous than the Sun, and usually these stars contribute most of the mass of a system without accounting for very much light. The mass-to-light ratio for low-mass stars is greater than 1 (you can verify this using the data in Table 18.3). Therefore, a galaxy's mass-to-light ratio is also generally greater than 1, with the exact value depending on the ratio of high-mass stars to low-mass stars.

Galaxies in which star formation is still occurring have many massive stars, and their mass-to-light ratios are usually in the range of 1 to 10. Galaxies consisting mostly of an older stellar population, such as ellipticals, in which the massive stars have already completed their evolution and have ceased to shine, have mass-to-light ratios of 10 to 20.

But these figures refer only to the inner, conspicuous parts of galaxies (Figure 26.10). In The Milky Way Galaxy and above, we discussed the evidence for dark matter in the outer regions of our own Galaxy, extending much farther from the galactic center than do the bright stars and gas. Recent measurements of the rotation speeds of the outer parts of nearby galaxies, such as the Andromeda galaxy we discussed earlier, suggest that they too have extended distributions of dark matter around the visible disk of stars and dust. This largely invisible matter adds to the mass of the galaxy while contributing nothing to its luminosity, thus increasing the mass-to-light ratio. If dark invisible matter is present in a galaxy, its mass-to-light ratio can be as high as 100. The two different mass-to-light ratios measured for various types of galaxies are given in Table 26.1.

Download for free at http://cnx.org/content/col11992/latest/

Figure 26.10. M101, the Pinwheel Galaxy. This galaxy is a face-on spiral at a distance of 21 million light-years. M101 is almost twice the diameter of the Milky Way, and it contains at least 1 trillion stars. (credit: NASA, ESA, K. Kuntz (Johns Hopkins University), F. Bresolin (University of Hawaii), J. Trauger (Jet Propulsion Lab), J. Mould (NOAO), Y.-H. Chu (University of Illinois, Urbana), and STScI)

These measurements of other galaxies support the conclusion already reached from studies of the rotation of our own Galaxy—namely, that most of the material in the universe cannot at present be observed directly in any part of the electromagnetic spectrum. An understanding of the properties and distribution of this invisible matter is crucial to our understanding of galaxies. It's becoming clearer and clearer that, through the gravitational force it exerts, dark matter plays a dominant role in galaxy formation and early evolution. There is an interesting parallel here between our time and the time during which Edwin Hubble was receiving his training in astronomy. By 1920, many scientists were aware that astronomy stood on the brink of important breakthroughs—if only the nature and behavior of the nebulae could be settled with better observations. In the same way, many astronomers today feel we may be closing in on a far more sophisticated understanding of the large-scale structure of the universe—if only we can learn more about the nature and properties of dark matter. If you follow astronomy articles in the news (as we hope you will), you should be hearing more about dark matter in the years to come.

26.4 | THE EXTRAGALACTIC DISTANCE SCALE

Learning Objectives

By the end of this section, you will be able to:

This OpenStax book is available for free at http://cnx.org/content/col11992/1.8

Download for free at http://cnx.org/content/col11992/latest/

> > Describe the use of variable stars to estimate distances to galaxies
> > Explain how standard bulbs and the Tully-Fisher relation can be used to estimate distances to galaxies

To determine many of the properties of a galaxy, such as its luminosity or size, we must first know how far away it is. If we know the distance to a galaxy, we can convert how bright the galaxy appears to us in the sky into its true luminosity because we know the precise way light is dimmed by distance. (The same galaxy 10 times farther away, for example, would look 100 times dimmer.) But the measurement of galaxy distances is one of the most difficult problems in modern astronomy: all galaxies are far away, and most are so distant that we cannot even make out individual stars in them.

For decades after Hubble's initial work, the techniques used to measure galaxy distances were relatively inaccurate, and different astronomers derived distances that differed by as much as a factor of two. (Imagine if the distance between your home or dorm and your astronomy class were this uncertain; it would be difficult to make sure you got to class on time.) In the past few decades, however, astronomers have devised new techniques for measuring distances to galaxies; most importantly, all of them give the same answer to within an accuracy of about 10%. As we will see, this means we may finally be able to make reliable estimates of the size of the universe.

Variable Stars

Before astronomers could measure distances to other galaxies, they first had to establish the scale of cosmic distances using objects in our own Galaxy. We described the chain of these distance methods in Celestial Distances (and we recommend that you review that chapter if it has been a while since you've read it). Astronomers were especially delighted when they discovered that they could measure distances using certain kinds of intrinsically luminous *variable stars*, such as cepheids, which can be seen at very large distances (Figure 26.11).

After the variables in nearby galaxies had been used to make distance measurements for a few decades, Walter Baade showed that there were actually two kinds of cepheids and that astronomers had been unwittingly mixing them up. As a result, in the early 1950s, the distances to all of the galaxies had to be increased by about a factor of two. We mention this because we want you to bear in mind, as you read on, that science is always a study in progress. Our first tentative steps in such difficult investigations are always subject to future revision as our techniques become more reliable.

The amount of work involved in finding cepheids and measuring their periods can be enormous. Hubble, for example, obtained 350 long-exposure photographs of the Andromeda galaxy over a period of 18 years and was able to identify only 40 cepheids. Even though cepheids are fairly luminous stars, they can be detected in only about 30 of the nearest galaxies with the world's largest ground-based telescopes.

As mentioned in Celestial Distances, one of the main projects carried out during the first years of operation of the Hubble Space Telescope was the measurement of cepheids in more distant galaxies to improve the accuracy of the extragalactic distance scale. Recently, astronomers working with the Hubble Space Telescope have extended such measurements out to 108 million light-years—a triumph of technology and determination.

Download for free at http://cnx.org/content/col11992/latest/

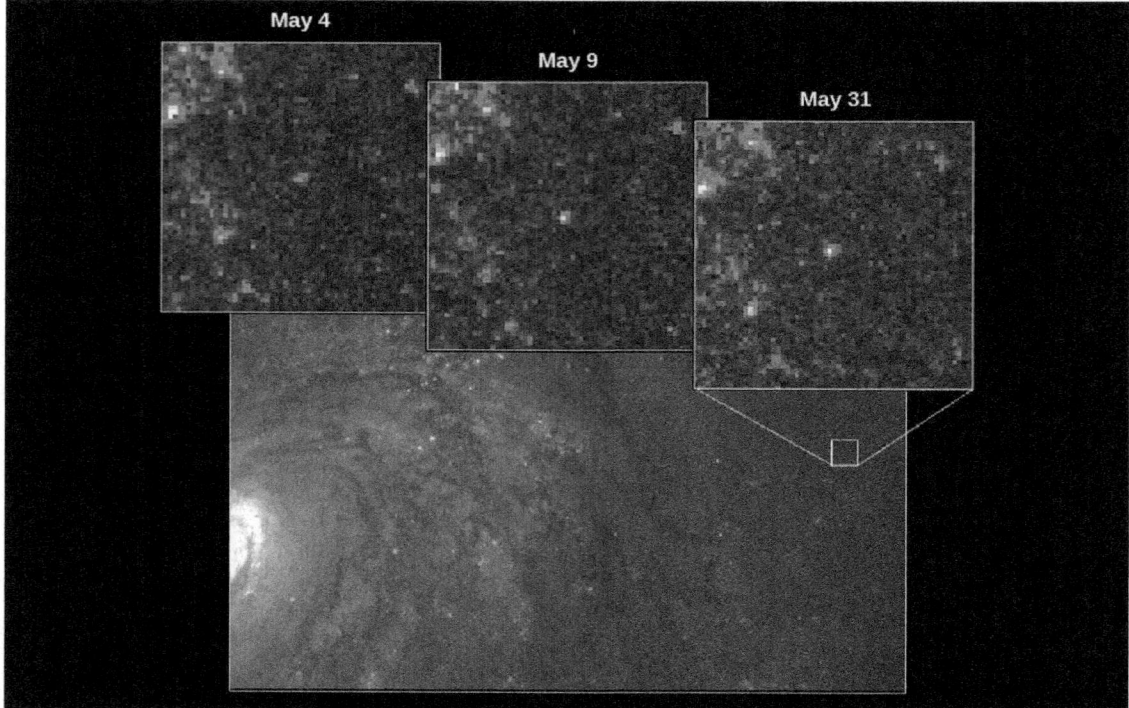

Figure 26.11. Cepheid Variable Star. In 1994, using the Hubble Space Telescope, astronomers were able to make out an individual cepheid variable star in the galaxy M100 and measure its distance to be 56 million light-years. The insets show the star on three different nights; you can see that its brightness is indeed variable. (credit: modification of work by Wendy L. Freedman, Observatories of the Carnegie Institution of Washington, and NASA/ESA)

Nevertheless, we can only use cepheids to measure distances within a small fraction of the universe of galaxies. After all, to use this method, we must be able to resolve single stars and follow their subtle variations. Beyond a certain distance, even our finest space telescopes cannot help us do this. Fortunately, there are other ways to measure the distances to galaxies.

Standard Bulbs

We discussed in Celestial Distances the great frustration that astronomers felt when they realized that the stars in general were not standard *bulbs*. If every light bulb in a huge auditorium is a standard 100-watt bulb, then bulbs that look brighter to us must be closer, whereas those that look dimmer must be farther away. If every star were a standard luminosity (or wattage), then we could similarly "read off" their distances based on how bright they appear to us. Alas, as we have learned, neither stars nor galaxies come in one standard-issue luminosity. Nonetheless, astronomers have been searching for objects out there that do act in some way like a standard bulb—that have the same intrinsic (built-in) brightness wherever they are.

A number of suggestions have been made for what sorts of objects might be effective standard bulbs, including the brightest supergiant stars, planetary nebulae (which give off a lot of ultraviolet radiation), and the average globular cluster in a galaxy. One object turns out to be particularly useful: the **type Ia supernova**. These supernovae involve the explosion of a white dwarf in a binary system (see The Evolution of Binary Star Systems) Observations show that supernovae of this type all reach nearly the same luminosity (about $4.5 \times 10^9 \; L_{Sun}$) at maximum light. With such tremendous luminosities, these supernovae have been detected out to

This OpenStax book is available for free at http://cnx.org/content/col11992/1.8

a distance of more than 8 billion light-years and are therefore especially attractive to astronomers as a way of determining distances on a large scale (Figure 26.12).

Figure 26.12. Type Ia Supernova. The bright object at the bottom left of center is a type Ia supernova near its peak intensity. The supernova easily outshines its host galaxy. This extreme increase and luminosity help astronomers use Ia supernova as standard bulbs. (credit: NASA, ESA, A. Riess (STScI))

Several other kinds of standard bulbs visible over great distances have also been suggested, including the overall brightness of, for example, giant ellipticals and the brightest member of a galaxy cluster. Type Ia supernovae, however, have proved to be the most accurate standard bulbs, and they can be seen in more distant galaxies than the other types of calibrators. As we will see in the chapter on The Big Bang, observations of this type of supernova have profoundly changed our understanding of the evolution of the universe.

Other Measuring Techniques

Another technique for measuring galactic distances makes use of an interesting relationship noticed in the late 1970s by Brent Tully of the University of Hawaii and Richard Fisher of the National Radio Astronomy Observatory. They discovered that the luminosity of a spiral galaxy is related to its rotational velocity (how fast it spins). Why would this be true?

The more mass a galaxy has, the faster the objects in its outer regions must orbit. A more massive galaxy has more stars in it and is thus more luminous (ignoring dark matter for a moment). Thinking back to our discussion from the previous section, we can say that if the mass-to-light ratios for various spiral galaxies are pretty similar, then we can estimate the luminosity of a spiral galaxy by measuring its mass, and we can estimate its mass by measuring its rotational velocity.

Tully and Fisher used the 21-cm line of cold hydrogen gas to determine how rapidly material in spiral galaxies is orbiting their centers (you can review our discussion of the 21-cm line in Between the Stars: Gas and Dust in Space). Since 21-cm radiation from stationary atoms comes in a nice narrow line, the width of the 21-cm line produced by a whole rotating galaxy tells us the range of orbital velocities of the galaxy's hydrogen gas. The broader the line, the faster the gas is orbiting in the galaxy, and the more massive and luminous the galaxy turns out to be.

Download for free at http://cnx.org/content/col11992/latest/

It is somewhat surprising that this technique works, since much of the mass associated with galaxies is dark matter, which does not contribute at all to the luminosity but does affect the rotation speed. There is also no obvious reason why the mass-to-light ratio should be similar for all spiral galaxies. Nevertheless, observations of nearer galaxies (where we have other ways of measuring distance) show that measuring the rotational velocity of a galaxy provides an accurate estimate of its intrinsic luminosity. Once we know how luminous the galaxy really is, we can compare the luminosity to the apparent brightness and use the difference to calculate its distance.

While the Tully-Fisher relation works well, it is limited—we can only use it to determine the distance to a spiral galaxy. There are other methods that can be used to estimate the distance to an elliptical galaxy; however, those methods are beyond the scope of our introductory astronomy course.

Table 26.2 lists the type of galaxy for which each of the distance techniques is useful, and the range of distances over which the technique can be applied.

Some Methods for Estimating Distance to Galaxies

Method	Galaxy Type	Approximate Distance Range (millions of light-years)
Planetary nebulae	All	0–70
Cepheid variables	Spiral, irregulars	0–110
Tully-Fisher relation	Spiral	0–300
Type Ia supernovae	All	0–11,000
Redshifts (Hubble's law)	All	300–13,000

Table 26.2

26.5 | THE EXPANDING UNIVERSE

Learning Objectives

By the end of this section, you will be able to:

> Describe the discovery that galaxies getting farther apart as the universe evolves
> Explain how to use Hubble's law to determine distances to remote galaxies
> Describe models for the nature of an expanding universe
> Explain the variation in Hubble's constant

We now come to one of the most important discoveries ever made in astronomy—the fact that the universe is expanding. Before we describe how the discovery was made, we should point out that the first steps in the study of galaxies came at a time when the techniques of spectroscopy were also making great strides. Astronomers using large telescopes could record the spectrum of a faint star or galaxy on photographic plates, guiding their telescopes so they remained pointed to the same object for many hours and collected more light. The resulting spectra of galaxies contained a wealth of information about the composition of the galaxy and the velocities of these great star systems.

This OpenStax book is available for free at http://cnx.org/content/col11992/1.8

Slipher's Pioneering Observations

Curiously, the discovery of the expansion of the universe began with the search for Martians and other solar systems. In 1894, the controversial (and wealthy) astronomer Percival Lowell established an observatory in Flagstaff, Arizona, to study the planets and search for life in the universe. Lowell thought that the spiral nebulae might be solar systems in the process of formation. He therefore asked one of the observatory's young astronomers, Vesto M. Slipher (Figure 26.13), to photograph the spectra of some of the spiral nebulae to see if their spectral lines might show chemical compositions like those expected for newly forming planets.

Figure 26.13. Vesto M. Slipher (1875–1969). Slipher spent his entire career at the Lowell Observatory, where he discovered the large radial velocities of galaxies. (credit: Lowell Observatory)

The Lowell Observatory's major instrument was a 24-inch refracting telescope, which was not at all well suited to observations of faint spiral nebulae. With the technology available in those days, photographic plates had to be exposed for 20 to 40 hours to produce a good spectrum (in which the positions of the lines could reveal a galaxy's motion). This often meant continuing to expose the same photograph over several nights. Beginning in 1912, and making heroic efforts over a period of about 20 years, Slipher managed to photograph the spectra of more than 40 of the spiral nebulae (which would all turn out to be galaxies).

To his surprise, the spectral lines of most galaxies showed an astounding **redshift**. By "redshift" we mean that the lines in the spectra are displaced toward longer wavelengths (toward the red end of the visible spectrum). Recall from the chapter on Radiation and Spectra that a redshift is seen when the source of the waves is moving away from us. Slipher's observations showed that most spirals are racing away at huge speeds; the highest velocity he measured was 1800 kilometers per second.

Only a few spirals—such as the Andromeda and Triangulum Galaxies and M81—all of which are now known to be our close neighbors, turned out to be approaching us. All the other galaxies were moving away. Slipher first announced this discovery in 1914, years before Hubble showed that these objects were other galaxies and before anyone knew how far away they were. No one at the time quite knew what to make of this discovery.

Hubble's Law

The profound implications of Slipher's work became apparent only during the 1920s. Georges Lemaître was a Belgian priest and a trained astronomer. In 1927, he published a paper in French in an obscure Belgian journal in which he suggested that we live in an expanding universe. The title of the paper (translated into English) is "A Homogenous Universe of Constant Mass and Growing Radius Accounting for the Radial Velocity of Extragalactic

Download for free at http://cnx.org/content/col11992/latest/

Nebulae." Lemaître had discovered that Einstein's equations of relativity were consistent with an expanding universe (as had the Russian scientist Alexander Friedmann independently in 1922). Lemaître then went on to use Slipher's data to support the hypothesis that the universe actually is expanding and to estimate the rate of expansion. Initially, scientists paid little attention to this paper, perhaps because the Belgian journal was not widely available.

In the meantime, Hubble was making observations of galaxies with the 2.5-meter telescope on Mt. Wilson, which was then the world's largest. Hubble carried out the key observations in collaboration with a remarkable man, Milton Humason, who dropped out of school in the eighth grade and began his astronomical career by driving a mule train up the trail on Mount Wilson to the observatory (Figure 26.14). In those early days, supplies had to be brought up that way; even astronomers hiked up to the mountaintop for their turns at the telescope. Humason became interested in the work of the astronomers and, after marrying the daughter of the observatory's electrician, took a job as janitor there. After a time, he became a night assistant, helping the astronomers run the telescope and record data. Eventually, he made such a mark that he became a full astronomer at the observatory.

Figure 26.14. Milton Humason (1891–1972). Humason was Hubble's collaborator on the great task of observing, measuring, and classifying the characteristics of many galaxies. (credit: Caltech Archives)

By the late 1920s, Humason was collaborating with Hubble by photographing the spectra of faint galaxies with the 2.5-meter telescope. (By then, there was no question that the spiral nebulae were in fact galaxies.) Hubble had found ways to improve the accuracy of the estimates of distances to spiral galaxies, and he was able to measure much fainter and more distant galaxies than Slipher could observe with his much-smaller telescope. When Hubble laid his own distance estimates next to measurements of the recession velocities (the speed with which the galaxies were moving away), he found something stunning: there was a relationship between distance and velocity for galaxies. *The more distant the galaxy, the faster it was receding from us.*

In 1931, Hubble and Humason jointly published the seminal paper where they compared distances and velocities of remote galaxies moving away from us at speeds as high as 20,000 kilometers per second and were able to show that the recession velocities of galaxies are directly proportional to their distances from us (Figure 26.15), just as Lemaître had suggested.

This OpenStax book is available for free at http://cnx.org/content/col11992/1.8

Download for free at http://cnx.org/content/col11992/latest/

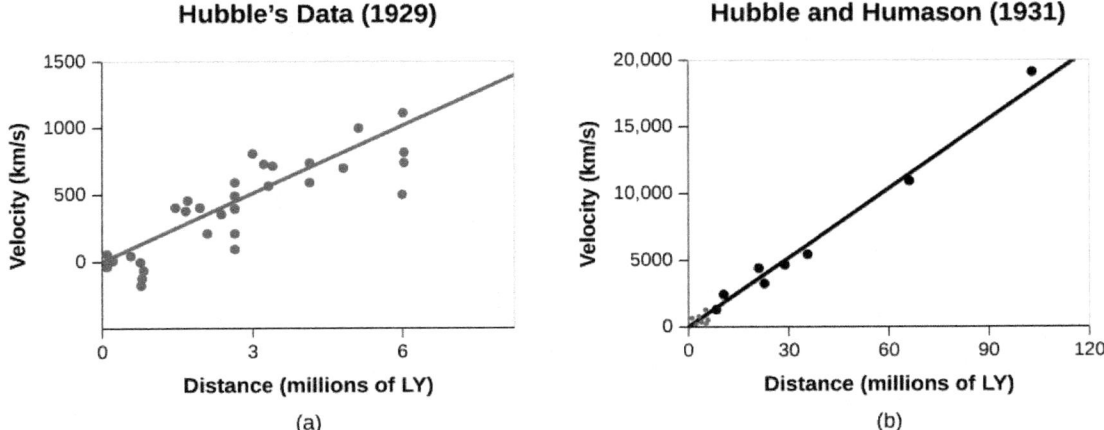

Figure 26.15. Hubble's Law. (a) These data show Hubble's original velocity-distance relation, adapted from his 1929 paper in the *Proceedings of the National Academy of Sciences*. (b) These data show Hubble and Humason's velocity-distance relation, adapted from their 1931 paper in *The Astrophysical Journal*. The red dots at the lower left are the points in the diagram in the 1929 paper. Comparison of the two graphs shows how rapidly the determination of galactic distances and redshifts progressed in the 2 years between these publications.

We now know that this relationship holds for every galaxy except a few of the nearest ones. Nearly all of the galaxies that are approaching us turn out to be part of the Milky Way's own group of galaxies, which have their own individual motions, just as birds flying in a group may fly in slightly different directions at slightly different speeds even though the entire flock travels through space together.

Written as a formula, the relationship between velocity and distance is

$$V = H \times d$$

where *v* is the recession speed, *d* is the distance, and *H* is a number called the **Hubble constant**. This equation is now known as **Hubble's law**.

ASTRONOMY BASICS

Constants of Proportionality

Mathematical relationships such as Hubble's law are pretty common in life. To take a simple example, suppose your college or university hires you to call rich alumni and ask for donations. You are paid $2.50 for each call; the more calls you can squeeze in between studying astronomy and other courses, the more money you take home. We can set up a formula that connects *p*, your pay, and *n*, the number of calls

$$p = A \times n$$

where *A* is the alumni constant, with a value of $2.50. If you make 20 calls, you will earn $2.50 times 20, or $50.

Suppose your boss forgets to tell you what you will get paid for each call. You can calculate the alumni constant that governs your pay by keeping track of how many calls you make and noting your gross pay each week. If you make 100 calls the first week and are paid $250, you can deduce that the constant is

Download for free at http://cnx.org/content/col11992/latest/

$2.50 (in units of dollars per call). Hubble, of course, had no "boss" to tell him what his constant would be—he *had* to calculate its value from the measurements of distance and velocity.

Astronomers express the value of Hubble's constant in units that relate to how they measure speed and velocity for galaxies. In this book, we will use kilometers per second per million light-years as that unit. For many years, estimates of the value of the Hubble constant have been in the range of 15 to 30 kilometers per second per million light-years The most recent work appears to be converging on a value near 22 kilometers per second per million light-years If *H* is 22 kilometers per second per million light-years, a galaxy moves away from us at a speed of 22 kilometers per second for every million light-years of its distance. As an example, a galaxy 100 million light-years away is moving away from us at a speed of 2200 kilometers per second.

Hubble's law tells us something fundamental about the universe. Since all but the nearest galaxies appear to be in motion away from us, with the most distant ones moving the fastest, we must be living in an expanding universe. We will explore the implications of this idea shortly, as well as in the final chapters of this text. For now, we will just say that Hubble's observation underlies all our theories about the origin and evolution of the universe.

Hubble's Law and Distances

The regularity expressed in Hubble's law has a built-in bonus: it gives us a new way to determine the distances to remote galaxies. First, we must reliably establish Hubble's constant by measuring both the distance and the velocity of many galaxies in many directions to be sure Hubble's law is truly a universal property of galaxies. But once we have calculated the value of this constant and are satisfied that it applies everywhere, much more of the universe opens up for distance determination. Basically, if we can obtain a spectrum of a galaxy, we can immediately tell how far away it is.

The procedure works like this. We use the spectrum to measure the speed with which the galaxy is moving away from us. If we then put this speed and the Hubble constant into Hubble's law equation, we can solve for the distance.

EXAMPLE 26.1

Hubble's Law

Hubble's law ($v = H \times d$) allows us to calculate the distance to any galaxy. Here is how we use it in practice.

We have measured Hubble's constant to be 22 km/s per million light-years. This means that if a galaxy is 1 million light-years farther away, it will move away 22 km/s faster. So, if we find a galaxy that is moving away at 18,000 km/s, what does Hubble's law tells us about the distance to the galaxy?

Solution

$$d = \frac{v}{H} = \frac{18,000 \text{ km/s}}{\frac{22 \text{ km/s}}{1 \text{ million light-years}}} = \frac{18,000}{22} \times \frac{1 \text{ million light-years}}{1} = 818 \text{ million light-years}$$

This OpenStax book is available for free at http://cnx.org/content/col11992/1.8

Download for free at http://cnx.org/content/col11992/latest/

Note how we handled the units here: the km/s in the numerator and denominator cancel, and the factor of million light-years in the denominator of the constant must be divided correctly before we get our distance of 818 million light-years.

Check Your Learning

Using 22 km/s/million light-years for Hubble's constant, what recessional velocity do we expect to find if we observe a galaxy at 500 million light-years?

Answer:

$$v = d \times H = 500 \text{ million light-years} \times \frac{22 \text{ km/s}}{1 \text{ million light-years}} = 11,000 \text{ km/s}$$

Variation of Hubble's Constant

The use of redshift is potentially a very important technique for determining distances because as we have seen, most of our methods for determining galaxy distances are limited to approximately the nearest few hundred million light-years (and they have large uncertainties at these distances). The use of Hubble's law as a distance indicator requires only a spectrum of a galaxy and a measurement of the Doppler shift, and with large telescopes and modern spectrographs, spectra can be taken of extremely faint galaxies.

But, as is often the case in science, things are not so simple. This technique works if, and only if, the Hubble constant has been truly constant throughout the entire life of the universe. When we observe galaxies billions of light-years away, we are seeing them as they were billions of years ago. What if the Hubble "constant" was different billions of years ago? Before 1998, astronomers thought that, although the universe is expanding, the expansion should be slowing down, or decelerating, because the overall gravitational pull of all matter in the universe would have a dominant, measureable effect. If the expansion is decelerating, then the Hubble constant should be decreasing over time.

The discovery that type Ia supernovae are standard bulbs gave astronomers the tool they needed to observe extremely distant galaxies and measure the rate of expansion billions of years ago. The results were completely unexpected. It turns out that the expansion of the universe is *accelerating* over time! What makes this result so astounding is that there is no way that existing physical theories can account for this observation. While a decelerating universe could easily be explained by gravity, there was no force or property in the universe known to astronomers that could account for the acceleration. In The Big Bang chapter, we will look in more detail at the observations that led to this totally unexpected result and explore its implications for the ultimate fate of the universe.

In any case, if the Hubble constant is not really a constant when we look over large spans of space and time, then the calculation of galaxy distances using the Hubble constant won't be accurate. As we shall see in the chapter on The Big Bang, the accurate calculation of distances requires a model for how the Hubble constant has changed over time. The farther away a galaxy is (and the longer ago we are seeing it), the more important it is to include the effects of the change in the Hubble constant. For galaxies within a few billion light-years, however, the assumption that the Hubble constant is indeed constant gives good estimates of distance.

Models for an Expanding Universe

At first, thinking about Hubble's law and being a fan of the work of Copernicus and Harlow Shapley, you might be shocked. Are all the galaxies really moving *away from us*? Is there, after all, something special about our

Download for free at http://cnx.org/content/col11992/latest/

position in the universe? Worry not; the fact that galaxies are receding from us and that more distant galaxies are moving away more rapidly than nearby ones shows only that the universe is expanding uniformly.

A uniformly expanding universe is one that is expanding at the same rate everywhere. In such a universe, we and all other observers, no matter where they are located, must observe a proportionality between the velocities and distances of equivalently remote galaxies. (Here, we are ignoring the fact that the Hubble constant is not constant over all time, but if at any given time in the evolution of the universe the Hubble constant has the same value everywhere, this argument still works.)

To see why, first imagine a ruler made of stretchable rubber, with the usual lines marked off at each centimeter. Now suppose someone with strong arms grabs each end of the ruler and slowly stretches it so that, say, it doubles in length in 1 minute (Figure 26.16). Consider an intelligent ant sitting on the mark at 2 centimeters—a point that is not at either end nor in the middle of the ruler. He measures how fast other ants, sitting at the 4-, 7-, and 12-centimeter marks, move away from him as the ruler stretches.

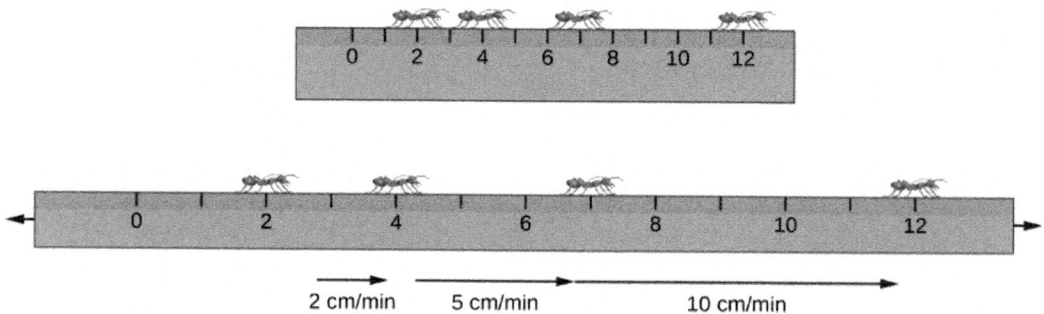

Figure 26.16. Stretching a Ruler. Ants on a stretching ruler see other ants move away from them. The speed with which another ant moves away is proportional to its distance.

The ant at 4 centimeters, originally 2 centimeters away from our ant, has doubled its distance in 1 minute; it therefore moved away at a speed of 2 centimeters per minute. The ant at the 7-centimeters mark, which was originally 5 centimeters away from our ant, is now 10 centimeters away; it thus had to move at 5 centimeters per minute. The one that started at the 12-centimeters mark, which was 10 centimeters away from the ant doing the counting, is now 20 centimeters away, meaning it must have raced away at a speed of 10 centimeters per minute. Ants at different distances move away at different speeds, and their speeds are proportional to their distances (just as Hubble's law indicates for galaxies). Yet, notice in our example that all the ruler was doing was stretching uniformly. Also, notice that none of the ants were actually moving of their own accord, it was the stretching of the ruler that moved them apart.

Now let's repeat the analysis, but put the intelligent ant on some other mark—say, on 7 or 12 centimeters. We discover that, as long as the ruler stretches uniformly, this ant also finds every other ant moving away at a speed proportional to its distance. In other words, the kind of relationship expressed by Hubble's law can be explained by a uniform stretching of the "world" of the ants. And all the ants in our simple diagram will see the other ants moving away from them as the ruler stretches.

For a three-dimensional analogy, let's look at the loaf of raisin bread in Figure 26.17. The chef has accidentally put too much yeast in the dough, and when she sets the bread out to rise, it doubles in size during the next hour, causing all the raisins to move farther apart. On the figure, we again pick a representative raisin (that is not at the edge or the center of the loaf) and show the distances from it to several others in the figure (before and after the loaf expands).

This OpenStax book is available for free at http://cnx.org/content/col11992/1.8

Download for free at http://cnx.org/content/col11992/latest/

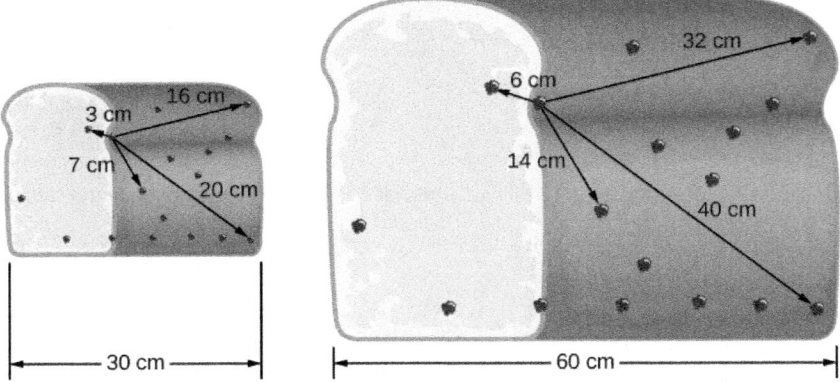

Figure 26.17. Expanding Raisin Bread. As the raisin bread rises, the raisins "see" other raisins moving away. More distant raisins move away faster in a uniformly expanding bread.

Measure the increases in distance and calculate the speeds for yourself on the raisin bread, just like we did for the ruler. You will see that, since each distance doubles during the hour, each raisin moves away from our selected raisin at a speed proportional to its distance. The same is true no matter which raisin you start with.

Our two analogies are useful for clarifying our thinking, but you must not take them literally. On both the ruler and the raisin bread, there are points that are at the end or edge. You can use these to pinpoint the middle of the ruler and the loaf. While our models of the universe have some resemblance to the properties of the ruler and the loaf, the universe has no boundaries, no edges, and no center (all mind-boggling ideas that we will discuss in a later chapter).

What is useful to notice about both the ants and the raisins is that they themselves did not "cause" their motion. It isn't as if the raisins decided to take a trip away from each other and then hopped on a hoverboard to get away. No, in both our analogies, it was the stretching of the medium (the ruler or the bread) that moved the ants or the raisins farther apart. In the same way, we will see in The Big Bang chapter that the galaxies don't have rocket motors propelling them away from each other. Instead, they are passive participants in the *expansion of space*. As space stretches, the galaxies are carried farther and farther apart much as the ants and the raisins were. (If this notion of the "stretching" of space surprises or bothers you, now would be a good time to review the information about spacetime in Black Holes and Curved Spacetime. We will discuss these ideas further as our discussion broadens from galaxies to the whole universe.)

The expansion of the universe, by the way, does not imply that the individual galaxies and clusters of galaxies themselves are expanding. Neither raisins nor the ants in our analogy grow in size as the loaf expands. Similarly, gravity holds galaxies and clusters of galaxies together, and they get farther away from each other—without themselves changing in size—as the universe expands.

Download for free at http://cnx.org/content/col11992/latest/

CHAPTER 26 REVIEW

 KEY TERMS

elliptical galaxy a galaxy whose shape is an ellipse and that contains no conspicuous interstellar material

Hubble constant a constant of proportionality in the law relating the velocities of remote galaxies to their distances

Hubble's law a rule that the radial velocities of remove galaxies are proportional to their distances from us

irregular galaxy a galaxy without any clear symmetry or pattern; neither a spiral nor an elliptical galaxy

mass-to-light ratio the ratio of the total mass of a galaxy to its total luminosity, usually expressed in units of solar mass and solar luminosity; the mass-to-light ratio gives a rough indication of the types of stars contained within a galaxy and whether or not substantial quantities of dark matter are present

redshift when lines in the spectra are displaced toward longer wavelengths (toward the red end of the visible spectrum)

spiral galaxy a flattened, rotating galaxy with pinwheel-like arms of interstellar material and young stars, winding out from its central bulge

type Ia supernova a supernova formed by the explosion of a white dwarf in a binary system and reach a luminosity of about $4.5 \times 10^9 \, L_{Sun}$; can be used to determine distances to galaxies on a large scale

 SUMMARY

26.1 The Discovery of Galaxies

Faint star clusters, clouds of glowing gas, and galaxies all appeared as faint patches of light (or nebulae) in the telescopes available at the beginning of the twentieth century. It was only when Hubble measured the distance to the Andromeda galaxy using cepheid variables with the giant 2.5-meter reflector on Mount Wilson in 1924 that the existence of other galaxies similar to the Milky Way in size and content was established.

26.2 Types of Galaxies

The majority of bright galaxies are either spirals or ellipticals. Spiral galaxies contain both old and young stars, as well as interstellar matter, and have typical masses in the range of 10^9 to $10^{12} \, M_{Sun}$. Our own Galaxy is a large spiral. Ellipticals are spheroidal or slightly elongated systems that consist almost entirely of old stars, with very little interstellar matter. Elliptical galaxies range in size from giants, more massive than any spiral, down to dwarfs, with masses of only about $10^6 \, M_{Sun}$. Dwarf ellipticals are probably the most common type of galaxy in the nearby universe. A small percentage of galaxies with more disorganized shapes are classified as irregulars. Galaxies may change their appearance over time due to collisions with other galaxies or by a change in the rate of star formation.

26.3 Properties of Galaxies

The masses of spiral galaxies are determined from measurements of their rates of rotation. The masses of elliptical galaxies are estimated from analyses of the motions of the stars within them. Galaxies can be characterized by their mass-to-light ratios. The luminous parts of galaxies with active star formation typically have mass-to-light ratios in the range of 1 to 10; the luminous parts of elliptical galaxies, which contain only old

This OpenStax book is available for free at http://cnx.org/content/col11992/1.8

stars, typically have mass-to-light ratios of 10 to 20. The mass-to-light ratios of whole galaxies, including their outer regions, are as high as 100, indicating the presence of a great deal of dark matter.

26.4 The Extragalactic Distance Scale

Astronomers determine the distances to galaxies using a variety of methods, including the period-luminosity relationship for cepheid variables; objects such as type Ia supernovae, which appear to be standard bulbs; and the Tully-Fisher relation, which connects the line broadening of 21-cm radiation to the luminosity of spiral galaxies. Each method has limitations in terms of its precision, the kinds of galaxies with which it can be used, and the range of distances over which it can be applied.

26.5 The Expanding Universe

The universe is expanding. Observations show that the spectral lines of distant galaxies are redshifted, and that their recession velocities are proportional to their distances from us, a relationship known as Hubble's law. The rate of recession, called the Hubble constant, is approximately 22 kilometers per second per million light-years. We are not at the center of this expansion: an observer in any other galaxy would see the same pattern of expansion that we do. The expansion described by Hubble's law is best understood as a stretching of space.

FOR FURTHER EXPLORATION

Articles

Andrews, B. "What Are Galaxies Trying to Tell Us?" *Astronomy* (February 2011): 24. Introduction to our understanding of the shapes and evolution of different types of galaxies.

Bothun, G. "Beyond the Hubble Sequence." *Sky & Telescope* (May 2000): 36. History and updating of Hubble's classification scheme.

Christianson, G. "Mastering the Universe." *Astronomy* (February 1999): 60. Brief introduction to Hubble's life and work.

Dalcanton, J. "The Overlooked Galaxies." *Sky & Telescope* (April 1998): 28. On low-brightness galaxies, which have been easy to miss.

Freedman, W. "The Expansion Rate and Size of the Universe." *Scientific American* (November 1992): 76.

Hodge, P. "The Extragalactic Distance Scale: Agreement at Last?" *Sky & Telescope* (October 1993): 16.

Jones, B. "The Legacy of Edwin Hubble." *Astronomy* (December 1989): 38.

Kaufmann, G. and van den Bosch, F. "The Life Cycle of Galaxies." *Scientific American* (June 2002): 46. On galaxy evolution and how it leads to the different types of galaxies.

Martin, P. and Friedli, D. "At the Hearts of Barred Galaxies." *Sky & Telescope* (March 1999): 32. On barred spirals.

Osterbrock, D. "Edwin Hubble and the Expanding Universe." *Scientific American* (July 1993): 84.

Russell, D. "Island Universes from Wright to Hubble." *Sky & Telescope* (January 1999) 56. A history of our discovery of galaxies.

Smith, R. "The Great Debate Revisited." *Sky & Telescope* (January 1983): 28. On the Shapley-Curtis debate concerning the extent of the Milky Way and the existence of other galaxies.

Download for free at http://cnx.org/content/col11992/latest/

Websites

ABC's of Distance: http://www.astro.ucla.edu/~wright/distance.htm. A concise summary by astronomer Ned Wright of all the different methods we use to get distances in astronomy.

Cosmic Times 1929: http://cosmictimes.gsfc.nasa.gov/online_edition/1929Cosmic/index.html. NASA project explaining Hubble's work and surrounding discoveries as if you were reading newspaper articles.

Edwin Hubble: The Man Behind the Name: https://www.spacetelescope.org/about/history/the_man_behind_the_name/. Concise biography from the people at the Hubble Space Telescope.

Edwin Hubble: http://apod.nasa.gov/diamond_jubilee/d_1996/sandage_hubble.html. An article on the life and work of Hubble by his student and successor, Allan Sandage. A bit technical in places, but giving a real picture of the man and the science.

NASA Science: Introduction to Galaxies: http://science.nasa.gov/astrophysics/focus-areas/what-are-galaxies/. A brief overview with links to other pages, and recent Hubble Space Telescope discoveries.

National Optical Astronomy Observatories Gallery of Galaxies: https://www.noao.edu/image_gallery/galaxies.html. A collection of images and information about galaxies and galaxy groups of different types. Another impressive archive can be found at the European Southern Observatory site: https://www.eso.org/public/images/archive/category/galaxies/.

Sloan Digital Sky Survey: Introduction to Galaxies: http://skyserver.sdss.org/dr1/en/astro/galaxies/galaxies.asp. Another brief overview.

Universe Expansion: http://hubblesite.org/newscenter/archive/releases/1999/19. The background material here provides a nice chronology of how we discovered and measured the expansion of the universe.

Videos

Edwin Hubble (Hubblecast Episode 89): http://www.spacetelescope.org/videos/hubblecast89a/. (5:59).

Galaxies: An Introduction: https://www.youtube.com/watch?v=HYYgangrkZg. A compilation of several short European videos that first describe galaxies in general and then focus on galaxies in Hubble telescope images (12:48).

Hubble's Views of the Deep Universe: https://www.youtube.com/watch?v=argR2U15w-M. A 2015 public talk by Brandon Lawton of the Space Telescope Science Institute about galaxies and beyond (1:26:20).

🚹 COLLABORATIVE GROUP ACTIVITIES

A. Throughout much of the last century, the 100-inch telescope on Mt. Wilson (completed in 1917) and the 200-inch telescope on Palomar Mountain (completed in 1948) were the only ones large enough to obtain spectra of faint galaxies. Only a handful of astronomers (all male—since, until the 1960s, women were not given time on these two telescopes) were allowed to use these facilities, and in general the observers did not compete with each other but worked on different problems. Now there are many other telescopes, and several different groups do often work on the same problem. For example, two different groups have independently developed the techniques for using supernovae to determine the distances to galaxies at high redshifts. Which approach do you think is better for the field of astronomy? Which is more cost effective? Why?

This OpenStax book is available for free at http://cnx.org/content/col11992/1.8

Download for free at http://cnx.org/content/col11992/latest/

B. A distant relative, whom you invite to dinner so you can share all the exciting things you have learned in your astronomy class, says he does not believe that other galaxies are made up of stars. You come back to your group and ask them to help you respond. What kinds of measurements would you make to show that other galaxies are composed of stars?

C. Look at Figure 26.1 with your group. What does the difference in color between the spiral arms and the bulge of Andromeda tell you about the difference in the types of stars that populate these two regions of the galaxy? Which side of the galaxy is closer to us? Why?

D. What is your reaction to reading about the discovery of the expanding universe? Discuss how the members of the group feel about a universe "in motion." Einstein was not comfortable with the notion of a universe that had some overall movement to it, instead of being at rest. He put a kind of "fudge factor" into his equations of general relativity for the universe as a whole to keep it from moving (although later, hearing about Hubble and Humason's work, he called it "the greatest blunder" he ever made). Do you share Einstein's original sense that this is not the kind of universe you feel comfortable with? What do you think could have caused space to be expanding?

E. In science fiction, characters sometimes talk about visiting other galaxies. Discuss with your group how realistic this idea is. Even if we had fast spaceships (traveling close to the speed of light, the speed limit of the universe) how likely are we to be able to reach another galaxy? Why?

F. Despite his son's fascination with astronomy in college, Edwin Hubble's father did not want him to go into astronomy as a profession. He really wanted his son to be a lawyer and pushed him hard to learn the law when he won a fellowship to study abroad. Hubble eventually defied his father and went into astronomy, becoming, as you learned in this chapter, one of the most important astronomers of all time. His dad didn't live to see his son's remarkable achievements. Do you think he would have reconciled himself to his son's career choice if he had? Do you or does anyone in your group or among your friends have to face a choice between the passion in your heart and what others want you to do? Discuss how people in college today are dealing with such choices.

⎙ EXERCISES

Review Questions

1. Describe the main distinguishing features of spiral, elliptical, and irregular galaxies.

2. Why did it take so long for the existence of other galaxies to be established?

3. Explain what the mass-to-light ratio is and why it is smaller in spiral galaxies with regions of star formation than in elliptical galaxies.

4. If we now realize dwarf ellipticals are the most common type of galaxy, why did they escape our notice for so long?

5. What are the two best ways to measure the distance to a nearby spiral galaxy, and how would it be measured?

6. What are the two best ways to measure the distance to a distant, isolated spiral galaxy, and how would it be measured?

7. Why is Hubble's law considered one of the most important discoveries in the history of astronomy?

Download for free at http://cnx.org/content/col11992/latest/

8. What does it mean to say that the universe is expanding? What is expanding? For example, is your astronomy classroom expanding? Is the solar system? Why or why not?

9. Was Hubble's original estimate of the distance to the Andromeda galaxy correct? Explain.

10. Does an elliptical galaxy rotate like a spiral galaxy? Explain.

11. Why does the disk of a spiral galaxy appear dark when viewed edge on?

12. What causes the largest mass-to-light ratio: gas and dust, dark matter, or stars that have burnt out?

13. What is the most useful standard bulb method for determining distances to galaxies?

14. When comparing two isolated spiral galaxies that have the same apparent brightness, but rotate at different rates, what can you say about their relative luminosity?

15. If all distant galaxies are expanding away from us, does this mean we're at the center of the universe?

16. Is the Hubble constant actually constant?

Thought Questions

17. Where might the gas and dust (if any) in an elliptical galaxy come from?

18. Why can we not determine distances to galaxies by the same method used to measure the parallaxes of stars?

19. Which is redder—a spiral galaxy or an elliptical galaxy?

20. Suppose the stars in an elliptical galaxy all formed within a few million years shortly after the universe began. Suppose these stars have a range of masses, just as the stars in our own galaxy do. How would the color of the elliptical change over the next several billion years? How would its luminosity change? Why?

21. Starting with the determination of the size of Earth, outline a sequence of steps necessary to obtain the distance to a remote cluster of galaxies. (Hint: Review the chapter on Celestial Distances.)

22. Suppose the Milky Way Galaxy were truly isolated and that no other galaxies existed within 100 million light-years. Suppose that galaxies were observed in larger numbers at distances greater than 100 million light-years. Why would it be more difficult to determine accurate distances to those galaxies than if there were also galaxies relatively close by?

23. Suppose you were Hubble and Humason, working on the distances and Doppler shifts of the galaxies. What sorts of things would you have to do to convince yourself (and others) that the relationship you were seeing between the two quantities was a real feature of the behavior of the universe? (For example, would data from two galaxies be enough to demonstrate Hubble's law? Would data from just the nearest galaxies—in what astronomers call "the Local Group"—suffice?)

24. What does it mean if one elliptical galaxy has broader spectrum lines than another elliptical galaxy?

25. Based on your analysis of galaxies in Table 26.1, is there a correlation between the population of stars and the quantity of gas or dust? Explain why this might be.

26. Can a higher mass-to-light ratio mean that there is gas and dust present in the system that is being analyzed?

Figuring For Yourself

27. According to Hubble's law, what is the recessional velocity of a galaxy that is 10^8 light-years away from us? (Assume a Hubble constant of 22 km/s per million light-years.)

This OpenStax book is available for free at http://cnx.org/content/col11992/1.8

28. A cluster of galaxies is observed to have a recessional velocity of 60,000 km/s. Find the distance to the cluster. (Assume a Hubble constant of 22 km/s per million light-years.)

29. Suppose we could measure the distance to a galaxy using one of the distance techniques listed in Table 26.2 and it turns out to be 200 million light-years. The galaxy's redshift tells us its recessional velocity is 5000 km/s. What is the Hubble constant?

30. Calculate the mass-to-light ratio for a globular cluster with a luminosity of 10^6 L_{Sun} and 10^5 stars. (Assume that the average mass of a star in such a cluster is 1 M_{Sun}.)

31. Calculate the mass-to-light ratio for a luminous star of 100 M_{Sun} having the luminosity of 10^6 L_{Sun}.

Download for free at http://cnx.org/content/col11992/latest/

This OpenStax book is available for free at http://cnx.org/content/col11992/1.8

27

ACTIVE GALAXIES, QUASARS, AND SUPERMASSIVE BLACK HOLES

Figure 27.1. Hubble Ultra-Deep Field. The deepest picture of the sky in visible light (left) shows huge numbers of galaxies in a tiny patch of sky, only 1/100 the area of the full Moon. In contrast, the deepest picture of the sky taken in X-rays (right) shows large numbers of point-like quasars, which astronomers have shown are supermassive black holes at the very centers of galaxies. (credit left: modification of work by NASA, ESA, H. Teplitz and M. Rafelski (IPAC/Caltech), A. Koekemoer (STScI), R. Windhorst (Arizona State University), and Z. Levay (STScI); credit right: modification of work by ESO/Mario Nonino, Piero Rosati, ESO GOODS Team)

Chapter Outline

27.1 Quasars

27.2 Supermassive Black Holes: What Quasars Really Are

27.3 Quasars as Probes of Evolution in the Universe

Thinking Ahead

During the first half of the twentieth century, astronomers viewed the universe of galaxies as a mostly peaceful place. They assumed that galaxies formed billions of years ago and then evolved slowly as the populations of stars within them formed, aged, and died. That placid picture completely changed in the last few decades of the twentieth century.

Today, astronomers can see that the universe is often shaped by violent events, including cataclysmic explosions of supernovae, collisions of whole galaxies, and the tremendous outpouring of energy as matter interacts in the environment surrounding very massive black holes. The key event that began to change our view of the universe was the discovery of a new class of objects: quasars.

27.1 QUASARS

Learning Objectives

By the end of this section, you will be able to:

> Describe how quasars were discovered
> Explain how astronomers determined that quasars are at the distances implied by their redshifts

Download for free at http://cnx.org/content/col11992/latest/

> Justify the statement that the enormous amount of energy produced by quasars is generated in a very small volume of space

The name " quasars" started out as short for "quasi-stellar radio sources" (here "quasi-stellar" means "sort of like stars"). The discovery of radio sources that appeared point-like, just like stars, came with the use of surplus World War II radar equipment in the 1950s. Although few astronomers would have predicted it, the sky turned out to be full of strong sources of radio waves. As they improved the images that their new radio telescopes could make, scientists discovered that some radio sources were in the same location as faint blue "stars." No known type of star in our Galaxy emits such powerful radio radiation. What then were these "quasi-stellar radio sources"?

Redshifts: The Key to Quasars

The answer came when astronomers obtained visible-light spectra of two of those faint "blue stars" that were strong sources of radio waves (Figure 27.2). Spectra of these radio "stars" only deepened the mystery: they had emission lines, but astronomers at first could not identify them with any known substance. By the 1960s, astronomers had a century of experience in identifying elements and compounds in the spectra of stars. Elaborate tables had been published showing the lines that each element would produce under a wide range of conditions. A "star" with unidentifiable lines in the ordinary visible light spectrum had to be something completely new.

Figure 27.2. Typical Quasar. The arrow in this image marks the quasar known by its catalog number, PKS 1117-248. Note that nothing in this image distinguishes the quasar from an ordinary star. Its spectrum, however, shows that it is moving away from us at a speed of 36% the speed of light, or 67,000 miles per second. In contrast, the maximum speed observed for any star is only a few hundred miles per second. (credit: modification of work by WIYN Telescope, Kitt Peak National Observatory, NOAO)

In 1963 at Caltech's Palomar Observatory, Maarten Schmidt (Figure 27.3) was puzzling over the spectrum of one of the radio stars, which was named 3C 273 because it was the 273rd entry in the third Cambridge catalog of radio sources (part (b) of Figure 27.3). There were strong emission lines in the spectrum, and Schmidt recognized that they had the same spacing between them as the Balmer lines of hydrogen (see Radiation and Spectra). But the lines in 3C 273 were shifted far to the red of the wavelengths at which the Balmer lines are normally located. Indeed, these lines were at such long wavelengths that if the redshifts were attributed to the Doppler effect, 3C 273 was receding from us at a speed of 45,000 kilometers per second, or about 15% the speed of light! Since stars don't show Doppler shifts this large, no one had thought of considering high redshifts to be the cause of the strange spectra.

This OpenStax book is available for free at http://cnx.org/content/col11992/1.8

(a)   (b)

Figure 27.3. Quasar Pioneers and Quasar 3C 273. (a) Maarten Schmidt (left), who solved the puzzle of the quasar spectra in 1963, shares a joke in this 1987 photo with Allan Sandage, who took the first spectrum of a quasar. Sandage was also instrumental in measuring the value of Hubble's constant. (b) This is the first quasar for which a redshift was measured. The redshift showed that the light from it took about 2.5 billion years to reach us. Despite this great distance, it is still one of the quasars closest to the Milky Way Galaxy. Note also the faint streak going toward the upper left from the quasar. Some quasars, like 3C 273, eject super-fast jets of material. The jet from 3C 273 is about 200,000 light-years long. (credit a: modification of work by Andrew Fraknoi; credit b: modification of work by ESA/Hubble/NASA)

The puzzling emission lines in other star-like radio sources were then reexamined to see if they, too, might be well-known lines with large redshifts. This proved to be the case, but the other objects were found to be receding from us at even greater speeds. Their astounding speeds showed that the radio "stars" could not possibly be stars in our own Galaxy. Any true star moving at more than a few hundred kilometers per second would be able to overcome the gravitational pull of the Galaxy and completely escape from it. (As we shall see later in this chapter, astronomers eventually discovered that there was also more to these "stars" than just a point of light.)

It turns out that these high-velocity objects only look like stars because they are compact and very far away. Later, astronomers discovered objects with large redshifts that appear star-like but have no radio emission. Observations also showed that quasars were bright in the infrared and X-ray bands too, and not all these X-ray or infrared-bright quasars could be seen in either the radio or the visible-light bands of the spectrum. Today, all these objects are referred to as *quasi-stellar objects* (*QSOs*), or, as they are more popularly known, **quasars**. (The name was also soon appropriated by a manufacturer of home electronics.)

LINK TO LEARNING

Read an interview (https://openstax.org/l/30SchmidtIntv) with Maarten Schmidt on the fiftieth anniversary of his insight about the spectrum of quasars and their redshifts.

Over a million quasars have now been discovered, and spectra are available for over a hundred thousand. All these spectra show redshifts, none show blueshifts, and their redshifts can be very large. Yet in a photo they look just like stars (Figure 27.4).

Download for free at http://cnx.org/content/col11992/latest/

Figure 27.4. Typical Quasar Imaged by the Hubble Space Telescope. One of these two bright "stars" in the middle is in our Galaxy, while the other is a quasar 9 billion light-years away. From this picture alone, there's no way to say which is which. (The quasar is the one in the center of the picture.) (credit: Charles Steidel (CIT)/NASA/ESA)

In the record-holding quasars, the first Lyman series line of hydrogen, with a laboratory wavelength of 121.5 nanometers in the ultraviolet portion of the spectrum, is shifted all the way through the visible region to the infrared. At such high redshifts, the simple formula for converting a Doppler shift to speed (Radiation and Spectra) must be modified to take into account the effects of the theory of relativity. If we apply the relativistic form of the Doppler shift formula, we find that these redshifts correspond to velocities of about 96% of the speed of light.

EXAMPLE 27.1

Recession Speed of a Quasar

The formula for the Doppler shift, which astronomers denote by the letters z, is

$$z = \frac{\Delta \lambda}{\lambda} = \frac{v}{c}$$

where λ is the wavelength emitted by a source of radiation that is not moving, $\Delta \lambda$ is the difference between that wavelength and the wavelength we measure, v is the speed with which the source moves away, and c (as usual) is the speed of light.

A line in the spectrum of a galaxy is at a wavelength of 393 nanometers (nm, or 10^{-9} m) when the source is at rest. Let's say the line is measured to be longer than this value (redshifted) by 7.86 nm. Then its redshift $z = \frac{7.86 \text{ nm}}{393 \text{ nm}} = 0.02$, so its speed away from us is 2% of the speed of light $\left(\frac{v}{c} = 0.02\right)$.

This formula is fine for galaxies that are relatively nearby and are moving away from us slowly in the expansion of the universe. But the quasars and distant galaxies we discuss in this chapter are moving

This OpenStax book is available for free at http://cnx.org/content/col11992/1.8

away at speeds close to the speed of light. In that case, converting a Doppler shift (redshift) to a distance must include the effects of the special theory of relativity, which explains how measurements of space and time change when we see things moving at high speeds. The details of how this is done are way beyond the level of this text, but we can share with you the relativistic formula for the Doppler shift:

$$\frac{v}{c} = \frac{(z+1)^2 - 1}{(z+1)^2 + 1}$$

Let's do an example. Suppose a distant quasar has a redshift of 5. At what fraction of the speed of light is the quasar moving away?

Solution

We calculate the following:

$$\frac{v}{c} = \frac{(5+1)^2 - 1}{(5+1)^2 + 1} = \frac{36-1}{36+1} = \frac{35}{37} = 0.946$$

The quasar is thus receding from us at about 95% the speed of light.

Check Your Learning

Several lines of hydrogen absorption in the visible spectrum have rest wavelengths of 410 nm, 434 nm, 486 nm, and 656 nm. In a spectrum of a distant galaxy, these same lines are observed to have wavelengths of 492 nm, 521 nm, 583 nm, and 787 nm respectively. What is the redshift of this galaxy? What is the recession speed of this galaxy?

Answer:

Because this is the same galaxy, we could pick any one of the four wavelengths and calculate how much it has shifted. If we use a rest wavelength of 410 nm and compare it to the shifted wavelength of 492 nm, we see that

$$z = \frac{\Delta\lambda}{\lambda} = \frac{(492 \text{ nm} - 410 \text{ nm})}{410 \text{ nm}} = \frac{82 \text{ nm}}{410 \text{ nm}} = 0.20$$

In the classical view, this galaxy is receding at 20% of the speed of light; however, at 20% of the speed of light, relativistic effects are starting to become important. So, using the relativistic Doppler equation, we compute the true recession rate as

$$\frac{v}{c} = \frac{(z+1)^2 - 1}{(z+1)^2 + 1} = \frac{(0.2+1)^2 - 1}{(0.2+1)^2 + 1} = \frac{1.44-1}{1.44+1} = \frac{0.44}{2.44} = 0.18$$

Therefore, the actual recession speed is only 18% of the speed of light. While this may not initially seem like a big difference from the classical measurement, there is already an 11% deviation between the classical and the relativistic solutions; and at greater recession speeds, the divergence between the classical and relativistic speeds increases rapidly!

Quasars Obey the Hubble Law

The first question astronomers asked was whether quasars obeyed the Hubble law and were really at the large distances implied by their redshifts. If they did not obey the rule that large redshift means large distance, then they could be much closer, and their luminosity could be a lot less. One straightforward way to show that

Download for free at http://cnx.org/content/col11992/latest/

quasars had to obey the Hubble law was to demonstrate that they were actually part of galaxies, and that their redshift was the same as the galaxy that hosted them. Since ordinary galaxies *do* obey the Hubble law, anything within them would be subject to the same rules.

Observations with the Hubble Space Telescope provided the strongest evidence showing that quasars are located at the centers of galaxies. Hints that this is true had been obtained with ground-based telescopes, but space observations were required to make a convincing case. The reason is that quasars can outshine their entire galaxies by factors of 10 to 100 or even more. When this light passes through Earth's atmosphere, it is blurred by turbulence and drowns out the faint light from the surrounding galaxy—much as the bright headlights from an oncoming car at night make it difficult to see anything close by.

The Hubble Space Telescope, however, is not affected by atmospheric turbulence and can detect the faint glow from some of the galaxies that host quasars (Figure 27.5). Quasars have been found in the cores of both spiral and elliptical galaxies, and each quasar has the same redshift as its host galaxy. A wide range of studies with the Hubble Space Telescope now clearly demonstrate that quasars are indeed far away. If so, they must be producing a truly impressive amount of energy to be detectable as points of light that are much brighter than their galaxy. Interestingly, many quasar host galaxies are found to be involved in a collision with a second galaxy, providing, as we shall see, an important clue to the source of their prodigious energy output.

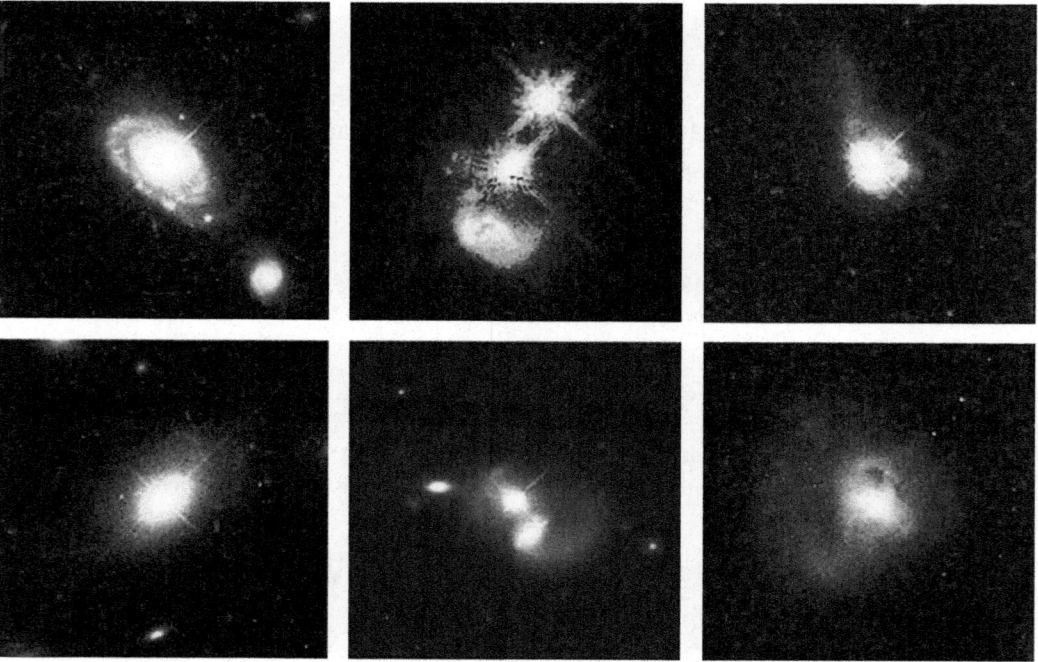

Figure 27.5. Quasar Host Galaxies. The Hubble Space Telescope reveals the much fainter "host" galaxies around quasars. The top left image shows a quasar that lies at the heart of a spiral galaxy 1.4 billion light-years from Earth. The bottom left image shows a quasar that lies at the center of an elliptical galaxy some 1.5 billion light-years from us. The middle images show remote pairs of interacting galaxies, one of which harbors a quasar. Each of the right images shows long tails of gas and dust streaming away from a galaxy that contains a quasar. Such tails are produced when one galaxy collides with another. (credit: modification of work by John Bahcall, Mike Disney, NASA)

The Size of the Energy Source

Given their large distances, quasars have to be extremely luminous to be visible to us at all—far brighter than any normal galaxy. In visible light alone, most are far more energetic than the brightest elliptical galaxies. But, as we saw, quasars also emit energy at X-ray and ultraviolet wavelengths, and some are radio sources as well.

This OpenStax book is available for free at http://cnx.org/content/col11992/1.8

Download for free at http://cnx.org/content/col11992/latest/

When all their radiation is added together, some QSOs have total luminosities as large as a hundred trillion Suns (10^{14} L_{Sun}), which is 10 to 100 times the brightness of luminous elliptical galaxies.

Finding a mechanism to produce the large amount of energy emitted by a quasar would be difficult under any circumstances. But there is an additional problem. When astronomers began monitoring quasars carefully, they found that some vary in luminosity on time scales of months, weeks, or even, in some cases, days. This variation is irregular and can change the brightness of a quasar by a few tens of percent in both its visible light and radio output.

Think about what such a change in luminosity means. A quasar at its dimmest is still more brilliant than any normal galaxy. Now imagine that the brightness increases by 30% in a few weeks. Whatever mechanism is responsible must be able to release new energy at rates that stagger our imaginations. The most dramatic changes in quasar brightness are equivalent to the energy released by 100,000 billion Suns. To produce this much energy we would have to convert the total mass of about ten Earths into energy every minute.

Moreover, because the fluctuations occur in such short times, the part of a quasar that is varying must be smaller than the distance light travels in the time it takes the variation to occur—typically a few months. To see why this must be so, let's consider a cluster of stars 10 light-years in diameter at a very large distance from Earth (see Figure 27.6, in which Earth is off to the right). Suppose every star in this cluster somehow brightens simultaneously and remains bright. When the light from this event arrives at Earth, we would first see the brighter light from stars on the near side; 5 years later we would see increased light from stars at the center. Ten years would pass before we detected more light from stars on the far side.

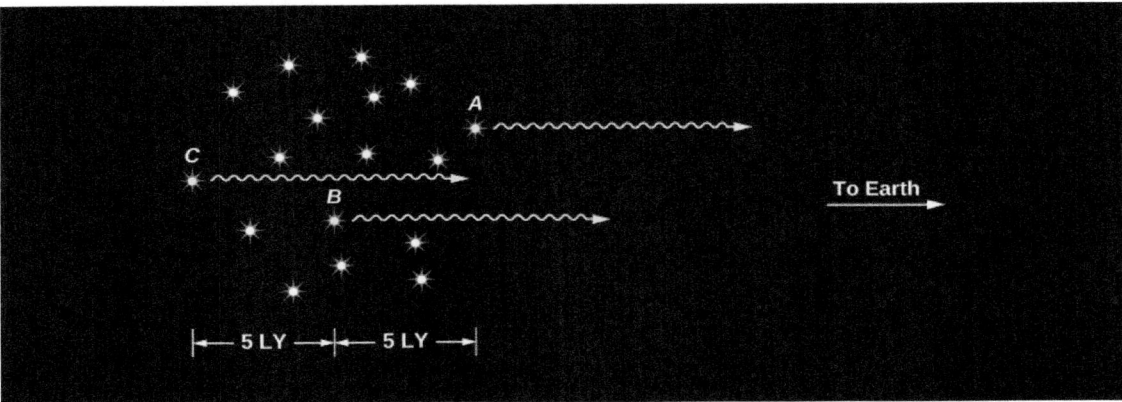

Figure 27.6. How the Size of a Source Affects the Timescale of Its Variability. This diagram shows why light variations from a large region in space appear to last for an extended period of time as viewed from Earth. Suppose all the stars in this cluster, which is 10 light-years across, brighten simultaneously and instantaneously. From Earth, star A will appear to brighten 5 years before star B, which in turn will appear to brighten 5 years earlier than star C. It will take 10 years for an Earth observer to get the full effect of the brightening.

Even though all stars in the cluster brightened at the same time, the fact that the cluster is 10 light-years wide means that 10 years must elapse before the increased light from every part of the cluster reaches us. From Earth we would see the cluster get brighter and brighter, as light from more and more stars began to reach us. Not until 10 years after the brightening began would we see the cluster reach maximum brightness. In other words, if an extended object suddenly flares up, it will seem to brighten over a period of time equal to the time it takes light to travel across the object from its far side.

We can apply this idea to brightness changes in quasars to estimate their diameters. Because quasars typically vary (get brighter and dimmer) over periods of a few months, the region where the energy is generated can be

Download for free at http://cnx.org/content/col11992/latest/

no larger than a few light-months across. If it were larger, it would take longer than a few months for the light from the far side to reach us.

How large is a region of a few light-months? Pluto, usually the outermost (dwarf) planet in our solar system, is about 5.5 light-hours from us, while the nearest star is 4 light-years away. Clearly a region a few light months across is tiny relative to the size of the entire Galaxy. And some quasars vary even more rapidly, which means their energy is generated in an even smaller region. Whatever mechanism powers the quasars must be able to generate more energy than that produced by an entire galaxy in a volume of space that, in some cases, is not much larger than our solar system.

Earlier Evidence

Even before the discovery of quasars, there had been hints that something very strange was going on in the centers of at least some galaxies. Back in 1918, American astronomer Heber Curtis used the large Lick Observatory telescope to photograph the galaxy Messier 87 in the constellation Virgo. On that photograph, he saw what we now call a jet coming from the center, or nucleus, of the galaxy (Figure 27.7). This jet literally and figuratively pointed to some strange activity going on in that galaxy nucleus. But he had no idea what it was. No one else knew what to do with this space oddity either.

The random factoid that such a central jet existed lay around for a quarter century, until Carl Seyfert, a young astronomer at Mount Wilson Observatory, also in California, found half a dozen galaxies with extremely bright nuclei that were almost stellar, rather than fuzzy in appearance like most galaxy nuclei. Using spectroscopy, he found that these nuclei contain gas moving at up to two percent the speed of light. That may not sound like much, but it is 6 million miles per hour, and more than 10 times faster than the typical motions of stars in galaxies.

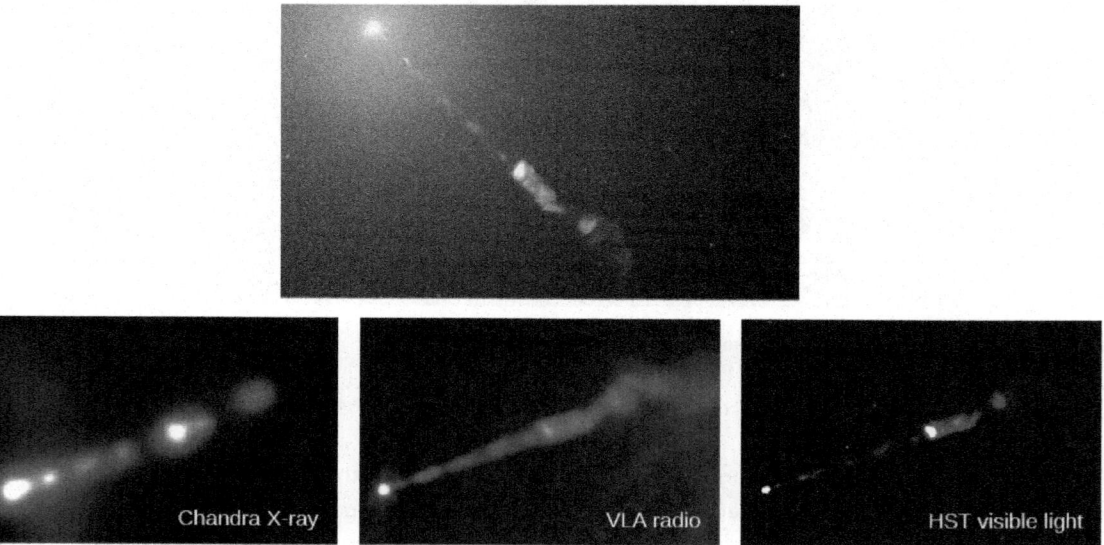

Figure 27.7. M87 Jet. Streaming out like a cosmic searchlight from the center of the galaxy, M87 is one of nature's most amazing phenomena, a huge jet of electrons and other particles traveling at nearly the speed of light. In this Hubble Space Telescope image, the blue of the jet contrasts with the yellow glow from the combined light of billions of unseen stars and yellow, point-like globular clusters that make up the galaxy (at the upper left). As we shall see later in this chapter, the jet, which is several thousand light-years long, originates in a disk of superheated gas swirling around a giant black hole at the center of M87. The light that we see is produced by electrons twisting along magnetic field lines in the jet, a process known as synchrotron radiation, which gives the jet its bluish tint. The jet in M87 can be observed in X-ray, radio, and visible light, as shown in the bottom three images. At the extreme left of each bottom image, we see the bright galactic nucleus harboring a supermassive black hole. (credit top: modification of work by NASA, The Hubble Heritage Team(STScI/AURA); credit bottom: modification of work by X-ray: H. Marshall (MIT), et al., CXC, NASA; Radio: F. Zhou, F. Owen (NRAO), J. Biretta (STScI); Optical: E. Perlman (UMBC), et al.)

This OpenStax book is available for free at http://cnx.org/content/col11992/1.8

Download for free at http://cnx.org/content/col11992/latest/

After decades of study, astronomers identified many other strange objects beyond our Milky Way Galaxy; they populate a whole "zoo" of what are now called **active galaxies** or **active galactic nuclei (AGN)**. Astronomers first called them by many different names, depending on what sorts of observations discovered each category, but now we know that we are always looking at the same basic mechanism. What all these galaxies have in common is some activity in their nuclei that produces an enormous amount of energy in a very small volume of space. In the next section, we describe a model that explains all these galaxies with strong central activity—both the AGNs and the QSOs.

LINK TO LEARNING

To see a jet for yourself, check out a time-lapse video (https://openstax.org/l/30timelapsejet) of the jet ejected from NGC 3862.

27.2 SUPERMASSIVE BLACK HOLES: WHAT QUASARS REALLY ARE

Learning Objectives

By the end of this section, you will be able to:

> Describe the characteristics common to all quasars
> Justify the claim that supermassive black holes are the source of the energy emitted by quasars (and AGNs)
> Explain how a quasar's energy is produced

In order to find a common model for quasars (and their cousins, the AGNs), let's first list the common characteristics we have been describing—and add some new ones:

- Quasars are hugely powerful, emitting more power in radiated light than all the stars in our Galaxy combined.

- Quasars are tiny, about the size of our solar system (to astronomers, that is really small!).

- Some quasars are observed to be shooting out pairs of straight jets at close to the speed of light, in a tight beam, to distances far beyond the galaxies they live in. These jets are themselves powerful sources of radio and gamma-ray radiation.

- Because quasars put out so much power from such a small region, they can't be powered by nuclear fusion the way stars are; they must use some process that is far more efficient.

- As we shall see later in this chapter, quasars were much more common when the universe was young than they are today. That means they must have been able to form in the first billion years or so after the universe began to expand.

The readers of this text are in a much better position than the astronomers who discovered quasars in the 1960s to guess what powers the quasars. That's because the key idea in solving the puzzle came from observations of the black holes. The discovery of the first stellar mass black hole in the binary system Cygnus X-1 was announced in 1971, several years after the discovery of quasars. Proof that there is a black hole at the center of our own Galaxy came even later. Back when astronomers first began trying to figure out what powered quasars,

Download for free at http://cnx.org/content/col11992/latest/

black holes were simply one of the more exotic predictions of the general theory of relativity that still waited to be connected to the real world.

It was only as proof of the existence of black holes accumulated over several decades that it became clearer that only supermassive black holes could account for all the observed properties of quasars and AGNs. As we saw in The Milky Way Galaxy, our own Galaxy has a black hole in its center, and the energy is emitted from a small central region. While our black hole doesn't have the mass or energy of the quasar black holes, the mechanism that powers them is similar. The evidence now shows that most—and probably all—elliptical galaxies and all spirals with nuclear bulges have black holes at their centers. The amount of energy emitted by material near the black hole depends on two things: the mass of the black hole and the amount of matter that is falling into it.

If a black hole with a billion Suns' worth of mass inside (10^9 M_{Sun}) accretes (gathers) even a relatively modest amount of additional material—say, about 10 M_{Sun} per year—then (as we shall see) it can, in the process, produce as much energy as a thousand normal galaxies. This is enough to account for the total energy of a quasar. If the mass of the black hole is smaller than a billion solar masses or the accretion rate is low, then the amount of energy emitted can be much smaller, as it is in the case of the Milky Way.

LINK TO LEARNING

Watch a video (https://openstax.org/l/30mataccrsupblh) of an artist's impression of matter accreting around a supermassive black hole.

Observational Evidence for Black Holes

In order to prove that a black hole is present at the center of a galaxy, we must demonstrate that so much mass is crammed into so small a volume that no normal objects—massive stars or clusters of stars—could possibly account for it (just as we did for the black hole in the Milky Way). We already know from observations (discussed in Black Holes and Curved Spacetime) that an accreting black hole is surrounded by a hot *accretion disk* with gas and dust that swirl around the black hole before it falls in.

If we assume that the energy emitted by quasars is also produced by a hot accretion disk, then, as we saw in the previous section, the size of the disk must be given by the time the quasar energy takes to vary. For quasars, the emission in visible light varies on typical time scales of 5 to 2000 days, limiting the size of the disk to that many light-days.

In the X-ray band, quasars vary even more rapidly, so the light travel time argument tells us that this more energetic radiation is generated in an even smaller region. Therefore, the mass around which the accretion disk is swirling must be confined to a space that is even smaller. If the quasar mechanism involves a great deal of mass, then the only astronomical object that can confine a lot of mass into a very small space is a black hole. In a few cases, it turns out that the X-rays are emitted from a region just a few times the size of the black hole event horizon.

The next challenge, then, is to "weigh" this central mass in a quasar. In the case of our own Galaxy, we used observations of the orbits of stars very close to the galactic center, along with Kepler's third law, to estimate the mass of the central black hole (The Milky Way Galaxy). In the case of distant galaxies, we cannot measure the orbits of individual stars, but we can measure the orbital speed of the gas in the rotating accretion disk. The Hubble Space Telescope is especially well suited to this task because it is above the blurring of Earth's

This OpenStax book is available for free at http://cnx.org/content/col11992/1.8

atmosphere and can obtain spectra very close to the bright central regions of active galaxies. The Doppler effect is then used to measure radial velocities of the orbiting material and so derive the speed with which it moves around.

One of the first galaxies to be studied with the Hubble Space Telescope is our old favorite, the giant elliptical M87. Hubble Space Telescope images showed that there is a disk of hot (10,000 K) gas swirling around the center of M87 (Figure 27.8). It was surprising to find hot gas in an elliptical galaxy because this type of galaxy is usually devoid of gas and dust. But the discovery was extremely useful for pinning down the existence of the black hole. Astronomers measured the Doppler shift of spectral lines emitted by this gas, found its speed of rotation, and then used the speed to derive the amount of mass inside the disk—applying Kepler's third law.

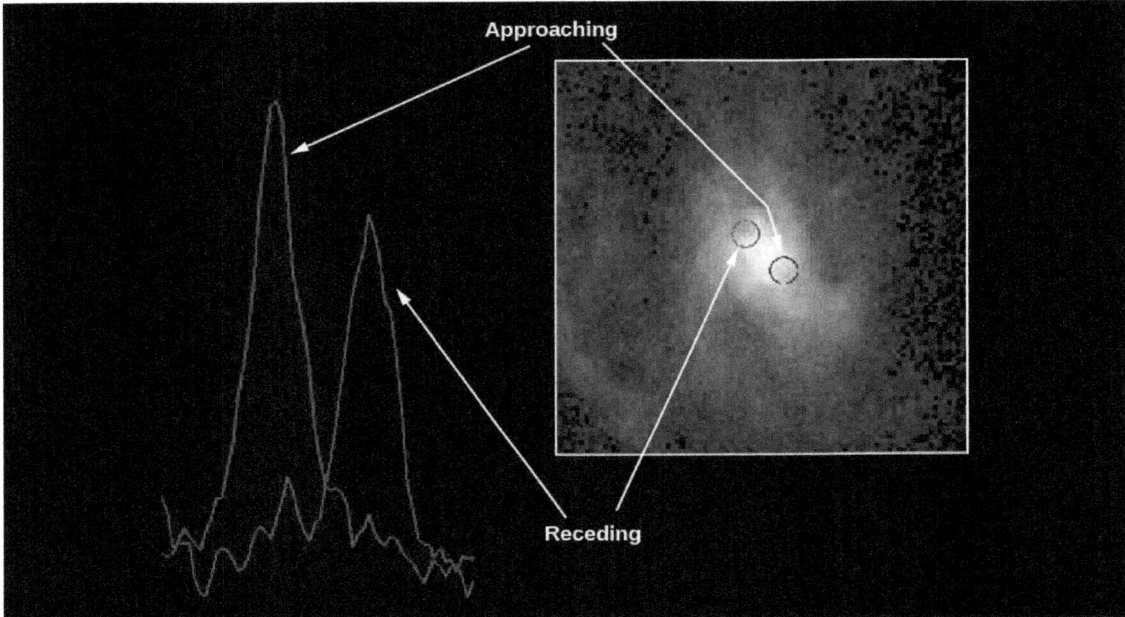

Figure 27.8. Evidence for a Black Hole at the Center of M87. The disk of whirling gas at right was discovered at the center of the giant elliptical galaxy M87 with the Hubble Space Telescope. Observations made on opposite sides of the disk show that one side is approaching us (the spectral lines are blueshifted by the Doppler effect) while the other is receding (lines redshifted), a clear indication that the disk is rotating. The rotation speed is about 550 kilometers per second or 1.2 million miles per hour. Such a high rotation speed is evidence that there is a very massive black hole at the center of M87. (credit: modification of work by Holland Ford, STScI/JHU; Richard Harms, Linda Dressel, Ajay K. Kochhar, Applied Research Corp.; Zlatan Tsvetanov, Arthur Davidsen, Gerard Kriss, Johns Hopkins; Ralph Bohlin, George Hartig, STScI; Bruce Margon, University of Washington in Seattle; NASA)

Modern estimates show that there is a mass of at least 3.5 billion M_{Sun} concentrated in a tiny region at the very center of M87. So much mass in such a small volume of space must be a black hole. Let's stop for a moment and take in this figure: a single black hole that has swallowed enough material to make 3.5 billion stars like the Sun. Few astronomical measurements have ever led to so mind-boggling a result. What a strange environment the neighborhood of such a supermassive black hole must be.

Another example is shown in Figure 27.9. Here, we see a disk of dust and gas that surrounds a 300-million-M_{Sun} black hole in the center of an elliptical galaxy. (The bright spot in the center is produced by the combined light of stars that have been pulled close together by the gravitational force of the black hole.) The mass of the black hole was again derived from measurements of the rotational speed of the disk. The gas in the disk is moving around at 155 kilometers per second at a distance of only 186 light-years from its center. Given the pull of the mass at the center, we expect that the whole dust disk should be swallowed by the black hole in several billion years.

Download for free at http://cnx.org/content/col11992/latest/

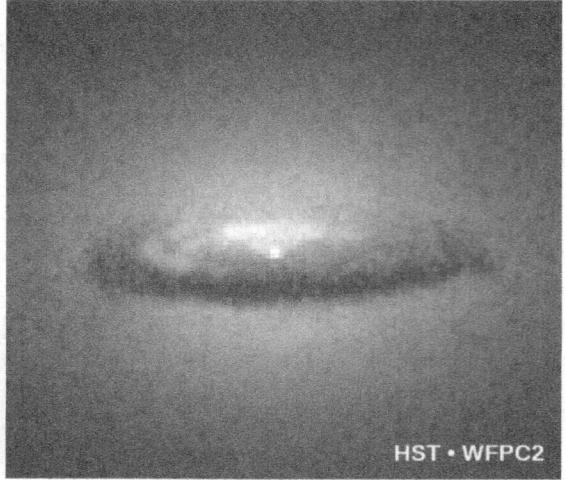

Figure 27.9. Another Galaxy with a Black-Hole Disk. The ground-based image shows an elliptical galaxy called NGC 7052 located in the constellation of Vulpecula, almost 200 million light-years from Earth. At the galaxy's center (right) is a dust disk roughly 3700 light-years in diameter. The disk rotates like a giant merry-go-round: gas in the inner part (186 light-years from the center) whirls around at a speed of 155 kilometers per second (341,000 miles per hour). From these measurements and Kepler's third law, it is possible to estimate that the disk is orbiting around a central black hole with a mass of 300 million Suns. (credit: modification of work by Roeland P. van der Marel (STScI), Frank C. van den Bosch (University of Washington), NASA)

But do we *have* to accept black holes as the only explanation of what lies at the center of these galaxies? What else could we put in such a small space other than a giant black hole? The alternative is stars. But to explain the masses in the centers of galaxies without a black hole we need to put at least a million stars in a region the size of the solar system. To fit, they would have be only 2 star diameters apart. Collisions between stars would happen all the time. And these collisions would lead to mergers of stars, and very soon the one giant star that they form would collapse into a black hole. So there is really no escape: only a black hole can fit so much mass into so small a space.

As we saw earlier, observations now show that all the galaxies with a spherical concentration of stars—either elliptical galaxies or spiral galaxies with nuclear bulges (see the chapter on Galaxies)—harbor one of these giant black holes at their centers. Among them is our neighbor spiral galaxy, the Andromeda galaxy, M31. The masses of these central black holes range from a just under a million up to at least 30 billion times the mass of the Sun. Several black holes may be even more massive, but the mass estimates have large uncertainties and need verification. We call these black holes "supermassive" to distinguish them from the much smaller black holes that form when some stars die (see The Death of Stars). So far, the most massive black holes from stars—those detected through gravitational waves detected by LIGO—have masses only a little over 30 solar masses.

Energy Production around a Black Hole

By now, you may be willing to entertain the idea that huge black holes lurk at the centers of active galaxies. But we still need to answer the question of how such a black hole can account for one of the most powerful sources of energy in the universe. As we saw in Black Holes and Curved Spacetime, a black hole itself can radiate no energy. Any energy we detect from it must come from material very close to the black hole, but not inside its event horizon.

In a galaxy, a central black hole (with its strong gravity) attracts matter—stars, dust, and gas—orbiting in the dense nuclear regions. This matter spirals in toward the spinning black hole and forms an accretion disk of material around it. As the material spirals ever closer to the black hole, it accelerates and becomes compressed,

This OpenStax book is available for free at http://cnx.org/content/col11992/1.8

heating up to temperatures of millions of degrees. Such hot matter can radiate prodigious amounts of energy as it falls in toward the black hole.

To convince yourself that falling into a region with strong gravity can release a great deal of energy, imagine dropping a printed version of your astronomy textbook out the window of the ground floor of the library. It will land with a thud, and maybe give a surprised pigeon a nasty bump, but the energy released by its fall will not be very great. Now take the same book up to the fifteenth floor of a tall building and drop it from there. For anyone below, astronomy could suddenly become a deadly subject; when the book hits, it does so with a great deal of energy.

Dropping things from far away into the much stronger gravity of a black hole is much more effective in turning the energy released by infall into other forms of energy. Just as the falling book can heat up the air, shake the ground, or produce sound energy that can be heard some distance away, so the energy of material falling toward a black hole can be converted to significant amounts of electromagnetic radiation.

What a black hole has to work with is not textbooks but streams of infalling gas. If a dense blob of gas moves through a thin gas at high speed, it heats up as it slows by friction. As it slows down, kinetic (motion) energy is turned into heat energy. Just like a spaceship reentering the atmosphere (Figure 27.10), gas approaching a black hole heats up and glows where it meets other gas. But this gas, as it approaches the event horizon, reaches speeds of 10% the speed of light and more. It therefore gets far, far hotter than a spaceship, which reaches no more than about 1500 K. Indeed, gas near a supermassive black hole reaches a temperature of about 150,000 K, about 100 times hotter than a spaceship returning to Earth. It can even get so hot—millions of degrees—that it radiates X-rays.

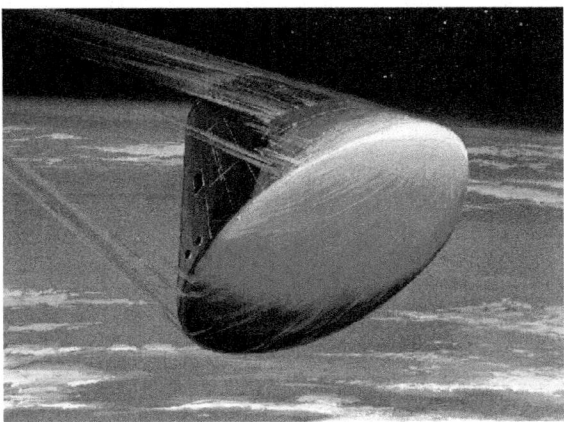

Figure 27.10. Friction in Earth's Atmosphere. In this artist's impression, the rapid motion of a spacecraft (the Apollo mission reentry capsule) through the atmosphere compresses and heats the air ahead of it, which heats the spacecraft in turn until it glows red hot. Pushing on the air slows down the spacecraft, turning the kinetic energy of the spacecraft into heat. Fast-moving gas falling into a quasar heats up in a similar way. (credit: modification of work by NASA)

The amount of energy that can be liberated this way is enormous. Einstein showed that mass and energy are interchangeable with his famous formula $E = mc^2$ (see The Sun: A Nuclear Powerhouse). A hydrogen bomb releases just 1% of that energy, as does a star. Quasars are much more efficient than that. The energy released falling to the event horizon of a black hole can easily reach 10% or, in the extreme theoretical limit, 32%, of that energy. (Unlike the hydrogen atoms in a bomb or a star, the gas falling into the black hole is not actually losing mass from its atoms to free up the energy; the energy is produced just because the gas is falling closer and closer to the black hole.) This huge energy release explains how a tiny volume like the region around a black hole can release as much power as a whole galaxy. But to radiate all that energy, instead of just falling inside

Download for free at http://cnx.org/content/col11992/latest/

the event horizon with barely a peep, the hot gas must take the time to swirl around the star in the accretion disk and emit some of its energy.

Most black holes don't show any signs of quasar emission. We call them "quiescent." But, like sleeping dragons, they can be woken up by being roused with a fresh supply of gas. Our own Milky Way black hole is currently quiescent, but it may have been a quasar just a few million years ago (Figure 27.11). Two giant bubbles that extend 25,000 light-years above and below the galactic center are emitting gamma rays. Were these produced a few million years ago when a significant amount of matter fell into the black hole at the center of the galaxy? Astronomers are still working to understand what remarkable event might have formed these enormous bubbles.

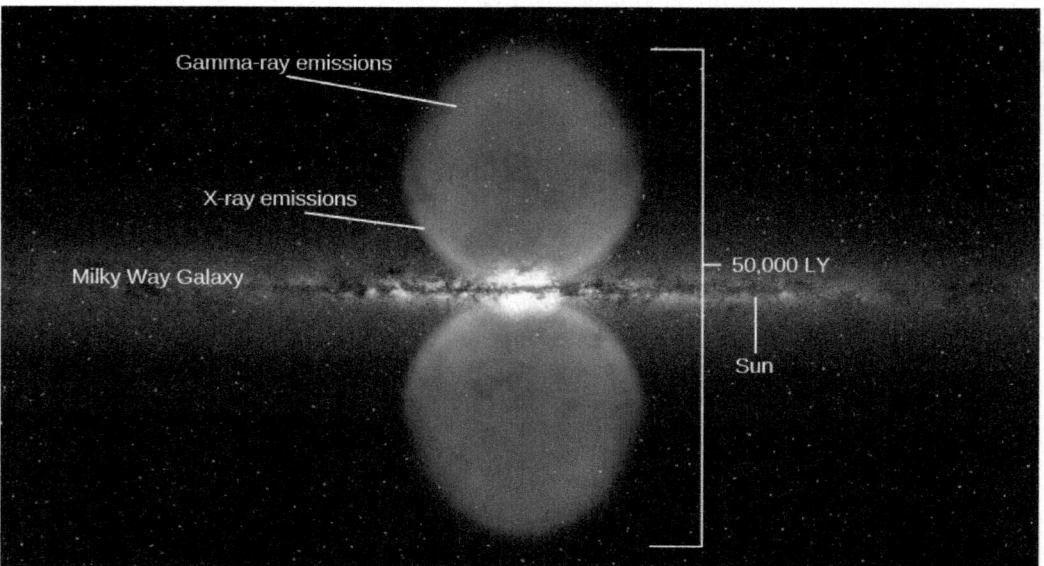

Figure 27.11. Fermi Bubbles in the Galaxy. Giant bubbles shining in gamma-ray light lie above and below the center of the Milky Way Galaxy, as seen by the Fermi satellite. (The gamma-ray and X-ray image is superimposed on a visible-light image of the inner parts of our Galaxy.) The bubbles may be evidence that the supermassive black hole at the center of our Galaxy was a quasar a few million years ago. (credit: modification of work by NASA's Goddard Space Flight Center)

The physics required to account for the exact way in which the energy of infalling material is converted to radiation near a black hole is far more complicated than our simple discussion suggests. To understand what happens in the "rough and tumble" region around a massive black hole, astronomers and physicists must resort to computer simulations (and they require supercomputers, fast machines capable of awesome numbers of calculations per second). The details of these models are beyond the scope of our book, but they support the basic description presented here.

Radio Jets

So far, our model seems to explain the central energy source in quasars and active galaxies. But, as we have seen, there is more to quasars and other active galaxies than the point-like energy source. They can also have long jets that glow with radio waves, light, and sometimes even X-rays, and that extend far beyond the limits of the parent galaxy. Can we find a way for our black hole and its accretion disk to produce these jets of energetic particles as well?

This OpenStax book is available for free at http://cnx.org/content/col11992/1.8

Download for free at http://cnx.org/content/col11992/latest/

Many different observations have now traced these jets to within 3 to 30 light-years of the parent quasar or galactic nucleus. While the black hole and accretion disk are typically smaller than 1 light-year, we nevertheless presume that if the jets come this close, they probably originate in the vicinity of the black hole. Another characteristic of the jets we need to explain is that they contain matter moving close to the speed of light.

Why are energetic electrons and other particles near a supermassive black hole ejected into jets, and often into two oppositely directed jets, rather than in all directions? Again, we must use theoretical models and supercomputer simulations of what happens when a lot of material whirls inward in a crowded black hole accretion disk. Although there is no agreement on exactly how jets form, it has become clear that any material escaping from the neighborhood of the black hole has an easier time doing so *perpendicular to* the disk.

In some ways, the inner regions of black hole accretion disks resemble a baby that is just learning to eat by herself. As much food as goes into the baby's mouth can sometimes wind up being spit out in various directions. In the same way, some of the material whirling inward toward a black hole finds itself under tremendous pressure and orbiting with tremendous speed. Under such conditions, simulations show that a significant amount of material can be flung outward—not back along the disk, where more material is crowding in, but above and below the disk. If the disk is thick (as it tends to be when a lot of material falls in quickly), it can channel the outrushing material into narrow beams perpendicular to the disk (Figure 27.12).

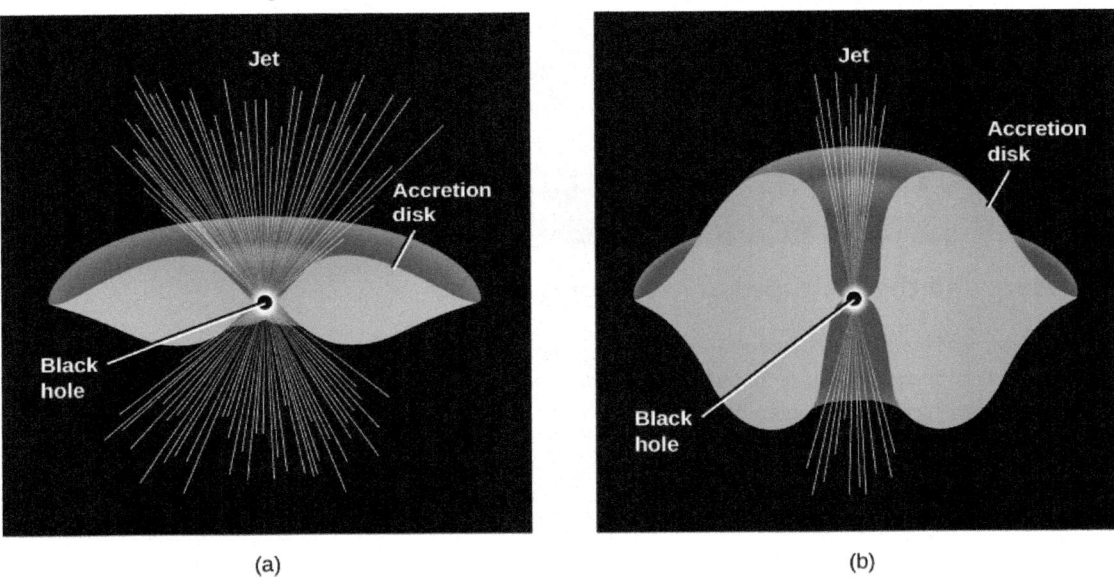

(a)   (b)

Figure 27.12. Models of Accretion Disks. These schematic drawings show what accretion disks might look like around large black holes for (a) a thin accretion disk and (b) a "fat" disk—the type needed to account for channeling the outflow of hot material into narrow jets oriented perpendicular to the disk.

Figure 27.13 shows observations of an elliptical galaxy that behaves in exactly this way. At the center of this active galaxy, there is a ring of dust and gas about 400 light-years in diameter, surrounding a 1.2-billion-M_{Sun} black hole. Radio observations show that two jets emerge in a direction perpendicular to the ring, just as the model predicts.

Download for free at http://cnx.org/content/col11992/latest/

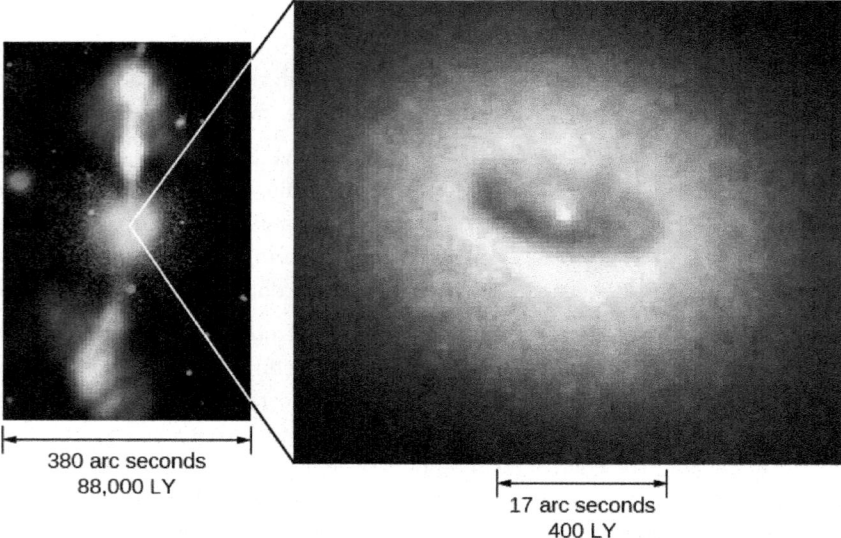

380 arc seconds
88,000 LY

17 arc seconds
400 LY

Figure 27.13. Jets and Disk in an Active Galaxy. The picture on the left shows the active elliptical galaxy NGC 4261, which is located in the Virgo Cluster at a distance of about 100 million light-years. The galaxy itself—the white circular region in the center—is shown the way it looks in visible light, while the jets are seen at radio wavelengths. A Hubble Space Telescope image of the central portion of the galaxy is shown on the right. It contains a ring of dust and gas about 800 light-years in diameter, surrounding a supermassive black hole. Note that the jets emerge from the galaxy in a direction perpendicular to the plane of the ring. (credit: modification of work by ESA/HST)

MAKING CONNECTIONS

Quasars and the Attitudes of Astronomers

The discovery of quasars in the early 1960s was the first in a series of surprises astronomers had in store. Within another decade they would find neutron stars (in the form of pulsars), the first hints of black holes (in binary X-ray sources), and even the radio echo of the Big Bang itself. Many more new discoveries lay ahead.

As Maarten Schmidt reminisced in 1988, "This had, I believe, a profound impact on the conduct of those practicing astronomy. Before the 1960s, there was much authoritarianism in the field. New ideas expressed at meetings would be instantly judged by senior astronomers and rejected if too far out." We saw a good example of this in the trouble Chandrasekhar had in finding acceptance for his ideas about the death of stars with cores greater than 1.4 M_{Sun} (see the feature box on Subrahmanyan Chandrasekhar).

"The discoveries of the 1960s," Schmidt continued, "were an embarrassment, in the sense that they were totally unexpected and could not be evaluated immediately. In reaction to these developments, an attitude has evolved where even outlandish ideas in astronomy are taken seriously. Given our lack of solid knowledge in extragalactic astronomy, this is probably to be preferred over authoritarianism."[1]

That is not to say that astronomers (being human) don't continue to have prejudices and preferences. For example, a small group of astronomers who thought that the redshifts of quasars were not connected with their distances (which was definitely a minority opinion) often felt excluded from meetings or from access to telescopes in the 1960s and 1970s. It's not so clear that they actually *were*

This OpenStax book is available for free at http://cnx.org/content/col11992/1.8

Download for free at http://cnx.org/content/col11992/latest/

excluded, as much as that they felt the very difficult pressure of knowing that most of their colleagues strongly disagreed with them. As it turned out, the evidence—which must ultimately decide all scientific questions—was not on their side either.

But today, as better instruments bring solutions to some problems and starkly illuminate our ignorance about others, the entire field of astronomy seems more open to discussing unusual ideas. Of course, before any hypotheses become accepted, they must be tested—again and again—against the evidence that nature itself reveals. Still, the many strange proposals published about what dark matter might be (see The Evolution and Distribution of Galaxies) attest to the new openness that Schmidt described.

With this black hole model, we have come a long way toward understanding the quasars and active galaxies that seemed very mysterious only a few decades ago. As often happens in astronomy, a combination of better instruments (making better observations) and improved theoretical models enabled us to make significant progress on a puzzling aspect of the cosmos.

27.3 QUASARS AS PROBES OF EVOLUTION IN THE UNIVERSE

Learning Objectives

By the end of this section, you will be able to:

> Trace the rise and fall of quasars over cosmic time
> Describe some of the ways in which galaxies and black holes influence each other's growth
> Describe some ways the first black holes may have formed
> Explain why some black holes are not producing quasar emission but rather are quiescent

The quasars' brilliance and large distance make them ideal probes of the far reaches of the universe and its remote past. Recall that when first introducing quasars, we mentioned that they generally tend to be far away. When we see extremely distant objects, we are seeing them as they were long ago. Radiation from a quasar 8 billion light-years away is telling us what that quasar and its environment were like 8 billion years ago, much closer to the time that the galaxy that surrounds it first formed. Astronomers have now detected light emitted from quasars that were already formed only a few hundred million years after the universe began its expansion 13.8 billion years ago. Thus, they give us a remarkable opportunity to learn about the time when large structures were first assembling in the cosmos.

The Evolution of Quasars

Quasars provide compelling evidence that we live in an evolving universe—one that changes with time. They tell us that astronomers living billions of years ago would have seen a universe that is very different from the universe of today. Counts of the number of quasars at different redshifts (and thus at different times in the evolution of the universe) show us how dramatic these changes are (Figure 27.14). We now know that the number of quasars was greatest at the time when the universe was only 20% of its present age.

---

1   M. Schmidt, "The Discovery of Quasars," in *Modern Cosmology in Retrospect*, ed. B. Bertotti et al. (Cambridge University Press, 1990).

Download for free at http://cnx.org/content/col11992/latest/

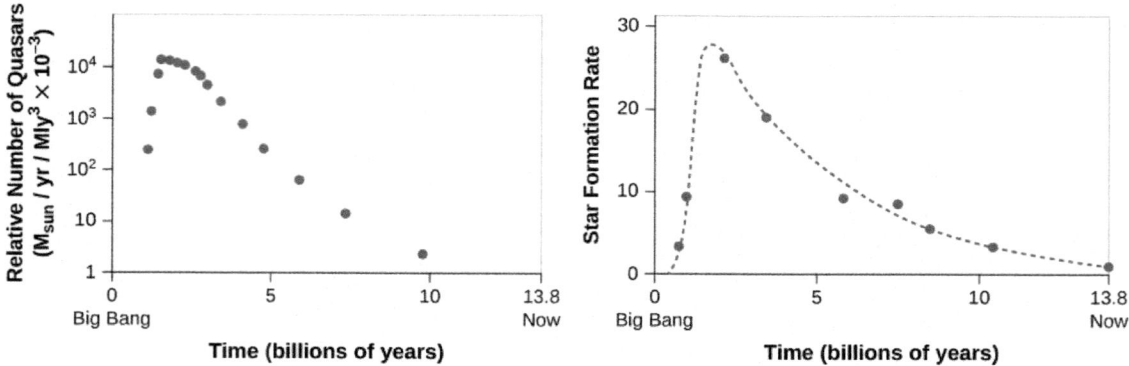

Figure 27.14. Relative Number of Quasars and Rate at Which Stars Formed as a Function of the Age of the Universe. An age of 0 on the plots corresponds to the beginning of the universe; an age of 13.8 corresponds to the present time. Both the number of quasars and the rate of star formation were at a peak when the universe was about 20% as old as it is now.

As you can see, the drop-off in the numbers of quasars as time gets nearer to the present day is quite abrupt. Observations also show that the emission from the accretion disks around the most massive black holes peaks early and then fades. The most powerful quasars are seen only at early times. In order to explain this result, we make use of our model of the energy source of the quasars—namely that quasars are black holes with enough fuel to make a brilliant accretion disk right around them.

The fact that there were more quasars long ago (far away) than there are today (nearby) could be explained if there was more material available to be accreted by black holes early in the history of the universe. You might say that the quasars were more active when their black holes had fuel for their "energy-producing engines." If that fuel was mostly consumed in the first few billion years after the universe began its expansion, then later in its life, a "hungry" black hole would have very little left with which to light up the galaxy's central regions.

In other words, if matter in the accretion disk is continually being depleted by falling into the black hole or being blown out from the galaxy in the form of jets, then a quasar can continue to radiate only as long as new gas is available to replenish the accretion disk.

In fact, there *was* more gas around to be accreted early in the history of the universe. Back then, most gas had not yet collapsed to form stars, so there was more fuel available for both the feeding of black holes and the forming of new stars. Much of that fuel was subsequently consumed in the formation of stars during the first few billion years after the universe began its expansion. Later in its life, a galaxy would have little left to feed a hungry black hole or to form more new stars. As we see from Figure 27.14, both star formation and black hole growth peaked together when the universe was about 2 billion years old. Ever since, both have been in sharp decline. We are late to the party of the galaxies and have missed some of the early excitement.

Observations of nearer galaxies (seen later in time) indicate that there is another source of fuel for the central black holes—the collision of galaxies. If two galaxies of similar mass collide and merge, or if a smaller galaxy is pulled into a larger one, then gas and dust from one may come close enough to the black hole in the other to be devoured by it and so provide the necessary fuel. Astronomers have found that collisions were also much more common early in the history of the universe than they are today. There were more small galaxies in those early times because over time, as we shall see (in The Evolution and Distribution of Galaxies), small galaxies tend to combine into larger ones. Again, this means that we would expect to see more quasars long ago (far away) than we do today (nearby)—as we in fact do.

This OpenStax book is available for free at http://cnx.org/content/col11992/1.8

Download for free at http://cnx.org/content/col11992/latest/

Codependence of Black Holes and Galaxies

Once black hole masses began to be measured reliably in the late 1990s, they posed an enigma. It looked as though the mass of the central black hole depended on the mass of the galaxy. The black holes in galaxies always seem to be just 1/200 the mass of the galaxy they live in. This result is shown schematically in Figure 27.15, and some of the observations are plotted in Figure 27.16.

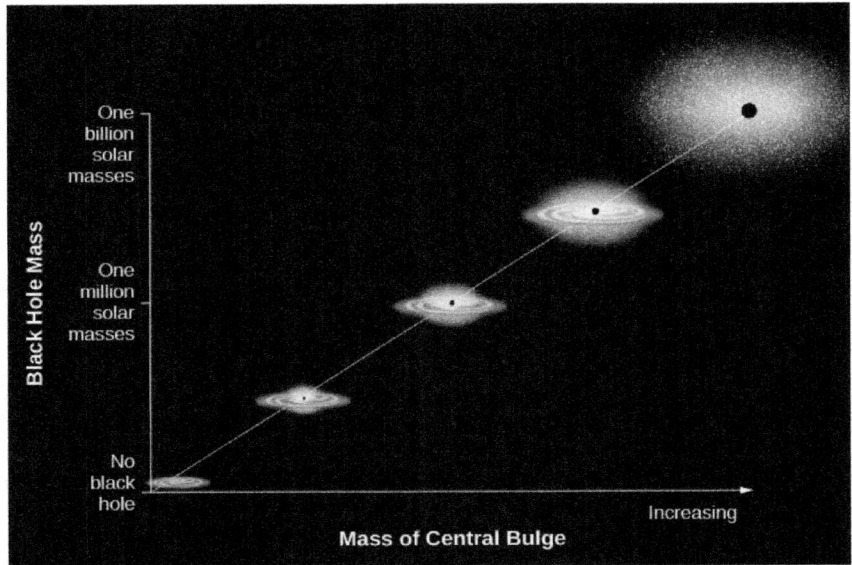

Figure 27.15. Relationship between Black Hole Mass and the Mass of the Host Galaxy. Observations show that there is a close correlation between the mass of the black hole at the center of a galaxy and the mass of the spherical distribution of stars that surrounds the black hole. That spherical distribution may be in the form of either an elliptical galaxy or the central bulge of a spiral galaxy. (credit: modification of work by K. Cordes, S. Brown (STScI))

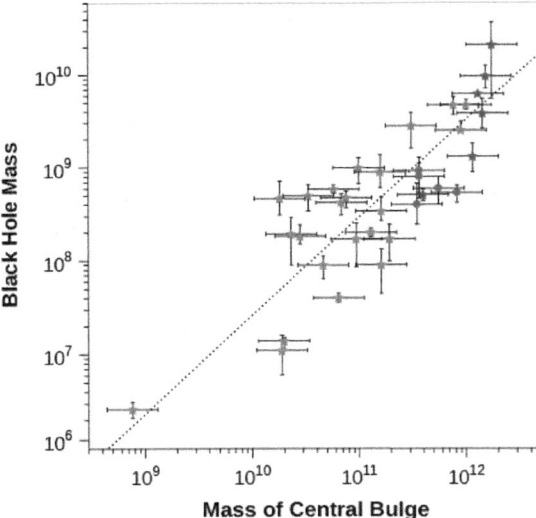

★ Stars/Early-type BCG
★ Stars/Early-type non-BCG
● Gas/Early-type BCG
● Gas/Early-type non-BCG

Figure 27.16. Correlation between the Mass of the Central Black Hole and the Mass Contained within the Bulge of Stars Surrounding the Black Hole, Using Data from Real Galaxies. The black hole always turns out to be about 1/200 the mass of the stars surrounding it. The horizontal and vertical bars surrounding each point show the uncertainty of the measurement. (credit: modification of work by Nicholas J. McConnell, Chung-Pei Ma, "Revisiting the Scaling Relations of Black Hole Masses and Host Galaxy Properties," *The Astrophysical Journal*, 764:184 (14 pp.), February 20, 2013.)

Download for free at http://cnx.org/content/col11992/latest/

Somehow black hole mass and the mass of the surrounding bulge of stars are connected. But why does this correlation exist? Unfortunately, astronomers do not yet know the answer to this question. We do know, however, that the black hole can influence the rate of star formation in the galaxy, and that the properties of the surrounding galaxy can influence how fast the black hole grows. Let's see how these processes work.

How a Galaxy Can Influence a Black Hole in Its Center

Let's look first at how the surrounding galaxy might influence the growth and size of the black hole. Without large quantities of fresh "food," the surroundings of black holes glow only weakly as bits of local material spiral inward toward the black hole. So somehow large amounts of gas have to find their way to the black hole from the galaxy in order to feed the quasar and make it grow and give off the energy to be noticed. Where does this "food" for the black hole come from originally and how might it be replenished? The jury is still out, but the options are pretty clear.

One obvious source of fuel for the black hole is matter from the host galaxy itself. Galaxies start out with large amounts of interstellar gas and dust, and at least some of this interstellar matter is gradually converted into stars as the galaxy evolves. On the other hand, as stars go through their lives and die, they lose mass all the time into the space between them, thereby returning some of the gas and dust to the interstellar medium. We expect to find more gas and dust in the central regions early in a galaxy's life than later on, when much of it has been converted into stars. Any of the interstellar matter that ventures too close to the black hole may be accreted by it. This means that we would expect that the number and luminosity of quasars powered in this way would decline with time. And as we have seen, that is just what we find.

Today both *elliptical galaxies* and the *nuclear bulges of spiral galaxies* have very little raw material left to serve as a source of fuel for the black hole. And most of the giant black holes in nearby galaxies, including the one in our own Milky Way, are now dark and relatively quiet—mere shadows of their former selves. So that fits with our observations.

We should note that even if you have a quiescent supermassive black hole, a star in the area could occasionally get close to it. Then the powerful tidal forces of the black hole can pull the whole star apart into a stream of gas. This stream quickly forms an accretion disk that gives off energy in the normal way and makes the black hole region into a temporary quasar. However, the material will fall into the black hole after only a few weeks or months. The black hole then goes back into its lurking, quiescent state, until another victim wanders by.

This sort of "cannibal" event happens only once every 100,000 years or so in a typical galaxy. But we can monitor millions of galaxies in the sky, so a few of these " tidal disruption events" are found each year (Figure 27.17). However, these individual events, dramatic as they are, are too rare to account for the huge masses of the central black holes.

This OpenStax book is available for free at http://cnx.org/content/col11992/1.8

Figure 27.17. A Black Hole Snacks on a Star. This artist's impression shows three stages of a star (red) swinging too close to a giant black hole (black circle). The star starts off (top left) in its normal spherical shape, then begins to be pulled into a long football shape by tides raised by the black hole (center). When the star gets closer still, the tides become stronger than the gravity holding the star together, and it breaks up into a streamer (right). Much of the star's matter forms a temporary accretion disk that lights up as a quasar for a few weeks or months. (credit: modification of work by NASA/CXC/M. Weiss)

Another source of fuel for the black hole is the collision of its host galaxy with another galaxy. Some of the brightest galaxies turn out, when a detailed picture is taken, to be pairs of colliding galaxies. And most of them have quasars inside them, not easily visible to us because they are buried by enormous amounts of dust and gas.

A collision between two cars creates quite a mess, pushing parts out of their regular place. In the same way, if two galaxies collide and merge, then gas and dust (though not so much the stars) can get pushed out of their regular orbits. Some may veer close enough to the black hole in one galaxy or the other to be devoured by it and so provide the necessary fuel to power a quasar. As we saw, galaxy collisions and mergers happened most frequently when the universe was young and probably help account for the fact that quasars were most common when the universe was only about 20% of its current age.

Collisions in today's universe are less frequent, but they do happen. Once a galaxy reaches the size of the Milky Way, most of the galaxies it merges with will be much smaller galaxies—*dwarf galaxies* (see the chapter on Galaxies). These don't disrupt the big galaxy much, but they can supply some additional gas to its black hole.

By the way, if two galaxies, each of which contains a black hole, collide, then the two black holes may merge and form an even larger black hole (Figure 27.18). In this process they will emit a burst of gravitational waves. One of the main goals of the European Space Agency's planned LISA (Laser Interferometer Space Antenna) mission is to detect the gravitational wave signals from the merging of supermassive black holes.

Download for free at http://cnx.org/content/col11992/latest/

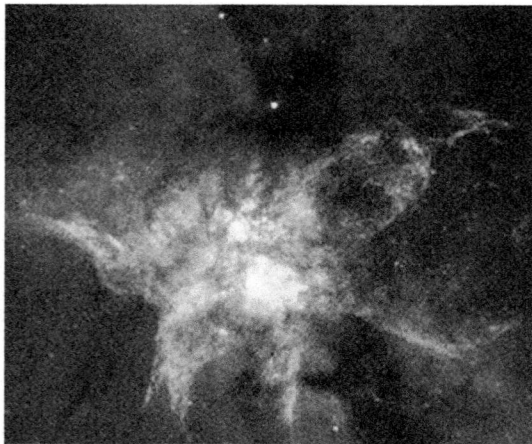 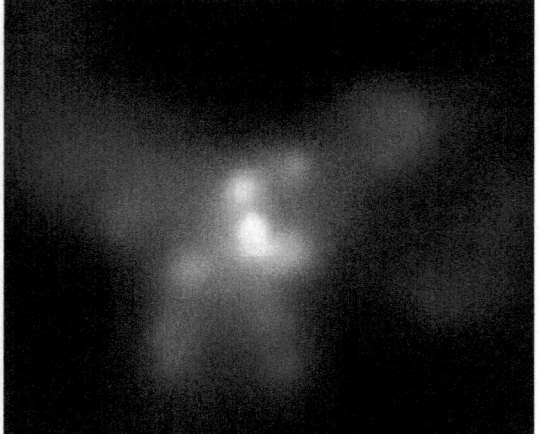

Figure 27.18. Colliding Galaxies with Two Black Holes. We compare Hubble Space Telescope visible-light (left) and Chandra X-ray (right) images of the central regions of NGC 6240, a galaxy about 400 million light-years away. It is a prime example of a galaxy in which stars are forming, evolving, and exploding at an exceptionally rapid rate due to a relatively recent merger (30 million years ago). The Chandra image shows two bright X-ray sources, each produced by hot gas surrounding a black hole. Over the course of the next few hundred million years, the two supermassive black holes, which are about 3000 light-years apart, will drift toward each other and merge to form an even larger black hole. This detection of a binary black hole supports the idea that black holes can grow to enormous masses in the centers of galaxies by merging with nearby galaxies. (credit left: modification of work by NASA/CXC/MPE/S.Komossa et al; credit right: NASA/STScI/R. P. van der Marel, J. Gerssen)

LINK TO LEARNING

Watch two galaxies collide (https://openstax.org/l/30galcolsuperma) to form a supermassive black hole.

How Does the Black Hole Influence the Formation of Stars in the Galaxy?

We have seen that the material in galaxies can influence the growth of the black hole. The black hole in turn can also influence the galaxy in which it resides. It can do so in three ways: through its jets, through winds of particles that manage to stream away from the accretion disk, and through radiation from the accretion disk. As they stream away from the black hole, all three can either promote star formation by compressing the surrounding gas and dust—or instead suppress star formation by heating the surrounding gas and shredding molecular clouds, thereby inhibiting or preventing star formation. The outflowing energy can even be enough to halt the accretion of new material and starve the black hole of fuel. Astronomers are still trying to evaluate the relative importance of these effects in determining the overall evolution of galactic bulges and the rates of star formation.

In summary, we have seen how galaxies and supermassive black holes can each influence the evolution of the other: the galaxy supplies fuel to the black hole, and the quasar can either support or suppress star formation. The balance of these processes probably helps account for the correlation between black hole and bulge masses, but there are as yet no theories that explain quantitatively and in detail why the correlation between black hole and bulge masses is as tight as it is or why the black hole mass is always about 1/200 times the mass of the bulge.

This OpenStax book is available for free at http://cnx.org/content/col11992/1.8

The Birth of Black Holes and Galaxies

While the connection between quasars and galaxies is increasingly clear, the biggest puzzle of all—namely, how the supermassive black holes in galaxies got started—remains unsolved. Observations show that they existed when the universe was very young. One dramatic example is the discovery of a quasar that was already shining when the universe was only 700 million years old. What does it take to create a large black hole so quickly? A related problem is that in order to eventually build black holes containing more than 2 billion solar masses, it is necessary to have giant "seed" black holes with masses at least 2000 times the mass of the Sun—and they must somehow have been created shortly after the expansion of the universe began.

Astronomers are now working actively to develop models for how these seed black holes might have formed. Theories suggest that galaxies formed from collapsing clouds of dark matter and gas. Some of the gas formed stars, but perhaps some of the gas settled to the center where it became so concentrated that it formed a black hole. If this happened, the black hole could form right away—although this requires that the gas should not be rotating very much initially.

A more likely scenario is that the gas will have some angular momentum (rotation) that will prevent direct collapse to a black hole. In that case, the very first generation of stars will form, and some of them, according to calculations, will have masses hundreds of times that of the Sun. When these stars finish burning hydrogen, just a few million years later, the supernovae they end with will create black holes a hundred or so times the mass of the Sun. These can then merge with others or accrete the rich gas supply available at these early times.

The challenge is growing these smaller black holes quickly enough to make the much larger black holes we see a few hundred million years later. It turns out to be difficult because there are limits on how fast they can accrete matter. These should make sense to you from what we discussed earlier in the chapter. If the rate of accretion becomes too high, then the energy streaming outward from the black hole's accretion disk will become so strong as to blow away the infalling matter.

What if, instead, a collapsing gas cloud doesn't form a black hole directly or break up and form a group of regular stars, but stays together and makes one fairly massive star embedded within a dense cluster of thousands of lower mass stars and large quantities of dense gas? The massive star will have a short lifetime and will soon collapse to become a black hole. It can then begin to attract the dense gas surrounding it. But calculations show that the gravitational attraction of the many nearby stars will cause the black hole to zigzag randomly within the cluster and will prevent the formation of an accretion disk. If there is no accretion disk, then matter can fall freely into the black hole from all directions. Calculations suggest that under these conditions, a black hole even as small as 10 times the mass of the Sun could grow to more than 10 billion times the mass of the Sun by the time the universe is a billion years old.

Scientists are exploring other ideas for how to form the seeds of supermassive black holes, and this remains a very active field of research. Whatever mechanism caused the rapid formation of these supermassive black holes, they do give us a way to observe the youthful universe when it was only about five percent as old as it is now.

LINK TO LEARNING

Take a look at some new results (https://openstax.org/l/30chanxrayobser) from the Chandra X-ray Observatory about the formation of supermassive black holes in the early universe.

Download for free at http://cnx.org/content/col11992/latest/

CHAPTER 27 REVIEW

 KEY TERMS

active galactic nuclei (AGN) galaxies that are almost as luminous as quasars and share many of their properties, although to a less spectacular degree; abnormal amounts of energy are produced in their centers

active galaxies galaxies that house active galactic nuclei

quasar an object of very high redshift that looks like a star but is extragalactic and highly luminous; also called a quasi-stellar object, or QSO

 SUMMARY

27.1 Quasars

The first quasars discovered looked like stars but had strong radio emission. Their visible-light spectra at first seemed confusing, but then astronomers realized that they had much larger redshifts than stars. The quasar spectra obtained so far show redshifts ranging from 15% to more than 96% the speed of light. Observations with the Hubble Space Telescope show that quasars lie at the centers of galaxies and that both spirals and ellipticals can harbor quasars. The redshifts of the underlying galaxies match the redshifts of the quasars embedded in their centers, thereby proving that quasars obey the Hubble law and are at the great distances implied by their redshifts. To be noticeable at such great distances, quasars must have 10 to 100 times the luminosity of the brighter normal galaxies. Their variations show that this tremendous energy output is generated in a small volume—in some cases, in a region not much larger than our own solar system. A number of galaxies closer to us also show strong activity at their centers—activity now known to be caused by the same mechanism as the quasars.

27.2 Supermassive Black Holes: What Quasars Really Are

Both active galactic nuclei and quasars derive their energy from material falling toward, and forming a hot accretion disk around, a massive black hole. This model can account for the large amount of energy emitted and for the fact that the energy is produced in a relatively small volume of space. It can also explain why jets coming from these objects are seen in two directions: those directions are perpendicular to the accretion disk.

27.3 Quasars as Probes of Evolution in the Universe

Quasars and galaxies affect each other: the galaxy supplies fuel to the black hole, and the quasar heats and disrupts the gas clouds in the galaxy. The balance between these two processes probably helps explain why the black hole seems always to be about 1/200 the mass of the spherical bulge of stars that surrounds the black hole.

Quasars were much more common billions of years ago than they are now, and astronomers speculate that they mark an early stage in the formation of galaxies. Quasars were more likely to be active when the universe was young and fuel for their accretion disk was more available.

Quasar activity can be re-triggered by a collision between two galaxies, which provides a new source of fuel to feed the black hole.

This OpenStax book is available for free at http://cnx.org/content/col11992/1.8

Download for free at http://cnx.org/content/col11992/latest/

FOR FURTHER EXPLORATION

Articles

Bartusiak, M. "A Beast in the Core." *Astronomy* (July 1998): 42. On supermassive black holes at the centers of galaxies.

Disney, M. "A New Look at Quasars." *Scientific American* (June 1998): 52.

Djorgovski, S. "Fires at Cosmic Dawn." *Astronomy* (September 1995): 36. On quasars and what we can learn from them.

Ford, H., & Tsvetanov, Z. "Massive Black Holes at the Hearts of Galaxies." *Sky & Telescope* (June 1996): 28. Nice overview.

Irion, R. "A Quasar in Every Galaxy?" *Sky & Telescope* (July 2006): 40. Discusses how supermassive black holes powering the centers of galaxies may be more common than thought.

Kormendy, J. "Why Are There so Many Black Holes?" *Astronomy* (August 2016): 26. Discussion of why supermassive black holes are so common in the universe.

Kruesi, L. "Secrets of the Brightest Objects in the Universe." *Astronomy* (July 2013): 24. Review of our current understanding of quasars and how they help us learn about black holes.

Miller, M., et al. "Supermassive Black Holes: Shaping their Surroundings." *Sky & Telescope* (April 2005): 42. Jets from black hole disks.

Nadis, S. "Exploring the Galaxy–Black Hole Connection." *Astronomy* (May 2010): 28. Overview.

Nadis, S. "Here, There, and Everywhere." *Astronomy* (February 2001): 34. On Hubble observations showing how common supermassive black holes are in galaxies.

Nadis, S. "Peering inside a Monster Galaxy." *Astronomy* (May 2014): 24. What X-ray observations tell us about the mechanism that powers the active galaxy M87.

Olson, S. "Black Hole Hunters." *Astronomy* (May 1999): 48. Profiles four astronomers who search for "hungry" black holes at the centers of active galaxies.

Peterson, B. "Solving the Quasar Puzzle." *Sky & Telescope* (September 2013): 24. A review article on how we figured out that black holes were the power source for quasars, and how we view them today.

Tucker, W., et al. "Black Hole Blowback." *Scientific American* (March 2007): 42. How supermassive black holes create giant bubbles in the intergalactic medium.

Voit, G. "The Rise and Fall of Quasars." *Sky & Telescope* (May 1999): 40. Good overview of how quasars fit into cosmic history.

Wanjek, C. "How Black Holes Helped Build the Universe." *Sky & Telescope* (January 2007): 42. On the energy and outflow from disks around supermassive black holes; nice introduction.

Websites

Monsters in Galactic Nuclei: http://chandra.as.utexas.edu/stardate.html. An article on supermassive black holes by John Kormendy, from *StarDate* magazine.

Quasar Astronomy Forty Years On: http://www.astr.ua.edu/keel/agn/quasar40.html. A 2003 popular article by William Keel.

Download for free at http://cnx.org/content/col11992/latest/

Quasars and Active Galactic Nuclei: www.astr.ua.edu/keel/agn/. An annotated gallery of images showing the wide range of activity in galaxies. There is also an introduction, a glossary, and background information. Also by William Keel.

Quasars: "The Light Fantastic": http://hubblesite.org/newscenter/archive/releases/1996/35/background/. This brief "backgrounder" from the public information office at the HubbleSite gives a bit of the history of the discovery and understanding of quasars.

Videos

Active Galaxies: https://www.youtube.com/watch?v=Y_HgsFmwCeg. Part of the *Astronomy: Observations and Theories* series; half-hour introduction to quasars and related objects (27:28).

Black Hole Chaos: The Environments of the Most Supermassive Black Holes in the Universe: https://www.youtube.com/watch?v=hzSgU-3d8QY. May 2013 lecture by Dr. Belinda Wilkes and Dr. Francesca Civano of the Center for Astrophysics in the CfA Observatory Nights Lecture Series (50:14).

Hubble and Black Holes: http://www.spacetelescope.org/videos/hubblecast43a/. Hubblecast on black holes and active galactic nuclei (9:10).

Monster Black Holes: https://www.youtube.com/watch?v=LN9oYjNKBm8. May 2013 lecture by Professor Chung-Pei Ma of the University of California, Berkeley; part of the Silicon Valley Astronomy Lecture Series (1:18:03).

⚍ COLLABORATIVE GROUP ACTIVITIES

A. When quasars were first discovered and the source of their great energy was unknown, some astronomers searched for evidence that quasars are much nearer to us than their redshifts imply. (That way, they would not have to produce so much energy to look as bright as they do.) One way was to find a "mismatched pair"—a quasar and a galaxy with different redshifts that lie in very nearly the same direction in the sky. Suppose you do find one and only one galaxy with a quasar very close by, and the redshift of the quasar is six times larger than that of the galaxy. Have your group discuss whether you could then conclude that the two objects are at the same distance and that redshift is *not* a reliable indicator of distance. Why? Suppose you found three such pairs, each with different mismatched redshifts? Suppose *every* galaxy has a nearby quasar with a different redshift. How would your answer change and why?

B. Large ground-based telescopes typically can grant time to only one out of every four astronomers who apply for observing time. One prominent astronomer tried for several years to establish that the redshifts of quasars do not indicate their distances. At first, he was given time on the world's largest telescope, but eventually it became clearer that quasars were just the centers of active galaxies and that their redshifts really did indicate distance. At that point, he was denied observing time by the committee of astronomers who reviewed such proposals. Suppose your group had been the committee. What decision would you have made? Why? (In general, what criteria should astronomers have for allowing astronomers whose views completely disagree with the prevailing opinion to be able to pursue their research?)

C. Based on the information in this chapter and in Black Holes and Curved Spacetime, have your group discuss what it would be like near the event horizon of a supermassive black hole in a quasar or active galaxy. Make a list of all the reasons a trip to that region would not be good for your health. Be specific.

This OpenStax book is available for free at http://cnx.org/content/col11992/1.8

D. Before we understood that the energy of quasars comes from supermassive black holes, astronomers were baffled by how such small regions could give off so much energy. A variety of models were suggested, some involving new physics or pretty "far out" ideas from current physics. Can your group come up with some areas of astronomy that you have studied in this course where we don't yet have an explanation for something happening in the cosmos?

🗒 EXERCISES

Review Questions

1. Describe some differences between quasars and normal galaxies.

2. Describe the arguments supporting the idea that quasars are at the distances indicated by their redshifts.

3. In what ways are active galaxies like quasars but different from normal galaxies?

4. Why could the concentration of matter at the center of an active galaxy like M87 not be made of stars?

5. Describe the process by which the action of a black hole can explain the energy radiated by quasars.

6. Describe the observations that convinced astronomers that M87 is an active galaxy.

7. Why do astronomers believe that quasars represent an early stage in the evolution of galaxies?

8. Why were quasars and active galaxies not initially recognized as being "special" in some way?

9. What do we now understand to be the primary difference between normal galaxies and active galaxies?

10. What is the typical structure we observe in a quasar at radio frequencies?

11. What evidence do we have that the luminous central region of a quasar is small and compact?

Thought Questions

12. Suppose you observe a star-like object in the sky. How can you determine whether it is actually a star or a quasar?

13. Why don't any of the methods for establishing distances to galaxies, described in Galaxies (other than Hubble's law itself), work for quasars?

14. One of the early hypotheses to explain the high redshifts of quasars was that these objects had been ejected at very high speeds from other galaxies. This idea was rejected, because no quasars with large blueshifts have been found. Explain why we would expect to see quasars with both blueshifted and redshifted lines if they were ejected from nearby galaxies.

15. A friend of yours who has watched many *Star Trek* episodes and movies says, "I thought that black holes pulled everything into them. Why then do astronomers think that black holes can explain the great *outpouring* of energy from quasars?" How would you respond?

16. Could the Milky Way ever become an active galaxy? Is it likely to ever be as luminous as a quasar?

17. Why are quasars generally so much more luminous (why do they put out so much more energy) than active galaxies?

18. Suppose we detect a powerful radio source with a radio telescope. How could we determine whether or not this was a newly discovered quasar and not some nearby radio transmission?

Download for free at http://cnx.org/content/col11992/latest/

19. A friend tries to convince you that she can easily see a quasar in her backyard telescope. Would you believe her claim?

Figuring For Yourself

20. Show that no matter how big a redshift (z) we measure, v/c will never be greater than 1. (In other words, no galaxy we observe can be moving away faster than the speed of light.)

21. If a quasar has a redshift of 3.3, at what fraction of the speed of light is it moving away from us?

22. If a quasar is moving away from us at $v/c = 0.8$, what is the measured redshift?

23. In the chapter, we discussed that the largest redshifts found so far are greater than 6. Suppose we find a quasar with a redshift of 6.1. With what fraction of the speed of light is it moving away from us?

24. Rapid variability in quasars indicates that the region in which the energy is generated must be small. You can show why this is true. Suppose, for example, that the region in which the energy is generated is a transparent sphere 1 light-year in diameter. Suppose that in 1 s this region brightens by a factor of 10 and remains bright for two years, after which it returns to its original luminosity. Draw its light curve (a graph of its brightness over time) as viewed from Earth.

25. Large redshifts move the positions of spectral lines to longer wavelengths and change what can be observed from the ground. For example, suppose a quasar has a redshift of $\frac{\Delta\lambda}{\lambda} = 4.1$. At what wavelength would you make observations in order to detect its Lyman line of hydrogen, which has a laboratory or rest wavelength of 121.6 nm? Would this line be observable with a ground-based telescope in a quasar with zero redshift? Would it be observable from the ground in a quasar with a redshift of $\frac{\Delta\lambda}{\lambda} = 4.1$?

26. Once again in this chapter, we see the use of Kepler's third law to estimate the mass of supermassive black holes. In the case of NGC 4261, this chapter supplied the result of the calculation of the mass of the black hole in NGC 4261. In order to get this answer, astronomers had to measure the velocity of particles in the ring of dust and gas that surrounds the black hole. How high were these velocities? Turn Kepler's third law around and use the information given in this chapter about the galaxy NGC 4261—the mass of the black hole at its center and the diameter of the surrounding ring of dust and gas—to calculate how long it would take a dust particle in the ring to complete a single orbit around the black hole. Assume that the only force acting on the dust particle is the gravitational force exerted by the black hole. Calculate the velocity of the dust particle in km/s.

27. In the Check Your Learning section of Example 27.1, you were told that several lines of hydrogen absorption in the visible spectrum have rest wavelengths of 410 nm, 434 nm, 486 nm, and 656 nm. In a spectrum of a distant galaxy, these same lines are observed to have wavelengths of 492 nm, 521 nm, 583 nm, and 787 nm, respectively. The example demonstrated that $z = 0.20$ for the 410 nm line. Show that you will obtain the same redshift regardless of which absorption line you measure.

28. In the Check Your Learning section of Example 27.1, the author commented that even at $z = 0.2$, there is already an 11% deviation between the relativistic and the classical solution. What is the percentage difference between the classical and relativistic results at $z = 0.1$? What is it for $z = 0.5$? What is it for $z = 1$?

29. The quasar that appears the brightest in our sky, 3C 273, is located at a distance of 2.4 billion light-years. The Sun would have to be viewed from a distance of 1300 light-years to have the same apparent magnitude as 3C 273. Using the inverse square law for light, estimate the luminosity of 3C 273 in solar units.

This OpenStax book is available for free at http://cnx.org/content/col11992/1.8

Download for free at http://cnx.org/content/col11992/latest/

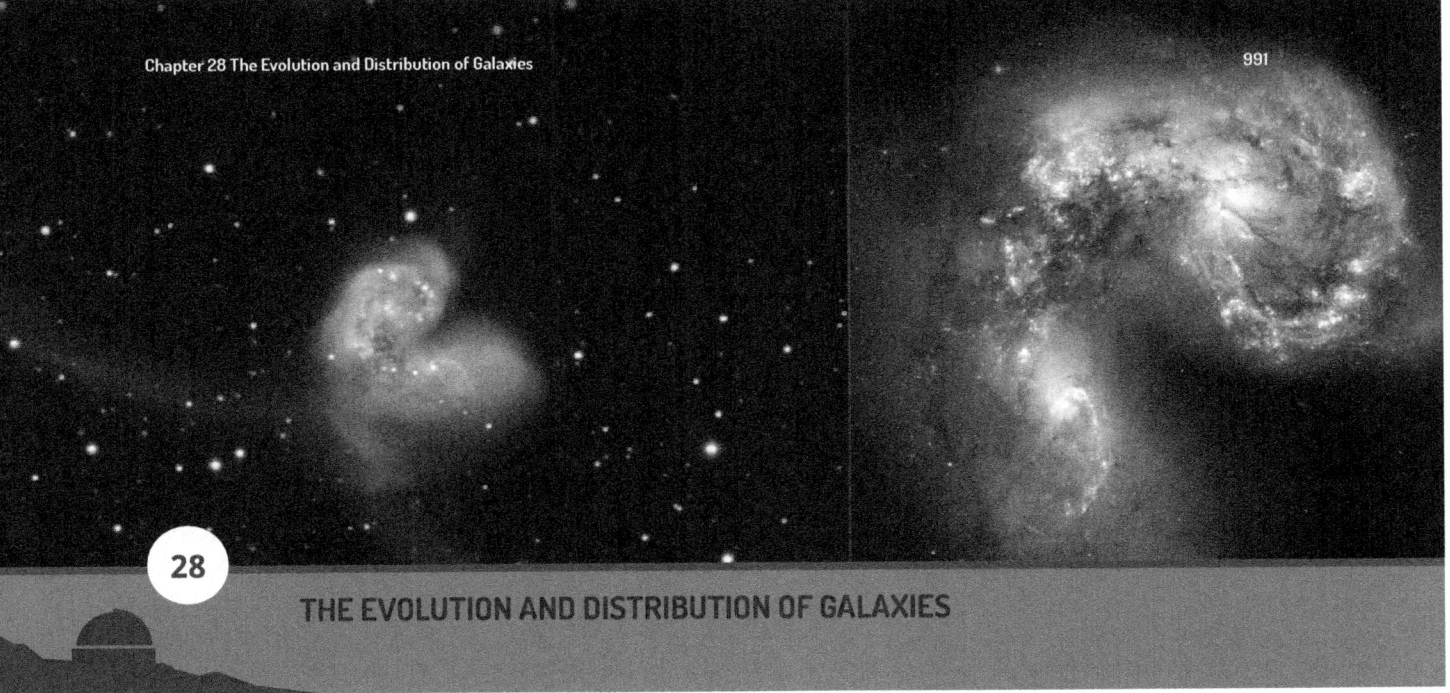

Figure 28.1. Colliding Galaxies. Collisions and mergers of galaxies strongly influence their evolution. On the left is a ground-based image of two colliding galaxies (NCG 4038 and 4039), sometimes nicknamed the Antennae galaxies. The long, luminous tails are material torn out of the galaxies by tidal forces during the collision. The right image shows the inner regions of these two galaxies, as taken by the Hubble Space Telescope. The cores of the twin galaxies are the orange blobs to the lower left and upper right of the center of the image. Note the dark lanes of dust crossing in front of the bright regions. The bright pink and blue star clusters are the result of a burst of star formation stimulated by the collision. (credit left: modification of work by Bob and Bill Twardy/Adam Block/NOAO/AURA/NSF; credit right: modification of work by NASA, ESA, and the Hubble Heritage Team (STScI/AURA)-ESA/Hubble Collaboration)

Chapter Outline

Thinking Ahead

How and when did galaxies like our Milky Way form? Which formed first: stars or galaxies? Can we see direct evidence of the changes galaxies undergo over their lifetimes? If so, what determines whether a galaxy will "grow up" to be spiral or elliptical? And what is the role of "nature versus nurture"? That is to say, how much of a galaxy's development is determined by what it looks like when it is born and how much is influenced by its environment?

Astronomers today have the tools needed to explore the universe almost back to the time it began. The huge new telescopes and sensitive detectors built in the last decades make it possible to obtain both images and spectra of galaxies so distant that their light has traveled to reach us for more than 13 billion years—more than 90% of the way back to the Big Bang: we can use the finite speed of light and the vast size of the universe as a cosmic time machine to peer back and observe how galaxies formed and evolved over time. Studying galaxies so far away in any detail is always a major challenge, largely because their distance makes them appear very faint. However, today's large telescopes on the ground and in space are finally making such a task possible.

Download for free at http://cnx.org/content/col11992/latest/

28.1 OBSERVATIONS OF DISTANT GALAXIES

Learning Objectives

By the end of this section, you will be able to:

> Explain how astronomers use light to learn about distant galaxies long ago
> Discuss the evidence showing that the first stars formed when the universe was less than 10% of its current age
> Describe the major differences observed between galaxies seen in the distant, early universe and galaxies seen in the nearby universe today

Let's begin by exploring some techniques astronomers use to study how galaxies are born and change over cosmic time. Suppose you wanted to understand how adult humans got to be the way they are. If you were very dedicated and patient, you could actually observe a sample of babies from birth, following them through childhood, adolescence, and into adulthood, and making basic measurements such as their heights, weights, and the proportional sizes of different parts of their bodies to understand how they change over time.

Unfortunately, we have no such possibility for understanding how galaxies grow and change over time: in a human lifetime—or even over the entire history of human civilization—individual galaxies change hardly at all. We need other tools than just patiently observing single galaxies in order to study and understand those long, slow changes.

We do, however, have one remarkable asset in studying galactic evolution. As we have seen, the universe itself is a kind of time machine that permits us to observe remote galaxies as they were long ago. For the closest galaxies, like the Andromeda galaxy, the time the light takes to reach us is on the order of a few hundred thousand to a few million years. Typically not much changes over times that short—individual stars in the galaxy may be born or die, but the overall structure and appearance of the galaxy will remain the same. But we have observed galaxies so far away that we are seeing them as they were when the light left them more than 10 billion years ago.

By observing more distant objects, we look further back toward a time when both galaxies and the universe were young (Figure 28.2). This is a bit like getting letters in the mail from several distant friends: the farther the friend was when she mailed the letter to you, the longer the letter must have been in transit, and so the older the news is when it arrives in your mailbox; you are learning something about her life at an earlier time than when you read the letter.

This OpenStax book is available for free at http://cnx.org/content/col11992/1.8

Download for free at http://cnx.org/content/col11992/latest/

Figure 28.2. Astronomical Time Travel. This true-color, long-exposure image, made during 70 orbits of Earth with the Hubble Space Telescope, shows a small area in the direction of the constellation Sculptor. The massive cluster of galaxies named Abell 2744 appears in the foreground of this image. It contains several hundred galaxies, and we are seeing them as they looked 3.5 billion years ago. The immense gravity in Abell 2744 acts as a gravitational lens (see the Astronomy Basics feature box on Gravitational Lensing later in this chapter) to warp space and brighten and magnify images of nearly 3000 distant background galaxies. The more distant galaxies (many of them quite blue) appear as they did more than 12 billion years ago, not long after the Big Bang. Blue galaxies were much more common in that earlier time than they are today. These galaxies appear blue because they are undergoing active star formation and making hot, bright blue stars. (credit: NASA, ESA, STScI)

If we can't directly detect the changes over time in individual galaxies because they happen too slowly, how then can we ever understand those changes and the origins of galaxies? The solution is to observe many galaxies at many different cosmic distances and, therefore, look-back times (how far back in time we are seeing the galaxy). If we can study a thousand very distant "baby" galaxies when the universe was 1 billion years old, and another thousand slightly closer "toddler" galaxies when it was 2 billion years old, and so on until the present 13.8-billion-year-old universe of mature "adult" galaxies near us today, then maybe we can piece together a coherent picture of how the whole ensemble of galaxies evolves over time. This allows us to reconstruct the "life story" of galaxies since the universe began, even though we can't follow a single galaxy from infancy to old age.

Fortunately, there is no shortage of galaxies to study. Hold up your pinky at arm's length: the part of the sky blocked by your fingernail contains about one million galaxies, layered farther and farther back in space and time. In fact, the sky is filled with galaxies, all of them, except for Andromeda and the Magellanic Clouds, too faint to see with the naked eye—more than 100 billion galaxies in the observable universe, each one with about 100 billion stars.

This cosmic time machine, then, lets us peer into the past to answer fundamental questions about where galaxies come from and how they got to be the way they are today. Astronomers call those galactic changes over cosmic time **evolution**, a word that recalls the work of Darwin and others on the development of life on Earth. But note that galaxy evolution refers to the changes in *individual* galaxies over time, while the kind of evolution biologists study is changes in *successive generations* of living organisms over time.

Download for free at http://cnx.org/content/col11992/latest/

Spectra, Colors, and Shapes

Astronomy is one of the few sciences in which all measurements must be made at a distance. Geologists can take samples of the objects they are studying; chemists can conduct experiments in their laboratories to determine what a substance is made of; archeologists can use carbon dating to determine how old something is. But astronomers can't pick up and play with a star or galaxy. As we have seen throughout this book, if they want to know what galaxies are made of and how they have changed over the lifetime of the universe, they must decode the messages carried by the small number of photons that reach Earth.

Fortunately (as you have learned) electromagnetic radiation is a rich source of information. The distance to a galaxy is derived from its *redshift* (how much the lines in its spectrum are shifted to the red because of the expansion of the universe). The conversion of redshift to a distance depends on certain properties of the universe, including the value of the Hubble constant and how much mass it contains. We will describe the currently accepted model of the universe in The Big Bang. For the purposes of this chapter, it is enough to know that the current best estimate for the age of the universe is 13.8 billion years. In that case, if we see an object that emitted its light 6 billion light-years ago, we are seeing it as it was when the universe was almost 8 billion years old. If we see something that emitted its light 13 billion years ago, we are seeing it as it was when the universe was less than a billion years old. So astronomers measure a galaxy's redshift from its spectrum, use the Hubble constant plus a model of the universe to turn the redshift into a distance, and use the distance and the constant speed of light to infer how far back in time they are seeing the galaxy—the look-back time.

In addition to distance and look-back time, studies of the Doppler shifts of a galaxy's spectral lines can tell us how fast the galaxy is rotating and hence how massive it is (as explained in Galaxies). Detailed analysis of such lines can also indicate the types of stars that inhabit a galaxy and whether it contains large amounts of interstellar matter.

Unfortunately, many galaxies are so faint that collecting enough light to produce a detailed spectrum is currently impossible. Astronomers thus have to use a much rougher guide to estimate what kinds of stars inhabit the faintest galaxies—their overall colors. Look again at Figure 28.2 and notice that some of the galaxies are very blue and others are reddish-orange. Now remember that hot, luminous blue stars are very massive and have lifetimes of only a few million years. If we see a galaxy where blue colors dominate, we know that it must have many hot, luminous blue stars, and that star formation must have taken place in the few million years before the light left the galaxy. In a yellow or red galaxy, on the other hand, the young, luminous blue stars that surely were made in the galaxy's early bursts of star formation must have died already; it must contain mostly old yellow and red stars that last a long time in their main-sequence stages and thus typically formed billions of years before the light that we now see was emitted.

Another important clue to the nature of a galaxy is its shape. Spiral galaxies can be distinguished from elliptical galaxies by shape. Observations show that spiral galaxies contain young stars and large amounts of interstellar matter, while elliptical galaxies have mostly old stars and very little or no star formation. Elliptical galaxies turned most of their interstellar matter into stars many billions of years ago, while star formation has continued until the present day in spiral galaxies.

If we can count the number of galaxies of each type during each epoch of the universe, it will help us understand how the pace of star formation changes with time. As we will see later in this chapter, galaxies in the distant universe—that is, young galaxies—look very different from the older galaxies that we see nearby in the present-day universe.

The First Generation of Stars

In addition to looking at the most distant galaxies we can find, astronomers look at the oldest stars (what we might call the fossil record) of our own Galaxy to probe what happened in the early universe. Since stars are the

This OpenStax book is available for free at http://cnx.org/content/col11992/1.8

Download for free at http://cnx.org/content/col11992/latest/

source of nearly all the light emitted by galaxies, we can learn a lot about the evolution of galaxies by studying the stars within them. What we find is that nearly all galaxies contain at least some very old stars. For example, our own Galaxy contains globular clusters with stars that are at least 13 billion years old, and some may be even older than that. Therefore, if we count the age of the Milky Way as the age of its oldest constituents, the Milky Way must have been born at least 13 billion years ago.

As we will discuss in The Big Bang, astronomers have discovered that the universe is expanding, and have traced the expansion backward in time. In this way, they have discovered that the universe itself is only about 13.8 billion years old. Thus, it appears that at least some of the globular-cluster stars in the Milky Way must have formed less than a billion years after the expansion began.

Several other observations also establish that star formation in the cosmos began very early. Astronomers have used spectra to determine the composition of some elliptical galaxies that are so far away that the light we see left them when the universe was only half as old as it is now. Yet these ellipticals contain old red stars, which must have formed billions of years earlier still.

When we make computer models of how such galaxies evolve with time, they tell us that star formation in elliptical galaxies began less than a billion years or so after the universe started its expansion, and new stars continued to form for a few billion years. But then star formation apparently stopped. When we compare distant elliptical galaxies with ones nearby, we find that ellipticals have not changed very much since the universe reached about half its current age. We'll return to this idea later in the chapter.

Observations of the most luminous galaxies take us even further back in time. Recently, as we have already noted, astronomers have discovered a few galaxies that are so far away that the light we see now left them less than a billion years or so after the beginning (Figure 28.3). Yet the spectra of some of these galaxies already contain lines of heavy elements, including carbon, silicon, aluminum, and sulfur. These elements were not present when the universe began but had to be manufactured in the interiors of stars. This means that when the light from these galaxies was emitted, an entire generation of stars had already been born, lived out their lives, and died—spewing out the new elements made in their interiors through supernova explosions—even before the universe was a billion years old. And it wasn't just a few stars in each galaxy that got started this way. Enough had to live and die to affect the overall composition of the galaxy, in a way that we can still measure in the spectrum from far away.

Download for free at http://cnx.org/content/col11992/latest/

Figure 28.3. Very Distant Galaxy. This image was made with the Hubble Space Telescope and shows the field around a luminous galaxy at a redshift z = 8.68, corresponding to a distance of about 13.2 billion light-years at the time when the light was emitted (indicated by the arrow and shown in the upper inset). Long exposures in the far-red and infrared wavelengths were combined to make the image, and additional infrared exposures with the Spitzer Space Telescope, which has lower spatial resolution than the Hubble (lower inset), show the redshifted light of normal stars. The very distant galaxy was detected because it has a strong emission line of hydrogen. This line is produced in regions where the formation of hot, young stars is taking place. (credit: modification of work by I. Labbé (Leiden University), NASA/ESA/JPL-Caltech)

Observations of *quasars* (galaxies whose centers contain a supermassive black hole) support this conclusion. We can measure the abundances of heavy elements in the gas near quasar black holes (explained in Active Galaxies, Quasars, and Supermassive Black Holes). The composition of this gas in quasars that emitted their light 12.5 billion light-years ago is very similar to that of the Sun. This means that a large portion of the gas surrounding the black holes must have already been cycled through stars during the first 1.3 billion years after the expansion of the universe began. If we allow time for this cycling, then their first stars must have formed when the universe was only a few hundred million years old.

A Changing Universe of Galaxies

Back in the middle decades of the twentieth century, the observation that all galaxies contain some old stars led astronomers to the hypothesis that galaxies were born fully formed near the time when the universe began its expansion. This hypothesis was similar to suggesting that human beings were born as adults and did not have to pass through the various stages of development from infancy through the teens. If this hypothesis were correct, the most distant galaxies should have shapes and sizes very much like the galaxies we see nearby. According to this old view, galaxies, after they formed, should then change only slowly, as successive generations of stars within them formed, evolved, and died. As the interstellar matter was slowly used up and fewer new stars formed, the galaxies would gradually become dominated by fainter, older stars and look dimmer and dimmer.

Thanks to the new generation of large ground- and space-based telescopes, we now know that this picture of galaxies evolving peacefully and in isolation from one another is completely wrong. As we will see later in this chapter, galaxies in the distant universe do not look like the Milky Way and nearby galaxies such as Andromeda, and the story of their development is more complex and involves far more interaction with their neighbors.

This OpenStax book is available for free at http://cnx.org/content/col11992/1.8

Why were astronomers so wrong? Up until the early 1990s, the most distant normal galaxy that had been observed emitted its light 8 billion years ago. Since that time, many galaxies—and particularly the giant ellipticals, which are the most luminous and therefore the easiest to see at large distances—did evolve peacefully and slowly. But the Hubble, Spitzer, Herschel, Keck, and other powerful new telescopes that have come on line since the 1990s make it possible to pierce the 8-billion-light-year barrier. We now have detailed views of many thousands of galaxies that emitted their light much earlier (some more than 13 billion years ago—see Figure 28.3).

Much of the recent work on the evolution of galaxies has progressed by studying a few specific small regions of the sky where the Hubble, Spitzer, and ground-based telescopes have taken extremely long exposure images. This allowed astronomers to detect very faint, very distant, and therefore very *young* galaxies (Figure 28.4). Our deep space telescope images show some galaxies that are 100 times fainter than the faintest objects that can be observed spectroscopically with today's giant ground-based telescopes. This turns out to mean that we can obtain the spectra needed to determine redshifts for only the very brightest five percent of the galaxies in these images.

Figure 28.4. Hubble Ultra-Deep Field. This image is the result of an 11-day-long observation with the Hubble Space Telescope of a tiny region of sky, located toward the constellation Fornax near the south celestial pole. This is an area that has only a handful of Milky Way stars. (Since the Hubble orbits Earth every 96 minutes, the telescope returned to view the same tiny piece of sky over and over again until enough light was collected and added together to make this very long exposure.) There are about 10,000 objects in this single image, nearly all of them galaxies, each with tens or hundreds of billions of stars. We can see some pinwheel-shaped spiral galaxies, which are like the Milky Way. But we also find a large variety of peculiar-shaped galaxies that are in collision with companion galaxies. Elliptical galaxies, which contain mostly old stars, appear as reddish blobs. (credit: modification of work by NASA, ESA, H. Teplitz and M. Rafelski (IPAC/Caltech), A. Koekemoer (STScI), R. Windhorst (Arizona State University), and Z. Levay (STScI))

Although we do not have spectra for most of the faint galaxies, the Hubble Space Telescope is especially well suited to studying their *shapes* because the images taken in space are not blurred by Earth's atmosphere. To the surprise of astronomers, the distant galaxies did not fit Hubble's classification scheme at all. Remember that Hubble found that nearly all nearby galaxies could be classified into a few categories, depending on whether they were ellipticals or spirals. The distant galaxies observed by the Hubble Space Telescope look very different from present-day galaxies, without identifiable spiral arms, disks, and bulges (Figure 28.5). They also tend to be much clumpier than most galaxies today. In other words, it's becoming clear that the shapes of galaxies have changed significantly over time. In fact, we now know that the Hubble scheme works well for only the last half of the age of the universe. Before then, galaxies were much more chaotic.

Download for free at http://cnx.org/content/col11992/latest/

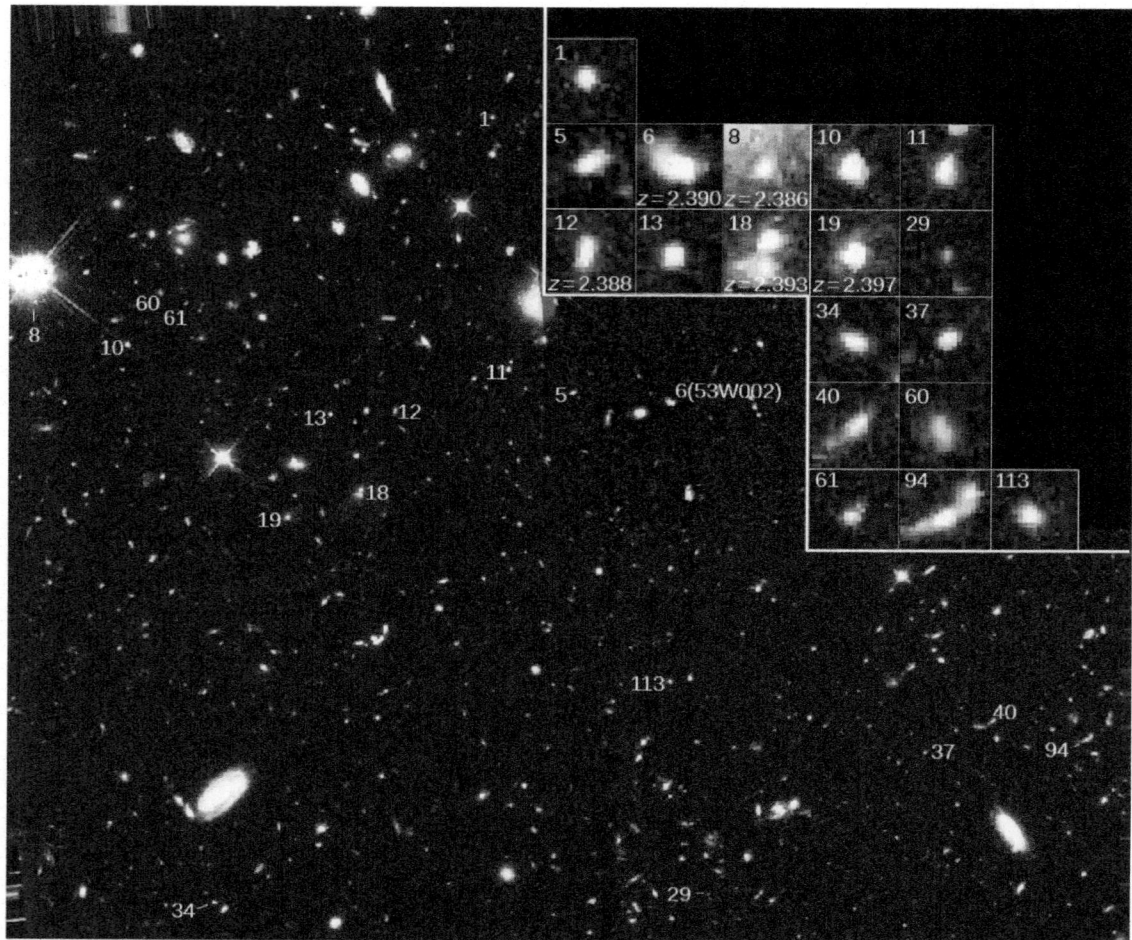

Figure 28.5. Early Galaxies. This Hubble Space Telescope image shows what are probably "galaxies under construction" in the early universe. The boxes in this color image show enlargements of 18 groups of stars smaller than galaxies as we know them. All these objects emitted their light about 11 billion years ago. They are typically only about 2,000 light-years across, which is much smaller than the Milky Way, with its diameter of 100,000 light-years. These 18 objects are found in a region only 2 million light-years across and are close enough together that they will probably collide and merge to build one or more normal galaxies. (credit: modification of work by Rogier Windhorst (Arizona State University) and NASA)

It's not just the shapes that are different. Nearly all the galaxies at distances greater than 11 billion light-years—that is, galaxies that we are seeing when they were less than 3 billion years old—are extremely blue, indicating that they contain a lot of young stars and that star formation in them is occurring at a higher rate than in nearby galaxies. Observations also show that very distant galaxies are systematically smaller on average than nearby galaxies. Relatively few galaxies present before the universe was about 8 billion years old have masses greater than 10^{11} M_{Sun}. That's 1/20 the mass of the Milky Way if we include its dark matter halo. Eleven billion years ago, there were only a few galaxies with masses greater than 10^{10} M_{Sun}. What we see instead seem to be small pieces or fragments of galactic material (Figure 28.6). When we look at galaxies that emitted their light 11 to 12 billion years ago, we now believe we are seeing the *seeds* of elliptical galaxies and of the central bulges of spirals. Over time, these smaller galaxies collided and merged to build up today's large galaxies.

This OpenStax book is available for free at http://cnx.org/content/col11992/1.8

Download for free at http://cnx.org/content/col11992/latest/

Bear in mind that stars that formed more than 11 billion years ago will be very old stars today. Indeed when we look nearby (at galaxies we see closer to our time), we find mostly old stars in the nuclear bulges of nearby spirals and in elliptical galaxies.

Figure 28.6. One of the Farthest, Faintest, and Smallest Galaxies Ever Seen. The small white boxes, labeled *a, b,* and *c,* mark the positions of three images of the same galaxy. These multiple images were produced by the massive cluster of galaxies known as Abell 2744, which is located between us and the galaxy and acts as a gravitational lens. The arrows in the enlarged insets at right point to the galaxy. Each magnified image makes the galaxy appear as much as 10 times larger and brighter than it would look without the intervening lens. This galaxy emitted the light we observe today when the universe was only about 500 million years old. When the light was emitted the galaxy was tiny—only 850 light-years across, or 500 times smaller than the Milky, and its mass was only 40 million times the mass of the Sun. Star formation is going on in this galaxy, but it appears red in the image because of its large redshift. (credit: modification of work by NASA, ESA, A. Zitrin (California Institute of Technology), and J. Lotz, M. Mountain, A. Koekemoer, and the HFF Team (STScI))

What such observations are showing us is that galaxies have grown in size as the universe has aged. Not only were galaxies smaller several billion years ago, but there were more of them; gas-rich galaxies, particularly the less luminous ones, were much more numerous then than they are today.

Those are some of the basic observations we can make of individual galaxies (and their evolution) looking back in cosmic time. Now we want to turn to the larger context. If stars are grouped into galaxies, are the galaxies also grouped in some way? In the third section of this chapter, we'll explore the largest structures known in the universe.

28.2 GALAXY MERGERS AND ACTIVE GALACTIC NUCLEI

Learning Objectives

By the end of this section, you will be able to:

Download for free at http://cnx.org/content/col11992/latest/

> Explain how galaxies grow by merging with other galaxies and by consuming smaller galaxies (for lunch)
> Describe the effects that supermassive black holes in the centers of most galaxies have on the fate of their host galaxies

One of the conclusions astronomers have reached from studying distant galaxies is that collisions and mergers of whole galaxies play a crucial role in determining how galaxies acquired the shapes and sizes we see today. Only a few of the nearby galaxies are currently involved in collisions, but detailed studies of those tell us what to look for when we seek evidence of mergers in very distant and very faint galaxies. These in turn give us important clues about the different evolutionary paths galaxies have taken over cosmic time. Let's examine in more detail what happens when two galaxies collide.

Mergers and Cannibalism

Figure 28.1 shows a dynamic view of two galaxies that are colliding. The stars themselves in this pair of galaxies will not be affected much by this cataclysmic event. (See the Astronomy Basics feature box Why Galaxies Collide but Stars Rarely Do.) Since there is a lot of space between the stars, a direct collision between two stars is very unlikely. However, the *orbits* of many of the stars will be changed as the two galaxies move through each other, and the change in orbits can totally alter the appearance of the interacting galaxies. A gallery of interesting colliding galaxies is shown in Figure 28.7. Great rings, huge tendrils of stars and gas, and other complex structures can form in such cosmic collisions. Indeed, these strange shapes are the signposts that astronomers use to identify colliding galaxies.

This OpenStax book is available for free at http://cnx.org/content/col11992/1.8

Download for free at http://cnx.org/content/col11992/latest/

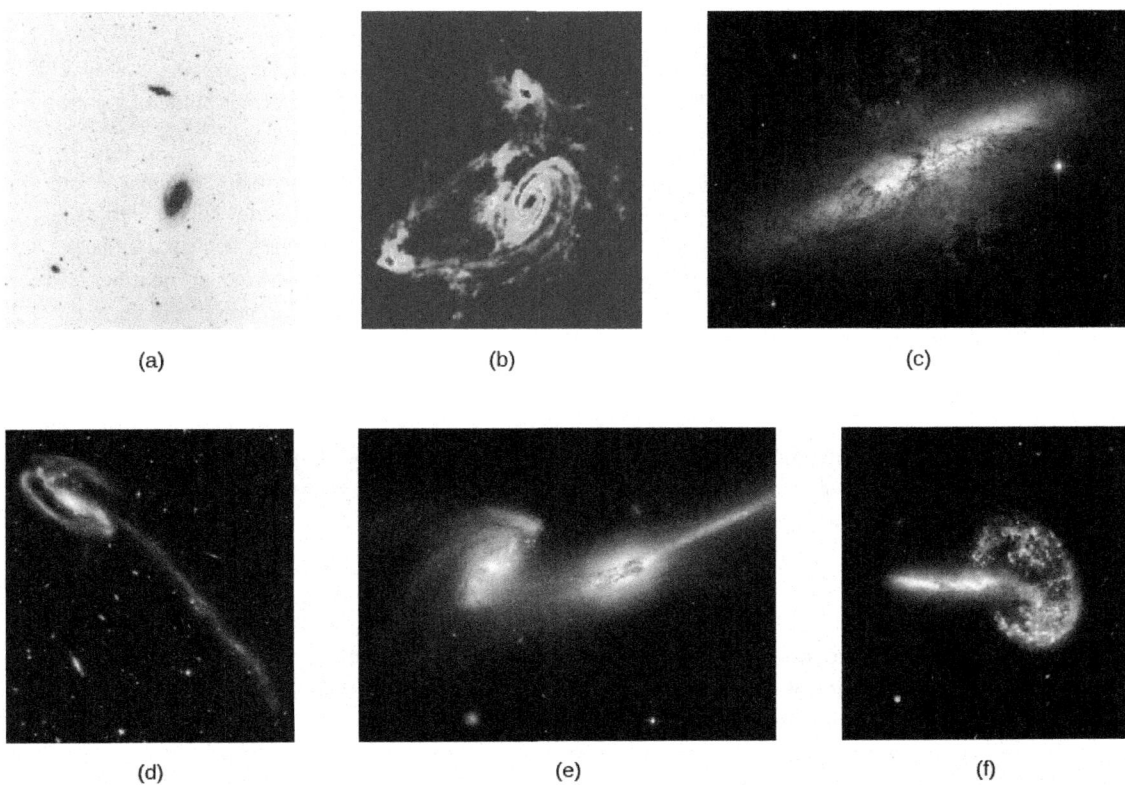

(a)   (b)   (c)

(d)   (e)   (f)

Figure 28.7. Gallery of Interacting Galaxies. (a and b) M82 (smaller galaxy at top) and M83 (spiral) are seen (a) in a black-and-white visible light image and (b) in radio waves given off by cold hydrogen gas. The hydrogen image shows that the two galaxies are wrapped in a common shroud of gas that is being tugged and stretched by the gravity of the two galaxies. (c) This close-up view by the Hubble Space Telescope shows some of the effects of this interaction on galaxy M82, including gas streaming outward (red tendrils) powered by supernovae explosions of massive stars formed in the burst of star formation that was a result of the collision. (d) Galaxy UGC 10214 ("The Tadpole") is a barred spiral galaxy 420 million light-years from the Milky Way that has been disrupted by the passage of a smaller galaxy. The interloper's gravity pulled out the long tidal tail, which is about 280,000 light-years long, and triggered bursts of star formation seen as blue clumps along the tail. (e) Galaxies NGC 4676 A and B are nicknamed "The Mice." In this Hubble Space Telescope image, you can see the long, narrow tails of stars pulled away from the galaxies by the interactions of the two spirals. (e) Arp 148 is a pair of galaxies that are caught in the act of merging to become one new galaxy. The two appear to have already passed through each other once, causing a shockwave that reformed one into a bright blue ring of star formation, like the ripples from a stone tossed into a pond. (credit a, b: modification of work by NRAO/AUI; credit c: modification of work by NASA, ESA, and The Hubble Heritage Team (STScI/AURA); credit d, e: modification of work by NASA, H. Ford (JHU), G. Illingworth (UCSC/LO), M.Clampin (STScI), G. Hartig (STScI), the ACS Science Team, and ESA; credit f: modification of work by NASA, ESA, the Hubble Heritage (STScI/AURA)-ESA/Hubble Collaboration, and A. Evans (University of Virginia, Charlottesville/NRAO/Stony Brook University))

ASTRONOMY BASICS

Why Galaxies Collide but Stars Rarely Do

Throughout this book we have emphasized the large distances between objects in space. You might therefore have been surprised to hear about collisions between galaxies. Yet (except at the very cores of galaxies) we have not worried at all about stars inside a galaxy colliding with each other. Let's see why there is a difference.

The reason is that stars are pitifully small compared to the distances between them. Let's use our Sun as an example. The Sun is about 1.4 million kilometers wide, but is separated from the closest other star by about 4 light-years, or about 38 trillion kilometers. In other words, the Sun is 27 million of its own

Download for free at http://cnx.org/content/col11992/latest/

diameters from its nearest neighbor. If the Sun were a grapefruit in New York City, the nearest star would be another grapefruit in San Francisco. This is typical of stars that are not in the nuclear bulge of a galaxy or inside star clusters. Let's contrast this with the separation of galaxies.

The visible disk of the Milky Way is about 100,000 light-years in diameter. We have three satellite galaxies that are just one or two Milky Way diameters away from us (and will probably someday collide with us). The closest major spiral is the Andromeda Galaxy (M31), about 2.4 million light-years away. If the Milky Way were a pancake at one end of a big breakfast table, M31 would be another pancake at the other end of the same table. Our nearest large galaxy neighbor is only 24 of our Galaxy's diameters from us, and it will begin to crash into the Milky Way in about 3 billion years.

Galaxies in rich clusters are even closer together than those in our neighborhood (see The Distribution of Galaxies in Space). Thus, the chances of galaxies colliding are far greater than the chances of stars in the disk of a galaxy colliding. And we should note that the difference between the separation of galaxies and stars also means that when galaxies do collide, their stars almost always pass right by each other like smoke passing through a screen door.

The details of galaxy collisions are complex, and the process can take hundreds of millions of years. Thus, collisions are best simulated on a computer (Figure 28.8), where astronomers can calculate the slow interactions of stars, and clouds of gas and dust, via gravity. These calculations show that if the collision is slow, the colliding galaxies may coalesce to form a single galaxy.

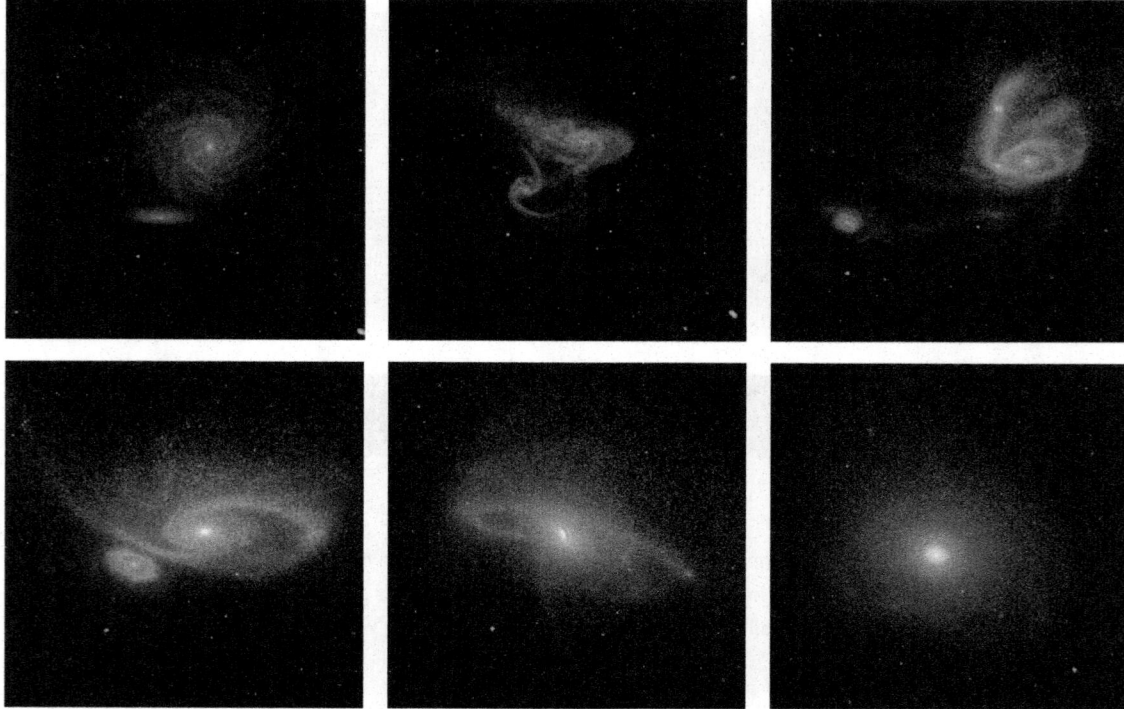

Figure 28.8. Computer Simulation of a Galaxy Collision. This computer simulation starts with two spiral galaxies merging and ends with a single elliptical galaxy. The colors show the colors of stars in the system; note the bursts of blue color as copious star formation gets triggered by the interaction. The timescale from start to finish in this sequence is about a billion years. (credit: modification of work by P. Jonsson (Harvard-Smithsonian Center for Astrophysics), G. Novak (Princeton University), and T. J. Cox (Carnegie Observatories))

This OpenStax book is available for free at http://cnx.org/content/col11992/1.8

Download for free at http://cnx.org/content/col11992/latest/

When two galaxies of equal size are involved in a collision, we call such an interaction a **merger** (the term applied in the business world to two equal companies that join forces). But small galaxies can also be swallowed by larger ones—a process astronomers have called, with some relish, **galactic cannibalism** (Figure 28.9).

LINK TO LEARNING

Modern personal computers are more than powerful enough to compute what happens when galaxies collide. Here's a website and Java applet (https://openstax.org/l/30whgalaxcoll) that will let you try your own hand at crashing two spiral galaxies together from the comfort of your own home or dorm room. By changing a few basic controls such as the relative masses, their separation, and the orientation of each galaxy's disk, you can create a wide range of resulting merger results. (You can also download a similar app (https://openstax.org/l/30iphoneapp) for your iPhone or iPad.)

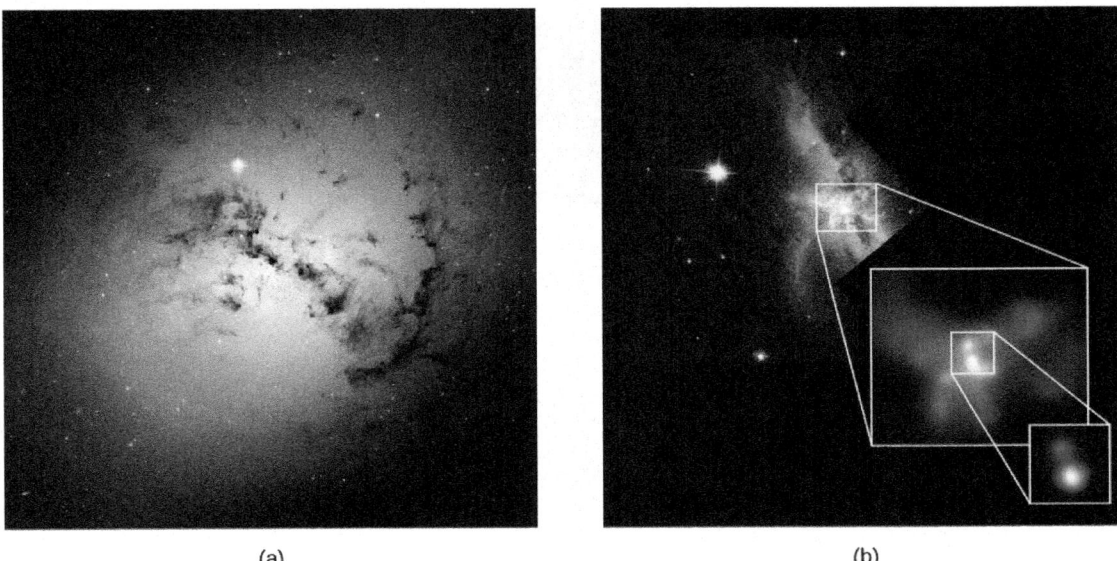

(a)   (b)

Figure 28.9. Galactic Cannibalism. (a) This Hubble image shows the eerie silhouette of dark dust clouds against the glowing nucleus of the elliptical galaxy NGC 1316. Elliptical galaxies normally contain very little dust. These clouds are probably the remnant of a small companion galaxy that was cannibalized (eaten) by NGC 1316 about 100 million years ago. (b) The highly disturbed galaxy NGC 6240, imaged by Hubble Space Telescope (background image) and Chandra X-ray Telescope (both insets) is apparently the product of a merger between two gas-rich spiral galaxies. The X-ray images show that there is not one but two nuclei, both glowing brightly in X-rays and separated by only 4000 light-years. These are likely the locations of two supermassive black holes that inhabited the cores of the two galaxies pre-merger; here they are participating in a kind of "death spiral," in which the two black holes themselves will merge to become one. (credit a: modification of work by NASA, ESA, and The Hubble Heritage Team (STScI/AURA); credit b: X-ray: NASA/CXC/MPE/S.Komossa et al.; Optical: NASA/STScI/R.P.van der Marel & J.Gerssen)

The very large elliptical galaxies we discussed in Galaxies probably form by cannibalizing a variety of smaller galaxies in their clusters. These "monster" galaxies frequently possess more than one nucleus and have probably acquired their unusually high luminosities by swallowing nearby galaxies. The multiple nuclei are the remnants of their victims (Figure 28.9). Many of the large, peculiar galaxies that we observe also owe their chaotic shapes to past interactions. Slow collisions and mergers can even transform two or more spiral galaxies into a single elliptical galaxy.

Download for free at http://cnx.org/content/col11992/latest/

A change in shape is not all that happens when galaxies collide. If either galaxy contains interstellar matter, the collision can compress the gas and trigger an increase in the rate at which stars are being formed—by as much as a factor of 100. Astronomers call this abrupt increase in the number of stars being formed a **starburst**, and the galaxies in which the increase occurs are termed starburst galaxies (Figure 28.10). In some interacting galaxies, star formation is so intense that all the available gas is exhausted in only a few million years; the burst of star formation is clearly only a temporary phenomenon. While a starburst is going on, however, the galaxy where it is taking place becomes much brighter and much easier to detect at large distances.

(a)   (b)

Figure 28.10. Starburst Associated with Colliding Galaxies. (a) Three of the galaxies in the small group known as Stephan's Quintet are interacting gravitationally with each other (the galaxy at upper left is actually much closer than the other three and is not part of this interaction), resulting in the distorted shapes seen here. Long strings of young, massive blue stars and hundreds of star formation regions glowing in the pink light of excited hydrogen gas are also results of the interaction. The ages of the star clusters range from 2 million to 1 billion years old, suggesting that there have been several different collisions within this group of galaxies, each leading to bursts of star formation. The three interacting members of Stephan's Quintet are located at a distance of 270 million light-years. (b) Most galaxies form new stars at a fairly slow rate, but members of a rare class known as starburst galaxies blaze with extremely active star formation. The galaxy II Zw 096 is one such starburst galaxy, and this combined image using both Hubble and Spitzer Space Telescope data shows that it is forming bright clusters of new stars at a prodigious rate. The blue colors show the merging galaxies in visible light, while the red colors show infrared radiation from the dusty region where star formation is happening. This galaxy is at a distance of 500 million light-years and has a diameter of about 50,000 light-years, about half the size of the Milky Way. (credit a: modification of work by NASA, ESA, and the Hubble SM4 ERO Team; credit b: modification of work by NASA/JPL-Caltech/STScI)

When astronomers finally had the tools to examine a significant number of galaxies that emitted their light 11 to 12 billion years ago, they found that these very young galaxies often resemble nearby starburst galaxies that are involved in mergers: they also have multiple nuclei and peculiar shapes, they are usually clumpier than normal galaxies today, with multiple intense knots and lumps of bright starlight, and they have higher rates of star formation than isolated galaxies. They also contain lots of blue, young, type O and B stars, as do nearby merging galaxies.

Galaxy mergers in today's universe are rare. Only about five percent of nearby galaxies are currently involved in interactions. Interactions were much more common billions of years ago (Figure 28.11) and helped build up the "more mature" galaxies we see in our time. Clearly, interactions of galaxies have played a crucial role in their evolution.

This OpenStax book is available for free at http://cnx.org/content/col11992/1.8

Download for free at http://cnx.org/content/col11992/latest/

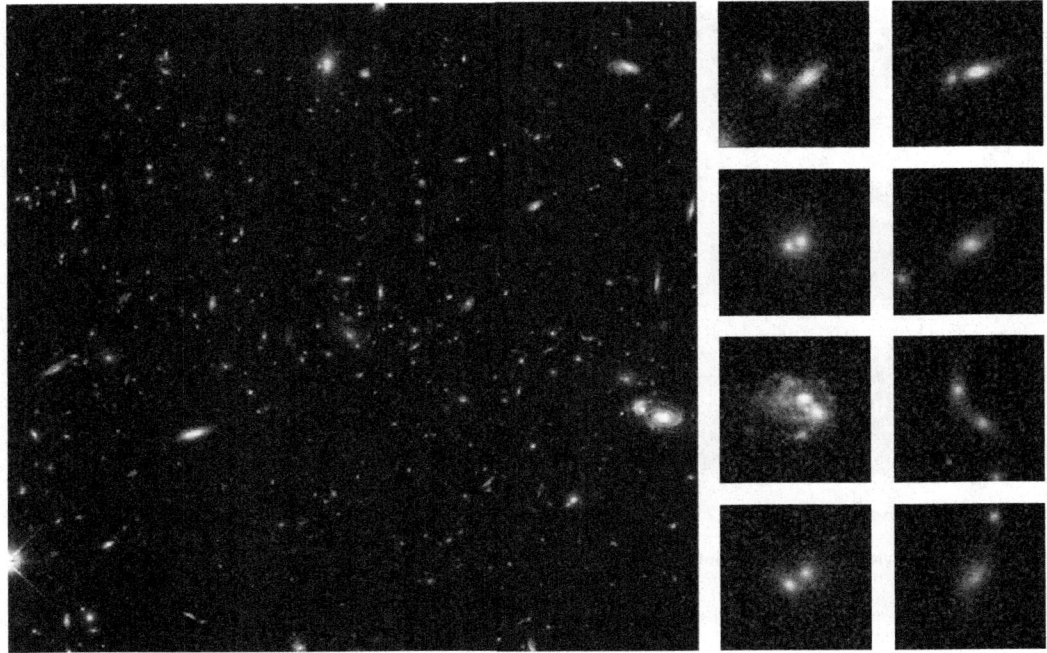

Figure 28.11. Collisions of Galaxies in a Distant Cluster. The large picture on the left shows the Hubble Space Telescope image of a cluster of galaxies at a distance of about 8 billion light-years. Among the 81 galaxies in the cluster that have been examined in some detail, 13 are the result of recent collisions of pairs of galaxies. The eight smaller images on the right are close-ups of some of the colliding galaxies. The merger process typically takes a billion years or so. (credit: modification of work by Pieter van Dokkum, Marijn Franx (University of Groningen/Leiden), ESA and NASA)

Active Galactic Nuclei and Galaxy Evolution

While galaxy mergers are huge, splashy events that completely reshape entire galaxies on scales of hundreds of thousands of light-years and can spark massive bursts of star formation, accreting black holes inside galaxies can also disturb and alter the evolution of their host galaxies. You learned in Active Galaxies, Quasars, and Supermassive Black Holes about a family of objects known as *active galactic nuclei* (AGN), all of them powered by supermassive black holes. If the black hole is surrounded by enough gas, some of the gas can fall into the black hole, getting swept up on the way into an accretion disk, a compact, swirling maelstrom perhaps only 100 AU across (the size of our solar system).

Within the disk the gas heats up until it shines brilliantly even in X-rays, often outshining the rest of the host galaxy with its billions of stars. Supermassive black holes and their accretion disks can be violent and powerful places, with some material getting sucked into the black hole but even more getting shot out along huge jets perpendicular to the disk. These powerful jets can extend far outside the starry edge of the galaxy.

AGN were much more common in the early universe, in part because frequent mergers provided a fresh gas supply for the black hole accretion disks. Examples of AGN in the nearby universe today include the one in galaxy M87 (see Figure 27.7), which sports a jet of material shooting out from its nucleus at speeds close to the speed of light, and the one in the bright galaxy NGC 5128, also known as Centaurus A (see Figure 28.12).

Download for free at http://cnx.org/content/col11992/latest/

Figure 28.12. Composite View of the Galaxy Centaurus A. This artificially colored image was made using data from three different telescopes: submillimeter radiation with a wavelength of 870 microns is shown in orange; X-rays are seen in blue; and visible light from stars is shown in its natural color. Centaurus A has an active galactic nucleus that is powering two jets, seen in blue and orange, reaching in opposite directions far outside the galaxy's stellar disk, and inflating two huge lobes, or clouds, of hot X-ray-emitting gas. Centaurus is at a distance of 13 million light-years, making it one of the closest active galaxies we know. (credit: modification of work by ESO/WFI (Optical); MPIfR/ESO/APEX/A. Weiss et al. (Submillimeter); NASA/CXC/CfA/R.Kraft et al. (X-ray))

Many highly accelerated particles move with the jets in such galaxies. Along the way, the particles in the jets can plow into gas clouds in the interstellar medium, breaking them apart and scattering them. Since denser clouds of gas and dust are required for material to clump together to make stars, the disruption of the clouds can halt star formation in the host galaxy or cut it off before it even begins.

In this way, quasars and other kinds of AGN can play a crucial role in the evolution of their galaxies. For example, there is growing evidence that the merger of two gas-rich galaxies not only produces a huge burst of star formation, but also triggers AGN activity in the core of the new galaxy. That activity, in turn, could then slow down or shut off the burst of star formation—which could have significant implications for the apparent shape, brightness, chemical content, and stellar components of the entire galaxy. Astronomers refer to that process as *AGN feedback*, and it is apparently an important factor in the evolution of most galaxies.

28.3 | THE DISTRIBUTION OF GALAXIES IN SPACE

Learning Objectives

By the end of this section, you will be able to:

> Explain the cosmological principle and summarize the evidence that it applies on the largest scales of the known universe
> Describe the contents of the Local Group of galaxies
> Distinguish among groups, clusters, and superclusters of galaxies
> Describe the largest structures seen in the universe, including voids

In the preceding section, we emphasized the role of mergers in shaping the evolution of galaxies. In order to collide, galaxies must be fairly close together. To estimate how often collisions occur and how they affect galaxy evolution, astronomers need to know how galaxies are distributed in space and over cosmic time. Are most of

This OpenStax book is available for free at http://cnx.org/content/col11992/1.8

Download for free at http://cnx.org/content/col11992/latest/

them isolated from one another or do they congregate in groups? If they congregate, how large are the groups and how and when did they form? And how, in general, are galaxies and their groups arranged in the cosmos? Are there as many in one direction of the sky as in any other, for example? How did galaxies get to be arranged the way we find them today?

Edwin Hubble found answers to some of these questions only a few years after he first showed that the spiral nebulae were galaxies and not part of our Milky Way. As he examined galaxies all over the sky, Hubble made two discoveries that turned out to be crucial for studies of the evolution of the universe.

The Cosmological Principle

Hubble made his observations with what were then the world's largest telescopes—the 100-inch and 60-inch reflectors on Mount Wilson. These telescopes have small fields of view: they can see only a small part of the heavens at a time. To photograph the entire sky with the 100-inch telescope, for example, would have taken longer than a human lifetime. So instead, Hubble sampled the sky in many regions, much as Herschel did with his star gauging (see The Architecture of the Galaxy). In the 1930s, Hubble photographed 1283 sample areas, and on each print, he carefully counted the numbers of galaxy images ((Figure 28.13).).

The first discovery Hubble made from his survey was that the number of galaxies visible in each area of the sky is about the same. (Strictly speaking, this is true only if the light from distant galaxies is not absorbed by dust in our own Galaxy, but Hubble made corrections for this absorption.) He also found that the numbers of galaxies increase with faintness, as we would expect if the density of galaxies is about the same at all distances from us.

To understand what we mean, imagine you are taking snapshots in a crowded stadium during a sold-out concert. The people sitting near you look big, so only a few of them will fit into a photo. But if you focus on the people sitting in seats way on the other side of the stadium, they look so small that many more will fit into your picture. If all parts of the stadium have the same seat arrangements, then as you look farther and farther away, your photo will get more and more crowded with people. In the same way, as Hubble looked at fainter and fainter galaxies, he saw more and more of them.

Figure 28.13. Hubble at Work. Edwin Hubble at the 100-inch telescope on Mount Wilson. (credit: NASA)

Hubble's findings are enormously important, for they indicate that the universe is both **isotropic** and **homogeneous**—it looks the same in all directions, and a large volume of space at any given redshift or distance is much like any other volume at that redshift. If that is so, it does not matter what section of the universe we observe (as long as it's a sizable portion): any section will look the same as any other.

Hubble's results—and many more that have followed in the nearly 100 years since then—imply not only that the universe is about the same everywhere (apart from changes with time) but also that aside from small-scale local differences, the part we can see around us is representative of the whole. The idea that the universe is

Download for free at http://cnx.org/content/col11992/latest/

the same everywhere is called the **cosmological principle** and is the starting assumption for nearly all theories that describe the entire universe (see The Big Bang).

Without the cosmological principle, we could make no progress at all in studying the universe. Suppose our own local neighborhood were unusual in some way. Then we could no more understand what the universe is like than if we were marooned on a warm south-sea island without outside communication and were trying to understand the geography of Earth. From our limited island vantage point, we could not know that some parts of the planet are covered with snow and ice, or that large continents exist with a much greater variety of terrain than that found on our island.

Hubble merely counted the numbers of galaxies in various directions without knowing how far away most of them were. With modern instruments, astronomers have measured the velocities and distances of hundreds of thousands of galaxies, and so built up a meaningful picture of the large-scale structure of the universe. In the rest of this section, we describe what we know about the distribution of galaxies, beginning with those that are nearby.

The Local Group

The region of the universe for which we have the most detailed information is, as you would expect, our own local neighborhood. It turns out that the Milky Way Galaxy is a member of a small group of galaxies called, not too imaginatively, the **Local Group**. It is spread over about 3 million light-years and contains more than 54 members. There are three large spiral galaxies (our own, the Andromeda galaxy, and M33), two intermediate ellipticals, and many dwarf ellipticals and irregular galaxies.

New members of the Local Group are still being discovered. We mentioned in The Milky Way Galaxy a dwarf galaxy only about 80,000 light-years from Earth and about 50,000 light-years from the center of the galaxy that was discovered in 1994 in the constellation of Sagittarius. (This dwarf is actually venturing too close to the much larger Milky Way and will eventually be consumed by it.)

Many of the recent discoveries have been made possible by the new generation of automated, sensitive, wide-field surveys, such as the Sloan Digital Sky Survey, that map the positions of millions of stars across most of the visible sky. By digging into the data with sophisticated computer programs, astronomers have turned up numerous tiny, faint dwarf galaxies that are all but invisible to the eye even in those deep telescopic images. These new findings may help solve a long-standing problem: the prevailing theories of how galaxies form predicted that there should be more dwarf galaxies around big galaxies like the Milky Way than had been observed—and only now do we have the tools to find these faint and tiny galaxies and begin to compare the numbers of them with theoretical predictions.

LINK TO LEARNING

You can read more about the Sloan survey (https://openstax.org/l/30sloansurvey) and its dramatic results. And check out this brief animation (https://openstax.org/l/30anifliarrgal) of a flight through the arrangement of the galaxies as revealed by the survey.

Several new dwarf galaxies have also been found near the Andromeda galaxy. Such dwarf galaxies are difficult to find because they typically contain relatively few stars, and it is hard to distinguish them from the foreground stars in our own Milky Way.

This OpenStax book is available for free at http://cnx.org/content/col11992/1.8

Figure 28.14 is a rough sketch showing where the brighter members of the Local Group are located. The average of the motions of all the galaxies in the Local Group indicates that its total mass is about $4 \times 10^{12}\ M_{Sun}$, and at least half of this mass is contained in the two giant spirals—the Andromeda galaxy and the Milky Way Galaxy. And bear in mind that a substantial amount of the mass in the Local Group is in the form of dark matter.

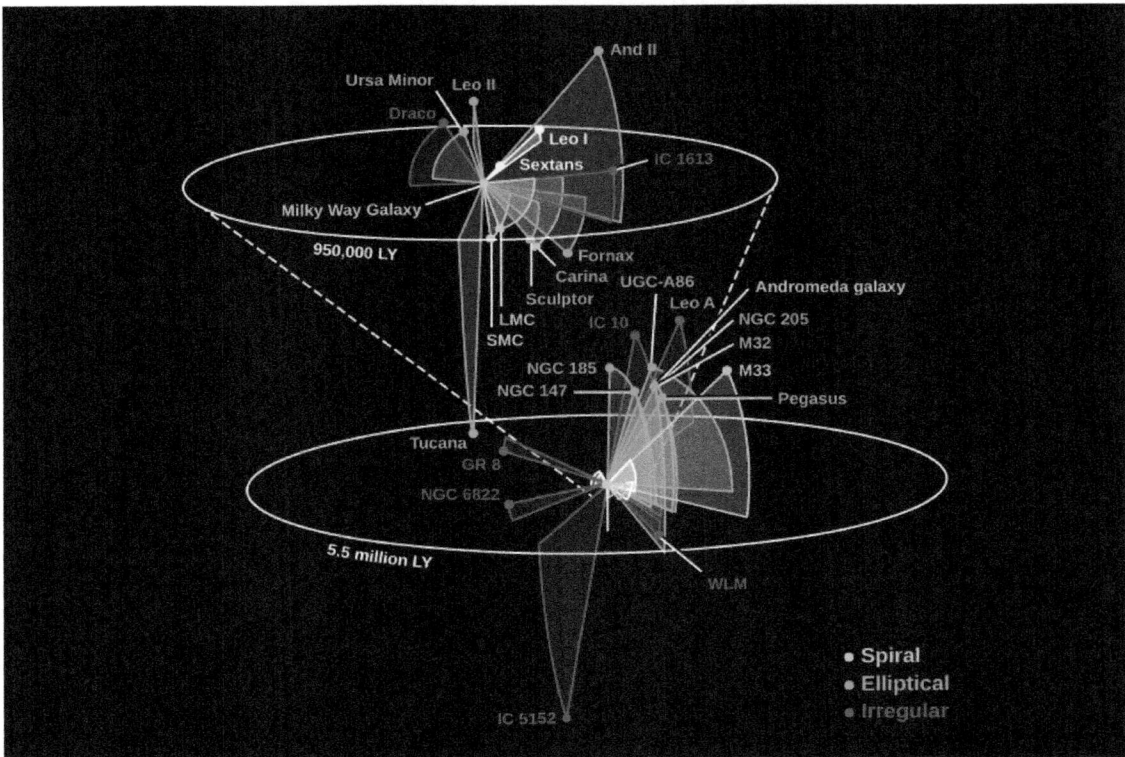

Figure 28.14. Local Group. This illustration shows some members of the Local Group of galaxies, with our Milky Way at the center. The exploded view at the top shows the region closest to the Milky Way and fits into the bigger view at the bottom as shown by the dashed lines. The three largest galaxies among the three dozen or so members of the Local Group are all spirals; the others are small irregular galaxies and dwarf ellipticals. A number of new members of the group have been found since this map was made.

Neighboring Groups and Clusters

Small galaxy groups like ours are hard to notice at larger distances. However, there are much more substantial groups called galaxy clusters that are easier to spot even many millions of light-years away. Such clusters are described as *poor* or *rich* depending on how many galaxies they contain. Rich clusters have thousands or even tens of thousands of galaxies, although many of the galaxies are quite faint and hard to detect.

The nearest moderately rich galaxy cluster is called the Virgo Cluster, after the constellation in which it is seen. It is about 50 million light-years away and contains thousands of members, of which a few are shown in Figure 28.15. The giant elliptical (and very active) galaxy M87, which you came to know and love in the chapter on Active Galaxies, Quasars, and Supermassive Black Holes, belongs to the Virgo Cluster.

Download for free at http://cnx.org/content/col11992/latest/

Figure 28.15. Central Region of the Virgo Cluster. Virgo is the nearest rich cluster and is at a distance of about 50 million light-years. It contains hundreds of bright galaxies. In this picture you can see only the central part of the cluster, including the giant elliptical galaxy M87, just below center. Other spirals and ellipticals are visible; the two galaxies to the top right are known as "The Eyes." (credit: modification of work by Chris Mihos (Case Western Reserve University)/ESO)

A good example of a cluster that is much larger than the Virgo complex is the Coma cluster, with a diameter of at least 10 million light-years (Figure 28.16). Some 250 to 300 million light-years distant, this cluster is centered on two giant ellipticals whose luminosities equal about 400 billion Suns each. Thousands of galaxies have been observed in Coma, but the galaxies we see are almost certainly only part of what is really there. Dwarf galaxies are too faint to be seen at the distance of Coma, but we expect they are part of this cluster just as they are part of nearer ones. If so, then Coma likely contains tens of thousands of galaxies. The total mass of this cluster is about 4×10^{15} M_{Sun} (enough mass to make 4 million billion stars like the Sun).

Let's pause here for a moment of perspective. We are now discussing numbers by which even astronomers sometimes feel overwhelmed. The Coma cluster may have 10, 20, or 30 thousand galaxies, and each galaxy has billions and billions of stars. If you were traveling at the speed of light, it would still take you more than 10 million years (longer than the history of the human species) to cross this giant swarm of galaxies. And if you lived on a planet on the outskirts of one of these galaxies, many other members of the cluster would be close enough to be noteworthy sights in your nighttime sky.

This OpenStax book is available for free at http://cnx.org/content/col11992/1.8

Download for free at http://cnx.org/content/col11992/latest/

Figure 28.16. Central Region of the Coma Cluster. This combined visible-light (from the Sloan Digital Sky Survey) and infrared (from the Spitzer Space Telescope) image has been color coded so that faint dwarf galaxies are seen as green. Note the number of little green smudges on the image. The cluster is roughly 320 million light-years away from us. (credit: modification of work by NASA/JPL-Caltech/L. Jenkins (GSFC))

Really rich clusters such as Coma usually have a high concentration of galaxies near the center. We can see giant elliptical galaxies in these central regions but few, if any, spiral galaxies. The spirals that do exist generally occur on the outskirts of clusters.

We might say that ellipticals are highly "social": they are often found in groups and very much enjoy "hanging out" with other ellipticals in crowded situations. It is precisely in such crowds that collisions are most likely and, as we discussed earlier, we think that most large ellipticals are built through mergers of smaller galaxies.

Spirals, on the other hand, are more "shy": they are more likely to be found in poor clusters or on the edges of rich clusters where collisions are less likely to disrupt the spiral arms or strip out the gas needed for continued star formation.

ASTRONOMY BASICS

Gravitational Lensing

As we saw in Black Holes and Curved Spacetime, spacetime is more strongly curved in regions where the gravitational field is strong. Light passing very near a concentration of matter appears to follow a curved path. In the case of starlight passing close to the Sun, we measure the position of the distant star to be slightly different from its true position.

Now let's consider the case of light from a distant galaxy or quasar that passes near a concentration of matter such as a cluster of galaxies on its journey to our telescopes. According to general relativity, the

Download for free at http://cnx.org/content/col11992/latest/

light path may be bent in a variety of ways; as a result we can observe distorted and even multiple images (Figure 28.17).

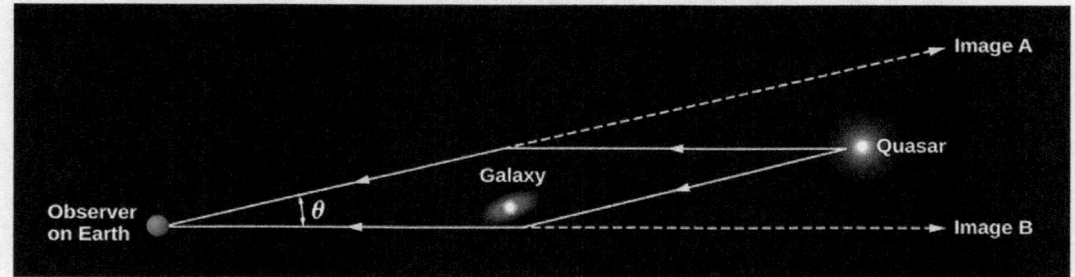

Figure 28.17. Gravitational Lensing. This drawing shows how a gravitational lens can make two images. Two light rays from a distant quasar are shown being bent while passing a foreground galaxy; they then arrive together at Earth. Although the two beams of light contain the same information, they now appear to come from two different points on the sky. This sketch is oversimplified and not to scale, but it gives a rough idea of the lensing phenomenon.

Gravitational lenses can produce not only double images, as shown in Figure 28.17, but also multiple images, arcs, or rings. The first gravitational lens discovered, in 1979, showed two images of the same distant object. Eventually, astronomers used the Hubble Space Telescope to capture remarkable images of the effects of gravitational lenses. One example is shown in Figure 28.18.

Figure 28.18. Multiple Images of a Gravitationally Lensed Supernova. Light from a supernova at a distance of 9 billion light-years passed near a galaxy in a cluster at a distance of about 5 billion light-years. In the enlarged inset view of the galaxy, the arrows point to the multiple images of the exploding star. The images are arranged around the galaxy in a cross-shaped pattern called an Einstein Cross. The blue streaks wrapping around the galaxy are the stretched images of the supernova's host spiral galaxy, which has been distorted by the warping of space. (credit: modification of work by NASA, ESA, and S. Rodney (JHU) and the FrontierSN team; T. Treu (UCLA), P. Kelly (UC Berkeley), and the GLASS team; J. Lotz (STScI) and the Frontier Fields team; M. Postman (STScI) and the CLASH team; and Z. Levay (STScI))

This OpenStax book is available for free at http://cnx.org/content/col11992/1.8

Download for free at http://cnx.org/content/col11992/latest/

General relativity predicts that the light from a distant object may also be amplified by the lensing effect, thereby making otherwise invisible objects bright enough to detect. This is particularly useful for probing the earliest stages of galaxy formation, when the universe was young. Figure 28.19 shows an example of a very distant faint galaxy that we can study in detail only because its light path passes through a large concentration of massive galaxies and we now see a brighter image of it.

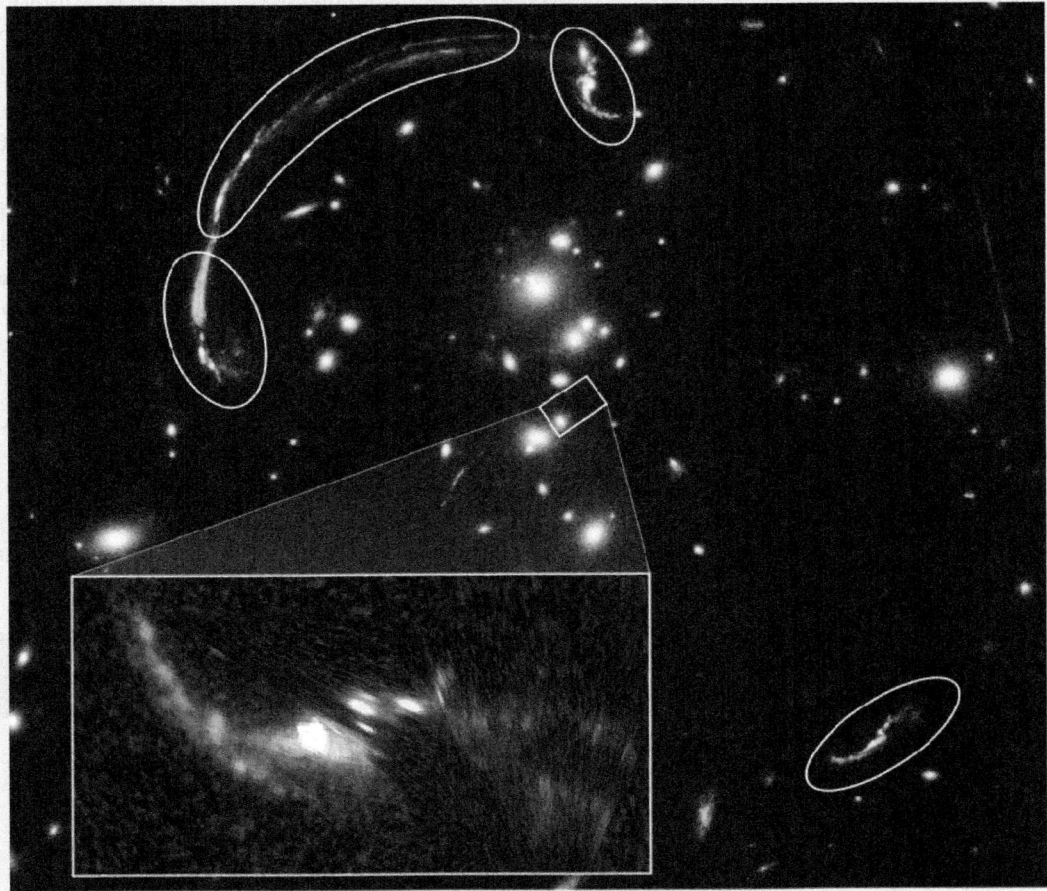

Figure 28.19. Distorted Images of a Distant Galaxy Produced by Gravitational Lensing in a Galaxy Cluster. The rounded outlines show the location of distinct, distorted images of the background galaxy resulting from lensing by the mass in the cluster. The image in the box at lower left is a reconstruction of what the lensed galaxy would look like in the absence of the cluster, based on a model of the cluster's mass distribution, which can be derived from studying the distorted galaxy images. The reconstruction shows far more detail about the galaxy than could have been seen in the absence of lensing. As the image shows, this galaxy contains regions of star formation glowing like bright Christmas tree bulbs. These are much brighter than any star-formation regions in our Milky Way Galaxy. (credit: modification of work by NASA, ESA, and Z. Levay (STScI))

We should note that the visible mass in a galaxy is not the only possible gravitational lens. Dark matter can also reveal itself by producing this effect. Astronomers are using lensed images from all over the sky to learn more about where dark matter is located and how much of it exists.

Superclusters and Voids

After astronomers discovered clusters of galaxies, they naturally wondered whether there were still larger structures in the universe. Do clusters of galaxies gather together? To answer this question, we must be able to

Download for free at http://cnx.org/content/col11992/latest/

map large parts of the universe in three dimensions. We must know not only the position of each galaxy on the sky (that's two dimensions) but also its distance from us (the third dimension).

This means we must be able to measure the redshift of each galaxy in our map. Taking a spectrum of each individual galaxy to do this is a much more time-consuming task than simply counting galaxies seen in different directions on the sky, as Hubble did. Today, astronomers have found ways to get the spectra of many galaxies in the same field of view (sometimes hundreds or even thousands at a time) to cut down the time it takes to finish their three-dimensional maps. Larger telescopes are also able to measure the redshifts—and therefore the distances—of much more distant galaxies and (again) to do so much more quickly than previously possible.

Another challenge astronomers faced in deciding how to go about constructing a map of the universe is similar to that confronted by the first team of explorers in a huge, uncharted territory on Earth. Since there is only one band of explorers and an enormous amount of land, they have to make choices about where to go first. One strategy might be to strike out in a straight line in order to get a sense of the terrain. They might, for example, cross some mostly empty prairies and then hit a dense forest. As they make their way through the forest, they learn how thick it is in the direction they are traveling, but not its width to their left or right. Then a river crosses their path; as they wade across, they can measure its width but learn nothing about its length. Still, as they go on in their straight line, they begin to get some sense of what the landscape is like and can make at least part of a map. Other explorers, striking out in other directions, will someday help fill in the remaining parts of that map.

Astronomers have traditionally had to make the same sort of choices. We cannot explore the universe in every direction to infinite "depth" or sensitivity: there are far too many galaxies and far too few telescopes to do the job. But we can pick a single direction or a small slice of the sky and start mapping the galaxies. Margaret Geller, the late John Huchra, and their students at the Harvard-Smithsonian Center for Astrophysics pioneered this technique, and several other groups have extended their work to cover larger volumes of space.

VOYAGERS IN ASTRONOMY

Margaret Geller: Cosmic Surveyor

Born in 1947, Margaret Geller is the daughter of a chemist who encouraged her interest in science and helped her visualize the three-dimensional structure of molecules as a child. (It was a skill that would later come in very handy for visualizing the three-dimensional structure of the universe.) She remembers being bored in elementary school, but she was encouraged to read on her own by her parents. Her recollections also include subtle messages from teachers that mathematics (her strong early interest) was not a field for girls, but she did not allow herself to be deterred.

Geller obtained a BA in physics from the University of California at Berkeley and became the second woman to receive a PhD in physics from Princeton. There, while working with James Peebles, one of the world's leading cosmologists, she became interested in problems relating to the large-scale structure of the universe. In 1980, she accepted a research position at the Harvard-Smithsonian Center for Astrophysics, one of the nation's most dynamic institutions for astronomy research. She saw that to make progress in understanding how galaxies and clusters are organized, a far more intensive series of surveys was required. Although it would not bear fruit for many years, Geller and her collaborators began the long, arduous task of mapping the galaxies ((Figure 28.20).).

This OpenStax book is available for free at http://cnx.org/content/col11992/1.8

Figure 28.20. Margaret Geller. Geller's work mapping and researching galaxies has helped us to better understand the structure of the universe. (credit: modification of work by Massimo Ramella)

Her team was fortunate to be given access to a telescope that could be dedicated to their project, the 60-inch reflector on Mount Hopkins, near Tucson, Arizona, where they and their assistants took spectra to determine galaxy distances. To get a slice of the universe, they pointed their telescope at a predetermined position in the sky and then let the rotation of Earth bring new galaxies into their field of view. In this way, they measured the positions and redshifts of over 18,000 galaxies and made a wide range of interesting maps to display their data. Their surveys now include "slices" in both the Northern and Southern Hemispheres.

As news of her important work spread beyond the community of astronomers, Geller received a MacArthur Foundation Fellowship in 1990. These fellowships, popularly called "genius awards," are designed to recognize truly creative work in a wide range of fields. Geller continues to have a strong interest in visualization and has (with filmmaker Boyd Estus) made several award-winning videos explaining her work to nonscientists (one is titled *So Many Galaxies . . . So Little Time*). She has appeared on a variety of national news and documentary programs, including the *MacNeil/Lehrer NewsHour*, *The Astronomers*, and *The Infinite Voyage*. Energetic and outspoken, she has given talks on her work to many audiences around the country, and works hard to find ways to explain the significance of her pioneering surveys to the public.

"It's exciting to discover something that nobody's seen before. [To be] one of the first three people to ever see that slice of the universe [was] sort of being like Columbus. . . . Nobody expected such a striking pattern!"—Margaret Geller

LINK TO LEARNING

Find out more about Geller and Huchra's work (including interviews with Geller) in this 4-minute NOVA (https://openstax.org/l/30gellhucwork) video. You can also learn more about their conclusions (https://openstax.org/l/30gellhucconc) and additional research it led to.

Download for free at http://cnx.org/content/col11992/latest/

The largest universe mapping project to date is the Sloan Digital Sky Survey (see the Making Connections feature box Astronomy and Technology: The Sloan Digital Sky Survey at the end of this section). A plot of the distribution of galaxies mapped by the Sloan survey is shown in Figure 28.21. To the surprise of astronomers, maps like the one in the figure showed that clusters of galaxies are not arranged uniformly throughout the universe, but are found in huge filamentary **superclusters** that look like great arcs of inkblots splattered across a page. The superclusters resemble an irregularly torn sheet of paper or a pancake in shape—they can extend for hundreds of millions of light-years in two dimensions, but are only 10 to 20 million light-years thick in the third dimension. Detailed study of some of these structures shows that their masses are a few times 10^{16} M_{Sun}, which is 10,000 times more massive than the Milky Way Galaxy.

LINK TO LEARNING

Check out this animated visualization (https://openstax.org/l/30anivisslosur) of large-scale structure from the Sloan survey.

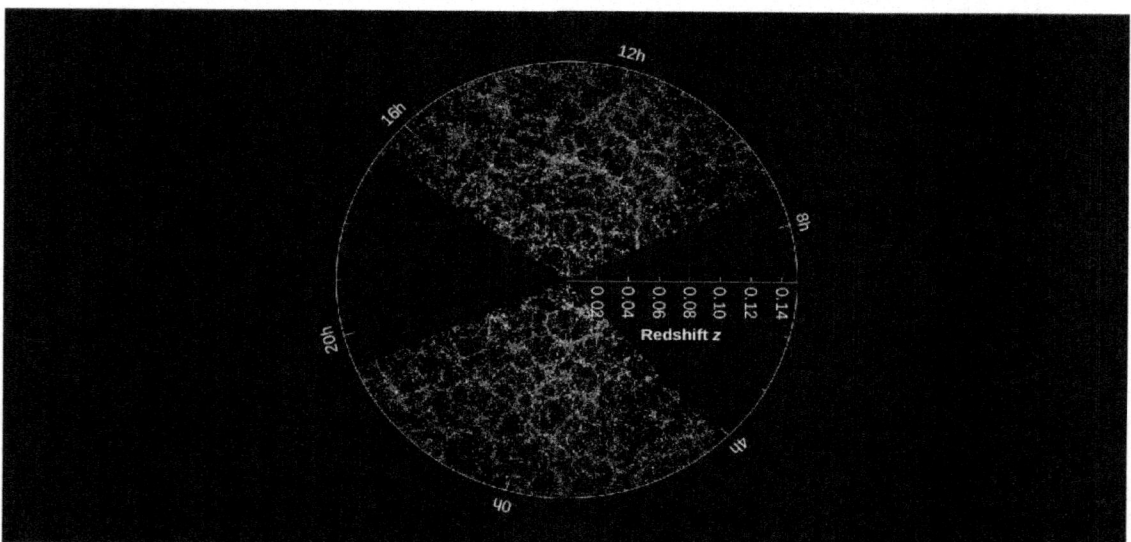

Figure 28.21. Sloan Digital Sky Survey Map of the Large-Scale Structure of the Universe. This image shows slices from the SDSS map. The point at the center corresponds to the Milky Way and might say "You Are Here!" Points on the map moving outward from the center are farther away. The distance to the galaxies is indicated by their redshifts (following Hubble's law), shown on the horizontal line going right from the center. The redshift $z = \Delta\lambda/\lambda$, where $\Delta\lambda$ is the difference between the observed wavelength and the wavelength λ emitted by a nonmoving source in the laboratory. Hour angle on the sky is shown around the circumference of the circular graph. The colors of the galaxies indicate the ages of their stars, with the redder color showing galaxies that are made of older stars. The outer circle is at a distance of two billion light-years from us. Note that red (older stars) galaxies are more strongly clustered than blue galaxies (young stars). The unmapped areas are where our view of the universe is obstructed by dust in our own Galaxy. (credit: modification of work by M. Blanton and the Sloan Digital Sky Survey)

Separating the filaments and sheets in a supercluster are **voids**, which look like huge empty bubbles walled in by the great arcs of galaxies. They have typical diameters of 150 million light-years, with the clusters of galaxies concentrated along their walls. The whole arrangement of filaments and voids reminds us of a sponge, the inside of a honeycomb, or a hunk of Swiss cheese with very large holes. If you take a good slice or cross-section through any of these, you will see something that looks roughly like Figure 28.21.

This OpenStax book is available for free at http://cnx.org/content/col11992/1.8

Before these voids were discovered, most astronomers would probably have predicted that the regions between giant clusters of galaxies were filled with many small groups of galaxies, or even with isolated individual galaxies. Careful searches within these voids have found few galaxies of any kind. Apparently, 90 percent of the galaxies occupy less than 10 percent of the volume of space.

EXAMPLE 28.1

Galaxy Distribution

To determine the distribution of galaxies in three-dimensional space, astronomers have to measure their positions and their redshifts. The larger the volume of space surveyed, the more likely the measurement is a fair sample of the universe as a whole. However, the work involved increases very rapidly as you increase the volume covered by the survey.

Let's do a quick calculation to see why this is so.

Suppose that you have completed a survey of all the galaxies within 30 million light-years and you now want to survey to 60 million light-years. What volume of space is covered by your second survey? How much larger is this volume than the volume of your first survey? Remember that the volume of a sphere, V, is given by the formula $V = 4/3\pi R^3$, where R is the radius of the sphere.

Solution

Since the volume of a sphere depends on R^3 and the second survey reaches twice as far in distance, it will cover a volume that is $2^3 = 8$ times larger. The total volume covered by the second survey will be $(4/3)\pi \times$ (60 million light-years)$^3 = 9 \times 10^{23}$ light-years^3.

Check Your Learning

Suppose you now want to expand your survey to 90 million light-years. What volume of space is covered, and how much larger is it than the volume of the second survey?

Answer:

The total volume covered is $(4/3)\pi \times$ (90 million light-years)$^3 = 3.05 \times 10^{24}$ light-years^3. The survey reaches 3 times as far in distance, so it will cover a volume that is $3^3 = 27$ times larger.

Even larger, more sensitive telescopes and surveys are currently being designed and built to peer farther and farther out in space and back in time. The new 50-meter Large Millimeter Telescope in Mexico and the Atacama Large Millimeter Array in Chile can detect far-infrared and millimeter-wave radiation from massive starbursting galaxies at redshifts and thus distances more than 90% of the way back to the Big Bang. These cannot be observed with visible light because their star formation regions are wrapped in clouds of thick dust. And in 2018, the 6.5-meter-diameter James Webb Space Telescope is scheduled to launch. It will be the first new major visible light and near-infrared telescope in space since Hubble was launched more than 25 years earlier. One of the major goals of this telescope is to observe directly the light of the first galaxies and even the first stars to shine, less than half a billion years after the Big Bang.

At this point, if you have been thinking about our discussions of the expanding universe in Galaxies, you may be wondering what exactly in Figure 28.21 is expanding. We know that the galaxies and clusters of galaxies are

Download for free at http://cnx.org/content/col11992/latest/

held together by their gravity and do not expand as the universe does. However, the voids do grow larger and the filaments move farther apart as space stretches (see The Big Bang).

MAKING CONNECTIONS

Astronomy and Technology: The Sloan Digital Sky Survey

In Edwin Hubble's day, spectra of galaxies had to be taken one at a time. The faint light of a distant galaxy gathered by a large telescope was put through a slit, and then a spectrometer (also called a spectrograph) was used to separate the colors and record the spectrum. This was a laborious process, ill suited to the demands of making large-scale maps that require the redshifts of many thousands of galaxies.

But new technology has come to the rescue of astronomers who seek three-dimensional maps of the universe of galaxies. One ambitious survey of the sky was produced using a special telescope, camera, and spectrograph atop the Sacramento Mountains of New Mexico. Called the Sloan Digital Sky Survey (SDSS), after the foundation that provided a large part of the funding, the program used a 2.5-meter telescope (about the same aperture as the Hubble) as a wide-angle astronomical camera. During a mapping program lasting more than ten years, astronomers used the SDSS's 30 charge-coupled devices (CCDs)—sensitive electronic light detectors similar to those used in many digital cameras and cell phones—to take images of over 500 million objects and spectra of over 3 million, covering more than one-quarter of the celestial sphere. Like many large projects in modern science, the Sloan Survey involved scientists and engineers from many different institutions, ranging from universities to national laboratories.

Every clear night for more than a decade, astronomers used the instrument to make images recording the position and brightness of celestial objects in long strips of the sky. The information in each strip was digitally recorded and preserved for future generations. When the seeing (recall this term from Astronomical Instruments) was only adequate, the telescope was used for taking spectra of galaxies and quasars—but it did so for up to *640 objects at a time*.

The key to the success of the project was a series of *optical fibers*, thin tubes of flexible glass that can transmit light from a source to the CCD that then records the spectrum. After taking images of a part of the sky and identifying which objects are galaxies, project scientists drilled an aluminum plate with holes for attaching fibers at the location of each galaxy. The telescope was then pointed at the right section of the sky, and the fibers led the light of each galaxy to the spectrometer for individual recording (Figure 28.22).

This OpenStax book is available for free at http://cnx.org/content/col11992/1.8

Download for free at http://cnx.org/content/col11992/latest/

(a) (b)

Figure 28.22. Sloan Digital Sky Survey. (a) The Sloan Digital Sky Survey telescope is seen here in front of the Sacramento Mountains in New Mexico. (b) Astronomer Richard Kron inserts some of the optical fibers into the pre-drilled plate to enable the instruments to make many spectra of galaxies at the same time. (credit a, b: modification of work by the Sloan Digital Sky Survey)

About an hour was sufficient for each set of spectra, and the pre-drilled aluminum plates could be switched quickly. Thus, it was possible to take as many as 5000 spectra in one night (provided the weather was good enough).

The galaxy survey led to a more comprehensive map of the sky than has ever before been possible, allowing astronomers to test their ideas about large-scale structure and the evolution of galaxies against an impressive array of real data.

The information recorded by the Sloan Survey staggers the imagination. The data came in at 8 megabytes per second (this means 8 million individual numbers or characters every second). Over the course of the project, scientists recorded over 15 terabytes, or 15 thousand billion bytes, which they estimate is comparable to the information contained in the Library of Congress. Organizing and sorting this volume of data and extracting the useful scientific results it contains is a formidable challenge, even in our information age. Like many other fields, astronomy has now entered an era of "Big Data," requiring supercomputers and advanced computer algorithms to sift through all those terabytes of data efficiently.

One very successful solution to the challenge of dealing with such large datasets is to turn to "citizen science," or crowd-sourcing, an approach the SDSS helped pioneer. The human eye is very good at recognizing subtle differences among shapes, such as between two different spiral galaxies, while computers often fail at such tasks. When Sloan project astronomers wanted to catalog the shapes of some of the millions of galaxies in their new images, they launched the "Galaxy Zoo" project: volunteers around the world were given a short training course online, then were provided with a few dozen galaxy images to classify by eye. The project was wildly successful, resulting in over 40 million galaxy classifications by more than 100,000 volunteers and the discovery of whole new types of galaxies.

Download for free at http://cnx.org/content/col11992/latest/

LINK TO LEARNING

Learn more about how you can be part of the project of classifying galaxies (https://openstax.org/l/30classgalax) in this citizen science effort. This program is part of a whole series of "citizen science" projects (https://openstax.org/l/30citizscien) that enable people in all walks of life to be part of the research that professional astronomers (and scholars in a growing number of fields) need help with.

28.4 THE CHALLENGE OF DARK MATTER

Learning Objectives

By the end of this section, you will be able to:

> Explain how astronomers know that the solar system contains very little dark matter
> Summarize the evidence for dark matter in most galaxies
> Explain how we know that galaxy clusters are dominated by dark matter
> Relate the presence of dark matter to the average mass-to-light ratio of huge volumes of space containing many galaxies

So far this chapter has focused almost entirely on matter that radiates electromagnetic energy—stars, planets, gas, and dust. But, as we have pointed out in several earlier chapters (especially The Milky Way Galaxy), it is now clear that galaxies contain large amounts of dark matter as well. There is much more dark matter, in fact, than matter we can see—which means it would be foolish to ignore the effect of this unseen material in our theories about the structure of the universe. (As many a ship captain in the polar seas found out too late, the part of the iceberg visible above the ocean's surface was not necessarily the only part he needed to pay attention to.) Dark matter turns out to be extremely important in determining the evolution of galaxies and of the universe as a whole.

The idea that much of the universe is filled with dark matter may seem like a bizarre concept, but we can cite a historical example of "dark matter" much closer to home. In the mid-nineteenth century, measurements showed that the planet Uranus did not follow exactly the orbit predicted from Newton's laws if one added up the gravitational forces of all the known objects in the solar system. Some people worried that Newton's laws may simply not work so far out in our solar system. But the more straightforward interpretation was to attribute Uranus' orbital deviations to the gravitational effects of a new planet that had not yet been seen. Calculations showed where that planet had to be, and Neptune was discovered just about in the predicted location.

In the same way, astronomers now routinely determine the location and amount of dark matter in galaxies by measuring its gravitational effects on objects we can see. And, by measuring the way that galaxies move in clusters, scientists have discovered that dark matter is also distributed among the galaxies in the clusters. Since the environment surrounding a galaxy is important in its development, dark matter must play a central role in galaxy evolution as well. Indeed, it appears that dark matter makes up most of the matter in the universe. But what *is* dark matter? What is it made of? We'll look next at the search for dark matter and the quest to determine its nature.

This OpenStax book is available for free at http://cnx.org/content/col11992/1.8

Dark Matter in the Local Neighborhood

Is there dark matter in our own solar system? Astronomers have examined the orbits of the known planets and of spacecraft as they journey to the outer planets and beyond. No deviations have been found from the orbits predicted on the basis of the masses of objects already discovered in our solar system and the theory of gravity. We therefore conclude that there is no evidence that there are large amounts of dark matter nearby.

Astronomers have also looked for evidence of dark matter in the region of the Milky Way Galaxy that lies within a few hundred light-years of the Sun. In this vicinity, most of the stars are restricted to a thin disk. It is possible to calculate how much mass the disk must contain in order to keep the stars from wandering far above or below it. The total matter that must be in the disk is less than twice the amount of luminous matter. This means that no more than half of the mass in the region near the Sun can be dark matter.

Dark Matter in and around Galaxies

In contrast to our local neighborhood near the Sun and solar system, there is (as we saw in The Milky Way Galaxy) ample evidence strongly suggesting that about 90% of the mass in the entire galaxy is in the form of a halo of dark matter. In other words, there is apparently about nine times more dark matter than visible matter. Astronomers have found some stars in the outer regions of the Milky Way beyond its bright disk, and these stars are revolving very rapidly around its center. The mass contained in all the stars and all the interstellar matter we can detect in the galaxy does not exert enough gravitational force to explain how those fast-moving stars remain in their orbits and do not fly away. Only by having large amounts of unseen matter could the galaxy be holding on to those fast-moving outer stars. The same result is found for other spiral galaxies as well.

Figure 28.23 is an example of the kinds of observations astronomers are making, for the Andromeda galaxy, a member of our Local Group. The observed rotation of spiral galaxies like Andromeda is usually seen in plots, known as *rotation curves,* that show velocity versus distance from the galaxy center. Such plots suggest that the dark matter is found in a large halo surrounding the luminous parts of each galaxy. The radius of the halos around the Milky Way and Andromeda may be as large as 300,000 light-years, much larger than the visible size of these galaxies.

Download for free at http://cnx.org/content/col11992/latest/

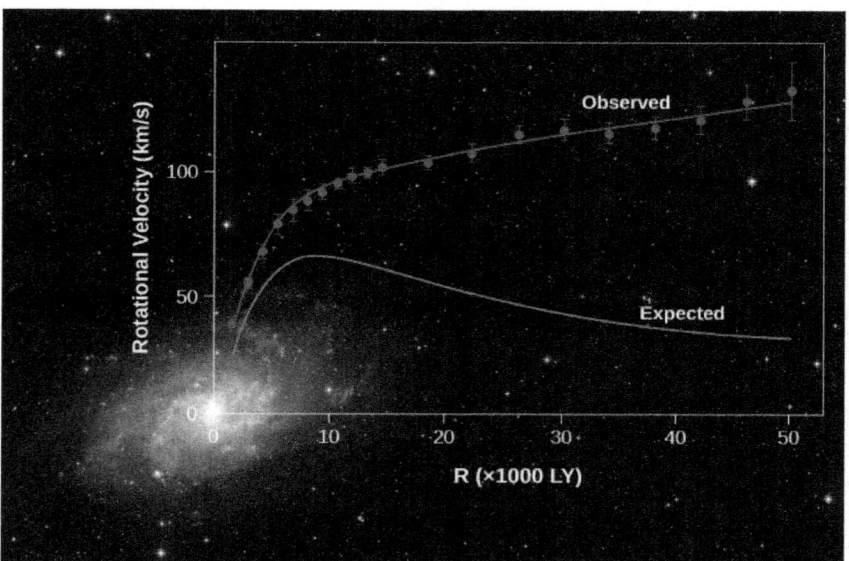

Figure 28.23. Rotation Indicates Dark Matter. We see the Milky Way's sister, the spiral Andromeda galaxy, with a graph that shows the velocity at which stars and clouds of gas orbit the galaxy at different distances from the center (red line). As is true of the Milky Way, the rotational velocity (or orbital speed) does not decrease with distance from the center, which is what you would expect if an assembly of objects rotates around a common center. A calculation (blue line) based on the total mass visible as stars, gas, and dust predicts that the velocity should be much lower at larger distances from the center. The discrepancy between the two curves implies the presence of a halo of massive dark matter extending outside the boundary of the luminous matter. This dark matter causes everything in the galaxy to orbit faster than the observed matter alone could explain. (credit background: modification of work by ESO)

Dark Matter in Clusters of Galaxies

Galaxies in clusters also move around: they orbit the cluster's center of mass. It is not possible for us to follow a galaxy around its entire orbit because that typically takes about a billion years. It is possible, however, to measure the velocities with which galaxies in a cluster are moving, and then estimate what the total mass in the cluster must be to keep the individual galaxies from flying out of the cluster. The observations indicate that the mass of the galaxies alone cannot keep the cluster together—some other gravity must again be present. The total amount of dark matter in clusters exceeds by more than ten times the luminous mass contained within the galaxies themselves, indicating that dark matter exists between galaxies as well as inside them.

There is another approach we can take to measuring the amount of dark matter in clusters of galaxies. As we saw, the universe is expanding, but this expansion is not perfectly uniform, thanks to the interfering hand of gravity. Suppose, for example, that a galaxy lies outside but relatively close to a rich cluster of galaxies. The gravitational force of the cluster will tug on that neighboring galaxy and slow down the rate at which it moves away from the cluster due to the expansion of the universe.

Consider the Local Group of galaxies, lying on the outskirts of the Virgo Supercluster. The mass concentrated at the center of the Virgo Cluster exerts a gravitational force on the Local Group. As a result, the Local Group is moving away from the center of the Virgo Cluster at a velocity a few hundred kilometers per second slower than the Hubble law predicts. By measuring such deviations from a smooth expansion, astronomers can estimate the total amount of mass contained in large clusters.

There are two other very useful methods for measuring the amount of dark matter in galaxy clusters, and both of them have produced results in general agreement with the method of measuring galaxy velocities: gravitational lensing and X-ray emission. Let's take a look at both.

This OpenStax book is available for free at http://cnx.org/content/col11992/1.8

Download for free at http://cnx.org/content/col11992/latest/

As Albert Einstein showed in his theory of general relativity, the presence of mass bends the surrounding fabric of spacetime. Light follows those bends, so very massive objects can bend light significantly. You saw examples of this in the Astronomy Basics feature box Gravitational Lensing in the previous section. Visible galaxies are not the only possible gravitational lenses. Dark matter can also reveal its presence by producing this effect. Figure 28.24 shows a galaxy cluster that is acting like a gravitational lens; the streaks and arcs you see on the picture are lensed images of more distant galaxies. Gravitational lensing is well enough understood that astronomers can use the many ovals and arcs seen in this image to calculate detailed maps of how much matter there is in the cluster and how that mass is distributed. The result from studies of many such gravitational lens clusters shows that, like individual galaxies, galaxy clusters contain more than ten times as much dark matter as luminous matter.

Figure 28.24. Cluster Abell 2218. This view from the Hubble Space Telescope shows the massive galaxy cluster Abell 2218 at a distance of about 2 billion light-years. Most of the yellowish objects are galaxies belonging to the cluster. But notice the numerous long, thin streaks, many of them blue; those are the distorted and magnified images of even more distant background galaxies, gravitationally lensed by the enormous mass of the intervening cluster. By carefully analyzing the lensed images, astronomers can construct a map of the dark matter that dominates the mass of the cluster. (credit: modification of work by NASA, ESA, and Johan Richard (Caltech))

The third method astronomers use to detect and measure dark matter in galaxy clusters is to image them in the light of X-rays. When the first sensitive X-ray telescopes were launched into orbit around Earth in the 1970s and trained on massive galaxy clusters, it was quickly discovered that the clusters emit copious X-ray radiation (see Figure 28.25). Most stars do not emit much X-ray radiation, and neither does most of the gas or dust between the stars inside galaxies. What could be emitting the X-rays seen from virtually all massive galaxy clusters?

It turns out that just as galaxies have gas distributed between their stars, clusters of galaxies have gas distributed between their galaxies. The particles in these huge reservoirs of gas are not just sitting still; rather, they are constantly moving, zooming around under the influence of the cluster's immense gravity like mini planets around a giant sun. As they move and bump against each other, the gas heats up hotter and hotter until, at temperatures as high as 100 million K, it shines brightly at X-ray wavelengths. The more mass the cluster has, the faster the motions, the hotter the gas, and the brighter the X-rays. Astronomers calculate that the mass present to induce those motions must be about ten times the mass they can see in the clusters, including all the galaxies and all the gas. Once again, this is evidence that the galaxy clusters are seen to be dominated by dark matter.

Download for free at http://cnx.org/content/col11992/latest/

Figure 28.25. X-Ray Image of a Galaxy Cluster. This composite image shows the galaxy cluster Abell 1689 at a distance of 2.3 billion light-years. The finely detailed views of the galaxies, most of them yellow, are in visible and near-infrared light from the Hubble Space Telescope, while the diffuse purple haze shows X-rays as seen by Chandra X-ray Observatory. The abundant X-rays, the gravitationally lensed images (thin curving arcs) of background galaxies, and the measured velocities of galaxies in the clusters all show that the total mass of Abell 1689—most of it dark matter—is about 10^{15} solar masses. (credit: modification of work by NASA/ESA/JPL-Caltech/Yale/CNRS)

Mass-to-Light Ratio

We described the use of the mass-to-light ratio to characterize the matter in galaxies or clusters of galaxies in Properties of Galaxies. For systems containing mostly old stars, the mass-to-light ratio is typically 10 to 20, where mass and light are measured in units of the Sun's mass and luminosity. A mass-to-light ratio of 100 or more is a signal that a substantial amount of dark matter is present. Table 28.1 summarizes the results of measurements of mass-to-light ratios for various classes of objects. Very large mass-to-light ratios are found for all systems of galaxy size and larger, indicating that dark matter is present in all these types of objects. This is why we say that dark matter apparently makes up most of the total mass of the universe.

Mass-To-Light Ratios

Type of Object	Mass-to-Light Ratio
Sun	1
Matter in vicinity of Sun	2
Total mass in Milky Way	10
Small groups of galaxies	50–150
Rich clusters of galaxies	250–300

Table 28.1

This OpenStax book is available for free at http://cnx.org/content/col11992/1.8

Download for free at http://cnx.org/content/col11992/latest/

The clustering of galaxies can be used to derive the total amount of mass in a given region of space, while visible radiation is a good indicator of where the luminous mass is. Studies show that the dark matter and luminous matter are very closely associated. The dark matter halos do extend beyond the luminous boundaries of the galaxies that they surround. However, where there are large clusters of galaxies, you will also find large amounts of dark matter. Voids in the galaxy distribution are also voids in the distribution of dark matter.

What Is the Dark Matter?

How do we go about figuring out what the dark matter consists of? The technique we might use depends on its composition. Let's consider the possibility that some of the dark matter is made up of normal particles: protons, neutrons, and electrons. Suppose these particles were assembled into black holes, brown dwarfs, or white dwarfs. If the black holes had no accretion disks, they would be invisible to us. White and brown dwarfs do emit some radiation but have such low luminosities that they cannot be seen at distances greater than a few thousand light-years.

We can, however, look for such compact objects because they can act as gravitational lenses. (See the Astronomy Basics feature box Gravitational Lensing.) Suppose the dark matter in the halo of the Milky Way were made up of black holes, brown dwarfs, and white dwarfs. These objects have been whimsically dubbed MACHOs (MAssive Compact Halo Objects). If an invisible MACHO passes directly between a distant star and Earth, it acts as a gravitational lens, focusing the light from the distant star. This causes the star to appear to brighten over a time interval of a few hours to several days before returning to its normal brightness. Since we can't predict when any given star might brighten this way, we have to monitor huge numbers of stars to catch one in the act. There are not enough astronomers to keep monitoring so many stars, but today's automated telescopes and computer systems can do it for us.

Research teams making observations of millions of stars in the nearby galaxy called the Large Magellanic Cloud have reported several examples of the type of brightening expected if MACHOs are present in the halo of the Milky Way (Figure 28.26). However, there are not enough MACHOs in the halo of the Milky Way to account for the mass of the dark matter in the halo.

Download for free at http://cnx.org/content/col11992/latest/

Figure 28.26. Large and Small Magellanic Clouds. Here, the two small galaxies we call the Large Magellanic Cloud and Small Magellanic Cloud can be seen above the auxiliary telescopes for the Very Large Telescope Array on Cerro Paranal in Chile. You can see from the number of stars that are visible that this is a very dark site for doing astronomy. (credit: ESO/J. Colosimo)

This result, along with a variety of other experiments, leads us to conclude that the types of matter we are familiar with can make up only a tiny portion of the dark matter. Another possibility is that dark matter is composed of some new type of particle—one that researchers are now trying to detect in laboratories here on Earth (see The Big Bang).

The kinds of dark matter particles that astronomers and physicists have proposed generally fall into two main categories: hot and cold dark matter. The terms *hot* and *cold* don't refer to true temperatures, but rather to the average velocities of the particles, analogous to how we might think of particles of air moving in your room right now. In a cold room, the air particles move more slowly on average than in a warm room.

In the early universe, if dark matter particles easily moved fast and far compared to the lumps and bumps of ordinary matter that eventually became galaxies and larger structures, we call those particles **hot dark matter**. In that case, smaller lumps and bumps would be smeared out by the particle motions, meaning fewer small galaxies would get made.

On the other hand, if the dark matter particles moved slowly and covered only small distances compared to the sizes of the lumps in the early universe, we call that **cold dark matter**. Their slow speeds and energy would mean that even the smaller lumps of ordinary matter would survive to grow into small galaxies. By looking at when galaxies formed and how they evolve, we can use observations to distinguish between the two kinds of dark matter. So far, observations seem most consistent with models based on cold dark matter.

Solving the dark matter problem is one of the biggest challenges facing astronomers. After all, we can hardly understand the evolution of galaxies and the long-term history of the universe without understanding what its most massive component is made of. For example, we need to know just what role dark matter played in starting the higher-density "seeds" that led to the formation of galaxies. And since many galaxies have large

This OpenStax book is available for free at http://cnx.org/content/col11992/1.8

halos made of dark matter, how does this affect their interactions with one another and the shapes and types of galaxies that their collisions create?

Astronomers armed with various theories are working hard to produce models of galaxy structure and evolution that take dark matter into account in just the right way. Even though we don't know what the dark matter is, we do have some clues about how it affected the formation of the very first galaxies. As we will see in The Big Bang, careful measurements of the microwave radiation left over after the Big Bang have allowed astronomers to set very tight limits on the actual sizes of those early seeds that led to the formation of the large galaxies that we see in today's universe. Astronomers have also measured the relative numbers and distances between galaxies and clusters of different sizes in the universe today. So far, most of the evidence seems to weigh heavily in favor of cold dark matter, and most current models of galaxy and large-scale structure formation use cold dark matter as their main ingredient.

As if the presence of dark matter—a mysterious substance that exerts gravity and outweighs all the known stars and galaxies in the universe but does not emit or absorb light—were not enough, there is an even more baffling and equally important constituent of the universe that has only recently been discovered: we have called it **dark energy** in parallel with dark matter. We will say more about it and explore its effects on the evolution of the universe in The Big Bang. For now, we can complete our inventory of the contents of the universe by noting that it appears that the entire universe contains some mysterious energy that pushes spacetime apart, taking galaxies and the larger structures made of galaxies along with it. Observations show that dark energy becomes more and more important relative to gravity as the universe ages. As a result, the expansion of the universe is accelerating, and this acceleration seems to be happening mostly since the universe was about half its current age.

What we see when we peer out into the universe—the light from trillions of stars in hundreds of billions of galaxies wrapped in intricate veils of gas and dust—is therefore actually only a sprinkling of icing on top of the cake: as we will see in The Big Bang, when we look outside galaxies and clusters of galaxies at the universe as a whole, astronomers find that for every gram of luminous normal matter, such as protons, neutrons, electrons, and atoms in the universe, there are about 4 grams of nonluminous normal matter, mainly intergalactic hydrogen and helium. There are about 27 grams of dark matter, and the energy equivalent (remember Einstein's famous $E = mc^2$) of about 68 grams of dark energy. Dark matter, and (as we will see) even more so dark energy, are dramatic demonstrations of what we have tried to emphasize throughout this book: science is always a "progress report," and we often encounter areas where we have more questions than answers.

Let's next put together all these clues to trace the life history of galaxies and large-scale structure in the universe. What follows is the current consensus, but research in this field is moving rapidly, and some of these ideas will probably be modified as new observations are made.

28.5 THE FORMATION AND EVOLUTION OF GALAXIES AND STRUCTURE IN THE UNIVERSE

Learning Objectives

By the end of this section, you will be able to:

> Summarize the main theories attempting to explain how individual galaxies formed
> Explain how tiny "seeds" of dark matter in the early universe grew by gravitational attraction over billions of years into the largest structures observed in the universe: galaxy clusters and superclusters, filaments, and voids

Download for free at http://cnx.org/content/col11992/latest/

As with most branches of natural science, astronomers and cosmologists always want to know the answer to the question, "How did it get that way?" What made galaxies and galaxy clusters, superclusters, voids, and filaments look the way they do? The existence of such large filaments of galaxies and voids is an interesting puzzle because we have evidence (to be discussed in The Big Bang) that the universe was extremely smooth even a few hundred thousand years after forming. The challenge for theoreticians is to understand how a nearly featureless universe changed into the complex and lumpy one that we see today. Armed with our observations and current understanding of galaxy evolution over cosmic time, dark matter, and large-scale structure, we are now prepared to try to answer that question on some of the largest possible scales in the universe. As we will see, the short answer to how the universe got this way is "dark matter + gravity + time."

How Galaxies Form and Grow

We've already seen that galaxies were more numerous, but smaller, bluer, and clumpier, in the distant past than they are today, and that galaxy mergers play a significant role in their evolution. At the same time, we have observed quasars and galaxies that emitted their light when the universe was less than a billion years old—so we know that large condensations of matter had begun to form at least that early. We also saw in Active Galaxies, Quasars, and Supermassive Black Holes that many quasars are found in the centers of elliptical galaxies. This means that some of the first large concentrations of matter must have evolved into the elliptical galaxies that we see in today's universe. It seems likely that the supermassive black holes in the centers of galaxies and the spherical distribution of ordinary matter around them formed at the same time and through related physical processes.

Dramatic confirmation of that picture arrived only in the last decade, when astronomers discovered a curious empirical relationship: as we saw in Active Galaxies, Quasars, and Supermassive Black Holes, the more massive a galaxy is, the more massive its central black hole is. Somehow, the black hole and the galaxy "know" enough about each other to match their growth rates.

There have been two main types of galaxy formation models to explain all those observations. The first asserts that massive elliptical galaxies formed in a single, rapid collapse of gas and dark matter, during which virtually all the gas was turned quickly into stars. Afterward the galaxies changed only slowly as the stars evolved. This is what astronomers call a "top-down" scenario.

The second model suggests that today's giant ellipticals were formed mostly through mergers of smaller galaxies that had already converted at least some of their gas into stars—a "bottom-up" scenario. In other words, astronomers have debated whether giant ellipticals formed most of their stars in the large galaxy that we see today or in separate small galaxies that subsequently merged.

Since we see some luminous quasars from when the universe was less than a billion years old, it is likely that at least some giant ellipticals began their evolution very early through the collapse of a single cloud. However, the best evidence also seems to show that mature *giant* elliptical galaxies like the ones we see nearby were rare before the universe was about 6 billion years old and that they are much more common today than they were when the universe was young. Observations also indicate that most of the gas in elliptical galaxies was converted to stars by the time the universe was about 3 billion years old, so it appears that elliptical galaxies have not formed many new stars since then. They are often said to be "red and dead"—that is, they mostly contain old, cool, red stars, and there is little or no new star formation going on.

These observations (when considered together) suggest that the giant elliptical galaxies that we see nearby formed from a combination of both top-down and bottom-up mechanisms, with the most massive galaxies forming in the densest clusters where both processes happened very early and quickly in the history of the universe.

This OpenStax book is available for free at http://cnx.org/content/col11992/1.8

The situation with spiral galaxies is apparently very different. The bulges of these galaxies formed early, like the elliptical galaxies (Figure 28.27). However, the disks formed later (remember that the stars in the disk of the Milky Way are younger than the stars in the bulge and the halo) and still contain gas and dust. However, the rate of star formation in spirals today is about ten times lower than it was 8 billion years ago. The number of stars being formed drops as the gas is used up. So spirals seem to form mostly "bottom up" but over a longer time than ellipticals and in a more complex way, with at least two distinct phases.

Rapid Collapse

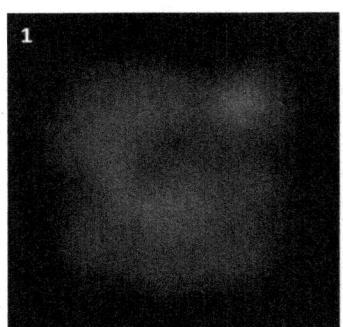 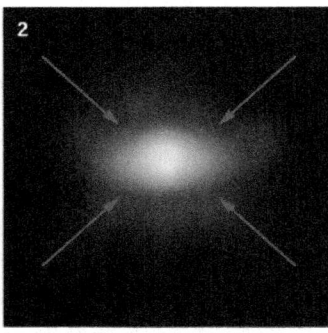 

Primordial hydrogen cloud.   Cloud collapses under gravity.   Large bulge of ancient stars dominates galaxy.

Environmental Effects

Disk galaxy and companion.   Smaller galaxy falls into disk galaxy.   Bulge inflates with addition of young stars and gas.

Figure 28.27. Growth of Spiral Bulges. The nuclear bulges of some spiral galaxies formed through the collapse of a single protogalactic cloud (top row). Others grew over time through mergers with other smaller galaxies (bottom row).

Hubble originally thought that elliptical galaxies were young and would eventually turn into spirals, an idea we now know is not true. In fact, as we saw above, it's more likely the other way around: two spirals that crash together under their mutual gravity can turn into an elliptical.

Despite these advances in our understanding of how galaxies form and evolve, many questions remain. For example, it's even possible, given current evidence, that spiral galaxies could lose their spiral arms and disks in a merger event, making them look more like an elliptical or irregular galaxy, and then regain the disk and arms again later if enough gas remains available. The story of how galaxies assume their final shapes is still being written as we learn more about galaxies and their environment.

Download for free at http://cnx.org/content/col11992/latest/

Forming Galaxy Clusters, Superclusters, Voids, and Filaments

If individual galaxies seem to grow mostly by assembling smaller pieces together gravitationally over cosmic time, what about the clusters of galaxies and larger structures such as those seen in Figure 28.21? How do we explain the large-scale maps that show galaxies distributed on the walls of huge sponge- or bubble-like structures spanning hundreds of millions of light-years?

As we saw, observations have found increasing evidence for concentrations, filaments, clusters, and superclusters of galaxies when the universe was less than 3 billion years old (Figure 28.28). This means that large concentrations of galaxies had already come together when the universe was less than a quarter as old as it is now.

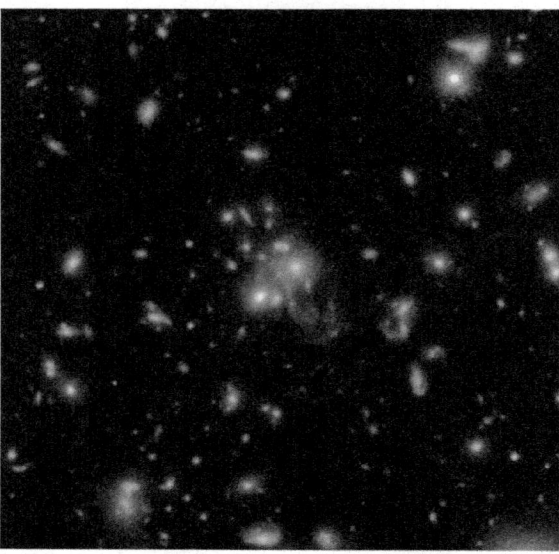

Figure 28.28. Merging Galaxies in a Distant Cluster. This Hubble image shows the core of one of the most distant galaxy clusters yet discovered, SpARCS 1049+56; we are seeing it as it was nearly 10 billion years ago. The surprise delivered by the image was the "train wreck" of chaotic galaxy shapes and blue tidal tails: apparently there are several galaxies right in the core that are merging together, the probable cause of a massive burst of star formation and bright infrared emission from the cluster. (credit: modification of work by NASA/STScI/ESA/JPL-Caltech/McGill)

Almost all the currently favored models of how large-scale structure formed in the universe tell a story similar to that for individual galaxies: tiny dark matter "seeds" in the hot cosmic soup after the Big Bang grew by gravity into larger and larger structures as cosmic time ticked on (Figure 28.29). The final models we construct will need to be able to explain the size, shape, age, number, and spatial distribution of galaxies, clusters, and filaments—not only today, but also far back in time. Therefore, astronomers are working hard to measure and then to model those features of large-scale structure as accurately as possible. So far, a mixture of 5% normal atoms, 27% cold dark matter, and 68% dark energy seems to be the best way to explain all the evidence currently available (see The Big Bang).

This OpenStax book is available for free at http://cnx.org/content/col11992/1.8

Download for free at http://cnx.org/content/col11992/latest/

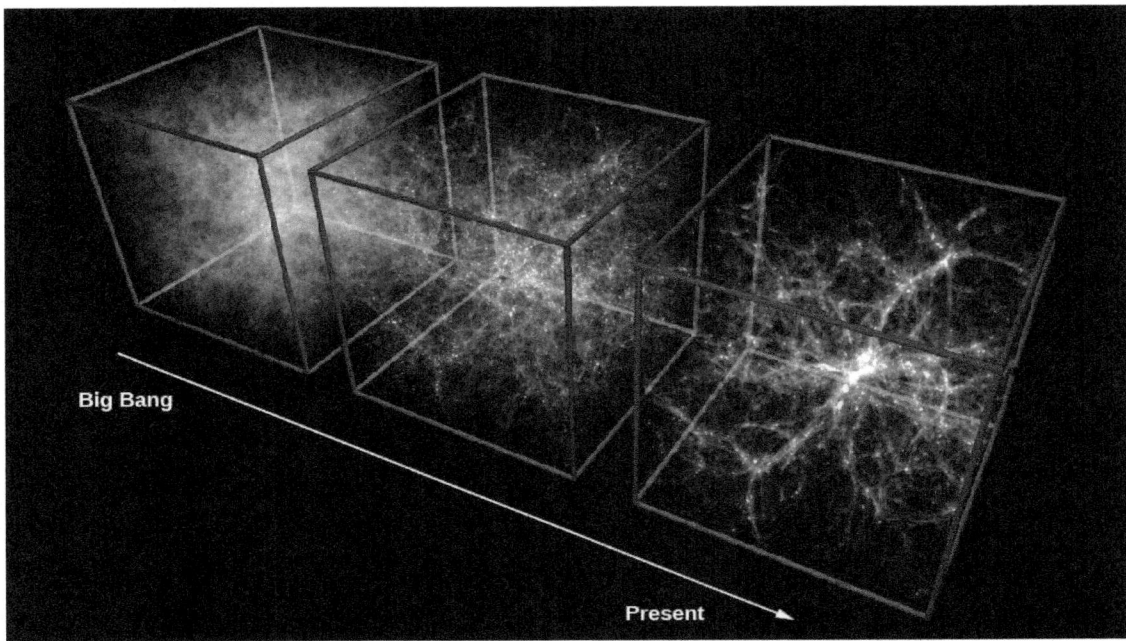

Figure 28.29. Growth of Large-Scale Structure as Calculated by Supercomputers. The boxes show how filaments and superclusters of galaxies grow over time, from a relatively smooth distribution of dark matter and gas, with few galaxies formed in the first 2 billion years after the Big Bang, to the very clumpy strings of galaxies with large voids today. Compare the last image in this sequence with the actual distribution of nearby galaxies shown in Figure 28.21. (credit: modification of work by CXC/MPE/V.Springel)

The box at left is labeled "Big Bang," the box at center is unlabeled and the box at right is labeled "Present". A white arrow points from left to right representing the direction of time.

Scientists even have a model to explain how a nearly uniform, hot "soup" of particles and energy at the beginning of time acquired the Swiss-cheese-like structure that we now see on the largest scales. As we will see in The Big Bang, when the universe was only a few hundred thousand years old, *everything* was at a temperature of a few thousand degrees. Theorists suggest that at that early time, all the hot gas was vibrating, much as sound waves vibrate the air of a nightclub with an especially loud band. This vibrating could have concentrated matter into high-density peaks and created emptier spaces between them. When the universe cooled, the concentrations of matter were "frozen in," and galaxies ultimately formed from the matter in these high-density regions.

The Big Picture

To finish this chapter, let's put all these ideas together to tell a coherent story of how the universe came to look the way it does. Initially, as we said, the distribution of matter (both luminous and dark) was nearly, but not quite exactly, smooth and uniform. That "not quite" is the key to everything. Here and there were lumps where the density of matter (both luminous and dark) was ever so slightly higher than average.

Initially, each individual lump expanded because the whole universe was expanding. However, as the universe continued to expand, the regions of higher density acquired still more mass because they exerted a slightly larger than average gravitational force on surrounding material. If the inward pull of gravity was high enough, the denser individual regions ultimately stopped expanding. They then began to collapse into irregularly shaped blobs (that's the technical term astronomers use!). In many regions the collapse was more rapid in one

Download for free at http://cnx.org/content/col11992/latest/

direction, so the concentrations of matter were not spherical but came to resemble giant clumps, pancakes, and rope-like filaments—each much larger than individual galaxies.

These elongated clumps existed throughout the early universe, oriented in different directions and collapsing at different rates. The clumps provided the framework for the large-scale filamentary and bubble-like structures that we see preserved in the universe today.

The universe then proceeded to "build itself" from the bottom up. Within the clumps, smaller structures formed first, then merged to build larger ones, like Lego pieces being put together one by one to create a giant Lego metropolis. The first dense concentrations of matter that collapsed were the size of small dwarf galaxies or globular clusters—which helps explain why globular clusters are the oldest things in the Milky Way and most other galaxies. These fragments then gradually assembled to build galaxies, galaxy clusters, and, ultimately, superclusters of galaxies.

According to this picture, small galaxies and large star clusters first formed in the highest density regions of all—the filaments and nodes where the pancakes intersect—when the universe was about two percent of its current age. Some stars may have formed even before the first star clusters and galaxies came into existence. Some galaxy-galaxy collisions triggered massive bursts of star formation, and some of these led to the formation of black holes. In that rich, crowded environment, black holes found constant food and grew in mass. The development of massive black holes then triggered quasars and other active galactic nuclei whose powerful outflows of energy and matter shut off the star formation in their host galaxies. The early universe must have been an exciting place!

Clusters of galaxies then formed as individual galaxies congregated, drawn together by their mutual gravitational attraction (Figure 28.30). First, a few galaxies came together to form groups, much like our own Local Group. Then the groups began combining to form clusters and, eventually, superclusters. This model predicts that clusters and superclusters should still be in the process of gathering together, and observations do in fact suggest that clusters are still gathering up their flocks of galaxies and collecting more gas as it flows in along filaments. In some instances we even see entire clusters of galaxies merging together.

Figure 28.30. Formation of Cluster of Galaxies. This schematic diagram shows how galaxies might have formed if small clouds formed first and then congregated to form galaxies and then clusters of galaxies.

This OpenStax book is available for free at http://cnx.org/content/col11992/1.8

Most giant elliptical galaxies formed through the collision and merger of many smaller fragments. Some spiral galaxies may have formed in relatively isolated regions from a single cloud of gas that collapsed to make a flattened disk, but others acquired additional stars, gas, and dark matter through collisions, and the stars acquired through these collisions now populate their halos and bulges. As we have seen, our Milky Way is still capturing small galaxies and adding them to its halo, and probably also pulling fresh gas from these galaxies into its disk.

Download for free at http://cnx.org/content/col11992/latest/

CHAPTER 28 REVIEW

 KEY TERMS

cold dark matter slow-moving massive particles, not yet identified, that don't absorb, emit, or reflect light or other electromagnetic radiation, and that make up most of the mass of galaxies and galaxy clusters

cosmological principle the assumption that, on the large scale, the universe at any given time is the same everywhere—isotropic and homogeneous

dark energy an energy that is causing the expansion of the universe to accelerate; the source of this energy is not yet understood

evolution (of galaxies) changes in individual galaxies over cosmic time, inferred by observing snapshots of many different galaxies at different times in their lives

galactic cannibalism a process by which a larger galaxy strips material from or completely swallows a smaller one

homogeneous having a consistent and even distribution of matter that is the same everywhere

hot dark matter massive particles, not yet identified, that don't absorb, emit, or reflect light or other electromagnetic radiation, and that make up most of the mass of galaxies and galaxy clusters; hot dark matter is faster-moving material than cold dark matter

isotropic the same in all directions

Local Group a small cluster of galaxies to which our Galaxy belongs

merger a collision between galaxies (of roughly comparable size) that combine to form a single new structure

starburst a galaxy or merger of multiple galaxies that turns gas into stars much faster than usual

supercluster a large region of space (more than 100 million light-years across) where groups and clusters of galaxies are more concentrated; a cluster of clusters of galaxies

void a region between clusters and superclusters of galaxies that appears relatively empty of galaxies

 SUMMARY

28.1 Observations of Distant Galaxies

When we look at distant galaxies, we are looking back in time. We have now seen galaxies as they were when the universe was about 500 million years old—only about five percent as old as it is now. The universe now is 13.8 billion years old. The color of a galaxy is an indicator of the age of the stars that populate it. Blue galaxies must contain a lot of hot, massive, young stars. Galaxies that contain only old stars tend to be yellowish red. The first generation of stars formed when the universe was only a few hundred million years old. Galaxies observed when the universe was only a few billion years old tend to be smaller than today's galaxies, to have more irregular shapes, and to have more rapid star formation than the galaxies we see nearby in today's universe. This shows that the smaller galaxy fragments assembled themselves into the larger galaxies we see today.

This OpenStax book is available for free at http://cnx.org/content/col11992/1.8

Download for free at http://cnx.org/content/col11992/latest/

28.2 Galaxy Mergers and Active Galactic Nuclei

When galaxies of comparable size collide and coalesce we call it a merger, but when a small galaxy is swallowed by a much larger one, we use the term galactic cannibalism. Collisions play an important role in the evolution of galaxies. If the collision involves at least one galaxy rich in interstellar matter, the resulting compression of the gas will result in a burst of star formation, leading to a starburst galaxy. Mergers were much more common when the universe was young, and many of the most distant galaxies that we see are starburst galaxies that are involved in collisions. Active galactic nuclei powered by supermassive black holes in the centers of most galaxies can have major effects on the host galaxy, including shutting off star formation.

28.3 The Distribution of Galaxies in Space

Counts of galaxies in various directions establish that the universe on the large scale is homogeneous and isotropic (the same everywhere and the same in all directions, apart from evolutionary changes with time). The sameness of the universe everywhere is referred to as the cosmological principle. Galaxies are grouped together in clusters. The Milky Way Galaxy is a member of the Local Group, which contains at least 54 member galaxies. Rich clusters (such as Virgo and Coma) contain thousands or tens of thousands of galaxies. Galaxy clusters often group together with other clusters to form large-scale structures called superclusters, which can extend over distances of several hundred million light-years. Clusters and superclusters are found in filamentary structures that are huge but fill only a small fraction of space. Most of space consists of large voids between superclusters, with nearly all galaxies confined to less than 10% of the total volume.

28.4 The Challenge of Dark Matter

Stars move much faster in their orbits around the centers of galaxies, and galaxies around centers of galaxy clusters, than they should according to the gravity of all the luminous matter (stars, gas, and dust) astronomers can detect. This discrepancy implies that galaxies and galaxy clusters are dominated by dark matter rather than normal luminous matter. Gravitational lensing and X-ray radiation from massive galaxy clusters confirm the presence of dark matter. Galaxies and clusters of galaxies contain about 10 times more dark matter than luminous matter. While some of the dark matter may be made up of ordinary matter (protons, neutrons, and electrons), perhaps in the form of very faint stars or black holes, most of it probably consists of some totally new type of particle not yet detected on Earth. Observations of gravitational lensing effects on distant objects have been used to look in the outer region of our Galaxy for any dark matter in the form of compact, dim stars or star remnants, but not enough such objects have been found to account for all the dark matter.

28.5 The Formation and Evolution of Galaxies and Structure in the Universe

Initially, luminous and dark matter in the universe was distributed almost—but not quite—uniformly. The challenge for galaxy formation theories is to show how this "not quite" smooth distribution of matter developed the structures—galaxies and galaxy clusters—that we see today. It is likely that the filamentary distribution of galaxies and voids was built in near the beginning, before stars and galaxies began to form. The first condensations of matter were about the mass of a large star cluster or a small galaxy. These smaller structures then merged over cosmic time to form large galaxies, clusters of galaxies, and superclusters of galaxies. Superclusters today are still gathering up more galaxies, gas, and dark matter. And spiral galaxies like the Milky Way are still acquiring material by capturing small galaxies near them.

Download for free at http://cnx.org/content/col11992/latest/

📐 FOR FURTHER EXPLORATION

Articles

Andrews, B. "What Are Galaxies Trying to Tell Us?" *Astronomy* (February 2011): 24. Introduction to our understanding of the shapes and evolution of different types of galaxies.

Barger, A. "The Midlife Crisis of the Cosmos." *Scientific American* (January 2005): 46. On how our time differs from the early universe in terms of what galaxies are doing, and what role supermassive black holes play.

Berman, B. "The Missing Universe." *Astronomy* (April 2014): 24. Brief review of dark matter, what it could be, and modified theories of gravity that can also explain it.

Faber, S., et al. "Staring Back to Cosmic Dawn." *Sky & Telescope* (June 2014): 18. Program to see the most distant and earliest galaxies with the Hubble.

Geller, M., & Huchra, J. "Mapping the Universe." *Sky & Telescope* (August 1991): 134. On their project mapping the location of galaxies in three dimensions.

Hooper, D. "Dark Matter in the Discovery Age." *Sky & Telescope* (January 2013): 26. On experiments looking for the nature of dark matter.

James, C. R. "The Hubble Deep Field: The Picture Worth a Trillion Stars." *Astronomy* (November 2015): 44. Detailed history and results, plus the Hubble Ultra-Deep Field.

Kaufmann, G., & van den Bosch, F. "The Life Cycle of Galaxies." *Scientific American* (June 2002): 46. On the evolution of galaxies and how the different shapes of galaxies develop.

Knapp, G. "Mining the Heavens: The Sloan Digital Sky Survey." *Sky & Telescope* (August 1997): 40.

Kron, R., & Butler, S. "Stars and Strips Forever." *Astronomy* (February 1999): 48. On the Sloan Digital Survey.

Kruesi, L. "What Do We Really Know about Dark Matter?" *Astronomy* (November 2009): 28. Focuses on what dark matter could be and experiments to find out.

Larson, R., & Bromm, V. "The First Stars in the Universe." *Scientific American* (December 2001): 64. On the dark ages and the birth of the first stars.

Nadis, S. "Exploring the Galaxy-Black Hole Connection." *Astronomy* (May 2010): 28. About the role of massive black holes in the evolution of galaxies.

Nadis, S. "Astronomers Reveal the Universe's Hidden Structure." *Astronomy* (September 2013): 44. How dark matter is the scaffolding on which the visible universe rests.

Schilling, G. "Hubble Goes the Distance." *Sky & Telescope* (January 2015): 20. Using gravitational lensing with HST to see the most distant galaxies.

Strauss, M. "Reading the Blueprints of Creation." *Scientific American* (February 2004): 54. On large-scale surveys of galaxies and what they tell us about the organization of the early universe.

Tytell, D. "A Wide Deep Field: Getting the Big Picture." *Sky & Telescope* (September 2001): 42. On the NOAO survey of deep sky objects.

Villard, R. "How Gravity's Grand Illusion Reveals the Universe." *Astronomy* (January 2013): 44. On gravitational lensing and what it teaches us.

This OpenStax book is available for free at http://cnx.org/content/col11992/1.8

Websites

Assembly of Galaxies: http://jwst.nasa.gov/galaxies.html. Introductory background information about galaxies: what we know and what we want to learn.

Brief History of Gravitational Lensing: http://www.einstein-online.info/spotlights/grav_lensing_history. From Einstein OnLine.

Cosmic Structures: http://skyserver.sdss.org/dr1/en/astro/structures/structures.asp. Brief review page on how galaxies are organized, from the Sloan Survey.

Discovery of the First Gravitational Lens: http://astrosociety.org/wp-content/uploads/2013/02/ab2009-33.pdf. By Ray Weymann, 2009.

Gravitational Lensing Discoveries from the Hubble Space Telescope: http://hubblesite.org/newscenter/archive/releases/exotic/gravitational-lens/. A chronological list of news releases and images.

Local Group of Galaxies: http://www.atlasoftheuniverse.com/localgr.html. Clickable map from the Atlas of the Universe project. See also their Virgo Cluster page: http://www.atlasoftheuniverse.com/galgrps/vir.html.

RotCurve: http://burro.astr.cwru.edu/JavaLab/RotcurveWeb/main.html. Try your hand at using real galaxy rotation curve data to measure dark matter halos using this Java applet simulation.

Sloan Digital Sky Survey Website: http://classic.sdss.org/. Includes nontechnical and technical parts.

Spyglasses into the Universe: http://www.spacetelescope.org/science/gravitational_lensing/. Hubble page on gravitational lensing; includes links to videos.

Virgo Cluster of Galaxies: http://messier.seds.org/more/virgo.html. A page with brief information and links to maps, images, etc.

Videos

Cosmic Simulations: http://www.tapir.caltech.edu/~phopkins/Site/Movies_cosmo.html. Beautiful videos with computer simulations of how galaxies form, from the FIRE group.

Cosmology of the Local Universe: http://irfu.cea.fr/cosmography. Narrated flythrough of maps of galaxies showing the closer regions of the universe (17:35).

Gravitational Lensing: https://www.youtube.com/watch?v=4Z71RtwoOas. Video from Fermilab, with Dr. Don Lincoln (7:14).

How Galaxies Were Cooked from the Primordial Soup: https://www.youtube.com/watch?v=wqNNCm7SNyw. A 2013 public talk by Dr. Sandra Faber of Lick Observatory about the evolution of galaxies; part of the Silicon Valley Astronomy Lecture Series (1:19:33).

Hubble Extreme Deep Field Pushes Back Frontiers of Time and Space: https://www.youtube.com/watch?v=gu_VhzhlqGw. Brief 2012 video (2:42).

Looking Deeply into the Universe in 3-D: https://www.eso.org/public/videos/eso1507a/. 2015 ESOCast video on how the Very Large Telescopes are used to explore the Hubble Ultra-Deep Field and learn more about the faintest and most distant galaxies (5:12).

Millennium Simulation: http://wwwmpa.mpa-garching.mpg.de/galform/virgo/millennium. A supercomputer in Germany follows the evolution of a representative large box as the universe evolves.

Movies of flying through the large-scale local structure: http://www.ifa.hawaii.edu/~tully/. By Brent Tully.

Download for free at http://cnx.org/content/col11992/latest/

Shedding Light on Dark Matter: https://www.youtube.com/watch?v=bZW_B9CC-gI. 2008 TED talk on galaxies and dark matter by physicist Patricia Burchat (17:08).

Sloan Digital Sky Survey overview movies: http://astro.uchicago.edu/cosmus/projects/sloanmovie/.

Virtual Universe: https://www.youtube.com/watch?v=SY0bKE10ZDM. An MIT model of a section of universe evolving, with dark matter included (4:11).

When Two Galaxies Collide: http://www.openculture.com/2009/04/when_galaxies_collide.html. Computer simulation, which stops at various points and shows a Hubble image of just such a system in nature (1:37).

🖧 COLLABORATIVE GROUP ACTIVITIES

A. Suppose you developed a theory to account for the evolution of New York City. Have your group discuss whether it would resemble the development of structure in the universe (as we have described it in this chapter). What elements of your model for NYC resemble the astronomers' model for the growth of structure in the universe? Which elements do not match?

B. Most astronomers believe that dark matter exists and is a large fraction of the total matter in the universe. At the same time, most astronomers do not believe that UFOs are evidence that we are being visited by aliens from another world. Yet astronomers have never actually seen either dark matter or a UFO. Why do you think one idea is widely accepted by scientists and the other is not? Which idea do you think is more believable? Give your reasoning.

C. Someone in your group describes the redshift surveys of galaxies to a friend, who says he's never heard of a bigger waste of effort. Who cares, he asks, about the large-scale structure of the universe? What is your group's reaction, and what reasons could you come up with for putting money into figuring out how the universe is organized?

D. The leader of a small but very wealthy country is obsessed by maps. She has put together a fabulous collection of Earth maps, purchased all the maps of other planets that astronomers have assembled, and now wants to commission the best possible map of the entire universe. Your group is selected to advise her. What sort of instruments and surveys should she invest in to produce a good map of the cosmos? Be as specific as you can.

E. Download a high-resolution image of a rich galaxy cluster from the Hubble Space Telescope (see the list of gravitational lens news stories in the "For Further Exploration" section). See if your group can work together to identify gravitational arcs, the images of distant background galaxies distorted by the mass of the cluster. How many can you find? Can you identify any multiple images of the same background galaxy? (If anyone in the group gets really interested, there is a Citizen Science project called Spacewarps, where you can help astronomers identify gravitational lenses on their images: https://spacewarps.org.)

F. You get so excited about gravitational lensing that you begin to talk about it with an intelligent friend who has not yet taken an astronomy course. After hearing you out, this friend starts to worry. He says, "If gravitational lenses can distort quasar images, sometimes creating multiple, or ghost, images of the same object, then how can we trust any point of light in the sky to be real? Maybe many of the stars we see are just ghost images or lensed images too!" Have your group discuss how to respond. (Hint: Think about the path that the light of a quasar took on its way to us and the path the light of a typical star takes.)

This OpenStax book is available for free at http://cnx.org/content/col11992/1.8

G. The 8.4-meter Large Synoptic Survey Telescope (LSST), currently under construction atop Cerro Pachón, a mountain in northern Chile, will survey the entire sky with its 3.2-gigapixel camera every few days, looking for transient, or temporary, objects that make a brief appearance in the sky before fading from view, including asteroids and Kuiper belt objects in our solar system, and supernovae and other explosive high-energy events in the distant universe. When it's fully operating sometime after 2021, the LSST will produce up to 30 terabytes of data *every night*. (A terabyte is 1000 gigabytes, which is the unit you probably use to rate your computer or memory stick capacity.) With your group, consider what you think might be some challenges of dealing with that quantity of data every night in a scientifically productive but efficient way. Can you propose any solutions to those challenges?

H. Quasars are rare now but were much more numerous when the universe was about one-quarter of its current age. The total star formation taking place in galaxies across the universe peaked at about the same redshift. Does your group think this is a coincidence? Why or why not?

I. One way to see how well the ideas in astronomy (like those in this chapter) have penetrated popular culture is to see whether you can find astronomical words in the marketplace. A short web search for the term "dark matter" turns up both a brand of coffee and a brand of "muscle growth accelerator" with that name. How many other terms used in this chapter can your group find in the world of products? (What's a really popular type of Android cell phone, for example?)

J. What's your complete address in the universe? Group members should write out their full address, based on the information in this chapter (and the rest of the book). After your postal code and country, you may want to add continent, planet, planetary system, galaxy, etc. Then each group member should explain this address to a family member or student not taking astronomy.

⬚ EXERCISES

Review Questions

1. How are distant (young) galaxies different from the galaxies that we see in the universe today?

2. What is the evidence that star formation began when the universe was only a few hundred million years old?

3. Describe the evolution of an elliptical galaxy. How does the evolution of a spiral galaxy differ from that of an elliptical?

4. Explain what we mean when we call the universe homogeneous and isotropic. Would you say that the distribution of elephants on Earth is homogeneous and isotropic? Why?

5. Describe the organization of galaxies into groupings, from the Local Group to superclusters.

6. What is the evidence that a large fraction of the matter in the universe is invisible?

7. When astronomers make maps of the structure of the universe on the largest scales, how do they find the superclusters of galaxies to be arranged?

8. How does the presence of an active galactic nucleus in a starburst galaxy affect the starburst process?

Download for free at http://cnx.org/content/col11992/latest/

Thought Questions

9. Describe how you might use the color of a galaxy to determine something about what kinds of stars it contains.

10. Suppose a galaxy formed stars for a few million years and then stopped (and no other galaxy merged or collided with it). What would be the most massive stars on the main sequence after 500 million years? After 10 billion years? How would the color of the galaxy change over this time span? (Refer to Evolution from the Main Sequence to Red Giants.)

11. Given the ideas presented here about how galaxies form, would you expect to find a giant elliptical galaxy in the Local Group? Why or why not? Is there in fact a giant elliptical in the Local Group?

12. Can an elliptical galaxy evolve into a spiral? Explain your answer. Can a spiral turn into an elliptical? How?

13. If we see a double image of a quasar produced by a gravitational lens and can obtain a spectrum of the galaxy that is acting as the gravitational lens, we can then put limits on the distance to the quasar. Explain how.

14. The left panel of Figure 27.1 shows a cluster of yellow galaxies that produces several images of blue galaxies through gravitational lensing. Which are more distant—the blue galaxies or the yellow galaxies? The light in the galaxies comes from stars. How do the temperatures of the stars that dominate the light of the cluster galaxies differ from the temperatures of the stars that dominate the light of the blue-lensed galaxy? Which galaxy's light is dominated by young stars?

15. Suppose you are standing in the center of a large, densely populated city that is exactly circular, surrounded by a ring of suburbs with lower-density population, surrounded in turn by a ring of farmland. From this specific location, would you say the population distribution is isotropic? Homogeneous?

16. Astronomers have been making maps by observing a slice of the universe and seeing where the galaxies lie within that slice. If the universe is isotropic and homogeneous, why do they need more than one slice? Suppose they now want to make each slice extend farther into the universe. What do they need to do?

17. Human civilization is about 10,000 years old as measured by the development of agriculture. If your telescope collects starlight tonight that has been traveling for 10,000 years, is that star inside or outside our Milky Way Galaxy? Is it likely that the star has changed much during that time?

18. Given that only about 5% of the galaxies visible in the Hubble Deep Field are bright enough for astronomers to study spectroscopically, they need to make the most of the other 95%. One technique is to use their colors and apparent brightnesses to try to roughly estimate their redshift. How do you think the inaccuracy of this redshift estimation technique (compared to actually measuring the redshift from a spectrum) might affect our ability to make maps of large-scale structures such as the filaments and voids shown in Figure 28.21?

Figuring For Yourself

19. Using the information from Example 28.1, how much fainter an object will you have to be able to measure in order to include the same kinds of galaxies in your second survey? Remember that the brightness of an object varies as the inverse square of the distance.

20. Using the information from Example 28.1, if galaxies are distributed homogeneously, how many times more of them would you expect to count on your second survey?

21. Using the information from Example 28.1, how much longer will it take you to do your second survey?

This OpenStax book is available for free at http://cnx.org/content/col11992/1.8

Download for free at http://cnx.org/content/col11992/latest/

22. Galaxies are found in the "walls" of huge voids; very few galaxies are found in the voids themselves. The text says that the structure of filaments and voids has been present in the universe since shortly after the expansion began 13.8 billion years ago. In science, we always have to check to see whether some conclusion is contradicted by any other information we have. In this case, we can ask whether the voids would have filled up with galaxies in roughly 14 billion years. Observations show that in addition to the motion associated with the expansion of the universe, the galaxies in the walls of the voids are moving in random directions at typical speeds of 300 km/s. At least some of them will be moving into the voids. How far into the void will a galaxy move in 14 billion years? Is it a reasonable hypothesis that the voids have existed for 14 billion years?

23. Calculate the velocity, the distance, and the look-back time of the most distant galaxies in Figure 28.21 using the Hubble constant given in this text and the redshift given in the diagram. Remember the Doppler formula for velocity $\left(v = c \times \frac{\Delta\lambda}{\lambda}\right)$ and the Hubble law ($v = H \times d$, where d is the distance to a galaxy). For these low velocities, you can neglect relativistic effects.

24. Assume that dark matter is uniformly distributed throughout the Milky Way, not just in the outer halo but also throughout the bulge and in the disk, where the solar system lives. How much dark matter would you expect there to be inside the solar system? Would you expect that to be easily detectable? Hint: For the radius of the Milky Way's dark matter halo, use $R = 300,000$ light-years; for the solar system's radius, use 100 AU; and start by calculating the ratio of the two volumes.

25. The simulated box of galaxy filaments and superclusters shown in Figure 28.29 stretches across 1 billion light-years. If you were to make a scale model where that box covered the core of a university campus, say 1 km, then how big would the Milky Way Galaxy be? How far away would the Andromeda galaxy be in the scale model?

26. The first objects to collapse gravitationally after the Big Bang might have been globular cluster-size galaxy pieces, with masses around 10^6 solar masses. Suppose you merge two of those together, then merge two larger pieces together, and so on, Lego-style, until you reach a Milky Way mass, about 10^{12} solar masses. How many merger generations would that take, and how many original pieces? (Hint: Think in powers of 2.)

Download for free at http://cnx.org/content/col11992/latest/

This OpenStax book is available for free at http://cnx.org/content/col11992/1.8

Download for free at http://cnx.org/content/col11992/latest/

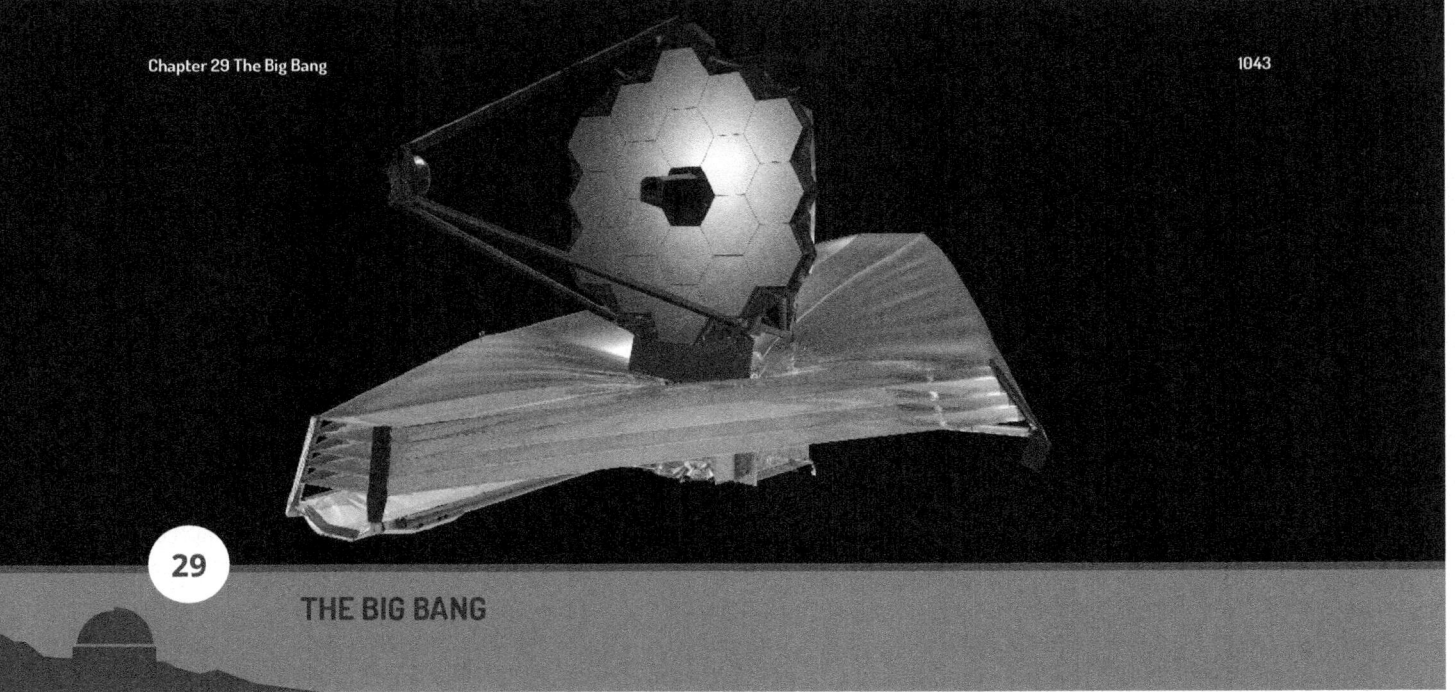

29

THE BIG BANG

Figure 29.1. Space Telescope of the Future. This drawing shows the James Webb Space Telescope, which is currently planned for launch in 2018. The silver sunshade shadows the primary mirror and science instruments. The primary mirror is 6.5 meters (21 feet) in diameter. Before and during launch, the mirror will be folded up. After the telescope is placed in its orbit, ground controllers will command it to unfold the mirror petals. To see distant galaxies whose light has been shifted to long wavelengths, the telescope will carry several instruments for taking infrared images and spectra. (credit: modification of work by NASA)

Chapter Outline

Thinking Ahead

In previous chapters, we explored the contents of the universe—planets, stars, and galaxies—and learned about how these objects change with time. But what about the universe as a whole? How old is it? What did it look like in the beginning? How has it changed since then? What will be its fate?

Cosmology is the study of the universe as a whole and is the subject of this chapter. The story of observational cosmology really begins in 1929 when Edwin Hubble published observations of redshifts and distances for a small sample of galaxies and showed the then-revolutionary result that we live in an expanding universe—one which in the past was denser, hotter, and smoother. From this early discovery, astronomers developed many predictions about the origin and evolution of the universe and then tested those predictions with observations. In this chapter, we will describe what we already know about the history of our dynamic universe and highlight some of the mysteries that remain.

Download for free at http://cnx.org/content/col11992/latest/

29.1 THE AGE OF THE UNIVERSE

Learning Objectives

By the end of this section, you will be able to:

> Describe how we estimate the age of the universe
> Explain how changes in the rate of expansion over time affect estimates of the age of the universe
> Describe the evidence that dark energy exists and that the rate of expansion is currently accelerating
> Describe some independent evidence for the age of the universe that is consistent with the age estimate based on the rate of expansion

To explore the history of the universe, we will follow the same path that astronomers followed historically—beginning with studies of the nearby universe and then probing ever-more-distant objects and looking further back in time.

The realization that the universe changes with time came in the 1920s and 1930s when measurements of the redshifts of a large sample of galaxies became available. With hindsight, it is surprising that scientists were so shocked to discover that the universe is expanding. In fact, our theories of gravity demand that the universe must be either expanding or contracting. To show what we mean, let's begin with a universe of finite size—say a giant ball of a thousand galaxies. All these galaxies attract each other because of their gravity. If they were initially stationary, they would inevitably begin to move closer together and eventually collide. They could avoid this collapse only if for some reason they happened to be moving away from each other at high speeds. In just the same way, only if a rocket is launched at high enough speed can it avoid falling back to Earth.

The problem of what happens in an infinite universe is harder to solve, but Einstein (and others) used his theory of general relativity (which we described in Black Holes and Curved Spacetime) to show that even infinite universes cannot be static. Since astronomers at that time did not yet know the universe was expanding (and Einstein himself was philosophically unwilling to accept a universe in motion), he changed his equations by introducing an arbitrary new term (we might call it a fudge factor) called the **cosmological constant**. This constant represented a hypothetical force of repulsion that could balance gravitational attraction on the largest scales and permit galaxies to remain at fixed distances from one another. That way, the universe could remain still.

This OpenStax book is available for free at http://cnx.org/content/col11992/1.8

Download for free at http://cnx.org/content/col11992/latest/

Figure 29.2. Einstein and Hubble. (a) Albert Einstein is shown in a 1921 photograph. (b) Edwin Hubble at work in the Mt. Wilson Observatory.

About a decade later, Hubble, and his coworkers reported that the universe is expanding, so that no mysterious balancing force is needed. (We discussed this in the chapter on Galaxies.) Einstein is reported to have said that the introduction of the cosmological constant was "the biggest blunder of my life." As we shall see later in this chapter, however, relatively recent observations indicate that the expansion is *accelerating*. Observations are now being carried out to determine whether this acceleration is consistent with a cosmological constant. In a way, it may turn out that Einstein was right after all.

LINK TO LEARNING

View this web exhibit (https://openstax.org/l/30exhcosmAIPCHP) on the history of our thinking about cosmology, with images and biographies, from the American Institute of Physics Center for the History of Physics.

The Hubble Time

If we had a movie of the expanding universe and ran the film *backward*, what would we see? The galaxies, instead of moving apart, would move *together* in our movie—getting closer and closer all the time. Eventually, we would find that all the matter we can see today was once concentrated in an infinitesimally small volume. Astronomers identify this time with the *beginning of the universe*. The explosion of that concentrated universe at the beginning of time is called the **Big Bang** (not a bad term, since you can't have a bigger bang than one that creates the entire universe). But when did this bang occur?

We can make a reasonable estimate of the time since the universal expansion began. To see how astronomers do this, let's begin with an analogy. Suppose your astronomy class decides to have a party (a kind of "Big Bang") at someone's home to celebrate the end of the semester. Unfortunately, everyone is celebrating with so much enthusiasm that the neighbors call the police, who arrive and send everyone away at the same moment. You

Download for free at http://cnx.org/content/col11992/latest/

get home at 2 a.m., still somewhat upset about the way the party ended, and realize you forgot to look at your watch to see what time the police got there. But you use a map to measure that the distance between the party and your house is 40 kilometers. And you also remember that you drove the whole trip at a steady speed of 80 kilometers/hour (since you were worried about the police cars following you). Therefore, the trip must have taken:

$$\text{time} = \frac{\text{distance}}{\text{velocity}} = \frac{40\,\text{kilometers}}{80\,\text{kilometers/hour}} = 0.5\,\text{hours}$$

So the party must have broken up at 1:30 a.m.

No humans were around to look at their watches when the universe began, but we can use the same technique to estimate when the galaxies began moving away from each other. (Remember that, in reality, it is space that is expanding, not the galaxies that are moving through static space.) If we can measure how far apart the galaxies are now, and how fast they are moving, we can figure out how long a trip it's been.

Let's call the age of the universe measured in this way T_0. Let's first do a simple case by assuming that the expansion has been at a constant rate ever since the expansion of the universe began. In this case, the time it has taken a galaxy to move a distance, d, away from the Milky Way (remember that at the beginning the galaxies were all together in a very tiny volume) is (as in our example)

$$T_0 = d/v$$

where v is the velocity of the galaxy. If we can measure the speed with which galaxies are moving away, and also the distances between them, we can establish how long ago the expansion began.

Making such measurements should sound very familiar. This is just what Hubble and many astronomers after him needed to do in order to establish the Hubble law and the Hubble constant. We learned in Galaxies that a galaxy's distance and its velocity in the expanding universe are related by

$$V = H \times d$$

where H is the Hubble constant. Combining these two expressions gives us

$$T_0 = \frac{d}{v} = \frac{d}{(H \times d)} = \frac{1}{H}$$

We see, then, that the work of calculating this time was already done for us when astronomers measured the Hubble constant. The age of the universe estimated in this way turns out to be just the *reciprocal of the Hubble constant* (that is, $1/H$). This age estimate is sometimes called the Hubble time. For a Hubble constant of 20 kilometers/second per million light-years, the Hubble time is about 15 billion years. The unit used by astronomers for the Hubble constant is kilometers/second per million parsecs. In these units, the Hubble constant is equal to about 70 kilometers/second per million parsecs, again with an uncertainty of about 5%.

To make numbers easier to remember, we have done some rounding here. Estimates for the Hubble constant are actually closer to 21 or 22 kilometers/second per million light-years, which would make the age closer to 14 billion years. But there is still about a 5% uncertainty in the Hubble constant, which means the age of the universe estimated in this way is also uncertain by about 5%.

To put these uncertainties in perspective, however, you should know that 50 years ago, the uncertainty was a factor of 2. Remarkable progress toward pinning down the Hubble constant has been made in the last couple of decades.

The Role of Deceleration

The Hubble time is the right age for the universe only if the expansion rate has been constant throughout the time since the expansion of the universe began. Continuing with our end-of-the-semester-party analogy, this is

This OpenStax book is available for free at http://cnx.org/content/col11992/1.8

Download for free at http://cnx.org/content/col11992/latest/

equivalent to assuming that you traveled home from the party at a constant rate, when in fact this may not have been the case. At first, mad about having to leave, you may have driven fast, but then as you calmed down—and thought about police cars on the highway—you may have begun to slow down until you were driving at a more socially acceptable speed (such as 80 kilometers/hour). In this case, given that you were driving faster at the beginning, the trip home would have taken less than a half-hour.

In the same way, in calculating the Hubble time, we have assumed that H has been constant throughout all of time. It turns out that this is not a good assumption. Earlier in their thinking about this, astronomers expected that the rate of expansion should be slowing down. We know that matter creates gravity, whereby all objects pull on all other objects. The mutual attraction between galaxies was expected to slow the expansion as time passed. This means that, if gravity were the only force acting (a big *if*, as we shall see in the next section), then the rate of expansion must have been faster in the past than it is today. In this case, we would say the universe has been *decelerating* since the beginning.

How much it has decelerated depends on the importance of gravity in slowing the expansion. If the universe were nearly empty, the role of gravity would be minor. Then the deceleration would be close to zero, and the universe would have been expanding at a constant rate. But in a universe with any significant density of matter, the pull of gravity means that the rate of expansion should be slower now than it used to be. If we use the current rate of expansion to estimate how long it took the galaxies to reach their current separations, we will overestimate the age of the universe—just as we may have overestimated the time it took for you to get home from the party.

A Universal Acceleration

Astronomers spent several decades looking for evidence that the expansion was decelerating, but they were not successful. What they needed were 1) larger telescopes so that they could measure the redshifts of more distant galaxies and 2) a very luminous *standard bulb* (or standard candle), that is, some astronomical object with known luminosity that produces an enormous amount of energy and can be observed at distances of a billion light-years or more.

Recall that we discussed standard bulbs in the chapter on Galaxies. If we compare how luminous a standard bulb is supposed to be and how dim it actually looks in our telescopes, the difference allows us to calculate its distance. The redshift of the galaxy such a bulb is in can tell us how fast it is moving in the universe. So we can measure its distance and motion independently.

These two requirements were finally met in the 1990s. Astronomers showed that supernovae of type Ia (see The Death of Stars), with some corrections based on the shapes of their light curves, are standard bulbs. This type of supernova occurs when a white dwarf accretes enough material from a companion star to exceed the Chandrasekhar limit and then collapses and explodes. At the time of maximum brightness, these dramatic supernovae can briefly outshine the galaxies that host them, and hence, they can be observed at very large distances. Large 8- to 10-meter telescopes can be used to obtain the spectra needed to measure the redshifts of the host galaxies (Figure 29.3).

Download for free at http://cnx.org/content/col11992/latest/

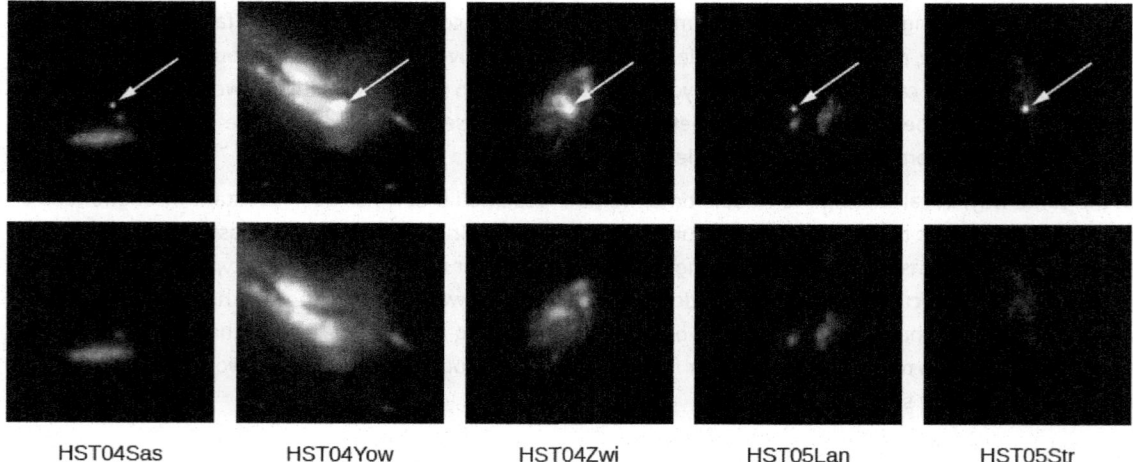

| HST04Sas | HST04Yow | HST04Zwi | HST05Lan | HST05Str |

Figure 29.3. Five Supernovae and Their Host Galaxies. The top row shows each galaxy and its supernova (arrow). The bottom row shows the same galaxies either before or after the supernovae exploded. (credit: modification of work by NASA, ESA, and A. Riess (STScI))

The result of painstaking, careful study of these supernovae in a range of galaxies, carried out by two groups of researchers, was published in 1998. It was shocking—and so revolutionary that their discovery received the 2011 Nobel Prize in Physics. What the researchers found was that these type Ia supernovae in distant galaxies were fainter than expected from Hubble's law, given the measured redshifts of their host galaxies. In other words, distances estimated from the supernovae used as standard bulbs disagreed with the distances measured from the redshifts.

If the universe were decelerating, we would expect the far-away supernovae to be *brighter* than expected. The slowing down would have kept them closer to us. Instead, they were *fainter*, which at first seemed to make no sense.

Before accepting this shocking development, astronomers first explored the possibility that the supernovae might not really be as useful as standard bulbs as they thought. Perhaps the supernovae appeared too faint because dust along our line of sight to them absorbed some of their light. Or perhaps the supernovae at large distances were for some reason intrinsically less luminous than nearby supernovae of type Ia.

A host of more detailed observations ruled out these possibilities. Scientists then had to consider the alternative that the distance estimated from the redshift was incorrect. Distances derived from redshifts assume that the Hubble constant has been truly constant for all time. We saw that one way it might not be constant is that the expansion is slowing down. But suppose neither assumption is right (steady speed or slowing down.)

Suppose, instead, that the universe is *accelerating*. If the universe is expanding faster now than it was billions of years ago, our motion away from the distant supernovae has sped up since the explosion occurred, sweeping us farther away from them. The light of the explosion has to travel a greater distance to reach us than if the expansion rate were constant. The farther the light travels, the fainter it appears. This conclusion would explain the supernova observations in a natural way, and this has now been substantiated by many additional observations over the last couple of decades. It really seems that *the expansion of the universe is accelerating*, a notion so unexpected that astronomers at first resisted considering it.

How can the expansion of the universe be speeding up? If you want to accelerate your car, you must supply energy by stepping on the gas. Similarly, energy must be supplied to accelerate the expansion of the universe. The discovery of the acceleration was shocking because scientists still have no idea what the source of the energy is. Scientists call whatever it is **dark energy**, which is a clear sign of how little we understand it.

This OpenStax book is available for free at http://cnx.org/content/col11992/1.8

Note that this new component of the universe is not the dark matter we talked about in earlier chapters. Dark energy is something else that we have also not yet detected in our laboratories on Earth.

What is dark energy? One possibility is that it is the cosmological constant, which is an energy associated with the vacuum of "empty" space itself. Quantum mechanics (the intriguing theory of how things behave at the atomic and subatomic levels) tells us that the source of this vacuum energy might be tiny elementary particles that flicker in and out of existence everywhere throughout the universe. Various attempts have been made to calculate how big the effects of this vacuum energy should be, but so far these attempts have been unsuccessful. In fact, the order of magnitude of theoretical estimates of the vacuum energy based on the quantum mechanics of matter and the value required to account for the acceleration of the expansion of the universe differ by an incredible factor of at least 10^{120} (that is a 1 followed by 120 zeros)! Various other theories have been suggested, but the bottom line is that, although there is compelling evidence that dark energy exists, we do not yet know the source of that energy.

Whatever the dark energy turns out to be, we should note that the discovery that the rate of expansion has not been constant since the beginning of the universe complicates the calculation of the age of the universe. Interestingly, the acceleration seems not to have started with the Big Bang. During the first several billion years after the Big Bang, when galaxies were close together, gravity was strong enough to slow the expansion. As galaxies moved farther apart, the effect of gravity weakened. Several billion years after the Big Bang, dark energy took over, and the expansion began to accelerate (Figure 29.4).

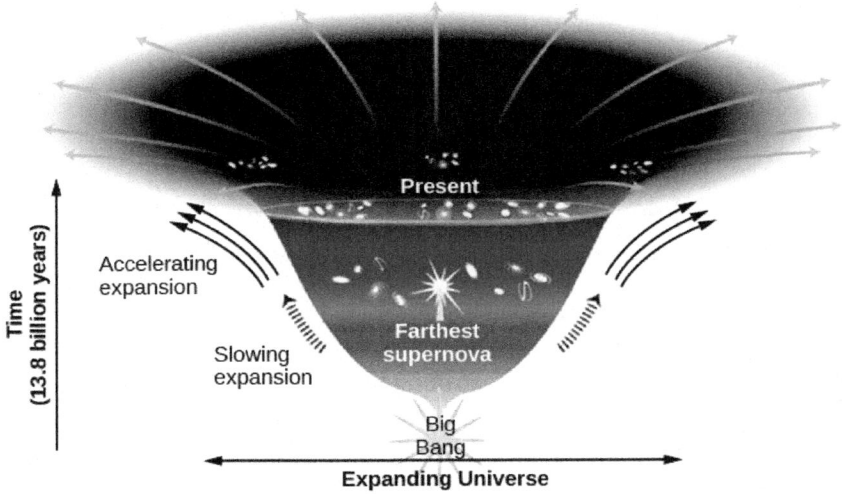

Figure 29.4. Changes in the Rate of Expansion of the Universe Since Its Beginning 13.8 Billion Years Ago. The more the diagram spreads out horizontally, the faster the change in the velocity of expansion. After a period of very rapid expansion at the beginning, which scientists call inflation and which we will discuss later in this chapter, the expansion began to decelerate. Galaxies were then close together, and their mutual gravitational attraction slowed the expansion. After a few billion years, when galaxies were farther apart, the influence of gravity began to weaken. Dark energy then took over and caused the expansion to accelerate. (credit: modification of work by Ann Feild (STScI))

Deceleration works to make the age of the universe estimated by the simple relation $T_0 = 1/H$ seem older than it really is, whereas acceleration works to make it seem younger. By happy coincidence, our best estimates of how much deceleration and acceleration occurred lead to an answer for the age very close to $T_0 = 1/H$. The best current estimate is that the universe is 13.8 billion years old with an uncertainty of only about 100 million years.

Throughout this chapter, we have referred to the Hubble *constant*. We now know that the Hubble constant does change with time. It is, however, constant everywhere in the universe at any given time. When we say the

Download for free at http://cnx.org/content/col11992/latest/

Hubble constant is about 70 kilometers/second/million parsecs, we mean that this is the value of the Hubble constant at the current time.

Comparing Ages

We now have one estimate for the age of the universe from its expansion. Is this estimate consistent with other observations? For example, are the oldest stars or other astronomical objects younger than 13.8 billion years? After all, the universe has to be at least as old as the oldest objects in it.

In our Galaxy and others, the oldest stars are found in the globular clusters (Figure 29.5), which can be dated using the models of stellar evolution described in the chapter Stars from Adolescence to Old Age.

Figure 29.5. Globular Cluster 47 Tucanae. This NASA/ESA Hubble Space Telescope image shows a globular cluster known as 47 Tucanae, since it is in the constellation of Tucana (The Toucan) in the southern sky. The second-brightest globular cluster in the night sky, it includes hundreds of thousands of stars. Globular clusters are among the oldest objects in our Galaxy and can be used to estimate its age. (credit: NASA, ESA, and the Hubble Heritage (STScI/AURA)-ESA/Hubble Collaboration)

The accuracy of the age estimates of the globular clusters has improved markedly in recent years for two reasons. First, models of interiors of globular cluster stars have been improved, mainly through better information about how atoms absorb radiation as they make their way from the center of a star out into space. Second, observations from satellites have improved the accuracy of our measurements of the distances to these clusters. The conclusion is that the oldest stars formed about 12–13 billion years ago.

This age estimate has recently been confirmed by the study of the spectrum of uranium in the stars. The isotope uranium-238 is radioactive and decays (changes into another element) over time. (Uranium-238 gets its designation because it has 92 protons and 146 neutrons.) We know (from how stars and supernovae make elements) how much uranium-238 is generally made compared to other elements. Suppose we measure the amount of uranium relative to nonradioactive elements in a very old star and in our own Sun, and compare the abundances. With those pieces of information, we can estimate how much longer the uranium has been decaying in the very old star because we know from our own Sun how much uranium decays in 4.5 billion years.

The line of uranium is very weak and hard to make out even in the Sun, but it has now been measured in one extremely old star using the European Very Large Telescope (Figure 29.6). Comparing the abundance with that in the solar system, whose age we know, astronomers estimate the star is 12.5 billion years old, with an uncertainty of about 3 billion years. While the uncertainty is large, this work is important confirmation of

This OpenStax book is available for free at http://cnx.org/content/col11992/1.8

the ages estimated by studies of the globular cluster stars. Note that the uranium age estimate is completely independent; it does not depend on either the measurement of distances or on models of the interiors of stars.

Figure 29.6. European Extremely Large Telescope, European Very Large Telescope, and the Colosseum. The European Extremely Large Telescope (E-ELT) is currently under construction in Chile. This image compares the size of the E-ELT (left) with the four 8-meter telescopes of the European Very Large Telescope (center) and with the Colosseum in Rome (right). The mirror of the E-ELT will be 39 meters in diameter. Astronomers are building a new generation of giant telescopes in order to observe very distant galaxies and understand what they were like when they were newly formed and the universe was young. (credit: modification of work by ESO)

As we shall see later in this chapter, the globular cluster stars probably did not form until the expansion of the universe had been underway for at least a few hundred million years. Accordingly, their ages are consistent with the 13.8 billion-year age estimated from the expansion rate.

29.2 | A MODEL OF THE UNIVERSE

Learning Objectives

By the end of this section, you will be able to:

- > Explain how the rate of expansion of the universe affects its evolution
- > Describe four possibilities for the evolution of the universe
- > Explain what is expanding when we say that the universe is expanding
- > Define critical density and the evidence that matter alone in the universe is much smaller than the critical density
- > Describe what the observations say about the likely long-term future of the universe

Let's now use the results about the expansion of the universe to look at how these ideas might be applied to develop a model for the evolution of the universe as a whole. With this model, astronomers can make predictions about how the universe has evolved so far and what will happen to it in the future.

The Expanding Universe

Every model of the universe must include the expansion we observe. Another key element of the models is that the cosmological principle (which we discussed in The Evolution and Distribution of Galaxies) is valid: on the large scale, the universe at any given time is the same everywhere (homogeneous and isotropic). As a result, the expansion rate must be the same everywhere during any epoch of cosmic time. If so, we don't need to think about the entire universe when we think about the expansion, we can just look at any sufficiently large portion

Download for free at http://cnx.org/content/col11992/latest/

of it. (Some models for dark energy would allow the expansion rate to be different in different directions, and scientists are designing experiments to test this idea. However, until such evidence is found, we will assume that the cosmological principle applies throughout the universe.)

In Galaxies, we hinted that when we think of the expansion of the universe, it is more correct to think of space itself stretching rather than of galaxies moving through static space. Nevertheless, we have since been discussing the redshifts of galaxies as if they resulted from the motion of the galaxies themselves.

Now, however, it is time to finally put such simplistic notions behind us and take a more sophisticated look at the cosmic expansion. Recall from our discussion of Einstein's theory of general relativity (in the chapter on Black Holes and Curved Spacetime) that space—or, more precisely, spacetime—is not a mere backdrop to the action of the universe, as Newton thought. Rather, it is an active participant—affected by and in turn affecting the matter and energy in the universe.

Since the expansion of the universe is the stretching of all spacetime, all points in the universe are stretching together. Thus, the expansion began *everywhere at once*. Unfortunately for tourist agencies of the future, there is no location you can visit where the stretching of space began or where we can say that the Big Bang happened.

To describe just how space stretches, we say the cosmic expansion causes the universe to undergo a uniform change in *scale* over time. By scale we mean, for example, the distance between two clusters of galaxies. It is customary to represent the scale by the factor R; if R doubles, then the distance between the clusters has doubled. Since the universe is expanding at the same rate everywhere, the change in R tells us how much it has expanded (or contracted) at any given time. For a static universe, R would be constant as time passes. In an expanding universe, R increases with time.

If it is space that is stretching rather than galaxies moving through space, then why do the galaxies show redshifts in their spectra? When you were young and naïve—a few chapters ago—it was fine to discuss the redshifts of distant galaxies as if they resulted from their motion away from us. But now that you are an older and wiser student of cosmology, this view will simply not do.

A more accurate view of the redshifts of galaxies is that the light waves are stretched by the stretching of the space they travel through. Think about the light from a remote galaxy. As it moves away from its source, the light has to travel through space. If space is stretching during all the time the light is traveling, the light waves will be stretched as well. A redshift is a stretching of waves—the wavelength of each wave increases (Figure 29.7). Light from more distant galaxies travels for more time than light from closer ones. This means that the light has stretched more than light from closer ones and thus shows a greater redshift.

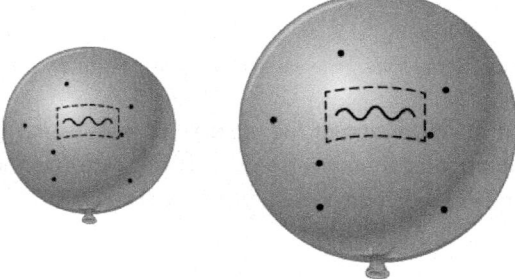

Figure 29.7. Expansion and Redshift. As an elastic surface expands, a wave on its surface stretches. For light waves, the increase in wavelength would be seen as a redshift.

This OpenStax book is available for free at http://cnx.org/content/col11992/1.8

Download for free at http://cnx.org/content/col11992/latest/

Thus, what the measured redshift of light from an object is telling us is how much the universe has expanded since the light left the object. If the universe has expanded by a factor of 2, then the wavelength of the light (and all electromagnetic waves from the same source) will have doubled.

Models of the Expansion

Before astronomers knew about dark energy or had a good measurement of how much matter exists in the universe, they made speculative models about how the universe might evolve over time. The four possible scenarios are shown in Figure 29.8. In this diagram, time moves forward from the bottom upward, and the scale of space increases by the horizontal circles becoming wider.

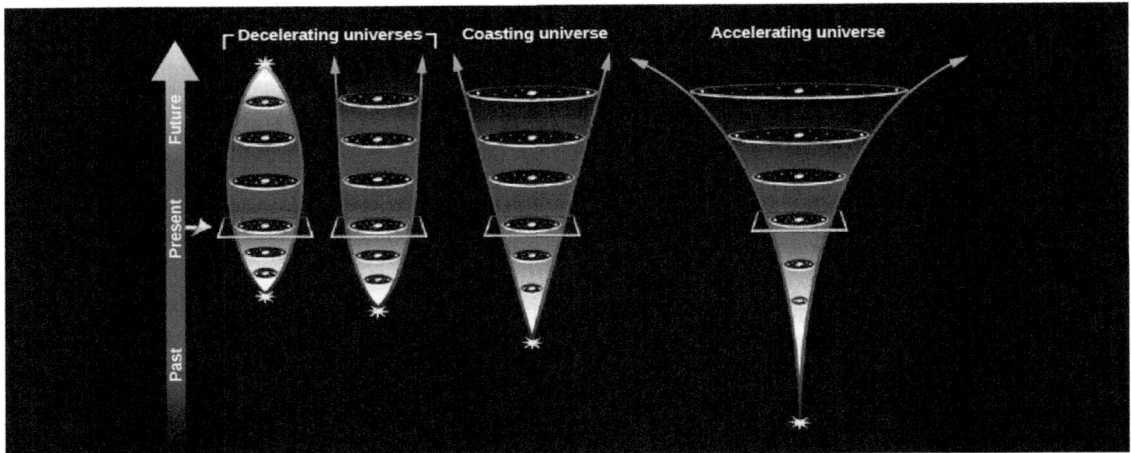

Figure 29.8. Four Possible Models of the Universe. The yellow square marks the present in all four cases, and for all four, the Hubble constant is equal to the same value at the present time. Time is measured in the vertical direction. The first two universes on the left are ones in which the rate of expansion slows over time. The one on the left will eventually slow, come to a stop and reverse, ending up in a "big crunch," while the one next to it will continue to expand forever, but ever-more slowly as time passes. The "coasting" universe is one that expands at a constant rate given by the Hubble constant throughout all of cosmic time. The accelerating universe on the right will continue to expand faster and faster forever. (credit: modification of work by NASA/ESA)

The simplest scenario of an expanding universe would be one in which R increases with time at a constant rate. But you already know that life is not so simple. The universe contains a great deal of mass and its gravity decelerates the expansion—by a large amount if the universe contains a lot of matter, or by a negligible amount if the universe is nearly empty. Then there is the observed acceleration, which astronomers blame on a kind of dark energy.

Let's first explore the range of possibilities with models for different amounts of mass in the universe and for different contributions by dark energy. In some models—as we shall see—the universe expands forever. In others, it stops expanding and starts to contract. After looking at the extreme possibilities, we will look at recent observations that allow us to choose the most likely scenario.

We should perhaps pause for a minute to note how remarkable it is that we can do this at all. Our understanding of the principles that underlie how the universe works on the large scale and our observations of how the objects in the universe change with time allow us to model the evolution of the entire cosmos these days. It is one of the loftiest achievements of the human mind.

What astronomers look at in practice, to determine the kind of universe we live in, is the *average density* of the universe. This is the mass of matter (including the equivalent mass of energy)[1] that would be contained

---

1   By equivalent mass we mean that which would result if the energy were turned into mass using Einstein's formula, $E = mc^2$.

Download for free at http://cnx.org/content/col11992/latest/

in each unit of volume (say, 1 cubic centimeter) if all the stars, galaxies, and other objects were taken apart, atom by atom, and if all those particles, along with the light and other energy, were distributed throughout all of space with absolute uniformity. If the average density is low, there is less mass and less gravity, and the universe will not decelerate very much. It can therefore expand forever. Higher average density, on the other hand, means there is more mass and more gravity and that the stretching of space might slow down enough that the expansion will eventually stop. An extremely high density might even cause the universe to collapse again.

For a given rate of expansion, there is a **critical density**—the mass per unit volume that will be just enough to slow the expansion to zero at some time infinitely far in the future. If the actual density is higher than this critical density, then the expansion will ultimately reverse and the universe will begin to contract. If the actual density is lower, then the universe will expand forever.

These various possibilities are illustrated in Figure 29.9. In this graph, one of the most comprehensive in all of science, we chart the development of the scale of space in the cosmos against the passage of time. Time increases to the right, and the scale of the universe, R, increases upward in the figure. Today, at the point marked "present" along the time axis, R is increasing in each model. We know that the galaxies are currently expanding away from each other, no matter which model is right. (The same situation holds for a baseball thrown high into the air. While it may eventually fall back down, near the beginning of the throw it moves upward most rapidly.)

The various lines moving across the graph correspond to different models of the universe. The straight dashed line corresponds to the empty universe with no deceleration; it intercepts the time axis at a time, T_0 (the Hubble time), in the past. This is not a realistic model but gives us a measure to compare other models to. The curves below the dashed line represent models with no dark energy and with varying amounts of deceleration, starting from the Big Bang at shorter times in the past. The curve above the dashed line shows what happens if the expansion is accelerating. Let's take a closer look at the future according to the different models.

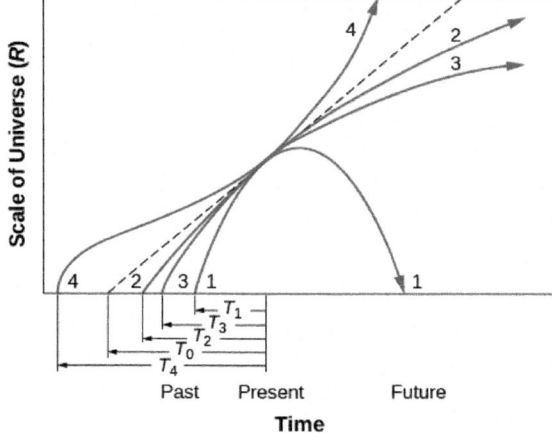

Figure 29.9. Models of the Universe. This graph plots R, the scale of the universe, against time for various cosmological models. Curve 1 represents a universe where the density is greater than the critical value; this model predicts that the universe will eventually collapse. Curve 2 represents a universe with a density lower than critical; the universe will continue to expand but at an ever-slower rate. Curve 3 is a critical-density universe; in this universe, the expansion will gradually slow to a stop infinitely far in the future. Curve 4 represents a universe that is accelerating because of the effects of dark energy. The dashed line is for an empty universe, one in which the expansion is not slowed by gravity or accelerated by dark energy. Time is very compressed on this graph.

Let's start with curve 1 in Figure 29.9. In this case, the actual density of the universe is higher than the critical density and there is no dark energy. This universe will stop expanding at some time in the future and begin

This OpenStax book is available for free at http://cnx.org/content/col11992/1.8

contracting. This model is called a **closed universe** and corresponds to the universe on the left in Figure 29.8. Eventually, the scale drops to zero, which means that space will have shrunk to an infinitely small size. The noted physicist John Wheeler called this the " big crunch," because matter, energy, space, and time would all be crushed out of existence. Note that the "big crunch" is the opposite of the Big Bang—it is an *implosion*. The universe is not expanding but rather collapsing in upon itself.

Some scientists speculated that another Big Bang might follow the crunch, giving rise to a new expansion phase, and then another contraction—perhaps oscillating between successive Big Bangs and big crunches indefinitely in the past and future. Such speculation was sometimes referred to as the oscillating theory of the universe. The challenge for theorists was how to describe the transition from collapse (when space and time themselves disappear into the big crunch) to expansion. With the discovery of dark energy, however, it does not appear that the universe will experience a big crunch, so we can put worrying about it on the back burner.

If the density of the universe is less than the critical density (curve 2 in Figure 29.9 and the universe second from the left in Figure 29.8), gravity is never important enough to stop the expansion, and so the universe expands forever. Such a universe is infinite and this model is called an **open universe**. Time and space begin with the Big Bang, but they have no end; the universe simply continues expanding, always a bit more slowly as time goes on. Groups of galaxies eventually get so far apart that it would be difficult for observers in any of them to see the others. (See the feature box on What Might the Universe Be Like in the Distant Future? for more about the distant future in the closed and open universe models.)

At the critical density (curve 3), the universe can just barely expand forever. The critical-density universe has an age of exactly two-thirds T_0, where T_0 is the age of the empty universe. Universes that will someday begin to contract have ages less than two-thirds T_0.

In an empty universe (the dashed line Figure 29.9 and the coasting universe in Figure 29.8), neither gravity nor dark energy is important enough to affect the expansion rate, which is therefore constant throughout all time.

In a universe with dark energy, the rate of the expansion will increase with time, and the expansion will continue at an ever-faster rate. Curve 4 in Figure 29.9, which represents this universe, has a complicated shape. In the beginning, when the matter is all very close together, the rate of expansion is most influenced by gravity. Dark energy appears to act only over large scales and thus becomes more important as the universe grows larger and the matter begins to thin out. In this model, at first the universe slows down, but as space stretches, the acceleration plays a greater role and the expansion speeds up.

The Cosmic Tug of War

We might summarize our discussion so far by saying that a "tug of war" is going on in the universe between the forces that push everything apart and the gravitational attraction of matter, which pulls everything together. If we can determine who will win this tug of war, we will learn the ultimate fate of the universe.

The first thing we need to know is the density of the universe. Is it greater than, less than, or equal to the critical density? The critical density today depends on the value of the expansion rate today, H_0. If the Hubble constant is around 20 kilometers/second per million light-years, the critical density is about 10^{-26} kg/m^3. Let's see how this value compares with the actual density of the universe.

EXAMPLE 29.1

Download for free at http://cnx.org/content/col11992/latest/

Critical Density of the Universe

As we discussed, the critical density is that combination of matter and energy that brings the universe coasting to a stop at time infinity. Einstein's equations lead to the following expression for the critical density (ρ_{crit}):

$$\rho_{crit} = \frac{3H^2}{8\pi G}$$

where H is the Hubble constant and G is the universal constant of gravity (6.67×10^{-11} Nm^2/kg^2).

Solution

Let's substitute our values and see what we get. Take an $H = 22$ km/s per million light-years. We need to convert both km and light-years into meters for consistency. A million light-years $= 10^6 \times 9.5 \times 10^{15}$ m $= 9.5 \times 10^{21}$ m. And 22 km/s $= 2.2 \times 10^4$ m/s. That makes $H = 2.3 \times 10^{-18}$ /s and $H^2 = 5.36 \times 10^{-36}$ /s^2. So,

$$\rho_{crit} = \frac{3 \times 5.36 \times 10^{-36}}{8 \times 3.14 \times 6.67 \times 10^{-11}} = 9.6 \times 10^{-27} \text{ kg/m}^3$$

which we can round off to the 10^{-26} kg/m^3. (To make the units work out, you have to know that N, the unit of force, is the same as kg $\times$ m/s^2.)

Now we can compare densities we measure in the universe to this critical value. Note that density is mass per unit volume, but energy has an equivalent mass of $m = E/c^2$ (from Einstein's equation $E = mc^2$).

Check Your Learning

a. A single grain of dust has a mass of about 1.1×10^{-13} kg. If the average mass-energy density of space is equal to the critical density on average, how much space would be required to produce a total mass-energy equal to a dust grain?

b. If the Hubble constant were twice what it actually is, how much would the critical density be?

Answer:

a. In this case, the average mass-energy in a volume V of space is $E = \rho_{crit} V$. Thus, for space with critical density, we require that

$$V = \frac{E_{grain}}{\rho_{crit}} = \frac{1.1 \times 10^{-13} \text{ kg}}{9.6 \times 10^{-26} \text{ kg/m}^3} = 1.15 \times 10^{12} \text{ m}^3 = (10{,}500 \text{ m})^3 \cong (10.5 \text{ km})^3$$

Thus, the sides of a cube of space with mass-energy density averaging that of the critical density would need to be slightly greater than 10 km to contain the total energy equal to a single grain of dust!

b. Since the critical density goes as the square of the Hubble constant, by doubling the Hubble parameter, the critical density would increase by a factor a four. So if the Hubble constant was 44 km/s per million light-years instead of 22 km/s per million light-years, the critical density would be $\rho_{crit} = 4 \times 9.6 \times 10^{-27}$ kg/m$^3 = 3.8 \times 10^{-26}$ kg/m^3.

We can start our survey of how dense the cosmos is by ignoring the dark energy and just estimating the density of all matter in the universe, including ordinary matter and dark matter. Here is where the cosmological principle really comes in handy. Since the universe is the same all over (at least on large scales), we only need to measure how much matter exists in a (large) representative sample of it. This is similar to the way a

This OpenStax book is available for free at http://cnx.org/content/col11992/1.8

representative survey of a few thousand people can tell us whom the millions of residents of the US prefer for president.

There are several methods by which we can try to determine the average density of matter in space. One way is to count all the galaxies out to a given distance and use estimates of their masses, including dark matter, to calculate the average density. Such estimates indicate a density of about 1 to 2 × 10^{-27} kg/m^3 (10 to 20% of critical), which by itself is too small to stop the expansion.

A lot of the dark matter lies outside the boundaries of galaxies, so this inventory is not yet complete. But even if we add an estimate of the dark matter outside galaxies, our total won't rise beyond about 30% of the critical density. We'll pin these numbers down more precisely later in this chapter, where we will also include the effects of dark energy.

In any case, even if we ignore dark energy, the evidence is that the universe will continue to expand forever. The discovery of dark energy that is causing the rate of expansion to speed up only strengthens this conclusion. Things definitely do not look good for fans of the closed universe (big crunch) model.

MAKING CONNECTIONS

What Might the Universe Be Like in the Distant Future?

> Some say the world will end in fire,
> Some say in ice.
> From what I've tasted of desire
> I hold with those who favor fire.
> —From the poem "Fire and Ice" by Robert Frost (1923)

Given the destructive power of impacting asteroids, expanding red giants, and nearby supernovae, our species may not be around in the remote future. Nevertheless, you might enjoy speculating about what it would be like to live in a much, much older universe.

The observed acceleration makes it likely that we will have continued expansion into the indefinite future. If the universe expands forever (R increases without limit), the clusters of galaxies will spread ever farther apart with time. As eons pass, the universe will get thinner, colder, and darker.

Within each galaxy, stars will continue to go through their lives, eventually becoming white dwarfs, neutron stars, and black holes. Low-mass stars might take a long time to finish their evolution, but in this model, we would literally have all the time in the world. Ultimately, even the white dwarfs will cool down to be black dwarfs, any neutron stars that reveal themselves as pulsars will slowly stop spinning, and black holes with accretion disks will one day complete their "meals." The remains of stars will all be dark and difficult to observe.

This means that the light that now reveals galaxies to us will eventually go out. Even if a small pocket of raw material were left in one unsung corner of a galaxy, ready to be turned into a fresh cluster of stars, we will only have to wait until the time that their evolution is also complete. And time is one thing this model of the universe has plenty of. There will surely come a time when all the stars are out, galaxies are as dark as space, and no source of heat remains to help living things survive. Then the lifeless galaxies will just continue to move apart in their lightless realm.

Download for free at http://cnx.org/content/col11992/latest/

If this view of the future seems discouraging (from a human perspective), keep in mind that we fundamentally do not understand why the expansion rate is currently accelerating. Thus, our speculations about the future are just that: speculations. You might take heart in the knowledge that science is always a progress report. The most advanced ideas about the universe from a hundred years ago now strike us as rather primitive. It may well be that our best models of today will in a hundred or a thousand years also seem rather simplistic and that there are other factors determining the ultimate fate of the universe of which we are still completely unaware.

Ages of Distant Galaxies

In the chapter on Galaxies, we discussed how we can use Hubble's law to measure the distance to a galaxy. But that simple method only works with galaxies that are not too far away. Once we get to large distances, we are looking so far into the past that we must take into account changes in the rate of the expansion of the universe. Since we cannot measure these changes directly, we must assume one of the models of the universe to be able to convert large redshifts into distances.

This is why astronomers squirm when reporters and students ask them exactly how far away some newly discovered distant quasar or galaxy is. We really can't give an answer without first explaining the model of the universe we are assuming in calculating it (by which time a reporter or student is long gone or asleep). Specifically, we must use a model that includes the change in the expansion rate with time. The key ingredients of the model are the amounts of matter, including dark matter, and the equivalent mass (according to $E = mc^2$) of the dark energy along with the Hubble constant.

Elsewhere in this book, we have estimated the mass density of ordinary matter plus dark matter as roughly 0.3 times the critical density, and the mass equivalent of dark energy as roughly 0.7 times the critical density. We will refer to these values as the "standard model of the universe." The latest (slightly improved) estimates for these values and the evidence for them will be given later in this chapter. Calculations also require the current value of the Hubble constant. For Table 29.1, we have adopted a Hubble constant of 67.3 kilometers/second/ million parsecs (rather than rounding it to 70 kilometers/second/million parsecs), which is consistent with the 13.8 billion-year age of the universe estimated by the latest observations.

Once we assume a model, we can use it to calculate the age of the universe at the time an object emitted the light we see. As an example, Table 29.1 lists the times that light was emitted by objects at different redshifts as fractions of the current age of the universe. The times are given for two very different models so you can get a feeling for the fact that the calculated ages are fairly similar. The first model assumes that the universe has a critical density of matter and no dark energy. The second model is the standard model described in the preceding paragraph. The first column in the table is the redshift, which is given by the equation $z = \Delta\lambda/\lambda_0$ and is a measure of how much the wavelength of light has been stretched by the expansion of the universe on its long journey to us.

This OpenStax book is available for free at http://cnx.org/content/col11992/1.8

Ages of the Universe at Different Redshifts

Redshift	Percent of Current Age of Universe When the Light Was Emitted (mass = critical density)	Percent of Current Age of Universe When the Light Was Emitted (mass = 0.3 critical density; dark energy = 0.7 critical density)
0	100 (now)	100 (now)
0.5	54	63
1.0	35	43
2.0	19	24
3.0	13	16
4.0	9	11
5.0	7	9
8.0	4	5
11.9	0.02	0.027
Infinite	0	0

Table 29.1

Notice that as we find objects with higher and higher redshifts, we are looking back to smaller and smaller fractions of the age of the universe. The highest observed redshifts as this book is being written are close to 12 (Figure 29.10). As Table 29.1 shows, we are seeing these galaxies as they were when the universe was only about 3% as old as it is now. They were already formed only about 700 million years after the Big Bang.

Download for free at http://cnx.org/content/col11992/latest/

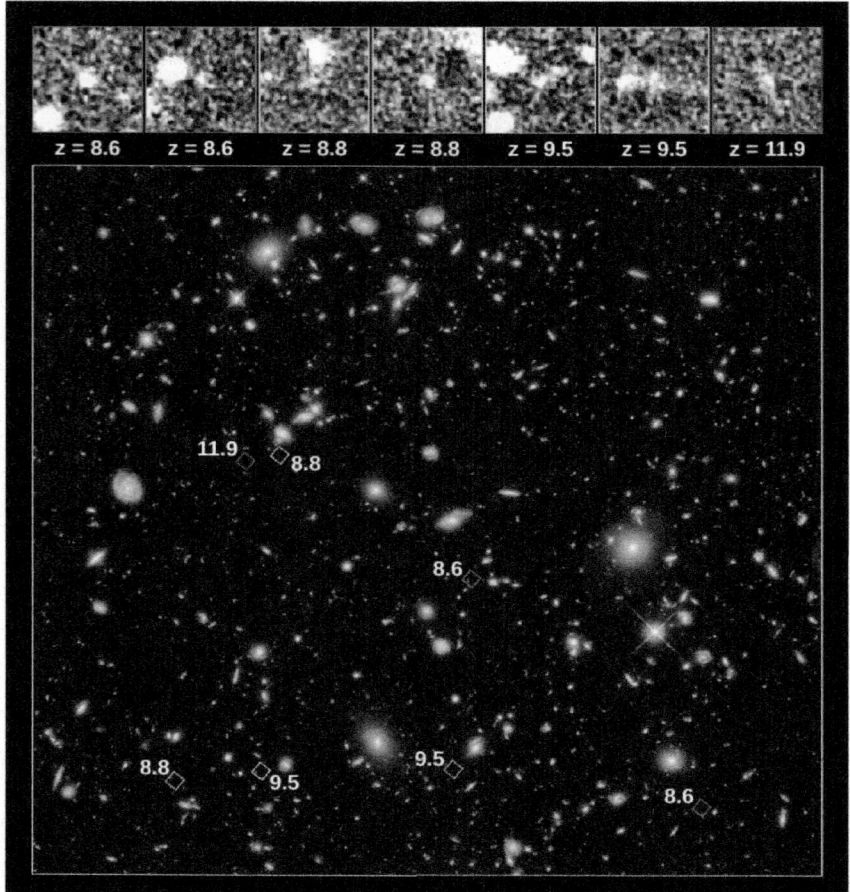

Figure 29.10. Hubble Ultra-Deep Field. This image, called the Hubble Ultra Deep Field, shows faint galaxies, seen very far away and therefore very far back in time. The colored squares in the main image outline the locations of the galaxies. Enlarged views of each galaxy are shown in the black-and-white images. The red lines mark each galaxy's location. The "redshift" of each galaxy is indicated below each box, denoted by the symbol "z." The redshift measures how much a galaxy's ultraviolet and visible light has been stretched to infrared wavelengths by the universe's expansion. The larger the redshift, the more distant the galaxy, and therefore the further astronomers are seeing back in time. One of the seven galaxies may be a distance breaker, observed at a redshift of 11.9. If this redshift is confirmed by additional measurements, the galaxy is seen as it appeared only 380 million years after the Big Bang, when the universe was less than 3% of its present age. (credit: modification of work by NASA, ESA, R. Ellis (Caltech), and the UDF 2012 Team)

29.3 | THE BEGINNING OF THE UNIVERSE

Learning Objectives

By the end of this section, you will be able to:

> Describe what the universe was like during the first few minutes after it began to expand
> Explain how the first new elements were formed during the first few minutes after the Big Bang
> Describe how the contents of the universe change as the temperature of the universe decreases

The best evidence we have today indicates that the first galaxies did not begin to form until a few hundred million years after the Big Bang. What were things like before there were galaxies and space had not yet stretched very significantly? Amazingly, scientists have been able to calculate in some detail what was happening in the universe in the first few minutes after the Big Bang.

This OpenStax book is available for free at http://cnx.org/content/col11992/1.8

Download for free at http://cnx.org/content/col11992/latest/

The History of the Idea

It is one thing to say the universe had a beginning (as the equations of general relativity imply) and quite another to describe that beginning. The Belgian priest and cosmologist Georges Lemaître was probably the first to propose a specific model for the Big Bang itself (Figure 29.11). He envisioned all the matter of the universe starting in one great bulk he called the *primeval atom*, which then broke into tremendous numbers of pieces. Each of these pieces continued to fragment further until they became the present atoms of the universe, created in a vast nuclear fission. In a popular account of his theory, Lemaître wrote, "The evolution of the world could be compared to a display of fireworks just ended—some few red wisps, ashes, and smoke. Standing on a well-cooled cinder, we see the slow fading of the suns and we try to recall the vanished brilliance of the origin of the worlds."

Figure 29.11. Abbé Georges Lemaître (1894–1966). This Belgian cosmologist studied theology at Mechelen and mathematics and physics at the University of Leuven. It was there that he began to explore the expansion of the universe and postulated its explosive beginning. He actually predicted Hubble's law 2 years before its verification, and he was the first to consider seriously the physical processes by which the universe began.

LINK TO LEARNING

View a short video (https://openstax.org/l/30Lemaitrevid) about the work of Lemaître, considered by some to be the father of the Big Bang theory.

Physicists today know much more about nuclear physics than was known in the 1920s, and they have shown that the primeval fission model cannot be correct. Yet Lemaître's vision was in some respects quite prophetic. We still believe that everything was together at the beginning; it was just not in the form of matter we now know. Basic physical principles tell us that when the universe was much denser, it was also much hotter, and that it cools as it expands, much as gas cools when sprayed from an aerosol can.

By the 1940s, scientists knew that fusion of hydrogen into helium was the source of the Sun's energy. Fusion requires high temperatures, and the early universe must have been hot. Based on these ideas, American physicist George Gamow (Figure 29.12) suggested a universe with a different kind of beginning that involved nuclear **fusion** instead of fission. Ralph Alpher worked out the details for his PhD thesis, and the results were

Download for free at http://cnx.org/content/col11992/latest/

published in 1948. (Gamow, who had a quirky sense of humor, decided at the last minute to add the name of physicist Hans Bethe to their paper, so that the coauthors on this paper about the beginning of things would be Alpher, Bethe, and Gamow, a pun on the first three letters of the Greek alphabet: alpha, beta, and gamma.) Gamow's universe started with fundamental particles that built up the heavy elements by fusion in the Big Bang.

Figure 29.12. George Gamow and Collaborators. This composite image shows George Gamow emerging like a genie from a bottle of ylem, a Greek term for the original substance from which the world formed. Gamow revived the term to describe the material of the hot Big Bang. Flanking him are Robert Herman (left) and Ralph Alpher (right), with whom he collaborated in working out the physics of the Big Bang. (The modern composer Karlheinz Stockhausen was inspired by Gamow's ideas to write a piece of music called *Ylem*, in which the players actually move away from the stage as they perform, simulating the expansion of the universe.)

Gamow's ideas were close to our modern view, except we now know that the early universe remained hot enough for fusion for only a short while. Thus, only the three lightest elements—hydrogen, helium, and a small amount of lithium—were formed in appreciable abundances at the beginning. The heavier elements formed later in stars. Since the 1940s, many astronomers and physicists have worked on a detailed theory of what happened in the early stages of the universe.

The First Few Minutes

Let's start with the first few minutes following the Big Bang. Three basic ideas hold the key to tracing the changes that occurred during the time just after the universe began. The first, as we have already mentioned, is that the universe cools as it expands. Figure 29.13 shows how the temperature changes with the passage of time. Note that a huge span of time, from a tiny fraction of a second to billions of years, is summarized in this diagram. In the first fraction of a second, the universe was unimaginably hot. By the time 0.01 second had elapsed, the temperature had dropped to 100 billion (10^{11}) K. After about 3 minutes, it had fallen to about 1 billion (10^9) K, still some 70 times hotter than the interior of the Sun. After a few hundred thousand years, the temperature was down to a mere 3000 K, and the universe has continued to cool since that time.

This OpenStax book is available for free at http://cnx.org/content/col11992/1.8

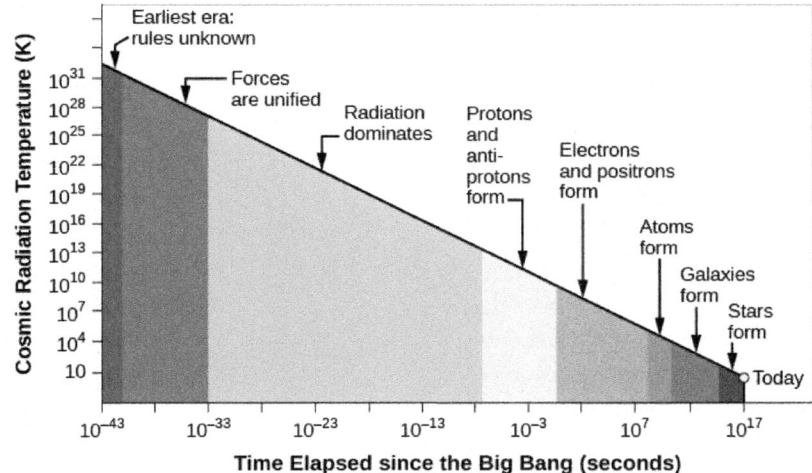

Figure 29.13. Temperature of the Universe. This graph shows how the temperature of the universe varies with time as predicted by the standard model of the Big Bang. Note that both the temperature (vertical axis) and the time in seconds (horizontal axis) change over vast scales on this compressed diagram.

All of these temperatures but the last are derived from theoretical calculations since (obviously) no one was there to measure them directly. As we shall see in the next section, however, we have actually detected the feeble glow of radiation emitted at a time when the universe was a few hundred thousand years old. We can measure the characteristics of that radiation to learn what things were like long ago. Indeed, the fact that we have found this ancient glow is one of the strongest arguments in favor of the Big Bang model.

The second step in understanding the evolution of the universe is to realize that at very early times, it was so hot that it contained mostly radiation (and not the matter that we see today). The photons that filled the universe could collide and produce material particles; that is, under the conditions just after the Big Bang, energy could turn into matter (and matter could turn into energy). We can calculate how much mass is produced from a given amount of energy by using Einstein's formula $E = mc^2$ (see the chapter on The Sun: A Nuclear Powerhouse).

The idea that energy could turn into matter in the universe at large is a new one for many students, since it is not part of our everyday experience. That's because, when we compare the universe today to what it was like right after the Big Bang, we live in cold, hard times. The photons in the universe today typically have far-less energy than the amount required to make new matter. In the discussion on the source of the Sun's energy in The Sun: A Nuclear Powerhouse, we briefly mentioned that when subatomic particles of matter and *antimatter* collide, they turn into pure energy. But the reverse, energy turning into matter and antimatter, is equally possible. This process has been observed in particle accelerators around the world. If we have enough energy, under the right circumstances, new particles of matter (and antimatter) are indeed created —and the conditions were right during the first few minutes after the expansion of the universe began.

Our third key point is that the hotter the universe was, the more energetic were the photons available to make matter and antimatter (see Figure 29.13). To take a specific example, at a temperature of 6 billion (6×10^9) K, the collision of two typical photons can create an electron and its antimatter counterpart, a positron. If the temperature exceeds 10^{14} K, much more massive protons and antiprotons can be created.

The Evolution of the Early Universe

Keeping these three ideas in mind, we can trace the evolution of the universe from the time it was about 0.01 second old and had a temperature of about 100 billion K. Why not begin at the very beginning? There are as yet

Download for free at http://cnx.org/content/col11992/latest/

no theories that allow us penetrate to a time before about 10^{-43} second (this number is a decimal point followed by 42 zeros and then a one). It is so small that we cannot relate it to anything in our everyday experience. When the universe was that young, its density was so high that the theory of general relativity is not adequate to describe it, and even the concept of time breaks down.

Scientists, by the way, have been somewhat more successful in describing the universe when it was older than 10^{-43} second but still less than about 0.01 second old. We will take a look at some of these ideas later in this chapter, but for now, we want to start with somewhat more familiar situations.

By the time the universe was 0.01 second old, it consisted of a soup of matter and radiation; the matter included protons and neutrons, leftovers from an even younger and hotter universe. Each particle collided rapidly with other particles. The temperature was no longer high enough to allow colliding photons to produce neutrons or protons, but it was sufficient for the production of electrons and positrons (Figure 29.14). There was probably also a sea of exotic subatomic particles that would later play a role as dark matter. All the particles jiggled about on their own; it was still much too hot for protons and neutrons to combine to form the nuclei of atoms.

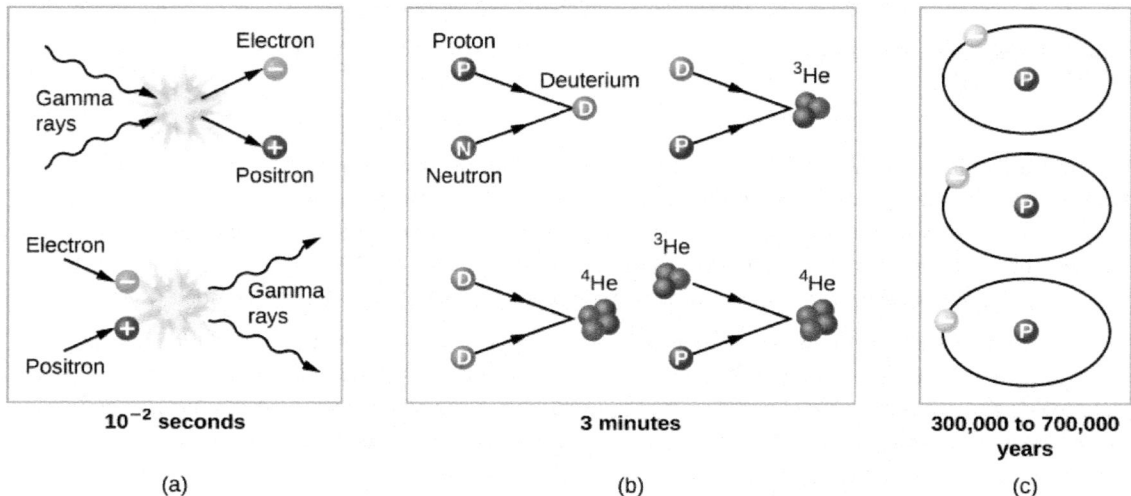

Figure 29.14. Particle Interactions in the Early Universe. (a) In the first fractions of a second, when the universe was very hot, energy was converted into particles and antiparticles. The reverse reaction also happened: a particle and antiparticle could collide and produce energy. (b) As the temperature of the universe decreased, the energy of typical photons became too low to create matter. Instead, existing particles fused to create such nuclei as deuterium and helium. (c) Later, it became cool enough for electrons to settle down with nuclei and make neutral atoms. Most of the universe was still hydrogen.

Think of the universe at this time as a seething cauldron, with photons colliding and interchanging energy, and sometimes being destroyed to create a pair of particles. The particles also collided with one another. Frequently, a matter particle and an antimatter particle met and turned each other into a burst of gamma-ray radiation.

Among the particles created in the early phases of the universe was the ghostly neutrino (see The Sun: A Nuclear Powerhouse), which today interacts only very rarely with ordinary matter. In the crowded conditions of the very early universe, however, neutrinos ran into so many electrons and positrons that they experienced frequent interactions despite their "antisocial" natures.

By the time the universe was a little more than 1 second old, the density had dropped to the point where neutrinos no longer interacted with matter but simply traveled freely through space. In fact, these neutrinos should now be all around us. Since they have been traveling through space unimpeded (and hence unchanged) since the universe was 1 second old, measurements of their properties would offer one of the best tests of the Big Bang model. Unfortunately, the very characteristic that makes them so useful—the fact that they interact

This OpenStax book is available for free at http://cnx.org/content/col11992/1.8

so weakly with matter that they have survived unaltered for all but the first second of time—also renders them unable to be measured, at least with present techniques. Perhaps someday someone will devise a way to capture these elusive messengers from the past.

Atomic Nuclei Form

When the universe was about 3 minutes old and its temperature was down to about 900 million K, protons and neutrons could combine. At higher temperatures, these atomic nuclei had immediately been blasted apart by interactions with high-energy photons and thus could not survive. But at the temperatures and densities reached between 3 and 4 minutes after the beginning, **deuterium** (a proton and neutron) lasted long enough that collisions could convert some of it into helium, (Figure 29.14). In essence, the entire universe was acting the way centers of stars do today—fusing new elements from simpler components. In addition, a little bit of element 3, **lithium**, could also form.

This burst of cosmic fusion was only a brief interlude, however. By 4 minutes after the Big Bang, more helium was having trouble forming. The universe was still expanding and cooling down. After the formation of helium and some lithium, the temperature had dropped so low that the fusion of helium nuclei into still-heavier elements could not occur. No elements beyond lithium could form in the first few minutes. That 4-minute period was the end of the time when the entire universe was a fusion factory. In the cool universe we know today, the fusion of new elements is limited to the centers of stars and the explosions of supernovae.

Still, the fact that the Big Bang model allows the creation of a good deal of helium is the answer to a long-standing mystery in astronomy. Put simply, there is just too much helium in the universe to be explained by what happens inside stars. All the generations of stars that have produced helium since the Big Bang cannot account for the quantity of helium we observe. Furthermore, even the oldest stars and the most distant galaxies show significant amounts of helium. These observations find a natural explanation in the synthesis of helium by the Big Bang itself during the first few minutes of time. We estimate that *10 times more helium* was manufactured in the first 4 minutes of the universe than in all the generations of stars during the succeeding 10 to 15 billion years.

LINK TO LEARNING

These nice animations (https://openstax.org/l/30origelemani) that explain the way in which different elements formed in the history of the universe are from the University of Chicago's *Origins of the Elements* site.

Learning from Deuterium

We can learn many things from the way the early universe made atomic nuclei. It turns out that all of the deuterium (a hydrogen nucleus with a neutron in it) in the universe was formed during the first 4 minutes. In stars, any region hot enough to fuse two protons to form a deuterium nucleus is also hot enough to change it further—either by destroying it through a collision with an energetic photon or by converting it into helium through nuclear reactions.

The amount of deuterium that can be produced in the first 4 minutes of creation depends on the density of the universe at the time deuterium was formed. If the density were relatively high, nearly all the deuterium would have been converted into helium through interactions with protons, just as it is in stars. If the density were relatively low, then the universe would have expanded and thinned out rapidly enough that some deuterium

Download for free at http://cnx.org/content/col11992/latest/

would have survived. The amount of deuterium we see today thus gives us a clue to the density of the universe when it was about 4 minutes old. Theoretical models can relate the density then to the density now; thus, measurements of the abundance of deuterium today can give us an estimate of the current density of the universe.

The measurements of deuterium indicate that the present-day density of ordinary matter—protons and neutrons—is about 5×10^{-28} kg/m^3. Deuterium can only provide an estimate of the density of ordinary matter because the abundance of deuterium is determined by the particles that interact to form it, namely protons and neutrons alone. From the abundance of deuterium, we know that not enough protons and neutrons are present, by a factor of about 20, to produce a critical-density universe.

We do know, however, that there are dark matter particles that add to the overall matter density of the universe, which is then higher than what is calculated for ordinary matter alone. Because dark matter particles do not affect the production of deuterium, measurement of the deuterium abundance cannot tell us how much dark matter exists. Dark matter is made of some exotic kind of particle, not yet detected in any earthbound laboratory. It is definitely not made of protons and neutrons like the readers of this book.

THE COSMIC MICROWAVE BACKGROUND

Learning Objectives

By the end of this section, you will be able to:

> Explain why we can observe the afterglow of the hot, early universe
> Discuss the properties of this afterglow as we see it today, including its average temperature and the size of its temperature fluctuations
> Describe open, flat, and curved universes and explain which type of universe is supported by observations
> Summarize our current knowledge of the basic properties of the universe including its age and contents

The description of the first few minutes of the universe is based on theoretical calculations. It is crucial, however, that a scientific theory should be testable. What predictions does it make? And do observations show those predictions to be accurate? One success of the theory of the first few minutes of the universe is the correct prediction of the amount of helium in the universe.

Another prediction is that a significant milestone in the history of the universe occurred about 380,000 years after the Big Bang. Scientists have directly observed what the universe was like at this early stage, and these observations offer some of the strongest support for the Big Bang theory. To find out what this milestone was, let's look at what theory tells us about what happened during the first few hundred thousand years after the Big Bang.

The fusion of helium and lithium was completed when the universe was about 4 minutes old. The universe then continued to resemble the interior of a star in some ways for a few hundred thousand years more. It remained hot and opaque, with radiation being scattered from one particle to another. It was still too hot for electrons to "settle down" and become associated with a particular nucleus; such free electrons are especially effective at scattering photons, thus ensuring that no radiation ever got very far in the early universe without having its path changed. In a way, the universe was like an enormous crowd right after a popular concert; if you get separated from a friend, even if he is wearing a flashing button, it is impossible to see through the dense crowd to spot him. Only after the crowd clears is there a path for the light from his button to reach you.

This OpenStax book is available for free at http://cnx.org/content/col11992/1.8

Download for free at http://cnx.org/content/col11992/latest/

The Universe Becomes Transparent

Not until a few hundred thousand years after the Big Bang, when the temperature had dropped to about 3000 K and the density of atomic nuclei to about 1000 per cubic centimeter, did the electrons and nuclei manage to combine to form stable atoms of hydrogen and helium (Figure 29.14). With no free electrons to scatter photons, the universe became transparent for the first time in cosmic history. From this point on, matter and radiation interacted much less frequently; we say that they *decoupled* from each other and evolved separately. Suddenly, electromagnetic radiation could really travel, and it has been traveling through the universe ever since.

Discovery of the Cosmic Background Radiation

If the model of the universe described in the previous section is correct, then—as we look far outward in the universe and thus far back in time—the first "afterglow" of the hot, early universe should still be detectable. Observations of it would be very strong evidence that our theoretical calculations about how the universe evolved are correct. As we shall see, we have indeed detected the radiation emitted at this **photon decoupling time**, when radiation began to stream freely through the universe without interacting with matter (Figure 29.15).

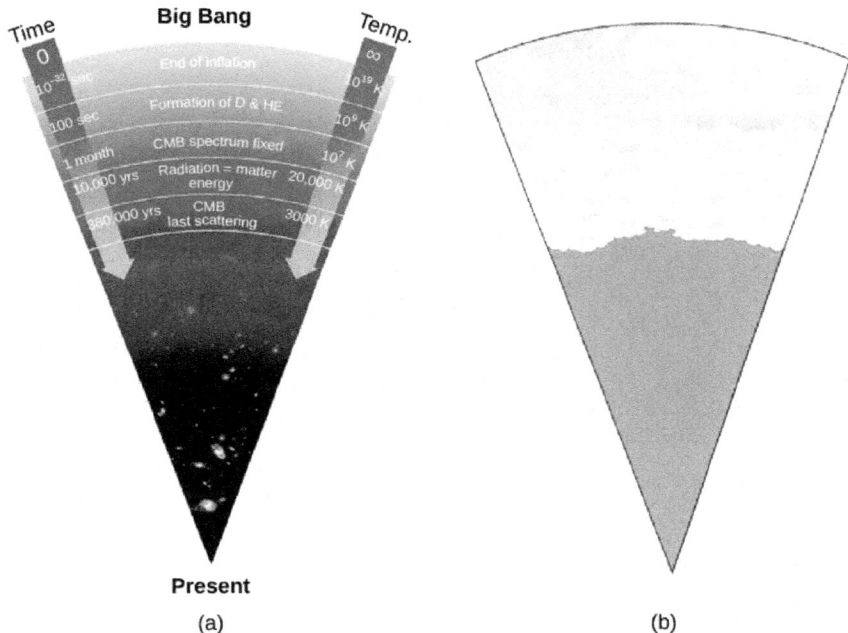

Figure 29.15. Cosmic Microwave Background and Clouds Compared. (a) Early in the universe, photons (electromagnetic energy) were scattering off the crowded, hot, charged particles and could not get very far without colliding with another particle. But after electrons and photons settled into neutral atoms, there was far less scattering, and photons could travel over vast distances. The universe became transparent. As we look out in space and back in time, we can't see back beyond this time. (b) This is similar to what happens when we see clouds in Earth's atmosphere. Water droplets in a cloud scatter light very efficiently, but clear air lets light travel over long distances. So as we look up into the atmosphere, our vision is blocked by the cloud layers and we can't see beyond them. (credit: modification of work by NASA)

The detection of this afterglow was initially an accident. In the late 1940s, Ralph Alpher and Robert Herman, working with George Gamow, realized that just before the universe became transparent, it must have been radiating like a blackbody at a temperature of about 3000 K—the temperature at which hydrogen atoms could begin to form. If we could have seen that radiation just after neutral atoms formed, it would have resembled radiation from a reddish star. It was as if a giant fireball filled the whole universe.

Download for free at http://cnx.org/content/col11992/latest/

But that was nearly 14 billion years ago, and, in the meantime, the scale of the universe has increased a thousand fold. This expansion has increased the wavelength of the radiation by a factor of 1000 (see Figure 29.7). According to Wien's law, which relates wavelength and temperature, the expansion has correspondingly lowered the temperature by a factor of 1000 (see the chapter on Radiation and Spectra).

Alpher and Herman predicted that the glow from the fireball should now be at radio wavelengths and should resemble the radiation from a blackbody at a temperature only a few degrees above absolute zero. Since the fireball was everywhere throughout the universe, the radiation left over from it should also be everywhere. If our eyes were sensitive to radio wavelengths, the whole sky would appear to glow very faintly. However, our eyes can't see at these wavelengths, and at the time Alpher and Herman made their prediction, there were no instruments that could detect the glow. Over the years, their prediction was forgotten.

In the mid-1960s, in Holmdel, New Jersey, Arno Penzias and Robert Wilson of AT&T's Bell Laboratories had built a delicate microwave antenna (Figure 29.16) to measure astronomical sources, including supernova remnants like Cassiopeia A (see the chapter on The Death of Stars). They were plagued with some unexpected background noise, just like faint static on a radio, which they could not get rid of. The puzzling thing about this radiation was that it seemed to be coming from all directions at once. This is very unusual in astronomy: after all, most radiation has a specific direction where it is strongest—the direction of the Sun, or a supernova remnant, or the disk of the Milky Way, for example.

Figure 29.16. Robert Wilson (left) and Arno Penzias (right). These two scientists are standing in front of the horn-shaped antenna with which they discovered the cosmic background radiation. The photo was taken in 1978, just after they received the Nobel Prize in physics.

Penzias and Wilson at first thought that any radiation appearing to come from all directions must originate from inside their telescope, so they took everything apart to look for the source of the noise. They even found that some pigeons had roosted inside the big horn-shaped antenna and had left (as Penzias delicately put it) "a layer of white, sticky, dielectric substance coating the inside of the antenna." However, nothing the scientists did could reduce the background radiation to zero, and they reluctantly came to accept that it must be real, and it must be coming from space.

Penzias and Wilson were not cosmologists, but as they began to discuss their puzzling discovery with other scientists, they were quickly put in touch with a group of astronomers and physicists at Princeton University (a short drive away). These astronomers had—as it happened—been redoing the calculations of Alpher and Herman from the 1940s and also realized that the radiation from the decoupling time should be detectable as a faint afterglow of radio waves. The different calculations of what the observed temperature would be for this **cosmic microwave background (CMB)**[2] were uncertain, but all predicted less than 40 K.

2   Recall that microwaves are in the radio region of the electromagnetic spectrum.

This OpenStax book is available for free at http://cnx.org/content/col11992/1.8

Download for free at http://cnx.org/content/col11992/latest/

Penzias and Wilson found the distribution of intensity at different radio wavelengths to correspond to a temperature of 3.5 K. This is very cold—closer to absolute zero than most other astronomical measurements—and a testament to how much space (and the waves within it) has stretched. Their measurements have been repeated with better instruments, which give us a reading of 2.73 K. So Penzias and Wilson came very close. Rounding this value, scientists often refer to "the 3-degree microwave background."

Many other experiments on Earth and in space soon confirmed the discovery by Penzias and Wilson: The radiation was indeed coming from all directions (it was isotropic) and matched the predictions of the Big Bang theory with remarkable precision. Penzias and Wilson had inadvertently observed the glow from the primeval fireball. They received the Nobel Prize for their work in 1978. And just before his death in 1966, Lemaître learned that his "vanished brilliance" had been discovered and confirmed.

LINK TO LEARNING

You may enjoy watching *Three Degrees*, a 26-minute video (https://openstax.org/l/30threedegvid) from Bell Labs about Penzias and Wilson's discovery of the cosmic background radiation (with interesting historical footage).

Properties of the Cosmic Microwave Background

One issue that worried astronomers is that Penzias and Wilson were measuring the background radiation filling space through Earth's atmosphere. What if that atmosphere is a source of radio waves or somehow affected their measurements? It would be better to measure something this important from space.

The first accurate measurements of the CMB were made with a satellite orbiting Earth. Named the Cosmic Background Explorer (COBE), it was launched by NASA in November 1989. The data it received quickly showed that the CMB closely matches that expected from a blackbody with a temperature of 2.73 K (Figure 29.17). This is exactly the result expected if the CMB was indeed redshifted radiation emitted by a hot gas that filled all of space shortly after the universe began.

Download for free at http://cnx.org/content/col11992/latest/

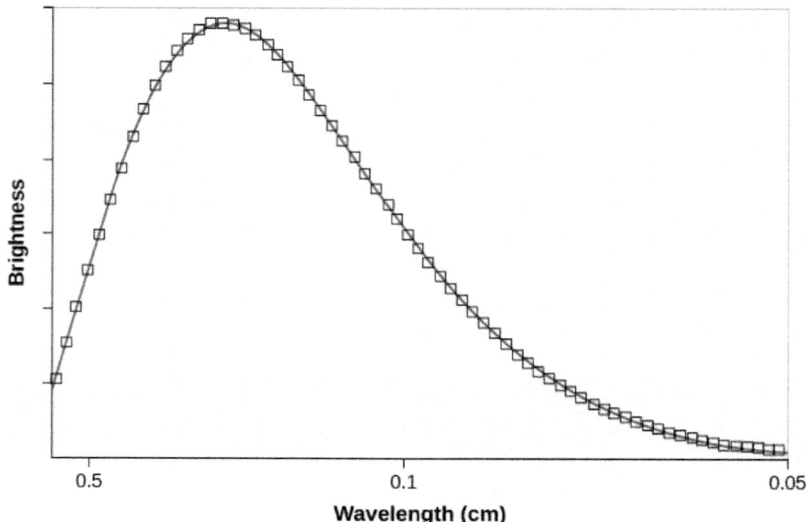

Figure 29.17. Cosmic Background Radiation. The solid line shows how the intensity of radiation should change with wavelength for a blackbody with a temperature of 2.73 K. The boxes show the intensity of the cosmic background radiation as measured at various wavelengths by COBE's instruments. The fit is perfect. When this graph was first shown at a meeting of astronomers, they gave it a standing ovation.

The first important conclusion from measurements of the CMB, therefore, is that the universe we have today has indeed evolved from a hot, uniform state. This observation also provides direct support for the general idea that we live in an evolving universe, since the universe is cooler today than it was in the beginning.

Small Differences in the CMB

It was known even before the launch of COBE that the CMB is extremely *isotropic*. In fact, its uniformity in every direction is one of the best confirmations of the cosmological principle— that the universe is homogenous and isotropic.

According to our theories, however, the temperature could not have been *perfectly* uniform when the CMB was emitted. After all, the CMB is radiation that was scattered from the particles in the universe at the time of decoupling. If the radiation were completely smooth, then all those particles must have been distributed through space absolutely evenly. Yet it is those particles that have become all the galaxies and stars (and astronomy students) that now inhabit the cosmos. Had the particles been completely smoothly distributed, they could not have formed all the large-scale structures now present in the universe—the clusters and superclusters of galaxies discussed in the last few chapters.

The early universe must have had tiny density fluctuations from which such structures could evolve. Regions of higher-than-average density would have attracted additional matter and eventually grown into the galaxies and clusters that we see today. It turned out that these denser regions would appear to us to be colder spots, that is, they would have lower-than-average temperatures.

The reason that temperature and density are related can be explained this way. At the time of decoupling, photons in a slightly denser portion of space had to expend some of their energy to escape the gravitational force exerted by the surrounding gas. In losing energy, the photons became slightly colder than the overall average temperature at the time of decoupling. Vice versa, photons that were located in a slightly less dense portion of space lost less energy upon leaving it than other photons, thus appearing slightly hotter than average. Therefore, if the seeds of present-day galaxies existed at the time that the CMB was emitted, we should see some slight variations in the CMB temperature as we look in different directions in the sky.

This OpenStax book is available for free at http://cnx.org/content/col11992/1.8

Download for free at http://cnx.org/content/col11992/latest/

Scientists working with the data from the COBE satellite did indeed detect very subtle temperature differences—about 1 part in 100,000—in the CMB. The regions of lower-than-average temperature come in a variety of sizes, but even the smallest of the colder areas detected by COBE is far too large to be the precursor of an individual galaxy, or even a supercluster of galaxies. This is because the COBE instrument had "blurry vision" (poor resolution) and could only measure large patches of the sky. We needed instruments with "sharper vision."

The most detailed measurements of the CMB have been obtained by two satellites launched more recently than COBE. The results from the first of these satellites, the Wilkinson Microwave Anisotropy Probe (WMAP) spacecraft, were published in 2003. In 2015, measurements from the Planck satellite extended the WMAP measurements to even-higher spatial resolution and lower noise (Figure 29.18).

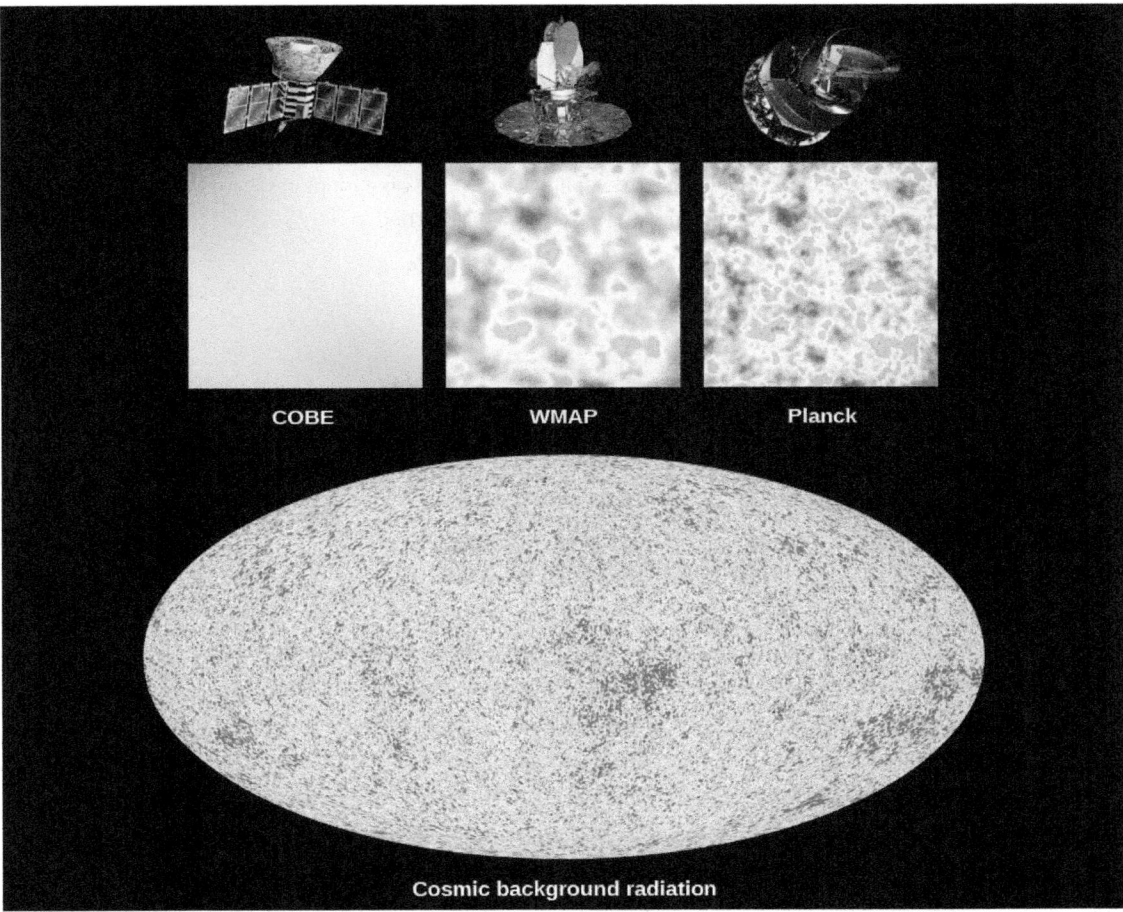

Figure 29.18. CMB Observations. This comparison shows how much detail can be seen in the observations of three satellites used to measure the CMB. The CMB is a snapshot of the oldest light in our universe, imprinted on the sky when the universe was just about 380,000 years old. The first spacecraft, launched in 1989, is NASA's Cosmic Background Explorer, or COBE. WMAP was launched in 2001, and Planck was launched in 2009. The three panels show 10-square-degree patches of all-sky maps. This cosmic background radiation image (bottom) is an all-sky map of the CMB as observed by the Planck mission. The colors in the map represent different temperatures: red for warmer and blue for cooler. These tiny temperature fluctuations correspond to regions of slightly different densities, representing the seeds of all future structures: the stars, galaxies, and galaxy clusters of today. (credit top: modification of work by NASA/JPL-Caltech/ESA; credit bottom: modification of work by ESA and the Planck Collaboration)

Theoretical calculations show that the sizes of the hot and cold spots in the CMB depend on the geometry of the universe and hence on its total density. (It's not at all obvious that it should do so, and it takes some pretty

Download for free at http://cnx.org/content/col11992/latest/

fancy calculations—way beyond the level of our text—to make the connection, but having such a dependence is very useful.) The total density we are discussing here includes both the amount of mass in the universe and the mass equivalent of the dark energy. That is, we must add together mass and energy: ordinary matter, dark matter, and the dark energy that is speeding up the expansion.

To see why this works, remember (from the chapter on Black Holes and Curved Spacetime) that with his theory of general relativity, Einstein showed that matter can curve space and that the amount of curvature depends on the amount of matter present. Therefore, the total amount of matter in the universe (including dark matter and the equivalent matter contribution by dark energy), determines the overall geometry of space. Just like the geometry of space around a black hole has a curvature to it, so the entire universe may have a curvature. Let's take a look at the possibilities (Figure 29.19).

If the density of matter is higher than the critical density, the universe will eventually collapse. In such a closed universe, two initially parallel rays of light will eventually meet. This kind of geometry is referred to as spherical geometry. If the density of matter is less than critical, the universe will expand forever. Two initially parallel rays of light will diverge, and this is referred to as hyperbolic geometry. In a critical-density universe, two parallel light rays never meet, and the expansion comes to a halt only at some time infinitely far in the future. We refer to this as a **flat universe**, and the kind of Euclidean geometry you learned in high school applies in this type of universe.

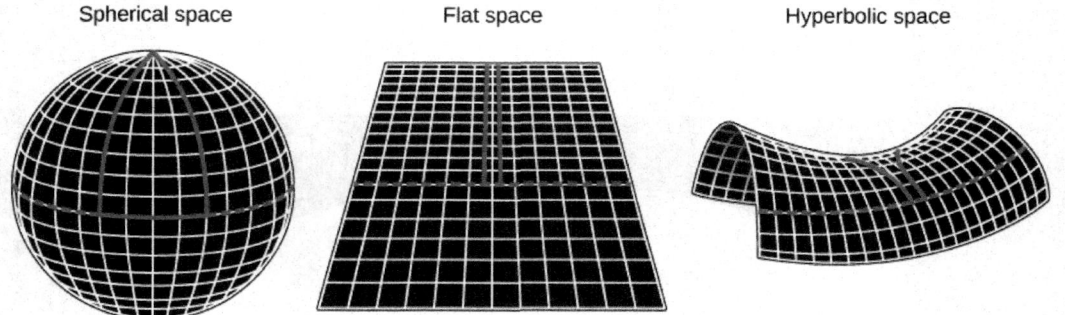

Spherical space   Flat space   Hyperbolic space

Figure 29.19. Picturing Space Curvature for the Entire Universe. The density of matter and energy determines the overall geometry of space. If the density of the universe is greater than the critical density, then the universe will ultimately collapse and space is said to be *closed* like the surface of a sphere. If the density exactly equals the critical density, then space is *flat* like a sheet of paper; the universe will expand forever, with the rate of expansion coming to a halt infinitely far in the future. If the density is less than critical, then the expansion will continue forever and space is said to be *open* and negatively curved like the surface of a saddle (where more space than you expect opens up as you move farther away). Note that the red lines in each diagram show what happens in each kind of space—they are initially parallel but follow different paths depending on the curvature of space. Remember that these drawings are trying to show how space for the entire universe is "warped"—this can't be seen locally in the small amount of space that we humans occupy.

If the density of the universe is equal to the critical density, then the hot and cold spots in the CMB should typically be about a degree in size. If the density is greater than critical, then the typical sizes will be larger than one degree. If the universe has a density less than critical, then the structures will appear smaller. In Figure 29.20, you can see the differences easily. WMAP and Planck observations of the CMB confirmed earlier experiments that we do indeed live in a flat, critical-density universe.

This OpenStax book is available for free at http://cnx.org/content/col11992/1.8

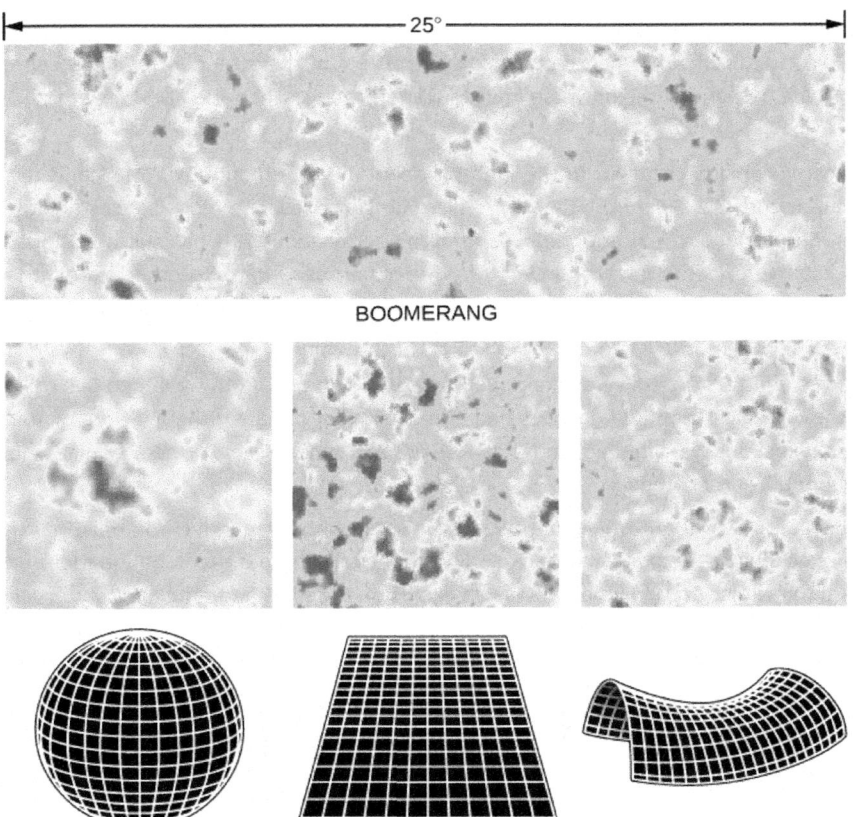

Figure 29.20. Comparison of CMB Observations with Possible Models of the Universe. Cosmological simulations predict that if our universe has critical density, then the CMB images will be dominated by hot and cold spots of around one degree in size (bottom center). If, on the other hand, the density is higher than critical (and the universe will ultimately collapse), then the images' hot and cold spots will appear larger than one degree (bottom left). If the density of the universe is less than critical (and the expansion will continue forever), then the structures will appear smaller (bottom right). As the measurements show, the universe is at critical density. The measurements shown were made by a balloon-borne instrument called BOOMERanG (Balloon Observations of Millimetric Extragalactic Radiation and Geophysics), which was flown in Antarctica. Subsequent satellite observations by WMAP and Planck confirm the BOOMERanG result. (credit: modification of work by NASA)

Key numbers from an analysis of the Planck data give us the best values currently available for some of the basic properties of the universe:

- Age of universe: 13.799 ± 0.038 billion years (Note: That means we know the age of the universe to within 38 million years. Amazing!)

- Hubble constant: 67.31 ± 0.96 kilometers/second/million parsecs

- Fraction of universe's content that is "dark energy": 68.5% ± 1.3%

- Fraction of the universe's content that is matter: 31.5% ± 1.3%

Note that this value for the Hubble constant is slightly smaller than the value of 70 kilometers/second/million parsecs that we have adopted in this book. In fact, the value derived from measurements of redshifts is 73 kilometers/second/million parsecs. So precise is modern cosmology these days that scientists are working hard to resolve this discrepancy. The fact that the difference between these two independent measurements is so small is actually a remarkable achievement. Only a few decades ago, astronomers were arguing about whether the Hubble constant was around 50 kilometers/second/million parsecs or 100 kilometers/second/million parsecs.

Download for free at http://cnx.org/content/col11992/latest/

Analysis of Planck data also shows that ordinary matter (mainly protons and neutrons) makes up 4.9% of the total density. Dark matter plus normal matter add up to 31.5% of the total density. Dark energy contributes the remaining 68.5%. The age of the universe at decoupling—that is, when the CMB was emitted—was 380,000 years.

Perhaps the most surprising result from the high-precision measurements by WMAP and the even higher-precision measurements from Planck is that there were no surprises. The model of cosmology with ordinary matter at about 5%, dark matter at about 25%, and dark energy about 70% has survived since the late 1990s when cosmologists were forced in that direction by the supernovae data. In other words, the very strange universe that we have been describing, with only about 5% of its contents being made up of the kinds of matter we are familiar with here on Earth, really seems to be the universe we live in.

After the CMB was emitted, the universe continued to expand and cool off. By 400 to 500 million years after the Big Bang, the very first stars and galaxies had already formed. Deep in the interiors of stars, matter was reheated, nuclear reactions were ignited, and the more gradual synthesis of the heavier elements that we have discussed throughout this book began.

We conclude this quick tour of our model of the early universe with a reminder. You must not think of the Big Bang as a *localized* explosion *in space*, like an exploding superstar. There were no boundaries and there was no single site where the explosion happened. It was an explosion *of space* (and time and matter and energy) that happened everywhere in the universe. All matter and energy that exist today, including the particles of which you are made, came from the Big Bang. We were, and still are, in the midst of a Big Bang; it is all around us.

29.5 WHAT IS THE UNIVERSE REALLY MADE OF?

Learning Objectives

By the end of this section, you will be able to:

> Specify what fraction of the density of the universe is contributed by stars and galaxies and how much ordinary matter (such as hydrogen, helium, and other elements we are familiar with here on Earth) makes up the overall density
> Describe how ideas about the contents of the universe have changed over the last 50 years
> Explain why it is so difficult to determine what dark matter really is
> Explain why dark matter helped galaxies form quickly in the early universe
> Summarize the evolution of the universe from the time the CMB was emitted to the present day

The model of the universe we described in the previous section is the simplest model that explains the observations. It assumes that general relativity is the correct theory of gravity throughout the universe. With this assumption, the model then accounts for the existence and structure of the CMB; the abundances of the light elements deuterium, helium, and lithium; and the acceleration of the expansion of the universe. All of the observations to date support the validity of the model, which is referred to as the standard (or concordance) model of cosmology.

Figure 29.21 and Table 29.2 summarize the current best estimates of the contents of the universe. Luminous matter in stars and galaxies and neutrinos contributes about 1% of the mass required to reach critical density. Another 4% is mainly in the form of hydrogen and helium in the space between stars and in intergalactic space. Dark matter accounts for about an additional 27% of the critical density. The mass equivalent of dark energy (according to $E = mc^2$) then supplies the remaining 68% of the critical density.

This OpenStax book is available for free at http://cnx.org/content/col11992/1.8

Composition of the Universe

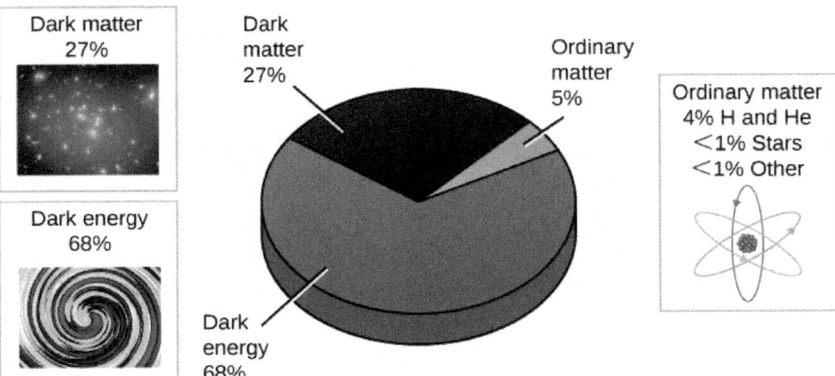

Figure 29.21. Composition of the Universe. Only about 5% of all the mass and energy in the universe is matter with which we are familiar here on Earth. Most ordinary matter consists of hydrogen and helium located in interstellar and intergalactic space. Only about one-half of 1% of the critical density of the universe is found in stars. Dark matter and dark energy, which have not yet been detected in earthbound laboratories, account for 95% of the contents of the universe.

What Different Kinds of Objects Contribute to the Density of the Universe

Object	Density as a Percent of Critical Density
Luminous matter (stars, etc.)	<1
Hydrogen and helium in interstellar and intergalactic space	4
Dark matter	27
Equivalent mass density of the dark energy	68

Table 29.2

This table should shock you. What we are saying is that 95% of the stuff of the universe is either dark matter or dark energy—neither of which has ever been detected in a laboratory here on Earth. This whole textbook, which has focused on objects that emit electromagnetic radiation, has generally been ignoring 95% of what is out there. Who says there aren't big mysteries yet to solve in science!

Figure 29.22 shows how our ideas of the composition of the universe have changed over just the past three decades. The fraction of the universe that we think is made of the same particles as astronomy students has been decreasing steadily.

Download for free at http://cnx.org/content/col11992/latest/

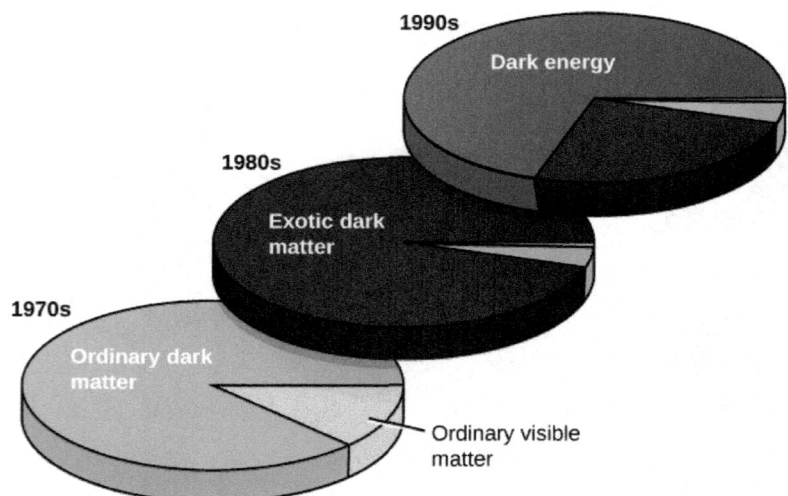

Figure 29.22. Changing Estimates of the Content of the Universe. This diagram shows the changes in our understanding of the contents of the universe over the past three decades. In the 1970s, we suspected that most of the matter in the universe was invisible, but we thought that this matter might be ordinary matter (protons, neutrons, etc.) that was simply not producing electromagnetic radiation. By the 1980s, it was becoming likely that most of the dark matter was made of something we had not yet detected on Earth. By the late 1990s, a variety of experiments had shown that we live in a critical-density universe and that dark energy contributes about 70% of what is required to reach critical density. Note how the estimate of the relative importance of ordinary luminous matter (shown in yellow) has diminished over time.

What Is Dark Matter?

Many astronomers find the situation we have described very satisfying. Several independent experiments now agree on the type of universe we live in and on the inventory of what it contains. We seem to be very close to having a cosmological model that explains nearly everything. Others are not yet ready to jump on the bandwagon. They say, "show me the 96% of the universe we can't detect directly—for example, find me some dark matter!"

At first, astronomers thought that **dark matter** might be hidden in objects that appear dark because they emit no light (e.g., black holes) or that are too faint to be observed at large distances (e.g., planets or white dwarfs). However, these objects would be made of ordinary matter, and the deuterium abundance tells us that no more than 5% of the critical density consists of ordinary matter.

Another possible form that dark matter can take is some type of elementary particle that we have not yet detected here on Earth—a particle that has mass and exists in sufficient abundance to contribute 23% of the critical density. Some physics theories predict the existence of such particles. One class of these particles has been given the name WIMPs, which stands for **weakly interacting massive particles**. Since these particles do not participate in nuclear reactions leading to the production of deuterium, the deuterium abundance puts no limits on how many WIMPs might be in the universe. (A number of other exotic particles have also been suggested as prime constituents of dark matter, but we will confine our discussion to WIMPs as a useful example.)

If large numbers of WIMPs do exist, then some of them should be passing through our physics laboratories right now. The trick is to catch them. Since by definition they interact only weakly (infrequently) with other matter, the chances that they will have a measurable effect are small. We don't know the mass of these particles, but various theories suggest that it might be a few to a few hundred times the mass of a proton. If WIMPs are 60 times the mass of a proton, there would be about 10 million of them passing through your outstretched hand every second—with absolutely no effect on you. If that seems too mind-boggling, bear in mind that neutrinos interact weakly with ordinary matter, and yet we were able to "catch" them eventually.

This OpenStax book is available for free at http://cnx.org/content/col11992/1.8

Download for free at http://cnx.org/content/col11992/latest/

Despite the challenges, more than 30 experiments designed to detect WIMPS are in operation or in the planning stages. Predictions of how many times WIMPs might actually collide with the nucleus of an atom in the instrument designed to detect them are in the range of 1 event per year to 1 event per 1000 years per kilogram of detector. The detector must therefore be large. It must be shielded from radioactivity or other types of particles, such as neutrons, passing through it, and hence these detectors are placed in deep mines. The energy imparted to an atomic nucleus in the detector by collision with a WIMP will be small, and so the detector must be cooled to a very low temperature.

The WIMP detectors are made out of crystals of germanium, silicon, or xenon. The detectors are cooled to a few thousandths of a degree—very close to absolute zero. That means that the atoms in the detector are so cold that they are scarcely vibrating at all. If a dark matter particle collides with one of the atoms, it will cause the whole crystal to vibrate and the temperature therefore to increase ever so slightly. Some other interactions may generate a detectable flash of light.

A different kind of search for WIMPs is being conducted at the Large Hadron Collider (LHC) at CERN, Europe's particle physics lab near Geneva, Switzerland. In this experiment, protons collide with enough energy potentially to produce WIMPs. The LHC detectors cannot detect the WIMPs directly, but if WIMPs are produced, they will pass through the detectors, carrying energy away with them. Experimenters will then add up all the energy that they detect as a result of the collisions of protons to determine if any energy is missing.

So far, none of these experiments has detected WIMPs. Will the newer experiments pay off? Or will scientists have to search for some other explanation for dark matter? Only time will tell (Figure 29.23).

Figure 29.23. Dark Matter. This cartoon from NASA takes a humorous look at how little we yet understand about dark matter. (credit: NASA)

Dark Matter and the Formation of Galaxies

As elusive as dark matter may be in the current-day universe, galaxies could not have formed quickly without it. Galaxies grew from density fluctuations in the early universe, and some had already formed only about

Download for free at http://cnx.org/content/col11992/latest/

400–500 million years after the Big Bang. The observations with WMAP, Planck, and other experiments give us information on the size of those density fluctuations. It turns out that the density variations we observe are too small to have formed galaxies so soon after the Big Bang. In the hot, early universe, energetic photons collided with hydrogen and helium, and kept them moving so rapidly that gravity was still not strong enough to cause the atoms to come together to form galaxies. How can we reconcile this with the fact that galaxies *did* form and are all around us?

Our instruments that measure the CMB give us information about density fluctuations only for *ordinary matter*, which interacts with radiation. Dark matter, as its name indicates, does not interact with photons at all. Dark matter could have had much greater variations in density and been able to come together to form gravitational "traps" that could then have begun to attract ordinary matter immediately after the universe became transparent. As ordinary matter became increasingly concentrated, it could have turned into galaxies quickly thanks to these dark matter traps.

For an analogy, imagine a boulevard with traffic lights every half mile or so. Suppose you are part of a motorcade of cars accompanied by police who lead you past each light, even if it is red. So, too, when the early universe was opaque, radiation interacted with ordinary matter, imparting energy to it and carrying it along, sweeping past the concentrations of dark matter. Now suppose the police leave the motorcade, which then encounters some red lights. The lights act as traffic traps; approaching cars now have to stop, and so they bunch up. Likewise, after the early universe became transparent, ordinary matter interacted with radiation only occasionally and so could fall into the dark matter traps.

The Universe in a Nutshell

In the previous sections of this chapter, we traced the evolution of the universe progressively further back in time. Astronomical discovery has followed this path historically, as new instruments and new techniques have allowed us to probe ever closer to the beginning of time. The rate of expansion of the universe was determined from measurements of nearby galaxies. Determinations of the abundances of deuterium, helium, and lithium based on nearby stars and galaxies were used to put limits on how much ordinary matter is in the universe. The motions of stars in galaxies and of galaxies within clusters of galaxies could only be explained if there were large quantities of dark matter. Measurements of supernovae that exploded when the universe was about half as old as it is now indicated that the rate of expansion of the universe has sped up since those explosions occurred. Observations of extremely faint galaxies show that galaxies had begun to form when the universe was only 400–500 million years old. And observations of the CMB confirmed early theories that the universe was initially very hot.

But all this moving further and further backward in time might have left you a bit dizzy. So now let's instead show how the universe evolves as time moves forward.

Figure 29.24 summarizes the entire history of the observable universe from the beginning in a single diagram. The universe was very hot when it began to expand. We have fossil remnants of the very early universe in the form of neutrons, protons, electrons, and neutrinos, and the atomic nuclei that formed when the universe was 3–4 minutes old: deuterium, helium, and a small amount of lithium. Dark matter also remains, but we do not yet know what form it is in.

This OpenStax book is available for free at http://cnx.org/content/col11992/1.8

Download for free at http://cnx.org/content/col11992/latest/

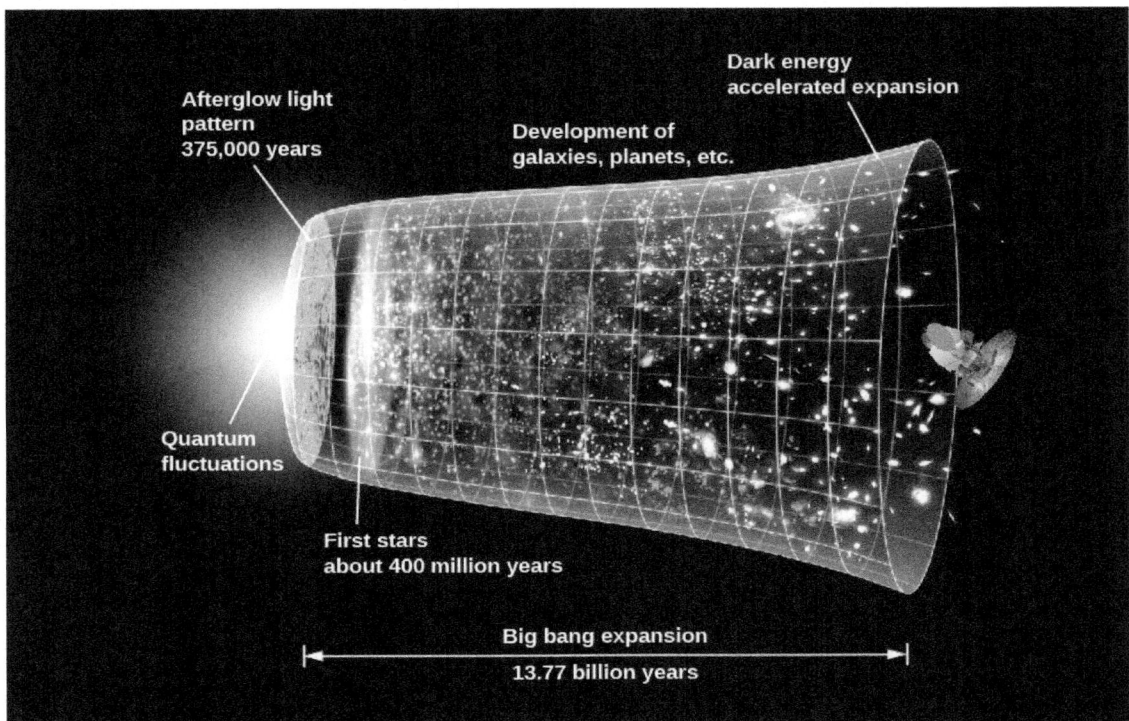

Figure 29.24. History of the Universe. This image summarizes the changes that have occurred in the universe during the last 13.8 billion years. Protons, deuterium, helium, and some lithium were produced in the initial fireball. About 380,000 years after the Big Bang, the universe became transparent to electromagnetic radiation for the first time. COBE, WMAP, Planck, and other instruments have been used to study the radiation that was emitted at that time and that is still visible today (the CMB). The universe was then dark (except for this background radiation) until the first stars and galaxies began to form only a few hundred million years after the Big Bang. Existing space and ground-based telescopes have made substantial progress in studying the subsequent evolution of galaxies. (credit: modification of work by NASA/WMAP Science Team)

The universe gradually cooled; when it was about 380,000 years old, and at a temperature of about 3000 K, electrons combined with protons to form hydrogen atoms. At this point, as we saw, the universe became transparent to light, and astronomers have detected the CMB emitted at this time. The universe still contained no stars or galaxies, and so it entered what astronomers call "the dark ages" (since stars were not lighting up the darkness). During the next several hundred million years, small fluctuations in the density of the dark matter grew, forming gravitational traps that concentrated the ordinary matter, which began to form galaxies about 400–500 million years after the Big Bang.

By the time the universe was about a billion years old, it had entered its own renaissance: it was again blazing with radiation, but this time from newly formed stars, star clusters, and small galaxies. Over the next several billion years, small galaxies merged to form the giants we see today. Clusters and superclusters of galaxies began to grow, and the universe eventually began to resemble what we see nearby.

During the next 20 years, astronomers plan to build giant new telescopes both in space and on the ground to explore even further back in time. In 2018, the James Webb Space Telescope, a 6.5-meter telescope that is the successor to the Hubble Space Telescope, will be launched and assembled in space. The predictions are that with this powerful instrument (see Figure 29.1) we should be able to look back far enough to analyze in detail the formation of the first galaxies.

Download for free at http://cnx.org/content/col11992/latest/

29.6 | THE INFLATIONARY UNIVERSE

Learning Objectives

By the end of this section, you will be able to:

> Describe two important properties of the universe that the simple Big Bang model cannot explain
> Explain why these two characteristics of the universe can be accounted for if there was a period of rapid expansion (inflation) of the universe just after the Big Bang
> Name the four forces that control all physical processes in the universe

The hot Big Bang model that we have been describing is remarkably successful. It accounts for the expansion of the universe, explains the observations of the CMB, and correctly predicts the abundances of the light elements. As it turns out, this model also predicts that there should be exactly three types of neutrinos in nature, and this prediction has been confirmed by experiments with high-energy accelerators. We can't relax just yet, however. This standard model of the universe doesn't explain *all* the observations we have made about the universe as a whole.

Problems with the Standard Big Bang Model

There are a number of characteristics of the universe that can only be explained by considering further what might have happened before the emission of the CMB. One problem with the standard Big Bang model is that it does not explain why the density of the universe is equal to the critical density. The mass density could have been, after all, so low and the effects of dark energy so high that the expansion would have been too rapid to form any galaxies at all. Alternatively, there could have been so much matter that the universe would have already begun to contract long before now. Why is the universe balanced so precisely on the knife edge of the critical density?

Another puzzle is the remarkable *uniformity* of the universe. The temperature of the CMB is the same to about 1 part in 100,000 everywhere we look. This sameness might be expected if all the parts of the visible universe were in contact at some point in time and had the time to come to the same temperature. In the same way, if we put some ice into a glass of lukewarm water and wait a while, the ice will melt and the water will cool down until they are the same temperature.

However, if we accept the standard Big Bang model, all parts of the visible universe were *not* in contact at any time. The fastest that information can go from one point to another is the speed of light. There is a maximum distance that light can have traveled from any point since the time the universe began—that's the distance light could have covered since then. This distance is called that point's *horizon distance* because anything farther away is "below its horizon"—unable to make contact with it. One region of space separated by more than the horizon distance from another has been completely isolated from it through the entire history of the universe.

If we measure the CMB in two opposite directions in the sky, we are observing regions that were significantly beyond each other's horizon distance at the time the CMB was emitted. We can see both regions, but *they* can never have seen each other. Why, then, are their temperatures so precisely the same? According to the standard Big Bang model, they have never been able to exchange information, and there is no reason they should have identical temperatures. (It's a little like seeing the clothes that all the students wear at two schools in different parts of the world become identical, without the students ever having been in contact.) The only explanation we could suggest was simply that the universe somehow *started out* being absolutely uniform (which is like saying all students were born liking the same clothes). Scientists are always uncomfortable when they must appeal to a special set of initial conditions to account for what they see.

This OpenStax book is available for free at http://cnx.org/content/col11992/1.8

The Inflationary Hypothesis

Some physicists suggested that these fundamental characteristics of the cosmos—its flatness and uniformity—can be explained if shortly after the Big Bang (and before the emission of the CMB), the universe experienced a sudden increase in size. A model universe in which this rapid, early expansion occurs is called an **inflationary universe**. The inflationary universe is identical to the Big Bang universe for all time after the first 10^{-30} second. Prior to that, the model suggests that there was a brief period of extraordinarily rapid expansion or inflation, during which the scale of the universe increased by a factor of about 10^{50} times more than predicted by standard Big Bang models (Figure 29.25).

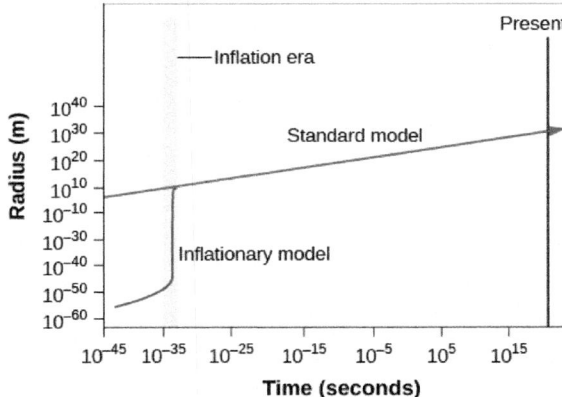

Figure 29.25. Expansion of the Universe. This graph shows how the scale factor of the observable universe changes with time for the standard Big Bang model (red line) and for the inflationary model (blue line). (Note that the time scale at the bottom is extremely compressed.) During inflation, regions that were very small and in contact with each other are suddenly blown up to be much larger and outside each other's horizon distance. The two models are the same for all times after 10–30 second.

Prior to (and during) inflation, all the parts of the universe that we can now see were so small and close to each other that they *could* exchange information, that is, the horizon distance included all of the universe that we can now observe. Before (and during) inflation, there was adequate time for the observable universe to homogenize itself and come to the same temperature. Then, inflation expanded those regions tremendously, so that many parts of the universe are now beyond each other's horizon.

Another appeal of the inflationary model is its prediction that the density of the universe should be exactly equal to the critical density. To see why this is so, remember that curvature of spacetime is intimately linked to the density of matter. If the universe began with some curvature of its spacetime, one analogy for it might be the skin of a balloon. The period of inflation was equivalent to blowing up the balloon to a tremendous size. The universe became so big that from our vantage point, no curvature should be visible (Figure 29.26). In the same way, Earth's surface is so big that it looks flat to us no matter where we are. Calculations show that a universe with no curvature is one that is at critical density. Universes with densities either higher or lower than the critical density would show marked curvature. But we saw that the observations of the CMB in Figure 29.18, which show that the universe has critical density, rule out the possibility that space is significantly curved.

Download for free at http://cnx.org/content/col11992/latest/

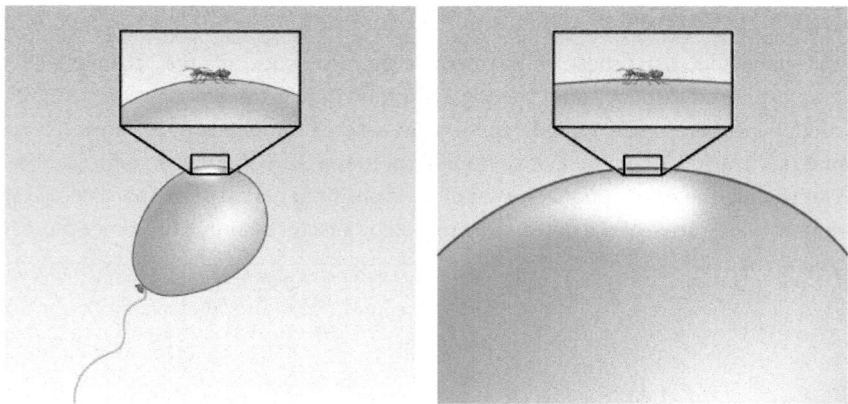

Figure 29.26. Analogy for Inflation. During a period of rapid inflation, a curved balloon grows so large that to any local observer it looks flat. The inset shows the geometry from the ant's point of view.

Grand Unified Theories

While inflation is an intriguing idea and widely accepted by researchers, we cannot directly observe events so early in the universe. The conditions at the time of inflation were so extreme that we cannot reproduce them in our laboratories or high-energy accelerators, but scientists have some ideas about what the universe might have been like. These ideas are called **grand unified theories** or GUTs.

In GUT models, the forces that we are familiar with here on Earth, including gravity and electromagnetism, behaved very differently in the extreme conditions of the early universe than they do today. In physical science, the term *force* is used to describe anything that can change the motion of a particle or body. One of the remarkable discoveries of modern science is that all known physical processes can be described through the action of just four forces: gravity, electromagnetism, the strong nuclear force, and the weak nuclear force (Table 29.3).

The Forces of Nature

Force	Relative Strength Today	Range of Action	Important Applications
Gravity	1	Whole universe	Motions of planets, stars, galaxies
Electromagnetism	10^{36}	Whole universe	Atoms, molecules, electricity, magnetic fields
Weak nuclear force	10^{33}	10^{-17} meters	Radioactive decay
Strong nuclear force	10^{38}	10^{-15} meters	The existence of atomic nuclei

Table 29.3

Gravity is perhaps the most familiar force, and certainly appears strong if you jump off a tall building. However, the force of gravity between two elementary particles—say two protons—is by far the weakest of the four

This OpenStax book is available for free at http://cnx.org/content/col11992/1.8

forces. Electromagnetism—which includes both magnetic and electrical forces, holds atoms together, and produces the electromagnetic radiation that we use to study the universe—is much stronger, as you can see in Table 29.3. The weak nuclear force is only weak in comparison to its strong "cousin," but it is in fact much stronger than gravity.

Both the weak and strong nuclear forces differ from the first two forces in that they act only over very small distances—those comparable to the size of an atomic nucleus or less. The weak force is involved in radioactive decay and in reactions that result in the production of neutrinos. The strong force holds protons and neutrons together in an atomic nucleus.

Physicists have wondered why there are four forces in the universe—why not 300 or, preferably, just one? An important hint comes from the name *electromagnetic force*. For a long time, scientists thought that the forces of electricity and magnetism were separate, but James Clerk Maxwell (see the chapter on Radiation and Spectra) was able to *unify* these forces—to show that they are aspects of the same phenomenon. In the same way, many scientists (including Einstein) have wondered if the four forces we now know could also be unified. Physicists have actually developed GUTs that unify three of the four forces (but not gravity).

In these theories, the strong, weak, and electromagnetic forces are not three independent forces but instead are different manifestations or aspects of what is, in fact, a single force. The theories predict that at high enough temperatures, there would be only one force. At lower temperatures (like the ones in the universe today), however, this single force has changed into three different forces (Figure 29.27). Just as different gases or liquids freeze at different temperatures, we can say that the different forces "froze out" of the unified force at different temperatures. Unfortunately, the temperatures at which the three forces acted as one force are so high that they cannot be reached in any laboratory on Earth. Only the early universe, at times prior to 10^{-35} second, was hot enough to unify these forces.

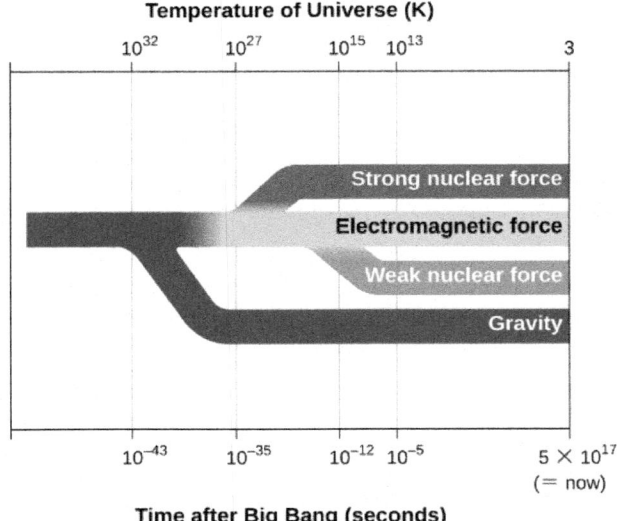

Figure 29.27. Four Forces That Govern the Universe. The behavior of the four forces depends on the temperature of the universe. This diagram (inspired by some grand unified theories) shows that at very early times when the temperature of the universe was very high, all four forces resembled one another and were indistinguishable. As the universe cooled, the forces took on separate and distinctive characteristics.

Many physicists think that gravity was also unified with the three other forces at still higher temperatures, and scientists have tried to develop a theory that combines all four forces. For example, in string theory, the point-like particles of matter that we have discussed in this book are replaced by one-dimensional objects called

Download for free at http://cnx.org/content/col11992/latest/

strings. In this theory, infinitesimal strings, which have length but not height or width, are the building blocks used to construct all the forms of matter and energy in the universe. These strings exist in 11-dimensional space (not the 4-dimensional spacetime with which we are familiar). The strings vibrate in the various dimensions, and depending on how they vibrate, they are seen in our world as matter or gravity or light. As you can imagine, the mathematics of string theory is very complex, and the theory remains untested by experiments. Even the largest particle accelerators on Earth do not achieve high enough energy to show whether string theory applies to the real world.

String theory is interesting to scientists because it is currently the only approach that seems to have the potential of combining all four forces to produce what physicists have termed the Theory of Everything.[3] Theories of the earliest phases of the universe must take both quantum mechanics and gravity into account, but at the simplest level, gravity and quantum mechanics are incompatible. General relativity, our best theory of gravity, says that the motions of objects can be predicted exactly. Quantum mechanics says you can only calculate the probability (chance) that an object will do something. String theory is an attempt to resolve this paradox. The mathematics that underpins string theory is elegant and beautiful, but it remains to be seen whether it will make predictions that can be tested by observations in yet-to-be-developed, high-energy accelerators on Earth or by observations of the early universe.

The earliest period in the history of the universe from time zero to 10^{-43} second is called the Planck time. The universe was unimaginably hot and dense, and theorists believe that at this time, quantum effects of gravity dominated physical interactions—and, as we have just discussed, we have no tested theory of quantum gravity. Inflation is hypothesized to have occurred somewhat later, when the universe was between perhaps 10^{-35} and 10^{-33} second old and the temperature was 10^{27} to 10^{28} K. This rapid expansion took place when three forces (electromagnetic, strong, and weak) are thought to have been unified, and this is when GUTs are applicable.

After inflation, the universe continued to expand (but more slowly) and to cool. An important milestone was reached when the temperature was down to 10^{15} K and the universe was 10^{-10} second old. Under these conditions, all four forces were separate and distinct. High-energy particle accelerators can achieve similar conditions, and so theories of the history of the universe from this point on have a sound basis in experiments.

As yet, we have no direct evidence of what the conditions were during the inflationary epoch, and the ideas presented here are speculative. Researchers are trying to devise some experimental tests. For example, the quantum fluctuations in the very early universe would have caused variations in density and produced gravitational waves that may have left a detectable imprint on the CMB. Detection of such an imprint will require observations with equipment whose sensitivity is improved from what we have today. Ultimately, however, it may provide confirmation that we live in a universe that once experienced an epoch of rapid inflation.

If you are typical of the students who read this book, you may have found this brief discussion of dark matter, inflation, and cosmology a bit frustrating. We have offered glimpses of theories and observations, but have raised more questions than we have answered. What is dark matter? What is dark energy? Inflation explains the observations of flatness and uniformity of the university, but did it actually happen? These ideas are at the forefront of modern science, where progress almost always leads to new puzzles, and much more work is needed before we can see clearly. Bear in mind that less than a century has passed since Hubble demonstrated the existence of other galaxies. The quest to understand just how the universe of galaxies came to be will keep astronomers busy for a long time to come.

---

3   This name became the title of a film about physicist Stephen Hawking in 2014.

This OpenStax book is available for free at http://cnx.org/content/col11992/1.8

Download for free at http://cnx.org/content/col11992/latest/

29.7 THE ANTHROPIC PRINCIPLE

Learning Objectives

By the end of this section, you will be able to:

> Name some properties of the universe that, if different, would have precluded the development of humans

Despite our uncertainties, we must admit that the picture we have developed about the evolution of our universe is a remarkable one. With new telescopes, we have begun to collect enough observational evidence that we can describe how the universe evolved from a mere fraction of a second after the expansion began. Although this is an impressive achievement, there are still some characteristics of the universe that we cannot explain. And yet, it turns out that if these characteristics were any different, we would not be here to ask about them. Let's look at some of these "lucky accidents," beginning with the observations of the cosmic microwave background (CMB).

Lucky Accidents

As we described in this chapter, the CMB is radiation that was emitted when the universe was a few hundred thousand years old. Observations show that the temperature of the radiation varies from one region to another, typically by about 10 millionths of a degree, and these temperature differences signal small differences in density. But suppose the tiny, early fluctuations in density had been much smaller. Then calculations show that the pull of gravity near them would have been so small that no galaxies would ever have formed.

What if the fluctuations in density had been much larger? Then it is possible that very dense regions would have condensed, and these would simply have collapsed directly to black holes without ever forming galaxies and stars. Even if galaxies had been able to form in such a universe, space would have been filled with intense X-rays and gamma rays, and it would have been difficult for life forms to develop and survive. The density of stars within galaxies would be so high that interactions and collisions among them would be frequent. In such a universe, any planetary systems could rarely survive long enough for life to develop.

So for us to be here, the density fluctuations need to be "just right"—not too big and not too small.

Another lucky accident is that the universe is finely balanced between expansion and contraction. It is expanding, but very slowly. If the expansion had been at a much higher rate, all of the matter would have thinned out before galaxies could form. If everything were expanding at a much slower rate, then gravity would have "won." The expansion would have reversed and all of the matter would have recollapsed, probably into a black hole—again, no stars, no planets, no life.

The development of life on Earth depends on still-luckier coincidences. Had matter and antimatter been present initially in exactly equal proportions, then all matter would have been annihilated and turned into pure energy. We owe our existence to the fact that there was slightly more matter than antimatter. (After most of the matter made contact with an equal amount of antimatter, turning into energy, a small amount of additional matter must have been present. We are all descendants of that bit of "unbalanced" matter.)

If nuclear fusion reactions occurred at a somewhat faster rate than they actually do, then at the time of the initial fireball, all of the matter would have been converted from hydrogen into helium into carbon and all the way into iron (the most stable nucleus). That would mean that no stars would have formed, since the existence of stars depends on there being light elements that can undergo fusion in the main-sequence stage and make the stars shine. In addition, the structure of atomic nuclei had to be just right to make it possible for three

Download for free at http://cnx.org/content/col11992/latest/

helium atoms to come together easily to fuse carbon, which is the basis of life. If the *triple-alpha process* we discussed in the chapter on Stars from Adolescence to Old Age were too unlikely, not enough carbon would have formed to lead to biology as we know it. At the same time, it had to be hard enough to fuse carbon into oxygen that a large amount of carbon survived for billions of years.

There are additional factors that have contributed to life like us being possible. Neutrinos have to interact with matter at just the right, albeit infrequent, rate. Supernova explosions occur when neutrinos escape from the cores of collapsing stars, deposit some of their energy in the surrounding stellar envelope, and cause it to blow out and away into space. The heavy elements that are ejected in such explosions are essential ingredients of life here on Earth. If neutrinos did not interact with matter at all, they would escape from the cores of collapsing stars without causing the explosion. If neutrinos interacted strongly with matter, they would remain trapped in the stellar core. In either case, the heavy elements would remain locked up inside the collapsing star.

If gravity were a much stronger force than it is, stars could form with much smaller masses, and their lifetimes would be measured in years rather than billions of years. Chemical processes, on the other hand, would not be sped up if gravity were a stronger force, and so there would be no time for life to develop while stars were so short-lived. Even if life did develop in a stronger-gravity universe, life forms would have to be tiny or they could not stand up or move around.

What Had to Be, Had to Be

In summary, we see that a specific set of rules and conditions in the universe has allowed complexity and life on Earth to develop. As yet, we have no theory to explain why this "right" set of conditions occurred. For this reason, many scientists are beginning to accept an idea we call the **anthropic principle**—namely, that the physical laws we observe must be what they are precisely because these are the only laws that allow for the existence of humans.

Some scientists speculate that our universe is but one of countless universes, each with a different set of physical laws—an idea that is sometimes referred to as the **multiverse**. Some of those universes might be stillborn, collapsing before any structure forms. Others may expand so quickly that they remain essentially featureless with no stars and galaxies. In other words, there may be a much larger multiverse that contains our own universe and many others. This multiverse (existing perhaps in more dimensions that we can become aware of) is infinite and eternal; it generates many, many inflating regions, each of which evolves into a separate universe, which may be completely unlike any of the other separate universes. Our universe is then the way it is because it is the only way it could be and have humans like ourselves in it to discover its properties and ask such questions.

LINK TO LEARNING

View the 2011 introductory talk (https://openstax.org/l/30mulcosinfla) on the Multiverse and Cosmic Inflation by Dr. Anthony Aguirre of the University of California, part of the Silicon Valley Astronomy Lecture Series.

It is difficult to know how to test these ideas since we can never make contact with any other universe. For most scientists, our discussion in this section borders on the philosophical and metaphysical. Perhaps in the future our understanding of physics will develop to the point that we can know why the gravitational constant is as strong as it is, why the universe is expanding at exactly the rate it is, and why all of the other "lucky accidents"

This OpenStax book is available for free at http://cnx.org/content/col11992/1.8

happened—why they were inevitable and could be no other way. Then this anthropic idea would no longer be necessary. No one knows, however, whether we will ever have an explanation for why this universe works the way it does.

We have come a long way in our voyage through the universe. We have learned a remarkable amount about *how* and *when* the cosmos came to be, but the question of *why* the universe is the way it is remains as elusive as ever.

Download for free at http://cnx.org/content/col11992/latest/

CHAPTER 29 REVIEW

 KEY TERMS

anthropic principle idea that physical laws must be the way they are because otherwise we could not be here to measure them

Big Bang the theory of cosmology in which the expansion of the universe began with a primeval explosion (of space, time, matter, and energy)

closed universe a model in which the universe expands from a Big Bang, stops, and then contracts to a big crunch

cosmic microwave background (CMB) microwave radiation coming from all directions that is the redshifted afterglow of the Big Bang

cosmological constant the term in the equations of general relativity that represents a repulsive force in the universe

cosmology the study of the organization and evolution of the universe

critical density in cosmology, the density that is just sufficient to bring the expansion of the universe to a stop after infinite time

dark energy the energy that is causing the expansion of the universe to accelerate; its existence is inferred from observations of distant supernovae

dark matter nonluminous material, whose nature we don't yet understand, but whose presence can be inferred because of its gravitational influence on luminous matter

deuterium a form of hydrogen in which the nucleus of each atom consists of one proton and one neutron

flat universe a model of the universe that has a critical density and in which the geometry of the universe is flat, like a sheet of paper

fusion the building of heavier atomic nuclei from lighter ones

grand unified theories (GUTs) physical theories that attempt to describe the four forces of nature as different manifestations of a single force

inflationary universe a theory of cosmology in which the universe is assumed to have undergone a phase of very rapid expansion when the universe was about 10^{-35} second old; after this period of rapid expansion, the standard Big Bang and inflationary models are identical

lithium the third element in the periodic table; lithium nuclei with three protons and four neutrons were manufactured during the first few minutes of the expansion of the universe

multiverse the speculative idea that our universe is just one of many universes, each with its own set of physical laws

open universe a model in which the density of the universe is not high enough to bring the expansion of the universe to a halt

This OpenStax book is available for free at http://cnx.org/content/col11992/1.8

photon decoupling time when radiation began to stream freely through the universe without interacting with matter

weakly interacting massive particles (WIMPs) weakly interacting massive particles are one of the candidates for the composition of dark matter

 # SUMMARY

29.1 The Age of the Universe

Cosmology is the study of the organization and evolution of the universe. The universe is expanding, and this is one of the key observational starting points for modern cosmological theories. Modern observations show that the rate of expansion has not been constant throughout the life of the universe. Initially, when galaxies were close together, the effects of gravity were stronger than the effects of dark energy, and the expansion rate gradually slowed. As galaxies moved farther apart, the influence of gravity on the expansion rate weakened. Measurements of distant supernovae show that when the universe was about half its current age, dark energy began to dominate the rate of expansion and caused it to speed up. In order to estimate the age of the universe, we must allow for changes in the rate of expansion. After allowing for these effects, astronomers estimate that all of the matter within the observable universe was concentrated in an extremely small volume 13.8 billion years ago, a time we call the Big Bang.

29.2 A Model of the Universe

For describing the large-scale properties of the universe, a model that is isotropic and homogeneous (same everywhere) is a pretty good approximation of reality. The universe is expanding, which means that the universe undergoes a change in scale with time; space stretches and distances grow larger by the same factor everywhere at a given time. Observations show that the mass density of the universe is less than the critical density. In other words, there is not enough matter in the universe to stop the expansion. With the discovery of dark energy, which is accelerating the rate of expansion, the observational evidence is strong that the universe will expand forever. Observations tell us that the expansion started about 13.8 billion years ago.

29.3 The Beginning of the Universe

Lemaître, Alpher, and Gamow first worked out the ideas that are today called the Big Bang theory. The universe cools as it expands. The energy of photons is determined by their temperature, and calculations show that in the hot, early universe, photons had so much energy that when they collided with one another, they could produce material particles. As the universe expanded and cooled, protons and neutrons formed first, then came electrons and positrons. Next, fusion reactions produced deuterium, helium, and lithium nuclei. Measurements of the deuterium abundance in today's universe show that the total amount of ordinary matter in the universe is only about 5% of the critical density.

29.4 The Cosmic Microwave Background

When the universe became cool enough to form neutral hydrogen atoms, the universe became transparent to radiation. Scientists have detected the cosmic microwave background (CMB) radiation from this time during the hot, early universe. Measurements with the COBE satellite show that the CMB acts like a blackbody with a temperature of 2.73 K. Tiny fluctuations in the CMB show us the seeds of large-scale structures in the universe. Detailed measurements of these fluctuations show that we live in a critical-density universe and that the critical density is composed of 31% matter, including dark matter, and 69% dark energy. Ordinary matter—the kinds of elementary particles we find on Earth—make up only about 5% of the critical density. CMB measurements also indicate that the universe is 13.8 billion years old.

Download for free at http://cnx.org/content/col11992/latest/

29.5 What Is the Universe Really Made Of?

Twenty-seven percent of the critical density of the universe is composed of dark matter. To explain so much dark matter, some physics theories predict that additional types of particles should exist. One type has been given the name of WIMPs (weakly interacting massive particles), and scientists are now conducting experiments to try to detect them in the laboratory. Dark matter plays an essential role in forming galaxies. Since, by definition, these particles interact only very weakly (if at all) with radiation, they could have congregated while the universe was still very hot and filled with radiation. They would thus have formed gravitational traps that quickly attracted and concentrated ordinary matter after the universe became transparent, and matter and radiation decoupled. This rapid concentration of matter enabled galaxies to form by the time the universe was only 400–500 million years old.

29.6 The Inflationary Universe

The Big Bang model does not explain why the CMB has the same temperature in all directions. Neither does it explain why the density of the universe is so close to critical density. These observations can be explained if the universe experienced a period of rapid expansion, which scientists call inflation, about 10^{-35} second after the Big Bang. New grand unified theories (GUTs) are being developed to describe physical processes in the universe before and at the time that inflation occurred.

29.7 The Anthropic Principle

Recently, many cosmologists have noted that the existence of humans depends on the fact that many properties of the universe—the size of density fluctuations in the early universe, the strength of gravity, the structure of atoms—were just right. The idea that physical laws must be the way they are because otherwise we could not be here to measure them is called the anthropic principle. Some scientists speculate that there may be a multiverse of universes, in which ours is just one.

 FOR FURTHER EXPLORATION

Articles

Kruesi, L. "Cosmology: 5 Things You Need to Know." *Astronomy* (May 2007): 28. Five questions students often ask, and how modern cosmologists answer them.

Kruesi, L. "How Planck Has Redefined the Universe." *Astronomy* (October 2013): 28. Good review of what this space mission has told us about the CMB and the universe.

Lineweaver, C. & Davis, T. "Misconceptions about the Big Bang." *Scientific American* (March 2005): 36. Some basic ideas about modern cosmology clarified, using general relativity.

Nadis, S. "Sizing Up Inflation." *Sky & Telescope* (November 2005): 32. Nice review of the origin and modern variants on the inflationary idea.

Nadis, S. "How We Could See Another Universe." *Astronomy* (June 2009): 24. On modern ideas about multiverses and how such bubbles of space-time might collide.

Nadis, S. "Dark Energy's New Face: How Exploding Stars Are Changing our View." *Astronomy* (July 2012): 45. About our improving understanding of the complexities of type Ia supernovae.

Naze, Y. "The Priest, the Universe, and the Big Bang." *Astronomy* (November 2007): 40. On the life and work of Georges Lemaître.

This OpenStax book is available for free at http://cnx.org/content/col11992/1.8

Panek, R. "Going Over to the Dark Side." *Sky & Telescope* (February 2009): 22. A history of the observations and theories about dark energy.

Pendrick, D. "Is the Big Bang in Trouble?" *Astronomy* (April 2009): 48. This sensationally titled article is really more of a quick review of how modern ideas and observations are fleshing out the Big Bang hypothesis (and raising questions.)

Reddy, F. "How the Universe Will End." *Astronomy* (September 2014): 38. Brief discussion of local and general future scenarios.

Riess, A. and Turner, M. "The Expanding Universe: From Slowdown to Speedup." *Scientific American* (September 2008): 62.

Turner, M. "The Origin of the Universe." *Scientific American* (September 2009): 36. An introduction to modern cosmology.

Websites

Cosmology Primer: https://preposterousuniverse.com/cosmologyprimer/. Caltech Astrophysicist Sean Carroll offers a non-technical site with brief overviews of many key topics in modern cosmology.

Everyday Cosmology: http://cosmology.carnegiescience.edu/. An educational website from the Carnegie Observatories with a timeline of cosmological discovery, background materials, and activities.

How Big Is the Universe?: http://www.pbs.org/wgbh/nova/space/how-big-universe.html. A clear essay by a noted astronomer Brent Tully summarizes some key ideas in cosmology and introduces the notion of the acceleration of the universe.

Universe 101: WMAP Mission Introduction to the Universe: http://map.gsfc.nasa.gov/universe/. Concise NASA primer on cosmological ideas from the WMAP mission team.

Cosmic Times Project: http://cosmictimes.gsfc.nasa.gov/. James Lochner and Barbara Mattson have compiled a rich resource of twentieth-century cosmology history in the form of news reports on key events, from NASA's Goddard Space Flight Center.

Videos

The Day We Found the Universe: http://www.cfa.harvard.edu/events/mon_video_archive09.html. Distinguished science writer Marcia Bartusiak discusses Hubble's work and the discovery of the expansion of the cosmos—one of the Observatory Night lectures at the Harvard-Smithsonian Center for Astrophysics (53:46).

Images of the Infant Universe: https://www.youtube.com/watch?v=x0AqCwElyUk. Lloyd Knox's public talk on the latest discoveries about the CMB and what they mean for cosmology (1:16:00).

Runaway Universe: https://www.youtube.com/watch?v=kNYVFrnmcOU. Roger Blandford (Stanford Linear Accelerator Center) public lecture on the discovery and meaning of cosmic acceleration and dark energy (1:08:08).

From the Big Bang to the Nobel Prize and on to the James Webb Space Telescope and the Discovery of Alien Life: http://svs.gsfc.nasa.gov/vis/a010000/a010300/a010370/index.html. John Mather, NASA Goddard (1:01:02). His Nobel Prize talk from Dec. 8, 2006 can be found at http://www.nobelprize.org/mediaplayer/index.php?id=74&view=1.

Dark Energy and the Fate of the Universe: https://webcast.stsci.edu/webcast/detail.xhtml?talkid=1961&parent=1. Adam Reiss (STScI), at the Space Telescope Science Institute (1:00:00).

Download for free at http://cnx.org/content/col11992/latest/

⚎ COLLABORATIVE GROUP ACTIVITIES

A. This chapter deals with some pretty big questions and ideas. Some belief systems teach us that there are questions to which "we were not meant to know" the answers. Other people feel that if our minds and instruments are capable of exploring a question, then it becomes part of our birthright as thinking human beings. Have your group discuss your personal reactions to discussing questions like the beginning of time and space, and the ultimate fate of the universe. Does it make you nervous to hear about scientists discussing these issues? Or is it exciting to know that we can now gather scientific evidence about the origin and fate of the cosmos? (In discussing this, you may find that members of your group strongly disagree; try to be respectful of others' points of view.)

B. A popular model of the universe in the 1950s and 1960s was the so-called steady-state cosmology. In this model, the universe was not only the same everywhere and in all directions (homogeneous and isotropic), but also the same *at all times*. We know the universe is expanding and the galaxies are thinning out, and so this model hypothesized that new matter was continually coming into existence to fill in the space between galaxies as they moved farther apart. If so, the infinite universe did not have to have a sudden beginning, but could simply exist forever in a steady state. Have your group discuss your reaction to this model. Do you find it more appealing philosophically than the Big Bang model? Can you cite some evidence that indicates that the universe was not the same billions of years ago as it is now—that it is not in a steady state?

C. One of the lucky accidents that characterizes our universe is the fact that the time scale for the development of intelligent life on Earth and the lifetime of the Sun are comparable. Have your group discuss what would happen if the two time scales were very different. Suppose, for example, that the time for intelligent life to evolve was 10 times greater than the main-sequence lifetime of the Sun. Would our civilization have ever developed? Now suppose the time for intelligent life to evolve is ten times shorter than the main-sequence lifetime of the Sun. Would we be around? (This latter discussion requires considerable thought, including such ideas as what the early stages in the Sun's life were like and how much the early Earth was bombarded by asteroids and comets.)

D. The grand ideas discussed in this chapter have a powerful effect on the human imagination, not just for scientists, but also for artists, composers, dramatists, and writers. Here we list just a few of these responses to cosmology. Each member of your group can select one of these, learn more about it, and then report back, either to the group or to the whole class.

- The California poet Robinson Jeffers was the brother of an astronomer who worked at the Lick Observatory. His poem "Margrave" is a meditation on cosmology and on the kidnap and murder of a child: http://www.poemhunter.com/best-poems/robinson-jeffers/margrave/.

- In the science fiction story "The Gravity Mine" by Stephen Baxter, the energy of evaporating supermassive black holes is the last hope of living beings in the far future in an ever-expanding universe. The story has poetic description of the ultimate fate of matter and life and is available online at: http://www.infinityplus.co.uk/stories/gravitymine.htm.

- The musical piece *YLEM* by Karlheinz Stockhausen takes its title from the ancient Greek term for primeval material revived by George Gamow. It tries to portray the oscillating universe in musical terms. Players actually expand through the concert hall, just as the universe does, and then return and expand again. See: http://www.karlheinzstockhausen.org/ylem_english.htm.

This OpenStax book is available for free at http://cnx.org/content/col11992/1.8

- The musical piece *Supernova Sonata* http://www.astro.uvic.ca/~alexhp/new/supernova_sonata.html by Alex Parker and Melissa Graham is based on the characteristics of 241 type Ia supernova explosions, the ones that have helped astronomers discover the acceleration of the expanding universe.

- Gregory Benford's short story "The Final Now" envisions the end of an accelerating open universe, and blends religious and scientific imagery in a very poetic way. Available free online at: http://www.tor.com/stories/2010/03/the-final-now.

E. When Einstein learned about Hubble's work showing that the universe of galaxies is expanding, he called his introduction of the cosmological constant into his general theory of relativity his "biggest blunder." Can your group think of other "big blunders" from the history of astronomy, where the thinking of astronomers was too conservative and the universe turned out to be more complicated or required more "outside-the-box" thinking?

EXERCISES

Review Questions

1. What are the basic observations about the universe that any theory of cosmology must explain?

2. Describe some possible futures for the universe that scientists have come up with. What property of the universe determines which of these possibilities is the correct one?

3. What does the term Hubble time mean in cosmology, and what is the current best calculation for the Hubble time?

4. Which formed first: hydrogen nuclei or hydrogen atoms? Explain the sequence of events that led to each.

5. Describe at least two characteristics of the universe that are explained by the standard Big Bang model.

6. Describe two properties of the universe that are not explained by the standard Big Bang model (without inflation). How does inflation explain these two properties?

7. Why do astronomers believe there must be dark matter that is not in the form of atoms with protons and neutrons?

8. What is dark energy and what evidence do astronomers have that it is an important component of the universe?

9. Thinking about the ideas of space and time in Einstein's general theory of relativity, how do we explain the fact that all galaxies outside our Local Group show a redshift?

10. Astronomers have found that there is more helium in the universe than stars could have made in the 13.8 billion years that the universe has been in existence. How does the Big Bang scenario solve this problem?

11. Describe the anthropic principle. What are some properties of the universe that make it "ready" to have life forms like you in it?

12. Describe the evidence that the expansion of the universe is accelerating.

Thought Questions

13. What is the most useful probe of the early evolution of the universe: a giant elliptical galaxy or an irregular galaxy such as the Large Magellanic Cloud? Why?

Download for free at http://cnx.org/content/col11992/latest/

14. What are the advantages and disadvantages of using quasars to probe the early history of the universe?

15. Would acceleration of the universe occur if it were composed entirely of matter (that is, if there were no dark energy)?

16. Suppose the universe expands forever. Describe what will become of the radiation from the primeval fireball. What will the future evolution of galaxies be like? Could life as we know it survive forever in such a universe? Why?

17. Some theorists expected that observations would show that the density of matter in the universe is just equal to the critical density. Do the current observations support this hypothesis?

18. There are a variety of ways of estimating the ages of various objects in the universe. Describe two of these ways, and indicate how well they agree with one another and with the age of the universe itself as estimated by its expansion.

19. Since the time of Copernicus, each revolution in astronomy has moved humans farther from the center of the universe. Now it appears that we may not even be made of the most common form of matter. Trace the changes in scientific thought about the central nature of Earth, the Sun, and our Galaxy on a cosmic scale. Explain how the notion that most of the universe is made of dark matter continues this "Copernican tradition."

20. The anthropic principle suggests that in some sense we are observing a special kind of universe; if the universe were different, we could never have come to exist. Comment on how this fits with the Copernican tradition described in Exercise 29.19.

21. Penzias and Wilson's discovery of the Cosmic Microwave Background (CMB) is a nice example of scientific *serendipity*—something that is found by chance but turns out to have a positive outcome. What were they looking for and what did they discover?

22. Construct a timeline for the universe and indicate when various significant events occurred, from the beginning of the expansion to the formation of the Sun to the appearance of humans on Earth.

Figuring For Yourself

23. Suppose the Hubble constant were not 22 but 33 km/s per million light-years. Then what would the critical density be?

24. Assume that the average galaxy contains 10^{11} M_{Sun} and that the average distance between galaxies is 10 million light-years. Calculate the average density of matter (mass per unit volume) in galaxies. What fraction is this of the critical density we calculated in the chapter?

25. The CMB contains roughly 400 million photons per m^3. The energy of each photon depends on its wavelength. Calculate the typical wavelength of a CMB photon. Hint: The CMB is blackbody radiation at a temperature of 2.73 K. According to Wien's law, the peak wave length in nanometers is given by $\lambda_{max} = \frac{3 \times 10^6}{T}$. Calculate the wavelength at which the CMB is a maximum and, to make the units consistent, convert this wavelength from nanometers to meters.

26. Following up on Exercise 29.27 calculate the energy of a typical photon. Assume for this approximate calculation that each photon has the wavelength calculated in Exercise 29.25. The energy of a photon is given by $E = \frac{hc}{\lambda}$, where h is Planck's constant and is equal to 6.626 × 10^{-34} J × s, c is the speed of light in m/s, and λ is the wavelength in m.

This OpenStax book is available for free at http://cnx.org/content/col11992/1.8

27. Continuing the thinking in Exercise 29.27 and Exercise 29.28, calculate the energy in a cubic meter of space, multiply the energy per photon calculated in Exercise 29.26 by the number of photons per cubic meter given above.

28. Continuing the thinking in the last three exercises, convert this energy to an equivalent in mass, use Einstein's equation $E = mc^2$. Hint: Divide the energy per m^3 calculated in Exercise 29.27 by the speed of light squared. Check your units; you should have an answer in kg/m^3. Now compare this answer with the critical density. Your answer should be several powers of 10 smaller than the critical density. In other words, you have found for yourself that the contribution of the CMB photons to the overall density of the universe is much, much smaller than the contribution made by stars and galaxies.

29. There is still some uncertainty in the Hubble constant. (a) Current estimates range from about 19.9 km/s per million light-years to 23 km/s per million light-years. Assume that the Hubble constant has been constant since the Big Bang. What is the possible range in the ages of the universe? Use the equation in the text, $T_0 = \frac{1}{H}$, and make sure you use consistent units. (b) Twenty years ago, estimates for the Hubble constant ranged from 50 to 100 km/s per Mps. What are the possible ages for the universe from those values? Can you rule out some of these possibilities on the basis of other evidence?

30. It is possible to derive the age of the universe given the value of the Hubble constant and the distance to a galaxy, again with the assumption that the value of the Hubble constant has not changed since the Big Bang. Consider a galaxy at a distance of 400 million light-years receding from us at a velocity, v. If the Hubble constant is 20 km/s per million light-years, what is its velocity? How long ago was that galaxy right next door to our own Galaxy if it has always been receding at its present rate? Express your answer in years. Since the universe began when all galaxies were very close together, this number is a rough estimate for the age of the universe.

Download for free at http://cnx.org/content/col11992/latest/

This OpenStax book is available for free at http://cnx.org/content/col11992/1.8

30

LIFE IN THE UNIVERSE

Figure 30.1. Astrobiology: The Road to Life in the Universe. In this fanciful montage produced by a NASA artist, we see one roadmap for discovering life in the universe. Learning more about the origin, evolution, and properties of life on Earth aids us in searching for evidence of life beyond our planet. Our neighbor world, Mars, had warmer, wetter conditions billions of years ago that might have helped life there begin. Farther out, Jupiter's moon Europa represents the icy moons of the outer solar system. Beneath their shells of solid ice may lie vast oceans of liquid water that could support biology. Beyond our solar system are stars that host their own planets, some of which might be similar to Earth in the ability to support liquid water—and a thriving biosphere—at the planet's surface. Research is pushing actively in all these directions with the goal of proving a scientific answer to the question, "Are we alone?" (credit: modification of work by NASA)

Chapter Outline

30.1 The Cosmic Context for Life

30.2 Astrobiology

30.3 Searching for Life beyond Earth

30.4 The Search for Extraterrestrial Intelligence

✎ Thinking Ahead

As we have learned more about the universe, we have naturally wondered whether there might be other forms of life out there. The ancient question, "Are we alone in the universe?" connects us to generations of humans before us. While in the past, this question was in the realm of philosophy or science fiction, today we have the means to seek an answer through scientific inquiry. In this chapter, we will consider how life began on Earth, whether the same processes could have led to life on other worlds, and how we might seek evidence of life elsewhere. This is the science of astrobiology.

The search for life on other planets is not the same as the search for *intelligent* life, which (if it exists) is surely much rarer. Learning more about the origin, evolution, and properties of life on Earth aids us in searching for evidence of all kinds of life beyond that on our planet.

Download for free at http://cnx.org/content/col11992/latest/

30.1 | THE COSMIC CONTEXT FOR LIFE

Learning Objectives

By the end of this section, you will be able to:

> Describe the chemical and environmental conditions that make Earth hospitable to life
> Discuss the assumption underlying the Copernican principle and outline its implications for modern-day astronomers
> Understand the questions underlying the Fermi paradox

We saw that the universe was born in the Big Bang about 14 billion years ago. After the initial hot, dense fireball of creation cooled sufficiently for atoms to exist, all matter consisted of hydrogen and helium (with a very small amount of lithium). As the universe aged, processes within stars created the other elements, including those that make up Earth (such as iron, silicon, magnesium, and oxygen) and those required for life as we know it, such as carbon, oxygen, and nitrogen. These and other elements combined in space to produce a wide variety of compounds that form the basis of life on Earth. In particular, life on Earth is based on the presence of a key unit known as an **organic molecule**, a molecule that contains carbon. Especially important are the hydrocarbons, chemical compounds made up entirely of hydrogen and carbon, which serve as the basis for our biological chemistry, or *biochemistry*. While we do not understand the details of how life on Earth began, it is clear that to make creatures like us possible, events like the ones we described must have occurred, resulting in what is called the *chemical evolution* of the universe.

What Made Earth Hospitable to Life?

About 5 billion years ago, a cloud of gas and dust in this cosmic neighborhood began to collapse under its own weight. Out of this cloud formed the Sun and its planets, together with all the smaller bodies, such as comets, that also orbit the Sun (Figure 30.2). The third planet from the Sun, as it cooled, eventually allowed the formation of large quantities of liquid water on its surface.

Figure 30.2. Comet Hyakutake. This image was captured in 1996 by NASA photographer Bill Ingalls. Comet impacts can deliver both water and a variety of interesting chemicals, including some organic chemicals, to Earth. (credit: NASA/Bill Ingalls)

The chemical variety and moderate conditions on Earth eventually led to the formation of molecules that could make copies of themselves (reproduce), which is essential for beginning life. Over the billions of years of Earth history, life evolved and became more complex. The course of evolution was punctuated by occasional planet-wide changes caused by collisions with some of the smaller bodies that did not make it into the Sun or one of its

This OpenStax book is available for free at http://cnx.org/content/col11992/1.8

accompanying worlds. As we saw in the chapter on Earth as a Planet, mammals may owe their domination of Earth's surface to just such a collision 65 million years ago, which led to the extinction of the dinosaurs (along with the majority of other living things). The details of such mass extinctions are currently the focus of a great deal of scientific interest.

Through many twisting turns, the course of evolution on Earth produced a creature with self-consciousness, able to ask questions about its own origins and place in the cosmos (Figure 30.3). Like most of Earth, this creature is composed of atoms that were forged in earlier generations of stars—in this case, assembled into both its body and brain. We might say that through the thoughts of human beings, the matter in the universe can become aware of itself.

Figure 30.3. Young Human. Human beings have the intellect to wonder about their planet and what lies beyond it. Through them (and perhaps other intelligent life), the universe becomes aware of itself. (credit: Andrew Fraknoi)

Think about those atoms in your body for a minute. They are merely on loan to you from the lending library of atoms that make up our local corner of the universe. Atoms of many kinds circulate through your body and then leave it—with each breath you inhale and exhale and the food you eat and excrete. Even the atoms that take up more permanent residence in your tissues will not be part of you much longer than you are alive. Ultimately, you will return your atoms to the vast reservoir of Earth, where they will be incorporated into other structures and even other living things in the millennia to come.

This picture of *cosmic evolution*, of our descent from the stars, has been obtained through the efforts of scientists in many fields over many decades. Some of its details are still tentative and incomplete, but we feel reasonably confident in its broad outlines. It is remarkable how much we have been able to learn in the short time we have had the instruments to probe the physical nature of the universe.

The Copernican Principle

Our study of astronomy has taught us that we have always been wrong in the past whenever we have claimed that Earth is somehow unique. Galileo, using the newly invented technology of the telescope, showed us that Earth is not the center of the solar system, but merely one of a number of objects orbiting the Sun. Our study of the stars has demonstrated that the Sun itself is a rather undistinguished star, halfway through its long main-sequence stage like so many billions of others. There seems nothing special about our position in the Milky Way Galaxy either, and nothing surprising about our Galaxy's position in either its own group or its supercluster.

The discovery of planets around other stars confirms our idea that the formation of planets is a natural consequence of the formation of stars. We have identified thousands of exoplanets—planets orbiting around

Download for free at http://cnx.org/content/col11992/latest/

other stars, from huge ones orbiting close to their stars (informally called "hot Jupiters") down to planets smaller than Earth. A steady stream of exoplanet discoveries is leading to the conclusion that earthlike planets occur frequently—enough that there are likely many billions of "exo-Earths" in our own Milky Way Galaxy alone. From a planetary perspective, smaller planets are not unique.

Philosophers of science sometimes call the idea that there is nothing special about our place in the universe the *Copernican principle*. Given all of the above, most scientists would be surprised if life were limited to our planet and had started nowhere else. There are billions of stars in our Galaxy old enough for life to have developed on a planet around them, and there are billions of other galaxies as well. Astronomers and biologists have long conjectured that a series of events similar to those on the early Earth probably led to living organisms on many planets around other stars, and possibly even on other planets in our solar system, such as Mars.

The real scientific issue (which we do not currently know the answer to) is whether organic biochemistry is likely or unlikely in the universe at large. Are we a fortunate and exceedingly rare outcome of chemical evolution, or is organic biochemistry a regular part of the chemical evolution of the cosmos? We do not yet know the answer to this question, but data, even an exceedingly small amount (like finding "unrelated to us" living systems on a world like Europa), will help us arrive at it.

So Where Are They?

If the Copernican principle is applied to life, then biology may be rather common among planets. Taken to its logical limit, the Copernican principle also suggests that intelligent life like us might be common. Intelligence like ours has some very special properties, including an ability to make progress through the application of technology. Organic life around other (older) stars may have started a billion years earlier than we did on Earth, so they may have had a lot more time to develop advanced technology such as sending information, probes, or even life-forms between stars.

Faced with such a prospect, physicist Enrico Fermi asked a question several decades ago that is now called the *Fermi paradox*: where are they? If life and intelligence are common and have such tremendous capacity for growth, why is there not a network of galactic civilizations whose presence extends even into a "latecomer" planetary system like ours?

Several solutions have been suggested to the Fermi paradox. Perhaps life is common but intelligence (or at least technological civilization) is rare. Perhaps such a network will come about in the future but has not yet had the time to develop. Maybe there are invisible streams of data flowing past us all the time that we are not advanced enough or sensitive enough to detect. Maybe advanced species make it a practice not to interfere with immature, developing consciousness such as our own. Or perhaps civilizations that reach a certain level of technology then self-destruct, meaning there are no other civilizations now existing in our Galaxy. We do not yet know whether any advanced life is out there and, if it is, why we are not aware of it. Still, you might want to keep these issues in mind as you read the rest of this chapter.

LINK TO LEARNING

Is there a network of galactic civilizations beyond our solar system? If so, why can't we see them? Explore the possibilities in the cartoon video (https://openstax.org/l/30fermparadox) "The Fermi Paradox—Where Are All the Aliens?"

This OpenStax book is available for free at http://cnx.org/content/col11992/1.8

30.2 ASTROBIOLOGY

Learning Objectives

By the end of this section, you will be able to:

> Describe the chemical building blocks required for life
> Describe the molecular systems and processes driving the origin and evolution of life
> Describe the characteristics of a habitable environment
> Describe some of the extreme conditions on Earth, and explain how certain organisms have adapted to these conditions

Scientists today take a multidisciplinary approach to studying the origin, evolution, distribution, and ultimate fate of life in the universe; this field of study is known as **astrobiology**. You may also sometimes hear this field referred to as *exobiology* or *bioastronomy*. Astrobiology brings together astronomers, planetary scientists, chemists, geologists, and biologists (among others) to work on the same problems from their various perspectives.

Among the issues that astrobiologists explore are the conditions in which life arose on Earth and the reasons for the extraordinary adaptability of life on our planet. They are also involved in identifying habitable worlds beyond Earth and in trying to understand in practical terms how to look for life on those worlds. Let's look at some of these issues in more detail.

The Building Blocks of Life

While no unambiguous evidence for life has yet been found anywhere beyond Earth, life's chemical building blocks have been detected in a wide range of extraterrestrial environments. Meteorites (which you learned about in Cosmic Samples and the Origin of the Solar System) have been found to contain two kinds of substances whose chemical structures mark them as having an extraterrestrial origin—amino acids and sugars. **Amino acids** are **organic compounds** that are the molecular building blocks of proteins. **Proteins** are key biological molecules that provide the structure and function of the body's tissues and organs and essentially carry out the "work" of the cell. When we examine the gas and dust around comets, we also find a number of organic molecules—compounds that on Earth are associated with the chemistry of life.

Expanding beyond our solar system, one of the most interesting results of modern radio astronomy has been the discovery of organic molecules in giant clouds of gas and dust between stars. More than 100 different molecules have been identified in these reservoirs of cosmic raw material, including formaldehyde, alcohol, and others we know as important stepping stones in the development of life on Earth. Using radio telescopes and radio spectrometers, astronomers can measure the abundances of various chemicals in these clouds. We find organic molecules most readily in regions where the interstellar dust is most abundant, and it turns out these are precisely the regions where star formation (and probably planet formation) happen most easily (Figure 30.4).

Download for free at http://cnx.org/content/col11992/latest/

Figure 30.4. Cloud of Gas and Dust. This cloud of gas and dust in the constellation of Scorpius is the sort of region where complex molecules are found. It is also the sort of cloud where new stars form from the reservoir of gas and dust in the cloud. Radiation from a group of hot stars (off the picture to the bottom left) called the Scorpius OB Association is "eating into" the cloud, sweeping it into an elongated shape and causing the reddish glow seen at its tip. (credit: Dr. Robert Gendler)

Clearly the early Earth itself produced some of the molecular building blocks of life. Since the early 1950s, scientists have tried to duplicate in their laboratories the chemical pathways that led to life on our planet. In a series of experiments known as the *Miller-Urey experiments*, pioneered by Stanley Miller and Harold Urey at the University of Chicago, biochemists have simulated conditions on early Earth and have been able to produce some of the fundamental building blocks of life, including those that form proteins and other large biological molecules known as nucleic acids (which we will discuss shortly).

Although these experiments produced encouraging results, there are some problems with them. The most interesting chemistry from a biological perspective takes place with hydrogen-rich or *reducing* gases, such as ammonia and methane. However, the early atmosphere of Earth was probably dominated by carbon dioxide (as Venus' and Mars' atmospheres still are today) and may not have contained an abundance of reducing gases comparable to that used in Miller-Urey type experiments. Hydrothermal vents—seafloor systems in which ocean water is superheated and circulated through crustal or mantle rocks before reemerging into the ocean—have also been suggested as potential contributors of organic compounds on the early Earth, and such sources would not require Earth to have an early reducing atmosphere.

Both earthly and extraterrestrial sources may have contributed to Earth's early supply of organic molecules, although we have more direct evidence for the latter. It is even conceivable that life itself originated elsewhere and was seeded onto our planet—although this, of course, does not solve the problem of how that life originated to begin with.

LINK TO LEARNING

Hydrothermal vents are beginning to seem more likely as early contributors to the organic compounds found on Earth. Read about hydrothermal vents, watch videos and slideshows on these and other deep-

This OpenStax book is available for free at http://cnx.org/content/col11992/1.8

Download for free at http://cnx.org/content/col11992/latest/

sea wonders, and try an interactive simulation of hydrothermal circulation at the Woods Hole Oceanographic Institution (https://openstax.org/l/30wohooceins) website.

The Origin and Early Evolution of Life

The carbon compounds that form the chemical basis of life may be common in the universe, but it is still a giant step from these building blocks to a living cell. Even the simplest molecules of the **genes** (the basic functional units that carry the genetic, or hereditary, material in a cell) contain millions of molecular units, each arranged in a precise sequence. Furthermore, even the most primitive life required two special capabilities: a means of extracting energy from its environment, and a means of encoding and replicating information in order to make faithful copies of itself. Biologists today can see ways that either of these capabilities might have formed in a natural environment, but we are still a long way from knowing how the two came together in the first life-forms.

We have no solid evidence for the pathway that led to the origin of life on our planet except for whatever early history may be retained in the biochemistry of modern life. Indeed, we have very little direct evidence of what Earth itself was like during its earliest history—our planet is so effective at resurfacing itself through plate tectonics (see the chapter on Earth as a Planet) that very few rocks remain from this early period. In the earlier chapter on Cratered Worlds, you learned that Earth was subjected to a heavy bombardment—a period of large impact events—some 3.8 to 4.1 billion years ago. Large impacts would have been energetic enough to heat-sterilize the surface layers of Earth, so that even if life had begun by this time, it might well have been wiped out.

When the large impacts ceased, the scene was set for a more peaceful environment on our planet. If the oceans of Earth contained accumulated organic material from any of the sources already mentioned, the ingredients were available to make living organisms. We do not understand in any detail the sequence of events that led from molecules to biology, but there is fossil evidence of microbial life in 3.5-billion-year-old rocks, and possible (debated) evidence for life as far back as 3.8 billion years.

Life as we know it employs two main molecular systems: the functional molecules known as proteins, which carry out the chemical work of the cell, and information-containing molecules of **DNA (deoxyribonucleic acid)** that store information about how to create the cell and its chemical and structural components. The origin of life is sometimes considered a "chicken and egg problem" because, in modern biology, neither of these systems works without the other. It is our proteins that assemble DNA strands in the precise order required to store information, but the proteins are created based on information stored in DNA. Which came first? Some origin of life researchers believe that prebiotic chemistry was based on molecules that could both store information and do the chemical work of the cell. It has been suggested that **RNA (ribonucleic acid)**, a molecule that aids in the flow of genetic information from DNA to proteins, might have served such a purpose. The idea of an early "RNA world" has become increasingly accepted, but a great deal remains to be understood about the origin of life.

Perhaps the most important innovation in the history of biology, apart from the origin of life itself, was the discovery of the process of **photosynthesis**, the complex sequence of chemical reactions through which some living things can use sunlight to manufacture products that store energy (such as carbohydrates), releasing oxygen as one by-product. Previously, life had to make do with sources of chemical energy available on Earth or delivered from space. But the abundant energy available in sunlight could support a larger and more productive biosphere, as well as some biochemical reactions not previously possible for life. One of these was the production of oxygen (as a waste product) from carbon dioxide, and the increase in atmospheric levels of oxygen about 2.4 billion years ago means that oxygen-producing photosynthesis must have emerged and

Download for free at http://cnx.org/content/col11992/latest/

become globally important by this time. In fact, it is likely that oxygen-producing photosynthesis emerged considerably earlier.

Some forms of chemical evidence contained in ancient rocks, such as the solid, layered rock formations known as **stromatolites**, are thought to be the fossils of oxygen-producing photosynthetic bacteria in rocks that are almost 3.5 billion years old (Figure 30.5). It is generally thought that a simpler form of photosynthesis that does not produce oxygen (and is still used by some bacteria today) probably preceded oxygen-producing photosynthesis, and there is strong fossil evidence that one or the other type of photosynthesis was functioning on Earth at least as far back as 3.4 billion years ago.

(a)   (b)

Figure 30.5. Stromatolites Preserve the Earliest Physical Representation of Life on Earth. In their reach for sunlight, the single-celled microbes formed mats that trapped sediments in the water above them. Such trapped sediments fell and formed layers on top of the mats. The microbes then climbed atop the sediment layers and trapped more sediment. What is found in the rock record are (a) the solidified, curved sedimentary layers that are signatures of biological activity. The earliest known stromatolite is 3.47 billion years old and is found in Western Australia. (b) This more recent example is in Lake Thetis, also in Western Australia. (credit a: modification of work by James St. John; credit b: modification of work by Ruth Ellison)

The free oxygen produced by photosynthesis began accumulating in our atmosphere about 2.4 billion years ago. The interaction of sunlight with oxygen can produce ozone (which has three atoms of oxygen per molecule, as compared to the two atoms per molecule in the oxygen we breathe), which accumulated in a layer high in Earth's atmosphere. As it does on Earth today, this ozone provided protection from the Sun's damaging ultraviolet radiation. This allowed life to colonize the landmasses of our planet instead of remaining only in the ocean.

The rise in oxygen levels was deadly to some microbes because, as a highly reactive chemical, it can irreversibly damage some of the biomolecules that early life had developed in the absence of oxygen. For other microbes, it was a boon: combining oxygen with organic matter or other reduced chemicals generates a lot of energy—you can see this when a log burns, for example—and many forms of life adopted this way of living. This new energy source made possible a great proliferation of organisms, which continued to evolve in an oxygen-rich environment.

The details of that evolution are properly the subject of biology courses, but the process of evolution by natural selection (survival of the fittest) provides a clear explanation for the development of Earth's remarkable variety of life-forms. It does not, however, directly solve the mystery of life's earliest beginnings. We hypothesize that life will arise whenever conditions are appropriate, but this hypothesis is just another form of the Copernican principle. We now have the potential to address this hypothesis with observations. If a second example of life is found in our solar system or a nearby star, it would imply that life emerges commonly enough that the universe is likely filled with biology. To make such observations, however, we must first decide where to focus our search.

This OpenStax book is available for free at http://cnx.org/content/col11992/1.8

LINK TO LEARNING

Just how did life arise in the first place? And could it have happened with a different type of chemistry? Watch the 15-minute video Making Matter Come Alive (https://openstax.org/l/30makmattcomali) in which a chemistry expert explores some answers to these questions, from a 2011 TED Talk.

Habitable Environments

Among the staggering number of objects in our solar system, Galaxy, and universe, some may have conditions suitable for life, while others do not. Understanding what conditions and features make a **habitable environment**—an environment capable of hosting life—is important both for understanding how widespread habitable environments may be in the universe and for focusing a search for life beyond Earth. Here, we discuss habitability from the perspective of the life we know. We will explore the basic requirements of life and, in the following section, consider the full range of environmental conditions on Earth where life is found. While we can't entirely rule out the possibility that other life-forms might have biochemistry based on alternatives to carbon and liquid water, such life "as we don't know it" is still completely speculative. In our discussion here, we are focusing on habitability for life that is chemically similar to that on Earth.

Life requires a solvent (a liquid in which chemicals can dissolve) that enables the construction of biomolecules and the interactions between them. For life as we know it, that solvent is water, which has a variety of properties that are critical to how our biochemistry works. Water is abundant in the universe, but life requires that water be in liquid form (rather than ice or gas) in order to properly fill its role in biochemistry. That is the case only within a certain range of temperatures and pressures—too high or too low in either variable, and water takes the form of a solid or a gas. Identifying environments where water is present within the appropriate range of temperature and pressure is thus an important first step in identifying habitable environments. Indeed, a "follow the water" strategy has been, and continues to be, a key driver in the exploration of planets both within and beyond our solar system.

Our biochemistry is based on molecules made of carbon, hydrogen, nitrogen, oxygen, phosphorus, and sulfur. Carbon is at the core of organic chemistry. Its ability to form four bonds, both with itself and with the other elements of life, allows for the formation of a vast number of potential molecules on which to base biochemistry. The remaining elements contribute structure and chemical reactivity to our biomolecules, and form the basis of many of the interactions among them. These "biogenic elements," sometimes referred to with the acronym CHNOPS (carbon, hydrogen, nitrogen, oxygen, phosphorus, and sulfur), are the raw materials from which life is assembled, and an accessible supply of them is a second requirement of habitability.

As we learned in previous chapters on nuclear fusion and the life story of the stars, carbon, nitrogen, oxygen, phosphorus, and sulfur are all formed by fusion within stars and then distributed out into their galaxy as those stars die. But how they are distributed among the planets that form within a new star system, in what form, and how chemical, physical, and geological processes on those planets cycle the elements into structures that are accessible to biology, can have significant impacts on the distribution of life. In Earth's oceans, for example, the abundance of phytoplankton (simple organisms that are the base of the ocean food chain) in surface waters can vary by a thousand-fold because the supply of nitrogen differs from place to place (Figure 30.6). Understanding what processes control the accessibility of elements at all scales is thus a critical part of identifying habitable environments.

Download for free at http://cnx.org/content/col11992/latest/

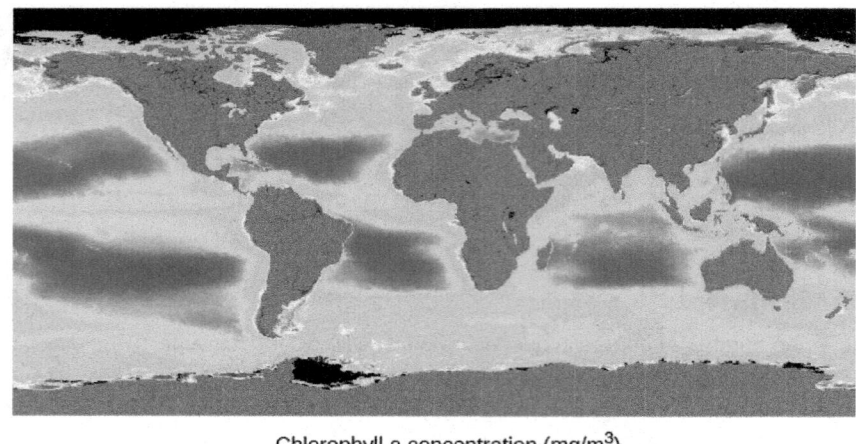

Chlorophyll a concentration (mg/m^3)

0.01   0.03   0.1   0.3   1   3   10

Figure 30.6. Chlorophyll Abundance. The abundance of chlorophyll (an indicator of photosynthetic bacteria and algae) varies by almost a thousand-fold across the ocean basins. That variation is almost entirely due to the availability of nitrogen—one of the major "biogenic elements" in forms that can be used by life. (credit: modification of work by NASA, Gene C. Feldman)

With these first two requirements, we have the elemental raw materials of life and a solvent in which to assemble them into the complicated molecules that drive our biochemistry. But carrying out that assembly and maintaining the complicated biochemical machinery of life takes energy. You fulfill your own requirement for energy every time you eat food or take a breath, and you would not live for long if you failed to do either on a regular basis. Life on Earth makes use of two main types of energy: for you, these are the oxygen in the air you breathe and the organic molecules in your food. But life overall can use a much wider array of chemicals and, while all animals require oxygen, many bacteria do not. One of the earliest known life processes, which still operates in some modern microorganisms, combines hydrogen and carbon dioxide to make methane, releasing energy in the process. There are microorganisms that "breathe" metals that would be toxic to us, and even some that breathe in sulfur and breathe out sulfuric acid. Plants and photosynthetic microorganisms have also evolved mechanisms to use the energy in light directly.

Water in the liquid phase, the biogenic elements, and energy are the fundamental requirements for habitability. But are there additional environmental constraints? We consider this in the next section.

This OpenStax book is available for free at http://cnx.org/content/col11992/1.8

Download for free at http://cnx.org/content/col11992/latest/

Figure 30.7. Grand Prismatic Spring in Yellowstone National Park. This hot spring, where water emerges from the bluish center at temperatures near the local boiling point (about 92 °C), supports a thriving array of microbial life. The green, yellow, and orange colors around the edges come from thick "mats" of photosynthetic bacteria. In fact, their coloration in part demonstrates their use of light energy—some wavelengths of incoming sunlight are selectively captured for energy; the rest are reflected back. Since it lacks the captured wavelengths, this light is now different in color than the sunlight that illuminates it. The blue part of the spring has temperatures too high to allow photosynthetic life (hence the lack of color except that supplied by water itself), but life is still present. Here, at nearly boiling temperatures, bacteria use the chemical energy supplied by the combination of hydrogen and other chemicals with oxygen. (credit: modification of work by Domenico Salvagnin)

Life in Extreme Conditions

At a chemical level, life consists of many types of molecules that interact with one another to carry out the processes of life. In addition to water, elemental raw materials, and energy, life also needs an environment in which those complicated molecules are stable (don't break down before they can do their jobs) and their interactions are possible. Your own biochemistry works properly only within a very narrow range of about 10 °C in body temperature and two-tenths of a unit in blood pH (pH is a numerical measure of acidity, or the amount of free hydrogen ions). Beyond those limits, you are in serious danger.

Life overall must also have limits to the conditions in which it can properly work but, as we will see, they are much broader than human limits. The resources that fuel life are distributed across a very wide range of conditions. For example, there is abundant chemical energy to be had in hot springs that are essentially boiling acid (see Figure 30.7). This provides ample incentive for evolution to fill as much of that range with life as is biochemically possible. An organism (usually a microbe) that tolerates or even thrives under conditions that most of the life around us would consider hostile, such as very high or low temperature or acidity, is known as an **extremophile** (where the suffix -*phile* means "lover of"). Let's have a look at some of the conditions that can challenge life and the organisms that have managed to carve out a niche at the far reaches of possibility.

Both high and low temperatures can cause a problem for life. As a large organism, you are able to maintain an almost constant body temperature whether it is colder or warmer in the environment around you. But this is not possible at the tiny size of microorganisms; whatever the temperature in the outside world is also the temperature of the microbe, and its biochemistry must be able to function at that temperature. High temperatures are the enemy of complexity—increasing thermal energy tends to break apart big molecules into smaller and smaller bits, and life needs to stabilize the molecules with stronger bonds and special proteins. But this approach has its limits.

Nevertheless, as noted earlier, high-temperature environments like hot springs and hydrothermal vents often offer abundant sources of chemical energy and therefore drive the evolution of organisms that can tolerate high temperatures (see Figure 30.8); such an organism is called a **thermophile**. Currently, the high temperature record holder is a methane-producing microorganism that can grow at 122 °C, where the pressure

Download for free at http://cnx.org/content/col11992/latest/

also is so high that water still does not boil. That's amazing when you think about it. We cook our food—meaning, we alter the chemistry and structure of its biomolecules—by boiling it at a temperature of 100 °C. In fact, food begins to cook at much lower temperatures than this. And yet, there are organisms whose biochemistry remains intact and operates just fine at temperatures 20 degrees higher.

Figure 30.8. Hydrothermal Vent on the Sea Floor. What appears to be black smoke is actually superheated water filled with minerals of metal sulfide. Hydrothermal vent fluid can represent a rich source of chemical energy, and therefore a driver for the evolution of microorganisms that can tolerate high temperatures. Bacteria feeding on this chemical energy form the base of a food chain that can support thriving communities of animals—in this case, a dense patch of red and white tubeworms growing around the base of the vent. (credit: modification of work by the University of Washington; NOAA/OAR/OER)

Cold can also be a problem, in part because it slows down metabolism to very low levels, but also because it can cause physical changes in biomolecules. Cell membranes—the molecular envelopes that surround cells and allow their exchange of chemicals with the world outside—are basically made of fatlike molecules. And just as fat congeals when it cools, membranes crystallize, changing how they function in the exchange of materials in and out of the cell. Some cold-adapted cells (called *psychrophiles*) have changed the chemical composition of their membranes in order to cope with this problem; but again, there are limits. Thus far, the coldest temperature at which any microbe has been shown to reproduce is about –25 °C.

Conditions that are very acidic or alkaline can also be problematic for life because many of our important molecules, like proteins and DNA, are broken down under such conditions. For example, household drain cleaner, which does its job by breaking down the chemical structure of things like hair clogs, is a very alkaline solution. The most acid-tolerant organisms (*acidophiles*) are capable of living at pH values near zero—about ten million times more acidic than your blood (Figure 30.9). At the other extreme, some *alkaliphiles* can grow at pH levels of about 13, which is comparable to the pH of household bleach and almost a million times more alkaline than your blood.

This OpenStax book is available for free at http://cnx.org/content/col11992/1.8

Download for free at http://cnx.org/content/col11992/latest/

Figure 30.9. Spain's Rio Tinto. With a pH close to 2, Rio Tinto is literally a river of acid. Acid-loving microorganisms (acidophiles) not only thrive in these waters, their metabolic activities help generate the acid in the first place. The rusty red color that gives the river its name comes from high levels of iron dissolved in the waters.

High levels of salts in the environment can also cause a problem for life because the salt blocks some cellular functions. Humans recognized this centuries ago and began to salt-cure food to keep it from spoiling—meaning, to keep it from being colonized by microorganisms. Yet some microbes have evolved to grow in water that is saturated in sodium chloride (table salt)—about ten times as salty as seawater (Figure 30.10).

Figure 30.10. Salt Ponds. The waters of an evaporative salt works near San Francisco are colored pink by thriving communities of photosynthetic organisms. These waters are about ten times as salty as seawater—enough for sodium chloride to begin to crystallize out—yet some organisms can survive and thrive in these conditions. (credit: modification of work by NASA)

Very high pressures can literally squeeze life's biomolecules, causing them to adopt more compact forms that do not work very well. But we still find life—not just microbial, but even animal life—at the bottoms of our ocean trenches, where pressures are more than 1000 times atmospheric pressure. Many other adaptions

Download for free at http://cnx.org/content/col11992/latest/

to environmental "extremes" are also known. There is even an organism, *Deinococcus radiodurans,* that can tolerate ionizing radiation (such as that released by radioactive elements) a thousand times more intense than you would be able to withstand. It is also very good at surviving extreme desiccation (drying out) and a variety of metals that would be toxic to humans.

From many such examples, we can conclude that life is capable of tolerating a wide range of environmental extremes—so much so that we have to work hard to identify places where life can't exist. A few such places are known—for example, the waters of hydrothermal vents at over 300 °C appear too hot to support any life—and finding these places helps define the possibility for life elsewhere. The study of extremophiles over the last few decades has expanded our sense of the range of conditions life can survive and, in doing so, has made many scientists more optimistic about the possibility that life might exist beyond Earth.

30.3 SEARCHING FOR LIFE BEYOND EARTH

Learning Objectives

By the end of this section, you will be able to:

> Outline what we have learned from exploration of the environment on Mars
> Identify where in the solar system life is most likely sustainable and why
> Describe some key missions and their findings in our search for life beyond our solar system
> Explain the use of biomarkers in the search for evidence of life beyond our solar system

Astronomers and planetary scientists continue to search for life in the solar system and the universe at large. In this section, we discuss two kinds of searches. First is the direct exploration of planets within our own solar system, especially Mars and some of the icy moons of the outer solar system. Second is the even more difficult task of searching for evidence of life—a **biomarker**—on planets circling other stars. In the next section, we will examine SETI, the *search for extraterrestrial intelligence.* As you will see, the approaches taken in these three cases are very different, even though the goal of each is the same: to determine if life on Earth is unique in the universe.

Life on Mars

The possibility that Mars hosts, or has hosted, life has a rich history dating back to the "canals" that some people claimed to see on the martian surface toward the end of the nineteenth century and the beginning of the twentieth. With the dawn of the space age came the possibility to address this question up close through a progression of missions to Mars that began with the first successful flyby of a robotic spacecraft in 1964 and have led to the deployment of NASA's *Curiosity* rover, which landed on Mars' surface in 2012.

The earliest missions to Mars provided some hints that liquid water—one of life's primary requirements—may once have flowed on the surface, and later missions have strengthened this conclusion. The NASA Viking landers, whose purpose was to search directly for evidence of life on Mars, arrived on Mars in 1976. Viking's onboard instruments found no organic molecules (the stuff of which life is made), and no evidence of biological activity in the martian soils it analyzed.

This result is not particularly surprising because, despite the evidence of flowing liquid water in the past, liquid water on the surface of Mars is generally not stable today. Over much of Mars, temperatures and pressures at the surface are so low that pure water would either freeze or boil away (under very low pressures, water will boil at a much lower temperature than usual). To make matters worse, unlike Earth, Mars does not have a magnetic field and ozone layer to protect the surface from harmful solar ultraviolet radiation and energetic particles.

This OpenStax book is available for free at http://cnx.org/content/col11992/1.8

However, Viking's analyses of the soil said nothing about whether life may have existed in Mars' distant past, when liquid water was more abundant. We do know that water in the form of ice exists in abundance on Mars, not so deep beneath its surface. Water vapor is also a constituent of the atmosphere of Mars.

Since the visit of Viking, our understanding of Mars has deepened spectacularly. Orbiting spacecraft have provided ever-more detailed images of the surface and detected the presence of minerals that could have formed only in the presence of liquid water. Two bold surface missions, the Mars Exploration Rovers *Spirit* and *Opportunity* (2004), followed by the much larger *Curiosity* Rover (2012), confirmed these remote-sensing data. All three rovers found abundant evidence for a past history of liquid water, revealed not only from the mineralogy of rocks they analyzed, but also from the unique layering of rock formations.

Curiosity has gone a step beyond evidence for water and confirmed the existence of habitable environments on ancient Mars. "Habitable" means not only that liquid water was present, but that life's requirements for energy and elemental raw materials could also have been met. The strongest evidence of an ancient habitable environment came from analyzing a very fine-grained rock called a mudstone—a rock type that is widespread on Earth but was unknown on Mars until *Curiosity* found it (see Figure 30.11). The mudstone can tell us a great deal about the wet environments in which they formed.

Figure 30.11. Mudstone. Shown are the first holes drilled by NASA's *Curiosity* Mars rover into a mudstone, with "fresh" drill-pilings around the holes. Notice the difference in color between the red ancient martian surface and the gray newly exposed rock powder that came from the drill holes. Each drill hole is about 0.6 inch (1.6 cm) in diameter. (credit: modification of work by NASA/JPL-Caltech/MSSS)

Five decades of robotic exploration have allowed us to develop a picture of how Mars evolved through time. Early Mars had epochs of warmer and wetter conditions that would have been conducive to life at the surface. However, Mars eventually lost much of its early atmosphere and the surface water began to dry up. As that happened, the ever-shrinking reservoirs of liquid water on the martian surface became saltier and more acidic, until the surface finally had no significant liquid water and was bathed in harsh solar radiation. The surface thus became uninhabitable, but this might not be the case for the planet overall.

Reservoirs of ice and liquid water could still exist underground, where pressure and temperature conditions make it stable. There is recent evidence to suggest that liquid water (probably very salty water) can occasionally (and briefly) flow on the surface even today. Thus, Mars might even have habitable conditions in the present day, but of a much different sort than we normally think of on Earth.

Our study of Mars reveals a planet with a fascinating history—one that saw its ability to host surface life dwindle billions of years ago, but perhaps allowing life to adapt and survive in favorable environmental niches. Even if life did not survive, we expect that we might find evidence of life if it ever took hold on Mars. If it is there, it is hidden in the crust, and we are still learning how best to decipher that evidence.

Download for free at http://cnx.org/content/col11992/latest/

Life in the Outer Solar System

The massive gas and ice giant planets of the outer solar system—Jupiter, Saturn, Uranus, and Neptune—are almost certainly not habitable for life as we know it, but some of their moons might be (see Figure 30.12). Although these worlds in the outer solar system contain abundant water, they receive so little warming sunlight in their distant orbits that it was long believed they would be "geologically dead" balls of hard-frozen ice and rock. But, as we saw in the chapter on Rings, Moons, and Pluto, missions to the outer solar system have found something much more interesting.

Jupiter's moon Europa revealed itself to the Voyager and Galileo missions as an active world whose icy surface apparently conceals an ocean with a depth of tens to perhaps a hundred kilometers. As the moon orbits Jupiter, the planet's massive gravity creates tides on Europa—just as our own Moon's gravity creates our ocean tides—and the friction of all that pushing and pulling generates enough heat to keep the water in liquid form (Figure 30.13). Similar tides act upon other moons if they orbit close to the planet. Scientists now think that six or more of the outer solar system's icy moons may harbor liquid water oceans for the same reason. Among these, Europa and Enceladus, a moon of Saturn, have thus far been of greatest interest to astrobiologists.

Figure 30.12. Jupiter's Moons. The Galilean moons of Jupiter are shown to relative scale and arranged in order of their orbital distance from Jupiter. At far left, Io orbits closest to Jupiter and so experiences the strongest tidal heating by Jupiter's massive gravity. This effect is so strong that Io is thought to be the most volcanically active body in our solar system. At far right, Callisto shows a surface scarred by billions of years' worth of craters—an indication that the moon's surface is old and that Callisto may be far less active than its sibling moons. Between these hot and cold extremes, Europa, second from left, orbits at a distance where Jupiter's tidal heating may be "just right" to sustain a liquid water ocean beneath its icy crust. (credit: modification of work by NASA/JPL/DLR)

Europa has probably had an ocean for most or all of its history, but habitability requires more than just liquid water. Life also requires energy, and because sunlight does not penetrate below the kilometers-thick ice crust of Europa, this would have to be chemical energy. One of Europa's key attributes from an astrobiology perspective is that its ocean is most likely in direct contact with an underlying rocky mantle, and the interaction of water and rocks—especially at high temperatures, as within Earth's hydrothermal vent systems—yields a *reducing chemistry* (where molecules tend to give up electrons readily) that is like one half of a chemical battery. To complete the battery and provide energy that could be used by life requires that an *oxidizing chemistry* (where molecules tend to accept electrons readily) also be available. On Earth, when chemically reducing vent fluids meet oxygen-containing seawater, the energy that becomes available often supports thriving communities of microorganisms and animals on the sea floor, far from the light of the Sun.

The Galileo mission found that Europa's icy surface does contain an abundance of oxidizing chemicals. This means that availability of energy to support life depends very much on whether the chemistry of the surface and the ocean can mix, despite the kilometers of ice in between. That Europa's ice crust appears geologically "young" (only tens of millions of years old, on average) and that it is active makes it tantalizing to think that such mixing might indeed occur. Understanding whether and how much exchange occurs between the surface

This OpenStax book is available for free at http://cnx.org/content/col11992/1.8

and ocean of Europa will be a key science objective of future missions to Europa, and a major step forward in understanding whether this moon could be a cradle of life.

Figure 30.13. Jupiter's Moon Europa, as Imaged by NASA's Galileo Mission. The relative scarcity of craters on Europa suggests a surface that is "geologically young," and the network of colored ridges and cracks suggests constant activity and motion. Galileo's instruments also strongly suggested the presence of a massive ocean of salty liquid water beneath the icy crust. (credit: modification of work by NASA/JPL-Caltech/SETI Institute)

In 2005, the Cassini mission performed a close flyby of a small (500-kilometer diameter) moon of Saturn, Enceladus (Figure 30.14), and made a remarkable discovery. Plumes of gas and icy material were venting from the moon's south polar region at a collective rate of about 250 kilograms of material per second. Several observations, including the discovery of salts associated with the icy material, suggest that their source is a liquid water ocean beneath tens of kilometers of ice. Although it remains to be shown definitively whether the ocean is local or global, transient or long-lived, it does appear to be in contact, and to have reacted, with a rocky interior. As on Europa, this is probably a necessary—though not sufficient—condition for habitability. What makes Enceladus so enticing to planetary scientists, though, are those plumes of material that seem to come directly from its ocean: samples of the interior are there for the taking by any spacecraft sent flying through. For a future mission, such samples could yield evidence not only of whether Enceladus is habitable but, indeed, of whether it is home to life.

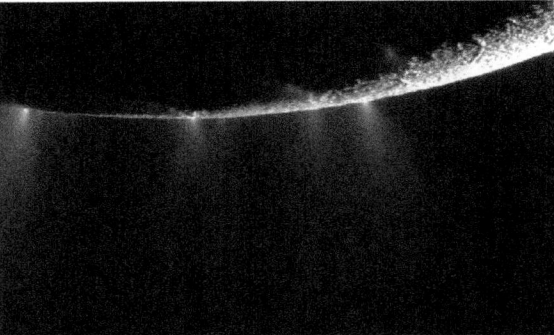

Figure 30.14. Image of Saturn's Moon Enceladus from NASA's Cassini Mission. The south polar region was found to have multiple plumes of ice and gas that, combined, are venting about 250 kilograms of material per second into space. Such features suggest that Enceladus, like Europa, has a sub-ice ocean. (credit: NASA/JPL/ SSI)

Saturn's big moon Titan is very different from both Enceladus and Europa (see Figure 30.15). Although it may host a liquid water layer deep within its interior, it is the surface of Titan and its unusual chemistry that

Download for free at http://cnx.org/content/col11992/latest/

makes this moon such an interesting place. Titan's thick atmosphere—the only one among moons in the solar system—is composed mostly of nitrogen but also of about 5% methane. In the upper atmosphere, the Sun's ultraviolet light breaks apart and recombines these molecules into more complex organic compounds that are collectively known as *tholins*. The tholins shroud Titan in an orange haze, and imagery from Cassini and from the Huygens probe that descended to Titan's surface show that heavier particles appear to accumulate on the surface, even forming "dunes" that are cut and sculpted by flows of liquid hydrocarbons (such as liquid methane). Some scientists see this organic chemical factory as a natural laboratory that may yield some clues about the solar system's early chemistry—perhaps even chemistry that could support the origin of life.

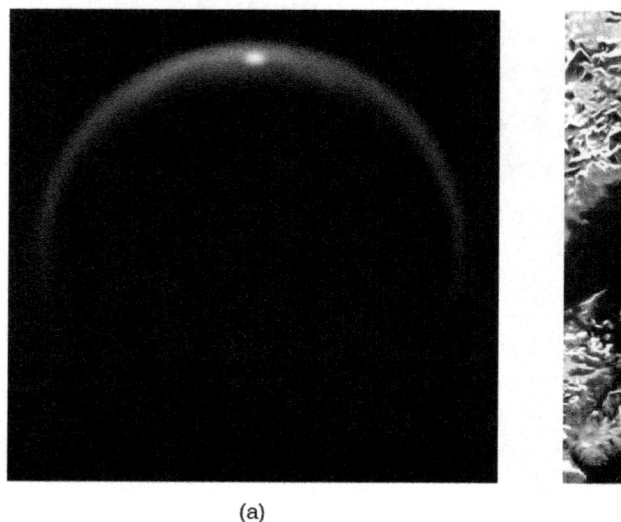
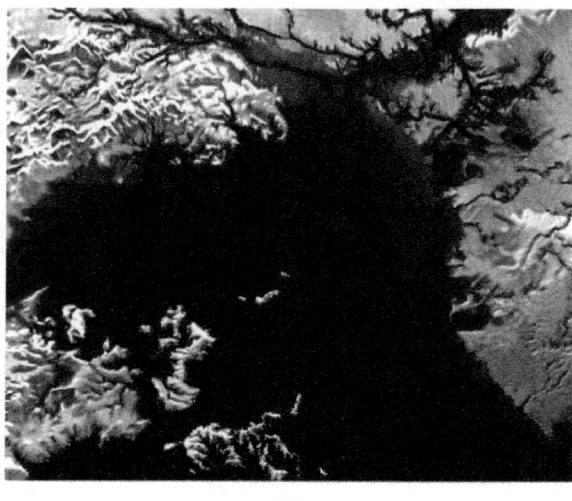

(a)   (b)

Figure 30.15. Image of Saturn's Moon Titan from NASA's Cassini Mission. (a) The hazy orange glow comes from Titan's thick atmosphere (the only one known among the moons of the solar system). That atmosphere is mostly nitrogen but also contains methane and potentially a variety of complex organic compounds. The bright spot near the top of the image is sunlight reflected from a very flat surface—almost certainly a liquid. We see this effect, called "glint," when sunlight reflects off the surface of a lake or ocean. (b) Cassini radar imagery shows what look very much like landforms and lakes on the surface of Titan. But the surface lakes and oceans of Titan are not water; they are probably made of liquid hydrocarbons like methane and ethane. (credit a: modification of work by NASA/JPL/University of Arizona/DLR; credit b: modification of work by NASA/JPL-Caltech/ASI)

LINK TO LEARNING

In January 2005, the Huygens probe descended to the surface of Titan and relayed data, including imagery of the landing site, for about 90 minutes. You can watch a video (https://openstax.org/l/30huytatsurf) about the descent of Huygens to Titan's surface.

Habitable Planets Orbiting Other Stars

One of the most exciting developments in astronomy during the last two decades is the ability to detect exoplanets—planets orbiting other stars. As we saw in the chapter on the formation of stars and planets, since the discovery of the first exoplanet in 1995, there have been thousands of confirmed detections, and many more candidates that are not yet confirmed. These include several dozen possibly habitable exoplanets. Such numbers finally allow us to make some predictions about exoplanets and their life-hosting potential. The

This OpenStax book is available for free at http://cnx.org/content/col11992/1.8

Download for free at http://cnx.org/content/col11992/latest/

majority of stars with mass similar to the Sun appear to host at least one planet, with multi-planet systems like our own not unusual. How many of these planets might be habitable, and how could we search for life there?

LINK TO LEARNING

The NASA Exoplanet Archive (https://openstax.org/l/30NASAexoarc) is an up-to-date searchable online source of data and tools on everything to do with exoplanets. Explore stellar and exoplanet parameters and characteristics, find the latest news on exoplanet discoveries, plot your own data interactively, and link to other related resources.

In evaluating the prospect for life in distant planetary systems, astrobiologists have developed the idea of a **habitable zone**—a region around a star where suitable conditions might exist for life. This concept focuses on life's requirement for liquid water, and the habitable zone is generally thought of as the range of distances from the central star in which water could be present in liquid form at a planet's surface. In our own solar system, for example, Venus has surface temperatures far above the boiling point of water and Mars has surface temperatures that are almost always below the freezing point of water. Earth, which orbits between the two, has a surface temperature that is "just right" to keep much of our surface water in liquid form.

Whether surface temperatures are suitable for maintaining liquid water depends on a planet's "radiation budget" —how much starlight energy it absorbs and retains—and whether or how processes like winds and ocean circulation distribute that energy around the planet. How much stellar energy a planet receives, in turn, depends on how much and what sort of light the star emits and how far the planet is from that star,[1] how much it reflects back to space, and how effectively the planet's atmosphere can retain heat through the greenhouse effect (see Earth as a Planet). All of these can vary substantially, and all matter a lot. For example, Venus receives about twice as much starlight per square meter as Earth but, because of its dense cloud cover, also reflects about twice as much of that light back to space as Earth does. Mars receives only about half as much starlight as Earth, but also reflects only about half as much. Thus, despite their differing orbital distances, the three planets actually absorb comparable amounts of sunlight energy. Why, then, are they so dramatically different?

As we learned in several chapters about the planets, some of the gases that make up planetary atmospheres are very effective at trapping infrared light—the very range of wavelengths at which planets radiate thermal energy back out to space—and this can raise the planet's surface temperature quite a bit more than would otherwise be the case. This is the same "greenhouse effect" that is of such concern for global warming on our planet. Earth's natural greenhouse effect, which comes mostly from water vapor and carbon dioxide in the atmosphere, raises our average surface temperature by about 33 °C over the value it would have if there were no greenhouse gases in the atmosphere. Mars has a very thin atmosphere and thus very little greenhouse warming (about 2 °C worth), while Venus has a massive carbon dioxide atmosphere that creates very strong greenhouse warming (about 510 °C worth). These worlds are much colder and much hotter, respectively, than Earth would be if moved into their orbits. Thus, we must consider the nature of any atmosphere as well as the distance from the star in evaluating the range of habitability.

---

1   The amount of starlight received per unit area of a planet's surface (per square meter, for example) decreases with the square of the distance from the star. Thus, when the orbital distance doubles, the illumination decreases by 4 times (2^2), and when the orbital distance increases tenfold, the illumination decreases by 100 times (10^2). Venus and Mars orbit the sun at about 72% and 152% of Earth's orbital distance, respectively, so Venus receives about $1/(0.72)^2 = 1.92$ (about twice) and Mars about $1/(1.52)^2 = 0.43$ (about half) as much light per square meter of planet surface as Earth does.

Download for free at http://cnx.org/content/col11992/latest/

Of course, as we have learned, stars also vary widely in the intensity and spectrum (the wavelengths of light) they emit. Some are much brighter and hotter (bluer), while others are significantly dimmer and cooler (redder), and the distance of the habitable zone varies accordingly. For example, the habitable zone around M-dwarf stars is 3 to 30 times closer in than for G-type (Sun-like) stars. There is a lot of interest in whether such systems could be habitable because—although they have some potential downsides for supporting life—M-dwarf stars are by far the most numerous and long-lived in our Galaxy.

The luminosity of stars like the Sun also increases over their main-sequence lifetime, and this means that the habitable zone migrates outward as a star system ages. Calculations indicate that the power output of the Sun, for example, has increased by at least 30% over the past 4 billion years. Thus, Venus was once within the habitable zone, while Earth received a level of solar energy insufficient to keep the modern Earth (with its present atmosphere) from freezing over. In spite of this, there is plenty of geological evidence that liquid water was present on Earth's surface billions of years ago. The phenomenon of increasing stellar output and an outwardly migrating habitable zone has led to another concept: the *continuously* habitable zone is defined by the range of orbits that would remain within the habitable zone during the entire lifetime of the star system. As you might imagine, the continuously habitable zone is quite a bit narrower than the habitable zone is at any one time in a star's history. The nearest star to the Sun, Proxima Centauri, is an M star that has a planet with a mass of at least 1.3 Earth masses, taking about 11 days to orbit. At the distance for such a quick orbit (0.05 AU), the planet may be in the habitable zone of its star, although whether conditions on such a planet near such a star are hospitable for life is a matter of great scientific debate.

Even when planets orbit within the habitable zone of their star, it is no guarantee that they are habitable. For example, Venus today has virtually no water, so even if it were suddenly moved to a "just right" orbit within the habitable zone, a critical requirement for life would still be lacking.

Scientists are working to understand all the factors that define the habitable zone and the habitability of planets orbiting within that zone because this will be our primary guide in targeting exoplanets on which to seek evidence of life. As technology for detecting exoplanets has advanced, so too has our potential to find Earth-size worlds within the habitable zones of their parent stars. Of the confirmed or candidate exoplanets known at the time of writing, nearly 300 are considered to be orbiting within the habitable zone and more than 10% of those are roughly Earth-size.

LINK TO LEARNING

Explore the habitable universe at the online Planetary Habitability Laboratory (https://openstax.org/l/30planhabitlab) created by the University of Puerto Rico at Arecibo. See the potentially habitable exoplanets and other interesting places in the universe, watch video clips, and link to numerous related resources on astrobiology.

Biomarkers

Our observations suggest increasingly that Earth-size planets orbiting within the habitable zone may be common in the Galaxy—current estimates suggest that more than 40% of stars have at least one. But are any of them inhabited? With no ability to send probes there to sample, we will have to derive the answer from the light and other radiation that come to us from these faraway systems (Figure 30.16). What types of observations might constitute good evidence for life?

This OpenStax book is available for free at http://cnx.org/content/col11992/1.8

Figure 30.16. Earth, as Seen by NASA's Voyager 1. In this image, taken from 4 billion miles away, Earth appears as a "pale blue dot" representing less than a pixel's worth of light. Would this light reveal Earth as a habitable and inhabited world? Our search for life on exoplanets will depend on an ability to extract information about life from the faint light of faraway worlds. (credit: modification of work by NASA/JPL-Caltech)

To be sure, we need to look for robust biospheres (atmospheres, surfaces, and/or oceans) capable of creating planet-scale change. Earth hosts such a biosphere: the composition of our atmosphere and the spectrum of light reflected from our planet differ considerably from what would be expected in the absence of life. Presently, Earth is the only body in our solar system for which this is true, despite the possibility that habitable conditions might prevail in the subsurface of Mars or inside the icy moons of the outer solar system. Even if life exists on these worlds, it is very unlikely that it could yield planet-scale changes that are both telescopically observable and clearly biological in origin.

What makes Earth "special" among the potentially habitable worlds in our solar system is that it has a photosynthetic biosphere. This requires the presence of liquid water at the planet's surface, where organisms have direct access to sunlight. The habitable zone concept focuses on this requirement for surface liquid water—even though we know that subsurface habitable conditions could prevail at more distant orbits—exactly because these worlds would have biospheres detectable at a distance.

Indeed, plants and photosynthetic microorganisms are so abundant at Earth's surface that they affect the color of the light that our planet reflects out into space—we appear greener in visible wavelengths and reflect more near-infrared light than we otherwise would. Moreover, photosynthesis has changed Earth's atmosphere at a large scale—more than 20% of our atmosphere comes from the photosynthetic waste product, oxygen. Such high levels would be very difficult to explain in the absence of life. Other gases, such as nitrous oxide and methane, when found simultaneously with oxygen, have also been suggested as possible indicators of life. When sufficiently abundant in an atmosphere, such gases could be detected by their effect on the spectrum of light that a planet emits or reflects. (As we saw in the chapter on exoplanets, astronomers today are beginning to have the capability of detecting the spectrum of the atmospheres of some planets orbiting other stars.)

Astronomers have thus concluded that, at least initially, a search for life outside our solar system should focus on exoplanets that are as much like Earth as possible—roughly Earth-size planets orbiting in the habitable zone—and look for the presence of gases in the atmosphere or colors in the visible spectrum that are hard to

Download for free at http://cnx.org/content/col11992/latest/

explain except by the presence of biology. Simple, right? In reality, the search for exoplanet life poses many challenges.

As you might imagine, this task is more challenging for planetary systems that are farther away and, in practical terms, this will limit our search to the habitable worlds closest to our own. Should we become limited to a very small number of nearby targets, it will also become important to consider the habitability of planets orbiting the M-dwarfs we discussed above.

If we manage to separate out a clean signal from the planet and find some features in the light spectrum that might be indicative of life, we will need to work hard to think of any nonbiological process that might account for them. "Life is the hypothesis of last resort," noted astronomer Carl Sagan—meaning that we must exhaust all other explanations for what we see before claiming to have found evidence of extraterrestrial biology. This requires some understanding of what processes might operate on worlds that we will know relatively little about; what we find on Earth can serve as a guide but also has potential to lead us astray (Figure 30.17).

Recall, for example, that it would be extremely difficult to account for the abundance of oxygen in Earth's atmosphere except by the presence of biology. But it has been hypothesized that oxygen could build up to substantial levels on planets orbiting M-dwarf stars through the action of ultraviolet radiation on the atmosphere—with no need for biology. It will be critical to understand where such "false positives" might exist in carrying out our search.

We need to understand that we might not be able to detect biospheres even if they exist. Life has flourished on Earth for perhaps 3.5 billion years, but the atmospheric "biosignatures" that, today, would supply good evidence for life to distant astronomers have not been present for all of that time. Oxygen, for example, accumulated to detectable levels in our atmosphere only a little over 2 billion years ago. Could life on Earth have been detected before that time? Scientists are working actively to understand what additional features might have provided evidence of life on Earth during that early history, and thereby help our chances of finding life beyond.

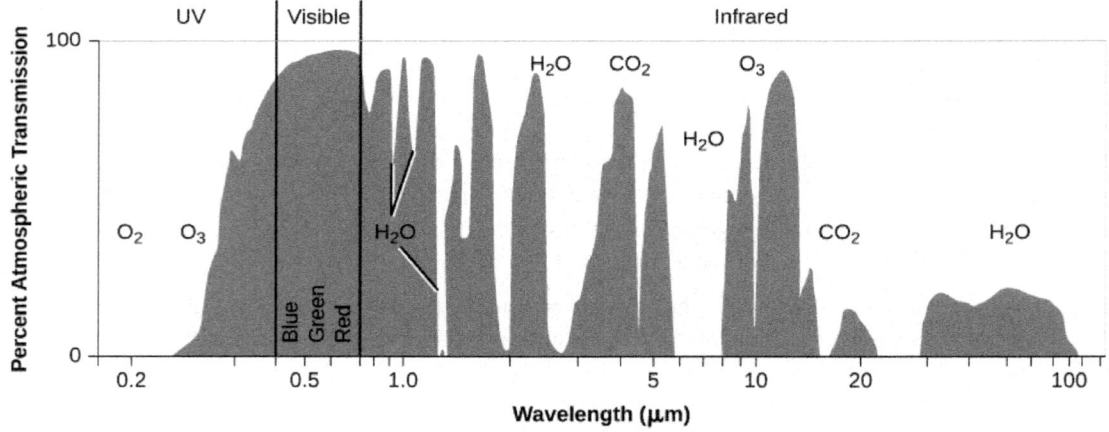

Figure 30.17. Spectrum of Light Transmitted through Earth's Atmosphere. This graph shows wavelengths ranging from ultraviolet (far left) to infrared. The many downward "spikes" come from absorption of particular wavelengths by molecules in Earth's atmosphere. Some of these compounds, like water and the combination oxygen/ozone and methane, might reveal Earth as both habitable and inhabited. We will have to rely on this sort of information to seek life on exoplanets, but our spectra will be of much poorer quality than this one, in part because we will receive so little light from the planet. (credit: modification of work by NASA)

This OpenStax book is available for free at http://cnx.org/content/col11992/1.8

Download for free at http://cnx.org/content/col11992/latest/

30.4 | THE SEARCH FOR EXTRATERRESTRIAL INTELLIGENCE

Learning Objectives

By the end of this section, you will be able to:

> Explain why spaceships from extraterrestrial civilizations are unlikely to have visited us
> List efforts by humankind to communicate with other civilizations via messages on spacecraft
> Understand the various SETI programs scientists are undertaking

Given all the developments discussed in this chapter, it seems likely that life could have developed on many planets around other stars. Even if that life is microbial, we saw that we may soon have ways to search for chemical biosignatures. This search is of fundamental importance for understanding biology, but it does not answer the question, "Are we alone?" that we raised at the beginning of this chapter. When we ask this question, many people think of other intelligent creatures, perhaps beings that have developed technology similar to our own. If any intelligent, technical civilizations have arisen, as has happened on Earth in the most recent blink of cosmic time, how could we make contact with them?

This problem is similar to making contact with people who live in a remote part of Earth. If students in the United States want to converse with students in Australia, for example, they have two choices. Either one group gets on an airplane and travels to meet the other, or they communicate by sending a message remotely. Given how expensive airline tickets are, most students would probably select the message route.

In the same way, if we want to get in touch with intelligent life around other stars, we can travel, or we can try to exchange messages. Because of the great distances involved, interstellar space travel would be very slow and prohibitively expensive. The fastest spacecraft the human species has built so far would take almost 80,000 years to get to the nearest star. While we could certainly design a faster craft, the more quickly we require it to travel, the greater the energy cost involved. To reach neighboring stars in less than a human life span, we would have to travel close to the speed of light. In that case, however, the expense would become truly astronomical.

Interstellar Travel

Bernard Oliver, an engineer with an abiding interest in life elsewhere, made a revealing calculation about the costs of rapid interstellar space travel. Since we do not know what sort of technology we (or other civilizations) might someday develop, Oliver considered a trip to the nearest star (and back again) in a spaceship with a "perfect engine"—one that would convert its fuel into energy with 100% efficiency. Even with a perfect engine, the energy cost of a single round-trip journey at 70% the speed of light turns out to be equivalent to several hundred thousand years' worth of total U.S. electrical energy consumption. The cost of such travel is literally out of this world.

This is one reason astronomers are so skeptical about claims that UFOs are spaceships from extraterrestrial civilizations. Given the distance and energy expense involved, it seems unlikely that the dozens of UFOs (and even UFO abductions) claimed each year could be visitors from other stars so fascinated by Earth civilization that they are willing to expend fantastically large amounts of energy or time to reach us. Nor does it seem credible that these visitors have made this long and expensive journey and then systematically avoided contacting our governments or political and intellectual leaders.

Not every UFO report has been explained (in many cases, the observations are sketchy or contradictory). But investigation almost always converts them to IFOs (identified flying objects) or NFOs (not-at-all flying objects). While some are hoaxes, others are natural phenomena, such as bright planets, ball lightning, fireballs (bright

Download for free at http://cnx.org/content/col11992/latest/

meteors), or even flocks of birds that landed in an oil slick to make their bellies reflective. Still others are human craft, such as private planes with some lights missing, or secret military aircraft. It is also interesting that the group of people who most avidly look at the night sky, the amateur astronomers, have never reported UFO sightings. Further, not a single UFO has ever left behind any physical evidence that can be tested in a laboratory and shown to be of nonterrestrial origin.

Another common aspect of belief that aliens are visiting Earth comes from people who have difficulty accepting human accomplishments. There are many books and TV shows, for example, that assert that humans could not have built the great pyramids of Egypt, and therefore they must have been built by aliens. The huge statues (called Moai) on Easter Island are also sometimes claimed to have been built by aliens. Some people even think that the accomplishments of space exploration today are based on alien technology.

However, the evidence from archaeology and history is clear: ancient monuments were built by ancient *people*, whose brains and ingenuity were every bit as capable as ours are today, even if they didn't have electronic textbooks like you do.

Messages on Spacecraft

While space travel by living creatures seems very difficult, robot probes can travel over long distances and over long periods of time. Five spacecraft—two Pioneers, two Voyagers, and New Horizons—are now leaving the solar system. At their coasting speeds, they will take hundreds of thousands or millions of years to get anywhere close to another star. On the other hand, they were the first products of human technology to go beyond our home system, so we wanted to put messages on board to show where they came from.

Each Pioneer carries a plaque with a pictorial message engraved on a gold-anodized aluminum plate (Figure 30.18). The Voyagers, launched in 1977, have audio and video records attached, which allowed the inclusion of over 100 photographs and a selection of music from around the world. Given the enormous space between stars in our section of the Galaxy, it is very unlikely that these messages will ever be received by anyone. They are more like a note in a bottle thrown into the sea by a shipwrecked sailor, with no realistic expectation of its being found soon but a slim hope that perhaps someday, somehow, someone will know of the sender's fate.

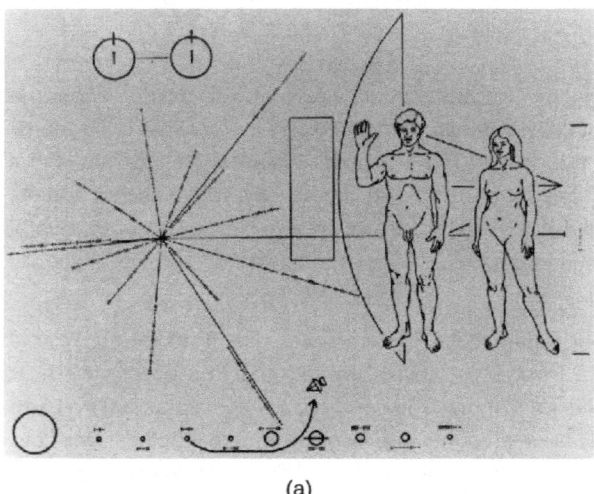
(a)

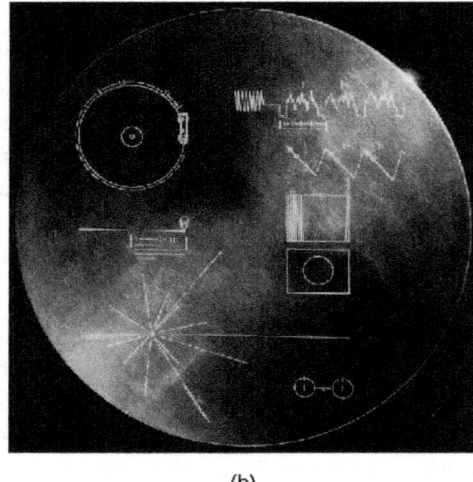
(b)

Figure 30.18. Interstellar Messages. (a) This is the image engraved on the plaques aboard the Pioneer 10 and 11 spacecraft. The human figures are drawn in proportion to the spacecraft, which is shown behind them. The Sun and planets in the solar system can be seen at the bottom, with the trajectory that the spacecraft followed. The lines and markings in the left center show the positions and pulse periods for a number of pulsars, which might help locate the spacecraft's origins in space and time. (b) Encoded onto a gold-coated copper disk, the Voyager record contains 118 photographs, 90 minutes of music from around the world, greetings in almost 60 languages, and other audio material. It is a summary of the sights and sounds of Earth. (credit a, b: modification of work by NASA)

This OpenStax book is available for free at http://cnx.org/content/col11992/1.8

The Voyager Message

An Excerpt from the Voyager Record:

"We cast this message into the cosmos. It is likely to survive a billion years into our future, when our civilization is profoundly altered. . . . If [another] civilization intercepts Voyager and can understand these recorded contents, here is our message:

This is a present from a small, distant world, a token of our sounds, our science, our images, our music, our thoughts, and our feelings. We are attempting to survive our time so we may live into yours. We hope, someday, having solved the problems we face, to join a community of galactic civilizations. This record represents our hope and our determination, and our goodwill in a vast and awesome universe."

—Jimmy Carter, President of the United States of America, June 16, 1977

Communicating with the Stars

If direct visits among stars are unlikely, we must turn to the alternative for making contact: exchanging messages. Here the news is a lot better. We already use a messenger—light or, more generally, electromagnetic waves—that moves through space at the fastest speed in the universe. Traveling at 300,000 kilometers per second, light reaches the nearest star in only 4 years and does so at a tiny fraction of the cost of sending material objects. These advantages are so clear and obvious that we assume they will occur to any other species of intelligent beings that develop technology.

However, we have access to a wide spectrum of electromagnetic radiation, ranging from the longest-wavelength radio waves to the shortest-wavelength gamma rays. Which would be the best for interstellar communication? It would not be smart to select a wavelength that is easily absorbed by interstellar gas and dust, or one that is unlikely to penetrate the atmosphere of a planet like ours. Nor would we want to pick a wavelength that has lots of competition for attention in our neighborhood.

One final criterion makes the selection easier: we want the radiation to be inexpensive enough to produce in large quantities. When we consider all these requirements, radio waves turn out to be the best answer. Being the lowest-frequency (and lowest-energy) band of the spectrum, they are not very expensive to produce, and we already use them extensively for communications on Earth. They are not significantly absorbed by interstellar dust and gas. With some exceptions, they easily pass through Earth's atmosphere and through the atmospheres of the other planets we are acquainted with.

The Cosmic Haystack

Having made the decision that radio is the most likely means of communication among intelligent civilizations, we still have many questions and a daunting task ahead of us. Shall we *send* a message, or try to *receive* one? Obviously, if every civilization decides to receive only, then no one will be sending, and everyone will be disappointed. On the other hand, it may be appropriate for us to *begin* by listening, since we are likely to be among the most primitive civilizations in the Galaxy who are interested in exchanging messages.

We do not make this statement to insult the human species (which, with certain exceptions, we are rather fond of). Instead, we base it on the fact that humans have had the ability to receive (or send) a radio message across interstellar distances for only a few decades. Compared to the ages of the stars and the Galaxy, this is a mere

Download for free at http://cnx.org/content/col11992/latest/

instant. If there are civilizations out there that are ahead of us in development by even a short time (in the cosmic sense), they are likely to have a technology head start of many, many years.

In other words, we, who have just started, may well be the "youngest" species in the Galaxy with this capability (see the discussion in Example 30.1). Just as the youngest members of a community are often told to be quiet and listen to their elders for a while before they say something foolish, so may we want to begin our exercise in extraterrestrial communication by listening.

Even restricting our activities to listening, however, leaves us with an array of challenging questions. For example, if an extraterrestrial civilization's signal is too weak to be detected by our present-day radio telescopes, we will not detect them. In addition, it would be very expensive for an extraterrestrial civilization to broadcast on a huge number of channels. Most likely, they select one or a few channels for their particular message. Communicating on a narrow band of channels also helps distinguish an artificial message from the radio static that comes from natural cosmic processes. But the radio band contains an astronomically large number of possible channels. How can we know in advance which one they have selected, and how they have coded their message into the signal?

Table 30.1 summarizes these and other factors that scientists must grapple with when trying to tune in to radio messages from distant civilizations. Because their success depends on either guessing right about so many factors or searching through all the possibilities for each factor, some scientists have compared their quest to looking for a needle in a haystack. Thus, they like to say that the list of factors in Table 30.1 defines the *cosmic haystack problem*.

Factors

From which direction (which star) is the message coming?
On what channels (or frequencies) is the message being broadcast?
How wide in frequency is the channel?
How strong is the signal (can our radio telescopes detect it)?
Is the signal continuous, or does it shut off at times (as, for example, a lighthouse beam does when it turns away from us)?
Does the signal drift (change) in frequency because of the changing relative motion of the source and the receiver?
How is the message encoded in the signal (how do we decipher it)?
Can we even recognize a message from a completely alien species? Might it take a form we don't at all expect?

Table 30.1 The Cosmic Haystack Problem: Some Questions about an Extraterrestrial Message

Radio Searches

Although the cosmic haystack problem seems daunting, many other research problems in astronomy also require a large investment of time, equipment, and patient effort. And, of course, if we don't search, we're sure not to find anything.

This OpenStax book is available for free at http://cnx.org/content/col11992/1.8

Download for free at http://cnx.org/content/col11992/latest/

The very first search was conducted by astronomer Frank Drake in 1960, using the 85-foot antenna at the National Radio Astronomy Observatory (Figure 30.19). Called Project Ozma, after the queen of the exotic Land of Oz in the children's stories of L. Frank Baum, his experiment involved looking at about 7200 channels and two nearby stars over a period of 200 hours. Although he found nothing, Drake demonstrated that we had the technology to do such a search, and set the stage for the more sophisticated projects that followed.

(a)   (b)

Figure 30.19. Project Ozma and the Allen Telescope Array. (a) This 25th anniversary photo shows some members of the Project Ozma team standing in front of the 85-foot radio telescope with which the 1960 search for extraterrestrial messages was performed. Frank Drake is in the back row, second from the right. (b) The Allen Telescope Array in California is made up of 42 small antennas linked together. This system allows simultaneous observations of multiple sources with millions of separate frequency channels. (credit a: modification of work by NRAO; credit b: modification of work by Colby Gutierrez-Kraybill)

Receivers are constantly improving, and the sensitivity of SETI programs— **SETI** stands for the search for extraterrestrial life—is advancing rapidly. Equally important, modern electronics and software allow simultaneous searches on millions of frequencies (channels). If we can thus cover a broad frequency range, the cosmic haystack problem of guessing the right frequency largely goes away. One powerful telescope array (funded with an initial contribution from Microsoft founder Paul Allen) that is built for SETI searches is the Allen Telescope in Northern California. Other radio telescopes being used for such searches include the giant Arecibo radio dish in Puerto Rico and the Green Bank Telescope in West Virginia, which is the largest steerable radio telescope in the world.

What kind of signals do we hope to pick up? We on Earth are inadvertently sending out a flood of radio signals, dominated by military radar systems. This is a kind of leakage signal, similar to the wasted light energy that is beamed upward by poorly designed streetlights and advertising signs. Could we detect a similar leakage of radio signals from another civilization? The answer is just barely, but only for the nearest stars. For the most part, therefore, current radio SETI searches are looking for beacons, assuming that civilizations might be intentionally drawing attention to themselves or perhaps sending a message to another world or outpost that lies in our direction. Our prospects for success depend on how often civilizations arise, how long they last, and how patient they are about broadcasting their locations to the cosmos.

Download for free at http://cnx.org/content/col11992/latest/

VOYAGERS IN ASTRONOMY

Jill Tarter: Trying to Make Contact

1997 was quite a year for Jill Cornell Tarter (Figure 30.20), one of the world's leading scientists in the SETI field. The SETI Institute announced that she would be the recipient of its first endowed chair (the equivalent of an endowed research professorship) named in honor of Bernard Oliver. The National Science Foundation approved a proposal by a group of scientists and educators she headed to develop an innovative hands-on high school curriculum based on the ideas of cosmic evolution (the topics of this chapter). And, at roughly the same time, she was being besieged with requests for media interviews as news reports identified her as the model for Ellie Arroway, the protagonist of *Contact*, Carl Sagan's best-selling novel about SETI. The book had been made into a high-budget science fiction film, starring Jodie Foster, who had talked with Tarter before taking the role.

Figure 30.20. Jill Tarter (credit: Christian Schidlowski)

Tarter is quick to point out, "Carl Sagan wrote a book about a woman who does what I do, not about me." Still, as the only woman in such a senior position in the small field of SETI, she was the center of a great deal of public attention. (However, colleagues and reporters pointed out that this was nothing compared to what would happen if her search for radio signals from other civilizations recorded a success.)

Being the only woman in a group is not a new situation to Tarter, who often found herself the only woman in her advanced science or math classes. Her father had encouraged her, both in her interest in science and her "tinkering." As an undergraduate at Cornell University, she majored in engineering physics. That training became key to putting together and maintaining the complex systems that automatically scan for signals from other civilizations.

Switching to astrophysics for her graduate studies, she wrote a PhD thesis that, among other topics, considered the formation of failed stars—those whose mass was not sufficient to ignite the nuclear reactions that power more massive stars like our own Sun. Tarter coined the term "brown dwarf" for these small, dim objects, and it has remained the name astronomers use ever since.

It was while she was still in graduate school that Stuart Bowyer, one of her professors at the University of California, Berkeley, asked her if she wanted to be involved in a small experiment to siphon off a bit of radiation from a radio telescope as astronomers used it year in and year out and see if there was any hint

This OpenStax book is available for free at http://cnx.org/content/col11992/1.8

Download for free at http://cnx.org/content/col11992/latest/

of an intelligently coded radio message buried in the radio noise. Her engineering and computer programming skills became essential to the project, and soon she was hooked on the search for life elsewhere.

Thus began an illustrious career working full time searching for extraterrestrial civilizations, leading Jill Tarter to receive many awards, including being elected fellow of the American Association for the Advancement of Science in 2002, the Adler Planetarium Women in Space Science Award in 2003, and a 2009 TED Prize, among others.

LINK TO LEARNING

Watch the TED talk (https://openstax.org/l/30TarterSETI) Jill Tarter gave on the fascination of the search for intelligence.

EXAMPLE 30.1

The Drake Equation

At the first scientific meeting devoted to SETI, Frank Drake wrote an equation on the blackboard that took the difficult question of estimating the number of civilizations in the Galaxy and broke it down into a series of smaller, more manageable questions. Ever since then, both astronomers and students have used this **Drake equation** as a means of approaching the most challenging question: How likely is it that we are alone? Since this is at present an unanswerable question, astronomer Jill Tarter has called the Drake equation a "way of organizing our ignorance." (See Figure 30.21.)

Figure 30.21. Drake Equation. A plaque at the National Radio Astronomy Observatory commemorates the conference where the equation was first discussed. (credit: NRAO/NSF/AUI)

The form of the Drake equation is very simple. To estimate the number of communicating civilizations that currently exist in the Galaxy (we will define these terms more carefully in a moment), we multiply the rate of formation of such civilizations (number per year) by their average lifetime (in years). In symbols,

Download for free at http://cnx.org/content/col11992/latest/

$$N = R_{\text{total}} \times L$$

To make this formula easier to use (and more interesting), however, Drake separated the rate of formation R_{total} into a series of probabilities:

$$R_{\text{total}} = R_{\text{star}} \times f_p \times f_e \times f_l \times f_i \times f_c$$

R_{star} is the rate of formation of stars like the Sun in our Galaxy, which is about 10 stars per year. Each of the other terms is a fraction or probability (less than or equal to 1.0), and the product of all these probabilities is itself the total probability that each star will have an intelligent, technological, communicating civilization that we might want to talk to. We have:

- f_p = the fraction of these stars with planets
- f_e = the fraction of the planetary systems that include habitable planets
- f_l = the fraction of habitable planets that actually support life
- f_i = the fraction of inhabited planets that develop advanced intelligence
- f_c = the fraction of these intelligent civilizations that develop science and the technology to build radio telescopes and transmitters

Each of these factors can be discussed and perhaps evaluated, but we must guess at many of the values. In particular, we don't know how to calculate the probability of something that happened once on Earth but has not been observed elsewhere—and these include the development of life, of intelligent life, and of technological life (the last three factors in the equation). One important advance in estimating the terms of the Drake equation comes from the recent discovery of exoplanets. When the Drake equation was first written, no one had any idea whether planets and planetary systems were common. Now we know they are—another example of the Copernican principle.

Solution

Even if we don't know the answers, we can make some guesses and calculate the resulting number N. Let's start with the optimism implicit in the Copernican principle and set the last three terms equal to 1.0. If R is 10 stars/year and if we measure the average lifetime of a technological civilization in years, the units of years cancel. If we also assume that f_p is 0.1, and f_e is 1.0, the equation becomes

$$N = R_{\text{total}} \times L = L$$

Now we see the importance of the term L, the lifetime of a communicating civilization (measured in years). We have had this capability (to communicate at the distances of the stars) for only a few decades.

Check Your Learning

Suppose we assume that this stage in our history lasts only one century.

Answer:

With our optimistic assumptions about the other factors, L = 100 years and N = 100 such civilizations in the entire Galaxy. In that case, there are so few other civilizations like ours that we are unlikely to detect any signals in a SETI search. But suppose the average lifetime is a million years; in that case, there are a million such civilizations in the Galaxy, and some of them may be within range for radio communication. The most important conclusion from this calculation is that even if we are extremely optimistic about the

This OpenStax book is available for free at http://cnx.org/content/col11992/1.8

Download for free at http://cnx.org/content/col11992/latest/

probabilities, the only way we can expect success from SETI is if other civilizations are much older (and hence probably much more advanced) than ours.

LINK TO LEARNING

Read Frank Drake's own account (https://openstax.org/l/30drakeequat) of how he came up with his "equation." And here is a recent interview (https://openstax.org/l/30frandrakinter) with Frank Drake by one of the authors of this textbook.

SETI outside the Radio Realm

For the reasons discussed above, most SETI programs search for signals at radio wavelengths. But in science, if there are other approaches to answering an unsolved question, we don't want to neglect them. So astronomers have been thinking about other ways we could pick up evidence for the existence of technologically advanced civilizations.

Recently, technology has allowed astronomers to expand the search into the domain of visible light. You might think that it would be hopeless to try to detect a flash of visible light from a planet given the brilliance of the star it orbits. This is why we usually cannot measure the reflected light of planets around other stars. The feeble light of the planet is simply swamped by the "big light" in the neighborhood. So another civilization would need a mighty strong beacon to compete with their star.

However, in recent years, human engineers have learned how to make flashes of light brighter than the Sun. The trick is to "turn on" the light for a very brief time, so that the costs are manageable. But ultra-bright, ultra-short laser pulses (operating for periods of a billionth of a second) can pack a lot of energy and can be coded to carry a message. We also have the technology to detect such short pulses—not with human senses, but with special detectors that can be "tuned" to hunt automatically for such short bursts of light from nearby stars.

Why would any civilization try to outshine its own star in this way? It turns out that the cost of sending an ultra-short laser pulse in the direction of a few promising stars can be less than the cost of sweeping a continuous radio message across the whole sky. Or perhaps they, too, have a special fondness for light messages because one of their senses evolved using light. Several programs are now experimenting with " optical SETI" searches, which can be done with only a modest telescope. (The term *optical* here means using visible light.)

If we let our imaginations expand, we might think of other possibilities. What if a truly advanced civilization should decide to (or need to) renovate its planetary system to maximize the area for life? It could do so by breaking apart some planets or moons and building a ring of solid material that surrounds or encloses the star and intercepts some or all of its light. This huge artificial ring or sphere might glow very brightly at infrared wavelengths, as the starlight it receives is eventually converted to heat and re-radiated into space. That infrared radiation could be detected by our instruments, and searches for such infrared sources are also underway (Figure 30.22).

Download for free at http://cnx.org/content/col11992/latest/

Figure 30.22. Wide-Field Infrared Survey Explorer (WISE). Astronomers have used this infrared satellite to search for infrared signatures of enormous construction projects by very advanced civilizations, but their first survey did not reveal any. (credit: modification of work by NASA/JPL-Caltech)

Should We Transmit in Addition to Listening?

Our planet has some leakage of radio waves into space, from FM radio, television, military radars, and communication between Earth and our orbiting spacecraft. However, such leakage radiation is still quite weak, and therefore difficult to detect at the distances of the stars, at least with the radio technology we have. So at the present time our attempts to communicate with other civilizations that may be out there mostly involve trying to receive messages, but not sending any ourselves.

Some scientists, however, think that it is inconsistent to search for beacons from other civilizations without announcing our presence in a similar way. (We discussed earlier the problem that if every other civilization confined itself to listening, no one would ever get in touch.) So, should we be making regular attempts at sending easily decoded messages into space? Some scientists warn that our civilization is too immature and defenseless to announce ourselves at this early point in our development. The decision whether to transmit or not turns out to be an interesting reflection of how we feel about ourselves and our place in the universe.

Discussions of transmission raise the question of who should speak for planet Earth. Today, anyone and everyone can broadcast radio signals, and many businesses, religious groups, and governments do. It would be a modest step for the same organizations to use or build large radio telescopes and begin intentional transmissions that are much stronger than the signals that leak from Earth today. And if we intercept a signal from an alien civilization, then the issue arises whether to reply.

Who should make the decision about whether, when, and how humanity announces itself to the cosmos? Is there freedom of speech when it comes to sending radio messages to other civilizations? Do all the nations of Earth have to agree before we send a signal strong enough that it has a serious chance of being received at the distances of the stars? How our species reaches a decision about these kinds of questions may well be a test of whether or not there is intelligent life on Earth.

This OpenStax book is available for free at http://cnx.org/content/col11992/1.8

Download for free at http://cnx.org/content/col11992/latest/

Conclusion

Whether or not we ultimately turn out to be the only intelligent species in our part of the Galaxy, our exploration of the cosmos will surely continue. An important part of that exploration will still be the search for biomarkers from inhabited planets that have not produced technological creatures that send out radio signals. After all, creatures like butterflies and dolphins may never build radio antennas, but we are happy to share our planet with them and would be delighted to find their counterparts on other worlds.

Whether or not life exists elsewhere is just one of the unsolved problems in astronomy that we have discussed in this book. A humble acknowledgment of how much we have left to learn about the universe is one of the fundamental hallmarks of science. This should not, however, prevent us from feeling exhilarated about how much we have already managed to discover, and feeling curious about what else we might find out in the years to come.

Our progress report on the ideas of astronomy ends here, but we hope that your interest in the universe does not. We hope you will keep up with developments in astronomy through media and online, or by going to an occasional public lecture by a local scientist. Who, after all, can even guess all the amazing things that future research projects will reveal about both the universe and our connection with it?

Download for free at http://cnx.org/content/col11992/latest/

CHAPTER 30 REVIEW

 KEY TERMS

amino acids organic compounds that are the molecular building blocks of proteins

astrobiology the multidisciplinary study of life in the universe: its origin, evolution, distribution, and fate; similar terms are *exobiology* and *bioastronomy*

biomarker evidence of the presence of life, especially a global indication of life on a planet that could be detected remotely (such as an unusual atmospheric composition)

DNA (deoxyribonucleic acid) a molecule that stores information about how to replicate a cell and its chemical and structural components

Drake equation a formula for estimating the number of intelligent, technological civilizations in our Galaxy, first suggested by Frank Drake

extremophile an organism (usually a microbe) that tolerates or even thrives under conditions that most of the life around us would consider hostile, such as very high or low temperature or acidity

gene the basic functional unit that carries the genetic (hereditary) material contained in a cell

habitable environment an environment capable of hosting life

habitable zone the region around a star in which liquid water could exist on the surface of terrestrial-sized planets, hence the most probable place to look for life in a star's planetary system

organic compound a compound containing carbon, especially a complex carbon compound; not necessarily produced by life

organic molecule a combination of carbon and other atoms—primarily hydrogen, oxygen, nitrogen, phosphorus, and sulfur—some of which serve as the basis for our biochemistry

photosynthesis a complex sequence of chemical reactions through which some living things can use sunlight to manufacture products that store energy (such as carbohydrates), releasing oxygen as one by-product

protein a key biological molecule that provides the structure and function of the body's tissues and organs, and essentially carries out the chemical work of the cell

RNA (ribonucleic acid) a molecule that aids in the flow of genetic information from DNA to proteins

SETI the search for extraterrestrial intelligence; usually applied to searches for radio signals from other civilizations

stromatolites solid, layered rock formations that are thought to be the fossils of oxygen-producing photosynthetic bacteria in rocks that are 3.5 billion years old

thermophile an organism that can tolerate high temperatures

This OpenStax book is available for free at http://cnx.org/content/col11992/1.8

Download for free at http://cnx.org/content/col11992/latest/

SUMMARY

30.1 The Cosmic Context for Life

Life on Earth is based on the presence of a key unit known as an organic molecule, a molecule that contains carbon, especially complex hydrocarbons. Our solar system formed about 5 billion years ago from a cloud of gas and dust enriched by several generations of heavier element production in stars. Life is made up of chemical combinations of these elements made by stars. The Copernican principle, which suggests that there is nothing special about our place in the universe, implies that if life could develop on Earth, it should be able to develop in other places as well. The Fermi paradox asks why, if life is common, more advanced life-forms have not contacted us.

30.2 Astrobiology

The study of life in the universe, including its origin on Earth, is called astrobiology. Life as we know it requires water, certain elemental raw materials (carbon, hydrogen, nitrogen, oxygen, phosphorus, and sulfur), energy, and an environment in which the complex chemistry of life is stable. Carbon-based (or organic) molecules are abundant in space and may also have been produced by processes on Earth. Life appears to have spread around our planet within 400 million years after the end of heavy bombardment, if not sooner. The actual origin of life—the processes leading from chemistry to biology—is not completely understood. Once life took hold, it evolved to use many energy sources, including first a range of different chemistries and later light, and diversified across a range of environmental conditions that humans consider "extreme." This proliferation of life into so many environmental niches, so relatively soon after our planet became habitable, has served to make many scientists optimistic about the chances that life could exist elsewhere.

30.3 Searching for Life beyond Earth

The search for life beyond Earth offers several intriguing targets. Mars appears to have been more similar to Earth during its early history than it is now, with evidence for liquid water on its ancient surface and perhaps even now below ground. The accessibility of the martian surface to our spacecraft offers the exciting potential to directly examine ancient and modern samples for evidence of life. In the outer solar system, the moons Europa and Enceladus likely host vast sub-ice oceans that may directly contact the underlying rocks—a good start in providing habitable conditions—while Titan offers a fascinating laboratory for understanding the sorts of organic chemistry that might ultimately provide materials for life. And the last decade of research on exoplanets leads us to believe that there may be billions of habitable planets in the Milky Way Galaxy. Study of these worlds offers the potential to find biomarkers indicating the presence of life.

30.4 The Search for Extraterrestrial Intelligence

Some astronomers are engaged in the search for extraterrestrial intelligent life (SETI). Because other planetary systems are so far away, traveling to the stars is either very slow or extremely expensive (in terms of energy required). Despite many UFO reports and tremendous media publicity, there is no evidence that any of these are related to extraterrestrial visits. Scientists have determined that the best way to communicate with any intelligent civilizations out there is by using electromagnetic waves, and radio waves seem best suited to the task. So far, they have only begun to comb the many different possible stars, frequencies, signal types, and other factors that make up what we call the cosmic haystack problem. Some astronomers are also undertaking searches for brief, bright pulses of visible light and infrared signatures of huge construction projects by advanced civilizations. If we do find a signal someday, deciding whether to answer and what to answer may be two of the greatest challenges humanity will face.

Download for free at http://cnx.org/content/col11992/latest/

🖋 FOR FURTHER EXPLORATION

Articles

Astrobiology

Chyba, C. "The New Search for Life in the Universe." *Astronomy* (May 2010): 34. An overview of astrobiology and the search for life out there in general, with a brief discussion of the search for intelligence.

Dorminey, B. "A New Way to Search for Life in Space." *Astronomy* (June 2014): 44. Finding evidence of photosynthesis on other worlds.

McKay, C., & Garcia, V. "How to Search for Life on Mars." *Scientific American* (June 2014): 44–49. Experiments future probes could perform.

Reed, N. "Why We Haven't Found Another Earth Yet." *Astronomy* (February 2016): 25. On the search for smaller earthlike planets in their star's habitable zones, and where we stand.

Shapiro, R. "A Simpler Origin of Life." *Scientific American* (June 2007): 46. New ideas about what kind of molecules formed first so life could begin.

Simpson, S. "Questioning the Oldest Signs of Life." *Scientific American* (April 2003): 70. On the difficulty of interpreting biosignatures in rocks and the implications for the search for life on other worlds.

SETI

Chandler, D. "The New Search for Alien Intelligence." *Astronomy* (September 2013): 28. Review of various ways of finding other civilizations out there, not just radio wave searches.

Crawford, I. "Where Are They?" *Scientific American* (July 2000): 38. On the Fermi paradox and its resolutions, and on galactic colonization models.

Folger, T. "Contact: The Day After." *Scientific American* (January 2011): 40–45. Journalist reports on efforts to prepare for ET signals; protocols and plans for interpreting messages; and discussions of active SETI.

Kuhn, J., et al. "How to Find ET with Infrared Light." *Astronomy* (June 2013): 30. On tracking alien civilizations by the heat they put out.

Lubick, N. "An Ear to the Stars." *Scientific American* (November 2002): 42. Profile of SETI researcher Jill Tarter.

Nadis, S. "How Many Civilizations Lurk in the Cosmos?" *Astronomy* (April 2010): 24. New estimates for the terms in the Drake equation.

Shostak, S. "Closing in on E.T." *Sky & Telescope* (November 2010): 22. Nice summary of current and proposed efforts to search for intelligent life out there.

Websites

Astrobiology

Astrobiology Web: http://astrobiology.com/. A news site with good information and lots of material.

Exploring Life's Origins: http://exploringorigins.org/index.html. A website for the Exploring Origins Project, part of the multimedia exhibit of the Boston Museum of Science. Explore the origin of life on Earth with an interactive timeline, gain a deeper knowledge of the role of RNA, "build" a cell, and explore links to learn more about astrobiology and other related information.

This OpenStax book is available for free at http://cnx.org/content/col11992/1.8

History of Astrobiology: https://astrobiology.nasa.gov/about/history-of-astrobiology/. By Marc Kaufman, on the NASA Astrobiology site.

Life, Here and Beyond: https://astrobiology.nasa.gov/about/. By Marc Kaufman, on the NASA Astrobiology site.

SETI

Berkeley SETI Research Center: https://seti.berkeley.edu/. The University of California group recently received a $100 million grant from a Russian billionaire to begin the Breakthrough: Listen project.

Fermi Paradox: http://www.seti.org/seti-institute/project/details/fermi-paradox. Could we be alone in our part of the Galaxy or, more dramatic still, could we be the only technological society in the universe? A useful discussion.

Planetary Society: http://www.planetary.org/explore/projects/seti/. This advocacy group for exploration has several pages devoted to the search for life.

SETI Institute: http://www.seti.org. A key organization in the search for life in the universe; the institute's website is full of information and videos about both astrobiology and SETI.

SETI: http://www.skyandtelescope.com/tag/seti/. *Sky & Telescope* magazine offers good articles on this topic.

Videos

Astrobiology

Copernicus Complex: Are We Special in the Cosmos?: https://www.youtube.com/watch?v=ERp0AHYRm_Q. A video of a popular-level talk by Caleb Scharf of Columbia University (1:18:54).

Life at the Edge: Life in Extreme Environments on Earth and the Search for Life in the Universe: https://www.youtube.com/watch?v=91JQmTn0SF0. A video of a 2009 nontechnical lecture by Lynn Rothschild of NASA Ames Research Center (1:31:21).

Saturn's Moon Titan: A World with Rivers, Lakes, and Possibly Even Life: https://www.youtube.com/watch?v=bbkTJeHoOKY. A video of a 2011 talk by Chris McKay of NASA Ames Research Center (1:23:33).

SETI

Allen Telescope Array: The Newest Pitchfork for Exploring the Cosmic Haystack: https://www.youtube.com/watch?v=aqsI1HZCgUM. A 2013 popular-level lecture by Jill Tarter of the SETI Institute (1:45:55).

Confessions of an Alien Hunter: http://fora.tv/2009/03/31/Seth_Shostak_Confessions_of_an_Alien_Hunter. 2009 interview with Seth Shostak on FORA TV (36:27).

Search for Extra-Terrestrial Intelligence: Necessarily a Long-Term Strategy: http://www.longnow.org/seminars/02004/jul/09/the-search-for-extra-terrestrial-intelligence-necessarily-a-long-term-strategy/. 2004 talk by Jill Tarter at the Long Now Foundation (1:21:13).

Search for Intelligent Life Among the Stars: New Strategies: https://www.youtube.com/watch?v=m9WxW2ktcKU. A 2010 nontechnical talk by Seth Shostak of the SETI Institute (1:29:58).

 ## COLLABORATIVE GROUP ACTIVITIES

A. If one of the rocks from Mars examined by a future mission to the red planet does turn out to have unambiguous signs of ancient life that formed on Mars, what does your group think would be the

Download for free at http://cnx.org/content/col11992/latest/

implications of such a discovery for science and for our view of life elsewhere? Would such a discovery have any long-term effects on your own thinking?

B. Suppose we receive a message from an intelligent civilization around another star. What does your group think the implications of this discovery would be? How would your own thinking or personal philsophy be affected by such a discovery?

C. A radio message has been received from a civilization around a star 40 light-years away, which contains (in pictures) quite a bit of information about the beings that sent the message. The president of the United States has appointed your group a high-level commission to advise whether humanity should answer the message (which was not particularly directed at us, but comes from a beacon that, like a lighthouse, sweeps out a circle in space). How would you advise the president? Does your group agree on your answer or do you also have a minority view to present?

D. If there is no evidence that UFOs are extraterrestrial visitors, why does your group think that television shows, newspapers, and movies spend so much time and effort publicizing the point of view that UFOs are craft from other worlds? Make a list of reasons. Who stands to gain by exaggerating stories of unknown lights in the sky or simply fabricating stories that alien visitors are already here?

E. Does your group think scientists should simply ignore all the media publicity about UFOs or should they try to respond? If so, how should they respond? Does everyone in the group agree?

F. Suppose your group is the team planning to select the most important sights and sounds of Earth to record and put on board the next interstellar spacecraft. What pictures (or videos) and sounds would you include to represent our planet to another civilization?

G. Let's suppose Earth civilization has decided to broadcast a message announcing our existence to other possible civilizations among the stars. Your group is part of a large task force of scientists, communications specialists, and people from the humanities charged with deciding the form and content of our message. What would you recommend? Make a list of ideas.

H. Think of examples of contact with aliens you have seen in movies and on TV. Discuss with your group how realistic these have been, given what you have learned in this class. Was the contact in person (through traveling) or using messages? Why do you think Hollywood does so many shows and films that are not based on our scientific understanding of the universe?

I. Go through the Drake equation with your group and decide on values for each factor in the estimate. (If you disagree on what a factor should be within the group, you can have a "minority report.") Based on the factors, how many intelligent, communicating civilizations do you estimate to be thriving in our Galaxy right now?

🖱 EXERCISES

Review Questions

1. What is the Copernican principle? Make a list of scientific discoveries that confirm it.

2. Where in the solar system (and beyond) have scientists found evidence of organic molecules?

This OpenStax book is available for free at http://cnx.org/content/col11992/1.8

Download for free at http://cnx.org/content/col11992/latest/

3. Give a short history of the atoms that are now in your little finger, going back to the beginning of the universe.

4. What is a biomarker? Give some possible examples of biomarkers we might look for beyond the solar system.

5. Why are Mars and Europa the top targets for the study of astrobiology?

6. Why is traveling between the stars (by creatures like us) difficult?

7. What are the advantages to using radio waves for communication between civilizations that live around different stars? List as many as you can.

8. What is the "cosmic haystack problem"? List as many of its components as you can think of.

9. What is a habitable zone?

10. Why is the simultaneous detection of methane and oxygen in an atmosphere a good indication of the existence of a biosphere on that planet?

11. What are two characteristic properties of life that distinguish it from nonliving things?

12. What are the three requirements that scientists believe an environment needs to supply life with in order to be considered habitable?

13. Can you name five environmental conditions that, in their extremes, microbial life has been challenged by and has learned to survive on Earth?

Thought Questions

14. Would a human have been possible during the first generation of stars that formed right after the Big Bang? Why or why not?

15. If we do find life on Mars, what might be some ways to check whether it formed separately from Earth life, or whether exchanges of material between the two planets meant that the two forms of life have a common origin?

16. What kind of evidence do you think would convince astronomers that an extraterrestrial spacecraft has landed on Earth?

17. What are some reasons that more advanced civilizations might want to send out messages to other star systems?

18. What are some answers to the Fermi paradox? Can you think of some that are not discussed in this chapter?

19. Why is there so little evidence of Earth's earliest history and therefore the period when life first began on our planet?

20. Why was the development of photosynthesis a major milestone in the evolution of life?

21. Does all life on Earth require sunshine?

22. Why is life unlikely to be found on the surface of Mars today?

23. In this chapter, we identify these characteristic properties of life: life extracts energy from its environment, and has a means of encoding and replicating information in order to make faithful copies of itself. Does this definition fully capture what we think of as "life"? How might our definition be biased by our terrestrial environment?

Download for free at http://cnx.org/content/col11992/latest/

24. Given that no sunlight can penetrate Europa's ice shell, what would be the type of energy that could make some form of europan life possible?

25. Why is Saturn's moon Enceladus such an exciting place to send a mission?

26. In addition to an atmosphere dominated by nitrogen, how else is Saturn's moon Titan similar to Earth?

27. How can a planet's atmosphere affect the width of the habitable zone in its planetary system?

28. Why are we limited to finding life on planets orbiting other stars to situations where the biosphere has created planet-scale changes?

Figuring For Yourself

29. Suppose astronomers discover a radio message from a civilization whose planet orbits a star 35 light-years away. Their message encourages us to send a radio answer, which we decide to do. Suppose our governing bodies take 2 years to decide whether and how to answer. When our answer arrives there, their governing bodies also take two of our years to frame an answer to us. How long after we get their first message can we hope to get their reply to ours? (A question for further thinking: Once communication gets going, should we continue to wait for a reply before we send the next message?)

30. The light a planet receives from the Sun (per square meter of planet surface) decreases with the square of the distance from the Sun. So a planet that is twice as far from the Sun as Earth receives $(1/2)^2 = 0.25$ times (25%) as much light and a planet that is three times as far from the Sun receives $(1/3)^2 = 0.11$ times (11%) as much light. How much light is received by the moons of Jupiter and Saturn (compared to Earth), worlds which orbit 5.2 and 9.5 times farther from the Sun than Earth?

31. Think of our Milky Way Galaxy as a flat disk of diameter 100,000 light-years. Suppose we are one of 1000 civilizations, randomly distributed through the disk, interested in communicating via radio waves. How far away would the nearest such civilization be from us (on average)?

This OpenStax book is available for free at http://cnx.org/content/col11992/1.8

Download for free at http://cnx.org/content/col11992/latest/

A | HOW TO STUDY FOR AN INTRODUCTORY ASTRONOMY CLASS

In this brief appendix, we want to give you some hints for the effective study of astronomy. These suggestions are based on ideas from good teachers and good students around the United States. Your professor will probably have other, more specific suggestions for doing well in your class.

Astronomy, the study of the universe beyond the borders of our planet, is one of the most exciting and rapidly changing branches of science. Even scientists from other fields often confess to having had a lifelong interest in astronomy, though they may now be doing something earthbound—like biology, chemistry, engineering, or writing software.

But some of the things that make astronomy so interesting also make it a challenge for the beginning student. The universe is a big place, full of objects and processes that do not have familiar counterparts here on Earth. Like a visitor to a new country, it will take you a while to feel familiar with the territory or the local customs. Astronomy, like other sciences, also has its own special vocabulary, some of which you will have to learn to communicate well with your professor and classmates.

Still, hundreds of thousands of non-science majors take an introductory astronomy course every year, and surveys show that students from a wide range of backgrounds have succeeded in (and even enjoyed) these classes. Astronomy is for everyone, not just those who are "science oriented."

So, here are some suggestions to help you increase your chances of doing well in your astronomy class.

1. The best advice we can give you is to be sure to leave enough time in your schedule to study the material in this class regularly. It sounds obvious, but it is not very easy to catch up with a subject like astronomy by trying to do everything just before an exam. (As astronomers like to put it, you can't learn the whole universe in one night!) Try to put aside some part of each day, or every other day, when you can have uninterrupted time for reading and studying astronomy.

2. In class, put your phone away and focus on the class activities. If you have to use a laptop or tablet in class, make a pact with yourself that you will *not* check email, get on social media, or play games during class. A number of careful studies of student behavior and grades have shown that students are not as good at such multi-tasking as they think they are, and that students who do *not* use screens during class get significantly better grades in the end.

3. Try to take careful notes during class. Many students start college without good note-taking habits. If you are not a good note-taker, try to get some help. Many colleges and universities have student learning centers that offer short courses, workbooks, tutors, or videos on developing good study habits. Good note-taking skills will also be useful for many jobs or activities you are likely get involved with after college.

4. Try to read each assignment in the textbook twice, once before it is discussed in class, and once afterward. Take notes as you read or use a highlighter to outline ideas that you may want to review later.

5. Form a small astronomy study group with people in your class. Get together with them regularly and discuss what you have been learning. Also, focus on the topics that may be giving group members trouble. Make up sample exam questions and make sure everyone in the group can answer them confidently. If you have always studied alone, you may at first resist this idea, but don't be too hasty to say no. Study groups are a very effective way to digest a large amount of new information.

6. Before each exam, create a concise outline of the main ideas discussed in class and presented in your text. Compare your outline with those of other students as a check on your own study habits.

7. If your professor suggests doing web-based sample quizzes, or looking at online apps, animations, or study guides, take advantage of these resources to enhance your studying.

Download for free at http://cnx.org/content/col11992/latest/

8. At the end of each chapter in this textbook you, will find four kinds of questions. The Collaborative Group Activities are designed to encourage you follow up on the material in the chapter as a group, rather than individually. Review Questions help you see if you have learned the material in the chapter. Thought Questions test deeper understanding by asking you to apply your knowledge to new situations. And Figuring for Yourself exercises test and extend some of the mathematical examples in the chapter. (Not all professors will use the math sections; if they don't, you may not have homework from this section.)

9. If you find a topic in the text or in class especially difficult or interesting, talk to your professor or teaching assistant. Many students are scared to show their ignorance in front of their teacher, but we can assure you that most professors and TA's *like* it when students come to office hours and show that they care enough about the course to ask for help.

10. Don't stay up all night before a test and then expect your mind to respond well. For the same reason, don't eat a big meal just before a test, since we all get a little sleepy and don't think as clearly after a big meal. Take many deep breaths and try to relax during the test itself.

11. Don't be too hard on yourself! If astronomy is new to you, many of the ideas and terms in this book may be unfamiliar. Astronomy is like any new language: it may take a while to become a good conversationalist. Practice as much as you can, but also realize that it is natural to feel overwhelmed by the vastness of the universe and the variety of things that are going on in it.

This OpenStax book is available for free at http://cnx.org/content/col11992/1.8

Download for free at http://cnx.org/content/col11992/latest/

B   ASTRONOMY WEBSITES, IMAGES, AND APPS

Throughout the textbook, we suggest useful resources for students on the specific topics in a given chapter. Here, we offer some websites for exploring astronomy in general, plus good sites for viewing and downloading the best astronomy images, and guides to astronomical apps for smartphones and tablets. This is not an exhaustive listing, but merely a series of suggestions to whet the appetite of those wanting to go beyond the textbook.

Websites For Exploring Astronomy In General

Astronomical Organizations

Amateur Astronomy Clubs. In most large cities and a number of rural areas, there are *amateur astronomy clubs*, where those interested in the hobby of astronomy gather to observe the sky, share telescopes, hear speakers, and help educate the public about the night sky. To find an astronomy club near you, you can try the following sites:

> - Night Sky Network club finder: http://nightsky.jpl.nasa.gov/club-map.cfm.
> - *Sky & Telescope* Magazine astronomy clubs and organizations: http://www.skyandtelescope.com/community/organizations.
> - *Astronomy* Magazine club finder: http://www.astronomy.com/groups.aspx.
> - Astronomical League astronomy clubs and societies: http://www.astroleague.org/societies/all.
> - Go-Astronomy club search: http://www.go-astronomy.com/astro-club-search.htm.

American Astronomical Society: http://www.aas.org. Composed mainly of professional astronomers. They have an active education office and various materials for students and the public on the education pages of their website.

Astronomical League: http://www.astroleague.org. The league is the umbrella organization of American astronomy clubs. They offer a newsletter, national observing programs, and support for how to form and support a club.

Astronomical Society of the Pacific: http://www.astrosociety.org. Founded in 1889, this international society is devoted to astronomy education and outreach. They have programs, publications, and materials for families, teachers, amateur astronomers, museum guides, and anyone interested in astronomy.

European Space Agency (ESA): http://www.esa.int/. Information on European space missions with an excellent gallery of images.

International Astronomical Union (IAU): http://www.iau.org/. International organization for professional astronomers; see the menu choice "IAU for the Public" for information on naming astronomical objects and other topics of interest to students.

International Dark-Sky Association: http://www.darksky.org. Dedicated to combating light pollution, the encroachment of stray light that wastes energy and washes out the glories of the night sky.

NASA: http://www.nasa.gov. NASA has a wide range of information on its many websites; the trick is to find what you need. Most space missions and NASA centers have their own sites.

Planetary Society: http://www.planetary.org. Founded by the late Carl Sagan and others, this group works to encourage planetary exploration and the search for life elsewhere. While much of their work is advocacy, they have some educational outreach too.

Download for free at http://cnx.org/content/col11992/latest/

Royal Astronomical Society of Canada: http://www.rasc.ca/. Unites professional and amateur astronomers around Canada; has 28 centers with local activities, plus national magazines and meetings.

Some Astronomical Publications Students Can Read

Astronomy Now: http://www.astronomynow.com/. A colorful British monthly, with excellent articles about astronomy, the history of astronomy, and stargazing.

Astronomy: http://www.astronomy.com. Has the largest circulation of any magazine devoted to the universe and is designed especially for astronomy hobbyists and armchair astronomers.

Free Astronomy: http://www.astropublishing.com/. A new web-based publication, with European roots.

Scientific American: http://www.sciam.com. Offers one astronomy article about every other issue. These articles, a number of which are reproduced on their website, are at a slightly higher level, but—often being written by the astronomers who have done the work—are authoritative and current.

Sky & Telescope: http://skyandtelescope.com. An older and somewhat higher-level magazine for astronomy hobbyists. Many noted astronomers write for this publication.

Sky News: http://www.skynews.ca/. A Canadian publication, featuring both astronomy and stargazing information. It also lists Canadian events for hobbyists.

StarDate: https://stardate.org/. Magazine that accompanies the brief radio program, with a useful website for beginners.

Sites that Cover Astronomy News

Exploring the Universe: http://fraknoi.blogspot.com. An astronomy news blog by one of the original authors of this textbook.

Portal to the Universe: http://www.portaltotheuniverse.org/. A site that gathers online astronomy and space news items, blogs, and pictures.

Science@NASA news stories and newscasts: http://science.nasa.gov/science-news/. Well-written stories with, of course, a NASA focus.

Space.com: http://www.space.com/news/. A commercial site, but with wide coverage of space and astronomy news.

Universe Today: http://www.universetoday.com/. Another commercial site, with good articles by science journalists, but a lot of ads.

Sites for Answering Astronomical Questions

Ask an Astrobiologist: http://astrobiology.nasa.gov/ask-an-astrobiologist/. On this site from the National Astrobiology Institute at NASA, astronomer David Morrison answered questions about the search for life on other planets, the origin of life on Earth, and many other topics.

Ask an Astronomer at Lick Observatory: http://www.ucolick.org/~mountain/AAA/. Graduate students and staff members at this California observatory answered selected astronomy questions, particularly from high school students.

Ask an Astrophysicist: http://imagine.gsfc.nasa.gov/docs/ask_astro/ask_an_astronomer.html. Questions and answers at NASA's Laboratory for High-Energy Astrophysics focus on X-ray and gamma-ray astronomy, and such objects as black holes, quasars, and supernovae.

This OpenStax book is available for free at http://cnx.org/content/col11992/1.8

Ask an Infrared Astronomer: http://coolcosmos.ipac.caltech.edu/cosmic_classroom/ask_astronomer/faq/index.shtml. A site from the California Institute of Technology, with an archive focusing on infrared (heat-ray) astronomy and the discoveries it makes about cool objects in the universe. No longer taking new questions.

Ask the Astronomer: http://www.astronomycafe.net/qadir/qanda.html. This site, run by astronomer Sten Odenwald, is no longer active, but lists 3001 answers to questions asked in the mid-1990s. They are nicely organized by topic.

Ask the Experts at PhysLink: http://www.physlink.com/Education/AskExperts/index.cfm. Lots of physics questions answered, with some astronomy as well, at this physics education site. Most answers are by physics teachers, not astronomers. Still taking new questions.

Ask the Space Scientist: http://image.gsfc.nasa.gov/poetry/ask/askmag.html. An archive of questions about the Sun and its interactions with Earth, answered by astronomer Sten Odenwald. Not accepting new questions.

Curious about Astronomy?: http://curious.astro.cornell.edu. An ask-an-astronomer site run by graduate students and professors of astronomy at Cornell University. Has searchable archives and is still answering new questions.

Miscellaneous Sites of Interest

A Guide to Careers in Astronomy: http://aas.org/files/resources/Careers-in-Astronomy.pdf. From the American Astronomical Society.

Astronomical Pseudo-Science: A Skeptic's Resource List: http://www.astrosociety.org/pseudo. Readings and websites that analyze such claims as astrology, UFOs, moon-landing denial, creationism, human faces on other worlds, astronomical disasters, and more.

Astronomy for Beginners: http://www.skyandtelescope.com/astronomy-information/. A page to find resources for getting into amateur astronomy.

Science Fiction Stories with Good Astronomy and Physics: http://www.astrosociety.org/scifi.

Space Calendar: http://www2.jpl.nasa.gov/calendar/. Ron Baalke at the Jet Propulsion Laboratory keeps a listing of what space events happened on each day of the year; great if you need a reason to have a space-theme party.

Unheard Voices: The Astronomy of Many Cultures: http://multiverse.ssl.berkeley.edu/multicultural. A guide to resources about the astronomy of native, African, Asian, and other non-Western groups.

Selected Websites For Viewing And Downloading Astronomical Images

The Top Image Sites

Astronomy Picture of the Day: http://antwrp.gsfc.nasa.gov/apod/lib/aptree.html. Two space scientists scour the internet and feature one interesting astronomy image each day.

European Southern Observatory Photo Gallery: http://www.eso.org/public/images/. Magnificent color images from ESO's largest telescopes. See the topical menu at the top.

Hubble Space Telescope Images: http://hubblesite.org/newscenter/archive/browse/images/. Starting at this page, you can select from among many hundreds of Hubble pictures by subject. Other ways to approach these images are through the more public-oriented Hubble Gallery (http://hubblesite.org/gallery/) or the European ESO site (http://www.spacetelescope.org/images/).

Download for free at http://cnx.org/content/col11992/latest/

National Optical Astronomy Observatories Image Gallery: http://www.noao.edu/image_gallery/. Growing archive of images from the many telescopes that are at the United States' National Observatories.

Planetary Photojournal: http://photojournal.jpl.nasa.gov/index.html. Features thousands of images from NASA's extensive set of planetary exploration missions with a good search menu. Does not include most of the missions from other countries.

The World at Night: http://www.twanight.org/newTWAN/index.asp. Dramatic night-sky images by professional photographers who are amateur astronomers. Note that while many of the astronomy sites allow free use of their images, these are copyrighted by photographers who make their living selling them.

Other Useful General Galleries

Anglo-Australian Observatory: http://203.15.109.22/images/. Soon at https://www.aao.gov.au/public/images. Great copyrighted color images by leading astro-photographer David Malin and others.

Canada-France-Hawaii Telescope: http://www.cfht.hawaii.edu/HawaiianStarlight/images.html. Remarkable color images from a major telescope on top of the Mauna Kea peak in Hawaii.

European Space Agency Gallery: http://www.esa.int/spaceinimages/Images. Access images from such missions as Mars Express, Rosetta, and Herschel.

Gemini Observatory Images: http://www.gemini.edu/index.php?option=com_gallery. Images from a pair of large telescopes in the northern and the southern hemispheres.

Isaac Newton Group of Telescopes Image Gallery: http://www.ing.iac.es/PR/images_index.html. Beautiful images from the Herschel, Newton, and Kapteyn telescopes on La Palma.

National Radio Astronomy Observatory Image Gallery: http://images.nrao.edu/. Organized by topic, the images show objects and processes that give off radio waves.

Our Infrared World Gallery: http://coolcosmos.ipac.caltech.edu/image_galleries/missions_gallery.html. Images from a variety of infrared astronomy telescopes and missions. See also their "Cool Cosmos" site for the public: http://coolcosmos.ipac.caltech.edu/.

Some Galleries on Specific Subjects

Astronaut Photography of Earth: http://eol.jsc.nasa.gov/.

Chandra X-Ray Observatory Images: http://chandra.harvard.edu/photo/category.html.

NASA Human Spaceflight Gallery: https://www.flickr.com/photos/nasa2explore or http://spaceflight1.nasa.gov/gallery/index.html. Astronaut images.

Robert Gendler: http://www.robgendlerastropics.com/. One of the amateur astro-photographers who comes closest to being professional.

Sloan Digital Sky Survey Images: http://www.sdss.org/gallery/.

Solar Dynamics Observatory Gallery: http://sdo.gsfc.nasa.gov/gallery/main. Sun images.

Spitzer Infrared Telescope Images: http://www.spitzer.caltech.edu/images.

Astronomy Apps For Smartphones And Tablets

A pretty comprehensive listing of such apps with brief descriptions and links to their websites can be found at: http://dx.doi.org/10.3847/AER2011036. The list is now a few years old, but most of the apps are still available.

This OpenStax book is available for free at http://cnx.org/content/col11992/1.8

Listings and Reviews of Apps

11 Best Astronomy Apps for Amateur Star Gazers: http://www.businessinsider.com/11-best-astronomy-apps-for-amateurs-2013-10. From Kelly Dickerson (2013).

14 Best Astronomy Apps for Stargazers and Space Lovers: http://nerdsmagazine.com/best-astronomy-apps-for-android/. Viney Dhiman's recommendations, part of *Nerd's Magazine* (2014).

15 Best Astronomy Applications for iPhone: http://www.iphoneness.com/iphone-apps/top-astronomy-applications-for-iphone/. From iPhoneness.

Apps for Stargazing: http://appadvice.com/appguides/show/astronomy-apps. App Advice site's reviews.

NASA Apps for Smartphones and Tablets: https://www.nasa.gov/connect/apps.html.

Phone/Tablet Apps and the Practical Astronomer: http://www.cloudynights.com/page/articles/cat/user-reviews/phonetablet-apps-and-the-practical-astronomer-r2925. Active amateur astronomer Tom Fowler reviews 22 apps (2014).

Sky & Telescope Mobile Apps: http://www.skyandtelescope.com/sky-and-stargazing-apps/. Apps from *Sky & Telescope* Magazine.

Smartphone Apps Can Make Astronomy as Easy as Point and Gaze: http://www.heraldnet.com/article/20140511/LIVING/140519988." Mike Lynch for HeraldNet (2014).

Download for free at http://cnx.org/content/col11992/latest/

This OpenStax book is available for free at http://cnx.org/content/col11992/1.8

Download for free at http://cnx.org/content/col11992/latest/

C | SCIENTIFIC NOTATION

In astronomy (and other sciences), it is often necessary to deal with very large or very small numbers. In fact, when numbers become truly large in everyday life, such as the national debt in the United States, we call them astronomical. Among the ideas astronomers must routinely deal with is that the Earth is 150,000,000,000 meters from the Sun, and the mass of the hydrogen atom is 0.0000000000000000000000000167 kilograms. No one in his or her right mind would want to continue writing so many zeros!

Instead, scientists have agreed on a kind of shorthand notation, which is not only easier to write, but (as we shall see) makes multiplication and division of large and small numbers much less difficult. If you have never used this powers-of-ten notation or scientific notation, it may take a bit of time to get used to it, but you will soon find it much easier than keeping track of all those zeros.

Writing Large Numbers

In scientific notation, we generally agree to have only one number to the left of the decimal point. If a number is not in this format, it must be changed. The number 6 is already in the right format, because for integers, we understand there to be a decimal point to the right of them. So 6 is really 6., and there is indeed only one number to the left of the decimal point. But the number 965 (which is 965.) has three numbers to the left of the decimal point, and is thus ripe for conversion.

To change 965 to proper form, we must make it 9.65 and then keep track of the change we have made. (Think of the number as a weekly salary and suddenly it makes a lot of difference whether we have $965 or $9.65.) We keep track of the number of places we moved the decimal point by expressing it as a power of ten. So 965 becomes 9.65×10^2 or 9.65 multiplied by ten to the second power. The small raised 2 is called an exponent, and it tells us how many times we moved the decimal point to the left.

Note that 10^2 also designates 10 squared, or 10×10, which equals 100. And 9.65×100 is just 965, the number we started with. Another way to look at scientific notation is that we separate out the messy numbers out front, and leave the smooth units of ten for the exponent to denote. So a number like 1,372,568 becomes 1.372568 times a million (10^6) or 1.372568 times 10 multiplied by itself 6 times. We had to move the decimal point six places to the left (from its place after the 8) to get the number into the form where there is only one digit to the left of the decimal point.

The reason we call this powers-of-ten notation is that our counting system is based on increases of ten; each place in our numbering system is ten times greater than the place to the right of it. As you have probably learned, this got started because human beings have ten fingers and we started counting with them. (It is interesting to speculate that if we ever meet intelligent life-forms with only eight fingers, their counting system would probably be a powers-of-eight notation!)

So, in the example we started with, the number of meters from Earth to the Sun is 1.5×10^{11}. Elsewhere in the book, we mention that a string 1 light-year long would fit around Earth's equator 236 million or 236,000,000 times. In scientific notation, this would become 2.36×10^8. Now if you like expressing things in millions, as the annual reports of successful companies do, you might like to write this number as 236×10^6. However, the usual convention is to have only one number to the left of the decimal point.

Writing Small Numbers

Now take a number like 0.00347, which is also not in the standard (agreed-to) form for scientific notation. To put it into that format, we must make the first part of it 3.47 by moving the decimal point three places *to the right*.

Download for free at http://cnx.org/content/col11992/latest/

Note that this motion to the right is the opposite of the motion to the left that we discussed above. To keep track, we call this change negative and put a minus sign in the exponent. Thus 0.00347 becomes 3.47×10^{-3}.

In the example we gave at the beginning, the mass of the hydrogen atom would then be written as 1.67×10^{-27} kg. In this system, one is written as 10^0, a tenth as 10^{-1}, a hundredth as 10^{-2}, and so on. Note that any number, no matter how large or how small, can be expressed in scientific notation.

Multiplication And Division

Scientific notation is not only compact and convenient, it also simplifies arithmetic. To multiply two numbers expressed as powers of ten, you need only multiply the numbers out front and then *add* the exponents. If there are no numbers out front, as in 100 × 100,000, then you just add the exponents (in our notation, $10^2 \times 10^5 = 10^7$). When there are numbers out front, you have to multiply them, but they are much easier to deal with than numbers with many zeros in them.

Here's an example:

$$\left(3 \times 10^5\right) \times \left(2 \times 10^9\right) = 6 \times 10^{14}$$

And here's another example:

$$\begin{aligned}
0.04 \times 6{,}000{,}000 &= \left(4 \times 10^{-2}\right) \times \left(6 \times 10^6\right) \\
&= 24 \times 10^4 \\
&= 2.4 \times 10^5
\end{aligned}$$

Note in the second example that when we added the exponents, we treated negative exponents as we do in regular arithmetic (−2 plus 6 equals 4). Also, notice that our first result had a 24 in it, which was not in the acceptable form, having two places to the left of the decimal point, and we therefore changed it to 2.4 and changed the exponent accordingly.

To divide, you divide the numbers out front and *subtract* the exponents. Here are several examples:

$$\frac{1{,}000{,}000}{1000} = \frac{10^6}{10^3} = 10^{(6-3)} = 10^3$$

$$\frac{9 \times 10^{12}}{2 \times 10^3} = 4.5 \times 10^9$$

$$\frac{2.8 \times 10^2}{6.2 \times 10^5} = 0.452 \times 10^{-3} = 4.52 \times 10^{-4}$$

In the last example, our first result was not in the standard form, so we had to change 0.452 into 4.52, and change the exponent accordingly.

If this is the first time that you have met scientific notation, we urge you to practice many examples using it. You might start by solving the exercises below. Like any new language, the notation looks complicated at first but gets easier as you practice it.

Exercises

1. At the end of September, 2015, the New Horizons spacecraft (which encountered Pluto for the first time in July 2015) was 4.898 billion km from Earth. Convert this number to scientific notation. How many astronomical units is this? (An astronomical unit is the distance from Earth to the Sun, or about 150 million km.)

This OpenStax book is available for free at http://cnx.org/content/col11992/1.8

2. During the first six years of its operation, the Hubble Space Telescope circled Earth 37,000 times, for a total of 1,280,000,000 km. Use scientific notation to find the number of km in one orbit.

3. In a large university cafeteria, a soybean-vegetable burger is offered as an alternative to regular hamburgers. If 889,875 burgers were eaten during the course of a school year, and 997 of them were veggie-burgers, what fraction and what percent of the burgers does this represent?

4. In a 2012 Kelton Research poll, 36 percent of adult Americans thought that alien beings have actually landed on Earth. The number of adults in the United States in 2012 was about 222,000,000. Use scientific notation to determine how many adults believe aliens have visited Earth.

5. In the school year 2009–2010, American colleges and universities awarded 2,354,678 degrees. Among these were 48,069 PhD degrees. What fraction of the degrees were PhDs? Express this number as a percent. (Now go and find a job for all those PhDs!)

6. A star 60 light-years away has been found to have a large planet orbiting it. Your uncle wants to know the distance to this planet in old-fashioned miles. Assume light travels 186,000 miles per second, and there are 60 seconds in a minute, 60 minutes in an hour, 24 hours in a day, and 365 days in a year. How many miles away is that star?

Answers

1. 4.898 billion is 4.898×10^9 km. One astronomical unit (AU) is 150 million km = 1.5×10^8 km. Dividing the first number by the second, we get $3.27 \times 10^{(9-8)} = 3.27 \times 10^1$ AU.

2. $\dfrac{1.28 \times 10^9 \text{ km}}{3.7 \times 10^4 \text{ orbits}} = 0.346 \times 10^{(9-4)} = 0.346 \times 10^5 = 3.46 \times 10^4$ km per orbit.

3. $\dfrac{9.97 \times 10^2 \text{ veggie burgers}}{8.90 \times 10^5 \text{ total burgers}} = 1.12 \times 10^{(2-5)} = 1.12 \times 10^{(2-5)} = 1.12 \times 10^{-3}$ (or roughly about one thousandth) of the burgers were vegetarian. Percent means per hundred. So $\dfrac{1.12 \times 10^{-3}}{10^{-2}} = 1.12 \times 10^{(-3-(-2))} = 1.12 \times 10^{-1}$ percent (which is roughly one tenth of one percent).

4. 36% is 36 hundredths or 0.36 or 3.6×10^{-1}. Multiply that by 2.22×10^8 and you get about $7.99 \times 10^{(-1+8)} = 7.99 \times 10^7$ or almost 80 million people who believe that aliens have landed on our planet. We need more astronomy courses to educate all those people.

5. $\dfrac{4.81 \times 10^4}{2.35 \times 10^6} = 2.05 \times 10^{(4-6)} = 2.05 \times 10^{-2} = $ about 2%. (Note that in these examples we are rounding off some of the numbers so that we don't have more than 2 places after the decimal point.)

6. One light-year is the distance that light travels in one year. (Usually, we use metric units and not the old British system that the United States is still using, but we are going to humor your uncle and stick with miles.) If light travels 186,000 miles every second, then it will travel 60 times that in a minute, and 60 times that in an hour, and 24 times that in a day, and 365 times that in a year. So we have $1.86 \times 10^5 \times 6.0 \times 10^1 \times 6.0 \times 10^1 \times 2.4 \times 10^1 \times 3.65 \times 10^2$. So we multiply all the numbers out front together and add all the exponents. We get $586.57 \times 10^{10} = 5.86 \times 10^{12}$ miles in a light year (which is roughly 6 trillion miles—a heck of a lot of miles). So if the star is 60 light-years away, its distance in miles is $6 \times 10^1 \times 5.86 \times 10^{12} = 35.16 \times 10^{13} = 3.516 \times 10^{14}$ miles.

Download for free at http://cnx.org/content/col11992/latest/

This OpenStax book is available for free at http://cnx.org/content/col11992/1.8

Download for free at http://cnx.org/content/col11992/latest/

D | UNITS USED IN SCIENCE

In the American system of measurement (originally developed in England), the fundamental units of length, weight, and time are the foot, pound, and second, respectively. There are also larger and smaller units, which include the ton (2240 lb), the mile (5280 ft), the rod (16 1/2 ft), the yard (3 ft), the inch (1/12 ft), the ounce (1/16 lb), and so on. Such units, whose origins in decisions by British royalty have been forgotten by most people, are quite inconvenient for conversion or doing calculations.

In science, therefore, it is more usual to use the *metric system,* which has been adopted in virtually all countries except the United States. Its great advantage is that every unit increases by a factor of ten, instead of the strange factors in the American system. The fundamental units of the metric system are:

> length: 1 meter (m)
> mass: 1 kilogram (kg)
> time: 1 second (s)

A meter was originally intended to be 1 ten-millionth of the distance from the equator to the North Pole along the surface of Earth. It is about 1.1 yd. A kilogram is the mass that on Earth results in a weight of about 2.2 lb. The second is the same in metric and American units.

Length

The most commonly used quantities of length of the metric system are the following.

Conversions

1 kilometer (km) = 1000 meters = 0.6214 mile
1 meter (m) = 0.001 km = 1.094 yards = 39.37 inches
1 centimeter (cm) = 0.01 meter = 0.3937 inch
1 millimeter (mm) = 0.001 meter = 0.1 cm
1 micrometer (μm) = 0.000001 meter = 0.0001 cm
1 nanometer (nm) = 10^{-9} meter = 10^{-7} cm

Table D1 Length

To convert from the American system, here are a few helpful factors:

> 1 mile = 1.61 km
> 1 inch = 2.54 cm

Mass

Although we don't make the distinction very carefully in everyday life on Earth, strictly speaking the kilogram is a unit of mass (measuring the quantity of matter in a body, roughly how many atoms it has,) while the pound is a unit of weight (measuring how strongly Earth's gravity pulls on a body).

Download for free at http://cnx.org/content/col11992/latest/

The most commonly used quantities of mass of the metric system are the following.

Conversions

1 metric ton = 10^6 grams = 1000 kg (and it produces a weight of 2.205×10^3 lb on Earth)
1 kg = 1000 grams (and it produces a weight of 2.2046 lb on Earth)
1 gram (g) = 0.0353 oz (and the equivalent weight is 0.002205 lb)
1 milligram (mg) = 0.001 g

Table D2 Mass

A weight of 1 lb is equivalent on Earth to a mass of 0.4536 kg, while a weight of 1 oz is produced by a mass of 28.35 g.

Temperature

Three temperature scales are in general use:

> Fahrenheit (F); water freezes at 32 °F and boils at 212 °F.

> Celsius or centigrade[1] (C); water freezes at 0 °C and boils at 100 °C.

> Kelvin or absolute (K); water freezes at 273 K and boils at 373 K.

All molecular motion ceases at about −459 °F = −273 °C = 0 K, a temperature called *absolute zero*. Kelvin temperature is measured from this lowest possible temperature, and it is the temperature scale most often used in astronomy. Kelvins have the same value as centigrade or Celsius degrees, since the difference between the freezing and boiling points of water is 100 degrees in each. (Note that we just say "kelvins," not kelvin degrees.)

On the Fahrenheit scale, the difference between the freezing and boiling points of water is 180 degrees. Thus, to convert Celsius degrees or kelvins to Fahrenheit degrees, it is necessary to multiply by 180/100 = 9/5. To convert from Fahrenheit degrees to Celsius degrees or kelvins, it is necessary to multiply by 100/180 = 5/9.

The full conversion formulas are:

> K = °C + 273

> °C = 0.555 × (°F − 32)

> °F = (1.8 × °C) + 32

---

1   Celsius is now the name used for centigrade temperature; it has a more modern standardization but differs from the old centigrade scale by less than 0.1°.

This OpenStax book is available for free at http://cnx.org/content/col11992/1.8

Download for free at http://cnx.org/content/col11992/latest/

E | SOME USEFUL CONSTANTS FOR ASTRONOMY

Physical Constants

Name	Value
speed of light (c)	2.9979×10^8 m/s
gravitational constant (G)	6.674×10^{-11} m^3/(kg s^2)
Planck's constant (h)	6.626×10^{-34} J-s
mass of a hydrogen atom (M_H)	1.673×10^{-27} kg
mass of an electron (M_e)	9.109×10^{-31} kg
Rydberg constant (R_∞)	1.0974×10^7 m^{-1}
Stefan-Boltzmann constant (σ)	5.670×10^{-8} J/(s·m^2 deg^4)[1]
Wien's law constant ($\lambda_{max}T$)	2.898×10^{-3} m K
electron volt (energy) (eV)	1.602×10^{-19} J
energy equivalent of 1 ton TNT	4.2×10^9 J

Table E1

Astronomical Constants

Name	Value
astronomical unit (AU)	1.496×10^{11} m
Light-year (ly)	9.461×10^{15} m
parsec (pc)	3.086×10^{16} m = 3.262 light-years
sidereal year (y)	3.156×10^7 s
mass of Earth (M_{Earth})	5.974×10^{24} kg
equatorial radius of Earth (R_{Earth})	6.378×10^6 m
obliquity of ecliptic	23.4° 26′

Table E2

---

1   deg stands for degrees Celsius or kelvins

Download for free at http://cnx.org/content/col11992/latest/

Astronomical Constants

Name	Value
surface gravity of Earth (g)	9.807 m/s^2
escape velocity of Earth (v_{Earth})	1.119×10^4 m/s
mass of Sun (M_{Sun})	1.989×10^{30} kg
equatorial radius of Sun (R_{Sun})	6.960×10^8 m
luminosity of Sun (L_{Sun})	3.85×10^{26} W
solar constant (flux of energy received at Earth) (S)	1.368×10^3 W/m^2
Hubble constant (H_0)	approximately 20 km/s per million light-years, or approximately 70 km/s per megaparsec

Table E2

This OpenStax book is available for free at http://cnx.org/content/col11992/1.8

Download for free at http://cnx.org/content/col11992/latest/

F | PHYSICAL AND ORBITAL DATA FOR THE PLANETS

Physical Data for the Major Planets

Major Planet	Mean Diameter (km)	Mean Diameter (Earth = 1)	Mass (Earth = 1)	Mean Density (g/cm³)	Rotation Period (d)	Inclination of Equator to Orbit (°)	Surface Gravity (Earth = 1[g])	Velocity of Escape (km/s)
Mercury	4879	0.38	0.055	5.43	58.	0.0	0.38	4.3
Venus	12,104	0.95	0.815	5.24	−243.	177	0.90	10.4
Earth	12,756	1.00	1.00	5.51	1.000	23.4	1.00	11.2
Mars	6779	0.53	0.11	3.93	1.026	25.2	0.38	5.0
Jupiter	140,000	10.9	318	1.33	0.414	3.1	2.53	60.
Saturn	117,000	9.13	95.2	0.69	0.440	26.7	1.07	36.
Uranus	50,700	3.98	14.5	1.27	−0.718	97.9	0.89	21.
Neptune	49,200	3.86	17.2	1.64	0.671	29.6	1.14	23.

Table F1

Physical Data for Well-Studied Dwarf Planets

Well-Studied Dwarf Planet	Diameter (km)	Diameter (Earth = 1)	Mass (Earth = 1)	Mean Density (g/cm³)	Rotation Period (d)	Inclination of Equator to Orbit (°)	Surface Gravity (Earth = 1[g])	Velocity of Escape (km/s)
Ceres	950	0.07	0.0002	2.2	0.378	3	0.03	0.5
Pluto	2470	0.18	0.0024	1.9	−6.387	122	0.06	1.3
Haumea	1700	0.13	0.0007	3	0.163	—	—	0.8
Makemake	1400	0.11	0.0005	2	0.321	—	—	0.8
Eris	2326	0.18	0.0028	2.5	1.25[1]	—	—	1.4

Table F2

---

1   This measurement is quite uncertain.

Download for free at http://cnx.org/content/col11992/latest/

Orbital Data for the Major Planets

Major Planet	Semimajor Axis (AU)	Semimajor Axis (10^6 km)	Sidereal Period (y)	Sidereal Period (d)	Mean Orbital Speed (km/s)	Orbital Eccentricity	Inclination of Orbit to Ecliptic (°)
Mercury	0.39	58	0.24	88.0	47.9	0.206	7.0
Venus	0.72	108	0.6	224.7	35.0	0.007	3.4
Earth	1.00	149	1.00	365.2	29.8	0.017	0.0
Mars	1.52	228	1.88	687.0	24.1	0.093	1.9
Jupiter	5.20	778	11.86	—	13.1	0.048	1.3
Saturn	9.54	1427	29.46	—	9.6	0.056	2.5
Uranus	19.19	2871	84.01	—	6.8	0.046	0.8
Neptune	30.06	4497	164.82	—	5.4	0.010	1.8

Table F3

Orbital Data for Well-Studied Dwarf Planets

Well-Studied Dwarf Planet	Semimajor Axis (AU)	Semimajor Axis (10^6 km)	Sidereal Period (y)	Mean Orbital Speed (km/s)	Orbital Eccentricity	Inclination of Orbit to Ecliptic (°)
Ceres	2.77	414.0	4.6	18	0.08	11
Pluto	39.5	5915	248.6	4.7	0.25	17
Haumea	43.1	6452	283.3	4.5	0.19	28
Makemake	45.8	6850	309.9	4.4	0.16	29
Eris	68.0	10,120	560.9	3.4	0.44	44

Table F4

This OpenStax book is available for free at http://cnx.org/content/col11992/1.8

Download for free at http://cnx.org/content/col11992/latest/

G | SELECTED MOONS OF THE PLANETS

Note: As this book goes to press, nearly two hundred moons are now known in the solar system and more are being discovered on a regular basis. Of the major planets, only Mercury and Venus do not have moons. In addition to moons of the planets, there are many moons of asteroids. In this appendix, we list only the largest and most interesting objects that orbit each planet (including dwarf planets). The number given for each planet is discoveries through 2015. For further information see https://solarsystem.nasa.gov/planets/solarsystem/moons and https://en.wikipedia.org/wiki/List_of_natural_satellites.

Selected Moons of the Planets

Planet (moons)	Satellite Name	Discovery	Semimajor Axis (km × 1000)	Period (d)	Diameter (km)	Mass (10^{20} kg)	Density (g/cm^3)
Earth (1)	Moon	—	384	27.32	3476	735	3.3
Mars (2)	Phobos	Hall (1877)	9.4	0.32	23	1×10^{-4}	2.0
	Deimos	Hall (1877)	23.5	1.26	13	2×10^{-5}	1.7
Jupiter (67)	Amalthea	Barnard (1892)	181	0.50	200	—	—
	Thebe	Voyager (1979)	222	0.67	90	—	—
	Io	Galileo (1610)	422	1.77	3630	894	3.6
	Europa	Galileo (1610)	671	3.55	3138	480	3.0
	Ganymede	Galileo (1610)	1070	7.16	5262	1482	1.9
	Callisto	Galileo (1610)	1883	16.69	4800	1077	1.9
	Himalia	Perrine (1904)	11,460	251	170	—	—
Saturn (62)	Pan	Voyager (1985)	133.6	0.58	20	3×10^{-5}	—
	Atlas	Voyager (1980)	137.7	0.60	40	—	—

Table G1

Download for free at http://cnx.org/content/col11992/latest/

Selected Moons of the Planets

Planet (moons)	Satellite Name	Discovery	Semimajor Axis (km × 1000)	Period (d)	Diameter (km)	Mass (10^{20} kg)	Density (g/cm^3)
	Prometheus	Voyager (1980)	139.4	0.61	80	—	—
	Pandora	Voyager (1980)	141.7	0.63	100	—	—
	Janus	Dollfus (1966)	151.4	0.69	190	—	—
	Epimetheus	Fountain, Larson (1980)	151.4	0.69	120	—	—
	Mimas	Herschel (1789)	186	0.94	394	0.4	1.2
	Enceladus	Herschel (1789)	238	1.37	502	0.8	1.2
	Tethys	Cassini (1684)	295	1.89	1048	7.5	1.3
	Dione	Cassini (1684)	377	2.74	1120	11	1.3
	Rhea	Cassini (1672)	527	4.52	1530	25	1.3
	Titan	Huygens (1655)	1222	15.95	5150	1346	1.9
	Hyperion	Bond, Lassell (1848)	1481	21.3	270	—	—
	Iapetus	Cassini (1671)	3561	79.3	1435	19	1.2
	Phoebe	Pickering (1898)	12,950	550 (R)[1]	220	—	—
Uranus (27)	Puck	Voyager (1985)	86.0	0.76	170	—	—

Table G1

---

1   R stands for retrograde rotation (backward from the direction that most objects in the solar system revolve and rotate).

This OpenStax book is available for free at http://cnx.org/content/col11992/1.8

Download for free at http://cnx.org/content/col11992/latest/

Selected Moons of the Planets

Planet (moons)	Satellite Name	Discovery	Semimajor Axis (km × 1000)	Period (d)	Diameter (km)	Mass (10²⁰ kg)	Density (g/cm³)
	Miranda	Kuiper (1948)	130	1.41	485	0.8	1.3
	Ariel	Lassell (1851)	191	2.52	1160	13	1.6
	Umbriel	Lassell (1851)	266	4.14	1190	13	1.4
	Titania	Herschel (1787)	436	8.71	1610	35	1.6
	Oberon	Herschel (1787)	583	13.5	1550	29	1.5
Neptune (14)	Despina	Voyager (1989)	53	0.33	150	—	—
	Galatea	Voyager (1989)	62	0.40	150	—	—
	Larissa	Voyager (1989)	118	1.12	400	—	—
	Triton	Lassell (1846)	355	5.88 (R)[2]	2720	220	2.1
	Nereid	Kuiper (1949)	5511	360	340	—	—
Pluto (5)	Charon	Christy (1978)	19.7	6.39	1200	—	1.7
	Styx	Showalter et al (2012)	42	20	20	—	—
	Nix	Weaver et al (2005)	48	24	46	—	2.1
	Kerberos	Showalter et al (2011)	58	24	28	—	1.4

Table G1

---

2   R stands for retrograde rotation (backward from the direction that most objects in the solar system revolve and rotate).

Download for free at http://cnx.org/content/col11992/latest/

Selected Moons of the Planets

Planet (moons)	Satellite Name	Discovery	Semimajor Axis (km × 1000)	Period (d)	Diameter (km)	Mass (10^{20} kg)	Density (g/cm^3)
	Hydra	Weaver et al (2005)	65	38	61	—	0.8
Eris (1)	Dysnomea	Brown et al (2005)	38	16	684	—	—
Makemake (1)	(MK2)	Parker et al (2016)	—	—	160	—	—
Haumea (2)	Hi'iaka	Brown et al (2005)	50	49	400	—	—
	Namaka	Brown et al (2005)	39	35	200	—	—

Table G1

This OpenStax book is available for free at http://cnx.org/content/col11992/1.8

Download for free at http://cnx.org/content/col11992/latest/

H   FUTURE TOTAL ECLIPSES

Future Total Solar Eclipses

We also include eclipses that are *annular*—where the Moon is directly in front of the Sun, but doesn't fully cover it—leaving a ring of light around the dark Moon's edges)

Future Total Solar Eclipses

Date	Type of Eclipse	Location on Earth[1]
September 1, 2016	Annular	S Atlantic Ocean, C Africa, Madagascar, Indian Ocean
February 26, 2017	Annular	SW Africa, S tip of South America
August 21, 2017	Total	U.S. and oceans on either side
July 2, 2019	Total	SW South America, Pacific Ocean
December 26, 2019	Annular	Saudi Arabia, S India, Malaysia
June 21, 2020	Annular	(very short) C Africa, Pakistan, India, China
December 14, 2020	Total	Chile, Argentina, and oceans on either side
June 10, 2021	Annular	N Canada, Greenland
December 4, 2021	Total	Only in Antarctica
April 20, 2023	Total[2]	Mostly in Indian and Pacific oceans, Indonesia
October 14, 2023	Annular	OR, NV, UT, NM, TX, C America, Colombia, Brazil
April 8, 2024	Total	N Mexico, U.S. (TX to ME), SE Canada and oceans on either side
October 2, 2024	Annular	S Chile, S Argentina, and oceans on either side
February 17, 2026	Annular	Only in Antarctica
August 12, 2026	Total	Greenland, Iceland, Spain
February 6, 2027	Annular	S Pacific, Argentina, Chile, Uruguay, S Atlantic
August 2, 2027	Total	Spain, Morocco, Egypt, Saudi Arabia, Yemen, Arabian Sea

Table H1

---

1   Remember that a total or annular eclipse is only visible on a narrow track. The same eclipse will be partial over a much larger area, but partial eclipses are not as spectacular as total ones.
2   This is a so-called hybrid eclipse, which is total in some places and annular in others.

Download for free at http://cnx.org/content/col11992/latest/

Future Total Solar Eclipses

Date	Type of Eclipse	Location on Earth
January 26, 2028	Annular	Ecuador, Peru, Brazil, North Atlantic Ocean, Portugal, Spain
July 22, 2028	Total	Indian Ocean, Australia, New Zealand, South Pacific Ocean

Table H1

Future Total Lunar Eclipses

Future Total Lunar Eclipses

Date	Location on Earth
January 31, 2018	Asia, Australia, W North America
July 27, 2018	S America, Asia, Africa, Australia, Indian Ocean
January 21, 2019	N America, S America, W Africa, W Europe
May 26, 2021	E Asia, Australia, Pacific Ocean, W North America, W South America
May 16, 2022	N America, S America, Europe, Africa
November 8, 2022	Asia, Australia, Pacific Ocean, N America, S America
March 14, 2025	Pacific Ocean, N America, S America, Atlantic Ocean, W Europe, W Africa
September 7, 2025	Europe, Africa, Asia, Australia, Indian Ocean
March 3, 2026	E Asia, Australia, Pacific Ocean, N America, C America
June 26, 2029	E North America, S America, Atlantic Ocean, W Europe, W Africa
December 20, 2029	E North America, E South America, Atlantic Ocean, Europe, Africa, Asia

Table H2

Additional Resources

For more information and detailed maps about eclipses, see these resources.

> - NASA's Eclipse Site: http://eclipse.gsfc.nasa.gov/
> - Mr. Eclipse site for beginners by Dr. Fred Espenak: http://www.mreclipse.com/
> - Eclipse Weather and Maps by Meteorologist Jay Anderson: http://home.cc.umanitoba.ca/~jander/

This OpenStax book is available for free at http://cnx.org/content/col11992/1.8

Download for free at http://cnx.org/content/col11992/latest/

> Eclipse Maps by Michael Zeiler: http://www.eclipse-maps.com/Eclipse-Maps/Welcome.html

> Eclipse Information and Maps by Xavier Jubier: http://xjubier.free.fr/en/site_pages/eclipses.html

Download for free at http://cnx.org/content/col11992/latest/

This OpenStax book is available for free at http://cnx.org/content/col11992/1.8

Download for free at http://cnx.org/content/col11992/latest/

I | THE NEAREST STARS, BROWN DWARFS, AND WHITE DWARFS

The Nearest Stars, Brown Dwarfs, and White Dwarfs

Star	System	Discovery Name	Distance (light-year)	Spectral Type	Location: RA[1]	Location: Dec[2]	Luminosity (Sun = 1)
		Sun	—	G2 V	—	—	1
1	1	Proxima Centauri	4.2	M5.5 V	14 29	−62 40	5×10^{-5}
2	2	Alpha Centauri A	4.4	G2 V	14 39	−60 50	1.5
3		Alpha Centauri B	4.4	K2 IV	14 39	−60 50	0.5
4	3	Barnard's Star	6.0	M4 V	17 57	+04 42	4.4×10^{-4}
5	4	Wolf 359	7.8	M6 V	10 56	+07 00	2×10^{-5}
6	5	Lalande 21 185	8.3	M2 V	11 03	+35 58	5.7×10^{-3}
7	6	Sirius A	8.6	A1 V	06 45	−16 42	23.1
8		Sirius B	8.6	DA2[3]	06 45	−16 43	2.5×10^{-3}
9	7	Luyten 726-8 A	8.7	M5.5 V	01 39	−17 57	6×10^{-5}
10		Luyten 726-8 B (UV Ceti)	8.7	M6 V	01 39	−17 57	4×10^{-5}
11	8	Ross 154	9.7	M.05 V	18 49	−23 50	5×10^{-4}
12	9	Ross 248 (HH Andromedae)	10.3	M5.5 V	23 41	+44 10	1.0×10^{-4}
13	10	Epsilon Eridani	10.5	K2 V	03 32	−09 27	0.29
14	11	Lacaille 9352	10.7	M0.5 V	23 05	−35 51	0.011
15	12	Ross 128 (FI Virginis)	10.9	M4 V	11 47	+00 48	3.4×10^{-4}
16	13	Luyten 789-6 A (EZ Aquarii A)	11.3	M5 V	22 38	−15 17	5×10^{-5}

Table I1

1   Location (right ascension) given for Epoch 2000.0
2   Location (declination) given for Epoch 2000.0
3   White dwarf stellar remnant

Download for free at http://cnx.org/content/col11992/latest/

The Nearest Stars, Brown Dwarfs, and White Dwarfs

Star	System	Discovery Name	Distance (light-year)	Spectral Type	Location: RA	Location: Dec	Luminosity (Sun = 1)
17		Luyten 789-6 B (EZ Aquarii B)	11.3	M5.5 V	22 38	−15 15	5×10^{-5}
18		Luyten 789-6 C (EZ Aquarii C)	11.3	M6.5 V	22 38	−15 17	2×10^{-5}
19	14	61 Cygni A	11.4	K5 V	21 06	+38 44	0.086
20		61 Cygni B	11.4	K7 V	21 06	+38 44	0.041
21	15	Procyon A	11.4	F51V	07 39	+05 13	7.38
22		Procyon B	11.4	wd[4]	07 39	+05 13	5.5×10^{-4}
23	16	Sigma 2398 A	11.5	M3 V	18 42	+59 37	0.003
24		Sigma 2398 B	11.5	M3.5 V	18 42	+59 37	1.4×10^{-3}
25	17	Groombridge 34 A (GX Andromedae)	11.6	M1.5 V	00 18	+44 01	6.4×10^{-3}
26		Groombridge 34 B (GQ Andromedae)	11.6	M3.5 V	00 18	+44 01	4.1×10^{-4}
27	18	Epsilon Indi A	11.8	K5 V	22 03	−56 46	0.150
28		Epsilon Indi Ba	11.7	T1[5]	22 04	−56 46	—
29		Epsilon Indi Bb	11.7	T6[6]	22 04	−56 46	—
30	19	G 51-15 (DX Cancri)	11.8	M6.5 V	08 29	+26 46	1×10^{-5}
31	20	Tau Ceti	11.9	G8.5 V	01 44	−15 56	0.458
32	21	Luyten 372-58	12.0	M5 V	03 35	−44 30	7×10^{-5}
33	22	Luyten 725-32 (YZ Ceti)	12.1	M4.5 V	01 12	−16 59	1.8×10^{-4}

Table I1

---

4   White dwarf stellar remnant
5   Brown dwarf
6   Brown dwarf

This OpenStax book is available for free at http://cnx.org/content/col11992/1.8

The Nearest Stars, Brown Dwarfs, and White Dwarfs

Star	System	Discovery Name	Distance (light-year)	Spectral Type	Location: RA	Location: Dec	Luminosity (Sun = 1)
34	23	Luyten's Star	12.4	M3.5 V	07 27	+05 13	1.4×10^{-3}
35	24	SCR J184-6357 A	12.6	M8.5 V	18 45	−63 57	1×10^{-6}
36		SCR J184-6357 B	12.7	T6[7]	18 45	−63 57	—
37	25	Teegarden's Star	12.5	M6 V	02 53	+16 52	1×10^{-5}
38	26	Kapteyn's Star	12.8	M1 V	05 11	−45 01	3.8×10^{-3}
39	27	Lacaille 8760 (AX Microscopium)	12.9	K7 V	21 17	−38 52	0.029

Table I1

---

7   Brown dwarf

Download for free at http://cnx.org/content/col11992/latest/

This OpenStax book is available for free at http://cnx.org/content/col11992/1.8

Download for free at http://cnx.org/content/col11992/latest/

J THE BRIGHTEST TWENTY STARS

Note: These are the stars that *appear* the brightest visually, as seen from our vantage point on Earth. They are not necessarily the stars that are intrinsically the most luminous.

The Brightest Twenty Stars										
Name					Proper Motion (arcsec/y)		Right Ascension		Declination	
Traditional	Bayer	Luminosity (Sun = 1)	Distance (light-years)	Spectral Type	RA	Dec	(h)	(m)	(deg)	(min)
Sirius	α Canis Majoris	22.5	8.6	A1 V	−0.5	−1.2	06	45.2	−16	43
Canopus	α Carinae	13,500	309	F0 II	+0.02	+0.02	06	24.0	−52	42
Rigil Kentaurus	α Centauri	1.94	4.32	G2 V + K IV	−3.7	+0.5	14	39.7	−60	50
Arcturus	α Bootis	120	36.72	K1.5 III	−1.1	−2.0	14	15.7	+19	11
Vega	α Lyrae	49	25.04	A0 V	+0.2	+0.3	18	36.9	+38	47
Capella	α Aurigae	140	42.80	G8 III + G0 III	+0.08	−0.4	05	16.7	+46	00
Rigel	β Orionis	50,600	863	B8 I	+0.00	+0.00	05	14.5	−08	12
Procyon	α Canis Minoris	7.31	11.46	F5 IV-V	−0.7	−1.0	07	39.3	+05	14
Achernar	α Eridani	1030	139	B3 V	+0.10	−0.04	01	37.7	−57	14
Betelgeuse	α Orionis	13,200	498	M2 I	+0.02	+0.01	05	55.2	+07	24
Hadar	β Centauri	7050	392	B1 III	−0.03	−0.02	14	03.8	−60	22
Altair	α Aquilae	11.2	16.73	A7 V	+0.5	+0.4	19	50.8	+08	52
Acrux	α Crucis	4090	322	B0.5 IV + B1V	−0.04	−0.01	12	26.6	−63	06
Aldebaran	α Tauri	160	66.64	K5 III	+0.1	−0.2	04	35.9	+16	31
Spica	α Virginis	2030	250	B1 III-IV + B2 V	−0.04	−0.03	13	25.2	−11	10
Antares	α Scorpii	9290	554	M1.5 I + B2.5 V	−0.01	−0.02	16	29.4	−26	26
Pollux	β Geminorum	31.6	33.78	K0 III	−0.6	−0.05	07	45.3	+28	02
Fomalhaut	α Piscis Austrini	17.2	25.13	A3 V	+0.03	−0.2	22	57.6	−29	37
Mimosa	β Crucis	1980	279	B0.5 III	−0.04	−0.02	12	47.7	−59	41
Deneb	α Cygni	50,600	1412	A2 I	+0.00	+0.00	20	41.4	+45	17

Figure J1 The brightest stars typically have names from antiquity. Next to each star's ancient name, we have added a column with its name in the system originated by Bayer (see the Naming Stars feature box.) The distances of the more remote stars are estimated from their spectral types and apparent brightnesses and are only approximate. The luminosities for those stars are approximate to the same degree. Right ascension and declination is given for Epoch 2000.0.

Download for free at http://cnx.org/content/col11992/latest/

This OpenStax book is available for free at http://cnx.org/content/col11992/1.8

K | THE CHEMICAL ELEMENTS

The Chemical Elements

Element	Symbol	Atomic Number	Atomic Weight[1]	Percentage of Naturally Occurring Elements in the Universe
Hydrogen	H	1	1.008	75
Helium	He	2	4.003	23
Lithium	Li	3	6.94	6×10^{-7}
Beryllium	Be	4	9.012	1×10^{-7}
Boron	B	5	10.821	1×10^{-7}
Carbon	C	6	12.011	0.5
Nitrogen	N	7	14.007	0.1
Oxygen	O	8	15.999	1
Fluorine	F	9	18.998	4×10^{-5}
Neon	Ne	10	20.180	0.13
Sodium	Na	11	22.990	0.002
Magnesium	Mg	12	24.305	0.06
Aluminum	Al	13	26.982	0.005
Silicon	Si	14	28.085	0.07
Phosphorus	P	15	30.974	7×10^{-4}
Sulfur	S	16	32.06	0.05
Chlorine	Cl	17	35.45	1×10^{-4}
Argon	Ar	18	39.948	0.02
Potassium	K	19	39.098	3×10^{-4}

Table K1

---

1   Where mean atomic weights have not been well determined, the atomic mass numbers of the most stable isotopes are given in parentheses.

Download for free at http://cnx.org/content/col11992/latest/

The Chemical Elements

Element	Symbol	Atomic Number	Atomic Weight	Percentage of Naturally Occurring Elements in the Universe
Calcium	Ca	20	40.078	0.007
Scandium	Sc	21	44.956	3×10^{-6}
Titanium	Ti	22	47.867	3×10^{-4}
Vanadium	V	23	50.942	3×10^{-4}
Chromium	Cr	24	51.996	0.0015
Manganese	Mn	25	54.938	8×10^{-4}
Iron	Fe	26	55.845	0.11
Cobalt	Co	27	58.933	3×10^{-4}
Nickel	Ni	28	58.693	0.006
Copper	Cu	29	63.546	6×10^{-6}
Zinc	Zn	30	65.38	3×10^{-5}
Gallium	Ga	31	69.723	1×10^{-6}
Germanium	Ge	32	72.630	2×10^{-5}
Arsenic	As	33	74.922	8×10^{-7}
Selenium	Se	34	78.971	3×10^{-6}
Bromine	Br	35	79.904	7×10^{-7}
Krypton	Kr	36	83.798	4×10^{-6}
Rubidium	Rb	37	85.468	1×10^{-6}
'Strontium	Sr	38	87.62	4×10^{-6}
Yttrium	Y	39	88.906	7×10^{-7}
Zirconium	Zr	40	91.224	5×10^{-6}
Niobium	Nb	41	92.906	2×10^{-7}

Table K1

This OpenStax book is available for free at http://cnx.org/content/col11992/1.8

Download for free at http://cnx.org/content/col11992/latest/

The Chemical Elements

Element	Symbol	Atomic Number	Atomic Weight	Percentage of Naturally Occurring Elements in the Universe
Molybdenum	Mo	42	95.95	5×10^{-7}
Technetium	Tc	43	(98)	—
Ruthenium	Ru	44	101.07	4×10^{-7}
Rhodium	Rh	45	102.906	6×10^{-8}
Palladium	Pd	46	106.42	2×10^{-7}
Silver	Ag	47	107.868	6×10^{-8}
Cadmium	Cd	48	112.414	2×10^{-7}
Indium	In	49	114.818	3×10^{-8}
Tin	Sn	50	118.710	4×10^{-7}
Antimony	Sb	51	121.760	4×10^{-8}
Tellurium	Te	52	127.60	9×10^{-7}
Iodine	I	53	126.904	1×10^{-7}
Xenon	Xe	54	131.293	1×10^{-6}
Cesium	Cs	55	132.905	8×10^{-8}
Barium	Ba	56	137.327	1×10^{-6}
Lanthanum	La	57	138.905	2×10^{-7}
Cerium	Ce	58	140.116	1×10^{-6}
Praseodymium	Pr	59	140.907	2×10^{-7}
Neodymium	Nd	60	144.242	1×10^{-6}
Promethium	Pm	61	(145)	—
Samarium	Sm	62	150.36	5×10^{-7}
Europium	Eu	63	151.964	5×10^{-8}

Table K1

Download for free at http://cnx.org/content/col11992/latest/

The Chemical Elements

Element	Symbol	Atomic Number	Atomic Weight	Percentage of Naturally Occurring Elements in the Universe
Gadolinium	Gd	64	157.25	2×10^{-7}
Terbium	Tb	65	158.925	5×10^{-8}
Dysprosium	Dy	66	162.500	2×10^{-7}
Holmium	Ho	67	164.930	5×10^{-8}
Erbium	Er	68	167.259	2×10^{-7}
Thulium	Tm	69	168.934	1×10^{-8}
Ytterbium	Yb	70	173.054	2×10^{-7}
Lutetium	Lu	71	174.967	1×10^{-8}
Hafnium	Hf	72	178.49	7×10^{-8}
Tantalum	Ta	73	180.948	8×10^{-9}
Tungsten	W	74	183.84	5×10^{-8}
Rhenium	Re	75	186.207	2×10^{-8}
Osmium	Os	76	190.23	3×10^{-7}
Iridium	Ir	77	192.217	2×10^{-7}
Platinum	Pt	78	195.084	5×10^{-7}
Gold	Au	79	196.967	6×10^{-8}
Mercury	Hg	80	200.592	1×10^{-7}
Thallium	TI	81	204.38	5×10^{-8}
Lead	Pb	82	207.2	1×10^{-6}
Bismuth	Bi	83	208.980	7×10^{-8}
Polonium	Po	84	(209)	—
Astatine	At	85	(210)	—

Table K1

This OpenStax book is available for free at http://cnx.org/content/col11992/1.8

Download for free at http://cnx.org/content/col11992/latest/

The Chemical Elements

Element	Symbol	Atomic Number	Atomic Weight	Percentage of Naturally Occurring Elements in the Universe
Radon	Rn	86	(222)	—
Francium	Fr	87	(223)	—
Radium	Ra	88	(226)	—
Actinium	Ac	89	(227)	—
Thorium	Th	90	232.038	4×10^{-8}
Protactinium	Pa	91	231.036	—
Uranium	U	92	238.029	2×10^{-8}
Neptunium	Np	93	(237)	—
Plutonium	Pu	94	(244)	—
Americium	Am	95	(243)	—
Curium	Cm	96	(247)	—
Berkelium	Bk	97	(247)	—
Californium	Cf	98	(251)	—
Einsteinium	Es	99	(252)	—
Fermium	Fm	100	(257)	—
Mendelevium	Md	101	(258)	—
Nobelium	No	102	(259)	—
Lawrencium	Lr	103	(262)	—
Rutherfordium	Rf	104	(267)	—
Dubnium	Db	105	(268)	—
Seaborgium	Sg	106	(271)	—
Bohrium	Bh	107	(272)	—

Table K1

Download for free at http://cnx.org/content/col11992/latest/

The Chemical Elements

Element	Symbol	Atomic Number	Atomic Weight	Percentage of Naturally Occurring Elements in the Universe
Hassium	Hs	108	(270)	—
Meitnerium	Mt	109	(276)	—
Darmstadtium	Ds	110	(281)	—
Roentgenium	Rg	111	(280)	—
Copernicium	Cn	112	(285)	—
Ununtrium	Uut	113	(284)	—
Flerovium	Fl	114	(289)	—
Ununpentium	Uup	115	(288)	—
Livermorium	Lv	116	(293)	—
Ununseptium	Uus	117	(294)	—
Ununoctium	Uuo	118	(294)	—

Table K1

Note: Some of the newest elements near the bottom of the table have suggested names that are still under review, and those names are not yet listed here. For example, Tennessine is suggested for element 117, after the state where the Oak Ridge National Laboratory is located.

This OpenStax book is available for free at http://cnx.org/content/col11992/1.8

Download for free at http://cnx.org/content/col11992/latest/

L THE CONSTELLATIONS

The Constellations

Constellation (Latin name)	Genitive Case Ending	English Name or Description	Abbreviation	Approximate Position: α (h)	Approximate Position: δ (°)
Andromeda	Andromedae	Princess of Ethiopia	And	1	+40
Antila	Antilae	Air pump	Ant	10	−35
Apus	Apodis	Bird of Paradise	Aps	16	−75
Aquarius	Aquarii	Water bearer	Aqr	23	−15
Aquila	Aquilae	Eagle	Aql	20	+5
Ara	Arae	Altar	Ara	17	−55
Aries	Arietis	Ram	Ari	3	+20
Auriga	Aurigae	Charioteer	Aur	6	+40
Boötes	Boötis	Herdsman	Boo	15	+30
Caelum	Cael	Graving tool	Cae	5	−40
Camelopardus	Camelopardis	Giraffe	Cam	6	+70
Cancer	Cancri	Crab	Cnc	9	+20
Canes Venatici	Canum Venaticorum	Hunting dogs	CVn	13	+40
Canis Major	Canis Majoris	Big dog	CMa	7	−20
Canis Minor	Canis Minoris	Little dog	CMi	8	+5
Capricornus	Capricorni	Sea goat	Cap	21	−20
Carina[1]	Carinae	Keel of Argonauts' ship	Car	9	−60
Cassiopeia	Cassiopeiae	Queen of Ethiopia	Cas	1	+60

Table L1

---

1   The four constellations Carina, Puppis, Pyxis, and Vela originally formed the single constellation Argo Navis.

Download for free at http://cnx.org/content/col11992/latest/

The Constellations

Constellation (Latin name)	Genitive Case Ending	English Name or Description	Abbreviation	Approximate Position: α (h)	Approximate Position: δ (°)
Centaurus	Centauri	Centaur	Cen	13	−50
Cepheus	Cephei	King of Ethiopia	Cep	22	+70
Cetus	Ceti	Sea monster (whale)	Cet	2	−10
Chamaeleon	Chamaeleontis	Chameleon	Cha	11	−80
Circinus	Circini	Compasses	Cir	15	−60
Columba	Columbae	Dove	Col	6	−35
Coma Berenices	Comae Berenices	Berenice's hair	Com	13	+20
Corona Australis	Coronae Australis	Southern crown	CrA	19	−40
Corona Borealis	Coronae Borealis	Northern crown	CrB	16	+30
Corvus	Corvi	Crow	Crv	12	−20
Crater	Crateris	Cup	Crt	11	−15
Crux	Crucis	Cross (southern)	Cru	12	−60
Cygnus	Cygni	Swan	Cyg	21	+40
Delphinus	Delphini	Porpoise	Del	21	+10
Dorado	Doradus	Swordfish	Dor	5	−65
Draco	Draconis	Dragon	Dra	17	+65
Equuleus	Equulei	Little horse	Equ	21	+10
Eridanus	Eridani	River	Eri	3	−20
Fornax	Fornacis	Furnace	For	3	−30
Gemini	Geminorum	Twins	Gem	7	+20

Table L1

This OpenStax book is available for free at http://cnx.org/content/col11992/1.8

Download for free at http://cnx.org/content/col11992/latest/

The Constellations

Constellation (Latin name)	Genitive Case Ending	English Name or Description	Abbreviation	Approximate Position: α (h)	Approximate Position: δ (°)
Grus	Gruis	Crane	Gru	22	−45
Hercules	Herculis	Hercules, son of Zeus	Her	17	+30
Horologium	Horologii	Clock	Hor	3	−60
Hydra	Hydrae	Sea serpent	Hya	10	−20
Hydrus	Hydri	Water snake	Hyi	2	−75
Indus	Indi	Indian	Ind	21	−55
Lacerta	Lacertae	Lizard	Lac	22	+45
Leo	Leonis	Lion	Leo	11	+15
Leo Minor	Leonis Minoris	Little lion	LMi	10	+35
Lepus	Leporis	Hare	Lep	6	−20
Libra	Librae	Balance	Lib	15	−15
Lupus	Lupi	Wolf	Lup	15	−45
Lynx	Lyncis	Lynx	Lyn	8	+45
Lyra	Lyrae	Lyre or harp	Lyr	19	+40
Mensa	Mensae	Table Mountain	Men	5	−80
Microscopium	Microscopii	Microscope	Mic	21	−35
Monoceros	Monocerotis	Unicorn	Mon	7	−5
Musca	Muscae	Fly	Mus	12	−70
Norma	Normae	Carpenter's level	Nor	16	−50
Octans	Octantis	Octant	Oct	22	−85
Ophiuchus	Ophiuchi	Holder of serpent	Oph	17	0

Table L1

Download for free at http://cnx.org/content/col11992/latest/

The Constellations

Constellation (Latin name)	Genitive Case Ending	English Name or Description	Abbreviation	Approximate Position: α (h)	Approximate Position: δ (°)
Orion	Orionis	Orion, the hunter	Ori	5	+5
Pavo	Pavonis	Peacock	Pav	20	−65
Pegasus	Pegasi	Pegasus, the winged horse	Peg	22	+20
Perseus	Persei	Perseus, hero who saved Andromeda	Per	3	+45
Phoenix	Phoenicis	Phoenix	Phe	1	−50
Pictor	Pictoris	Easel	Pic	6	−55
Pisces	Piscium	Fishes	Psc	1	+15
Piscis Austrinus	Piscis Austrini	Southern fish	PsA	22	−30
Puppis[2]	Puppis	Stern of the Argonauts' ship	Pup	8	−40
Pyxis[3] (=Malus)	Pyxidus	Compass of the Argonauts' ship	Pyx	9	−30
Reticulum	Reticuli	Net	Ret	4	−60
Sagitta	Sagittae	Arrow	Sge	20	+10
Sagittarius	Sagittarii	Archer	Sgr	19	−25
Scorpius	Scorpii	Scorpion	Sco	17	−40
Sculptor	Sculptoris	Sculptor's tools	Scl	0	−30
Scutum	Scuti	Shield	Sct	19	−10
Serpens	Serpentis	Serpent	Ser	17	0

Table L1

2   The four constellations Carina, Puppis, Pyxis, and Vela originally formed the single constellation Argo Navis.
3   The four constellations Carina, Puppis, Pyxis, and Vela originally formed the single constellation Argo Navis.

This OpenStax book is available for free at http://cnx.org/content/col11992/1.8

Download for free at http://cnx.org/content/col11992/latest/

The Constellations

Constellation (Latin name)	Genitive Case Ending	English Name or Description	Abbreviation	Approximate Position: α (h)	Approximate Position: δ (°)
Sextans	Sextantis	Sextant	Sex	10	0
Taurus	Tauri	Bull	Tau	4	+15
Telescopium	Telescopii	Telescope	Tel	19	−50
Triangulum	Trianguli	Triangle	Tri	2	+30
Triangulum Australe	Trianguli Australis	Southern triangle	TrA	16	−65
Tucana	Tucanae	Toucan	Tuc	0	−65
Ursa Major	Ursae Majoris	Big bear	UMa	11	+50
Ursa Minor	Ursae Minoris	Little bear	UMi	15	+70
Vela[4]	Velorum	Sail of the Argonauts' ship	Vel	9	−50
Virgo	Virginis	Virgin	Vir	13	0
Volans	Volantis	Flying fish	Vol	8	−70
Vulpecula	Vulpeculae	Fox	Vul	20	+25

Table L1

---

4   The four constellations Carina, Puppis, Pyxis, and Vela originally formed the single constellation Argo Navis.

Download for free at http://cnx.org/content/col11992/latest/

This OpenStax book is available for free at http://cnx.org/content/col11992/1.8

Download for free at http://cnx.org/content/col11992/latest/

STAR CHART AND SKY EVENT RESOURCES

Star Charts

To obtain graphic charts of the sky over your head tonight, there are a number of free online resources.

> One of the easiest is at the Skymaps website: http://www.skymaps.com/downloads.html. Here you can print out a free PDF version of the northern sky (for roughly latitude 40°, which is reasonable for much of the United States). Maps of the sky from the equator and the southern hemisphere can also be printed.

> Sky charts and other summaries of astronomical information can also be found at http://www.heavens-above.com/.

> A free, open-source computer application that shows the sky at any time from any place is called Stellarium. You can find it at http://www.stellarium.org/.

> Appendix B provides a section of information for finding astronomy apps for cell phones and tablets. Many of these also provide star charts. If you have a smartphone, you can find a variety of inexpensive apps that allow you to simply hold your phone upward to see what is in the sky behind your phone.

> A *planisphere* is a sky chart that turns inside a round frame and can show you the night sky at your latitude on any date and time of the year. You can buy them at science supply and telescope stores or online. Or you can construct your own from templates at these two websites:

>> Dennis Schatz's AstroAdventures Star Finder: http://dennisschatz.org/activities/Star%20Finder.pdf

>> Uncle Al's Star Wheel: http://www.lawrencehallofscience.org/do_science_now/science_apps_and_activities/starwheels

Calendars Of Night Sky Events

The following resources offer calendars of night sky events.

> Sea and Sky: http://www.seasky.org/astronomy/astronomy.html (click on the fourth menu button)

> *Sky & Telescope's* This Week's Sky at a Glance: http://www.skyandtelescope.com/observing/sky-at-a-glance/

> *Astronomy* Magazine's The Sky This Week: http://www.astronomy.com/observing/sky-this-week

> NASA Sky Cal: http://eclipse.gsfc.nasa.gov/SKYCAL/SKYCAL.html

> Night Sky Network Sky Planner: https://nightsky.jpl.nasa.gov/planner.cfm

Download for free at http://cnx.org/content/col11992/latest/

This OpenStax book is available for free at http://cnx.org/content/col11992/1.8

Download for free at http://cnx.org/content/col11992/latest/

INDEX

Download for free at http://cnx.org/content/col11992/latest/

This OpenStax book is available for free at http://cnx.org/content/col11992/1.8

Download for free at http://cnx.org/content/col11992/latest/

This OpenStax book is available for free at http://cnx.org/content/col11992/1.8

Download for free at http://cnx.org/content/col11992/latest/

Download for free at http://cnx.org/content/col11992/latest/

This OpenStax book is available for free at http://cnx.org/content/col11992/1.8

Download for free at http://cnx.org/content/col11992/latest/

Z

This OpenStax book is available for free at http://cnx.org/content/col11992/1.8